Geometry of String Theory Compactifications

String theory is a leading candidate for the unification of universal forces and matter, and one of its most striking predictions is the existence of small additional dimensions that have escaped detection so far. This book focuses on the geometry of these dimensions, beginning with the basics of the theory, the mathematical properties of spinors, and differential geometry. It further explores advanced techniques at the core of current research, such as G-structures and generalized complex geometry. Many significant classes of solutions to the theory's equations are studied in detail, from special holonomy and Sasaki–Einstein manifolds to their more recent generalizations involving fluxes for form fields. Various explicit examples are discussed, of interest to graduates and researchers.

Alessandro Tomasiello is Professor of Physics at the University of Milano–Bicocca. He has held various positions in Harvard University, Stanford University, and École Polytechnique, Paris, during the early stages of his career, and has been a plenary speaker at the annual Strings Conference several times. His research applies modern mathematical techniques to problems of string theory and modern high-energy physics.

Geometry of String Theory Compactifications

ALESSANDRO TOMASIELLO
University of Milano–Bicocca

CAMBRIDGE
UNIVERSITY PRESS

University Printing House, Cambridge CB2 8BS, United Kingdom

One Liberty Plaza, 20th Floor, New York, NY 10006, USA

477 Williamstown Road, Port Melbourne, VIC 3207, Australia

314–321, 3rd Floor, Plot 3, Splendor Forum, Jasola District Centre, New Delhi – 110025, India

103 Penang Road, #05–06/07, Visioncrest Commercial, Singapore 238467

Cambridge University Press is part of the University of Cambridge.

It furthers the University's mission by disseminating knowledge in the pursuit of education, learning, and research at the highest international levels of excellence.

www.cambridge.org
Information on this title: www.cambridge.org/9781108473736
DOI: 10.1017/9781108635745

First published 2022

A catalogue record for this publication is available from the British Library.

Library of Congress Cataloging-in-Publication Data
Names: Tomasiello, Alessandro, 1974- author.
Title: Geometry of string theory compactifications / Alessandro Tomasiello.
Description: Cambridge : Cambridge University Press, 2021. | Includes bibliographical references and index.
Identifiers: LCCN 2021029912 (print) | LCCN 2021029913 (ebook) | ISBN 9781108473736 (hardback) | ISBN 9781108635745 (epub)
Subjects: LCSH: String models–Mathematics.
Classification: LCC QC794.6.S85 T66 2021 (print) | LCC QC794.6.S85 (ebook) | DDC 539.7/258–dc23
LC record available at https://lccn.loc.gov/2021029912
LC ebook record available at https://lccn.loc.gov/2021029913

ISBN 978-1-108-47373-6 Hardback

Contents

Preface

There are already several excellent references on string theory [1–11]. This book focuses on one particular aspect: the geometry of the extra dimensions. Many interesting techniques have been developed over the years to find and classify string theory vacuum solutions, such as G-structures and pure spinors; I felt it would be useful to collect these ideas in a single place.

The intended audience is mostly advanced graduate students, but I tried to make the book interesting also to more experienced researchers who are not already working on this subject. I assume the reader has basic knowledge of general relativity, Lie groups, and algebras, and nodding acquaintance of the main ideas of supersymmetry. Proficiency in quantum field theory is very welcome but not heavily used. The basics of string theory and supergravity are recalled in Chapter 1, but in a presentation skewed toward the needs of the rest of the book, and not meant to give a complete picture of the field. Many details, such as the supersymmetry transformations, are postponed to a later stage, after a long mathematical detour in Chapters 2–7 allows us to present them with the appropriate level of sophistication.

Chapters 2 and 3 focus on the algebraic properties of spinors, and their deep relationship with forms. Here spacetime is taken to be flat. With respect to other introductions to these topics, I have emphasized the relation between a spinor and its bilinear tensors, and reviewed how forms can be considered as spinors for a doubled Clifford algebra; these are central to efforts in later chapters to rewrite supersymmetry in terms of exterior algebra. I have also considered a wide range of dimensions, both in Lorentzian and Euclidean signatures; this is perhaps a bit more than is really needed in later chapters, but it might be useful for readers who intend to go beyond the topics covered in the book. These chapters are also the most technically detailed; the aim was to teach how to carry out these computations as painlessly as possible – I have taken care to describe most steps, without resorting too often to the magical sentence "it can be shown that" to hide inaccessible derivations.

Chapters 4–7 are dedicated to geometry, but I have tried to keep them focused on physics needs. Chapter 4 is an introduction to differential geometry; these are standard topics, but sometimes they give an occasion to put the techniques of the earlier chapters to good use. In Chapter 5, we encounter G-structures, a well-known geometrical concept that has become very useful in supersymmetry. It is a very general framework, and I have discussed complex, Kähler, and Calabi–Yau geometry from this point of view. Kähler manifolds are those where computations are easiest, and so the entire Chapter 6 is devoted to them. Chapter 7 is devoted to manifolds with special holonomy where the Ricci tensor vanishes, such as Calabi–Yau's. This includes a lengthy close-up on conical manifolds, which are important later for AdS compactifications.

We get back to physics with Chapter 8. This is an elementary introduction to compactifications with relatively little mathematics, in the simple settings of pure gravity and string theory without flux fields. I have also provided here a quick review of four-dimensional supergravity, for later use. Chapter 9 starts with Calabi–Yau compactifications; these are not too realistic but are still the field's gold standard for rigor and depth. We later modify them by including D-branes and fluxes, covering in particular the important F-theory and conformal Kähler classes of Minkowski vacua.

Chapter 10 is a more systematic investigation of vacuum solutions. Here we finally introduce the supersymmetry transformation in full generality, and rewrite them in terms of forms using G-structures, and more precisely their *doubled* variants using pure forms from Chapter 3. This chapter is again rather technical at times, but the result is a very general system of form equations, which we can then use to look for supersymmetric solutions without having to consider spinors any more. Here and elsewhere, parts marked by an asterisk are harder and can be skipped on first reading. At the end of the chapter, we give a geometrical interpretation to this system in terms of so-called generalized complex geometry. We then proceed in Chapter 11 to a more detailed review of AdS_d solutions in $d \geq 4$, focusing on supersymmetric ones but mentioning supersymmetry-breaking in various instances. In some cases, we can give a complete classification of explicit solutions. We end in Chapter 12 with a quicker review of efforts to obtain dS vacua and of the swampland program, and with some final thoughts.

The book is not meant to be comprehensive. Notably, I have mostly focused on vacuum solutions, perhaps not paying enough attention to the broader geometry of reductions. After Chapter 1, I have devoted most attention to type II supergravity, and perhaps not enough to M-theory and heterotic strings. In general, I almost always avoided $d < 4$ vacua; and I have covered holography only superficially, in Chapter 11. Several other important topics have not been given the space they deserved. I hope readers disappointed by such omissions will forgive me after checking the total page count, which is already testing my editors' patience; I was also wary of the danger of producing a soulless encyclopedia. In general, the number of pages dedicated to a subject should not be construed as a judgment of its importance. I have been lengthier on topics that I feel are less thoroughly covered in other books, and sketchier on those where lots of great material is available already, and which I am including for context and completeness. On controversial issues, I have tried to represent all sides as fairly as I could; I have not tried to hide my opinion, but I believe the proper place to articulate it is in research articles.

I have learned these topics from my teachers, my collaborators, and my students. I am especially grateful to Loriano Bonora, Michael Douglas, Davide Gaiotto, Mariana Graña, Shamit Kachru, Dario Martelli, Ruben Minasian, Michela Petrini, and Alberto Zaffaroni. During the writing phase, I was helped by Bruno De Luca, Suvendu Giri, Andrea Legramandi, Gabriele Lo Monaco, Luca Martucci, Achilleas Passias, Vivek Saxena, and Riccardo Villa, and by the great staff at CUP. Special thanks go to Francesca Baviera, Concetta Fratantonio, and Luciano Tomasiello, although I am sorry the latter could not wait to see this finished. In spite of all this help, I am of course aware that the final product will turn out to have lots of typos, imprecisions, and outright mistakes; I will maintain a list of corrections on my personal website.

Conventions

- Lorentzian signature is "mostly plus."
- The word "generic" means "for any choice except for a set of measure zero."
- The antisymmetrizer of k indices is denoted by square brackets and includes a $1/k!$; for example, $v_{[m}w_{n]} = \frac{1}{2}(v_m w_n - v_n w_m)$. The symmetrizer is denoted by round brackets, so $v_{(m}w_{n)} = \frac{1}{2}(v_m w_n + v_n w_m)$. A vertical slash | is used to exclude indices from these operations.
- The floor function $\lfloor x \rfloor$ is the integer part of x or, in other words, the largest integer n such that $n \leq x$.
- The chiral matrix is $\gamma = c\gamma^0 \dots \gamma^{d-1}$ ($c\gamma^1 \dots \gamma^d$) for Lorentzian (Euclidean) signature. The constant c is constrained by (2.1.20) so that $\gamma^2 = 1$: notably, we take $c = \mathrm{i}$ for $d = 4$ Lorentzian, $c = -\mathrm{i}$ for $d = 6$ Euclidean, $c = 1$ for $d = 10$ Lorentzian.
- The identity matrix in d dimensions is denoted by 1_d or often simply by 1.
- d is most often the real dimension of a manifold; occasionally it denotes degree of a polynomial. Complex dimension is sometimes denoted by N.
- When working in an index-free notation, we use the same symbol v for a vector field with components v^μ and its associated one-form $g_{\mu\nu}v^\nu$.
- Indices $\mu, \nu \dots$ are for Lorentzian signature; $m, n, \dots$ for Euclidean signature; i, $j \dots$ are holomorphic indices. An exception is $d = 10$ (and $d = 26$) Lorentzian, where we use $M, N \dots$. Flat (vielbein) indices are $a, b \dots$. Indices $\alpha, \beta \dots$ are usually spinorial.
- The vielbein $e^a = e^a_m \mathrm{d}x^m$ is an orthonormal basis of one-forms, so the line element is $\mathrm{d}s^2 = g_{mn}\mathrm{d}x^m \mathrm{d}x^n = e^a e_a$. Its inverse is denoted by $E_a = E_a^m \partial_m$, an orthonormal basis of vector fields. We also often need a holomorphic vielbein, defined by (5.1.22), (5.1.35), and hence $\mathrm{d}s^2 = \sum_{a=1}^{d/2} h^a \bar{h}^{\bar{a}}$.
- The spinor covariant derivative is $D_m = \partial_m + \frac{1}{4}\omega_m^{ab}\gamma_{ab}$ (Section 4.3.3).
- The components of a k-form are defined by $\alpha_k = \frac{1}{k!}\alpha_{m_1 \dots m_k} \mathrm{d}x^{m_1} \wedge \dots \wedge \mathrm{d}x^{m_k}$. The Clifford map associates to it a bispinor $\alpha\!\!\!/_k \equiv \frac{1}{k!}\alpha_{m_1 \dots m_k}\gamma^{m_1 \dots m_k}$, with $\gamma^{m_1 \dots m_k} \equiv \gamma^{[m_1} \dots \gamma^{m_k]}$. For lengthy expressions, we also use the notation $(\alpha_k)_/$ $\equiv \alpha\!\!\!/_k$, but often we don't use any symbol at all and denote by α_k both a form and the associated bispinor, with an abuse of language.
- A vector field acts on a spinor as $v \cdot \eta \equiv v^\mu \gamma_\mu \eta = v_\mu \gamma^\mu \eta = v\!\!\!/\eta$, or also just $v\eta$ by the previous point.
- $\mathrm{vol} = e^1 \wedge \dots \wedge e^d$ is the volume form, while the volume of a manifold M is denoted by $\mathrm{Vol}(M)$.
- A chiral spinor η_+ is said to be pure if it is annihilated by $d/2$ gamma matrices; in flat space, this defines a notion of a (anti-)holomorphic index, for which we use the convention $\gamma_{\bar{\imath}}\eta_+ = 0$

- The complex conjugate of a complex number z is denoted by z^* or $\bar{z}$. For a complex matrix M, $M^\dagger \equiv (M^*)^t$. The conjugate of a spinor is $\zeta^c \equiv B\zeta^*$ (Section 2.3.1). In Lorentzian signature, we also define $\overline{\zeta} = \zeta^\dagger \gamma_0$.
- We typically use the letter ζ for spinors in Lorentzian signature, and η for Euclidean signature.
- Span$\{v_1, \ldots, v_k\}$ is the vector subspace of linear combinations of the v_a.

Introduction

The idea that spacetime might have additional dimensions might seem preposterous at first. It has come to the fore of current research in theoretical physics by two strands of thought.

0.1 String theory

One comes from attempts at quantizing gravity. The problems one encounters in general relativity at high energies suggest that it is superseded in that regime by a different theory. A prominent candidate is *string theory* (to be reviewed in Chapter 1). It describes interacting strings, which at low energies are seen as particles, some of which behave as gravitons, thus reducing at low energies to a version of general relativity (GR). The other particles behave in ways that look complicated enough to accommodate the phenomena we see in particle physics. So string theory solves the high-energy problems of GR, and gives a possible strategy to unify not only all forces but also all matter. Remarkably, the theory is so constrained by various anomalies that it has no free parameters. This is perhaps what one should expect from a unified theory of all physics.

String theory does come, however, with a heavy conceptual framework. This includes supersymmetry, which plays an important role in the theory's internal consistency; it could be broken spontaneously at Planck energies and be hidden from observations for a long time. More importantly for us, string theory only works in more than four dimensions. In its best-understood phase, *six* additional dimensions are needed, with a seventh also sometimes emerging. To avoid conflict with observations, we need to postulate that the compact space M_6 they span is small enough that current experiments have not revealed it yet. A *compactification* is a spacetime that looks four-dimensional macroscopically, even if it actually has a larger number of dimensions.

0.2 Kaluza–Klein reduction

This idea is natural enough that it had been considered long before string theory [12, 13]. The reason is that it gives a simpler, independent way to unify gravity with other elementary forces. This was first noticed in GR with a single additional dimension, wound up on a circle S^1 with an extra coordinate x^4. The various components of the metric are viewed by a four-dimensional observer as fields of different spin: $g_{\mu\nu}$ as a four-dimensional metric, $g_{\mu 4}$ as a vector field, and g_{44} as a scalar. The first two can be interpreted as describing gravity and electromagnetism in four dimensions.

The field dependence on the extra dimension x^4 also gives rise to a "tower" of massive spin-two fields with masses

$$m_k = \frac{2\pi k}{L}, \tag{0.2.1}$$

where L is the size of the S^1. A similar phenomenon can be seen already with a free scalar σ on $\mathbb{R}^4 \times S^1$: if we expand σ in a Fourier series with respect to x^4 and plug the expansion in the Klein–Gordon equation $(\partial_\mu \partial^\mu + \partial_4^2)\sigma = 0$, the term ∂_4^2 gives a mass (0.2.1) to the kth Fourier mode. With gravity, $g_{\mu\nu}$, $g_{\mu 4}$, and g_{44} all undergo the same phenomenon: the massive spin-two fields then "eat" the massive modes of the other components, in a version of the Brout–Englert–Higgs (BEH) mechanism. The infinite sequence of masses (0.2.1) is called a *KK tower*, and the corresponding fields are called *KK modes*.

As expected by dimensional analysis, (0.2.1) are inversely proportional to L; so when L is small, these masses are large and might have avoided detection so far. Even $L \sim 10^{-19} m$ leads to $m_k \sim O(\text{TeV})$. We will review in more depth the physics of this five-dimensional model in Section 8.1.1.

With $d > 1$ additional dimensions, one can consider more complicated spaces M_d, which can now also realize Yang–Mills (YM) theories. The symmetry group of M_d becomes the YM gauge group. From this point of view, the idea of extra dimensions is an evolution of that of "internal symmetry" in the world of elementary particles. It is the postulate that those symmetries have a geometrical origin.

0.3 String compactifications

The topic of this book is the study of the "internal space" of string theory. While the theory itself has no free parameters, the choice of M_6 introduces a lot of freedom. As we will see in Section 4.2.5, the possible topologies for a six-dimensional compact space are classified by a few algebraic data (the dimensions of two vector spaces and two polynomial functions). But the space of possible metrics for each given topology is infinite dimensional.

Perhaps the simplest question we can ask is whether by compactifying on M_6 we can find at least a *vacuum solution*: namely, one where the macroscopic spacetime is empty. This means that the stress energy tensor is zero, or consists at most of a cosmological constant. Such a space should locally have as many symmetries as flat Minkowski space, and is said to be *maximally symmetric (MS)* (Section 4.5); the possibilities are Minkowski space itself, de Sitter (dS) space, and anti-de Sitter (AdS) space, with a positive, zero, or negative cosmological constant Λ. We will argue in Chapter 8 that for such a solution the line element for the ten-dimensional metric reads

$$ds_{10}^2 = e^{2A} ds_{\text{MS}_4}^2 + ds_{M_6}^2 \,. \tag{0.3.1}$$

$A = A(y)$ is called the *warping function* of the coordinates y^m on M_6, and this form of the metric is called *warped product*.

The question of which M_6 lead to vacuum solutions is already rather hard: it involves solving the equations of motion, which reduce to partial differential equations on M_6. An easy case is when all of string theory's fields have zero expectation value except the metric; the equations of motion then say that $R_{MN} = 0$. These imply that the maximally symmetric $\mathrm{MS}_4 = \mathrm{Mink}_4$, and that M_6 is Ricci-flat (Chapter 7). Many such spaces are the so-called *Calabi–Yau manifolds*. When the other fields are also present, finding and classifying solutions is much harder. This is one of the main topics of this book.

After finding a vacuum solution, one would like a description of the physics one would observe in it. Here one faces a choice between precision and broadness. At one extreme, the *KK spectrum* is the information about all the particle masses and spins for a single given vacuum, but without any information about their interactions.

At the other extreme, one focuses on a small subset of fields, but with complete information about their interactions, and in particular about the potential for the scalars. Such an action S_4 might describe many vacua at once. As usual, we call it an effective theory if the scalars we kept span a "valley" with a relatively mild potential V_{eff}, much smaller than the potential for the scalars we discarded. Sometimes in string theory we are forced to work without such a scale separation; one calls this a *nonlinear reduction*, and we then want at least that its vacua correspond to vacuum solutions of fully fledged string theory. We will give a longer introduction to these ideas in Section 8.3, and then see several examples in later parts of the book.

Sometimes one reverses this procedure, and one uses an effective theory or nonlinear reduction to find new vacuum solutions rather than to describe the physics of one that was previously found. Some of the theory's most celebrated solutions have been found this way. However, our focus will be on techniques to find vacua directly in ten (or eleven) dimensions.

0.4 Supersymmetric vacua and geometry

To find vacuum solutions, it proves easier to start from those where supersymmetry is partially preserved and to break it later, rather than trying to solve string theory's equations of motion in general. Technically, preserved supersymmetry gives a first-order system of equations, which partially implies the second-order equations of motion. More importantly, supersymmetry helps finding solutions because it naturally invokes several deep geometrical ideas.

We cannot cover the internal M_6 with a single coordinate system; we have to use several, related by coordinate changes called transition functions, in general valued in the group $\mathrm{GL}(6, \mathbb{R})$ of invertible matrices. A *G-structure* on M_6 is a choice of transition functions valued in a smaller group G (Chapter 5). The infinitesimal parameters for supersymmetry are spinors η; they naturally define a G-structure, with $G = \mathrm{Stab}(\eta)$ their little group (or *stabilizer*), the group of rotations that keep η invariant. This helps trading η with other, nonspinorial geometrical objects on M_6; often antisymmetric tensors, or *forms*.

For example, a single η on M_6 defines an SU(3)-structure, which can also be defined by the metric and a complex three-form (an antisymmetric tensor with three indices) Ω. It is further possible to trade even the metric for a real two-form J. The G-structure techniques allow one to recast the supersymmetry equations, originally involving η, directly in terms of J and Ω. Often one then recognizes a well-known mathematical concept, and this helps finding solutions. This procedure is also natural because most of the string theory fields beyond the metric are themselves forms, analogues with many indices of the electromagnetic field-strength $F_{\mu\nu}$.

Some of the most interesting vacua are in type II, where there are two η^a. In this case, it proves more fruitful to work with a doubled, or *generalized*, version of the rotation group. Again, the upshot is that we may trade the data of the η^a and of the metric with forms, this time a pair $\Phi_\pm$ of them with an algebraic property called *purity* (Chapters 2 and 3). In this language, the supersymmetry equations become particularly elegant, making contact with *generalized complex geometry* (Chapter 10).

The fact that the metric is included in this trade-off with forms is particularly intriguing. It is reminiscent of previous attempts to reformulate GR such as [14, 15].

0.5 The cosmological constant

The observed cosmological constant is positive, so one would like to focus on the case $\text{MS}_4 = \text{dS}_4$. These are actually the hardest solutions to obtain. To see why, consider a general gravitational theory with an Einstein–Hilbert (EH) kinetic term [16–18]. The equations of motion read $R_{MN} - \frac{1}{2}g_{MN}R = 8\pi G_{\text{N}}T_{MN}$, where as usual G_{N} is Newton's constant, R_{MN} is the Ricci tensor, R its trace, and T_{MN} the stress–energy tensor of various matter fields; the indices $M, N = 0, \ldots, 3 + d$. The "trace-reversed" Einstein equations are

$$R_{MN} = 8\pi G_{\text{N}} \left(T_{MN} - \frac{1}{2+d} g_{MN} T_P{}^P \right) . \tag{0.5.1}$$

For a warped product metric as in (0.3.1) (generalizing $M_6 \to M_d$),

$$R_{\mu\nu} = \left(\Lambda - \frac{1}{4} \text{e}^{-2A} \nabla^2 \text{e}^{4A} \right) g^4_{\mu\nu} , \tag{0.5.2}$$

where $g^4_{\mu\nu}$ are the components of the metric of the external MS_4 space, Λ its cosmological constant (normalized as $R^4_{\mu\nu} = \Lambda g^4_{\mu\nu}$), and ∇ the internal covariant derivative. (We will derive (0.5.2) in an exercise in Chapter 4.) In coordinates where $g^4_{00} = -1$, the time components of (0.5.1) give

$$- \text{e}^{2A}\Lambda + \frac{1}{4}\nabla^2 \text{e}^{4A} = 8\pi G_{\text{N}} \text{e}^{2A} \left(T_{00} - \frac{1}{2+d} g_{00} T_P{}^P \right) . \tag{0.5.3}$$

The parenthesis on the right-hand side is nonnegative if the higher-dimensional theory obeys the *strong energy condition*. This is an assumption often made in general relativity, for example in proving singularity theorems for black holes and cosmology; see, for example, the discussions in [19, 4.3;8.2] and [20, chap. 9].

Supposing it holds, we integrate (0.5.3) on the compact M_6; the second term on the left-hand side is a total derivative and gives no contribution, so

$$\Lambda \leq 0\,. \tag{0.5.4}$$

We conclude that a theory with an EH kinetic term obeying the strong energy condition has *no de Sitter compactifications*. Even Minkowski compactifications are only marginally allowed, requiring the parenthesis in (0.5.3) to vanish.

Does this apply to string theory? We mentioned that it reduces at low energies to a version of GR, which due to supersymmetry is called *supergravity*. In this regime, the graviton has an EH kinetic term, coupled to various other fields and to certain localized sources. All the fields satisfy the strong energy condition, except for a term called *Romans mass*. But this possible loophole in the argument was closed in [18] (Section 10.3.1).

For localized sources, violating the strong energy condition requires negative tension, leading to repulsive gravity, and usually to instabilities for dynamical objects. The sources in string theory are of two types, called *D-branes* and *O-planes* (Sections 1.3 and 1.4.4). The first are defined as spacetime defects on which strings can end; this makes them dynamical. The second arise after quotienting string theory by a parity-like symmetry, and arise at its fixed loci; so they are not dynamical. Some of them have indeed negative tension, and so they do invalidate (0.5.3).

The conclusion is that, in the regime where string theory is described by supergravity, de Sitter compactifications require O-planes. Minkowski compactifications also need them, unless all the fields except gravity are turned off.

0.6 Beyond supergravity

Let us consider now a $d = 10$ effective field theory S_{10}. This is useful in a regime of energies high enough to see the extra dimensions, but low enough to see strings as particles; it is not to be confused with the $d = 4$ effective action S_4, relevant at lower energies where we cannot resolve the extra dimensions. In S_{10}, supergravity is the collection of the most relevant operators at energies well below the Planck scale, which in turn is related to a new fundamental length scale l_s, the "typical length" of strings. But supergravity is not renormalizable; this manifests itself in the presence of higher-derivative corrections, which become relevant at high energies, or when the curvature gets large. These are not known completely, but there is no reason to expect that the result (0.5.4) still holds when they are introduced. For example, one famous leading correction has the form $\int \mathrm{d}^{10}x(\text{Riemann})^4$. So not even the metric appears simply through an EH kinetic term, as in supergravity. Unfortunately, only the first few terms have been computed.

This introduces other challenges. These corrections will contribute to the potential of the $d = 4$ effective action S_4 (Chapter 8), whose vacua should approximate string theory's vacuum solutions as defined earlier. Very naively, suppressing the dependence on other fields, we will see that the EH term and the $(\text{Riemann})^4$ term give two contributions to the potential:

$$V_4 \sim ar^{-2} + br^{-8}\,, \tag{0.6.1}$$

where $r \equiv R/l_s$ is the length scale of the internal space M_6, in units of the string length. If a and b have opposite sign, this has an extremum at $r = (-4b/a)^{1/6}$; but if $a, b \sim O(1)$, also $r \sim O(1)$, and $R \sim l_s$. But in this regime, other (Riemann)^k, $k > 4$ corrections would also be relevant, contributing further terms r^{-2k} to (0.6.1); so we cannot trust our extremum. This illustrates a general issue: if we find a solution by using one string correction to supergravity, we can expect it to be in a regime where *all* string corrections are relevant, where we in fact cannot compute anything. This is the *Dine–Seiberg problem* [21].

Fortunately, not all the terms in the $d = 4$ effective potential are $\sim O(1)$. The aforementioned form fields have to satisfy a certain Dirac quantization; this introduces integers, which can be taken to be large, introducing a hierarchy that eventually makes $R \gg l_s$. From the point of view of S_4, this is behind the existence of most vacua, but usually the terms that compete originate from the leading supergravity approximation. (For the Calabi–Yau vacua, the leading supergravity contribution vanishes; we will see in Chapter 9 a more delicate argument to show that the corrections don't destroy the vacuum.)

0.7 Overview of vacua

The argument (0.5.4) indicates that finding vacua is easiest when the cosmological constant is negative, which as we mentioned is contrary to observations. These AdS vacua have found applications in *holography*, which relates them to quantum field theory models with conformal invariance, or conformal field theories (CFTs). Another reason not to discard them is that the supersymmetry-breaking procedure sometimes also changes Λ. For these reasons, we dedicate Chapter 11 to a survey of such vacua. Several classification results are available here, and several more are likely to emerge in the near future, as techniques improve. For example, a list of all supersymmetric AdS_6 and AdS_7 solutions has been achieved relatively recently.

Minkowski vacua are more tightly constrained. Relative to AdS, this is expected, if nothing else because $\Lambda = 0$ is an equality, not an inequality. A priori the general equations seem to allow for M_6 of any curvature; but so far the vast majority of known supersymmetric vacua are related to Calabi–Yau manifolds, one way or another. For example, in a famous class in Chapter 9 the Calabi–Yau metric is only modified by an overall function. Until recently, one might have thought this to be an artifact of technical limitations in including O-planes, which as we have already argued are necessary. However, many AdS vacua with O-planes have now been constructed and seem to allow for a far greater variety of internal spaces.

Finally, for the de Sitter case, the situation gets even less clear (Chapter 12). One additional complication is that supersymmetry is necessarily broken and cannot guide us any more. Most models evade (0.5.4) by involving both O-planes and quantum effects, which are harder to control. As a result, all of them have attracted some objections.

The first and most successful proposal, the KKLT model [22], again obtained by modifying a Calabi–Yau metric on M_6, generates a large quantity of dS vacua, with numbers such as 10^{hundreds} or even $10^{\text{thousands}}$ often quoted. The resulting picture has been dubbed the *string theory landscape* [23], borrowing a metaphor from protein-folding research. Leaving aside for now any criticism of this and other models, the possibility that a seemingly formidable obstacle such as the cosmological constant could be overcome so easily has suggested that the number of vacua reproducing *all* other observed features of our Universe could still be very large. This has created much confusion in causal observers of the field. Perhaps string theory has no predictive power?

This question appears misguided. First of all, the large numbers 10^N arise from discretizing an N-dimensional continuous space; in other words, from allowing several discrete possibilities to a set of N free parameters. Conceptually, this is not that different from the 19 (or more) free parameters in the standard model, which prior to experiment has an ∞^{19} of possibilities. Rightly, no one complains about the latter large number because, given enough experiments to fix the parameters within a certain range, the standard model makes testable predictions about other experiments. This would be true as well for string theory; the fact that such experiments are beyond human capabilities for the foreseeable future is of course unfortunate, but is in the nature of the problem of quantum gravity, whose characteristic scale is after all m_{Planck}.

Even more importantly, string *theory* is a framework, within which there are *models* with free parameters, such as the KKLT model. It would of course be senseless to criticize quantum field theory because it cannot predict the standard model of particle physics from first principles, or criticize quantum mechanics because it does not predict the potential in the Schrödinger equation. Of course, quantum field theory is a scientific *theory*, in the appropriate sense: given enough experimental data, it can provide a *model*, such as the Standard Model, which makes new experimental predictions.

An alternative point of view is to focus on the vacua that *cannot* be found in string theory. The *swampland* program [24] looks for models that look consistent in field theory but cannot be coupled to quantum gravity. While the inspiration often comes from string theory, the aim is to find universal properties that are valid beyond it. In recent years, this program has started to clash with many of the predictions of the effective field theory approach, including the existence of dS vacua. Some of this debate is covered in the final Chapter 12, with the unfortunate result of ending with more questions than answers.

1 String theory and supergravity

As stated in the Preface, this book assumes some rudimentary knowledge of string theory, but it is a good idea to recall the basics. The field is notoriously vast and complex, so this chapter should not be understood as a replacement for serious study on one of the many great introductions [1–11]. In most of the book, we will approximate string theory by *supergravity*, an effective theory of gravitons and other fields; the presentation will be biased toward that.

In this chapter, we also assume knowledge of general relativity (GR) and some acquaintance with spinors, but we will try to keep mathematical sophistication at a minimum. We will develop some ideas, such as spinors and differential geometry, in much greater detail in the next few chapters before we return to physics. Still, already in this chapter we will pepper our presentation with occasional forward references to those mathematically more advanced treatments, to whet the reader's appetite.

1.1 Perturbative strings

A quantum field collects creation (and annihilation) operators for a representation of the Poincaré group. Once one fixes the value of the momentum p of the created state, the remaining degrees of freedom are a representation of the *little group*, or *stabilizer*, of p, namely the subgroup $\mathrm{Stab}(p) \subset \mathrm{SO}(d)$ of elements that leave p invariant. This is

$$\mathrm{Stab}(p) = \mathrm{SO}(d-1)\ (p^2 < 0)\,, \qquad \mathrm{Stab}(p) = \mathrm{SO}(d-2) \ltimes \mathbb{R}^{d-2}\ (p^2 = 0)\,, \quad (1.1.1)$$

for the massive or massless case. (We will review the $p^2 = 0$ case in Section 3.3.6.) In the massless case, we would also have the possibility of selecting an infinite-dimensional representation, but this is usually regarded as exotic; so we select a finite-dimensional representation, ignoring the $\mathbb{R}^{d-2}$ factor. Ordinary fields then represent objects with finitely many degrees of freedom, which we call spin and helicity for $m^2 > 0$ and $= 0$, respectively. Moreover, we usually take these objects to interact via terms of the type $\int \mathrm{d}^d x \phi_1(x) \ldots \phi_2(x)$: these allow the value of a field to influence directly that of another only at the same point.

All these reasons make us think of the quanta of a field as point particles. To describe a quantum theory of interacting extended objects, we need to change this picture somehow. First of all, a string can have infinitely many vibration modes, so a field that creates a string must be somehow a collection of infinitely many ordinary fields. Second, extended objects can interact when their centers of mass are not superimposed. So the interaction terms should be nonlocal.

Such a *string field theory* (SFT) is fascinating but also just as complicated as our description suggests. So in fact most studies of interacting strings focus on an approach that is *first-quantized*: one first decides the Feynman diagram one wants to consider, and then computes the amplitude associated with it. (A similar approach is used sometimes in quantum field theory too, under the name of *world-line formalism*.)

In this section, we will review quickly some aspects of this perturbative treatment of string interactions. There are five possible consistent string models:

- Type IIA
- Type IIB
- Heterotic with gauge group $E_8 \times E_8$
- Heterotic with gauge group SO(32)
- Type I

All these select $d = 10$ as spacetime dimension, in a sense we will clarify later in this chapter. The last case, type I, can be viewed as a certain quotient procedure from IIB strings, which we will introduce in Section 1.4.4. So in this section we will discuss the other four. We will actually start our discussion from a model that has a *tachyon*, namely a scalar with a negative mass, but whose discussion is simpler: the *bosonic* string.

1.1.1 Bosonic strings

The action for a particle moving in a curved background is proportional to its "length in spacetime," namely, to the proper time measured along its world-line (its trajectory γ in spacetime):

$$S_{\text{part}} = -m \int_\gamma \mathrm{d}\sigma_0 \sqrt{-g_{\mu\nu}\dot{x}^\mu \dot{x}^\nu}, \tag{1.1.2}$$

where $x^\mu(\sigma_0)$ are the coordinates of the point in spacetime as a function of the world-line coordinate σ_0, and $\dot{x}^\mu \equiv \partial_0 x^\mu$. In flat space, this is indeed minimized on straight lines in spacetime, which maximize proper time. For curved $g_{\mu\nu}$, (1.1.2) is minimized on geodesics. If we also have a Maxwell field and our particle is charged, we have to add a term

$$S_{\text{part,EM}} = q \int_\gamma \mathrm{d}\sigma_0 A_\mu \partial_0 x^\mu, \tag{1.1.3}$$

where q is the charge, and A_μ is the vector potential. In Section 4.1.4, we will see that the integrand is an example of a natural operation called *pull-back*.

String action

By analogy with (1.1.2), the natural action for a string would seem to be the volume of its two-dimensional world-sheet in spacetime. However, it is classically equivalent to the *Polyakov action*, which is easier to quantize:

$$S_{\text{F1},g} = -\frac{1}{2}T_{\text{F1}} \int_\Sigma \mathrm{d}^2\sigma h^{\alpha\beta}\sqrt{-h}g_{MN}\partial_\alpha x^M \partial_\beta x^N . \tag{1.1.4}$$

This type of action is also called a *sigma model*, for reasons going back to four-dimensional models of mesons, or sometimes *nonlinear* sigma model when g_{MN} is not flat. The $x^M(\sigma_0, \sigma_1)$, $M = 0, \ldots, d-1$, describe the embedding of Σ in physical spacetime (often called *target space*), and h is a metric on Σ. The mass m in (1.1.2) has been replaced by the mass/length ratio, or *tension*:

$$T_{\mathrm{F1}} = \frac{1}{2\pi l_s^2} \,. \tag{1.1.5}$$

"F" stands for *fundamental*, to distinguish this string from other extended objects that will appear later; 1 denotes the space extension of the string. The constant l_s is called *string length*. (We will always keep it explicit in this chapter, but later we will often work in string units and set $l_s = 1$.)

In this section, we are going to focus on strings that are *closed* or, in other words, that have no boundary. A generic[1] time slice is then a collection of several copies of the circle S^1. The time evolution of each of these for a finite time will be a cylinder; then σ^1 is a periodic coordinate, $\sigma^1 \sim \sigma^1 + \pi$. These cylinders are then glued together at some values of σ^0 to obtain a general Σ.

Spectrum in flat space

Quantizing (1.1.4) is challenging for general g_{MN} but relatively easy in Minkowski space $g_{MN} = \eta_{MN}$: superficially (1.1.4) then becomes a collection of free bosons, with equations of motion $\partial^2 x^M = 0$. For a closed string, the slice at $\sigma^0 =$ constant is an S^1; there are then discrete Fourier modes for each x^M. Since the equation of motion is of second order, the states are in correspondence to the values of these Fourier modes and their derivatives. Alternatively, we can write a solution of the world-sheet equations of motion as $x^M = x^M(\sigma^+) + x^M(\sigma^-)$, where $\sigma^\pm = \sigma^1 \pm \sigma^0$, and introduce Fourier modes α_i^M, $\tilde{\alpha}_i^M$ for the left- and right-movers $x^M(\sigma^\pm)$. The only subtlety is that the world-sheet metric $h_{\alpha\beta}$ is a Lagrange multiplier, which gives a constraint. This can be taken care of in many ways: by solving the constraint, or by introducing Faddeev–Popov ghosts and the Becchi–Rouet–Stora–Tyutin (BRST) method (the so-called covariant quantization). Skipping many interesting details, here we will just give the results.

Even for a fixed momentum, the spectrum has infinitely many states, of the form

$$\alpha_{-i_1}^{N_1} \ldots \alpha_{-i_n}^{N_n} \tilde{\alpha}_{-\tilde{\imath}_1}^{\tilde{N}_1} \ldots \tilde{\alpha}_{-\tilde{\imath}_n}^{\tilde{N}_n} |0\rangle, \tag{1.1.6}$$

where $|0\rangle$ is the world-sheet vacuum, and i_k, $j_k \geq 0$ (possibly repeated). As we mentioned, these correspond to the vibration modes of the string, and in a spacetime picture they would require infinitely many ordinary quantum fields to create them. Their masses are

$$m^2 = \frac{4}{l_s^2}\left(\frac{2-d}{24} + N\right), \tag{1.1.7}$$

where $N = \sum i_k = \sum \tilde{\imath}_k$ is a nonnegative integer. The identity between these two expressions is called *level matching* and is the link between the left- and right-moving sectors, which otherwise proceed on parallel tracks. If $d > 2$, we see that the lowest

[1] The mathematical meaning of the word *generically*, which we will use in this book, is "for any choice except for a set of measure zero."

value of m^2, for $N = 0$, is actually *negative*. Such a mode is usually called a *tachyon* and signals an instability. For this reason, the bosonic string we are discussing in this section is usually only considered a toy model.

Nevertheless, it already displays a very interesting feature. For the *critical dimension* $d = 26$, the modes with $n = 1$ in (1.1.6) and (1.1.7) are massless. They read

$$\alpha^M_{-1}\tilde{\alpha}^N_{-1}|0\rangle\,, \tag{1.1.8}$$

and so they correspond to fields with two indices. Among these we thus find a *massless spin-two field* $h_{MN} = \delta g_{MN}$. The action (1.1.4) can then be thought of as a string moving in a condensate of such a field. This is a bit similar to expanding a quantum field theory (QFT) around a vacuum where a field has acquired a nonzero expectation value.

So we have found that in $d = 26$, the string modes include those that would normally be associated with a graviton. Remarkably, the scattering amplitudes one obtains with this formalism are *finite*. The string tension acts as a regulator: Taking the limit $l_s \to 0$, the scattering amplitudes become divergent again. In this limit, the theory becomes a local QFT model again, and a local theory of gravity has divergent amplitudes.

Coupling to condensates of other fields

Among (1.1.8), we find other massless modes. Following (1.1.1), we need to consider only the components of (1.1.8) in the $d - 2 = 24$ dimensions transverse to the momentum p, which are 24^2. The physical components h_{MN} of a graviton are represented by a traceless 24×24 matrix; this is the generalization of the transverse traceless (TT) gauge familiar from the treatment of gravitational waves in four dimensions. The remaining modes are thus the antisymmetric part of (1.1.8) and its trace. The fields that create these states are an antisymmetric *Kalb–Ramond* field $B_{MN} = -B_{NM}$, and a scalar field ϕ called *dilaton*. So in total the massless fields of the bosonic string are

$$g_{MN}\,, \qquad B_{MN}\,, \qquad \phi\,. \tag{1.1.9}$$

We can consider condensates of B_{MN} and ϕ, too; this leads to the extra terms in the action:

$$S_{\mathrm{F1},B,\phi} = -\frac{1}{2}T_{\mathrm{F1}}\int_\Sigma \mathrm{d}^2\sigma\left[\epsilon^{\alpha\beta}B_{MN}\partial_\alpha x^M\partial_\beta x^N + l_s^2\sqrt{-h}R_{(2)}\phi\right]. \tag{1.1.10}$$

Here $R_{(2)}$ is the scalar curvature of the world-sheet metric $h_{\alpha\beta}$, and $\epsilon = \left(\begin{smallmatrix}0 & -1\\ 1 & 0\end{smallmatrix}\right)$. The coupling with B is the natural generalization of the coupling (1.1.3). The coupling with the dilaton is peculiar in that

$$\frac{1}{4\pi}\int_\Sigma \sqrt{-h}R_{(2)} = 2 - 2g\,, \tag{1.1.11}$$

where g is the *genus* of the world-sheet Σ. This is the stringy analogue of the number of loops, and can be intuitively described (when Σ has no boundary) as the *number of handles*; a more formal definition will be given in Section 4.1.10. Because of this, the computation of all scattering amplitudes is organized in powers:

$$g_s^{2g-2}\,, \qquad g_s \equiv e^{\phi}\,. \tag{1.1.12}$$

We can think of g_s as a *string coupling constant*: when it is small, the powers (1.1.12) are smaller for Riemann surfaces Σ of increasing g, which can be thought of as the stringy analogue of Feynman diagrams of increasing complexity.

The action

$$S_{\text{bos}} = S_{\text{F1},g} + S_{\text{F1},B,\phi} \tag{1.1.13}$$

is classically invariant under general coordinate transformation $\sigma_\alpha \to \sigma'_\alpha(\sigma_0, \sigma_1)$, if we also take care to transform the world-sheet metric $h_{\alpha\beta}$. This is a gauge invariance, in that it doesn't affect the physical configuration, the image of the world-sheet embedding $x^\mu(\sigma)$, but only how we parameterize it. Equation (1.1.13) is also invariant under Weyl rescaling $h_{\alpha\beta} \to e^f h_{\alpha\beta}$. In two dimensions, one can fix the coordinate-change freedom by taking, for example, $h_{\alpha\beta}$ to have constant scalar curvature. Even so, a residual invariance remains: coordinate transformations that leave the metric invariant up to a Weyl transformation. These are called *conformal transformations*.

Conformal invariance and effective action

It is crucial that this residual gauge invariance remains at the quantum level. It decouples potentially harmful negative-norm states that would come from the fact that x^0 in (1.1.4) has a wrong-sign kinetic term. This is similar to what happens in the quantization of the electromagnetic field, for example. Conformal invariance is also behind the absence of high-energy divergences. Usually scattering amplitudes become problematic when two particles collide at a small impact parameter. The world-sheet of a string scattering is a non-compact Riemann surface with several spikes s_i corresponding to the incoming and outgoing strings. Conformal invariance means that the distance between two points on the world-sheet has no intrinsic meaning: only ratios of distances do. So a small impact parameter might seem to correspond to two such spikes s_1 and s_2 getting close, but that only means that they are close relative to their distance from other external strings s_i. This corresponds to a Riemann surface that develops a long neck, where the two s_i are both attached, far from the others.

The Noether current associated to dilatations in a field theory is $T^{\mu\nu}x_\nu$, where $T_{\mu\nu}$ is the stress–energy tensor. This is conserved if $0 = \partial_\mu(T^{\mu\nu}x_\nu) = T^{\mu\nu}g_{\mu\nu} = T^\mu_\mu$. Evaluating the expectation value $\langle T^\mu_\mu\rangle$ of this trace is thus a way to check if there is a *Weyl anomaly*.

From the point of view of the world-sheet, the spacetime fields (1.1.9) are really couplings for the action of the fields $x^M(\sigma)$. So a Weyl anomaly can also be detected by computing the beta functions of the action (1.1.13) for the couplings (1.1.9). This can be obtained by the usual perturbative methods; the coupling for this computation is given by l_s^2, or rather the dimensionless combination $l_s^2\times$ (spacetime curvature). This results in the following three conditions:

$$R_{MN} + 2\nabla_M \partial_N \phi - \frac{1}{4} H_{MPQ} H_N{}^{PQ} + O(l_s^2) = 0\,, \tag{1.1.14a}$$

$$\nabla_M (\mathrm{e}^{-2\phi} H^M{}_{NP}) + O(l_s^2) = 0\,, \tag{1.1.14b}$$

$$\frac{2}{3l_s^2}(26 - d) + R - \frac{1}{2}|H|^2 - 4\mathrm{e}^{\phi}\nabla^2 \mathrm{e}^{-\phi} + O(l_s^2) = 0\,. \tag{1.1.14c}$$

We have introduced

$$H_{MNP} = \partial_M B_{NP} + \partial_N B_{PM} + \partial_P B_{MN}\,, \qquad |H|^2 \equiv \frac{1}{6} H_{MNP} H^{MNP}\,. \tag{1.1.15}$$

This can be considered as a *field-strength* for the *potential* B_{MN}, similar to the relation between $F_{MN} = \partial_M A_N - \partial_N A_M$ and A_M in electromagnetism. Indeed, there is also a gauge transformation

$$B_{MN} \to B_{MN} + \partial_M \hat{\lambda}_N - \partial_N \hat{\lambda}_M\,, \tag{1.1.16}$$

under which (1.1.15) is invariant. The world-sheet action (1.1.10) is invariant too under this, because the transformation adds a total derivative term.

From spacetime point of view, where (1.1.9) are fields, (1.1.14) are to be interpreted as equations of motion. They can be obtained by extremizing[2]

$$S_{\rm bos} = \frac{1}{2\kappa_{\rm b}^2} \int \mathrm{d}^d x \sqrt{-g}\, \mathrm{e}^{-2\phi} \left(\frac{2}{3l_s^2}(26 - d) + R + 4\partial_M \phi \partial^M \phi - \frac{1}{2}|H|^2 + O(l_s^2) \right) \tag{1.1.17}$$

with respect to (1.1.9). By dimensional reasons, $\kappa_{\rm b}$ has dimension l_s^{12}. (The metric coefficients have no dimension, while R contains two derivatives and has mass dimension two.) In general, the *Planck mass* $m_{\rm P}$ is defined as the mass scale entering the Einstein–Hilbert action; the *Planck length* $l_{\rm P}$ is its inverse, and (1.1.17) tells us that it is proportional to l_s.

As a consistency check, we see that flat space is a solution of (1.1.14) only if we set $d = 26$, which is the value where we found the massless fields (1.1.9) in the first place. More generally, to trust (1.1.14) we have to make sure that the expansion parameter $l_s^2\times$ (curvature) is small, so we better solve those equations of motion separately at every order. This leads again to taking

$$d = 26\,. \tag{1.1.18}$$

It is conceptually possible to consider solutions where $d \neq 26$, and the first term in (1.1.14c) competes with the others, but in that case we have to worry that the other terms in the l_s expansion become relevant too, and we have not given them in (1.1.14). If, on the other hand, one is able to prove that a certain world-sheet model is conformal exactly, without using the l_s expansion at all, then $d = 26$ is not necessary. There are not many such cases: one is the *linear dilaton* background, where ϕ is linear in one of the coordinates. This leads to *noncritical string theories*, which historically have been important toy models.

Another point of view on the critical dimension is this. We observed that (1.1.13) has conformal invariance. Conformal transformations form a group; for flat space it is SO$(d-2, 2)$ for $d > 2$, but for $d = 2$ it becomes infinite dimensional. Indeed,

[2] This variation is a little more involved than the usual Einstein–Hilbert action variation because of the prefactor $\mathrm{e}^{-2\phi}$. More details will be given in Section 10.1.2.

any transformation $x^\pm \to x^{\pm\prime}(x^\pm)$ is conformal for any metric of the type $ds^2 = e^f dx^+ dx^-$. The generators L_m, $m \in \mathbb{Z}$, of such transformations on the x^+ obey the Lie algebra

$$[L_m, L_n] = (m-n)L_{m+n} + \frac{c}{12} m(m^2-1)\, \delta_{n+m,0}\,, \tag{1.1.19}$$

called *Virasoro algebra*. The L_0, $L_{\pm 1}$ form an SO(1, 2) subalgebra where c does not appear. As usual, spacetime transformations are generated by the stress–energy tensor, so these L_m are related to it. After a Wick rotation, $x_+ \to z = \sigma^1 + i\sigma^0$, and we can collect all the generators in

$$T_{zz}(z) = \sum_n L_n z^{-n-2}\,. \tag{1.1.20}$$

In a Lie algebra, the commutation relations should always be linear, so we need to think of the second term in (1.1.19) as containing a new generator c, which commutes with all the others, and thus lies in the center of the algebra; so c in (1.1.19) is called *central charge*. The $\tilde{L}_m$ on the x_- variable generate a second copy of the same algebra (1.1.19), and they are collected in $T_{\bar{z}\bar{z}}$.

This c is also a measure of the Weyl anomaly: for any QFT model that is conformal on a flat (world-sheet) metric $h_{\alpha\beta} = \eta_{\alpha\beta}$, a nonzero c tells us that conformal invariance is broken for more general $h_{\alpha\beta} \neq \eta_{\alpha\beta}$. A free boson contributes $c = 1$, while the ghosts give −26. Thus if we quantize *around flat space*, where the $x^M(\sigma)$ bosons are free, for quantum conformal invariance we need to take $d = 26$.

The fact that the action (1.1.17) exists at all is nontrivial from the point of view of the world-sheet derivation we described. We can think of it as being an approximation to the string field theory action S_{SFT}, which would also contain the massive fields creating all the states (1.1.6). We can call it an effective action, in the usual quantum field theory sense: It reproduces the results one would obtain from S_{SFT}, at energies that are *low*, namely much smaller than l_s^{-1}. Indeed, another way to compute (1.1.17) is to compute string scattering amplitudes using the world-sheet approach, and then guessing what spacetime action would reproduce them.

The diagrams leading to (1.1.17) have $g = 0$ in (1.1.12), leading to $g_s^{-2} = e^{-2\phi}$, thus explaining the presence of that exponential. The higher powers of l_s hidden in (1.1.17) also receive contributions from higher values of g (and thus from more complicated Feynman diagrams). So the effective action will have a double expansion in powers of both:

$$S = \sum_{j,k} S_{j,k} l_s^j e^{k\phi}\,. \tag{1.1.21}$$

These higher-order corrections can in principle be computed; we will see some examples for superstrings. When we first discover that GR is non-renormalizable and needs (curvature)^2 counterterms [25], we might perhaps hope that by adding more and more such counterterms, with arbitrary powers (curvature)^k, we might eventually find a theory that has no divergence. Finding such a renormalizable theory of gravity would be very hard without some sort of guidance: not only would we have to find a fixed point of the renormalization group (RG) flow by going backward in energy, but we would also have to worry about modes with wrong kinetic energy, which in such theories generically abound. (Adding operators with higher numbers of derivatives to a Lagrangian also adds propagating modes, each of which might be a ghost.) String theory is renormalizable, and in principle we can reexpress it precisely as such a sum of infinitely many corrections to (1.1.17).

This discussion seems, however, to assume that the effective action is analytic in the parameters l_s, e^ϕ, or, in other words, that it coincides with its Taylor expansion (1.1.21). In mathematics, we know many functions that are not analytic, and they might also appear here. This is the reason we have put the word "perturbative" in the title of this section; we will make amends in Section 1.4.

Some critics of string theory complain that the theory has not been proven to be *background independent*. What they mean is that in the world-sheet approach based on (1.1.13), we first have to fix a background configuration for the spacetime fields (1.1.9), and then we can compute an action for the small fluctuations around it. A priori, it might even be unclear if this procedure is describing a single theory or a collection of theories that have nothing to do with each other. The emergence of (1.1.17) should be reassuring in this respect: that effective action can be expanded around any background, and matches the result of the world-sheet method around it. A more satisfactory rebuttal is the proof at the level of string field theory in [26].

Torus compactification

Finally, let us have a first taste of string compactifications, by supposing that the theory lives on $\mathbb{R}^{25} \times S^1$. Thus we declare one of the coordinates to be periodically identified, say $x^{25} \equiv x^{25} + 2\pi R$. Now $x^{25}(\sigma^0, \sigma^1)$ is no longer necessarily periodic as a function of σ^1, even for a closed string: rather, if we take $\sigma^1 \sim \sigma^1 + \pi$, we demand

$$x^{25}(\sigma^0, \sigma^1 + \pi) = x^{25}(\sigma^0, \sigma^1) + 2\pi w R\,. \tag{1.1.22}$$

This represents a string that winds $w \in \mathbb{Z}$ times around the S^1. Another new effect is familiar from quantum mechanics: the overall momentum of the string in the S^1 direction is now not continuous but quantized: $p^{25} = \frac{q}{R}$, $q \in \mathbb{Z}$.

The mass spectrum in $\mathbb{R}^{25}$ is now modified from (1.1.7) to

$$m^2 = \frac{4}{l_s^2}(-1 + N) + \left(\frac{q}{R} - w\frac{R}{l_s^2}\right)^2 = \frac{4}{l_s^2}(-1 + \tilde{N}) + \left(\frac{q}{R} + w\frac{R}{l_s^2}\right)^2 , \tag{1.1.23}$$

where now $N = \sum i_k$ and $\tilde{N} = \sum_k \tilde{\imath}_k$ are no longer necessarily equal (as they were in (1.1.7)); comparing the two expressions, we have $N - \tilde{N} = wq$.

For a generic value of R, the massless spectrum is still (1.1.8) and (1.1.9); but now it should be reinterpreted. The components

$$g_{M\,25}\,, \qquad B_{M\,25} \tag{1.1.24}$$

are now two vector fields in $\mathbb{R}^{25}$; $g_{25\,25}$ is a scalar. The remaining components of (1.1.9) then give a metric, a Kalb–Ramond field, and a scalar in $\mathbb{R}^{25}$.

From (1.1.23), however, we also see another option: if $\frac{q}{R} - w\frac{R}{l_s^2} = \pm\frac{2}{l_s}$, then we have a new massless state for $N = 0$. This is possible for

$$R = l_s\,, \tag{1.1.25}$$

taking $q = -w = \pm 1$; then $\tilde{N} = \sum_k \tilde{\imath}_k = 1$. This state $\tilde{\alpha}^M_{-1}|0\rangle$ has a single index, and so it is created by a vector field. At this value of R, we also have the possibility of using the same trick with the other expression in (1.1.23), this time leading to

$q = w = \pm 1$, $\tilde{N} = 0$, $N = 1$. So we have a total of four more vector fields in $\mathbb{R}^{25}$. It turns out that these combine with the previous two (1.1.24) to give a nonabelian gauge group

$$\mathrm{SU}(2) \times \mathrm{SU}(2) \,. \tag{1.1.26}$$

This compactification was rather nice in that the string could be quantized exactly, at least perturbatively in g_s. In more complicated cases, we won't be so lucky, and we will have to limit ourselves to the less powerful effective field theory methods, potentially missing phenomena such as this non-abelian gauge group enhancement.

1.1.2 Type II superstrings

Supersymmetric world-sheet action

The world-sheet action (1.1.13) can be made supersymmetric. At the most basic level, this means that we promote the $x^M(\sigma)$ to a function of σ and of new formal coordinates $\theta^\pm$ that anticommute: $\theta^+\theta^- = -\theta^-\theta^+$, $(\theta^\pm)^2 = 0$. The Taylor expansion in the new coordinates truncates:

$$X^M = x^M + \theta^+ \psi^M_+ + \theta^- \psi^M_- + \theta^+\theta^- F^M \,. \tag{1.1.27}$$

We can also introduce the derivative operators

$$D_\pm = \partial_{\theta^\pm} + \mathrm{i}\theta^\pm \partial_\pm \,, \qquad \partial_\pm \equiv \partial_{\sigma^\pm} \,. \tag{1.1.28}$$

Then, (1.1.4), for example, is replaced by

$$S^{1,1}_{\mathrm{F1},g} = -\frac{1}{2} T_{\mathrm{F1}} \int_\Sigma \mathrm{d}^2\sigma \mathrm{d}^2\theta (g + B)_{MN}(X) D_+ X^M D_- X^N \,, \tag{1.1.29}$$

with the integration rule $\int \mathrm{d}\theta^\pm \theta^\pm = 1$, $\int \mathrm{d}\theta^\pm 1 = \int \mathrm{d}\theta^\pm \theta^\mp = 0$. We also added the contribution from B. The terms (1.1.10) can also be supersymmetrized in this way. The final result is quite messy for a general background where g_{MN} and B_{MN} are arbitrary; it can be found, for example, in [27, sec. 6.3.1]. For example, it contains a kinetic term

$$g_{MN}(\psi^M_+ \partial_- \psi^N_+ + \psi^M_- \partial_+ \psi^N_-) \tag{1.1.30}$$

for the world-sheet fermions $\psi^M_\pm$.[3] The F^M in (1.1.27) are auxiliary fields: they have no kinetic term, and can be replaced with the solutions of their equations of motion.

Since we have introduced a single θ^+ and a single θ^-, the resulting model is said to have $\mathcal{N} = (1, 1)$ supersymmetry. Any two-dimensional bosonic model can be promoted to such a model. In the context of compactifications, one often needs to separate external and internal dimensions, and the supersymmetrization of the world-sheet model in the latter has more supercharges; a common case one needs is $\mathcal{N} = (2, 2)$. This is more challenging to achieve, because such *extended* supersymmetry requires that one combine the x^M with each other in pairs. Such a pairing is reminiscent of the idea of complex coordinates, and is at the root of why differential geometry is useful for compactifications. This idea will return in Chapter 9.

[3] The bosonic world-sheet indices $\pm$ are conceptually not the same as the $\pm$ on the fermions, denoting chirality. To emphasize the difference, some authors change the world-sheet indices to $+ \to \#$ and $- \to =$. This also has the benefit that every term in a world-sheet will then have an equal number of pluses and minuses; see for example [28].

Equation (1.1.29) is called the *Neveu–Schwarz–Ramond* (NSR) model. While we introduced it by supersymmetrizing the world-sheet action, we will see later that the resulting spacetime theory also has the much more nontrivial property of *spacetime* supersymmetry.

Spectrum

Even around flat space, the spectrum of (1.1.29) is now more complicated because it depends on what we impose on the fermionic $\psi^M_\pm$. Since a fermion should only get back to itself after a 4π rotation, under 2π we can impose either periodic or antiperiodic boundary conditions, called Neveu–Schwarz (NS) and Ramond (R) respectively. These can be imposed independently on the $\psi^M_\pm$, leading to four sectors: NSNS, NSR, RNS, and RR. The spectrum has to be analyzed in each sector separately, because the Fourier modes for the $\psi^M_\pm$ behave differently in each.

In the NS sector, the fermionic Fourier modes are $b^M_{-i-1/2}$, $i \geq 0$. The two lowest-lying states are

$$|0\rangle_{\rm NS}\,, \qquad \mathbf{1}\,, \qquad m^2 = \frac{1}{8l_s^2}(2-d)\,; \tag{1.1.31a}$$

$$b^M_{-1/2}|0\rangle_{\rm NS}\,, \qquad \mathbf{8}_{\rm V}\,, \qquad m^2 = \frac{1}{l_s^2}\left(\frac{(2-d)}{8}+1\right)\,. \tag{1.1.31b}$$

We have also indicated what representation these states form under the compact part $SO(d-2) = SO(8)$ of the massless little group (1.1.1). For (1.1.31b), the subscript "V" is because there are two more dimension-eight representations of SO(8), which will soon play a role too.

In the R sector, the fermionic Fourier modes are d^M_{-i}, $i \geq 0$. In this case, the vacuum has already $m^2 = 0$, but in fact it is not unique: the modes d^M_0 now don't raise the energy, and they act on the space of vacua. These d^M_0 satisfy a *Clifford algebra* $\{d^M_0, d^N_0\} = 2g^{MN}1$, and as a consequence the space of R vacua transforms as a spinor under spacetime symmetries. In Section 2.1, we will attack Clifford algebras and spinors systematically in every dimension; for now, we only state the main features we need, which are quite similar to the properties of gamma matrices in four dimensions.

- Gamma matrices Γ^M can be defined in every dimension as matrices that satisfy $\{\Gamma^M, \Gamma^N\} = 2g^{MN}1$.
- In $d = 10$ dimensions, they are 32×32 matrices; in $d = 8$, they are 16×16.
- The space of spinors on which the Γ^M act is a representation for the Lorentz group; in $d =$ even, it decomposes in two chiralities, for which we introduce indices α, $\dot\alpha$. Multiplication by a single Γ^M changes chirality, so the nonzero blocks are $\Gamma^M_{\alpha\dot\beta}$ and $\Gamma^M_{\dot\alpha\beta}$.
- In both $d = 10$ with Lorentzian signature and $d = 8$ with Euclidean signature, there is a choice of Γ^M that are all real. (This aspect will be treated more specifically in Sections 2.2.3 and 2.3.)

As a representation of the transverse SO(8) in the little group (1.1.1), the R states then form a reducible representation of dimension 16, which further splits in two representation of dimension eight, traditionally called $\mathbf{8}_S$ and $\mathbf{8}_C$. We summarize all this by writing

$$|0, \alpha\rangle_R\,, \qquad \mathbf{8}_S\,, \qquad m^2 = 0\,; \tag{1.1.32a}$$

$$|0, \dot\alpha\rangle_R\,, \qquad \mathbf{8}_C\,, \qquad m^2 = 0\,. \tag{1.1.32b}$$

The closed-string spectrum is obtained by taking tensor products of these two sectors. For example, if we impose NS conditions for both $\psi^M_\pm$, we have the NSNS sector, whose low-lying states are obtained by taking tensor products of two copies of (1.1.31). Here we see once again the presence of a tachyon, $|0\rangle_{NS} \otimes |0\rangle_{NS}$. In the critical dimension

$$d = 10\,, \tag{1.1.33}$$

this sector contains massless states $b^M_{-1/2}\tilde b^M_{-1/2}|0\rangle_{NS} \otimes |0\rangle_{NS}$, which would be created by the fields (1.1.9) we encountered for the bosonic string, so we are going to stick to $d = 10$ from now on.

Quite nontrivially, just like in the bosonic string, this is also the dimension where the *Weyl anomaly vanishes*. This can be checked in any of the methods we saw for the bosonic string. In terms of the central charge of the Virasoro algebra, for example, there are now ghosts and superghosts, which give a total contribution -15. Quantizing around flat space, each boson gives a contribution $+1$, but now each fermion gives an extra $+1/2$. Since (1.1.29) has an equal number of bosons of fermions, we get $(1 + 1/2)d = 15$, which leads again to $d = 10$.

A less welcome similarity with the bosonic string is the presence of a tachyon. Fortunately, this can be eliminated by the so-called *Gliozzi–Scherk–Olive (GSO) projection* on the spectrum. Initially an ad hoc prescription, it later emerged to be required by consistency when one takes the world-sheet Σ to be a torus, an invariance under coordinate changes called *modular invariance*. It projects out $|0\rangle_{NS}$ in the NS sector, and also eliminates one of the two sets in (1.1.32). For the left-moving ψ^M_+, it is immaterial which one we choose; we keep $\mathbf{8}_S$. However, now the choice for the right-moving ψ^M_- does matter; this leads to two different theories.

If we keep $\mathbf{8}_C$ for the ψ^M_-, we obtain the theory called *IIA string theory*; its spectrum is given in Table 1.1. We use SO(8) group theory to decompose the tensor products of representations as direct sums. In the last column, we have named the corresponding spacetime fields.

- In the NSNS sector, the decomposition is that of an 8×8 matrix in its symmetric traceless, antisymmetric and trace, with the same logic that led us to the fields (1.1.9) for the bosonic string.

Table 1.1. Massless IIA spectrum.

NSNS	$\mathbf{8}_V \otimes \mathbf{8}_V = \mathbf{35}_V \oplus \mathbf{28}_V \oplus \mathbf{1}_V$	g_{MN}, B_{MN}, ϕ
RNS	$\mathbf{8}_S \otimes \mathbf{8}_V = \mathbf{56}_S \oplus \mathbf{8}_C$	$\psi^1_{M\alpha}, \lambda^1_{\dot\alpha}$
NSR	$\mathbf{8}_V \otimes \mathbf{8}_C = \mathbf{56}_C \oplus \mathbf{8}_S$	$\psi^2_{M\dot\alpha}, \lambda^2_{\alpha}$
RR	$\mathbf{8}_S \otimes \mathbf{8}_C = \mathbf{56}_V \oplus \mathbf{8}_V$	C_{MNP}, C_M

- In the RNS sector, the $\mathbf{8}_S \otimes \mathbf{8}_V$ representation has a vector and a spinor index, leading to an object $\Psi_{M\alpha}$. This is not irreducible: we can use an eight-dimensional gamma matrix to extract a sort of trace $\lambda_{\dot\alpha} \equiv (\Gamma^M)_\alpha{}^{\dot\beta}\Psi_{M\dot\beta}$, transforming as $\mathbf{8}_C$; the remaining traceless part of $\Psi_{M\alpha}$ is the $\mathbf{56}_S$. Both have one spinor index, and thus are spacetime fermions. Since they only have an index of one kind (either α or $\dot\alpha$), they are chiral, or *Weyl*. In the aforementioned real basis for the Γ^M, these fermions are also all real. The basis-independent notion is called the *Majorana* property (Section 2.3.1).
- The NSR sector is similar, but with $\mathbf{8}_S \leftrightarrow \mathbf{8}_C$. These are also spacetime fermions. The $\psi^1_{\alpha M}$, $\psi^2_{\dot\alpha M}$ are related by supersymmetry to the metric field, and hence are called *gravitinos*. The $\lambda^1_{\dot\alpha}$, ψ^2_{α} are called *dilatinos*.
- In the RR sector, we have an object with two spinorial indices:

$$C_{\alpha\dot\beta} \,. \tag{1.1.34}$$

  These have two spinor indices, and thus are sometimes called *bispinors*; they are spacetime *bosons*.

A more familiar description for the RR (1.1.34) is obtained by expanding them in a basis for bispinors. This is familiar from QFT in $d = 4$: it consists of antisymmetrized products

$$\Gamma^{MN} \equiv \frac{1}{2}(\Gamma^M\Gamma^N - \Gamma^N\Gamma^M)\,, \qquad \Gamma^{MNP} = \frac{1}{6}(\Gamma^M\Gamma^N\Gamma^P \pm \text{perm.})\,, \tag{1.1.35}$$

and so on. The expansion on this basis is called a *Fierz identity* and will be analyzed systematically in Section 3.4; for now, we sketch the result. If we contract (1.1.34) with a single eight-dimensional gamma matrix, we obtain the vector

$$C_M = (\Gamma_M)^{\alpha\dot\beta} C_{\alpha\dot\beta}\,, \tag{1.1.36}$$

which is the $\mathbf{8}_V$ there. One can further contract with products of gamma matrices; with two of them, we obtain $\Gamma^{MN}_{\alpha\beta}$ or $\Gamma^{MN}_{\dot\alpha\dot\beta}$, which cannot be contracted with (1.1.34), but with three we do obtain $C_{MNP} = (\Gamma_{MNP})^{\alpha\dot\beta} C_{\alpha\dot\beta}$. This explains the entries in the bottom-right corner of Table 1.1. By construction, this is completely antisymmetric:

$$C_{MNP} = -C_{NMP} = -C_{PNM} = -C_{MPN}\,. \tag{1.1.37}$$

The number of independent components of a completely antisymmetric tensor with k indices in d (transverse) dimensions is $\binom{d}{k}$; for (1.1.37) this gives $\binom{8}{3} = 56$, confirming the last row of Table 1.1. One might think that we also need to consider $\Gamma^{M_1 \ldots M_k}$ with higher odd k, but these are actually related to the ones already present in Table 1.1, as shown by the matching dimension count in the representation. Sometimes it is convenient to work with a redundant set of RR potentials; this *democratic* formalism will be presented in Section 10.1.

The second choice is to keep $\mathbf{8}_S$ for the ψ^M_-. The resulting theory is now called *IIB string*, and its spectrum is shown in Table 1.2. Most differences with Table 1.1 are straightforward; the one worthy of most attention is that now the RR sector consists of a bispinor

$$C_{\alpha\beta} \,. \tag{1.1.38}$$

Table 1.2. Massless IIB spectrum.

NSNS	$\mathbf{8}_V \otimes \mathbf{8}_V = \mathbf{35}_V \oplus \mathbf{28}_V \oplus \mathbf{1}_V$	g_{MN}, B_{MN}, ϕ
RNS	$\mathbf{8}_S \otimes \mathbf{8}_V = \mathbf{56}_S \oplus \mathbf{8}_C$	$\psi^1_{M\alpha}, \lambda^1_{\dot\alpha}$
NSR	$\mathbf{8}_V \otimes \mathbf{8}_S = \mathbf{56}_S \oplus \mathbf{8}_C$	$\psi^2_{M\alpha}, \lambda^2_{\dot\alpha}$
RR	$\mathbf{8}_S \otimes \mathbf{8}_S = \mathbf{35}_C \oplus \mathbf{28}_C \oplus \mathbf{1}_C$	C^+_{MNPQ}, C_{MN}, C_0

In the Fierz expansion, now the Γ^{MN} has the correct index structure, leading to $C_{MN} = (\Gamma_{MN})^{\alpha\beta} C_{\alpha\beta}$; a similar projection can be defined for all the products $\Gamma^{M_1 \ldots M_k}$ with even k, but in fact $k > 6$ are redundant, and even the one for $k = 4$ has a "self-duality property" that halves its degrees of freedom, and whose consequences we will see soon. This is the reason of the superscript $+$ on C^+_{MNPQ}. Just like for (1.1.37), by construction these tensors are all completely antisymmetric.

Spacetime supersymmetry

In both IIA and IIB, the spacetime bosons arise from the NSNS and RR sectors, while the fermions arise from the NSR and RNS sectors. They have the same total number (128) of degrees of freedom. This is a symptom of the aforementioned *spacetime* supersymmetry. These mix the bosonic fields (NSNS, RR) with the fermionic ones (NSR, RNS). They are rather complicated, and at this stage their expression would not look very informative. We will have a first look at them in Section 8.2 for the case where only the metric is present, and in Section 10.1 in full. For now, we just comment about their infinitesimal parameters, which are *two* Majorana–Weyl spinors:

$$\epsilon^1, \qquad \epsilon^2 . \tag{1.1.39}$$

(This is the original reason these theories are called "type II.") They have the same chirality as the gravitino: thus in IIA ϵ^1_α has positive chirality, and $\epsilon^2_{\dot\alpha}$ has negative chirality, while in IIB both ϵ^a_α have positive chirality. Altogether this gives 32 supercharges, which is the highest number for any supersymmetric theory.

This property is made manifest in the alternative *Green–Schwarz* model. A more recent formulation is the *Berkovits*, or *pure spinor*, model [29, 30]. (Pure spinors will play an important role in this book, but for somewhat different reasons.) These alternative formulations are more complicated, but are better at describing strings in condensates of RR fields, while with the NSR model (1.1.29) this is difficult.

1.1.3 Heterotic strings

Definition

In both the bosonic string and the superstring, quantization of the left- and right-movers seems to proceed almost independently. This is because they are almost free as two-dimensional field theories, apart from the constraint associated to the Lagrange multiplier $h_{\alpha\beta}$. For example, in the bosonic string spectrum (1.1.6), the constraint is only visible in the level matching condition mentioned after (1.1.7).

This creates an opportunity: we can try to define a hybrid, or *heterotic* theory which looks like the bosonic string for the left-movers, and like the superstring for the right-movers [31]. This might look impossible: is the spacetime dimension going to be 26 or 10? The dimension of spacetime, however, is a macroscopic concept, irrelevant at length scales below l_P, where quantum gravity sets in. So a solution to the puzzling mismatch in dimension is to compactify 16 of the 26 bosonic coordinates. Provided we manage to satisfy the level matching condition relating the left- and right-moving sectors, all should be well.

The easiest way to compactify the 16 left-moving bosons x_+^I is to take them to belong to a *torus*. In Section 1.1.1, we saw a circle compactification of the bosonic string: we took one of the coordinates $x^{25} \sim x^{25} + 2\pi R$. The simplest torus would be obtained as $T^{16} \equiv (S^1)^{16}$, with the same identification for all the x_+^I. A straightforward generalization is to take a basis $\{R_a^I\}$, $a = 1, \ldots, 16$ of $\mathbb{R}^{16}$, and to introduce equivalence relations

$$x^I = x^I + \pi R_a^I \quad \forall a\,. \tag{1.1.40}$$

The set of integer multiples of the $\{R_a^I\}$ is called a *lattice* Γ. The space defined by (1.1.40) is topologically still T^{16}; the choice of Γ affects its size and shape. The momenta are quantized according to

$$p^I R_a^I \in \mathbb{Z}\,. \tag{1.1.41}$$

The solutions to this equation are the elements of a *colattice*: there is a dual set of vectors p_a^I of which all solutions to (1.1.41) are integer multiples.

Spectrum

As usual, now the spectrum is computed independently for the left-movers and for the right-movers, imposing level matching. For the left-movers, $(\partial_0 - \partial_1)x^I = 0$ relates p^I to the R_a^I: this generalizes setting $q/R = wR/l_s^2$ in (1.1.23). So in fact the lattice and colattice coincide,[4] and Γ is said to be *self-dual*. Moreover, the analogue of (1.1.23) now reads

$$m^2 = \frac{4}{l_s^2}\left(-1 + N + \frac{1}{2}\hat{p}^2\right) = \frac{4}{l_s^2}\tilde{N}\,. \tag{1.1.42}$$

From this we see that $\frac{1}{2}\hat{p}^2 \in \mathbb{Z}$: the lattice is said to be *even*.

The massless spectrum is again obtained as a tensor product of left- and right-movers. The latter have both an NS and an R sector, just as in the superstring (recall (1.1.31) and (1.1.32)). For the left-movers, (1.1.42) gives us two possibilities. We can take $\hat{p}^2 = 0$ and $N = 1$: a single Fourier mode acting on the vacuum (compare with the left-moving part of (1.1.8)). This can be α_{-1}^M, with $M = 0, \ldots, 9$, or α_{-1}^I, with $I = 1, \ldots, 16$ one of the extra 16 bosonic directions. So far, this gives us the massless states

$$\alpha_{-1}^M|0\rangle_{\rm bos} \otimes b_{-1/2}^M|0\rangle_{\rm NS}\,, \qquad \alpha_{-1}^M|0\rangle_{\rm bos} \otimes |0,\alpha\rangle_{\rm R}\,; \tag{1.1.43a}$$

$$\alpha_{-1}^I|0\rangle_{\rm bos} \otimes b_{-1/2}^M|0\rangle_{\rm NS}\,, \qquad \alpha_{-1}^I|0\rangle_{\rm bos} \otimes |0,\alpha\rangle_{\rm R}\,. \tag{1.1.43b}$$

The first in (1.1.43a) gives the familiar (g_{MN}, B_{MN}, ϕ) in the NSNS sector of the superstring; the second is identical to the RNS sector of the superstring. These are

[4] At this point, one might also take more generally the lattice to be a sublattice of the colattice, but modular invariance (already mentioned for the superstring) eliminates this possibility.

Table 1.3. Massless heterotic spectrum.

NS	$\mathbf{8}_V \otimes \mathbf{8}_V = \mathbf{35}_V \oplus \mathbf{28}_V \oplus \mathbf{1}_V$	g_{MN}, B_{MN}, ϕ
R	$\mathbf{8}_V \otimes \mathbf{8}_S = \mathbf{56}_S \oplus \mathbf{8}_C$	$\psi_{M\alpha}, \lambda_{\dot\alpha}$
NS	$\mathbf{8}_V$	A^a_M
R	$\mathbf{8}_S$	χ^a_α

the first two lines in Table 1.3. The (1.1.43b) are 16 abelian vector fields A^I_M and spinors χ^I_α; they are the Cartan subalgebra part of the second two lines in Table 1.3.

Indeed, we are not done yet, because (1.1.42) allows us a second possibility: we can take $N = 0$ and look for $\hat{p}^2 = 2$, namely vectors in Γ of length two. There exist only *two* lattices in $\mathbb{R}^{16}$ that are both self-dual and even; both are the root lattices $\Gamma_\mathfrak{g}$ for a Lie algebra $\mathfrak{g}$ of a Lie group that can be either

$$\mathrm{SO}(32) \quad \text{or} \quad E_8 \times E_8 \,. \tag{1.1.44}$$

The elements of Γ of length two are the nonzero roots of $\mathfrak{g}$. A tensor product of all these states with $b^M_{-1/2}|0\rangle_{\mathrm{NS}}$ and $|0, \alpha\rangle_{\mathrm{R}}$, as in (1.1.43b), produces one vector field and one spinor for each nonzero root of $\mathfrak{g}$. Both (1.1.44) have rank 16, which is just the number of the A^I_M we found previously. So in total we have A^a_M, χ^a_α with $a = 1, \ldots, \dim(\mathfrak{g})$, and we reproduce the missing part of the last two lines in Table 1.3. The A^a_M are the gauge vectors of a nonabelian gauge algebra $\mathfrak{g}$. This enhancement is similar to (1.1.26) for the bosonic string.

The heterotic string has the advantage of having a built-in nonabelian gauge symmetry. We will see in Section 1.3 that D-branes give an alternative way of obtaining nonabelian gauge groups G in type II, but only of the type $G = \mathrm{U}(N)$; later in Section 1.4.4 we will see $G = \mathrm{SO}(N)$ and $\mathrm{Sp}(N)$. Obtaining a more interesting gauge group such as E_8 is in fact possible in IIB, but requires more sophisticated techniques that we will only study much later, in Section 9.4.[5]

1.2 Supergravity

An effective spacetime action for superstrings can be found using the same methods that led us to (1.1.17). Because of its spacetime supersymmetry, it is called 10-*dimensional supergravity*.

1.2.1 RR fields

In this book, we will actually only see the action for the *bosonic* fields. The superstring NSNS fields are the same as those (1.1.9) of the bosonic string. The new

[5] It might seem that a unitary gauge group is quite enough to accommodate the Standard Model's $\mathrm{SU}(3) \times \mathrm{SU}(2) \times \mathrm{U}(1)$. Even if these three factors unify at higher energies, the gauge groups that seem most promising are $G = \mathrm{SU}(5)$ and $\mathrm{SO}(10)$. However, the particular representations that one needs for those *grand unified theories (GUT)* are easier to obtain by starting from a group such as E_8 or its subgroup E_6.

bosonic fields are those in the RR sector. In this subsection, we focus on them before writing the effective actions for IIA and IIB in the next subsections.

Completely antisymmetric tensors

The RR fields are all completely antisymmetric tensors with k-indices. Such tensors are also called k-forms; we give here a minimal introduction to their properties, leaving a deeper treatment to Sections 3.1 and 4.1.

The *antisymmetrized derivative* or *exterior differential*

$$(\mathrm{d}A)_{M_1\ldots M_k} \equiv k\partial_{[M_1}C_{M_2\ldots M_k]} \equiv \partial_{M_1}C_{M_2\ldots M_k} - \partial_{M_2}C_{M_1\ldots M_k} \pm \ldots\,, \tag{1.2.1}$$

takes a $(k-1)$-form to a k-form. The index antisymmetrizer $[M_1 \ldots M_k]$ sums over all permutations with a ± 1 equal to the sign $\sigma(i_1 \ldots i_k)$ of the permutation taking $1 \ldots k$ to $i_1 \ldots i_k$, and divides by a $k!$. So, for example, $(\mathrm{d}A)_{MN} = \partial_M C_N - \partial_N C_M$ is the Maxwell field-strength of C_M. As another example, (1.1.15) can be written as $H_{MNP} = (\mathrm{d}B)_{MNP} = 3\partial_{[M}B_{NP]}$, with only three terms instead of six because of antisymmetry of B. To avoid index proliferation, often we will write $C_{M_1\ldots M_k}$ symbolically simply as C_k, and denote (1.2.1) by $\mathrm{d}C_k$. An important property of (1.2.1) is

$$\mathrm{d}\,\mathrm{d}\,\alpha_k = 0\,. \tag{1.2.2}$$

Explicitly,

$$(\mathrm{d}\mathrm{d}\alpha)_{M_1\ldots M_k} = k(k-1)\partial_{[M_1}\partial_{M_2}\alpha_{M_3\ldots M_k]} = 0\,, \tag{1.2.3}$$

because $\partial_{[M_1}\partial_{M_2]} = 0$ on smooth functions. For an electromagnetic potential A_M, this gives $\partial_{[M}F_{NP]} = \partial_{[M}\partial_N A_{P]} = 0$, one of the Maxwell equations. As another example, from (1.1.15) it now follows $\partial_{[M}H_{NPQ]} = 0$, or in other words,

$$\mathrm{d}H = 0\,. \tag{1.2.4}$$

The antisymmetrized, or *wedge*, product of two forms α_k, $\alpha'_{k'}$ is defined as

$$(\alpha \wedge \alpha')_{M_1\ldots M_k N_1\ldots N_{k'}} \equiv \frac{(k+k')!}{k!\,k'!}\alpha_{[M_1\ldots M_k}\alpha'_{N_1\ldots N_{k'}]}\,. \tag{1.2.5}$$

This satisfies $\alpha_k \wedge \alpha_{k'} = (-1)^{kk'}\alpha_{k'} \wedge \alpha_k$. In particular,

$$H \wedge H = 0\,. \tag{1.2.6}$$

We now also have the Leibniz identity:

$$\mathrm{d}(\alpha_k \wedge \alpha_{k'}) = \mathrm{d}\alpha_k \wedge \alpha_{k'} + (-1)^k \alpha_k \wedge \mathrm{d}\alpha_{k'}\,. \tag{1.2.7}$$

In $d = 10$, the largest possible number of indices of a k-form is $k = 10$. Moreover, a ten-form α_{10} is unique up to rescaling: it must be proportional to the Levi-Civita tensor $\epsilon^{(0)}_{M_1\ldots M_{10}} \equiv \sigma(M_1 \ldots M_{10})$, the sign of the permutation taking $M_1 \ldots M_{10}$ to $0 \ldots 9$. (The (0) label is used here because a more mathematically sophisticated version of this tensor will enter the scene later.) Writing then $\alpha_{0\ldots 9} = f$, we define

$$\int \alpha_{10} \equiv \int \mathrm{d}^{10}x\, f\,. \tag{1.2.8}$$

A similar definition holds in any dimension d.

Twisted field-strengths

RR forms usually appear in the action through a *twisted* field-strength[6]

$$F_k \equiv dC_{k-1} - H \wedge C_{k-3} . \tag{1.2.9}$$

For example, we will soon encounter $F_{MNPQ} = 4\partial_{[M}C_{NPQ]} - 4H_{[MNP}C_{Q]}$. (The factors come from (1.2.1) and (1.2.5).) For $k < 3$, the second term is absent; for example, $F_2 \equiv dC_1$. The F_k satisfy a *Bianchi identity*:

$$dF_k = H \wedge F_{k-2}, \tag{1.2.10}$$

as one sees using (1.2.2), (1.2.4), and (1.2.6). In a sense, this is more fundamental than (1.2.9): sometimes we will modify (1.2.9), with (1.2.10) remaining true. Later in this chapter, we will consider sources, which will violate (1.2.10) on some spacetime defects. A form field H or F_k satisfying (1.2.10) with no source is often called a *flux*. By an abuse of language, one sometimes calls by this name any form field, sourced or not.

Gauge transformations

In electromagnetism, the field-strength $F_{MN} = \partial_M A_N - \partial_N A_M$ is invariant under the gauge transformation $A_M \to \partial_M \lambda_0$. We already observed after (1.1.15) that the Kalb–Ramond field has a gauge transformation $B \to B + d\hat{\lambda}_1$, which leaves invariant its field-strength $H = dB$. For the RR fields, (1.2.9) are invariant under the gauge transformations:

$$\delta C_k = d\lambda_{k-1} - H \wedge \lambda_{k-3} . \tag{1.2.11}$$

So, for example, under λ_1 we have $\delta C_2 = d\lambda_1$, $\delta C_4 = -H \wedge \lambda_1$, while under λ_3 we have $\delta C_4 = d\lambda_3$, $\delta C_2 = 0$.

In electromagnetism, λ_0 need not be a function: it can be multivalued, as long as

$$e^{iq_e\lambda_0} \tag{1.2.12}$$

is single-valued, where q_e is the elementary electric charge. Multivalued λ_0 are called a *large* gauge transformation. They exist because the gauge group is compact, which in turn comes from the requirement that the wave function should be single-valued. This is relevant when there is a nontrivial loop in spacetime, either because of an excluded region (as in the Aharanov–Bohm effect) or because spacetime has a nontrivial topology. Consider for example an S^1 with a periodic coordinate $x \sim x+L$. Single-valuedness of (1.2.12) requires $\lambda_0 \to \lambda_0 + 2\pi N/q_e$, $N \in \mathbb{Z}$. The gauge transformation $A_M \to A_M + \partial_M \lambda_0$ is more correctly rewritten as

$$A_M \to A_M + \Lambda_M , \tag{1.2.13}$$

where Λ_M is required to have $\int dx \Lambda_x = 2\pi N/q_e$.

Suppose we want to give a *constant* expectation value to a component, $A_x = A_x^0$. This has no physical meaning if the direction x is noncompact, because we can gauge it away with a gauge transformation $\lambda = xA_x^0$. But if $x \sim x + L$ is a coordinate on a

[6] Many authors use the symbol F_k for dC_{k-1} alone, and then put a tilde on our (1.2.9).

circle, then with a periodic λ we cannot gauge away A^0_x any more; with a large gauge transformation, (1.2.13) gives us the identification

$$A^0_x \cong A^0_x + \frac{2\pi}{Lq_e} . \tag{1.2.14}$$

In the limit $L \to \infty$ we see again that all constant values can be gauged away. Similar large generalizations of (1.2.11) exist; we will discuss them after we introduce the analogue of the elementary electric charge for the C_k in Section 1.3.

1.2.2 IIA supergravity

We now come to the effective action for IIA string theory. As we anticipated, we will only show the action for the bosonic fields.

Bosonic action of IIA supergravity

The leading bosonic action for type IIA superstrings is called type IIA supergravity. It reads

$$S_{\rm IIA} = S_{\rm kNS} + \frac{1}{4\kappa^2}\left[-\int d^{10}x\sqrt{-g}(|F_2|^2 + |F_4|^2) + \int B \wedge dC_3 \wedge dC_3\right] , \tag{1.2.15}$$

where $|F_k|^2 \equiv \frac{1}{k!}F_{M_1\ldots M_k}F^{M_1\ldots M_k}$, extending the definition in (1.1.15);

$$2\kappa^2 = (2\pi)^7 l_s^8 ; \tag{1.2.16}$$

and the kinetic NSNS term

$$S_{\rm kNS} = \frac{1}{2\kappa^2}\int d^{10}x\sqrt{-g}e^{-2\phi}\left(R + 4\partial_\mu\phi\partial^\mu\phi - \frac{1}{2}|H|^2\right) \tag{1.2.17}$$

is the same Lagrangian as in (1.1.17), only now integrated in $d = 10$ dimensions rather than 26. This term will also appear in all the supergravity effective actions we will see later. The $e^{-2\phi}$ prefactor signals its origin from $g = 0$ string diagrams, as we remarked following (1.1.17). In a region where the dilaton is constant, the Planck length is now

$$l_{\rm P} = g_s^{1/4} l_s . \tag{1.2.18}$$

The string scattering amplitudes originally give a prefactor $e^{-2\phi}$ for the rest of (1.2.15) as well, but the action happens to look nicer if one rescales it away by redefining the C_k.

RR terms

The first parenthesis in (1.2.15) can be regarded as the kinetic term for $F_2 = dC_1$, and for $F_4 = dC_3 - H \wedge C_1$ from (1.2.9). The last term is more peculiar in (1.2.15). Unpacking the form notation (1.2.1), (1.2.5), and (1.2.8), one obtains:

$$\int B \wedge dC_3 \wedge dC_3 = \frac{10!}{2\cdot(3!)^2}\int d^{10}x B_{[01}\partial_2 C_{345}\partial_6 C_{789]} . \tag{1.2.19}$$

It does not involve the metric, and it contains a potential B_2 without an exterior derivative. For this reason, it is called the *Chern–Simons* term, after the Chern–Simons action in three dimensions $S_{CS} = \int_{M_3} \mathrm{CS}_A$, where

$$\mathrm{CS}_A \equiv \mathrm{Tr}\left(A \wedge \mathrm{d}A + \frac{2}{3} A \wedge A \wedge A\right) \tag{1.2.20}$$

is a three-form with the property $\mathrm{dCS}_A = \mathrm{Tr}(F \wedge F)$; the two-form $F = \mathrm{d}A + A \wedge A$ is the nonabelian field-strength in the form notation of Section 1.2.1. Alternative expressions for (1.2.19), such as $-\int H \wedge C_3 \wedge \mathrm{d}C_3$, can be obtained by integration by parts. Note that if we worked in $d = 11$ rather than $d = 10$, we could compute[7]

$$\mathrm{d}(B \wedge \mathrm{d}C_3 \wedge \mathrm{d}C_3) = H \wedge \mathrm{d}C_3 \wedge \mathrm{d}C_3 \overset{(1.2.9),(1.2.6)}{=} H \wedge F_4 \wedge F_4 \,. \tag{1.2.21}$$

Frame change

Superficially, the sign of the kinetic term for ϕ in (1.2.17) looks wrong: it is not of the usual form "kinetic energy minus potential energy," since $\dot{\phi}^2$ appears with a minus sign. (This issue of course was already present in (1.1.17).) But ϕ also appears multiplying the Einstein–Hilbert term R; so its dynamics is less simple than it looks. To make it more transparent, one can define an alternative metric

$$g^{\mathrm{E}}_{MN} \equiv \mathrm{e}^{-\phi/2} g_{MN} \,. \tag{1.2.22}$$

This is called *Einstein frame* metric, because in terms of it (1.2.17) becomes

$$S_{\mathrm{E,\,kNS}} = \frac{1}{2\kappa^2} \int \mathrm{d}^{10}x \sqrt{-g_{\mathrm{E}}} \left(R_{\mathrm{E}} - \frac{1}{2}\partial_\mu \phi \partial^\mu \phi - \frac{1}{2}\mathrm{e}^{-\phi}|H|^2\right), \tag{1.2.23}$$

with the Einstein–Hilbert term now appearing without the dilaton prefactor; indices are also now contracted with the Einstein frame metric. Here the dilaton's kinetic term has the conventional sign. In most of the book, we will use the original *string frame* metric g_{MN}.

Supersymmetry

As we mentioned in Section 1.1.2, type II superstrings are symmetric under the 32 supercharges ϵ^a; so the fermionic completion of the supergravity action (1.2.15) will enjoy such symmetry too. Theories with 32 supercharges are relatively rare: besides IIB, an important example we will see later in this chapter is eleven-dimensional supergravity, of which IIA is a *dimensional reduction*.[8] This large amount of supersymmetry has consequences on the structure of the l_s and g_s corrections; for example, it fixes the two-derivative action completely.

This is a symmetry of the *theory*: not all supercharges will leave invariant a particular field configuration. The generic field configuration breaks supersymmetry

[7] Equation (1.2.21), together with a generalization of Stokes's theorem that we will study in Section 4.1.10, allows us to reinterpret the CS term as an integral over an eleven-dimensional space of which the ten-dimensional one is a boundary.

[8] Most other theories with 32 supercharges are further dimensional reductions of IIB or eleven-dimensional supergravity, some of which we'll see in Chapter 11; but there now exist examples which are not thought to arise this way [32].

completely, just like the generic metric g_{MN} has no Killing vectors. One of the topics of interest for this book will be the study of how many supercharges leave invariant given field configurations, or vice versa of which field configurations are invariant under a certain number of supercharges.

String corrections

Corrections to the supergravity approximation first occur at the eight-derivative level. The curvature terms read

$$S_{\text{IIA},R^4} = \frac{1}{2\kappa^2}\frac{l_s^6\zeta(3)}{3\cdot 2^{11}}\int \mathrm{d}^{10}x\sqrt{-g}\mathrm{e}^{-2\phi}\left(t_{M_1\ldots M_8}t^{N_1\ldots N_8} + \frac{1}{8}\epsilon_{PQM_1\ldots M_8}\epsilon^{PQN_1\ldots N_8}\right)$$
$$\cdot R^{M_1M_2}{}_{N_1N_2}R^{M_3M_4}{}_{N_3N_4}R^{M_5M_6}{}_{N_5N_6}R^{M_7M_8}{}_{N_7N_8} \tag{1.2.24}$$

at the leading order in g_s. Here ζ is Riemann's zeta function, and the tensor t is defined by

$$t_{M_1\ldots M_8}M^{M_1M_2}M^{M_3M_4}M^{M_4M_5}M^{M_7M_8} = 24\mathrm{Tr}M^4 - 6(\mathrm{Tr}M^2)^2\,. \tag{1.2.25}$$

Just like for supergravity, (1.2.24) can be obtained either by computing string amplitudes [33], or by computing the world-sheet beta functions beyond the leading approximation [34]. At $g = 1$, or in other words $\mathrm{e}^{0\cdot\phi}$ according to (1.1.12), there is a similar term, where the combination $t_8t_8 - \epsilon\epsilon/8$ appears.

Even at this l_s^8 level, the complete structure of the action is not completely established beyond (1.2.24). For example, one expects couplings to the form field strengths. There is a famous coupling of the type $\int B_2R^4$, related to anomalies [35, 36]. One can try to infer the remaining terms by using dualities [37] or supersymmetry [38]. A complementary approach is via string amplitudes in the four-field approximation (which gives a contribution to (1.2.24), when one linearizes it in the metric fluctuation $g_{MN} \sim \eta_{MN} + h_{MN}$); with this restriction, one can go to arbitrary precision in l_s in the aforementioned Berkovits formalism [39].

Massive IIA

IIA supergravity has a deformation by a parameter F_0 called *Romans mass* [40], still preserving 32 supercharges.[9] This theory was originally obtained by realizing a *Stückelberg mechanism*, by which the Kalb–Ramond field B acquires a mass, at the price of absorbing (or "eating") the degrees of freedom of C_1.

The original Stückelberg mechanism [43] is a variant of (and predates) the more familiar Brout–Englert–Higgs (BEH) mechanism. In both cases, a vector field A_μ acquires a mass by eating the degrees of freedom of a scalar a. In BEH, we gauge a rotation in field space, leading to a covariant derivative schematically of the form $\partial_\mu a + A_\mu a$. In the Stückelberg case, we instead introduce

$$D_\mu a \equiv \partial_\mu a + mA_\mu\,, \tag{1.2.26}$$

where m is a mass. This is invariant under a translation: $a \to a - m\lambda$, if also $A_\mu \to A_\mu + \partial_\mu\lambda$. This transformation can of course be used to set $a = 0$; in this sense, a

[9] There is in fact also a second deformation of IIA with 32 supercharges [41]; this arises as a dimensional reduction of eleven-dimensional supergravity where one identifies spacetime with itself up to an overall rescaling. These are the only maximally supersymmetric deformations [42].

is "eaten" by A_μ. The covariant kinetic term $D_\mu a D^\mu a$, when expanded, is now seen to contain a mass term $m^2 A_\mu A^\mu$ for the vector field. Equation (1.2.26) appears often in string compactifications with fluxes. We will comment further on the difference between BEH and Stückelberg, and generalize both, in Section 4.2.2.

The variant of (1.2.26) we need in IIA is obtained by adding an index to both participant fields, thus introducing

$$F_2 \equiv \mathrm{d}C_1 + F_0 B. \tag{1.2.27}$$

This modifies (1.2.9), which would have given $F_2 = \mathrm{d}C_1$. Now C_1 and B plays the role of a and A_μ in (1.2.26), respectively. We have now used the symbol F_0 for the mass parameter. The reason for this name becomes apparent when we notice that

$$\mathrm{d}F_2 = HF_0\,. \tag{1.2.28}$$

This is of the form (1.2.10), suggesting that F_0 should be regarded as a new, non-dynamical RR "field-strength" with zero indices. We will see later more cogent reasons for this interpretation.

We also have to change F_4 with respect to (1.2.9):

$$F_4 = \mathrm{d}C_3 - H \wedge C_1 + \frac{1}{2} F_0 B \wedge B\,; \tag{1.2.29}$$

(1.2.10) still holds. The gauge transformations (1.2.11) still leave these invariant, but that of B does not, unless we modify it:

$$\delta B = \mathrm{d}\hat{\lambda}_1\,, \qquad \delta C_1 = -F_0 \hat{\lambda}_1\,, \qquad \delta C_3 = -F_0 \hat{\lambda}_1 \wedge B\,. \tag{1.2.30}$$

With these new definitions for the F_k, the IIA action is now modified as

$$S_{\mathrm{IIA}} = S_{\mathrm{kNS}} + \frac{1}{4\kappa^2}\left[-\int \mathrm{d}^{10}x\sqrt{-g}(F_0^2 + |F_2|^2 + |F_4|^2) + \int \mathrm{mCS}_{10}\right], \tag{1.2.31}$$

where mCS_{10} is a ten-form that again has the formal property

$$\mathrm{d}(\mathrm{mCS}_{10}) = H \wedge F_4 \wedge F_4 \tag{1.2.32}$$

as in (1.2.21). Viewed in this way the modification with respect to (1.2.15) is minimal, although an explicit expression is less nice:

$$\mathrm{mCS}_{10} = B \wedge \left(\mathrm{d}C_3^2 + \frac{1}{3} F_0 \mathrm{d}C_3 \wedge B^2 + \frac{1}{20} F_0^2 B^4\right), \tag{1.2.33}$$

where $B^k \equiv B \wedge \ldots \wedge B$.

We will see a more elegant way of understanding both (1.2.15) and (1.2.31) in Section 10.1, at the cost of a slightly more sophisticated mathematical apparatus. On that occasion, we will also see the supersymmetry transformations.

1.2.3 IIB supergravity

IIB supergravity is very similar to IIA [44–46]. The most important difference is that it is a *chiral* theory: in Table 1.2, we see that the two ψ^a and the two λ^a have the same chirality. Usually the presence of chiral spinors threatens an anomaly.

While there are no gauge fields, in a gravitational theory we have to worry about potential anomalies for *diffeomorphisms*, which are just as lethal as their gauge counterparts. The chiral ψ^a and λ_a indeed give a nonvanishing contribution to such a diffeomorphism anomaly. Fortunately, however, the RR field C^+_{MNPQ} also gives a contribution, which exactly cancels the fermionic ones [47], avoiding a potential catastrophe in a nontrivial way.

C^+_{MNPQ} gives a contribution to an anomaly in spite of being bosonic because of its self-duality property:

$$F_{M_1\ldots M_5} = \frac{1}{5!}\sqrt{-g}\,\epsilon^{(0)}_{M_1\ldots M_5}{}^{M_6\ldots M_{10}}F_{M_6\ldots M_{10}}, \qquad (1.2.34a)$$

where again $\epsilon^{(0)}$ is the completely antisymmetric tensor defined later in (1.2.6). In the condensed notation of the previous subsection, (1.2.34a) is written as

$$F_5 = *F_5\,. \qquad (1.2.34b)$$

Just like in (1.2.9), we can take $F_5 = \mathrm{d}C_4^+ - H \wedge C_2$. There are alternatives, such as $F_5' = \mathrm{d}C_4^+ + \frac{1}{2}(B \wedge \mathrm{d}C_2 - H \wedge C_2)$. Both F_5 and F_5' satisfy the Bianchi identities (1.2.10); one can bring one into the other by redefining C_4^+ (with no ill effect on the action (1.2.36)).

Pseudoaction

While self-duality of F_5 saves the theory, it also makes it hard to write down an action. A self-duality property similar to (1.2.34) can be introduced for any k-form potential a_k with self-dual field-strength $\mathrm{d}a_k = *\mathrm{d}a_k$ in $d = 2(k+1)$ dimensions. The simplest example is for $k = 0$: a_0 is then a scalar in $d = 2$ dimensions, and self-duality implies

$$\partial_M a_0\, \partial^M a_0 = \sqrt{-g}\,\epsilon^{MN}_{(0)}\,\partial_M a_0\, \partial_N a_0 = 0\,, \qquad (1.2.35)$$

so the usual Lagrangian density vanishes. More generally, this happens for any even k, and in particular for our (1.2.34): the naive Lagrangian density $|F_5|^2 = 0$. There are strategies to cope with this issue; see [48, 49] for a recent proposal. In the following, we will simply write a *pseudoaction*: all equations of motion can be derived by varying it, *except* for the constraint (1.2.34). From now on, we will drop the superscript and call $C_4^+ \to C_4$, for uniformity with the other potentials in both type II theories.

A pseudoaction for IIB is then

$$S_{\rm IIB} = S_{\rm kNS} + \frac{1}{4\kappa^2}\left[-\int \mathrm{d}^{10}x\sqrt{-g}\left(|F_1|^2 + |F_3|^2 + \frac{1}{2}|F_5|^2\right) + \int B \wedge \mathrm{d}C_2 \wedge \mathrm{d}C_4\right], \qquad (1.2.36)$$

where $S_{\rm kNS}$ was given in (1.2.17). As in IIA, the last term is of Chern–Simons type, and has the formal property that in eleven dimensions $\mathrm{d}(B\wedge \mathrm{d}C_2 \wedge \mathrm{d}C_4) = H \wedge F_3 \wedge F_5$, similar to (1.2.21). Alternative expressions can be obtained by integration by parts, such as $-\int H \wedge C_2 \wedge \mathrm{d}C_4$. The fermionic completion of (1.2.36) is supersymmetric with 32 supercharges, just like IIA.

The leading l_s corrections are the same as in IIA, (1.2.24). At the $g = 1$ level, the same combination $t_8 t_8 - \epsilon\epsilon/8$ is present. Knowledge of the full eight-derivative corrections is as incomplete as in IIA, but in this case dualities are more powerful and determine the dependence on ϕ of the terms in (1.2.24) beyond the $g = 0$ and $g = 1$ terms [50]. This is based on an important additional symmetry of IIB that is already present at the level of supergravity, to which we now turn.

SL(2, ℝ) symmetry

Given a 2×2 matrix with unit determinant,

$$m = \begin{pmatrix} a & b \\ c & d \end{pmatrix} \in \mathrm{SL}(2, \mathbb{R}), \tag{1.2.37}$$

we define the transformation law

$$\begin{aligned} \tau \to m \cdot \tau \equiv \frac{a\tau + b}{c\tau + d}, \qquad & g_{MN} \to |c\tau + d|\, g_{MN} \\ F_5 \to F_5, \qquad \begin{pmatrix} C_2 \\ B \end{pmatrix} \to & \, m \begin{pmatrix} C_2 \\ B \end{pmatrix}, \end{aligned} \tag{1.2.38}$$

where

$$\tau \equiv C_0 + \mathrm{i}e^{-\phi} \tag{1.2.39}$$

is called the *axiodilaton*.

Equation (1.2.38) is a symmetry of the action (1.2.36). This type of nonlinear action on τ is called a *Möbius transformation*; it will appear in several other contexts. The string-frame metric transforms, but the Einstein frame metric (1.2.22) does not:

$$g^{\mathrm{E}}_{MN} \to g^{\mathrm{E}}_{MN}\,. \tag{1.2.40}$$

There is also an alternative expression for the transformation law of the two-form potentials, in terms of

$$G_3 \equiv \mathrm{d}C_2 - \tau H = \mathrm{d}C_2 - HC_0 - \mathrm{i}e^{-\phi} H = F_3 - \mathrm{i}e^{-\phi} H\,. \tag{1.2.41}$$

From (1.2.38):

$$G_3 \to (a\,\mathrm{d}C_2 + bH) - \frac{a\tau + b}{c\tau + d}(c\,\mathrm{d}C_2 + dH) \overset{(1.2.37)}{=} \frac{1}{c\tau + d} G_3\,. \tag{1.2.42}$$

To see why this symmetry is remarkable, consider for example $s = \left(\begin{smallmatrix} 0 & 1 \\ -1 & 0 \end{smallmatrix}\right)$. Starting from a configuration where $C_0 = 0$, we see that

$$e^{\phi} \to e^{-\phi}\,. \tag{1.2.43}$$

The string coupling $g_s \to 1/g_s$: strong coupling is mapped to weak coupling.

We hasten to add that this is a symmetry of the *supergravity* approximation; it does not fully survive in string theory. One reason is the general expectation that theories of quantum gravity should have no continuous symmetries, which we will review in Chapter 12. A perhaps more concrete argument is that the fundamental string couples to B; an arbitrary transformation in (1.2.38) transforms $B \to cC_2 + dB$, and if this were a symmetry for arbitrary c and d there would have to exist a continuum of string-like objects, coupling to any such linear combination. We will see in Section 1.4.3 that generalizations of the fundamental string do exist, but only a *discrete* infinity of them. This argument suggests that $SL(2,\mathbb{R})$ gets discretized; we will see this in Section 1.4.3.

1.2.4 Heterotic supergravity

Action

The bosonic part of the action for heterotic supergravity is

$$S_{\text{het}} = S_{\text{kNS}} + \frac{l_s^2}{8\kappa^2} \int_{M_{10}} d^{10}x \sqrt{-g_{10}} e^{-2\phi} \text{Tr}|F^2| . \tag{1.2.44}$$

(The positive sign is because $\text{Tr}(T^a T^b)$ is negative-definite; we do not include i's in the gauge fields.) The term S_{kNS} is the usual (1.2.17), but now

$$H \equiv dB - \frac{l_s^2}{4} \text{CS}_A , \tag{1.2.45}$$

where CS_A was given in (1.2.20). As in type II supergravity, there are l_s corrections; some of these will play a role in anomaly cancellation, to which we now turn.

Anomaly cancellation

Since we only supersymmetrized the right-movers, (1.2.44) only has half the supersymmetry of the type II theories: 16 supercharges. Each line marked R in Table 1.3 is related by supersymmetry to the line marked NS above it. As in IIB, the spectrum is chiral: the gravitino $\psi_{M\alpha}$ and the gaugino χ_α have positive chirality, while the $\lambda_{\dot\alpha}$ have negative chirality. Again, this creates the danger of an anomaly, this time not only for diffeomorphisms but also for gauge transformations. Superficially, this does not appear to vanish, but exactly for the two gauge groups (1.1.44) it takes the factorized form

$$\delta\Gamma = \int \omega_2^1 \wedge Y_8 . \tag{1.2.46}$$

Here the symbol δ denotes variation under both diffeomorphisms and gauge transformations; as usual in QFT, Γ is the quantum effective action (the action whose tree-level amplitudes equal the full quantum amplitudes of S). Y_8 is an eight-form, quartic in curvature and in the Yang–Mills field-strength F^a. The two-form ω_2^1 is such that

$$d\omega_2^1 = \delta(\text{CS}_g - \text{CS}_A) , \tag{1.2.47}$$

where now CS_g is a three-form with the property $(d\text{CS}_g)_{MNPQ} = 6R^{AB}{}_{[MN}R_{PQ]BA}$, similar to (1.2.20).

Equation (1.2.46) is nonzero, but it is not the full story; there are contributions from the higher-derivative corrections. First of all, (1.2.45) should be modified to

$$H \equiv dB + \frac{l_s^2}{4}(\text{CS}_g - \text{CS}_A) . \tag{1.2.48}$$

CS_g contains up to three derivatives, since its derivative contains by definition two Riemann tensors. Now (1.2.47) implies $\delta H \neq 0$, unless we also make B transform as $\delta B = -\frac{l_s^2}{4}\omega_2^1$. Another l_s correction is the term

$$\int B \wedge Y_8 \,; \qquad (1.2.49)$$

its transformation now cancels the anomaly (1.2.46). This only works for the two choices (1.1.44) of the gauge group, which allowed us to write (1.2.46) in the first place. For more general gauge groups, $\delta\Gamma$ does not factorize.

Exercise 1.2.1 Compute the equations of motion for the RR fluxes in IIA and IIB by varying the actions (1.2.15), (1.2.31), and (1.2.36) with respect to the C_p. Check that they are formally identical to the Bianchi identities (1.2.10), if we extend them to $k \geq 6$ by defining

$$F_6 = *F_4\,, \qquad F_7 = -*F_3\,, \qquad F_8 = -*F_2\,, \qquad F_9 = *F_1\,, \qquad F_{10} = *F_0\,, \qquad (1.2.50)$$

in the notation of (1.2.34b).

Exercise 1.2.2 Use (1.2.38) and (1.2.39) to show

$$e^{\phi} \to |c\tau + d|^2 e^{\phi}\,. \qquad (1.2.51)$$

1.3 D-branes

We saw that string theory has two expansion parameters: l_s (or rather the dimensionless l_s/r, with r a curvature radius), and e^{ϕ}, which when constant is also called g_s. In (1.1.21), we wrote the expansion as a power series, but we wondered whether this really captures the whole dependence. Recall that a function is called (real) analytic if it coincides with its Taylor series around every point. A famous nonanalytic function is

$$f(g) = e^{-1/g^2}\,. \qquad (1.3.1)$$

Both f and all its derivatives vanish at $g = 0$, so its Taylor series vanishes identically, but $f(g) \neq 0$. In physics, we call an effect with such a dependence on the coupling *nonperturbative*. In attempts at describing nonperturbative aspects of field theory, two types of objects often come up: solitons, which are large and stable field configurations, localized in space; and instantons, which occur in the Euclidean path integral and are localized in time as well. In string theory, the most prominent objects that play these roles are *D-branes*; this section is devoted to them.

1.3.1 Solitons and instantons

We begin with a quick reminder of solitons and instantons in field theory.

Solitons

In perturbative field theory, we often deal with quanta of small field perturbations around the vacuum, which we call elementary particles. However, many field theories also contain other objects. A *soliton* is a "localized" solution of the equations of motion that does not dissipate. This is in contrast with plane waves, which are delocalized. Already in a linear system we find both types of solutions. For example, for a free scalar ϕ in $d = 2$, the equation of motion $(-\partial_t^2 + \partial_x^2)\phi = 0$ is solved by $\phi = \phi_0(t - x)$ for any ϕ_0; this includes the plane wave solutions $\phi = e^{i(t-x)}$, but we can also take ϕ_0 to be any localized function, such as $\phi = (\cosh(x - t))^{-2}$, which is very small everywhere except for a narrow band of order one.

More notably, solitons exist in nonlinear systems also: the venerable Korteweg–De Vries (KdV) equation

$$\partial_t \phi + \partial_x^3 \phi + 6\phi \partial_x \phi = 0\,, \tag{1.3.2}$$

describing water waves in shallow channels, has many solitonic solutions, such as

$$\phi = \frac{c}{2}\left(\cosh\left(\frac{\sqrt{c}}{2}(x - ct)\right)\right)^{-2} \tag{1.3.3}$$

for any c. Once again, this is localized in a narrow band at any t, and velocity c. There also exist *multisoliton* solutions, where many waves similar to (1.3.3) coexist and interact, just like particles. The dissipative and nonlinear effects from the second and third term of (1.3.2) compete in exactly such a way that these waves do not dissipate. (At a deeper level, this phenomenon is really due to the presence of infinitely many conserved quantities for (1.3.2); see, for example, [51].)

Going back to relativistic field theories, many nonlinear theories also have such stable solutions. Any $d = 2$ theory of a single scalar ϕ with a potential $V(\phi)$ with two or more vacua $\phi_\pm$ will have solitons: in this case, they are field configurations where $\phi \to \phi_\pm$ at $x \to \pm\infty$. This is protected from dissipation by topological reasons: one cannot take such a configuration to a vacuum without an infinite energy expenditure. Unlike (1.3.3), these solitons can be at rest. Solitons also exist for $d > 2$, and again they are usually stable for topological reasons. They are usually heavy when the theory is weakly coupled. Nevertheless, just like the soliton waves of the KdV equation, they behave almost as particles, and we should be able to compute their dynamics, for example, their scattering amplitudes.

Monopoles

A *magnetic monopole* is a soliton around which the flux integral of the magnetic field is nonzero. In $\mathbb{R}^3$, this means that

$$\int_{S_2} \mathrm{d}a_i B_i = q_{\mathrm{m}} \neq 0 \tag{1.3.4}$$

on a sphere S_2. $\mathrm{d}a_i$ is an outward-directed vector whose norm $\mathrm{d}a$ is the infinitesimal area. Usually in electromagnetism $q_{\mathrm{m}} = 0$, because of the Maxwell equation $\partial_i B_i = 0$, the space part of $\partial_{[\mu} F_{\nu\rho]} = 0$. So to introduce such objects, we should change this equation to $\partial_i B_i = \Pi_{i=1}^3 \delta^i(x^i)$. A striking feature is their *charge quantization*. Suppose we have a monopole localized at the origin. Consider a potential A_μ^{N}, and integrate it on the equator E of S_2. Since $B_i = \epsilon_{ijk}\partial_j A_k^{\mathrm{N}}$, we can apply Stokes's theorem to the semisphere $\mathrm{U_N}$ of S_2 bounded by E:

$$\oint_{\mathrm{E}} \mathrm{d}x_i A_i^{\mathrm{N}} = \int_{U_{\mathrm{N}}} \mathrm{d}a_i B_i\,. \tag{1.3.5}$$

However, we could do the same with the other portion of the sphere U_S; because of orientation, there is a minus sign, so this time we would have $\oint_E dx_i A_i^N \stackrel{?}{=} -\int_{U_S} da_i B_i$. Subtracting this from (1.3.5), we would find $\int_{S^2} da_i B_i \stackrel{?}{=} 0$, a contradiction. The way out is to have a second potential A_μ^S for the region U_S; then $\oint_E dx_i A_i^S = -\int_{US} da_i B_i$, and subtraction from (1.3.5) now just gives

$$\int_{S_2} da_i B_i = \oint_E dx_i (A^N - A^S)_i = q_m . \tag{1.3.6}$$

Two potentials are related by a gauge transformation, so $(A^N - A^S)_i = \partial_i \lambda_0$; now (1.3.6) tells us that λ_0 is not periodic, but undergoes a shift q_m after a turn around E. Now, a gauge transformation acts on the wave function of a particle with electric charge q_e by $\psi \to e^{-iq_e\lambda_0}\psi$. This should be periodic, so $e^{-iq_e q_m} = 1$, or in other words,

$$\frac{1}{2\pi} q_e q_m \in \mathbb{Z} . \tag{1.3.7}$$

This is called *Dirac quantization*, and will play many roles in this book.[10]

Monopole solutions can be found in the Yang–Mills–Higgs model

$$S_{YMH} = -\int d^4x \,\mathrm{Tr}\left(\frac{1}{2g_{YM}^2}|F|^2 + D_\mu a D^\mu a + \lambda(a^2 - a_0^2)^2\right), \tag{1.3.8}$$

where F is an SU(2) gauge field, a an adjoint scalar, and $D_\mu a \equiv \partial_\mu a + [A_\mu, a]$ the gauge covariant derivative. For the *'t Hooft–Polyakov monopole* solutions, q_m is defined as the magnetic charge under $\mathrm{Tr}(aF)$, and all fields are nonsingular. One can prove the *Bogomolnyi–Prasad–Sommerfield (BPS) bound* for the mass of any monopole solution:

$$m_{mon} \geq a_0 q_m = \frac{2\pi a_0}{g_{YM}} . \tag{1.3.9}$$

As anticipated, the mass is large at weak coupling. Conversely, at strong coupling they may become light.

These effects are under better control in supersymmetric theories. A famous example is the $\mathcal{N} = 2$-supersymmetric version of Yang–Mills (*super-YM*), which is a bit similar to (1.3.8). The $\mathcal{N} = 1$ super-YM involves the gauge field A_μ and a gaugino λ_α. The $\mathcal{N} = 2$ version involves *two* gauginos, and a scalar a (now complex), all in the adjoint representation of the gauge group. The bosonic Lagrangian is[11]

$$S_{SYM} = \int d^4x \mathrm{Tr}\left(-\frac{1}{2g^2}|F|^2 + \frac{\theta}{64\pi^2}\epsilon_{(0)}^{\mu\nu\rho\sigma} F_{\mu\nu}F_{\rho\sigma} - D_\mu a^\dagger D^\mu a - \frac{1}{2}\mathrm{Tr}([a, a^\dagger])^2\right) . \tag{1.3.10}$$

[10] Monopoles in electromagnetism have never been detected; but in quantum chromodynamics (QCD), the theory of strong interactions, their condensation is believed to play an important role in confinement. The mechanism is similar to the Meissner effect in superconductors, where the electric field effectively acquires a mass and the magnetic field is zero almost everywhere, except in thin tubes. The same should happen in QCD, but now with the electric field confined in thin tubes; since its flux lines no longer disperse, this leads to a potential that grows with distance.

[11] The θ term is a total derivative, but it does have physical effects on instantons and on monopoles. In QCD, it is allowed but observed to be $< 10^{-9}$ for unknown reasons (the *strong CP* problem). In supersymmetric theories, it appears naturally.

The anticommutators of the $\mathcal{N} = 2$ algebra read

$$\{Q^I, \bar{Q}_J\} = P_\mu \gamma^\mu \delta^I_J, \qquad \{Q^I, Q^J\} = \epsilon^{IJ} Z. \tag{1.3.11}$$

The generator Z commutes with all other generators in the algebra, and as such it is called *central* charge, just like c in (1.1.19). The BPS bound is now reinterpreted as $m_{\text{mon}} \geq |Z|$. The representation theory of (1.3.11) shows that BPS states form a special short representation; because of this, they are protected against time evolution and against deformations of the theory, so they cannot just disappear as the coupling is changed.

There is a low-energy effective description where we only keep the "abelian" part of the fields, in the Cartan subalgebra. For $G = \text{SU}(2)$, this is U(1), which we can take along the σ^3 generator of su(2); so we keep A^3_μ and its supersymmetric partners. The bosonic part of the effective action is now an Abelian version of (1.3.10), but with both $g = g(a_3)$ and $\theta = \theta(a_3)$ depending on the vacuum expectation value for the scalar a_3. $\mathcal{N} = 2$ supersymmetry determines this *Seiberg–Witten (SW) effective theory* [52] exactly as $\tau(a_3) = \partial^3_{a_3} \mathcal{F}$, where the *prepotential* $\mathcal{F}$ is a holomorphic function of a_3, and we defined it:

$$\tau \equiv \frac{4\pi \mathrm{i}}{g^2_{\text{YM}}} + \frac{\theta}{2\pi}. \tag{1.3.12}$$

In this solution, monopoles do become light at strong coupling, and there is a "dual" description of the theory, of which they are the elementary photons, described by a vector $\tilde{A}_\mu$.

In conclusion, solitons behave a lot like particles; at weak coupling, they are collective excitations, but at strong coupling they may become the fundamental degrees of freedom. So in quantum field theory it is not always clear which objects are made of which others.

Instantons

In cases where exact results are available, such as in the $\mathcal{N} = 2$ SW theory, the effective action is not analytic: it does not coincide with its perturbative expansion. The prepotential $\mathcal{F}$ depends on a_3, but after reexpressing it in terms of the high-energy g_{YM}, we find a sum of contributions of the type

$$\mathrm{e}^{-s/g^2_{\text{YM}}}, \tag{1.3.13}$$

with $s = 8\pi^2 k$, for k an integer.

Such effects are ubiquitous: they appear even in quantum mechanics, in the dependence on $\hbar$ of the energy spectrum, or of the tunnel effect probability (see, for example, [53–55]). The easiest example is perhaps the energy of the vacuum in a double-well potential $V_{\text{dw}} = \lambda(x^2 - x_0^2)^2$. Near each vacuum $x = \pm x_0$, it is approximately harmonic, $V_{\text{dw}} \sim \frac{1}{2}\omega^2(x \mp x_0)^2$, $\omega^2 = 8x_0^2\lambda$; so we expect the lowest energies to be $\sim \frac{\hbar}{2}\omega$, with a small splitting due to tunneling between the two. To find this, one computes the probability $\langle x_0 | \mathrm{e}^{-\mathrm{i}HT/\hbar} | -x_0 \rangle$, which becomes an integral over histories:

$$\int Dx(t) \mathrm{e}^{-\mathrm{i}S/\hbar}. \tag{1.3.14}$$

After a Wick rotation $T \to -\mathrm{i}T_{\mathrm{E}}$, the integral is dominated by the Euclidean-time history $x(t_{\mathrm{E}})$ that solves the Euclidean equations of motion, which are obtained from the ordinary ones by an overall sign of the potential, $V_{\mathrm{E}} = -V$. This classical solution x_{cl} needs to asymptote to $\pm x_0$ for $t_{\mathrm{E}} \to \pm\infty$; it is called an *instanton* because most of its action comes from a particular time t_0, when x switches from one vacuum to the other. Evaluating the integrand on this solution gives already a good approximation to the integral; a better one is found by performing the integral over fluctuations $(x - x_{\mathrm{cl}})(t)$ around it. For the double-well V_{dw}, both these steps can be carried out exactly, and the energy of the lowest eigenstate is

$$E_0 \sim \frac{\hbar}{2}\omega - \hbar\omega\sqrt{\frac{6s}{\pi\hbar}}\mathrm{e}^{-s/\hbar}\,, \qquad s = \frac{2}{3}\omega x_0^2\,. \tag{1.3.15}$$

The splitting of the levels contains the nonanalytic $\mathrm{e}^{-s/\hbar}$. The coefficient $s = S(x_{\mathrm{cl}})$ is the action evaluated on the classical solution; the coefficient multiplying it comes from the integral over small fluctuations.

In field theory, one again Wick-rotates the path integral, looking for classical Euclidean solutions ϕ_{cl} with finite action $s = S(\phi_{\mathrm{cl}})$, so that their contribution $\mathrm{e}^{-s} \neq 0$. So a field theory instanton should be localized in space and time. A famous example occurs for YM theories in $d = 4$, where instantons are solutions of the *self-duality equation*

$$F_{\mu\nu} = \frac{1}{2}\epsilon_{\mu\nu}{}^{\rho\sigma}F_{\rho\sigma}\,, \tag{1.3.16}$$

which in the notation of (1.2.34b) reads

$$F = *F\,. \tag{1.3.17}$$

Computing these effects exactly is more challenging, but sometimes it can be done; again, supersymmetry helps (see, for example, [56] for a review). While the SW $\mathcal{N} = 2$ solution was found with other methods, one can in fact reproduce it exactly by counting gauge instantons [57].

World-sheet instantons in string theory

In string theory, we have two expansion parameters, l_s and $g_s = \mathrm{e}^{\phi}$. For l_s, we can apply the preceding discussion to the two-dimensional QFT on the world-sheet. An instanton for this model is a finite-action solution to the Euclidean equations of motion. Equation (1.1.4) is not directly the world-sheet area, but it is classically equivalent to it, so we can look for maps $x(\sigma)\colon\ \Sigma \to M_{10}$, where the area $A(\Sigma)$ is minimized. In flat space, this would give the degenerate situation where Σ is shrunk to a point. But for spacetimes with nontrivial topology, such as compactifications, the internal space may contain a two-dimensional subspace S_2 that cannot be continuously shrunk to a point; so its minimal area $A(\Sigma) \neq 0$, and this embedding is a *world-sheet instanton*, which, recalling (1.1.4) and (1.1.5), contributes

$$\mathrm{e}^{-A(\Sigma)/4\pi l_s^2} \tag{1.3.18}$$

to the path integral. Remarkably, these effects can sometimes be computed exactly, as we will see in Section 7.1.

Effects that are nonperturbative in g_s are a different matter altogether. The world-sheet approach is intrinsically perturbative in g_s; so we need a new ingredient.

1.3.2 Open string definition

A *D-brane* is an additional extended object in string theory; so far we have not encountered it because we limited ourselves to a perturbative approach.

In general relativity, no solitons exist, if defined as fully regular solutions with localized energy and flat asymptotics ([58]; for a recent account, see, for example, [59, IV.8]). In a general gravitational theory, we might be tempted to say that black holes are the analogue of solitons: they have localized energy, and they are stable. They do have a singularity, unlike the solitons we have studied in the previous section. It is natural to think, however, that this singularity signals the breakdown of general relativity, and that in a full theory of quantum gravity it disappears.

Besides black holes, string theory has solutions whose energy is localized along a subspace, and not just a point in space; so they are called *p-branes*, a generalization of the word "membrane," where p denotes the number of space dimensions they span. As solutions of the supergravity equations, they have a singularity at their core; but string theory does provide an alternative, smooth understanding, in terms of *open strings*.

An open string is one whose time slice is an interval rather than a circle. It requires choosing boundary conditions for the world-sheet fields; Dirichlet boundary conditions for some of the $x^M(\sigma)$ correspond to open strings that end on a subspace of spacetime, which is accordingly called a Dirichlet-brane, or more commonly a *D-brane*.

These D-branes are the fundamental definition of the p-brane gravity solutions [60]. Consider, for example, a D-brane in flat space. We have both closed and open strings, interacting with one another, both described by a world-sheet model with a flat background metric $g_{MN} = \eta_{MN}$. Suppose we now perform the path integral over the open string degrees of freedom – in QFT jargon, we "integrate them out." As in QFT, the action for the remaining degrees of freedom, those of closed strings, is now modified: the background metric is distorted to a different metric g_{MN}. So we obtain an effective description, with only closed strings in a curved background metric, which is nothing but the p-brane metric, or in other words the "back-reaction" of the D-brane on spacetime. Because of the difficulties with the world-sheet description of RR fields that we mentioned at the end of Section 1.1.2, this identification is a bit difficult to demonstrate explicitly, but several indirect arguments point toward it.

We will now review the definition of D-branes as loci where open strings end, and then their gravitational description.

To describe an open string, we take $\sigma^1 \in [0, \pi]$, rather than a periodic coordinate as in Section 1.1. When we vary the action, we now have to pay attention to boundary terms. For example, let us consider flat space $g_{MN} = \eta_{MN}$ and $h_{\alpha\beta} = \eta_{\alpha\beta}$, and focus on a single coordinate x. Then the variation gives

$$\begin{aligned}\frac{1}{2}\delta \int_\Sigma d^2\sigma \partial^\alpha x \partial_\alpha x &= \int d^2\sigma \partial^\alpha x \partial_\alpha \delta x = \int d^2\sigma \left(\partial_\alpha(\delta x \partial^\alpha x) - \delta x \partial^2 x\right) \\ &= \int_{\partial\Sigma} d\sigma^0 \delta x \partial_{\sigma^1} x - \int_\Sigma d^2\sigma \delta x \partial^2 x \,.\end{aligned} \tag{1.3.19}$$

The second term gives us the free equation of motion $\partial^2 x = 0$, familiar from closed strings; the first is new. To set it to zero, we can either set $\partial_{\sigma^1} x = 0$ or $\delta x = 0$. The first is called *Neumann (N)* boundary condition; the second sets to zero the variation of this world-sheet scalar at the boundary, and is called *Dirichlet (D)* boundary condition.

All this was for a single x; still remaining in flat space, we can select N or D boundary conditions for each of the x^M independently. Each x^M for which we are choosing, a D boundary condition is fixed, $x^M = x_0^M$, at the boundary $\partial\Sigma$, or in other words at the endpoints of the open string. So if we choose N boundary conditions for $p+1$ fields x^M (including time), and D for the remaining $9-p$, we have a theory of open strings that end on a $p+1$-dimensional object; by definition, this is called a Dp-brane.

Clearly, the presence of a Dp breaks some of the symmetries of the background; in flat space, a flat Dp breaks the ten-dimensional Poincaré group ISO(1, 9) to a ISO$(1, p+1) \times$ SO$(9-p)$. For applications to compactifications, we will want Dp-branes that are completely extended along $\mathbb{R}^4$ and localized along some of the internal dimensions. This preserves the four-dimensional Poincaré group.

The analysis in (1.3.19) is enough for D-branes in bosonic string theory; for type II superstrings, we also need to give boundary conditions for world-sheet spinors. The procedure (1.3.19) applied to the spinorial action (1.1.30) shows that at the boundary $\partial\Sigma$ the combination $\psi_+\delta\psi_+ - \psi_-\delta\psi_-$ should vanish; this can be arranged by having

$$(\psi_+ = \pm\psi_-)|_{\partial\Sigma} \,. \tag{1.3.20}$$

Both for the x^M and ψ^M, the boundary conditions now relate left- and right-movers, and the open string spectrum has only one set of oscillators. Just like in the closed string sector, massless states come from the fermionic oscillators; the sign in (1.3.20) gives two different sectors, which again we call NS and R. After GSO projection, we have the massless spectrum

$$b^M_{-1/2}|0\rangle_{\rm NS}\,, \qquad \mathbf{8}_{\rm V}\,; \qquad |0,\alpha\rangle_{\rm R}\,, \qquad \mathbf{8}_{\rm S}\,. \tag{1.3.21}$$

If we are considering N boundary conditions, the endpoints are not fixed to lie anywhere, and we have a D9-brane; in this case (1.3.21) is interpreted as a vector field a_M and a gaugino λ_α. If we have D directions $i = p+1, \ldots, 9$, then the components $b^i_{-1/2}|0\rangle_{\rm NS}$ are transverse to the Dp-brane, and behave as scalars under ISO$(1, p+1)$; the parallel components $b^a_{-1/2}|0\rangle_{\rm NS}$, $a = 0, \ldots, p$ still represent a vector field. As for the $|0,\alpha\rangle_{\rm R}$, they represent a spinor under both the parallel and transverse rotations. So in total we have

$$a^a\,, \qquad x^i\,, \qquad \lambda^\alpha\,. \tag{1.3.22}$$

Calling the scalars x^i might seem confusing, given that we also called x^M the world-sheet scalars. But we will see shortly that the scalars x^i on a D-brane parameterize its transverse fluctuations, and thus give an embedding of Dp into spacetime.

Just as closed (bosonic) strings can couple to condensates of their massless fields (1.1.9), we can couple open strings to condensates of (1.3.22). For example, the coupling to a^a reads

$$\int_{\partial\Sigma} d\sigma^0 a_a \partial_0 x^a\,; \tag{1.3.23}$$

recalling (1.1.3), this means that the endpoints behave as charged particles under a. Rewriting this as $-\int_\Sigma d^2\sigma \partial_1(a_a\partial_0 x^a)$ and adding the total derivative

$0 = \int d^2\sigma \partial_0(a_a \partial_1 x^a)$, this becomes $\int_\Sigma d^2\sigma f_{ab}\partial_0 x^a \partial_1 x^b$, where the two-form $f = da$ is the field-strength of the vector a. This has the same form as in (1.1.10); in the parallel directions, the two can be collected together, and the combination $(B + 2\pi l_s^2 f)_{ab}$ appears.

1.3.3 Effective action

So far, we considered a Dp-brane extended along a flat subspace $\mathbb{R}^{p+1} \subset \mathbb{R}^{10}$. We can try to generalize this to arbitrary subspaces in arbitrary background metrics, but we still need to impose conformal invariance. For closed strings, this led to the beta functions (1.1.14), which we reinterpreted as spacetime equations of motion for the effective action (1.1.17); for open strings, it leads to equations of motion for the brane, which again come from an effective action $S_{\mathrm{D}p}$. The fields appearing in this action are those of the closed string spectrum in Tables 1.1 and 1.2, plus the open string fields (1.3.22). Once again, we only give here the bosonic part:

$$S_{\mathrm{D}p} = \tau_{\mathrm{D}p}\left[-\int d^{p+1}\sigma e^{-\phi}\sqrt{-\det(g|_{\mathrm{D}p} + \mathcal{F})} \pm \sum_k (-1)^k \frac{1}{k!}\int_{\mathrm{D}p} C_{p+1-2k} \wedge \mathcal{F}^k\right] \tag{1.3.24}$$

where

$$\tau_{\mathrm{D}p} = \frac{1}{(2\pi)^p l_s^{p+1}} . \tag{1.3.25}$$

The coordinates σ^a, $a = 0, \ldots, p$ parameterize the world-volume[12] Dp. The restriction or pull-back operation $|_{\mathrm{D}p}$ consists in contracting each index M with $\partial_a x^M$, as in (1.1.3) and (1.1.10); the $x^M(\sigma^0, \ldots, \sigma^p)$ gives the embedding of Dp into spacetime. In the second term of (1.3.24), the spacetime forms are integrated on the Dp world-volume by first pulling them back; so, for example, we should really write $\int_{\mathrm{D}p} C_p \equiv \int_{\mathrm{D}p} C_p|_{\mathrm{D}p}$. The sign $\pm$ is related to the "orientation" on Dp, namely to the order of the coordinates in the world-volume measure; we will see this in more detail later, and fix this convention more precisely in Chapter 10.

Recall that $f^k = f \wedge \ldots \wedge f$; we also defined

$$\mathcal{F}_{ab} \equiv (2\pi l_s^2 f + B|_{\mathrm{D}p})_{ab} = 2\pi l_s^2 f_{ab} + B_{MN}\partial_a x^M \partial_b x^N , \tag{1.3.26}$$

in line with the remark that follows (1.3.23). The gauge transformation for B no longer leaves the action invariant, unless it also acts on a:

$$B \to B + d\hat{\lambda}_1 , \qquad a \to a - 2\pi l_s^2 \hat{\lambda}_1 . \tag{1.3.27}$$

For open strings, (1.1.12) has to be modified as $g_s^{2g-2+\#b}$, where $\#b$ is the number of boundaries added to Σ; the prefactor $e^{-\phi}$ in (1.3.29) indicates then that this action originates from a sphere to which we add a single boundary, which is topologically a disk. This is the simplest open string diagram: an open string that is created and later disappears. The term containing C_p does not contain this prefactor because of the

[12] We will call "world-volume" the subspace swept in spacetime by a D-brane, and keep using "world-sheet" for a fundamental string.

customary RR rescaling noted following (1.2.17). For constant $e^{\phi} = g_s$ effectively the tension of a D-brane is not really τ_{Dp}, but rather

$$T_{Dp} = \frac{\tau_{Dp}}{g_s} . \tag{1.3.28}$$

So at small g_s, D-branes are heavy, similar to the bound on monopole mass in (1.3.9) in terms of g_{YM}.

The DBI term

If we set to zero all fields except g_{MN}, (1.3.24) reduces to

$$-\tau_{Dp} \int d^{p+1}\sigma\, e^{-\phi} \sqrt{-\det g_{MN}\partial_a x^M \partial_b x^N} . \tag{1.3.29}$$

This is the natural generalization of the particle action (1.1.2) to an extended object; it measures the volume of the $(p+1)$-dimensional object Dp in spacetime, relative to the background metric g_{MN}. In the flat background metric $g_{MN} = \eta_{MN}$, a subspace that is itself flat extremizes (1.3.29); a curved Dp will tend to relax to such a flat subspace, or shrink to a point. Without loss of generality, we can locally take the first $p+1$ coordinates of spacetime x^M to coincide with the σ^a; then the remaining $9-p$ can be identified with the transverse scalars x^i in (1.3.22). If we choose the latter so that $g_{aj} = 0$, then

$$(g|_{Dp})_{ab} = g_{ab} + g_{ij}\partial_a x^i \partial_b x^j . \tag{1.3.30}$$

For small x^i, (1.3.29) contains then the usual kinetic term, and terms with more than two derivatives.

The full first term $\sqrt{-\det(g|_{Dp} + \mathcal{F})}$ is called *(Dirac–)Born–Infeld (DBI)* because it is reminiscent of early proposals to improve the short-distance behavior of the electromagnetic field, inspired by the resemblance to the relativistic contraction factor $\gamma^{-1} = \sqrt{1-v^2}$. While this combination might seem odd, we will see that it has many natural properties (including a duality with Pythagoras' theorem in Section 1.4.2). We can expand it for small open string fields. Recall that $\log\det M = \mathrm{Tr}\log(M)$; then for a matrix $M \sim 1 + m$ near the identity, $m \ll 1$:

$$\begin{aligned}\det(1+m) &= \exp\left[\mathrm{Tr}\log(1+m)\right] = \exp\left[\mathrm{Tr}\, m - \frac{1}{2}\mathrm{Tr}(m^2) + \dots\right] \\ &= 1 + \mathrm{Tr}\, m + \frac{1}{2}((\mathrm{Tr}\, m)^2 - \mathrm{Tr}(m^2)) + \dots .\end{aligned} \tag{1.3.31}$$

Using this and (1.3.30), for $B = 0$:

$$\sqrt{-\det(g|_{Dp} + \mathcal{F})} \sim \sqrt{-\det_{ab} g_{ab}}\left(1 + \frac{1}{2}g_{ij}\partial_a x^i \partial^a x^j + \frac{1}{2}|2\pi l_s^2 f|^2\right) + \dots , \tag{1.3.32}$$

where we kept only terms with two derivatives, and indices are raised and contracted with g_{ab}. Taking into account the prefactors in (1.3.24), the YM coupling is given by

$$g_{YM}^2 = (2\pi)^{p-2} l_s^{p-3} g_s . \tag{1.3.33}$$

In particular, it is dimensionless for $p = 3$, as expected.

The Wess–Zumino term

We now turn to a discussion of the term involving C_{p+1-2k} in (1.3.24). This is variously called *Chern–Simons* or *Wess–Zumino* (WZ) term, inspired by the names of two famous actions that don't involve the metric. When $\mathcal{F} = 0$, the integral on Dp is defined as in (1.2.8), after the pull-back operation described following (1.3.24). Explicitly,

$$\int_{\mathrm{D}p} C_{p+1} = \int C_{M_0 \ldots M_p} \partial_0 x^{M_0} \ldots \partial_p x^{M_p} \,. \tag{1.3.34}$$

For $k = 1$ this has the form (1.1.3); for $k = 2$, it looks like the coupling (1.1.10) of the fundamental string to B. So the coefficient of (1.3.34) is a charge density; we see from (1.3.24) that it equals the brane tension $T_{\mathrm{D}p}$.

Recall from Section 1.1.2 that the C_{p+1} exist with p = even in IIA, and p = odd in IIB; so there are

$$\mathrm{D}p\text{-branes:} \quad p = \begin{cases} \text{even} & \text{IIA}\,, \\ \text{odd} & \text{IIB}\,. \end{cases} \tag{1.3.35}$$

In IIA, the smallest object is a point-like soliton in IIA, the D0-brane. In IIB we also have the possibility of a D(-1)-brane or *D-instanton*, with a D boundary condition for the time coordinate as well, so that it is localized also in time. In the context of compactifications, there are additional brane instanton that wrap the internal directions and are completely localized in the noncompact directions.

A subtlety mentioned following (1.3.24) is that the overall sign of (1.3.34) can change if we embed the world-volume differently: if we flip the sign of one of the σ^a,

$$x^M(\sigma^0, \sigma^1, \ldots, \sigma^p) \to x^M(\sigma^0, -\sigma^1, \ldots, \sigma^p)\,, \tag{1.3.36}$$

the image of the embedding remains the same, but (1.3.34) changes sign, because $\partial_{\sigma^1} x^M$ appears only a single time in it. A similar sign change happens upon exchanging two coordinates. Since the charge density changes sign, we call this an *anti-Dp-brane*. The difference between a brane and antibrane is conventional.

Effect of world-volume flux

When $\mathcal{F} \neq 0$, a Dp-brane also couples to RR fields C_k with $k < p$. For example, the WZ term for a D2-brane becomes

$$\int_{\mathrm{D}2} (C_3 - \mathcal{F} \wedge C_1)\,. \tag{1.3.37}$$

In (1.3.34), C_1 would couple to a D0; so we interpret the second term in (1.3.37) as the presence of a distribution of D0-branes on the D2. In other words, we have a D2 that also has D0 charge, or a *D2/D0 bound state*.

A final subtlety about the WZ term regards the Romans mass F_0 [61, 62]. Since this flux has no potential, it might seem that no brane couples to it. In fact, for $F_0 \neq 0$ the WZ term in (1.3.24) needs to be modified to

$$\int_{\mathrm{D}p} \mathrm{wz}_{p+1}\,, \qquad d\mathrm{wz}_{p+1} = \sum \frac{1}{k!}(-1)^k F_{p+2-2k} \wedge \mathcal{F}^k\,. \tag{1.3.38}$$

For $F_0 = 0$, $\mathrm{wz}_{p+1} = \sum_k \frac{1}{k!}(-1)^k C_{p+1-2k} \wedge \mathcal{F}^k$ as in (1.3.24); for $F_0 \neq 0$, we can use $\sum_k F_{p+2-2k}\mathrm{cs}_k$, where cs_k generalizes the Chern–Simons form (1.2.20). For example, for $p = 0$, taking (1.3.24) literally would give a term $\int C_1 - C_{-1} \wedge \mathcal{F}$; the correct coupling can be obtained by formally integrating the second term by part:

$$\int_{\text{D0}} (C_1 - F_0 a) \,. \tag{1.3.39}$$

Magnetic dual potentials

In (1.3.34), RR potentials C_p appear for any p, while in Section 1.2 we only saw $p \leq 4$. We need to extend the definition to higher p by introducing the *magnetic dual* potentials, defined through

$$\mathrm{d}C_p - H \wedge C_{p-2} \equiv F_{p+1} \equiv (-1)^{p(p-1)/2} * F_{9-p} \quad (p > 4) \tag{1.3.40}$$

in the notation of (1.2.34b). These new C_p, $p \geq 5$ are the higher-dimensional analogue of the magnetic dual gauge fields we mentioned for the SW solution that follows (1.3.12). For example, a D6-brane couples to C_7, which when $H = 0$ is defined by $\mathrm{d}C_7 = F_8$, or more explicitly

$$F_{M_1 \dots M_8} = -\sqrt{-g}\,\epsilon^{(0)}_{M_1 \dots M_8}{}^{M_9 M_{10}} \partial_{M_9} C_{M_{10}} \,. \tag{1.3.41}$$

In Exercise 1.2.1, it was noted that the RR equations of motion look like an extension of the Bianchi identities in terms of the magnetic duals. This allows to rewrite the action by swapping one or more of the original potentials with their magnetic duals. In presence of a Dp with $p \geq 4$, it is convenient to use this reformulation to vary with respect to C_{p+1}. The combined action, (1.3.24) plus the supergravity action in (1.2.31) or (1.2.36), can be rewritten as an integral over ten dimensions by using delta functions. In the language of forms, recalling our notation x^i for the directions transverse to the Dp, we can introduce a form

$$\delta_{\text{D}p} \equiv \delta(x^1) \dots \delta(x^{9-p}) \mathrm{d}x^1 \wedge \dots \mathrm{d}x^{9-p} \,. \tag{1.3.42}$$

After the variation, we then obtain (for $\mathcal{F} = 0$)

$$\mathrm{d}F_p - H \wedge F_{p-2} = \mp 2\kappa^2 \tau_{\text{D}(8-p)} \wedge \delta_{\text{D}(8-p)} \,; \tag{1.3.43}$$

the source term on the right-hand side is the effect of D-branes on the RR field strengths. A similar source term appears in the equations of motion for the other closed string fields.

1.3.4 Flux quantization

The coupling (1.3.34) is the natural generalization of (1.1.3) to extended objects. In this sense, a Dp-brane is the elementary charge for the RR field C_{p+1}, much as the electron for the electric field. The generalization of the large gauge transformation (1.2.13) is

$$C_{p+1} \to C_{p+1} + \Lambda_{p+1} \,, \qquad \frac{\tau_{\text{D}p}}{2\pi} \int_{S_{p+1}} \Lambda_{p+1} \in \mathbb{Z} \tag{1.3.44}$$

for any subspace S_{p+1}. This includes our old "small" RR gauge transformations (1.2.11): $\Lambda_{p+1} = \mathrm{d}\lambda_p$ is a total derivative, and the integral (1.3.44) is zero.

The non-single-valuedness of a gauge transformation λ_0 was used in the Dirac quantization argument (1.3.7). This suggests a generalization of that result to RR fields. Surround a Dp-brane with a sphere S^{8-p}, consider an "equator" E on it (itself a sphere S^{7-p}), and the two semispheres U_N, U_S into which it divides S^{8-p}. The integral $\tau_{D(6-p)} \int_{U_N} (dC)_{8-p}$ can be rewritten using an analogue of Stokes's theorem (which we will formalize in Section 4.1.10) as

$$\tau_{D(6-p)} \int_E C_{7-p} \,. \tag{1.3.45}$$

Repeating this over U_S yields a competing result for (1.3.45); the two results in general disagree, their difference being exactly the integral of F_{8-p} over all of S^{8-p}. But there is no inconsistency if we use two different C_{p+1} on U_N and U_S, differing by a large gauge transformation (1.3.44). This leads to the condition

$$\tau_{D(6-p)} \int_{S^k} (dC)_{8-p} = 2\pi n \,, \tag{1.3.46}$$

where $n \in \mathbb{Z}$. Recalling (1.3.25),

$$\frac{1}{(2\pi l_s)^{k-1}} \int_{S^k} (dC)_k \in \mathbb{Z} \,. \tag{1.3.47}$$

Although for concreteness we presented our argument with spheres, (1.3.47) holds for any k-dimensional subspace S_k. Often such a subspace can be continuously deformed to a point, and in that case the integral in (1.3.47) just vanishes. The integral can be nonzero in two types of situations:

- If S_k surrounds a brane, we cannot continuously deform it to zero without crossing it. The integral $\int F_{8-p}$ can be taken as the definition of the charge of a Dp; one could use this to rederive (1.3.25). In this case, flux quantization can also be derived by suitably integrating (1.3.43).
- For compactifications, some S_k in the internal space cannot be continuously deformed to a point for *topological* reasons (Section 4.1.10). In this case, (1.3.47) can be nonzero even if there are no D-branes.

The logic behind (1.3.47) can also be applied to the magnetic dual potentials (1.3.40); one concludes that (1.3.47) is valid for all $k \leq 10$, and not just for the F_k with $k \leq 5$ that appear in the supergravity actions. Notice that the $*$ in (1.3.40) contains the metric, as we see explicitly in (1.3.41). This seems to create a paradox: if we impose flux quantization both for F_3 and for its dual $*F_3$ (say), it seems we are imposing some quantization conditions on the metric itself. In practice, however, there are always some noncompact directions in spacetime, and only one of the two flux integrals makes sense. For example, for $M_{10} = \mathbb{R}^4 \times M_6$, if we impose flux quantization for F_3, the one for $*F_3$ would involve an integral over $\mathbb{R}^4$, which would diverge.

The case $k = 0$ in (1.3.47) deserves a separate discussion. There is no potential C_{-1}, and $S^0 = \{x_1^2 = 1\}$ is just the union of two points. But integrating (1.3.43) for $p = 0$ on a segment crossing a D8, we still obtain

$$2\pi l_s \Delta F_0 = N_{D8} \in \mathbb{Z} \,. \tag{1.3.48}$$

Strictly speaking, in this case we only quantized the jumps of F_0 rather than F_0 itself. It would now be possible to entertain the notion that $2\pi l_s F_0 = n_0 + \theta$, with $n_0 \in \mathbb{Z}$ and $\theta \in [0, 1)$, fixed for any background. This looks unlikely; we will see that T-duality relates the RR fluxes to one another, and if $\theta \neq 0$ were allowed it would eventually violate one of the (1.3.47) for $k > 0$. So we will take

$$2\pi l_s F_0 \in \mathbb{Z}. \tag{1.3.49}$$

Our discussion can also be easily applied to B, since its coupling to the fundamental string is identical to the one of C_1 to the D1. Looking at (1.3.44) and (1.3.47), we conclude

$$B \to B + \tilde{\Lambda}_2, \qquad \frac{1}{(2\pi l_s)^2} \int_{S^2} \tilde{\Lambda}_2 \in \mathbb{Z}; \tag{1.3.50a}$$

$$\frac{1}{(2\pi l_s)^2} \int H \in \mathbb{Z}. \tag{1.3.50b}$$

1.3.5 Supersymmetry

D-branes break some of the bosonic symmetries of a background; perhaps more importantly, D-branes also partially break some of its supersymmetry. Recall that both type II string theories are invariant under 32 supercharges, whose infinitesimal parameters are two spacetime fermions ϵ^a. A Dp extended along a flat $\mathbb{R}^{p+1} \subset \mathbb{R}^{10}$ is invariant only if the two ϵ^a are related by

$$\epsilon^1 = \Gamma_{\|} \epsilon^2, \tag{1.3.51}$$

where

$$\Gamma_{\|} = \frac{1}{p! \sqrt{-\det(g|_{\mathrm{D}p})}} \epsilon_{(0)}^{a_0 \ldots a_p} \Gamma_{a_0 \ldots a_p}, \qquad \Gamma_a \equiv \Gamma_M \partial_a x^M \qquad (\mathcal{F} = 0). \tag{1.3.52}$$

The flat space background is invariant under all 32 supercharges. If we have a Dp-brane extended along a $\mathbb{R}^{p+1} \subset \mathbb{R}^{10}$ subspace, without loss of generality we can take it to be $\{x^{p+1} = \ldots = x^9 = 0\}$; (1.3.52) then becomes

$$\Gamma_{\|} = \Gamma_0 \ldots \Gamma_p. \tag{1.3.53}$$

Equation (1.3.51) gives a relation among the two ϵ^a: only ϵ^1 is now independent, and thus we are left with 16 supercharges. This preserved supersymmetry makes these flat D-branes the analogue of the field-theory BPS states in Section 1.3.1. In particular, they are stable against time evolution and deformations of the theory.[13] This is the reason they are useful: if we make the string coupling g_s large, we lose perturbative control, but we know that D-branes have to remain in the spectrum. Since their tension (1.3.28) is the inverse of the string coupling, they become light at strong coupling and might become new fundamental objects, similar to the SW solution of $\mathcal{N} = 2$-supersymmetric Yang–Mills (Section 1.3.1).

We saw in Section 1.1.2 that the ϵ^a have opposite chiralities in IIA, and equal in IIB, and that multiplication by a single Γ^M changes chirality. It follows that

[13] A D-brane with a more general shape is not guaranteed to preserve supersymmetry, and hence to be stable under time evolution. In Section 9.2, we will consider this further, and also generalize (1.3.52) to $\mathcal{F} \neq 0$.

(1.3.51) can only be solved for p = even in IIA, and for p = odd in IIB; this confirms (1.3.35).[14]

The origin of the constraint (1.3.51) is roughly the following. The fermionic completion of the effective action (1.3.24) includes two world-sheet fermions θ^a. Closed string supersymmetry acts on these as a spinorial translation:

$$\delta\theta^a = \epsilon^a \, . \tag{1.3.54}$$

Imposing supersymmetry would then set both $\epsilon^a = 0$. But there is also a gauge equivalence among the θ^a, called *κ-symmetry*, acting as

$$\delta\theta^1 = \Gamma_{\parallel}\kappa \, , \qquad \delta\theta^2 = \kappa \, . \tag{1.3.55}$$

Imposing invariance under a combined supersymmetry and κ-symmetry gives (1.3.51).

1.3.6 Multiple D-branes

D-brane stacks and nonabelian gauge groups

To each endpoint of an open string, it is possible to add an extra discrete quantum number $I = 1, \ldots, N$, called the *Chan–Paton (CP) label*. It can be interpreted as the presence of $N > 1$ superimposed D-branes, or in other words a stack of N D-branes.

Let us see why. Since open strings have two endpoints, the states (1.3.21) and the fields (1.3.22) acquire two extra labels IJ, and so they are now promoted to Hermitian matrices. Now a becomes a nonabelian U(N) gauge field, and x a scalar in the adjoint representation. In the action (1.3.24), we should turn all fields into matrices with an overall trace, but there are potential ordering ambiguities.

The situation is much clearer for the two-derivative approximation (1.3.32), whose nonabelian extension is dictated by supersymmetry. We know from (1.3.51) that a brane extended along a flat $\mathbb{R}^{p+1} \subset \mathbb{R}^{10}$ subspace preserves 16 supercharges; we expect its effective action to have this invariance. This is the maximum number of supercharges for a QFT model (not including gravity), and models with this property are quite constrained. For example, for $p = 9$, when we have a D9 extended along all of spacetime, there are no scalars x^i, and in (1.3.32) only the $|f|^2$ term remains. Its nonabelian version is the YM Lagrangian density $\mathrm{Tr}|f|^2$; supersymmetry involves also the gaugino λ^α of (1.3.22), and the requirement of 16 supercharges fixes the action uniquely.

For lower p, some components of the gauge field now become x^i, because of their common origin in (1.3.57). The resulting supersymmetric YM theory with 16 supercharges is again uniquely fixed: it is the dimensional reduction of the action for $p = 9$ along the directions to the Dp. The components of the gauge field along the transverse directions become the transverse scalars x_i; the world-volume field strengths then become

$$F_{ai} = D_a x_i = \partial_a x_i + [A_a, x_i] \, , \qquad F_{ij} = [x_i, x_j] \, . \tag{1.3.56}$$

[14] There also exist non-BPS branes that violate (1.3.35) and are of course not described by the effective action (1.3.24). These branes are unstable, but they become stable after some quotients, including the orientifolds we will introduce in Section 1.4.4 [63].

In particular, the YM term generates a potential for the scalars

$$V \propto \mathrm{Tr}\left([x_i, x_j][x_i, x_j]\right), \tag{1.3.57}$$

which vanishes in the Abelian case. The x_i are diagonalizable because Hermitian; the vacua of (1.3.57) are given by configurations where $[x_i, x_j] = 0$, which means that the x_i are simultaneously diagonalizable. In the generic vacuum where they are all different, the BEH mechanism gives a mass to all fields except the diagonal ones; the action of these massless modes is again the Abelian (1.3.32). The D-brane positions are the eigenvalues λ_i^I. This picture motivates the proposed interpretation of the CP label as N superimposed D-branes.

An additional check of this conclusion is that the off-diagonal modes receive a mass that is proportional to the $\sum_i(\lambda^I - \lambda^J)_i^2$; these can be interpreted as the lengths of the strings going from the Ith to the Jth D-brane. Another is supersymmetry: the condition for (1.3.51) is the same for parallel Dp, since $\partial_a x^M$ is insensitive to translations $x^M \to x^M + x_0^M$. So having parallel Dp still preserves 16 supercharges, which is what we assumed for the nonabelian super-YM theory in presence of the CP label.

Brane–antibrane system

Recall from (1.3.36) that parity in one of the embedding coordinates changes the WZ coupling (1.3.34) by a sign; we called this an anti-Dp. Now in the condition for supersymmetry (1.3.51) the matrix $\Gamma_{\parallel}$ changes sign, too. If we have a Dp and an anti-Dp together, we have to solve

$$\epsilon^1 = \Gamma_{\parallel}\epsilon^2, \qquad \epsilon^1 = -\Gamma_{\parallel}\epsilon^2, \tag{1.3.58}$$

which is impossible. So a brane–antibrane system breaks supersymmetry.

Moreover, an analysis of the open string modes reveals the presence of a tachyon. It has been shown [64, 65] that in fact this tachyon has a nontrivial potential, with a stable vacuum, which represents the closed-string vacuum without any branes; in other words, the tachyon can condense (get a nonzero expectation value) and become stable. The spacetime interpretation is that the Dp–anti-Dp pair annihilates, much as a particle–antiparticle pair. The story becomes even more interesting when one of the two has a nontrivial flux of the world-volume field-strength f; in this case, after the annihilation we are left with a lower-dimensional brane instead of the vacuum. This motivates the *K-theory* interpretation of D-branes [66], also suggested by the higher-derivative corrections to (1.3.24) [67]. (This will reappear in Section 9.2.3.)

1.3.7 Gravity solutions

We anticipated at the beginning of this section that a D-brane back-reacts on closed string fields, distorting the flat space metric to a so-called p-brane solution. These can be found by solving the closed string equations of motion with the symmetries we expect the brane to preserve, namely ISO(p+1)×SO($d-p$) (parallel Poincaré and transverse rotations), and 16 supercharges. Here we give the result; we will check in later chapters that they are supersymmetric solutions, in various ways.

The solutions for N superimposed Dp-branes, extended along a flat $\{x^{p+1} = \ldots = x^9 = 0\} = \mathbb{R}^{p+1} \subset \mathbb{R}^{10}$, is

$$ds^2_{\text{D}p} = h^{-1/2}ds^2_{\parallel} + h^{1/2}ds^2_{\perp}\,, \qquad e^{\phi} = g_s h^{\frac{3-p}{4}}\,, \tag{1.3.59a}$$

$$F_{i_1\ldots i_{8-p}} = \frac{f_{8-p}}{r^{9-p}} x^i \epsilon^{\perp}_{ii_1\ldots i_{8-p}}\,, \qquad f_{8-p} \equiv \frac{(2\pi l_s)^{7-p}N}{v_{8-p}}\,. \tag{1.3.59b}$$

$ds^2_{\parallel} = -(dx^0)^2 + \sum_{a=1}^{p}(dx^a)^2$, and $ds^2_{\perp} = \sum_{i=p+1}^{9}(dx^i)^2$ are the metrics respectively of the $1+p$ parallel and $9-p$ transverse dimensions prior to introducing the D-branes; recall that a and i are our names in this section for the parallel and transverse indices respectively. $\epsilon^{\perp}$ is the completely antisymmetric tensor in the transverse directions (such that $\epsilon^{\perp}_{p+1\ldots 9} = 1$); $v_d \equiv \text{Vol}(S^d) = 2\frac{\pi^{(d+1)/2}}{\Gamma((d+1)/2)}$ is the volume of the unit-radius sphere S^d; in form notation, as we will see in Exercise 4.1.9, $F = f_p \text{vol}_{S^d}$, the volume form of S^d.

h is a function of the radial direction $r = (\sum_{i=p+1}^{9}(x^i)^2)^{1/2}$, which satisfies

$$\sum_{i=p+1}^{9} \partial_i^2 h = 0 \tag{1.3.60}$$

away from $r = 0$. For $p < 7$, we can take

$$h = 1 + N\frac{r_0^{7-p}}{r^{7-p}}\,, \qquad r_0^{7-p} \equiv \frac{g_s(2\pi l_s)^{7-p}}{(7-p)v_{8-p}}\,. \tag{1.3.61}$$

The fields not mentioned in (1.3.59) are zero, and in particular so is $H = dB$.

For the sake of a unified description, in (1.3.59) we have given the transverse RR field-strength F_{8-p}. For $p \geq 4$, this is one of the RR forms that appear in the type II supergravity actions of Section 1.2. For $p \leq 2$, this is one of the magnetic duals defined in (1.3.40); using that definition backward on (1.3.59) gives the original RR form F_{p+2}, which has components $F_{a_0\ldots a_p r}$. For $p = 3$, the flux F_5 is self-dual, and both the transverse and parallel directions appear.

The $x^i\epsilon^{\perp}_{ii_1\ldots i_{8-p}}$ in (1.3.59b) can be interpreted as the angular part of the transverse directions, as we will see better in Section 4.1.3; its integral is simply v_{8-p}, and the prefactor is fixed so that (1.3.46) and (1.3.47) holds with $n = 1$. The equations of motion then determine the coefficient of r^{7-p} in (1.3.61), but only up to a sign; the correct choice can be decided by examining the long-distance behavior of the gravitational field, which should be attractive given that we want positive tension as in (1.3.28).

Generalizations

The solutions (1.3.59) are part of a more general family, where there are two horizons; (1.3.59) is the extremal limit where the two horizons have coalesced, and have been set to $r = 0$ by a coordinate change.[15] This extremality is related to the equality of the tension and the charge density of a D-brane, or in other words the coefficients of the DBI and WZ terms in (1.3.24). It is quite common for the extremal limit to saturate the BPS bound, and vice versa.

[15] See, for example, [68, sec. 1] or [69, chap. 19] for more details; the change of variables is similar to the one we will see in Section 11.1 for charged black holes.

Another generalization, which will be more important for us, consists in considering parallel branes. This can be achieved by solving (1.3.60) with a harmonic function, which has several point-like sources in the transverse space rather than just one. For $p < 7$, (1.3.61) is replaced by

$$h = 1 + r_0^{7-p} \sum_{\alpha=1} \frac{1}{|x - x_I|^{7-p}} , \tag{1.3.62}$$

where x_I^i is the position of the Ith D-brane in transverse space. The fact that this solution is static signals that *parallel Dp-branes don't exert any force on each other*. We could also have expected this from the analysis in Section 1.3.6, where we concluded that parallel Dp preserve the same 16 supercharges as a single one.

Asymptotics and singularities

Returning to a single stack (1.3.61), at large r we would expect the solution to be asymptotic to flat space. The radius r_0 is interpreted as the region where the gravitation field is strong. For $g_s \to 0$, it shrinks to zero; this might seem strange, since the D-brane's tension (1.3.28) gets large in this limit. The reason is the prefactor $e^{-2\phi}$ in the NSNS sector action (1.2.17), so that the effective Newton constant is $2\kappa^2 g_s^2$; this overcomes the g_s^{-1} in the D-brane tension, so that their product goes like g_s, which is the power observed in r_0^{7-p}. (Indeed, $(2\pi l_s)^{7-p} = 2\kappa^2 \tau_{\mathrm{D}p}$.)

In fact, as we will now see, $r = 0$ is in most cases a singularity; either curvature, or dilaton, diverge there. As we mentioned, this singularity is expected to be resolved in fully fledged string theory. For example, in Section 7.2.3 we are going to see how this happens for the D6-brane solution.

The behavior of this solution as $r \to 0$ depends on p:

- If $p < 3$, the curvature invariants remain finite as $r \to 0$; in particular, the Ricci scalar $R \to 0$. But the string coupling diverges. $r = 0$ is at infinite distance.
- If $p = 3$, again the curvature invariants are finite, and moreover ϕ is constant. In fact, an analytic continuation beyond $r = 0$ exists [70], but once again $r = 0$ is at infinite distance.
- If $3 < p < 7$, the curvature diverges as $r \to 0$, but the string coupling $e^{\phi} \to 0$.
- For $p = 7$, there are two transverse directions, and a function satisfying (1.3.60) is a logarithm:

$$h = -\frac{g_s N}{2\pi} \log(r/r_0) . \tag{1.3.63}$$

  $r = 0$ is again a singularity. The metric is no longer asymptotic to flat space at large distance: for $r > r_0$ the function h becomes negative, and the solution loses meaning. In practice, this is seldom a problem, because in the context of compactifications the transverse directions to a D7 are compact anyway.
- For $p = 8$, there is only one transverse direction x^9, and a function satisfying (1.3.60) is piecewise-linear:

$$h = h_0 + \frac{g_s N}{2\pi l_s} |x_9| . \tag{1.3.64}$$

  Again, the solution is no longer asymptotic to flat space, and has a critical distance, dependent on the integration constant h_0, which is related to the value of the dilaton at $x_9 = 0$.

NS5-branes

Similar solutions to (1.3.59) exist that are supersymmetric and charged under the Kalb–Ramond field B.

One represents the back-reaction of a stack of superimposed fundamental strings F1; we will see it later in an exercise. Another describes a new object, extended along five space dimensions and hence called *NS5-brane*. The solution is rather similar to (1.3.59) for $p = 5$:

$$\mathrm{d}s^2_{\mathrm{NS5}} = \mathrm{d}s^2_{\parallel} + h\mathrm{d}s^2_{\perp}\,, \qquad e^{\phi} = g_s h^{1/2}\,, \tag{1.3.65a}$$

$$H_{ijk} = 2Nx^l \epsilon^{\perp}_{lijk}\,. \tag{1.3.65b}$$

$\mathrm{d}s^2_{\parallel}$, $\mathrm{d}s^2_{\perp}$ describe the six parallel and four transverse directions; $h = 1 + Nl_s^2/r^2$. (In form notation, (1.3.65b) reads $H = 2N\mathrm{vol}_{S^3}$.) A definition of this solitonic object from open strings is not available, but we will see later that it plays an important role in string dualities, which give indirect information. One can infer from this an effective action in the style of (1.3.24). In particular, one obtains that it couples to the dual potential B_6, defined similar to (1.3.40) as

$$* H = -\mathrm{d}B_6\,. \tag{1.3.66}$$

The tension is

$$T_{\mathrm{NS5}} \equiv \frac{\tau_{\mathrm{NS5}}}{g_s^2} \equiv \frac{1}{(2\pi)^5 l_s^6 g_s^2}\,, \tag{1.3.67}$$

which differs from $T_{\mathrm{D5}} = 1/((2\pi)^5 l_s^6 g_s)$ from (1.3.28) in the power of g_s, signaling that the NS5 is not a D-brane. Its effect on the Bianchi identity is

$$\mathrm{d}H = -2\kappa^2 \tau_{\mathrm{NS5}} \delta_{\mathrm{NS5}}\,, \tag{1.3.68}$$

in the spirit of (1.3.43).

Exercise 1.3.1 In flat ten-dimensional space $\mathbb{R}^{10}$, consider a D-brane on $S^p \times \mathbb{R}$, where $S^p = \{\sum_{m=1}^{p}(x^m)^2 = R^2\}$. Does it satisfy (1.3.51)?

Exercise 1.3.2 Obtain (1.3.47) from (1.3.43).

Exercise 1.3.3 Show that the metric of the $D8$ solution can be put in the *conformally flat* form

$$\mathrm{d}s^2_{10} = f\mathrm{d}s^2_{\mathrm{Mink}_{10}}\,, \tag{1.3.69}$$

where f is a function and Mink_{10} denotes as usual *flat* Minkowski space.

Exercise 1.3.4 Check the sign of the coefficient of r^{7-p} in (1.3.61) by considering the potential of a p-brane with DBI action (1.3.29) in the solution (1.3.59): the force should be attractive. Next, consider a Dp-brane probe, with its full action (1.3.24); check that the total force in this case is zero.

1.4 Dualities

We have learned that in flat space D-branes extended along flat subspaces are BPS solitons, the analogue of monopoles in supersymmetric YM theories; in this section, we will use them to find information on the strong-coupling behavior of string theory.

1.4.1 From IIA to eleven dimensions

D0-branes as KK-modes

D0-branes are particle states. When $g_s = e^{\phi}$ is constant, from (1.3.28) we see that they have a mass

$$m_{D0} = \frac{1}{l_s g_s}\,. \tag{1.4.1}$$

States made of k coincident D0-branes have masses $k/(l_s g_s)$. They are all multiple of the same fundamental value (1.4.1), just like a tower of KK states (0.2.1). Taking this analogy seriously suggests that a new dimension has been generated dynamically, of size

$$L_{10} = 2\pi l_s g_s\,. \tag{1.4.2}$$

At weak coupling, (1.2.18) implies $L_{10} \ll l_{\rm Pl}$: the size of this extra circle is sub-Planckian, and has unclear physical meaning. But at strong coupling $L_{10} \gg l_{\rm Pl}$: the new dimension is macroscopic, and cannot be ignored. So if the KK interpretation of D0-branes is correct, we should expect the strong-coupling limit of IIA string theory to have an eleven-dimensional interpretation.

Since these states are BPS, supersymmetry should not be broken in the process, and we would expect the eleven-dimensional theory to still have 32 supercharges. Such a theory exists and is unique.

Eleven-dimensional supergravity

The fields are the metric, a three-form potential, and a single gravitino:

$$g_{MN}\,, \qquad A_{MNP}\,; \qquad \psi_{\alpha M}\,. \tag{1.4.3}$$

The field-strength of the three-form is $G_4 = dA_3$, in the form notation of Section 1.2.1. The bosonic action is [71]

$$S_{11} = \frac{1}{2\kappa_{11}^2}\left[\int d^{11}x\sqrt{-g_{11}}\left(R_{11} - \frac{1}{2}|G_4|^2\right) - \frac{1}{6}\int A_3 \wedge G_4 \wedge G_4\right]. \tag{1.4.4}$$

The supersymmetry transformations contain a single Majorana ϵ, which has 32 independent components.

On $\text{Mink}_{10} \times S^1$, if the circle is small we can ignore dependence on it, and describe the theory by ten-dimensional fields. The metric g^{11}_{MN} generates in $d = 10$ a metric, a vector field $g_{M\,10}$ and a scalar $g_{10\,10}$. The latter two can be identified with C_M and the dilaton in IIA. The three-form potential A_{MNP} generates a three-form and two-form $A_{MN\,10}$; these can be identified with C_{MNP} and B_{MN} respectively. To reduce the gravitino, we notice that a spinor in $d = 11$ becomes a pair of spinors of both chiralities in $d = 10$; so $\psi_{M\alpha}$ generates $\psi_{M\alpha}$, $\psi_{M\dot\alpha}$, $\psi_{10\,\alpha}$, $\psi_{10\,\dot\alpha}$, which can be identified with the gravitinos and dilatinos in Table 1.1.

More precisely, with the field identification

$$ds_{11}^2 = e^{-\frac{2}{3}\phi}(ds_{10}^2 + e^{2\phi}(dx^{10} - C_1)^2), \quad (1.4.5a)$$

$$A_3 = C_3 - B \wedge dx^{10}, \quad (1.4.5b)$$

the action (1.4.4) turns into (1.2.15), the IIA action for $F_0 = 0$. The four-form field-strength is

$$G_4 = dC_3 - H \wedge dx^{10} \overset{(1.2.9)}{=} F_4 - e^{-2/3\phi} H \wedge e^{10}. \quad (1.4.5c)$$

To identify κ_{11}, notice that in (1.4.4) the integral over dx^{10} gives an overall L_{10}; this would give $\kappa_{11}^2 = \kappa^2 L_{10}$. For the physical Planck length, however, we also want to include the $e^{-2\phi} = g_s^{-2}$ factor in (1.2.17), as we did in (1.2.18). Defining $2\kappa_{11}^2 = (2\pi)^8 l_{P11}^9$ by analogy with (1.2.16):

$$l_{P11} = g_s^{1/3} l_s. \quad (1.4.6)$$

L_{10} is macroscopic with respect to this length again when $g_s \gg 1$.

The dimensional reduction we have just described might look dangerous, since in this regime IIA is strongly coupled, and strictly speaking we don't know its action. However, as we mentioned after (1.2.23), the two-derivative action is protected by supersymmetry, and so for slowly varying fields (1.2.15) is still appropriate.

Still (1.4.4) cannot be the end of the story: it is nonrenormalizable, so it cannot be itself the strong-coupling definition of IIA string theory. Rather, we should think of it as another effective action, useful in the regime where $g_s \gg 1$, just like (1.2.15) is useful in the regime $g_s \ll 1$. Of what theory is it an effective action? It cannot be one of the perturbative string theories we know, since it is defined on an eleven-dimensional spacetime. Moreover, in (1.4.4) there is no coupling parameter that can be made small, like g_s was for perturbative strings. So the chances that we can find a weakly coupled description at this point appear slim. We will proceed anyway, and see what we can learn about it. This mysterious theory in $d = 11$ is called *M-theory*, and we conclude [72]

$$\text{M-theory on } S^1 \quad \cong \quad \text{IIA}. \quad (1.4.7)$$

Membranes

The relation to IIA strings suggests that M-theory should also have extended objects.

All BPS objects are charged under some potential. Dp-branes are charged under C_{p+1}; the M-theory potential A_3 suggests the existence of a 2-brane, also called a membrane. One can indeed find a gravitational solution charged under A_3, along the lines of (1.3.59) (Exercise 1.4.6). Perhaps more impressively, one can write a supersymmetric theory [73] for a membrane embedded in spacetime $\mathbb{R}^d$, *only for* $d = 11$. This is strongly reminiscent of string theory, where the one-dimensional object F1 is embedded in $d = 10$. Unfortunately this theory is difficult to quantize; we will find a better version in Chapter 11. Still we can postulate this object's existence and see what follows from it. This membrane is called *M2-brane*, and is one of the inspirations for the name "M-theory."

It is natural to think that the M2 becomes the D2 when the eleventh dimension becomes small, $L_{10} \to 0$. In particular, the two tensions should be the same when viewed by a ten-dimensional observer:

$$T_{\rm M2} = T_{\rm D2} \overset{(1.3.28)}{=} \frac{1}{4\pi^2 l_s^3 g_s} \overset{(1.4.6)}{=} \frac{1}{4\pi^2 l_{\rm P11}^3} . \tag{1.4.8}$$

We can also consider a D2 extended along x^{10}. In the limit $L_{10} \to 0$, in $d = 10$ this will look like a $p = 1$-brane. The only such object is the fundamental string F1. So we are led to also postulate

$$L_{10} T_{\rm M2} = T_{\rm F1} \overset{(1.1.5)}{=} \frac{1}{2\pi l_s^2} . \tag{1.4.9}$$

Indeed, this follows from (1.4.2) and (1.4.8); we can regard this as an alternative derivation for (1.4.2).

When several M2-branes are brought together, one expects a nonabelian theory to arise, since this happens for D2-branes. As we noted earlier, for M-theory we don't have direct control over the membrane dynamics, so we cannot just derive such a theory directly. Moreover, in IIA the YM parameter on the D-brane world-volume is (1.3.33), and we could make it small by changing g_s; in M-theory we don't have a similar parameter. In spite of these difficulties, the non-abelian M2 theory was found [74–77]. The trick was essentially to consider M2-branes on a geometry obtained as an *orbifold*, namely by identifying some space directions by the action of a discrete group $\mathbb{Z}_k$; the geometry is still flat, but the M2 dynamics now contains factors of $1/k$ that provide perturbative control. We will comment on this further in Section 11.3.

M5-branes

D-branes can couple to the magnetic dual potentials defined in (1.3.40). In the same spirit we can define a potential A_6 by

$$* G_4 = {\rm d}A_6 . \tag{1.4.10}$$

A brane coupling to it would be extended along $p = 5$ space dimensions; hence we call it *M5-brane*.

Its properties can again be inferred from its dimensional reductions. It is expected to have a two-form potential b_2 on its world-volume, whose IIA reduction becomes the gauge field a. To avoid it from also generating an unwanted two-form, one imposes a duality condition $h_3 = *h_3$, $h_3 = {\rm d}b_2$. As for IIB supergravity, this creates a problem in writing an action, although a proposal for overcoming it is known [78]. This world-volume theory has $\mathcal{N} = (2,0)$ in $d = 6$. For coincident M5-branes, the world-volume theory is expected to have $\sim N^3$ degrees of freedom, rather than the familiar gauge theory $\sim N^2$. This can be inferred from anomalies [79], holography [80], field theory [81–83], and conformal bootstrap [84]. This theory remains mysterious to this day; for recent efforts, see, for example, [85, 86].

We can reduce an M5 to IIA in two ways, by wrapping it along the S^1 or not; this produces branes with $p = 4$ and $p = 5$. Given the list of available BPS objects in IIA, these must be the D4 and the NS5 respectively. This leads to

$$T_{\rm M5} = T_{\rm NS5} \overset{(1.3.67)}{=} \frac{1}{(2\pi)^5 l_s^6 g_s^2} \overset{(1.4.6)}{=} \frac{1}{(2\pi)^5 l_{\rm P11}^6}\,,$$
$$L_{10} T_{\rm M5} = T_{\rm D4} \overset{(1.3.28)}{=} \frac{1}{(2\pi)^4 l_s^5 g_s}\,. \tag{1.4.11}$$

Again the two are in agreement once we recall (1.4.2). We could also regard the first relation as a derivation of $T_{\rm NS5}$.

There are still some objects in IIA that we have not recovered from M-theory: D6- and D8-branes. D6-branes are a source for the flux F_2; we see from (1.4.5a) that its potential C_1 becomes part of the metric in eleven dimensions. So we expect a D6-brane to become a purely geometrical object in $d = 11$; we will see how this comes about in Section 9.2.

Massive IIA

D8-branes are more troubling. They a source for the Romans mass F_0; but the dimensional reduction (1.4.5) of eleven-dimensional supergravity only reproduced the $F_0 = 0$ version (1.2.15) of IIA. Even our original argument for the emergence of an eleventh dimension no longer works with $F_0 \neq 0$, because D0-branes have a term $\int F_0 a$ from (1.3.39), and the equation of motion on the $0+1$-dimensional world-sheet gives $F_0 = 0$, a contradiction. This is known as a "tadpole" problem because the field a has a nonzero one-point function.

The nonperturbative definition of massive IIA cannot be conventional M-theory; it is an interesting open question. We will see in Section 10.1 that there are no classical supergravity solutions with $F_0 \neq 0$ and $g_s \gg 1$ [87]; but from a quantum-mechanical point of view, the path integral should be over all configurations, so the nonperturbative formulation is still needed. Applying the formulas (1.4.5) to the D8-brane solution, one obtains a metric in eleven dimensions [88], but this does not clarify what the action is. There are in fact arguments against the existence of deformations of eleven-dimensional supergravity that preserve 32 supercharges [89–91]. More recently, it has been suggested that these can be overcome by a "dual" formulation for the graviton [92].

This might make us doubt that massive IIA is really part of string theory. There are several reasons to believe that it is. Anticipating a bit some later discussions:

- We will see in Section 1.4.2 that *T-duality* relates F_0 to the other RR fluxes, and the D8 to the other Dp-branes.
- Related to this, one can get to M-theory from massive IIA with a *sequence* of dualities. At the end of Section 1.4.5, a particular quotient of massive IIA will be related to heterotic string theories, and to M-theory on spaces with boundaries. There are other related duality sequences [93, 94].
- Several AdS solutions exist with $F_0 \neq 0$, some of which we will see in Chapter 11. These can be tested quantitatively using the AdS/CFT correspondence [87, 95–97].

1.4.2 T-duality

T-duality can be obtained from the world-sheet; we could have discussed it already in Section 1.1. But it can also be interpreted in terms of KK states, similar to our motivation for eleven-dimensional supergravity.

T-duality can be used when at least one of the directions is an S^1. For definiteness and simplicity, in this section we consider a spacetime

$$\mathbb{R}^{10-k} \times (S^1)^k , \tag{1.4.12}$$

with a line element $ds^2_{10} = ds^2_{\mathbb{R}^{10-k}} + \sum_{M=k+1^9} (dx^M)^2$. The only data of the torus are the periodicities L_k of the S^1s. In Section 4.2.6 we will deal with more complicated situations, which we will cast in the language of fiber bundles.

We first consider $k = 1$, so that there is a single periodic coordinate $x^9 \sim x^9 + L$. In this case, T-duality is the equivalence $L \cong \tilde{L}$,

$$\tilde{L} = \frac{4\pi^2 l_s^2}{L} . \tag{1.4.13}$$

For superstrings, the duality also exchanges IIA and IIB.

KK argument

We first show (1.4.13) in the bosonic string. A first piece of evidence comes from the formula for the spectrum: (1.1.23) is invariant under

$$R \to \tilde{R} = \frac{l_s^2}{R} , \qquad w \leftrightarrow q , \tag{1.4.14}$$

matching (1.4.13) for $L = 2\pi R$, $\tilde{L} = 2\pi\tilde{R}$. We did not derive (1.1.23), so here is a perhaps more intuitive version of the same statement, modeled on the D0 argument leading to (1.4.2). Consider a state with a string winding $w \in \mathbb{Z}$ times around the S^1. From the nine-dimensional point of view, it looks like a particle, with mass

$$L T_{\text{F1}} = \frac{Rw}{l_s^2} . \tag{1.4.15}$$

Comparison with (0.2.1) suggests that such states form a KK tower for an extra S^1 dimension of size $\tilde{L} = 2\pi\tilde{R} = 2\pi l_s^2/R$, reproducing (1.4.14). This version of the argument also applies directly to superstrings.

Dualization of world-sheet scalars

We now show that the world-sheet action is invariant under (1.4.13) (see, for example, [98, 99]). We focus on the periodic scalar $x \equiv x^9$, ignoring the other directions; for simplicity, we also set $B = 0$ and the dilaton to a constant. The relevant part of (1.1.4) is then

$$S_x = -\frac{1}{4\pi l_s^2} \int d^2\sigma \partial_\alpha x \partial^\alpha x . \tag{1.4.16}$$

Introduce a vector field b_α and consider the alternative action

$$S_{b,x} = -\frac{1}{4\pi l_s^2} \int d^2\sigma (b_\alpha b^\alpha - 2\epsilon^{\alpha\beta}\partial_\alpha x b_\beta) . \tag{1.4.17}$$

Since b_α has no kinetic term, we can substitute it with the solution of its equation of motion, $b_\alpha = \epsilon_{\alpha\beta}\partial^\beta x$; then $S_{b,x}$ reduces to (1.4.16). On the other hand, the x equations of motion are

$$\epsilon^{\alpha\beta}\partial_\alpha b_\beta = 0 . \tag{1.4.18}$$

Naively, the solution is simply $b_\alpha = \partial_\alpha \tilde{x}$ for a new scalar $\tilde{x}$. Substituting this in (1.4.17) gives an action identical to (1.4.16), but with $x \to \tilde{x}$.

This does not seem very interesting: the action for a free scalar is equivalent to that of another free scalar. Our real interest is how the periodicities are mapped. Let us Wick-rotate σ^0 and consider the Euclidean model for (1.4.17). For definiteness, we take $\sigma^i \sim \sigma^i + 2\pi$. Since $x \sim x + 2\pi R$, we can parameterize

$$\partial_\alpha x = \partial_\alpha x_{\rm per} + 2\pi n_\alpha R \,, \tag{1.4.19}$$

where $x_{\rm per}$ is periodic. $\partial_\alpha x$ is only locally the gradient of a function: its integral along the σ^i can be nonzero. Likewise, the solution to (1.4.18) is not just $b_\alpha = \partial_\alpha \tilde{x}$, but more generally

$$b_\alpha = \partial_\alpha \tilde{x}_{\rm per} + 2\pi \lambda_\alpha \,, \tag{1.4.20}$$

where λ_α are constant. If we replace this and (1.4.19) in (1.4.17), we get a term $-\frac{2\pi}{l_s^2}\epsilon^{\alpha\beta} n_\alpha R \lambda_\beta$. The n_α are part of the degrees of freedom of x; since they are discrete we cannot vary them, but in the path integral we should still sum over them. But now $\sum_m e^{\mathrm{i}xm} = 2\pi\delta(x - 2\pi m)$, the Fourier series analogue of the Fourier transform of the delta function; this implies $\lambda_\alpha = (l_s^2/R) m_\alpha$ with $m_\alpha \in \mathbb{Z}$. Comparing (1.4.20) with (1.4.19), the periodicity of $\tilde{x}$ is determined as $2\pi\tilde{R}$, with $\tilde{R} = l_s^2/R$. This reproduces (1.4.14).

Returning to Lorentzian signature, the equation of motion (1.4.18) is $\partial_\alpha \tilde{x} = \epsilon_\alpha{}^\beta \partial_\beta x$; so

$$\partial_\pm \tilde{x}^9 = \pm \partial_\pm x^9 \,, \tag{1.4.21}$$

while of course the x^M with $M \neq 9$ are unchanged. In other words, T-duality is equivalent to a parity transformation acting on the right-movers only. (In the bosonic string, this can also be seen from the spectrum (1.1.23).) As familiar from QFT (and reviewed in Section 2.1.1), a parity acts on spinors by a gamma matrix. For type II superstrings, this changes the chirality of the GSO projection in such a way that IIA is mapped to IIB. To summarize:

$$\text{IIA on } S^1 \quad \cong \quad \text{IIB on } \tilde{S}^1 \,. \tag{1.4.22}$$

Action on D-branes

Another consequence of (1.4.21) is that N and D boundary conditions are exchanged. Indeed, the N condition is $\partial_{\sigma^1} x^9 = 0$, which becomes $\partial_{\sigma^0}\tilde{x}^9 = 0$; this fixes $\tilde{x}^9$ to be a constant, so it is the D condition. More precisely, since the a_a and x^i have a common origin from the $8_{\rm V}$ in (1.3.21), the component a_9 is exchanged by T-duality with the scalar x^9. By (1.2.14) and (1.3.23), $a_9 \sim a_9 + 2\pi/L_9$. We conclude

$$\text{D}p \text{ along } x^9 \text{ with } a_9 = a \quad \overset{T_9}{\longleftrightarrow} \quad \text{D}(p-1) \text{ at } \tilde{x}^9 = 2\pi l_s^2 a \,. \tag{1.4.23}$$

We know from (1.3.59) that a Dp sources an RR field along the angular directions surrounding it. Since in (1.4.23) a Dp loses a parallel direction, the RR flux gains an index:

$$F_{m_1\ldots m_k} \overset{T_9}{\longleftrightarrow} F_{m_1\ldots m_k 9} \,. \tag{1.4.24}$$

We are going to give a more complete description of the T-duality's action on fields in Section 4.2.6.

Let us compare D-brane tensions on both sides of (1.4.23), in the spirit of (1.4.15). From a nine-dimensional point of view, both objects in (1.4.23) look $(p-1)$-dimensional. T-duality states that physics should be the same on both sides, so the tensions should be equal:

$$L_9 T_{\mathrm{D}p} = T_{\mathrm{D}(p-1)} \quad \Rightarrow \quad \frac{L_9}{(2\pi)^p g_s l_s^{p+1}} = \frac{1}{(2\pi)^{p-1} \tilde{g}_s l_s^p}\,; \tag{1.4.25}$$

so the value of g_s after T-duality is

$$\tilde{g}_s = g_s \frac{2\pi l_s}{L_9}\,. \tag{1.4.26}$$

Thus weak coupling is mapped to weak coupling, unlike the SL(2, $\mathbb{R}$) of Section 1.2.3. This is consistent with our use on both sides of the world-sheet approach, which is perturbative in g_s.

Tori and bound states

In presence of a torus (1.4.12), we can choose to T-dualize along any subset of the k compact directions. If $k = 2$, T-duality along a single direction exchanges IIA and IIB, so T-duality T_{89} along both maps IIA to IIA, and IIB to IIB. The action on D-branes is also a straightforward iteration of the rule (1.4.23). For example, a D2 extended along both x^8 and x^9, or in other words "wrapping" the torus $T^2 = S^1 \times S^1$, is mapped by a single T-duality T_9 to a D1, and by T_{89} to a D0. In the top part of Figure 1.1, the T^2 is represented as a rectangle; the opposite sides are meant to be identified. The D1-branes along x^9 are mapped by T_9 to D0-branes; and a D1 along x^8 is mapped to a D2. But what is the T-dual of an *oblique* D1-brane, which winds w_8 times along S^1_8 and w_9 times along S^1_9, such as the one on the bottom-left of Figure 1.1?

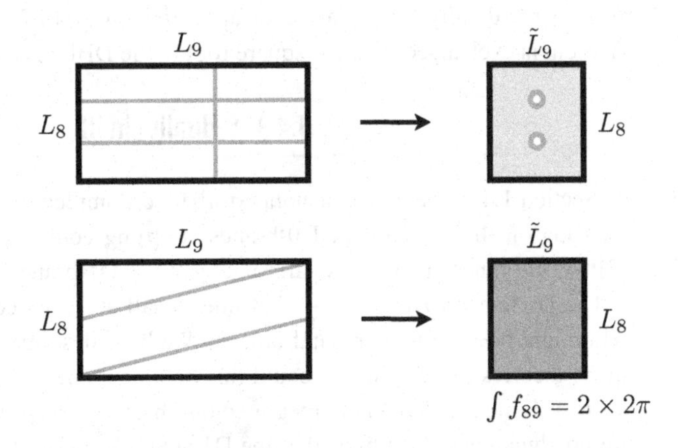

Figure 1.1 T-duality acting on D1-branes on a T^2. It maps horizontal and vertical D1-branes to D0- and D2-branes respectively; it maps oblique D1s to D2/D0 bound states.

Such a D1 is still straight: so it should be BPS. From the point of view of an observer in the noncompact eight dimensions, it is a particle with mass

$$m_{\text{obD1}} = T_{\text{D1}}\,\text{Length(obD1)} = \frac{1}{2\pi l_s^2 g_s}\sqrt{w_8^2 L_8^2 + w_9^2 L_9^2}\,. \tag{1.4.27}$$

We may imagine a process where w_8 D1-branes wrapping S^1_8 and w_9 D1s wrapping S^1_9 snap together to create a single oblique D1. In Figure 1.1, this process would connect the top-left and bottom-left situations. In this process, the total mass decreases: because of the triangular inequality, (1.4.27) is smaller than $T_{\text{D1}} \times (w_8 L_8 + w_9 L_9)$. In a sense, the oblique D1 is a bound state of the horizontal and vertical ones. Now, the latter are mapped under T_9 to w_8 D2s and w_9 D0s respectively; so we expect the oblique D1 to be mapped to a bound state of these objects on the IIA side.

This is confirmed by the duality between world-volume gauge fields and transverse scalars. To avoid having to deal with nonabelian branes, we take $w_8 = 1$. We describe an oblique D1 by $x^9 = \frac{w_9 L_9}{L_8} x^8$, so (1.4.23) gives

$$a_9 = \frac{1}{2\pi l_s^2}\frac{w_9 L_9}{L_8} x^8\,. \tag{1.4.28}$$

As a cross-check, $\frac{1}{2\pi}\int_{\text{D2}} f_{89} \text{d}x^8 \text{d}x^9 = w_9 \in \mathbb{Z}$, in agreement with Dirac quantization. Now recall from the discussion in (1.3.37) that the presence of f turns the D2 into a bound state of a D2 and w_9 D0s, confirming our expectations.

This D2 is viewed as a particle in eight dimensions, but to compute its mass we cannot just use the formula $T_{\text{D2}} \times L_8 \tilde{L}_9$; because of the presence of $f \neq 0$, we have to use the DBI term. The potential energy is obtained by integrating the Lagrangian density over the space dimensions; we need to evaluate the determinant of the 2×2 matrix $\begin{pmatrix} 1 & \mathcal{F}_{89} \\ -\mathcal{F}_{89} & 1 \end{pmatrix}$. Recalling also (1.3.26) with $B = 0$:

$$\begin{aligned} m_{\text{D2/D0}} &= T_{\text{D2}} L_8 \tilde{L}_9 \sqrt{1 + (2\pi l_s^2 f_{89})^2} \overset{(1.4.28)}{=} \frac{L_8 \tilde{L}_9}{(2\pi)^2 l_s^3 \tilde{g}_s}\sqrt{1 + w_9^2 \frac{L_9^2}{L_8^2}} \\ &\overset{(1.4.13),(1.4.26)}{=} \frac{1}{2\pi l_s^2 g_s}\sqrt{L_8^2 + w_9^2 L_9^2} \overset{(1.4.27)}{=} m_{\text{obD1}}\,. \end{aligned} \tag{1.4.29}$$

So under T-duality the square root appearing in (1.4.27) because of Pythagoras' theorem is exchanged with the square root in the DBI term of (1.3.24).

1.4.3 S-duality in IIB

In Section 1.4.1, we found a nonperturbative completion to IIA string theory by focusing on the behavior of D0-branes at strong coupling. In IIB, we don't have (BPS) D0-branes; the next "smallest" object is a D1-brane.

The D1 tension $T_{\text{D1}} = \frac{1}{2\pi l_s^2 g_s}$ becomes small at strong coupling, suggesting that D1s might become fundamental objects. We have described all known perturbative string theories in Section 1.1; could this be a new one? The fields are (1.3.22); the world-volume a field can be integrated out, because gauge fields in two dimensions are nondynamical. One finds that the D1 action becomes at strong coupling simply the IIB perturbative string again, in the Green–Schwarz formulation. So the strong-coupling limit of IIB is IIB, but with F1 $\leftrightarrow$ D1.

The SL(2, ℝ) symmetry (1.2.38) can relate the weak and strong coupling regimes of IIB supergravity; but in Section 1.2.3 we gave arguments against it being a symmetry of IIB *strings*. One argument was that it would have required a continuum of string states, coupling to any linear combination $cC_2 + dB$. Now we know at least one such a string state: the D1, which couples to C_2. So it is possible that at least the element

$$s = \begin{pmatrix} 0 & -1 \\ 1 & 0 \end{pmatrix} \tag{1.4.30}$$

is a symmetry of the full IIB string theory: it exchanges B and C_2, and hence F1s and D1s. This is called *S-duality*.

D-brane duals

As usual, let us check how the various extended objects transform, setting for simplicity $C_0 = 0$. From (1.2.38) and (1.4.30):

$$\phi \to -\phi\,, \qquad g_{MN} \to e^{-\phi} g_{MN}\,, \qquad F_3 \leftrightarrow -H\,, \qquad F_5 \to F_5\,. \tag{1.4.31}$$

Because of how the metric transforms, a p-brane's tension picks up a factor

$$g_s^{-(p+1)/2}\,. \tag{1.4.32}$$

So, for example,

$$T_{\rm F1} = \frac{1}{2\pi l_s^2} \to \frac{1}{2\pi l_s^2 g_s} = T_{\rm D1}\,. \tag{1.4.33}$$

On the other hand, $T_{\rm D1}$ picks up a g_s^{-1}, but at the same time the g_s in the denominator gets inverted; so the two factors cancel out and we obtain $T_{\rm F1}$. This confirms that (1.4.30) indeed exchanges F1s and D1s.

We next transform the D3:

$$T_{\rm D3} = \frac{1}{(2\pi)^3 l_s^4 g_s} \overset{(1.4.32)}{\to} \frac{g_s^{-2}}{(2\pi)^3 l_s^4 g_s^{-1}} = T_{\rm D3}\,. \tag{1.4.34}$$

So the D3-brane should be invariant under (1.4.30), consistent with the invariance of F_5 in (1.4.31), of which it is a source. Next we have

$$T_{\rm D5} = \frac{1}{(2\pi)^5 l_s^6 g_s} \overset{(1.4.32)}{\to} \frac{g_s^{-3}}{(2\pi)^3 l_s^4 g_s^{-1}} \overset{(1.3.67)}{=} T_{\rm NS5}\,. \tag{1.4.35}$$

Again this is consistent with the transformation in (1.4.31) of the fields they create, B and C_2.

The next object would be the D7-brane. The field it creates is C_0, which we have chosen to set to zero. More importantly, we have seen after (1.3.63) that its back-reaction on the metric is not localized, but grows logarithmically; so it is not really a soliton. We will discuss its S-dual in Section 9.4. Finally, we would have the D9-brane, which fills all of spacetime. In fact, a D9 by itself is inconsistent. We discussed its world-volume theory preceding (1.3.57), where we argued it to be a supersymmetric YM with 16 supercharges. This theory is chiral, and suffers from a gauge anomaly. This can be cured by introducing more ingredients, as we will do in Section 1.4.4. It is still possible, but subtler, to define an S-dual to the D9, dubbed NS9 [100, sec. 6].

Other S-dualities

We now would like to know whether there are other elements of $SL(2, \mathbb{R})$ besides (1.4.30) that survive in full string theory. A natural candidate is

$$t = \begin{pmatrix} 1 & 1 \\ 0 & 1 \end{pmatrix} . \tag{1.4.36}$$

Since $|c\tau + d| = 1$, the metric does not transform under (1.2.38). Moreover $\tau \to t \cdot \tau = \tau + 1$; taking real and imaginary parts, $\phi \to \phi$, and

$$C_0 \to C_0 + 1 . \tag{1.4.37}$$

By (1.3.44), this is a large gauge transformation, suggesting that t is indeed a symmetry of string theory.

Equations (1.2.38) and (1.4.36) also give $B \to B$, $C_2 \to C_2 + B$. The latter implies that the D1 turns into an object that couples to both C_2 and B: a *D1/F1 bound state*. As a cross-check, consider placing a D1 and an F1 next to each other, both wrapping an S^1 [101]. The F1 can break in two open strings that end on the D1. Recall from (1.3.23) that the endpoints of an open string behave on a D-brane as electric charges for the world-sheet potential a. Now the endpoints of the F1 can recombine, making the open F1 disappear altogether. This leaves behind a flux $f_{01} \neq 0$ on the D1, constant because of the Maxwell equation. Equation (1.3.32) contains a term $|\mathcal{F}|^2$, which in turn contains $-2\pi l_s^2 f_{01} B_{01}|_{\text{D1}}$. This is a coupling to B; so we have indeed obtained a D1/F1 bound state.

Since s and t, (1.4.30), (1.4.36) are both symmetries, so are all the elements of the group they generate:

$$SL(2, \mathbb{Z}) . \tag{1.4.38}$$

This is called *S-duality group* to distinguish it from the particular element (1.4.30).

Acting on D1s and F1s, (1.4.38) generates more general bound states of p F1s and q D1s called *(p, q)-strings*. More precisely, it acts on $\binom{p}{q}$ in the vector representation, with an F1 represented by the vector $\binom{1}{0}$, and a D1 by $\binom{0}{1}$.

Since F_5 is invariant under (1.2.38), D3-branes are invariant under the whole $SL(2, \mathbb{Z})$. As we discussed around (1.3.57), the field theory living on a D3 stack has 16 supercharges; for $p = 3$, this is $\mathcal{N} = 4$-supersymmetric YM in $d = 4$. Equation (1.4.38) implies a QFT duality that maps weak to strong coupling, because $g_{\text{YM}}^2 = 2\pi g_s$, by (1.3.33). This was indeed discovered long before it resurfaced in string theory [102]:

$$\tau_{\text{YM}} \to \frac{a\tau_{\text{YM}} + b}{c\tau_{\text{YM}} + d} , \tag{1.4.39}$$

with τ_{YM} as in (1.3.12); this is a strong check of S-duality. (The θ angle is created by the C_0 contribution to the WZ term, similar to the discussion leading to (1.3.37).)

Next we consider D5- and NS5-branes. They couple to the magnetic potentials C_6 and B_6 respectively, which were defined in (1.3.40) and (1.3.66) from F_3 and H_3. This implies the existence of bound states of D5s and NS5s, called *(p, q)-five-branes*. We expect (p, q)-seven-branes for similar reasons, but we postpone their discussion to Section 9.4.

From M-theory to IIB

Consider now M-theory on $\mathbb{R}^9 \times T^2$ [103, 104]; let the two sides of the T^2 have lengths L_A and L_{10}. From Section 1.4.1, this is equivalent to IIA on $\mathbb{R}^9 \times S^1$, with $g_s \equiv g_A = L_{10}/2\pi l_s$. Further T-dualizing along the remaining S^1 gives IIB on the same space; from (1.4.13) and (1.4.26), the length of the S^1 and the string coupling are

$$L_B = \frac{4\pi^2 l_s^2}{L_A}, \qquad g_B = g_A \frac{2\pi l_s}{L_A} = \frac{L_{10}}{L_A}. \tag{1.4.40}$$

There is also an equivalent but more direct logic taking M-theory on $\mathbb{R}^9 \times T^2$ to IIB. Naively, shrinking the T^2 results in a nine-dimensional spacetime. However, M2s wrapping the T^2 w times give particle states of mass

$$T_{M2}\,\mathrm{Area(M2)} = \frac{w L_A L_{10}}{(2\pi)^2 l_{P11}^3}. \tag{1.4.41}$$

Interpreting this once again as a KK tower (0.2.1) gives

$$\frac{L_A L_{10}}{(2\pi)^2 l_{P11}^3} = \frac{2\pi}{L_B}, \tag{1.4.42}$$

suggesting the emergence of a new S^1 of length L_B, that becomes large as the T^2 gets small; so spacetime is ten-dimensional after all. Wrapping the M2s along the two circles of the torus also gives rise to two types of string states; so the ten-dimensional theory is IIB, which has the F1 and D1. Comparing the tensions:

$$L_A T_{M2} = T_{D1} \quad \Rightarrow \quad \frac{L_A}{(2\pi)^2 l_{P11}^3} = \frac{1}{2\pi l_s^2 g_B}, \tag{1.4.43a}$$

$$L_{10} T_{M2} = T_{F1} \quad \Rightarrow \quad \frac{L_{10}}{(2\pi)^2 l_{P11}^3} = \frac{1}{2\pi l_s^2}. \tag{1.4.43b}$$

Combining (1.4.42) with (1.4.43), we recover (1.4.40). We can generalize (1.4.43) by wrapping one direction of the M2 along an oblique direction (k, l) direction, winding p times along the first circle and q along the second (as in the bottom-left panel of Figure 1.1). This gives the (p, q)-strings, the bound states of p D1s and q F1s.

It was a bit arbitrary to decide in (1.4.43) that wrapping the M2s on the first circle gives a D1, and on the second an F1. Had we done the opposite, we would have obtained (1.4.40) with $L_{10} \leftrightarrow L_A$, and hence with $g_B \to 1/g_B$. So exchanging the role of the two directions of the torus in M-theory gives rise to the S-duality (1.4.30) in IIB. To see what generates the other elements, we consider a more general geometry for the torus. We define the two-dimensional torus as a quotient

$$T^2 \equiv \mathbb{C}/\mathbb{Z}^2, \tag{1.4.44}$$

much as we did in (1.1.40) when we compactified 16 of the bosonic coordinates to define the heterotic string. A fundamental region for this equivalence relation is now a parallelogram, with the two opposite sides identified. The metric on $\mathbb{C} \cong \mathbb{R}^2$ is taken to be the Euclidean one. Up to rescalings and rotations, we can always make one of the generators of the $\mathbb{Z}^2$ lattice to be 1; but the second is then an arbitrary complex number τ, called *modular parameter*. So the two identifications in (1.4.44) are $z \sim z+1 \sim z+\tau$. All this is illustrated in Figure 1.2. The tori we have discussed so far, in particular (1.4.12) for $k = 2$, were "rectangular": their τ was purely imaginary.

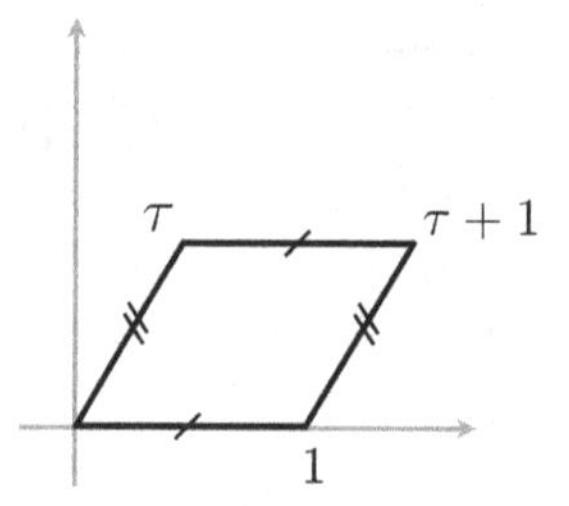

Figure 1.2 Fundamental region for the identification (5.3.64) defining a torus. The sides marked by the same symbol are to be considered as identified.

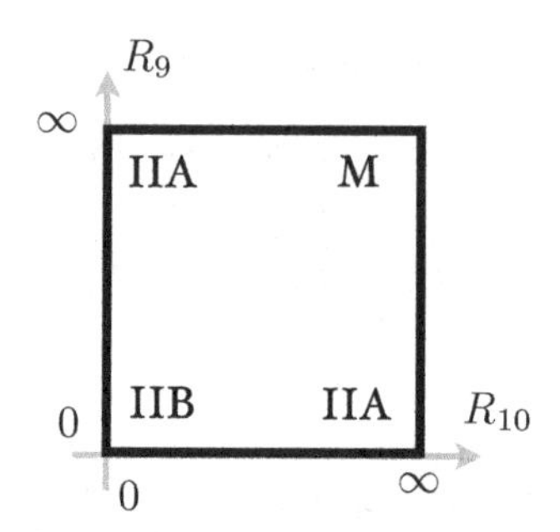

Figure 1.3 M-theory on $\mathbb{R}^9 \times T^2$.

Sometimes, different τ correspond to the same lattice: for example, τ and $\tau + 1$. More generally, any Möbius transformation $\tau' = \frac{a\tau+b}{c\tau+d}$, with $a, b, c, d \in \mathbb{Z}$, also defines the same torus as τ. This is how the $SL(2,\mathbb{Z})$ symmetry of IIB manifests itself in M-theory. An M2 wrapping one of the two sides of the fundamental region in Figure 1.2 is now mapped under a Möbius transformation to one wrapping an oblique line, which we argued corresponds to a (p, q)-string.

We summarize the relation among M-theory, IIA, and IIB with the parameter space in Figure 1.3. When both radii are large, spacetime is eleven-dimensional, and the most appropriate description is the mysterious M-theory. When one of the radii is small and the other is large, there are ten macroscopic dimensions; the theory becomes IIA on $\mathbb{R}^9 \times S^1$. When both radii are small, spacetime naively becomes nine-dimensional, but in fact there is a dual IIB description where it is ten-dimensional.

1.4.4 Orientifolds

It is common to define a space by means of an equivalence relation, as for example in the definition (1.4.44) of the torus. In string theory, one can also quotient by symmetries that act on the world-sheet.

A time slice of the world-sheet is one-dimensional; the only possible nontrivial action on it is world-sheet parity Ω: $\sigma^1 \to -\sigma^1$. This squares to the identity, so the action R on spacetime should be an *involution*, a map $x^M \to R^M(x)$ such that $R^2 = 1$. We define an *orientifold* to be the quotient of a string theory by a simultaneous action of Ω and such a spacetime involution [105–107]. (For more thorough discussions, see [108, 109].)

R acts on fermions as a product of the gamma matrices in the transverse directions, as we mentioned after (1.4.21) and will review in Section 2.1.1. This sometimes squares to -1, and one then has to include a factor $(-1)^{F_L}$, defined as -1 on fields that are fermionic in the left-mover Hilbert space, or in other words on the RNS and RR sectors. If R_p is locally the reflection of $9-p$ coordinates, the orientifold action is

$$\begin{array}{ll} \Omega R_p & \text{if } p = 0,\,1,\,4,\,5,\,8,\,9\,; \\ \Omega(-1)^{F_L} R_p & \text{if } p = 2,\,3,\,6,\,7\,. \end{array} \tag{1.4.45}$$

The action of Ω in the NSNS sector states (1.1.8) is obtained by noting that exchanging left- and right-movers is equivalent to exchanging the two indices MN. So g_{MN} and ϕ remain invariant, while B_{MN} picks up a sign. The Ω action on the RR sector can be inferred likewise. The overall conclusion is that the bosonic fields should satisfy

$$R_p^*\phi = \phi\,, \qquad R_p^* g = g\,, \qquad R_p^* B = -B\,, \qquad R_p^* F_k = -(-1)^{\lfloor p/2 \rfloor + \lfloor k/2 \rfloor} F_k\,, \tag{1.4.46}$$

where $R_p^* \alpha_{M_1 \ldots M_k}(x) = \partial_{M_1} R_p^{N_1} \ldots \partial_{M_k} R_p^{N_k} \alpha_{N_1 \ldots N_k}(x)$ (a definition that will find its proper place in Section 4.1.4). So an orientifold acts a bit like a mirror, dictating that the physics in one half of spacetime should be a specular copy of that in the other half.

R_p has a $(p+1)$-dimensional fixed locus, which is called an *orientifold plane* and denoted by Op, in analogy with the notation for D-branes. Op-planes have a tension and RR charge, as Dp-branes do; but as we will see, they can have *negative tension*, so they are in a sense a source of antigravity. Usually objects with negative mass can generate instabilities,[16] but this is avoided here because Op-planes are stuck at the fixed locus and are not dynamical. Since open strings do not end on an O-plane (unless D-branes also happen to be located on top of it), the open string modes (1.3.22) are absent, and in particular there is no transverse fluctuation x^i.

Op-planes also preserve half-supersymmetry in flat space, as D-branes do; this makes them stable under time evolution. The preserved supercharges are again those that satisfy (1.3.51).

O$p_\pm$-planes

The action on open strings needs to be specified separately. World-sheet parity still reverses the orientation of the string, and now exchanges the two endpoints: $\Omega\colon \sigma^1 \to \pi - \sigma^1$. Since Ω exchanges the endpoints, it should also reverse the order of CP labels (Section 1.3.6). Omitting other quantum numbers, this would lead to $\Omega|IJ\rangle = |JI\rangle$, or more generally to

$$\Omega|IJ\rangle = \sum_{JK} M_{IK} M^{-1}_{LJ} |LK\rangle\,. \tag{1.4.47}$$

Imposing $\Omega^2 = 1$ gives

$$M^{-1} M^t = \mp 1\,. \tag{1.4.48}$$

[16] Already in Newtonian mechanics, a naive argument for instability is the following: if we have two objects with mass m, $-m$, the first will feel a repulsive force, the second an attractive one, so they will both start accelerating in the same direction.

The choice of sign has important consequences on the O-plane, which is then called O$p_\pm$.[17] Most notably, the tension is

$$\tau_{\mathrm{O}p_\pm} = \mp 2^{p-5}\tau_{\mathrm{D}p} = \mp\frac{1}{32\pi^p l_s^{p+1}}\,. \tag{1.4.49}$$

Both O$p_\pm$ are charged under C_{p+1} with a charge density equaling the tension. In other words, their effective action is

$$S_{\mathrm{O}p} = \tau_{\mathrm{O}p}\left[-\int \mathrm{d}^{p+1}\sigma\, \mathrm{e}^{-\phi}\sqrt{-\det g|_{\mathrm{O}p}} + \int_{\mathrm{O}p} C_{p+1}\right], \tag{1.4.50}$$

to be compared with (1.3.24). Notice that the terms involving $\mathcal{F}$ are absent: $B=0$ vanishes on the Op because of (1.4.46), while f vanishes because there are no dynamical fields, as already observed. Other orientifold projections are possible, but we will not need them in this book. See, for example, [111].

Dp-branes on Op-planes

Consider now Dp-branes parallel to an Op-plane, for simplicity in flat space, along $\{x^{p+1} = 0, \ldots, x^9 = 0\}$. This configuration is BPS, just like parallel Dps. If the branes are not atop the O-plane, then the orientifold action dictates that they should come in mirror pairs, related by the involution R_p. So for a stack of N Dps at $\{x^{p+1} = x_0^{p+1}, \ldots, x^9 = x_0^9\}$, there should be a mirror stack at $\{x^{p+1} = -x_0^{p+1}, \ldots, x^9 = -x_0^9\}$. In this case, the gauge group on each stack is still U(N) (see Section 1.3.6). But for a Dp stack on an Op-plane, the orientifold action relates the open string states to themselves, projecting some of them out. Consider, for example, the states that give rise to the vector field, which are the parallel components of (1.3.21) with the additional CP labels; a linear combination of such states is of the form

$$\sum_{IJ}\lambda_{IJ}\, b^a_{-1/2}|0, IJ\rangle_{\mathrm{NS}}\,. \tag{1.4.51}$$

The action of Ω on $b^a_{-1/2}|0, IJ\rangle_{\mathrm{NS}}$ gives a minus sign. Recalling (1.4.47), the states that survive are those such that

$$\lambda = -M\lambda^t M^{-1}\,. \tag{1.4.52}$$

Now the gauge group depends on whether we have an Op_+ or Op_-:

- For an Op_--plane, the sign in (1.4.48) is -1. Then M is symmetric; it can be brought to the identity by a change of basis ($\lambda \to C^{-1}\lambda C$, $M \to CMC^t$). So (1.4.52) says $\lambda \in \mathrm{so}(N)$, and the gauge group is SO(N).
- For an Op_+-plane, the sign in (1.4.48) is -1, and M is antisymmetric; by a change of basis, it can be taken to be $\begin{pmatrix} 0 & 1_N \\ -1_N & 0\end{pmatrix}$. By definition, (1.4.52) then says that $\lambda \in \mathrm{sp}(N)$. The gauge group is Sp($N$), also known as USp($2N$).

If we start from the earlier situation with N Dp-branes parallel to the Op, and move them on top of it, the N mirror images will also do so, resulting in $2N$ Dp-branes and a gauge group SO($2N$) or Sp(N). Sometimes one calls a pair of two mirror images a "full" brane. A single Dp cannot be moved away from an Op and is called a "half" brane.

[17] In older literature, such as [110], this convention is sometimes reversed.

If a Dp-brane is instead orthogonal to an Op, it is mapped to itself by the orientifold action; the gauge group is U(N) away from the orientifold, with some boundary conditions on the gauge fields that depend on how many directions the Op-plane and the Dp-brane have in common.

T-duality

T-duality acts on orientifolds as in (1.4.23) for D-branes. Under T_9, an orientifold whose spacetime involution is a reflection R_p becomes a reflection $R_p R_9$; a factor $(-1)^{F_L}$ is added or erased according to (1.4.45). For example, on $\mathbb{R}^9 \times S^1$,

$$T_9 : \Omega \to \Omega R_9 . \tag{1.4.53}$$

The left-hand side Ω has no spacetime involution. The O-plane extends over all of spacetime, so we call it an O9. On the right-hand side, on the periodic direction, $x^9 \sim x^9 + L_9$, the involution $R_9 : x^9 \to x^9$ has two fixed loci, at $\{x^9 = 0\}$ and $\{x^9 = L_9/2\}$. So there are *two* O8-planes. If we compactify p directions and T-dualize along all of them, we get 2^p O$(9-p)$-planes. The $\pm$ type is preserved by this operation, so an O9_- will generate 2^p O$(9-p)_-$-planes. Notice that the total tension and RR charge is unchanged, because of (1.4.49).

It is possible to include Op-planes of different types in the same model [110]. If we have the ΩR_9 orientifold above, and choose one of the planes to be O8_+ and the other O8_-, and T-dualize back along x^9, the result is no longer an O9, but a *shift-orientifold*, defined by $\Omega\,\text{sh}$, where

$$\text{sh} : x_9 \to x_9 + L_9/2 \tag{1.4.54}$$

is the translation by half a period. This has no fixed loci, and so there are no O-planes. This is consistent with the O$8_\pm$ having opposite tensions and RR charges.

Gravitational description

Even though Op-planes have a fixed position, they have a tension and charge, and thus have an effect on the gravitational and RR fields. Given the similarity between (1.4.50) and (1.3.24), this Op gravitational solution is formally identical to that for Dp-branes, (1.3.59); but the constant in the harmonic function h changes, because of the tension (1.4.49). For the Op_-, (1.3.61) changes to

$$h = 1 - 2^{p-5} \frac{r_0^{7-p}}{r^{7-p}} , \tag{1.4.55}$$

again valid for $p < 7$. (For the Op_+, the relative sign is +.) The negative sign changes the behavior of the solution: now in the region $r < 2^{p-5} r_0$ the function h is negative, and the solution (1.3.59) is meaningless.[18] This is in fact not worrisome because the supergravity approximation breaks down already well outside the hole $r < r_0$, due to large curvature and sometimes string coupling. This unphysical hole should be resolved in full-fledged string theory, much like what we expect for the singularity at $r = 0$ for Dp-branes; we will see examples in Sections 7.2.3 and 9.4.3.

[18] This should not be confused with the region inside a Schwarzschild black hole horizon, where $g_{tt} < 0$ but the metric is real.

Equation (1.4.55) can also be combined with Dp-branes, similar to (1.3.62). This confirms that parallel Op-planes and Dp-branes are also simultaneously BPS, just like parallel Dp.

Here are some more details:

- For $p < 3$, at the hole boundary $r = 2^{p-5} r_0$ the scalar curvature diverges,while the string coupling goes to zero.
- For $p = 3$, $R = 0$, but other curvature invariant diverge at $r = 2^{p-5} r_0$. The string coupling is constant.
- For $3 < p < 7$, both curvature and string coupling diverge at $r = 2^{p-5} r_0$.
- For $p = 7$, the harmonic function is

$$h = 4 \frac{g_s}{2\pi} \log(r/r_0) , \tag{1.4.56}$$

the -4 of difference with (1.3.63) for D7-branes again due to (1.4.49). The unphysical hole is $r < r_0$.

- For $p = 8$,

$$h = h_0 - 8 \frac{g_s}{2\pi l_s} |x_9| . \tag{1.4.57}$$

For $h_0 > 0$, there is no hole: the dilaton approaches $e^\phi \sim g_s h_0^{-5/4}$. The curvature is finite except at $x_9 = 0$, where the second derivatives of h generate a delta function that can be interpreted as the effect of the O8 source. This is a much milder singularity than all the others we have encountered, especially for large h_0. However, for small h_0 the string coupling is large; for $h_0 = 0$, both dilaton and curvature diverge at $x_9 = 0$.

O9-plane and type I superstrings

We now focus on the O9, where the spacetime involution is the identity, which we already encountered after (1.4.53). The fields that survive the projection (1.4.46) for $p = 9$ are

$$\phi , \qquad g_{MN} , \qquad C_2 . \tag{1.4.58}$$

(The magnetic dual C_6 (1.3.40) is not independent.) The fermionic fields are also projected: only one gravitino and one dilatino remain. The branes that are still BPS should be

$$\text{D1} , \qquad \text{D5} , \qquad \text{D9} . \tag{1.4.59}$$

The F1 and the NS5 should not be BPS, because they would be charged under B, which has been projected out in (1.4.58). Indeed, a closed F1 is unstable toward breaking in open strings. Ordinarily this can happen only near D-branes, but in type I we have D9-branes, which are everywhere.

The O9 by itself has a problem: the chiral fermions have an anomaly. In IIB, this was canceled by the contribution of C_4 (Section 1.2.3), but now that field has been projected out. A curious alternative way of noticing the problem is to focus on the WZ term in (1.4.50), which for $p = 9$ gives

$$T_{\text{O9}} \int C_{10} . \tag{1.4.60}$$

Since C_{10} only appears here, its equation of motion reads $1 = 0$. (This "tadpole" is reminiscent of the one on D0s in massive IIA (Section 1.4.1).) This point of view suggests a way out: D9-branes also have a WZ coupling to C_{10}, and adding N of them results in a total $(N T_{D9} + T_{O9}) \int C_{10}$. In view of (1.4.49), we need $N = 16$ full D9-branes, leading to a gauge group

$$\mathrm{SO}(2N) = \mathrm{SO}(32)\,. \tag{1.4.61}$$

This is the gauge group for one of the heterotic theories, (1.1.44); the anomaly is now canceled in a similar fashion as in that theory. This O9 plus 16 (full) D9 is called a *type I superstring*, because the O9 breaks supersymmetry by half. The spacetime action S is obtained from the IIB (1.2.36) by setting to zero the fields not included in (1.4.58), $B_{MN} = C_0 = C_4 = 0$, and adding an action for the D9–O9 system. By Section 1.3.6, this is supersymmetric Yang–Mills with gauge group SO(32).

This finally completes the list of perturbative string theories at the beginning of Section 1.1.

1.4.5 Heterotic dualities

In Sections 1.4.1–1.4.3, we have seen that IIA, IIB and eleven-dimensional supergravity are all connected by a web of dualities, which can be explained by postulating a theory in $d = 11$ of extended objects called M-theory. We will now see how to obtain from M-theory the remaining perturbative string theories, namely type I and the two heterotic theories [112, 113].

M-theory on an interval and heterotic strings

We begin by considering M-theory on a spacetime

$$\mathbb{R}^{10} \times I\,, \tag{1.4.62}$$

with I an interval. In general, even in QFT, introducing space boundaries requires care with boundary conditions; in a theory of gravity, where geometry is dynamical, one might even worry that a boundary can be unstable and expand. In string theory, so far we have only seen one possible spacetime boundary, the O8-plane. We can try to introduce a boundary in a similar way in M-theory, by quotienting it.

We then view $I = S^1/\mathbb{Z}_2$: if S^1 is defined by a periodic coordinate $x^{10} \sim x^{10} + 2\pi R_{10}$, then $\mathbb{Z}_2$ identifies opposite points according to $x^{10} \to -x^{10}$. The two fixed loci of this action, $\{x^{10} = 0\}$ and $\{x^{10} = \pi R_{10}\}$, are the endpoints of the interval. Once we take the product with $\mathbb{R}^{10}$, these loci become ten-dimensional boundaries. Unfortunately, this $x^{10} \to -x^{10}$ is not a symmetry: in the $d = 11$ action (1.4.4), the CS term $A_3 \wedge G_4 \wedge G_4$ is not invariant, because its definition contains an ϵ tensor (similar to (1.2.19)), which picks up a minus sign under a parity. Said differently, $A_3 \wedge G_4 \wedge G_4$ only contains the index $M = 10$ a single time. We can cure this problem by also reversing the sign of A_3:

$$x^{10} \to -x^{10}\,, \qquad A_3 \to -A_3\,. \tag{1.4.63}$$

This is now a symmetry of (1.4.4), and we can quotient M-theory by it.

In the limit where the interval is small, $R_{10} \ll l_{P11}$, the resulting quotient theory will look ten-dimensional. In a KK reduction, in first approximation we can keep only the zero modes in x^{10}. (1.4.63) implies that the components A_{MNP}, $M, N, P \neq 10$, are odd functions of x^{10}; so the zero KK mode, which would be the constant, is projected out. The components $A_{MN\,10}$ are even functions, and the zero KK mode remains. So the ten-dimensional theory we obtain has a two-form B_{MN}, but no three-form. There are also zero modes for the metric and dilaton; so the $d = 10$ bosonic fields are those of the common NSNS sector or of the bosonic string, (1.1.9).

Equation (1.4.63) should be supplemented by the spinorial parity action, multiplication by Γ_{10}. The result for the zero modes is that we keep a positive-chirality gravitino $\psi_{M\alpha}$, and a negative-chirality dilatino $\lambda_{\dot\alpha}$. The supersymmetry parameter ϵ becomes a single Majorana spinor ϵ of positive chirality; so there are 16 supercharges.

The massless fields we have found in ten dimensions so far are the same as the first two rows of Table 1.3 for the heterotic theories. This is problematic, because as we saw there the fermions $\psi_{M\alpha}$ and $\lambda_{\dot\alpha}$ give rise to an anomaly. Before taking the limit $R_{10} \ll l_{\mathrm{P}11}$, we might think we can have no anomaly because we are in $d = 11$ dimension, which is odd; but there can be anomalies localized on the boundaries. So on a spacetime boundary in M-theory, there should be degrees of freedom that cancel such anomalies. There are two identical boundaries, and it is natural to expect that they give the same contribution. Given that in the heterotic theory the anomaly is canceled by the χ^a_α, it must then be that each of the two boundaries gives half that contribution.

One of the heterotic gauge groups (1.1.44) is indeed a product of two copies of E_8; so the anomaly is canceled if on each $d = 10$ spacetime boundary there is a χ^a_α in the adjoint of E_8. Supersymmetry then demands that a bosonic vector field should also be present. In other words, the remaining two lines of Table 1.3 are also present in the spectrum, but are localized on the two spacetime boundaries, each with an E_8 gauge group.

We conclude then that the $R_{10} \ll l_{\mathrm{P}11}$ limit of M-theory on $\mathbb{R}^{10} \times I$ is the $E_8 \times E_8$ heterotic string:

$$\text{M-theory on } I \quad \cong \quad E_8 \times E_8 \text{ heterotic} . \tag{1.4.64}$$

M-theory on (1.4.62) can be regarded as a nonperturbative completion of heterotic $E_8 \times E_8$, much as M-theory on $\mathbb{R}^{10} \times S^1$ can be regarded as a nonperturbative completion of IIA.

Heterotic T-duality

The two heterotic theories are *T-dual*. To see this, we put both theories on $\mathbb{R}^9 \times S^1$. This is more complicated than IIA or IIB on such space: the length of the S^1 is now no longer the only parameter. There is also A_9, which is physical because of (1.2.14). By a gauge transformation, we can make it lie in the Cartan subalgebra, which for both heterotic gauge groups (1.1.44) is 16-dimensional. So we have 17 physical parameters:

$$(A_9^I, L_9) . \tag{1.4.65}$$

The gauge group is broken to the commutant of A_9^I. On generic points of the parameter space, this is $U(1)^{16}$; moreover, there is an extra $U(1)^2$ coming from g_{M9} and B_{M9}, much as in (1.1.24) for the bosonic string. At the special points where $A_9^I = 0$ in either heterotic theory, this is enhanced back to

$$E_8 \times E_8 \times U(1)^2 , \qquad SO(32) \times U(1)^2 . \tag{1.4.66}$$

There is also a world-sheet description of such S^1 compactifications [114, 115]. In Section 1.1.3, the requirements of level matching constrained the T^{16} for the left-moving bosons so much that only two possibilities remained, leading to the two heterotic theories. Compactifying one spacetime direction, we have 17 compact left-movers and one compact right-mover; the spectrum analysis is a mix of those for the heterotic theory and for the compactified bosonic string, (1.1.23). We now need an $(1 + 17)$-dimensional lattice. These are all mathematically equivalent under rotations that mix all 18 bosons, but physically we should regard as equivalent only those that are related by rotations of the 17 left-movers. Another possible perspective is that we should choose a single linear combination of the 17 compact left-movers to pair with the single compact right-mover. This again gives 17 physical parameters, as in (1.4.65). As for the gauge group, a generic $(1 + 17)$-lattice won't have elements of length two, so the generic gauge group is Abelian. For special values of the lattice, one can obtain (1.4.66).

From the world-sheet point of view, we didn't have to specify which of the two heterotic theories we were discussing. Before compactifications, there are two possibilities; after compactification, there is a 17-dimensional space of possibilities, and each point in this space can be thought of as arising from either of the two heterotic theories. In other words, a configuration obtained from the $E_8 \times E_8$ theory for one choice of (1.4.65) also arises from the SO(32) theory for a different choice of parameters. In this identification, the length of the S^1 gets inverted, as in type II T-duality. In this sense, the two heterotic theories are T-dual.

Heterotic-type I duality

Earlier in this subsection, we have argued that the strong-coupling limit of the $E_8 \times E_8$ heterotic theory is M-theory on (1.4.62). What is the strong-coupling limit of the SO(32) heterotic theory?

The SO(32) gauge group suggests that it might be type I. Indeed, this works at the level of the effective action. The map is (1.4.31), applied to the fields (1.4.58) that survive the O9 orientifold projection, and leaving the SO(32) gauge field unchanged. In particular, the C_2 potential of type I becomes the heterotic B; the D1-brane of type I becomes the heterotic F1. We also argued after (1.4.59) that the F1 in type I is unstable, and this is dual to the lack of D-branes in the heterotic theory.

M-theory on a cylinder

The dualities of this subsection can be summarized conveniently by considering M-theory on

$$\mathbb{R}^9 \times S^1 \times I , \tag{1.4.67}$$

sketched in Figure 1.4. This is similar to how M-theory on $\mathbb{R}^9 \times T^2$ summarizes the dualities among M-theory, IIA, and IIB (see Figure 1.3).

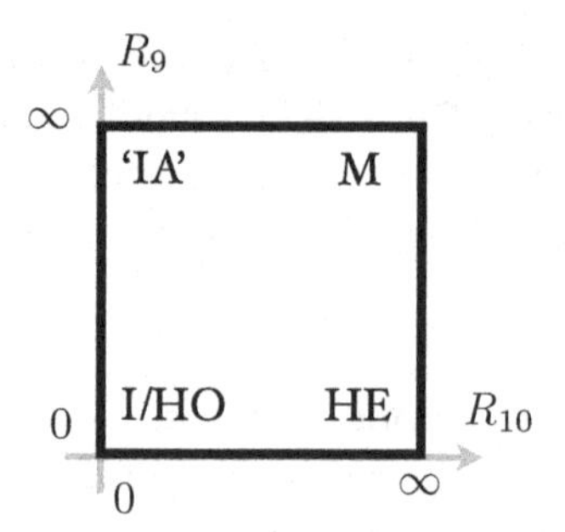

Figure 1.4 M-theory on $\mathbb{R}^9 \times S^1 \times I$.

We call R_9 and R_{10} the radii of S^1 and I respectively. When both are large, spacetime is eleven-dimensional, and the best description is M-theory. When R_9 is large and R_{10} is small, we recover the $E_8 \times E_8$ heterotic string by (1.4.63). In the figure, we have gone from the top-right corner of Figure 1.4 to the bottom-right (from "M" to "HE"). If we now also make R_{10} small, naively spacetime becomes nine-dimensional, but we argued that a T-duality can make it ten-dimensional again, turning the theory into SO(32) heterotic (bottom-left corner "HO" in the figure).

For M-theory on $\mathbb{R}^9 \times T^2$, the exchange of the two radii was interpreted as the S-duality (1.4.31). This exchanges the two axes in Figure 1.3 and leaves the bottom-left IIB corner invariant, consistent with S-duality being a symmetry of IIB. For M-theory on (1.4.67), the same operation

$$R_9 \leftrightarrow R_{10} \tag{1.4.68}$$

is again interpreted as an S-duality, this time relating the SO(32) heterotic theory to type I. This is why the bottom-left corner is labeled with both theories.

The two bottom corners in the figure are related by T-duality; (1.4.68) then predicts that the duality taking type I to the top-left corner should also be a T-duality. By (1.4.53), this is a configuration with two O8$_-$-planes in IIA. It is called *type IA*, although it is simply an orientifold of IIA (just like type I is just an orientifold of IIB). The 16 full D9-branes in type I are T-dualized to 16 full D8-branes. Reducing directly from M-theory on the S^1, we would have expected to find IIA on $\mathbb{R}^9 \times I$; the only boundaries we know in both type II theories are indeed O8-planes.

In type I, there are also 16 full D9-branes, which are T-dualized to 16 full D8-branes parallel to the boundaries. By (1.4.23), their positions in the interval are dual to the values of the gauge field a_9 on the D9s (in the Cartan subalgebra). Going further back in the duality chain, this originates from the values of a_9 in the two copies of E_8 that live on the M-theory boundaries.

D8-branes couple to F_{10}, the dual of the Romans mass parameter F_0. When the D8 positions are generic, there will thus be a nonzero value for F_0 in the spacetime on the IA side, thus achieving a lift of sorts of massive IIA to eleven dimensions. This map from IA to M-theory looks quite nonlocal: the position of the D8s is encoded in the values of two gauge fields that live on the boundaries of the M-theory spacetime. It is not obvious how to generalize this to cases without O8-planes.

Exercise 1.4.1 Show with an argument similar to (1.3.47) that in eleven-dimensional supergravity

$$\frac{1}{(2\pi l_{\mathrm{P}11})^3}\int G_4 \in \mathbb{Z}\,, \qquad \frac{1}{(2\pi l_{\mathrm{P}11})^6}\int *G_4 \in \mathbb{Z}\,. \tag{1.4.69}$$

Exercise 1.4.2 The M5 solution is

$$\mathrm{d}s^2_{\mathrm{M5}} = h^{-1/3}\mathrm{d}s^2_{\parallel} + h^{2/3}\mathrm{d}s^2_{\perp}\,, \qquad G_{ijkl} = 3\pi l^3_{\mathrm{P}11} x^m \epsilon^{\perp}_{mijkl} r^{-5}\,, \tag{1.4.70}$$

in the notation of (1.3.59): the parallel and transverse directions are six and five, and $h = 1 + r_0^3/r^3$, $r_0^3 = \pi N l^3_{\mathrm{P}11}$.

Identify periodically one of the transverse directions, $\mathbb{R}^5 \to \mathbb{R}^4 \times S^1$ and reduce along it: namely, calling the compact direction x^{11}, write (1.4.70) in the form (1.4.5a). Check that you obtain the NS5 solution from (1.3.65).

Exercise 1.4.3 Now reduce (1.4.70) along one of the parallel directions, and check that you obtain the D4 solution from (1.3.59).

Exercise 1.4.4 Use S-duality (1.2.38) and (1.4.30) to derive the NS5 solution (1.3.65) from the D5 solution (1.3.59).

Exercise 1.4.5 Now use the same S-duality to derive the back-reaction of an F1 from the D1 solution.

Exercise 1.4.6 Finally, use the F1 solution to obtain that of an M2 whose metric reads

$$\mathrm{d}s^2_{\mathrm{M2}} = h^{-2/3}\mathrm{d}s^2_{\parallel} + h^{1/3}\mathrm{d}s^2_{\perp}\,, \qquad h = 1 + \frac{r_0^6}{r^6}\,, \tag{1.4.71}$$

with $r_0^6 = 32\pi^2 N l^6_{\mathrm{P}11}$. What happens if we try to obtain it by lifting a D2?

Exercise 1.4.7 Check that (1.4.46) is a symmetry of the type II actions (1.2.15) and (1.2.36).

2 Spinors

Since in this chapter we are interested in the algebraic properties of spinors, we will work in *flat space*. Both in this chapter and in the next, we will consider a wider range of dimensions than we really need for the later developments.

2.1 Clifford algebra

In this first section, we will focus on Lorentzian signature, commenting occasionally on what changes for the Euclidean case; later we will change focus as needed. Since in this chapter we only consider flat space, the metric is $g_{\mu\nu} = \eta_{\mu\nu}$ in Lorentzian signature, and $g_{mn} = \delta_{mn}$ in the Euclidean.

2.1.1 Gamma matrices and spin representation

Recall that the Lie algebra of the Lorentz group $\mathrm{O}(1, d-1)$ in d dimensions is the space of matrices[1]

$$\mathrm{so}(1, d-1) = \{\lambda \text{ such that } g\lambda = -(g\lambda)^t\} . \tag{2.1.1}$$

In components, λ has index structure[2] $\lambda^\mu{}_\nu$, and $g\lambda$ has index structure $\lambda_{\mu\nu}$; so the condition in (2.1.1) reads $\lambda_{\mu\nu} = -\lambda_{\nu\mu}$. A basis consists of matrices $J_{\mu\nu}$ with components $(J_{\mu\nu})^{\alpha\beta} = -\delta^\alpha_\mu \delta^\beta_\nu + \delta^\alpha_\nu \delta^\beta_\mu$, and commutation relations

$$[J_{\mu\nu}, J^{\rho\sigma}] = 4\delta^{[\rho}_{[\mu} J_{\nu]}{}^{\sigma]} = \delta^\rho_\mu J_\nu{}^\sigma - \delta^\rho_\nu J_\mu{}^\sigma - \delta^\sigma_\mu J_\nu{}^\rho + \delta^\sigma_\nu J_\mu{}^\rho . \tag{2.1.2}$$

The gamma matrices γ_μ are defined as obeying the Clifford algebra

$$\{\gamma_\mu, \gamma_\nu\} = 2g_{\mu\nu}\, 1 . \tag{2.1.3}$$

(In Chapter 1, we used the symbol Γ_M; we use a lower case γ_μ for $d \neq 10$ and for general discussions valid in any d.) There are many possible choices, or *bases*, of such matrices, but we will see later that they are almost unique, and that their minimal dimension is

$$2^{\lfloor \frac{d}{2} \rfloor} . \tag{2.1.4}$$

[1] In this chapter, both the Lorentzian metric η and the Euclidean metric δ will be denoted by g.

[2] In this book, we use Greek indices $\mu, \nu, \ldots$ for Lorentzian signature and Latin indices $m, n, \ldots$ for Euclidean signature. Indices $i, j, \ldots$ will appear for holomorphic indices. The capitals $M, N, \ldots$ are reserved for $d = 10$ Lorentzian dimensions.

Spinor representation

To see why (2.1.3) is useful, consider

$$\gamma_{\mu\nu} \equiv \gamma_{[\mu}\gamma_{\nu]} = \frac{1}{2}[\gamma_\mu, \gamma_\nu]. \tag{2.1.5}$$

One can compute the commutator of two such matrices from (2.1.3); we will soon learn several ways to do it effectively. For now, we just give the result:

$$[\gamma_{\mu\nu}, \gamma^{\rho\sigma}] = -8\delta^{[\rho}_{[\mu}\gamma_{\nu]}{}^{\sigma]}. \tag{2.1.6}$$

Its similarity to (2.1.2) signals that $-\frac{1}{2}\gamma_{\mu\nu}$ is a representation ρ_s of the so$(1, d-1)$ Lie algebra. Spinors are defined as elements of the vector space S on which ρ_s acts. In other words, an infinitesimal transformation on a vector v is represented on a spinor ζ as follows:

$$\delta v^\mu = \Lambda^\mu{}_\nu v^\nu, \qquad \delta\zeta = \rho_s(\Lambda)\zeta \equiv -\frac{1}{2}\mathbb{\Lambda}\zeta, \qquad \mathbb{\Lambda} \equiv \frac{1}{2}\Lambda_{\mu\nu}\gamma^{\mu\nu}. \tag{2.1.7}$$

At the level of finite transformations:

$$v^\mu \to (e^\Lambda)^\mu{}_\nu v^\nu, \qquad \zeta \to \rho_s(\exp[\Lambda])\,\zeta \equiv \exp\left[-\frac{1}{2}\mathbb{\Lambda}\right]\zeta. \tag{2.1.8}$$

This suffers from a famous sign ambiguity. Consider, for example, a rotation in the plane 12 by an angle ϕ; Λ is a block-diagonal matrix with $\phi\left(\begin{smallmatrix} 0 & -1 \\ 1 & 0 \end{smallmatrix}\right)$ in the 12 block, and the identity elsewhere. Thus $\mathbb{\Lambda} = -\phi\gamma_{12} = -\phi\gamma_1\gamma_2$, and

$$\exp\left[-\frac{1}{2}\mathbb{\Lambda}\right] = \exp[(\phi/2)\gamma_{12}] = \cos(\phi/2)1 + \sin(\phi/2)\gamma_{12}. \tag{2.1.9}$$

(In resuming the exponential, one uses $\gamma_{12}^2 = -1$.) We see that $\phi = 2\pi$ corresponds in Spin$(1, d-1)$ to minus the identity -1, while in SO$(1, d-1)$ it corresponds to 1. This is the familiar physical fact that a rotation by 2π changes a spinor by a sign. More generally, a Lorentz transformation $\Lambda = e^\Lambda$ determines its spinorial counterpart $e^{-\frac{1}{2}\mathbb{\Lambda}}$ only up to a sign. One way to take care of this subtlety is to introduce the *spin group* Spin$(1, d-1)$, whose elements are all the expressions of the form $e^{-\frac{1}{2}\mathbb{\Lambda}}$. This comes with a homomorphism:

$$\text{Spin}(1, d-1) \to \text{SO}(1, d-1), \tag{2.1.10}$$

which is $2:1$, in that the counterimage of any SO$(1, d-1)$ consists of two elements of Spin$(1, d-1)$, which differ by an overall sign.

Parity

There also exist elements in O$(1, d-1)$ that cannot be written in the form e^Λ,[3] notably any reflection (or parity):

$$R_\mu : x^\nu \mapsto \begin{cases} -x^\nu & \text{if } \mu = \nu, \\ x^\nu & \text{if } \mu \neq \nu. \end{cases} \tag{2.1.11}$$

[3] O$(1, d-1)$ breaks up in four connected components, characterized by the determinant $\det(\Lambda)$ and by the sign of $\Lambda^0{}_0$. One defines SO$(1, d-1) \equiv \{\Lambda \in \text{O}(1, d-1) | \det(\Lambda) = 1\}$, and further SO$^+(1, d-1) \equiv \{\Lambda \in \text{SO}(1, d-1) | \Lambda^0{}_0 > 0\}$.

It is easy to guess how to extend ρ_s to such elements by noting that $R_\mu R_\nu$ is a rotation of angle π in the $\mu\nu$ plane. For example, with $\mu = 1$, $\nu = 2$, $e^\lambda = \text{diag}(1, -1, -1, 1, \ldots, 1)$. From (2.1.9) for $\phi = \pi$, $\rho_s(R_1 R_2) = \gamma_{12} = \gamma_1\gamma_2$. More generally, $\rho_s(R_\mu R_\nu) = \gamma_\mu\gamma_\nu$. This leads to

$$\rho_s(R_\mu) = \gamma_\mu \,. \tag{2.1.12}$$

(Some people prefer calling *pinors* the representation of $O(1, d-1)$ obtained this way, but we will not use this name.) We referred to this result several times in Chapter 1.

Gamma matrices as invariants

Another notable identity that follows from (2.1.3) is

$$[\gamma_{\mu\nu}, \gamma^\rho] = -2\delta^\rho_\mu \gamma_\nu + 2\delta^\rho_\nu \gamma_\mu = 4\gamma_{[\mu}\delta^\rho_{\nu]}\,; \tag{2.1.13}$$

this too will be derived later in this section. It says that gamma matrices are a Lorentz invariant, in the following sense. A matrix α acting on spinors has two spinorial indices, and is hence called a *bispinor*. By (2.1.8), such an α should transform as $\alpha \to \exp[-\frac{1}{2}\lambda]\alpha \exp[\frac{1}{2}\lambda]$; infinitesimally, $\delta\alpha = [-\frac{1}{2}\lambda, \alpha]$. A γ_μ is a bispinor, but also has a vector index. Multiplying (2.1.13) by $\lambda^{\mu\nu}$ gives

$$\left[-\frac{1}{2}\lambda, \gamma^\mu\right] = \lambda^\mu{}_\nu \gamma^\nu\,; \tag{2.1.14}$$

comparing with (2.1.7), we see that it converts the bispinor transformation into the vector transformation. We can also see this at the finite action. The *Hadamard lemma* says[4]

$$\begin{aligned} e^A B e^{-A} &= B + [A, B] + \frac{1}{2}[A, [A, B]] + \ldots \\ &= B + \text{ad}_A(B) + \frac{1}{2}\text{ad}_A(\text{ad}_A(B)) + \ldots = \exp[\text{ad}_A](B)\,. \end{aligned} \tag{2.1.15}$$

Using this and (2.1.14), resuming the exponential we get

$$\exp\left[-\frac{1}{2}\lambda\right]\gamma^\mu \exp\left[\frac{1}{2}\lambda\right] = \left(e^\lambda\right)^\mu{}_\nu \gamma^\nu\,, \tag{2.1.16}$$

to be compared with (2.1.8).

2.1.2 Chirality

When d is even, the *chirality matrix* is

$$\gamma \equiv c\,\gamma^0\gamma^1 \ldots \gamma^{d-1} \tag{2.1.17}$$

[4] This formula is often used in quantum mechanics under this name. Other results in analysis are also called in the same way. For $A = f(x)$, $B = \partial_x$, (2.1.15) reduces to the Taylor expansion of f.

in Lorentzian signature, and $\gamma \equiv c\,\gamma^1 \ldots \gamma^d$ in Euclidean signature. It anticommutes with all of the γ_μ:

$$\{\gamma, \gamma_\mu\} = 0\,. \tag{2.1.18}$$

(In $d = 4$, (2.1.17) is often called γ_5; this would be confusing when dealing with higher dimensions.) The constant c in (2.1.17) is chosen so that

$$\gamma^2 = 1\,. \tag{2.1.19}$$

This imposes

$$c^2 = \begin{cases} -(-1)^{\lfloor \frac{d}{2} \rfloor} & \text{(Lor.)}\,; \\ (-1)^{\lfloor \frac{d}{2} \rfloor} & \text{(Eucl.)}\,. \end{cases} \tag{2.1.20}$$

Some important cases for us are $d = 4$ in Lorentzian signature, where we will take $c = \mathrm{i}$, and $d = 6$ Euclidean, where we will take $c = -\mathrm{i}$. When $c^2 = 1$, we will always take $c = 1$; notably this is the case for $d = 10$ in Lorentzian signature.

It follows from (2.1.18) that γ commutes with products of even numbers of γ_μ's, such as the exponential in (2.1.8). Thus, if a spinor ζ_+ obeys $\gamma\zeta_+ = \zeta_+$, its transformed $\zeta'_+ = \exp[-\frac{1}{2}\lambda]\zeta_+$ obeys the same constraint. In other words, the space $S_+ = \{\zeta_+ \in S | \gamma\zeta_+ = \zeta_+\}$ of spinors of *positive chirality* is an invariant subspace for the representation ρ_s. Its orthogonal complement $S_- = \{\zeta_- \in S | \gamma\zeta_- = -\zeta_-\}$ is also invariant; ρ_s decomposes as the direct sum of two irreducible representations $\rho_\pm$ acting on $S_\pm$. Spinors of definite chirality are sometimes also called *Weyl* spinors. The projectors on $S_\pm$ read

$$P_\pm \equiv \frac{1}{2}(1 \pm \gamma)\,. \tag{2.1.21}$$

To denote the various components of a spinor, one often uses indices from the beginning of the Greek alphabet. One uses the index $\alpha, \beta, \ldots = 1, \ldots, 2^{\frac{d}{2}-1}$ for positive-chirality spinors and $\dot\alpha, \dot\beta, \ldots = 1, \ldots, 2^{\frac{d}{2}-1}$ for negative chirality. The gamma matrices act on the space of spinors S, so they have two spinorial indices. They change chirality: if $\zeta_\pm \in S_\pm$, $\gamma_\mu\zeta_\pm \in S_\mp$:

$$\gamma\gamma_\mu\zeta_\pm = \mp\gamma_\mu\zeta_\pm\,. \tag{2.1.22}$$

Thus the only nonzero components of the gamma matrices are the $(\gamma_\mu)^\alpha{}_{\dot\beta}$ and $(\gamma_\mu)^{\dot\alpha}{}_\beta$.

Odd dimensions

Given a choice of γ_μ in d = even, one can obtain one in $d+1$ dimensions by adding the chirality matrix: $\{\gamma_0, \gamma_1, \ldots, \gamma_{d-1}, \gamma_d \equiv \gamma\}$.

On the other hand, applying (2.1.17) to odd d now gives a matrix that *commutes* with all γ_μ, rather than satisfying (2.1.18). With (2.1.20), $\gamma^2 = 1$ still holds; if the gamma matrices have the smallest possible dimension (2.1.4), $\gamma = 1$ or -1. Given a choice of γ_μ that obey the Clifford algebra, the $\gamma'_\mu = -\gamma_\mu$ also do. Now $\gamma' = -\gamma = \mp 1$, so this new choice cannot be related to the γ_μ by a basis change M: $\gamma'_\mu = -\gamma_\mu \neq M\gamma_\mu M^{-1}$, and one says it is an *inequivalent* choice. In other words, a matrix satisfying (2.1.18) does not exist, so one cannot define chirality and ρ_s is irreducible. In this case, one uses a single type of index $\alpha, \beta, \ldots = 1, \ldots, 2^{\frac{d-1}{2}}$.

The same argument does not apply to d = even, because there $M = \gamma$ does relate $-\gamma_\mu = \gamma\gamma_\mu\gamma^{-1}$. Indeed, we will see later that all choices of gamma matrices of minimal dimension (2.1.4) are equivalent.

2.1.3 A more abstract point of view

Equation (2.1.3) can also be viewed more abstractly, as a relation between formal symbols γ_μ rather than as a matrix equation.

Associative algebras

An *associative algebra* (often just *algebra*) is a vector space endowed with a product $\cdot$ that is associative ($v_1 \cdot (v_2 \cdot v_3) = (v_1 \cdot v_2) \cdot v_3$) and that has the distributive property $v_1 \cdot (v_2 + v_3) = v_1 \cdot v_2 + v_1 \cdot v_3$, $(v_1 + v_2) \cdot v_3 = v_1 \cdot v_3 + v_2 \cdot v_3$. Some examples follow:

- The space $\mathrm{Mat}(N, \mathbb{R})$ of real $N \times N$ matrices.
- The space $\mathbb{R}[x_1, \ldots, x_n]$ of real polynomials in n variables.
- The algebra of *quaternions* $\mathbb{H}$ is defined as the real $\mathrm{Span}\{1, i_1, i_2, i_3\}$, where the i_a are "imaginary units" such that

$$i_a \cdot i_b = -\delta_{ab} + \epsilon_{abc} i_c \,. \tag{2.1.23}$$

  They are also often called $\mathrm{i} \equiv i_1$, $\mathrm{j} \equiv i_2$, $\mathrm{k} \equiv i_3$.
- The *direct sum* $A_1 \oplus A_2$ of two associative algebras A_1, A_2 is the space of pairs (the cartesian product), with the pairwise product $(v_1, v_2) \cdot (w_1, w_2) \equiv (v_1 \cdot v_2, w_1 \cdot w_2)$.

Abstract Clifford algebra

As a generalization of $\mathbb{R}[x_1, \ldots, x_n]$, one can consider an algebra of polynomials T_d with d abstract variables that don't commute, which we also call γ_μ. We can view (2.1.3) as a relation in such an algebra: this means that whenever we encounter the combination $\gamma_\mu\gamma_\nu + \gamma_\nu\gamma_\mu$ in a polynomial, we can replace it with $2g_{\mu\nu}1$. The resulting associative algebra is the abstract Clifford algebra

$$\mathrm{Cl}_{1,d-1} \equiv \frac{T_d}{\{\gamma_\mu, \gamma_\nu\} = 2g_{\mu\nu}1} \,. \tag{2.1.24}$$

As a vector space, it is generated by all monomials in the formal variables γ_μ; due to the relation (2.1.3), the number of such monomials is not infinite. For example, with the flat metric we are considering in this chapter, the square of each γ_μ is proportional to the identity, so each γ_μ can appear in a monomial with power at most one. Equation (2.1.3) also allows us to write the γ_μ appearing in any monomial in a given order, up to monomials of lower degree. Thus a possible basis for the algebra is given by ordered monomials $\gamma_{\mu_1}\gamma_{\mu_2}\cdots\gamma_{\mu_k}$, $\mu_1 < \mu_2 < \ldots < \mu_k$. We can also introduce the antisymmetrized products[5]

$$\gamma_{\mu_1\ldots\mu_k} \equiv \gamma_{[\mu_1} \cdots \gamma_{\mu_k]} \,, \tag{2.1.25}$$

[5] In our conventions, the antisymmetrizer has a $\frac{1}{k!}$; so, for example, $\gamma_{123} = \frac{1}{6}(\gamma_1\gamma_2\gamma_3 - \gamma_2\gamma_1\gamma_3 + \ldots) = \gamma_1\gamma_2\gamma_3$.

generalizing (2.1.5). There are $\binom{d}{k}$ such monomials of degree k; so

$$\dim \mathrm{Cl}_{1,d-1} = \sum_{k=0}^{d} \binom{d}{k} = (1+1)^d = 2^d . \tag{2.1.26}$$

In the d = even case, this coincides with (2.1.4), indicating that (2.1.25) is a basis for the space of all matrices acting on spinors. In the d = odd case, (2.1.4) and (2.1.26) differ by a factor of 2, indicating some redundancies in (2.1.25); for example, as already mentioned, the would-be chirality matrix (2.1.17) is in fact proportional to the identity. We will see this more clearly in Section 3.1.2.

Isomorphisms to matrix algebras

A *representation* ρ of an associative algebra A is a linear map to a matrix algebra that is a homomorphism, namely $\rho(v_1) \cdot \rho(v_2) = \rho(v_1 \cdot v_2)$. Often, when no confusion should arise, one skips the symbol ρ to avoid a proliferation of symbols. For a Clifford algebra $\mathrm{Cl}(1, d-1)$, a representation is often even *isomorphic* to a matrix algebra, namely the linear map ρ is one-to-one.

For example, for $d = 1$ the generators are 1 and γ_0, so a general element can be written as $x + \gamma_0 y$. Since γ_0 squares to -1, it formally behaves as the imaginary unit i; so we have an isomorphism $\mathrm{Cl}_{1,0} \cong \mathbb{C}$. For $d = 2$, the generators are $1, \gamma_0, \gamma_1, \gamma_{01}$ (in agreement with (2.1.26)). In the two-dimensional representation where

$$\gamma_0 = \mathrm{i}\sigma_2 \equiv \epsilon = \begin{pmatrix} 0 & 1 \\ -1 & 0 \end{pmatrix}, \qquad \gamma_1 = \sigma_1 = \begin{pmatrix} 0 & 1 \\ 1 & 0 \end{pmatrix}, \tag{2.1.27}$$

we also have $\gamma_{01} = \sigma_3 = \left(\begin{smallmatrix} 1 & 0 \\ 0 & -1 \end{smallmatrix}\right)$, and of course 1 is the identity. So we see that

$$\mathrm{Cl}_{1,1} \cong \mathrm{Mat}(2, \mathbb{R}) , \tag{2.1.28}$$

the associative algebra of real 2×2 matrices.

There are similar results for higher dimensions [116]:

$$\begin{aligned}
\mathrm{Cl}_{1,2} &\cong \mathrm{Mat}(2,\mathbb{R}) \oplus \mathrm{Mat}(2,\mathbb{R}) , & \mathrm{Cl}_{1,3} &\cong \mathrm{Mat}(4,\mathbb{R}) , \\
\mathrm{Cl}_{1,4} &\cong \mathrm{Mat}(4,\mathbb{C}) , & \mathrm{Cl}_{1,5} &\cong \mathrm{Mat}(4,\mathbb{H}) , \\
\mathrm{Cl}_{1,6} &\cong \mathrm{Mat}(4,\mathbb{H}) \oplus \mathrm{Mat}(4,\mathbb{H}) , & \mathrm{Cl}_{1,7} &\cong \mathrm{Mat}(8,\mathbb{H}) , \\
\mathrm{Cl}_{1,8} &\cong \mathrm{Mat}(16,\mathbb{C}) , & \mathrm{Cl}_{1,9} &\cong \mathrm{Mat}(32,\mathbb{R}) , \\
\mathrm{Cl}_{1,10} &\cong \mathrm{Mat}(32,\mathbb{R}) \oplus \mathrm{Mat}(32,\mathbb{R}) .
\end{aligned} \tag{2.1.29a}$$

For Euclidean signature, g in (2.1.24) is the Euclidean metric, and we call the resulting algebra $\mathrm{Cl}_{0,d}$ or simply Cl_d. One can still use (2.1.25) as a basis, and the dimension is thus $\dim \mathrm{Cl}_d = 2^d$, similarly to (2.1.26). For $d = 1$, $\gamma_1^2 = 1$, and there is no isomorphism with $\mathbb{C}$ as there was with $\mathrm{Cl}_{1,0}$. The elements $P_{1\pm} = \frac{1}{\sqrt{2}}(1 \pm \gamma_1)$ are orthogonal projectors: $P_{1\pm}^2 = P_{1\pm}$, $P_{1+}P_{1-} = 0$. The algebra thus decomposes as a vector sum of two copies of $\mathbb{R}$ that ignore each other: $(x_+ P_{1+} + x_- P_{1-})(y_+ P_{1+} + y_- P_{1-}) = x_+ y_+ P_{1+} + x_- y_- P_{1-}$; we have $\mathrm{Cl}_1 \cong \mathbb{R} \oplus \mathbb{R}$. Here is a table including higher dimensions [116]:

$$
\begin{aligned}
&\mathrm{Cl}_1 \cong \mathrm{Mat}(1,\mathbb{R}) \oplus \mathrm{Mat}(1,\mathbb{R})\,, && \mathrm{Cl}_2 \cong \mathrm{Mat}(2,\mathbb{R})\,, \\
&\mathrm{Cl}_3 \cong \mathrm{Mat}(2,\mathbb{C})\,, && \mathrm{Cl}_4 \cong \mathrm{Mat}(2,\mathbb{H})\,, \\
&\mathrm{Cl}_5 \cong \mathrm{Mat}(2,\mathbb{H}) \oplus \mathrm{Mat}(2,\mathbb{H})\,, && \mathrm{Cl}_6 \cong \mathrm{Mat}(4,\mathbb{H})\,, \\
&\mathrm{Cl}_7 \cong \mathrm{Mat}(8,\mathbb{C})\,, && \mathrm{Cl}_8 \cong \mathrm{Mat}(16,\mathbb{R})\,, \\
&\mathrm{Cl}_9 \cong \mathrm{Mat}(16,\mathbb{R}) \oplus \mathrm{Mat}(16,\mathbb{R})\,.
\end{aligned}
\tag{2.1.29b}
$$

In principle, applying (2.1.29) to the abstract symbols γ_μ in the abstract Clifford algebra (2.1.24) yields concrete matrices satisfying $\{\gamma_\mu, \gamma_\nu\} = 2g_{\mu\nu}1$. Since we did not provide the explicit isomorphisms, this is not a practical way to find such matrices; we will see some methods for this in Section 2.2. Still, we can use (2.1.29) to obtain some general properties.

We are interested in gamma matrices with real or complex entries. The entries $\mathrm{Mat}(N,\mathbb{R})$, $\mathrm{Mat}(N,\mathbb{C})$ provide such matrices directly. The elements of $\mathrm{Mat}(N,\mathbb{H})$ can be converted to complex $2N \times 2N$ matrices by applying to each quaternionic entry the famous map

$$
\mathrm{i} \mapsto -\mathrm{i}\sigma_1\,, \qquad \mathrm{j} \mapsto -\mathrm{i}\sigma_2\,, \qquad \mathrm{k} \mapsto -\mathrm{i}\sigma_3\,. \tag{2.1.30}
$$

Applying this trick to every case in (2.1.29) gives the matrix dimension $2^{\lfloor d/2 \rfloor}$, as anticipated in (2.1.4).

Moreover, we see that the choice of gamma matrices is essentially unique. For example, the algebra $\mathrm{Mat}(N,\mathbb{R})$ has only one representation of minimal dimension N, the fundamental. For $\mathrm{Mat}(N,\mathbb{C})$ we have the fundamental and antifundamental, related by $\mathrm{i} \to -\mathrm{i}$ (even if over the reals these are considered equivalent). Taking this into account, (2.1.29) for d = even shows that a Clifford representation (and hence a choice of gamma matrices) is unique up to equivalence, and that for d = odd there are two possibilities. This is consistent with our remarks at the end of Section 2.1.2.

Finally, (2.1.29) shows that in some d we can represent the γ_μ as real matrices; in Lorentzian signature, (2.1.29a) indicates that this is possible in $d = 2, 3, 4, 10, 11$. In Section 2.3.1, we will see a more general type of reality condition, which works in other dimensions as well.

2.1.4 Commutators

There are several ways to compute a commutator such as (2.1.6). The most naive method would be to write the commutator using the definition and to use (2.1.3) many times to shuffle things around. A slightly more sophisticated procedure is to use matrix identities relating commutators and anticommutators. For later use, we give here a few:

$$
[B_1, B_2B_3] - [B_1, B_2]B_3 + B_2[B_1, B_3]\,, \tag{2.1.31a}
$$

$$
[B_1, F_2B_3] = [B_1, F_2]B_3 + F_2[B_1, B_3]\,, \tag{2.1.31b}
$$

$$
[B_1, F_2F_3] = [B_1, F_2]F_3 + F_2[B_1, F_3]\,; \tag{2.1.31c}
$$

$$
[F_1, B_2B_3] = [F_1, B_2]B_3 + B_2[F_1, B_3]\,, \tag{2.1.31d}
$$

$$
\{F_1, F_2B_3\} = \{F_1, F_2\}B_3 - F_2[F_1, B_3]\,, \tag{2.1.31e}
$$

$$
[F_1, F_2F_3] = \{F_1, F_2\}F_3 - F_2\{F_1, F_3\}\,. \tag{2.1.31f}
$$

While the B_a and F_a can be any matrices, a physicist's way to remember these identities is to think of them respectively as bosons and fermions. Indeed, in all (2.1.31) we have an anticommutator between two fermions, and a commutator otherwise; a minus sign appears only when two fermions change order. Equation (2.1.31a) is a Leibniz identity for the "derivation" operator $[B_1, \cdot]$; (2.1.31e) and (2.1.31f) are fermionic analogues. Using these, one can also obtain boson–fermion analogues of the Jacobi identity:

$$[F_1, [B_2, B_3]] = [[F_1, B_2], B_3] + [B_2, [F_1, B_3]], \quad (2.1.32a)$$

$$\{F_1, [F_2, B_3]\} = [\{F_1, F_2\}, B_3] - \{F_2, [F_1, B_3]\}, \quad (2.1.32b)$$

$$[F_1, \{F_2, F_3\}] = [\{F_1, F_2\}, F_3] - [F_2, \{F_1, F_3\}]. \quad (2.1.32c)$$

This method is enough to obtain some simple identities. For example, (2.1.13) can be obtained from (2.1.31f), thinking of the gamma matrices as fermions, and from

$$\gamma_\mu \gamma_\nu = \frac{1}{2}[\gamma_\mu, \gamma_\nu] + \frac{1}{2}\{\gamma_\mu, \gamma_\nu\} = \gamma_{\mu\nu} + g_{\mu\nu}. \quad (2.1.33)$$

From (2.1.13) and also (2.1.31a) and (2.1.31e), we then also obtain (2.1.6).

Computing (anti-)commutators of $\gamma_{\mu_1 \ldots \mu_k}$ with large k, however, quickly gets hard in this way. So we describe here a second method. The idea is to consider the product of gamma matrices we want to compute as an element α of the Clifford algebra $\mathrm{Cl}_{1,d-1}$ (see (2.1.24)). It can then be expanded in the basis (2.1.25):

$$\alpha = \alpha_0 + \alpha^1_\mu \gamma^\mu + \frac{1}{2}\alpha^2_{\mu\nu}\gamma^{\mu\nu} + \ldots = \sum_{k=1} \frac{1}{k!}\alpha^k_{\mu_1 \ldots \mu_k}\gamma^{\mu_1 \ldots \mu_k}. \quad (2.1.34)$$

We then determine the coefficients α^k using Lorentz invariance and some examples. To illustrate the idea, we compute the product $\gamma_{\mu\nu}\gamma^{\rho\sigma}$.

- Because of Lorentz invariance, an expansion of the type (2.1.34) for $\gamma_{\mu\nu}\gamma^{\rho\sigma}$ must have four free indices. For example, we can consider $\gamma_{\mu\nu}{}^{\rho\sigma}$ (the totally antisymmetric $\gamma_{\mu\nu\rho\sigma}$, with two indices raised by the metric). Or we can consider elements of the basis (2.1.25) with fewer indices, leaving the others to be provided by Lorentz-invariant objects. These would be the metric $g_{\mu\nu} = \eta_{\mu\nu}$ and the totally antisymmetric $\epsilon_{\mu_1 \ldots \mu_d}$, but the latter does not appear in the Clifford algebra. Thus we can consider $\gamma_\mu{}^\rho \delta^\sigma_\nu$, with the indices in various positions, or $\delta^\rho_\mu \delta^\sigma_\nu$.

  In general, a product $\gamma_{\mu_1 \ldots \mu_k}\gamma^{\nu_1 \ldots \nu_l}$ will involve elements $\gamma_{\rho_1 \ldots \rho_p}$ with $p = k + l, k + l - 2, \ldots, |k - l|$.
- We know that $\gamma_{\mu\nu}\gamma^{\rho\sigma}$ is antisymmetric under exchange $\mu \leftrightarrow \nu$ and $\rho \leftrightarrow \sigma$. Thus we have to consider properly antisymmetrized combinations of the elements identified so far. In our example, this leaves us with

$$\gamma_{\mu\nu}\gamma^{\rho\sigma} = a_1 \gamma_{\mu\nu}{}^{\rho\sigma} + a_2 \gamma_{[\mu}{}^{[\rho}\delta^{\sigma]}_{\nu]} + a_3 \delta^{[\rho}_{[\mu}\delta^{\sigma]}_{\nu]}. \quad (2.1.35)$$

  If one is interested only in the commutator $[\gamma_{\mu\nu}, \gamma_{\rho\sigma}]$, one can also observe that it is antisymmetric under the simultaneous exchange $(\mu \leftrightarrow \rho, \nu \leftrightarrow \sigma)$; in (2.1.35), only the a_2 term has this property, while the a_1 and a_3 terms are symmetric (and would be relevant for the anticommutator $\{\gamma_{\mu\nu}, \gamma_{\rho\sigma}\}$).
- At this point, we can identify the coefficients a_i in (2.1.35) by examining some well-chosen examples for the free indices. Taking $\mu = 1, \nu = 2, \rho = 1, \sigma = 4$,

the left-hand side gives $\gamma_{12}\gamma^{14} = -\gamma_2\gamma^4$, and the right-hand side $\frac{a_2}{4}\delta^1_1\gamma_2\gamma^4$; hence $a_2 = -4$. One can similarly obtain $a_1 = 1$ and $a_3 = -2$: thus

$$\gamma_{\mu\nu}\gamma^{\rho\sigma} = \gamma_{\mu\nu}{}^{\rho\sigma} - 4\gamma_{[\mu}{}^{[\rho}\delta^{\sigma]}_{\nu]} - 2\delta^{[\rho}_{[\mu}\delta^{\sigma]}_{\nu]} . \tag{2.1.36}$$

Equation (2.1.6) follows from this.

In Section 3.2.1 we will see a third, even more powerful method, which is in a sense an evolution of the second.

Exercise 2.1.1 Show some of the isomorphisms in (2.1.29).

Exercise 2.1.2 Show the following gamma matrix identities:

$$[\gamma_{\mu\nu}, \gamma^{\rho_1\rho_2\rho_3}] = -12\delta^{[\rho_1}_{[\mu}\gamma_{\nu]}{}^{\rho_2\rho_3]} , \tag{2.1.37a}$$

$$\{\gamma_{\mu\nu}, \gamma^{\rho_1\rho_2\rho_3}\} = 2\gamma_{\mu\nu}{}^{\rho_1\rho_2\rho_3} - 12\delta^{[\rho_1}_{\mu}\delta^{\rho_2}_{\nu}\gamma^{\rho_3]} ; \tag{2.1.37b}$$

$$[\gamma_{\mu\nu}, \gamma^{\rho_1\rho_2\rho_3\rho_4}] = -16\delta^{[\rho_1}_{[\mu}\gamma_{\nu]}{}^{\rho_2\rho_3\rho_4]} , \tag{2.1.37c}$$

$$\{\gamma_{\mu\nu}, \gamma^{\rho_1\rho_2\rho_3\rho_4}\} = 2\gamma_{\mu\nu}{}^{\rho_1\rho_2\rho_3\rho_4} - 24\delta^{[\rho_1}_{\mu}\delta^{\rho_2}_{\nu}\gamma^{\rho_3\rho_4]} ; \tag{2.1.37d}$$

$$[\gamma_{\mu\nu\rho}, \gamma^{\sigma_1\sigma_2\sigma_3}] = 2\gamma_{\mu\nu\rho}{}^{\sigma_1\sigma_2\sigma_3} - 36\delta^{[\sigma_1}_{[\mu}\delta^{\sigma_2}_{\nu}\gamma_{\rho]}{}^{\sigma_3]} , \tag{2.1.37e}$$

$$\{\gamma_{\mu\nu\rho}, \gamma^{\sigma_1\sigma_2\sigma_3}\} = 18\delta^{[\sigma_1}_{[\mu}\gamma_{\nu\rho]}{}^{\sigma_2\sigma_3]} - 12\delta^{[\sigma_1}_{[\mu}\delta^{\sigma_2}_{\nu}\delta^{\sigma_3]}_{\rho]} . \tag{2.1.37f}$$

Exercise 2.1.3 Generalize (2.1.37a) and (2.1.37c) to $[\gamma_{\mu\nu}, \gamma^{\rho_1\cdots\rho_k}]$.

2.2 Bases

2.2.1 Creators and annihilators

In this subsection, we will see the so-called *creator basis* of gamma matrices, which will help us gain some intuition on the space of spinors. This time we start with the Euclidean case.

$d = 2$ Euclidean dimensions

It is useful to work in complex coordinates, defining $z \equiv x + \mathrm{i}y$, $\bar{z} = x - \mathrm{i}y$. In these coordinates, the line element $\mathrm{d}x^2 + \mathrm{d}y^2$ is $\mathrm{d}z\mathrm{d}\bar{z}$. Thus the metric reads $\frac{1}{2}\begin{pmatrix} 0 & 1 \\ 1 & 0 \end{pmatrix}$; in other words, $g_{z\bar{z}} = \frac{1}{2}$. Equation (2.1.3) then reduces to

$$\{\gamma_z, \gamma_{\bar{z}}\} = 1 , \qquad \gamma_z^2 = \gamma_{\bar{z}}^2 = 0 . \tag{2.2.1}$$

This relation is familiar from the quantization of a free fermion. It is often represented by a two-state space, generated by a state $|+\rangle$ and a state $|-\rangle$, following a notation familiar from quantum mechanics. We can take γ_z to be a lowering operator, and $\gamma_z^\dagger$ to be a raising operator:

$$\gamma_z|+\rangle = |-\rangle , \qquad \gamma_{\bar{z}}|-\rangle = |+\rangle . \tag{2.2.2}$$

From (2.2.1), it also follows $\gamma_z|-\rangle = 0$, $\gamma_{\bar{z}}|+\rangle = 0$. We can translate all this in matrices by taking $|+\rangle = \binom{1}{0}$, $|-\rangle = \binom{0}{1}$:

$$\gamma_z = b \equiv \frac{1}{2}(\sigma_1 - \mathrm{i}\sigma_2) = \begin{pmatrix} 0 & 0 \\ 1 & 0 \end{pmatrix}, \qquad \gamma_{\bar{z}} = b^\dagger \equiv \frac{1}{2}(\sigma_1 + \mathrm{i}\sigma_2) = \begin{pmatrix} 0 & 1 \\ 0 & 0 \end{pmatrix}, \tag{2.2.3}$$

with a notation meant to evoke spinorial creators and annihilators. Since $\gamma_z = \frac{1}{2}(\gamma_x - \mathrm{i}\gamma_y)$, we can easily go back to real indices:

$$\gamma_1 = \sigma_1, \qquad \gamma_2 = \sigma_2. \tag{2.2.4}$$

This is related to (2.1.27) by Wick rotation and relabeling.

d = even Euclidean dimensions

For $d > 2$ but still even, we can similarly define

$$z^1 \equiv x_1 + \mathrm{i}x_{d/2+1}, \qquad z^2 \equiv x_2 + \mathrm{i}x_{d/2+2}, \qquad \ldots \qquad z^{d/2} \equiv x_{d/2} + \mathrm{i}x_d \tag{2.2.5}$$

and their conjugates $\overline{z^i} \equiv z^{\bar{i}}$. The line element is $\sum_{i=1}^{d/2} \mathrm{d}z^i \mathrm{d}z^{\bar{i}}$; so $g_{i\bar{j}} = \frac{1}{2}\delta_{i\bar{j}}$, leading to

$$\{\gamma_i, \gamma_{\bar{j}}\} = \delta_{i\bar{j}}, \qquad \{\gamma_i, \gamma_j\} = \{\gamma_{\bar{i}}, \gamma_{\bar{j}}\} = 0 \tag{2.2.6}$$

with $\gamma_i \equiv \gamma_{z^i}$ for short. This again looks familiar, this time from $d/2$ fermions. This suggests to take a representation space made of $d/2$ variables s_i, each assuming $\pm$ values:

$$|s_1 \ldots s_{d/2}\rangle, \qquad s_i \in \{+,-\}. \tag{2.2.7}$$

For example, for $d = 4$ we have

$$|++\rangle, \qquad |+-\rangle, \qquad |-+\rangle, \qquad |--\rangle. \tag{2.2.8}$$

In general, the number of vectors in (2.2.7) is $2^{d/2}$, in agreement with (2.1.4).

In (2.2.8), we are borrowing notation from quantum mechanics. We are realizing $\mathbb{C}^4$ as a tensor product of two copies $\mathbb{C}^2$, each with a basis $|\pm\rangle$. Given two vector spaces V and W with bases $\{v_i\}$ and $\{w_j\}$, their tensor product $V \otimes W$ is a vector space with a basis $\{v_i \otimes w_j\}$. A matrix $M_1 \otimes M_2$ acts as $(M_1 \otimes M_2) v_i \otimes w_j = (M_1 v_i) \otimes (M_2 w_j)$, as in quantum mechanics when an operator acts on several separate quantum systems. An alternative, basis-free definition of $V \otimes W$ is as the quotient $(V \times W)/\sim$ of the cartesian product (the space of pairs) by the equivalence relation $(\lambda v, w) \sim (v, \lambda w)$ for any $\lambda \in \mathbb{C}$.

We now want to guess a choice of gamma matrices acting on (2.2.8). Following (2.2.3), a possibility might seem to be writing each γ_i as a copy of b acting on the ith sign s_i in (2.2.7):

$$\gamma_{z^1} \stackrel{?}{=} b \otimes 1 \otimes \ldots \otimes 1, \qquad \gamma_{z^2} \stackrel{?}{=} 1 \otimes b \otimes 1 \ldots \otimes 1. \tag{2.2.9}$$

In this notation, the ith matrix is meant to act on the ith sign s_i. The notation establishes that 1 is the 2×2 identity matrix, and the $\gamma_{\bar{i}}$ their complex conjugates. However, in (2.2.9) the γ_i would commute with each other, rather than anticommute as (2.2.6) prescribes. Fortunately, there is a way to cure this naive guess:

$$
\begin{aligned}
\gamma_{z^1} &= b \otimes 1 \otimes 1 \otimes \ldots \otimes 1\,, \\
\gamma_{z^2} &= \sigma_3 \otimes b \otimes 1 \otimes \ldots \otimes 1\,, \\
&\ \vdots \\
\gamma_{z^{d/2}} &= \sigma_3 \otimes \sigma_3 \otimes \sigma_3 \otimes \ldots \otimes b\,,
\end{aligned}
\tag{2.2.10}
$$

and $\gamma_{\bar{i}} = \gamma_{\bar{z}^{\bar{i}}} = \gamma^\dagger_{z^i}$. This now does realize (2.2.6), because the third Pauli matrix σ_3 anticommutes with both b and $b^\dagger$. We can go back to real indices using (2.2.5):

$$
\begin{aligned}
\gamma_1 &= \sigma_1 \otimes 1 \otimes 1 \otimes \ldots \otimes 1\,, & \gamma_{d/2+1} &= \sigma_2 \otimes 1 \otimes 1 \otimes \ldots \otimes 1\,, \\
\gamma_2 &= \sigma_3 \otimes \sigma_1 \otimes 1 \otimes \ldots \otimes 1\,, & \gamma_{d/2+2} &= \sigma_3 \otimes \sigma_2 \otimes 1 \otimes \ldots \otimes 1\,, \\
&\ \vdots \\
\gamma_{d/2} &= \sigma_3 \otimes \sigma_3 \otimes \sigma_3 \otimes \ldots \otimes \sigma_1\,, & \gamma_d &= \sigma_3 \otimes \sigma_3 \otimes \sigma_3 \otimes \ldots \otimes \sigma_2\,.
\end{aligned}
\tag{2.2.11}
$$

The chirality matrix (2.1.17) can be taken to be, choosing c appropriately,

$$
\gamma = \sigma_3 \otimes \sigma_3 \otimes \ldots \otimes \sigma_3\,. \tag{2.2.12}
$$

So we have realized the space S_d of spinors in even dimensions as a tensor product of $d/2$ copies of the space S_2 of spinors in two dimensions: $S_d \cong S_2 \otimes \ldots \otimes S_2$.

d = even Lorentzian dimensions

To adapt our discussion in this section to the Lorentzian case, again we Wick-rotate and relabel:

$$
\begin{aligned}
\gamma^0 &= \mathrm{i}\sigma_2 \otimes 1 \otimes 1 \otimes \ldots \otimes 1\,, & \gamma_1 &= \sigma_1 \otimes 1 \otimes 1 \otimes \ldots \otimes 1\,, \\
\gamma_2 &= \sigma_3 \otimes \sigma_1 \otimes 1 \otimes \ldots \otimes 1\,, & \gamma_{d/2+1} &= \sigma_3 \otimes \sigma_2 \otimes 1 \otimes \ldots \otimes 1\,, \\
&\ \vdots \\
\gamma_{d/2} &= \sigma_3 \otimes \sigma_3 \otimes \sigma_3 \otimes \ldots \otimes \sigma_1\,, & \gamma_{d-1} &= \sigma_3 \otimes \sigma_3 \otimes \sigma_3 \otimes \ldots \otimes \sigma_2\,.
\end{aligned}
\tag{2.2.13}
$$

This is equivalent to defining light-cone coordinates $x^\pm \equiv x^1 \pm x^0$, and then $z^i \equiv x_i + \mathrm{i}x_{d/2+i-1}$ for $i \geq 2$. In this basis, the Clifford algebra is then the same as (2.2.6), with $\gamma_{z^1} \to \gamma_{x^+}$ and $\gamma_{\bar{z}^1} \to \gamma_{x^-}$. The chiral γ again reads (2.2.12).

d = odd dimensions

It is also easy to extend the discussion to odd d. As we mentioned in Section 2.1.2, we can simply take the basis (2.2.11) for $d-1$ and add to it the chiral γ in (2.2.12):

$$
\gamma_\mu = \gamma_\mu^{(d)}\,, \qquad \gamma_d = \gamma\,. \tag{2.2.14}
$$

The product of all gammas, which in even dimension would give the chirality matrix (2.1.17), now becomes proportional to the identity, as we anticipated.

2.2.2 Quaternions and octonions

We will now see a couple of notable sets of gamma matrices based on important algebraic structures.

Quaternions and gamma matrices

In $d = 3$ Euclidean dimensions, we can use the Pauli matrices σ_i, which famously satisfy

$$\sigma_i \sigma_j = \delta_{ij} + \mathrm{i}\epsilon_{ijk}\sigma_k \,, \tag{2.2.15}$$

from which it follows that $\{\sigma_i, \sigma_j\} = 2\delta_{ij}$, the Euclidean version of (2.1.3). (This is also what one gets if one specializes the Euclidean basis of the previous subsection to $d = 3$.) The infinitesimal spinor transformation matrix in this case reads $\lambda = \frac{1}{2}\lambda_{mn}\gamma^{mn} = \frac{\mathrm{i}}{2}\lambda_{ij}\epsilon^{ijk}\sigma_k$. The σ_i are all traceless and Hermitian, so λ is a general element of the Lie algebra su(2). Exponentiating this, one recovers the well-known fact that spinor transformations form the group SU(2); in other words, the spin group (defined after (2.1.9)) is Spin(3) $\cong$ SU(2). The 2 : 1 homomorphism SU(2) $\to$ SO(3), and is a Euclidean counterpart of (2.1.10).

For $d = 3$ Lorentzian dimensions, one can of course simply multiply one of the σ_i by an i, so that it now squares to -1. A natural choice is to do so for σ_2, which is purely imaginary, so that we end up with a real basis:

$$\gamma^0 = \mathrm{i}\sigma_2 \,, \qquad \gamma^1 = \sigma_1 \,, \qquad \gamma^2 = \sigma_3 \,. \tag{2.2.16}$$

(We had already predicted the existence of a real basis from (2.1.29a).) This time $\gamma^{\mu\nu} = \epsilon^{\mu\nu}{}_{\rho}\gamma^{\rho}$, and it follows that λ is a linear combination of the γ^{μ}. Since they are all real and traceless, we see that $\lambda \in \mathrm{sl}(2,\mathbb{R})$. The spin group is then SL(2, $\mathbb{R}$), which again has a 2:1 homomorphism to $\mathrm{SO}^{+}(1,2)$.

Octonions

As we already mentioned in (2.1.23), (2.2.15) is also related to quaternions.[6] Quaternions are a *division algebra*, namely an algebra such that if $a \cdot b = 0$, then either $a = 0$ or $b = 0$. Other well-known division algebras are $\mathbb{R}$ and $\mathbb{C}$. All three are also *normed* division algebra: they have a Euclidean norm q compatible with the product, $q(a \cdot b) = q(a)q(b)$. We defined algebras in Section 2.1.3, including a condition on associativity of the product. If we drop this requirement but we keep the division property, other examples exist. If we also insist on the normed property, there is only one more example: the so-called *octonions* $\mathbb{O}$. In other words, the only normed division algebras are $\mathbb{R}$, $\mathbb{C}$, $\mathbb{H}$, and $\mathbb{O}$. (There is also a more exotic property, that of *alternativity*, that singles these four out.)

$\mathbb{O}$ has dimension eight, and is defined as the real Span$\{1, i_1, \ldots, i_7\}$, where i_m are seven imaginary units (rather than one for $\mathbb{C}$ and three for $\mathbb{H}$) with multiplication

$$i_m \cdot i_n = -\delta_{mn} + \phi_{mnp} i_p \,, \tag{2.2.17}$$

similar to relations (2.1.23) for the quaternionic units; ϕ is completely antisymmetric and has nonzero entries:

$$\phi_{123} = \phi_{435} = \phi_{624} = \phi_{516} = \phi_{471} = \phi_{572} = \phi_{673} = 1 \,, \tag{2.2.18}$$

with all the other entries not related to these by antisymmetry being zero. These are summarized in Figure 2.1. This particular choice might seem mysterious right now;

[6] With opposite signature, we could represent the gamma matrices directly with $\mathbb{H}$ itself. The full Clifford algebra would then be isomorphic to $\mathbb{H} \oplus \mathbb{H}$.

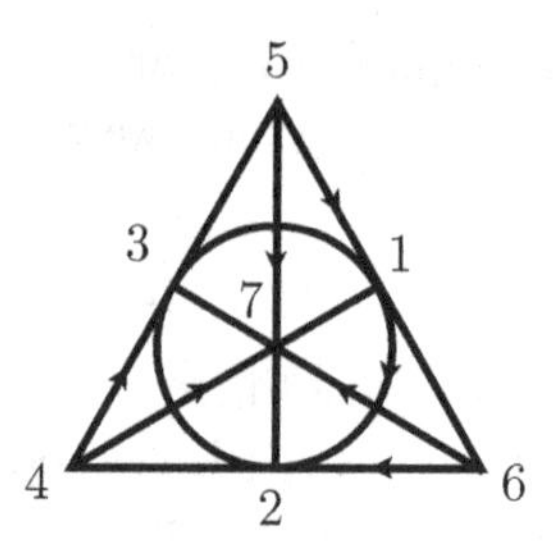

Figure 2.1 Octonion multiplication table.

it will look much more natural later on, when we will see its relation to the G_2 and SU(3) Lie groups.

Given the role of the other normed division algebras $\mathbb{R}$, $\mathbb{C}$, and $\mathbb{H}$ in the study of Clifford algebras, one can try to use the octonions $\mathbb{O}$ as well to define gamma matrices. Indeed, (2.2.17) already implies that $\{i_i, i_j\} = -\delta_{ij}$; so the octonionic imaginary units give a representation of the Clifford algebra whose metric is minus the Euclidean one. To obtain a representation for the usual Euclidean metric, we can proceed as follows. On a generic quaternion, left multiplication by an imaginary unit $i_m : x_8 1 + x_n i_n \mapsto x_8 i_m + x_n(-\delta_{mn} + \phi_{mnp} i_p)$ induces an action on the coefficients $x_1, \ldots x_8$ that can be summarized by a real 8×8 matrix $\hat{\gamma}_m$. These real matrices satisfy the negative-signature Clifford algebra $\{\hat{\gamma}_m, \hat{\gamma}_n\} = -2\delta_{mn}$, again because of (2.2.17); now define $\gamma_m = -\mathrm{i}\hat{\gamma}_m$ to obtain a set of seven 8×8 imaginary matrices that satisfy the Euclidean $d = 7$ Clifford algebra. This can be summarized by writing

$$\gamma^m_{\alpha\beta} = \mathrm{i}\phi^m{}_{\alpha\beta} + 2\mathrm{i}\delta^m_{[\alpha}\delta^8_{\beta]} \qquad \text{(Eucl. } d = 7\text{)}. \tag{2.2.19}$$

Here $\alpha, \beta = 1, \ldots, 8$ are spinorial indices, and ϕ has been extended by declaring it to be zero whenever one of its indices is 8.

The basis (2.2.19) has the property that all γ^m are purely imaginary. It follows that there is also a purely imaginary basis in $d = 6$: it is enough to discard γ^7. (The minimal dimension $2^{\lfloor d/2 \rfloor}$ for a Clifford representation is equal to 8 both for $d = 7$ and $d = 6$.) We will see later that there is no basis in either $d = 7$ or $d = 6$ where they are all real.

2.2.3 Tensor products

We will now see a simple way to produce gamma matrices in $d = d_1 + d_2$ once one has a basis in both d_1 and d_2 dimensions, and correspondingly to write the space of spinors in $d_1 + d_2$ dimensions as a tensor product: $S_d \cong S_{d_1} \otimes S_{d_2}$. This is a generalization of the procedure we followed to get the creator basis in Section 2.2.1, which took us, for example, to (2.2.11).

One of the d_i is even

Rather than writing this out in full generality, we will present the idea in some examples. First we consider the important case $d_1 = 4$ and $d_2 = 6$, in Lorentzian and Euclidean signature respectively, with bases γ_μ and γ_m.

We cannot quite take $\gamma_\mu \otimes 1$ and $1 \otimes \gamma_m$: these would satisfy the Clifford algebra for the indices μ and m separately, but they would not anticommute, as the Clifford algebra for mixed indices would prescribe. This is a similar problem as the one in (2.2.9), and it has a similar solution:

$$\Gamma_\mu = \gamma_\mu \otimes 1 \,, \qquad \Gamma_{m+3} = \gamma \otimes \gamma_m \,, \qquad (d = 10) \tag{2.2.20}$$

γ being the chiral matrix in $d = 4$. We noted at the end of the previous subsection that there exists a basis in $d = 6$ Euclidean dimensions where all the gamma matrices are purely imaginary. We will see in (2.2.25) a real basis in $d = 4$ Lorentzian dimensions; then the chiral matrix for these, $\gamma = \mathrm{i}\gamma^{0123}$, is purely imaginary. With this choice, (2.2.20) is a real basis in $d = 10$.

As another example, notice that our Lorentzian creator basis (2.2.13) in any d can be thought of as a tensor product of a $d_1 = 2$ Lorentzian basis $\{\gamma_0 = \mathrm{i}\sigma_2, \gamma_1 = \sigma_1\}$ and of a $d_2 = d - 2$ Euclidean creator basis (2.2.11):

$$\Gamma_\mu = \gamma_\mu \otimes 1 \,, \qquad \Gamma_{m+1} = \sigma_3 \otimes \gamma_m \qquad (\text{any Lor. } d). \tag{2.2.21}$$

Because of this, several properties of spinors in d Lorentzian dimensions parallel those of spinors in $d - 2$ Euclidean dimensions.

In (2.2.20), we could have used a chiral matrix in $d_2 = 6$ in the Γ_μ, rather than using the chirality matrix in $d_1 = 4$ in the Γ_m. When either d_1 or d_2 is odd, only one choice is available. When both d_1 and d_2 are odd, one cannot proceed as in (2.2.20).

Chiral basis in $d = 4$ Lorentzian dimensions

Take $d_1 = 1$, $d_2 = 3$, in Lorentzian and Euclidean signature respectively. The single gamma matrix in $d_1 = 1$ can be represented by a 1×1 matrix i, and the $d_2 = 3$ gamma matrices can be represented by the Pauli matrices σ_i. There is no chirality matrix in either d_1 or d_2. To resolve this issue, we can proceed as follows:[7]

$$\gamma^0 = \mathrm{i}\sigma^1 \otimes 1_2 = \mathrm{i}\begin{pmatrix} 0 & 1_2 \\ 1_2 & 0 \end{pmatrix}, \quad \gamma^i = \sigma^2 \otimes \sigma^i = \mathrm{i}\begin{pmatrix} 0 & -\sigma^i \\ \sigma^i & 0 \end{pmatrix} \qquad (d = 4). \tag{2.2.22}$$

We have introduced an "auxiliary space" along which the matrices read either σ^1 or σ^2, which makes the gamma's anticommute; it is auxiliary in the sense that it was not present either in the original $d_1 = 1$ or $d_2 = 3$ gamma matrices. The basis (2.2.22) is called *chiral basis* in four dimensions, since the chirality matrix reads $\gamma = \mathrm{i}\gamma^0\gamma^1\gamma^2\gamma^3 = \sigma_3 \otimes 1_2 = \begin{pmatrix} 1_2 & 0 \\ 0 & -1_2 \end{pmatrix}$. In this basis, we also find

$$\lambda = \begin{pmatrix} \lambda_i^+ \sigma_i & 0 \\ 0 & \lambda_i^- \sigma_i \end{pmatrix}, \qquad \lambda_i^\pm \equiv \frac{\mathrm{i}}{2}\epsilon_{ijk}\lambda^{jk} \mp \lambda_{0i} \,. \tag{2.2.23}$$

We saw that $\mathrm{i}\sigma_i$ form a basis for the Lie algebra su(2); the block $\lambda_i^+\sigma^i$ is a linear combination of them but with complex coefficients, so by definition it spans the *complexification* $\mathrm{su}(2)_\mathbb{C}$, which is isomorphic to $\mathrm{sl}(2,\mathbb{C})$. Alternatively, we saw that (2.2.16) is a basis for $\mathrm{sl}(2,\mathbb{R})$, so $\lambda_i^+\sigma^i \in \mathrm{sl}(2,\mathbb{R})_\mathbb{C} \cong \mathrm{sl}(2,\mathbb{C})$. The block $\lambda_i^-\sigma_i$ in (2.2.23) equals $(\lambda_i^+\sigma_i)^\dagger$, so it does not add any new information. So in fact the λ form a Lie algebra isomorphic to $\mathrm{sl}(2,\mathbb{C})$; the spin group is $\mathrm{SL}(2,\mathbb{C})$.

[7] In quantum field theory books, a $(+ - - -)$ signature is often thought to be more convenient, and the basis (2.2.22) will appear multiplied by an overall i.

There is an amusing alternative way of seeing this $\mathrm{SL}(2,\mathbb{C})$. The space of all possible directions in $\mathbb{R}^3$ is a copy of S^2, which for obvious reasons is called the *celestial sphere*. Performing a Lorentz transformation acts on this sphere. (Think, for example, of performing a boost and watching the positions of the fixed stars change.) If we put on the S^2 a coordinate z with a point at infinity (identifying it with a Riemann sphere $\mathbb{C}\cup\{\infty\}$), this action turns out to be a Möbius transformation

$$z \mapsto \frac{az+b}{cz+d} \tag{2.2.24}$$

for some $a, b, c, d \in \mathbb{C}$ such that $ad-bc = 1$. These transformations form a group that is isomorphic to $\mathrm{SL}(2,\mathbb{C})$ (up to a $\mathbb{Z}_2$ identifying elements with minus themselves). Recall that this is the same action we saw in (1.2.38) for the $\mathrm{SL}(2,\mathbb{R})$ symmetry on the axiodilaton τ.

Other cases with both d_i = odd

Alternatively to (2.2.22), again in $d = 4$ Lorentzian dimensions, one can take $d_1 = 3$, $d_2 = 1$, and use the real basis (2.2.16) in $d_1 = 3$ to produce a real basis:

$$\gamma^\mu = \gamma^\mu_{(3)} \otimes \sigma_1\,, \qquad \gamma^4 = 1 \otimes \sigma_3 \qquad (d=4). \tag{2.2.25}$$

For a more challenging example, we return to $d = 10$, but now we combine gamma matrices γ_μ and γ_m in $d_1 = 3$ and $d_2 = 7$. Again we cannot use the chirality matrix trick in (2.2.20); so we have to introduce an auxiliary space, which we again write first:

$$\Gamma_\mu = \sigma_1 \otimes \gamma_\mu \otimes 1\,, \qquad \Gamma_{m+2} = \sigma_2 \otimes 1 \otimes \gamma_m \qquad (d=10). \tag{2.2.26}$$

If we take the γ_μ to the real basis in (2.2.16) and γ_m to be the purely imaginary basis in (2.2.19), (2.2.26) is again a real basis, like (2.2.20). The chirality matrix reads $\gamma = -\sigma_3 \otimes 1 \otimes 1$. We can again evaluate λ; the block acting on chirality + spinors reads

$$\lambda|_{\text{++block}} = -\gamma_\mu \otimes \left(\frac{1}{2}\epsilon^{\mu\nu\rho}\lambda_{\nu\rho}1 + \mathrm{i}\lambda^{\mu m}\gamma_m\right) + \frac{1}{2}\lambda_{mn} 1 \otimes \gamma^{mn}\,. \tag{2.2.27}$$

The first term of this is similar to $\lambda_i^+\sigma_i$ in (2.2.23), and can be interpreted as $\mathbb{O}$-valued traceless matrices. Unlike with $\mathbb{R}$ and $\mathbb{C}$, such matrices don't close under commutator, but generate the second term $1\otimes\gamma^{mn}$. One can take (2.2.27) as a *definition* of $\mathrm{sl}(2,\mathbb{O})$, although this feels less compelling than the $\mathrm{sl}(2,\mathbb{R})$ and $\mathrm{sl}(2,\mathbb{C})$ we saw for three and four dimensions after (2.2.16) and (2.2.23).

As a final application, one can define a basis in eight Euclidean dimensions by combining gamma matrices in $d_1 = 7$ and $d_2 = 1$, similar to (2.2.22):

$$\Gamma_m = \sigma_2 \otimes \gamma_m \quad (m = 1, \ldots, 7)\,, \qquad \Gamma_8 = \sigma_1 \otimes 1 \qquad (d=8); \tag{2.2.28}$$

these are real if the γ_m are taken to be the purely imaginary basis defined in (2.2.19). (Flipping $\sigma_1 \leftrightarrow \sigma_2$ in (2.2.28), one instead obtains a purely imaginary basis.)

Exercise 2.2.1 Compute c in (2.1.17) for (2.2.12).

Exercise 2.2.2 Use (2.2.18) or Figure 2.1 to evaluate $\phi^{mnp}\phi_{mnp}$ and $\phi_m{}^{pq}\phi_{npq}$.

Exercise 2.2.3 With the same method, check the identity

$$\phi^{mnr}\phi_{pqr} = 2\delta_p^{[m}\delta_q^{n]} - \tilde{\phi}^{mn}{}_{pq}\,, \qquad \tilde{\phi}_{mnpq} \equiv \frac{1}{6}\epsilon^{mnpqrst}\phi_{rst}\,. \tag{2.2.29}$$

Exercise 2.2.4 Finally check

$$\phi_{mpq}\phi_{nrs}\tilde{\phi}^{prqs} = 24\delta_{mn}\,. \tag{2.2.30}$$

A long list of identities similar to (2.2.29) and (2.2.30) are given in [117, sec. 7].

Exercise 2.2.5 Check that (2.2.19) satisfy the Clifford algebra by using (2.2.29).

2.3 Reality

2.3.1 Conjugation

We now would like to define a real spinor. The naive definition would be

$$\zeta = \zeta^*\,, \tag{2.3.1}$$

but in general this is spoiled by a Lorentz transformation. Already at the infinitesimal level (2.1.7), the transformed spinor $\delta\zeta = -\frac{1}{2}\lambda\zeta$ is not real unless the $\gamma^{\mu\nu}$ in λ are real. Thus saying that a spinor is real is a frame-dependent statement and makes no physical sense.

As we commented after (2.1.29), in some dimensions one can in fact take the gamma matrices to be real, and (2.3.1) has no problem. But there is an other possibility: as we saw in Section 2.1.3, for d = even an irreducible representation of the Clifford algebra is unique up to change of basis in the space of spinors. Given a representation γ_μ, the conjugate matrices γ_μ^* are also a representation, as one can see by taking the complex conjugate of (2.1.3). Thus there exists a B_+ such that

$$B_+\gamma_\mu^* B_+^{-1} = \gamma_\mu\,. \tag{2.3.2a}$$

The same logic holds for $-\gamma_\mu^*$. This is realized by $B_- \equiv \gamma B_+$:

$$B_-\gamma_\mu^* B_-^{-1} = -\gamma_\mu\,. \tag{2.3.2b}$$

With either $B = B_+$ or $B = B_-$, we have

$$B\gamma_{\mu\nu}^* B^{-1} = \gamma_{\mu\nu}\,. \tag{2.3.3}$$

Now the *Majorana* condition

$$\zeta = \zeta^c \equiv B\zeta^* \tag{2.3.4}$$

is no longer spoiled by a Lorentz transformation:

$$\delta\zeta = \delta(\zeta^c) = -\frac{1}{2}\lambda B\zeta^* \overset{(2.3.3)}{=} -\frac{1}{2}B\lambda^*\zeta^* = (\delta\zeta)^c\,. \tag{2.3.5}$$

So (2.3.4) is compatible with Lorentz, while in general (2.3.1) is not.

Consistency

However, for (2.3.4) to be consistent, we have to make sure that the operation $(\ )^c$ squares to the identity. Taking its conjugate and multiplying by B, we obtain $\zeta = BB^*\zeta$. From (2.3.2), we see that BB^* commutes with the γ_μ, and thus commutes with a representation of the Lorentz group (including parities; recall (2.1.12)). By Schur's lemma, it should be proportional to the identity, $BB^* = b1$. However, this does not mean that $b = 1$; so (2.3.4) might still be inconsistent after all. We need to check explicitly in each dimension whether $b = 1$ for $B = B_+$ or B_-.

We can do so in the creator basis (2.2.11), which has a useful feature: all gamma matrices are either real or purely imaginary. For example, in the $d =$ even Lorentzian case $\gamma_0, \gamma_1, \ldots \gamma_{d/2}$ are real, while $\gamma_{d/2+1}, \ldots \gamma_{d-1}$ are purely imaginary. This makes it is easy to find $B_\pm$ as the product of either all real or all purely imaginary gamma's: we get

$$\begin{aligned} d = 4k: &\quad B_+ = \gamma_0\gamma_1 \ldots \gamma_{d/2}\,, \qquad B_- = \gamma_{d/2+1} \ldots \gamma_{d-1}\,;\\ d = 4k+1: &\quad B_- = \gamma_{(d+1)/2} \ldots \gamma_{d-2}\,;\\ d = 4k+2: &\quad B_+ = \gamma_{d/2+1} \ldots \gamma_{d-1}\,, \qquad B_- = \gamma_0\gamma_1 \ldots \gamma_{d/2}\,;\\ d = 4k+3: &\quad B_+ = \gamma_{(d+1)/2} \ldots \gamma_{d-2}\,. \end{aligned} \tag{2.3.6a}$$

(For $d =$ odd, only one of $B_\pm$ can be found; this is related to the absence of the chirality matrix.) We can now check directly:

$$B_+B_+^* = 1 \text{ for } d = 2, 3, 4 \bmod 8\,; \qquad B_-B_-^* = 1 \text{ for } d = 2, 8, 9 \bmod 8\,. \tag{2.3.6b}$$

Thus the Majorana condition (2.3.4) can only be imposed when

$$d = 2, 3, 4, 8, 9 \bmod 8\,, \qquad \text{(Lorentzian)}. \tag{2.3.6c}$$

We are free to change any B in (2.3.6) by an overall phase if we find it convenient; this does not spoil (2.3.2) or (2.3.6b).

For the Euclidean case, the computation is similar, and gives:

$$\begin{aligned} d = 4k: &\quad B_+ = \gamma_{d/2+1} \ldots \gamma_d\,, \qquad B_- = \gamma_1 \ldots \gamma_{d/2}\,;\\ d = 4k+1: &\quad B_+ = \gamma_{(d+1)/2} \ldots \gamma_{d-1}\,;\\ d = 4k+2: &\quad B_+ = \gamma_1 \ldots \gamma_{d/2}\,, \qquad B_- = \gamma_{d/2+1} \ldots \gamma_d\,;\\ d = 4k+3: &\quad B_- = \gamma_{(d+1)/2} \ldots \gamma_{d-1}\,. \end{aligned} \tag{2.3.7a}$$

We then have

$$B_+B_+^* = 1 \text{ for } d = 1, 2, 8 \bmod 8\,; \qquad B_-B_-^* = 1 \text{ for } d = 6, 7, 8 \bmod 8 \tag{2.3.7b}$$

and thus (2.3.4) can be imposed when

$$d = 1, 2, 6, 7, 8 \bmod 8\,, \qquad \text{(Euclidean)}. \tag{2.3.7c}$$

We summarize all this in Tables 2.1 and 2.2. The B such that $BB^* = +1$ are those that can be used to impose a Majorana condition. We see that it *cannot* be imposed for $d = 5, 6, 7$ in Lorentzian signature, and in $d = 3, 4, 5$ in Euclidean signature. Indeed, the results for Euclidean d match with those for Lorentzian $d+2$, compatible with our comment following (2.2.21). The value of B^2 is not shown, because one can always arrange $B^2 = 1$ by multiplying our B in (2.3.6) and (2.3.7) by an i if needed.

Table 2.1. Majorana matrices in various Lorentzian dimensions.

Lor. d	2	3	4	5	6	7	8	9
$B_+B_+^*$	+	+	+		−	−	−	
$B_-B_-^*$	+		−	−	−		+	+

Table 2.2. Majorana matrices in various Euclidean dimensions.

Eucl. d	1	2	3	4	5	6	7	8
$B_+B_+^*$	+	+		−	−	−		+
$B_-B_-^*$		−	−	−		+	+	+

When the Majorana condition can be imposed, there is a basis where the γ_μ are real (if B_+ works) or purely imaginary (if B_- works). Indeed, for the Euclidean case we saw in (2.2.19) a purely imaginary basis for $d = 7$, commenting there that it also implied the existence of such a basis in $d = 6$. We then saw in (2.2.28) a real basis for $d = 8$, commenting there that it could also be made purely imaginary. A real basis for $d = 2$ is $\gamma_1 = \sigma_1$, $\gamma_2 = \sigma_3$; a real basis for $d = 1$ is obvious. Together, these remarks reproduce all the possibilities in (2.3.7b) or Table 2.2.

When (2.3.7b) is not satisfied and the Majorana condition cannot be imposed, then $BB^* = -1$, and it follows that $(\zeta^c)^c = -\zeta$. In other words, we have two spinors such that

$$\zeta^c = \zeta' , \qquad (\zeta')^c = -\zeta . \tag{2.3.8}$$

This is called a *symplectic-Majorana pair*.

Majorana condition and chirality

If $\zeta = \zeta_+$ is a chiral spinor, the conjugate $(\zeta_+)^c$ is also chiral.[8] Indeed, since $\gamma\zeta_+ = \zeta_+$, for the Lorentzian case

$$\gamma(\zeta_+)^c = \gamma B(\zeta_+)^* \overset{(2.3.6)}{=} -(-1)^{d/2} B\gamma(\zeta_+)^* = -(-1)^{d/2} B(\gamma\zeta_+)^* = -(-1)^{d/2}(\zeta_+)^c ; \tag{2.3.9}$$

Euclidean signature differs by a sign. (We have used the reality of γ in (2.2.12)). So

$$(\zeta_+)^c \text{ has chirality } \begin{cases} -(-1)^{d/2} & \text{(Lorentzian);} \\ (-1)^{d/2} & \text{(Euclidean).} \end{cases} \tag{2.3.10}$$

In other words, conjugation $(\cdot)^c$ keeps the same chirality in $d = 2, 6 \bmod 8$ in Lorentzian signature, and in $d = 0, 4 \bmod 8$ in Euclidean signature, and changes it otherwise.

In the dimensions where conjugation does not change chirality, it is possible to impose both the Majorana (2.3.4) and the Weyl condition together. Naturally these are then called Majorana–Weyl spinors.

[8] Dropping the parenthesis and writing ζ_+^c can sometimes be ambiguous. If $\zeta_+ = P_+\zeta$ by (2.1.21), then $(\zeta_+)^c = B(P_+\zeta)^*$, while $(\zeta^c)_+ = P_+\zeta^c = P_+B\zeta^*$.

In the dimensions where conjugation does change chirality, such as in $d = 4$ in Lorentzian signature, we can parameterize Majorana spinors with a Weyl one as $\zeta = \zeta_+ + (\zeta_+)^c$, or vice versa a Weyl spinor with a Majorana one as $\zeta_+ = \frac{1}{2}(1+\gamma)\zeta$.

2.3.2 Inner products

Given two spinors ζ_1, ζ_2, one might want to define their inner product as $\zeta_1^\dagger \zeta_2$, inspired by the Hermitian inner product in $\mathbb{C}^d$. However, these definitions would in general not be Lorentz invariant: for example, we see that

$$\delta(\zeta_1^\dagger \zeta_2) = -\frac{1}{2}\zeta_1^\dagger(\lambda^\dagger + \lambda)\zeta_2, \tag{2.3.11}$$

which need not be zero if $\gamma_\mu^\dagger \neq \gamma_\mu$, i.e. if the gamma matrices are not Hermitian.

To overcome this issue, one might be tempted to take all the gamma matrices to be Hermitian, $\gamma_\mu^\dagger = \gamma_\mu$. This is possible in Euclidean signature, and in fact the creator basis (2.2.11) already has this feature; so in the Euclidean case (2.3.11) is zero, and $\eta_1^\dagger \eta_2$ is a perfectly fine SO(d)-invariant inner product.

For Lorentzian signature, however, (2.1.3) implies $\gamma_0^2 = -1$; so the eigenvalues of γ_0 are $\pm$i, and γ_0 cannot possibly be Hermitian. The next best thing is to have γ_0 anti-Hermitian, and the others Hermitian; this is also what one obtains by Wick-rotating one of the gamma's in the Euclidean creator basis (2.2.11). One then has

$$\gamma_0 \gamma_\mu^\dagger \gamma_0^{-1} = -\gamma_\mu\,. \tag{2.3.12}$$

This is similar to (2.3.2): $-\gamma_\mu^\dagger$ satisfy the Clifford algebra just like the γ_μ, and γ_0 is the spinor change of basis, or *intertwiner*, that relates them. (If d = even, one can also consider $\gamma_0\gamma$, which relates γ_μ to $\gamma_\mu^\dagger$.) It is now possible to define a Lorentz-invariant "inner product":[9]

$$\overline{\zeta_1}\zeta_2 \equiv \zeta_1^\dagger \gamma_0 \zeta_2\,. \tag{2.3.13}$$

Given (2.3.12), one can also obtain intertwiners that relate the γ_μ to $\pm\gamma_\mu^t$:

$$C_\pm \gamma_\mu^t C_\pm^{-1} = \pm\gamma_\mu\,, \qquad C_\pm = \gamma_0 (B_\mp^\dagger)^{-1}\,. \tag{2.3.14}$$

This makes it now possible to also define an invariant version of $\zeta_1^t \zeta_2$, namely $\zeta_1^t c \zeta_2$ (with $C = C_+$ or C_-, depending on which of $B_\pm$ is allowed, recalling (2.3.7b)). In the Euclidean case, where the inner product $\eta_1^\dagger \eta_2$ is already invariant, we will sometimes use the notation $\overline{\eta_1} \equiv \eta_1^t C$; in the Lorentzian case, where we need the bar for (2.3.13), we will use the slightly clumsier notation $\overline{\zeta_1^c}$ (which is proportional to $\zeta_1^t C$).

2.4 Orbits

Two spinors are said to be *equivalent* if one can be mapped into the other by Lorentz action. A set of equivalent spinors is called an *orbit*.

[9] We put this in quotes because this expression is usually reserved for bilinear products that are either symmetric (in real vector spaces) or Hermitian (for complex vector spaces). In contrast, (2.3.13) obeys $(\overline{\zeta_1}\zeta_2)^* = (\zeta_1^\dagger\gamma_0\zeta_2)^\dagger = -\zeta_2^\dagger\gamma_0\zeta_1 = -\overline{\zeta_2}\zeta_1$.

More precisely, in the Euclidean case we will consider the action of SO(d); in the Lorentzian case, the action of the connected component of the identity, $\mathrm{SO}^+(1, d-1)$ (see footnote 3 preceding (2.1.11)); the action of the parities (2.1.11) is then easy to work out.

If d = even, spinors of opposite chirality cannot be related by the action of SO(d) or $\mathrm{SO}^+(1, d-1)$. Another obvious obstacle to equivalence is the norm. In the Euclidean case, the inner product $\eta_1^\dagger \eta_2$ is invariant under rotations (see Section 2.3.2); so two spinors with unequal norms are not in the same orbit.

When d is low, any two spinors of the same chirality and norm are equivalent. For $d = 2$, all unit-norm chiral spinors are of the form $\eta_+ = \mathrm{e}^{\mathrm{i}\varphi}|+\rangle$, in the basis (2.2.4). The only rotation is $\exp[-\frac{1}{2}\gamma_{12}] = \exp[-\frac{\mathrm{i}}{2}\sigma_3]$, which shifts φ; so this orbit is topologically an S^1. We will now try to classify orbits in higher d.

2.4.1 Four Euclidean dimensions

In the basis (2.2.8), the chirality matrix in (2.2.12) shows that a spinor of chirality $+1$ is generated by $|++\rangle$ and $|--\rangle$. We now ask whether a general chiral spinor of unit norm can be related by SO(4) action to a given such spinor, say $|++\rangle$:

$$\mathrm{e}^{-\frac{1}{2}\lambda}|++\rangle = \eta_+ = a|++\rangle + b|--\rangle , \qquad |a|^2 + |b|^2 = 1 . \tag{2.4.1}$$

This can be answered directly by computing $\mathrm{e}^{-\frac{1}{2}\lambda}$ for the most general $\lambda \in \mathrm{so}(4)$. Given that $\mathrm{so}(4)_{\mathbb{C}} \cong \mathrm{sl}(2,\mathbb{R}) \oplus \mathrm{sl}(2,\mathbb{R})$, one concludes easily that for any a and b in (2.4.1) there is an λ that satisfies it. So all unit-norm spinors belong to the same orbit.

It is more instructive to use a different logic that generalizes better to higher dimensions. From (2.2.3) and (2.2.10), we see that $|++\rangle$ is annihilated by two gammas:

$$\gamma_{\bar{\imath}}|++\rangle = 0 . \tag{2.4.2}$$

We call a four-dimensional spinor *pure* if it is annihilated by two independent complex gamma matrices.

It sometimes proves useful to rewrite this in terms of real indices m rather than antiholomorphic indices (see footnote 2 after (2.1.1)). For that, we can introduce the *holomorphic projector*:

$$\Pi_n{}^p \equiv \frac{1}{2}\left(\delta_n^p - \mathrm{i}I^{0}{}_n{}^p\right) , \qquad I^0 = \begin{pmatrix} 0 & 1_2 \\ -1_2 & 0 \end{pmatrix} . \tag{2.4.3}$$

When applied to any real vector, it produces a linear combination of the z^i, the complex coordinates in (2.2.5). Indeed, its image is spanned by the two vectors in (2.4.2), which are the eigenvectors of I^0. In holomorphic indices,

$$I^{0j}_i = \mathrm{i}\delta_i^j , \qquad I^{0\bar{\jmath}}_{\bar{\imath}} = -\mathrm{i}\delta_{\bar{\imath}}^{\bar{\jmath}} , \qquad I^{0\bar{\jmath}}_i = I^{0j}_{\bar{\imath}} = 0 , \tag{2.4.4a}$$

$$\Pi_i{}^j = \delta_i^j , \qquad \Pi_i{}^{\bar{\jmath}} = \Pi_{\bar{\imath}}{}^j = \Pi_{\bar{\imath}}{}^{\bar{\jmath}} = 0 . \tag{2.4.4b}$$

Equation (2.4.2) then reads

$$\bar{\Pi}_n{}^m \gamma_m|++\rangle = 0 . \tag{2.4.5}$$

Now, if another spinor η_+ is related by a rotation to $|++\rangle$ as in (2.4.1), by acting on (2.4.2) by $\mathrm{e}^{-\frac{1}{2}\lambda}$ and using (2.1.16) we see that

$$\bar{\Pi}_m{}^n O_n{}^p \gamma_p \eta_+ = 0 \,, \qquad O \equiv \mathrm{e}^{\lambda} \,, \tag{2.4.6}$$

so that again η_+ is annihilated by two complex linear combinations of gamma matrices and thus is pure.

Conversely, if a spinor is annihilated by two complex linear combinations of gamma matrices, writing them as in (2.4.6) determines a rotation e^{λ} that takes them to the $\gamma_{\bar{\imath}}$. The spinor η_+ is then mapped to a spinor that is annihilated by the $\gamma_{\bar{\imath}}$, which is necessarily of the form $\mathrm{e}^{\mathrm{i}\varphi}|++\rangle$. This is then taken to $|++\rangle$ by a further rotation of the form $\exp[-\frac{1}{2}\varphi(\gamma_{13}+\gamma_{24})]$, since $(\gamma_{13}+\gamma_{24})|++\rangle = -2\mathrm{i}\gamma_{i\bar{\imath}}|++\rangle = 2\mathrm{i}|++\rangle$.

We then conclude that all pure spinors with the same chirality and norm belong to the same orbit.

Our problem is then reduced to checking whether all chiral unit-norm spinors are pure. Let us look for the gamma matrices that annihilate (2.4.1):

$$\begin{aligned}(d_i\gamma_{z^i} + \tilde{d}_i\gamma_{\bar{z}^i})(a|++\rangle + b|--\rangle) = \\ (d_2 a + \tilde{d}_1 b)|+-\rangle + (d_1 a - \tilde{d}_2 b)|-+\rangle = 0 \,.\end{aligned} \tag{2.4.7}$$

We get two conditions on four coefficients; the space of solutions is two dimensional, and can be written as a span of two generators:

$$a\gamma_{\bar{z}^2} + b\gamma_{z^1} \,, \qquad a\gamma_{\bar{z}^1} - b\gamma_{z^2} \,. \tag{2.4.8}$$

Thus any η_+ of chirality $+1$ is annihilated by two linear combinations of gammas and is thus pure. We earlier concluded that all pure spinors belong to a single orbit; so all unit-norm chiral spinors belong to a single orbit.

There is nothing special about our choice to take $z^1 = x^1 + \mathrm{i}x^3$, $z^2 = x^2 + \mathrm{i}x^4$. We could define different complex linear combinations z'_i of the x_m, in such a way that (2.4.8) actually read $\gamma_{\bar{z}'_1} \equiv \gamma'_{\bar{1}}$, $\gamma_{\bar{z}'_2} \equiv \gamma'_{\bar{2}}$. Equivalently, we can multiply (2.4.6) by an $O^{-1} = O^t$, so that it reads

$$\bar{\Pi}'_n{}^m \gamma_m \eta_+ = 0 \,, \qquad \Pi'_n{}^p \equiv \frac{1}{2}\left(\delta_n^p - \mathrm{i}I_n{}^p\right) , \qquad I \equiv O^t I^0 O, \tag{2.4.9}$$

which is formally exactly like (2.4.5), but with an $I \neq I^0$. By definition, I will still square to -1

$$I^2 = -1 \tag{2.4.10}$$

like I^0 did. But it will now single out the z' coordinates, just like I^0 singled out the z coordinates, in the sense of (2.4.4b). In terms of these new complex coordinates, one can also reexpress (2.4.6) as

$$\gamma'_{\bar{\imath}}\eta_+ = 0 \,, \tag{2.4.11}$$

which generalizes (2.4.2).

In general, a choice of complex coordinates in a real space, or equivalently of a matrix I such that $I^2 = -1$, is called a *complex structure*. A consequence of our discussion is that a chiral spinor in four dimensions determines a complex structure.

2.4.2 Pure spinors

We can generalize some aspects of the $d = 4$ discussion to any $d =$ even, still in Euclidean signature.

Definition

In general, we define the *annihilator* of a spinor as the space of all vectors that annihilate it:

$$\mathrm{Ann}(\eta) \equiv \left\{ v \in \mathbb{C}^d \mid v \cdot \eta = 0 \right\}, \tag{2.4.12}$$

where

$$v \cdot \eta \equiv v^m \gamma_m \eta \,. \tag{2.4.13}$$

(With a slight abuse of notation, we will sometimes also consider the annihilator directly as the space of the elements $v^m \gamma_m$ of the Clifford algebra.) Generalizing the $d = 4$ case, a *pure spinor* η is one such that $\mathrm{Ann}(\eta)$ has complex dimension $d/2$. For example, (2.4.2) tells us $\mathrm{Ann}(|++\rangle) = \mathrm{Span}\{\gamma_{\bar{\imath}}\}$.

The annihilator is maximally isotropic

Suppose v_1, v_2 are two elements of $\mathrm{Ann}(\eta)$, with η not necessarily pure. Then[10]

$$0 = \{v_1 \cdot, v_2 \cdot\}\eta = v_1^m v_2^n \{\gamma_m, \gamma_n\}\eta = 2 g_{mn} v_1^m v_2^n \eta = 2(v_1 \cdot v_2)\eta \,. \tag{2.4.14}$$

It follows that $v_1 \cdot v_2 = 0$. So the metric is identically zero when restricted to $\mathrm{Ann}(\eta)$ for any η; one says that $\mathrm{Ann}(\eta)$ is *isotropic*. (We are in Euclidean signature, so this can, of course, only be true because we are considering *complex* vectors.) Hence its dimension cannot be too large: for example, it cannot be the whole space, or else the metric would be zero. The largest possible zero block is $d/2$; this is achieved precisely when η is pure. So the annihilator $\mathrm{Ann}(\eta)$ of a pure η has the largest possible dimension for an isotropic subspace: it is *maximally isotropic*.

As an example, the spinor

$$\eta_+^0 = |\underbrace{+ + \cdots +}_{d/2}\rangle \tag{2.4.15}$$

is annihilated by all $\gamma_{\bar{z}^i}$ just as in (2.4.2), and thus by $d/2$ gamma matrices; so it is pure. In complex coordinates, the line element $\sum_{i=1}^{d/2} \mathrm{d}z^i \mathrm{d}\bar{z}^i$, and the metric is $\begin{pmatrix} 0 & 1_{d/2} \\ 1_{d/2} & 0 \end{pmatrix}$. Its restriction to $\mathrm{Ann}(\eta_+^0)$ is the upper-left block, which is indeed zero, as predicted earlier.

Pure spinor orbit

As in $d = 4$, all pure spinors with the same chirality and norm belong to the same orbit. The logic is the same as in Section 2.4.1: we can find a rotation that takes the $d/2$ complex generators of $\mathrm{Ann}(\eta_+)$ to the standard $d/2$ generators γ_{z^i} of

[10] We use the symbol $\cdot$ for several "natural" actions. In (2.4.13), we defined the action of a vector on a spinor; in (2.4.14) we see first this, and then at the end the action of a vector on another vector, i.e., their contraction.

Ann($|+\cdots+\rangle$). This rotation is then guaranteed to take η_+ to (a multiple of) $|+\cdots+\rangle$. In other words, a positive chirality spinor is pure if it is related to $|+\cdots+\rangle$ by a rotation. In particular, a pure spinor is always chiral.

Unlike for $d = 4$, however, for general d not all chiral spinors are pure. Intuitively, a pure spinor is annihilated by as many gamma matrices as possible, and thus acting by rotations, $\delta\eta = -\frac{1}{2}\lambda\eta$, will produce fewer new spinors than it generically would. So we expect the orbit of a pure spinor to be small, which is a justification of sorts for its name. We will quantify this later in this section.

Complex structure (in flat space)

Also generalizing the $d = 4$ logic at the end of Section 2.4.1, for any given η_+ it is possible to define a new choice of complex coordinates z'_i such that the generators of Ann(η_+) read

$$\gamma_{\bar{z}'_i} \, . \tag{2.4.16}$$

Again, a choice of complex coordinates is called a complex structure. In any d, a pure spinor determines a complex structure.

Pure Spinors as Clifford vacua

Pure spinors can also be thought of as *Clifford vacua*. This means that starting from the spinor $|+\cdots+\rangle$, by acting with γ_{z^i} in (2.2.10) one will generate all other spinors. (This is different from the Lorentz or orthogonal action, which infinitesimally consists of the γ_{mn}.) For example, in $d = 4$:

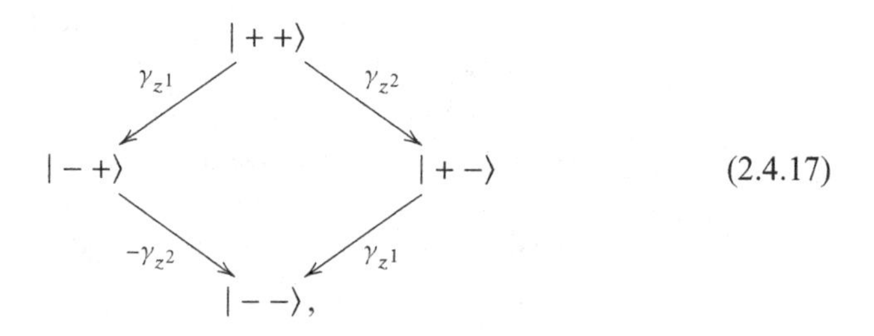

(2.4.17)

while in $d = 6$:

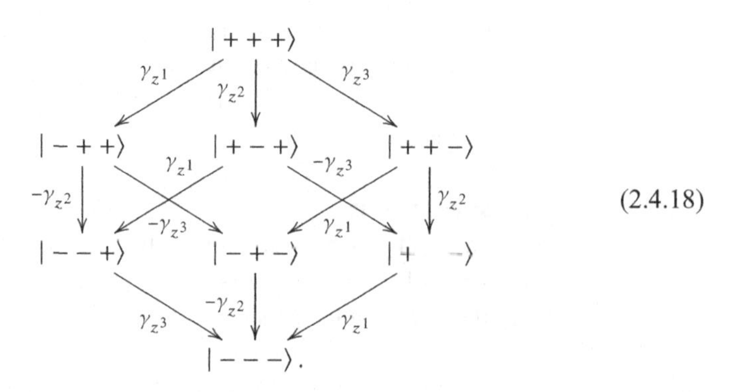

(2.4.18)

More generally, one generates the whole space of spinors by acting repeatedly on any pure η_+ with the complex conjugates of Ann(η_+).

2.4.3 Even Euclidean dimensions

We have seen that all $d = 4$ spinors are pure; we now argue that this is still the case for $d = 6$, but not for $d = 8$.

$d = 6$ **Euclidean dimensions**

Working again in the creator basis of Section 2.2.1, the most general positive-chirality spinor can be written as

$$\eta_+ = a|+++\rangle + b^1|+--\rangle - b^2|-+-\rangle + b^3|--+\rangle\,, \tag{2.4.19}$$

from the first and third line in (2.4.18). By imposing $(d^i\gamma_i + \tilde{d}^{i}\gamma_{\bar{\imath}})\eta_+ = 0$, we find the conditions

$$d^i a = \epsilon^{ijk}\tilde{d}^{j} b^k\,, \qquad d^i b^i = 0\,, \tag{2.4.20}$$

which has a three-dimensional space of solutions: if $a \neq 0$, the first determines the d^i and the second is redundant; if $a = 0$, the solution is $\tilde{d}^{i} \propto b^i$, $d^i \perp b^i$. In other words, the annihilator can be written as the span of three gamma matrices:

$$\mathrm{Ann}(\eta_+) = \begin{cases} \mathrm{Span}\{a\gamma_{\bar{\imath}} + \epsilon_{\bar{\imath}}^{jk} b^j \gamma_k\} & a \neq 0, \\ \mathrm{Span}\{b^i\gamma_{\bar{\imath}},\, d^i\gamma_i | d^i b^i = 0\} & a = 0\,. \end{cases} \tag{2.4.21}$$

This demonstrates that, just like in $d = 4$, in $d = 6$ all chiral spinors are pure; thus all positive-chirality unit-norm spinors belong to a single orbit. (We could have simplified this computation by using a $SU(3) \subset SO(6)$ transformation to set the vector b_i to be along a particular direction, say 3.)

$d = 8$ **Euclidean dimensions**

In $d = 8$, there do exist nonpure chiral spinors. Consider the following example:

$$\eta_+ = |++++\rangle + |----\rangle\,. \tag{2.4.22}$$

Acting with the γ_{z^i} produces a spinor with one – and three +; e.g. $\gamma_{z^1}\eta_+ = |-+++\rangle$. On the other hand, the $\gamma_{\bar{z}^{\bar{\imath}}}$ produce a spinor with one + and three –, e.g., $\gamma_{\bar{z}^{\bar{1}}}\eta_+ = |+---\rangle$. So there is no linear combination of the γ_i and $\gamma_{\bar{\imath}}$ that gives zero:

$$\mathrm{Ann}\,(|++++\rangle + |----\rangle) = \{0\}\,. \tag{2.4.23}$$

With either choice $B_\pm$ in (2.3.7), $|++++\rangle^c \propto |----\rangle$. This suggests that the condition in $d = 8$ for a chiral spinor η_+ to be pure is

$$(\eta_+^c)^\dagger \eta_+ = 0\,. \tag{2.4.24}$$

We will confirm this later with a different approach.

Comparing with $d = 6$, we see that (2.4.22) is nonpure because the two terms $|++++\rangle$ and $|----\rangle$ differ by four signs rather than just two. This is barely possible in $d = 8$, but it becomes more and more likely as d increases. Thus we expect pure spinors to become even rarer for $d > 8$. We will now make this expectation more precise.

2.4.4 Stabilizers

We will compute the dimension of the pure spinor orbit for any d. The methods we will use are in fact more important than the dimension formula itself.

A *group action* is a map $G \times M \to M$, $(g, m) \mapsto g \cdot m$ on a space M, such that $e \cdot m = m$ and $(g_1 g_2) \cdot m = g_1 \cdot (g_2 \cdot m)$. A representation is an example of group action, where M is a vector space. We say that an action is *transitive* if for any two points $m_1, m_2 \in M$ there exists a g that relates them: $g \cdot m_1 = m_2$. In this case, we can write

$$M \cong G/H\,, \tag{2.4.25}$$

where H is the *stabilizer* (or *little group*)

$$H = \mathrm{Stab}(m) \equiv \{h \in G \mid h \cdot m = m\} \tag{2.4.26}$$

of any point $m \in M$. (Given two points $m, m' \in M$ and a g such that $m' = g \cdot m$, $\mathrm{Stab}(m) \cong \mathrm{Stab}(m')$, under the isomorphism $h \mapsto g^{-1} h g$.)

Spaces such as (2.4.25) will play a role as potential factors of internal space in later chapters. Here we will use (2.4.25) to study the space of spinors in a given dimension.

Pure spinors in even d.
Let us apply (2.4.25) to the pure spinor orbit. We know $G = \mathrm{SO}(d)$; we need to find H. The stabilizers of all points in the orbit are isomorphic, so we can take $\eta_+ = |+\ldots+\rangle$. The Lie algebra of $\mathrm{Stab}(\eta_+)$ is given by all λ_{mn} such that

$$\lambda\eta_+ = \frac{1}{2}\lambda_{mn}\gamma^{mn}\eta_+ = 0 \tag{2.4.27}$$

(because then the exponentiated action $\exp[-\frac{1}{2}\lambda]\eta_+ = \eta_+$). Since we know $\gamma_{\bar\imath}\eta_+ = 0$, it is convenient to work in complex coordinates. Here $\lambda_{\bar\imath\bar\jmath} = \overline{\lambda_{ij}}$ and $\lambda_{\bar\imath j} = \overline{\lambda_{i\bar\jmath}}$, by reality. We see then (using (2.1.33)) that

$$\gamma_{\bar\imath\bar\jmath}\eta_+ = 0\,, \qquad \gamma_{i\bar\jmath}\eta_+ = -\frac{1}{2}\delta_{i\bar\jmath}\eta_+\,. \tag{2.4.28}$$

$\gamma^\dagger_{mn} = \gamma_{nm} = -\gamma_{mn}$; so λ is anti-Hermitian. Hence $\gamma_{\bar\imath\bar\jmath}$ cannot appear by itself in λ, but only with γ_{ij}, which does not annihilate η_+. As for $\gamma_{i\bar\jmath}$, it sends η_+ to itself; so we need to subtract its trace, which can be done after computing $\gamma_{k\bar k}\eta_+ = -\frac{d}{4}\eta_+$. All in all, the Lie algebra of the stabilizer is

$$\mathrm{stab}(\eta_+) = \mathrm{Span}\left\{\gamma_{i\bar\jmath} - \frac{2}{d}\delta_{i\bar\jmath}\gamma_{k\bar k}\right\}\,. \qquad (d = \text{even Eucl.};\ \eta_+ \text{ pure.}) \tag{2.4.29}$$

For example, for $d = 4$ this consists of $\gamma_{z^1\bar z^1} - \gamma_{z^2\bar z^2}$ and of $\gamma_{z^1\bar z^2}$. Strictly speaking, the latter is complex and counts for two; if one prefers, one can consider $\gamma_{z^1\bar z^2} - \gamma_{z^2\bar z^1}$ and $\mathrm{i}(\gamma_{z^1\bar z^2} + \gamma_{z^2\bar z^1})$, which are both anti-Hermitian as they should be.

To write (2.4.29) in real indices, we can introduce two antisymmetric tensors. One is I from Section 2.4.1, generalized to any even d. We lower its second index using the metric; we could call the resulting object simply I_{mn}, but for later convenience we call it

$$J_{mn} \equiv I_m{}^p g_{pn}\,. \tag{2.4.30}$$

This distinction will prove useful in later chapters, when g will no longer be the Euclidean metric. By (2.4.4a),

$$J_{ij} = 0\,, \qquad J_{i\bar{j}} = \mathrm{i} g_{i\bar{j}} = \frac{\mathrm{i}}{2}\delta_{i\bar{j}}\,. \tag{2.4.31}$$

Since J is in fact real (see (2.4.3)), we also have $J_{\bar{i}\bar{j}} = 0$, $J_{\bar{i}j} = -\frac{\mathrm{i}}{2}\delta_{\bar{i}j}$. The second tensor is complex, and is a sort of holomorphic Levi–Civita tensor:

$$\Omega_{i_1\ldots i_{d/2}} = \epsilon_{i_1\ldots i_{d/2}}\,, \tag{2.4.32}$$

with all other entries being zero. Now (2.4.29) is equivalent to

$$\mathrm{stab}(\eta_+) = \left\{\lambda \mid \lambda^{m_1 m_2}\Omega_{m_1\ldots m_{d/2}} = 0 = \lambda^{mn} J_{mn}\right\}. \qquad (d = \text{even Eucl.};\ \eta_+ \text{ pure.}) \tag{2.4.33}$$

The first condition enforces the fact that the allowed elements in (2.4.29) only have mixed indices; the second subtracts the trace.

What is this Lie algebra? Let us first work out that of the $\gamma_{i\bar{j}}$, without the second term in (2.4.29). $\lambda_{ij} = \lambda_{\bar{i}\bar{j}} = 0$, so λ does not mix holomorphic indices with anti-holomorphic ones. Since $z^1 = x_1 + \mathrm{i}x_{d/2+1}$, $z^2 = x_2 + \mathrm{i}x_{d/2+2}\ldots$, in real indices λ preserves the space of vectors of the type $\binom{v}{\mathrm{i}v}$ (where v is any $d/2$-dimensional vector). We can impose this on the antisymmetric matrix λ by decomposing it in $d/2 \times d/2$ blocks $\begin{pmatrix} A_1 & M \\ -M^t & A_2 \end{pmatrix}$; this results in $A_1 = A_2 \equiv A$ and $M = M^t \equiv S$. In this block decomposition then,

$$\lambda = \begin{pmatrix} A & S \\ -S & A \end{pmatrix}, \tag{2.4.34}$$

where A is antisymmetric and S is symmetric, and both are $d/2 \times d/2$. Matrices of the form (2.4.34) close under taking commutators, and thus define a Lie subalgebra of so(d), as they should; their Lie algebra is easily seen to be isomorphic to that of matrices $A + \mathrm{i}S$, which are anti-Hermitian and thus generate the Lie algebra u($d/2$). Finally, in (2.4.29) we also subtracted a trace $\gamma_{k\bar{k}}$; this modifies the Lie algebra to su($d/2$). (For example, the three generators we found for $d = 4$ after (2.4.29) generate su(2).) We thus conclude the following:

$$\mathrm{Stab}(\eta_+) = \mathrm{SU}(d/2)\,. \qquad (d = \text{even Eucl.};\ \eta_+ \text{ pure.}) \tag{2.4.35}$$

This determines the dimension of the pure spinor orbit O_{pure}. Using (2.4.25) for $G = \mathrm{SO}(d)$ and $H = \mathrm{Stab}(\eta_+) = \mathrm{SU}(d/2)$, we get

$$\dim(O_{\mathrm{pure}}) = \frac{1}{2}d(d-1) - \left(\frac{d^2}{4} - 1\right) = \frac{d}{2}\left(\frac{d}{2} - 1\right) + 1\,. \tag{2.4.36}$$

In Table 2.3, we list this dimension for the first few even d, along with the dimension of the space of all chiral spinors S_+. (Recall that the orbit O_{pure} only contains spinors of unit norm; the space of all pure spinors then has one additional dimension.) For $d = 2, 4, 6$, the two coincide, as expected from previous subsections. For $d = 8$, the codimension is two, as anticipated after (2.4.24). For higher d, one grows quadratically, the other exponentially. In $d = 10$, the codimension is 10; this coincidence is relevant for the pure spinor formalism for the superstring world-sheet action [29, 30]. In $d = 12$, the space of pure spinors has half the dimension of S_+;

Table 2.3. Dimensions of the orbit of pure spinors and of the space of chiral spinors.

d	2	4	6	8	10	12
$\dim(O_{\text{pure}})+1$	2	4	8	14	22	32
$\dim(S_+)$	2	4	8	16	32	64

this will be relevant later, when we will consider a certain "doubled" Clifford algebra with twice as many gamma matrices as the dimension of spacetime.

Nonpure spinors in $d = 8$

We now give more details about $d = 8$. We saw that O_{pure} is singled out by the constraint (2.4.24). To see what happens in the other orbits, let us first write η in terms of two Majorana–Weyl spinors η_1, η_2:

$$\eta = \eta_1 + \mathrm{i}\eta_2 , \qquad \eta_1 \equiv \frac{\eta + \eta^c}{2} , \qquad \eta_2 \equiv \frac{\eta - \eta^c}{2\mathrm{i}} . \tag{2.4.37}$$

In the real basis (2.2.28), $B_+ = 1$ and the η_i are real; (2.4.37) then simply means decomposing η in its real and imaginary parts. In this basis, the constraint (2.4.24) reads now $\eta_1^t\eta_2 = 0$, $\eta_1^t\eta_1 = \eta_2^t\eta_2$. Even if η_1 and η_2 do not satisfy this, as long as they are linearly independent there exist two real parameters ψ, A such that $\eta_1' \equiv \cos\psi\eta_1 + \sin\psi\eta_2$, $\eta_2' \equiv A(\cos\psi\eta_1 - \sin\psi\eta_2)$ do satisfy

$$(\eta_1')^t\eta_2' = 0 , \qquad (\eta_1')^t\eta_1 = (\eta_2')^t\eta_2 ; \tag{2.4.38}$$

so $\eta' = \eta_1' + \mathrm{i}\eta_2'$ is pure. Hence we can parameterize $\eta = \eta_1 + \mathrm{i}\eta_2$ in terms of two parameters A, ψ and of a new spinor $\eta' \equiv \eta_1' + \mathrm{i}\eta_2'$ which is pure. So the stabilizer is still SU(4).

But if η_1 and η_2 are linearly dependent, the situation changes drastically. Multiplying by an overall phase, we can set $\eta = \eta_1$ to be Majorana–Weyl itself. An example is (2.4.22). To compute the stabilizer, we can invoke a property of SO(8) called *triality*: it has automorphisms (isomorphisms that map it to itself) that permute the vector representation $\mathbf{8}_{\text{V}}$ with the two Majorana–Weyl spinor representations of chirality ± 1, $\mathbf{8}_{\text{S}}$ and $\mathbf{8}_{\text{C}}$ (which appeared in Section 1.1.2). The stabilizer under SO(8) of a vector in $\mathbb{R}^8$ is SO(7). Triality then implies that the same should be true for a Majorana–Weyl spinor. More precisely, the group of spinorial transformations is Spin(8), which has a 2:1 homomorphism to SO(8) (see Section 2.1.1). The stabilizer of a Majorana–Weyl spinor in $d = 8$ is then

$$\text{Stab}(\eta_{\text{MW}}) = \text{Spin}(7) . \qquad (d = 8.) \tag{2.4.39}$$

2.4.5 Lorentzian case

We now turn to the Lorentzian case.

Here we need to use the inner product $\overline{\zeta_1}\zeta_2 = \zeta_1^\dagger\gamma_0\zeta_2$ (see Section 2.3.2). Since γ_0 changes chirality, $\overline{\zeta_+}\zeta_+ = 0$ for a chiral spinor ζ_+; so we cannot use the notion of norm to separate orbits.

Pure spinors, on the other hand, still exist. In the creator basis (2.2.13), one of the $\gamma_{\bar{\imath}}$ is replaced by γ_-. So $|+\cdots+\rangle$ is still annihilated by $d/2$ gamma matrices:

$$\gamma_- \zeta_+ = 0\,, \qquad \gamma_{\bar{\imath}} \zeta_+ = 0\,, \qquad i = 1, \ldots, d/2-1\,. \tag{2.4.40}$$

Pure spinors in $d = 4$

Let us compute the Lie algebra of the stabilizer, starting from simplicity from the case $d = 4$. In the Euclidean, the generators were listed after (2.4.29) as $\gamma_{z^1\bar{z}^1} - \gamma_{z^2\bar{z}^2}$ and $\gamma_{z^1\bar{z}^2}$. Equation (2.4.40) now suggests that $\gamma_{-z^1} \in \mathrm{stab}(\zeta_+)$ in Lorentzian signature. It would also suggest looking at a real linear combination of $\gamma_{+-} = -\frac{1}{2}\gamma_{01}$ and $\gamma_{z^1\bar{z}^1} = \frac{\mathrm{i}}{2}\gamma_{23}$. But there is none, because by (2.2.13)

$$\gamma_{01}|++\rangle = -|++\rangle\,, \qquad \gamma_{23}|++\rangle = \mathrm{i}|++\rangle\,. \tag{2.4.41}$$

So we are only left with a single complex generator:

$$\mathrm{stab}(|++\rangle) = \mathrm{Stab}\{\gamma_{-z^1}\}\,. \tag{2.4.42}$$

In real coordinates, it reads $\frac{1}{4}(\gamma_0 - \gamma_1)(\gamma_2 - \mathrm{i}\gamma_3)$, so two generators directly of the form λ would be $\gamma_{02} - \gamma_{12}$ and $\gamma_{03} - \gamma_{13}$. Recalling (2.1.6), these two commute, and the Lie algebra is simply abelian:

$$\mathrm{Stab}(\zeta_+) = \mathbb{R}^2. \qquad (d = 4 \text{ Lor.}) \tag{2.4.43}$$

The orbit $\mathrm{SO}^+(3,1)/\mathbb{R}^2$ has real dimension 4, the same as the real dimension of the space of all chiral spinors. Indeed, the arguments in Section 2.4.1 can be adapted to the Lorentzian case to show that all chiral spinors are pure.

Pure spinors in any d

For higher even d, all the γ_{-z^i} are in $\mathrm{stab}(\zeta_+)$. Similar to (2.4.41), we have $\gamma_{01}\zeta_+ = -\zeta_+$, $\gamma_{i\bar{\jmath}}\zeta_+ = -\frac{1}{2}\delta_{i\bar{\jmath}}\zeta_+$. Again we cannot find a real linear combination of γ_{01} and $\gamma_{i\bar{\jmath}}$ that leaves ζ_+ invariant, but we can consider a traceless combination of the latter ones as in (2.4.29). Thus

$$\mathrm{stab}(\zeta_+) = \mathrm{Span}\left\{\gamma_{-i}\,,\ \gamma_{i\bar{\jmath}} - \frac{2}{d-1}\delta_{i\bar{\jmath}}\gamma_{k\bar{k}}\right\}. \qquad (d = \text{even Lor.};\ \zeta_+ \text{ pure.}) \tag{2.4.44}$$

Recall that now $i = 1,\ldots,d/2-1$. Just like in the Euclidean case, $\gamma_{i\bar{\jmath}} - \frac{2}{d-1}\delta_{i\bar{\jmath}}\gamma_{k\bar{k}}$ generate a Lie algebra $\mathrm{su}(d/2-1)$. They don't commute with the γ_{-i}; the commutators express the fact that the γ_{-i} transform in the fundamental of $\mathrm{SU}(d/2-1)$. The γ_{-i} commute with one another, so they generate an invariant abelian Lie subalgebra of (2.4.44). All in all,

$$\mathrm{Stab}(\zeta_+) = \mathrm{SU}(d/2-1) \ltimes \mathbb{R}^{d-2}\,. \qquad (d = \text{even Lor.};\ \zeta_+ \text{ pure.}) \tag{2.4.45}$$

Its dimension is $(d/2-1)^2-1+d-2 = (d/2)^2-2 = \dim(\mathrm{SU}(d/2))-1$. Comparing with (2.4.36), we see that the pure spinor orbit in the Lorentzian case has one dimension more than in the Euclidean case. But the dimension of the space of pure spinors is the same in the two cases; the difference is only that there is no Lorentz-invariant norm in the Lorentzian case, as pointed out at the beginning of this subsection.

(Nonpure) MW spinors in $d = 10$

Just like in the Euclidean case, for large d not all spinors are pure. Let us, however, analyze MW spinors ζ_{MW} in $d = 10$, of special interest for string theory. We will also comment at the end about the Weyl case.

ζ_{MW} is not pure. A positive-chirality pure spinor can be put by a rotation in the form $|+++++\rangle$; but this is not Majorana, as we see by picking $B = -B_+ = -\sigma_3 \otimes \sigma_1 \otimes \sigma_2 \otimes \sigma_1 \otimes \sigma_2$ from (2.3.6) (remembering our freedom to change an overall sign for convenience). To make it Majorana, we can add to it its conjugate: this produces

$$\zeta_{\text{MW}} = |+++++\rangle + |+----\rangle\,, \tag{2.4.46}$$

a Lorentzian analogue of our nonpure example (2.4.22) in $d = 8$, and is not pure by the same arguments we gave there. Its annihilator is given by the span of Γ_- and nothing else:

$$\text{Ann}(\zeta_{\text{MW}}) = \text{Span}\{\Gamma_-\}. \qquad (d = 10 \text{ Lor.}) \tag{2.4.47}$$

The stabilizer has two types of generators. The first consists of Γ_{-m}, where $m = 2,\ldots,9$. For the second, we see our creator basis (2.2.13) as a tensor product, as in (2.2.21), of a $d_1 = 2$ basis acting on the first $\pm$ sign, and a $d_2 = 8$ basis on the remaining four signs. In this spirit, (2.4.46) is

$$|+\rangle \otimes \eta^8_{\text{MW}}\,, \qquad \eta^8_{\text{MW}} = |++++\rangle + |----\rangle\,. \tag{2.4.48}$$

The stabilizer of ι_{MW} should then also contain the stabilizer of η^8_{MW}, which is Spin(7) by (2.4.22) and (2.4.39). This second type of generator doesn't commute with the Γ_{-m} because of the presence of the m index, and we get

$$\text{Stab}(\zeta_{\text{MW}}) = \text{Span}\{\Gamma_{-m}\} \oplus \text{Stab}(\eta^8_{\text{MW}}) \cong \text{Spin}(7) \ltimes \mathbb{R}^8. \qquad (d = 10 \text{ Lor.}) \tag{2.4.49}$$

We now check that there is a single orbit of positive-chirality MW spinors. We do this at the level of the infinitesimal action, by checking that $\Gamma_{MN}\zeta_{\text{MW}}$ span all possible such spinors. We split again the spacetime indices in $\pm$ in two dimensions, and $m = 2, \ldots, 9$ in eight dimensions, using the basis (2.2.21). Now we are interested in the γ_{MN} that have a nonzero action on ζ_{MW}, rather than in those that annihilate it. We find the following:

- The single $\Gamma_{+-}\zeta_{\text{MW}} = -2\zeta_{\text{MW}}$.
- $\Gamma_{+m} = |-\rangle \otimes \gamma_m \eta^8_{\text{MW}}$; since by (2.4.23) there is no annihilator in $d = 8$, these are eight independent spinors.
- Finally, $\Gamma_{mn}\zeta_{\text{MW}} = |+\rangle \otimes \gamma_{mn}\zeta_{\text{MW}}$. Among the γ_{mn}, the 21 generators of $\text{Stab}(\zeta_{\text{MW}}) = \text{Spin}(7)$ give zero, so only $28 - 21 = 7$ spinors get generated.

The total is $1 + 8 + 7 = 16$, which equals the dimension of the space S_+ of chiral spinors.

Weyl spinors in $d = 10$

Still in $d = 10$, we may also consider an η_+ which is Weyl but not Majorana. This is equivalent to having two MW spinors:

$$\eta_1 \equiv \frac{1}{2}(\eta_+ + (\eta_+)^c)\,, \qquad \eta_2 \equiv \frac{1}{2i}(\eta_+ - (\eta_+)^c)\,, \tag{2.4.50}$$

in terms of which $\eta_+ = \eta_1 + i\eta_2$. Now η_+ can be pure again, in which case the stabilizer is (2.4.45) for $d = 10$, namely $\text{SU}(4) \ltimes \mathbb{R}^8$. There are other possibilities as well, but they become a lot easier to analyze with the methods of Section 3.3, to which we postpone this discussion.

2.4.6 Odd dimensions

For odd d, the definition of pure spinors does not apply. But we do have some distinguished orbits; we will describe some in this subsection. We start from Euclidean signature, and then more briefly look at the Lorentzian case.

$d = 3$ Euclidean dimensions

The case of $d = 3$ Euclidean dimensions is not very interesting. The creator basis of Section 2.2.1 here just becomes the Pauli-matrix basis σ_i. The spinor $|+\rangle$ has an annihilator $\gamma_{\bar{z}}$, but the stabilizer is empty. The orbit then has the same dimension as SO(3), which is 3; this equals the dimension of the space of unit-norm spinors.

$d = 5$ Euclidean dimensions

The $d = 5$ case is more interesting. The creator basis is obtained by taking (2.2.11) for $d = 4$ and defining $\gamma_5 = \sigma_3 \otimes \sigma_3$, by (2.2.12). We can consider the orbit of $|++\rangle$, now understood as a $d = 5$ spinor. The stabilizer certainly contains the stabilizer in $d = 4$, which we showed after (2.4.29) to be su(2). The generators of so(5) not in so(4) are of the form γ_{5m}, $m = 1, \ldots, 4$. While $\gamma_{5\bar{\imath}}$ does annihilate $|++\rangle$ (recall (2.4.2), which still applies), its complex conjugate does not; since the stabilizer should be real, we have no new elements in stab($|++\rangle$) in $d = 5$ that we did not already have in $d = 4$.

A general $d = 5$ spinor, when considered as a $d = 4$ spinor (or "reduced" to $d = 4$), has no reason to be chiral; in general, it will contain both chiralities. So it might seem $|++\rangle$ is a very special $d = 5$ spinor. But for any spinor η, one can reduce along a direction v^m such that η chiral. This is identified by imposing $v^m\gamma_m\eta = \eta$. If we impose this explicitly, we get four conditions (from the four components of η) for the five components v^m. For example, for the spinor $|++\rangle + |+-\rangle$ (which is not chiral in $d = 4$) we find $v^m\gamma_m = \gamma_2 + \gamma_{\bar{2}}$. (We will see a more practical method to identify v^m in Section 3.3.) If we perform a rotation so that v^m becomes direction 5, η will become proportional to η. Thus, in general we can say

$$\text{Stab}(\eta) = \text{SU}(2)\,. \qquad (d = 5 \text{ Eucl.}) \tag{2.4.51}$$

In particular, the orbit is $10 - 3 = 7$-dimensional, which is indeed the dimension of the space of unit-norm spinors.

$d = 7$ Euclidean dimensions

The next nontrivial case is $d = 7$. Repeating the previous $d = 5$ arguments leads to an SU(3) stabilizer, with a $21 - 8 = 13$-dimensional orbit. This equals the dimension of the space of spinors with fixed values for the norm $\eta^\dagger\eta$ and for the complex $(\eta^c)^\dagger\eta$: $16 - 3 = 13$.

We can also consider a Majorana spinor η_M, recalling (2.3.7b); this is a more interesting case. The quickest way to analyze Majorana spinors in $d = 7$ is to work in the basis we defined using octonions. From (2.2.19), taking for example the spinor $\eta_\alpha = \delta^8_\alpha$, we get

$$\gamma_{mn}\eta_M = -i\phi_{mnp}\gamma^p\eta_M\,. \tag{2.4.52}$$

Perhaps more intuitively, recall that (2.2.19) is obtained as left multiplication of an octonionic imaginary unit i^m on a general octonion $x_8 1 + x_n i_n$, to be thought of as a spinor. If we consider the spinor η_M defined by the element $1 \in \mathbb{O}$ (in other words, we take $x_n = 0$, $x_8 = 1$), then the action of the imaginary octonions on it can be immediately read off from (2.2.17) upon antisymmetrizing m and n.[11] In any case, (2.4.52) implies that the infinitesimal stabilizer is given by λ_{mn} such that

$$\text{stab}(\eta_M) = \{\lambda \mid \lambda_{mn}\phi^{mnp} = 0\}. \qquad (d = 7 \text{ Eucl.}) \tag{2.4.53}$$

This gives seven constraints on the 21 independent λ_{mn}, resulting in a 14-dimensional stabilizer. These λ close under commutator, so they form a Lie algebra, as they should. The only 14-dimensional Lie subalgebra of so(7) is g_2, one of the exceptional Lie algebras. So we have the following:

$$\text{Stab}(\eta_M) = G_2. \qquad (d = 7 \text{ Eucl.}) \tag{2.4.54}$$

It is instructive to rederive this in the creator basis of Section 2.2.1. We again extend the creator basis (2.2.11) by adding $\gamma_7 = \sigma_3 \otimes \sigma_3 \otimes \sigma_3$. From (2.3.7), we see $B = B_- = \gamma_4\gamma_5\gamma_6 = \gamma_6 = \mathrm{i}\sigma_2 \otimes \sigma_1 \otimes \sigma_2$, so a possible Majorana spinor is $\eta_M \equiv |{+}{+}{+}\rangle - \mathrm{i}|{-}{-}{-}\rangle$. The annihilator of η_M still contains the elements of (2.4.29) or (2.4.33) for $d = 6$, but now we also get more generators: it is possible for the effect of two annihilators on $|+++\rangle$ to be compensated by the effect of one creator on $|---\rangle$. We can obtain this way three complex new elements in the stabilizer:

$$\left(\frac{1}{2}\gamma_{z^i z^j}\epsilon^{ijk} - \mathrm{i}\gamma_{\bar{z}^k x^7}\right)(|+++\rangle - \mathrm{i}|---\rangle) = 0\,. \tag{2.4.55}$$

Together with the eight generators of su(3), this adds up to a total of 14, the dimension of G_2. Another way of thinking about this is that there is a Lie algebra inclusion $g_2 \supset \text{su}(3)$, under which the adjoint decomposes as $\mathbf{14} \to \mathbf{8} \oplus \mathbf{3} \oplus \bar{\mathbf{3}}$; the $\mathbf{3}$ is spanned by the (2.4.55), and the $\bar{\mathbf{3}}$ by their complex conjugates.

$d = 3$ **Lorentzian dimensions**

Just like in the Euclidean case, there is not much to discuss. In the real basis (2.2.16), where the B matrix is 1, consider $|+\rangle$, which is Majorana. It is annihilated by γ_-, but the stabilizer is empty. So the orbit has dimension 3 (the dimension of SO(2, 1)), which is the space of spinors with $\bar{\zeta}\zeta = 1$.

$d = 5$ **Lorentzian dimensions**

In $d = 5$, we find two different orbits, distinguished by whether $\overline{\zeta^c}\zeta = (B\zeta^*)^\dagger\gamma^0\zeta = \zeta^t B^\dagger \gamma^0\zeta$ is zero or not. For a case where it is nonzero, pick $\zeta = |++\rangle$ in the usual creator basis (2.2.13) augmented by $\gamma^4 = \sigma_3 \otimes \sigma_3$. By (2.3.6), $B = B_- = \gamma_3 = \sigma_3 \otimes \sigma_2$ and $\zeta^c = \mathrm{i}|+-\rangle$. The stabilizer contains the elements it had in $d = 4$ from (2.4.42), and a further γ_{-4}. Together these give

$$\text{Stab}(\zeta) = \mathbb{R}^3 \qquad (d = 5 \text{ Lor.}, \overline{\zeta^c}\zeta \neq 0)\,. \tag{2.4.56}$$

For the other type of orbit, it is best to use a slightly different basis: start from the $d = 4$ Euclidean creator basis (2.2.11) and add $\gamma^0 = \mathrm{i}\sigma_3 \otimes \sigma_3$ to it. Now $B =$

[11] Equation (2.2.17) might even suggest a gamma matrix identity of the form $\gamma_{mn} \stackrel{?}{=} -\mathrm{i}\phi_{mnp}\gamma^p$, in other words, an extension of (2.4.52) valid on every spinor; this does not work because the octonion multiplication is not associative.

$\gamma^0\gamma^2\gamma^3 = -\sigma_2 \otimes \sigma_1$; if we take $\zeta = |++\rangle$ in this basis, now $\zeta^c = -\mathrm{i}|--\rangle$, and $\overline{\zeta^c}\zeta = 0$. The stabilizer is now the same as in $d = 4$ Euclidean dimensions:

$$\mathrm{Stab}(\zeta) = \mathrm{SU}(2). \qquad (d = 5 \text{ Lor.}, \overline{\zeta^c}\zeta = 0). \tag{2.4.57}$$

Summary

To summarize the discussion in this section, we present Tables 2.4 and 2.5 with some notable spinor orbits in low dimensions. The column "Orbit dimension" shows that $\dim\mathrm{SO}(d) - \dim\mathrm{Stab}$ equals the space of spinors minus the number of invariants. We indicate which invariants are complex with a subscript $_\mathbb{C}$.

Exercise 2.4.1 Show the last entry in Table 2.4: the stabilizer of a spinor η_M that is Majorana but not Weyl is G_2.

Exercise 2.4.2 Work out the two $d = 7$ entries in Table 2.5. (The logic is similar to $d = 5$.)

Exercise 2.4.3 How many linearly independent spinors η_I do we need in d dimensions so that their common stabilizer is the identity? Pay attention to whether you are considering spinors that are Majorana, Weyl, MW, or unconstrained.

Table 2.4. Some spinor orbits in Euclidean signature.

Type	d	Stab	Orbit dim.	Invariants
Weyl ≅ Maj.	2		$1-0=2-1$	$\|\eta\|^2$
	3		$3-0=4-1$	$\|\eta\|^2$
Weyl	4	SU(2)	$6-3=4-1$	$\|\eta_+\|^2$
	4		$6-0=8-2$	$(\overline{\eta}\eta)_\mathbb{C}$
	5	SU(2)	$10-3=8-1$	$\|\eta\|^2$
Weyl ≅ Maj.	6	SU(3)	$15-8=8-1$	$\|\eta\|^2$
	6	SU(2)	$15-3=16-4$	$\|\eta_a\|^2$, $(\eta_1^\dagger\eta_2)_\mathbb{C}$ $(\eta=\eta_1+\mathrm{i}\eta_2)$
Maj.	7	G_2	$21-14=8-1$	$\|\eta\|^2$
	7	SU(3)	$21-8=16-3$	$\|\eta\|^2$, $(\overline{\eta}\eta)_\mathbb{C}$
MW	8	Spin(7)	$28-21=8-1$	$\|\eta\|^2$
Weyl	8	SU(4)	$28-15=16-3$	$\|\eta_a\|^2$, $\eta_1^\dagger\eta_2$ $(\eta=\eta_1+\mathrm{i}\eta_2)$
Maj.	8	G_2	$28-14=16-2$	$\|\eta_\pm\|^2$ $(\eta=\eta_++\eta_-)$

Table 2.5. Some spinor orbits in Lorentzian signature; $K_\mu \equiv \overline{\zeta}\gamma_\mu\zeta$.

Type	d	Stab	Orbit dim.	Invariants
MW	2		$1-0=1$	
Maj.	3		$3-0=4-1$	$\overline{\zeta}\eta$
Maj. ≅ Weyl	4	$\mathbb{R}^2$	$6-2=4-1$	
	4		$6=8-2$	$(\overline{\zeta}\eta)_\mathbb{C}$
	5	$\mathbb{R}^3$	$10-3=8-1$	$\overline{\zeta}\zeta$; $K^2=0$
	5	SU(2)	$10-3=8-1$	$\overline{\zeta}\zeta$; $K^2<0$
Weyl	6	$\mathrm{SU}(2)\ltimes\mathbb{R}^4$	$15-7=8$	
	7	$\mathrm{SU}(2)\ltimes\mathbb{R}^5$	$21-8=16-3$	$\overline{\zeta}\zeta$, $\overline{\zeta^c}\zeta$; $K^2=0$
	7	SU(3)	$21-8=16-3$	$\overline{\zeta}\zeta$, $\overline{\zeta^c}\zeta$; $K^2<0$

3 From spinors to forms

In this chapter, we get closer to traditional geometrical ideas, introducing forms and showing how they are related to spinors. We will still work in flat space.

3.1 Forms and gamma matrices

An object with many indices is often called a *tensor*. For example, a tensor with two indices is a matrix. The reason for this name is that a basis of the tensor product of several copies of the vector space V with itself is given by objects with many indices (see the discussion leading to (2.2.9)). Tensors with k indices that are antisymmetric under exchange of any two indices are called *k-forms*. They form an algebra, whose relation to the Clifford algebra will be important for us.

3.1.1 Exterior algebra

We begin with a more detailed introduction to the algebraic aspect of forms, expanding some remarks in Chapter 1. The vector space of k-forms $\alpha_{m_1 \dots m_k}$ is called Λ^k. It has dimension

$$\dim \Lambda^k = \binom{d}{k} = \frac{d!}{k!\,(d-k)!} \,. \tag{3.1.1}$$

In particular, we see that there is only one d-form up to a factor, and that there are no k-forms for $k > d$. Although in flat space indices can be lowered and raised with the metric, by convention forms will be taken to have lower indices. When we use an index-free notation, we will use the same letter v both for a contravariant vector v^μ and for the associated one-form (or covariant vector) $g_{\mu\nu} v^\nu$.

We already saw in (1.2.5) a natural product $\Lambda^{k_1} \times \Lambda^{k_2} \to \Lambda^{k_1+k_2}$, obtained by antisymmetrizing all indices:

$$(\alpha_{m_1 \dots m_{k_1}}, \alpha'_{n_1 \dots n_{k_2}}) \mapsto \frac{(k+k')!}{k!\,k'!} \alpha_{[m_1 \dots m_k} \alpha'_{n_1 \dots n_{k_2}]} \,. \tag{3.1.2}$$

So, for example, when $k_1 = k_2 = 1$, this takes two one-forms (i.e., covariant vectors) to their antisymmetrized product, a two-form:

$$(\alpha_m, \alpha'_n) \mapsto 2\alpha_{[m} \alpha'_{n]} = \alpha_m \alpha'_n - \alpha_n \alpha'_m \,. \tag{3.1.3}$$

A convenient index-free notation for one-forms is obtained by considering them as components of a differential: $\alpha \equiv \alpha_m \mathrm{d}x^m$. For higher k-forms, we define

$$\mathrm{d}x^m \wedge \mathrm{d}x^n \equiv \mathrm{d}x^m \otimes \mathrm{d}x^n - \mathrm{d}x^n \otimes \mathrm{d}x^m \,. \tag{3.1.4}$$

This can be extended by linearity to any two one-forms:

$$\alpha \wedge \alpha' \equiv \alpha_m \alpha'_n \mathrm{d}x^m \wedge \mathrm{d}x^n = \alpha_{[m}\alpha'_{n]} \mathrm{d}x^m \wedge \mathrm{d}x^n \; ; \tag{3.1.5}$$

we used the property $\mathrm{d}x^m \wedge \mathrm{d}x^n = -\mathrm{d}x^n \wedge \mathrm{d}x^m$, which follows from the definition (3.1.4). The components of (3.1.5) then reproduce (3.1.3). If intuitively we think of a vector in $\mathbb{R}^d$ as pointing in a direction, we can think of (3.1.5) as describing a plane.

Iterating (3.1.4), we can define an index-free notation for a k-form:

$$\alpha \equiv \frac{1}{k!} \alpha_{m_1 \ldots m_k} \mathrm{d}x^{m_1} \wedge \ldots \wedge \mathrm{d}x^{m_k} \, , \tag{3.1.6a}$$

$$\mathrm{d}x^{m_1} \wedge \mathrm{d}x^{m_2} \wedge \ldots \wedge \mathrm{d}x^{m_k} = k! \, \mathrm{d}x^{[m_1} \otimes \mathrm{d}x^{m_2} \otimes \cdots \otimes \mathrm{d}x^{m_k]} \, . \tag{3.1.6b}$$

The factor $1/k!$ proves useful in taking care of repetitions. For example, a two-form whose only nonvanishing components are $\alpha_{12} = -\alpha_{21} = 1$ will read, according to (3.1.6a), $\frac{1}{2}(\mathrm{d}x^1 \wedge \mathrm{d}x^2 + \mathrm{d}x^2 \wedge \mathrm{d}x^1) = \mathrm{d}x^1 \wedge \mathrm{d}x^2$. Another example is the so-called *flat-space volume form*:

$$\mathrm{vol} \equiv \frac{1}{d!} \epsilon_{m_1 \ldots m_d} \mathrm{d}x^{m_1} \wedge \ldots \wedge \mathrm{d}x^{m_d} = \mathrm{d}x^1 \wedge \ldots \wedge \mathrm{d}x^d \, . \tag{3.1.7}$$

As we mentioned earlier, the space of d-forms is one dimensional, so they are in fact all proportional to vol. In this chapter, $\epsilon_{m_1 \ldots m_d} = \epsilon^{(0)}_{m_1 \ldots m_d} \equiv \sigma(m_1 \ldots m_d)$, a permutation sign. (This flat space definition will need to change in the next chapter.)

One often appends a label to a form to denote its number of indices (or *degree*) k: so in (3.1.6a), we might write the left-hand side as α_k. Now

$$\alpha_{k_1} \wedge \alpha'_{k_2} = \frac{1}{k_1! \, k_2!} \alpha_{[m_1 \ldots m_{k_1}} \alpha'_{n_1 \ldots n_{k_2}]} \mathrm{d}x^{m_1} \wedge \ldots \wedge \mathrm{d}x^{m_k} \wedge \mathrm{d}x^{n_1} \wedge \ldots \wedge \mathrm{d}x^{n_{k_2}} \, , \tag{3.1.8}$$

reproducing (3.1.2).

The dimension of Λ^k is equal to that of Λ^{d-k}, since $\binom{d}{k} = \binom{d}{d-k}$. There is a natural bijection between the two spaces:

$$* : \mathrm{d}x^{m_1} \wedge \ldots \wedge \mathrm{d}x^{m_k} \mapsto \frac{1}{(d-k)!} \epsilon_{m_{k+1} \ldots m_d}{}^{m_1 \ldots m_k} \mathrm{d}x^{m_{k+1}} \wedge \ldots \wedge \mathrm{d}x^{m_d} \, , \tag{3.1.9}$$

called $*$ *map* or *Hodge dual*, already used in Chapter 1 starting from (1.2.34b). Once again, the prefactor is meant to eliminate redundancies: for example,

$$* \, \mathrm{d}x^1 \wedge \ldots \wedge \mathrm{d}x^k = (-1)^{k(d-k)} \mathrm{d}x^{k+1} \wedge \ldots \wedge \mathrm{d}x^d \tag{3.1.10}$$

(again this will be modified in curved space). In general, it is good practice to include in definitions a factor of $1/k!$ whenever two sets of k antisymmetric indices are contracted. Intuitively, $*$ takes off any $\mathrm{d}x^m$ that is already present, and inserts any that is absent. So applying $*$ twice gives back the same form up to a factor:

$$*^2 \, \alpha_k = (-1)^{k(d-k)} \alpha_k \, . \tag{3.1.11}$$

This formula has an additional minus sign in Lorentzian signature.

We can also introduce $\Lambda \equiv \oplus_{k=0}^d \Lambda^k$; its elements are formal sums of forms of all degrees, $\alpha_1 + \alpha_2 + \cdots + \alpha_d$, or *polyforms*. The wedge product $\wedge : \Lambda \times \Lambda \to \Lambda$ makes Λ an algebra, called *exterior algebra*. Its dimension is

$$\sum_{k=0}^{d} \binom{d}{k} = 2^d \, . \tag{3.1.12}$$

3.1.2 Clifford map

Equation (3.1.12) is equal to the Clifford algebra dimension (2.1.26). Moreover, the basis (2.1.25) is rather similar to the basis in (3.1.6b). Let us define the map

$$dx^{m_1} \wedge \ldots \wedge dx^{m_k} \mapsto \gamma^{m_1 \ldots m_k} . \tag{3.1.13}$$

Extending it by linearity, this defines the *Clifford map* $\Lambda \to \mathrm{Cl}$, usually denoted by a slash:

$$\alpha_k = \frac{1}{k!} \alpha_{m_1 \ldots m_k} dx^{m_1} \wedge \ldots \wedge dx^{m_k} \mapsto \not{\alpha}_k = \frac{1}{k!} \alpha_{m_1 \ldots m_k} \gamma^{m_1 \ldots m_k} . \tag{3.1.14}$$

We already used this notation for infinitesimal spinor transformations, since back in (2.1.7). Using the notation / can be clumsy on long expressions; we will sometimes use the alternative notation

$$\not{\alpha}_k \equiv [\alpha_k]_/ . \tag{3.1.15}$$

Eventually, we will drop the / altogether and confuse a form α_k and its associated $\not{\alpha}_k$, when no confusion is likely to arise.

At the formal level, one can even define Λ in a way reminiscent of (2.1.24), calling dx^m rather than γ^m the generators of T_d, and quotienting similarly:

$$\Lambda \equiv T_d / \{dx^m \wedge dx^n + dx^n \wedge dx^m = 0\} . \tag{3.1.16}$$

So in a sense Λ is a Clifford algebra relative to the zero metric. The map (3.1.13) is not an isomorphism of algebras. Nevertheless, it will be quite useful for us.

As defined in (3.1.13), the map is bijective, as we can see from the fact that both Λ and the Clifford algebra Cl_d have dimension 2^d. In practice, however, we usually consider the version of this map where the $\gamma^{m_1 \ldots m_k}$ are in a matrix representation. Such matrices are often called *bispinors* because they have two spinor indices; we know from Section 2.1.3 that this space has dimension $2^{\lfloor d/2 \rfloor} \times 2^{\lfloor d/2 \rfloor}$. This equals to 2^d when d is even, but $2^{(d-1)}$ when d is odd. So when d is even, the matrices $\gamma^{m_1 \ldots m_k}$ are a basis for bispinors; when d is odd, they are twice as many as needed for a basis. For example, when d is odd, the would-be chiral gamma (2.1.17) is in fact proportional to the identity.

3.1.3 Clifford product and exterior algebra

The difference between the exterior algebra Λ and the Clifford algebra Cl is obviously that the product in the latter algebra is much more complicated. For example, in Λ the product between dx^m and dx^n is simply $dx^m \wedge dx^n$, while in Cl the product between γ^m and γ^n is $\gamma^{mn} + g^{mn}$, (2.1.33). The counterimage of the right-hand side under the Clifford map (3.1.13) is $dx^m \wedge dx^n + g^{mn}$, a sum of a two- and a zero-form.

More generally, we can compute with the methods described in Section 2.1.4

$$\gamma^m \gamma^{n_1 \ldots n_k} = \gamma^{m n_1 \ldots n_k} + k\, g^{m[n_1} \gamma^{n_2 \ldots n_k]}\,; \tag{3.1.17}$$

the counterimage of the right-hand side under (3.1.13) is

$$\mathrm{d}x^m \wedge \mathrm{d}x^{n_1} \wedge \ldots \wedge \mathrm{d}x^{n_k} + k g^{m[n_1} \mathrm{d}x^{n_2} \wedge \ldots \wedge \mathrm{d}x^{n_k]}\,, \tag{3.1.18}$$

a sum of a $(k+1)$- and a $(k-1)$-form.

We introduce the *contraction* operator ι_m by[1]

$$\iota_m(\mathrm{d}x^{n_1} \wedge \ldots \wedge \mathrm{d}x^{n_k}) \equiv k\delta_m^{[n_1} \mathrm{d}x^{n_2} \wedge \ldots \wedge \mathrm{d}x^{n_k]} = \tag{3.1.19a}$$

$$= \delta_m^{n_1} \mathrm{d}x^{n_2} \wedge \ldots \wedge \mathrm{d}x^{n_k} - \mathrm{d}x^{n_1} \wedge \delta_m^{n_2} \mathrm{d}x^{n_3} \wedge \ldots \wedge \mathrm{d}x^{n_k} + \cdots$$
$$+ (-1)^{k-1} \mathrm{d}x^{n_1} \wedge \ldots \wedge \mathrm{d}x^{n_{k-1}} \delta_m^{n_k}\,. \tag{3.1.19b}$$

For example, $\iota_m \mathrm{d}x^n = \delta_m^n$, and $\iota_m \mathrm{d}x^n \wedge \mathrm{d}x^p = \delta_m^n \mathrm{d}x^p - \delta_m^p \mathrm{d}x^n$. We can then summarize (3.1.17) and (3.1.18) as

$$\gamma^m \not{\alpha}_k = [(\mathrm{d}x^m \wedge + g^{mn} \iota_n)\alpha_k]_{/} \tag{3.1.20}$$

in the notation of (3.1.15). We can also write it as an "operator" equation, abstracting from the "test form" α_k:

$$\overrightarrow{\gamma^m} = \mathrm{d}x^m \wedge + g^{mn} \iota_n\,. \tag{3.1.21}$$

The $\rightarrow$ emphasizes that we multiply from the left. As anticipated, we are now deliberately confusing bispinors on the left-hand side with forms on the right-hand side. We will use a similar short-hand notation in several subsequent form identities.

As another example, another Clifford algebra identity is

$$\gamma^{n_1 \ldots n_k} \gamma^m = (-1)^k \left(\gamma^{m n_1 \ldots n_k} - k\, g^{m[n_1} \gamma^{n_2 \ldots n_k]}\right)\,. \tag{3.1.22}$$

To write it in the operator spirit of (3.1.21), we can introduce the slightly funny-looking right action

$$\overleftarrow{\gamma^m} \not{\alpha}_k \equiv \not{\alpha}_k \gamma^m\,. \tag{3.1.23}$$

Then (3.1.22) can be written as

$$\overleftarrow{\gamma^m} = (\mathrm{d}x^m \wedge - g^{mn} \iota_n)(-1)^{\deg}\,. \tag{3.1.24}$$

Now deg denotes the operator whose eigenvalue is the form degree, $\deg \alpha_k = k\alpha_k$. We note for later use that

$$\{(-1)^{\deg}, \mathrm{d}x^m \wedge\} = \{(-1)^{\deg}, \iota_m\} = 0\,. \tag{3.1.25}$$

[1] The factor of k is again natural to avoid repetitions: for example, $\iota_1(\mathrm{d}x^1 \wedge \ldots \wedge \mathrm{d}x^k) = k\delta_1^{[1} \mathrm{d}x^2 \wedge \ldots \wedge \mathrm{d}x^{k]} = \frac{k(k-1)!}{k!} \delta_1^1 \mathrm{d}x^2 \wedge \ldots \wedge \mathrm{d}x^k = \mathrm{d}x^2 \wedge \ldots \wedge \mathrm{d}x^k$.

3.2 Forms as spinors for a doubled Clifford algebra

3.2.1 Generalized Lorentz group

In using (3.1.21) and (3.1.24), it is useful to know how the two operators ι_m and $dx^n\wedge$ interact. The version (3.1.19b) of the definition of ι_m makes it clear that it obeys a Leibniz identity:

$$\iota_m(\alpha_{k_1} \wedge \alpha'_{k_2}) = (\iota_m \alpha_{k_1}) \wedge \alpha'_{k_2} + (-1)^{k_1} \alpha_{k_1} \wedge (\iota_m \alpha'_{k_2}) . \qquad (3.2.1)$$

In other words, we can write an operator equation

$$\iota_m \alpha_k \wedge -(-1)^k \alpha_k \wedge \iota_m = \begin{cases} [\iota_m, \alpha_k \wedge] & \text{even } k \\ \{\iota_m, \alpha_k \wedge\} & \text{odd } k \end{cases} = (\iota_m \alpha_k) \wedge . \qquad (3.2.2)$$

In particular,

$$\{\iota_m, dx^n \wedge\} = \delta^n_m \, 1 . \qquad (3.2.3a)$$

Using (3.1.19a), we see $\iota_m \iota_n (dx^{p_1} \wedge \ldots \wedge dx^{p_k}) = k \delta^{[p_1}_n \iota_m (dx^{p_2} \wedge \ldots \wedge dx^{p_k]}) = k(k-1)\delta^{[p_1}_n \delta^{p_2}_m dx^{p_3} \wedge \ldots \wedge dx^{p_k]}$, which is antisymmetric in m and n; thus

$$\{\iota_m, \iota_n\} = 0 . \qquad (3.2.3b)$$

Finally, it is also clear that

$$\{dx^m \wedge, dx^n \wedge\} = 0 . \qquad (3.2.3c)$$

Together, (3.2.3) are the anticommutation relations of a Clifford algebra. Define indeed an index $A = 1, \ldots, 2d$, and the gamma matrices

$$\Gamma_A = \left\{ dx^1 \wedge, \, dx^2 \wedge, \, \ldots, \, dx^d \wedge, \, \iota_1, \, \iota_2, \, \ldots, \, \iota_d \right\} . \qquad (3.2.4)$$

We see then that

$$\{\Gamma_A, \Gamma_B\} = \mathcal{I}_{AB} \, 1 . \qquad (3.2.5)$$

Comparison with (2.1.3) shows that the relevant metric is $\frac{1}{2}\mathcal{I}$, where

$$\mathcal{I} = \begin{pmatrix} 0 & 1_d \\ 1_d & 0 \end{pmatrix} . \qquad (3.2.6)$$

This metric has signature (d, d); so (3.2.5) is the Clifford algebra $\mathrm{Cl}_{d,d}$. Since the operators (3.2.4) that span it act on forms, we see that the algebra Λ of forms can be viewed as the space of spinors for this $\mathrm{Cl}_{d,d}$.

In retrospect, the emergence of the "doubled" Clifford algebra (3.2.5) for operators acting on forms should come as no surprise. After all, there are two copies of an ordinary Clifford algebra acting on bispinors: gamma matrices acting from the left and from the right, $\overrightarrow{\gamma}_m$ and $\overleftarrow{\gamma}_m$. These would commute, but $\overrightarrow{\gamma}_m$ and $\overleftarrow{\gamma}_m \, (-1)^{\deg}$ anticommute; moreover, $\{\overleftarrow{\gamma}_m \, (-1)^{\deg}, \overleftarrow{\gamma}_m \, (-1)^{\deg}\} = -g_{mn}$. Thus $\overrightarrow{\gamma}_m$ and $\overleftarrow{\gamma}_m \, (-1)^{\deg}$ together satisfy a Clifford algebra with the metric

$$\begin{pmatrix} g_{mn} & 0 \\ 0 & -g_{mn} \end{pmatrix} . \qquad (3.2.7)$$

This is related to $\mathcal{I}$ in (3.2.6) by a change of basis, which is indeed the one in (3.1.21) and (3.1.24). We can also invert it by taking sums and differences:

$$dx^m \wedge = \frac{1}{2}\left(\overrightarrow{\gamma^m} + \overleftarrow{\gamma^m}\,(-1)^{\deg}\right), \qquad (3.2.8a)$$

$$\iota_m = \frac{1}{2}\left(\overrightarrow{\gamma_m} - \overleftarrow{\gamma_m}\,(-1)^{\deg}\right). \qquad (3.2.8b)$$

More explicitly:

$$(dx^m \wedge \alpha_k)_{/} = \begin{cases} \frac{1}{2}\{\gamma^m, \alpha\!\!\!/_k\} & \text{even } k \\ \frac{1}{2}[\gamma^m, \alpha\!\!\!/_k] & \text{odd } k \end{cases}; \qquad (3.2.9a)$$

$$(\iota_m \alpha_k)_{/} = \begin{cases} \frac{1}{2}[\gamma_m, \alpha\!\!\!/_k] & \text{even } k \\ \frac{1}{2}\{\gamma_m, \alpha\!\!\!/_k\} & \text{odd } k \end{cases}. \qquad (3.2.9b)$$

There is also a notion of chirality for the Clifford algebra (3.2.5). The expression (2.1.17) is appropriate for a flat metric in diagonal form, but not for (3.2.6). Fortunately, it is immediate to check that $(-1)^{\deg}$ anticommutes with all (3.2.4); so the space of positive (negative) chirality forms is the space of even (odd) forms:

$$\Lambda_+ \equiv \oplus_{k=\text{even}}\Lambda^k, \qquad \Lambda_- \equiv \oplus_{k=\text{odd}}\Lambda^k. \qquad (3.2.10)$$

Similar to (2.1.8), we can also consider a rotation group $O(d,d)$ acting on this space of spinors, whose infinitesimal generators are $\Gamma_{AB} \equiv \Gamma_{[A}\Gamma_{B]}$. We will call this $O(d,d)$ the *generalized* (or *doubled*) *Lorentz group*, with Lie algebra

$$o(d,d) = \{o \in \text{Mat}(2d) \text{ such that } \mathcal{I}o = -(\mathcal{I}o)^t\}, \qquad (3.2.11)$$

analogous to (2.1.1). To make this more explicit, we can decompose O in four $d \times d$ blocks; the condition on o in (3.2.11) then reads

$$o = \begin{pmatrix} \lambda & \beta \\ b & -\lambda^t \end{pmatrix}, \qquad b = -b^t, \qquad \beta = -\beta^t. \qquad (3.2.12)$$

In terms of these blocks, the infinitesimal action then reads

$$o\cdot \equiv \frac{1}{2}o^{AB}\Gamma_{AB} = \frac{1}{2}\left(b_{mn}dx^m \wedge dx^n + 2\lambda^m{}_n dx^n \wedge \iota_m + \beta^{mn}\iota_m\iota_n\right). \qquad (3.2.13)$$

The λ term is simply how an infinitesimal rotation acts on forms. The b term at the finite level gets exponentiated to the *b-transform*:

$$\alpha \to e^b \wedge \alpha \equiv \left(1 + b + \frac{1}{2}b\wedge b + \frac{1}{6}b\wedge b\wedge b + \cdots\right)\wedge\alpha. \qquad (3.2.14)$$

Recalling $b^k \equiv \underbrace{b\wedge\ldots\wedge b}_{k}$, we have introduced the *form exponential* of b.

3.2.2 Gamma matrix identities from operators

The remarks of Section 3.2.1, especially (3.1.21) and (3.1.24), provide a method to compute gamma matrix identities that is in many ways deeper and more powerful than those of Section 2.1.4.

As a first example, let us rederive (2.1.36). The left action of γ_{mn} is given by

$$\begin{aligned}\overrightarrow{\gamma}_{mn}=\overrightarrow{\gamma}_{[m}\overrightarrow{\gamma}_{n]} &= \mathrm{d}x_m \wedge \mathrm{d}x_n \wedge + \mathrm{d}x_{[m} \wedge \iota_{n]} + \iota_{[m}\mathrm{d}x_{n]} \wedge + \iota_m\iota_n \\ &\overset{(3.2.3a)}{=} \mathrm{d}x_m \wedge \mathrm{d}x_n \wedge + 2\mathrm{d}x_{[m} \wedge \iota_{n]} + \iota_m\iota_n \\ &\equiv (\mathrm{d}x \wedge + \iota)^2_{[mn]} \,.\end{aligned} \tag{3.2.15}$$

In the last step, we have introduced a condensed notation that will be useful later. The terms from (3.1.19) act on a two-form as

$$\begin{aligned}\mathrm{d}x_m \wedge \iota_n(\mathrm{d}x^p \wedge \mathrm{d}x^q) &= 2\delta_n^{[p}\mathrm{d}x_m \wedge \mathrm{d}x^{q]}\,, \\ \iota_m\iota_n(\mathrm{d}x^p \wedge \mathrm{d}x^q) &= 2\delta_n^{[p}\iota_m(\mathrm{d}x^{q]}) = 2\delta_n^{[p}\delta_m^{q]}\,.\end{aligned} \tag{3.2.16}$$

Putting together (3.2.15), (3.2.16), and the Clifford map (3.1.13), we reproduce immediately (2.1.36). The advantage is that (3.2.16) are trivial to generalize to the action on a k-form:

$$\mathrm{d}x_m \wedge \iota_n(\mathrm{d}x^{p_1} \wedge \ldots \wedge \mathrm{d}x^{p_k}) = k\delta_n^{[p_1}\mathrm{d}x_m \wedge \mathrm{d}x^{p_2} \wedge \ldots \wedge \mathrm{d}x^{p_k]}\,, \tag{3.2.17a}$$

$$\iota_m\iota_n(\mathrm{d}x^{p_1} \wedge \ldots \wedge \mathrm{d}x^{p_k}) = k(k-1)\delta_n^{[p_1}\delta_m^{p_2}\mathrm{d}x^{p_3} \wedge \ldots \wedge \mathrm{d}x^{p_k]}\,, \tag{3.2.17b}$$

thus providing also an identity for $\gamma_{mn}\gamma^{p_1\cdots p_k}$ extending (2.1.36).

As another example, let us consider (2.1.6). We need the right-acting version of (3.2.15):

$$\begin{aligned}\overleftarrow{\gamma}_{mn} = \overleftarrow{\gamma}_{[n}\overleftarrow{\gamma}_{m]} &= -(\mathrm{d}x_{[m} - \iota_{[m})(-1)^{\deg}(\mathrm{d}x_{n]} - \iota_{n]})(-1)^{\deg} \\ &\overset{(3.1.25)}{=} \mathrm{d}x_m \wedge \mathrm{d}x_n \wedge - 2\mathrm{d}x_{[m} \wedge \iota_{n]} + \iota_m\iota_n \\ &\equiv (\mathrm{d}x \wedge - \iota)^2_{[mn]} \,.\end{aligned} \tag{3.2.18}$$

Again the last line is in a condensed notation, similar to that in (3.2.15). Combining this with (3.2.15), we see

$$[\gamma_{mn},\,\cdot\,] = 4\mathrm{d}x_{[m} \wedge \iota_{n]}\,. \tag{3.2.19}$$

With (3.2.16), we recover (2.1.6). Again, the advantage of (3.2.19) is that it generates immediately the generalization

$$[\gamma_{mn}, \gamma^{p_1\cdots p_k}] = -4k\delta_{[m}{}^{[p_1}\gamma_{n]}{}^{p_2\cdots p_k]}\,. \tag{3.2.20}$$

This reproduces (2.1.13), (2.1.37a), and (2.1.37c).

We end by considering a more curious operator: $\overrightarrow{\gamma}_m\overleftarrow{\gamma}^m$. When it acts on a bispinor α_k, it gives $\gamma_m\alpha_k\gamma^m$. Using (3.1.21), (3.1.24), and (3.2.3), this time we get

$$\begin{aligned}&\overleftarrow{\gamma}_m\overrightarrow{\gamma}^m\alpha_k = (\mathrm{d}x_m \wedge + \iota_m)(\mathrm{d}x^m \wedge - \iota^m)(-1)^k\alpha_k \\ &= (\iota_m\mathrm{d}x^m \wedge - \mathrm{d}x^m \wedge \iota_m)(-1)^k\alpha_k = (d - 2\mathrm{d}x^m \wedge \iota_m)(-1)^k\alpha_k = (d-2k)(-1)^k\alpha_k\,.\end{aligned} \tag{3.2.21}$$

For the last step, notice that $\mathrm{d}x^m \wedge \iota_m$ simply takes off every $\mathrm{d}x$ from a form and replaces it, giving a 1 for each $\mathrm{d}x$ present in the form; in other words,

$$\mathrm{d}x^m \wedge \iota_m\alpha_k = k\alpha_k = \deg\alpha_k\,, \tag{3.2.22}$$

recalling the definition of deg that follows (3.1.24). Summing up, we have

$$\gamma_m \alpha_k \gamma^m = (d - 2k)(-1)^k \alpha_k . \tag{3.2.23}$$

See also Exercise 3.2.4 for a more pedestrian derivation.

3.2.3 Hodge dual and chiral gamma

The chiral gamma (2.1.18) for spinors has an interesting relation to the Hodge star (3.1.9).

Consider first d = even. The Clifford map applied to the volume form (3.1.7) is proportional to the chiral gamma (2.1.17):

$$\gamma = c \not{\text{vol}} ; \tag{3.2.24}$$

c was discussed in (2.1.17). More interestingly, multiplying by γ has a similar effect than the $*$ map in (3.1.9). Consider for example $d = 4$ Euclidean dimensions: recalling the comment that follows (2.1.20), $c = 1$, and $\gamma = \gamma^1\gamma^2\gamma^3\gamma^4$. Multiplying it from the left gives $\gamma\gamma^1 = -\gamma^2\gamma^3\gamma^4 = -\gamma^{234}$, $\gamma\gamma^2 = \gamma^{134}$ and so on. This resembles the action of (3.1.9): $*\mathrm{d}x^1 = -\mathrm{d}x^2 \wedge \mathrm{d}x^3 \wedge \mathrm{d}x^4$, $*\mathrm{d}x^2 = \mathrm{d}x^1 \wedge \mathrm{d}x^3 \wedge \mathrm{d}x^4$. In other words, in this case

$$\gamma \not{\alpha}_1 = \gamma\, \alpha_m \gamma^m = \frac{1}{6} \alpha_m \epsilon_{npq}{}^m \gamma^{npq} = \not{*\alpha_1} , \tag{3.2.25}$$

where at the end the slash acts on the whole $*\alpha_1$. We can easily generalize this example for arbitrary form degree, dimension, and signature. First we can work out

$$\gamma\, \gamma^{m_1 \ldots m_k} = \frac{c(-1)^{\lfloor \frac{k}{2} \rfloor}}{(d-k)!} \epsilon_{m_{k+1} \ldots m_d}{}^{m_1 \ldots m_k} \gamma^{m_{k+1} \ldots m_d} , \tag{3.2.26}$$

where c is the constant back in (2.1.17) and (2.1.20). We can summarize this more succinctly as follows:

$$\gamma \not{\alpha} = c[*\lambda\alpha]_{/} \qquad (d = \text{even}), \tag{3.2.27}$$

where the λ operator multiplies a k-form by a k-dependent sign:

$$\lambda\alpha_k \equiv (-1)^{\lfloor \frac{k}{2} \rfloor} \alpha_k = (-1)^{\frac{k(k-1)}{2}} \alpha_k . \tag{3.2.28}$$

Recall that the notation in (3.2.27) means that the slash acts on the whole $*\lambda\alpha$.

We now consider d = odd. Equation (3.2.27) is in fact still valid. But we now have to recall that γ is proportional to the identity; we can choose c so that $\gamma = 1$. So (3.1.13) is not bijective, when the $\gamma^{m_1 \ldots m_k}$ are understood as bispinors. For example, both forms 1 and vol are mapped to the identity. More generally, (3.2.27) means that both forms α and $*\lambda\alpha$ are mapped by (3.1.13) to the same bispinor:

$$\not{\alpha} = c[*\lambda\alpha]_{/} . \qquad (d = \text{odd.}) \tag{3.2.29}$$

This explains concretely why the dimension of the space of bispinors is half the dimension of the space of forms when d is odd, as we saw in Section 3.1.2. (For the creator basis in Section 2.2.1, in the Euclidean case we take $c = \mathrm{i}^{(d-1)/2}$.)

As an application of (3.2.27), we can easily show how the Hodge dual $*$ relates wedge products and contractions. Intuitively, this happens because $\mathrm{d}x^m\wedge$ adds an index to a form, while ι_m takes it off. In $d =$ even,

$$\begin{aligned} \vec{\gamma}\, \mathrm{d}x^m\wedge &\overset{(3.2.8\mathrm{a})}{=} \vec{\gamma}\, \frac{1}{2}\left(\vec{\gamma}^m + \overleftarrow{\gamma}^m (-1)^{\deg}\right) \\ &= \frac{1}{2}\left(-\vec{\gamma}^m + \overleftarrow{\gamma}^m (-1)^{\deg}\right) \vec{\gamma} \overset{(3.2.9\mathrm{b})}{=} -\iota^m\, \vec{\gamma} \end{aligned} \tag{3.2.30}$$

and similarly $\vec{\gamma}\ \iota^m = -\mathrm{d}x^m\wedge\, \vec{\gamma}$. In $d =$ odd, $\vec{\gamma}$ is proportional to the identity and commutes with both the left and right γ^m actions. Viewed as the map from the left- to the right-hand side of (3.2.29), it maps even forms to odd and vice versa; in this sense, $\{\vec{\gamma}, (-1)^{\deg}\} = 0$. Changing the signs in (3.2.30), one obtains now $\vec{\gamma}\ \mathrm{d}x^m\wedge = \iota^m\ \vec{\gamma}$. Several similar computations relate the action of γ to $\mathrm{d}x^m$ and ι^m, also using Exercise 3.2.12. We summarize them here:

$$\begin{aligned} &*\lambda\, \mathrm{d}x^m\wedge = (-1)^{d-1}\iota^m * \lambda\,, \qquad &&\lambda *\mathrm{d}x^m\wedge = \iota^m\lambda *\,; \\ &*\lambda\, \iota^m = (-1)^{d-1}\mathrm{d}x^m \wedge *\lambda\,, \qquad &&\lambda *\iota^m = \mathrm{d}x^m \wedge \lambda *\,. \end{aligned} \tag{3.2.31}$$

3.2.4 Inner products

Bispinors are matrices, and on the space of bispinors there is then a natural, positive-definite inner product:

$$\mathrm{Tr}(A^\dagger B)\,. \tag{3.2.32}$$

(To see that it is positive definite, notice that for $B = A$ it becomes the sum of the norms of the entries.) We can use the Clifford map (3.1.13) to translate this inner product into one among forms. In Euclidean signature, the adjoint of our basis gives $(\gamma^{m_1\ldots m_k})^\dagger = \gamma^{m_k\ldots m_1} = (-1)^{\lfloor k/2\rfloor}\gamma^{m_1\ldots m_k}$; if α is a complex form,

$$(\not{\alpha})^\dagger = (\lambda\bar{\alpha})_/ \qquad \text{(Eucl.)} \tag{3.2.33}$$

taking into account that the coefficients of α also get conjugated by the †. When $d =$ even, the basis is also orthonormal with respect to (3.2.32):

$$\mathrm{Tr}((\gamma^{m_1\ldots m_k})^\dagger\gamma_{n_1\ldots n_l}) = \mathrm{Tr}(\gamma^{m_k\ldots m_1}\gamma_{n_1\ldots n_l}) = \begin{cases} 0 & \text{if } k \neq l\,, \\ 2^{\lfloor d/2\rfloor} k!\, \delta^{m_1}_{[n_1} \cdots \delta^{m_k}_{n_k]} & \text{if } k = l\,. \end{cases} \tag{3.2.34}$$

So for two k-forms α_k, β_k, with coefficients defined as in (3.1.6a), we have the following:

$$\mathrm{Tr}\left((\not{\alpha}_k)^\dagger \not{\beta}_k\right) = \frac{1}{(k!)^2}\bar{\alpha}^{m_1\ldots m_k}\beta_{n_1\ldots n_k}\mathrm{Tr}(\gamma^\dagger_{m_1\ldots m_k}\gamma^{n_1\ldots n_k}) = \frac{2^{\lfloor d/2\rfloor}}{k!}\bar{\alpha}^{m_1\ldots m_k}\beta_{m_1\ldots m_k}\,. \tag{3.2.35}$$

Now we notice $\mathrm{d}x^{m_1} \wedge \ldots \wedge \mathrm{d}x^{m_d} = \epsilon^{m_1 \ldots m_d} \mathrm{vol}$, using which

$$(*\mathrm{d}x^{m_1} \wedge \ldots \wedge \mathrm{d}x^{m_k}) \wedge \mathrm{d}x_{n_1} \wedge \ldots \wedge \mathrm{d}x_{n_k} \tag{3.2.36}$$
$$= \frac{1}{(d-k)!} \epsilon_{m_{k+1} \ldots m_d}{}^{m_1 \ldots m_k} \mathrm{d}x^{m_{k+1}} \wedge \ldots \wedge \mathrm{d}x^{m_d} \wedge \mathrm{d}x_{n_1} \wedge \ldots \wedge \mathrm{d}x_{n_k}$$
$$= \frac{1}{(d-k)!} \epsilon_{m_{k+1} \ldots m_d}{}^{m_1 \ldots m_k} \epsilon^{m_{k+1} \ldots m_d}{}_{n_1 \ldots n_k} \mathrm{vol} = k!\, \delta^{[m_1}_{n_1} \ldots \delta^{m_k]}_{n_k} \mathrm{vol}\,.$$

On general forms, this becomes

$$* \alpha_k \wedge \beta_k = \frac{1}{k!} \alpha^{m_1 \ldots m_k} \beta_{m_1 \ldots m_k} \mathrm{vol} \equiv \alpha_k \cdot \beta_k \mathrm{vol}\,. \tag{3.2.37}$$

If we repeat the computation for two forms with unequal degrees, we will get zero, by (3.2.34). So for two general polyforms α and β, we can write (3.2.35) as

$$\frac{1}{2^{\lfloor d/2 \rfloor}} \mathrm{Tr}\left((\not{\alpha})^\dagger \not{\beta} \right) \mathrm{vol} = \bar{\alpha} \cdot \beta \mathrm{vol} = (*\bar{\alpha} \wedge \beta)_d, \qquad \text{(Eucl.)} \tag{3.2.38}$$

where $(\ldots)_d$ denotes keeping only the d-form part.

In Lorentzian signature, gamma matrices are not all Hermitian; we should use (2.3.12). Retracing our steps, we obtain

$$\frac{(-1)^k}{2^{\lfloor d/2 \rfloor}} \mathrm{Tr}(\gamma_0 \not{\alpha}_k^\dagger \gamma^0 \not{\beta}_k) = \bar{\alpha}_k \cdot \beta_k\,, \qquad \gamma_0 \not{\alpha}_k^\dagger \gamma^0 = (-1)^k (\lambda(\bar{\alpha}_k))_{/}. \qquad \text{(Lor.)} \tag{3.2.39}$$

The inner product (3.2.32) is not invariant under $\mathrm{O}(d, d)$, for the same reason as for (2.3.11). In that case, the solution was to add an intertwiner that took care of the fact that not all gamma matrices are Hermitian, thus arriving for Lorentzian signature at the invariant (2.3.13). We could use the same procedure here, but we just give the answer:

$$(\alpha, \beta) \equiv \frac{(\alpha \wedge \lambda(\beta))_d}{\mathrm{vol}}\,, \tag{3.2.40}$$

where again $_d$ means keeping the d-form part only, so the quotient makes sense. It pairs a k-form and a $(d-k)$-form, and is invariant $\mathrm{O}(d, d)$-invariant (Exercise 3.2.13). This is called *Chevalley–Mukai pairing*; it is not quite correct to call it an inner product, for reasons similar to those in footnote 9 prior to (2.3.13). For example, in Euclidean $d = 6$ it is antisymmetric rather than symmetric:

$$(\alpha, \beta) = -(\beta, \alpha) \qquad (d = 6)\,. \tag{3.2.41}$$

(See Exercise 3.2.17.) Sometimes it is useful to also consider more generally

$$(\alpha, \beta)_k \equiv (\alpha \wedge \lambda(\beta))_k \tag{3.2.42}$$

even when $k < d$. In this case, of course, we do not divide by the volume form as in (3.2.40). This is not an inner product and is thus less mathematically significant, but it will occasionally also play a role.

Equation (3.2.40) does not involve the metric, as one can see from the fact that one need not raise any indices to compute it. For now, this only means that its definition does not depend on the signature of spacetime, but it will become even more important in later chapters, where we will deal with curved spaces.

3.2.5 Pure forms

As we saw in Section 3.2.1, the contraction and wedge operators ι_m and $dx^m\wedge$ obey a "doubled" Clifford algebra. Forms are then spinors for this "split-signature" $\mathrm{Cl}_{d,d}$, and for a generalized Lorentz group $\mathrm{O}(d,d)$. We can then consider the orbits of this group on Λ, as in Section 2.4. We will limit ourselves to the pure orbit, which is particularly important. The definition is the same as usual: a form Φ is pure if $\mathrm{Ann}(\Phi)$ has half as the number of gamma matrices, which is now $\frac{1}{2}(d+d) = d$.

The most trivial example of a pure form is simply $\Phi = 1$. In this case, the annihilator is simply given by all contraction operators:

$$\mathrm{Ann}(1) = \mathrm{Span}\{\iota_m\}, \tag{3.2.43}$$

which clearly has dimension d. Another easy example is given by the volume form vol, which is annihilated instead by all the wedge operators: $\mathrm{Ann}(\mathrm{vol}) = \mathrm{Span}\{dx^m\wedge\}$.

An easy generalization of these two is $dx^1 \wedge \ldots \wedge dx^k$. It is annihilated by $dx^1\wedge, \ldots dx^k\wedge$, and by the contraction operators $\iota_{k+1}, \ldots \iota_d$, for a total of d gamma matrices; so it is also pure. Even more generally, we can consider any k-form that is a wedge of one-forms:

$$\Phi = \alpha_1 \wedge \ldots \wedge \alpha_k\,. \tag{3.2.44}$$

Such a k-form is called *decomposable* (or *simple*). It is pure because its annihilator consists of the wedges of the k one-forms, plus any contraction that is orthogonal to the one-forms:

$$\mathrm{Ann}(\alpha_1 \wedge \ldots \wedge \alpha_k) = \mathrm{Span}(\{\alpha_1\wedge, \ldots, \alpha_k\wedge\} \cup \{v\,|\iota_v\alpha_1 = \cdots = \iota_v\alpha_k = 0\})\,, \tag{3.2.45}$$

where

$$\iota_v \equiv v^m \iota_m\,. \tag{3.2.46}$$

A k-form Φ is decomposable if and only if it satisfies the *Plücker relations*:

$$(\iota_{m_1} \ldots \iota_{m_{k-2}}\Phi) \wedge \Phi = 0\,. \tag{3.2.47}$$

To find pure forms that are not decomposable, we should impose that $\mathrm{Ann}(\Phi)$ is not generated by contractions and wedges separately as in (3.2.45), but rather by a set of linear combinations, such as

$$\iota_m + \beta_{mn} dx^n \wedge\,. \tag{3.2.48}$$

Since m here can get any value from 1 to d, the span of (3.2.48) is d-dimensional. We know from Section 2.4.2 that an annihilator must be isotropic: the inner product of any two of its generators must be zero. In our case, the inner product of two operators of the type (3.2.48) should vanish, with respect to the inner product of the metric (3.2.6), which pairs a ι with a $dx\wedge$. Alternatively, one can simply compute the anticommutator of two elements using (3.2.5):

$$\left\{\iota_m + \beta_{mn}dx^n\wedge,\, \iota_p + \beta_{pq}dx^q\wedge\right\} = \delta^q_m\beta_{pq} + \delta^n_p\beta_{mn} = \beta_{pm} + \beta_{mp}\,. \tag{3.2.49}$$

Thus β in (3.2.48) should be antisymmetric. With this single condition, the space of gamma matrices of the form (3.2.48) is isotropic. Since it is also maximal (it has dimension d), we know from Section 2.4.2 that it should be the annihilator of some pure spinor Φ. A pure spinor should be chiral; we try with the positive chirality space Λ_+, namely even forms (recalling (3.2.10)). So $\Phi = \Phi_0 + \Phi_2 + \cdots \Phi_d$, where Φ_k is a k-form. Imposing that this is annihilated by (3.2.48) gives

$$\beta_{mn} dx^n \Phi_0 + \iota_m \Phi_2 = 0\,, \qquad \beta_{mn} dx^n \wedge \Phi_2 + \iota_m \Phi_4 = 0\,, \qquad \ldots\,. \tag{3.2.50}$$

The one-form equation is solved by $\Phi_2 = -\frac{1}{2}\beta_{mn} dx^m \wedge dx^n \Phi_0 \equiv -\beta\Phi_0$. This solution is unique, since there are no forms that are annihilated by all ι_m. The three-form equation then has solution $\Phi_4 = \frac{1}{2}\beta \wedge \beta \Phi_0$. Continuing in this vein, one obtains

$$\Phi = \left(1 - \beta + \frac{1}{2}\beta \wedge \beta - \frac{1}{6}\beta \wedge \beta \wedge \beta + \ldots\right)\Phi_0 = e^{-\beta}\Phi_0\,, \tag{3.2.51}$$

where Φ_0 is a number. We have thus found a new kind of pure form, different from our earlier example (3.2.45), with annihilator (3.2.48):

$$\text{Ann}(e^{-\beta}) = \text{Span}\left(\{\iota_m + \beta_{mn} dx^n \wedge\}\right)\,. \tag{3.2.52}$$

These two classes can be combined: a form

$$\Phi = \alpha_1 \wedge \ldots \wedge \alpha_k \wedge e^{-\beta}\,, \tag{3.2.53}$$

where α_i are one-forms and β is a two-form, is always pure. In fact, *any* pure form can be written as (3.2.53). Notice that the b-transform (3.2.14) maps a form of this type into another ($\beta \to \beta - b$); so it preserves purity. The integer k, which is the lowest form degree present in (3.2.53), is called the *type* of Φ.

Exercise 3.2.1 In dimension $d = 2$, consider a two-form α_2. Compute $\exp[\alpha_2]$ and $(\exp[\alpha_2])_{/}$.

Exercise 3.2.2 As noted in the main text, (3.1.21) and (3.1.24) diagonalize $\mathcal{I}$. Using this and (2.1.17), find $\Gamma = (1 - 2dx^1 \wedge \iota_1) \ldots (1 - 2dx^d \wedge \iota_d)$ for the chirality operator for the doubled Clifford algebra (3.2.5); by acting on a test form, show $\Gamma = (-1)^{\deg}$, as claimed in the text preceding (3.2.10).

Exercise 3.2.3 Consider a mixed wedge-contraction operator $M\cdot \equiv M_m^n dx^m \wedge \iota_n$. Show that

$$[M\cdot, \iota_v] = -\iota_{M^t v}\,, \qquad [M\cdot, \alpha\wedge] = (M\alpha)\wedge\,, \tag{3.2.54}$$

with v a vector, $\iota_v \equiv v^m \iota_m$ as in (3.2.46), and α is a one-form.

Exercise 3.2.4 Rederive (3.2.23) more directly by checking it on a test form $\alpha_k = \gamma^{1\ldots k}$. (Hint: the γ^m, $m = 1, \ldots, k$ on the right can be moved to the left by paying a sign $(-1)^{k-1}$, while the γ^m, $m = k+1, \ldots, d$ by paying $(-1)^k$.)

Exercise 3.2.5 Use (3.2.15) and (3.2.18) to find an analogue of (3.2.20) for $\{\gamma_{mn}, \gamma^{p_1 \ldots p_k}\}$. Check that it agrees with (2.1.37b) and (2.1.37d).

Exercise 3.2.6 Similar to (3.2.15) and (3.2.18), show

$$\overrightarrow{\gamma}_{mnp} = \mathrm{d}x_m \wedge \mathrm{d}x_n \wedge \mathrm{d}x_p \wedge + 3\mathrm{d}x_{[m} \wedge \mathrm{d}x_n \wedge \iota_{p]} + 3\mathrm{d}x_{[m} \wedge \iota_n \iota_{p]} + \iota_m \iota_n \iota_p \equiv (\mathrm{d}x + \iota)^3_{[mnp]}\,; \tag{3.2.55a}$$

$$\overleftarrow{\gamma}_{mnp} = (\mathrm{d}x_m \wedge \mathrm{d}x_n \wedge \mathrm{d}x_p \wedge - 3\mathrm{d}x_{[m} \wedge \mathrm{d}x_n \wedge \iota_{p]} + 3\mathrm{d}x_{[m} \wedge \iota_n \iota_{p]} - \iota_m \iota_n \iota_p)(-1)^{\deg} \equiv (\mathrm{d}x - \iota)^3_{[mnp]}(-1)^{\deg}\,. \tag{3.2.55b}$$

Exercise 3.2.7 Use (3.2.55) to compute the (anti)commutator of γ_{mnp} with $\gamma^{r_1 \ldots r_k}$. Check that it agrees with (2.1.37e) and (2.1.37f).

Exercise 3.2.8 Consider a two-form α_2 in $d = 2$. Show

$$(\mathrm{e}^{-\alpha_2})_/ (\mathrm{e}^{\alpha_2})_/ = \det_{(mn)} (\delta_{mn} + \alpha_{mn})\,. \tag{3.2.56}$$

Now generalize (3.2.56) to $d > 2$, by using the fact that every antisymmetric matrix can be put in the block-diagonal form $\mathrm{diag}(\alpha_{12}\epsilon, \alpha_{34}\epsilon, \ldots)$, where $\epsilon = \begin{pmatrix} 0 & 1 \\ -1 & 0 \end{pmatrix}$.

Exercise 3.2.9 Show

$$\gamma_{mn}\, \alpha_k\, \gamma^{mn} = (-(d - 2k)^2 + d)\alpha_k\,. \tag{3.2.57}$$

You might want to use (2.1.33) and (3.2.23). Alternatively, check it in a basis, as described after (3.2.23).

Exercise 3.2.10 Show the following identities:

$$\begin{aligned} \lambda\, \alpha_1 \wedge &= \alpha_1 \wedge \lambda\, (-1)^{\deg}\,, & \lambda\, \alpha_2 \wedge &= -\alpha_2 \wedge \lambda\,, \\ \lambda\, \alpha_3 \wedge &= -\alpha_3 \wedge \lambda\, (-1)^{\deg}\,, & \lambda\, \alpha_4 \wedge &= \alpha_4 \wedge \lambda\,, \end{aligned} \tag{3.2.58}$$

where α_k is a k-form. (Hint: check them on a test j-form for any j. You might find it easier with the last expression in (3.2.28); alternatively, use $(-1)^{\lfloor \frac{j+1}{2} \rfloor} = (-1)^{j + \lfloor \frac{j}{2} \rfloor}$.)

Exercise 3.2.11 Show in a similar way

$$* \lambda\, \alpha_k = (-1)^{(d+1)k + \lfloor \frac{d}{2} \rfloor} \lambda * \alpha_k \tag{3.2.59}$$

(in both Euclidean and Lorentzian signatures). So, for example, $*\lambda = -\lambda * (-1)^{\deg}$ for $d = 6$ and $d = 10$.

Exercise 3.2.12 Similar to (3.2.27) and (3.2.29), show

$$\not{\alpha}\, \gamma = (-1)^{\lfloor d/2 \rfloor} c[\lambda * \alpha]_/\,. \tag{3.2.60}$$

Exercise 3.2.13 Use (3.2.58) to show

$$\begin{aligned} (\alpha, \mathrm{d}x^m \wedge \beta) &= (-1)^{d+1} (\mathrm{d}x^m \wedge \alpha, \beta)\,, \\ (\alpha, \iota_m \beta) &= (-1)^{d+1} (\iota_m \alpha, \beta)\,. \end{aligned} \tag{3.2.61}$$

In other words, $\mathrm{d}x^m \wedge$ and ι_m are (anti-)self-adjoint with respect to the Chevalley–Mukai pairing (3.2.40). The same equations also hold for (3.2.42). Use this to check the Chevalley–Mukai pairing (3.2.40) is $\mathrm{O}(d,d)$-invariant at the infinitesimal level.

Exercise 3.2.14 Check that (3.2.40) is $\mathrm{O}(d,d)$-invariant at the finite level for the b-transform (3.2.14):

$$(\mathrm{e}^b \wedge \alpha, \mathrm{e}^b \wedge \beta) = (\alpha, \beta)\,. \tag{3.2.62}$$

Exercise 3.2.15 Using (3.2.27) and (3.2.31), show

$$\iota_v \lambda * \alpha_k = (-1)^k \iota_{\alpha_k} \lambda * v, \tag{3.2.63}$$

where v is a one-form and

$$\iota_{\alpha_k} \equiv \frac{1}{k!} \alpha_{m_1 \ldots m_k} \iota^{m_1} \ldots \iota^{m_k} \tag{3.2.64}$$

is contraction by a k-form.

Exercise 3.2.16 Again as a one-line application of (3.2.31), show

$$\iota_{\alpha_k} \text{vol} = (-1)^k * \lambda \alpha_k . \tag{3.2.65}$$

Exercise 3.2.17 In d dimensions, show that (3.2.40) obeys

$$(\alpha, \beta) = (-1)^{\lfloor \frac{d}{2} \rfloor} (\beta, \alpha) . \tag{3.2.66}$$

Exercise 3.2.18 For $\alpha_\pm$, an even/odd real form in d = even, show

$$(\iota_m \alpha_\pm) \cdot (\iota_n \alpha_\pm) - \frac{1}{2} g_{mn} \alpha_\pm \cdot \alpha_\pm = \mp \frac{1}{2^{\lfloor d/2 \rfloor + 1}} \text{Tr}(\lambda(\alpha_\pm) \gamma_{(m} \alpha_\pm \gamma_{n)}) . \tag{3.2.67}$$

(Hint: use (3.2.38), (3.2.9b), $(\lambda \alpha)_/ = \alpha\!\!\!/^\dagger$, and the cyclicity of the trace. Alternatively, test it on the basis of forms (3.1.6b).) Check that contracting this with g^{mn} leads back to (3.2.38) upon using (3.2.23). Equation (3.2.67) is also valid in Lorentzian signature, although there $(\lambda \alpha)_/ = (-1)^k \gamma_0 \alpha\!\!\!/_k^\dagger \gamma^0$, (3.2.39).

Exercise 3.2.19 Again in d = even, show

$$(\iota_m \alpha_\pm) \cdot (dx_n \wedge \alpha_\pm) = \pm \frac{1}{2^{\lfloor d/2 \rfloor + 1}} \text{Tr}(\lambda(\alpha_\pm) \gamma_{[m} \alpha_\pm \gamma_{n]}) . \tag{3.2.68}$$

(Hint: use (3.2.38), (3.2.9a), and that γ_{mn} is anti-Hermitian.) Again, this is also valid in Lorentzian signature, with the same comment as in the previous exercise.

3.3 From bilinears to bispinors

A spinor *bilinear* is a tensor associated to a spinor via a quadratic map. Of particular importance are bilinear k-forms: for example:[2]

$$\eta^\dagger \gamma_{m_1 \ldots m_k} \eta . \tag{3.3.1}$$

If enough bilinears $\{B_1, \ldots, B_n\}$ are considered, η can be fully reconstructed from them. In other words, there is an inverse map

$$\{B_1, \ldots, B_n\} \mapsto \eta . \tag{3.3.2}$$

To check this, one may compute their common stabilizer Stab$\{B_1, \ldots, B_n\}$, namely the subgroup of SO(d) that leaves all the B_i invariant. A priori, Stab$\{B_1, \ldots, B_n\} \supset$ Stab(η), but if an inverse map (3.3.2) exists, Stab$\{B_1, \ldots, B_n\} \cong$ Stab(η).

[2] Physical spinor fields are fermionic, and one may wonder whether one should take this into account in considering bilinears. But we will consider bilinears of spinorial parameters of the supersymmetry transformations, not propagating fields. More importantly, the algebraic properties we are concerned with here can be analyzed with any mathematical device we see fit, independent of the eventual physics application.

3.3.1 Two Euclidean dimensions

To parallel our discussion in Section 2.4, we will start with chiral spinors $\eta = \eta_+$ in low Euclidean dimensions, beginning with $d = 2$.

If $\eta = \eta_+$ is chiral, among the bilinears (3.3.1), the case $k = 1$ vanishes because it is an inner product between two spinors of different chiralities, and γ_m changes chirality, by (2.1.22). The cases $k = 0, 2$ are not very informative: $k = 0$ is simply the norm $||\eta_+||^2$, and for $k = 2$ the only nonzero entry is $\eta_+^\dagger \gamma_{12} \gamma_+ = -\mathrm{i}\eta_+^\dagger \gamma \eta_+ = -\mathrm{i}||\eta_+||^2$ (recalling (2.1.20) and the comment that follows it).

We can obtain more interesting results using an inner product with the conjugate spinor. This has chirality -1 by (2.3.10), so we can call it $\eta_- \equiv (\eta_+)^{\mathrm{c}}$. This time the only surviving bilinear is $v_m \equiv \eta_-^\dagger \gamma_m \eta_+$.

We can be more explicit in our usual basis (2.2.4). All positive-chirality spinors are proportional to $\eta_+ = |+\rangle$. Recalling (2.3.7), we have $\eta_- = |-\rangle$. With a complex coordinate $z = x + \mathrm{i}y$, (2.2.2) gives us $v_z = 1$, $v_{\bar{z}} = 0$. In other words

$$\eta_-^\dagger \gamma_m \eta_+ \, \mathrm{d}x^m = \mathrm{d}z = \mathrm{d}x + \mathrm{i}\mathrm{d}y\,. \tag{3.3.3}$$

The stabilizer of this is trivial; this is in agreement with (2.4.35), which tells us that $\mathrm{Stab}(\eta_+)$ is trivial too. In this case, however, the issue of reconstructing η_+ from its bilinears is moot, since there is only one anyway up to a proportionality constant.

3.3.2 Four Euclidean dimensions

The case $d = 4$ is more interesting. This time, (3.3.1) vanishes for $k = 1$ and 3. Again the $k = 0$ case is simply the norm $||\eta_+||^2$. For $k = 2$:

$$J_{mn} \equiv -\mathrm{i}\eta_+^\dagger \gamma_{mn} \eta_+\,. \tag{3.3.4}$$

This two-form is real:

$$J^*_{mn} = \mathrm{i}(\eta_+^\dagger \gamma_{mn} \eta_+)^\dagger = \mathrm{i}\eta_+^\dagger \gamma_{nm} \eta_+ = J_{mn}\,. \tag{3.3.5}$$

Again we can also consider inner products with η_+^{c}; but in this case it has positive chirality, and we cannot call it η_-. Among these new bilinears,

$$(\eta_+^{\mathrm{c}})^\dagger \gamma_{m_1 \ldots m_k} \eta_+ = \eta_+^t B^\dagger \gamma_{m_1 \ldots m_k} \eta_+\,, \tag{3.3.6}$$

again $k = 1$ and 3 vanish by chirality. The $k = 0$ case in fact also vanishes: both $B_\pm$ in (2.3.7), for example $B_+ = \mathrm{i}\sigma_1 \otimes \sigma_2$, are antisymmetric. The $k = 4$ case then also vanishes, because $\gamma_{mnpq} = \epsilon_{mnpq}\gamma$ (the $k = 0$ case of (3.2.26)). So only the $k = 2$ case survives:

$$\Omega_{mn} \equiv -(\eta_+^{\mathrm{c}})^\dagger \gamma_{mn} \eta_+\,. \tag{3.3.7}$$

This two-form is complex, in contrast to (3.3.4).

A canonical example

We can get more intuition about the two-forms J and Ω by picking a particular spinor, say $\eta_+ = |++\rangle$. After all, we know from Section 2.4.1 that any chiral spinor can be related to this by a rotation. From (2.4.28), we obtain

$$J_{i\bar{j}} = \frac{\mathrm{i}}{2}\delta_{i\bar{j}} \tag{3.3.8}$$

and $J_{\bar{i}\bar{j}} = 0$. (We are using here holomorphic indices; recall footnote 2 following (2.1.1).) Then we also have $J_{ij} = 0$ by reality, or directly by $\gamma_{ij}|++\rangle = \epsilon_{ij}|--\rangle$. We recognize this tensor as the one we introduced back in (2.4.31). In the wedge formalism:

$$J \equiv \frac{1}{2}J_{mn}\mathrm{d}x^m \wedge \mathrm{d}x^n = J_{i\bar{j}}\mathrm{d}z^i \wedge \mathrm{d}\bar{z}^{\bar{j}} = \frac{\mathrm{i}}{2}\mathrm{d}z^i \wedge \mathrm{d}\bar{z}^{\bar{i}}. \tag{3.3.9}$$

Going back to real indices,

$$J = \mathrm{d}x^1 \wedge \mathrm{d}x^3 + \mathrm{d}x^2 \wedge \mathrm{d}x^4. \tag{3.3.10}$$

The components of (3.3.7) are also most easily evaluated in holomorphic indices. The conjugate of η_+ is $\eta_- = \mathrm{i}\sigma_1 \otimes \sigma_2|++\rangle = -|--\rangle$. The nonzero components are those where γ_{mn} contain two annihilators, turning $|{+}{+}\rangle$ into $|{-}{-}\rangle$. So $\Omega_{\bar{i}\bar{j}} = \Omega_{\bar{i}j} = 0$, $\Omega_{ij} = \epsilon_{ij}$; in other words,

$$\Omega = \mathrm{d}z^1 \wedge \mathrm{d}z^2. \tag{3.3.11}$$

This tensor also already appeared, in (2.4.32). It is sometimes called a *holomorphic volume form*: it is similar in spirit to (3.1.7), but it is a wedge of the differentials of all complex coordinates z^i rather than of all real coordinates x_m. Applying the Clifford map to the $\mathrm{d}z^i$ gives the $\gamma^i = g^{i\bar{j}}\gamma_{\bar{j}}$; remembering that the Euclidean metric reads $\frac{1}{2}\delta_{i\bar{j}}$ in complex coordinates, this is $2\gamma_{\bar{i}}$. These are nothing but the two generators of $\mathrm{Ann}(|++\rangle)$ in (2.4.2). So Ω gives a way to recover the annihilator.

Characterization of bilinears

The spaces of *(p,q)-forms* are defined as

$$\Lambda^{p,q} \equiv \mathrm{Span}\{\mathrm{d}z^{i_1} \wedge \ldots \wedge \mathrm{d}z^{i_p} \wedge \mathrm{d}\bar{z}^{\bar{j}_1} \wedge \ldots \wedge \mathrm{d}\bar{z}^{\bar{j}_q}\} \subset \Lambda^{p+q}. \tag{3.3.12}$$

For example, (3.3.11) is a $(2,0)$-form, while from (3.3.9) we see that J is a $(1,1)$-form.

While there are several $(1,1)$-forms, Ω is the only $(2,0)$-form, up to a proportionality constant. We also see from (3.3.11) that it is decomposable: it is the wedge of the two independent $(1,0)$-forms. Recalling (3.2.45), we see that Ω is a pure form; we will soon see a deeper reason for this.

The two-forms J and Ω are both eigenvectors of the $*$-operator defined in (3.1.9). The space Λ^2 of two-forms in $d = 4$ is six dimensional. The $*$ operator squares to the identity on Λ^2 (as can be seen from (3.2.27)); so its eigenvalues are ± 1. It is easy to see that the trace of the $*$ operator is zero; so the eigenvalues ± 1 must both occur with multiplicity 3. In other words, the eigenspaces of *self-dual* and *anti-self-dual* two-forms

$$\Lambda^2_\pm \equiv \{\alpha | *\alpha = \pm\alpha\} \tag{3.3.13}$$

both have dimension 3. The forms

$$J_a \equiv \{J\,,\ \mathrm{Re}\Omega\,,\ \mathrm{Im}\Omega\} \tag{3.3.14}$$

are all anti-self-dual:

$$* J_a = -J_a\,, \tag{3.3.15}$$

and thus are a basis for Λ^2_-. We will soon see how to obtain (3.3.15) with almost no computation at all, but let us see it now in a more pedestrian fashion using holomorphic indices, as an exercise. For example, $*dz^1 \wedge dz^2 = \epsilon_{12}{}^{12} dz^1 \wedge dz^2$. (In the computation in this paragraph, the indices on the ϵ tensor refer to complex coordinates rather than real: so $\epsilon_{12}{}^{12} \equiv \epsilon_{z^1 z^2}{}^{z^1 z^2}$.) Then

$$\epsilon_{12}{}^{12} = 4\epsilon_{12\bar{1}\bar{2}} = -4\epsilon_{1\bar{1}2\bar{2}}\,; \tag{3.3.16}$$

we used $g_{i\bar{j}} = \frac{1}{2}\delta_{i\bar{j}}$ to lower indices (see the comment that precedes (2.2.6)). Now $\mathrm{vol} = dx^1 \wedge dx^2 \wedge dx^3 \wedge dx^4 = \epsilon_{1\bar{1}2\bar{2}} dz^1 \wedge d\bar{z}^1 \wedge dz^2 \wedge d\bar{z}^2 = \epsilon_{1\bar{1}2\bar{2}}(-2i dx_1 \wedge dx_3) \wedge (-2i dx_2 \wedge dx_4)$, so $\epsilon_{1\bar{1}2\bar{2}} = \frac{1}{4}$ and $\epsilon_{12}{}^{12} = -1$. So in the end $*\Omega = -\Omega$. In a similar way, one also sees $*J = -J$.

Equation (3.3.14) are also an orthonormal basis, with respect to the inner product $\alpha \cdot \beta = \frac{1}{2}\alpha_{mn}\beta^{mn}$. This can also be expressed as

$$J_a \wedge J_b = -2\delta_{ab}\mathrm{vol}\,, \tag{3.3.17}$$

or perhaps more explicitly as

$$J \wedge \Omega = \Omega \wedge \Omega = 0\,, \qquad \frac{1}{4}\Omega \wedge \bar{\Omega} = \frac{1}{2}J^2 = -\mathrm{vol}\,. \tag{3.3.18}$$

A basis for Λ^2_+ can instead be obtained by considering $(1,1)$-forms β that are orthogonal to J:

$$\Lambda^2_+ = \{\beta \in \Lambda^{1,1} \mid J^{mn}\beta_{mn} = 0\}\,. \tag{3.3.19}$$

Such $(1,1)$-forms are also called *primitive*. Notice that Λ^2_+ and Λ^2_- are orthogonal under the inner product (3.2.37).

η_+ from Ω

We have obtained the properties (3.3.15) and (3.3.17) from the explicit expressions (3.3.10) and (3.3.11), valid for the spinor $|++\rangle$. But as we anticipated, any spinor $\eta_+ = R|++\rangle$ for some rotation R; the J and Ω for η_+ are related to (3.3.10) and (3.3.11) by R. Since the properties (3.3.15) and (3.3.17) are covariant under rotations, we see that they are still valid for any η_+.

Explicitly, if $R = e^\lambda$, from (2.1.16) we see that $\Omega = (e^\lambda)^{z^1}{}_m (e^\lambda)^{z^2}{}_n dx^m \wedge dx^n$, and similarly for J. In particular, Ω will still be a wedge of two one-forms: by (2.4.6) we recognize these explicitly as two generators of $\mathrm{Ann}(\eta_+)$. So the statement we made after (3.3.11) is true in general: Ω decomposes as the wedge product of two generators of $\mathrm{Ann}(\eta_+)$. This shows that one can reconstruct η_+ from Ω alone. (Indeed, J is also determined by Ω: it can be determined as the third element of the orthonormal basis (3.3.17).)

To see this even more explicitly, consider the general positive-chirality spinor (2.4.1). The conjugate $(\eta_+)^c = -\bar{a}|--\rangle + \bar{b}|++\rangle$, and

$$\begin{aligned}\Omega &= \frac{1}{2}\left(a\langle --| - b\langle ++|\right)\gamma_{mn}\left(a|++\rangle + b|--\rangle\right)\mathrm{d}x^m \wedge \mathrm{d}x^n \\ &= a^2\mathrm{d}z^1 \wedge \mathrm{d}z^2 + b^2\mathrm{d}\bar{z}^1 \wedge \mathrm{d}\bar{z}^2 + a\,b(\mathrm{d}z^1 \wedge \mathrm{d}\bar{z}^1 + \mathrm{d}z^2 \wedge \mathrm{d}\bar{z}^2) \\ &= (a\mathrm{d}z^1 - b\mathrm{d}\bar{z}^2) \wedge (a\mathrm{d}z^2 + b\mathrm{d}\bar{z}^1)\,.\end{aligned} \tag{3.3.20}$$

This confirms our expectations: the two one-forms in the last line are proportional to the two generators in (2.4.8), after applying the Clifford map and lowering indices with the metric $g_{i\bar{j}} = \frac{1}{2}\delta_{i\bar{j}}$.

At the end of Section 2.4.1, we also commented that the annihilator of any η_+ can be brought into the form $\mathrm{Span}\{\gamma_{\bar{z}'_i}\}$ by choosing new appropriate complex coordinates z'_i, complex linear combinations of the x_m; we called such a choice a complex structure. In these coordinates, the bilinears read $J = \frac{\mathrm{i}}{2}\mathrm{d}z'^i \wedge \mathrm{d}\bar{z}'^{\bar{i}}$, $\Omega = \mathrm{d}z'^1 \wedge \mathrm{d}z'^2$. For lighter notation, from now on we will drop the prime, $z'_i \to z_i$, with the understanding that the choice of complex coordinates depends on the particular spinor η_+ we are considering.

A property that is unique to $d = 4$ is that the J and Ω associated to any spinor η_+ are a linear combination of the J and Ω associated to the reference spinor $|++\rangle$. This is because all these forms are in Λ^2_-, which is a three-dimensional space. The definition (3.3.19) is then in fact independent of the choice of the spinor (as long as it is of positive chirality).

Stabilizers

If a rotation $R = \mathrm{e}^\lambda$ leaves Ω invariant, it maps each $\mathrm{d}z^i$ into a linear combination of the $\mathrm{d}z^i$, without any $\mathrm{d}\bar{z}^i$; otherwise, $R\Omega$ would also have components outside $\Lambda^{2,0}$. In other words, R preserves $\Lambda^{1,0}$. λ is then of the form (2.4.34), and that discussion shows that it is in one-to-one correspondence with a matrix in u(2), which is nothing but $\lambda_i{}^j$ (in holomorphic indices). This condition is necessary but not sufficient: the infinitesimal action of $\lambda_i{}^j \in \mathrm{u}(2)$ on $\mathrm{d}z^1 \wedge \mathrm{d}z^2$ gives $\lambda_i{}^i\mathrm{d}z^1 \wedge \mathrm{d}z^2$, so we also need to impose $\lambda_i{}^i = 0$, which implies $\lambda \in \mathrm{su}(2)$. Summing up:

$$\mathrm{Stab}(\Omega) = \mathrm{SU}(2)\,. \tag{3.3.21}$$

This coincides with (2.4.35) for $d = 4$; this confirms our discussion around (3.3.2), since we saw earlier that Ω determines η_+. If we instead compute the stabilizer of J, R should now only preserve $\Lambda^{1,0}$:

$$\mathrm{Stab}(J) = \mathrm{U}(2)\,. \tag{3.3.22}$$

$\mathrm{Stab}(\Omega) \subset \mathrm{Stab}(J)$ is compatible with Ω determining J.

3.3.3 Fierz identities

To analyze $d > 4$, it proves useful to introduce an alternative point of view.

We know that a bispinor is an object with two spinor indices; in other words, a linear map from the space of spinors S to itself. In this subsection, we will focus on bispinors of the type

$$\eta \otimes \eta^\dagger\,. \tag{3.3.23}$$

The tensor-product symbol $\otimes$ is often used in physics. For example, when considering Hilbert spaces for composite systems, we often write states such as $|\psi\rangle \otimes |\psi'\rangle$, perhaps sometimes omitting the $\otimes$ for brevity. In (3.3.23), we are considering instead a tensor product of the space of spinors S with its dual. We also do that in quantum mechanics, when writing an expression such as

$$|\psi\rangle\langle\psi|, \tag{3.3.24}$$

which should read more correctly (or pedantically) $|\psi\rangle \otimes \langle\psi|$. This is an operator from the Hilbert space $\mathcal{H}$ to itself; its action on a state $|\chi\rangle$ gives $|\psi\rangle\langle\psi|\chi\rangle$. Its image consists then of states proportional to $|\psi\rangle$; its kernel consists of states orthogonal to $|\psi\rangle$. In particular, it is a rank-one operator.

Similar to this, (3.3.23) maps a spinor η' to $\eta(\eta^\dagger\eta')$; so it is a rank-one matrix with image $=$ Span(η) and kernel $= \{\eta\}^\perp$.

We now want to compute the preimage of (3.3.23) under the Clifford map (3.1.14); we expect this form to be related to the bilinears (3.3.1). We will treat the d = even and odd cases separately, because of the subtleties discussed in Section 3.1.2.

Even dimensions

To find the preimage under (3.1.14) of a bispinor α, we need to write it as $\sum_{k=0}^{d} \alpha_{m_1\dots m_k}\gamma^{m_1\dots m_k}$ and find the coefficients $\alpha_{m_1\dots m_k}$. For this, we can project on the basis $\gamma_{m_1\dots m_k}$, which by (3.2.34) is orthonormal with respect to the inner product $\mathrm{Tr}(\alpha_1^\dagger\alpha_2)$:

$$\mathrm{Tr}(\alpha\gamma_{n_l\dots n_1}) = \sum_{k=0}^{d} \mathrm{Tr}(\alpha_{m_1\dots m_k}\gamma^{m_1\dots m_k}\gamma_{n_l\dots n_1}) \overset{(3.2.34)}{=} \delta_{lk}2^{d/2}l!\,\alpha_{n_1\dots n_l}\,. \tag{3.3.25}$$

So we have obtained

$$\alpha = \sum_{k=0}^{d} \frac{1}{2^{d/2}k!}\mathrm{Tr}\left(\alpha\gamma_{n_k\dots n_1}\right)\gamma^{n_1\dots n_k}\,. \tag{3.3.26}$$

We already used this in Chapter 1: the RR potentials originally arise in the spectrum as bispinors, and are then recast as forms by using (3.3.26), which we described informally after (1.1.34) and (1.1.38).

We can now apply this to the bispinor (3.3.23). In this case, the trace $\mathrm{Tr}(\alpha\gamma_{n_k\dots n_1}) = \mathrm{Tr}(\eta\otimes\eta^\dagger\gamma_{n_k\dots n_1}) = \eta^\dagger\gamma_{n_k\dots n_1}\eta$, where in the last step we used cyclicity of the trace in the spinorial indices. (We also often do this in quantum mechanics, when we write $\mathrm{Tr}(|\psi\rangle\langle\psi|O) = \langle\psi|O|\psi\rangle$.) Then (3.3.26) becomes the *Fierz identity*:

$$\eta\otimes\eta^\dagger = \sum_{k=0}^{d} \frac{1}{2^{d/2}k!}\left(\eta^\dagger\gamma_{n_k\dots n_1}\eta\right)\gamma^{n_1\dots n_k}\,. \tag{3.3.27}$$

The parenthesis has no free spinorial indices; it is one of the bilinear forms (3.3.1). Now it is easy to find the counterimage under the Clifford map (3.1.13):

$$\eta \otimes \eta^{\dagger} = \left(\sum_{k=0}^{d} \frac{1}{2^{d/2} k!} (\eta^{\dagger} \gamma_{n_k \ldots n_1} \eta) \, dx^{n_1} \wedge \ldots \wedge dx^{n_k} \right)_{/} . \qquad (3.3.28)$$

So we see that $\eta \otimes \eta^{\dagger}$ collects together all the bilinear forms.

The same logic can, of course, be applied to other bispinors, such as $\eta \otimes (\eta^{c})^{\dagger}$, for which (3.3.26) becomes

$$\eta \otimes (\eta^{c})^{\dagger} = \sum_{k=0}^{d} \frac{1}{2^{d/2} k!} \left((\eta^{c})^{\dagger} \gamma_{n_k \ldots n_1} \eta \right) \gamma^{n_1 \ldots n_k} . \qquad (d = \text{even}). \qquad (3.3.29)$$

Odd dimensions

We now also consider d = odd. Here we have the familiar subtlety that the $\gamma_{m_1 \ldots m_k}$ are twice as many as would be needed for a basis. To obtain a basis, we can pick all even k or all odd k; indeed, we saw in (3.2.29) that an even form is equivalent to its $*$ odd form as a bispinor. So the analogue of (3.3.27) can read

$$\begin{aligned} \eta \otimes \eta^{\dagger} &= \left(\sum_{k \text{even}} \frac{1}{2^{(d-1)/2} k!} (\eta^{\dagger} \gamma_{n_k \ldots n_1} \eta) \, dx^{n_1} \wedge \ldots \wedge dx^{n_k} \right)_{/} \\ &= \left(\sum_{k \text{odd}} \frac{1}{2^{(d-1)/2} k!} (\eta^{\dagger} \gamma_{n_k \ldots n_1} \eta) \, dx^{n_1} \wedge \ldots \wedge dx^{n_k} \right)_{/} . \end{aligned} \qquad (d = \text{odd}). \qquad (3.3.30)$$

In fact, we will also use other possibilities, such as keeping all forms of degree 0 to $(d-1)/2$. One can, of course, also take the average of the two expressions in (3.3.30), obtaining an expression where all form degrees are present, at the cost of some redundance.

For later use, we also record the behavior (in any d) of the terms in (3.3.27) and (3.3.29) under the adjoint †. For the left-hand sides, we need to point out

$$(\eta \otimes \chi^{\dagger})^{\dagger} = \chi \otimes \eta^{\dagger} . \qquad (3.3.31)$$

For the right-hand sides, we can just recall (3.2.33).

3.3.4 Bispinors in two dimensions

Let us see how (3.3.27) works in a simple example: we pick $\eta = \eta_{+} = |+\rangle$ in $d = 2$. On the left-hand side, we have $|+\rangle \otimes \langle +|$. We can write this as an explicit matrix as follows. When we introduced the creator basis (2.2.4) in Section 2.2.1, we thought of $|+\rangle$ as the vector $\binom{1}{0}$; see our comment that precedes (2.2.3). We can then write

$$|+\rangle \otimes \langle +| = \begin{pmatrix} 1 \\ 0 \end{pmatrix} \otimes (1\ 0) = \begin{pmatrix} 1 & 0 \\ 0 & 0 \end{pmatrix} . \qquad (3.3.32a)$$

Indeed, this matrix has rank one, and behaves as it should: when applied to any vector $|v\rangle = \binom{v_1}{v_2}$, it gives $\binom{v_1}{0} = \binom{1}{0}((1\ 0) \cdot \binom{v_1}{v_2}) = |+\rangle\langle +|v\rangle$.[3] As for the right-hand side of (3.3.27), we know from Section 3.3.1 that the nonzero bilinear forms

[3] More generally, the tensor product $v \otimes w^{\dagger}$ can be represented by a matrix with entries $v_i \overline{w_j}$.

are $\eta_+^\dagger\eta_+ = ||\eta_+||^2$ and $\eta_+^\dagger\gamma_{12}\eta_+ = \mathrm{i}||\eta_+||^2$. So the right-hand side of (3.3.27) gives

$$\frac{1}{2}\left(||\eta_+||^2 1 - \mathrm{i}||\eta_+||^2\gamma^{12}\right) = \frac{1}{2}(1+\gamma) = \frac{1}{2}(1+\sigma_3) = \begin{pmatrix} 1 & 0 \\ 0 & 0 \end{pmatrix} \qquad (3.3.32b)$$

in agreement with (3.3.32a). Even without fixing the basis to be (2.2.4), we could have noticed that (3.3.27) reduces to

$$|+\rangle \otimes \langle +| = \frac{1}{2}(1+\gamma) = \frac{1}{2}(1+\mathrm{i}dx^1\wedge dx^2)_/\,. \qquad (3.3.33)$$

This is as it should be: the left-hand side is the projector on positive-chirality spinors.

We can check (3.3.29) in the same spirit. $(\eta_+)^c = |-\rangle$ can be identified with $\binom{0}{1}$. With a similar logic as (3.3.32a), we find

$$|+\rangle \otimes \langle -| = \begin{pmatrix} 1 \\ 0 \end{pmatrix} \otimes (0\ 1) = \begin{pmatrix} 0 & 1 \\ 0 & 0 \end{pmatrix}. \qquad (3.3.34a)$$

The only bilinear contributing to the right-hand side of (3.3.29) is (3.3.3), which upon applying the Clifford map becomes

$$\frac{1}{2}(\gamma^1 + \mathrm{i}\gamma^2) = \frac{1}{2}(\sigma_1 + \mathrm{i}\sigma_2) = \begin{pmatrix} 0 & 1 \\ 0 & 0 \end{pmatrix}, \qquad (3.3.34b)$$

again in agreement with (3.3.34a). In other words

$$|+\rangle \otimes \langle -| = \frac{1}{2}(dx^1 + \mathrm{i}dx^2)_/\,. \qquad (3.3.35)$$

3.3.5 Bispinors in four dimensions

As another example, we now show how the bispinor formalism works in $d = 4$ Euclidean dimensions, again with a chiral spinor $\eta = \eta_+$. This time we will be more ambitious: rather than just checking (3.3.27) and (3.3.29), we will use them to rederive the four-dimensional results of Section 3.3.2. In $d > 4$, these methods are much quicker than those of Section 3.3.2.

We start from (3.3.28). Since we are taking $\eta = \eta_+$ to be chiral, only even k contribute to the sum on the right-hand side. The $k = 0$ term is simply $||\eta_+||^2$; the $k = 4$ term is a four-form, and as such it is proportional to vol_4. So we can parameterize

$$\Phi_+ \equiv \eta_+ \otimes \eta_+^\dagger = \frac{1}{4}||\eta_+||^2(1 + \beta_2 + a\mathrm{vol}_4) \qquad (3.3.36)$$

with a two-form β_2. (As we warned earlier, we have dropped the slash.) Let us take the dagger of this equation. Because of (3.3.31), the left-hand side remains invariant; (3.2.33) turns the right-hand side into $\frac{1}{4}||\eta_+||^2(1 - \bar\beta_2 + \bar a\mathrm{vol}_4)$. So a is real, while β_2 is purely imaginary and we parameterize it as $-\mathrm{i}J$, with J real.

We now study the behavior under the $*$ operator. Applying (3.2.27) with $c = 1$:

$$\gamma\eta_+ \otimes \eta_+^\dagger = \eta_+ \otimes \eta_+^\dagger \quad \Rightarrow \quad 1 - \mathrm{i}J + a\mathrm{vol}_4 = *(1 + \mathrm{i}J + a\mathrm{vol}_4)\,. \qquad (3.3.37)$$

The zero-form part of this gives $1 = a * \mathrm{vol}_4$; from the definition (3.1.9), $*\mathrm{vol}_4 = \epsilon^{1234} = 1$, so $a = 1$. The four-form part gives the same result. The two-form part of (3.3.37) gives $*J = -J$, recovering one of (3.3.15). Summing up, so far we have

$$\Phi_+ = \eta_+ \otimes \eta_+^\dagger = \frac{1}{4}||\eta_+||^2(1 - \mathrm{i}J + \mathrm{vol}_4)\,. \tag{3.3.38}$$

We now turn to (3.3.29). Here again only $k = 0, 2, 4$ survive on the right-hand side, but we also know that the zero-form part $(\eta_+^c)^\dagger\eta_+ = 0$.[4] Since $\gamma\eta_+ \otimes (\eta_+^c)^\dagger = \eta_+ \otimes (\eta_+^c)^\dagger$, the same logic as in (3.3.37) implies that the four-form part vanishes, and that the two-form part, which we called Ω in (3.3.7), satisfies $*\Omega = -\Omega$. This reproduces the remaining two of (3.3.15). We are left with

$$\tilde{\Phi}_+ = \eta_+ \otimes (\eta_+^c)^\dagger = \frac{1}{4}||\eta_+||^2\Omega\,. \tag{3.3.39}$$

To reproduce (3.3.18), we can use the inner-product formula (3.2.38). If we take $A = \tilde{\Phi}_+$, $B = \Phi_+$, the left-hand side reads $\mathrm{Tr}(\tilde{\Phi}_+^\dagger\Phi_+) = \mathrm{Tr}((\eta_+)^c\eta_+^\dagger\eta_+\eta_+^\dagger) = \eta_+^\dagger(\eta_+)^c\eta_+^\dagger\eta_+ = 0$; the right-hand side is proportional to

$$(*\bar{\Omega} \wedge (1 - \mathrm{i}J + \mathrm{vol}))_4 = \mathrm{i}\bar{\Omega} \wedge J\,, \tag{3.3.40}$$

where we used $*\Omega = -\Omega$. Thus we conclude $\bar{\Omega} \wedge J = 0$, which is one of (3.3.18). To obtain the other equations there, we can consider (3.2.38) with $A = B = \Phi_+$ to find $-\frac{1}{2}J^2 = \mathrm{vol}$; with $A = B = \tilde{\Phi}_+$ to find $\bar{\Omega} \wedge \Omega = -4\mathrm{vol}$; and with $A = \overline{\tilde{\Phi}_+}$, $B = \tilde{\Phi}_+$ to obtain $\Omega \wedge \Omega = 0$.

We will later retrieve some of these equations once again using the theory of pure spinors and pure forms.

3.3.6 Four Lorentzian dimensions

We now turn to Lorentzian signature. We will consider a Weyl spinor, like we did in the Euclidean case. We will focus on $d = 4$; later in the next section, we will also consider $d = 10$.

In four Lorentzian dimensions, we can impose a Majorana condition instead of Weyl, but not both Majorana and Weyl simultaneously (see Section 2.3.1). We also saw that the set of Weyl spinors and the set of Majorana spinors are in one-to-one correspondence: given ζ_+, the conjugate

$$\zeta_- \equiv \zeta_+^c \tag{3.3.41}$$

has negative chirality. It also follows $\zeta_+ = \zeta_-^c$. Then $\zeta_+ + \zeta_-$ is Majorana. Conversely, given a Majorana ζ, $P_+\zeta = \frac{1}{2}(1+\gamma)\zeta$ has positive chirality. These two maps are each other's inverse. Recall also from (2.2.25) that there exists a real gamma matrix basis, in which $B = B_+ = 1$ and $\zeta_\pm = \zeta_\mp^c = \zeta_\mp^*$. We will often work in this basis, and then phrase our conclusions in a basis-invariant way.

Recalling our definition (2.3.13), we can consider bilinears of the type

$$\overline{\zeta_+}\gamma_{\mu_1\ldots\mu_k}\zeta_+\,, \qquad \overline{\zeta_-}\gamma_{\mu_1\ldots\mu_k}\zeta_+\,. \tag{3.3.42}$$

These are collected in bispinors $\zeta_+ \otimes \overline{\zeta_+}$ and $\zeta_+ \otimes \overline{\zeta_-}$, by a Lorentzian analogue of (3.3.27) and (3.3.29):

$$\zeta \otimes \overline{\zeta'} = \sum_{k=0}^{d} \frac{1}{2^{d/2}k!}\left(\overline{\zeta'}\gamma_{\mu_k\ldots\mu_1}\zeta\right)\gamma^{\mu_1\ldots\mu_k}\,. \tag{3.3.43}$$

[4] In this section, we let $\eta_+^c \equiv (\eta_+)^c$; the ambiguity we warned about in footnote 8 that precedes (2.3.9) should not arise here.

Bilinears with ζ_- on the right can be obtained from (3.3.42) by conjugation: in the real basis (2.2.25), $(\overline{\zeta_\pm}\gamma_{\mu_1\ldots\mu_k}\zeta_+)^* = \overline{\zeta_\mp}\gamma_{\mu_1\ldots\mu_k}\zeta_-$. Just this once, let us also do it in a basis-invariant way; from (2.3.6), we only need to know that $B = B_+ = 1 \otimes \sigma_1$ satisfies $B = B^\dagger = B^{-1} = B^*$. So we can compute

$$\begin{aligned}(\overline{\zeta_\pm}\gamma_{\mu_1\ldots\mu_k}\zeta_+)^* &= (\zeta_\pm^*)^\dagger \gamma_0^* \gamma^*_{\mu_1\ldots\mu_k}\zeta_+^* \\ &= \zeta_\mp^\dagger B \gamma_0^* \gamma^*_{\mu_1\ldots\mu_k} B \zeta_- = \overline{\zeta_\mp}\gamma_{\mu_1\ldots\mu_k}\zeta_- \,.\end{aligned} \tag{3.3.44}$$

The one-form

By chirality, the bilinears $\overline{\zeta_+}\gamma_{\mu_1\ldots\mu_k}\zeta_+$ are only nonzero when k is odd. (Recall $\overline{\zeta_+} = \zeta_+^\dagger\gamma_0$.) The one-form

$$v_\mu \equiv \overline{\zeta_+}\gamma_\mu\zeta_+ \tag{3.3.45}$$

is real: using (2.3.12), $v_\mu^* = (\overline{\zeta_+}\gamma_\mu\zeta_+)^\dagger = \zeta_+^\dagger\gamma_\mu^\dagger\gamma_0^\dagger\zeta_+ = \zeta_+^\dagger\gamma_0\gamma_\mu\zeta_+ = v_\mu$. Let us now consider the bispinor $\zeta_+ \otimes \overline{\zeta_+}$: just as in (3.3.37), we notice $\gamma\zeta_+ \otimes \overline{\zeta_+} = \zeta_+ \otimes \overline{\zeta_+}$. Now (3.3.43) with $\zeta = \zeta' = \zeta_+$ contains the one-form $v \equiv v_\mu \mathrm{d}x^\mu$ and a three-form, which can be reexpressed as the $*$ of v:

$$\begin{aligned}\zeta_+ \otimes \overline{\zeta_+} &= \frac{1}{4}\left((\overline{\zeta_+}\gamma_\mu\zeta_+)\gamma^\mu + \frac{1}{6}(\overline{\zeta_+}\gamma_{\rho\nu\mu}\zeta_+)\gamma^{\mu\nu\rho}\right) \\ &= \frac{1}{4}(1+\gamma)v \overset{(3.2.27)}{=} \frac{1}{4}(v + \mathrm{i} * v)\,.\end{aligned} \tag{3.3.46}$$

We knew this had to be of the form $(1+\gamma)(\ldots)$, since it is invariant under multiplication by γ from the left. We took $c = \mathrm{i}$, in agreement with (2.1.20).

v is light-like

We can compute the action of v on ζ with a trick. Start as follows:[5]

$$v\zeta_+ = v_\mu\gamma^\mu\zeta_+ = \gamma^\mu\zeta_+\overline{\zeta_+}\gamma_\mu\zeta_+\,. \tag{3.3.47a}$$

So far we just used the definition (3.3.45) of v and moved it to the right. Now we can interpret the right-hand side as $\gamma^\mu(\zeta_+ \otimes \overline{\zeta_+})\gamma_\mu\zeta_+$; we can then use (3.3.46):

$$\gamma^\mu\zeta_+\overline{\zeta_+}\gamma_\mu\zeta_+ = \gamma^\mu(\zeta_+ \otimes \overline{\zeta_+})\gamma_\mu\zeta_+ \overset{(3.3.46)}{=} \frac{1}{4}\gamma^\mu(1+\gamma)v\gamma_\mu\zeta_+ = \frac{1}{4}(1-\gamma)\gamma^\mu v\gamma_\mu\zeta_+\,. \tag{3.3.47b}$$

Finally we recall (3.2.23):

$$\frac{1}{4}(1-\gamma)\gamma^\mu v\gamma_\mu\zeta_+ = -\frac{1}{2}(1-\gamma)v\zeta_+ = -v\zeta_+\,. \tag{3.3.47c}$$

Following all of (3.3.47) from the beginning to the end, we see that $v\zeta_+ = -v\zeta_+$, or in other words

$$v\zeta_+ = 0\,. \tag{3.3.48}$$

[5] As stated at the beginning of the chapter, we use the same symbol v both for the vector with components v^μ and for the one-form with components $v_\mu = g_{\mu\nu}v^\nu$, with an abuse of language. The natural action (2.4.13) on a spinor, $\zeta \to v \cdot \zeta = v^\mu\gamma_\mu\eta$, is also equal to $v_\mu\gamma^\mu\zeta = \not{v}\zeta$, and recall that by another abuse of language we sometimes drop the slash; so $v.\zeta = v\zeta$.

Thus ζ_+ is annihilated by a *real* vector, unlike in the Euclidean case, where a chiral spinor is annihilated by complex vectors. We can also multiply (3.3.48) by $\overline{\zeta_+}$ from the left: we get $\overline{\zeta_+} v^\mu \gamma_\mu \zeta_+ = 0$, or in other words

$$v_\mu v^\mu = 0 : \tag{3.3.49}$$

v is light-like. All this is not terribly surprising if we recall our analysis in (2.4.40), where we saw that $|++\rangle$ (for $d = 4$) is annihilated by $\gamma_- = \frac{1}{2}(\gamma_0 - \gamma_1)$, which can be written as $v_\mu \gamma^\mu$ for a real light-like $v^\mu = (1, 1, 0, 0)$.

The space of light-like vectors is a three-dimensional cone in $\mathbb{R}^4$; so v determines ζ up to a single parameter. We can be more precise if we normalize ζ so that the component $v^0 = 1$. The map

$$\{\zeta_+ | \overline{\zeta_+} \gamma^0 \zeta_+ = 1\} \cong S^3 \to \{v \,|\, v^2 = 0, v^0 = 1\} \cong S^2 \tag{3.3.50}$$

is known as *Hopf map*; the counterimage of any point is $S^1 \cong \mathrm{U}(1)$. So v determines ζ up to an overall phase.

The two-form

Now we consider $\zeta_+ \otimes \overline{\zeta_-}$. By chirality, it contains only even forms. The zero-form bilinear is $\overline{\zeta_-} \zeta_+ = \zeta_-^\dagger \gamma_0 \zeta_+ = 0$; in the last step, we have used the fact that γ_0 is anti-Hermitian. Again we notice $\gamma \zeta_+ \otimes \overline{\zeta_-} = \zeta_+ \otimes \overline{\zeta_-}$; this implies that the four-form part is proportional to the zero-form, and hence vanishes. Thus

$$\zeta_+ \otimes \overline{\zeta_-} = \frac{1}{4} \omega \tag{3.3.51}$$

is a two-form, which by (3.2.27) satisfies $\omega = -\mathrm{i} * \omega$; in particular, it is complex. From (3.3.48), we also get

$$v \zeta_+ \otimes \overline{\zeta_-} = 0 \quad \overset{(3.1.21)}{\Rightarrow} \quad v \wedge \omega + \iota_v \omega = 0\,, \tag{3.3.52}$$

where remember $v \cdot = v^m \iota_m$. The two terms in this equation are a one-form and a three-form, so in fact they have to vanish separately. $v \wedge \omega = 0$ implies that there exists[6] a one-form w such that $\omega = v \wedge w$. Since ω is complex, w is complex. Equation (3.3.52) also contains $\iota_v \omega = 0$, so it also follows that $\iota_v w = 0$. It is not unique: one can add to w a component along v, and all the conditions on it are still met. Summarizing:

$$\zeta_+ \otimes \overline{\zeta_+} = \frac{1}{4}(v + \mathrm{i} * v)\,, \qquad \zeta_+ \otimes \overline{\zeta_-} = \frac{1}{4}\omega = \frac{1}{4} v \wedge w\,. \tag{3.3.53}$$

Reproducing the stabilizer

In (2.4.43), we concluded that the stabilizer of a chiral spinor in four Lorentzian dimensions is the abelian group $\mathbb{R}^2$. From the form perspective, we should look for the stabilizer of v and ω. The stabilizer in SO(3, 1) of a light-like vector v can be computed from (2.2.24): the celestial sphere defined there can be thought of as the

[6] In general, if v is a one-form, $v \wedge \alpha_k = 0$ implies that there exists a $(k-1)$-form α_{k-1} such that $\alpha_k = v \wedge \alpha_{k-1}$.

intersection of the light cone through the origin with a plane at fixed x^0. So we need to compute the stabilizer of a point in the celestial sphere, say $z = \infty$. This is the subgroup of (2.2.24) of maps $z \mapsto az + b$ with $|a| = 1$, or in other words, rotations and translations in the complex plane. So

$$\mathrm{Stab}(v) \cong \mathrm{SO}(2) \ltimes \mathbb{R}^2 \qquad (d = 4). \tag{3.3.54}$$

This is familiar from quantum field theory books, where it appears as the little group of a light-like momentum vector. (The integer parameterizing SO(2) representations is helicity.) For later use, we note that the stabilizer of a light-like vector in any dimension is a natural generalization of (3.3.54):

$$\mathrm{Stab}(v) \cong \mathrm{SO}(d-2) \ltimes \mathbb{R}^{d-2}\,. \tag{3.3.55}$$

Now w is a complex one-form orthogonal to v: the condition that it too should be invariant kills the SO(2) in (3.3.54). Thus

$$\mathrm{Stab}\{v, \omega\} = \mathbb{R}^2\,, \tag{3.3.56}$$

in agreement with (2.4.43).

Exercise 3.3.1 With a computation similar to (3.3.16), show that $*\colon \Lambda^{p,q} \to \Lambda^{d/2-q,d/2-p}$.

Exercise 3.3.2 Let ζ^I_+ be a set of chiral spinors in $d = 4$ Lorentzian dimensions. Show that

$$\overline{\zeta^I_-}\zeta^J_+ = -\overline{\zeta^J_-}\zeta^I_+\,, \qquad (\overline{\zeta^I_-}\zeta^J_+)^* = \overline{\zeta^I_-}\zeta^J_+\,; \tag{3.3.57a}$$

$$\overline{\zeta^I_-}\gamma_\mu\zeta^J_- = \overline{\zeta^J_+}\gamma_\mu\zeta^I_+\,, \qquad (\overline{\zeta^I_-}\gamma_\mu\zeta^J_-)^* = \overline{\zeta^I_+}\zeta^J_+\,. \tag{3.3.57b}$$

(Hint: use $v^t A w = w^t A^t v$. Work in the real basis as in Section 3.3.6, or use a basis-independent approach recalling Section 2.3.1.)

Exercise 3.3.3 Repeat the previous exercise in Euclidean signature.

Exercise 3.3.4 Compute (3.3.55), at least at the Lie algebra level, picking a particular v, say $(1, 1, 0, \ldots, 0)$.

3.4 Bispinors in higher dimensions

3.4.1 Pure spinor bilinears

We will now consider pure spinors in $d > 4$. We will use both the more classic bilinear methods of Sections 3.3.1 and 3.3.2, and the bispinor method introduced in later subsections. For the rest of this section, we will take our spinors to have unit norm, $||\eta_+||^2 = 1$, unless otherwise stated.

Bispinors as pure forms

We begin with a general remark, which will spare us many lengthy computations.

If η_+ is pure, by definition (discussed after (2.4.12)) it is annihilated by $d/2$ complex linear combinations of gamma matrices:

$$\gamma_{\bar{\imath}}\eta_+ = 0\,. \tag{3.4.1}$$

The bispinor $\Phi_+ = \eta_+ \otimes \eta_+^\dagger$ is then annihilated by $d/2$ left-acting gamma matrices, and by $d/2$ right-acting gamma matrices:

$$\gamma_{\bar{\imath}}\eta_+ \otimes \eta_+^\dagger = 0\,, \qquad \eta_+ \otimes \eta_+^\dagger \gamma_i = 0. \tag{3.4.2}$$

Remember, however, that left- and right-acting gamma matrices are related to the wedge and contraction operators $dx^m\wedge$ and ι_m by (3.1.21) and (3.1.24). So there are $d/2 + d/2 = d$ complex linear combinations of the Γ_Λ, the gamma matrices of $\text{Cl}_{d,d}$ in (3.2.4), that annihilate $\Phi_+ = \eta_+ \otimes \eta_+^\dagger$. Since there are a total of $2d$ such gamma matrices, we conclude that the form Φ_+ is pure as a $\text{Cl}_{d,d}$ spinor.

The same logic can be applied to $\tilde{\Phi}_+ = \eta_+ \otimes (\eta_+^c)^\dagger$, which is also annihilated by $d/2 + d/2 = d$ linear combinations of gamma matrices:

$$\gamma_{\bar{\imath}}\eta_+ \otimes (\eta_+^c)^\dagger = 0\,, \qquad \eta_+ \otimes (\eta_+^c)^\dagger \gamma_{\bar{\imath}} = 0\,. \tag{3.4.3}$$

More generally, the same idea applies to *two* pure spinors η, η':

$$\eta\,,\ \eta' \text{ pure spinors} \quad \Rightarrow \quad \eta \otimes \eta'^\dagger \text{ pure form.} \tag{3.4.4}$$

This will be useful several times in the future, especially when combined with our general theory of pure forms in Section 3.2.1.

For example, in $d = 4$, Φ_+ in (3.3.38) has type zero (its lowest form degree is zero, as discussed in Section 3.2.5); so it should be a form exponential (3.2.51). That implies $\text{vol}_4 = -\frac{1}{2}J^2$, one of (3.3.18). Likewise, for $\tilde{\Phi}_+$ in (3.3.39) to be a pure spinor, it should be decomposable, as in (3.2.45) for $k = 2$. This implies again $\Omega \wedge \Omega = 0$, another equation in (3.3.18).

Bispinors

We now turn to a detailed study of the bilinears and bispinors defined by a pure spinor η_+.

Bilinears of the type (3.3.1) again vanish for odd k by chirality. The $k = 0$ gives the norm $||\eta_+||^2$; the $k = 2$ case can be called J_{mn} as for $d = 4$ in (3.3.4):

$$J_{mn} \equiv -i\eta_+^\dagger \gamma_{mn} \eta_+\,. \tag{3.4.5}$$

$\Phi_+ \equiv \eta_+ \otimes \eta_+^\dagger$ is a pure form, because of the general principle (3.4.4), and we already noted that its zero-form part is nonzero; so it must be of the form (3.2.51), which fixes it to be

$$\begin{aligned}\Phi_+ &= \eta_+ \otimes \eta_+^\dagger \\ &= \frac{1}{2^{d/2}}\left(1 - iJ - \frac{1}{2}J\wedge J + \frac{i}{6}J\wedge J\wedge J + \cdots\right) = \frac{1}{2^{d/2}}e^{-iJ}\,.\end{aligned} \tag{3.4.6}$$

This agrees with both (3.3.33) in $d = 2$ and (3.3.38) in $d = 4$. From this, we see immediately that the higher k bilinears in (3.4.6) are all in fact wedges of J; for example, $\eta_+^\dagger \gamma_{mnpq}\eta_+$ is proportional to $(J\wedge J)_{mnpq} = 6J_{[mn}J_{pq]}$, and so on.

There is a clumsier, more traditional way of showing that the higher k bilinears are powers of J. We mention it anyway because in some other situations, where purity does not help, it can be a weapon of last resort. One starts with

$$J_{[mn}J_{pq]} = -\eta_+^\dagger \gamma_{[mn}\eta_+\eta_+^\dagger \gamma_{pq]}\eta_+ \,. \tag{3.4.7}$$

Noticing the presence of $\eta_+\eta_+^\dagger$ in the middle of this expression, one can replace it using (3.3.27). The $k = 0$ term in that expansion makes $\eta_+^\dagger \gamma_{mnpq}\eta_+$ appear in (3.4.7); the other terms make higher k bilinear forms appear. One then considers products similar to (3.4.7) but with higher k bilinear forms on the left-hand side. This gives a system of relations that when solved with care gives again (3.4.6).

We now turn to bilinears of the type

$$(\eta_+^c)^\dagger \gamma_{m_1\ldots m_k}\eta_+ \,. \tag{3.4.8}$$

By (2.3.10), the spinor η_+^c has positive chirality for $d = 4$, but negative for $d = 6$, and so on. So only even k survive for $d/2$ = even, and only odd k survive for $d/2$ = odd. In other words, the bispinor $\tilde{\Phi} \equiv \eta_+ \otimes (\eta_+^c)^\dagger$, which collects these bilinears because of (3.3.29), is even for $d/2$ = even and odd for $d/2$ = odd. We also know that $\tilde{\Phi}$ is a pure form, by the same logic surrounding (3.4.3). To determine $\tilde{\Phi}$, we need to know the form of lowest degree.

Here we can use an important property of pure spinors, one we have not needed so far. Consider, for example, $\eta_+ = |+\cdots+\rangle$. The spinor $\gamma_{m_1\ldots m_k}|+\cdots+\rangle$ has at most k minus signs; but it needs to have $d/2$ to make (3.4.8) nonzero, since $\eta_+^c \propto |-\cdots-\rangle$ (for both $B_\pm$ in (2.3.7b)). We thus conclude that (3.4.8) vanishes unless $k \geq d/2$. All pure spinors are related to $|+\cdots+\rangle$ by rotation, so this is in fact true for any pure spinor. Finally, the (3.4.8) are related to those with $k \to d-k$, because of (3.2.26); so in conclusion:

$$(\eta_+^c)^\dagger \gamma_{m_1\ldots m_k}\eta_+ = 0 \qquad \text{unless } k = d/2 \qquad (\eta_+ \text{ pure}). \tag{3.4.9}$$

This property is in fact *equivalent* to purity, and in some contexts it is taken as a definition of it.

In analogy with (3.3.7), we call Ω the $k = d/2$ bilinear:

$$\Omega_{m_1\ldots m_k} \equiv (\eta_+^c)^\dagger \gamma_{m_k\ldots m_1}\eta_+ = (-1)^{\lfloor \frac{k}{2} \rfloor}(\eta_+^c)^\dagger \gamma_{m_1\ldots m_k}\eta_+ \,. \tag{3.4.10}$$

$\tilde{\Phi}$ is a $d/2$-form:

$$\tilde{\Phi} = \eta_+ \otimes (\eta_+^c)^\dagger = \frac{1}{2^{d/2}}\Omega \,. \tag{3.4.11}$$

(Notice that $(\eta_+)^c$ has chirality $(-1)^{d/2}$, from (2.3.10); this is consistent with Ω having even degree for $d/2$ = even, and odd for $d/2$ = odd.) This agrees with (3.3.35) in $d = 2$, and (3.3.39) in $d = 4$.

Since Ω is a pure form, we know from (3.2.45) that it should be decomposable, namely a wedge product of $d/2$ one-forms. Generalizing our $d = 4$ discussion in Section 3.3.2, there is always a choice of complex coordinates z_i such that those one-forms are $\mathrm{d}z_i$:

$$\Omega \propto \mathrm{d}z_1 \wedge \ldots \wedge \mathrm{d}z_{d/2} \,, \tag{3.4.12}$$

again with the understanding that the complex coordinates z_i depend on the η_+. (The proportionality coefficient depends on d and on our choice of B.) Again Ω is the holomorphic volume form associated to the complex structure defined by the pure spinor η_+; in the language of (3.3.12), it is a $(d/2, 0)$-form, and the dz_i are $(1, 0)$-forms. The z_i are independent from the $\bar{z}_i$, so Ω is *nondegenerate*:

$$\Omega \wedge \bar{\Omega} \neq 0\,. \tag{3.4.13}$$

Stabilizer

The same logic that led to (3.3.21) for $d = 4$ now gives

$$\mathrm{Stab}(\Omega) = \mathrm{SU}(d/2), \tag{3.4.14}$$

which matches with (2.4.35); so again Ω determines η_+:

$$\eta_+ \leftrightarrow \Omega\ : \tag{3.4.15}$$

Just like in $d = 4$, η_+ can be determined explicitly by $\mathrm{Ann}(\eta_+) = \{\gamma_{\bar{z}_i}\}$, with the z_i defined by (3.4.12).

One can also extract the complex structure tensor I from

$$\frac{1}{2^{d/2-1}(d/2-1)!}\Omega_{mm_2\ldots m_{d/2}}\bar{\Omega}^{nm_2\ldots m_{d/2}} = \Pi_m{}^n = \frac{1}{2}(\delta_m^n - \mathrm{i}I_m{}^n)\,, \tag{3.4.16}$$

which one can derive from (2.4.32) and (2.4.4b). Then (2.4.30) tells us this tensor I is equivalent to J. (Recall that in this chapter we are still working in flat space.)

The generalization of (3.3.22) is $\mathrm{Stab}(J) = \mathrm{U}(d/2)$, so again J is redundant. Explicitly, this is because Ω is the holomorphic volume form: its expression (3.4.12) singles out the expression of the complex coordinates z_i, and J then has the expression

$$J = J_{i\bar{j}}dz^i \wedge d\bar{z}^{\bar{j}} = \frac{\mathrm{i}}{2}dz^i \wedge d\bar{z}^{\bar{i}}\,, \tag{3.4.17}$$

formally identical to (3.3.9). So again it is a $(1, 1)$-form, as it was in four dimensions.

We could now use the coordinate expressions (3.4.12) and (3.4.17) to work out the properties of J and Ω that generalize (3.3.15) and (3.3.18); these will be useful later, especially in Chapter 5. However, it is more instructive to do this using bispinors. For simplicity, we will do this in $d = 6$, in the next subsection.

3.4.2 Pure spinor bilinears in six dimensions

The six-dimensional case is of particular interest in string compactifications. Recall from Section 2.4.3 that in $d = 6$ all spinors are pure.

The conjugate

$$\eta_+^c \equiv \eta_- \tag{3.4.18}$$

has negative chirality, and indeed $\tilde{\Phi} = \eta_+ \otimes (\eta_+^c)^\dagger = \eta_+ \otimes \eta_-$ is a three-form; we rename it Φ_- to reflect that in particular it is odd. The statement (3.4.9) in our case means that there is no one-form bilinear:

$$\eta_-^\dagger \gamma_m \eta_+ = 0 \tag{3.4.19}$$

nor a five-form bilinear. This is also immediate in the basis where γ_m are purely imaginary and antisymmetric, found at the end of Section 2.2.2

The pure spinors from (3.4.6) and (3.4.11) thus now read

$$\Phi_+ = \eta_+ \otimes \eta_+^\dagger = \frac{1}{8} e^{-iJ}, \tag{3.4.20a}$$

$$\Phi_- = \eta_+ \otimes \eta_-^\dagger = \frac{1}{8} \Omega. \tag{3.4.20b}$$

Since we are in $d = 6$, the exponential in (3.4.20a) truncates at $J^3 = J \wedge J \wedge J$: the further ... terms in (3.4.6) are not present.

As a canonical example, we can take $\eta_+ = |+++\rangle$; its bilinears are then given by (3.4.17) and (3.4.12):

$$J = \frac{i}{2}(dz^1 \wedge d\bar{z}^1 + dz^2 \wedge d\bar{z}^2 + dz^3 \wedge d\bar{z}^3), \qquad \Omega = dz^1 \wedge dz^2 \wedge dz^3. \tag{3.4.21}$$

One can derive the properties of an SU(3) structure directly by working on this example, since all chiral spinors are in the same orbit in $d = 6$. But in what follows we will choose a more formal route.

To derive the behavior under $*$, we can use (3.2.27), (3.4.6), and (3.4.11). By (2.1.20), $c^2 = -1$; here and later, we will take $c = -i$, so $*\lambda = i\,\vec{\gamma}$. On (3.4.6), we get

$$* \lambda\, e^{-iJ} = i\, e^{-iJ}. \tag{3.4.22}$$

Comparing form degrees as in (3.3.37) now gives

$$* J = -\frac{1}{2} J^2. \tag{3.4.23}$$

On (3.4.11), (3.2.27) gives

$$* \Omega = -i\, \Omega. \tag{3.4.24}$$

To find properties similar to (3.3.18), we can proceed as in (3.3.40). For $d = 6$, the inner product of $\Phi_- = \tilde{\Phi}$ and Φ_+ would be automatically zero and would not give any information. We should rather consider (3.2.38) with $A = \Phi_-$ and $B = \{\gamma^m, \Phi_+\}$, which as a form reads $2dx^m \wedge \Phi_+$ (recall (3.1.21) and (3.1.24)). Now the left-hand side of (3.2.38) reads

$$\mathrm{Tr}(\Phi_-^\dagger \{\gamma^m, \Phi_+\}) = \mathrm{Tr}(\eta_- \eta_+^\dagger \{\gamma^m, \eta_+ \eta_+^\dagger\}); \tag{3.4.25}$$

both terms in the anticommutator are zero, because there are no one-form bilinears. Using (3.4.24), the right-hand side is proportional to

$$(*\bar{\Omega} \wedge dx^m \wedge e^{-iJ})_6 = \bar{\Omega} \wedge dx^m \wedge J. \tag{3.4.26}$$

Since this has to be true for any m, it follows that

$$J \wedge \Omega = 0. \tag{3.4.27a}$$

With similar computations (taking, for example, $A = B = \Phi_+$, and so on), one also finds the normalizations

$$\mathrm{vol} = -\frac{1}{6} J^3 = -\frac{i}{8} \Omega \wedge \bar{\Omega}. \tag{3.4.27b}$$

We stress once again that Ω is not any old complex form: as we found in the previous subsection, it should be decomposable, namely it should be possible to write it as a wedge of one-forms, which are the $(1,0)$-forms of the complex structure associated to η_+. Sometimes this constraint is not spelled out, with the understanding that it is implicit in the use of the letter Ω.

It is also useful to have expressions for the complex conjugates $\bar{\Phi}_\pm$. We now work in the basis where all the gammas are purely imaginary, B is proportional to the identity, and $\eta^c = \eta^*$. Taking the complex conjugate of (3.4.20a) turns the bispinor $\eta_+ \otimes \eta_+^\dagger$ into $\eta_+^* \otimes (\eta_+^*)^\dagger = \eta_+^c \otimes (\eta_+^c)^\dagger = \eta_- \otimes \eta_-^\dagger$. Reinstating the slash, the right-hand side is proportional to $(e^{-iJ})_{/}$. On each term $\alpha_{m_1 \ldots m_k} \gamma^{m_k \ldots m_1}$ of that sum, complex conjugation acts both on the coefficients $\alpha_{m_1 \ldots m_k}$ and on the $\gamma^{m_k \ldots m_1}$. But the latter are all real, because all the terms in Φ_+ have even k, and the product of an even number of purely imaginary gammas is real. So in the end,

$$\bar{\Phi}_+ = \eta_- \otimes \eta_-^\dagger = \frac{1}{8} e^{iJ} . \tag{3.4.28a}$$

Coming now to (3.4.20b), $\eta_+ \otimes \eta_-^\dagger$ is turned into $\eta_+^* \otimes (\eta_-^*)^\dagger = \eta_+^c \otimes \eta_+^\dagger = \eta_- \otimes \eta_+^\dagger$. The expression Ω should again be read as $\Omega_{/} = \not{\Omega}$. This now contains the product of three gamma matrices, which is purely imaginary; so in fact $\overline{\Omega_{mnp}\gamma^{mnp}} = \bar{\Omega}_{mnp}\overline{\gamma^{mnp}} = -\bar{\Omega}_{mnp}\gamma^{mnp}$. All in all,

$$(\bar{\Phi}_-)_{/} = \eta_- \otimes \eta_+^\dagger = -\frac{1}{8}(\bar{\Omega})_{/} . \tag{3.4.28b}$$

We have kept the slash here because a misunderstanding might arise: as we just saw,

$$\overline{\not{\Omega}} = -\not{\bar{\Omega}} . \tag{3.4.29}$$

Our convention from now on will be to keep the slash out as in (3.4.28b).

For later use, we also derive here a formula for a gamma action on a bispinor:

$$\begin{aligned} \gamma^m \eta_+ \otimes \eta_+^\dagger \gamma_m &\overset{(3.2.23),(3.4.20a)}{=} \frac{1}{8}\left(6 - 2iJ + \frac{2}{2}J^2 - \frac{6i}{6}J^3\right) \\ &\overset{(3.4.20b)}{=} -2\eta_- \otimes \eta_-^\dagger + 1 - \frac{i}{6}J^3 \\ &\overset{(3.2.24),(3.4.27b)}{=} -2\eta_- \otimes \eta_-^\dagger + 1 - \gamma . \end{aligned} \tag{3.4.30}$$

In the second line, we rewrote $6 = 8 - 2$. As a cross-check, acting with this on η_- on the left-hand side we obtain zero because of (3.4.19), and on the right-hand side we obtain $-2\eta_- + 2\eta_- = 0$ because of the unit-norm assumption and chirality.

3.4.3 Spinor pair in six dimensions

For applications, in $d = 6$ it will also be important to deal with two chiral spinors η_+^a, $a = 1, 2$. For simplicity, we will also assume here that the two are orthogonal:

$$\eta_+^{1\dagger}\eta_+^2 = 0 , \tag{3.4.31}$$

and that they have norm one, as previously stipulated in this section. Also, just as in (3.4.18), we will define $(\eta_+^a)^c \equiv \eta_-^a$. The bispinors $\eta_\mp^a \otimes \eta_\pm^{a\dagger}$ are covered by the discussion in the previous subsection, so we only really need to analyze the bispinors obtained with two different spinors, $\eta_+^1 \otimes \eta_\pm^{2\dagger}$.

To build intuition, it is also helpful to have an example in mind. We know that one of the two spinors, say the first, can be put in the form $\eta^1_+ = |+++\rangle$ by a rotation. The second has still positive chirality, so we can write it in the form (2.4.19). Taking $B = B_+$ in (2.3.7) and (2.2.11), we see that $\eta^1_- = |---\rangle$. So three of the terms in (2.4.19) are of the form $\gamma_{\bar{z}^{\bar{\imath}}}\eta^1_-$. We find it convenient to redefine here $b^i \to v^{\bar{\imath}}$, and (2.4.19) becomes

$$\eta^2_+ = a\eta^1_+ + \frac{1}{2} v \cdot \eta^1_- . \tag{3.4.32}$$

Here $v\cdot = v^m \gamma_m = \not{v}$ (recall (2.4.13)), but in fact $\gamma_{z_i}\eta^1_- = 0$, so only the $v^{\bar{\imath}}$ appear in this expression; this means we are taking v such that

$$v \cdot \eta^1_+ = 0 . \tag{3.4.33}$$

Our simplifying orthogonality assumption (3.4.31) demands now $a = 0$, so

$$\eta^2_+ = \frac{1}{2} v \cdot \eta^1_- , \qquad \eta^2_- = -\frac{1}{2}\bar{v} \cdot \eta^1_+ , \tag{3.4.34}$$

the second arising by conjugating the first. The minus sign is easy to see in a basis where the γ^m are purely imaginary, so $\bar{v} = (v_m \gamma^m)^* = -\bar{v}_m \gamma^m$. (Recall from Section 2.3.1 that a basis where they are all real does not exist.)

At this point, it might be possible to use an SU(3) $\subset$ SO(6) transformation, leaving $\eta^1_+ = |+++\rangle$ invariant while setting $\bar{v}$ to be say along the direction 3, thus also picking a particular choice for the second spinor, $\eta^2_+ = |--+\rangle$. However, in what follows we will use a more formal logic, without assuming a particular choice for the spinors. The relations (3.4.33) and (3.4.34) are still true, because every positive-chirality spinor is in the same orbit. We will now see we can derive all the properties of the bispinors from these properties alone.

First we reduce the new bispinors to the ones of η^1 alone, which we know from (3.4.20); we will call (J_a, Ω_a), $a = 1,2$ the forms appearing in $\eta^a_+ \otimes \eta^{a\,\dagger}_\pm$. As a preliminary step, we consider

$$0 \overset{(3.4.33)}{=} v \cdot \eta^1_+ \otimes \eta^{1\,\dagger}_+ \overset{(3.1.21),(3.4.20a)}{=} \frac{1}{16}(v \wedge + \iota_v)\mathrm{e}^{-\mathrm{i}J_1} . \tag{3.4.35}$$

The one-form part of (3.4.35) is

$$\iota_v J_1 = -\mathrm{i}v . \tag{3.4.36}$$

We can also rewrite this (raising an index) as $v^m (I^1)_m{}^n = -\mathrm{i}v^n$; in other words, v is a $(1,0)$-form with respect to the complex structure I_1 associated to η^1_+. This also implies that $v^2 = 0$.

We now consider the odd-form bispinor:

$$\begin{aligned} \eta^1_+ \otimes \eta^{2\,\dagger}_- &\overset{(3.4.34)}{=} -\frac{1}{2}\eta^1_+ \otimes \eta^{1\,\dagger}_+ \cdot v \\ &\overset{(3.1.24),(3.4.20a)}{=} -\frac{1}{16}(v \wedge - \iota_v)\mathrm{e}^{-\mathrm{i}J_1} \overset{(3.4.35)}{=} -\frac{1}{8} v \wedge \mathrm{e}^{-\mathrm{i}J_1} . \end{aligned} \tag{3.4.37}$$

In particular, its one-form part is, recalling (3.3.28),

$$v_m = -\eta^{2\,\dagger}_- \gamma_m \eta^1_+ , \tag{3.4.38}$$

which can be taken as an alternative definition of v. This one-form bilinear vanishes for $\eta^2 = \eta^1$, as we found in (3.4.19). A similar computation also gives

$$\eta^1_+ \otimes \eta^{2\dagger}_+ = \frac{1}{2}\eta^1_+ \otimes \eta^{1\dagger}_- \cdot \bar{v} = \frac{1}{16}(-\bar{v} \wedge + \iota_{\bar{v}})\Omega_1 \,. \tag{3.4.39}$$

We can refine our results by decomposing J_1 and Ω_1 in parts parallel and transverse to v. We first compute the norm of v by using

$$\begin{aligned} 1 = \eta^{2\dagger}_+ \eta^2_+ &\overset{(3.4.34)}{=} \frac{1}{4}\eta^{1\dagger}_- \cdot \bar{v} \cdot v \cdot \eta^1_- = \frac{1}{4}\eta^{1\dagger}_- \cdot (\{\bar{v}\cdot, v\cdot\} - v \cdot \bar{v}\cdot)\eta^1_- \\ &\overset{(3.4.33),(2.1.3)}{=} \frac{1}{2}|v|^2 \eta^{1\dagger}_- \eta^1_- = \frac{1}{2}|v|^2 \quad \Rightarrow \quad |v|^2 = 2 \,. \end{aligned} \tag{3.4.40}$$

Now we observe that

$$(P_\perp)_m{}^n \equiv \delta^n_m - \frac{1}{2} v_m \bar{v}^n \tag{3.4.41}$$

is a projector: using (3.4.40) we can check that it squares to itself. It has the property $(P_\perp)_m{}^n v_n = 0$. So the two-form

$$j \equiv \frac{1}{2}(P_\perp)_m{}^p (J_1)_{pn} \mathrm{d}x^m \wedge \mathrm{d}x^n = J_1 - \frac{\mathrm{i}}{2} v \wedge \bar{v} \tag{3.4.42}$$

has rank four rather than six: it satisfies $\iota_v j = \iota_{\bar{v}} j = 0$. Intuitively, it only spans the directions that do not include v and $\bar{v}$.

Moreover, recall from previous subsections that Ω is a $(3,0)$-form, unique up to rescaling; and from (3.4.36) that v is a $(1,0)$-form. So we can write

$$\Omega_1 = v \wedge \omega \tag{3.4.43}$$

for some $(2,0)$-form ω, which can be chosen such that $\iota_{\bar{v}}\omega = 0$.

Using this in (3.4.37) and (3.4.44a), we end up with

$$\Phi^{12}_+ = \eta^1_+ \otimes \eta^{2\dagger}_+ = \frac{1}{8} \mathrm{e}^{\frac{1}{2} v \wedge \bar{v}} \wedge \omega \,, \tag{3.4.44a}$$

$$\Phi^{12}_- = \eta^1_+ \otimes \eta^{2\dagger}_- = -\frac{1}{8} v \wedge \mathrm{e}^{-\mathrm{i}j} \,. \tag{3.4.44b}$$

We see that the bispinors "factorize" into a part along the v, $\bar{v}$ directions and a part along the remaining four directions. In each part, one of the $\Phi^{12}_\pm$ has an exponential, and the other has a $(k,0)$-form, similar to (3.4.20).

One can also reverse the role of η^1 and η^2. A computation similar to (3.4.40) gives an opposite of sorts to (3.4.34):

$$\eta^1_- = \frac{1}{2}\bar{v} \cdot \eta^2_+ \,. \tag{3.4.45}$$

This can also be obtained by multiplying (3.4.30) (with $\eta \to \eta^1$) by η^1_- from the right. Also, (3.4.34) and $v^2 = 0$ implies $v \cdot \eta^2_+ = 0$. A logic similar to (3.4.35) then tells us that v is $(1,0)$ with respect to the complex structure I_2 associated to η^2_+. The results for $\Phi^{12}_\pm$ can now be expressed in terms of (J_2, Ω_2) associated to η^2_+. Alternatively, we can use compute $\eta^2_+ \otimes \eta^{2\dagger}_\pm$ using (3.4.34) and compare them to (3.4.20). The result is similar to (3.4.42) and (3.4.43), and we give them here together:

$$J_1 = j + \frac{\mathrm{i}}{2} v \wedge \bar{v} \,, \qquad \Omega_1 = v \wedge \omega \,,$$
$$J_2 = -j + \frac{\mathrm{i}}{2} v \wedge \bar{v} \,, \qquad \Omega_2 = v \wedge \bar{\omega} \,; \tag{3.4.46}$$

recalling also

$$\iota_v j = \iota_{\bar{v}} j = 0 \,, \qquad \iota_v \omega = \iota_{\bar{v}} \omega = 0 \,. \tag{3.4.47}$$

We can think of j as being obtained from either J_1 or J_2 by projecting away the components along v and $\bar{v}$, so that the orthogonality (3.4.47) holds.

Applying (3.4.27) to either (J_1, Ω_1) or (J_2, Ω_2), we can work out the algebraic constraints satisfied by j and ω. The result contains v and $\bar{v}$, but we can strip them off using (3.4.47), arriving at

$$j \wedge \omega = \omega \wedge \omega = 0 \,, \qquad \omega \wedge \bar{\omega} = 2 j^2 \,, \tag{3.4.48}$$

which are the same as (3.3.18) in four dimensions.

The stabilizer of the spinor pair η^a_+ should leave invariant all bilinears; these can in turn be all expressed in terms of v, j, and ω. Leaving v and $\bar{v}$ invariant restricts the action to only happen in the four directions orthogonal to them; leaving j and ω invariant then forces such rotations to be in SU(2). So we conclude

$$\mathrm{Stab}(\eta^a_+) = \mathrm{SU}(2) \,. \tag{3.4.49}$$

As we mentioned earlier, by orthogonal transformations we could have reduced our analysis to the case $\eta^1_+ = |+++\rangle$, $\eta^2_+ = |--+\rangle$. Here the forms would read

$$v = \mathrm{d}z^3 \,, \qquad j = \frac{\mathrm{i}}{2}(\mathrm{d}z^1 \wedge \mathrm{d}\bar{z}^1 + \mathrm{d}z^2 \wedge \mathrm{d}\bar{z}^2) \,, \qquad \omega = \mathrm{d}z^1 \wedge \mathrm{d}z^2 \,. \tag{3.4.50}$$

3.4.4 Nonpure spinor bilinears in eight dimensions

$d = 8$ is the lowest dimension where we find nonpure spinors (see Section 2.4.3).

Nonpure MW spinors in $d = 8$.
In particular, let us consider a Majorana–Weyl $\eta = \eta_{\mathrm{MW}}$; by (2.4.39), it is nonpure and has stabilizer Spin(7). We saw an example in (2.4.22); we can generalize it by taking

$$\eta_{\mathrm{MW}} = \eta_+ + (\eta_+)^{\mathrm{c}} \,, \tag{3.4.51}$$

where η_+ is pure. We can compute $\eta_{\mathrm{MW}} \otimes \eta^\dagger_{\mathrm{MW}}$ using (3.4.51) and the bispinors involving η_+, $(\eta_+)^{\mathrm{c}}$. Two of these we know already from (3.4.6) and (3.4.11); we also need their conjugates. To simplify computations, we can work in the real gamma basis (2.2.28). (Eventually all the results we derive about bilinears are independent of the basis we choose.) There, $B = 1$, and $(\eta_+)^{\mathrm{c}} = \eta^*_+$. It then follows that $(\eta_+)^{\mathrm{c}} \otimes (\eta_+)^{\mathrm{c}\dagger}$ is just the complex conjugate as a form of $\eta_+ \otimes \eta^\dagger_+$; and likewise $((\eta_+)^{\mathrm{c}} \otimes \eta^\dagger_+)^* = \eta_+ \otimes (\eta_+)^{\mathrm{c}\dagger}$. Putting all this together:

$$\eta_{\rm MW} \otimes \eta_{\rm MW}^\dagger = \eta_+ \otimes \eta_+^\dagger + (\eta_+)^c \otimes (\eta_+)^{c\dagger} + \eta_+ \otimes (\eta_+)^{c\dagger} + (\eta_+)^c \otimes \eta_+^\dagger$$
$$= \frac{1}{16}\left(e^{-iJ} + e^{iJ} + \Omega + \bar{\Omega}\right) = \frac{1}{8}(1 + \Psi_4 + {\rm vol})\,, \tag{3.4.52}$$

where the four-form

$$\Psi_4 = -\frac{1}{2}J^2 + {\rm Re}\Omega \tag{3.4.53}$$

is self-dual, because $\gamma\eta_{\rm MW} \otimes \eta_{\rm MW}^\dagger = \eta_{\rm MW} \otimes \eta_{\rm MW}^\dagger$. An explicit coordinate expression could be obtained by using (3.4.12) and (3.4.17).

Nonpure Majorana spinors in $d = 8$.
Let us also consider a spinor η that is Majorana but not Weyl. We can decompose it on the two chiralities as $\eta = \eta_+ + \eta_-$, where both $\eta_\pm$ are MW. Now both $\eta_\pm$ define bispinors along the lines of (3.4.52), with some changes due to chirality: $\eta_\pm \otimes \eta_\pm^\dagger = \frac{1}{8}(1 + \Psi_{4\pm} \pm {\rm vol})$. One can also define the vector

$$w_m \equiv \eta_+^\dagger \gamma_m \eta_- = \eta_-^\dagger \gamma_m \eta_+\,; \tag{3.4.54}$$

the equality is clear in the basis (2.2.28), where the gamma matrices are real and symmetric. This basis also makes it manifest that w is real. (Both properties then also hold in all bases; they can also be inferred in a basis-independent way.) Its action exchanges $\eta_\pm$ (Exercise 3.4.5):

$$w\eta_\pm = w_m \gamma^m \eta_\pm = \eta_\mp\,. \tag{3.4.55}$$

So η_- is determined by w and η_+. The independent bilinears of a spinor that is Majorana but not Weyl are then a real one-form w, and the four-form Ψ_4 (3.4.53). The common stabilizer ${\rm Stab}(\Psi_4, w) = {\rm Spin}(7) \cap {\rm SO}(7)$. We will see in (3.4.84) that this is G_2, confirming the last entry of Table 2.4 and Exercise 2.4.1. (The idea there was to compute the stabilizer more directly and explicitly using an adapted basis.)

3.4.5 Ten Lorentzian dimensions

(Nonpure) MW spinor
We again begin by considering a Majorana–Weyl spinor $\epsilon_{\rm MW}$. The only bispinor we have to compute here is $\epsilon_{\rm MW} \otimes \overline{\epsilon_{\rm MW}}$.

The bilinears $\overline{\epsilon_{\rm MW}}\gamma_{\mu_1\ldots\mu_k}\epsilon_{\rm MW}$ as usual vanish for k = even by chirality. For simplicity, we will work in a real basis ((2.2.20) or (2.2.26)). Then $\epsilon_{\rm MW}^c = \epsilon_{\rm MW}^* = \epsilon_{\rm MW}$, and $\overline{\epsilon_{\rm MW}} = \epsilon_{\rm MW}^t\gamma_0$. One can also check that

$$(\gamma_0\gamma_{\mu_1\ldots\mu_k})^t = -(-1)^{\lfloor\frac{k+1}{2}\rfloor}\gamma_0\gamma_{\mu_1\ldots\mu_k}\,. \tag{3.4.56}$$

So $\gamma_0\gamma_{\mu_1\ldots\mu_k}$ is antisymmetric for $k = 0, 3, 4, 7, 8$, in which cases the bilinears $\overline{\epsilon_{\rm MW}}\gamma_{\mu_1\ldots\mu_k}\epsilon_{\rm MW} = \epsilon_{\rm MW}^t\gamma_0\gamma_{\mu_1\ldots\mu_k}\epsilon_{\rm MW}$ vanish. (It would also be possible to obtain this result in a basis-independent fashion.) The remaining nonzero bilinears occur for $k = 1, 5, 9$; (3.3.43) now gives

$$\epsilon_{\rm MW} \otimes \overline{\epsilon_{\rm MW}} = \frac{1}{32}\left((1+\gamma)V + \Omega\right) = \frac{1}{32}(V + \Omega + *V)\,, \tag{3.4.57}$$

where $V_\mu \equiv \overline{\epsilon_{\rm MW}}\gamma_\mu\epsilon_{\rm MW}$ is again the one-form part, Ω is a five-form, and $\gamma V = *V$ is the nine-form; all of them are real. In the first step, we determined the nine-form by using $\gamma\epsilon_{\text{MW}} \otimes \overline{\epsilon_{\text{MW}}} = \epsilon_{\text{MW}} \otimes \overline{\epsilon_{\text{MW}}}$ as usual. In the second step, we used (3.2.27), which in $d = 10$ Lorentzian dimensions reads $\overrightarrow{\gamma} = *\lambda$. The same arguments show

$$\gamma\Omega = *\Omega = \Omega\,. \tag{3.4.58}$$

By applying (3.2.23), we also have

$$\gamma^\mu \epsilon_{\text{MW}} \otimes \overline{\epsilon_{\text{MW}}}\gamma_\mu = -\frac{1}{4}(1-\gamma)V\,; \tag{3.4.59}$$

notice that Ω drops out, since it has $k = d/2 = 5$. We can then carry out a computation similar to (3.3.47):

$$\begin{aligned} V\epsilon_{\text{MW}} = V_\mu\gamma^\mu\epsilon_{\text{MW}} = \gamma^\mu\epsilon_{\text{MW}}\overline{\epsilon_{\text{MW}}}\gamma_\mu\epsilon_{\text{MW}} &= -\frac{1}{4}(1-\gamma)V\epsilon_{\text{MW}} = -\frac{1}{2}V\epsilon_{\text{MW}} \\ \Rightarrow \quad V\epsilon_{\text{MW}} = 0\,. \end{aligned} \tag{3.4.60}$$

Multiplying this from the right by $\overline{\epsilon_{\text{MW}}}$,

$$(V\wedge + \iota_V)(V + \Omega + *V) = 0\,. \tag{3.4.61}$$

Every form-degree in this equation vanishes independently. The zero-form part says that V is light-like, as in $d = 4$:

$$V^2 = 0\,. \tag{3.4.62}$$

The six-form part of (3.4.61) gives $V \wedge \Omega = 0$. It follows that there exists a four-form Ψ_4 such that

$$\Omega = V \wedge \Psi_4\,. \tag{3.4.63}$$

One can now rewrite the whole $\epsilon_{\text{MW}} \otimes \overline{\epsilon_{\text{MW}}}$ as $V \wedge (1 + \Psi_4 + \text{vol}_8)$, where vol_8 is an eight-form orthogonal to V. The parenthesis looks like our result in (3.4.52). Thus we have factorized our bispinor in $d_1 = 2$ and $d_2 = 8$ bispinors; this is related to our $2 + 8$ factorization of ϵ_{MW} back in (2.4.48).[7]

In fact, we saw in (2.4.47) that the annihilator of ϵ_{MW} has dimension one. Thus V is the only vector that annihilates ϵ_{MW} (up to proportionality):

$$\text{Ann}(\epsilon_{\text{MW}}) = \text{Span}\{V\} \qquad (d = 10 \text{ Lor.}). \tag{3.4.64}$$

We can see that ϵ_{MW} is not pure also in another way. If it were, $\epsilon_{\text{MW}} \otimes \overline{\epsilon_{\text{MW}}}$ would be a pure form because of (3.4.4). But (3.4.57) is not of the general form (3.2.53), since it doesn't contain any three- and seven-forms.

We now compute the common stabilizer of V and Ω. The stabilizer of the light-like V is (3.3.55), which for $d = 10$ is $\text{SO}(8) \ltimes \mathbb{R}^8$. The stabilizer of Ω equals that of Ψ_4 in (3.4.63); so SO(8) gets broken to Spin(7), and

$$\text{Stab}(V, \Omega) = \text{Spin}(7) \ltimes \mathbb{R}^8 \qquad (d = 10 \text{ Lor.}) \tag{3.4.65}$$

in agreement with (2.4.49).

[7] One subtlety is that the forms Ψ_4 and vol_8 do not quite identify an $\mathbb{R}^8 \subset \mathbb{R}^{1,9}$ subspace unambiguously. Indeed, one can change $\Psi_4 \to \Psi_4 + V \wedge (\ldots)$, and (3.4.63) is still satisfied. Ψ_4 and vol_8 live in the quotient space $V^\perp/\text{Span}(V)$: the space $V^\perp$ orthogonal to V is an $\mathbb{R}^9$, which includes V itself because it is light-like, and one defines a quotient by identifying vectors whose difference is proportional to V.

Weyl spinor

We now consider a spinor ϵ_+ that is only Weyl, and not Majorana. As noted at the end of Section 2.4.5, this is equivalent to having two MW spinors $\epsilon_{1,2}$, in terms of which $\epsilon_+ = \epsilon_1 + \mathrm{i}\epsilon_2$. The common stabilizer of the ϵ_i is the intersection of two copies of (3.4.65). There are several ways these can intersect. Each ϵ_i defines a one-form V_i; the stabilizer depends on whether they coincide or not:

- If $V_1 \neq V_2$, by Lorentz action we can make these two light-like vectors to be $\mathrm{d}x^+$ and $\mathrm{d}x^-$. The common stabilizer can then only act on the eight-dimensional space orthogonal to both, $V_i^\perp$. (This is now naturally defined; compare with footnote 7.) In the tensor-product basis (2.2.21) for $d = 10$, the spinors factorize as

$$\epsilon_1 = |+\rangle \otimes \eta_+ \,, \qquad \epsilon_2 = |-\rangle \otimes \eta_- \,, \tag{3.4.66}$$

  where $\eta_\pm$ are eight-dimensional MW spinors of chiralities ± 1. The common stabilizer of these two was found in Exercise 2.4.1 to be G_2. (It will be justified again in (3.4.85).)
- If $V_1 = V_2 = V$, setting again $V = \mathrm{d}x^+$, infinitesimal transformations of the type γ_{-m} are in the common stabilizer. In the factorized basis (2.2.21), we can write $\epsilon_i = |+\rangle \otimes \eta_i$. We are now down to computing the common stabilizer of two MW spinors with the same chirality, which we know from Section 2.4.4. If $\epsilon_1 = \epsilon_2$, clearly the common stabilizer is Spin(7). If $\epsilon_1 \neq \epsilon_2$ we have a common stabilizer of two MW spinors of two self-dual Ψ_{4i} in eight dimensions, which corresponds to that of a pure spinor, and hence equals SU(4).

Summarizing, we have found three possibilities:

$$\mathrm{Stab}(\epsilon_+) = \begin{cases} G_2 & V_1 \neq V_2\,; \\ \mathrm{SU}(4) \ltimes \mathbb{R}^8 & V_1 = V_2, \epsilon_1 \neq \epsilon_2\,; \\ \mathrm{Spin}(7) \ltimes \mathbb{R}^8 & \epsilon_1 = \epsilon_2\,. \end{cases} \tag{3.4.67}$$

3.4.6 Odd Euclidean dimensions

We now look at $d =$ odd Euclidean dimensions. Again we will consider our spinors to have unit norm, for simplicity.

$d = 3$ Euclidean dimensions

We can no longer use chirality to deduce that some of these are zero. Thus, for example, $\eta \otimes \eta^\dagger$ contains forms of all possible degrees. Recall, however, that half of these are related to the other half by (3.2.29), which here reads $\not{\alpha} = -\mathrm{i}(*\lambda\alpha)_{/}$. In (3.3.30), we showed two possible ways of resolving this redundance. Defining $w_m \equiv \eta^\dagger \gamma_m \eta$, we can write

$$\eta \otimes \eta^\dagger = \frac{1}{2}(w - \mathrm{i}\mathrm{vol})_{/} = \frac{1}{2}(1 + w)_{/}. \tag{3.4.68}$$

In the first equality, we have used the odd basis (the second line in (3.3.30)). In the second expression, we have used (3.2.29) to infer $-\mathrm{i}\not{\mathrm{vol}} = 1$, keeping only the zero- and one-form parts. The vector w is real: $w_m^* = (\eta^\dagger \gamma_m \eta)^\dagger = \eta^\dagger \gamma_m^\dagger \eta = \eta^\dagger \gamma_m \eta = w_m$. We can also consider

$$\eta \otimes (\eta^c)^\dagger = \frac{1}{2}(\omega_1)_/ \, . \tag{3.4.69}$$

Its zero-form part vanishes by $(\eta^c)^\dagger \eta = \eta^t B \eta = 0$, with $B = \sigma_2$ from (2.3.7) and (2.2.11).

Following the method in (3.3.47),

$$w\eta = \gamma_m \eta\eta^\dagger \gamma^m \eta \overset{(3.4.68)}{=} \frac{1}{2}\gamma_m(1+w)\gamma^m \eta \overset{(3.2.23)}{=} \frac{1}{2}(3-w)\eta \qquad \Rightarrow \qquad w\eta = \eta \, . \tag{3.4.70}$$

In a similar way,

$$\omega_1 \eta = 0 \, , \qquad \omega_1 \eta^c = \eta \, . \tag{3.4.71}$$

Together these imply that $w_i \equiv \{w, \mathrm{Re}\omega_1, \mathrm{Im}\omega_1\}$ are an orthonormal basis: $w_i \cdot w_j = \delta_{ij}$.

As usual, we can gain some intuition with a model spinor, such as $\eta = |+\rangle$. We can borrow the $d = 2$ computations in Section 3.3.4, basically because of (2.2.14): the $d = 3$ gamma basis can be obtained from the $d = 2$ one by simply declaring γ_3 to be the $d = 2$ chiral gamma. We can choose $\not{w}$ to be this third direction, and (3.4.70) now says that η is chiral in the two directions orthogonal to w. With this trick, we can import the formulas of Section 3.3.4 to $d = 3$. The main difference is that we can now trade a form with components along x^3 with one without, using (3.2.29). For example, $(\mathrm{d}x^1 \wedge \mathrm{d}x^2)_/ = \mathrm{i}(\mathrm{d}x^3)_/$. For example, (3.3.32b) is also already written as $1/2(1+\sigma_3)$, which we can reinterpret in $d = 3$ as $1/2(1 + \mathrm{d}x^3)_/$. So (3.3.32b) and (3.3.34b) give

$$w = \mathrm{d}x^3 \, , \qquad \omega_1 = \mathrm{d}x^1 + \mathrm{i}\mathrm{d}x^2 \, . \tag{3.4.72}$$

(We could also have identified ω_1 by recalling from the beginning of Section 2.4.6 that η was annihilated by $\gamma_{\bar z} = \frac{1}{2}\gamma^z$.) For a more general spinor η, there still exist coordinates x'_i such that (3.4.72) holds with $x_i \to x'_i$.

The common stabilizer of w and ω_1 is now obviously just the identity, since $\{w, \mathrm{Re}\omega_1, \mathrm{Im}\omega_1\}$ are a basis of $\mathbb{R}^3$, and there is no nontrivial rotation that keeps all the vectors of a basis invariant. This agrees with our result in Section 2.4.6.

$d = 5$ Euclidean dimensions

For $d = 5$, one can start from the known bispinors for $d = 4$ (say those for a chiral spinor, which we know already) and reinterpret them as $d = 5$ forms. Equation (3.2.29) now reads $\not{\alpha} = (*\lambda\alpha)_/$. So, for example, (3.3.38) naively does not appear to have a one-form, but in fact it does because $(J^2)_/ = (*J^2)_/$, and $*J^2$ is a one-form.

We can define a real $w_m \equiv \eta^\dagger \gamma_m \eta$. It is also still true, as it was in $d = 3$, that $w\eta = \eta$; intuitively, our η is chiral in a four-dimensional subspace orthogonal to the fifth dimension w. All in all, we can now write (3.3.38) and (3.3.39) as

$$\begin{aligned} \eta \otimes \eta^\dagger &= \frac{1}{4}(e^{-\mathrm{i}J})_/ \overset{(3.2.29)}{=} \frac{1}{4}(1 + w - \mathrm{i}J)_/ \, , \\ \eta \otimes (\eta^c)^\dagger &= \frac{1}{4}(\Omega)_/ \, . \end{aligned} \tag{3.4.73}$$

From $w\eta = \eta$, recalling (3.1.21) and (3.1.24), we see $\iota_w \eta \otimes \eta^\dagger$; using the first expression for $\eta \otimes \eta^\dagger$ in (3.4.73), we obtain $\iota_w J = 0$. Similarly we obtain $\iota_w \Omega = 0$.

For $\eta = |++\rangle$, we have $w = dx^5$, while J and Ω can be written in terms of the first four coordinates as (3.4.17) and (3.4.12) for $d = 4$. For a more general spinor, w identifies a subspace $w^\perp \cong \mathbb{R}^4$ on which J and Ω live. The common stabilizer of w, J, Ω is now easy to work out: if we impose that a rotation does not change w, it only acts on $w^\perp$; we are then left with the stabilizer of J and Ω, which is SU(2) because of (3.3.21). Thus

$$\text{Stab}\{w, J, \Omega\} = \text{SU}(2) \qquad (d = 5) \tag{3.4.74}$$

in agreement with (2.4.51).

$d = 7$ Euclidean dimensions

In $d = 7$, a spinor η gives rise to bilinears w, J, Ω, respectively a real one-form, real two-form, and complex three-form.

Again $w\eta = \eta$, from which it follows w identifies a subspace $w^\perp \cong \mathbb{R}^6$ on which J and Ω live, with properties inherited from $d = 6$, namely (3.4.27). In other words,

$$\iota_w J = \iota_w \Omega = 0\,, \qquad J \wedge \Omega = 0 = \Omega^2\,, \qquad -\frac{1}{6}J^3 = -\frac{\mathrm{i}}{8}\Omega \wedge \bar{\Omega}\,. \tag{3.4.75}$$

The bispinors are inherited from $d = 6$, via (3.4.6) and (3.4.11). From $w\eta = \eta$, it follows that

$$\begin{aligned} &(w \wedge e^{-\mathrm{i}J})_{/} = w \wedge \eta \otimes \eta^\dagger = \eta \otimes \eta^\dagger = (e^{-\mathrm{i}J})_{/} \\ \Rightarrow \quad & \not{w} = \frac{\mathrm{i}}{6}(J^3)_{/}\,, \qquad -\mathrm{i}(w \wedge J)_{/} = -\frac{1}{2}(J^2)_{/}\,. \end{aligned} \tag{3.4.76}$$

We can then rewrite

$$\begin{aligned} \eta \otimes \eta^\dagger &= \frac{1}{8}(e^{-\mathrm{i}J})_{/} = \frac{1}{8}(1 + w)(1 - \mathrm{i}J)_{/}\,, \\ \eta \otimes (\eta^c)^\dagger &= \frac{1}{8}(-\Omega)_{/}\,. \end{aligned} \tag{3.4.77}$$

(The minus sign in front of Ω is a matter of definition, and is chosen for later convenience.) The computation of the stabilizer works as in $d = 5$, and this time produces an SU(3).

As we saw in Section 2.4.6, in $d = 7$ we also have a more interesting possibility: a Majorana spinor $\eta_{\rm M}$. (We could not do this in $d = 3, 5$: recall (2.3.7c)). We can parameterize it as $\eta_{\rm M} = \eta + \eta^c$. In the basis (2.2.19), all gammas are purely imaginary; there is no basis where they are real. So the bilinears

$$\eta_{\rm M}^\dagger \gamma_{m_1 \ldots m_k} \eta_{\rm M} \tag{3.4.78}$$

are real when k is even, but purely imaginary when k is odd. The gammas are also antisymmetric, so $\gamma^t_{m_1\ldots m_k} = (-1)^k \gamma_{m_k \ldots m_1} = (-1)^{k + \lfloor k/2 \rfloor}\gamma_{m_1 \ldots m_k}$, and (3.4.78) vanishes for $k = 1, 2, 5, 6$. The remaining cases $k = 0, 3, 4, 7$ are related by (3.2.29), namely $\not{\alpha} = -\mathrm{i}(*\lambda\alpha)_{/}$. So we have derived

$$\eta_{\rm M} \otimes \eta_{\rm M}^\dagger = \frac{1}{8}(1 + *\phi)_{/} = -\frac{\mathrm{i}}{8}(\phi + \text{vol})_{/}\,. \tag{3.4.79}$$

By (3.3.30), $\mathrm{i}\phi_{mnp} = \eta_{\rm M}^\dagger \gamma_{mnp} \eta_{\rm M}$, $*\phi_{mnpq} = \eta_{\rm M}^\dagger \gamma_{mnpq}\eta_{\rm M}$. This three-form ϕ should presumably be related to the one appearing in (2.2.18). To see this, it is convenient to relate it to $d = 6$ forms, whose explicit expressions we know already. With a computation similar to (3.4.52):

$$\eta_M \otimes \eta_M^\dagger \overset{(3.4.77),(3.4.29)}{=} \frac{1}{8}(e^{-iJ} + e^{iJ} - \Omega + \bar{\Omega})_/ \\ = \frac{1}{4}\left(1 - \frac{1}{2}J^2 - i\mathrm{Im}\Omega\right)_/ \overset{(3.2.29),(3.4.76)}{=} -\frac{i}{4}(\mathrm{Im}\Omega + w \wedge J + \mathrm{vol})_/ \,. \tag{3.4.80}$$

Thus we have identified

$$\phi = \mathrm{Im}\Omega + w \wedge J \,. \tag{3.4.81}$$

If we now use

$$w = dx^7 \,, \qquad J = dx^1 \wedge dx^4 + dx^2 \wedge dx^5 + dx^3 \wedge dx^6 \,, \\ \Omega = i(dx^1 + i dx^4) \wedge (dx^2 + i dx^5) \wedge (dx^3 + i dx^6) \tag{3.4.82}$$

(see (3.4.12) and (3.4.17)), we reproduce the components of (2.2.18). The stabilizer is not SU(3), because it need not leave w, J, Ω separately invariant, but only their combination (3.4.81). Compared to the logic that precedes (3.3.21), imposing $\delta\phi_{ijk} = 0$ still imposes $\lambda_i{}^i = 0$, but $\delta\phi_{ij7} = \delta\phi_{ij7} = 0$ only relate λ_{ij} to $\epsilon_{ijk}\lambda_7{}^k$; so there are three complex new free entries in λ, leading to a Lie algebra of dimension $8 + 6 = 14$, which turns out to be g_2. In conclusion,

$$\mathrm{Stab}(\phi) = G_2 \,, \tag{3.4.83}$$

in agreement with (2.4.53).

G_2 **and** Spin(7)

We can use this result to get a different view on Majorana spinors in $d = 8$. Take $w = dx^8$ and work in the basis (2.2.28). The chirality matrix reads $\Gamma = \sigma_3 \otimes 1$, so we can write $\eta_+ = \binom{\eta_7}{0}$ and $\eta_- = \binom{0}{\eta_7'}$, with η_7 and η_7' spinors in $d = 7$. The condition (3.4.55) now gives $\eta_7 = \eta_7'$, so $\eta = \binom{\eta_7}{\eta_7}$. We can now use this to compare the bilinears of η, which according to Section 3.4.4 is stabilized by Spin(7), with those of η_7 in $d = 7$, which are stabilized by G_2. This leads to

$$\Psi_4 = \phi \wedge w + \tilde{\phi} \,, \tag{3.4.84}$$

where ϕ has $\mathrm{Stab}(\phi) = G_2$, and $\tilde{\phi} = *_7\phi$ is its Hodge dual in the $\mathbb{R}^7$ transverse to w. After some reshuffling of coordinates, one can see that the components ϕ_{mnp} are of the form (2.2.18). So we find

$$\mathrm{Stab}\{\Psi_4, w\} = \mathrm{Spin}(7) \cap \mathrm{SO}(7) = G_2 \,. \tag{3.4.85}$$

3.4.7 Basis associated to a spinor

Often, a given spinor η defines naturally a basis for the space of all spinors, via a gamma matrix action. As we will see now, such a basis can be used to write explicitly the action of the bilinears $\gamma_{m_1 \ldots m_k}$ on η.

$d = 6$ **Euclidean dimensions**

For a pure spinor η in any (even) d, the $\gamma_{m_1 \ldots m_k} \eta$ are a basis for the space of spinors S. For $d = 6$, this was illustrated in (2.4.18), with $\eta_+ = |+++\rangle$, $|+++\rangle^c \propto |---\rangle$. The generalization to any η_+ is

$$\eta_+ , \qquad \gamma_i \eta_+ , \qquad \gamma_{\bar{\imath}} \eta_- , \qquad \eta_- , \tag{3.4.86}$$

where $\eta_- \equiv (\eta_+)^c$.

The four groups in (3.4.86) can also be seen in terms of representation theory. The spaces of chiral spinors $S_\pm$ form two irreducible representations of SO(6) of dimension four. Under Stab(η_+) = SU(3), these both reduce into a singlet and a triplet: so (3.4.86) expresses the split of $S = S_+ \oplus S_-$ as $\mathbf{1} \oplus \mathbf{3} \oplus \bar{\mathbf{3}} \oplus \mathbf{1}$.

If in (3.4.86) we prefer not to use holomorphic and antiholomorphic indices i, $\bar{\imath}$, we can replace them by real indices m by allowing a little redundancy: $\gamma_m \eta$ now also contain some spinors that are zero, by (2.4.5), which we rewrite here as

$$\gamma_m \eta_+ = -\mathrm{i} I_m{}^n \gamma_n \eta_+ \, ; \tag{3.4.87}$$

I is the complex structure tensor first introduced in Section 2.4.1, and related to J by (2.4.30).

The basis (3.4.86) is *orthonormal*; this can be shown using (3.4.19), and

$$\begin{aligned} \eta_+^\dagger \gamma_q \gamma^p \eta_+ &= ||\eta_+||^2 (\delta_q{}^p + \mathrm{i} I_p{}^q) = 2||\eta_+||^2 \bar{\Pi}_p{}^q \, , \\ \eta_-^\dagger \gamma_q \gamma^p \eta_- &= ||\eta_+||^2 (\delta_q{}^p - \mathrm{i} I_p{}^q) = 2||\eta_+||^2 \Pi_p{}^q \, , \end{aligned} \tag{3.4.88}$$

which in turn follow from (2.1.33) and (3.4.5).

As a first application of (3.4.86), we write the projector on the space of chiral spinors by summing over an orthonormal basis of chiral spinors:[8]

$$\begin{aligned} \frac{1}{2}(1-\gamma) &= \sum_{\psi_a \text{ basis of } S_-} \psi_a \otimes \psi_a^\dagger = \eta_- \otimes \eta_-^\dagger + \frac{1}{2} \gamma_i \eta_+ \otimes \eta_+^\dagger \gamma^i \\ &= \eta_- \otimes \eta_-^\dagger + \frac{1}{2} \gamma_m \eta_+ \otimes \eta_+^\dagger \gamma^m \, . \end{aligned} \tag{3.4.89}$$

This reproduces (3.4.30) from a different point of view. In the last step, we have turned the sum over holomorphic indices to one over all indices using (3.4.87).

A more common application is to the action of multiple gamma matrices. For example, let us focus on $\gamma_{mn} \eta_+$. For any m, n this is a positive-chirality spinor; so it will have nonzero coefficients only on the elements η_+, $\gamma_m \eta_-$ in (3.4.86):

$$\gamma_{mn} \eta_+ = C_{mn} \eta_+ + D_{mnp} \gamma^p \eta_- \, , \tag{3.4.90}$$

for some coefficients C_{mn} and D_{mnp}. To identify these coefficients, we project on the basis. Multiplying from the left by $\eta_+^\dagger$ and recalling (3.4.5), we obtain $\mathrm{i} J_{mn} = C_{mn}$. To identify D_{mnp}, we multiply by $\eta_-^\dagger \gamma_q$. By (3.4.19), C_{mn} drops out, and we get

$$\eta_-^\dagger \gamma_q \gamma_{mn} \eta_+ = D_{mnp} \eta_-^\dagger \gamma_q \gamma^p \eta_- \, . \tag{3.4.91}$$

[8] This is similar to when in quantum mechanics we write a projector on a subspace $V \subset \mathcal{H}$ as $\sum_a |\psi_a\rangle\langle\psi_a|$, for $|\psi_a\rangle$ a basis for V. The factor 1/2 can be checked by acting on $\gamma^j \eta_+$ and using (3.4.88).

By (3.1.17) for $k = 2$ and (3.4.19), the left-hand side is $\eta_-^\dagger \gamma_{qmn}\eta_+$, which by (3.4.11) is $-||\eta_+||^2\Omega_{qmn}$. By (3.4.88), the right-hand side is $2||\eta_+||^2 D_{mnp}\Pi_q{}^p$. So we find $-\Omega_{qmn} = 2D_{mnp}\Pi_q{}^p$. We have not determined D completely. But we could not hope to do so: in (3.4.90) it appears multiplying $\gamma^p\eta_-$, which vanishes when the real index p is projected to become antihilomorphic: $\gamma^{\bar{\imath}}\eta_- = 0$. So the components $D_{mn\bar{\imath}}$ don't even appear anywhere; it is enough to determine D_{mni}, or in terms of purely real indices $D_{mnp}\Pi_q{}^p$. A solution is then $D_{mnp} = -\frac{1}{2}\Omega_{mnp}$. All in all, (3.4.90) becomes

$$\gamma_{mn}\eta_+ = \mathrm{i}J_{mn}\eta_+ - \frac{1}{2}\Omega_{mnp}\gamma^p\eta_- \,. \tag{3.4.92a}$$

As an immediate application of this formula, we can now reproduce quickly the expression (2.4.33) for the stabilizer of η_+ (for $d = 6$).

To obtain a similar equation for the action on η_-, we can simply use conjugation (recalling $\eta_- = (\eta_+)^c$). As usual in $d = 6$, it is easiest to work in a basis where all gammas are purely imaginary; here $B = 1$, and $\eta_- = (\eta_+)^c = \eta_+^*$. The conjugate of (3.4.92a) is then

$$\gamma_{mn}\eta_- = -\mathrm{i}J_{mn}\eta_- + \frac{1}{2}\bar{\Omega}_{mnp}\gamma^p\eta_+ \,. \tag{3.4.92b}$$

One can use this method to obtain similar equations with more indices. For example,

$$\gamma_{mnp}\eta_+ = -\Omega_{mnp}\eta_- + 3J_{[mn}J_{pq]}\gamma^q\eta_+ \,, \qquad \gamma_{mnp}\eta_- = \bar{\Omega}_{mnp}\eta_+ + 3J_{[mn}J_{pq]}\gamma^q\eta_- \,. \tag{3.4.93}$$

Basis for bispinors in $d = 6$

As a byproduct of our analysis, we can also produce a basis for bispinors (still for $d = 6$ Euclidean dimensions). This is obtained by taking a tensor product of two copies of the basis (3.4.86) for spinors, acting with a † on the second copy:

$$\begin{array}{ccccccc}
 & & & \eta_+\otimes\eta_+^\dagger & & & \\
 & & \eta_+\otimes\eta_+^\dagger\gamma_{\bar{\imath}} & & \gamma_i\eta_+\otimes\eta_+^\dagger & & \\
 & \eta_+\otimes\eta_-^\dagger\gamma_i & & \gamma_i\eta_+\otimes\eta_+^\dagger\gamma_{\bar{\jmath}} & & \gamma_{\bar{\imath}}\eta_-\otimes\eta_+^\dagger & \\
\eta_+\otimes\eta_-^\dagger & & \gamma_i\eta_+\otimes\eta_-^\dagger\gamma_j & & \gamma_{\bar{\imath}}\eta_-\otimes\eta_+^\dagger\gamma_{\bar{\jmath}} & & \eta_-\otimes\eta_+^\dagger \\
 & \gamma_i\eta_+\otimes\eta_-^\dagger & & \gamma_{\bar{\imath}}\eta_-\otimes\eta_-^\dagger\gamma_j & & \eta_-\otimes\eta_+^\dagger\gamma_{\bar{\imath}} & \\
 & & \gamma_{\bar{\imath}}\eta_-\otimes\eta_-^\dagger & & \eta_-\otimes\eta_-^\dagger\gamma_i & & \\
 & & & \eta_-\otimes\eta_-^\dagger & & &
\end{array} \,. \tag{3.4.94}$$

This is alternative to the bispinor basis $\gamma_{m_1 \ldots m_k}$, and is useful in other contexts.

It can be rewritten as a basis for forms, recalling the Clifford map (3.1.13) and (3.4.20) and (3.4.28):

$$\begin{array}{ccccccc}
 & & & \Phi_+ & & & \\
 & & \Phi_+\gamma_{\bar{\imath}} & & \gamma_i\Phi_+ & & \\
 & \Phi_-\gamma_i & & \gamma_i\Phi_+\gamma_{\bar{\jmath}} & & \gamma_{\bar{\imath}}\bar{\Phi}_- & \\
\Phi_- & & \gamma_i\Phi_-\gamma_j & & \gamma_{\bar{\imath}}\bar{\Phi}_-\gamma_{\bar{\jmath}} & & \bar{\Phi}_- \\
 & \gamma_i\Phi_- & & \gamma_{\bar{\imath}}\bar{\Phi}_+\gamma_j & & \bar{\Phi}_-\gamma_{\bar{\imath}} & \\
 & & \gamma_{\bar{\imath}}\bar{\Phi}_+ & & \bar{\Phi}_+\gamma_i & & \\
 & & & \bar{\Phi}_+ & & &
\end{array} \,, \tag{3.4.95}$$

where now the gamma matrices have to be understood via (3.1.21) and (3.1.24).

Since (3.4.86) is orthonormal, an orthogonality property holds for (3.4.95) as well, with respect to the inner product (3.2.40). If we denote by $\alpha_{j,k}$ a form in position (j,k) in (3.4.95), then

$$\left(\alpha_{j,k}, \bar{\beta}_{j',k'}\right) = 0 \quad \text{unless } j + j' = k + k' = 3\,. \tag{3.4.96}$$

The computations are similar to those in (3.4.25).

$d = 7$ Euclidean dimensions

For a slightly different example, let us look at $d = 7$ Euclidean dimensions, focusing on a Majorana spinor η_{M}. We already saw a formula for the action of two gammas in (2.4.52), using octonions. Now we can rederive it following (3.4.92). The basis for the space of spinors is simply

$$\eta_{\text{M}}\,, \qquad \gamma_m \eta_{\text{M}}\,; \tag{3.4.97}$$

the index m goes from 1 to 7, and so we have a total of eight spinors, as appropriate for a basis. Again this has a group-theoretical interpretation: the space of spinors is the **8** of SO(7), but it decomposes under $\text{Stab}(\eta_{\text{M}}) = G_2$ as $\mathbf{1} \oplus \mathbf{7}$. Now a priori $\gamma_{mn}\eta_{\text{M}} = A_{mn}\eta_{\text{M}} + B_{mnp}\gamma^p\eta_{\text{M}}$; but from (3.4.79) we know that η_{M} has no bilinears with two indices, so upon multiplying by $\eta_{\text{M}}^\dagger$ we find $A_{mn} = 0$. Multiplying by $\eta_{\text{M}}^\dagger\gamma_p$ then reproduces (2.4.52).

$d = 10$ Lorentzian dimensions

We finally consider a MW spinor ζ_{MW} in $d = 10$, say of positive chirality. The spinors

$$\Gamma_{MN}\zeta_{\text{MW}}\,. \tag{3.4.98}$$

were shown after (2.4.49) to form a basis for S_+; this was used to argue that there is only one orbit of MW spinors of a given chirality.

It is harder to exhibit a basis for S_-. The $\Gamma_M\zeta_{\text{MW}}$ are only nine spinors (since $\Gamma_-\zeta_{\text{MW}} = 0$); a natural possibility would seem to add the $\Gamma_{MNP}\zeta_{\text{MW}}$, but that is still not enough. A somewhat silly way out is to consider

$$\Gamma_{MN}\Gamma_+\zeta_{\text{MW}}\,, \tag{3.4.99}$$

which works simply because $\Gamma_+\zeta_{\text{MW}}$ is again MW. It is less natural than the other bases introduced in this subsection, because Γ_+ is not intrinsically defined; the annihilator is Γ_-, and there is no particularly natural choice of the direction Γ_+ – there are infinitely many possibilities, related by Lorentz transformations.

Exercise 3.4.1 Consider a set of chiral spinors η_+^I in $d = 6$ Euclidean dimensions, and let $\eta_-^I = (\eta_+^I)^{\text{c}}$. Show that

$$\eta_-^{I\dagger}\eta_-^J = \eta_+^{J\dagger}\eta_+^I\,, \qquad (\eta_-^{I\dagger}\eta_-^J)^* = \eta_+^{I\dagger}\eta_+^J\,; \tag{3.4.100a}$$

$$\eta_-^{I\dagger}\gamma_m\eta_+^J = -\eta_-^{J\dagger}\gamma_m\eta_+^I\,, \qquad (\eta_-^{I\dagger}\gamma_m\eta_+^J)^* = -\eta_+^{I\dagger}\gamma_m\eta_-^J\,. \tag{3.4.100b}$$

Exercise 3.4.2 Rederive (3.4.44) from the explicit choice of spinors $\eta_+^1 = |+{+}+\rangle$, $\eta_+^2 = |+{+}-\rangle$ and the results of Sections 3.3.4 and 3.3.5.

Exercise 3.4.3 Starting from (3.4.44), use complex conjugation and (3.2.33) to compute the following bispinors:

$$\eta^1_\pm \otimes \eta^{2\,\dagger}_- , \qquad \eta^2_+ \otimes \eta^{1\,\dagger}_\pm , \qquad \eta^2_- \otimes \eta^{1\,\dagger}_\pm . \tag{3.4.101}$$

(Pay attention to the sign issue in (3.4.29).)

Exercise 3.4.4 Using the methods of Section 3.4.3, compute

$$\eta^2_+ \otimes \eta^{2\,\dagger}_\pm . \tag{3.4.102}$$

Exercise 3.4.5 Show (3.4.55) using similar steps as (3.3.47).

Exercise 3.4.6 Show (3.4.56), again working in the real ten-dimensional basis (2.2.20) or (2.2.26).

Exercise 3.4.7 Work out the analogue of Section 3.4.3 in $d = 4$.

Exercise 3.4.8 Using (3.2.38) and (3.4.19), show in $d = 6$ that the forms $\Phi_\pm$ defined via $\Phi_\pm \equiv \eta^1_+ \otimes \eta^{2\,\dagger}_\pm$ satisfy

$$(\bar\Phi_+, \Phi_+) = (\bar\Phi_-, \Phi_-) = \frac{\mathrm{i}}{8} , \qquad (\Gamma_A \Phi_+, \Phi_-) = (\Gamma_A \bar\Phi_+, \Phi_-) = 0 , \tag{3.4.103}$$

in the notation of Section 3.2.

Exercise 3.4.9 Work out the analogue of (3.4.67) when ϵ is Majorana but not Weyl. It is useful to decompose it in its two chiralities, $\epsilon = \epsilon_+ + \epsilon_-$. You should find the following three possibilities:

$$G_2 \ltimes \mathbb{R}^8 , \qquad \mathrm{Spin}(7) , \qquad \mathrm{SU}(4) . \tag{3.4.104}$$

Exercise 3.4.10 Verify (3.4.84), either using bilinears or bispinors.

Exercise 3.4.11 Compute the analogue of (3.4.92) and (3.4.93) for $\gamma^{mnpq}\eta_\pm$.

Exercise 3.4.12 Compute the analogue of (2.4.52) for $\gamma^{mnp}\eta_{\mathrm{M}}$.

Exercise 3.4.13 Check that $\Gamma_M \zeta_{\mathrm{MW}}$ and $\Gamma_{MNP}\zeta_{\mathrm{MW}}$ don't give a basis of chiral spinors in $d = 10$ Lorentzian dimensions, but only span a 15-dimensional space. Which spinor is missing?

4 Differential geometry

The previous two chapters explored various properties of spinors and forms. We were interested in algebraic aspects, which did not involve derivatives; so we just worked in flat space. In this chapter, we start considering nontrivial spaces.

4.1 Tensors on manifolds

One of the main features of general relativity is that it is invariant under general coordinate transformations, where the new coordinate system is allowed to depend on the old in any C^∞ way

$$x^m \to x'^m(x^0, \ldots, x^{d-1}) \tag{4.1.1}$$

and not necessarily linearly, as for nongravitational theories on flat space. In differential geometry, we use this fact to glue pieces of space; the resulting object is called a *manifold*. In this section, we will introduce this concept, and populate it with geometric objects such as vector fields and metrics. We begin with some aspects of general relativity; even though it is a prerequisite for this book, it is good to go through the basics again before plunging in to more advanced material.

4.1.1 Vector and tensor fields

General coordinate transformations (4.1.1) can also be viewed in an "active" way as defining a map of spacetime to itself; in this context, one often calls (4.1.1) *diffeomorphisms*. The requirement that the objects in the theory should transform well under (4.1.1) is called general covariance.

Vector fields

Let us consider in particular contravariant vectors v^m, which from now on we will simply call vectors. (In Section 3.1, we started calling covariant vectors one-forms.) Already on $\mathbb{R}^d$, this needs to be a more general concept than in flat space: we must allow for space-dependent vectors, often called vector fields. Indeed, a vector transforms under a general coordinate change (4.1.1) as

$$v^m \to v'^m = \frac{\partial x'^m}{\partial x^n} v^n \equiv J^m{}_n v^n \, . \tag{4.1.2}$$

So a constant vector becomes nonconstant upon a general coordinate change. In other words, it assigns a vector in $\mathbb{R}^d$ to each point in space. This is a new entity, which we call a *vector field*, even though for simplicity we will often slip into calling it simply a vector. It can be visualized as the choice of an arrow at each point.

To free oneself from the explicit appearance of indices, one often introduces a *coordinate basis* of vector fields. In particular,

$$\partial_m \equiv \frac{\partial}{\partial x^m} \tag{4.1.3}$$

is defined as having components all zero except the mth, which is equal to one. Intuitively, it is represented at every point by an arrow pointing in the direction of the mth axis. Such a vector transforms under coordinate change as $\partial_m \to \partial'_m = \frac{\partial x^n}{\partial x'^m}\partial_n$. Now a vector field with components v^m can be written as

$$v = v^m \partial_m \,, \tag{4.1.4}$$

and now the transformation law (4.1.2) simply becomes the statement $v = v'$. In other words, the vector remains the same, even if its components along a given coordinate grid change.

Forms and tensors

Let us now consider one-form α_m (in other words, covariant vectors). Its transformation law is

$$\alpha_m \to \alpha'_m = \frac{\partial x^n}{\partial x'^m}\alpha_n = \tilde{J}_m{}^n \alpha_n \,, \qquad \tilde{J} = (J^t)^{-1} \,. \tag{4.1.5}$$

Just like for vectors, such a coordinate transformation turns a form into a new entity, a map that assigns a form to each point; we should perhaps call this a one-form field, but again we will drop the word "field" for simplicity, just as we often do for "vector fields." Also, as we did for vectors, we can avoid the excessive use of indices by thinking of α_m as the components of an object $\alpha \equiv \alpha_m \mathrm{d}x^m$, extending to curved space a notation we introduced in Section 3.1.1.

Another way of viewing one-forms is as linear maps from the space of vector fields to the space of functions. For example, the $\mathrm{d}x^m$ can now be viewed as maps such that

$$\mathrm{d}x^m(\partial_n) = \delta^m_n \,. \tag{4.1.6}$$

One can follow a similar logic to introduce tensor fields. Recall that a tensor is simply an object with many indices (see beginning of Section 3.1). A tensor field of (k, l)-type, i.e., with k upper and l lower indices, transforms as

$$T^{m_1 \ldots m_k}_{n_1 \ldots n_l} \to T'^{m_1 \ldots m_k}_{n_1 \ldots n_l} = J^{m_1}{}_{p_1} \ldots J^{m_k}{}_{p_k} \tilde{J}_{n_1}{}^{r_1} \ldots J_{n_l}{}^{r_l} T^{p_1 \ldots p_k}_{r_1 \ldots r_l} \tag{4.1.7}$$

in a mix of (4.1.2) and (4.1.5).

A notable example is the central object of general relativity, the metric g_{mn}, which is a tensor field of $(0, 2)$-type, symmetric under the exchange of two indices. The infinitesimal distance between a point with coordinates x^m and a nearby point with coordinates $x^m + \mathrm{d}x^m$ is given by the line element $\mathrm{d}s$, where

$$\mathrm{d}s^2 = g_{mn}\mathrm{d}x^m \mathrm{d}x^n \,, \tag{4.1.8}$$

with g_{mn} depending on the x^m. The expression on the right-hand side is a symmetrized tensor product, which we denote by simple juxtaposition rather than by introducing a dedicated symbol (unlike for antisymmetric products, which we will denote by $\wedge$ as in flat space). We use the metric to lower indices: we associate a form $v_m \equiv g_{mn}v^n$ to any vector field v. Vice versa, we use the inverse metric $g^{mn} \equiv (g^{-1})^{mn}$ to raise indices: we associate a vector field $\alpha^m \equiv g^{mn}\alpha_n$ to any one-form α_m.[1]

4.1.2 Covariant derivatives and curvature

Covariant derivative

Indices can be deceiving. A derivative $\partial_m v^n$ looks like a $(1,1)$-tensor, but in fact transforms to $\tilde{J}_m{}^p\partial_p(J^n{}_r v^r)$, which is not of the type (4.1.7). One overcomes this problem in a similar way as for fields coupled to electromagnetism or to Yang–Mills fields, by introducing a *covariant derivative*

$$\nabla_m v^n \equiv \partial_m v^n + \Gamma^n_{mp} v^p \,, \tag{4.1.9}$$

with the *connection* Γ^n_{mp} also not being a tensor, but transforming so as to make the full $\nabla_m v^n$ a tensor:

$$\Gamma^m_{np} \to \left(J^m{}_q \Gamma^q_{rs} - \partial_r J^m{}_s\right) \tilde{J}^r{}_n \tilde{J}^s{}_p \,. \tag{4.1.10}$$

Alternatively, we can define the covariant derivative directly on the coordinate basis (4.1.3). Since ∂_m are a basis for vector fields, $\nabla_m(\partial_n)$, which is a vector field, can be expanded in the basis, and we can define the Γ^p_{mn} as the coefficients of this expansion:

$$\nabla_m(\partial_n) \equiv \Gamma^p_{mn}\partial_p \,. \tag{4.1.11}$$

Then

$$\nabla_m(v^n\partial_n) = (\partial_m v^n)\partial_n + v^n(\nabla_m\partial_n) = (\partial_m v^n)\partial_n + v^n\Gamma^p_{mn}\partial_p = (\partial_m v^n + \Gamma^n_{mp}v^p)\partial_n \tag{4.1.12}$$

reproducing (4.1.9).

∇_m gives a way of comparing the components of a vector field at two nearby points. A vector is said to be *parallel transported* along a curve γ when $\nabla_m v^p = 0$ in directions tangent to γ.

Riemann tensor

While Γ does not transform as a tensor, the commutator of covariant derivatives[2]

$$[\nabla_m, \nabla_n]v^p = 2\nabla_{[m}\nabla_{n]}v^p = R^p{}_{qmn}v^q \,, \qquad R^p{}_{qmn} \equiv 2(\partial_{[m}\Gamma^p_{n]q} + \Gamma^p_{[m|r}\Gamma^r_{n]q}) \tag{4.1.13}$$

defines the *Riemann tensor* $R^q{}_{pmn}$. (This is similar to an identity for the electromagnetic or YM covariant derivative, to be reviewed in Section 4.2.2.) Intuitively, if we take a vector v_0^p at a point and try to extend it to a vector field that is parallel transported along a small rectangular path γ in the x^m–x^n plane of sides ϵ, we will find that its value v_γ^p at the end of γ is rotated by the Riemann tensor:

$$v_\gamma^p = v_0^p + \epsilon^2 R^p{}_{qmn} v_0^p \,. \tag{4.1.14}$$

[1] As already stated in the previous chapter, when using an index-free notation, it then becomes natural to use the same symbol α for $\alpha_m dx^m$ and $\alpha^m\partial_m$, with a slight abuse of language. Some authors, especially in the mathematical literature, prefer introducing so-called *musical maps* ♭ and ♯ to distinguish the two, but we will not do that.

[2] An index enclosed in vertical | is excluded from antisymmetrization.

One usually takes the connection ∇_m to be the *Levi-Civita* one, namely to satisfy

$$\nabla_m g_{np} = 0 \,, \qquad T^p_{mn} \equiv \Gamma^p_{[mn]} = 0 \qquad \text{(Levi-Civita connection)}\,. \tag{4.1.15}$$

So the metric is covariantly constant and the *torsion tensor* T^p_{mn} vanishes. From the two conditions (4.1.15), one can find the unique explicit expression

$$\Gamma^p_{mn} = g^{pq}\left(\partial_{(m}g_{n)q} - \frac{1}{2}\partial_q g_{mn}\right) . \tag{4.1.16}$$

(The round brackets denote symmetrized indices.) Other connections will become useful in the next chapter, but from now on we will assume ∇_m to be the Levi-Civita connection unless we explicitly state otherwise.

For one-forms, again we can take the covariant derivatives $\nabla_m \mathrm{d}x^n$ of the basis elements and expand them on the basis itself. Imposing the Leibniz identity and using (4.1.6), the coefficients are fixed as

$$\nabla_m(\mathrm{d}x^n) = -\Gamma^n_{mp}(\mathrm{d}x^p)\,. \tag{4.1.17}$$

Now $\nabla_m(\alpha_n \mathrm{d}x^n) = (\partial_m \alpha_n)\mathrm{d}x^n - \alpha_n \Gamma^n_{mp}\mathrm{d}x^p = (\partial_m\alpha_n - \Gamma^p_{mn}\alpha_p)\mathrm{d}x^n \equiv (\nabla_m\alpha_n)\mathrm{d}x^n$, similar to (4.1.12). There is also an analogue of (4.1.13):

$$[\nabla_m, \nabla_n]\alpha_p = -R^q{}_{pmn}\alpha_q\,. \tag{4.1.18}$$

Similar definitions can be given for other tensors, using (4.1.11) and (4.1.17) to determine the action of ∇_m on them. In general, the action of ∇ on tensor components that generalizes (4.1.9) can be written as

$$\nabla_m \equiv \partial_m + \Gamma^n_{mp}\rho^p{}_n\,, \tag{4.1.19}$$

where $\rho^p{}_n$ is the Lie algebra representation on the space of tensors, which satisfies[3]

$$[\rho^p{}_q, \rho^r{}_s] = \rho^p{}_s\delta^r_q - \rho^r{}_q\delta^p_s\,. \tag{4.1.20}$$

Using this, one can also prove an analogue of (4.1.13) for general tensors:

$$[\nabla_m, \nabla_n] = R^q{}_{pmn}\rho^p{}_q\,. \tag{4.1.21}$$

Explicitly, $\rho^m{}_n$ can be obtained by expanding $J^m{}_p \sim \delta^m_p + \epsilon\rho^m{}_p$ near the identity in (4.1.7); for example, for vectors $\rho^p{}_n v^q = v^p\delta^q_n$, giving back (4.1.9), while for one forms $\rho^p{}_n\alpha_q = -\alpha_n\delta^p_q$. More generally on k-forms

$$\rho^m{}_n = -\mathrm{d}x^m \wedge \iota_n\,, \tag{4.1.22}$$

so that (4.1.19), (4.1.21) become

$$\nabla_m = \partial_m - \Gamma^n_{mp}\mathrm{d}x^p \wedge \iota_n\,, \tag{4.1.23}$$

$$[\nabla_m, \nabla_n] = -R^q{}_{pmn}\mathrm{d}x^p \wedge \iota_q\,. \tag{4.1.24}$$

On one-forms, these reproduce (4.1.17) and (4.1.18) respectively.

[3] This is similar to (2.1.2); the latter is in fact an antisymmetrization of (4.1.20). The reason is that the Jacobian matrices we are dealing with here are in $\mathrm{GL}(d,\mathbb{R})$ rather than $\mathrm{SO}(d)$; this will play a role in Section 4.3.

Various properties of the Riemann tensor follow from (4.1.21):

$$R_{mnpq} = -R_{mnqp} \tag{4.1.25a}$$
$$R_{mnpq} = -R_{nmpq} \tag{4.1.25b}$$
$$R^m{}_{[npq]} = 0 \tag{4.1.25c}$$
$$R_{mnpq} = R_{pqmn} \, , \tag{4.1.25d}$$

as well as the differential property

$$\nabla_{[m} R_{np]qr} = 0 \, . \tag{4.1.26}$$

(4.1.25a) is an immediate consequence of (4.1.13). (4.1.25b), (4.1.25c), and (4.1.26) come from applying (4.1.21). (4.1.25d) is a consequence of the previous ones in (4.1.25).

Other curvature tensors and scalars

One can introduce various other tensors related to Riemann. The *Ricci tensor* is

$$R_{mn} \equiv R^p{}_{mpn} \, . \tag{4.1.27}$$

It follows from (4.1.25d) that it is symmetric. Going in the opposite direction, we can also subtract the trace from the Riemann tensor, defining a tensor W_{mnpq} such that the trace in (4.1.27) vanishes, $g^{mp} W_{mnpq} = 0$. If we want to keep the symmetry properties (4.1.25a), (4.1.25b), (4.1.25d) of the Riemann tensor, the only option is the *Weyl tensor*

$$W_{mnpq} = R_{mnpq} - \frac{2}{d-2}\left(g_{m[p}R_{q]n} - g_{n[p}R_{q]m}\right) + \frac{2}{(d-1)(d-2)} g_{m[p}g_{q]n}R \, . \tag{4.1.28}$$

The symmetry properties (4.1.25a), (4.1.25b), (4.1.25d) define an SO(d) representation. This is not reducible: in $d > 4$ it is the sum of three irreducible representations. The idea is to view the Riemann tensor as a symmetric matrix $R_{\Lambda\Sigma}$, where each index is now a group of two antisymmetrized ordinary indices: $\Lambda \sim [mn]$. In other words, Riemann can be viewed as a map

$$\Lambda^2 \to \Lambda^2 \tag{4.1.29}$$

from the space of two-forms Λ^2 into itself. So now we need to take the symmetric part of the tensor product of the antisymmetric representation of SO(d) (the adjoint) with itself. This gives three representations, which correspond to

- The Weyl tensor (4.1.28)
- The *Ricci scalar* (or *scalar curvature*) $R \equiv g^{mn} R_{mn}$
- The traceless part of the Ricci tensor, $R^0_{mn} \equiv R_{mn} - \frac{1}{d} g_{mn} R$

In $d < 4$, the Weyl tensor vanishes identically, giving a relation between Ricci and Riemann. Physically, this means that gravity does not propagate below $d = 4$.

For $d = 4$, the only change is that we can further decompose $\Lambda^2 = \Lambda^2_+ \oplus \Lambda^2_-$ as we did in (3.3.13) in flat space. (In Lorentzian signature, that definition would require an extra i.) This gives then a 2×2 block decomposition. The block mapping $\Lambda^2_+ \to \Lambda^2_-$ (and vice versa, since R_{MN} is symmetric) turns out to be the traceless Ricci tensor. The blocks mapping $\Lambda^2_+ \to \Lambda^2_+$ and $\Lambda^2_- \to \Lambda^2_-$ are two matrices whose traceless parts

$W^{\pm}$ are (anti-)self-dual projections of (4.1.28); the two traces are both related to the Ricci scalar R. So schematically, the Riemann tensor in $d = 4$ decomposes as

$$\begin{pmatrix} W^+ + R & R^0 \\ R^0 & W^- + R \end{pmatrix} . \tag{4.1.30}$$

(More details will be given in Exercise 4.3.11.)

Other scalars related to the Riemann tensor can also be useful. In physics, we often encounter loci where curvature diverges. Simply computing components of Ricci or Riemann might not be too meaningful, since individual components may become finite upon multiplication by a Jacobian matrix with some vanishing elements. This issue is avoided by considering scalars built out of curvature, such as

$$R\,, \qquad R_{\mu\nu}R^{\mu\nu}\,, \qquad R_{\mu\nu\rho\sigma}R^{\mu\nu\rho\sigma}\,. \tag{4.1.31}$$

If any of these diverges at a point in spacetime, it diverges in any coordinate system. These also appear in the leading quantum corrections to general relativity [25].

4.1.3 Manifolds

Having built a formalism where general coordinate transformations, or diffeomorphisms, are a gauge invariance, we can use this large symmetry to glue pieces of space together. The resulting object is called a manifold.

Definitions

A *topological* d-dimensional manifold is a (connected) space which is locally homeomorphic to $\mathbb{R}^d$: namely, around every point there is a neighborhood that can be mapped to $\mathbb{R}^d$ with a map that is continuous, and whose inverse is continuous. However, we are more interested in *smooth* manifolds. These are obtained by gluing together open sets $U_\alpha \subset \mathbb{R}^d$, called *charts* or *patches*, with respectively a geographic or sartorial metaphor. Namely, we consider a union of the U_α, where some subsets $U_{\alpha\beta} \subset U_\alpha$ are identified:

$$M = \cup_\alpha U_\alpha / \sim\,, \qquad x_{(\alpha)} \in U_{\alpha\beta} \sim x_{(\beta)} \in U_{\beta\alpha} \;\text{ if }\; x_{(\alpha)} = g_{\alpha\beta}(x_{(\beta)})\,; \tag{4.1.32}$$

the identifying maps $g_{\alpha\beta}$ are called *transition functions*, and are supposed to be smooth (C^∞). The U_α are then also open sets in M, and have a coordinate system $x_{(\alpha)}$, by virtue of their being subsets of $\mathbb{R}^d$. The two open subsets $U_{\alpha\beta}$ and $U_{\beta\alpha}$ are identified and become the same open set, on which two coordinate systems $x_{(\alpha)}$, $x_{(\beta)}$ are present; the $g_{\alpha\beta}$ can be thought of as a change of coordinates from one to the other. All manifolds in this book will be smooth unless otherwise stated.

Of course, even the local coordinate systems $x_{(\alpha)}$ are not unique, and one may decide to change them in favor of new ones, $x'_{(\alpha)} = g_\alpha(x_{(\alpha)})$, functions of the old ones. If one performs such local coordinate changes, the transition functions change too:

$$g_{\alpha\beta} \to g'_{\alpha\beta} = g_\alpha \circ g_{\alpha\beta} \circ g_\beta^{-1}\,, \tag{4.1.33}$$

where $\circ$ denotes function composition.

Many of our manifolds will also be *compact*. In topology, a compact space is closed and bounded; since we have required neighborhoods to be homeomorphic to

$\mathbb{R}^d$, for us a compact manifold has no boundary. Also, most of our manifolds will be *orientable*: namely, the transition functions can be chosen so that their Jacobian matrices have positive determinant, $\det \frac{\partial x^m_{(\alpha)}}{\partial x^n_{(\beta)}} > 1$. A famous example of nonorientable manifold is the *Möbius strip* $\frac{([0,1]\times(0,1))}{\sim}$, with the identification $(0, x^2) \sim (1, 1 - x^2)$. Further gluing antipodal point on the boundary, we can define the *Klein bottle* $K \equiv M_2/\{(x^1, 0) \sim (x^1, 1)\}$, nonorientable but also compact.

Tensors on manifolds

A function on M is defined as a collection of functions on the U_α that respect the identification (4.1.32). Namely, a function $f\colon M \to \mathbb{R}$ is a collection of functions $f_{(\alpha)}\colon U_\alpha \to \mathbb{R}$ such that $f_{(\alpha)}(x_{(\alpha)}) = f_{(\alpha)}(g_{\alpha\beta}(x_{(\beta)})) = f_{(\beta)}(x_{(\beta)})$ on $U_{\alpha\beta}$.

Defining tensors requires more care. We saw in Section 4.1.1 that general covariance requires us to generalize a vector to a vector field, even in $\mathbb{R}^d$, because of the transformation law (4.1.2). It is now easy to define a vector field also on a manifold M, which is after all just a gluing of charts by coordinate transformations. The definition is similar to the one we saw for functions on M: a vector field on M is a collection of vector fields $v_{(\alpha)}$ on each U_α, such that now

$$v^m_{(\alpha)} = \frac{\partial x^m_{(\alpha)}}{\partial x^n_{(\beta)}} v^n_{(\beta)} \equiv (J_{(\alpha\beta)})^m{}_n v^n_{(\beta)} \,. \tag{4.1.34}$$

Alternatively, in terms of the index-free notation in (4.1.4), the transformation from one U_α to a U_β reads more simply $v_{(\alpha)} = v_{(\beta)}$.

A function on a manifold M unambiguously assigns a real value to each point in M. One might want to think of a vector field in a similar way. At first, one might try to think of it as assigning d real numbers to each point of M. This would be wrong: (4.1.2) shows that the numbers get mixed under change of coordinates. A better way is to introduce a new manifold called *tangent bundle*. This is a larger manifold TM associated to another manifold M, whose charts are of the form $\hat{U}_\alpha = U_\alpha \times \mathbb{R}^d$, where U_α are the charts of M. Similar to (4.1.32), we can define

$$\begin{aligned} TM = \cup_\alpha \hat{U}_\alpha/\sim \qquad & (x_{(\alpha)}, v_{(\alpha)}) \in \hat{U}_{\alpha\beta} \sim (x_{(\beta)}, v_{(\beta)}) \in \hat{U}_{\beta\alpha} \text{ if} \\ x_{(\alpha)} = g_{\alpha\beta}(x_{(\beta)}) \,, \qquad & v^m_{(\alpha)} = \frac{\partial x^m_{(\alpha)}}{\partial x^n_{(\beta)}} v^n_{(\beta)} = (J_{(\alpha\beta)})^m{}_n v^n_{(\beta)} \,. \end{aligned} \tag{4.1.35}$$

Thus TM is again a manifold, but one with a peculiar type of chart and transition function; the latter, in particular, acts linearly on the components of $\hat{U}_\alpha$ in $\mathbb{R}^d$. This is of course simply our vector transformation law (4.1.34). There is a natural *projection map*

$$\pi\colon TM \to M \,, \tag{4.1.36}$$

defined in each chart by $\pi\colon (x_{(\alpha)}, v_{(\alpha)}) \mapsto x_{(\alpha)}$. The counterimage of any point is $\mathbb{R}^d$ and is called the *fiber*. A vector field can now be thought of as assigning to any point $x \in M$ a point TM whose projection π gives x. One also says that a vector field is a *section* of TM.

We can introduce likewise other tensor fields on M. A one-form on a manifold M can be defined similar to (4.1.34), as a collection of one-forms $C_{(\alpha)}$ on the open sets U_α such that $C_{(\alpha)} = C_{(\beta)}$ (in index-free notation). There is also a bundle of one-forms, called the *cotangent* bundle T^*M. The fiber is still $\mathbb{R}^d$ as in (4.1.35), but the transition functions now act on the fiber with the inverse Jacobian matrix:

$$\alpha_{m\,(\alpha)} = \frac{\partial x^n_{(\beta)}}{\partial x^m_{(\alpha)}} \alpha_{n\,(\beta)} \,. \tag{4.1.37}$$

All this can also be generalized to tensors with upper and lower indices, introducing more complicated bundles that generalize TM and T^*M. The Jacobian matrix will simply act on each index as appropriate for an upper or lower index. For example, the bundle of tensors of type $(1,1)$ (one index up and one down) is denoted by

$$(T \otimes T^*)M \,. \tag{4.1.38}$$

Another notable example is the bundle of tensors with two lower indices $T^* \otimes T^*$. A metric on M is a symmetric section of this bundle, $g_{mn} = g_{nm}$. A Riemannian metric is also required to be positive-definite at every point, and a Lorentzian metric to have signature $(- + \cdots +)$ at every point. At least one Riemannian metric can be proven to exist on any manifold, satisfying a very mild topological condition (called paracompactness).

The bundle of k-forms is denoted by $\Lambda^k T$, with a terminology inspired by exterior algebra (Section 3.1.1). Their fiber is a copy of the space Λ^k of forms in $\mathbb{R}^d$. Thus we have the same algebraic structure, and there is a wedge product that takes two form fields of degree k_1, k_2 to a third of degree $k_1 + k_2$. Its component expression is just the same we gave in (3.1.8), but now with the understanding that the α_{k_i} are no longer constant.

One can also introduce a bundle ΛT^* whose fiber is the total exterior algebra $\Lambda = \oplus_{k=0}^d \Lambda^k$. The transition functions are acting in a block-diagonal fashion on each summand Λ^k. This is a particular case of a direct sum of bundles, which is obtained by taking a fiber $F = F_1 \oplus F_2$, a direct sum of the fibers F_i of two other bundles, and letting the transition functions act block-diagonally. Another example is the *generalized tangent bundle*

$$(T \oplus T^*)M \,, \tag{4.1.39}$$

of which are sections the $\mathrm{Cl}(d,d)$ gamma matrices (3.2.4). This will play an important role in Chapter 10.

So far, we have considered *vector* bundles: the fiber was a vector space, and the transition functions were acting linearly on it. In other words, the transition functions in all our examples so far are really always $J_{(\alpha\beta)}$, but acting in different $G = \mathrm{GL}(d,\mathbb{R})$ representations. One can view T as arising from having selected the fundamental representation of G, T^* as arising from the antifundamental representation (where we use $(J^t_{(\alpha\beta)})^{-1}$ instead of $J_{(\alpha\beta)}$), (4.1.38) from the representation obtained as a tensor product of the previous two, and so on.

In general, the fiber F need not be a vector space. One may consider it to be any space on which $G = \mathrm{GL}(d,\mathbb{R})$ has a *group action*. By this we mean a map that associates to a $g \in G$ and to $f \in F$ an element $g \cdot f \in F$, such that

$$(g_1 g_2) \cdot f = g_1 \cdot (g_2 f) \tag{4.1.40}$$

for any $g_i \in G$ and $f \in F$, and $e \cdot f = f$ for e the identity element of G.

An example of this more general type of bundle is obtained by taking F to be the space of *frames*, namely of ordered bases in $\mathbb{R}^d$. One can also view F as a copy of $G = \mathrm{GL}(d, \mathbb{R})$ itself. We take the action of G on F to be the left action $f \to gf$. The resulting bundle is called the *frame bundle* FM.

Two-sphere S^2

We will illustrate some of the preceding ideas on the two-sphere

$$S^2 \equiv \{y_1^2 + y_2^2 + y_3^2 = 1\} \subset \mathbb{R}^3 . \qquad (4.1.41)$$

We can cover it with two charts $U_\mathrm{N} \equiv S^2 - \{s\}$, $U_\mathrm{S} \equiv S^2 - \{n\}$, where $n \equiv (0, 0, 1)$ and $s \equiv (0, 0, -1)$ are the North and South poles. One often uses spherical coordinates θ and ϕ such that

$$y_1 = r \sin\theta \sin\phi , \qquad y_2 = r \sin\theta \cos\phi , \qquad y_3 = r\cos\theta \qquad (4.1.42)$$

and the locus in (4.1.41) becomes $\{r = 1\}$. Equation (4.1.42), however, provides only good coordinates on $U_\mathrm{NS} \equiv U_\mathrm{N} \cap U_\mathrm{S}$, in the sense that ϕ is ambiguous at $\theta = 0, \pi$ (namely at n and s). Good coordinates on U_N and U_S may be obtained by the so-called *stereographic projections* $\pi_{\mathrm{N,S}}$. For example, $\pi_\mathrm{N}(p)$, $p \in S^2$, is defined by tracing a line in R^3 through $s = (0, 0, -1)$ and p, and intersecting it with the plane $\{z = 0\}$. This map identifies U_N with $T_n \cong \mathbb{R}^2$; the coordinate system x^m_N can be obtained by using ordinary cartesian coordinates on $\mathbb{R}^2$. The other projection π_S can be defined similarly. Making all this elementary geometry explicit,

$$x^1_\mathrm{N} + \mathrm{i}x^2_\mathrm{N} \equiv z_\mathrm{N} = \tan(\theta/2)\, \mathrm{e}^{\mathrm{i}\phi} , \qquad x^1_\mathrm{S} + \mathrm{i}x^2_\mathrm{S} \equiv z_\mathrm{S} = \cot(\theta/2)\, \mathrm{e}^{-\mathrm{i}\phi} . \qquad (4.1.43)$$

This gives maps connecting the coordinate systems x^m_N, x^m_S and (θ, ϕ) on the three open sets U_N, U_S, and U_NS. The latter is of course redundant; one can eliminate (θ, ϕ) from (4.1.43) to obtain

$$z_\mathrm{N} = z_\mathrm{S}^{-1} , \qquad (4.1.44)$$

whose real and imaginary parts give the transition function g_NS. We can also think of the S^2 as the Riemann sphere

$$\mathbb{C} \cup \{z = \infty\} . \qquad (4.1.45)$$

Equation (4.1.44) is holomorphic in $U_\mathrm{NS} = U_\mathrm{N} \cap U_\mathrm{S}$; as we will discuss later, this indicates that S^2 is a *complex manifold*.

A simple metric on S^2 is the so-called *round* metric, with a line element that on $\mathrm{U_{NS}}$ reads

$$\mathrm{d}s^2_{S^2} = \mathrm{d}\theta^2 + \sin^2\theta \mathrm{d}\phi^2 . \qquad (4.1.46)$$

From (4.1.46) and (4.1.8), one reads off $g = \mathrm{diag}(1, \sin^2\theta)$, which is singular at $\theta = 0, \pi$, namely at the poles. Nevertheless, one can extend (4.1.46) to U_N and U_S:

$$\mathrm{d}s^2_{S^2} = 4\frac{\mathrm{d}z_\mathrm{N}\mathrm{d}\bar{z}_\mathrm{N}}{(1 + |z_\mathrm{N}|^2)^2} = 4\frac{\mathrm{d}z_\mathrm{S}\mathrm{d}\bar{z}_\mathrm{S}}{(1 + |z_\mathrm{S}|^2)^2} . \qquad (4.1.47)$$

The expressions for g^N_{mn} and g^S_{mn} that can be read off from (4.1.8) are now smooth at the poles. So $\mathrm{d}s^2_{S^2}$ is in fact a metric on S^2, as defined in this section.

4.1.4 Maps

A map $\phi\colon M \to N$ connecting two topological spaces is called a *homeomorphism* if it is continuous, bijective, and its inverse ϕ^{-1} is also continuous. M and N are then said to be homeomorphic. This is what we most commonly think of when we say that two spaces are "topologically equivalent."

In physics, however, we also want maps to be differentiable. If M and N are manifolds of dimensions d and d', we can decide whether a map f is differentiable by using their local coordinate systems. We take the charts $U \subset M$ and $V \subset N$ to which p and $f(p)$ belong, with their respective coordinate systems x^m and y^n, and express the map near p in coordinates as the d' functions

$$y^n(x^1, \ldots, x^d)\,. \tag{4.1.48}$$

Now these are maps in $\mathbb{R}^d$, and we know how to check if they are smooth in p. If that is the case, by definition f is said to be smooth in p. From now on we will consider maps that are smooth everywhere.

When $N = \mathbb{R}$, a map $f\colon M \to \mathbb{R}$ should be understood as a function $f_{(\alpha)}$ on every U_α, consistent with our definition in Section 4.1.3; f is smooth if all the $f_{(\alpha)}$ are.

Another notable case is when M and N have the same dimension, $d' = d$. If ϕ is bijective and smooth, and ϕ^{-1} is also smooth, ϕ is called a *diffeomorphism*, extending our earlier definition on $\mathbb{R}^d$. In effect, ϕ says that M and N are the same smooth manifold. This is similar to our earlier statement that topological spaces are considered equivalent if they are homeomorphic.

A diffeomorphism that is very close to the identity

$$y^m = x^m + \epsilon\, \delta x^m \tag{4.1.49}$$

produces a vector field δx^m. Again we can intuitively imagine the components of the vector field as assigning to every point a little arrow, which in turn can be thought of as an infinitesimal map to a nearby point.

Given a smooth map $\phi\colon M \to N$, the Jacobian matrix $\frac{\partial y^n}{\partial x^m}$ of (4.1.48) defines a map $\phi_*\colon T_pM \to T_{\phi(p)}N$ for every point p, called the *differential* of ϕ: it maps a vector field v on M to a vector field $\phi_* v$ on N, with local components

$$(\phi_* v)^n = \frac{\partial y^n}{\partial x^m} v^m\,. \tag{4.1.50}$$

This is also called the *push-forward* of v. A similar definition applies to multivector fields, namely tensors with many upper indices. Given instead a tensor on N with lower indices T, we can define one on M whose local components are

$$(\phi^* T)_{m_1 \ldots m_k} \equiv \frac{\partial y^{n_1}}{\partial x^{m_1}} \cdots \frac{\partial y^{n_k}}{\partial x^{m_k}} T_{n_1 \ldots n_k}\,; \tag{4.1.51}$$

this is the *pull-back* of T. In particular, when ϕ is an inclusion map we denote this by $T|_\Sigma$; we recover the operation that we used several times in Chapter 1, for example in (1.1.3), (1.1.10), and (1.3.24). Notice the opposite behavior depending on index position: tensors with upper indices (such as vectors) are pushed from M to N, the same direction as the map ϕ; tensors with lower indices (such as one-forms) are pulled back from N to M. This is a natural consequence of the index structure $\frac{\partial y^n}{\partial x^m}$ of the Jacobian matrix of the map.

Specializing now (4.1.51) to the metric, a map $\phi: M \to M$ that leaves it invariant,

$$(\phi^* g)_{mn} = g_{mn}, \tag{4.1.52}$$

is called an *isometry*. The composition of two isometries is still an isometry; so they form a group ISO(g), whose dimension can be proven to be bounded:

$$\dim(\mathrm{ISO}(g)) \le d(d-1)/2. \tag{4.1.53}$$

4.1.5 Lie derivatives

The partial derivative $\partial_m f$ of a function measures the variation of f in direction m. More generally, to a vector field v^m one can associate the *Lie derivative*

$$L_v f \equiv v^m \partial_m f, \tag{4.1.54}$$

which measures the variation of f in the direction of the vector field v, or in other words compares $f(x + \epsilon v)$ with $f(x)$. We will now extend this idea to vectors and other tensors.

Lie derivatives of vectors

Naively one might just copy (4.1.54) and define, for example, the Lie derivative of a vector field w as $L_v w \stackrel{?}{=} v^m \partial_m w^\nu$. This, however, would not transform as a vector. The reason is once again that a vector field is not just a d-uple of functions; we cannot compare directly $w(x+\epsilon v)$ with $w(x)$, because they belong to different vector spaces. To cure this, we should view the vector field v as an infinitesimal diffeomorphism $x^m \to x^m + \epsilon\, v^m(x)$, as in (4.1.49). Applying this map to the vector field w, from (4.1.2) we obtain $(\delta^m_n + \epsilon \partial_n v^m) w^n$. This can now be compared with the value of the original vector field evaluated at in the point with coordinates $x^m + \epsilon v^m$:

$$w^n(x + \epsilon v) - (\delta^m_n + \epsilon \partial_n v^m) w^n(x) \sim \epsilon (v^m \partial_m w^n - \partial_n v^m w^n)(x). \tag{4.1.55}$$

This suggests the definition

$$L_v w \equiv (v^m \partial_m w^n - w^m \partial_m v^n)\partial_n . \tag{4.1.56}$$

This indeed now transforms as a vector field. The first term comes from evaluating w at the displaced point; the second term comes from the Jacobian matrix in the vector field transformation law (4.1.2). So L_v is the infinitesimal version of the diffeomorphism action on vectors (4.1.50).

Equation (4.1.56) also arises by computing the commutator $[v, w]$, where now both v and w are treated as operator, just like in quantum mechanics. Recall first the identity $[\partial_x, x] = 1$, which can be checked quickly on a test function; this generalizes to

$$[\partial_m, f] = \partial_m f, \tag{4.1.57}$$

where by f we mean the multiplication operator by a function f. Now, we use (4.1.57) and (2.1.31a) repeatedly:

$$\begin{aligned} [v, w] &= [v^m \partial_m, w^n \partial_n] = [v^m \partial_m, w^n]\partial_n + w^n [v^m \partial_m, \partial_n] \\ &= v^m [\partial_m, w^n]\partial_n - w^n [v^m, \partial_n]\partial_m = (v^m \partial_m w^n - w^m \partial_m v^n)\partial_n . \end{aligned} \tag{4.1.58}$$

This is called the *Lie bracket* of two vector fields; it happens to be equal to $L_v w$. The bracket is antisymmetric, linear in both entries, and satisfies the Jacobi identity; by definition, it turns the vector space of vector fields into a Lie algebra (of infinite dimension). We will see an intuitive interpretation in Section 4.1.8.

Lie derivatives of tensors

The Lie derivative for a one-form can be obtained by a logic similar to (4.1.55):

$$L_v \alpha \equiv (v^m \partial_m \alpha_n + \alpha_m \partial_n v^m) dx^n ; \tag{4.1.59}$$

the different sign and different index structure in the second term relative to (4.1.56) reflects the presence of $(J^t)^{-1}$ rather than J in (4.1.5). Another way to obtain (4.1.59) is to impose the Leibniz identity $L_v(w^n \alpha_n) = (L_v w)^n \alpha_n + w^n (L_v \alpha)_n$.

Both of these approaches can be applied to more general tensors; the general result can be summarized as

$$L_v = v^n \partial_n - \partial_m v^n \rho^m{}_n , \tag{4.1.60}$$

where again $\rho^m{}_n$ is the Lie algebra representation of the tensor (4.1.19). One can equivalently write this as

$$L_v = v^n \nabla_n - \nabla_m v^n \rho^m{}_n : \tag{4.1.61}$$

indeed, using (4.1.19) and the torsion-free property of the connection, we can see that the Γ terms cancel out. For connections more general than the Levi-Civita one, (4.1.61) and (4.1.60) are unequal; acting with their difference on a vector w gives $([v, w] - \nabla_v w + \nabla_w v)^p = v^m w^n T^p_{mn}$, where $\nabla_v \equiv v^m \nabla_m$. This is sometimes taken as a definition of the torsion tensor T^p_{mn}.

Using (4.1.60) and (4.1.20), we can also show

$$[L_v, L_w] = L_{[v,w]} . \tag{4.1.62}$$

This says that L provides a morphism from the Lie algebra of vector fields on M to the Lie algebra of operators on forms. (Both algebras are infinite dimensional.)

When a tensor T is such that $L_v T = 0$, we say that it is *invariant* under L_v; if it is such $L_v T = nT$, we say that it has *weight*, or *degree*, or *charge* n.

On forms, using (4.1.22), (4.1.60), and (4.1.61) simplify to

$$L_v = v^n \partial_n + \partial_m v^n dx^m \wedge \iota_n = v^n \nabla_n + \nabla_m v^n dx^m \wedge \iota_n . \tag{4.1.63}$$

Killing vectors

Another important case is when L acts on the metric. Using (4.1.60),

$$(L_v g)_{mn} = v^p \partial_p g_{mn} + \partial_m v^p g_{pn} + \partial_n v^p g_{mp} \overset{(4.1.61),(4.1.16)}{=} 2\nabla_{(m} v_{n)} . \tag{4.1.64}$$

A K such that

$$(L_K g)_{mn} = 2\nabla_{(m} K_{n)} = 0 \tag{4.1.65}$$

is called a *Killing vector*. We remarked in (4.1.49) that a vector represents an infinitesimal diffeomorphism; in the same spirit, a Killing vector represents an infinitesimal isometry (recalling (4.1.52)). From (4.1.62), we see that if K_1 and K_2 are two Killing vectors, their Lie bracket $[K_1, K_2]$ is Killing too. So the vector space of Killing vectors is a Lie algebra iso(g), whose associated Lie group is ISO(g). Unlike the space of *all* vector fields, this is finite dimensional, because of the bound (4.1.53).

We also derive a refinement of (4.1.18). Applying that equation to K_p, and using (4.1.65),

$$\nabla_m \nabla_n K_p + \nabla_n \nabla_p K_m = -R^q{}_{pmn} K_q \,. \tag{4.1.66}$$

Summing over cyclic permutations on both sides and using (4.1.25c), we obtain $\nabla_m \nabla_n K_p + \text{cycl} = 0$. Returning to (4.1.66), we get

$$\nabla_m \nabla_n K_p = R^q{}_{mnp} K_q \,. \tag{4.1.67}$$

Conformal Killing vectors

A more general concept is that of *conformal Killing vectors* (CKV), namely those such that

$$(L_K g)_{mn} = f g_{mn} \,. \tag{4.1.68}$$

for some function f. They again form a Lie algebra conf(g); the corresponding group is called the *conformal group*, or *group of conformal transformations*, of a metric.

For example, the conformal group of the flat Lorentzian metric η is

$$\text{Conf}(\eta) = \text{SO}(d-2, 2) \,. \tag{4.1.69}$$

As another example, define the round sphere metric on $S^d = \{\sum_{M=1}^{d+1} (y^M)^2 = 1\} \subset \mathbb{R}^{d+1}$ by generalizing (4.1.88):

$$\mathrm{d}s^2_{\mathbb{R}^{d+1}} = \mathrm{d}r^2 + r^2 \mathrm{d}s^2_{S^d} \,. \tag{4.1.70}$$

The conformal group of S^d is

$$\text{Conf}(g_{S^d}) = \text{SO}(d+1, 1) \,, \tag{4.1.71}$$

generalizing the celestial sphere argument leading to Möbius transformations (2.2.24) for $d = 2$.

4.1.6 Differential operators as bosons and fermions

We now focus on forms. We discussed their algebraic properties in Section 3.1; we will now consider some differential aspects.

The first observation is that for taking derivatives of forms, we don't really need to introduce a connection. A partial derivative of a tensor is not a tensor: from the transformation law (4.1.7), one sees that the derivatives of a tensor picks up derivatives $\partial_m J_n{}^p$. This is what motivated introducing Γ^m_{np}. But the antisymmetrized derivative of a k-form $\partial_{[m_1} \alpha_{m_2 \ldots m_{k+1}]}$ transforms already as the components of a $(k+1)$-form: one can check that the $\partial_m J_n{}^p$ drop out. Related to this, with the Levi-Civita connection,

$$\nabla_{[m_1} \alpha_{m_2 \ldots m_{k+1}]} = \partial_{[m_1} \alpha_{m_2 \ldots m_{k+1}]} \, : \tag{4.1.72}$$

all the Γ terms cancel out because of the torsion-free condition in (4.1.15).

So we have a new differential operation that creates a $(k+1)$-form out of a k-form. This is the *exterior differential* that we already used in Chapter 1:

$$\mathrm{d} \equiv \mathrm{d}x^m \wedge \partial_m \overset{(4.1.72),(4.1.23)}{=} \mathrm{d}x^m \wedge \nabla_m\,. \tag{4.1.73}$$

Given a k-form $\alpha = \alpha_k$ as in (3.1.6a),

$$\begin{aligned} \mathrm{d}\alpha_k &= \frac{1}{k!}\mathrm{d}x^m \wedge \partial_m(\alpha_{m_1\ldots m_k}\mathrm{d}x^{m_1}\wedge\ldots\wedge\mathrm{d}x^{m_k}) \\ &= \frac{1}{k!}\partial_m\alpha_{m_1\ldots m_k}\mathrm{d}x^m\wedge\mathrm{d}x^{m_1}\wedge\ldots\wedge\mathrm{d}x^{m_k} \\ &= \frac{1}{k!}\partial_{[m}\alpha_{m_1\ldots m_k]}\mathrm{d}x^m\wedge\mathrm{d}x^{m_1}\wedge\ldots\wedge\mathrm{d}x^{m_k}\,; \end{aligned} \tag{4.1.74}$$

so for the components we have $(\mathrm{d}\alpha_k)_{mm_1\ldots m_k} = (k+1)\partial_{[m}\alpha_{m_1\ldots m_k]}$. Importantly, then, $\mathrm{d}(\mathrm{d}\alpha_k) = (k+2)\partial_{[m}(\mathrm{d}\alpha_k)_{nm_1\ldots m_k]} = (k+2)(k+1)\partial_{[m}\partial_n\alpha_{m_1\ldots m_k]}$, which vanishes because $\partial_{[m}\partial_{n]} = 0$. So

$$\mathrm{d}^2 = 0\,. \tag{4.1.75}$$

Since d contains a derivative, one expects it to obey a Leibniz identity. But the $\mathrm{d}x^m\wedge$ introduces some extra signs:

$$\mathrm{d}(\alpha_k\wedge\alpha'_{k'}) = (\mathrm{d}\alpha_k)\wedge\alpha'_{k'} + (-1)^k\alpha_k\wedge(\mathrm{d}\alpha'_{k'})\,. \tag{4.1.76}$$

This can also be read as an operator identity acting on a test form $\alpha'_{k'}$:

$$\mathrm{d}(\alpha_k\wedge) - (-1)^k\alpha_k\wedge\mathrm{d} = \begin{cases} [\mathrm{d},\alpha_k\wedge] & \text{even } k \\ \{\mathrm{d},\alpha_k\wedge\} & \text{odd } k \end{cases} = (\mathrm{d}\alpha_k)\wedge\,. \tag{4.1.77}$$

For more complicated identities, it is more convenient to use the "operatorial" point of view we already used in Section 3.1 (for example, in (3.1.21) and (3.1.24)), and more recently in (4.1.58). In that perspective, the wedges $\mathrm{d}x^m\wedge$ and the contractions ι_m were treated as gamma matrices for a "doubled" Clifford algebra $\mathrm{Cl}_{d,d}$. We now extend those methods to include derivatives. We assign a fictitious "fermion number" to the various operators we need, which guides us quickly to which identity of the type in (2.1.31) to use:

- $\mathrm{d}x^m\wedge$, ι_m are to be considered fermionic.
- ∂_m is to be considered bosonic.

As an application, let us compute the anticommutator $\{\iota_v, \mathrm{d}\}$, which is similar to (4.1.77) for $k = 1$, but with a contraction instead of a wedge. (Recall from (3.2.46) that $\iota_v \equiv v^m\iota_m$.) We can use

$$\begin{aligned} \{\iota_v,\mathrm{d}\} = \{v^m\iota_m, \mathrm{d}x^n\wedge\partial_n\} &\overset{(2.1.31e)}{=} \{v^m\iota_m,\mathrm{d}x^n\wedge\}\partial_n - \mathrm{d}x^n\wedge[v^m\iota_m,\partial_n] \\ &\overset{(2.1.31e),(2.1.31b)}{=} v^m\{\iota_m,\mathrm{d}x^n\wedge\}\partial_n - \mathrm{d}x^n\wedge[v^m,\partial_n]\iota_m \\ &\overset{(4.1.57),(3.2.3a)}{=} v^m\partial_m + \partial_n v^m\mathrm{d}x^n\wedge\iota_m \overset{(4.1.63)}{=} L_v\,. \end{aligned} \tag{4.1.78}$$

So we have

$$L_v = \{\iota_v, \mathrm{d}\} = \iota_v \mathrm{d} + \mathrm{d}\iota_v \tag{4.1.79}$$

on forms. This is called *Cartan's magic formula* because of its simplicity.

As a nice consequence of (4.1.79):

$$[\mathrm{d}, L_v] = 0\,. \tag{4.1.80}$$

Using the expression of L_v in (4.1.78), we can also compute

$$\begin{aligned}[\{\iota_v, \mathrm{d}\}, \iota_w] &= [L_v, \iota_w] = [v^m\partial_m + \partial_n v^m \mathrm{d}x^n \wedge \iota_m, w^p \iota_p] \\ &= v^m \partial_m w^p \iota_p - \partial_n v^m w^n \iota_m = [v, w]^p \iota_p = \iota_{[v,w]}\,.\end{aligned} \tag{4.1.81}$$

Finally we can take an anticommutator of this equation with d and get

$$\begin{aligned}L_{[v,w]} &\overset{(4.1.79)}{=} \{\mathrm{d}, \iota_{[v,w]}\} \overset{(4.1.81)}{=} \{\mathrm{d}, [L_v, \iota_w]\} \\ &\overset{(2.1.32b)}{=} \{[\mathrm{d}, L_v], \iota_w\} + [L_v, \{\mathrm{d}, \iota_w\}] \overset{(4.1.79),\,(4.1.80)}{=} [L_v, L_w]\,,\end{aligned} \tag{4.1.82}$$

which recovers (4.1.62) for forms.

4.1.7 Submanifolds

A *submanifold* is a subspace S of a manifold M that has itself the structure of a manifold. We will usually require submanifolds to be *embedded*: namely, the inclusion map,

$$i : S \hookrightarrow M, \tag{4.1.83}$$

should be a homeomorphism to its image $i(S)$, and its differential i_* (4.1.50) should be injective.

The origins of differential geometry lie in the study of submanifolds in $\mathbb{R}^d$. In fact, *Whitney's theorem* asserts that a smooth manifold can always be realized in this way, for large enough d. A key idea, however, is that one may study its properties (such as curvature) without any reference to such an ambient space; this led to the standard definition of manifolds with open sets and transition functions of Section 4.1.3. Still, it can sometimes be useful to be able to relate the geometry of a manifold and that of a submanifold.

In general, given a submanifold $S \subset M$, one may define a metric on S by a pull-back under (4.1.83) of the metric on M. Sometimes it is useful to represent this by a projector that annihilates the components orthogonal to S. To illustrate the idea, let us consider the codimension-one case, where $\dim S = \dim M - 1$. Then at every point on S, we define a normal vector n such that

$$n^2 = g^M_{mn} n^m n^n = \pm 1\,. \tag{4.1.84}$$

(The case $n^2 = 1$ is relevant to Euclidean signature; the case $n^2 = -1$ is relevant to Lorentzian signature.) Then at every point on S, we can also define the tensor

$$P_{mn} \equiv g_{mn} \mp n_m n_n\,, \tag{4.1.85}$$

called *first fundamental form*. It is orthogonal to n: $P_{mn}n^n = 0$. It is a projector once we raise an index: $P^m{}_n P^n{}_p = P^m{}_p$. We can also define the *second fundamental form* as a Lie derivative under the normal vector field:

$$K_{mn} = \frac{1}{2} L_n P_{mn} \,. \tag{4.1.86}$$

Now the curvatures of the metrics on S and M are related by the *Gauss–Codazzi equation*:

$$\begin{aligned} R^S_{m_1m_2m_3m_4} &= P_{m_1}{}^{n_1} P_{m_2}{}^{n_2} P_{m_3}{}^{n_3} P_{m_4}{}^{n_4} R^M_{n_1n_2n_3n_4} \\ &\pm (K_{m_1m_3}K_{m_2m_4} - K_{m_1m_4}K_{m_2m_3}) \,. \end{aligned} \tag{4.1.87}$$

A simple example is the two-sphere S^2: the metric (4.1.46) is the pull-back of the flat $\mathbb{R}^3$ metric by the map $i : S^2 \subset \mathbb{R}^3$: $i^* g_{\mathbb{R}^3} = g_{S^2}$. We can also relate the two as

$$ds^2_{\mathbb{R}^3} = dy^m dy^m \overset{(4.1.42)}{=} dr^2 + r^2 ds^2_{S^2} \,. \tag{4.1.88}$$

One can also write

$$dr = \frac{1}{r} y^m dy^m \,. \tag{4.1.89}$$

In this case, the normal vector can be taken to be $n = \partial_r$. The submanifold S^2 is $\{r = 1\}$, so that the first fundamental form is $g^{S^2}_{mn}$, and the second is $\frac{1}{2}\partial_r(r^2 g^{S^2})_{mn}|_{r=1} = g^{S^2}_{mn}$. Applying (4.1.87), we get the Riemann tensor:

$$R_{mnpq} = g_{mp}g_{nq} - g_{mq}g_{np} \,. \tag{4.1.90}$$

4.1.8 Foliations

Sometimes a manifold can be *foliated*, or subdivided in submanifolds (called "leaves") that are all homeomorphic to each other. As a regularity requirement, one usually demands that around each point there exist a coordinate system where the leaves can be written as $\{x^1 = x^1_0, \ldots, x^k = x^k_0\}$, with $x^1_0, \ldots, x^k_0$ some real numbers that are different for each leaf.

This is a bit similar to the notion of a bundle we encountered in Section 4.1.3, except that now we have no requirement on the parameter space of leaves; a fiber bundle is foliated by its fibers, and the parameter space of these leaves is the base B, a manifold. Both for foliations and for fiber bundles, one sometimes drops the requirement that the leaves or fibers should all be homeomorphic to each other. This looser concept is often called a *fibration*.

Submanifolds and foliations often arise by "integrating" vector fields. Given a single vector field v on M, we can define its *integral curves* as the one-dimensional submanifolds S_1 to which v is everywhere tangent; they can be obtained by solving the differential equation

$$\dot{x}^m(t) = v^m(x^1(t), \ldots, x^d(t)) \,. \tag{4.1.91}$$

Following an integral curve for a certain amount of time t gives a map $M \to M$, a finite counterpart of the infinitesimal map (4.1.49), which is often called its *exponential*. If v and M are smooth, solutions to (4.1.91) always exist; by varying the

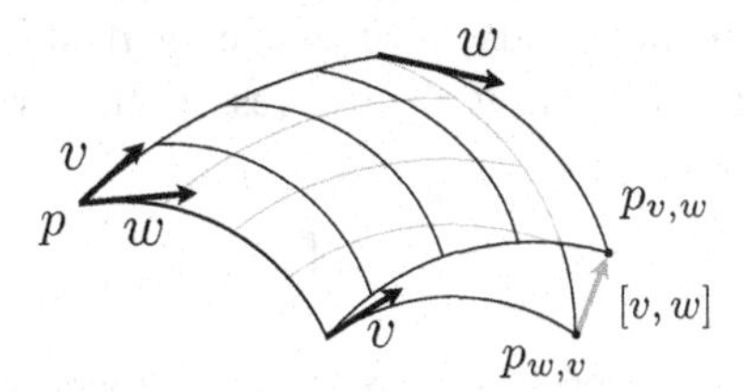

Figure 4.1 Lie bracket of two vector fields v and w.

point $x^m(t = 0)$, we obtain many such S_1, which foliate M. (However, the dimension of these leaves can degenerate to 0, as we can see with $v = x^2\partial_1 - x^1\partial_2$ in $\mathbb{R}^2$.)

A similar attempt at defining foliations with higher-dimensional leaves does not always succeed. Let us take $k > 1$ vector fields $v_1, \ldots, v_k$ and look for submanifolds S_k to which they are everywhere tangent. The data of the subspace $D_p \subset T_pM$ spanned at every point p by the vector fields are called a *tangent distribution.*

A strategy to integrate this distribution and find S_k might be to use each v_i separately to define one-dimensional submanifolds and then put them together to obtain S_k. This, however, often fails, already for $k = 2$ (two vector fields). This can be seen already infinitesimally. Suppose from a point p we follow v for a time ϵ by applying (4.1.49), and then do the same with w, landing at a point $p_{v,w}$. Then we do the same, but with w and v exchanged; this will land at a point $p_{w,v} \neq p_{v,w}$, and the mismatch has coordinates $\sim \epsilon^2[v,w]^\mu$ (Exercise 4.1.7). If this is linearly independent from v and w, a two-dimensional integral submanifold S_2 cannot exist; see Figure 4.1. For example, this is the case for the vector fields

$$v = \partial_1 , \qquad w = \partial_2 + x^1\partial_3 \tag{4.1.92}$$

in $\mathbb{R}^3$. Their Lie bracket is $[v, w] = \partial_3$, which cannot be obtained as a linear combination of v and w, not even a point-dependent one. On the other hand, suppose $[v, w]$ is pointwise linearly dependent from v and w; in other words, suppose there exist two *functions* f, g such that

$$[v, w] = fv + gw . \tag{4.1.93}$$

Then the mismatch can be "repaired" and a two-dimensional integral submanifold S_2 does exist. For example, again in $\mathbb{R}^3$, the two vector fields

$$v = x^1\partial_2 - x^2\partial_1 , \qquad w = x^2\partial_3 - x^3\partial_2 \tag{4.1.94}$$

have a nonzero Lie bracket $[v, w] = x^1\partial_3 - x^3\partial_1$; this can be written a $[v, w] = \frac{x^3}{x^2}v + \frac{x^1}{x^2}w$ as in (4.1.93).[4] So our argument predicts that integral submanifolds S_2 should exist; indeed, they are the spheres $\{\sum_{i=1}^3 (x^i)^2 = r^2\}$.

Generalizing these observations to a higher number k of vector fields, we arrive at the *Frobenius theorem*: given k vector fields $v_1, \ldots, v_k$, there exist k-dimensional

[4] However, v, w, and $[v, w]$ are considered linearly independent if we only allow *constant* coefficients; they are of course the three generators of the rotation group SO(3). In most contexts, we indeed want to take linear combinations of vectors with coefficients that are constant rather than point dependent; for example, a linear combination $f_1K_1 + f_2K_2$ of Killing vectors is in general not Killing if f_1 and f_2 are functions, while it is if they are constant. Indeed, the Lie algebra iso(g) of Killing vectors is finite dimensional.

integral submanifolds S_k if and only if all the Lie brackets $[v_a, v_b]$ are linear combinations of the v_i themselves; in other words, functions $f^c{}_{ab}$ exist such that

$$[v_a, v_b] = \sum_c f^c{}_{ab} v_c \,. \tag{4.1.95}$$

If this is true everywhere on M, then a foliation with k-dimensional leaves exists to which the v_a are everywhere tangent. One then says that the distribution is *integrable*, or that it can be integrated to a foliation.

An alternative characterization exists using one-forms rather than vector fields. We now characterize the D_p as being the spaces of vectors v whose contractions ι_v annihilates the $d-k$ one-forms $\omega_1, \ldots, \omega_{d-k}$. Suppose ω_a is one such form; then $\iota_{v_b}\omega_a = 0$ for all the v_b, $b = 1, \ldots, k$ generators of D_p. Because of (4.1.95), $\iota_{[v_b,v_c]}\omega_a = 0$ also. Now from (4.1.81)

$$0 = \iota_{[v_b,v_c]}\omega_a = [\{\iota_{v_b}, \mathrm{d}\}, \iota_{v_c}]\omega_a = -\iota_{v_c}\{\iota_{v_b}, \mathrm{d}\}\omega_a = -\iota_{v_c}\iota_{v_b}\mathrm{d}\omega_a \,. \tag{4.1.96}$$

Since the space of one-forms annihilated by ι_{v_a} is precisely the span of the ω_a, we obtain that

$$\mathrm{d}\omega_a = \sum_{b=1}^{d-k} \xi_{ab} \wedge \omega_b \tag{4.1.97}$$

for some one-forms ξ_{ab}. So (4.1.97) is again equivalent to integrability of the distribution, or in other words to the existence of a foliation whose leaves S_k are orthogonal to the one-forms ω_a. This is called the *dual form* of the Frobenius theorem.

An example that often occurs is $k = d - 1$, when we have a single equation of the type $\mathrm{d}\omega = \xi \wedge \omega$ for ω a real one-form. In this case the Frobenius theorem implies the existence of a codimension-one foliation. We can then introduce a variable x to label the leaves. Since $\mathrm{d}x$ is annihilated by the vectors tangent to the leaves, ω and $\mathrm{d}x$ must be proportional, and we conclude

$$\omega = f\mathrm{d}x \tag{4.1.98}$$

for some real function f.

4.1.9 Hodge duality

We have already defined the Hodge dual of forms in flat space in (3.1.9). We now adapt it to curved space.

Invariant measure and Hodge operator

The symbol $\epsilon_{m_1\ldots m_k} = \epsilon^{(0)}_{m_1\ldots m_k} = \sigma(m_1 \ldots m_k)$, used, for example, in (3.1.9), is not quite a tensor in curved space. Indeed, if this expression is valid in one coordinate system, in another it becomes

$$\epsilon^{(0)}_{1\ldots d} = 1 \to \tilde{J}_1{}^{m_1} \ldots \tilde{J}_d{}^{m_d} \epsilon^{(0)}_{m_1\ldots m_d} = \det(\tilde{J})\epsilon^{(0)}_{m_1\ldots m_d} \,. \tag{4.1.99}$$

We used the linear algebra formula

$$\det(M) = \epsilon^{(0)}_{m_1\ldots m_d} M_1{}^{m_1} \ldots M_d{}^{m_d} \,. \tag{4.1.100}$$

From the metric transformation $g_{mn} \to \tilde{J}_m{}^p \tilde{J}_n{}^q g_{pq}$, one sees $g \equiv \det(g_{mn}) \to \det(\tilde{J})^2 g$. So the alternative

$$\epsilon_{m_1 \ldots m_d} \equiv \sqrt{g}\, \epsilon^{(0)}_{m_1 \ldots m_d} \tag{4.1.101}$$

transforms instead correctly as a tensor.[5] So, for example, in Euclidean signature $\epsilon_{1\ldots d} = \sqrt{g}\epsilon^{(0)}_{1\ldots d} = \sqrt{g}$. It is now also natural to introduce the associated *volume form*

$$\mathrm{vol} = \frac{1}{d!}\epsilon_{m_1 \ldots m_d} \mathrm{d}x^{m_1} \wedge \ldots \mathrm{d}x^{m_d} = \sqrt{g}\mathrm{d}x^1 \wedge \ldots \mathrm{d}x^d\,, \tag{4.1.102}$$

which is invariant under coordinate changes. The Hodge duality operator (3.1.9) should now be written

$$*\colon \mathrm{d}x^{m_1} \wedge \ldots \wedge \mathrm{d}x^{m_k} \mapsto \frac{1}{(d-k)!}\epsilon_{m_{k+1}\ldots m_d}{}^{m_1 \ldots m_k} \mathrm{d}x^{m_{k+1}} \wedge \ldots \wedge \mathrm{d}x^{m_d}\,, \tag{4.1.103}$$

with (4.1.101); writing it in terms of $\epsilon^{(0)}$, we need then to add a $\sqrt{g}$. Both (4.1.101) and (4.1.103) will become a lot more natural in the vielbein formalism of Section 4.3.1. In particular, it will become clear there that the flat space properties still hold, such as the relation to the chiral gamma in (3.2.27) and the formula (3.1.11) for $*^2$.

Equation (4.1.102) is also used to define the integral of a d-form ω_d over a d-dimensional manifold M_d. Since Λ^d has dimension 1 (Section 3.1.1), all d-forms are proportional, and $\omega_d = f_0 \mathrm{d}x^1 \wedge \ldots \wedge \mathrm{d}x^d$. Now we define

$$\int_{M_d} \omega_d \equiv \int_{M_d} \mathrm{d}^d x f_0\,. \tag{4.1.104}$$

(This is basically the definition we gave back in (1.2.8).) Alternatively, it might look more intrinsic to write $\omega_d = f\mathrm{vol}$, where $f = f_0/\sqrt{g}$. This also implies $*\omega_d = f$. We can then rewrite

$$\int_{M_d} \omega_d = \int_{M_d} \mathrm{d}^d x \sqrt{g} f = \int_{M_d} \mathrm{d}^d x \sqrt{g}(*\omega_d)\,. \tag{4.1.105}$$

In a sense, we just regard the $\mathrm{d}x^1 \wedge \ldots \wedge \mathrm{d}x^d$ as the coordinate measure $\mathrm{d}^d x$. From this point of view, it is more natural to integrate d-forms than functions on M_d.

Inner products and adjoints

We saw back in Section 3.2.4 that $*\alpha_k \wedge \beta_k$ defines a positive-definite inner product among forms in flat space. For forms on manifolds, positivity is true at every point, but $*\alpha_k \wedge \beta_k$ is now a form on M_d; an inner product should associate a number to two forms. So we also take an integral over M_d, recalling also (3.2.37):[6]

$$\langle \alpha_k, \beta_k \rangle \equiv \int_{M_d} *\alpha_k \wedge \beta_k = \int_{M_d} \mathrm{d}^d x \sqrt{g}\, \alpha_k \cdot \beta_k\,, \tag{4.1.106a}$$

$$\alpha_k \cdot \beta_k \equiv \frac{1}{k!}\alpha_{m_1 \ldots m_k} \beta^{m_1 \ldots m_k}\,. \tag{4.1.106b}$$

[5] We can also define instead a *tensor density* $t^{m_1 \ldots m_k}_{n_1 \ldots n_l}$ of weight w, where (4.1.7) has an additional factor $(\det(\tilde{J}))^w$ on the right-hand side. Then $\epsilon^{(0)}$ is a tensor density of weight 1.

[6] This pairing is not to be confused with the Chevalley–Mukai pairing (,) defined as (3.2.40), which is not integrated and not positive-definite.

From the definition, (4.1.106a) is clearly symmetric and positive-definite. We also define a positive-definite norm:

$$||\alpha||^2 \equiv \langle \alpha, \alpha \rangle = \int_{M_d} \mathrm{d}^d x \sqrt{g}\, |\alpha_k|^2 , \tag{4.1.107a}$$

$$|\alpha_k|^2 \equiv \alpha_k \cdot \alpha_k = \frac{1}{k!} \alpha_{m_1 \ldots m_k} \alpha^{m_1 \ldots m_k} . \tag{4.1.107b}$$

The adjoint of the exterior differential is defined with respect to (4.1.106a):

$$\langle \mathrm{d}\alpha_k, \beta_{k+1} \rangle \equiv \langle \alpha_k, \mathrm{d}^\dagger \beta_{k+1} \rangle . \tag{4.1.108}$$

Integrating by parts and using (3.1.11), we get

$$\mathrm{d}^\dagger \alpha_k = (-1)^{kd} * \mathrm{d} * \alpha_k . \tag{4.1.109}$$

In a similar way, one can also check that

$$(\mathrm{d}x^m \wedge)^\dagger = \iota^m , \tag{4.1.110}$$

again with respect to the inner product (4.1.106b). So, recalling $\mathrm{d} = \mathrm{d}x^m \wedge \partial_m = \mathrm{d}x^m \wedge \nabla_m$, we have an alternative expression for (4.1.109):

$$\mathrm{d}^\dagger = -\iota^m \nabla_m . \tag{4.1.111}$$

(The minus sign comes from integrating by parts.)

Laplace–de Rham operator

The definition of $\mathrm{d}^\dagger$ suggests a natural self-adjoint operator:

$$\Delta \equiv \{\mathrm{d}, \mathrm{d}^\dagger\} = \mathrm{d}\mathrm{d}^\dagger + \mathrm{d}^\dagger \mathrm{d} , \tag{4.1.112}$$

which is known under various names, such as *Laplace–de Rham* operator; we will simply call it "Laplacian." We can evaluate (4.1.112) in more detail in terms of ∇, using our operatorial methods from Section 4.1.6. Applying (2.1.31) repeatedly,

$$\begin{aligned} \Delta &= -\{\mathrm{d}x^m \wedge \nabla_m, \iota^n \nabla_n\} = -\mathrm{d}x^m \wedge \iota^n [\nabla_m, \nabla_n] - \{\mathrm{d}x^m \wedge, \iota^n\} \nabla_n \nabla_m = \\ &\overset{(4.1.24),(3.2.3a)}{=} -\mathrm{d}x^m \wedge \iota^n \mathrm{d}x^p \wedge \iota^q R_{mnpq} - \nabla^2 , \end{aligned} \tag{4.1.113}$$

where we have shortened $\nabla^2 \equiv g^{mn} \nabla_m \nabla_n$. Using (3.2.3a) again, the other term on the right-hand side can also be written as $\mathrm{d}x^m \wedge \mathrm{d}x^p \wedge \iota^n \iota^q R_{mnpq} + \mathrm{d}x^m \iota^n R_{mn}$. Equation (4.1.113) is known as the *Weitzenböck formula*. On functions, Δ acts as follows:

$$-\Delta f = \nabla^2 f \overset{(4.1.15)}{=} \frac{1}{\sqrt{g}} \partial_m (\sqrt{g} g^{mn} \partial_n f) . \tag{4.1.114}$$

On one-forms, (4.1.113) gives

$$\Delta \alpha_m = -\nabla^n \nabla_n \alpha_m + R_{mn} \alpha^n . \tag{4.1.115}$$

In flat space, $\Delta = -\partial^2 = -\partial_m \partial^m$. This minus sign might look annoying, but it is included so that the spectrum is positive. Indeed, on a compact M_d, assume $\Delta\alpha = \lambda\alpha$; then

$$\lambda||\alpha||^2 = \langle\alpha, \Delta\alpha\rangle = \langle \mathrm{d}^\dagger\alpha, \mathrm{d}^\dagger\alpha\rangle + \langle \mathrm{d}\alpha, \mathrm{d}\alpha\rangle = ||\mathrm{d}^\dagger\alpha||^2 + ||\mathrm{d}\alpha||^2 \geq 0\,. \qquad (4.1.116)$$

4.1.10 Cohomology

For most of this subsection we will assume our manifolds to be orientable and compact; at the very end we will comment on the non-compact case.

Homology

It is often useful to know if a subspace $S_k \subset M_d$ is a boundary: namely, if there exists another subspace S_{k+1} such that $\partial S_{k+1} = S_k$.[7] Crucially, the boundary of a boundary is empty:

$$\partial^2 = 0\,. \qquad (4.1.117)$$

On the other hand, an S_k that has no boundary (which is often called a *cycle*) is not necessarily a boundary. This motivates introducing the kth *homology group*:

$$H_k(M_d, \mathbb{Z}) = \frac{\{S_k \mid \partial S_k = 0\}}{\{\partial S_{k+1} = 0\}}\,. \qquad (4.1.118)$$

We denote by $[S_k]$ the equivalence class of S_k in this quotient. The $\mathbb{Z}$ in the notation denotes that we are allowing formal linear sums over $\mathbb{Z}$, so, for example, $2S_k$ or $S_k - S'_k$ belong to (4.1.118). We could even take real linear combinations; in this case, we would denote the resulting space by $H_k(M_d, \mathbb{R})$. The latter are real vector spaces, and thus we only care about their dimensions,

$$h_k(M_d) \equiv \dim H_k(M_d, \mathbb{R})\,, \qquad (4.1.119)$$

also called *Betti numbers*. On the other hand, $H_k(M_d, \mathbb{Z})$ is an abelian group that can have h_k summands of $\mathbb{Z}$, but also some cyclic groups: $H_k(M_d, \mathbb{Z}) = \mathbb{Z}^{h_k} \oplus T_k$, where the T_k, called *torsion*, consists of a sum of $\mathbb{Z}_p$'s.

On compact manifolds, $H_k(M_d)$ is finite dimensional. It vanishes for $k < 0$ and $k > d$. $H_0(M_d, \mathbb{Z}) = \mathbb{Z}$: the generator of this $\mathbb{Z}$ is simply a point $p \in M_d$. Indeed, all points are identified: a line γ joining p and p' then is such that $p - p' = \partial\gamma$, which is set to zero in (4.1.118). In $H_d(M_d, \mathbb{Z}) = \mathbb{Z}$, the only d-dimensional cycle can be $S_d = M_d$ itself. (In the nonorientable case, this would be absent.)

For example, on S^d any k-cycle S_k is a boundary; so $H_k(S^d, \mathbb{Z})$ are zero except for $k = 0$ and d, where $H_0(S^d, \mathbb{Z}) = H_d(S^d, \mathbb{Z}) = \mathbb{Z}$.

Another simple and useful result is the *Künneth theorem*:

$$H_k(M \times M', \mathbb{Z}) = \oplus_{i+j=k} H_i(M, \mathbb{Z}) \otimes H_j(M', \mathbb{Z})\,. \qquad (4.1.120)$$

[7] The symbol ∂ for the boundary of a space is traditional; the fact that it is the same as the partial derivative symbol will hopefully generate no confusion.

A different way to classify subspaces in M is given by *homotopy*, which denotes a continuous deformation of an object into another. The *homotopy groups*

$$\pi_i(M) \tag{4.1.121}$$

are the equivalence classes of closed subspaces of dimension i under homotopy. While this sounds superficially related to homology, it is in general different; in particular, homotopy groups can be nonabelian. But the *Hurewicz theorem* says that if $\pi_i(M)$ are trivial for $i < k$, then $H_i(M)$ are trivial as well, and $H_k(M)$ is the abelianization of $\pi_k(M)$.

A space with $\pi_1 = 0$ is said to be *simply-connected*. For a space M with $\pi_1(M) \neq 0$, the *universal covering* $\hat{M}$ is a simply-connected space such that

$$M = \hat{M}/\pi_1(M)\,. \tag{4.1.122}$$

It can be constructed explicitly as the space of endpoints that can be reached from a path starting from a fixed point.

Intersections

It is useful to consider intersections of cycles. The intersection of two generic representatives $S_k \in H_k$ and $S_{d-k} \in H_{d-k}$ consists of a collection of points. The number of such points can change, but there is a way to weigh the points with appropriate signs[8] σ_p in such a way that their sum,

$$\#(S_k \cdot S_{d-k}) = \sum_{p \in S_k \cap S_{d-k}} \sigma_p, \tag{4.1.123}$$

doesn't depend on the representatives, and is thus a pairing between the two classes $[S_k]$ and $[S_{d-k}]$.

This intersection pairing implies a duality between H_k and H_{d-k}. We first define the *cohomology* as the space of linear functions from H_k to $\mathbb{Z}$:

$$H^k(M_d, \mathbb{Z}) = \{\varphi \colon H_k(M_d, \mathbb{Z}) \to \mathbb{Z}\}\,. \tag{4.1.124}$$

One type $\varphi_{S_{d-k}}$ of such a function is obtained by fixing an S_{d-k} and taking intersections with it: $\varphi_{S_{d-k}}(S_k) \equiv \#(S_k \cdot S_{d-k})$. The *Poincaré duality theorem* says that on a compact M the pairing is nondegenerate, and all elements of H^k are of this type: in other words,

$$H^k(M_d, \mathbb{Z}) \cong H_{d-k}(M_d, \mathbb{Z})\,. \tag{4.1.125}$$

A separate *universal coefficient theorem* then says $H^k(M_d, \mathbb{Z}) = \mathbb{Z}^{h_k} \oplus T_{k-1}$, where recall T_k was the torsion of H_k. As a consequence, we can conclude

$$h_k(M_d) = h_{d-k}(M_d)\,. \tag{4.1.126}$$

This is also often called Poincaré duality. It is reminiscent of the way the Hodge $*$ maps k-forms to $(d-k)$-forms.

Poincaré duality also implies that (4.1.123) is nondegenerate. In particular, when $d = 2k$,

$$\#(S_k \cdot S'_k) \tag{4.1.127}$$

[8] One takes bases of vector fields $\{v_1, \ldots, v_k\}$, $\{v_{k_1}, \ldots, v_d\}$ tangent to S_k and S_{d-k}; then σ_p is the sign of $\iota_{i_* v_1} \ldots \iota_{i_* v_d}$vol, where i denotes both inclusion maps in M_d.

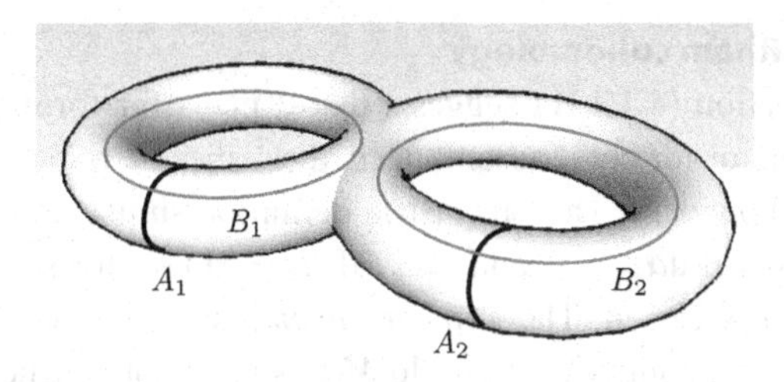

Figure 4.2 A and B cycles for a Riemann surface of genus $g = 2$.

becomes a nondegenerate pairing in the middle homology H_k. This is symmetric for k = even; in other words, it defines a quadratic form. Its *signature* $\sigma(M_{2k})$ (the difference between positive and negative eigenvalues) is an important topological invariant, especially useful for $d = 4$, $k = 2$. On the other hand, it is antisymmetric for k = odd. A nondegenerate antisymmetric matrix with integer coefficients can always be put in the block-antidiagonal form $\left(\begin{smallmatrix} 0 & -1 \\ 1 & 0 \end{smallmatrix}\right)$, so there exists a basis $\{A^A, B_A\}_{A=1,\ldots,h_k/2}$ of H_k such that

$$\#(A^A \cdot A^B) = 0\,, \qquad \#(A^A \cdot B_B) = -\#(B_B \cdot A^A) = \delta^A_B\,, \qquad \#(B_A \cdot B_B) = 0\,. \tag{4.1.128}$$

In particular, we see that when k is odd, the middle homology H_k has even dimension.

As an example, consider (orientable, compact) Riemann surfaces. Since $d = 2k$ and $k = 1$ is odd, the middle homology is even-dimensional,

$$h_1 = 2g \quad \text{(Riemann surfaces)}. \tag{4.1.129}$$

This g is called the *genus* and is visualized as the "number of handles" of this $M = \Sigma_g$. For $g = 0$, we have $\Sigma_0 = S^2$; for $g = 1$, $\Sigma_1 = T^2$. The cases with higher g, $\Sigma_{g\geq 2}$, can be constructed topologically by taking a polygon with $4g$ sides and identifying them in pairs in an appropriate way. We show the A and B cycles for $g = 2$ in Figure 4.2.

The introduction of cohomology in (4.1.124) as a space of linear functions might look a little artificial. A more geometrical alternative is provided by forms, which have a natural pairing with cycles, given by

$$\int_{S_k} \alpha_k \equiv \int_{S_k} i^* \alpha_k, \tag{4.1.130}$$

where $i\colon S_k \to M_d$ is the inclusion map. The operation of taking boundaries ∂ of subspaces is now "dual" to the exterior differential:

$$\int_{\partial S_k} \alpha_{k-1} = \int_{S_k} \mathrm{d}\alpha_{k-1}\,. \tag{4.1.131}$$

This is called *Stokes's theorem*, just like (1.3.5), because it reduces to (1.3.5) for $k = 2$ in $\mathbb{R}^3$. With $k = d$ in $\mathbb{R}^d$, it also reproduces the divergence theorem $\int_V \mathrm{d}^d x \nabla_m v^m = \int_{\partial V} v^m (\mathrm{d}a)_m$. Another important consequence of (4.1.131) is that the integral of a form $\mathrm{d}(\ldots)$ on a compact manifold vanishes, as usual for integrals of total derivatives.

De Rham cohomology

Equation (4.1.131) suggests that d plays for forms a role similar to the boundary operator for subspaces. Indeed, d satisfies (4.1.75), which is similar to (4.1.117) for the boundary. This suggests a definition similar to (4.1.118). A k-form α_k is called *closed* if $\mathrm{d}\alpha_k = 0$, and *exact* if $\alpha_k = \mathrm{d}\alpha_{k-1}$ for some α_{k-1}. Since $\mathrm{d}^2 = 0$, an exact form is closed. The *Poincaré lemma* says that locally a closed form is also exact, but on a nontrivial manifold M this need not be true. So we define the kth *de Rham cohomology group* on a manifold M:

$$H^k(M_d) = \frac{\{\alpha_k \,|\, \mathrm{d}\alpha_k = 0\}}{\{\mathrm{d}\alpha_{k-1}\}} . \tag{4.1.132}$$

In other words, H^k is the space of closed forms modulo exact forms. The *de Rham theorem* says that $H^k(M_d)$ is equal to $H^k(M_d, \mathbb{R})$, the analogue of (4.1.124) with real coefficients. In other words,

$$H^k(M_d) = \mathbb{R}^{h_k} , \tag{4.1.133}$$

where h_k was the dimension of homology we introduced earlier in (4.1.119). From our earlier results on homology, it follows that on compact manifolds H^k has finite dimension for all k, and that it vanishes for $k < 0$ and $k > d$. Moreover, $H^0(M_d)$ and $H^d(M_d)$ both have dimension 1; they are generated by the function 1 and the top-form vol respectively.

For example, for S^d we have

$$H^k(S^d) = 0 \tag{4.1.134}$$

except for $k = 0$ and d, where $H^0(S^d) = H^d(S^d) = \mathbb{R}$.

Equation (4.1.133) shows that de Rham cohomology forgets some of the information in homology and in $\mathbb{Z}$-valued cohomology. But some classes of de Rham cohomology can be viewed as belonging to $H^k(M, \mathbb{Z})$: if we define $\varphi_{\alpha_k} : S_k \mapsto \int_{S_k} \alpha_k$, we obtain an element of $H^k(M, \mathbb{Z})$ as defined in (4.1.124) if the integrals of α_k over a basis of homology (also called its *periods*) are integer. For example, the Poincaré dual of an $S_k \in H_k$ can now be viewed as a $(d-k)$-form

$$\sigma_{d-k} \equiv \mathrm{PD}[S_k] \;\text{ such that }\; \#(S_k \cdot S_{d-k}) = \int_{S_{d-k}} \sigma_{d-k} \quad \forall S_{d-k} \in H_{d-k}(M) . \tag{4.1.135}$$

We can think of this property as a generalization of a delta function, $\int \mathrm{d}x f(x)\delta(x) = f(0)$, $\forall f$. Inspired by this, we can write locally $S_k = \{x^1 = \cdots = x^{d-k} = 0\}$ for some choice of coordinates, and then $\mathrm{PD}[S_k] = \delta(x^1)\ldots\delta(x^{d-k})\mathrm{d}x^1 \wedge \ldots \mathrm{d}x^{d-k}$; this is the source form we were writing in (1.3.42). However, there are also smooth representatives of the class $\mathrm{PD}[S_k]$, and in most mathematical applications those are preferable. Conversely, the Poincaré dual of a form $\sigma_k \in H^k$ is a $(d-k)$-cycle:

$$S_{d-k} \equiv \mathrm{PD}[\sigma_k] \;\text{ such that }\; \int_M \sigma_k \wedge \sigma_{d-k} = \int_{S_{d-k}} \sigma_{d-k} \quad \forall \sigma_{d-k} \in H^{d-k}(M) . \tag{4.1.136}$$

It follows from these definitions that

$$\int_M \sigma_k \wedge \sigma_{d-k} = \int_{S_k} \sigma_k = \#(S_k \cdot S_{d-k}), \tag{4.1.137}$$

where $S_k = \mathrm{PD}[\sigma_{d-k}]$, $S_{d-k} = \mathrm{PD}[\sigma_k]$. In other words, the integrated wedge product has a topological meaning: it is related to the intersection pairing (4.1.123).

A natural generalization in $d = 6$ is the *triple intersection form*:

$$D(\alpha_2, \beta_2, \gamma_2) \equiv \int_{M_6} \alpha_2 \wedge \beta_2 \wedge \gamma_2 \,. \tag{4.1.138}$$

Upon choosing a basis for H^2, this produces a symmetric tensor with three indices D_{abc}, with $a, b, c = 1, \ldots, h^2(M)$.

Harmonic forms

A distinguished representative for a cohomology class in H^k is given by a *harmonic* form:[9] namely, one which is not only closed (as it should be, as an element of H^k) but also *co-closed*, namely closed under $\mathrm{d}^\dagger$:

$$\mathrm{d}\alpha = 0\,, \qquad \mathrm{d}^\dagger \alpha = 0\,. \tag{4.1.139}$$

Clearly a harmonic function satisfies $\Delta\alpha = 0$. On a compact manifold, the opposite is also true: by (4.1.116), $\langle \alpha, \Delta\alpha \rangle = ||\mathrm{d}^\dagger \alpha||^2 + ||\mathrm{d}\alpha||^2$, so if $\Delta\alpha = 0$, then (4.1.139) also holds.

Let $\mathcal{H}$ be the projector on the space of harmonic forms. The Laplacian is invertible on the orthogonal complement $(\mathrm{Im}\mathcal{H})^\perp$. Indeed, given a complete set of eigenforms $\Delta\alpha_i = \lambda_i \alpha_i$, any form can be written as $\alpha = \sum c_i \alpha_i$; the operator G defined by $G\alpha \equiv \sum_{\lambda_i \neq 0} \frac{c_i}{\lambda_i} \alpha_i$ is such that $\Delta G\alpha = \sum_{\lambda_i \neq 0} c_i \alpha_i = (1 - \mathcal{H})\alpha$; it commutes with d and $\mathrm{d}^\dagger$. We can rewrite this as

$$\alpha = \mathcal{H}\,\alpha + \mathrm{d}(\mathrm{d}^\dagger G\alpha) + \mathrm{d}^\dagger(\mathrm{d}G\alpha)\,. \tag{4.1.140}$$

In other words, every form can be decomposed into a harmonic form, an exact form, and a coexact form. This is known as *Hodge decomposition*. For a closed form, the third term is absent, and $\alpha - \mathrm{d}(\mathrm{d}^\dagger G\alpha)$ is a form in the same class of $\mathcal{H}\alpha$, which is harmonic. We can also write this schematically as a direct sum:

$$\Lambda^k = \mathcal{H}^k \oplus \mathrm{cE}^k \oplus \mathrm{E}^k \,, \tag{4.1.141}$$

where cE^k and E^k denote the space of coexact and exact k-forms.

So a harmonic representative exists in every cohomology class; the decomposition also shows that it is unique. With the exception of the form $\alpha = 0$, a harmonic form is a nontrivial element of cohomology. So the space of harmonic forms has finite dimension h_k.

[9] In particular, harmonic zero-forms are of course constant; unfortunately, it is also common to call a function "harmonic" if it is the eigenfunction of $\Delta = -\nabla^2$.

Noncompact case

Some of the results in this subsection need modification on a noncompact M_d. We will focus here on the case where M_d has a boundary; the case where it is not bounded has a similar treatment, after adding points at infinity.

The main novelty is that there are now several relevant notions of homology and cohomology. For homology, besides the usual definition (4.1.118), we have a new *relative homology*:

$$H_k(M_d, \partial M_d) = \frac{\{S_k \mid \partial S_k \subset \partial M_d\}}{\{\partial S_{k+1} = 0\}} . \qquad (4.1.142)$$

The difference with (4.1.118) is that the numerator is now the larger space of subspaces whose boundary is contained in ∂M_d, rather than those that have no boundary at all. (In the noncompact case, this means that we include cycles that extend all the way to infinity.) For cohomology, an alternative to (4.1.132) is

$$H^k(M_d, \partial M_d) = \frac{\{\alpha_k \mid \mathrm{d}\alpha_k = 0,\ \alpha_k|_{\partial M_d} = \mathrm{d}\lambda_{k-1}\}}{\{\mathrm{d}\alpha_{k-1}\}} . \qquad (4.1.143)$$

Here the numerator is the smaller space of closed forms that are exact when restricted to the boundary. The definitions (4.1.142) and (4.1.143) imply the obvious inclusions

$$H_k(M_d) \subset H_k(M_d, \partial M_d) , \qquad H^k(M_d) \supset H^k(M_d, \partial M_d) . \qquad (4.1.144)$$

Poincaré duality now is replaced by

$$H^k(M_d) \cong H_{d-k}(M, \partial M_d) , \qquad H_k(M_d) \cong H^{d-k}(M, \partial M_d) . \qquad (4.1.145)$$

For example, consider a ball $B_d = \{\sum_{i=m}^d x_m^2 \leq 1\} \subset \mathbb{R}^d$. (The case of $\mathbb{R}^d$ itself can be reduced to B_d by adding a sphere S^d of points at infinity.) In ordinary homology H_k, only a point is nontrivial, so

$$h_0(B_d) = 1 , \qquad h_i(B_d) = 0 \quad i \neq 0 . \qquad (4.1.146)$$

In particular, we don't have a d-dimensional cycle as would be predicted by (4.1.126) in the compact case, since B_d itself has a boundary. In relative homology (4.1.142), B_d is allowed, since its boundary is all on ∂B_d. But a point p is now trivial, because it is the boundary of a segment that joins p to ∂B_d. So here

$$h_d(B_d) = 1 , \qquad h_i(B_d) = 0 \quad i \neq d . \qquad (4.1.147)$$

A more interesting example will be given in Section 7.2.2.

Another subtlety regards the Hodge decomposition (4.1.141), which separates the space of forms in harmonic, exact, and coexact subspaces. In the noncompact case, it is replaced by the following [118]

$$\Lambda^k = \mathcal{H}^k \oplus \mathrm{cE}^k_{\mathrm{N}} \oplus \mathrm{E}^k_{\mathrm{D}} , \qquad \mathcal{H}^k = \mathcal{H}^k_{\mathrm{N}} \oplus \mathrm{EcC}^k = \mathcal{H}^k_{\mathrm{D}} \oplus \mathrm{CcE}^k . \qquad (4.1.148)$$

Here $\mathcal{H}^k$ is the space of k-forms that are closed and coclosed, which does not necessarily coincide with the space of forms that are annihilated by the Laplacian. cE^k, E^k, EcC^k, CcE^k denote the space of k-forms that are coexact, exact, exact and co-closed, and closed and coexact. The subscripts $_{\mathrm{N}}$ and $_{\mathrm{D}}$ stand for *Neumann* or *Dirichlet* forms, in the following sense. Choosing a function ρ such that ∂M_d is a level set of ρ, a form $\alpha_{\rm N}$ is Neumann if $\iota_{\partial_\rho}\alpha_{\rm N} = 0$; Dirichlet forms $\alpha_{\rm D}$ are of the form $d\rho\wedge$ a Neumann form. Finally we have

$$H^k(M_d) \cong \mathcal{H}^k_{\rm N}\,, \qquad H^k(M_d, \partial M_d) \cong \mathcal{H}^k_{\rm D}\,. \tag{4.1.149}$$

Exercise 4.1.1 Show (4.1.21) and (4.1.62) with the methods of Section 4.1.6, using (4.1.20) and (4.1.57).

Exercise 4.1.2 If a vector field v has components v_0^m at a point p with coordinates x^n, check that it is covariantly constant in p if at the point p' with coordinates $x^n + \delta x^n$ it has components $v_0^m - \delta x^n \Gamma^m_{np} v_0^p$.

Exercise 4.1.3 Apply the result of the previous exercise to a small displacement of the mth coordinate, $\delta x^p = \epsilon \delta^p_m$, and then of the nth coordinate, $\delta x^p = \epsilon\delta^p_n$. Now do the same in the opposite order. Subtract the two results to obtain (4.1.14).

Exercise 4.1.4 Show (4.1.25b) and (4.1.25c) by applying (4.1.21) respectively to $\nabla_{[m}\nabla_{n]}g_{pq} = 0$ and to $\nabla_{[m}\nabla_n\alpha_{p]} = 0$ (for α any one-form).

Exercise 4.1.5 Show (4.1.26) by applying (4.1.21) in two different ways to

$$\nabla_{[m}\nabla_n\nabla_{p]}\alpha_q\,, \tag{4.1.150}$$

again for α any one-form.

Exercise 4.1.6 Use (4.1.25a–4.1.25c) to show that the number of independent components of the Riemann tensor in d dimensions is $\frac{1}{12}d^2(d^2-1)$. (Equation (4.1.25d) is a consequence of these three, and can be ignored.)

Exercise 4.1.7 Check the intuitive interpretation of $[v, w]$ of Figure 4.1. To a point p with coordinates x^m, apply the infinitesimal map (4.1.49) associated to v, and then the one associated to w: you should get a point $p_{v,w}$ with coordinates

$$x^m + \epsilon v^m + \epsilon w^m + \epsilon^2 v^n \partial_n w^m\,. \tag{4.1.151}$$

Now apply first the map associated to w, followed by that associated to v. Subtract the two results and check that you get $\sim \epsilon^2[v, w]$.

Exercise 4.1.8 Show that the volume form of the round metric on S^d (defined in (4.1.70)) is given by

$$\mathrm{vol}_{S^d} = \frac{1}{(k-1)!\,r^k}\epsilon_{m_1\ldots m_k} y^{m_1} \mathrm{d}y^{m_2}\wedge\ldots\wedge \mathrm{d}y^{m_k}\,. \tag{4.1.152}$$

In particular, for $d = 2$

$$\mathrm{vol}_{S^2} = \frac{1}{2r^3}\epsilon_{mnp}y^m\mathrm{d}y^n\wedge\mathrm{d}y^p = \sin\theta\mathrm{d}\theta\wedge\mathrm{d}\phi\,. \tag{4.1.153}$$

Exercise 4.1.9 Use (4.1.152) to rewrite (1.3.59b) as

$$F_{8-p} = \frac{(2\pi l_s)^{7-p}}{v_{8-p}}\mathrm{vol}_{S^{8-p}}\,. \tag{4.1.154}$$

Use this to check that (1.3.47) is satisfied.

Exercise 4.1.10 Show that the pull-back and exterior differential commute:

$$\mathrm{d}f^* = f^*\mathrm{d}\,. \tag{4.1.155}$$

Use this and (4.1.131) to reinterpret the integration by parts below (1.3.23) as $\int_{\partial\Sigma}\iota^* a = \int_\Sigma \iota^* f$.

Exercise 4.1.11 Is the equality in (4.1.106a) still valid in Lorentzian signature? (It is enough to check it for $\alpha_k = \beta_k = e^0 \wedge \ldots \wedge e^{k-1}$ and $\alpha_k = \beta_k = e^1 \wedge \ldots \wedge e^k$.)

Exercise 4.1.12 Show that

$$(*\alpha)^2 = \alpha^2 \qquad \text{(Eucl.)} \tag{4.1.156a}$$

$$(*\alpha)^2 = -\alpha^2 \qquad \text{(Lor.)} \tag{4.1.156b}$$

(Once again, it is enough to do so for α chosen as in the previous exercise.)

Exercise 4.1.13 Show that the reduction (1.4.5) turns the Bianchi identity $\mathrm{d}G_4 = 0$ and the equation of motion $\mathrm{d} * G_4 + \frac{1}{2} G_4 \wedge G_4 = 0$ into the RR Bianchi identities and equations of motion.

Exercise 4.1.14 Show that on one-forms $\alpha_1 = \alpha_m \mathrm{d}x^m$

$$\mathrm{d}^\dagger \alpha_1 = \frac{1}{\sqrt{g}} \partial_m (\sqrt{g} g^{mn} \alpha_n) \,. \tag{4.1.157}$$

Exercise 4.1.15 Consider the Weyl transformations

$$g_{mn} \to \tilde{g}_{mn} = \mathrm{e}^{2\lambda} g_{mn} \,. \tag{4.1.158}$$

Show that the Laplacian transforms as

$$\nabla^2 f \to \tilde{\nabla}^2 f = \mathrm{e}^{-2\lambda} (\nabla^2 f - (2-d) \mathrm{d}\lambda \cdot \mathrm{d}f). \tag{4.1.159}$$

Exercise 4.1.16 Consider a warped product line element

$$\mathrm{d}s^2 = \mathrm{e}^{2A} \mathrm{d}s^2_{d_1} + \mathrm{d}s^2_{d_2} \,, \tag{4.1.160}$$

where A is a function of the coordinates of $\mathrm{d}s^2_{d_2}$. Show that the nonvanishing components of the connection are

$$\begin{aligned} &\Gamma^\mu_{\nu\rho} = \Gamma^\mu_{(d_1)\,\nu\rho} \,, \qquad \Gamma^\mu_{\nu p} = \partial_p A \delta^\mu_\nu \,, \\ &\Gamma^m_{\nu\rho} = -\mathrm{e}^{2A} g^{mn}_{(d_2)} \partial_n A g^{(d_1)}_{\nu\rho} \,, \qquad \Gamma^m_{np} = \Gamma^m_{(d_2)\,np} \,, \end{aligned} \tag{4.1.161}$$

where $\mu, \nu \ldots$ are the indices along $\mathrm{d}s^2_{d_1}$, and $m, n \ldots$ along $\mathrm{d}s^2_{d_2}$.

Exercise 4.1.17 Show that the Hodge dual for a rescaled metric $\mathrm{d}\tilde{s}^2 = \mathrm{e}^{2\lambda} \mathrm{d}s^2$ obeys

$$\tilde{*}\alpha_k = \mathrm{e}^{2(d-k)\lambda} * \alpha_k \,. \tag{4.1.162}$$

Exercise 4.1.18 Rederive (3.1.11) by using (2.1.20) and (3.2.27).

Exercise 4.1.19 For all the homology classes $h_i(T^d)$, find representatives for the Poincaré duals that are smooth and have support everywhere.

Exercise 4.1.20 Compute the ordinary and relative cohomology of a ball $B_d \subset \mathbb{R}^d$, and check (4.1.145).

4.2 Bundles

4.2.1 Definitions

We saw already some examples of bundles in Section 4.1.3. In all those examples, the transition functions $t_{\alpha\beta}$ acting on the fiber are related to the transition functions

$g_{\alpha\beta}$ acting on the coordinates by being their partial derivatives $\frac{\partial x^n_{(\alpha)}}{\partial x^m_{(\beta)}} \in \mathrm{GL}(d,\mathbb{R})$. The idea of bundle can be generalized by taking the two to be unrelated, and the $t_{\alpha\beta}$ to belong to any group G, called the *structure group*. (Thus the structure group of the tangent bundle is $\mathrm{GL}(d,\mathbb{R})$.)

This has several applications. One is that several important geometric objects are sections of this more general type of bundle. More importantly for a physicist, bundles are the mathematical way to make sense of nonabelian gauge fields on manifolds; this is in fact the origin of the idea. Finally, bundles are also a particular type of manifold themselves that we might want to use as part of the internal space of string theory.

General fiber

Once we have selected the transition functions $t_{\alpha\beta}$ valued in G, we can choose the fiber F on which they will act. As in Section 4.1.3, F can be any space on which G has a group action, (4.1.40). Now a bundle E can be defined similar to (4.1.35):

$$\begin{aligned} E = \cup_\alpha \hat{U}_\alpha / \sim \qquad & (x_{(\alpha)}, f_{(\alpha)}) \in \hat{U}_{\alpha\beta} \sim (x_{(\beta)}, f_{(\beta)}) \in \hat{U}_{\beta\alpha} \ \text{ if} \\ x_{(\alpha)} = g_{\alpha\beta}(x_{(\beta)}) , \qquad & f_{(\alpha)} = t_{\alpha\beta} \cdot f_{(\beta)} . \end{aligned} \tag{4.2.1}$$

To summarize all this, one sometimes writes

$$\begin{array}{ccc} F & \overset{\iota}{\hookrightarrow} & E \\ & & \downarrow \pi \\ & & M . \end{array} \tag{4.2.2}$$

ι represents the inclusion of F in the bundle E; π represents the projection map $\pi\colon E \to M$, defined in each chart by $(x_{(\alpha)}, f_{(\alpha)}) \to x_{(\alpha)}$.

If F is a vector space, and the action is a group representation, we call E a *vector bundle*; $\dim F$ is called the *rank* of E. We call the bundle real if $G \subset \mathrm{GL}(n,\mathbb{R})$ and complex if $G \subset \mathrm{GL}(n,\mathbb{C})$. Another possibility is taking $F = G$, the structure group itself, and the group action to be the left multiplication in G; in this case, the bundle is called *principal*. All the bundles obtained from the same choice of $t_{\alpha\beta}$ acting on different fibers are said to be *associated* to one another. So TM, T^*M, and the frame bundle FM are all associated to one another; FM in particular is a $\mathrm{GL}(d,\mathbb{R})$ principal bundle.

For rank-one real vector bundles, the structure group is $\mathrm{GL}(1,\mathbb{R}) \cong \mathbb{R}^* \equiv \mathbb{R} - \{0\}$. For rank-one complex vector bundles, the structure group is $\mathrm{GL}(1,\mathbb{C}) \cong \mathbb{C}^* \equiv \mathbb{C} - \{0\}$. One usually calls the latter *line bundles*, because in algebraic geometry $\mathbb{C}$ is the complex analogue of a line. Given a line bundle $\mathcal{L}$ with transition functions $l_{\alpha\beta}$, one can define the inverse $\mathcal{L}^{-1}$, whose transition functions are $l^{-1}_{\alpha\beta}$.

Similar to our definition for the tangent bundle, a *section* s of (4.2.2) is a map from M to the fiber F, such that the projection $\pi(s(x)) = x$ for any $x \in M$. In practice, this is a collection of functions $s_{(\alpha)}(x_{(\alpha)})$ on every U_α, which on every $U_{\alpha\beta}$ satisfies the gluing condition

$$s_{(\alpha)} = t_{\alpha\beta} \cdot s_{(\beta)}, \tag{4.2.3}$$

just as in (4.2.1). One sometimes also calls a section a *global* section, to distinguish it from a single $s_{(\alpha)}$, which one might want to call instead a *local* section.

Structure group reduction

The transition functions $t_{\alpha\beta}$ are not uniquely defined. We know that under local coordinate changes (4.1.33). By the chain rule and recalling $x_{(\alpha)} = g_{\alpha\beta}(x_{(\beta)})$, we see that they also act on TM as $(J_{\alpha\beta})^m{}_n \to \frac{\partial x'^{\,m}_{(\alpha)}}{\partial x^p_{(\alpha)}} (J_{\alpha\beta})^p{}_q \frac{\partial x^q_{(\beta)}}{\partial x'^{\,n}_{(\beta)}}$. More generally, when the $t_{\alpha\beta}$ are not related to the $J_{\alpha\beta}$, we can change coordinates on the fiber, $f_{(\alpha)} \to f'_{(\alpha)} = t_\alpha \cdot f_{(\alpha)}$, where t_α now takes values in the structure group G and $\cdot$ is the group action. The transition functions then change as

$$t_{\alpha\beta} \to t_\alpha \cdot t_{\alpha\beta} \cdot t_\beta^{-1} . \tag{4.2.4}$$

If two bundles can be made to have the same transition function under such a coordinate change, they are called *equivalent.*

Sometimes one can use (4.2.4) to make all the $t_{\alpha\beta}$ take value in a subgroup H of the original structure group G. In such a case, we say that there is a *reduction of the structure group*. When G can be continuously deformed into H, then a G-bundle always admits a reduction to H. An example is

$$\mathrm{U}(1) \subset \mathbb{C}^* \cong \mathrm{GL}(1, \mathbb{C}) : \tag{4.2.5}$$

Here is a continuous deformation: $\mathbb{C}^* \ni z \mapsto (1-t)z + t\frac{z}{|z|}$ gives z for $t = 0$, and its projection $\frac{z}{|z|} \in \mathrm{U}(1)$ for $t = 1$. (The proper word for such a map is *retraction.*) As another example, a reduction is always possible from a vector bundle, with fiber $\mathbb{R}^N$, to a *sphere bundle* with fiber $S^{N-1} \subset \mathbb{R}^N$. In this case, the reduction is from $\mathrm{GL}(N, \mathbb{R})$ to $\mathrm{O}(N)$. When G cannot be continuously deformed into H, the reduction might still be possible, depending on M on the choice of $t_{\alpha\beta}$; this is revealed in part by the theory of characteristic classes, to which we will give a quick introduction in Section 4.2.5.

An extreme example is when the transition functions can all be made to be the identity, $t_{\alpha\beta} = 1$; the bundle is then called (topologically) *trivial*. For some types of bundles, there are criteria for triviality involving sections. A principal bundle is trivial if and only if it has a section. Another important example is a line bundle $\mathcal{L}$: a section s of $\mathcal{L}$ with no zeros can also be viewed as a section of a $\mathbb{C}^*$-bundle, which is a principal bundle. So $\mathcal{L}$ is trivial if and only if it has a never-vanishing global section.

Products and sums

More or less every operation we know in linear algebra inspires a counterpart on bundles.

As a first example, let us consider the *tensor product* of two vector bundles. We have already seen an example in (4.1.38). If one has two vector bundles E_1 and E_2, with transition functions $t_{\alpha\beta}$ acting in the representations ρ_1 and ρ_2, then by definition $E_1 \otimes E_2$ has $\theta_{\alpha\beta}$ acting in the representation $\rho_1 \otimes \rho_2$. More generally, we can take tensor products of vector bundles E_i with different structure groups G_i: calling F^i and $t^i_{\alpha\beta}$ their fibers and transition functions, the fiber of $E_1 \otimes E_2$ will be $F_1 \otimes F_2$, with transition functions $t^1_{\alpha\beta} \otimes t^2_{\alpha\beta}$ acting as $f^1_\alpha \otimes f^2_\alpha = (t^1_{\alpha\beta} \otimes t^2_{\alpha\beta}) \cdot f^1_\beta \otimes f^2_\beta \equiv (t^1_{\alpha\beta} \cdot f^1_\beta) \otimes (t^2_{\alpha\beta} \cdot f^2_\beta)$. An example is

$$\Lambda^p T^* \otimes E , \tag{4.2.6}$$

whose sections are called *E-valued*, or *twisted*, forms.

We have also seen around (4.1.39) that vector bundles can be summed. The general definition follows the lines explained there: $E_1 \oplus E_2$ is defined as having transition functions in the block-diagonal form $\left(\begin{smallmatrix}\rho_1 & 0 \\ 0 & \rho_2\end{smallmatrix}\right)$.

The *determinant bundle* $\det(E)$ of a vector bundle E is a line bundle, whose transition functions are given by

$$\det t_{\alpha\beta} \tag{4.2.7}$$

of the $t_{\alpha\beta}$ of E.

A *subbundle* $E' \hookrightarrow E$ is one whose fiber F' is a vector subspace of the fiber F of E. The *quotient* E/E' is then obtained by quotienting fiber by fiber. A notable example is given by the *normal bundle* to a submanifold $S \hookrightarrow M$:

$$N(S, M) \equiv TM|_S/TS\,. \tag{4.2.8}$$

The fiber of the bundle $TM|_S$ on S consists of all vectors in M, while that of TS consists of vectors tangent to S.

Vectors and forms on the total space

We now consider vectors and forms on the total space E itself of a bundle, thought of as a manifold. A vector field v on E is said to be *vertical* if $\pi_* v = 0$; intuitively, such a vector is tangent to the fiber. One might perhaps expect a similar definition of horizontal vector, which would be tangent to the base, but we don't have another map similar to π to help us define this; in fact, such a definition requires additional data, as we will see in the next subsection. On the other hand, we can define a horizontal form α as one such that $\iota_v \alpha = 0$ for any vertical v. If α is also invariant under the vertical Lie derivative L_v, a theorem shows that it is the pull-back to E of a form α_B on the base B:

$$\iota_v \alpha = 0 = L_v \alpha \quad \forall \text{ vertical } v \qquad \Leftrightarrow \qquad \alpha = \pi^* \alpha_B\,. \tag{4.2.9}$$

(An equivalent condition is that both α and $d\alpha$ should be horizontal.) Such a form is called *basic*. For all intents and purposes, basic forms on E can be thought of as forms living on B.

A generalization involves forms with a definite charge under vertical vectors. Take for example a U(1)-bundle $\mathcal{L}$, and consider a form α such that $L_v \alpha = n\alpha$. In each local patch $\hat{U}$, α can be written as $e^{in\psi}\alpha_0$, where $\psi \sim \psi + 2\pi$ is the periodic coordinate on the fiber and α_0 is now invariant under L_v. Following the definitions, we see that α is the product of a form on B and of a section of $\mathcal{L}^n$:

$$\alpha = \mathrm{e}^{in\psi}\alpha_0, \tag{4.2.10}$$

or in other words a bundle-valued form.

4.2.2 Connections

The derivative of a vector field is usually not a tensor; in (4.1.9), one remedies this by combining it with a connection Γ, also not a tensor. A similar problem is that the derivative of a section of a bundle is not usually a section: $\partial_m f_{(\alpha)} = \partial_m (t_{\alpha\beta} \cdot f_{(\beta)}) \neq t_{\alpha\beta}\partial_m f_{(\beta)}$. To fix this, we again introduce a connection

$A = A_m \mathrm{d}x^m$, a collection of local one-forms $A_{m(\alpha)} \mathrm{d}x^m_{(\alpha)}$ valued in the Lie algebra $\mathfrak{g}$ of G, such that

$$A_{(\alpha)} = t_{\alpha\beta}(\mathrm{d} + A_{(\beta)}) t^{-1}_{\alpha\beta} \tag{4.2.11}$$

on $U_{\alpha\beta}$. This is more complicated than (4.1.37); so a connection A is globally not a one-form, despite having a single lower index. We recognize instead the gauge transformations of a nonabelian Yang–Mills (YM) gauge field. So a connection is simply a Yang–Mills field on a manifold. The $t_{\alpha\beta}$ are G-valued; (4.2.11) makes sense because

$$g \in G, \qquad x \in \mathfrak{g} \qquad \Rightarrow \qquad g x g^{-1} \in \mathfrak{g}. \tag{4.2.12}$$

It is sometimes useful to expand in a basis of generators $T^a \in \mathfrak{g}$ and write $A = A^a T^a = A^a_m T^a \mathrm{d}x^m$.

We now show how to use (4.2.11) to define a notion of derivative of sections of bundles. We begin with vector bundles, and then generalize to more complicated fibers.

Covariant derivative on vector bundles

Given a section s of a vector bundle E, its *covariant derivative*

$$D_m s = \partial_m s + A_m s \tag{4.2.13}$$

is again a section of E: indeed from (4.2.3) and (4.2.11),

$$D_m s_{(\alpha)} = t_{\alpha\beta} D_m s_{(\beta)} \, . \tag{4.2.14}$$

At every point, A_m is a matrix and s a vector in the fiber V of E; $A_m s$ is obtained by matrix multiplication. (This is why we dropped the $\cdot$ that we had in (4.2.3), to denote a more general action of G on the fiber F.) In a more condensed notation, we can introduce $Ds \equiv \mathrm{d}x^m D_m s$, and

$$Ds \equiv \mathrm{d}s + A\, s \tag{4.2.15}$$

is now a section of $T^*M \otimes E$.

We have seen that on every chart we can perform a local gauge transformation (4.2.4), which is just as immaterial as a coordinate transformation. Via (4.2.11), this implies that locally the two connections A and $A' = t(\mathrm{d} + A)t^{-1}$ are really to be considered equivalent. At the infinitesimal level, this reads

$$\delta A = \mathrm{d}\lambda + A\lambda - \lambda A \equiv \mathrm{d}\lambda + [A, \lambda] \, . \tag{4.2.16}$$

Here now λ is valued in the Lie algebra $\mathfrak{g}$.

The gauge covariant derivative (4.2.15) is similar to the geometric covariant derivative in (4.1.9). We can try to define an analogue of the Riemann tensor with a logic similar to (4.1.13):

$$[D_m, D_n] s = 2 D_{[m} D_{n]} s = 2(\partial_{[m} A_{n]} + A_{[m} A_{n]})\, s \equiv F_{mn}\, s \, . \tag{4.2.17}$$

In the more compact notation (4.2.15), we write $D^2 = \mathrm{d}x^m \wedge \mathrm{d}x^n D_{[m} D_{n]}$, and

$$D^2 s = (\mathrm{d}A + A \wedge A)\, s \equiv F\, s \, . \tag{4.2.18}$$

One should recall in this expression that $A = A_a T^a$ is valued in the Lie algebra $\mathfrak{g}$ of G; more explicitly,

$$F_a T^a = \mathrm{d}A_a T^a + A_a \wedge A_b T^a T^b = \mathrm{d}A_a T^a + \frac{1}{2} A_a \wedge A_b [T^a, T^b]$$
$$= \left(\mathrm{d}A_a + \frac{1}{2} f^{bc}{}_a A_b \wedge A_c \right) T^a , \tag{4.2.19}$$

where now we have introduced the structure constants of $\mathfrak{g}$:

$$[T^a, T^b] \equiv f^{ab}{}_c T^c . \tag{4.2.20}$$

In a notation where the wedge product is left implicit, (4.2.19) reads

$$F = \mathrm{d}A + \frac{1}{2}[A, A]. \tag{4.2.21}$$

This $F = \frac{1}{2} F_{mn} \mathrm{d}x^m \wedge \mathrm{d}x^n$ is called the *curvature* of the connection A. If we compute $F_{(\alpha)}$ in every patch, then from (4.2.11) it follows that

$$F_{(\alpha)} = t_{\alpha\beta} F_{(\beta)} t_{\alpha\beta}^{-1} . \tag{4.2.22}$$

In particular, for a bundle with $G = \mathrm{U}(1)$, the curvature is really a two-form. More generally, for a vector bundle we see from (4.2.22) that $\mathrm{Tr}F$ is a two-form. We can view it as the curvature on the determinant line bundle $\det(E)$ defined in (4.2.7). Similar to (4.2.22), under local gauge transformations (4.2.4) we have $F \to t^{-1} F t$; the infinitesimal version is that under (4.2.16) we have

$$\delta F = [F, \lambda] . \tag{4.2.23}$$

We also have the *nonabelian Bianchi identity*

$$\mathrm{d}F + [A, F] = 0 , \tag{4.2.24}$$

generalizing $\mathrm{d}F = 0$ for the abelian case.

Gauge theories

As we mentioned earlier, the idea of bundle is the proper framework to define gauge theories on manifolds. A Yang–Mills (YM) field with gauge group G can be thought of as a connection on a G-principal bundle P; indeed, the transformation (4.2.11) is of course the one of a Yang–Mills gauge field. Fields charged under a YM field can be viewed as sections of bundles associated to P. Their kinetic terms can be written in terms of their covariant derivatives. The most common cases are (4.2.15) and (4.2.28), which occur when we want to gauge a translation or a rotation isometry of field theory space. Equation (4.2.15) appears in a Higgs-type covariant derivative, while (4.2.28) in the Stückelberg mechanism (1.2.26).

General fiber

Let us now consider an arbitrary fiber F. The group action of G on F has an infinitesimal counterpart: each element in the Lie algebra $\mathfrak{g}$ of G is represented by a vector field

$$K^a = K^{ai} \partial_i , \tag{4.2.25}$$

where as usual $\partial_i = \partial_{y^i}$, in terms of coordinates y^i on F. The *horizontal* vector fields

$$D_m \equiv \partial_m - A_{am}K^a = \partial_m - A_{am}K^{aj}\partial_j \tag{4.2.26}$$

give a possible definition of covariant derivative; as anticipated, this notion depends on the choice of a connection. We can also define

$$Dy^i \equiv dy^i + A_a K^{ai}, \tag{4.2.27}$$

which are one-forms on the total space E and satisfy $\iota_{D_m} Dy^i = 0$.

For example, if E is a real vector bundle, we reproduce (4.2.15) by taking $K^{ai} = T^{ai}{}_j y^j$. Under the Lie bracket (4.1.58), $[K^a, K^b] = (K^{aj}\partial_j K^{bi} - K^{bj}\partial_j K^{ai})\partial_i = (T^{aj}{}_k T^{bi}{}_j - T^{bj}{}_k T^{ai}{}_j) y^k \partial_i = -[T^a, T^b]^i{}_k y^k \partial_i = -f^{ab}_c K^c$. For consistency with this particular case, we define $[K^a, K^b] = -f^{ab}{}_c K^c$ in general.

As another example, if E is a U(1) principal bundle, and we call ψ the coordinate on the fiber $U(1) \cong S^1$, then the vertical vector field is ∂_ψ, and (4.2.27) simply gives

$$D\psi = d\psi + A. \tag{4.2.28}$$

Generalizing (4.2.27), we can define

$$D \equiv d + A_a \wedge L_{K^a}; \tag{4.2.29}$$

L_{K^a} is the Lie derivative under the vector field K^a. Indeed, on the function y^i, the definition (4.1.54) gives $L_{K^a} y^i = K^{aj}\partial_j y^i = K^{ai}$, in agreement with (4.2.27). Computing the square is instructive:

$$\begin{aligned} D^2 &= \frac{1}{2}\{D, D\} = \frac{1}{2}\{d + A_a \wedge L_{K^a}, d + A_b \wedge L_{K^b}\} \\ &\overset{(4.1.75)}{=} (dA_a) \wedge L_{K^a} + \frac{1}{2} A_a \wedge A_b [L_{K^a}, L_{K^b}] \\ &\overset{(4.1.82)}{=} (dA_a) \wedge L_{K^a} + \frac{1}{2} A_a \wedge A_b L_{[K^a, K^b]} = \left(dA_a - \frac{1}{2} A_b \wedge A_c f^{bc}{}_a \right) L_{K^a} \\ &\equiv \hat{F}_a L_{K^a}. \end{aligned} \tag{4.2.30}$$

(The sign in the last step is due to $[K^a, K^b] = -f^{ab}{}_c K^c$, as motivated previously.)

Making forms covariant*

The operator

$$K\cdot \equiv A_a \wedge \iota_{K^a} \tag{4.2.31}$$

maps forms on F to forms on E; for example, $e^{K\cdot} dy^i = Dy^i = dy^i + A_a K^i_a$. More generally, $e^{K\cdot} dy^i \wedge e^{-K\cdot} = Dy^i \wedge$, because the series in (2.1.15) truncates at the first step. This implies

$$e^{K\cdot} \frac{1}{k!} \alpha_{i_1 \ldots i_k} dy^{i_1} \wedge \ldots \wedge dy^{i_k} = \frac{1}{k!} \alpha_{i_1 \ldots i_k} Dy^{i_1} \wedge \ldots \wedge Dy^{i_k}. \tag{4.2.32}$$

We can use (2.1.15) also to compute $e^{-K\cdot} d e^{K\cdot}$. Note the first two nontrivial terms in the series:

$$\begin{aligned} [d, K\cdot] &= [d_B, A_a \wedge \iota_{K^a}] + [d_F, A_a \wedge \iota_{K^a}] = d_B A_a \wedge \iota_{K^a} - A_a \wedge L_{K^a}, \\ -[[d, K\cdot], K\cdot] &= [A_a \wedge L_{K^a}, A_b \wedge \iota_{K^b}] = A_a \wedge A_b \wedge [L_{K^a}, \iota_{K^b}] \\ &= A_a \wedge A_b \wedge \iota_{[K^a, K^b]}. \end{aligned} \tag{4.2.33}$$

The next term $[[[\mathrm{d}, K\cdot], K\cdot], K\cdot]$ vanishes. The series (2.1.15) then reassembles as

$$\mathrm{e}^{-K\cdot}\mathrm{d}\,\mathrm{e}^{K\cdot} = \mathrm{d} - A_a \wedge L_{K^a} + \hat{F}_a \wedge \iota_{K^a}\,. \tag{4.2.34}$$

4.2.3 Bundle metric

One usually takes the G action to be isometric; the vector fields K^a in (4.2.27) are then *Killing vectors* of F. In fact, (4.2.27) now suggests a natural class of metrics on the total space E:

$$\mathrm{d}s^2_E \equiv g^B_{mn}\mathrm{d}x^m\mathrm{d}x^n + g^F_{ij}Dy^iDy^j\,, \tag{4.2.35}$$

where g^B and g^F are metrics on B and F respectively.

Typically one also makes some additional assumptions. Just like the A_{am} are forms on B and don't depend on the coordinates of F, it is natural to assume that the fiber metric only depends on the B coordinates, and that g^B only depend on F:

$$\partial_m g^F_{ij} = 0\,, \qquad \partial_i g^B_{mn} = 0\,. \tag{4.2.36}$$

Both these assumptions are sometimes slightly relaxed, as we will see later. If they do hold, we will now see that (4.2.35) inherits the isometries of g^B and g^F to some extent.

The most notable example is when the fiber is S^1, in which case (4.2.35) reads

$$\mathrm{d}s^2 = \mathrm{d}s^2_B + \mathrm{e}^{2\sigma}D\psi^2\,, \qquad D\psi \equiv \mathrm{d}\psi + A\,. \tag{4.2.37}$$

We recognize that the relation (1.4.5a) between the M-theory and IIA metrics is of this form. In this sense, we say that the M-theory circle is fibered over M_{10}.

Fiber isometries

First we consider a Killing vector v of g^F. It is not clear that it is still an isometry of (4.2.35), because of the presence of K^{ai}. For this reason, we will assume

$$[v, K^a] = 0\,. \tag{4.2.38}$$

Let us check if this is enough. Act with (4.2.30) on y^i, and rewrite the result as

$$\mathrm{d}Dy^i = -A_a \wedge L_{K^a}Dy^i + \hat{F}_aK^{ai}\,. \tag{4.2.39}$$

(We can also obtain this acting with (4.2.34) on $\mathrm{d}y^i$.) Now

$$\begin{aligned} L_vDy^i &\overset{(4.1.79)}{=} \mathrm{d}\iota_vDy^i + \iota_v\mathrm{d}Dy^i \overset{(4.2.27),(4.2.39)}{=} \mathrm{d}(v^i) + A_a \wedge \iota_vL_{K^a}Dy^i = \\ &= \mathrm{d}(v^i) + A_a \wedge (L_{K^a}\iota_v + [\iota_v, L_{K^a}])Dy^i \overset{(4.1.81),(4.2.38)}{=} \partial_jv^iDy^j\,. \end{aligned} \tag{4.2.40}$$

Now the first term in (4.2.40) assembles with the derivative of g^F_{ij}:

$$\begin{aligned} L_v(\mathrm{d}s^2_E) &\overset{(4.2.35)}{=} v^k\partial_kg^F_{ij}Dy^iDy^j + g^F_{ij}(\partial_kv^iDy^kDy^j + \partial_kv^jDy^iDy^k) \\ &\overset{(4.1.64)}{=} (L_vg^F_{ij})Dy^iDy^j = 0\,. \end{aligned} \tag{4.2.41}$$

So the Killing vectors of g^F that commute with the K^a are also Killing vectors of the bundle metric (4.2.35). For example, in (4.2.37), where the fiber is an S^1, the vertical isometry $v = \partial_\psi$ commutes with $K = \partial_\psi$ and is an isometry of the total space E.

Lifting base isometries

We now consider a Killing vector h of g^B. Again, it is not obvious that it should be a Killing vector of the full metric (4.2.35), this time because of the presence of the A_{am}. One might imagine we have to impose $L_h A_a = 0$, but this criterion would almost never be satisfied. A more reasonable idea is to impose that $L_h A_a$ is a gauge transformation; we have seen after all that (4.2.16) should be considered an equivalence of connections. So we impose

$$L_h A_a = \mathrm{d}\lambda_a + f^{bc}{}_a A_b \lambda_c \tag{4.2.42}$$

for some λ_a. We also have to compensate for the gauge transformation of A by adding a vertical piece: we define the *lift* of h to be

$$\hat{h} \equiv h - \lambda_a K^a \, . \tag{4.2.43}$$

Let us check that this works:

$$\begin{aligned} L_{\hat{h}} D y^i &\overset{(4.2.27)}{=} L_{-\lambda_a K^a} \mathrm{d}y^i + (L_h A_a) K^{ai} + A_a L_{-\lambda_b K^b} K^{ai} \\ &\overset{(4.1.79)}{=} -\mathrm{d}(\lambda_a K^{ai}) + (\mathrm{d}\lambda_a + f^{bc}{}_a A_b \lambda_c) K^{ai} - A_a \lambda_b K^{bj} \partial_j K^{ai} \\ &\overset{(4.2.20),(4.1.58)}{=} -\lambda_a \mathrm{d}K^{ai} - A_a \lambda_b K^{aj} \partial_j K^{bi} \\ &= -\lambda_a \partial_j K^{ai} D y^j \, . \end{aligned} \tag{4.2.44}$$

Now the same computation as in (4.2.41) shows that $L_{\hat{h}} \mathrm{d}s^2_E = 0$.

We will see a concrete example of the lift (4.2.43) in Section 4.4.1.

4.2.4 Bundles on S^2

Vector bundles on S^2 have the peculiar property that they can be written as direct sums of line bundles (*Birkhoff–Grothendieck theorem*); this doesn't happen on more general manifolds. So we only need to study line bundles (bundles whose fiber is $\mathbb{C}$) and their close cousins the U(1)-bundles.

Line bundles

We have seen in Section 4.1.3 the coordinate systems $x^m_{\mathrm{N,S}}$ on its two charts $U_{\mathrm{N,S}}$, and a third coordinate system (θ, ϕ) on $U_{\mathrm{NS}} = U_{\mathrm{N}} \cap U_{\mathrm{S}}$. We describe a line bundle $O(n)$ over S^2 as obtained by gluing $U_{\mathrm{N}} \times \mathrm{U}(1)$ and $U_{\mathrm{S}} \times \mathrm{U}(1)$ with the transition function

$$(z_{\mathrm{N}}, f_{\mathrm{N}}) \sim (z_{\mathrm{S}}, f_{\mathrm{S}}) \, , \qquad z_{\mathrm{S}} = z_{\mathrm{N}}^{-1} \, , \qquad f_{\mathrm{S}} = z_{\mathrm{N}}^{-n} f_{\mathrm{N}} \, . \tag{4.2.45}$$

The notation is as in (4.1.43) and (4.1.44). The single transition function is $t_{\mathrm{SN}} = z^n_{\mathrm{S}} = z_{\mathrm{N}}^{-n}$, acting by multiplication, along with its inverse $t_{\mathrm{NS}} = t^{-1}_{\mathrm{SN}}$. It depends holomorphically on the complex coordinate on U_{N}; we say that the bundle $O(n)$ is *holomorphic*. (This is an analogue for bundles of the notion of complex manifold, mentioned after (4.1.44) and to be developed in Section 5.3.1.)

Since the structure group is abelian, (4.2.4) acts trivially; thus the only trivial $O(n)$ is $O(0)$. The tensor product of two such bundles is

$$O(n) \otimes O(m) = O(n+m)\,. \tag{4.2.46}$$

Let us look for global sections. Since the transition functions are holomorphic, we can even look for holomorphic global sections. On $\mathrm{U_N}$, a possible basis of holomorphic functions is given by the monomials z_N^k, $k \geq 0$. According to (4.2.45), on U_S these become $t_\mathrm{SN} z_\mathrm{N}^k = z_\mathrm{N}^{-n} z_\mathrm{N}^k = z_\mathrm{S}^{n-k}$. This is holomorphic if $k \leq n$. So we have the $n+1$ sections

$$s_\mathrm{N} = z_\mathrm{N}^k\,, \qquad s_\mathrm{S} = z_\mathrm{S}^{n-k}\,. \tag{4.2.47}$$

For $n = 0$, we have the trivial bundle $O(0)$, whose sections are holomorphic functions; so we find that the only such function on S^2 is a constant. (This is a particular case of a general result we will mention in Section 5.3.1: the only globally holomorphic function on a compact manifold is constant.) If we again think of our S^2 as the Riemann sphere (4.1.45), any nontrivial holomorphic function on $\mathbb{C}$ will develop a singularity at $z = \infty$. If $n > 0$, the transition function z_N^{-n} "saves" some monomials from developing a pole at infinity. If $n < 0$, one has the opposite effect and no global holomorphic sections exist.

Recall that a holomorphic function with a set of isolated poles is a called a *meromorphic* function. Without introducing transition functions, the monomials z_N^k would have a pole of order k at $z_\mathrm{N} = \infty$ (or $z_\mathrm{S} = 0$). This is a general idea in algebraic geometry: there is a correspondence between meromorphic functions and holomorphic sections of nontrivial line bundles.

Another observation is that the sum of the order of the poles of a meromorphic function is equal to the sum of the order of the zeros. This is trivially so for a monomial z_N^k, which have a single zero and a single pole both of order k, but it is true more generally for sums of such sections, as one can see by standard complex analysis (for example, by considering the integral $\oint \mathrm{d}f/f$ over a closed path). A concrete consequence is that a section of $O(n)$ always has n zeros (taking multiplicities into account). This also has a generalization to higher dimensions, which we will mention in Section 4.2.5.

U(1)-**bundles**

The transition function $t_\mathrm{SN} = z_\mathrm{N}^{-n}$ from (4.2.45) is in $\mathrm{GL}(1,\mathbb{C}) = \mathbb{C}^*$, as appropriate for a line bundle. By a local gauge transformation (4.2.4), it can be brought to belong to $\mathrm{U}(1) \subset \mathbb{C}^*$. Indeed, take

$$t_\mathrm{N} = \frac{1}{1+|z_\mathrm{N}|^n}\,, \qquad t_\mathrm{S} = \frac{1}{1+|z_\mathrm{S}|^n}\,, \tag{4.2.48}$$

which are continuous in their domains U_N, and U_S. From (4.2.4), we get a new $t_\mathrm{SN}^{\mathrm{U}(1)} = t_\mathrm{S} t_\mathrm{SN} t_\mathrm{N}^{-1}$ given by

$$t_\mathrm{SN}^{\mathrm{U}(1)} = \left(\frac{z_\mathrm{N}}{|z_\mathrm{N}|}\right)^{-n} = \mathrm{e}^{-in\phi}\,. \tag{4.2.49}$$

This is an example of the structure group reduction we mentioned in (4.2.5). From now on, we call U_n the U(1)-bundles defined by (4.2.49), but one should keep in mind that they are simply obtained from the bundles $O(n)$ of (4.2.45), at the cost of losing their holomorphicity properties.

It turns out that any line bundle over S^2 is isomorphic to one of the $O(n)$, and any U(1)-bundle to one of the U_n. Here is the argument for the latter case: whatever the transition function t_{NS} is on U_{NS}, it can be put in the form $e^{-in\phi}$ by a change of coordinates. Intuitively, all that matters in t_{NS} is how many times it winds around the origin as one goes around the equator; this is the number n. Our result (4.2.47) has a topological counterpart: the bundle U_n for $n < 0$ has no global sections.

Hopf fibration

The total space of the bundle U_{-1} is topologically a three-sphere S^3. The homeomorphism identifying the two is obtained by parameterizing $S^3 \subset \mathbb{R}^4 \cong \mathbb{C}^2$ with two complex coordinates w_1, w_2, so that $S^3 = \{|w_1|^2 + |w_2|^2 = 1\}$. Taking

$$w_1 = \cos(\theta/2)e^{i\left(\psi+\frac{\phi}{2}\right)}, \qquad w_2 = \sin(\theta/2)e^{i\left(\psi-\frac{\phi}{2}\right)}, \tag{4.2.50}$$

the coordinates (θ, ϕ) become the coordinates on S^2 (valid on its open set U_{NS}), while ψ becomes the coordinate on the $S^1 \cong U(1)$ fiber of the bundle U_{-1}. To cover all values of w_i, the periodicities of ψ and ϕ should be $\Delta\psi = \Delta\phi = 2\pi$. In these coordinates, the round metric on S^3 from (4.1.70) becomes an example of a bundle metric (4.2.35):

$$ds^2_{S^3} = dw_1 d\bar{w}_1 + dw_2 d\bar{w}_2 = (d\psi + A)^2 + \frac{1}{4} ds^2_{S^2}. \tag{4.2.51}$$

Here $ds^2_{S^2}$ is the usual round metric in (4.1.46); the connection can be written as

$$A = \frac{1}{2}\cos\theta d\phi. \tag{4.2.52}$$

This expression of course only makes sense on $U_N \cap U_S$, where the (θ, ϕ) coordinates make sense. Indeed, (4.2.52) becomes $\frac{1}{2}d\phi$ at the poles, and ϕ is not defined there. To work on U_N and U_S, one should rewrite (4.2.51) in terms of

$$\begin{aligned} \psi_N &= \psi + \frac{1}{2}\phi, \qquad & A_N &= \frac{1}{2}(\cos\theta - 1)d\phi; \\ \psi_S &= \psi - \frac{1}{2}\phi, \qquad & A_S &= \frac{1}{2}(\cos\theta + 1)d\phi. \end{aligned} \tag{4.2.53}$$

In particular we see $e^{i\psi_S} = e^{i\phi}e^{i\psi_N}$; comparing with (4.2.49), we see that this is the transition function for U_{-1}, as promised. For the connection, we can write $A_S = A_N - d\phi$, which agrees with (4.2.11).

The projection $\pi : S^3 \to S^2$ is called the *Hopf map*; we already encountered it in (3.3.50). In the notation of (4.2.2), we can write

$$\begin{array}{ccc} S^1 & \hookrightarrow & S^3 \\ & & \downarrow \\ & & S^2. \end{array} \tag{4.2.54}$$

For U_{-n}, one obtains the same metric (4.2.51) but with a periodicity $2\pi/n$ for the angle ψ. The total space is now no longer S^3 but its quotient $S^3/\mathbb{Z}_n$, obtained from S^3 by the identification $(w_1, w_2) \sim (\lambda w_1, \lambda w_2)$, $\lambda = e^{2\pi i/n}$.

4.2.5 Characteristic classes

A *characteristic class* is an element of the cohomology of M associated to a bundle $E \to M$, which is invariant under bundle equivalence.

First Chern class c_1

The simplest example, and the one we will need most often, is the so-called *first Chern class* c_1 and is associated to any complex vector bundle E. Consider a connection A and its curvature $F = dA + A \wedge A = dA + \frac{1}{2}[A, A]$. We observe from (4.2.19) that $\mathrm{Tr}F = dA_a \mathrm{Tr}T^a$, so

$$d\mathrm{Tr}F = 0\,. \tag{4.2.55}$$

Even if F is not a two-form, $\mathrm{Tr}F$ is, as remarked after (4.2.22). Its class in cohomology $[\mathrm{Tr}F] \in H^2(M, \mathbb{R})$ is by definition $c_1(E)$. We will now show that

$$\frac{1}{2\pi i}\int_{S_2} \mathrm{Tr}F \in \mathbb{Z} \tag{4.2.56}$$

for any two-cycle $S_2 \in H_2(M, \mathbb{Z})$. By our remark that follows (4.1.133), we can regard $\frac{1}{2\pi i}\mathrm{Tr}F$ as a representative of a class in integer-valued cohomology:

$$c_1(E) \equiv \left[\frac{1}{2\pi i}\mathrm{Tr}F\right] \in H^2(M, \mathbb{Z})\,. \tag{4.2.57}$$

Rather than showing (4.2.56) in general, we will now see how it works for $M = S^2$. This is essentially the same argument as the one leading to Dirac quantization (1.3.7).

The S^2 case

By the Birkhoff–Grothendieck theorem (Section 4.2.4), all vector bundles on S^2 are direct sums of the line bundles $O(n)$. In this case, F is already abelian, and we can drop the trace in (4.2.55). Clearly $H^2(S^2) = \mathbb{R}$, so (4.2.57) reduces to the statement that the integral $\frac{1}{2\pi i}\int F = n \in \mathbb{Z} = H^2(S^2, \mathbb{Z})$. We write $S^2 = \mathrm{N} \cup \mathrm{S}$, where $\mathrm{N} = \{\theta \in [0, \pi/2)\}$ and $\mathrm{S} = \{\theta \in (\pi/2, \pi]\}$ are the northern and southern hemisphere; on each, $F = F_\mathrm{N} = dA_\mathrm{N}$ and $F = F_\mathrm{S} = dA_\mathrm{S}$. So

$$\int_{S^2} F = \int_\mathrm{N} F_\mathrm{N} + \int_\mathrm{S} F_\mathrm{S} = \int_\mathrm{N} dA_\mathrm{N} + \int_\mathrm{S} dA_\mathrm{S} \overset{(4.1.131)}{=} \int_\mathrm{E} (A_\mathrm{N} - A_\mathrm{S})\,. \tag{4.2.58a}$$

The equator $\mathrm{E} = \{\theta = \pi/2\}$ satisfies $\partial\mathrm{N} = \mathrm{E}$, $\partial\mathrm{S} = -\mathrm{E}$; this is the origin of the relative sign in the last step of (4.2.58a). Now we use the transformation law (4.2.11) of a connection, which in the abelian case simply reads $A_{(\alpha)} = A_{(\beta)} + d\log t_{\alpha\beta}$. The transition function is in (4.2.45), from which $A_\mathrm{N} = A_\mathrm{S} + \frac{n}{z_\mathrm{N}} dz_\mathrm{N}$. Then (4.2.58a) continues as

$$\int_\mathrm{E} (A_\mathrm{N} - A_\mathrm{S}) = n\int_\mathrm{E} \frac{dz_\mathrm{N}}{z_\mathrm{N}} = 2\pi i n\,. \tag{4.2.58b}$$

The last step is from Cauchy's theorem in complex analysis, or from noticing that at the equator $z_N = e^{i\phi}$. We conclude that in this case,

$$c_1(O(n)) = n\,, \tag{4.2.59}$$

in agreement with (4.2.57). The general proof of (4.2.57) goes along similar lines: one uses the abelian transformation $A_{(\alpha)} = A_{(\beta)} + \mathrm{d}\log t_{\alpha\beta}$ on all open set intersections.

The case of U(1)-bundles U_n is almost identical, because of their close relation to the $O(n)$ bundles. A subtlety is that the quantization law depends on the periodicity $\Delta\psi$ of the S^1-fiber, $\psi \sim \psi + \Delta\psi$. If the U(1)-bundle originates from the reduction of a line bundle, it is natural to take $\Delta\psi = 2\pi$, but more generally $\frac{1}{\Delta\psi}\int F \in \mathbb{Z}$ for the curvature.[10] So for a U(1)-bundle U, we define

$$c_1(U) = \left[\frac{1}{\Delta\psi}\int F\right] \in H^2(M,\mathbb{Z})\,. \tag{4.2.60}$$

For example, in (4.2.51) the fiber coordinate ψ has periodicity 2π; the connection A has $F = \mathrm{d}A = -\frac{1}{2}\sin\theta\mathrm{d}\theta\wedge\mathrm{d}\phi$, and $c_1 = \frac{1}{2\pi}\int F = -1$. Indeed, (4.2.51) was a metric on S^3, the total space of U_{-1}.

It is also interesting to note that the Poincaré dual (4.1.136) of $c_1(O(n))$ lives in $H_0(S^2)$; so it is a collection of n points. One can think of these points as the zeros s_i of a section of $O(n)$. For our basis (4.2.47), these were all collected at $z = 0$ or ∞, but by taking linear combinations, the s_i can be anywhere on S^2.

For an alternative perspective, still in $d = 2$, consider for example $a_z = s^{-1}\partial_z s$, $a_{\bar z} = 0$, where s is a holomorphic section; using (4.2.11), we see that this transforms as a connection. Its curvature is

$$f_{z\bar z} = -\partial_z\partial_{\bar z}\log s\,. \tag{4.2.61}$$

This might seem to be zero, but it is in fact a sum of delta forms with support on the zeros of s, ultimately because the log is flat Laplacian's Green's function in two dimensions. Indeed, from the residue theorem and Stokes's theorem (4.1.131)

$$2\pi i = \oint_{S^1}\frac{\mathrm{d}z}{z} = -\int_D \partial_{\bar z}\frac{1}{z}\mathrm{d}z\wedge\mathrm{d}\bar z \quad\Rightarrow\quad \partial_{\bar z}\frac{1}{z} = \pi\delta(x)\delta(y)\,, \tag{4.2.62}$$

where D is a disk containing the origin, and $S^1 = \partial D$ its boundary. Near each zero of s, $\log s \sim \log z$, and $f_{z\bar z}$ contains a delta. Thus integrating this curvature to compute c_1 is equivalent to counting the zeros of s.

All this has a generalization to higher dimensions. The Poincaré dual of $c_1(\mathcal{L})$ of a line bundle lives in $H_{d-2}(M_d)$; a representative of this class is the zero set of a holomorphic section s of a $\mathcal{L}$:

$$\{s = 0\} \cong \mathrm{PD}[c_1(\mathcal{L})]\,. \tag{4.2.63}$$

A generalization of the argument around (4.2.62) also exists, although it would require a definition of complex coordinates in a manifold, which we will see in Section 5.3.

[10] The loss of i in the denominator with respect of (4.2.57) is a convention: the curvature of a u(1)-bundle is $F = F^1T^1$ with only one generator $T^1 = i$.

Higher Chern classes

There are generalizations of c_1 that are more intrinsically nonabelian. The $2k$-forms $\mathrm{Tr}(F^k) \equiv \mathrm{Tr}\,\underbrace{F \wedge \ldots \wedge F}_{k}$ are also closed:

$$\mathrm{dTr}(F^k) = 0\,. \tag{4.2.64}$$

This offers an opportunity to define new characteristic classes in higher cohomology groups. One defines the polyform

$$\det\left(1 + \frac{1}{2\pi \mathrm{i}} F\right) \equiv 1 + C_1 + C_2 + \cdots\,. \tag{4.2.65}$$

The $2k$-form C_k is a linear combination of $\mathrm{Tr}(F^k)$, and is thus closed. Just like for c_1, its periods are integer; so the cohomology class $c_k \equiv [C_k] \in H^{2k}(M, \mathbb{Z})$. The formal sum $c = \sum_k c_k$, the total cohomology class of (4.2.65), is called the *total Chern class*. It has the property that

$$c(E \oplus F) = c(E)c(F)\,. \tag{4.2.66}$$

This can be proven with the *splitting principle*: in checking properties of characteristic classes, we may assume that a vector bundle is a sum of line bundles,

$$c_k(E) = c_k(\oplus_i \mathcal{L}_i)\,. \tag{4.2.67}$$

By (4.2.65) we see that $c(E) = \Pi_i(1 + x_i)$, $x_i \equiv c_1(\mathcal{L}_i)$, from which (4.2.66) follows. On the other hand, the *Chern character* $\mathrm{ch}(E)$ is defined as the cohomology representative of the polyform

$$\mathrm{Ch}(E) = \mathrm{Tr}\exp\left(\frac{1}{2\pi \mathrm{i}} F\right)\,. \tag{4.2.68}$$

The splitting principle tells us now $\mathrm{ch}(E) = \sum_i \mathrm{e}^{x_i}$, from which

$$\mathrm{ch}(E \oplus F) = \mathrm{ch}(E) + \mathrm{ch}(F)\,, \qquad \mathrm{ch}(E \otimes F) = \mathrm{ch}(E)\mathrm{ch}(F)\,. \tag{4.2.69}$$

Brane charges

As we anticipated, the argument in (4.2.58) and (4.2.59) for Chern class quantization has a physics origin in the Dirac quantization argument (1.3.7), with periodicity of the fiber playing the role of the particle's wavefunction; the main difference is that here the subspace on which we integrated is topologically nontrivial, rather than surrounding a monopole. So in a way we are giving mathematical names to ideas we already developed in Section 1.3.4.

Brane charges can also be interpreted in this spirit. For example, we saw in (1.3.37) that a D2 can couple to C_1, and thus has a D0 charge, in presence of world-volume flux. We now see that this is interpreted as the c_1 of the gauge bundle on the D2. More generally, the sum of the WZ terms in (1.3.24) reconstructs the exponential $\mathrm{e}^{\mathcal{F}}$; recalling (4.2.68), we see that the D$(p-2k)$-charge on a Dp-brane is interpreted in terms of higher Chern classes.

Pontryagin, Euler, Stiefel–Whitney

There are other notable characteristic classes:

- For real vector bundles, the structure group is $\mathrm{GL}(n, \mathbb{R})$. Similar to the discussion that follows (4.2.49), it is possible to reduce it to the smaller $\mathrm{O}(n)$, and hence to take the connection and curvature to take values in the Lie algebra so(n). So the generators T^a are antisymmetric matrices, and the C_k in (4.2.65) vanish for odd k; for example, Tr$F = 0$. The remaining forms define the so-called *Pontryagin classes* $p_i(E) \equiv (-1)^i c_{2i}(E \otimes \mathbb{C}) \in H^{4i}(M, \mathbb{Z})$, the first nontrivial being $p_1 = \frac{1}{8\pi^2}[\mathrm{Tr}F^2] \in H^4(M, \mathbb{Z})$.

- Also for real vector bundles, when the rank n is even, one introduces another characteristic class by using a cousin of the determinant called the *Pfaffian*, which is defined for an antisymmetric matrix A and satisfies Pf$A \equiv \sqrt{\det A}$. The corresponding *Euler class* $e \in H^n(M, \mathbb{Z})$ is then such that $p_{n/2} = e^2$; for a complex bundle, $c_n = e$.
- Not all characteristic classes live in de Rham cohomology. The *Stiefel–Whitney classes* w_i of a real bundle are in $H^i(M, \mathbb{Z}_2)$, defined as in (4.1.124) replacing $\mathbb{Z} \to \mathbb{Z}_2$. Their definition is most natural within *Čech cohomology*, which is defined using objects with indices α relative to the charts U_α of a manifold (see, for example, [119]). The first class $w_1(TM) = 0$ if and only if M is orientable (Section 4.1.3).

Characteristic classes as a classification tool

Characteristic classes help recognize whether a bundle is trivial or not. For example, a line bundle is trivial if and only if its c_1 vanishes:[11]

$$c_1(\mathcal{L}) = 0 \quad \Leftrightarrow \quad \mathcal{L} \text{ is trivial}. \tag{4.2.70}$$

In other words, line bundles are *characterized* by their c_1. On S^2, all bundles $O(n)$ and U_n were indeed characterized by the single integer n, which vanishes only when the bundle is trivial.

For bundles with arbitrary fiber, characteristic classes are usually not enough by themselves to classify bundles topologically. A relatively simple case is given by bundles on a sphere S^d, for which we only need a single transition function $f_{\mathrm{NS}} : U_{\mathrm{NS}} \to G$. Since topologically $U_{\mathrm{NS}} \cong S^{d-1} \times I$, with I an interval, the bundle is classified topologically by the class of f_{NS} in the homotopy group (Section 4.1.10):

$$\pi_{d-1}(G) \quad (G\text{-bundles over } S^d). \tag{4.2.71}$$

On an arbitrary manifold, a more sophisticated *obstruction theory* aims to give criteria to recognize that a bundle is trivial. The idea is to triangulate the manifold, and to analyze the bundle on loci of the triangulation of growing dimensions – first on its vertices, then on the lines connecting the vertices, and so on. This leads to the definition of invariants living in the cohomology groups $H^{i+1}(M, \pi_i(G))$. For a $G = \mathrm{U}(1)$ bundle, $\pi_1(\mathrm{U}(1)) = \mathbb{Z}$, and the invariant in $H^2(M, \mathbb{Z})$ is simply c_1.

Index theorems

The *Gauss–Bonnet theorem* states that the *Euler characteristic* of a compact M, defined as the alternate sum of its Betti numbers, is equal to the integral over M of its Euler class:[12]

[11] We will mention in Section 5.3.1 a different notion of triviality, which is not guaranteed by the vanishing of c_1.

[12] Famously, it can be computed in many other interesting ways; for example, if we write M as a union of simpler spaces, $\cup_i M_i$, $\chi(M) = \sum_i (-1)^{\dim(M_i)} \chi(M_i)$.

$$\chi(M) \equiv \sum_{k=1} (-1)^k h_k(M) = \int_M e(TM)\,. \tag{4.2.72}$$

This is an example of an *index*, counting with alternating signs the zero-modes of an operator, in this case the exterior differential d. For example, for a Riemann surface Σ_g we have (4.1.129) and $e = c_1$; (4.2.72) reads

$$2 - 2g = \int_\Sigma e(T\Sigma_g)\,. \tag{4.2.73}$$

This is in fact just (1.1.11). We will see more explicitly why e is the Ricci scalar in Section 4.3.3; for now, let us stress that TM is a real bundle, so at this stage we haven't defined its Chern classes yet. This will happen in the next chapter, when we will learn that sometimes we can split TM as a sum of two complex vector bundles.

Another example is the *Hirzebruch signature theorem* for the signature $\sigma(M_{2k})$ (defined after (4.1.127)):

$$\sigma(M_{2k}) = \int_{M_{2k}} L(TM)\,, \tag{4.2.74}$$

with L defined by the splitting principle (4.2.67) as $2^k \Pi_i (x_i/2)(\tanh(x_i/2))^{-1}$.

Topology of $d = 6$ manifolds

We are now ready to state an important classification result for manifolds in $d = 6$. *Wall's theorem* [120] characterizes all six-dimensional manifolds that are simply connected and whose cohomology has no torsion (no cyclic groups $\mathbb{Z}_p$). They are in one-to-one correspondence with the following data:

- Two abelian groups: $H^2 = \mathbb{Z}^{h_2}$, $H^3 = \mathbb{Z}^{h_3}$ (the latter with even dimension)
- A trilinear map $D\colon H^2 \times H^2 \times H^2 \to \mathbb{Z}$
- A homomorphism $p_1 : H^2 \to \mathbb{Z}$

such that

$$D(x,x,y) = D(x,y,y) \ (\mathrm{mod}\ 2)\,, \qquad p_1(x) = 4D(x,x,x) \ (\mathrm{mod}\ 24)\,. \tag{4.2.75}$$

Given one set of such data, there always exists a simply connected torsionless six-manifold M_6 whose second and third cohomology groups are H^2 and H^3, whose triple intersection form (4.1.138) is D, and whose Pontryagin class is p_1 (upon using the isomorphism $H^2 \cong H^4$).

4.2.6 T-duality on torus bundles

In Section 1.4.2, we discussed T-duality for a simple product geometry (1.4.12); among other results, we saw that it inverts the size of the circles it acts on, and that it exchanges winding and momentum modes. We are now ready to consider a generalization where spacetime is a fibration with $F = T^k$; so there are $y^i \sim y^i + 2\pi$ periodic coordinates. The metric is of the bundle metric form (4.2.35).

In the supergravity approximation, T-duality is taken to act on a geometry where momentum modes are switched off, so that they don't generate winding modes, which are massive. This means that we consider only spaces with an abelian group $\mathrm{U}(1)^k$ of continuous symmetries: the Lie derivative $L_{\partial_{y_i}}$ annihilates all fields, which then have no dependence on the y^i. T-duality on these T^k-bundles can be worked out with a generalization of the world-sheet argument in Section 1.4.2 [98].

O(d,d) action

We begin by describing the action on a torus T^d. This is most conveniently done in terms of the group $\mathrm{O}(d,d) = \{O|O^t \mathcal{I} O = \mathcal{I}\}$, where $\mathcal{I} = \begin{pmatrix} 0 & 1_d \\ 1_d & 0 \end{pmatrix}$ [121–124]. We already encountered this in Section 3.2.1. $\mathcal{I}$ appeared there as the natural pairing of vector fields and one-forms, which we can now interpret as the natural metric on the generalized tangent bundle $T \oplus T^*$; the Lie algebra $\mathrm{o}(d,d)$ was acting on forms in the spinorial representation (3.2.13). Here we need its subgroup

$$\mathrm{O}(d,d,\mathbb{Z}) \subset \mathrm{O}(d,d)\,, \tag{4.2.76}$$

defined by imposing all entries to be integer. Its action on the metric g and B field can be summarized as

$$\mathcal{G} \to \tilde{\mathcal{G}} = O\mathcal{G}O^t\,, \qquad \mathcal{G} \equiv \begin{pmatrix} g - Bg^{-1}B & Bg^{-1} \\ -g^{-1}B & g^{-1} \end{pmatrix}. \tag{4.2.77}$$

Equivalently, it acts on $E_{mn} \equiv (g+B)_{mn}$ with a nonlinear action reminiscent of the Möbius transformations in (1.2.38):

$$E \to \tilde{E} = (\alpha_{11}E + \alpha_{12})\frac{1}{\alpha_{21}E + \alpha_{22}}\,, \qquad O = \begin{pmatrix} \alpha_{11} & \alpha_{12} \\ \alpha_{21} & \alpha_{22} \end{pmatrix}. \tag{4.2.78}$$

(The equivalence of (4.2.77) and (4.2.78) is shown, for example, in [125, sec. 2.4].) The action on the dilaton is

$$\phi \to \tilde{\phi} = \phi + \frac{1}{2}\log(g/\tilde{g})\,, \tag{4.2.79}$$

where as usual $g = \det(g_{mn})$. The action on the RR fields is more complicated: it is best written in terms of the bispinor [126]

$$\mathcal{C} \to \sqrt{\det(\alpha_{21}E + \alpha_{22})}\,\Lambda\,\mathcal{C}\,, \tag{4.2.80}$$

where Λ is the spinor representation (Section 2.1.1) of $e^a_M((\alpha_{21}E+\alpha_{22})^{-1}(-\alpha_{21}E^t + \alpha_{22}))^{MN}e^b_N$. (We will consider spinors in curved space in Section 4.3, but T^d is flat.)

T-duality along some directions, say the first k, can be realized in this language by

$$\alpha_{11} = \alpha_{22} = \begin{pmatrix} 0 & 0 \\ 0 & 1_{d-k} \end{pmatrix}, \qquad \alpha_{12} = \alpha_{21} = \begin{pmatrix} 1_k & 0 \\ 0 & 0 \end{pmatrix}; \tag{4.2.81}$$

the corresponding O belongs to $\mathrm{O}(d,d,\mathbb{Z})$. Other elements belonging to it are $O = \begin{pmatrix} M & 0 \\ 0 & (M^t)^{-1} \end{pmatrix}$, $M \in \mathrm{GL}(d,\mathbb{Z})$, which represents modular transformations such as the $\mathrm{SL}(2,\mathbb{Z})$ acting on τ in Section 1.4.3; and $O = \begin{pmatrix} 1 & \delta B \\ 0 & 1 \end{pmatrix}$, which are large gauge transformations for B (Section 1.3.4). Together with (4.2.81), these generate (4.2.76), which is sometimes just called *T-duality group*.

For the "full" T-duality, along all the directions in T^d, (4.2.81) instructs us to take $\alpha_{11} = \alpha_{22} = 0$, $\alpha_{12} = \alpha_{21} = 1$. Equation (4.2.78) then gives $E \to \tilde{E} = E^{-1}$; decomposing it in its symmetric and antisymmetric part, we get the NSNS fields on the dual torus $\tilde{T}^d$:

$$g \to \tilde{g} = \frac{1}{g+B}\,g\,\frac{1}{g-B}\,, \qquad B \to \tilde{B} = -\frac{1}{g+B}\,B\,\frac{1}{g-B}\,. \tag{4.2.82}$$

In particular for $d = 1$, this reproduces (1.4.14); (4.2.80) reproduces (1.4.24). More generally, for $B = 0$, $\tilde{g} = g^{-1}$. Writing this out in indices can look confusing: a lower index is mapped to an upper index along the dualized directions. But this is natural if we recall that $\mathrm{O}(d,d)$ acts on $T \oplus T^*$.

Field space

The space of metrics g and B fields on T^d can be understood as the quotient

$$\frac{\mathrm{O}(d,d)}{\mathrm{O}(d) \times \mathrm{O}(d) \times \mathrm{O}(d,d,\mathbb{Z})} . \tag{4.2.83}$$

To see this, we use the logic in (2.4.25). The action of $\mathrm{O}(d,d)$ on the space of g and B is *transitive*, namely any value can be obtained from any other (Exercise 4.3.1). The stabilizers are then all isomorphic; that of $E = g + B = 1$ is

$$O = \frac{1}{2}\begin{pmatrix} O_1 + O_2 & O_1 - O_2 \\ O_1 - O_2 & O_1 + O_2 \end{pmatrix}, \qquad O_1, O_2 \in \mathrm{O}(d) . \tag{4.2.84}$$

These satisfy also $O^t O = 1_{2d}$, so they are in $\mathrm{O}(2d) \cap \mathrm{O}(d,d) = \mathrm{O}(d) \times \mathrm{O}(d)$. Physically we should also consider equivalent two E's related by an element of the T-duality group (4.2.76), which is then part of the stabilizer. This proves (4.2.83).

Action on bundle metrics

The preceding results have a simple extension to bundle metrics (4.2.35). Formally, we still use $\mathrm{O}(d,d)$, and decompose the $d \times d$ matrices α_{ab} in blocks acting on the $F = T^k$ fiber and the $(d-k)$-dimensional base B, so that the blocks acting along B are the identity:

$$\alpha_{ab} = \begin{pmatrix} \alpha^F_{ab} & 0 \\ 0 & \delta_{ab} 1_{d-k} \end{pmatrix} . \tag{4.2.85}$$

If, for example, we apply the full T-duality, along all the directions of T^k, (4.2.81) now tells us to take $\alpha^F_{11} = \alpha^F_{22} = 0$, $\alpha^F_{12} = \alpha^F_{21} = 1_k$. Using either (4.2.77) or (4.2.78), this gives

$$\begin{aligned} &E_{mn} \mathrm{d}x^m \otimes \mathrm{d}x^n + E_{im} \mathrm{d}y^i \otimes \mathrm{d}x^m + E_{mi} \mathrm{d}x^m \otimes \mathrm{d}y^i + E_{ij} \mathrm{d}y^i \otimes \mathrm{d}y^j \\ &E_{mn} \mathrm{d}x^m \otimes \mathrm{d}x^n + E^{ij} (\mathrm{d}\tilde{y}_i + E_{mi} \mathrm{d}x^m) \otimes (\mathrm{d}\tilde{y}_j - E_{jn} \mathrm{d}x^n) . \end{aligned} \tag{4.2.86}$$

Here the tensor products are neither symmetrized nor antisymmetrized. E^{mn} is the inverse of E_{mn}. So if $k = d$, (4.2.86) becomes $E \to E^{-1}$, in agreement with (4.2.82).

If we parametrize [127]

$$B = \frac{1}{2} B_{mn} \mathrm{d}x^m \wedge \mathrm{d}x^n + b_i \wedge \left(\mathrm{d}y^i + \frac{1}{2} A^i \right) + \frac{1}{2} B_{ij} D y^i \wedge D y^j , \tag{4.2.87}$$

then (4.2.86) gives that g^F and B transform as in (4.2.82), and

$$B_i \leftrightarrow A^i , \qquad g^B_{mn} \to g^B_{mn} , \qquad B_{mn} \to B_{mn} . \tag{4.2.88}$$

(The factor of 1/2 in the b_i term in (4.2.87) can be eliminated at the cost of making (4.2.88) slightly more complicated.) The raised i index in (4.2.88) again has to do with the $\mathrm{O}(d,d)$ exchanging $T \leftrightarrow T^*$ along the fiber.

T-duality of D-brane solutions

As an application, we can T-dualize the Dp-brane solution (1.3.59). We take x^9 to be one of the parallel directions and make it compact by declaring $x^9 \sim x^9 + 2\pi$. Then we apply (4.2.86) for $d = 1$; the coefficient $g_{99} = h^{-1/2} \to h^{1/2}$, so now there are p directions (including time) multiplied by $h^{-1/2}$, and $10 - p$ multiplied by $h^{1/2}$. Superficially, this looks like the metric of a D$(p-1)$-brane, seemingly reproducing the rule (1.4.23). However, the function h is still the one appropriate for the original Dp-brane; it still depends on $9-p$ directions rather than $10-p$. To fix ideas, consider a D7 stack, transverse to directions 7, 8; (1.4.23) would predict a localized D6 stack, say at $x^9 = 0$. Recall from (1.3.61) that $h_{D6} = 1 + N r_{0\,D6}(x_7^2 + x_8^2 + x_9^2)^{-1/2}$; but this has to be changed to account for direction 9 being compact. The method of images gives

$$h = 1 + N r_{0\,D6} \sum_{k\in\mathbb{Z}} (r^2 + (x_9 - 2\pi k))^{-1/2}\,, \qquad r^2 = x_7^2 + x_8^2\,. \tag{4.2.89}$$

But T-duality produces a solution with $h = h_{D7} = -\frac{g_s N}{2\pi}\log(r/r_{0\,D7})$, from (1.3.63).

To find agreement, we should not localize all the N D6 at $x^9 = 0$, but at various locations along direction 9. When N is large, this configuration can be approximated by a continuous distribution. This is much like the many point-like electric charges on a charged electric wire, whose electric potential is indeed a logarithm. Such a brane distribution is also called a *smeared* D-brane. This is an artifact of supergravity; it is believed that string corrections would unsmear the dualized branes. This was demonstrated for NS5-branes, where a similar issue presents itself, but we have better control over the world-sheet description, since there are no RR fields [128].

Topology change

The connection A^i has a topological meaning; so (4.2.88) implies that the dual bundle $\tilde{M}_d$ has in general a different topology from the original M_d. For $F = S^1$, the c_1 of the bundle is exchanged with the cohomology class of $\iota_{\partial_y} H$ [129].

To illustrate this phenomenon, we consider a simple example [130]: we begin with $M_3 = T^3$ with flat metric $ds^2 = dx_i dx_i$, and $B = a x^1 dx^2 \wedge dx^3$. Flux quantization (1.3.50b) imposes $2\pi a \equiv N \in \mathbb{Z}$. If we T-dualize along direction 3, we can use (4.2.86) or (4.2.88) to obtain

$$d\tilde{s}^2 = dx_1^2 + dx_2^2 + (d\tilde{x}_3 + a x_1 dx_2)^2\,, \qquad \tilde{B} = 0\,. \tag{4.2.90}$$

This is a bundle metric (4.2.35) $S^1 \to \tilde{M}_3 \hookrightarrow T^2$, with $A = a x_1 dx_2$. The definition (4.2.57) reveals $c_1 = n$. The explicit appearance of x_1 in the metric might raise concerns that the metric is not periodic in that direction, but in fact this is just the usual nontrivial gauge transformation (4.2.11) of a connection. Equivalently, the metric at $x_1 + 2\pi$ is glued to that at x_1 by a diffeomorphism of the total space. We will analyze this geometry in more detail in Section 4.4.4, where we will interpret it as a *nilmanifold*.

T-folds

Since ∂_2 is an isometry of (4.2.90), it is natural to wonder what happens if we T-dualize in that direction as well [131]. This gives

$$d\hat{s}^2 = dx_1^2 + \frac{1}{1 + a^2 x_1^2}(d\tilde{x}_2^2 + d\tilde{x}_3^2)\,, \qquad B = -\frac{a x_1}{1 + a^2 x_1^2} d\tilde{x}_2 \wedge d\tilde{x}_3\,. \tag{4.2.91}$$

This time the fields are really no longer periodic under $x_1 \to x_1 + 2\pi$. This would not make sense in ordinary differential geometry. It does make sense in string theory, precisely because of T-duality: the fields at $x_1 + 2\pi$ can be glued to those at x_1 by an element of SO(3, 3):

$$\alpha_{ab} = \begin{pmatrix} \delta_{ab} & 0 \\ 0 & \alpha^{(2)}_{ab} \end{pmatrix}, \qquad \alpha^{(2)}_{11} = \alpha^{(2)}_{22} = 1_2\,, \qquad \alpha^{(2)}_{12} = 0\,, \qquad \alpha^{(2)}_{21} = N\left(\begin{smallmatrix} 0 & 1 \\ -1 & 0 \end{smallmatrix}\right). \tag{4.2.92}$$

(For more intuition behind this, see Exercise 4.2.9.) This is called a *T-fold*: a field configuration obtained using transition functions in the T-duality group rather than diffeomorphisms alone.

String theory solutions that include T-folds rather than manifolds are a very interesting extension of the usual geometrical ideas in this chapter. A general problem is that a candidate solution often has regions where some directions are very small. This is not an issue if one has a world-sheet model [132], but it can prevent us from using the supergravity approximation.

Exercise 4.2.1 Compute the Lie bracket $[D_m, D_n]$ of two horizontal vector fields. Use the Frobenius theorem (4.1.95) to show that a covariantly constant section exists if and only if the bundle curvature vanishes.

Exercise 4.2.2 Show that TM is trivial if and only there exists a set of vector fields $\{v_1, \ldots, v_d\}$ that are a basis at every point. (Hint: consider the frame bundle FM associated to TM, and recall the triviality criterion for principal bundles.) M is said to be *parallelizable*.

Exercise 4.2.3 Show (4.2.64) for $k = 2$.

Exercise 4.2.4 Consider S^3, written as the total space of the Hopf bundle (4.2.54). We know that $H_2(S^3)$ is trivial (Section 4.1.10). But why is the base of (4.2.54) not a nontrivial generator of this homology group? (Hint: recall the criterion for a principal bundle to be nontrivial.)

Exercise 4.2.5 Show that $c_1(E) = c_1(\det E)$ for a complex vector bundle E.

Exercise 4.2.6 Using the splitting principle, compute the first few degrees of the Chern character $\mathrm{ch}(E)$ defined by (4.2.68). For example, you should get $\mathrm{ch}_0 =$ the rank of E, $\mathrm{ch}_1 = c_1$, $\mathrm{ch}_2 = (c_1^2 - 2c_2)/2$.

Exercise 4.2.7 Show that (4.2.78) implies

$$g \to \frac{1}{E^t\alpha^t_{21} + \alpha^t_{22}}\, g\, \frac{1}{\alpha_{21}E + \alpha_{22}}\,. \tag{4.2.93}$$

(Hint: $g = (E + E^t)/2$.) This generalizes (4.2.82).

Exercise 4.2.8 Show that

$$\mathrm{SO}(2,2,\mathbb{Z}) \cong \mathrm{SL}(2,\mathbb{Z}) \times \mathrm{SL}(2,\mathbb{Z}) \tag{4.2.94}$$

by finding the explicit isomorphism. (Hint: the modular transformations described after (4.2.76) are one $\mathrm{SL}(2,\mathbb{Z})$.) Show that the two factors are acting by Möbius transformations on the τ of Section 1.4.3, and $\rho \equiv \sqrt{g} + iB_{12}$. Finally, show that the two $\mathrm{SL}(2,\mathbb{Z})$ factors are interchanged by a T-duality along a single direction.

Exercise 4.2.9 Using the previous exercise, show that (4.2.92) can be represented as the action on the coordinates $\tilde{x}^2$, $\tilde{x}^3$ of

$$\rho^{-1} \to \rho^{-1} + iN\,. \tag{4.2.95}$$

4.3 Spinors on manifolds

We now turn to spinors on curved spaces, which require a subtler treatment than tensors.

4.3.1 Spinors and diffeomorphisms

Coordinate changes (or diffeomorphism) motivated us to consider vector and tensor fields (Section 4.1.3). We likewise expect to be forced to consider point-dependent spinors. We might want to write a spinorial analogue of the transformation laws (4.1.2) and (4.1.7). The problem is that the spinorial action $e^{-\lambda/2}$ of a flat-space rotation (or Lorentz transformation) e^{λ} was only defined for an element of $O(d)$; indeed, λ_{mn} was taken to be antisymmetric in (2.1.8). But the Jacobian matrix $J^m{}_n = \frac{\partial x'^m}{\partial x^n}$ is in general in $GL(d, \mathbb{R})$, not in $O(d)$.

Vielbein formalism

To find a solution to this problem, we have to back up and consider gamma matrices first. With a general Riemannian metric g_{mn}, it is natural to generalize the Clifford algebra (2.1.3) to

$$\{\gamma_m, \gamma_n\} = 2g_{mn}\, 1\,. \tag{4.3.1}$$

Since in general g_{mn} can be point dependent, the gamma matrices also have to depend on the point. There is a natural way to find such "curved" γ_m. We call *frame* a basis for the space of one-forms, or vectors. (We used this word already while defining the frame bundle in Section 4.1.3.) An *orthonormal* frame, or *vielbein*,[13] is a frame

$$e^a \equiv e^a_m \mathrm{d}x^m \tag{4.3.2}$$

such that

$$e^a_m e^b_n g^{mn} = e^a \cdot e^b = \delta^{ab}\,. \tag{4.3.3}$$

The new indices a, b appear in a flat metric, and for this reason are sometimes called *flat indices*; the usual space indices m are sometimes called *curved* or *tangent*. Flat indices are raised and lowered with the flat metric. (In the Lorentzian case, the vielbein satisfies $e^a_\mu e^b_\nu g^{\mu\nu} = \eta^{ab}$; so in that case, flat indices are raised and lowered with the Lorentzian flat metric η.) Now

$$\gamma_m \equiv \gamma_a e^a_m \tag{4.3.4}$$

satisfies (4.3.1) if the γ_a satisfy the flat Clifford algebra $\{\gamma_a, \gamma_b\} = 2\delta_{ab}\, 1$.

It is also useful to introduce the vector fields

$$E_a \equiv E^m_a \partial_m\,, \qquad E^m_a \equiv g^{mn} e_{an}\,; \tag{4.3.5}$$

in terms of these, (4.3.3) reads $e^a_m E^m_b = \delta^a_b$; in other words, E is the inverse of e. From (4.3.3), we also see that it is an orthonormal basis of vectors: $E^m_a E^n_b g_{mn} = \delta_{ab}$.

[13] This is "multileg" in German; plural *vielbeine*. The variant *vierbein* ("four-leg") is also in use in $d = 4$; the use of other German numerals is fun but less frequent.

Reading differently the statement that E is the inverse of e, we have $e^a_m E^n_a = \delta^n_m$, or in other words

$$e^a_m e_{an} = g_{mn} . \tag{4.3.6}$$

Local orthogonal (or Lorentz) transformations

A vielbein is far from unique. Indeed, we can transform

$$e^a_m \to e'^a_m = \Lambda^a{}_b e^b_m , \tag{4.3.7}$$

and if $\Lambda^a{}_b$ is an orthogonal matrix at every point, then e'^a_m is also a vielbein: it satisfies (4.3.3) and (4.3.6). Equation (4.3.7) may depend on the point, and thus are called *local orthogonal* transformations in Euclidean signature, or *local Lorentz* in Lorentzian signature.

So we have defined gamma matrices in curved space, at the cost of introducing a large degree of arbitrariness in the choice of the vielbein. One would like this choice not to appear in any physical observables; in other words, we would like local orthogonal transformations to be a gauge freedom. This is not the gauge freedom of coordinate changes; indeed, we had no need to talk about the concept of vielbein until we introduced spinors.

Consider, for example, the bilinears $\eta^\dagger \gamma_{m_1 \ldots m_k} \eta$; we defined them in flat space, and we would like them to make sense in curved spaces too, as coefficients of a k-form. Thus they should not depend on the choice of vielbein. But the curved gamma matrices (4.3.4) do change under (4.3.7). To make up for this, we have to postulate that spinors η also transform under local orthogonal transformations, so that $\eta^\dagger \gamma_{m_1 \ldots m_k} \eta$ is invariant. Since Λ is in $\mathrm{O}(d)$, we now have a natural candidate for this transformation from (2.1.8):

$$\eta \to \exp\left[-\frac{1}{2}\lambda\right] \eta , \tag{4.3.8}$$

where λ is defined by $\Lambda = \mathrm{e}^\lambda$. With this definition and the by now familiar application (2.1.16) of the Hadamard lemma, it is easy to check that the bilinears $\eta^\dagger \gamma_{m_1 \ldots m_k} \eta$ do not change under the combined (4.3.7) and (4.3.8), thus realizing our wish that the local orthogonal transformations should be a gauge freedom.

Alternatively, we can write bilinears in flat indices, e.g., $\eta^\dagger \gamma_{a_1 \ldots a_k} \eta$. These are formally identical to the bilinears in flat space we have studied in Section 3.3, since the γ_a obey the usual flat Clifford algebra. One can then multiply by $e^{a_1}_{m_1} \ldots e^{a_k}_{m_k}$ to obtain the curved-space form coefficients $\eta^\dagger \gamma_{m_1 \ldots m_k} \eta$.

Hodge dual

In terms of a vielbein, the definition of the Hodge operator is natural:

$$* : e^{a_1} \wedge \ldots \wedge e^{a_k} \mapsto \frac{1}{(d-k)!} \epsilon_{a_{k+1} \ldots a_d}{}^{a_1 \ldots a_k} e^{a_{k+1}} \wedge \ldots \wedge e^{a_d} . \tag{4.3.9}$$

This is the easiest definition both in practice and conceptually, and is equivalent to (4.1.103). Here ϵ is equal to the flat space definition:

$$\epsilon_{m_1 \ldots m_d} = \epsilon_{a_1 \ldots a_d} e^{a_1}_{m_1} \ldots e^{a_d}_{m_d} \overset{(4.1.100)}{=} \det(e)\, \epsilon^{(0)}_{m_1 \ldots m_d} , \tag{4.3.10}$$

which leads us again to (4.1.101), because (4.3.3) implies $\det g = \det(e)^2$. The volume form can also be introduced more naturally:

$$\mathrm{vol} = \frac{1}{d!}\epsilon_{a_1 \ldots a_d} e^{a_1} \wedge \ldots e^{a_d} = e^1 \wedge \ldots \wedge e^d \,. \tag{4.3.11}$$

This is indeed equivalent to (4.1.103):

$$\begin{aligned} e^1 \wedge \ldots \wedge e^d &= e^1_{m_1} \cdots e^d_{m_d} \mathrm{d}x^{m_1} \wedge \ldots \wedge \mathrm{d}x^{m_d} \\ &= e^1_{m_1} \cdots e^d_{m_d} \epsilon^{m_1 \ldots m_d}_{(0)} \mathrm{d}x^1 \wedge \ldots \wedge \mathrm{d}x^d \overset{(4.1.100)}{=} \sqrt{g} \mathrm{d}x^1 \wedge \ldots \wedge \mathrm{d}x^d \,. \end{aligned} \tag{4.3.12}$$

4.3.2 Spin structures

Following the same logic as in Sections 4.1.1–4.1.3, now that we have extended the concept of spinor to curved spaces, we want to see how to define it on manifolds.

On most manifolds, we cannot use a single vielbein everywhere. Indeed, this would mean that there are d vector fields that are linearly independent (and in particular nonzero) at each point. Manifolds where this happens are called *parallelizable* and are rather rare (Exercise 4.2.2). So in general, we have to work with a different vielbein $e^a_{(\alpha)}$ on each open set U_α. On an intersection $U_\alpha \cap U_\beta$, both $e^a_{(\alpha)}$ and $e^b_{(\beta)}$ are vielbeine for the same metric g; thus there is a local orthogonal transformation $(\Lambda_{(\alpha\beta)})^a{}_b$ such that

$$e^a_{(\alpha)} = (\Lambda_{(\alpha\beta)})^a{}_b e^b_{(\beta)} \,. \tag{4.3.13}$$

(Let us stress once again that this is separate from the need to use different coordinate systems on each U_α, which has to do with diffeomorphism invariance and not with local orthogonal transformations.) We now define a spinor field on M as a collection of spinors $\eta_{(\alpha)}$ on each U_α, such that

$$\eta_{(\alpha)} = \rho_\mathrm{s}(\Lambda_{(\alpha\beta)})\eta_{(\beta)} \,. \tag{4.3.14}$$

This condition ensures that the $\eta_{(\alpha)}$ on different patches represent the same physical object.

Recall, however, that $\rho_s(\mathrm{e}^\lambda) = \mathrm{e}^{-\frac{1}{2}\lambda}$ is determined only up to a sign ambiguity, as noted in (2.1.9) and (2.1.10). This sign choice is restricted by a consistency condition. By using (4.3.14) together with the version with α and β exchanged, one sees that $\rho_\mathrm{s}(\Lambda_{(\alpha\beta)})\rho_\mathrm{s}(\Lambda_{(\beta\alpha)}) = 1$. On a triple intersection $U_\alpha \cap U_\beta \cap U_\gamma$ we get instead

$$\rho_\mathrm{s}(\Lambda_{(\alpha\beta)})\rho_\mathrm{s}(\Lambda_{(\beta\gamma)})\rho_\mathrm{s}(\Lambda_{(\gamma\alpha)}) = 1 \,. \tag{4.3.15}$$

We can apply the same logic also to (4.3.13), obtaining

$$\Lambda_{(\alpha\beta)}\Lambda_{(\beta\gamma)}\Lambda_{(\gamma\alpha)} = 1 \,. \tag{4.3.16}$$

We might then try to derive (4.3.15) from (4.3.16) by acting with ρ_s on the latter. This indeed works for the left-hand side, since ρ_s is a representation and in particular a homomorphism, $\rho_\mathrm{s}(\Lambda\Lambda') = \rho_\mathrm{s}(\Lambda)\rho_\mathrm{s}(\Lambda')$. But the right-hand side becomes $\rho_\mathrm{s}(1) \equiv s_{\alpha\beta\gamma}$, which, according to our discussion around (2.1.9), (2.1.10) can be either ± 1. So we have not quite obtained the desired (4.3.15). We can try to fix this by using the sign ambiguity in $\rho_\mathrm{s}(\Lambda_{(\alpha\beta)})$, flipping its sign by some $s_{\alpha\beta}$ such that $s_{\alpha\beta}s_{\beta\gamma}s_{\gamma\alpha} = s_{\alpha\beta\gamma}$.

On an orientable manifold, this can be solved if and only if the first and second Stiefel–Whitney classes (Section 4.2.5) vanish:

$$w_1(TM) = w_2(TM) = 0\,. \tag{4.3.17}$$

The upshot of this discussion is that (4.3.17) is equivalent to the consistency condition (4.3.15), which is needed to define spinors. Such an M is said to be a *spin manifold*. The $\rho_s(\Lambda_{(\alpha\beta)})$ then become the transition function of a bundle that we call the *spinor bundle* SM, of which spinors are sections.

4.3.3 Spin connection

Our discussion was so far limited to the algebraic properties of spinors. We now wonder how derivatives act on a vielbein. One possible definition of ∇ was in terms of its action (4.1.17) on the one-form basis dx^m. The vielbein e^a is an alternative one-form basis, and we can also use it to define the action of ∇ in a similar way:

$$\nabla_m e^a = -\omega_m^{ab} e_b\,, \tag{4.3.18}$$

with some new coefficients ω_m^{ab} collectively called *spin connection*, for a reason we will see shortly. (Recall that the flat a, b indices are raised and lowered by the flat metric; the e_b are still one-forms.)

Spinorial covariant derivative

Under the infinitesimal analogue of (4.3.7), $\delta e^a = \lambda^a{}_b e^b$, consistency with (4.3.18) demands

$$\delta\omega^{ab} = -d\lambda^{ab} + [\lambda, \omega]^{ab}\,; \qquad \omega^{ab} \equiv \omega_m^{ab} dx^m\,, \qquad [\lambda, \omega]^{ab} = \lambda^{ac}\omega_c{}^b - \omega^{ac}\lambda_c{}^b\,. \tag{4.3.19}$$

This resembles the transformation law (4.1.10) of the connection, in that it is not just linear in ω, but also involves a derivative of λ. It can be used to solve a similar problem that (4.1.10) solved. If under the local orthogonal transformations (4.3.7) a spinor transforms as in (4.3.8), its derivative $\partial_m\psi$ does not transform in the same way, because the term $(\partial_m\Lambda)\psi$ appears. But the *spinorial covariant derivative*

$$D_m\eta \equiv \left(\partial_m + \frac{1}{4}\omega_m^{ab}\gamma_{ab}\right)\eta \tag{4.3.20}$$

does transform well. (This is easiest to check infinitesimally, using (4.3.19), the infinitesimal analogue of (4.3.8), and (2.1.6).)

We can also define the *Dirac operator*:

$$D \equiv \gamma^m D_m\,. \tag{4.3.21}$$

Its index is defined as the difference between its positive- and negative-chirality zero modes; a famous result gives it as [133]

$$\operatorname{ind}(D) = \int_M \hat{A}(M)\,, \tag{4.3.22}$$

where the $\hat{A}$ class is defined via the splitting principle (4.2.67) as $\hat{A} = \Pi_i(x_i/2)(\sinh(x_i/2))^{-1}$; it can be expanded as a sum of Pontryagin classes p_i. Not coincidentally, this class also appears in gravitational anomalies [134].

Structure equations

Recalling (4.3.3),

$$-\nabla_m(e^a \cdot e^b) = \omega_m^{ac} e_c \cdot e^b + e^a \cdot \omega_m^{bc} e_c = \omega_m^{ab} + \omega_m^{ba} = 0\,. \tag{4.3.23}$$

So ω_m^{ab} is antisymmetric in its two flat indices, and we have to pay attention to their order, especially when one of them is lowered. In contrast, the curved index m is to be thought of as independent with respect to a and b, and their relative order will be immaterial.

By linearity and (4.1.6),

$$e^a(E_b) = e^a_m E^n_b \mathrm{d}x^m(\partial_n) = e^a_m E^n_b \delta^m_n = e^a_m E^m_b = \delta^a_b\,, \tag{4.3.24}$$

so the action of ∇_m on the E_a is again determined by demanding that the Leibniz identity holds:

$$\nabla_m E^a = -\omega_m^{ab} E_b\,. \tag{4.3.25}$$

Equation (4.3.18) can be taken as a definition of ∇_m, just as good as (4.1.17); but they better be consistent. Writing $e^a = e^a_m \mathrm{d}x^m$ and using (4.1.17):

$$\nabla_m(e^a_n \mathrm{d}x^n) = (\partial_m e^a_n - \Gamma^p_{mn} e^a_p)\mathrm{d}x^n \tag{4.3.26}$$

and comparing with (4.3.18), we see

$$\partial_m e^a_n - \Gamma^p_{mn} e^a_p + \omega_m^{ab} e_{bn} = 0\,. \tag{4.3.27}$$

This consistency condition relates the Γ^p_{mn} to the ω_m^{ab}; it expresses the fact that they are the coefficients of the same connection ∇_m, acting on two different bases of one-forms. Another interpretation of (4.3.27) is that, if we define ∇ to be covariant both with respect to the m index and to the a index, e^a_m is covariantly constant with respect to it.

If we multiply (4.3.27) by $\mathrm{d}x^m \wedge \mathrm{d}x^n$, which is by definition antisymmetric, the Γ^p_{mn} term disappears, and we obtain

$$\mathrm{d}e^a + \omega^{ab} \wedge e_b = 0\,. \tag{4.3.28}$$

Often, solving this equation (recalling the constraint that $\omega^{ab} = -\omega^{ba}$) is the best way to find ω. Sometimes it can also be practical to have an explicit formula. For that, multiply (4.3.27) by a further e_{ap}, using (4.1.16) and (4.3.3). This results in

$$\omega_m^{ab} e_{an} e_{bp} \equiv \omega_{m\,np} = (\partial_m e_{a[n}) e^a_{p]} - \partial_{[n} g_{p]m}\,. \tag{4.3.29}$$

This is often even faster to compute than (4.1.16) itself. It is particularly useful for the spinorial covariant derivative (4.3.20), since $\omega_{m\,np}\gamma^{np} = \omega_m^{ab}\gamma_{ab}$.

Consider for a moment a connection $\tilde{\nabla}_m$, which is still compatible with the metric, but with nonzero torsion, $\tilde{T}^p_{mn} = \tilde{\Gamma}^m_{[np]} \neq 0$. We can still define coefficients $\tilde{\omega}_m^{ab}$

via (4.3.18), and the preceding discussion is almost unchanged; only, when we antisymmetrize (4.3.27), now we get $\partial_{[m}e^a_{n]} - \tilde{\Gamma}^p_{[mn]}e^a_p + \tilde{\omega}^{ab}_{[m}e_{b|n]} = 0$, or

$$de^a + \tilde{\omega}^{ab} \wedge e_b = \tilde{T}^a , \tag{4.3.30}$$

where $\tilde{T}^a \equiv dx^m \wedge dx^n \tilde{\Gamma}^p_{mn} e^a_p = dx^m \wedge dx^n \tilde{T}^p_{mn} e^a_p$. So (4.3.28) is another way of saying that the torsion of the Levi-Civita connection ∇_m is zero.

Let us now take a commutator of derivatives from (4.3.18):

$$\begin{aligned} \frac{1}{2}[\nabla_m, \nabla_n]E^a &= \nabla_{[m}\nabla_{n]}E^a = -\nabla_{[m}(\omega^{ab}_{n]}E_b) \\ &= -(\partial_{[m}\omega^{ab}_{n]})E_b - \omega^{ab}_{[n}(\nabla_{m]}E_b) = \left(-\partial_{[m}\omega^{ac}_{n]} + \omega^{ab}_{[n}\omega_{m]b}{}^c\right) E_c . \end{aligned} \tag{4.3.31}$$

We know from (4.1.13) that $[\nabla_m, \nabla_n]$ acts linearly and algebraically on vectors; multiplying by e^a and by $dx^m \wedge dx^n$, we then get[14]

$$d\omega^{ab} + \omega^{ac} \wedge \omega_c{}^b = \hat{R}^{ab} \equiv \frac{1}{2} e^a_p e^b_q R^{pq}{}_{mn} dx^m \wedge dx^n . \tag{4.3.32}$$

Together, (4.3.28) and (4.3.32) are called the *Cartan structure equations*.

Equation (4.3.32) shows that (4.1.21) also holds for the spinorial representation. Indeed

$$\begin{aligned} [D_m, D_n] &= \left[\partial_m + \frac{1}{4}\omega^{ab}_m \gamma_{ab}, \partial_n + \frac{1}{4}\omega^{cd}_n \gamma_{cd}\right] \\ &= \frac{1}{4}(\partial_m \omega^{ab}_n - \partial_n \omega^{ab}_m)\gamma_{ab} + \frac{1}{16}\omega^{ab}_m \omega_{ncd}[\gamma_{ab}, \gamma^{cd}] \\ &\overset{(2.1.6)}{=} \frac{1}{2}\left(\partial_{[m}\omega^{ab}_{n]} + \omega^{[a|c}_m \omega_{nc}{}^{b]}\right)\gamma_{ab} \overset{(4.3.32)}{=} \frac{1}{4}R^{ab}{}_{mn}\gamma_{ab} . \end{aligned} \tag{4.3.33}$$

In (4.3.32), the Riemann tensor is viewed as a two-form (the curved mn indices) valued in the Lie algebra so(d) (the flat ab indices). This offers a way to represent the Pontryagin classes of TM, which are defined in general for bundles with structure group O(n) (Section 4.2.5). So, for example, p_1 is represented by Tr$F^2/8\pi^2$, which for TM is then

$$R^{ab} \wedge R_{ab} . \tag{4.3.34}$$

As another example, the Euler class $e(TM)$ is obtained as the Pfaffian on flat indices; in $d = 2$, this is $R^{12} = \frac{1}{2}R^{12}{}_{mn}dx^m \wedge dx^n$. This is in fact just proportional to the Ricci scalar, since all nonzero components of the Riemann tensor are proportional: $R = R^{ab}{}_{mn}E^m_a E^n_b = R^{12}{}_{12}\det E$. Using this, (4.2.73) reduces to (1.1.11).

A Leibniz identity

As a cross-check, we now return to (4.3.20) to show that it plays well with the covariant derivative of tensors. Consider, for example, the bilinear $\eta^\dagger \gamma_m \eta$. It is a one-form, so its covariant derivative reads

[14] The hat on $\hat{R}^{ab}$ is meant to dispel possible confusion with the Ricci tensor.

$$
\begin{aligned}
\nabla_m(\eta^\dagger\gamma_n\eta) &= \partial_m(\eta^\dagger\gamma_n\eta) - \Gamma^p_{mn}(\eta^\dagger\gamma_p\eta) \\
&= (\partial_m\eta)^\dagger\gamma_n\eta + \eta^\dagger(\partial_m e^a_n)\gamma_a\eta + \eta^\dagger\gamma_n\partial_m\eta - \Gamma^p_{mn}\eta^\dagger e^a_p\gamma_a\eta \\
&\overset{(4.3.27)}{=} (\partial_m\eta)^\dagger\gamma_n\eta + \eta^\dagger\gamma_n\partial_m\eta - \omega^{ab}_m e_{bn}\eta^\dagger\gamma_a\eta \qquad (4.3.35)\\
&\overset{(2.1.13)}{=} (\partial_m\eta)^\dagger\gamma_n\eta + \eta^\dagger\gamma_n\partial_m\eta + \frac{1}{4}\omega^{ab}_m\eta^\dagger[\gamma_n,\gamma_{ab}]\eta \\
&\overset{(4.3.20)}{=} (D_m\eta)^\dagger\gamma_n\eta + \eta^\dagger\gamma_n(D_m\eta)\,.
\end{aligned}
$$

This is a sort of Leibniz identity connecting the spinorial and ordinary (tensorial) covariant derivatives; it will be very useful. It can be extended to more general bilinears $\eta^\dagger\gamma^{m_1\ldots m_k}\eta$ with the same method, but in Section 4.3.5, we will present a result on bispinors (Section 3.4) that extends (4.3.35) to all bilinears.

4.3.4 Lie derivatives and spinors

In Section 4.1.5, we have considered Lie derivatives of tensor fields. What about spinors? In Section 4.3.1, we saw that making spinors transform under diffeomorphisms would not be obvious, since the Jacobian matrix $\frac{\partial x'^m}{\partial x^n}$ is in $\mathrm{GL}(d,\mathbb{R})$ rather than $\mathrm{O}(d)$. Defining gamma matrices involves local orthogonal transformations (4.3.7), a second type of gauge freedom whose gauge group is instead naturally $\mathrm{O}(d)$. Spinors need to transform under this, but under diffeomorphisms they could transform perfectly well as functions. In a sense, spinorial indices should be thought of as having to do with the flat a indices, not with the curved m indices. $\rho^m{}_n$ in (4.1.60) should then be zero, and the Lie derivative of spinors should simply be the partial derivative $L^0_v\eta \equiv v^n\partial_n\eta$ that we had for functions in (4.1.54). This satisfies Leibniz identities with respect to the tensor Lie derivatives: for example,

$$
L_v(\eta^\dagger\gamma_n\eta) = (L^0_v\eta)^\dagger\gamma_n\eta + \eta^\dagger\gamma_n(L^0_v\eta) + \eta^\dagger(L_v e^a_n)\gamma_a\eta\,. \qquad (4.3.36)
$$

A first sign that this definition is not fully satisfactory is the flat space example: there we know that spinors do transform under rotations, and rotations are particular cases of diffeomorphisms. The source of the problem is that the vielbein itself transforms under diffeomorphisms; this should be reflected in an additional variation of the spinor. The variations $\delta e^a_m = L_v e^a_m$ are d vectors that can be expanded in the vielbein basis; so we can also find coefficients λ^{ab} such that

$$
L_v e^a_m = \lambda^{ab} e_{bm}\,. \qquad (4.3.37)
$$

Since $L_v g_{mn} = L_v(e^a_m e_{an}) = 2\lambda_{(ab)}e^a_m e^b_n$, these coefficients are antisymmetric when v is Killing. We can compute them explicitly:

$$
\lambda^{ab} = (L_v e^a_m)E^{bm} \overset{(4.1.61)}{=} (v^n\nabla_n e^a_m + \nabla_m v^n e^a_n)E^{bm} \overset{(4.3.18)}{=} -v^n\omega^{ab}_n + \nabla_m v_n e^{bm}e^{an}\,. \qquad (4.3.38)
$$

As a cross-check, symmetrizing we get $\lambda^{(ab)} = \nabla_{(m}v_{n)}e^{am}e^{bn}$, which gives $L_v g_{mn} = 2\nabla_{(m}v_{n)}$, in agreement with (4.1.64).

In view of (4.3.8), this vielbein variation should be included in the Lie derivative; this leads to the *Lie–Kosmann* derivative [135]:

$$L_v\eta = v^m\partial_m\eta - \frac{1}{4}\lambda_{ab}\gamma^{ab}\eta = v^m D_m\eta + \frac{1}{4}\nabla_m v_n\gamma^{mn}\eta \\ \stackrel{(4.1.72)}{=} v^m D_m\eta + \frac{1}{4}\partial_m v_n\gamma^{mn}\eta. \tag{4.3.39}$$

Since (4.3.39) now takes into account the vielbein transformation, we expect that the last term in (4.3.36) might be dropped. Indeed, substituting there $L_v^0 = L_v + \frac{1}{4}\lambda_{ab}\gamma^{ab}$ generates two terms with λ, which together form a commutator $-\frac{1}{4}\lambda_{ab}[\gamma^{ab},\gamma_n] = \lambda_{ab}e_n^{[a}\gamma^{b]} = \lambda_{[nb]}\gamma^b$. This reassembles nicely with the third term in (4.3.36). If v is Killing, λ_{ab} is antisymmetric, and the two terms cancel out; more generally, the symmetric $\lambda_{(mn)} = \frac{1}{2}L_v g_{mn}$ remains. All in all, (4.3.36) becomes

$$L_v(\eta^\dagger\gamma_n\eta) = (L_v\eta)^\dagger\gamma_n\eta + \eta^\dagger\gamma_n(L_v\eta) + \frac{1}{2}(L_v g_{mn})\eta^\dagger\gamma^m\eta\,. \tag{4.3.40}$$

The reason our definition (4.3.39) did not reabsorb also the transformation of the metric is that the $\lambda\eta$ term was conceived to represent an element of $\mathrm{o}(d)$ on η, not of $\mathrm{gl}(d,\mathbb{R})$; indeed, the symmetric part of λ^{ab} cancels from it.

When v is Killing, the last term in (4.3.40) vanishes and we have a Leibniz identity simpler than (4.3.36). Even when v is not Killing, (4.3.40) is more convenient. The slogan is that to transform a bilinear of a spinor, we need to transform the spinor only, and additionally take into account the transformation of the metric if the vector is not Killing.

Equation (4.3.40) could be extended to higher-form bilinears $\eta\gamma^{m_1\ldots m_k}\eta$, but we will do so using bispinors in the next subsection.

4.3.5 Bispinors on manifolds

In Section 3.4, we saw that spinor bilinears are often handled most conveniently by using bispinors. It is then useful to be able to use bispinors on manifolds as well.

Perhaps the most important subtlety with respect to flat space is that the Clifford map (3.1.13) relating forms to bispinors now depends on the metric, because the Clifford algebra (4.3.1) does. Raising an index to (4.3.4), $\gamma^m = E_a^m\gamma^a$. Under a vielbein variation,

$$\delta \not{dx}^m = \delta\gamma^m = \delta E_a^m\gamma^a = \delta E_a^m e_n^a\gamma^n = (\delta E_a^p e_n^a \mathrm{d}x^n \wedge \iota_p \mathrm{d}x^m)_/\,. \tag{4.3.41}$$

Extending this to antisymmetrized products $\gamma^{m_1\ldots m_k}$ leads to a general formula for the variation of a bispinor under change of vielbein:

$$\delta\not{\alpha} = (\delta E_a^m e_n^a \mathrm{d}x^n \wedge \iota_m\alpha)_/ = (e^a \wedge \iota_{\delta E_a}\alpha)_/\,. \tag{4.3.42}$$

(As in flat space, here α is a form, while $\not{\alpha}$ or $(\alpha)_/$ is the bispinor associated to it by the Clifford map (3.1.13).)

This plays a role for example when deriving $\not{\alpha}$. We have a contribution from the variation of the coefficients of the form α, and a contribution from (4.3.42):

$$\partial_m\not{\alpha} = (\partial_m\alpha + e_n^a\partial_m E_a^p \mathrm{d}x^n \wedge \iota_p\alpha)_/\,. \tag{4.3.43}$$

To define a covariant derivative for bispinors, we can start from a tensor product $\eta^1 \otimes \eta^{2\,\dagger}$. In this case, to enforce Leibniz it is natural to define

$$D_m(\eta^1 \otimes \eta^{2\,\dagger}) \equiv (D_m\eta^1) \otimes \eta^{2\,\dagger} + \eta^1 \otimes (D_m\eta^2)^\dagger = \partial_m(\eta^1 \otimes \eta^{2\,\dagger}) + \frac{1}{4}\omega_m^{ab}[\gamma_{ab}, \eta^1 \otimes \eta^{2\,\dagger}]\,. \tag{4.3.44}$$

But any bispinor can be obtained as a sum of tensor products of this type; so we define in general

$$D_m \slashed{\alpha} \equiv [D_m, \slashed{\alpha}] = \partial_m \slashed{\alpha} + \frac{1}{4}\omega_m^{ab}[\gamma_{ab}, \slashed{\alpha}]. \tag{4.3.45}$$

We wonder how this is related to the covariant derivative $\nabla_m \alpha$ of the form. We can compare the two explicitly:

$$\begin{aligned} D_m \slashed{\alpha} &\overset{(3.2.19)}{=} (\partial_m + \omega_m^{ab} e_a \wedge \iota_{e_b})\slashed{\alpha} \\ &\overset{(4.3.43),(4.3.5)}{=} (\partial_m \alpha - (\omega_m^{ab} e_{bn} + \partial_m e^a_n) e_{ap} \mathrm{d}x^n \wedge \iota^p \alpha)_/ \\ &\overset{(4.3.27)}{=} (\partial_m \alpha - \Gamma^p_{mn} \mathrm{d}x^n \wedge \iota_p \alpha)_/ \\ &\overset{(4.1.23)}{=} \slashed{\nabla_m \alpha}\,. \end{aligned} \tag{4.3.46}$$

So in an appropriate sense the covariant derivative operator commutes with the Clifford map $\alpha \mapsto \slashed{\alpha}$. This extends to arbitrary form degree the one-form result in (4.3.35). An important example of (4.3.46) is

$$[D_m, \gamma^n] = -\Gamma^n_{mp}\gamma^p\,, \tag{4.3.47}$$

which agrees with (4.1.17); see also Exercise 4.3.12.

A similar issue presents itself with Lie derivatives. Once again, we start by enforcing Leibniz on a bispinor of the type $\eta^1 \otimes \eta^{2\,\dagger}$:

$$\begin{aligned} L_v(\eta^1 \otimes \eta^{2\,\dagger}) &\equiv (L_v\eta^1) \otimes \eta^{2\,\dagger} + \eta^1 \otimes (L_v\eta^2)^\dagger \\ &\overset{(4.3.39)}{=} v^m D_m(\eta^1 \otimes \eta^{2\,\dagger}) + \frac{1}{4}\nabla_m v_n[\gamma^{mn}, \eta^1 \otimes \eta^{2\,\dagger}]\,. \end{aligned} \tag{4.3.48}$$

Hence by linearity we again define in general

$$L_v \slashed{\alpha} = v^m D_m \slashed{\alpha} + \frac{1}{4}\nabla_m v_n[\gamma^{mn}, \slashed{\alpha}]\,. \tag{4.3.49}$$

We now compare with the Lie derivative for forms:

$$\begin{aligned} L_v \slashed{\alpha} &\overset{(3.2.19)}{=} v^m D_m \slashed{\alpha} + (\nabla_{[m} v_{n]} \mathrm{d}x^m \wedge \iota^n \alpha)_/ \\ &\overset{(4.3.46)}{=} (v^m \nabla_m \alpha + (\nabla_m v_n - \nabla_{(m} v_{n)}) \mathrm{d}x^m \wedge \iota^n \alpha)_/ \\ &\overset{(4.1.63),(4.1.64)}{=} \left(L_v \alpha - \frac{1}{2}(L_v g_{mn}) \mathrm{d}x^m \wedge \iota^n \alpha\right)_/\,. \end{aligned} \tag{4.3.50}$$

This extends to arbitrary forms the result in (4.3.40).

As a simple example, let us consider constant spinors η^a in flat space $\mathbb{R}^d$, with the vector

$$r\partial_r = x^m \partial_m \,. \tag{4.3.51}$$

This is known as *Euler vector field* and will resurface later on with more interesting manifolds. It is a conformal Killing vector for the flat metric: $L_{r\partial_r} g_{mn} = 2g_{mn}$. It is immediate to check that $L_v \eta^a = 0$ in this case. Defining $\phi^{ab} \equiv \eta^a \otimes \eta^{b\dagger}$,

$$0 \overset{(4.3.48)}{=} L_v(\phi^{ab})_/ \overset{(4.3.50)}{=} (L_v \phi^{ab} - g_{mn} dx^m \iota^n \phi^{ab})_/ \overset{(3.2.22)}{=} \sum_k (L_v \phi^{ab}_k - k\phi^{ab}_k)_/, \tag{4.3.52}$$

where as usual $_k$ denotes the k-form part. So in this case,

$$L_v \phi^{ab}_k = k \phi^{ab}_k \,. \tag{4.3.53}$$

Intuitively, the spinors don't transform at all under v, and the only reason the bilinears transform is that the vielbein has degree one under rescaling, and so a k-form has degree k.

As a subexample, consider a single spinor η_+ in flat space $\mathbb{R}^6$. This is familiar from Sections 2.4.2 and 3.4.2, where we saw that its bilinears consist of a constant two-form $J^0_{mn} = -i\eta_+^\dagger \gamma_{mn} \eta_+$ and $\Omega^0_{mnp} = \eta_-^\dagger \gamma_{mnp} \eta_+$, with $\eta_- = \eta_+^c$. Now (4.3.53) gives

$$(L_v J^0)_{mn} = 2J^0_{mn} \,, \qquad (L_v \Omega^0)_{mnp} = 3\Omega^0_{mnp} \,. \tag{4.3.54}$$

These can indeed be checked directly from (3.4.12) and (3.4.17).

4.3.6 Anholonomy coefficients

We have given an explicit expression for the spin connection in (4.3.29). We will now see an alternative formula, which has perhaps a clearer geometrical meaning.

Since the e^a are a basis for one-forms at every point, the $e^a \wedge e^b$ are a basis for two-forms. The exterior derivatives de^a can be expanded in this basis:

$$de^a = \frac{1}{2} F^a{}_{bc} e^b \wedge e^c \,. \tag{4.3.55}$$

The $F^a{}_{bc}$ are called *anholonomy coefficients*. They also have a dual interpretation in terms of the vector fields E_a. To see this:

$$\iota_{[E_b,E_c]} e^d \overset{(4.1.81)}{=} [\{\iota_{E_b}, d\}, \iota_{E_c}] e^d = \{\iota_{E_b}, d\} \delta^d_c - \iota_{E_c} \{\iota_{E_b}, d\} e^d = -\iota_{E_c} \iota_{E_b} de^d \,, \tag{4.3.56}$$

where we have used $\iota_{E_a} e^b = \delta^b_a$ (which are annihilated both by d and by ι_{E_c}). We can now compute $\iota_{E_c} \iota_{E_d} e^e \wedge e^f = \iota_{E_c} \left(-e^e \wedge \iota_{E_d} + \{\iota_{E_d}, e^e \wedge\}\right) e^f = -\delta^e_c \delta^f_d + \delta^f_c \delta^e_d$, and using (4.3.55), we arrive at

$$[E_b, E_c] = -F^a{}_{bc} E_a \,. \tag{4.3.57}$$

In other words, the $F^a{}_{bc}$ are the "structure constants" for the algebra of the vector fields on M. The reason for the quotes is that actually in general the $F^a{}_{bc}$ depend on M; in other words, they are structure constants for an infinite-dimensional Lie algebra. The $F^a{}_{bc}$ become actually constant only when there is some group-theoretical reason for that to happen, such as when M is a Lie group, a case we will analyze in Section 4.4.1.

We now use (4.3.55) in (4.3.28). Both terms are linear combinations of the basis $e^b \wedge e^c$, and so

$$\frac{1}{2}F^a{}_{bc} + \omega^a{}_{[bc]} = 0\,, \qquad \omega^a{}_b{}^c \equiv \omega^{ac}_m E^m_b\,. \tag{4.3.58}$$

We also know $\omega^a{}_b{}^c = -\omega^c{}_b{}^a$, so (4.3.58) becomes $F^a{}_b{}^c = -\omega^a{}_b{}^c + \omega^{ac}{}_b = -\omega^a{}_b{}^c - \omega_b{}^{ca}$; antisymmetrizing in ac,

$$F^{[a}{}_b{}^{c]} = -\omega^a{}_b{}^c - \omega_b{}^{[ca]} = -\omega^a{}_b{}^c + \frac{1}{2}F_b{}^{ca}\,; \tag{4.3.59}$$

in other words,

$$\omega^{ab}_m = -\left(\frac{1}{2}F_c{}^{ab} + F^{[a}{}_c{}^{b]}\right) e^c_m\,. \tag{4.3.60}$$

Together, (4.3.55) and (4.3.60) give an algorithm to compute the spin connection ω directly from the vielbein.

Exercise 4.3.1 Show that the $\mathrm{O}(d,d)$ action on the space of metrics and B fields is transitive, by considering the element

$$O = \begin{pmatrix} e & B(e^t)^{-1} \\ 0 & (e^t)^{-1} \end{pmatrix}. \tag{4.3.61}$$

Exercise 4.3.2 Compute the spin connection for the metric (4.1.46) just by solving (4.3.28).

Exercise 4.3.3 Consider the line element $\mathrm{d}r^2 + a^2 \mathrm{d}s^2_{(d)}$ in $d+1$ dimensions, where $\mathrm{d}s^2_{(d)}$ is the line element of any d-dimensional metric and $a = a(r)$. You can choose a vielbein

$$e^{\hat{r}} = \mathrm{d}r\,, \qquad e^a = a\, e^a_{(d)}, \tag{4.3.62}$$

where $\hat{r}$ denotes the extra flat index. Check that the spin connection in this frame reads

$$\omega^{ab} = \omega^{ab}_{(d)}\,, \qquad \omega^{a\hat{r}} = a' e^a_{(d)}\,. \tag{4.3.63}$$

Compute now the Riemann and Ricci tensor. You should get in particular

$$R_{mn} = R^{(d)}_{mn} + \left((1-d)(a')^2 - aa''\right) g^{(d)}_{mn}\,, \qquad R_{rr} = -d\frac{a''}{a}\,. \tag{4.3.64}$$

Exercise 4.3.4 Using the operator methods of Section 4.1.6, show

$$[\nabla_m, L_K] = \partial_m K^n \nabla_n \tag{4.3.65}$$

if K is a Killing vector field. You will probably find (4.1.24) and (4.1.67) useful. Equation (4.3.65) shows that covariantly constant forms are a representation of the algebra of isometries.

Exercise 4.3.5 Show from (4.3.32) that the Ricci tensor and scalar satisfy

$$R^b{}_q \mathrm{d}x^q = \iota_{E^a} \hat{R}^{ab}\,, \qquad (4.3.66a)$$
$$R = -\iota_{E^a} \iota_{E^b} \hat{R}^{ab}\,. \qquad (4.3.66b)$$

Exercise 4.3.6 Consider the bundle metric (4.2.37) for $F = S^1$. A natural vielbein is $e^a = e^a_B$, $a = 1, \ldots, \dim(B) \equiv D-1$, $e^D = \mathrm{e}^\sigma D\psi$.

(i) Check that the spin connection is

$$\omega^{ab} = \omega^{ab}_B - \frac{1}{2} e^\sigma F^{ab} e^D\,, \qquad \omega^{aD} = -\frac{1}{2}\mathrm{e}^\sigma F^{ab} e_b + \partial^a \sigma\, e^D\,. \qquad (4.3.67)$$

(ii) Using (4.3.66b), show that

$$R = R_B - \frac{1}{4}\mathrm{e}^{2\sigma} F^{ab} F_{ab} - 2\left(\nabla^2\sigma + |\mathrm{d}\sigma|^2\right). \qquad (4.3.68)$$

Exercise 4.3.7 Show that under a Weyl transformation (4.1.158), the spinor covariant derivative transforms as

$$D_m \to \tilde{D}_m = D_m - \frac{1}{2}\partial_n \lambda \gamma^n{}_m\,. \qquad (4.3.69)$$

Using (4.3.66b), show also that

$$\begin{aligned} R \to \tilde{R} &= \mathrm{e}^{-2\lambda}(R - 2(d-1)\nabla^2\lambda - (d-2)(d-1)|\mathrm{d}\lambda|^2) \\ &= \mathrm{e}^{-2\lambda} R - 4\frac{d-1}{d-2}\mathrm{e}^{-\frac{d+2}{2}\lambda}\nabla^2 \mathrm{e}^{\frac{d-2}{2}\lambda}, \end{aligned} \qquad (4.3.70)$$

where as usual d is the space(time) dimension. (More details can be found in [20, app. D].)

Exercise 4.3.8 Show using (4.1.159) that the Ricci tensor for (4.1.160) is

$$\begin{aligned} R_{\mu\nu} &= R^{(d_1)}_{\mu\nu} - \mathrm{e}^{2A} g^{(d_1)}_{\mu\nu}(\nabla^2_{(d_2)} A + d_1 |\mathrm{d}A|^2)\,, \\ R_{mn} &= R^{(d_2)}_{mn} - d_1(\nabla^{(d_2)}_m \nabla^{(d_2)}_n A + \partial_m A \partial_n A)\,. \end{aligned} \qquad (4.3.71)$$

Exercise 4.3.9 Use (4.3.68) and (4.3.70) to check that the M-theory action (1.4.4) with (1.4.5) indeed reduces to the IIA action (1.2.15).

Exercise 4.3.10 Use (4.3.70) to show that (1.2.23) indeed follows from (1.2.17), (1.2.22), discarding total derivative terms.

Exercise 4.3.11 Give more details about (4.1.30), using the spin connection formalism. Begin by decomposing the flat indices of ω^{ab} in self-dual and anti-self-dual parts, using the self-duality matrices (4.4.22). (You might want to rename the index $a \in \{1,2,3\}$ occurring there to $\alpha \in \{1,2,3\}$.) Then decompose both the flat and curved indices of R^{ab}_{mn} in the same way, to obtain four tensors $R^{\alpha\beta}_{\pm\pm}$, $R^{\alpha\beta}_{\pm\mp}$. Now check that the traceless part of the Ricci tensor R^0_{mn} is related to

$$R^{\alpha\beta}_{+-} = R^{\beta\alpha}_{-+}\,, \qquad (4.3.72)$$

and that the Ricci scalar is related to $R^\alpha_{\alpha++} = R^\alpha_{\alpha--}$.

Exercise 4.3.12 Show that (4.3.47) implies

$$[D_m, \gamma_n] = \Gamma^p_{mn}\gamma_p\,. \qquad (4.3.73)$$

Exercise 4.3.13 Use (4.3.46) to show

$$[D_m, \alpha_{np}\gamma^{np}] = (\nabla_m \alpha_{np})\gamma^{np}\,. \qquad (4.3.74)$$

Exercise 4.3.14 Show that the Lie–Kosmann derivative (4.3.39) satisfies the Lie algebra property (4.1.62), assuming that v, w are Killing for simplicity. Useful equations for this are (4.1.61) applied on vectors; (4.1.18); (4.1.61); and (4.1.65).

Exercise 4.3.15 Show the *Bismut identity*

$$D^2 - D_m D^m = -\frac{1}{8} R \,, \tag{4.3.75}$$

where D is the Dirac operator (4.3.21). (Hint: write $2D^2 = \{D, D\}$ and use (4.3.33). Notice that D^m acts on a spinor η, but D_m acts on the vector-spinor $D^m \eta$.)

4.4 Homogeneous spaces

A space is called *homogeneous* if it has a group action (defined in (4.1.40)) that is transitive: any two points can be related by at least one g. We have already seen in (2.4.25) that such a space can be written as a coset G/H. The formalism in Section 4.3.6 is particularly useful in the study of such spaces. In this section, we will first study the geometry of a Lie group G itself, viewed as a manifold, and then cosets G/H.

4.4.1 Group manifolds

Frames

A Lie group G is by definition a group that is also a manifold, where both the group product and the group inverse are smooth maps as defined in Section 4.1.4. An important consequence of the group structure is the existence of a set of d vector fields and d one-forms that are everywhere independent. These imply that G is a parallelizable manifold: TM and T^*M are trivial (Exercise 4.2.2).

We can define such one-forms by considering

$$g^{-1} \mathrm{d} g \,. \tag{4.4.1}$$

At every $g \in G$, this gives a Lie algebra-valued one-form. The Lie algebra associated to a Lie group G is originally defined as the tangent space at the identity, $\mathfrak{g} \equiv T_e G$. The *exponential map* $\exp : T_e G \to G$ has an abstract definition in terms of integral curves, but for groups that can be realized in terms of matrices (which in finite dimension is always the case), it is simply the matrix exponential. We now write (4.4.1) as $\exp[-T_a x^a] \mathrm{d} \exp[T_a x^a] 1$. (Without the final 1, we would have the operator $g^{-1} \circ \mathrm{d} \circ g$.) Using (2.1.15), the first term $\mathrm{d}1 = 0$, and the second is $[\mathrm{d}, x]1 = [\mathrm{d}x^a \wedge \partial_a, T_b x^b]1 = \mathrm{d}x^a T_a$. The next terms all consist of commutators of $T_a x^a$ with this; so they can all be collected in an expression of the form

$$g^{-1} \mathrm{d} g = \mathrm{d}x^a T_a + \frac{1}{2} x^b \mathrm{d}x^a [T_a, T_b] + \cdots = \left(\delta_a^b + \frac{1}{2} f^b{}_{ac} x^c + \cdots \right) \mathrm{d}x^a T_b \,. \tag{4.4.2}$$

So we see that it is indeed Lie algebra valued. A similar result holds for $\mathrm{d}g g^{-1}$, which is not the same for nonabelian groups.

Now we can just expand in $\mathfrak{g}$ and obtain two sets of d one-forms:

$$g^{-1}\mathrm{d}g = \lambda^a T_a \,, \qquad \mathrm{d}g g^{-1} = \rho^a T_a \,. \tag{4.4.3}$$

Both of these are linearly independent at every point.

If G is compact, we can normalize the T_a so that the Killing form is equal to the identity:[15]

$$K^{ab} \equiv -\mathrm{Tr}(T^a T^b) = \delta^{ab} \,, \tag{4.4.4}$$

and we can use this to raise and lower Lie algebra indices. Then (4.4.3) can also be written as

$$\lambda^a \equiv -\mathrm{Tr}(g^{-1}\mathrm{d}g T^a) \,, \qquad \rho^a \equiv -\mathrm{Tr}(\mathrm{d}g g^{-1} T^a) \,. \tag{4.4.5}$$

The λ^a are called *left-invariant* forms because they are invariant under the left multiplication map $L_{g_0}: g \mapsto g_0 g$, for g_0 a constant $\in G$. Likewise, the ρ^a are *right-invariant* in the sense that they do not change under the right multiplication map $R_{g_0}: g \mapsto g g_0$.

The action of the exterior differential on these forms is particularly simple. From $0 = \mathrm{d}(g^{-1}g) = (\mathrm{d}g^{-1})g + g^{-1}\mathrm{d}g$, we have

$$\mathrm{d}(g^{-1}\mathrm{d}g) = -(g^{-1}\mathrm{d}g) \wedge (g^{-1}\mathrm{d}g) \,, \qquad \mathrm{d}(\mathrm{d}g g^{-1}) = (\mathrm{d}g g^{-1}) \wedge (\mathrm{d}g g^{-1}) \,. \tag{4.4.6a}$$

With (4.4.3) and using the definition of structure constants (4.2.20), we can rewrite these as

$$\mathrm{d}\lambda^a = -\frac{1}{2} f^a{}_{bc} \lambda^b \wedge \lambda^c \,, \qquad \mathrm{d}\rho^a = \frac{1}{2} f^a{}_{bc} \rho^b \wedge \rho^c \,. \tag{4.4.6b}$$

Equation (4.4.6) are called the *Maurer–Cartan equations*.

We define *Lie algebra cohomology* H^k_{LI} just like de Rham cohomology (4.1.132), but with the forms there restricted to being left invariant:

$$\alpha_k = \frac{1}{k!} \alpha_{a_1 \ldots a_k} \lambda^{a_1} \wedge \ldots \wedge \lambda^{a_k} \,. \tag{4.4.7}$$

Thanks to (4.4.6b),

$$\mathrm{d}\alpha_k = -\frac{1}{2\,k!} f^a{}_{[a_1 a_2|} \alpha_{a|a_3 \ldots a_{k+1}]} \lambda^{a_1} \wedge \ldots \wedge \lambda^{a_{k+1}} \,, \tag{4.4.8}$$

so d reduces to an algebraic computation. If G is compact,

$$H^k_{\mathrm{LI}}(G) \cong H^k(G) \,. \tag{4.4.9}$$

Let us now consider a basis of vector fields. The families of diffeomorphisms L_{g_0} induce vectors ℓ_a, by taking g_0 near the identity with the usual logic (4.1.49). In the

[15] In this book, we avoid introducing imaginary units in the structure constants; so, for example, the generators of the Lie algebra $\mathrm{u}(N)$ are anti-Hermitian. With this convention, $\mathrm{Tr}(T^a T^b)$ is semi-negative-definite.

same way, the diffeomorphisms R_{g_0} induce vector fields r_a. These are dual to the forms (4.4.5) in the sense that

$$\ell_a \cdot \rho^b = \delta^b_a\,, \qquad r_a \cdot \lambda^b = \delta^b_a\,. \tag{4.4.10}$$

From their definition, we know that they should commute:

$$[\ell_a, r_b] = 0\,. \tag{4.4.11a}$$

The ℓ_a are the infinitesimal version of the L_{g_0}, but (4.4.11) shows that they are right invariant; vice versa, the r_a are left invariant.[16]

From the Maurer–Cartan equations (4.4.6) and the computation leading from (4.3.55) to (4.3.57), we see that these satisfy the algebra

$$[\ell_a, \ell_b] = -f^c{}_{ab}\ell_c\,, \qquad [r_a, r_b] = f^c{}_{ab} r_c\,. \tag{4.4.11b}$$

We already know that (4.4.5) are respectively left and right invariant, so

$$L_{\ell_a}\lambda^b = 0\,, \qquad L_{r_a}\rho^b = 0\,. \tag{4.4.12a}$$

Moreover, from (4.1.79) and (4.4.6) we see

$$L_{\ell_a}\rho^b = f^b{}_{ac}\rho^c\,, \qquad L_{r_a}\lambda^b = -f^b{}_{ac}\lambda^c\,. \tag{4.4.12b}$$

Metrics

It is tempting to take one of the frames we introduced to be the orthonormal frame, or vielbein, $e^a = e^a_m \mathrm{d}x^m$. For example, we can take

$$e^a = \rho^a\,. \tag{4.4.13}$$

It is important to stress that (4.4.13) is not necessary; it is only a natural choice on a group manifold. It leads to the metric

$$\mathrm{d}s^2_{\mathrm{G}} = \delta_{ab}\rho^a\rho^b = -\mathrm{Tr}(\mathrm{d}g g^{-1}\mathrm{d}g g^{-1}) = -\mathrm{Tr}(g^{-1}\mathrm{d}g g^{-1}\mathrm{d}g) = \delta_{ab}\lambda^a\lambda^b\,. \tag{4.4.14}$$

Comparing (4.4.6b) with (4.3.55), we see that with this choice of metric the anholonomy coefficients are actually proportional to the structure constants of the Lie group:

$$F^a{}_{bc} = f^a_{bc}\,. \tag{4.4.15}$$

Since λ^a and the ρ^a are invariant under the actions of the R_{g_0} and of the L_{g_0} diffeomorphisms, (4.4.14) has two sets of isometries, each in one-to-one correspondence with the elements of G. In other words, the isometry group of (4.4.14) is

$$G \times G\,. \tag{4.4.16}$$

The corresponding Killing vectors will be (4.4.11).

If we take a more general linear combination of the ρ instead of (4.4.13), the metric will be of the form

$$\mathrm{d}s^2 = R_{ab}\rho^a\rho^b\,. \tag{4.4.17}$$

[16] From the point of view of naming conventions, we faced here two unpleasant alternatives: the generators of left multiplications are right invariant, so they might have been called with either the letter ℓ or r. We chose the former.

This is still invariant under the R_{g_0} but not under all the L_{g_0}; for a generic R_{ab}, the isometry group is only G. Only for some special R_{ab} some of the L_{g_0} are isometries. Vice versa,

$$\mathrm{d}s^2 = L_{ab}\lambda^a\lambda^b\,, \tag{4.4.18}$$

is invariant under the L_{g_0}, and possibly some of the R_{g_0}. For $L_{ab} = R_{ab} = \delta_{ab}$, both (4.4.17) and (4.4.18) will coincide with (4.4.14). We stress once again that it is also possible to take a choice of metric that is unrelated to either the ρ^a or the λ^a.

The group $\mathrm{SU}(2) \cong S^3$

As an example of this formalism, consider S^3. This is homeomorphic to the group SU(2), since we can parameterize

$$g = x_4 1_2 + \mathrm{i} x_i\sigma_i\,, \qquad \sum_{m=1}^{4} x_m^2 = 1\,. \tag{4.4.19}$$

Indeed, using (2.2.15), we can check that $g^\dagger g = 1$ and $\det g = 1$ automatically.[17] Now

$$\begin{aligned} g^{-1}\mathrm{d}g = g^\dagger \mathrm{d}g &= (x_4 1_2 - \mathrm{i}x_i\sigma_i)\mathrm{d}(x_4 1_2 + \mathrm{i}x_i\sigma_i) \\ &\overset{(2.2.15)}{=} \mathrm{i}\sigma_i(x_4\mathrm{d}x_i - x_i\mathrm{d}x_4) + \mathrm{i}\epsilon_{ijk}\sigma_k(x_i\mathrm{d}x_j)\,, \end{aligned} \tag{4.4.20}$$

and so the definition (4.4.3) gives, with $T_a = -\frac{\mathrm{i}}{2}\sigma_a$,[18]

$$\lambda_a = 2x_m j_{+a}^{mn}\mathrm{d}x_n\,. \tag{4.4.21}$$

The matrices

$$(j_{\pm a})_{mn} \equiv -\delta_{m4}\delta_{na} + \delta_{n4}\delta_{ma} \mp \epsilon_{mna} \tag{4.4.22}$$

represent the two su(2) subalgebras of so(4) $\cong$ su(2) $\oplus$ su(2); indeed, they satisfy $[j_{\pm a}, j_{\pm b}] = \pm 2\epsilon_{abc} j_{\pm c}$, $[j_{a+}, j_{b-}] = 0$. They also obey

$$j_{\pm a} j_{\pm b} = -\delta_{ab}1 \pm \epsilon_{abc} j_{\pm c}\,; \tag{4.4.23}$$

moreover, the matrices $j_{+a}j_{-b}$ are a basis for the space of 4×4 symmetric traceless matrices.

Another possibility is to view g as a quaternion using the map (2.1.30); then $g^\dagger \mathrm{d}g = \frac{1}{2} i_a\lambda^a$. To see the link with (4.4.22), we define $i_4 = 1$: then

$$i_a(x_4 + i_b x_b) = i_a(x_m i_m) = x^4 i_a - x_a + \epsilon_{abc}x_c = -2j_{+a}^{mn} x_m i_n\,. \tag{4.4.24}$$

The constraint $\sum_{m=1}^4 x_m^2 = 1$ in (4.4.19) can also be solved in terms of angles. Defining complex coordinates

$$w_1 \equiv x_4 - \mathrm{i}x_3\,, \qquad w_2 \equiv -x_1 + \mathrm{i}x_2\,, \tag{4.4.25}$$

and with the Hopf parameterization (4.2.50), (4.4.19) becomes

$$\begin{aligned} g &= \begin{pmatrix} \bar w_1 & -\mathrm{i}w_2 \\ -\mathrm{i}\bar w_2 & w_1 \end{pmatrix} = \begin{pmatrix} \mathrm{e}^{-\mathrm{i}(\psi+\phi/2)}\cos(\theta/2) & -\mathrm{i}\,\mathrm{e}^{\mathrm{i}(\psi-\phi/2)}\sin(\theta/2) \\ -\mathrm{i}\,\mathrm{e}^{\mathrm{i}(-\psi+\phi/2)}\sin(\theta/2) & \mathrm{e}^{\mathrm{i}(\psi+\phi/2)}\cos(\theta/2) \end{pmatrix} \\ &= \mathrm{e}^{-\frac{\mathrm{i}}{2}\phi\sigma_3}\mathrm{e}^{-\frac{\mathrm{i}}{2}\theta\sigma_1}\mathrm{e}^{-\mathrm{i}\psi\sigma_3}\,. \end{aligned} \tag{4.4.26}$$

[17] The fact that we are able to expand the group element in terms of the $\mathrm{i}\sigma_i$, which are generators of the Lie algebra, is an accident due to the σ_i and 1_2 being closed under matrix product and not only under commutator, by (2.2.15). This allows us to resume the exponential map exp: su(2) $\to$ SU(2).

[18] For this particular group, we bow to tradition and give up the normalization (4.4.4).

We recognize the Euler parameterization for SU(2). Now (4.4.5) gives the left- and right-invariant one-forms as

$$\begin{aligned} \lambda^1 &= \cos 2\psi d\theta + \sin 2\psi \sin\theta d\phi\,, & \rho^1 &= \cos\phi d\theta + 2\sin\phi\sin\theta d\psi \\ \lambda^2 &= -\sin 2\psi d\theta + \cos 2\psi \sin\theta d\phi\,, & \rho^2 &= \sin\phi d\theta - 2\cos\phi\sin\theta d\psi\,, \\ \lambda^3 &= 2d\psi + \cos\theta d\phi\,, & \rho^3 &= d\phi + 2\cos\theta d\psi\,. \end{aligned} \qquad (4.4.27)$$

It is easy to check (4.4.6b) with $f^a{}_{bc} = \epsilon^a{}_{bc}$, the structure constant for su(2). Equation (4.4.14) now gives $ds^2_{\mathrm{SU}(2)} = 4ds^2_{S^3}$ in the Hopf parameterization (4.2.51). According to (4.4.16), its isometry group should be at least $\mathrm{SU}(2)\times\mathrm{SU}(2) \cong \mathrm{SO}(4)/\mathbb{Z}_2$. Indeed this SO(4) can be seen more directly from the embedding $S^3 \subset \mathbb{R}^4$ that defines the three-sphere.

The vectors ℓ_a and r_a can be obtained by imposing (4.4.10). (Recall that the ℓ_a, the generators of left multiplications, are right invariant, and vice versa.) So

$$\begin{aligned} \ell_1 &= \cos\phi\partial_\theta - \frac{\sin\phi}{\tan\theta}\partial_\phi + \frac{\sin\phi}{2\sin\theta}\partial_\psi\,, & r_1 &= \cos 2\psi\partial_\theta + \frac{\sin 2\psi}{\sin\theta}\partial_\phi - \frac{1}{2}\frac{\sin 2\psi}{\tan\theta}\partial_\psi\,, \\ \ell_2 &= \sin\phi\partial_\theta + \frac{\cos\phi}{\tan\theta}\partial_\phi - \frac{\cos\phi}{2\sin\theta}\partial_\psi\,, & r_2 &= -\sin 2\psi\partial_\theta + \frac{\cos 2\psi}{\sin\theta}\partial_\phi - \frac{1}{2}\frac{\cos 2\psi}{\tan\theta}\partial_\psi, \\ \ell_3 &= \partial_\phi\,, & r_3 &= \frac{1}{2}\partial_\psi\,. \end{aligned} \qquad (4.4.28)$$

The components along θ and ϕ of the ℓ_a are the three Killing vectors l_a of S^2 in coordinates (θ,ϕ); the ℓ_a are a lift (4.2.43) of the $\ell_a^{S^2}$ to the total space of the Hopf bundle (4.2.54). The r_a have a similar interpretation as lifts of the *conformal* Killing vectors $r_a^{S^2}$ of S^2. (Recall from (4.1.71) that the conformal algebra of S^2 has dimension six; the generators are the $\ell_a^{S^2}$ and the $r_a^{S^2}$.)

We can also consider an example of metric with smaller isometry group. Taking $L_{ab} = \mathrm{diag}(1,1,\sigma/2)$, (4.4.18) becomes

$$ds^2_{\mathrm{sq}S^3} = ds^2_{S^2} + \sigma^2(d\psi + A)^2\,. \qquad (4.4.29)$$

This is traditionally called *squashed* S^3;[19] since the L_{ab} we have chosen has an identity block, (4.4.29) retains some of its right invariance, and the isometry group is $\mathrm{SU}(2)\times\mathrm{U}(1)$, with generators ℓ_a and r_3.

4.4.2 Cosets

We already introduced group quotients G/H long ago in (2.4.25), to study spaces of spinors in a given dimension. From now on, we will more interested in considering them as potential parts of spacetime, and for this reason we will now focus more on their differential-geometric properties. The formal definition is that G/H (or $\frac{G}{H}$) is the set of equivalence classes in G under the relation $\sim$ defined by

$$g \sim g' \quad \Leftrightarrow \quad \exists\, h \in H \mid g = g'h\,. \qquad (4.4.30)$$

[19] Some authors prefer using this name more generally for any metric obtained from (4.4.18) with $G = \mathrm{SU}(2)$ and any L_{ab}.

The equivalence class of g can be denoted by gH. Sometimes the notation G/H can be ambiguous, if there is more than one way to embed H in G; we will call attention to this subtlety when needed.

If H is a normal subgroup of G, i.e. if $ghg^{-1} \in H$ for any $g \in G$ and $h \in H$, then one can show that the product in G induces a product in G/H, under which G/H is still a group. For an H that is not normal, however, this is not the case, and the group structure is lost.

The definition given so far applies to any H. When H is itself a Lie group, we call G/H a *coset*. It is useful to view it as the base of a principal fiber bundle:

$$\begin{array}{ccc} H & \hookrightarrow & G \\ & & \downarrow \pi \\ & & G/H\,. \end{array} \tag{4.4.31}$$

Invariant forms

We split the generators T_a of the Lie algebra $\mathfrak{g}$ into those of $\mathfrak{h}$ and those of a vector subspace $\mathfrak{k}$ (usually *not* a Lie algebra) such that $\mathfrak{h} \oplus \mathfrak{k} = \mathfrak{g}$:

$$T_\alpha \in \mathfrak{h}\,, \qquad T_i \in \mathfrak{k}\,. \tag{4.4.32}$$

There is some freedom in how we choose $\mathfrak{k}$, but for semisimple or compact H one can always choose this complement such that $[T_\alpha, T_i]$ is a linear combination of the T_i:

$$[T_\alpha, T_i] = f^j{}_{\alpha i} T_j\,. \tag{4.4.33}$$

In other words, $f^\beta{}_{\alpha i} = 0$. We will assume this from now on. The Jacobi identities in $\mathfrak{g}$ now imply that the $f^j{}_{\alpha i}$ define a representation of $\mathfrak{h}$ on $\mathfrak{k}$, called the *isotropy representation*. In general, this is nontrivial; it is often irreducible, but sometimes it decomposes:

$$\mathfrak{k} = \mathfrak{k}_0 \oplus \ldots \oplus \mathfrak{k}_p\,. \tag{4.4.34}$$

By definition $\mathfrak{k}_0$ we mean the singlet, which might or might not be present. It consists of elements T_i^0 such that $[T_\alpha, T_i] = 0$; these belong to the Lie algebra of the *normalizer* $N(G, H) \equiv \{g \in G \,|\, gHg^{-1} \subset H\}$. (More generally, the Lie algebra of $N(G, H)$ might also contain elements such that $[T_\alpha, T_i]$ is a linear combination of the T_α, but these are not present if (4.4.33) holds.) More precisely, $\mathfrak{k}_0$ is the infinitesimal counterpart of the coset

$$N(G, H)/H\,. \tag{4.4.35}$$

We now want to define an analogue of the left-invariant forms (4.4.3). The first step is to split that formula according to (4.4.32):

$$g^{-1}\mathrm{d}g = \lambda^i T_i + \lambda^\alpha T_\alpha\,. \tag{4.4.36}$$

(Practically speaking, it can be useful to write this by parameterizing G/H as $g = kh$, where k is the exponential of a linear combination of the T_i and $h \in T_\alpha$.)

Recall our discussion in (4.2.9): for a bundle $F \hookrightarrow E \to B$, forms on B are in one-to-one correspondence with forms on E that are horizontal and also invariant under vertical vectors v. In the present case, the vertical v are the vectors r_α representing the infinitesimal right action of H; so the λ^i in (4.4.36) are horizontal. They are not invariant in general under the action of H: the representation (4.4.33) is nontrivial. Another way of seeing this is that a horizontal form α is invariant if and only if $\mathrm{d}\alpha$ is also horizontal; but the $\mathrm{d}\lambda^i$, which we can compute from (4.4.6b), may contain the λ^α, and in that case are not invariant. Infinitesimally, (4.4.12) here becomes

$$L_{r_\alpha}\lambda^i = -f^i{}_{\alpha j}\lambda^j \,. \tag{4.4.37}$$

So the one-forms λ_0^i that are invariant are those in the singlet $\mathfrak{k}_0$ of the isotropy representation (4.4.34).

As we stressed after (4.4.34), the singlet $\mathfrak{k}_0$ does not always exist. In such case, no λ^i is invariant; this might look disheartening. But in higher-form degrees, a k-form

$$\alpha = \frac{1}{k!}\alpha_{i_1\ldots i_k}\lambda^{i_1}\wedge\ldots\wedge\lambda^{i_k} \tag{4.4.38}$$

can be invariant even if its constituent λ^{i_a} are not. Indeed, a k-form transforms in the kth antisymmetric product of the isotropy representation (4.4.34), and that might contain a singlet even if (4.4.34) does not. Concretely, this happens if

$$f^j{}_{\alpha[i_1}\alpha_{i_2\ldots i_k]j} = 0 \,. \tag{4.4.39}$$

As a cross-check, since α is horizontal, we have to require that $\mathrm{d}\alpha$ be horizontal as well, or in other words that it is a wedge of the λ^i only, without any λ^α. From (4.4.6b), we see that this is the case exactly if (4.4.39) holds. (As an example, the top-form built from all the λ^i will always be invariant, since it is automatically closed.)

To summarize, forms of the type (4.4.38) and (4.4.39) can be thought of as forms on G/H. We still call them left invariant, because they are still left unchanged by the action of $L_{g_0}:\ gH \to g_0 gH$.

Metrics on cosets

We would now like to write a metric on a coset, similar to (4.4.14), (4.4.17) and (4.4.18) for group manifolds. We saw that the λ^i do not descend in general to forms on G/H, but their combination

$$\mathrm{d}s^2 = \delta_{ij}\lambda^i\lambda^j = -\mathrm{Tr}(P_{\mathfrak{k}}k^{-1}\mathrm{d}k P_{\mathfrak{k}}k^{-1}\mathrm{d}k) \tag{4.4.40}$$

does: the Killing form of $\mathfrak{g}$ restricted to $\mathfrak{k}$, which in our normalization reads δ_{ij}, builds an invariant. In the second equality, we have used the parameterization $g = kh$ in (4.4.36), and inserted a projector $P_{\mathfrak{k}}$ to keep only the λ^i. (Recall that $\mathfrak{k}$ is not a subalgebra, so $k^{-1}\mathrm{d}k$ is not necessarily automatically valued in $\mathfrak{k}$.)

A more general symmetric product $L_{ij}\lambda^i\lambda^j$ would not descend to a metric on G/H. If the isotropy representation (4.4.34) contains a singlet, the corresponding one-forms λ_0^i are invariant and can be included in the metric as $L_{ij}\lambda_0^i\lambda_0^j$ with any matrix L_{ij}. The nonsinglet representations in (4.4.34) can still be included, with care: the tensor product of a representation with itself contains a singlet $\delta_{ij}\lambda^i\lambda^j$. So a more general expression for a G-invariant metric on a coset is

$$\mathrm{d}s^2 = L_{ij}\lambda_0^i\lambda_0^j + a_2\delta_{ij}\lambda_1^i\lambda_1^j + \cdots + a_p\delta_{ij}\lambda_p^i\lambda_p^j, \tag{4.4.41}$$

where λ_1^i correspond to generators in $\mathfrak{k}_1$, and so on.

Under left-multiplication, all the λ^i are left invariant, because they are components of $g^{-1}\mathrm{d}g$; so (4.4.40) has at least the left G-action in its isometry group. Under right-multiplication, elements of the normalizer $N(G,H)$ act as $kh \to khn = knh'$, and so their action can be summarized by $k \to kn$, which leaves (4.4.40) invariant. However, the right action of H does not lead to isometries, because they do not act at all on G/H. A little more explicitly, we can consider the vector field r_i corresponding to the right action of an element in $\mathfrak{k}$. From (4.4.12),

$$L_{r_i} \lambda^j = -f^j{}_{ik} \lambda^k - f^j{}_{i\alpha} \lambda^\alpha \,. \tag{4.4.42}$$

If r_i corresponds to a generator in the singlet $\mathfrak{k}_0$ of the decomposition (4.4.34), then by definition $f^j{}_{i\alpha} = 0$, so the second term in (4.4.42) cancels out. Now acting on (4.4.40), we obtain $L_{r_i} \lambda^j \lambda^j = -2 f^i{}_{jk} \lambda^j \lambda^k$, which is zero because of antisymmetry of the structure constants. So the isometry group of (4.4.40) contains

$$G \times N(G, H)/H \,. \tag{4.4.43}$$

(There might be "accidental" isometries not predicted by this logic.) For the more general metric (4.4.41), the G factor is still present, while $N(G, H)/H$ might not.

4.4.3 Some examples

A very simple example of coset is provided by

$$\mathrm{SU}(2)/\mathrm{U}(1) \cong S^2 \,. \tag{4.4.44}$$

One of the generators R_i of the infinitesimal right action is $R_3 = \partial_\psi$; for simplicity, we can take this to be the generator of $H = \mathrm{U}(1)$. Then (4.4.31) becomes simply the Hopf bundle (4.2.54). In the decomposition in (4.4.36), the T_i are T_1 and T_2, while the only T_α is T_3. So from the λ^a in (4.4.27), the λ^i are the first two, λ^1 and λ^2. Applying the condition (4.4.39), we find that the only left-invariant form is

$$\lambda^1 \wedge \lambda^2 = \sin\theta \mathrm{d}\theta \wedge \mathrm{d}\phi \,. \tag{4.4.45}$$

The metric $\mathrm{d}s^2 = (\lambda^1)^2 + (\lambda^2)^2 = \mathrm{d}s^2_{S^2}$, the usual round metric (4.1.46) on S^2, whose volume form is (4.4.45).

We will now list some other prominent examples of right quotients G/H. In all these cases, we will apply the criterion we gave long ago in (2.4.25): if M has a transitive G-action and H is its stabilizer, then $M \cong G/H$.

- Consider a higher-dimensional sphere:

$$S^d \cong \mathrm{SO}(d+1)/\mathrm{SO}(d) \,. \tag{4.4.46}$$

  Here the metric is unique, because the isotropy representation (4.4.34) consists of the fundamental representation alone.
- Odd-dimensional spheres can be realized as cosets in another way:

$$S^{2N+1} \cong \mathrm{SU}(N+1)/\mathrm{SU}(N) \,. \tag{4.4.47}$$

  Here (4.4.34) contains a singlet, and as a consequence the metric is not unique. We saw an example in (4.4.29).
- Yet another homogeneous metric on a sphere is obtained as

$$S^{4n+3} \cong \mathrm{Sp}(n+1)/\mathrm{Sp}(n) \,. \tag{4.4.48}$$

Again this depends on a parameter. Both (4.4.47) and (4.4.48) are sometimes called *squashed spheres*. See [136] for a more complete list of homogeneous metrics on spheres and projective spaces.

- The space of complex one-dimensional subspaces $\mathbb{C} \subset \mathbb{C}^{N+1}$ is called *complex projective space*:

$$\mathbb{CP}^N \equiv \mathrm{SU}(N+1)/\mathrm{U}(N)\,. \tag{4.4.49}$$

(The embedding of $\mathrm{U}(N)$ is described slightly more accurately by writing it as $\mathrm{S}(\mathrm{U}(N) \times \mathrm{U}(1))$.) As in (4.4.46), the metric is unique. The real dimension is $(N+1)^2 - 1 - N^2 = 2N$, or complex dimension N. We will discuss this thoroughly in Section 6.1.

- An alternative metric on $\mathbb{CP}^N$ is obtained as

$$\mathbb{CP}^{2N+1} \cong \mathrm{Sp}(N+1)/(\mathrm{Sp}(N) \times \mathrm{U}(1)). \tag{4.4.50}$$

The metric is not unique, and in general differs from the one for (4.4.49). By analogy with (4.4.48), this can be called a squashed $\mathbb{CP}^N$.

- The space of k-dimensional vector subspaces (k-planes through the origin) in $\mathbb{R}^d$ is called the *Grassmannian*:

$$\mathrm{Gr}(k,d) \cong \mathrm{O}(d)/\left(\mathrm{O}(k) \times \mathrm{O}(d-k)\right)\,. \tag{4.4.51}$$

We already saw an occurrence of this in (4.2.83). The complex analogue is

$$\mathrm{Gr}_{\mathbb{C}}(k,d) \cong \mathrm{U}(d)/\left(\mathrm{U}(k) \times \mathrm{U}(d-k)\right)\,, \tag{4.4.52}$$

describing subspaces $\mathbb{C}^k \subset \mathbb{C}^d$. This generalizes (4.4.49), which is $\mathbb{CP}^N = \mathrm{Gr}_{\mathbb{C}}(1, N+1)$. Its dimension is $k(d-k)$, respectively real and complex.

- Generalizing even further, we can consider *flags*, namely nested subspaces $\mathbb{C}^{d_1} \subset \mathbb{C}^{d_2} \subset \ldots \subset \mathbb{C}^d$. The space of such flags is called a *flag manifold*:

$$\mathbb{F}(d_1, d_2, \ldots; d) \cong \mathrm{U}(d)/\left(\mathrm{U}(d_1) \times \mathrm{U}(d_2 - d_1) \ldots \times \mathrm{U}(d - d_{k-1})\right)\,. \tag{4.4.53}$$

In some cases, the nonuniqueness of the metric can be understood in terms of a bundle metric (4.2.35). A typical case is when $G \supset H' \supset H$; then

$$H'/H \hookrightarrow G/H \to G/H'\,, \tag{4.4.54}$$

generalizing (4.4.31). For example, applying this to (4.4.47) and (4.4.49), we get $S^1 \hookrightarrow S^{2N+1} \to \mathbb{CP}^N$; then the parameter mentioned after (4.4.47) is related to the volumes of the fiber and base. Explicitly, $\mathrm{d}s^2 = \lambda(\mathrm{d}\psi + A)^2 + \mathrm{d}s^2_{\mathbb{CP}^N}$, for $\lambda \in \mathbb{R}$.

4.4.4 Nil- and sol-manifolds

We will now see some examples of right quotients where H is discrete, rather than a Lie group. In such cases, we usually write $H = \Gamma$, and call G/Γ a *discrete quotient*. When G is noncompact, there exists sometimes a Γ such that G/Γ is compact.

A notable example is when G is a *nilpotent* group, in which case G/Γ is called *nilmanifold*. This can be characterized in two ways. One is that G is a subgroup of the group of upper-triangular matrices, excluding the diagonal. Another, more abstract, is in terms of its Lie algebra $\mathfrak{g}$. We define

$$\mathfrak{g}^1 \equiv [\mathfrak{g}, \mathfrak{g}] = \{x \in \mathfrak{g} \mid \exists x_1,\, x_2 \in \mathfrak{g}\; x = [x_1, x_2]\}\,. \tag{4.4.55}$$

In other words, this is the subset of elements that can arise as the result of a Lie bracket. It is a normal subalgebra. (This is the Lie algebra counterpart of a normal subgroup: $\mathfrak{n} \subset \mathfrak{g}$ is normal if $[\mathfrak{n}, \mathfrak{g}] \subset \mathfrak{n}$.) A semisimple algebra has no nontrivial normal subalgebras by definition, so $\mathfrak{g}^1 = \mathfrak{g}$. But more generally, $\mathfrak{g}^1 \subset \mathfrak{g}$ will be smaller. Defining then $\mathfrak{g}^k \equiv [\mathfrak{g}, \mathfrak{g}^{k-1}]$, $\mathfrak{g}$ is nilpotent if there is a k for which $\mathfrak{g}^k = \{0\}$.

A nilpotent Lie group is noncompact, but *Malcev's theorem* guarantees that there is a discrete subgroup Γ such that G/Γ is compact; it is enough to work in a basis where the structure constants are rational. A list of nilmanifolds in $d = 6$ was obtained in [137, 138]; see, for example, in [139, table 1], [140, table 4].

Even if (4.4.9) does not apply to G, a theorem by Nomizu guarantees that

$$H^k(G/\Gamma) \cong H^k_{\mathrm{LI}}(G) . \tag{4.4.56}$$

So again, one can use (4.4.8) to compute the de Rham cohomology algebraically.

The simplest nontrivial example Nil_3 of nilmanifold is three dimensional and is obtained by considering the structure constants

$$f^3{}_{12} = -f^3{}_{21} = -1 , \qquad f^b{}_{ac} = 0 \text{ otherwise} . \tag{4.4.57}$$

So (4.4.6b) read

$$\mathrm{d}\lambda^1 = \mathrm{d}\lambda^2 = 0 , \qquad \mathrm{d}\lambda^3 = \lambda^1 \wedge \lambda^2 . \tag{4.4.58}$$

(For more complicated examples, it is common to give the structure constants indirectly through the analogue of (4.4.58). A common way to summarize this equation is also via a table; for (4.4.58), we would write $(0, 0, 12)$.) We can realize this concretely by taking three coordinates x^i, and defining

$$\lambda^1 = \mathrm{d}x^1 , \qquad \lambda^2 = \mathrm{d}x^2 , \qquad \lambda^3 = \mathrm{d}x^3 + x^1 \mathrm{d}x^2 . \tag{4.4.59}$$

This turns $\mathbb{R}^3$ into a copy of the nilpotent group G with the structure constants (4.4.57). The subgroup Γ promised by Malcev's theorem is then defined by

$$(x^1, x^2, x^3) \cong (x^1, x^2 + 1, x^3) \cong (x^1, x^2, x^3 + 1) \cong (x^1 + 1, x^2, x^3 - x^2) . \tag{4.4.60}$$

The third is chosen so as to be consistent with the expression for λ^3 in (4.4.59).

If we choose a vielbein $e^a = \lambda^a$, we recognize from (4.4.59) the metric in (4.2.90) (for $n = 1$, and after a coordinate rescaling). As we observed there, the metric suggests that this space should be an S^1-bundle; we now see it more explicitly. Consider the map $\pi\colon (x^1, x^2, x^3) \to (x^1, x^2)$, where now $(x^1, x^2) \in T^2$, so $(x^1, x^2) \cong (x^1 + 1, x^2) \cong (x^1, x^2 + 1)$; π is well-defined, meaning that it maps equivalence classes to equivalence classes. The counterimage of every point is an S^1, realizing the expected fiber bundle

$$\begin{array}{ccc} S^1 & \hookrightarrow & \mathrm{Nil}_3 \\ & & \downarrow \\ & & T^2 . \end{array} \tag{4.4.61}$$

Generalizing this example, every nilmanifold can be written as an iterated torus bundle: a bundle Nil_d whose fiber is a torus T^{k_1}, whose base Nil_{d-k_1} is itself a T^{k_2}-bundle over another nilmanifold, and so on, with the last base $\mathrm{Nil}_{d-\sum k_i}$ being a torus itself.

More generally, we can consider *solvable* groups G, in which case G/Γ is called *solmanifold*. Again there are two definition; one is that G is solvable if it is a subgroup of the group of upper-triangular matrices, this time including the diagonal. A more abstract definition is obtained by defining $\mathfrak{g}_1 \equiv \mathfrak{g}^1$ as before, but now $\mathfrak{g}_k \equiv [\mathfrak{g}_{k-1}, \mathfrak{g}_{k-1}]$. We now say $\mathfrak{g}$ is solvable if there is a k such that $\mathfrak{g}_k = \{0\}$; this is clearly a laxer condition, and indeed a nilpotent algebra is also solvable. The famous *Levi decomposition theorem* says that one can write any Lie algebra as a sum of a semisimple and a solvable algebra.

Unfortunately, there is no complete classification of what solvable groups admit a Γ such that G/Γ is compact, as in Malcev's theorem. Some criteria, however, are known [141–143]; see, for example, [140, 144] for discussions.

Exercise 4.4.1 Show that a metric $\mathrm{Tr}(g^{-1}\mathrm{d}g + A)^2 + \mathrm{d}s^2_{\mathrm{B}}$ is of bundle type (4.2.35), where the Killing vectors T^a are the left-invariant vector fields r_a.

Exercise 4.4.2 Check that ℓ_1 and ℓ_2 in (4.4.28) can be obtained by applying the lift formula (4.2.43) to the Hopf metric (4.2.51) and (4.2.52).

Exercise 4.4.3 Consider the coset $\mathbb{F}(1,2;3) \cong \mathrm{SU}(3)/\mathrm{U}(1)^2$ from (4.4.53). In an appropriate basis, the structure constants of $\mathfrak{g} = \mathrm{su}(3)$ read

$$\begin{aligned} &f^1{}_{54} = f^1{}_{36} = f^2{}_{46} = f^2{}_{35} = f^3{}_{47} = f^5{}_{76} = \frac{1}{2}\,, \\ &f^1{}_{27} = 1\,, \qquad f^3{}_{48} = f^5{}_{68} = \frac{\sqrt{3}}{2}\,, \end{aligned} \tag{4.4.62}$$

plus cyclic permutations (so, for example, we also have $f^5{}_{41} = \frac{1}{2}$). Take $\mathfrak{h} = \mathrm{u}(1) \oplus \mathrm{u}(1)$ to be the span of the generators T_7, T_8. Show that the left-invariant two-forms and three-forms are (with $\lambda^{i_1 \ldots i_k} \equiv \lambda^{i_1} \wedge \ldots \wedge \lambda^{i_k}$)

$$\begin{aligned} &\lambda^{12}\,, \qquad \lambda^{34}\,, \qquad \lambda^{56}\,, \\ &\lambda^{245} + \lambda^{135} + \lambda^{146} - \lambda^{236}\,, \qquad \lambda^{235} + \lambda^{136} + \lambda^{246} - \lambda^{145}\,. \end{aligned} \tag{4.4.63}$$

Exercise 4.4.4 Using (4.4.54), give an interpretation to the squashing parameter for (4.4.48).

Exercise 4.4.5 Use MW spinors in $d = 10$ to define a bundle

$$S^7 \hookrightarrow S^{15} \xrightarrow{\pi} S^8\,. \tag{4.4.64}$$

(Hint: define π similar to (3.3.50), recalling Section 3.4.5. To compute the fiber, recall also that Ψ_4 defines a Spin(7)-structure, and use (4.4.46).)

Exercise 4.4.6 Similar to the previous exercise, use Weyl spinors in $d = 6$ to define a bundle

$$S^3 \hookrightarrow S^7 \xrightarrow{\pi} S^4\,. \tag{4.4.65}$$

Both this and (4.4.64) are generalizations of the Hopf bundle (4.2.54), related to octonions and quaternions respectively.

4.5 Maximally symmetric spaces

We conclude this section with a brief overview of *maximally symmetric* spaces. These are defined as manifolds M_d having the maximal number of Killing vectors, namely $d(d+1)/2$, that of flat space. Their treatment doesn't require particularly sophisticated geometrical ideas, but they are important in this book, especially in Lorentzian signature, for reasons made clear already in the introduction.

The Riemann tensor of a maximally symmetric space is severely constrained by its isometries:[20]

$$R_{mnpq} = k(g_{mp}g_{nq} - g_{mq}g_{np}) \tag{4.5.1}$$

for a constant k. (For example, we saw this to be true in (4.1.90) for S^2, for $k = 1$.) It follows from (4.5.1) that

$$R_{mn} = k(d-1)g_{mn} \,. \tag{4.5.2}$$

A metric proportional to its Ricci tensor is said to be *Einstein*. This is obviously a less stringent condition than (4.5.1).

Maximally symmetric spaces are so constrained that they are locally unique for a given k. For $k = 0$, we have flat space. The cases $k \neq 0$ are discussed later for both Euclidean and Lorentzian signature. Since under rescaling $g_{mn} \to R^2 g_{mn}$, the Ricci tensor remains invariant, we can always set a nonzero k to ± 1; the metrics we present in this section will all have this value.

4.5.1 Spheres

For $k > 0$ in Euclidean signature, a maximally symmetric space is locally isometric to a sphere $S^d = \{\sum_{M=1}^{d+1}(y^M)^2 = 1\} \subset \mathbb{R}^{d+1}$, which we considered earlier in (4.1.70) and (4.4.46). It follows from the latter that its isometry group is $\mathrm{SO}(d+1)$, which indeed has dimension $d(d+1)/2$.

Its line element can be written recursively by considering polar coordinates in $\mathbb{R}^{d+1}$ of the form

$$y^m = r \sin\theta_d \, \hat{y}^m \quad (m = 1, \ldots, d) \,, \qquad y^{d+1} = r\cos\theta_d \,, \tag{4.5.3}$$

with $\hat{y}^m \hat{y}^m = 1$, restricting to $r = 1$. This gives

$$\mathrm{d}s^2_{S^d} = \mathrm{d}\theta_d^2 + \sin^2\theta_d \mathrm{d}s^2_{S^{d-1}} \,, \tag{4.5.4}$$

generalizing (4.1.46) for S^2. If we define $R \equiv \tan(\theta_d/2)$,

$$\mathrm{d}s^2_{S^d} = 4\frac{\mathrm{d}R^2 + R^2 \mathrm{d}s^2_{S^{d-1}}}{(1+R^2)^2} \,, \tag{4.5.5}$$

generalizing (4.1.47).

[20] See, for example, [145, sec. 13.2] for a proof of this, and of the uniqueness of the metrics in this section.

A generalization of the coordinate system (4.5.3) is

$$y^M = r \sin\sigma\, \hat{y}_1^M \ (M = 1, \ldots, k{+}1)\,, \qquad y^M = r\cos\sigma\, \hat{y}_2^{M-k} \ (M = k{+}2, \ldots, d{+}1), \tag{4.5.6}$$

where again $\sum_{m=1}^{k+1}(\hat{y}_1^m)^2 = 1$, $\sum_{m=1}^{d-k}(\hat{y}_2^m)^2 = 1$. Both of these parametrize spheres, of dimensions k and $d-k-1$ respectively. The resulting line element is now

$$\mathrm{d}s^2_{S^d} = \mathrm{d}\sigma^2 + \sin^2\sigma \mathrm{d}s^2_{S^k} + \cos^2\sigma \mathrm{d}s^2_{S^{d-k-1}}\,. \tag{4.5.7}$$

This is called the *join* form of the metric. The name comes from the *topological* join of two spaces A and B, defined as

$$A * B \equiv \frac{A \times B \times [0,1]}{(a,b,0) \sim (a',b,0),\ (a,b,1) \sim (a,b',1)}\,. \tag{4.5.8}$$

The two relations in the denominator identify all points of A to one point at the left extremum 0 of the interval, and all points of B to one point at the right-extremum. So (4.5.7) expresses the fact that

$$S^k * S^{d-k-1} \cong S^d\,. \tag{4.5.9}$$

A famous example is $S^1 * S^1 \cong S^3$.

4.5.2 Hyperbolic space

For $k < 0$ in Euclidean space, we have *hyperbolic space*:

$$H_d \equiv \left\{ -\eta^{(d+1)}_{\mu\nu} X^\mu X^\nu \equiv X_0^2 - \sum_{m=1}^{d} X_m^2 = 1 \right\} \subset \mathbb{R}^{1,d}\,. \tag{4.5.10}$$

The isometry group $\mathrm{ISO}(H_d) = \mathrm{SO}(1,d)$ can be thought of as the group of linear transformations in $\mathbb{R}^{1,d}$ leaving H_d invariant. It acts transitively; the isotropy group of a point, for example $(1, 0, \ldots, 0)$, is $\mathrm{SO}(d)$, so we can also write H_d as a coset:

$$H_d \cong \mathrm{SO}(1,d)/\mathrm{SO}(d)\,. \tag{4.5.11}$$

To obtain an explicit metric, we can proceed similar to S^d by introducing embedding coordinates

$$X_0 = r\cosh\rho\,, \qquad X_m = r\sinh\rho\, \hat{x}_m, \tag{4.5.12}$$

where $\hat{x}_m \hat{x}_m = 1$ describes an S^{d-1}, and the Minkowski metric:

$$\eta^{(d+1)}_{\mu\nu} \mathrm{d}X^\mu \mathrm{d}X^\nu \equiv -\mathrm{d}X_0^2 + \sum_{m=1}^{d} \mathrm{d}X_i^2\,. \tag{4.5.13}$$

Plugging (4.5.12) into (4.5.13) and setting $r = 1$, we obtain the line element

$$\mathrm{d}s^2_{H_d} - \mathrm{d}\rho^2 + \sinh^2\rho\, \mathrm{d}s^2_{S^{d-1}}\,. \tag{4.5.14}$$

This is to be compared with (4.5.4) for S^d. Similar to (4.5.5), it is also possible to define $R \equiv \tanh(\rho/2)$, obtaining

$$\mathrm{d}s^2_{H_d} = 4\frac{\mathrm{d}R^2 + R^2 \mathrm{d}s^2_{S^{d-1}}}{(1-R^2)^2}\,, \tag{4.5.15}$$

which now covers a ball $B_d = \{R \leq 1\} \subset \mathbb{R}^{d+1}$. This can also be obtained from a hyperbolic analogue of the stereographic projection from the hyperboloid (4.5.10) to B_d.

Another famous parameterization is obtained by slicing (4.5.10) with planes. Define

$$u_\pm \equiv X_0 \pm X_d\,, \qquad X_m = u_+ x_m \; (m = 1, \ldots, d-1)\,, \tag{4.5.16}$$

where now the x_m are unconstrained. Equation (4.5.10) then becomes $u_- = u_+^{-1} + u_+ x_m x_m$. Defining $y \equiv u_+^{-1}$ leads to the *Poincaré metric*

$$\mathrm{d}s^2_{H_d} = \frac{1}{y^2}(\mathrm{d}y^2 + \mathrm{d}s^2_{\mathbb{R}^{d-1}})\,. \tag{4.5.17}$$

4.5.3 De Sitter

For $k > 0$ in Lorentzian signature, we have *de Sitter space*:

$$\mathrm{dS}_d \equiv \left\{ \eta^{\mu\nu}_{(d+1)} X_\mu X_\nu \equiv -X_0^2 + \sum_{m=1}^{d} X_m^2 = 1 \right\} \subset \mathbb{R}^{1,d}\,, \tag{4.5.18}$$

where the metric on $\mathbb{R}^{1,d}$ is $\eta^{\mu\nu}_{(d+1)} = -\mathrm{d}X_0^2 + \sum_{m=1}^{d} \mathrm{d}X_m^2$. Notice the sign difference with respect to (4.5.10). This definition makes it manifest that the isometry group is again $\mathrm{ISO}(\mathrm{dS}_d) = \mathrm{SO}(1,d)$, as for H_d. However, now all points are below the light cone; the isotropy group of a point, for example $(0,1,0,\ldots,0)$, is $\mathrm{SO}(1,d-1)$, so

$$\mathrm{dS}_d \equiv \mathrm{SO}(1,d)/\mathrm{SO}(1,d-1)\,. \tag{4.5.19}$$

FLRW coordinates

There are three coordinate systems that put the metric into *Friedmann–Lemaître–Robertson–Walker (FLRW)* type

$$\mathrm{d}s^2 = -\mathrm{d}t^2 + a^2(t)\mathrm{d}s^2_{\mathrm{MS}_d}\,, \tag{4.5.20}$$

where MS_d is one of the maximally symmetric d-dimensional Riemannian spaces discussed earlier. They are

$$X_0 = \sinh t_\mathrm{S}\,, \qquad X_m = \cosh t_\mathrm{S} \hat{x}_m \;\; (m = 1, \ldots, d) \quad \hat{x}_m \hat{x}_m = 1\,; \tag{4.5.21a}$$

$$\begin{gathered} X_0 + X_d = \mathrm{e}^{t_\mathrm{F}}\,, \qquad X_0 - X_d = \mathrm{e}^{t_\mathrm{F}} x_m x_m - \mathrm{e}^{-t_\mathrm{F}}\,, \\ X_m = \mathrm{e}^{t_\mathrm{F}} x_m \;\; (m = 1, \ldots, d-1)\,; \end{gathered} \tag{4.5.21b}$$

$$\begin{aligned} X_\mu &= \sinh t_\mathrm{H} \hat{x}_\mu \;\; (\mu = 0, \ldots, d-1) \;\; \eta^{\mu\nu}_{(d+1)} \hat{x}^\mu \hat{x}^\nu = -1\,, \\ X_d &= \cosh t_\mathrm{H}\,. \end{aligned} \tag{4.5.21c}$$

Equations (4.5.21a) and (4.5.21c) are similar to (4.5.6): they correspond to grouping the squares in (4.5.18) in different ways. Equation (4.5.21b) comes from a computation similar to the one leading to (4.5.17). These three lead to three metrics of FLRW

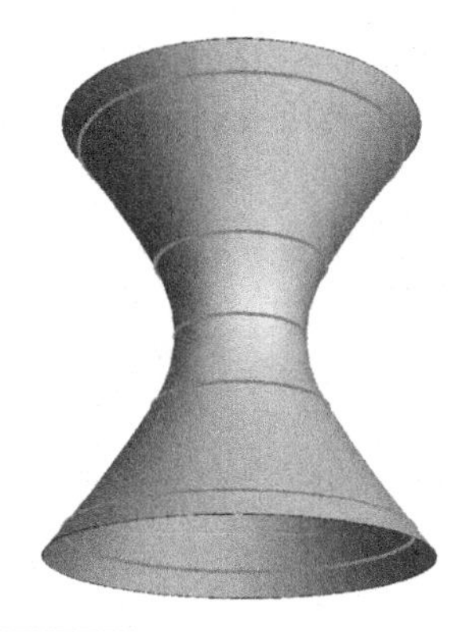
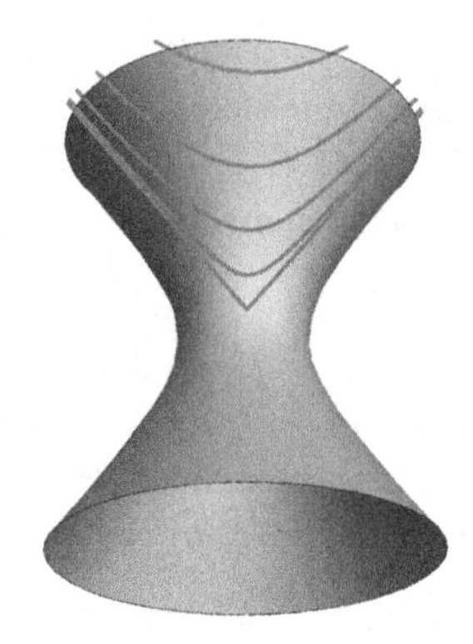
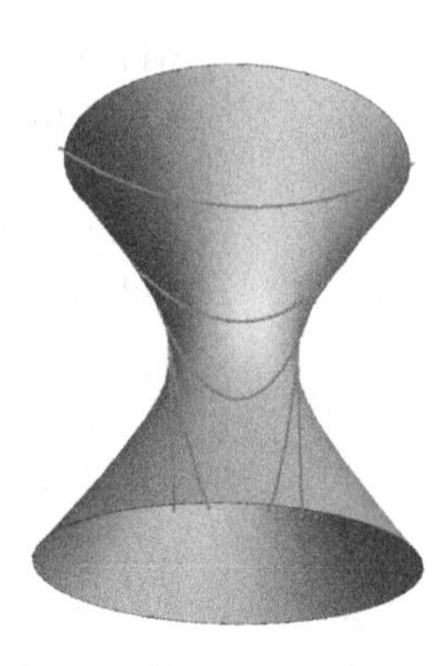

Figure 4.3 Hyperboloids (4.5.18) for $d = 2$, with constant-t loci corresponding to the three coordinate systems in (4.5.21).

type:

$$\begin{aligned} ds^2_{\mathrm{dS}_d} &= -dt_S^2 + \cosh^2 t_S ds^2_{S^{d-1}} & (4.5.22a)\\ &= -dt_F^2 + e^{2t_F} ds^2_{\mathbb{R}^{d-1}} & (4.5.22b)\\ &= -dt_H^2 + \sinh^2 t_H ds^2_{H_{d-1}} . & (4.5.22c)\end{aligned}$$

We show in Figure 4.3 the curves at constant t in all three cases. The three coordinate systems correspond to slicing the hyperboloid in three different ways; but only the first covers all of dS_d.

The situation becomes clearer if we introduce coordinates with a finite range. We start from (4.5.21a), which covers the whole space, and define $\tan(\eta/2) = \tanh(t_S/2)$, turning it into

$$ds^2_{\mathrm{dS}_d} = \frac{1}{\cos^2\eta}\left(-d\eta^2 + ds^2_{S^{d-1}}\right) . \qquad (4.5.23)$$

Writing in turn S^{d-1} as in (4.5.4), now dS_d is an S^{d-2}-fibration over a square with coordinates

$$\eta \in [-\pi/2, \pi/2] , \qquad \theta_{d-1} \in [0, \pi] \qquad (4.5.24)$$

(see Figure 4.4c). In general, such a diagram, the compact base of a sphere fibration, is called a *Penrose diagram*. The vertical sides of the square are not a boundary, because S^{d-2} shrinks there; these are the poles of the S^{d-1}. The horizontal sides represent the infinite past and future $t_S \to \pm\infty$. In Figure 4.4, we show the same square again, but shading the regions covered by the coordinate systems (4.5.21b) and (4.5.21c). These are the same regions covered by the constant-t lines in Figure 4.3, after turning the hyperboloid and changing coordinates to η.

Static coordinates

Another notable coordinate system is again of join type (4.5.6):

$$\begin{aligned} &X_0 = \sin\psi \sinh t , \qquad X_d = \sin\psi \cosh t ,\\ &X_m = \cos\psi\, \hat{x}_m \ (m = 1, \ldots , d-1) , \qquad \hat{x}_m \hat{x}_m = 1 . \end{aligned} \qquad (4.5.25)$$

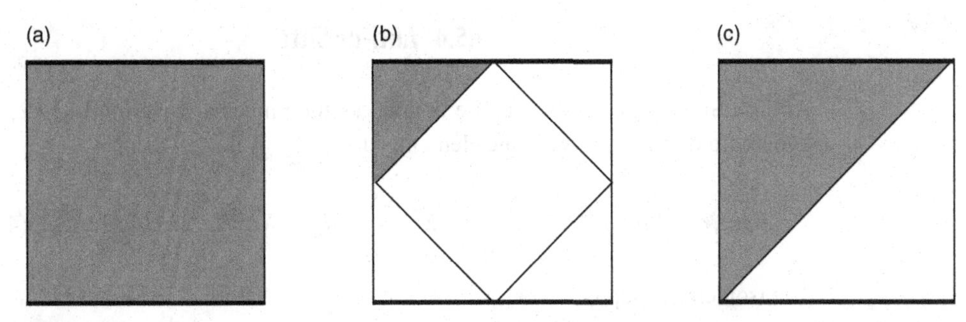

Figure 4.4 The Penrose diagram of de Sitter space is a square. We see shaded here the loci covered by the coordinate systems in (4.5.21).

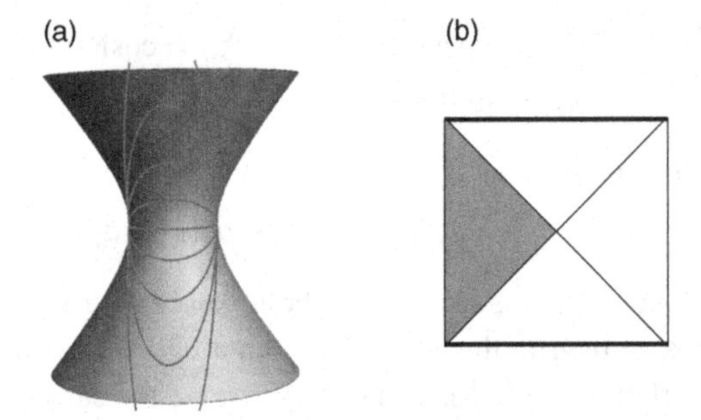

Figure 4.5 (a) Constant-t loci for dS in coordinates (4.5.26); (b) region covered by those coordinates.

The corresponding metric is

$$ds^2_{\mathrm{dS}_d} = d\psi^2 - \sin^2\psi\, dt^2 + \cos^2\psi\, ds^2_{S^{d-2}}\,, \tag{4.5.26}$$

which is similar to (4.5.7). No metric component depends on t, so ∂_t is a Killing vector. This coordinate system again doesn't cover all of dS, but only the shaded region in Figure 4.5b, called "static patch."

Amusingly, (4.5.26) can also be put in Schwarzschild-type form, with the redefinition $\hat{r} = \cos\psi$:[21]

$$ds^2_{\mathrm{dS}_d} = -e^{2U}dt^2 + e^{-2U}d\hat{r}^2 + \hat{r}^2 ds^2_{S^{d-2}} \qquad (e^{2U} = 1 - \hat{r}^2)\,. \tag{4.5.27}$$

The boundary of the static patch is at $\{\hat{r} = 1\}$, called *cosmological horizon*: just like for a black hole horizon, g_{tt} becomes positive for $\hat{r} > 1$, so t is no longer a time coordinate there. This is, however, an observer-dependent phenomenon: the action of the isometry group is transitive, so any point can be brought to lie at $\{\hat{r} = 0\}$. So there is a cosmological horizon surrounding any observer. The dynamics of free-falling particles can be understood by considering the gravitational potential

$$V_{\mathrm{grav}} \equiv \sqrt{-g_{00}}\,, \tag{4.5.28}$$

which in this case gives $V_{\mathrm{grav}} = e^U = \sqrt{1-\hat{r}^2}$. Particles tend to diverge from any observer, until they reach the horizon $\hat{r} = 1$, past which they are out of sight forever.

[21] In fact, modifying (4.5.27) to $e^{2U} = 1 + 2\frac{GM}{\hat{r}} - \hat{r}^2$ describes a Schwarzschild black hole in de Sitter.

4.5.4 Anti-de Sitter

Anti-de Sitter space AdS_d can be defined as the universal covering (4.1.122) of a hyperboloid in $\mathbb{R}^{2,d-1}$, with line element $-\mathrm{d}X_0^2 + \sum_{m=1}^{d-1} \mathrm{d}X_m^2 - \mathrm{d}X_d^2$:

$$\mathrm{SO}(2, d-1)/\mathrm{SO}(1, d-1) \cong \left\{ -X_0^2 + \sum_{m=1}^{d-1} X_m^2 - X_d^2 = -1 \right\} \subset \mathbb{R}^{2,d-1} . \tag{4.5.29}$$

The isometry group is

$$\mathrm{ISO}(\mathrm{AdS}_d) = \mathrm{SO}(2, d-1) . \tag{4.5.30}$$

We start with a set of coordinates covering the whole space, hence called *global*:

$$\begin{aligned} X_0 &= \cosh\rho \cos\tau , \qquad X_d = \cosh\rho \sin\tau , \\ X_m &= \sinh\rho\, \hat{x}_m \quad (m = 1, \ldots, d-1) , \qquad \hat{x}_m \hat{x}_m = 1 . \end{aligned} \tag{4.5.31}$$

This gives

$$\mathrm{d}s^2_{\mathrm{AdS}_d} = -\cosh^2\rho\, \mathrm{d}\tau^2 + \mathrm{d}\rho^2 + \sinh^2\rho\, \mathrm{d}s^2_{S^{d-2}} . \tag{4.5.32}$$

A constant-τ slice is hyperbolic space (4.5.14); as we know from (4.5.15), we can also rewrite this as a metric on a ball B_d, with a coordinate $R \in [0, 1)$. In the hyperboloid (4.5.29), τ and $\tau + 2\pi$ give the same point, signaling that the space is not simply connected, but as we mentioned, AdS_d is the universal covering, which means that we "unwind" this time coordinate and make it take values in $\mathbb{R}$ rather than S^1. This avoids the presence of closed time-like curves. So the Penrose diagram in this case is simply an infinite strip, with coordinate range

$$R \in [0, 1) , \qquad \tau \in \mathbb{R} . \tag{4.5.33}$$

It is also possible to rewrite this in Schwarzschild form by taking $\hat{r} \equiv \sinh\rho$:

$$\mathrm{d}s^2_{\mathrm{AdS}_d} = -\mathrm{e}^{2U}\mathrm{d}\tau^2 + \mathrm{e}^{-2U}\mathrm{d}\hat{r}^2 + \hat{r}^2 \mathrm{d}s^2_{S^{d-2}} \qquad (\mathrm{e}^{2U} = 1 + \hat{r}^2) . \tag{4.5.34}$$

This is to be contrasted to (4.5.27). The gravitational potential

$$V_{\mathrm{grav}} = \sqrt{-g_{00}} = \mathrm{e}^U = \sqrt{1 + \hat{r}^2} \tag{4.5.35}$$

now attracts all particles to the center. Once again, since the isometry group acts transitively, any point can be brought to lie at $\{\hat{r} = 0\}$. If from any point we shoot particles in all directions, they refocus at the same point after a time $\Delta\tau = 2\pi$.

Another famous coordinate system, useful in the AdS/CFT correspondence, is

$$\begin{aligned} X_{d-1} + X_d &= z^{-1} , \qquad X_{d-1} - X_d = z^{-1} x^\mu x_\mu - z , \\ X_\mu &= z^{-1} x_\mu \quad (\mu = 0, \ldots, d-2) , \end{aligned} \tag{4.5.36}$$

again inspired by the Poincaré coordinates (4.5.17) for H_d. This leads to

$$\mathrm{d}s^2_{\mathrm{AdS}_d} = \frac{1}{z^2}(\mathrm{d}z^2 + \mathrm{d}s^2_{\mathbb{R}^{1,d-2}}) = \frac{1}{r^2}\mathrm{d}r^2 + r^2 \mathrm{d}s^2_{\mathbb{R}^{1,d-2}} , \tag{4.5.37}$$

where we have also shown a popular rewriting using $r \equiv 1/z$. Equation (4.5.37) only covers the so-called *Poincaré patch*. This is shown in Figure 4.6: the cylinder

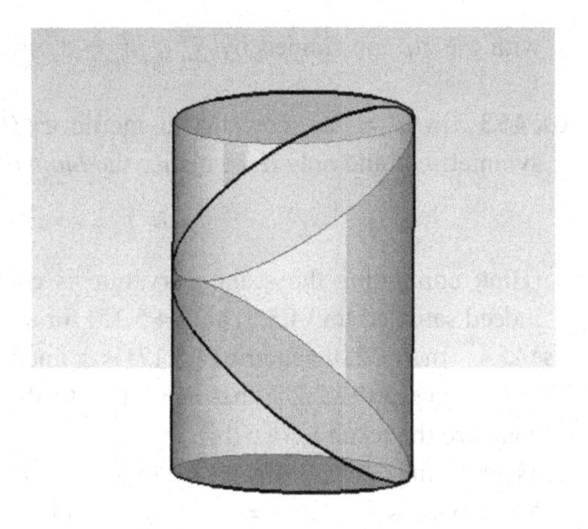

Figure 4.6 AdS_d can be represented as a cylinder, showing only one of the angular coordinates $\hat{x}_m$ in (4.5.31). The region inside the two oblique disks is the Poincaré patch.

is the strip (4.5.33) with one of the angular coordinates, say $\hat{x}_{d-1} = \cos\phi$, depicted as an angular coordinate; $\sigma \equiv \arctan\sinh\rho$ as radial coordinate; and τ as vertical coordinate. With this choice, the region $X_{d-1} + X_d > 0$ is contained inside the two shown diagonal slices inside the cylinder. Forgetting the single angular coordinate in Figure 4.6, we recover the Penrose diagram.

Another slightly less known coordinate system, but sometimes also useful in AdS/CFT, is

$$X_0 = \sinh\lambda\sinh t\,, \qquad X_d = \cosh\lambda$$
$$X_m = \sinh\lambda\cosh t\hat{x}_m \;\; (m = 1, \ldots, d-1)\,, \qquad \hat{x}_m\hat{x}_m = 1\,, \tag{4.5.38}$$

leading to

$$ds^2_{AdS_d} = -\sinh^2\lambda\, dt^2 + d\lambda^2 + \cosh^2\lambda\, ds^2_{S^{d-2}}\,, \tag{4.5.39}$$

again of join type. The three coordinate systems (4.5.32), (4.5.37), and (4.5.39) in a sense belong to a sequence: a radial slice, at large ρ, z^{-1} and λ respectively, is asymptotical to

$$-dt^2 + ds^2_{MS_{d-2}}, \tag{4.5.40}$$

with MS_{d-2} a maximally symmetric Riemannian space.

Finally it is also possible to write AdS in FLRW coordinates, although not in three ways but only one way, by taking

$$X_\mu = \sin t\,\hat{x}_\mu \;\; (\mu = 0, \ldots, d-1)\,, \qquad \eta^{\mu\nu}_{(d)}\hat{x}_\mu\hat{x}_\nu = -1\,; \qquad X_d = \cos t\,;$$
$$ds^2_{AdS_4} = -dt^2 + \sin^2 t\, ds^2_{H_{d-1}}\,. \tag{4.5.41}$$

Exercise 4.5.1 Evaluate the Weyl tensor (4.1.28) of a maximally symmetric space.

Exercise 4.5.2 In the spirit of (4.5.7), find coordinates on S^{2k-1} such that

$$ds^2_{S^{2k-1}} = \sum_{i=1}^{k}(d\mu_i^2 + \mu_i^2 d\phi_i^2)\,, \tag{4.5.42}$$

with the μ_i constrained by $\sum_{i=1}^k \mu_i^2 = 1$. Indeed, this reduces to (4.5.7) for $k = 2$.

Exercise 4.5.3 In $d = 2$, show that a metric $e^{2f}((dx^1)^2 + (dx^2)^2)$ is maximally symmetric if and only if it satisfies the *Liouville equation*:

$$\partial^2 f = -k e^{2f} . \tag{4.5.43}$$

(Hint: computing the scalar curvature is enough. Why?) Check that this is indeed satisfied for (4.5.5) and (4.5.15) for $d = 2$.

Exercise 4.5.4 In $d = 2$, the metric (4.5.17) is defined over a half-plane. Find a Möbius transformation (2.2.24) that maps this to the unit disk $\{|z| < 1\} \subset \mathbb{C}$, and compare the result to (4.5.14).

Exercise 4.5.5 Find a coordinate system for dS_d where it is conformal to Minkowski space, $ds^2 = f ds^2_{\text{Mink}_d}$. (Hint: start from one of the coordinate systems in the main text.)

Exercise 4.5.6 Find a coordinate system where AdS_d has line element $ds^2 = f(\theta)(-d\eta^2 + d\theta^2 + \sin^2\theta ds^2_{S^{d-2}})$. Use this to show that a light ray arrives at the boundary in finite time $\Delta\eta$. (Hint: start from (4.5.32).)

Exercise 4.5.7 By comparing (4.5.18) and (4.5.29), find coordinate systems where

$$ds^2_{\text{AdS}_d} = d\psi_1^2 + \cosh^2\psi_1 ds^2_{\text{AdS}_{d-1}} = d\psi_2^2 + \sinh^2\psi_2 ds^2_{\text{dS}_{d-1}} . \tag{4.5.44}$$

These are similar to (4.5.37): AdS_d is sliced in copies of AdS_{d-1} and dS_{d-1} rather than Mink_{d-1}.

5 Geometry of forms

5.1 G-structures

5.1.1 Overview

A *G-structure* is a tensor or spinor field whose stabilizer is everywhere G. The reason for this name is that it reduces the structure group (Section 4.2.1): rather than in $GL(d,\mathbb{R})$, the transition functions take values in a subgroup G. Equivalently, we can say that a G-structure defines a subbundle of the frame bundle FM, whose fiber is the space of frames (or ordered bases of vectors: Sections 4.1.3 and 4.2.1).

Metric = $O(d)$ structure

Already a metric g gives an example of a G-structure. Indeed, we can use it to define the subbundle $OFM \subset FM$, whose fibers are the frames that are orthonormal with respect to g, or in other words the choices of vielbein e^a. Pick a $e^a_{(\alpha)}$ in each patch U_α; from the gluing condition on the metric on intersections $U_\alpha \cap U_\beta$, it follows that the local vielbeine are glued as in (4.3.13). In other words, at every point:

$$\Lambda_{(\alpha\beta)} \in O(d) \subset GL(d,\mathbb{R}) \tag{5.1.1}$$

(or $O(d-1,1)$ in Lorentz signature). Combining (4.3.13) with (4.1.37), we also see that $(J_{(\alpha\beta)})_m{}^n = E^a_{m\,(\alpha)}(\Lambda_a^b)_{(\alpha\beta)} e^n_{b\,(\beta)}$. In other words, the transition functions on TM take values in a group isomorphic to $O(d)$, with the vielbein being the isomorphism.

Vector fields

Even just a humble vector field v defines a G-structure, if it never vanishes. Take it to be the first vector in a vielbein, after by dividing it by its norm: $E_1 = v/|v|$. If we do this on every patch, then on $U_{\alpha\beta}$

$$\Lambda_{(\alpha\beta)} E_1 = E_1 \,. \tag{5.1.2}$$

Together with (5.1.1), we conclude that the transition functions are now in $O(d-1)$. In other words, a never-vanishing vector and a Riemannian metric define an $O(d-1)$-structure.

Existence of such a structure imposes a topological constraint on M. The *Poincaré–Hopf theorem* gives an alternative characterization of the Euler characteristic $\chi(M)$ (4.2.72) in terms of the zeros of a vector field; in particular, on a compact M the existence of a vector field without zeros implies

$$\chi(M) = 0 \,. \tag{5.1.3}$$

More generally, one can consider a tangent distribution, which we defined in Section 4.1.7 as the span $D \subset TM$ of k vector fields. If they are everywhere independent, the stabilizer is the group of all matrices of block-triangular form $\left(\begin{smallmatrix} a & b \\ 0 & d \end{smallmatrix}\right) \in \mathrm{GL}(d, \mathbb{R})$; adding a metric reduces this further to

$$\mathrm{O}(d-k) \times \mathrm{O}(k)\,. \tag{5.1.4}$$

A different generalization arises from considering k vectors $\{v^1, \ldots, v^k\}$, again independent. The difference with the previous case is that each individual vector should now be stabilized, not just their collective span. In this case, with a metric we get an $\mathrm{O}(d-k)$-structure. In the $k = d$ case, we get a basis for the space of vectors, or equivalently of one-forms; for example, a vielbein, and

$$\mathrm{Stab}(\{e^a\}) = \{1\}\,, \tag{5.1.5}$$

the trivial group whose only element is the identity. So a vielbein defines an *identity structure*. In this case, the tangent bundle is trivial (Exercise 4.2.2).

Orientation

A manifold is orientable if the transition functions on TM_d have positive determinant, $\det \frac{\partial x^m_{(\alpha)}}{\partial x^n_{(\beta)}} > 1$ (Section 4.1.3). By definition, this reduces the structure group to $\mathrm{GL}^+(d, \mathbb{R})$. It is easy to see that orientability is equivalent to existence of a globally defined d-form, also called an *orientation*. In presence of a metric, the reduction is from $\mathrm{O}(d)$ to $\mathrm{SO}(d)$, and the orientation can be taken to be the volume form (4.3.11). Orientability imposes that the Stiefel–Whitney class $w_1(TM) = 0$.

$\mathrm{Stab}(\eta)$-structure

To obtain more interesting examples, suppose we have, besides the metric, also a spinor field on M, whose orbit is everywhere the same: the stabilizer at every point is isomorphic to a fixed group

$$\mathrm{Stab}(\eta) \subset \mathrm{SO}(d)\,. \tag{5.1.6}$$

A spinor field was defined in Section 4.3.2 as the choice of local spinors η_α on each chart U_α, with the gluing condition (4.3.14). On each U_α, we still have the freedom to change vielbein using local orthogonal transformations, $e^a_{(\alpha)} \to (\Lambda_{(\alpha)})^a{}_b e^b_{(\alpha)}$; this has the effect that the transition functions change as $\Lambda_{(\alpha\beta)} \to \Lambda_{(\alpha)}\Lambda_{(\alpha\beta)}\Lambda^{-1}_{(\beta)}$; the local spinors change according to the spinorial representation, namely $\eta_\alpha \to \rho_s(\Lambda_{(\alpha)})\eta_\alpha$. We can use this freedom of local gauge transformations to set the local spinors to be constant; given that we assumed the spinor to have the same orbit everywhere, this can be taken to be the same in every chart:

$$\eta_\alpha = \eta^0\,. \tag{5.1.7}$$

After doing this, the new transition functions satisfy

$$\rho_s(\Lambda_{(\alpha\beta)})\eta^0 = \eta^0\,. \tag{5.1.8}$$

In other words, the transition functions belong to the stabilizer of η^0:

$$\Lambda_{(\alpha\beta)} \in \mathrm{Stab}(\eta^0)\,. \tag{5.1.9}$$

So the structure group of TM has been further reduced from $\mathrm{SO}(d)$ to $\mathrm{Stab}(\eta)$.

Equation (5.1.6) suggests that a Stab(η)-structure defines more data than those of a metric, but also that there is a natural metric associated to it, since the latter defines an O(d)-structure. We will see this explicitly in various examples.

The theme of Sections 3.3 and 3.4 was that the data of a spinor can often be encoded in those of its bilinears; recall the discussion around (3.3.2). So we should be able to define a Stab(η) structure also by considering tensor bilinears of the spinor, promoting the correspondence (3.3.2) to curved space. We will see this explicitly later in various examples. Many of the G-structures we will see in this book arise in this way. A list of stabilizers was discussed in Section 2.4 (see Tables 2.4 and 2.5), and in Sections 3.3 and 3.4 from the point of view of bilinears. Of particular note will be the following:

- SU($d/2$)-structures in even d, defined by a pure spinor
- G_2-structures in $d = 7$, defined by a Majorana spinor
- Spin(7)-structures in $d = 8$, defined by a Majorana–Weyl spinor

SU($d/2$)-structure

As a particular example of Stab(η)-structure, let us consider d = even and take the spinor η_+ to be pure, so that Stab(η_+) = SU($d/2$) (Section 2.4.4).

Alternatively, we can define this SU($d/2$) structure via the bilinears of η^0 (Section 3.4.1). We saw in (3.4.15) that in flat space the spinor η_+ is in one-to-one correspondence with a $d/2$-form Ω, which had the nondegeneracy property (3.4.13). The metric is not flat; but the whole idea of the vielbein formalism of Section 4.3.1 is that we relate the curved-space gamma matrices γ_m to flat-space ones γ_a. One can then import flat space results about bilinears to results in curved space about bilinears with the γ_a. Indeed, we already argued that there is a preferred vielbein such that the $(\eta_+^c)^\dagger \gamma_{m_1 \ldots m_k} \eta_+$ bilinear can be written as (5.1.22).

Applying this to every patch U_α produces a $d/2$-form Ω_α on each; this defines a global $d/2$-form Ω on the manifold M, which is in one-to-one correspondence with the spinor η_+. The transition functions $\Lambda_{(\alpha\beta)}$ now have to satisfy $(\Lambda_{(\alpha\beta)})^{a_1}{}_{b_1} \ldots (\Lambda_{(\alpha\beta)})^{a_{d/2}}{}_{b_{d/2}} \Omega_{b_1 \ldots b_{d/2}} = \Omega_{a_1 \ldots a_{d/2}}$; this requires again

$$\Lambda_{(\alpha\beta)} \in \mathrm{SU}(d/2) \,. \tag{5.1.10}$$

In flat space, Ω defines a tensor I of type $(1, 1)$, for example via (3.4.16). In the z_i' coordinates, this had an expression like the one in (2.4.3): $I^0 = \begin{pmatrix} 0 & 1_{d/2} \\ -1_{d/2} & 0 \end{pmatrix}$. In curved space, the same expression (3.4.16) holds, but in flat indices; the logic is the same that brought us to (3.4.12). So we have an $(I^0)_a{}^b$ such that $(I^0)_a{}^c (I^0)_c{}^b = -\delta_a^b$. To go back to curved indices, we can multiply by a vielbein and an inverse vielbein; the resulting object

$$I_m{}^n \equiv e_m^a (I^0)_a{}^b E_b^n \tag{5.1.11}$$

still squares to -1:

$$I_m{}^p I_p{}^n = -\delta_m^n \,. \tag{5.1.12}$$

This $I_m{}^n$ is still related to the curved space Ω introduced by (3.4.16), so it is still determined by it:

$$\Omega \to I \,. \tag{5.1.13}$$

We will learn how to reformulate this further in Section 5.1.2.

A tensor satisfying (5.1.12) is called an *almost complex structure*. The "almost" refers to the fact that the curved indices $I_m{}^n$ might not be as simple as the I^0 in (2.4.3), even if the flat-index expression $(I^0)_a{}^b$ does. We will return to this in Section 5.3.

Almost complex structure = $\mathrm{GL}(d/2, \mathbb{C})$**-structure (or** $\mathrm{U}(d/2)$**-structure)**

We can also consider directly I rather than Ω. Without a metric, we have to compute the stabilizer of $I^0 = \begin{pmatrix} 0 & 1_{d/2} \\ -1_{d/2} & 0 \end{pmatrix}$ in $\mathrm{GL}(d, \mathbb{R})$. This proceeds along similar lines as (2.4.34). The difference is that we don't impose the Lie algebra matrix to be antisymmetric. We obtain a Lie algebra element of the type $\begin{pmatrix} M_1 & M_2 \\ -M_2 & M_1 \end{pmatrix}$; the map taking this to $M_1 + \mathrm{i}M_2 \in \mathrm{gl}(d/2, \mathbb{C})$ is a Lie algebra isomorphism. We conclude

$$\mathrm{Stab}_{\mathrm{GL}(d,\mathbb{R})}(I) = \mathrm{GL}(d/2, \mathbb{C})\,. \tag{5.1.14}$$

Such an I exists on a manifold M only if the third Stiefel–Whitney class $w_3(TM)$ vanishes.

When we also add a metric, we intersect (5.1.14) with $\mathrm{O}(d)$ and obtain

$$\mathrm{Stab}(I) = \mathrm{O}(d) \cap \mathrm{GL}(d/2, \mathbb{C}) \cong \mathrm{U}(d/2) \tag{5.1.15}$$

rather than $\mathrm{SU}(d/2)$ as in (5.1.10). This indicates that Ω determines I, but not the other way around. See also Exercise 5.1.1.

But not all almost complex structures are *compatible* with a given metric g: they cannot necessarily be obtained as (5.1.11) for e a vielbein of g. An equivalent way of stating this condition is as follows:

$$J_{mn} = -J_{nm}\,, \qquad J_{mn} = I_m{}^p g_{pn}\,. \tag{5.1.16}$$

This property already appeared in (2.4.30) (where, however, g was still the Euclidean metric). We can call the pair (I, g) an *almost Hermitian structure*. In fact, a metric compatible with an almost complex structure I always exists. This is because $\mathrm{GL}(d/2, \mathbb{C})$ can be continuously deformed to $\mathrm{U}(d/2)$ (as we saw for $d = 2$ after (4.2.49)).

We will discuss $\mathrm{U}(d/2)$- and $\mathrm{SU}(d/2)$-structures further in Section 5.1.2.

Almost symplectic structure = $\mathrm{Sp}(d, \mathbb{R})$**-structure**

The pure spinor η_+ defining an $\mathrm{SU}(d/2)$ has also another bilinear, the two-form J. It can also be defined by (5.1.16), with I associated to Ω and the latter to η_+. But let us first study here the properties of J alone, without a metric and an associated spinor.

A two-form satisfying $J^{d/2} \neq 0$ is said to be nondegenerate. Indeed,

$$J^{d/2} = \mathrm{Pf}(J)\mathrm{d}x^1 \wedge \ldots \wedge \mathrm{d}x^d\,, \tag{5.1.17}$$

where again the Pfaffian of an antisymmetric matrix A is defined by $\det A = \mathrm{Pf}^2 A$. So nondegeneracy is equivalent to the matrix of components J_{mn} having nonzero determinant. A nondegenerate real two-form is also called an *almost symplectic structure* for reasons that will become clear later.

To work out the stabilizer of J, again we can look at a simple example. In flat space, g_{mn} was the Euclidean metric δ_{mn}; so, in (2.4.30), J_{mn} had the same entries as $I_m{}^n$. The index structure, however, tells us that the two have a different $\mathrm{GL}(d, \mathbb{R})$ action: the one on J_{mn} is $J \to MJM^t$, which maps antisymmetric matrices to antisymmetric matrices, while the one on $I_m{}^n$ was $I \to MIM^{-1}$. The prototype for I, and hence for J, was I^0 in (2.4.3), namely

$$J_0 = \begin{pmatrix} 0 & 1_{d/2} \\ -1_{d/2} & 0 \end{pmatrix} . \tag{5.1.18}$$

Its stabilizer under $J \to MJM^t$ is then by definition the Lie group:

$$\mathrm{Sp}(d, \mathbb{R}) . \tag{5.1.19}$$

Once again, we can now add a metric. Equation (5.1.16) can be read as a compatibility condition between J and g, as we will see more explicitly in the next section. Their common stabilizer now becomes

$$\mathrm{Sp}(d, \mathbb{R}) \cap \mathrm{O}(d) \cong \mathrm{U}(d/2) . \tag{5.1.20}$$

$\mathbb{R}^2$-structures in $d = 4$ Lorentzian dimensions

So far, we only gave examples in Euclidean signature, but the same ideas can also be used in Lorentzian signature. As an example, in $d = 4$ a never-vanishing Weyl spinor ζ_+ defines a $\mathrm{Stab}(\zeta_+) = \mathbb{R}^2$-structure, recalling (2.4.43). This can also be reformulated in terms of a real light-like one-form v and a complex one-form w (Section 3.3.6). In the rest of the book, a G-structure will be assumed to be in Euclidean signature unless otherwise stated.

5.1.2 More on unitary structures

We will now see more details about $\mathrm{U}(d/2)$- and $\mathrm{SU}(d/2)$-structures. First we will compare the information encoded in the almost complex structure I with that in the associated holomorphic volume form Ω. Then we will reformulate $\mathrm{SU}(d/2)$-structures in terms of Ω and the two-form J.

***I* versus Ω**

Let us see more explicitly why Ω has more information than I.

If we have Ω, we can define a holomorphic basis of one-forms[1]

$$h^a = e^a + \mathrm{i} e^{a+d/2} , \tag{5.1.21}$$

such that

$$\Omega = h^1 \wedge \ldots \wedge h^{d/2} , \tag{5.1.22}$$

inspired by (3.4.12) in flat space. If we only have I, we can determine the h^a as the i-eigenvectors of $I_a{}^b$, or in other words as solutions of

$$I \cdot h^a = \mathrm{i} h^a , \tag{5.1.23}$$

where

$$I \cdot \equiv I_a{}^b e^a \wedge \iota_{E_b} = I_m{}^n \mathrm{d}x^m \wedge \iota_{\partial_n} \tag{5.1.24}$$

[1] We keep using the same index as for the usual vielbein e^a, for lighter notation. We do put bars on the indices on the complex conjugates, as we will see shortly.

is the natural action of I, which in components acts on a form as $(I \cdot \alpha)_m = I_m{}^n \alpha_n$. But (5.1.23) only determines the h^a up to rescaling, which instead Ω fixes. In other words, if we define a holomorphic basis h^a using (5.1.23), we are only able to reconstruct Ω only up to an overall function.

We define the *holomorphic cotangent bundle*[2]

$$T^*_{1,0}M \subset (T^*M)_{\mathbb{C}} \tag{5.1.25}$$

to have a fiber spanned by the h^a at every point. One can similarly define the holomorphic tangent bundle

$$T_{1,0}M \subset (TM)_{\mathbb{C}} . \tag{5.1.26}$$

(In some contexts, especially in algebraic geometry, this might be directly called tangent bundle, omitting the qualifier "holomorphic.")

More generally, we can divide forms of higher degree according to how many of the h^a they contain and how many of their complex conjugates $\overline{h^a} \equiv \bar{h}^{\bar{a}}$; this generalizes to curved space the decomposition (3.3.12). We define the bundle

$$\Lambda^{p,q}T^*M \subset (\Lambda^{p+q}T^*M)_{\mathbb{C}} \tag{5.1.27}$$

of (p, q)-forms, whose sections are

$$\text{Span}\{h^{a_1} \wedge \ldots \wedge h^{a_p} \wedge \bar{h}^{\bar{b}_1} \wedge \ldots \wedge \bar{h}^{\bar{b}_q}\} . \tag{5.1.28}$$

(In particular, the h^a themselves are defined to be $(1, 0)$-forms, and $T^*_{1,0} = \Lambda^{1,0}T^*$.) For later use, we also note a practical way of finding p and q if a form is not already expressed as in (5.1.28). In the spirit of Sections 3.1 and 4.1.6, (5.1.23) can be generalized to an operator identity

$$[I\cdot, h^a \wedge] = (I \cdot h^a)\wedge = \mathrm{i}h^a \wedge . \tag{5.1.29}$$

Acting with this operator identity on 1 gives back (5.1.23). (Recall that $\iota_{\partial_m} 1 = 0$.) Using this and its complex conjugate, we can see that on a p, q-form $\alpha_{p,q}$

$$I \cdot \alpha_{p,q} = (p - q)\, \mathrm{i}\, \alpha_{p,q} . \tag{5.1.30}$$

So the sections of $\Lambda^{p,q}T^*$ are eigenforms of $I\cdot$.

Among the bundles (5.1.27), of particular note for us is the *canonical bundle*

$$K \equiv \Lambda^{d/2,0}T^* , \tag{5.1.31}$$

of which Ω should be a section. The space of $(d/2, 0)$-forms in flat $\mathbb{C}^{d/2} = \mathbb{R}^d$ has dimension one: all such forms are proportional to $h^1 \wedge \ldots \wedge h^{d/2}$ as in (5.1.22). So K is a line bundle; we saw in Section 4.2.1 that such a bundle is trivial if and only if it has a never-vanishing section. In other words, a never-vanishing Ω exists if and only if the manifold has trivial canonical bundle.

A useful reformulation of this criterion is via characteristic classes. We define the Chern classes of the almost complex manifold M to be

$$c_i(M) \equiv c_i(T_{1,0}M) . \tag{5.1.32}$$

[2] The symbol $_{\mathbb{C}}$ here denotes complexification, which simply means that one is allowed to consider complex rather than real linear combinations; in other words, the sections of $(T^*M)_{\mathbb{C}}$ are complex one-forms; indeed the h^a are complex.

Notice that this notion depends on the choice of I; as we commented already in Section 4.2.5, we cannot talk of Chern classes of TM without it. It would be more correct to denote (5.1.32) by $c_i(M, I)$, but one often has in mind a definite I for a given M. With (4.2.70) and Exercise 4.2.5, we find that

$$c_1(M) = 0 \quad \Leftrightarrow \quad K \text{ is topologically trivial}. \tag{5.1.33}$$

When K is not trivial, there is a more exotic possibility. We can take Ω to be not a $d/2$-form, but a section of

$$\Lambda^{d/2,0}T^* \otimes K^{-1}, \tag{5.1.34}$$

namely a $(d/2, 0)$-form valued in K^{-1}; this is a trivial bundle because of (5.1.31), so in this sense Ω does exist.

Metric determined by forms

In flat space, the two-form J_{mn} was the same as the almost complex structure $I_m{}^n$. In curved space, we can still assume this in flat indices. In other words, $J_{ab} = \begin{pmatrix} 0 & 1_{d/2} \\ -1_{d/2} & 0 \end{pmatrix}$. The holomorphic basis h^a can then be chosen such that

$$J = \frac{1}{2} J_{ab} e^a \wedge e^b = \frac{1}{2} (I^0)_a{}^b e^a \wedge e_b = \frac{\mathrm{i}}{2} \sum_{a=1}^{d/2} h^a \wedge \bar{h}^{\bar{a}}, \tag{5.1.35}$$

extending the flat-space (3.4.17). For example, for $d = 6$, this and (5.1.22) read

$$J = e^1 \wedge e^4 + e^2 \wedge e^5 + e^3 \wedge e^6 = \frac{\mathrm{i}}{2}(h^1 \wedge \bar{h}^{\bar{1}} + h^2 \wedge \bar{h}^{\bar{2}} + h^3 \wedge \bar{h}^{\bar{3}}), \tag{5.1.36a}$$

$$\Omega = (e^1 + \mathrm{i}e^4) \wedge (e^2 + \mathrm{i}e^5) \wedge (e^3 + \mathrm{i}e^6) = h^1 \wedge h^2 \wedge h^3. \tag{5.1.36b}$$

So a basis h^a such that (5.1.22) and (5.1.35) hold is called a *holomorphic vielbein*, because the metric reads

$$g = e^a e^a = \sum_{a=1}^{d/2} h^a \bar{h}^{\bar{a}}. \tag{5.1.37}$$

The curved-indices components of J,

$$J_{mn} = (I^0)_a{}^b e^a_m e_{bn}, \qquad J = \frac{1}{2} J_{mn} \mathrm{d}x^m \wedge \mathrm{d}x^n, \tag{5.1.38}$$

are now no longer equal to $I_m{}^n$ in (5.1.11). Rather, we have a compatibility condition (5.1.16). So we have in fact three tensors, which are all different: J, I, and g. Given two of them, the third is determined. For example, multiplying (5.1.16) by I on both sides, one determines g in terms of I and J:

$$g_{mn} = J_{mp} I_n{}^p = -I_m{}^p J_{pn}. \tag{5.1.39}$$

Given an almost complex structure I and a nondegenerate two-form J, if g defined by (5.1.39) is a Riemannian metric (and in particular it is symmetric) we say that I and J are *compatible*. Recall from (5.1.16) that in this case I and g form an almost Hermitian structure.

Since $I_m{}^p$ is in turn determined by Ω, we can say that the metric g is determined by J and Ω.[3] So an SU(3)-structure can also be thought of as the data of J and Ω, without even specifying g. In other words, an SU(3)-structure has more data than a metric, and in particular defines one. This is in agreement with the general expectation we anticipated after (5.1.9).

To be sure, J and Ω cannot just be any old forms. As we stated earlier, Ω has to be decomposable and nondegenerate, $\Omega \wedge \bar{\Omega} \neq 0$. J also has to be nondegenerate, $J^{d/2} \neq 0$. The compatibility (5.1.39) between J and I can be reformulated directly in terms of J and Ω. In $d = 6$, we can take this from flat space in (3.4.27), which we repeat here:

$$J \wedge \Omega = 0\,, \tag{5.1.40a}$$

$$\mathrm{vol}_6 = -\frac{1}{6} J^3 = -\frac{\mathrm{i}}{8} \Omega \wedge \bar{\Omega}\,. \tag{5.1.40b}$$

A simple example of these was given in flat $\mathbb{R}^6$ in (3.4.21). The generalization of (5.1.40b) to any d can be obtained from the vielbein expression (5.1.35):

$$\mathrm{vol}_d = \frac{(-1)^{\lfloor d/4 \rfloor}}{(d/2)!} J^{d/2} = \left(\frac{\mathrm{i}}{2}\right)^{d/2} \Omega \wedge \bar{\Omega}\,. \tag{5.1.40c}$$

We now show that these are enough for compatibility. For clarity and concreteness, we will focus on $d = 6$, commenting occasionally on the generalization to any d, which is mostly straightforward.

SU(3)-structures in $d = 6$

We fix the holomorphic vielbein so that Ω reads as in (5.1.36b) and take at first J to be any two-form. It can be decomposed according to (5.1.28) as a sum of a $(2,0)$-, a $(1,1)$-, and a $(0,2)$-form:

$$J = J_{2,0} + J_{1,1} + J_{0,2}\,. \tag{5.1.41}$$

But J is real, $J^* = J$; and complex conjugation takes a (p,q)-form to a (q,p)-form. So we also have $J^*_{2,0} = J_{0,2}$.

Now, since Ω is a $(3,0)$-form, $J \wedge \Omega = J_{0,2} \wedge \Omega$. The wedge product of a $(0,2)$-form and of a $(3,0)$-form is never zero unless one of the two is; so (5.1.40a) imposes $J_{0,2} = 0$, and by reality $J_{2,0} = 0$ too. So J is a $(1,1)$-form. As such, it can be written as

$$J = J_{a\bar{b}} h^a \wedge \bar{h}^{\bar{b}}\,. \tag{5.1.42}$$

Reality now requires $J^*_{a\bar{b}} = -J_{b\bar{a}}$; in other words, the matrix $J_{a\bar{b}}$ is anti-Hermitian. The holomorphic vielbein h^a is not completely fixed by (5.1.36b). We can still redefine $h^a \to M^a_b h^b$, as long as Ω remains the same; that means $M \in \mathrm{SL}(3,\mathbb{C})$. Under this transformation, $J_{a\bar{b}} \to M_a{}^c J_{c\bar{d}} M_{\bar{b}}{}^{\bar{d}} = (MJM^\dagger)_{a\bar{b}}$. We know an anti-Hermitian matrix can be diagonalized by unitary transformations, so by taking M unitary we can bring J into diagonal form; in a second step, with a diagonal M we

[3] Recall that the opposite is not quite true, from our discussion in Section 5.1.1, in particular following (5.1.15). It was noticed there that the data of g and I give a $\mathrm{U}(d/2)$-structure.

can make J proportional to the identity. Equation (5.1.40b) fixes the proportionality coefficient so that we get

$$J_{a\bar{b}} = \frac{\mathrm{i}}{2}\delta_{a\bar{b}} \,. \tag{5.1.43}$$

This reproduces (5.1.35). We have thus shown that (5.1.40) imply (5.1.36a). Since that determines a vielbein, we have also reconstructed the metric g.

To summarize, we can then define an SU(3)-structure as one of the following:

(1) A metric g and a pure chiral spinor η_+. (Recall from Section 2.4.3 that in fact all chiral spinors in $d = 6$ are pure.)
(2) A metric g and a three-form Ω that is decomposable and nondegenerate, as defined in (3.2.44) and (3.4.13). (Replacing Ω with the slightly weaker almost complex structure I results in a U(3)-structure.)
(3) A real two-form J and a decomposable three-form Ω that satisfy (5.1.40). (Nondegeneracy is included in (5.1.40b).) From this point of view, the metric g is determined by J and Ω.

Let us also recall how one can go from one definition to another. To map definition (1) to (2), we can define Ω as a bilinear of η_+ as in (3.4.10):

$$\Omega_{mnp} = -\eta_-^\dagger \gamma_{mnp}\eta_+ \,, \qquad \eta_- = (\eta_+)^{\mathrm{c}} \,. \tag{5.1.44a}$$

One can also extract the (slightly weaker) almost complex structure I by the annihilator property of η_+, which we first introduced way back in (2.4.2) and that in present language can be variously written as

$$\gamma_{\bar{a}}\eta_+ = 0 \,, \qquad \bar{\Pi}_m{}^n\gamma_n\eta_+ = \frac{1}{2}(\delta^n_m + \mathrm{i}I_m{}^n)\gamma_n\eta_+ = 0 \,. \tag{5.1.44b}$$

Recall from Section 2.4.3 that these are valid because any chiral spinor in $d = 6$ Euclidean dimensions is pure.

To map (1) to (3), we can define

$$J_{mn} = -\mathrm{i}\,\eta_+^\dagger\gamma_{mn}\eta_+ \,. \tag{5.1.44c}$$

as in (3.4.5), assuming unit norm for simplicity:[4]

To go from (2) back to (1), we can identify $\mathrm{Ann}(\eta_+)$ from the decomposition of Ω as a wedge of one-forms, as we saw in flat space in Section 3.4.1. (Reconstructing a spinor from its bilinears was more broadly the idea behind Section 3.3.) Finally, the logic in this section shows that g itself can be reconstructed from J and Ω, thus allowing one to go from the point of view (3) back to (2) or (1).

The formulation (3) is the one we will use most often: it has the advantage of only relying on forms and exterior algebra. It is not surprising that one can reformulate the data of a spinor in terms of forms: this was one of the themes of Section 3.3. Here, however, we see that we can even reformulate the metric in terms of forms. This will become a recurring theme in this book.

SU(2)-structures in $d = 6$

So far, we have considered SU(d/2)-structures in d dimensions. However, sometimes there are additional tensors that reduce the structure group further. A notable example

[4] Recall also that there is no one-form nor five-form bilinear, as we saw back in (3.4.19) and earlier.

is given by two chiral spinors η^a_+, $a = 1, 2$ in six dimensions. We already performed the relevant computations in Section 3.4.3, assuming them to be orthogonal; in particular, (3.4.44) and (3.4.46) show that the bilinears are all determined in terms of forms

$$v\,, \qquad j\,, \qquad \omega\,. \tag{5.1.45}$$

v is a one-form, and j, ω are two-forms; they satisfy the algebraic constraints (3.4.47) and (3.4.48). We concluded in (3.4.49) that the stabilizer of these forms is SU(2). A simple example in flat space was given in (3.4.50).

So on a manifold M_6, we reinterpret the results of Section 3.4.3 as saying that two chiral spinors $\eta^a_\pm$, or equivalently the forms (5.1.45) satisfying (3.4.47) and (3.4.48), define an SU(2)-structure.

This structure can also arise when we add an SU(3) structure to a single vector or one-form that is never vanishing. Acting with the holomorphic projector $\Pi_m{}^n$ of the SU(3)-structure, we obtain a $(1, 0)$-form v. One can then define j and ω via

$$j = J - \frac{\mathrm{i}}{2} v \wedge \bar{v}\,, \qquad \omega = \frac{1}{2}\iota_{\bar{v}}\Omega\,. \tag{5.1.46}$$

This is equivalent to defining a second spinor $\eta^2_+ = \frac{1}{2} v \cdot \eta^1_-$.

Just like for SU(3)-structures in $d = 6$, the forms (5.1.45) can be taken to define the metric. We define it so that $\iota_v j = \iota_v \omega = 0$ as in (3.4.47); in other words,

$$\mathrm{d}s^2_6 = v\bar{v} + \mathrm{d}s^2_4\,. \tag{5.1.47}$$

Then $\mathrm{d}s^2_4$ is defined by (j, ω) with the procedure explained earlier for SU(d/2)-structures in d dimensions.

SU(3)-**structures in** $d = 7$

We can also define structures of SU type in odd dimension by introducing additional vector fields or one-forms.

The case most relevant to us is that of SU(3)-structures in $d = 7$, defined, for example, by a non-Majorana spinor η. This was discussed in flat space in Section 3.4.6, and we can model our discussion on those results. In particular, we found the relevant bilinears are

$$w\,, \qquad J\,, \qquad \Omega, \tag{5.1.48}$$

where this time w is a real one-form, and (J, Ω) satisfy the same algebraic constraints of an SU(3)-structure in six dimensions. The metric is once again determined as

$$\mathrm{d}s^2_7 = w^2 + \mathrm{d}s^2_6 \tag{5.1.49}$$

so that $\iota_w J = \iota_w \Omega = 0$, and $\mathrm{d}s^2_6$ is determined by (J, Ω) as explained earlier in this subsection.

Another point of view is that we can write $\eta = \eta_{\mathrm{M}1} + \mathrm{i}\eta_{\mathrm{M}2}$, with $\eta_{\mathrm{M}a}$ Majorana. Both of these define G_2-structures, as noted in Section 5.1.3. The intersection of these two G_2 stabilizers is SU(3). The relation of $\eta_{\mathrm{M}1} = \frac{1}{2}(\eta + \eta^{\mathrm{c}})$ to η was explained around (3.4.79) in terms of bilinears.

5.1.3 Other spinorial examples

G_2-structure

We have seen in (2.4.53) and again in (3.4.83) that the stabilizer of a Majorana spinor $\eta = \eta_{\mathrm{M}}$ in $d = 7$ Euclidean dimensions is G_2. Thus a spinor on a Riemannian seven-manifold defines a G_2 structure, as a particular case of Stab(η)-structure. We can also reformulate it in terms of the three-form $\phi_{mnp} = -\mathrm{i}\eta^\dagger \gamma_{mnp}\eta$ from (3.4.79).

Once again, the properties we know from flat space will be valid in flat vielbein indices, and can be promoted to identities with curved indices upon multiplication by vielbeine: by (2.2.18),

$$\begin{aligned} \phi &= e^{147} + e^{257} + e^{367} + e^{123} - e^{453} - e^{426} - e^{156} , \\ \tilde{\phi} &= *\phi = e^{2356} + e^{1346} + e^{1245} + e^{4567} - e^{1267} + e^{1357} - e^{2347} ; \\ e^{a_1 \ldots a_k} &\equiv e^{a_1} \wedge \ldots \wedge e^{a_k} . \end{aligned} \tag{5.1.50}$$

The identity (2.2.30) becomes

$$g_{mn} = \frac{1}{24}\phi_{mpq}\,\phi_{nrs}\,\tilde{\phi}^{prqs} . \tag{5.1.51}$$

According to the general expectation explained after (5.1.9), ϕ should define a metric. We can see this explicitly as follows. In flat space, we also have the dual four-form $\tilde{\phi}$. In curved space, we cannot define this until we have defined a metric. But we can define a four-vector $\tilde{\phi}^{mnpq} = \epsilon^{mnpqrst}\phi_{rst}$: the ϵ tensor is a section of $\Lambda^7 TM$, which defines an orientation and can be defined without a metric (Section 5.1.1). Now we can define the metric using (5.1.51). If we now lower the indices of $\tilde{\phi}$ to define a four-form, it will coincide with the Hodge dual of ϕ.

It is also instructive to see what happens if we have two Majorana spinors η^a_{M}, or equivalently a single Dirac spinor η. In this case, we have learned in Section 3.4.6 that the stabilizer is reduced to SU(3) $\subset G_2$. At the level of forms, this reduction is reflected by writing

$$\phi = \mathrm{Im}\Omega + J \wedge w , \qquad *\phi = -\frac{1}{2}J^2 + \mathrm{Re}\Omega \wedge w ; \tag{5.1.52}$$

the forms w, J, Ω describe an SU(3)-structure in $d = 7$, as in (5.1.48). We derived (5.1.52) already in (3.4.81) in flat space; it agrees with (5.1.50) if we adapt (3.4.82) to curved space as $w = e^7$, $J = e^{14} + e^{25} + e^{36}$, $\Omega = \mathrm{i}(e^1 + \mathrm{i}e^4) \wedge (e^2 + \mathrm{i}e^5) \wedge (e^3 + \mathrm{i}e^6)$.

We can decompose ϕ further under an SU(2)-structure. This we can do by decomposing an SU(3)-structure using the first line of (3.4.46):

$$\begin{aligned} \phi &= \mathrm{Im}\omega \wedge \mathrm{Re}v + \mathrm{Re}\omega \wedge \mathrm{Im}v + j \wedge w + \mathrm{Re}v \wedge \mathrm{Im}v \wedge w \\ &= j_i \wedge w_i + w_1 \wedge w_2 \wedge w_3 , \end{aligned} \tag{5.1.53}$$

where in the second line we defined $j_i \equiv (\mathrm{Im}\omega, \mathrm{Re}\omega, j)$, $w_i = (\mathrm{Re}v, \mathrm{Im}v, w)$. So we can describe an SU(2)-structure in $d = 7$ with a triplet w_i of one forms, and a triplet j_i of two-forms.

Spin(7)-structure

Using the results of Section 3.4.4, an MW spinor $\eta = \eta_{\rm MW}$ in eight dimensions defines a $\text{Stab}(\eta_{\rm MW}) = \text{Spin}(7)$-structure, which can also be reformulated in terms of a certain real four-form Ψ_4.

Not just any four-form will do; a possible parameterization was given in (3.4.53), (3.4.84) in terms of an SU(4) and of a G_2 structure respectively. In particular, we can be more explicit with $\Psi_4 = \phi \wedge e^8 + \tilde{\phi}$ (3.4.84) and the vielbein expression (5.1.50):

$$\begin{aligned}\Psi = {} & e^{1478} + e^{2578} + e^{3678} + e^{1238} - e^{4538} - e^{4268} - e^{1568} \\ & + e^{2356} + e^{1346} + e^{1245} + e^{4567} - e^{1267} + e^{1357} - e^{2347} ,\end{aligned} \tag{5.1.54}$$

again in the notation where $e^{a_1 \ldots a_k} \equiv e^{a_1} \wedge \ldots \wedge e^{a_k}$. So there should be a vielbein such that Ψ has this expression.

If we consider a spinor $\eta_{\rm M}$ which is Majorana but not Weyl, we know according to Table 2.4 and to (3.4.85) that we obtain a G_2-structure in $d = 8$. The more explicit (3.4.84) shows that we can also describe it by adding a one-form w to a Spin(7)-structure.

5.1.4 Quaternionic examples

One can also define a quaternionic analogue of the $U(d/2)$-structures. This requires a little more work, since we have not encountered yet its flat-space avatar, for reasons that will become clear.

Almost hyper-Kähler structure = $\text{Sp}(d/4)$-structure

The analogue of unitary matrices over quaternions is the group

$$\text{Sp}(n) \equiv \{ S \in \text{Mat}(n, \mathbb{H}) \mid S^\dagger S = 1 \} . \tag{5.1.55}$$

Morally speaking, in the case of $SU(d/2)$-structures the imaginary unit i is realized geometrically as an almost complex structure I. Quaternions have three imaginary units i_a satisfying $i_a i_b = -\delta_{ab} 1 + \epsilon_{abc} i_c$. So we define an *almost quaternionic structure* to be a triple of tensors I_a with one lower and one upper index, such that

$$I_a I_b = -\delta_{ab} 1 + \epsilon_{abc} I_c . \tag{5.1.56}$$

These would describe a $GL(d, \mathbb{H})$-structure, but that becomes $\text{Sp}(d/4)$ upon adding a metric. Just like in the $SU(d/2)$ case, lowering one index of the I_a leads to a triplet of two-forms J_a.

The four-dimensional case

For $d = 4$,

$$\text{Sp}(1) \cong SU(2) , \tag{5.1.57}$$

so we simply recover an SU(2)-structure in $d = 4$, discussed in Section 5.1.2.

A manifestation of (5.1.57) was noticed already while studying spinors in $\mathbb{R}^4$ in Section 3.3.2. The algebraic conditions (5.1.40) for $SU(d/2)$-structures in $d = 4$ read

$$J \wedge \Omega = 0 , \qquad -\frac{1}{4} \Omega \wedge \bar{\Omega} = -\frac{1}{2} J^2 = \text{vol}_4 , \tag{5.1.58}$$

which in terms of the triplet $J_a = \{J, \mathrm{Re}\Omega, \mathrm{Im}\Omega\}$ can be rewritten as

$$J_a \wedge J_b = -2\delta_{ab}\mathrm{vol}_4\,; \tag{5.1.59}$$

these are invariant under rotations acting on the $_a$ index.

It is useful to keep in mind a flat-space prototype: for example, we can take

$$J_a = \frac{1}{2} j_{a+}^{mn} \mathrm{d}x_m \wedge \mathrm{d}x_n, \tag{5.1.60}$$

where j_{a+} are the su(2) generators in (4.4.22), which as we saw there have a quaternionic interpretation. In complex coordinates, these reproduce (3.3.10) and (3.3.11).

Almost holomorphic symplectic form

In $d > 4$, instead of the isomorphism (5.1.57) we have the inclusion

$$\mathrm{Sp}(d/4) \subset \mathrm{SU}(d/2)\,. \tag{5.1.61}$$

We then expect an Sp(d/4)-structure to define an SU(d/2)-structure. Pick one of the almost complex structures, say I_1; using the metric, it defines a U(d/2)-structure together with the corresponding two-form J_1. Combining the other two-forms,

$$\omega \equiv J_2 - \mathrm{i}J_3 \tag{5.1.62}$$

defines a two-form that is of type $(2,0)$ for I_1, as one can see from (5.1.56) and (5.1.39). Moreover,

$$\Omega_1 \equiv \frac{1}{(d/4)!}\omega^{d/4} \tag{5.1.63}$$

is of type $(d/2, 0)$; together with J_1, it defines the expected SU(d/2)-structure. (This can of course be repeated to define Ω_2 and Ω_3 as well.) Equation (5.1.62) is said to be an *almost holomorphic symplectic form*. The reason for the name is that (5.1.63) is similar to how the (d/2)th power of an almost symplectic form J is the volume form, (5.1.40c).

Conversely, given an SU(d/2)-structure (J, Ω) with $d/2$ even, suppose we can find an almost holomorphic symplectic form ω satisfying (5.1.62). Then in terms of a holomorphic vielbein, we can write $\omega = \omega_{ab} h^a \wedge h^b$; the antisymmetric ω_{ab} is nondegenerate, and so by vielbein rotation and rescalings it can be put in the form

$$\omega_0 = \begin{pmatrix} 0 & 1_{d/4} \\ -1_{d/4} & 0 \end{pmatrix}; \tag{5.1.64}$$

so $\omega = h^1 \wedge h^2 + \cdots$ Defining $J_1 \equiv J$, $J_2 \equiv \mathrm{Re}\omega$, $J_3 \equiv -\mathrm{Im}\omega$, (5.1.56) follow.

In $d = 4$, $\omega^{d/4} = \omega$, so the holomorphic symplectic form is already part of the SU(2)-structures; this is another manifestation of the difference between (5.1.57) and (5.1.61).

The definitions (5.1.62) and (5.1.63) leading to an SU(d/2)-structure can be repeated in an S^2-worth of ways: any linear combination

$$I = x_a I_a \tag{5.1.65}$$

with $x_a x_a = 1$ is an almost complex structure, and it defines an SU(d/2)-structure similar to the preceding one.

As a flat-space example, in $d = 8$ we can consider, for example, the usual $J = \frac{\mathrm{i}}{2}\mathrm{d}z^i \wedge \mathrm{d}\bar{z}^{\bar{i}}$; a holomorphic volume form is $\Omega = \mathrm{d}z^1 \wedge \mathrm{d}z^2 \wedge \mathrm{d}z^3 \wedge \mathrm{d}z^4$, and a possible holomorphic symplectic form is

$$\omega = \mathrm{d}z^1 \wedge \mathrm{d}z^2 + \mathrm{d}z^3 \wedge \mathrm{d}z^4 . \tag{5.1.66}$$

Indeed, $\frac{1}{2}\omega \wedge \omega = \Omega$. (5.1.62) gives

$$J_a = \frac{1}{2} j^{mn}_{+a} \mathrm{d}x^I_m \wedge \mathrm{d}x^I_n, \tag{5.1.67}$$

where the j_{+a} are the matrices in (4.4.22), in an appropriate choice of real coordinates x^I_m, $I = 1, 2$. This can be generalized to higher d.

Stabilizer

Let us now show that the stabilizer of the I_a is indeed $\mathrm{Sp}(d/4)$. The stabilizer of a single I_1 and of the metric is $\mathrm{U}(d/2)$. The almost holomorphic symplectic form ω is $(2,0)$ with respect to I_1. In flat space, this would be $\omega = \frac{1}{2}\omega_{ij}\mathrm{d}z^i \wedge \mathrm{d}z^j$. The stabilizer of the antisymmetric ω_{ij} can be computed by putting it in the form (5.1.64). The stabilizer of this is by definition $\mathrm{Sp}(d/2, \mathbb{C})$, similar to (5.1.19). So we have reduced our problem to computing the intersection of $\mathrm{U}(d/2)$ with $\mathrm{Sp}(d/2, \mathbb{C})$. At the Lie algebra level, this computation is similar to (2.4.34). We take an anti-Hermitian matrix (an element of $\mathrm{u}(d/2)$) and split its indices in two sets, so that in block form it reads $\lambda = \begin{pmatrix} A_1 & M \\ -M^\dagger & A_2 \end{pmatrix}$, for A_i anti-Hermitian. Now we impose the Lie algebra condition for $\mathrm{sp}(d/2, \mathbb{C})$, namely that

$$\lambda \begin{pmatrix} 0 & 1_{d/4} \\ -1_{d/4} & 0 \end{pmatrix} + \begin{pmatrix} 0 & 1_{d/4} \\ -1_{d/4} & 0 \end{pmatrix} \lambda^t = 0 . \tag{5.1.68}$$

This imposes $A_1 = -A_2^t$, $M = M^t$. We can now define $A_1 = a + \mathrm{i}s_3$, $M = s_2 + \mathrm{i}s_1$, where a is real antisymmetric and s_i are real symmetric. We end up with

$$\lambda = \begin{pmatrix} a + \mathrm{i}s_3 & s_2 + \mathrm{i}s_1 \\ -s_2 + \mathrm{i}s_1 & a - \mathrm{i}s_3 \end{pmatrix} = a \otimes 1 + \mathrm{i}s_a \otimes \sigma_a, \tag{5.1.69}$$

which is isomorphic to the Lie algebra of (5.1.55), upon mapping $\mathrm{i}\sigma_a \mapsto i_a$ as in (2.1.30).

Spinors

There is also a spinorial point of view on $\mathrm{Sp}(d/4)$-structures, but this time it is clumsier. We pick an $\mathrm{SU}(d/2)$-structure (J_1, Ω_1); to this, we associate a pure spinor η^1_+. Acting on it with the holomorphic symplectic form ω_1 (via the Clifford map) produces the $(d/4 + 1)$ spinors

$$\eta^1_+, \qquad \not{\omega}\eta^1_+, \qquad (\not{\omega})^2\eta^1_+, \qquad \ldots, \qquad (\not{\omega})^{d/4}\eta^1_+ \propto \eta^1_- . \tag{5.1.70}$$

Starting the process from another $x_a J_a$, one obtains the same spinors; the $\mathrm{SU}(2)$ rotation acts in the representation of spin $\ell = d/8$. For example, if $d = 4$ and $\eta^1_+ = |++\rangle$, (5.1.70) are $|++\rangle$ and $|--\rangle$. If $d = 8$ and $\eta^1_+ = |++++\rangle$, with ω_1 as in (5.1.64), then (5.1.70) produces the three spinors:

$$|++++\rangle, \qquad |++--\rangle + |--++\rangle, \qquad |----\rangle . \tag{5.1.71}$$

Almost quaternionic-Kähler structures

A slightly different G-structure is obtained by considering the span of the three I_a rather than the triple. This means that we don't demand the transition functions to keep all three I_a separately invariant, but rather that they leave invariant the set of almost complex structures (5.1.65).

Since the set (5.1.65) is an S^2, the generators that shuffle the I_a form an so(3) $\cong$ sp(1) and commute with the generators of sp($d/4$) we already had. The corresponding Lie group is not exactly the direct product of the two, but

$$\mathrm{Sp}(d/4)\cdot\mathrm{Sp}(1)\equiv(\mathrm{Sp}(d/4)\times\mathrm{Sp}(1))/\mathbb{Z}_2\,. \tag{5.1.72}$$

Exercise 5.1.1 Show that the stabilizer of a decomposable, nondegenerate form Ω, without considering any metric, is

$$\mathrm{Stab}_{\mathrm{GL}(d,\mathbb{R})}(I^0)=\mathrm{SL}(d/2,\mathbb{C})\,. \tag{5.1.73}$$

Compare with (5.1.15), and conclude again that the map (5.1.13) is not one-to-one.

Exercise 5.1.2 Check that the Möbius strip (Section 4.1.3) does not admit a globally defined two-form.

Exercise 5.1.3 Describe the space of G-structures at every point as a coset.

Exercise 5.1.4 Using (3.2.31) and (5.1.39), show

$$[*\lambda, I\cdot]=0\,. \tag{5.1.74}$$

(This can also be inferred from (5.1.30) and Exercise 3.3.1.)

Exercise 5.1.5 Using (2.4.25), show that at every point a G_2-structure is an open subset of the space Λ^3 of all three-forms.

Exercise 5.1.6 Show using (3.3.50) that two Lorentzian-signature Weyl spinors ζ_+^A in $d=4$ define an identity structure.

Exercise 5.1.7 Generalize (5.1.48) to an SU(k)-structure in $d=2k+1$.

Exercise 5.1.8 Show that two spinors in $d=5$ define an identity structure. (Hint: start from (5.1.45) in $d=6$, and reduce along a new vector ξ, paying attention to its relation to v.)

Exercise 5.1.9 Find the appropriate coordinates x_m^I for (5.1.67).

5.2 Decomposing forms as representations

We have seen in (5.1.27) how an almost complex structure I can be used to decompose the bundle of forms in subbundles. Here we will generalize this idea to more complicated G-structures, by decomposing forms in irreducible representations of G. This will be a useful tool in many applications, such as for the action of the Hodge $*$ operator.

5.2.1 Almost symplectic structures and Lefschetz operators

In Section 5.1.1, we mentioned that a nondegenerate two-form J defines an Sp($d,\mathbb{R}$)-structure, which we also called an almost symplectic structure.

Given that J is nondegenerate, by the comment that follows (5.1.17), the matrix J_{mn} of components of J is invertible; so the inverse $(J^{-1})^{mn}$ exists. When J comes from a Hermitian metric as in (2.4.30), we can define an almost complex structure $I_m{}^n = J_{mp}g^{pn}$; then $J^{mn} \equiv g^{mp}J_{pq}g^{qn}$ satisfies $J^{mp}J_{pn} = g^{mq}I_q{}^pI_p{}^rg_{rn} = -g^{mq}g_{qn} = -\delta^m_n$, and

$$(J^{-1})^{mn} = -J^{mn} = -g^{mp}J_{pq}g^{qn}. \tag{5.2.1}$$

In this section, we will usually not make any reference to a metric or an almost complex structure, but we are going to *define* $J^{mn} \equiv -(J^{-1})^{mn}$ so as to agree with (5.2.1).[5]

We can now define two natural operators, wedging by J or contracting with its inverse:

$$J\wedge = \frac{1}{2}J_{mn}\mathrm{d}x^m \wedge \mathrm{d}x^n\wedge : \; \Lambda^k \to \Lambda^{k+2}, \tag{5.2.2a}$$

$$J\cdot \equiv -\frac{1}{2}J^{mn}\iota_m\iota_n : \; \Lambda^{k+2} \to \Lambda^k. \tag{5.2.2b}$$

(The operator in (5.2.2b) could also be called $-\iota_J$, in the notation (3.2.64).) We can compute their commutator as usual with (3.2.3). We need

$$\begin{aligned}[\iota_m\iota_n, \mathrm{d}x^p \wedge \mathrm{d}x^q\wedge] &\overset{(2.1.31a)}{=} \iota_m[\iota_n, \mathrm{d}x^p \wedge \mathrm{d}x^q\wedge] + [\iota_m, \mathrm{d}x^p \wedge \mathrm{d}x^q\wedge]\iota_n \\ &\overset{(2.1.31f),(3.2.3a)}{=} 2\iota_m\delta^{[p}_n\mathrm{d}x^{q]}\wedge + 2\delta^{[p}_m\mathrm{d}x^{q]}\wedge\iota_n \overset{(3.2.3a)}{=} 4\delta^{[p}_{[m}\mathrm{d}x^{q]}\wedge\iota_{n]} + 2\delta^{[p}_{[n}\delta^{q]}_{m]}.\end{aligned} \tag{5.2.3}$$

From this, using (5.2.1),

$$\begin{aligned}[J\cdot, J\wedge] &= \frac{1}{4}(J^{-1})^{mn}J_{pq}[\iota_m\iota_n, \mathrm{d}x^p \wedge \mathrm{d}x^q\wedge] \\ &= -\mathrm{d}x^m \wedge \iota_m + \frac{d}{2} \overset{(3.2.22)}{=} \frac{d}{2} - \deg \equiv h.\end{aligned} \tag{5.2.4}$$

For example, we can act with it on the form 1 and get

$$J\cdot J = \frac{d}{2}. \tag{5.2.5}$$

We can confirm this in indices:

$$\begin{aligned}-\frac{1}{2}J^{mn}\iota_m\iota_nJ &= -\frac{1}{4}J^{mn}J_{pq}\iota_m 2\delta^p_n\mathrm{d}x^q = -\frac{1}{2}J^{mn}J_{pq}\delta^p_n\delta^q_m \\ &= -\frac{1}{2}J^{mn}J_{nm} \overset{(5.2.1)}{=} \frac{1}{2}(J^{-1})^{mn}J_{nm} = \frac{1}{2}\delta^m_m = \frac{d}{2}.\end{aligned} \tag{5.2.6}$$

Even more explicitly, we can work in flat indices: in $d = 6$, for example, J is given by (5.1.36a) and $J\cdot = -\iota_{E_1}\iota_{E_4} - \iota_{E_2}\iota_{E_5} - \iota_{E_3}\iota_{E_6}$; since $\iota_{E_a}e^b = \delta^b_a$ we obtain $J\cdot J = 1 + 1 + 1 = 3$.

Acting on a k-form, we can also check easily

[5] For a metric, we denote the inverse by simply raising the indices; for J, we should resist the temptation to do the same, since we just saw J^{mn} defined by raising indices with the metric is in fact *minus* the inverse.

$$[J\wedge, h] = 2J\wedge\,, \qquad [J\cdot, h] = -2J\cdot\,, \tag{5.2.7}$$

simply because they raise and lower the degree by two. Together, (5.2.4) and (5.2.7) are the commutation relations of the Lie algebra $\mathrm{sl}(2,\mathbb{R})$. If we define

$$L_+ = J\wedge\,, \qquad L_- = J\cdot\,, \qquad h = \frac{d}{2} - \deg\,, \tag{5.2.8}$$

under the isomorphism $\mathrm{sl}(2,\mathbb{R})_\mathbb{C} \cong \mathrm{su}(2)_\mathbb{C}$ with the complexification of su(2), we have

$$L_\pm \cong L_1 \pm \mathrm{i}L_2\,, \qquad h \cong -2iL_3\,, \tag{5.2.9}$$

with L_i the generators of su(2) obeying $[L_i, L_j] = \epsilon_{ijk}L_k$. These L_i are collectively known as *Lefschetz operators*.

So an almost symplectic structure organizes forms in an $\mathrm{sl}(2,\mathbb{R})$ representation with generators L_i. This is not irreducible: even and odd forms $\Lambda^\pm$ are not mixed by the three generators. In Λ^+, the forms 1 and vol are lowest- and highest-weight states, since they are automatically annihilated by L_- and L_+ respectively. But there exist other such forms ω_k such that

$$J\cdot\omega_k = 0\,, \tag{5.2.10a}$$

which are lowest-weight states in their representation, or

$$J \wedge \omega_k = 0\,, \tag{5.2.10b}$$

highest-weight states. Such forms signal that $\Lambda^\pm$ are not an irreducible representation, since that would only have one lowest- and one highest-weight state. Both (5.2.10) are called *primitive* forms, extending a term already introduced after (3.3.19) in flat space. One can build an irreducible $\mathrm{sl}(2,\mathbb{R})$ representation by acting with powers of $L_+ = J\wedge$ on a primitive form of the type (5.2.10a) until one arrives at a highest-weight state, of the type (5.2.10b).

In most of the analysis in this subsection, we did not use the metric. If we do consider it, then the Hodge dual (4.3.9) takes primitive forms of the type (5.2.10a) into the type (5.2.10b) and vice versa. More generally it exchanges the raising and lowering operators:

$$J\cdot * \lambda = - * \lambda\, J\wedge\,, \tag{5.2.11}$$

which follows easily from (3.2.31) and from the definitions (5.2.1) and (5.2.2).

5.2.2 Unitary structures and Hodge operator

SU(3)-structures and (p, q)-forms

Since an SU(3)-structure has both an I and a J (see the summary at the end of Section 5.1.2), we can now use the operators associated to both to decompose forms in SU(3) representations.

Many of the $\Lambda^{p,q}T^*M$ bundles defined in (5.1.27) are already irreducible SU(3) representations. For example, the one $(0, 0)$-form (up to rescaling) at every point is obviously a singlet, and so is the only $(3, 3)$-form vol_6. Similarly, the space of $(3, 0)$-forms is at every point generated by Ω, which is also a singlet. The space of $(1,0)$-forms transforms in the fundamental representation **3**. The space of $(2,0)$-forms transforms in the antifundamental $\bar{\mathbf{3}}$, as can be seen by using the map

$$\alpha_{ij} \to \alpha_{\bar{k}} \equiv \frac{1}{2}\bar{\Omega}_{\bar{k}}{}^{ij}\alpha_{ij}\,. \tag{5.2.12}$$

Things become more interesting for $(1,1)$-forms $\alpha_{1,1}$. Each element of a basis

$$\{dz^i \wedge d\bar{z}^{\bar{j}}\} \tag{5.2.13}$$

is a wedge product of a $(1,0)$ with a $(0,1)$-form; thus they transform in the $\mathbf{3}\otimes\bar{\mathbf{3}}$ of SU(3), which decomposes as $\mathbf{8}\oplus\mathbf{1}$. This decomposition can be seen in terms of $J\cdot$ in (5.2.2b): the singlet **1** is the space of forms of the type $J(J\cdot\alpha_{1,1})$, while the **8** is given by the forms such that $J\cdot\alpha_{1,1}=0$, called primitive in (5.2.10a). Dually to this, $(2,2)$-forms also transform in the $\bar{\mathbf{3}}\otimes\mathbf{3}=\mathbf{8}\oplus\mathbf{1}$; the **8** now consists of forms such that $J\wedge\alpha_{2,2}=0$, also defined to be primitive in (5.2.10b).

A similar story holds for $(2,1)$-forms $\alpha_{2,1}$. A basis

$$\{dz^i \wedge dz^j \wedge d\bar{z}^{\bar{k}}\} \tag{5.2.14}$$

is a wedge of a $(2,0)$-form with a $(0,1)$-form, and thus transforms as $\bar{\mathbf{3}}\otimes\bar{\mathbf{3}}\cong\bar{\mathbf{6}}\oplus\mathbf{3}$. The $\bar{\mathbf{6}}$ is annihilated by both $J\cdot$ and $J\wedge$, and is thus primitive in both senses (5.2.10) (so it is a Lefschetz singlet). The nonprimitive part in the **3** is $J\wedge(J\cdot\alpha_{2,1})$; indeed, $J\cdot\alpha_{2,1}$ is a $(1,0)$-form and thus transforms in the **3**.

We can summarize all this with the following diagram. Following a common custom, we have arranged the degrees of the forms in a diamond.

$$\begin{matrix} & & & \Lambda^{0,0} & & & \\ & & \Lambda^{1,0} & & \Lambda^{0,1} & & \\ & \Lambda^{2,0} & & \Lambda^{1,1} & & \Lambda^{0,2} & \\ \Lambda^{3,0} & & \Lambda^{2,1} & & \Lambda^{1,2} & & \Lambda^{0,3} \\ & \Lambda^{3,1} & & \Lambda^{2,2} & & \Lambda^{1,3} & \\ & & \Lambda^{3,2} & & \Lambda^{2,3} & & \\ & & & \Lambda^{3,3} & & & \end{matrix} \quad = \quad \begin{matrix} & & & \mathbf{1} & & & \\ & & \mathbf{3} & & \bar{\mathbf{3}} & & \\ & \bar{\mathbf{3}} & & \mathbf{1}\oplus\mathbf{8} & & \mathbf{3} & \\ \mathbf{1} & & \bar{\mathbf{6}}\oplus\mathbf{3} & & \mathbf{6}\oplus\bar{\mathbf{3}} & & \mathbf{1} \\ & \bar{\mathbf{3}} & & \mathbf{1}\oplus\mathbf{8} & & \mathbf{3} & \\ & & \mathbf{3} & & \bar{\mathbf{3}} & & \\ & & & \mathbf{1} & & & \end{matrix}\,. \tag{5.2.15}$$

Action of Hodge $*$

We now apply this decomposition to the Hodge $*$ operator. Since it is defined using the metric, which is an SU(3) singlet, its action maps each representation to an isomorphic one.

Using the vielbein point of view, the flat space formulas (3.4.23) and (3.4.24) still apply for their curved-space counterparts:

$$*J = -\frac{1}{2}J^2\,, \qquad *\Omega = \mathrm{i}\Omega\,. \tag{5.2.16}$$

So in this case $*$ keeps the singlets in the third row of (5.2.15) invariant, and exchanges the ones in the second and fourth rows.

Consider next $(2,1)$-forms. The $*$ commutes with SU(3) (as follows, for example, from (5.1.74)). So by Schur's lemma, it is proportional to the identity on each SU(3) irreducible representation. We then have to determine its eigenvalue on the **6** and $\bar{\mathbf{3}}$ separately. The computation can be done similarly to the flat-space ones that follow (3.3.15), using vielbeine and in particular the holomorphic ones defined in (5.1.21). First we pick a convenient primitive form, for example $h^1 \wedge h^2 \wedge \bar{h}^{\bar{3}}$. Just as discussed after (3.3.16), in four dimensions, we have $\mathrm{vol}_6 = \mathrm{d}x^1 \wedge \ldots \wedge \mathrm{d}x^6 = \epsilon_{1\bar{1}2\bar{2}3\bar{3}}\mathrm{d}z^1 \wedge \mathrm{d}\bar{z}^{\bar{1}} \wedge \mathrm{d}z^2 \wedge \mathrm{d}\bar{z}^{\bar{2}} \wedge \mathrm{d}z^3 \wedge \mathrm{d}\bar{z}^{\bar{3}}$, from which

$$\epsilon_{1\bar{1}2\bar{2}3\bar{3}} = -\frac{1}{(-2\mathrm{i})^3} = \frac{\mathrm{i}}{8} . \tag{5.2.17}$$

Now we can compute

$$* h^1 \wedge h^2 \wedge \bar{h}^{\bar{3}} \overset{(4.3.9)}{=} \epsilon_{12\bar{3}}{}^{12\bar{3}} h^1 \wedge h^2 \wedge \bar{h}^{\bar{3}} = 8\epsilon_{123\bar{1}\bar{2}3} h^1 \wedge h^2 \wedge \bar{h}^{\bar{3}} = \mathrm{i} h^1 \wedge h^2 \wedge \bar{h}^{\bar{3}} . \tag{5.2.18}$$

So we conclude

$$* \, \alpha^0_{2,1} = \mathrm{i}\alpha^0_{2,1} \tag{5.2.19}$$

on every primitive (2, 1)-form $\alpha^0_{2,1}$. We can use a similar logic for (2, 1)-forms in the **3**, namely forms of the type $J \wedge \alpha_{1,0}$. But this time, we will compute this with our fancier pure-spinor techniques. From (3.4.22) in flat space, we have

$$\begin{aligned} *\lambda\left(\alpha_{1,0} \wedge \mathrm{e}^{-\mathrm{i}J}\right) &\overset{(3.2.31)}{=} -\iota_{\alpha_{1,0}} * \lambda \mathrm{e}^{-\mathrm{i}J} \overset{(3.4.22)}{=} -\mathrm{i}\, \iota_{\alpha_{1,0}} \mathrm{e}^{-\mathrm{i}J} \\ &\overset{(3.2.52)}{=} I \cdot \alpha_{1,0} \wedge \mathrm{e}^{-\mathrm{i}J} = \mathrm{i}\alpha_{1,0} \wedge \mathrm{e}^{-\mathrm{i}J} . \end{aligned} \tag{5.2.20}$$

Decomposing this in form degrees, we get

$$* \, \alpha_{1,0} = -\frac{\mathrm{i}}{2}\alpha_{1,0} J^2 , \qquad *(\alpha_{1,0} \wedge J) = -\mathrm{i}\, \alpha_{1,0} \wedge J , \qquad * \left(\alpha_{1,0} \wedge \frac{J^2}{2}\right) = -\mathrm{i}\alpha_{1,0} . \tag{5.2.21}$$

So the $*$ action on forms $J \wedge \alpha_{1,0}$ has an opposite eigenvalue with respect to (5.2.19). We summarize the action of $*$ on three-forms as follows:

$$* : \begin{array}{cccc} \mathbf{1} & \bar{\mathbf{6}} \oplus \mathbf{3} & \mathbf{6} \oplus \bar{\mathbf{3}} & \mathbf{1} \\ -\mathrm{i} & (\mathrm{i}, -\mathrm{i}) & (-\mathrm{i}, \mathrm{i}) & \mathrm{i} \end{array} . \tag{5.2.22}$$

We now consider even forms. Expanding (3.4.22), we have

$$* \, 1 = -\frac{1}{6}J^3 , \qquad *J = -\frac{1}{2}J^2 , \qquad *\frac{1}{2}J^2 = -J , \qquad *\frac{1}{6}J^3 = -1 . \tag{5.2.23}$$

Next we have the **8** of primitive two-forms and four-forms. This can be fixed in a similar way as for primitive three-forms, by picking, for example, the form $h^1 \wedge \bar{h}^{\bar{2}}$. We find

$$* \, \alpha^0_{1,1} = J \wedge \alpha^0_{1,1} . \tag{5.2.24}$$

So the action on the two representations **8** and **1** has a different factor. The action on a general (1, 1)-form can be obtained by projecting on the two representations. Recalling (5.2.5), we see that for any (1, 1)-form $\alpha_{1,1}$ the projection $\alpha_{1,1} - \frac{1}{3} J \, J \cdot \alpha_{1,1}$ is primitive. So

$$\begin{aligned} *\alpha_{1,1} &= *\left(\alpha_{1,1} - \frac{1}{3} J\, J{\cdot}\alpha_{1,1}\right) + *\frac{1}{3} J\, J{\cdot}\alpha_{1,1} \\ &\overset{(5.2.23),(5.2.24)}{=} J \wedge \left(\alpha_{1,1} - \frac{1}{3} J\, J{\cdot}\alpha_{1,1}\right) - \frac{1}{6} J^2 J{\cdot}\alpha_{1,1} \\ &= J \wedge \alpha_{1,1} - \frac{1}{2} J^2 J{\cdot}\alpha_{1,1}\,. \end{aligned} \tag{5.2.25}$$

For (2, 2)-forms, we can use the relation (5.2.11). Using this and (5.2.24), we get

$$*\,\alpha^0_{2,2} = J{\cdot}\alpha^0_{2,2}\,. \tag{5.2.26}$$

Finally we have two-forms and four-forms in the $\mathbf{3}$ and $\bar{\mathbf{3}}$:

$$\begin{aligned} &*\,\alpha_{2,0} = -J \wedge \alpha_{2,0}\,, \\ &*\,(\alpha_{0,1} \wedge \Omega) = *\lambda(\alpha_{0,1} \wedge \Omega) \overset{(3.2.31)}{=} -\iota_{\alpha_{0,1}} * \lambda\Omega \overset{(5.2.16)}{=} -\mathrm{i}\iota_{\alpha_{0,1}}\Omega\,, \end{aligned} \tag{5.2.27}$$

SU(2)-structures

The story is very similar for SU(d/2)-structures in other dimensions; let us look for example at SU(2)-structures. The form decomposition in representations is much simpler than in (5.2.15):

$$\begin{matrix} & & \Lambda^{0,0} & & \\ & \Lambda^{1,0} & & \Lambda^{0,1} & \\ \Lambda^{2,0} & & \Lambda^{1,1} & & \Lambda^{0,2} \\ & \Lambda^{2,1} & & \Lambda^{2,1} & \\ & & \Lambda^{2,2} & & \end{matrix} \;=\; \begin{matrix} & & \mathbf{1} & & \\ & \mathbf{2} & & \mathbf{2} & \\ \mathbf{1} & & \mathbf{1} \oplus \mathbf{3} & & \mathbf{1} \\ & \mathbf{2} & & \mathbf{2} & \\ & & \mathbf{1} & & \end{matrix}\,. \tag{5.2.28}$$

We have used the fact that the $\mathbf{2}$ is real, in the sense that the $\bar{\mathbf{2}}$ is equivalent to it. The three singlets in the middle row are Ω, J, and $\bar{\Omega}$.

The analogue of (5.2.22) in $d = 6$ is the action on two-forms. The action on the three singlet two-forms was given in (3.3.15), and it is still valid in curved space:

$$*\,J = -J\,, \qquad *\Omega = -\Omega\,. \tag{5.2.29}$$

The $\mathbf{3}$, namely the primitive (1, 1)-forms, are then self-dual. (This too was anticipated in flat space, in (3.3.19).) So we have

$$*: \begin{matrix} \mathbf{1} & \mathbf{1} \oplus \mathbf{3} & \mathbf{1} \\ -1 & (-1, 1) & -1 \end{matrix}\,. \tag{5.2.30}$$

We will not describe the same results in arbitrary d, but for future reference we do note the analogue of (5.2.16):

$$*\,J = (-1)^{\lfloor d/4 \rfloor} \frac{1}{(d/2-1)!} J^{d/2-1}\,, \qquad *\Omega = \mathrm{i}^{d/2}\Omega\,. \tag{5.2.31}$$

The signs and factors can be worked out from the vielbein expression (5.1.35).

5.2.3 Representations for G_2-structures

In the G_2 case, we cannot define $\Lambda^{p,q}$; we can only use ϕ and $\tilde{\phi}$ to decompose forms. The space of one-forms is irreducible: it transforms as the fundamental representation $\mathbf{7}$ of G_2. But that of two-forms can be reduced. Consider the equation

$$\phi^{mnp}\alpha^0_{mn} = 0\,. \tag{5.2.32}$$

These are seven constraints (one for each choice of the free index p), which can be easily seen to be independent; so the space of solutions to (5.2.32) is $\binom{7}{2}-7 = 21-7 =$ 14-dimensional. (One could try to use $\tilde{\phi}_{mnpq}$ to further decompose this space, but (2.2.29) shows that $\tilde{\phi}^{mnpq}\alpha_{mn} = 0$ follows from (5.2.32).) So (5.2.32) form the **14** of G_2, the adjoint representation. We will call these *primitive*, by analogy with the SU(d/2)-structure case. The orthogonal complement is the space of forms, which can be written as

$$\phi_{mnp} w^p \,, \tag{5.2.33}$$

for some w^p, which is seven dimensional and forms the fundamental **7** of G_2.

For three-forms, we can write two independent conditions using ϕ and $\tilde{\phi}$:

$$\phi^{mnp}\alpha^0_{mnp} = 0 \,, \qquad \tilde{\phi}^{mnpq}\alpha^0_{mnp} = 0 \,. \tag{5.2.34}$$

These are now 7 + 1 constraints; so the space of forms obeying them is $\binom{7}{3} - 7 - 1 = 35 - 8 = 27$-dimensional. There is an irreducible representation of G_2 with that dimension; again we define these three-forms to be primitive. The complement consists of two representations: those that can be written as $\tilde{\phi}_{mnpq}w^q$, a **7**; and those that are proportional to ϕ_{mnp}, a **1**.

Higher forms can be related to this analysis by using the $*$ operator. To summarize:

$$\begin{array}{cccccccc} \Lambda^0 & \Lambda^1 & \Lambda^2 & \Lambda^3 & \Lambda^4 & \Lambda^5 & \Lambda^6 & \Lambda^7 \\ \mathbf{1} & \mathbf{7} & \mathbf{14}\oplus\mathbf{7} & \mathbf{27}\oplus\mathbf{7}\oplus\mathbf{1} & \mathbf{27}\oplus\mathbf{7}\oplus\mathbf{1} & \mathbf{14}\oplus\mathbf{7} & \mathbf{7} & \mathbf{1} \end{array} \,. \tag{5.2.35}$$

The action of $*$ maps each G_2 representation in an isomorphic one. To work out the proportionality factors, we recall (5.1.50) and consider simple examples; for example, for two-forms, $e^1 \wedge e^4 - e^2 \wedge e^5$ in the **14**, and $e^1 \wedge e^4 + e^2 \wedge e^5 + e^3 \wedge e^6$ in the **7**. We obtain

$$* \,\alpha^0_2 = \phi \wedge \alpha^0_2 \,, \qquad *(\iota_v\phi) = -\frac{1}{2}\phi \wedge (\iota_v\phi) \,. \tag{5.2.36}$$

Finally, we can also work out how the G_2 representations in (5.2.35) decompose if the G_2-structure is reduced to SU(3) as in (5.1.52):

$$\mathbf{7} \to \mathbf{3}\oplus\bar{\mathbf{3}}\oplus\mathbf{1} \,, \qquad \mathbf{14} \to \mathbf{8}\oplus\mathbf{3}\oplus\bar{\mathbf{3}}\oplus\mathbf{1} \,, \qquad \mathbf{27} \to \mathbf{8}\oplus\mathbf{6}\oplus\bar{\mathbf{6}}\oplus\mathbf{3}\oplus\bar{\mathbf{3}}\oplus\mathbf{1} \,. \tag{5.2.37}$$

Exercise 5.2.1 Use (5.2.4) to compute $J \cdot J^k$.

Exercise 5.2.2 Rederive (5.2.21) and (5.2.24) with an explicit computation similar to the one we followed to obtain (5.2.19).

Exercise 5.2.3 Again with this explicit method, show (5.2.27).

Exercise 5.2.4 Compute the analogue of (5.2.36) on three-forms.

5.3 Complex geometry

So far, we have only considered G-structures at the algebraic level; we did not act with any derivatives. In this section, we will study the action of various differential operators, such as the exterior differential d and the covariant derivative ∇_m. A general theme of this section will be the notion of *integrable* G-structure, which means that there is a coordinate system in which the tensors T_a defining it are constant.

5.3.1 Complex structures

Local complex structure

In flat space, a complex structure was defined in Section 2.4 to be a choice of complex coordinates in $\mathbb{R}^d$ (for d even). On curved space, we introduced in (5.1.12) the concept of almost complex structure I, which is a tensor that squares to -1 at every point. It is a priori not clear that one can define complex coordinates from such an I. The closest we have gotten is the concept of holomorphic basis of one-forms h^a, defined from I in (5.1.23) or from a $(d/2, 0)$-form Ω in (5.1.22). Notice that at this point we don't need to mention a metric yet, and hence the h^a are not necessarily a holomorphic vielbein.

The h^a cannot necessarily be written in terms of complex coordinates: in general, there need not exist coordinates z^i such that

$$h^a = h^a{}_i \mathrm{d}z^i \,. \tag{5.3.1}$$

Of course, the $h^a = e^a + \mathrm{i} e^{a+d/2}$ are linear combinations of the real coordinates $\mathrm{d}x^m$; (5.3.1) demands that complex coordinates z^i exist, so that the h^a are linear combinations of the $\mathrm{d}z^i$ but don't involve the $\mathrm{d}\bar{z}^{\bar{\imath}}$. If (5.3.1) holds, the almost complex structure I is said to be *integrable*; alternatively, it is called a *complex structure.* In indices relative to these complex coordinates, I has an expression identical to the flat space expression (2.4.4a):

$$I_i{}^j = \mathrm{i}\delta^j_i \,, \qquad I_{\bar{\imath}}{}^{\bar{\jmath}} = -\mathrm{i}\delta_{\bar{\imath}}{}^{\bar{\jmath}} \,, \qquad I_i{}^{\bar{\jmath}} = I_{\bar{\imath}}{}^j = 0 \,. \tag{5.3.2}$$

So this agrees with the general definition of integrability we gave at the beginning of this section.

A metric g is compatible with I if (5.1.16) holds; as observed there, a metric always exists. One sometimes calls (I, g) a *Hermitian structure* (now without the "almost"). In terms of holomorphic indices, (5.1.16) or (5.1.39) give

$$g_{i\bar{\jmath}} = -\mathrm{i} J_{i\bar{\jmath}} \,, \qquad g_{ij} = 0 = g_{\bar{\imath}\bar{\jmath}} \,. \tag{5.3.3}$$

Recall that our basis h^a above is called a holomorphic vielbein if $\mathrm{d}s^2 = \sum_{a=1}^{d/2} h^a \bar{h}^a$ as in (5.1.37).

For future use, we note here what the Levi-Civita connection (4.1.16) becomes in this language:

$$\begin{aligned}
\Gamma^i_{jk} &= g^{i\bar{l}} \partial_{(j} g_{k)\bar{l}} \,, & \Gamma^{\bar{\imath}}_{jk} &= 0 \,, \\
\Gamma^i_{j\bar{k}} &= g^{i\bar{l}} \partial_{[\bar{k}} g_{\bar{l}]j} \,, & \Gamma^{\bar{\imath}}_{j\bar{k}} &= g^{\bar{\imath}l} \partial_{[j} g_{l]\bar{k}} \,, \\
\Gamma^i_{\bar{\jmath}\bar{k}} &= 0 \,, & \Gamma^{\bar{\imath}}_{\bar{\jmath}\bar{k}} &= g^{\bar{\imath}l} \partial_{(\bar{\jmath}} g_{\bar{k})l} \,.
\end{aligned} \tag{5.3.4}$$

Local example of nonintegrable almost complex structure

Equation (5.3.1) is a local condition, which can fail even in the neighborhood of a point. In $d = 2$, we will see that (5.3.1) is always met: any one-form h such that $h \wedge \bar{h} \neq 0$ can be written as $h = f\mathrm{d}z$, for some function z of the real coordinates. But already in $d = 4$, (5.3.1) can fail. A simple example that does not satisfy it is

$$h^1 = \mathrm{d}x^1 + \mathrm{i}\, x^4 \mathrm{d}x^2 \,, \qquad h^2 = \mathrm{d}x^3 + \mathrm{i}\, x^2 \mathrm{d}x^4 \,, \tag{5.3.5}$$

which is a basis of $(1, 0)$-forms for the almost complex structure

$$I = \begin{pmatrix} 0 & 1/x^4 & 0 & 0 \\ -x^4 & 0 & 0 & 0 \\ 0 & 0 & 0 & 1/x^2 \\ 0 & 0 & -x^2 & 0 \end{pmatrix} . \tag{5.3.6}$$

Suppose a solution to (5.3.1) did exist: then $\mathrm{d}h^1 = \mathrm{d}(h^1{}_a) \wedge \mathrm{d}z^a$ would have no component proportional to $\mathrm{d}\bar{z}^{\bar{1}} \wedge \mathrm{d}\bar{z}^{\bar{2}}$, and thus no $(0, 2)$-part. But (5.3.5) gives

$$\mathrm{d}h^1 = \mathrm{i}\mathrm{d}x^4 \wedge \mathrm{d}x^2 = \frac{\mathrm{i}}{4x^2 x^4}(h^1 - \bar{h}^{\bar{1}}) \wedge (h^2 - \bar{h}^{\bar{2}}) \,, \tag{5.3.7}$$

which does have a $(0, 2)$-form part $\frac{\mathrm{i}}{4x^2x^4}\bar{h}^{\bar{1}} \wedge \bar{h}^{\bar{2}}$. So (5.3.1) cannot hold for (5.3.5), and (5.3.6) is not integrable.

Complex manifolds

Consider now an almost complex structure I on a smooth manifold M, which is integrable everywhere. M is now called a *complex manifold*. A holomorphic vielbein $h^a_{(\alpha)}$ on U_α is a linear combination of the $h^a_{(\beta)}$ on another patch U_β, but not of its complex conjugates: this is a consequence of (5.1.10). So if the integrability condition (5.3.1) is met on every patch, the $\mathrm{d}z^i_{(\alpha)}$ are linear combinations of the $\mathrm{d}z^i_{(\beta)}$, but not of its complex conjugates. In other words, the transition functions are *holomorphic*:

$$z^i_{(\alpha)} = g_{\alpha\beta}(z^i_{(\beta)}) \,. \tag{5.3.8}$$

We can now define a function to be holomorphic on M if it is holomorphic on every patch. Without (5.3.8), such a definition would not make sense. Unfortunately, when M is *compact*, holomorphic functions don't actually exist: this is *Liouville's theorem*, but generalizations such as meromorphic ones do.

Of course, each U_α is homeomorphic to a subset of $\mathbb{R}^d$, by the definition of manifold; so if we want, we can always choose some local way of pairing the local real coordinates $x^m_{(\alpha)}$ into complex coordinates $z^i_{(\alpha)}$. However, in general such a choice will have nothing to do with the h^a defined in (5.1.22) or (5.1.23), and (5.3.1) will not hold. If we try to ignore the initially assigned h^a, and simply use the local $z^i_{(\alpha)}$ on each U_α, we can no longer use to our advantage the fact that the $h^i_{(\alpha)}$ don't mix with the $\bar{h}^{\bar{i}}_{(\beta)}$ on $U_\alpha \cap U_\beta$. So in general the transition functions $g_{\alpha\beta}$ would not be holomorphic: the $z^i_{(\alpha)}$ would be functions not only of the $z^i_{(\beta)}$, but also of the $\bar{z}^i_{(\alpha)}$. One could then not consistently define holomorphic functions on M.

Just like two topological spaces are considered equivalent if they are homeomorphic, and two smooth manifolds are considered equivalent if they are diffeomorphic, two complex manifolds are considered equivalent if they are connected by a *biholomorphic transformation*: a holomorphic map ϕ that is invertible, and whose inverse is also holomorphic.

As we mentioned earlier, a similar notion for bundles also exists. A *holomorphic bundle* is one whose transition functions are holomorphic on M.

Complex structure on S^2

As a simple example, consider S^2. We already introduced a complex structure when we considered complex coordinates $z_{\rm N}$ and $z_{\rm S}$ in Section 4.1.3; the transition function is (4.1.44), which is holomorphic.

At first, we ignore the metric and consider h to be just a basis of $(1,0)$-forms. The simplest is defined by $h_{\rm N} = {\rm d}z_{\rm N}$ and $h_{\rm S} = -{\rm d}z_{\rm S}$ on the two charts. The two are related by $h_{\rm S} = z_{\rm N}^{-2} h_{\rm N}$; comparison with (4.2.45) shows that h is not a section of the bundle $K = T^*_{1,0}$, but rather of $K \otimes O(2)$. Since h has no zeros, the line bundle $K \otimes O(2)$ must be trivial; in other words,

$$K = T^*_{1,0} = \Lambda^{1,0} T^* = O(-2)\,. \tag{5.3.9}$$

On the other hand,

$$T_{1,0} = O(2)\,. \tag{5.3.10}$$

By our discussion in Section 4.2.4, this is related the famous fact that a vector field on an S^2 always has two zeros.

If we also consider the metric (4.1.47), then we want h to be not just a basis but a *holomorphic vielbein*, that is a form that satisfies ${\rm d}s^2 = h\bar{h}$ in both patches. One choice is

$$h_{\rm N} = 2\frac{{\rm d}z_{\rm N}}{1+|z_N|^2}\,, \qquad h_{\rm S} = -2\frac{{\rm d}z_{\rm S}}{1+|z_S|^2}\,. \tag{5.3.11}$$

The expressions on the two patches are related by $h_{\rm S} = \frac{|z_{\rm N}^2|}{z_{\rm N}^2} h_{\rm N}$; from (4.2.49), we see that h is a section of $K \otimes U_2$. Indeed,

$${\rm d}h = -2{\rm i}A \wedge h\,. \tag{5.3.12}$$

In $U_{\rm N} \cap U_{\rm S}$, we can use the (θ, ϕ) coordinates, where we have

$$h = {\rm d}\theta + {\rm i}\sin\theta {\rm d}\phi\,. \tag{5.3.13}$$

If we view S^2 as the Riemann sphere (4.1.45), $\mathbb{C} \cup \{z = \infty\}$, the group of biholomorphic transformations $S^2 \to S^2$ is PSL$(2,\mathbb{C})$, acting by Möbius transformations (2.2.24).

5.3.2 Various equivalent criteria for integrability

From the holomorphic volume form

The example in (5.3.6) already suggests a necessary criterion for (5.3.1): ${\rm d}h^a$ should contain terms of the type $h^b \wedge h^c$ and $h^b \wedge \bar{h}^{\bar{c}}$, but no terms $\bar{h}^{\bar{b}} \wedge \bar{h}^{\bar{c}}$. In other words, there should exist one-forms $\xi^a{}_b$ such that

$${\rm d}h^a = \xi^a{}_b \wedge h^b\,. \tag{5.3.14}$$

The resemblance with (4.1.97) is not accidental: integrability of a complex structure is a sort of complex analogue of the dual Frobenius theorem. We will see soon what sort of distribution is involved here. (We saw around (5.1.4) that tangent distributions are also G-structures.)

We can also write (5.3.14) more succinctly by projecting on the $(0,2)$ part:

$$(dh^a)_{0,2} = 0\,. \tag{5.3.15}$$

One can also reformulate this criterion in terms of Ω. From (5.1.22), we compute

$$\begin{aligned} d\Omega = (dh^1) \wedge h^2 \wedge \ldots \wedge h^{d/2} - h^1 \wedge (dh^2) \wedge h^3 \wedge \ldots \wedge h^{d/2} \\ + \cdots + (-1)^{d/2-1} h^1 \wedge \ldots \wedge (dh^{d/2}). \end{aligned} \tag{5.3.16}$$

If (5.3.15) holds, each dh^a in (5.3.16) only generates a $(2,0)$- and a $(1,1)$-term; the former cancels out in $d\Omega$, while the latter gives rise to $(d/2,1)$-forms. If on the other hand (5.3.15) doesn't hold, then dh^a can also generate $(0,2)$-terms, and (5.3.16) will have a $(d/2-1,2)$-part as well. So (5.3.15) can be reformulated as

$$(d\Omega)_{d/2-1,2} = 0, \tag{5.3.17a}$$

or equivalently

$$d\Omega = (d\Omega)_{d/2,1}\,. \tag{5.3.17b}$$

Finally, a $(d/2,1)$-form can always be written as a wedge of a $(0,1)$-form with Ω. So yet another reformulation is

$$d\Omega = W \wedge \Omega \quad \Leftrightarrow \quad I \text{ integrable}\,. \tag{5.3.18}$$

As we will see later, (5.3.17) or (5.3.18) are also *sufficient* for (5.3.1), and so for integrability of the almost complex structure I, and thus equivalent to it. These conditions on $d\Omega$ are often the most convenient way to check integrability in practice.

For example, in $d = 2$ (5.3.17b) says that $d\Omega$ should be a $(1,1)$-form; but in $d = 2$ all two-forms are proportional to $h^1 \wedge \bar{h}^{\bar{1}}$, which is a $(1,1)$-form. So in $d = 2$, (5.3.17b) is automatic and any complex structure is integrable, as we mentioned before (5.3.5). Already in $d = 4$, this logic fails: now (5.3.17b) says that $d\Omega$ should be a $(2,1)$-form; but there now exist also $(1,2)$-forms $h^a \wedge \bar{h}^{\bar{1}} \wedge \bar{h}^{\bar{2}}$. In the example (5.3.5), $d\Omega = d(h^1 \wedge h^2)$ does have a $(1,2)$-form part $\frac{i}{4x^2x^4}(h^1 + h^2) \wedge \bar{h}^{\bar{1}} \wedge \bar{h}^{\bar{2}}$, confirming I in (5.3.6) is nonintegrable.

As we remarked after (5.1.31), Ω does not always exist globally as a $d/2$-form: if K is not topologically trivial, Ω only exists as a twisted $d/2$-form, a section of (5.1.34). In this case, we can read (5.3.18) as the vanishing of a covariant derivative $D\Omega \equiv (d - W\wedge)\Omega = 0$, in the spirit of (4.2.15). W is then interpreted as a connection for the canonical bundle K.

Nijenhuis tensor

There are other ways to check the integrability of an almost complex structure I, which don't rely on Ω. One possibility is to work with vector fields rather than with forms. If (5.3.1) holds, we can define the vector fields

$$\partial_i \equiv \partial_{z^i}\,. \tag{5.3.19}$$

These are sections of $T_{1,0}M$, and in fact a basis; so every section of $T_{1,0}M$ can be written as $v^i\partial_i$. The Lie bracket of two such vector fields reads

$$[v^i\partial_i, w^j\partial_j] = (v^i\partial_i w^j - w^i\partial_i v^j)\partial_j \tag{5.3.20}$$

and thus is again a section of $T_{1,0}M$. With a slight abuse of notation, we can write this symbolically as

$$[T_{1,0}M, T_{1,0}M] \subset T_{1,0}M. \tag{5.3.21}$$

The equivalence of (5.3.21) to integrability is called the *Newlander–Nirenberg theorem*. The resemblance with (4.1.95) is again not accidental; (5.3.21) can be seen as a complex extension of the Frobenius theorem, dual to (5.3.14).

Let us see (5.3.21) more explicitly in terms of the almost complex structure I. Working now in real indices, $T_{1,0}M$ is the image of the projector

$$\Pi_n{}^p \equiv \frac{1}{2}\left(\delta_n^p - \mathrm{i}I_n{}^p\right), \tag{5.3.22}$$

a generalization of the flat-space (2.4.3). In other words, $(1,0)$-vectors are of the form $\Pi_m{}^n\partial_n$. The Lie bracket of two such vectors reads

$$[\Pi_m{}^p\partial_p, \Pi_n{}^q\partial_q] = \left(\Pi_m{}^p\partial_p\Pi_n{}^q - \Pi_n{}^p\partial_p\Pi_m{}^q\right)\partial_q = 2\Pi_{[m|}{}^p\partial_p\Pi_{|n]}{}^q\partial_q\,. \tag{5.3.23}$$

To impose (5.3.21), we can demand that the $(0,1)$-part of this vector field vanishes, which can be done using the $\bar\Pi$ projector, $\bar\Pi(v^m\partial_m) \equiv v^m\bar\Pi_m{}^n\partial_n$. This gives

$$\begin{aligned} 0 &= \bar\Pi[\Pi_m{}^p\partial_p, \Pi_n{}^q\partial_q] = 2\bar\Pi_q{}^r\Pi_{[m|}{}^p\partial_p\Pi_{|n]}{}^q\partial_r \\ &= 2(-I_{[m|}{}^p\partial_p I_{|n]}{}^r + I_q{}^r\partial_{[m}I_{n]}{}^q)\partial_r - 2\mathrm{i}(\partial_{[m}I_{n]}{}^r + I_q{}^r I_{[m|}{}^p\partial_p I_{|n]}{}^q)\partial_r \\ &= \bar\Pi_q{}^r N_{mn}{}^q\partial_r \end{aligned} \tag{5.3.24}$$

where the *Nijenhuis tensor* is

$$N_{mn}{}^r \equiv -2I_{[m|}{}^p\partial_p I_{|n]}{}^r + 2I_q{}^r\partial_{[m}I_{n]}{}^q\,. \tag{5.3.25}$$

Thus setting $N_{mn}{}^r = 0$ is equivalent to (5.3.21), and is a necessary condition for I to be integrable.

Let us show that the $N_{mn}{}^r = 0$ condition (and thus (5.3.21)) is equivalent to (5.3.17b). We will actually use the complex conjugate of the latter. The strategy is very similar to that in (4.3.56) of "dualizing" the definition of the anholonomy coefficients in terms of Lie brackets of vector fields. Consider two $(1,0)$ vectors v, w. They contract to zero on $\bar\Omega$:

$$\iota_v\bar\Omega = \iota_w\bar\Omega = 0\,. \tag{5.3.26}$$

By (4.1.81) and the definition of (anti-)commutator, it follows that

$$\iota_{[v,w]}\bar\Omega = [\{\iota_v, \mathrm{d}\}, \iota_w]\bar\Omega = -\iota_w\{\iota_v, \mathrm{d}\}\bar\Omega = -\iota_w\iota_v\mathrm{d}\bar\Omega\,. \tag{5.3.27}$$

If (5.3.21) holds, $[v,w]$ is $(1,0)$ and also annihilates $\bar\Omega$, so the left-hand side is zero; it then follows that $\iota_w\iota_v\mathrm{d}\bar\Omega = 0$. But if (5.3.17b) did not hold, $\mathrm{d}\bar\Omega$ would have at least one component with two or more h^a, and $\iota_w\iota_v\mathrm{d}\bar\Omega$ could not possibly be zero; so (5.3.17b) does in fact hold. This logic can also be inverted.

Equation (5.3.17b) implies $d\bar{\Omega}$ is a $(1, d/2)$-form, and has only one h^a. Then $\iota_w \iota_v$ annihilate it, and (5.3.27) implies $\iota_{[v,w]}\bar{\Omega} = 0$; in other words, $[v, w]$ is $(1, 0)$, and (5.3.21) is proven. We have thus shown equivalence of the criteria for integrability in terms of Ω, (5.3.17b), and in terms of $T_{1,0}M$, (5.3.21).

Dolbeault differential

When I is a complex structure, we can introduce a version of the exterior differential d in (4.1.73), using the complex coordinates z^i:

$$\partial \equiv dz^i \wedge \partial_i \; : \Lambda^{p,q}T^* \to \Lambda^{p+1,q}T^* . \tag{5.3.28}$$

This is called *Dolbeault differential.* It satisfies

$$\partial^2 = 0 , \tag{5.3.29}$$

by the same logic that gave us (4.1.75). We will also use the complex conjugate $\bar{\partial} = d\bar{z}^{\bar{\imath}} \wedge \bar{\partial}_{\bar{\imath}}$. The exterior differential can be written as a real combination of these two:

$$d = \partial + \bar{\partial} . \tag{5.3.30}$$

Another real combination we will sometimes use is

$$d^c \equiv i(\bar{\partial} - \partial) . \tag{5.3.31}$$

This can also be written as

$$d^c = [d, I\cdot] , \tag{5.3.32}$$

where $I\cdot$ was defined in (5.1.24). We can see this easily on a test form:

$$\begin{aligned} d^c\alpha_{p,q} &= [d, I\cdot]\alpha_{p,q} = d(i(p-q)\alpha_{p,q}) - I\cdot(d\alpha_{p,q}) \\ &= i(p-q)(\partial\alpha_{p,q} + \bar{\partial}\alpha_{p,q}) - (i(p+1-q)\partial\alpha_{p,q} + i(p-q-1)\bar{\partial}\alpha_{p,q}) \\ &= i(-\partial + \bar{\partial})\alpha_{p,q} . \end{aligned} \tag{5.3.33}$$

The Dolbeault differential can be defined even if I is not integrable, and only an almost complex structure. For this, we go back to working in real indices and use (5.3.32) as a definition of d^c, and then

$$\partial = \frac{1}{2}(d + i\, d^c) = \frac{1}{2}(d + i[d, I\cdot]) \tag{5.3.34}$$

as a definition of ∂. We will now see that this satisfies (5.3.29) if and only if I is integrable.

In the spirit of the computations of Section 4.1.6 (see, for example, (4.1.78)), explicitly we have

$$d^c = [d, I\cdot] = [dx^p \wedge \partial_p, I_m{}^n dx^m \wedge \iota_n] \overset{(2.1.31),\,(3.2.3)}{=} \partial_p I_m{}^n dx^p \wedge dx^m \iota_n - I_m{}^n dx^m \wedge \partial_n . \tag{5.3.35}$$

A longer but similar computation now gives

$$[d^c, I\cdot] = [[d, I\cdot], I\cdot] = -d + N_{mn}{}^p dx^m \wedge dx^n \wedge \iota_p . \tag{5.3.36}$$

So $N_{mn}{}^p = 0$ (and thus (5.3.21)) is equivalent to $[\mathrm{d}^c, I\cdot] = -\mathrm{d}$. This already gives us an alternative formulation of integrability. To reformulate it in terms of ∂, notice that the anticommutator $\{\mathrm{d}, \mathrm{d}^c\}$ vanishes for algebraic reasons:

$$\{\mathrm{d}, \mathrm{d}^c\} = \{\mathrm{d}, [\mathrm{d}, I\cdot]\} \overset{(2.1.32b)}{=} [\{\mathrm{d}, \mathrm{d}\}, I\cdot] - \{\mathrm{d}, [\mathrm{d}, I\cdot]\} \overset{(4.1.75)}{=} -\{\mathrm{d}, \mathrm{d}^c\}$$
$$\Rightarrow \quad \{\mathrm{d}, \mathrm{d}^c\} = 0\,. \tag{5.3.37}$$

In the language of Section 4.1.6, recall that d and d^c are to be thought of as fermionic, since they contain an odd number of $\mathrm{d}x^m\wedge$ and ι_m operators, while $I\cdot$ contains an even number of them and is to be considered bosonic. Now

$$\begin{aligned}\{\mathrm{d}^c, \mathrm{d}^c\} &= \{\mathrm{d}^c, [\mathrm{d}, I\cdot]\} = [\{\mathrm{d}^c, \mathrm{d}\}, I\cdot] - \{\mathrm{d}, [\mathrm{d}^c, I\cdot]\} \\ &\overset{(5.3.37)}{=} -\{\mathrm{d}, [\mathrm{d}^c, I\cdot]\} \overset{(5.3.36)}{=} -\{\mathrm{d}, N_{mn}{}^p \mathrm{d}x^m \wedge \mathrm{d}x^n \wedge \iota_p\}\,;\end{aligned} \tag{5.3.38}$$

the right-hand side now contains the term $N_{mn}{}^p \mathrm{d}x^m \wedge \mathrm{d}x^n \wedge \partial_p$, which is zero if and only if $N_{mn}{}^p = 0$. In other words,

$$(\mathrm{d}^c)^2 = \frac{1}{2}\{\mathrm{d}^c, \mathrm{d}^c\} = 0 \quad \Leftrightarrow \quad N_{mn}{}^p = 0\,. \tag{5.3.39}$$

Finally then

$$\partial^2 = \frac{1}{2}\{\partial, \partial\} \overset{(5.3.34),\,(4.1.75)}{=} -\frac{1}{8}\{\mathrm{d}^c, \mathrm{d}^c\} = 0 \quad \Leftrightarrow \quad N_{mn}{}^p = 0\,. \tag{5.3.40}$$

Thus $\partial^2 = 0$, with ∂ defined as (5.3.34), is equivalent to integrability of I. If that is the case, ∂ can then be written as in (5.3.28).

Summary of integrability criteria

To summarize, we have seen that the integrability of an almost complex structure I can be formulated in the following equivalent ways:

(1) As the existence of complex coordinates z^i such that (5.3.1)
(2) As the differential condition (5.3.15) on the h^a, or (5.3.17b) on the holomorphic volume form Ω
(3) As the closure of $T_{1,0}M$ under Lie brackets, (5.3.21), which explicitly becomes the vanishing of the Nijenhuis tensor (5.3.25)
(4) As the Dolbeault differential (5.3.34) squaring to zero, $\partial^2 = 0$

5.3.3 Dolbeault cohomology

The property (5.3.29) is similar to (4.1.75), which motivated in part our definition of de Rham cohomology in (4.1.132). So it is natural to also define the *Dolbeault cohomology*

$$H^{p,q}(M) = \frac{\{\alpha_{p,q} \,|\, \bar\partial \alpha_{p,q} = 0\}}{\{\bar\partial \alpha_{p,q-1}\}}\,, \tag{5.3.41}$$

where $\alpha_{p,q}$ is a p, q-form. (The use of the complex conjugate $\bar\partial$ instead of ∂ itself is of course conventional.)

Bundle-valued version

Dolbeault cohomology also has an analogue for bundle-valued forms. Recalling the bundle covariant derivative (4.2.15), we can introduce

$$\bar{\partial}_E \equiv \bar{\partial} + A_{0,1} \tag{5.3.42}$$

where A is a connection on a vector bundle E. If

$$F_{0,2} = \bar{\partial} A_{0,1} + A_{0,1} \wedge A_{0,1} = 0\,, \tag{5.3.43}$$

recalling (4.2.18) we see that the operator $\bar{\partial}_E$ squares to zero and is called a *twisted Dolbeault differential.* We can define with it a cohomology similar to (5.3.41):

$$H^{p,q}(M, E)\,. \tag{5.3.44}$$

The existence of a twisted Dolbeault differential $\bar{\partial}_E$ is equivalent to E having holomorphic transition functions. This is the analogue for bundles of the equivalence (5.3.40) of $\bar{\partial}^2 = 0$ and integrability.

Since the differential in (5.3.41) only acts on the $\mathrm{d}\bar{z}$ components, we can view the bundles $\Lambda^{p,q}T^*$ as $\Lambda^{0,q}T^* \otimes \Lambda^{p,0}T^*$, and then (5.3.41) as a bundle-valued cohomology:

$$H^{p,q}(M) = H^{0,q}(M, \Lambda^{p,0}T^*) \equiv H^q_{\bar{\partial}}(M, \Omega^p)\,. \tag{5.3.45}$$

This is called *Dolbeault theorem.*

In general, it is not clear how (5.3.41) is related to de Rham cohomology. If a form A is closed, we can split it in (p, q)-forms, $A = \sum_{p,q} A_{p,q}$, but closure becomes $\partial A_{p-1,q} + \bar{\partial} A_{p,q-1} = 0$, and it is not clear that the $\bar{\partial} A_{p,q} = 0$ follows from this. Conversely, if a form is annihilated by $\bar{\partial}$, it might not be annihilated by $\bar{\partial}$. More information may be obtained through a technique called *Frölicher spectral sequence.* This is a procedure that takes from Dolbeault to de Rham cohomology in a finite set of steps. Describing it would be a long detour for us, but it leads to the inequality

$$h_k \leq \sum_{p+q=k} h^{p,q}\,, \tag{5.3.46}$$

where $h_{p,q} = \dim H^{p,q}$ and $h_k = \dim H^k$ are the Betti numbers (Section 4.1.10). This relation will become more precise for Kähler manifolds (Chapter 6).

We now give a few more properties of bundle cohomology:

- The particular case

$$H^{0,0}(M, E) \tag{5.3.47}$$

is simply the space of global holomorphic sections of E.

- *Serre duality* relates

$$h^{p,q}(M, E) = h^{d/2-p,d/2-q}(M, E^*)\,. \tag{5.3.48}$$

For $E = \mathcal{O}$, this becomes a Dolbeault analogue of (4.1.126).

- A $(1, 1)$-form α is called *positive* if $\alpha_{mp} I_n{}^p$ is a positive-definite quadratic form. A *positive line bundle* is one for which the cohomology class $c_1(\mathcal{L})$ has a positive $(1, 1)$-form representative. Now the *Kodaira vanishing theorem* says

$$H^{p,q}(M, \mathcal{L}^{-1}) = 0 \quad p + q < d/2 \tag{5.3.49}$$

if $\mathcal{L}$ is positive (and as usual $d/2 = \dim_{\mathbb{C}} M$). In particular, only a positive line bundle has holomorphic sections.

- Bundle-valued cohomology groups can sometimes be reduced to those for smaller bundles. For example, given a quotient bundle E/E' (Section 4.2.1), there is a *(long) exact sequence*:

$$\begin{aligned} \ldots &\to H^{0,q}(M,E') \to H^{0,q}(M,E) \to H^{0,q}(M,E/E') \\ &\to H^{0,q+1}(M,E') \to H^{0,q+1}(M,E) \to H^{0,q+1}(M,E/E') \to \ldots \end{aligned} \tag{5.3.50}$$

Here the image of each map is equal to the kernel of the next map. This is useful when many of the groups in the sequence happen to vanish; for example, if a piece of the sequence reads $0 \to H^{0,q}(E) \to 0$, then $H^{0,q}(E)$ is forced to vanish.

- An alternating sum of Dolbeault groups can be computed by the *Hirzebruch–Riemann–Roch* index theorem:

$$\chi(E) \equiv \sum_q (-1)^q h^{0,q}(M,E) = \int_M \mathrm{ch}(E)\mathrm{Td}(TM)\,, \tag{5.3.51}$$

where $\mathrm{ch}(E)$ is the Chern character defined by (4.2.68), and the *Todd class* is defined by the splitting principle (4.2.67) as $\mathrm{Td} \equiv \Pi_i x_i(1 - \mathrm{e}^{-x_i})^{-1} = \mathrm{e}^{c_1/2}\hat{A}$, with $\hat{A}$ defined after (4.3.22). This is a particular case of the even more powerful Grothendieck–Riemann–Roch theorem, whose formulation would require sheaves (to be discussed briefly in Section 9.2.3).

Holomorphic triviality

Recall that a bundle is topologically trivial if its transition functions can be brought to be the identity by using local gauge transformations t_α acting as (4.2.4). On complex manifolds, a bundle is called *holomorphically* trivial if the t_α can be taken to be holomorphic. One writes this as

$$K \cong \mathcal{O}\,, \tag{5.3.52}$$

since $\mathcal{O}$, the bundle of holomorphic functions, is holomorphically trivial. In Section 5.3.5, we will see many examples of bundles that are trivial topologically but not holomorphically.

Topological triviality is equivalent to the existence of a global section; holomorphic triviality, to the existence of a global *holomorphic* section. Rewriting (5.3.18) in terms of Dolbeault, we see $\bar{\partial}\Omega = W \wedge \Omega$; using $\bar{\partial}$ again leads to $\bar{\partial} W = 0$, so

$$W \in H^{0,1}(M)\,. \tag{5.3.53}$$

If $K \cong \mathcal{O}$, this class is trivial and $W = \bar{\partial}\varphi$. The form $\Omega' = \mathrm{e}^{-\varphi}\Omega$ is then closed. So when (5.3.52) holds, we can set $W = 0$ in (5.3.18). Also, when $h^{0,1} = 0$, we know that holomorphic triviality is automatic.

5.3.4 Complex deformations

We now ask on how many parameters a complex structure can depend, with special emphasis on the continuous ones, often called *moduli*.[6]

[6] This word is often used interchangeably with "parameters" in many branches of mathematics, but one tends to use it for continuous ones. It has percolated into physics, as we will see starting in Chapter 8.

Let $I = I_0 + \delta I$ be a deformation of a complex structure I_0, and correspondingly $\Omega = \Omega_0 + \delta\Omega$ be the deformation of the three-form. The forms that are $(1,0)$ with respect to I will not be $(1,0)$ with respect to I_0; they will pick up a small $(0,1)$ piece. So the deformed Ω will be of type $(d/2, 0)$ with respect to I, but with respect to I_0 it will have a $(d/2-1, 1)$ piece, which without loss of generality we can assume to be $\delta\Omega$.

We now want to impose that the deformed complex structure is integrable too. Deforming the condition (5.3.18) gives $d\delta\Omega = \delta W \wedge \Omega_0 + W_0 \wedge \delta\Omega$, whose $(d/2-1, 2)$ part reads

$$(\bar\partial_0 - W_0 \wedge)\delta\Omega = 0\,, \tag{5.3.54}$$

where ∂_0 is the Dolbeault differential associated to the undeformed I_0. According to (5.1.34), in general $\delta\Omega$ is a form valued in the bundle K^{-1}, and W is a connection on K. By (5.3.44), $\delta\Omega$ is then a section of

$$H^{d/2-1,1}(M, K^{-1})\,. \tag{5.3.55}$$

By Cartan's magic formula (4.1.79), the $(d/2-1, 1)$ part of a Lie derivative $L_\xi \Omega_0$ is $(\bar\partial_0 - W_0\wedge)\iota_\xi\Omega_0$; so exact elements of the cohomology (5.3.55) are mere coordinate transformations of Ω_0 under the infinitesimal diffeomorphism ξ.

This gives a cohomological description of the space of infinitesimal deformations that respect the integrability condition. It is particularly useful when K is trivial.

Beltrami differentials

As an alternative, it is also customary to parameterize

$$\delta\Omega = \mu \cdot \Omega_0\,, \tag{5.3.56}$$

with a tensor $\mu_{\bar\imath}^{j}$ whose lower index is antiholomorphic and whose upper index is holomorphic, which we can think of directly as the components $\delta I_{\bar\imath}^{j}$ of the complex structure deformations. Indeed, the action $\mu\,\cdot\, \equiv \mu_{\bar\imath}^{j} d\bar z^{\bar\imath} \wedge \iota_j$ on Ω_0 results in a $(d/2-1, 1)$-form. Now we can rewrite (5.3.54) as

$$(\bar\partial_0 - W_0\wedge)\mu \cdot \Omega_0 = \bar\partial_0(\mu \cdot \Omega_0) - \mu \cdot W_0 \wedge \Omega_0 = [\bar\partial_0, \mu\cdot]\Omega_0\,. \tag{5.3.57}$$

The commutator on the right-hand side can be evaluated with the usual operator methods from Section 4.1.6:

$$[\bar\partial_0, \mu\cdot] = [dz^{\bar\imath} \wedge \partial_{\bar\imath}, \mu_{\bar\jmath}^{k} dz^{\bar\jmath} \wedge \iota_k] = \partial_{\bar\imath}\mu_{\bar\jmath}^{k} dz^{\bar\imath} \wedge dz^{\bar\jmath} \wedge \iota_k\,. \tag{5.3.58}$$

So the infinitesimally deformed Ω corresponds to an integrable complex structure if it is a *Beltrami differential*:

$$\partial_{[\bar\imath}\mu_{\bar\jmath]}^{k} = 0\,. \tag{5.3.59}$$

If the index k was not present, this would just be the condition that a $(0,1)$-form is closed. The presence of the index k can be interpreted as $\mu_{\bar\jmath}^{k}$ being a form valued in the holomorphic tangent bundle $T_{1,0}$. So (5.3.59) can be interpreted as μ being a section of

$$H^{0,1}(M, T_{1,0}) \equiv H^1_{\bar\partial_0}(M, T_{1,0})\,. \tag{5.3.60}$$

Equations (5.3.55) and (5.3.60) are isomorphic, the map being simply (5.3.56). Given a complex structure I, these are two possible descriptions of the space of its infinitesimal complex deformations, namely of $I + \delta I$, which are still complex structures.

Higher-order deformations

The preceding analysis is valid at first order in μ. A finite deformation extending the infinitesimal (5.3.56) is

$$\Omega = \mathrm{e}^{\mu\cdot}\Omega_0 \,. \tag{5.3.61}$$

(Given the index structure of $\mu^j_{\bar{\imath}}$, the exponential in fact truncates.) The condition (5.3.18) can be rewritten by multiplying by $\mathrm{e}^{-\mu\cdot}$ and using (2.1.15):

$$\begin{aligned} \mathrm{e}^{-\mu\cdot}\mathrm{d}\mathrm{e}^{\mu\cdot}\Omega_0 &= \left(\mathrm{d} + [\mathrm{d}, \mu\cdot] + \frac{1}{2}[[\mathrm{d}, \mu\cdot]\mu\cdot] + \ldots\right)\Omega_0 = \\ &= \left(W_0 \wedge +[\partial_0, \mu\cdot]\right)\Omega_0 + \left([\bar{\partial}_0, \mu\cdot] + \frac{1}{2}[[\partial_0, \mu\cdot], \mu\cdot]\right)\Omega_0 \,. \end{aligned} \tag{5.3.62}$$

In the second line, we separated the $(d/2, 1)$-part from the $(d/2 - 1, 2)$-part. The former just tells us how W_0 is deformed; the operator in the latter is algebraic, and has to vanish even before we apply it to Ω_0:

$$[\bar{\partial}_0, \mu\cdot] + \frac{1}{2}[[\partial_0, \mu\cdot], \mu\cdot] = 0 \,. \tag{5.3.63a}$$

Evaluating this explicitly,

$$\partial_{[\bar{\imath}}\mu^k_{\bar{\jmath}]} - \mu^l_{[\bar{\imath}}\partial_l\mu^k_{\bar{\jmath}]} = 0 \,. \tag{5.3.63b}$$

Equation (5.3.63) is called *Kodaira–Spencer equation.*

A μ might satisfy the first-order condition (5.3.59) and fail to satisfy (5.3.63) at second order, because of the second term. In this case, we say that the deformation has an *obstruction*. The potential obstruction term in (5.3.63b) has two lower antiholomorphic indices and an upper holomorphic one; so it is a section of $H^{0,2}(M, T_{1,0}) \equiv H^2_{\bar{\partial}_0}(M, T_{1,0})$.

5.3.5 Riemann surfaces

We have analyzed S^2 in detail earlier. More generally, an almost complex structure I is always integrable in $d = 2$, so all Riemann surfaces Σ_g are complex.

Torus

For $g = 1$, we can realize $T^2 = \Sigma_1$ as a quotient $\mathbb{C}/\mathbb{Z}^2$, as we did back in (1.4.44) (see also Figure 1.2), leading to the equivalence relations

$$z \sim z + 1 \sim z + \tau \,. \tag{5.3.64}$$

The complex structure is then inherited by that of $\mathbb{C}$. As we discussed following (1.4.44), two τ related by an $\mathrm{SL}(2, \mathbb{Z})$ Möbius transformation

$$\tau \to \frac{a\tau + b}{c\tau + d} \tag{5.3.65}$$

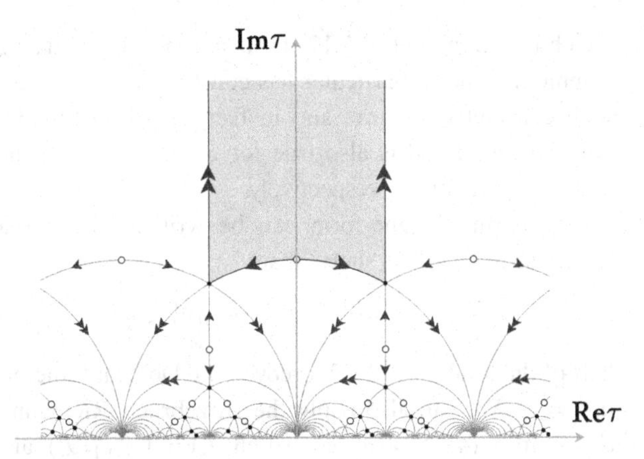

Figure 5.1 Moduli space of complex structures on a T^2.

define the same lattice and hence the same complex structure. On the other hand, two τ that are not related by such a transformation lead to a different complex structure. So the space of complex structures is the upper half-plane $H_\tau = \{\tau \in \mathbb{C}, \tau > 0\}$ parameterized by τ, up to these Möbius transformations (5.3.65). A fundamental region for this equivalence relation can be taken to be

$$\mathcal{M}_\tau = H_\tau/\mathrm{SL}(2,\mathbb{Z}) = \left\{-\frac{1}{2} \leq \mathrm{Re}\tau \leq \frac{1}{2}, \, |\tau|^2 \geq 1\right\} / \sim, \tag{5.3.66}$$

where $\sim$ denotes further identification of its boundary under $\tau \to -\bar{\tau}$. We show this region in Figure 5.1. The two arcs with a single arrow, and the two lines with double arrows, have to be identified. There are three special points on the boundary:

$$\tau = \mathrm{i}, \mathrm{e}^{\mathrm{i}\pi/6}, \infty . \tag{5.3.67}$$

The point $\tau = \mathrm{i}$ sits at the confluence of the two identified arcs; its neighborhood is not really $\mathbb{C}$ but $\mathbb{C}/\mathbb{Z}_2$. The point $\tau = \mathrm{e}^{\mathrm{i}\pi/6}$ has a neighborhood $\mathbb{C}/\mathbb{Z}_3$; the same is true for $\mathrm{e}^{\mathrm{i}\pi/3}$, which we have not listed in (5.3.67) since it is identified with $\mathrm{e}^{\mathrm{i}\pi/6}$. Finally, $\tau = \infty$ is a limiting point, but it is identified by (5.3.65) with several other finite points in the τ plane.

In this case, we can take $\Omega = \mathrm{d}z$; this is a holomorphic section of $K \cong T^*_{1,0}$, so $K \cong O$. By (5.3.55), the complex moduli are in $H^{0,1}$. The only representative in this space is the form $\mathrm{d}\bar{z}$, so $h^{0,1} = 1$. Recalling (5.3.60), this agrees with $\dim M_\tau = 1$.

Higher genus

For $g \geq 2$, Σ_g can also be represented as a quotient, as

$$H_z/\Gamma, \tag{5.3.68}$$

where now $H_z = \{z \in \mathbb{C}, \mathrm{Im}z > 0\}$ is the complex upper half-plane parameterized by z, and Γ is a discrete group of $\mathrm{SL}(2,\mathbb{R})$, the subgroup of (2.2.24) that maps $H \to H$. The fundamental region for this is a polygon with $4g$ sides, as we know from our discussion following (4.1.129).

It is useful to put on H the Hermitian Poincaré metric

$$\mathrm{d}s^2 = \frac{\mathrm{d}z\mathrm{d}\bar{z}}{(\mathrm{Im}z)^2}, \tag{5.3.69}$$

which is nothing but (4.5.17) for $d = 2$ (for $ds^2_{\mathbb{R}} = dx^2$, $z = x + iy$). It is maximally symmetric, and in particular has constant scalar curvature. So any Σ_g has a metric with constant curvature, and in fact locally maximally symmetric; this is called *uniformization* and is also true for $g = 0, 1$, taking there the round metric on S^2 and flat metric on T^2, respectively.

Any harmonic one-form can be written as the real part of a form in $H^{0,1}$ (Exercise 5.3.4). This shows that

$$h^{0,1} = g\,, \tag{5.3.70}$$

half of the total $h^1 = 2g$. Moreover, the Dolbeault theorem (5.3.45) gives $h^{0,1}(K) = h^{1,1} = 1$. We can also apply the Hirzebruch–Riemann–Roch theorem (5.3.51). In $d = 2$, from (4.2.68) we see that $\mathrm{ch}(\mathcal{L}) = 1 + c_1(\mathcal{L})$, and the definition that follows (5.3.51) gives $\mathrm{Td(TM)} = 1 + \frac{1}{2}c_1(\mathrm{TM})$. Using (4.2.73) and $c_1 = e$ from Section 4.2.5, we arrive at the classic *Riemann–Roch theorem*:

$$h^{0,0}(\mathcal{L}) - h^{0,1}(\mathcal{L}) = 1 - g + c_1(\mathcal{L})\,. \tag{5.3.71}$$

As a cross-check, for $\mathcal{L} = O$ we get agreement with (5.3.70). For $\mathcal{L} = K$, we obtain instead

$$c_1(K) = 2g - 2\,. \tag{5.3.72}$$

The derivation in this paragraph also works in $g = 0, 1$, and indeed (5.3.72) agrees with our results for S^2 ($g = 0$, $K = O(-2)$; (5.3.9)) and for T^2 ($g = 1$, $K = O$).

Returning to $g \geq 2$, (5.3.55) and (5.3.71) give us the number of complex moduli:

$$h^{0,1}(K^{-1}) = 3g - 3\,. \tag{5.3.73}$$

In this $d = 2$ case, we can also solve quite explicitly the problem of distinguishing topological and holomorphic triviality of bundles, posed after (5.3.52). The space of line bundles with $c_1 = 0$ (and hence topologically trivial, by (4.2.70)) but not holomorphically trivial is a torus $T^{2g} \cong (S^1)^{2g}$, called *Jacobian variety of* Σ. It can be parameterized by the integrals $\int a$ on the $2g$ generators of $H_1(\Sigma_g)$.

Exercise 5.3.1 Show that when I is not integrable, the Dolbeault operator defined by (5.3.34) maps $\Lambda^{p,q} \to \Lambda^{p+2,q-1} \oplus \Lambda^{p+1,q} \oplus \Lambda^{p,q+1} \oplus \Lambda^{p-1,q+2}$.

Exercise 5.3.2 Show (5.3.36).

Exercise 5.3.3 Obtain (5.3.63b) from (5.3.63a), using the operator methods in Section 4.1.6.

Exercise 5.3.4 Taking the metric to be Hermitian, show that any harmonic one-form α_1 can be written as the real part of a form in $H^{0,1}$.

5.4 Symplectic geometry

We now consider a notion of integrability for a nondegenerate two-form J, which we also called an almost symplectic structure. We draw inspiration from almost complex structures, where we defined integrability by demanding that the decomposition (5.1.27) in (p, q) forms should agree with differential operations such as Lie brackets or the exterior differential. For J, we will impose a similar condition on the Lefschetz decomposition of Section 5.2.1.

5.4.1 Integrability of two-form

In Section 5.3.2, one of the various ways to impose integrability for an almost complex structure I was by imposing $\bar{\partial}^2 = 0$ for the Dolbeault differential, or $(\mathrm{d}^c)^2 = 0$ for the operator $\mathrm{d}^c = [\mathrm{d}, I\cdot]$. We can try to proceed similarly by considering a commutator of d with either $J\wedge$ or $J\cdot$. By (4.1.77), $[\mathrm{d}, J\wedge] = (\mathrm{d}J)\wedge$, which is not a differential operator and whose square is zero automatically. So we consider instead

$$\mathrm{d}_J \equiv [\mathrm{d}, J\cdot] \overset{(2.1.31),(3.2.3)}{=} -\frac{1}{2}[\mathrm{d}x^m \wedge \partial_m, J^{np}\iota_n\iota_p] = -\frac{1}{2}\partial_m J^{np}\mathrm{d}x^m \wedge \iota_n\iota_p + J^{mn}\iota_m\partial_n, \tag{5.4.1}$$

which is a differential operator, since it contains the partial derivative ∂_n. A lengthier computation gives

$$[\mathrm{d}_J, J\cdot] = [[\mathrm{d}, J\cdot], J\cdot] = \frac{1}{6}P^{mnp}\iota_m\iota_n\iota_p \equiv \iota_P\,, \tag{5.4.2}$$

where

$$\frac{1}{6}P^{mnp} \equiv J^{q[m}\partial_q J^{np]} = J^{q[m}J^{n|r}J^{p]s}\partial_q J_{rs} = -J^{mq}J^{nr}J^{ps}\partial_{[q}J_{rs]}\,. \tag{5.4.3}$$

We used the familiar trick $\partial_m(J_{np}J^{pq}) = 0 \Rightarrow \partial_m J^{pq} = -J^{pn}\partial_m J_{nr}J^{rq}$. Notice the appearance of the components $\partial_{[q}J_{rs]}$ of the form $\mathrm{d}J$.

The same steps as in (5.3.37) give $\{\mathrm{d}, \mathrm{d}_J\} = 0$. The square of d_J can now be computed:

$$2\mathrm{d}_J^2 = \{\mathrm{d}_J, \mathrm{d}_J\} = \{\mathrm{d}_J, [\mathrm{d}, J\cdot]\} \overset{(2.1.31e)}{=} [\{\mathrm{d}_J, \mathrm{d}\}, J\cdot] - \{\mathrm{d}, [\mathrm{d}_J, J\cdot]\} = -\{\mathrm{d}, \iota_P\}\,. \tag{5.4.4}$$

So we conclude

$$\mathrm{d}_J^2 = 0 \quad \Leftrightarrow \quad P^{mnp} = 0 \quad \Leftrightarrow \quad \mathrm{d}J = 0\,. \tag{5.4.5}$$

Inspired by this, we say that a nondegenerate two-form J is a *symplectic structure* (dropping the "almost") if it is closed, $\mathrm{d}J = 0$; a manifold on which one exists is called a symplectic manifold.

Infinitesimal deformations $J_0 \to J_0 + \delta J$ lead to a new symplectic structure if and only if δJ is closed, and hence lives in the de Rham cohomology H^2. Moreover, from Cartan's magic formula we see that an infinitesimal diffeomorphism $\delta J = L_\xi J_0 = \mathrm{d}\iota_\xi J_0$; this is similar to our treatment of complex deformations around (5.3.55). So deformations of symplectic structures are parameterized by

$$H^2(M)\,. \tag{5.4.6}$$

Symplectic structures appear in physics in the context of classical Hamiltonian mechanics. One can interpret phase space as a manifold of even dimension M_d. For example, for the motion of a single particle on a manifold $M_{d/2}$, positions x^m live in $M_{d/2}$ itself, while momenta p_m live in the cotangent space $T^*_x M_{d/2}$ at the point described by the positions; so phase space in this case is the total space of the bundle

$M_d = T^* M_{d/2}$. In this case, Poisson brackets are defined by $\{x^m, p_n\}_{\rm PB} = \delta^m_n$ (where now $m, n = 1, \ldots, d/2$), which then implies $\{f, g\}_{\rm PB} = \partial_{x^m} f \partial_{p_m} g - \partial_{p_m} f \partial_{x^m} g$ for two functions f, g of positions and momenta. Going back to the more general case where the phase space is just a manifold M_d of even dimension, a Poisson bracket can be defined by J via the formula

$$\{f, g\}_{\rm PB} = J^{mn} \partial_m f \partial_n g \,. \tag{5.4.7}$$

Imposing the Jacobi identity results in $P^{mnp} = J^{q[m} \partial_q J^{np]} = 0$, and thus says that phase space is a symplectic manifold. The *Darboux theorem* says that around each point there is a neighborhood with coordinates x^m, p_n such that

$$J = {\rm d}x^m \wedge {\rm d}p_m \tag{5.4.8}$$

(where now $m, n = 1, \ldots, d/2$). So again our notion of integrability agrees with the general definition for G-structures at the beginning of this section.

Even more generally, one can define phase space by starting directly with the tensor J^{mn}; namely, as a manifold M_d endowed with a tensor J^{mn}, not necessarily invertible, which obeys

$$P^{mnp} = J^{q[m} \partial_q J^{np]} = 0 \,, \tag{5.4.9}$$

namely setting (5.4.3) to zero. Such a J^{mn} is called a *Poisson tensor*, and M_d is called a Poisson manifold. One can use this definition even in odd dimension, where a symplectic structure cannot exist.

5.4.2 Moment maps

A *symplectomorphism* is a diffeomorphism under which J is invariant. At the infinitesimal level, this is generated by a vector field ξ such that

$$L_\xi J = 0 \,. \tag{5.4.10}$$

From (4.1.79) and the fact that J is closed we obtain ${\rm d}(\iota_\xi J) = 0$. So *locally*, there exists a function μ_ξ such that

$$\iota_\xi J = {\rm d}\mu_\xi \,. \tag{5.4.11}$$

So for every symplectomorphism generator, locally on M there exists a function μ_ξ, defined up to a constant. The map $\mu \colon \xi \mapsto \mu_\xi$ is called *moment map* (or *momentum map*). It is invariant under the action of ξ, since

$$L_\xi \mu_\xi = \iota_\xi {\rm d}\mu_\xi = \iota_\xi^2 J = 0 \,. \tag{5.4.12}$$

More generally, the moment map is said to be *equivariant* if

$$L_{\xi_1} \mu_{\xi_2} = \mu_{[\xi_1, \xi_2]} \,. \tag{5.4.13}$$

This can always be arranged by fixing the constants left undetermined in each μ_ξ by the definition (5.4.11).

In applications to classical mechanics, time evolution is a symplectomorphism, and the μ_ξ associated to its generator ξ is the Hamiltonian. More generally, given ξ the generator of any physical symmetry, μ_ξ is the associated conserved quantity. In classical mechanics, sometimes it is also important to identify different points of the phase space under action of a group G. Geometrically, this would seem to suggest a quotient M/G as the phase space. But this would not be a symplectic manifold: if, for example, G is one-dimensional, then M/G would have odd dimension. However, if the generator ξ of G is a symmetry, it should be a symplectomorphism; because of (5.4.12), the action of ξ is along the level sets $\{\mu_\xi = c\}$, or in other words $\mu_\xi^{-1}(c)$, for c any constant. All this suggests considering

$$\mu_\xi^{-1}(c)/G \tag{5.4.14}$$

as the new phase space. Now this has dimension $\dim(M) - 2\dim(G)$: one $\dim(G)$ is lost because of considering the level set, the other by the G quotient. Moreover, the definition (5.4.11) tells us that the two lost directions are related by J: the generator ξ is mapped by J to the form $d\mu_\xi$, which is normal to the level set. In the language of classical mechanics, in (5.4.14) we are losing two conjugate variables. Eliminating two conjugate directions from a symplectic form J is expected to lead to a smaller symplectic form. For example, if in (5.4.8) we imagine losing two coordinates q_1, p_1, we obtain a two-form that is again of that type.

Formalizing these observations, one can prove that (5.4.14) is in fact again a symplectic manifold, variously called *reduced phase space* or *symplectic quotient*; the procedure is sometimes also called *Marsden–Weinstein reduction* [146]. We will see many examples in Section 6.4.2.

5.5 Intrinsic torsion

Given an $O(d)$-structure, or in other words a metric, we know that the Levi-Civita connection exists, which preserves it and which has zero torsion, (4.1.15). For other G-structures, one cannot satisfy both conditions. There are connections with zero torsion, such as the Levi-Civita ∇, but they do not preserve the G-structure; and several connection $\tilde{\nabla}$ that preserve the G-structure, but have nonzero torsion.

This gives rise to two points of view:

(1) We can keep working with the Levi-Civita connection ∇, which is torsion-free. Applying it on the tensor T or spinor η defining the G-structure gives rise to new tensors W_i.
(2) We can find a connection ∇_G under which the G-structure is covariantly constant. In general, this connection has torsion. Comparing ∇_G to the Levi-Civita ∇, one sees that some components of the torsion are related to the W_i, and thus cannot be set to zero.

The second point of view motivates calling the W_i "intrinsic torsion." As we will see, they transform in the representation

$$(\text{adjoint})^\perp \otimes (\text{fundamental}) \tag{5.5.1}$$

of G. Here by $(\text{adjoint})^\perp$ we mean the complement of the adjoint of G in the adjoint of $SO(d)$.

We will show how this works for G-structures defined by a tensor η, first in the case $G = \mathrm{SU}(3)$ to fix ideas, and then for an almost complex structure $I_m{}^n$, which defines a $\mathrm{GL}(d/2, \mathbb{C})$-structure (Section 5.1.1).

5.5.1 Unitary structures

Levi-Civita action on the spinor

We start with the spinorial point of view, regarding an SU(3)-structure as defined by a metric g and a chiral spinor η_+.

The spinorial Levi-Civita connection D_m does not always preserve η: in general, $D_m\eta \neq 0$. This is a vector-spinor, a section of the bundle $(T \otimes S)M$. If we evaluate it at a point and choose a value of the index m, we find a spinor; we can expand it in the basis defined by η_+ itself, (3.4.86). The positive-chirality spinors there are η_+ itself and $\gamma^m\eta_-$. Giving names to the coefficients, we can write

$$D_m\eta_+ = p_m\eta_+ + q_{mn}\gamma^n\eta_- \,. \tag{5.5.2}$$

The p_m are the q_{mn} are the intrinsic torsion tensors in this case. Since $\gamma^{\bar{\imath}}\eta_- = 0$ (from (5.1.44b)), there is a possible ambiguity in the definition of q_{mn}, because some of its components do not act; we fix it by imposing

$$q_{mn} = \Pi_n{}^r q_{mr} \tag{5.5.3}$$

so that the second index is holomorphic. For later use, let us also decompose q_{mn} in SU(3) representations. We split it according to the type of the first:

$$q_{mn} = q^{++}_{mn} + q^{-+}_{mn}\,, \qquad q^{++}_{mn} \equiv \Pi_m{}^p\Pi_n{}^r q_{pr}\,, \qquad q^{-+}_{mn} = \bar{\Pi}_m{}^p\Pi_n{}^r q_{pr}\,. \tag{5.5.4}$$

Now q^{++}_{mn} transforms in the $\mathbf{3} \otimes \mathbf{3} = \mathbf{6} \oplus \bar{\mathbf{3}}$; the $\mathbf{6}$ and $\bar{\mathbf{3}}$ are the symmetric and antisymmetric parts:

$$q^{\mathbf{6}}_{mn} \equiv q^{++}_{(mn)}\,, \qquad q^{\bar{\mathbf{3}}}_{mn} \equiv q^{++}_{[mn]} \equiv \Omega_{mnp}\hat{q}^p\,. \tag{5.5.5a}$$

The one-form $\hat{q} = \hat{q}_m \mathrm{d}x^m$ is of type (0, 1). Also, q^{-+}_{mn} transforms in the $\bar{\mathbf{3}}\otimes\mathbf{3} = \mathbf{8}\oplus\mathbf{1}$, which are simply the traceless and trace part:

$$J^{mn}q^{\mathbf{8}}_{mn} = 0\,, \qquad q^{\mathbf{1}}_{mn} \equiv q^{\mathbf{1}}\bar{\Pi}_{mn}\,. \tag{5.5.5b}$$

Recall from Section 5.1.2 (see the summary at its end) that an SU(3)-structure can also be defined as a pair of forms (J, Ω). In particular, from (5.1.44) we see that they are bilinears of η. We can try to define some tensors analogous to the p_m, q_{mn} by following a logic similar to (5.5.2). The three-form $\mathrm{d}J$ and the four-form $\mathrm{d}\Omega$ can be decomposed in representations following (5.2.15):

$$\mathrm{d}J = -\frac{3}{2}\mathrm{Im}(W_1\bar{\Omega}) + W_3 + W_4 \wedge J\,, \tag{5.5.6a}$$

$$\mathrm{d}\Omega = \bar{W}_5 \wedge \Omega + W_2 \wedge J + W_1 J^2\,. \tag{5.5.6b}$$

The names for the various forms W_i are traditional [147, 148].

- W_1 is a function; it appears in two spots because the **1** in $d\Omega$ and in dJ are proportional, as one sees by taking the exterior derivative of (5.1.40a) and using (5.1.40b). The term $-\frac{3}{2}\text{Im}(W_1\bar{\Omega})$ is the $(3,0)$ part of dJ.
- The $(2,1)$ part of dJ is divided into primitive and nonprimitive parts as $W_3+W_4\wedge J$; so W_3 is in the **6**, while W_4 is a $(1,0)$-form.
- The term $\bar{W}_5\wedge\Omega$ in $d\Omega$ is its $(3,1)$ part; its $(2,2)$ part is again divided into primitive and nonprimitive parts as $W_2\wedge J + W_1 J^2$; so the $(1,1)$-form W_2 is in the **8** and is primitive, $J\cdot W_2 = 0$.
- W_5 has an ambiguity similar to the one for q_{mn}: only its $(1,0)$ part really enters in (5.5.6). We resolve this by declaring $(W_5)_{0,1} = 0 = (\bar{W}_5)_{1,0}$.[7]
- Finally, notice that (5.2.15) would have allowed a $(1,3)$ part in $d\Omega$, but this vanishes automatically; this is a consequence for $d = 6$ of the discussion in (5.3.16) and (5.3.17a).

Summarizing, the W_i are in the following representations:

$$\begin{array}{ccccc} W_1 & W_2 & W_3 & W_4 & W_5 \\ \mathbf{1} & \mathbf{8} & \mathbf{6}\oplus\bar{\mathbf{6}} & \mathbf{3}\oplus\bar{\mathbf{3}} & \mathbf{3} \end{array} . \tag{5.5.7}$$

Some particular cases of SU(3)-structures can be characterized in terms of these W_i. Comparing (5.3.18) with (5.5.6), we see that the almost complex structure I associated to Ω is integrable if and only if W_1 and W_2 vanish:

$$I_\Omega \text{ integrable} \quad \Leftrightarrow \quad W_1 = W_2 = 0\,. \tag{5.5.8}$$

More obviously, J is a symplectic structure if it is closed, so

$$J \text{ symplectic} \quad \Leftrightarrow \quad W_1 = W_3 = W_4 = 0\,. \tag{5.5.9}$$

Because η_+ and (J,Ω) are two equivalent points of view on an SU(3)-structure, the information carried by the (p_m, q_{mn}) in (5.5.2) and by the W_i in (5.5.6) should be the same. This can be made explicit by using (5.1.44); this computation is also a good warm-up for similar ones we will see later on. Either generalizing (4.3.35) or applying (4.3.46):

$$i\nabla_m J_{np} = \nabla_m(\eta_+^\dagger\gamma_{np}\eta_+) = (D_m\eta_+^\dagger)\gamma_{np}\eta_+ + \eta_+^\dagger\gamma_{np}D_m\eta_+\,. \tag{5.5.10a}$$

We now plug in (5.5.2) and its dagger:

$$\begin{aligned} i\nabla_m J_{np} &= (\eta_+^\dagger\bar{p}_m + \eta_-^\dagger\gamma^q\bar{q}_{mq})\gamma_{np}\eta_+ + \eta_+^\dagger\gamma_{np}(p_m\eta_+ + q_{mq}\gamma^q\eta_-) \\ &= 2i\text{Re}p_m J_{np} - \bar{q}_{mq}\Omega_{np}{}^q + q_{mq}\bar{\Omega}_{np}{}^q\,, \end{aligned} \tag{5.5.10b}$$

recalling the definitions (5.1.44). If we now antisymmetrize the three indices,

$$dJ = 2\text{Re}p\wedge J + 2\text{Im}(q\cdot\bar{\Omega}), \tag{5.5.11a}$$

where $p \equiv p_m dx^m$ and $q\cdot \equiv q_{mn}dx^m\wedge\iota^n$. A similar computation for Ω yields

$$d\Omega = 2p\wedge\Omega + 8iq_{(2)}\wedge J\,, \tag{5.5.11b}$$

where now $q_{(2)} \equiv \frac{1}{2}q_{mn}dx^m\wedge dx^n$. Comparing (5.5.11) with (5.5.6) and recalling the SU(3) decompositions (5.5.5), we get

[7] Another possibility would be to make it real: adding $W_5\wedge\Omega = 0$, we find $d\Omega = 2\text{Re}W_5\wedge\Omega + \cdots$

$$W_1 = -4q^1 \,, \qquad W_2 = 8\mathrm{i} q^8_{(2)} \,, \qquad W_3 = 2\mathrm{Im}(q^6 \cdot \bar{\Omega}) \,,$$
$$W_4 = \mathrm{Re}\,(2p + 16\hat{q}) \,, \qquad \bar{W}_5 = 2p_{0,1} + 8\hat{q} \,. \tag{5.5.12}$$

Torsion of compatible connection

We now turn to the second point of view we explained at the beginning of the section.

We first need to find a connection $\tilde{D}_m$ that keeps η_+ invariant. We can try to reabsorb the terms on the right-hand side of (5.5.2) by a change in ω_m^{ab}. Recalling (3.4.92a), we indeed see that we can obtain η_+ and $\gamma^m \eta_-$ by an action of some appropriate object on η_+. As remarked after (5.2.4), $J_{mn}J^{mn} = -6$. Moreover, from (3.4.24), (5.1.40a), and (5.2.11) we have $J \cdot \Omega = 0$; in components, $J^{mn}\Omega_{mnp} = 0$. (Together with $J \wedge \Omega = 0$, this expresses the fact that Ω is in a Lefschetz singlet.) Using these, (3.4.16) and (3.4.92a), we obtain

$$J_{mn}\gamma^{mn}\eta_+ = 6\mathrm{i}\eta_+ \,, \qquad \bar{\Omega}_{mnp}\gamma^{np}\eta_+ = -4\gamma_m \eta_- \,. \tag{5.5.13}$$

It follows that

$$\tilde{\omega}_m^{ab} = \omega_m^{ab} + \frac{2}{3}\mathrm{i} p_m J^{ab} + q_{mn}\bar{\Omega}^{nab} + n_m^{ab} \,, \qquad n_m^{ab} J_{ab} = n_m^{ab}\Omega_{abn} = 0 \,, \tag{5.5.14}$$

defines a connection $\tilde{D}_m = \partial_m + \frac{1}{4}\tilde{\omega}_m^{ab}\gamma_{ab}$ that indeed leaves the SU(3)-structure invariant: $\tilde{D}_m \eta_+ = 0$.

The torsion of this new connection (5.5.14) can now be computed by (4.3.30); applying it to the Levi-Civita ω_m^{ab}, which has zero torsion, and to the torsionful $\tilde{\omega}_m^{ab}$, in components, it gives $T^a_{mn} = (\tilde{\omega} - \omega)^{ab}_{[m} e_{b|n]}$. By splitting the indices in holomorphic and antiholomorphic ones, one can check that the p_m and all the representations (5.5.5) appearing in q_{mn} appear in some components of T^a_{mn}. Thus it is justified to think of p_m and q_{mn} as torsion coefficients. Moreover, they are "intrinsic" in that their presence in (5.5.14) is necessary: without including them, the connection $\tilde{D}_m$ would not preserve η_+. The coefficients n_m^{ab}, on the other hand, are allowed but not necessary, or "nonintrinsic." Taking these observations together, we can see why calling the p_m and q_{mn} "intrinsic torsion" is justified: they appear in the torsion of the η_+-preserving connection $\tilde{\nabla}_m$, and they cannot be eliminated from it.

Finally let us justify (5.5.1). Once we know that an SU(3)-preserving connection exists, we can write

$$D_m \eta_+ = (D - \tilde{D})_m \eta_+ = \frac{1}{4}(\omega - \tilde{\omega})_m^{ab}\gamma_{ab}\eta_+ \,. \tag{5.5.15}$$

The right-hand side parameterizes the failure of ∇_m to leave η_+ invariant. From (2.4.33) and (5.5.14), we see that the nonintrinsic n_m^{ab} disappear from it. These are the coefficients along the generators of su(3). In other words, the only components of $(\omega - \tilde{\omega})_m^{ab}$ that really enter in (5.5.15) are those in the orthogonal complement $\mathrm{su}(3)^\perp$. This logic gives (5.5.1) for $G = \mathrm{SU}(3)$. To be more explicit: the adjoint of so(6) decomposes under su(3) as $\mathbf{8} \oplus \mathbf{3} \oplus \bar{\mathbf{3}} \oplus \mathbf{1}$. The adjoint of su(3) is the $\mathbf{8}$, so $\mathrm{su}(3)^\perp = \mathbf{3} \oplus \bar{\mathbf{3}} \oplus \mathbf{1}$. So (5.5.1) gives

$$\begin{aligned}(\text{adjoint})^\perp \otimes (\text{fundamental}) &= (\mathbf{3} \oplus \bar{\mathbf{3}} \oplus \mathbf{1}) \otimes (\mathbf{3} \oplus \bar{\mathbf{3}}) \\ &= \mathbf{6} \oplus \bar{\mathbf{3}} \oplus \bar{\mathbf{6}} \oplus \mathbf{3} \oplus \mathbf{8} \oplus \mathbf{1} \oplus \mathbf{8} \oplus \mathbf{1} \oplus \mathbf{3} \oplus \bar{\mathbf{3}} \,.\end{aligned} \tag{5.5.16}$$

Since W_1, W_2, and W_5 are complex, this is in agreement with (5.5.7).

U(3) structures
An SU(3)-structure requires the existence of a three-form Ω, which is not guaranteed in general if the canonical bundle K is nontrivial. Fortunately, we can still find one as a bundle-valued form as in (5.1.34). In this sense, we can apply the formalism of SU(3) structures to U(3) structures as well. From this point of view, W_5 is a connection on K, in keeping with our earlier (more general) observation following (5.3.18). Since $\Omega_{mnp} = \eta_-^\dagger \gamma_{mnp}\eta_+$, the spinor η_+ is now a section of $K^{-1/2} \otimes \Sigma_+$. ($\Sigma_+$ requires a spin structure, but $K^{-1/2} \otimes \Sigma_+$ always exists.)

5.5.2 Other spinorial examples

G_2-structures
We defined a G_2-structure in Section 5.1.3 by a Majorana spinor $\eta = \eta_{\rm M}$ or by the associated three-form $\phi_{mnp} = -{\rm i}\eta^\dagger \gamma_{mnp}\eta$, recalling (3.4.79). The basis (3.4.97) suggests that we take

$$D_m \eta = {\rm i} q_{mn} \gamma^n \eta \,, \tag{5.5.17}$$

similar to (5.5.2). We could have included a term $p_m \eta$, but this turns out to be zero if we impose that η has norm one. The i was included for consistency with the Majorana condition; this is clearest in the purely imaginary basis (2.2.19).

In terms of G_2-representations, the symmetric part $q_{(mn)}$ decomposes as $\mathbf{27} \oplus \mathbf{1}$: the traceless part, and the part proportional to the identity:

$$g^{mn} q^{\mathbf{27}}_{mn} = 0 \,, \qquad q^{\mathbf{1}}_{mn} = q^{\mathbf{1}} \delta_{mn} \,. \tag{5.5.18a}$$

The antisymmetric part $q_{[mn]} = q^{\mathbf{21}}_{mn} + q^{\mathbf{7}}_{mn}$, where

$$q^{\mathbf{21}}_{mn} \phi^{mnp} = 0 \,, \qquad q^{\mathbf{7}}_{mn} = \phi_{mnp} q^p_{\mathbf{7}} \,. \tag{5.5.18b}$$

From the form point of view, recall from Section 5.1.3 that a G_2-structure can be defined by a three-form ϕ. We can decompose ${\rm d}\phi$ and ${\rm d} * \phi$ using (5.2.35) and the discussion preceding it [149]:

$$\begin{aligned} {\rm d}\phi &= X_1 * \phi + X_4 \wedge \phi + X_3 \,, \\ {\rm d} * \phi &= \frac{4}{3} X_4 \wedge * \phi + X_2 \wedge \phi \,, \end{aligned} \tag{5.5.19}$$

where the X_i are in the representations

$$\begin{array}{cccc} X_1 & X_2 & X_3 & X_4 \\ \mathbf{1} & \mathbf{14} & \mathbf{27} & \mathbf{7} \end{array} . \tag{5.5.20}$$

These can be compared with (5.5.18) by computing ${\rm d}\phi$ and ${\rm d} * \phi$ from (5.5.17):

$$X_1 = -8q^{\mathbf{1}} \,, \qquad X_2 = 4q^{\mathbf{14}}_{(2)} \,, \qquad X_3 = -2q^{\mathbf{27}} \cdot * \phi \,, \qquad X_4 = -6q^{\mathbf{7}} \,, \tag{5.5.21}$$

where as earlier $q\cdot = q_{mn} {\rm d}x^m \wedge \iota^n$, $q_{(2)} = q_{[mn]} {\rm d}x^m \wedge {\rm d}x^n$. The relative factor between the two occurrences of X_4 in (5.5.19) can also be determined as in Exercise 5.5.7.

Finally, let us compute intrinsic torsions also from the point of view of (5.5.1). The adjoint of so(7) decomposes under g_2 as $\mathbf{14} \oplus \mathbf{7}$, so $g_2^\perp = \mathbf{7}$. Now (5.5.1) gives

$$(\text{adjoint})^\perp \otimes (\text{fundamental}) = \mathbf{7} \otimes \mathbf{7} = \mathbf{27} \oplus \mathbf{14} \oplus \mathbf{7} \oplus \mathbf{1} \tag{5.5.22}$$

which matches (5.5.18) and (5.5.20).

Spin(7)-structures

We will be brief in this case. Spin(7)-structure can be defined by a four-form Ψ which can locally be put in the form (5.1.54). Its exterior derivative is a five-form; in terms of Spin(7)-representations, Λ^5 can be decomposed as the space of forms that can be written as $\alpha_1 \wedge \Psi_4$, forming a **8**; and those that can't, which satisfy $\alpha_{m_1 \dots m_5} \Psi^{m_2 \dots m_5} = 0$ and form a 48. So we can decompose

$$d\Psi_4 = X_1 \wedge \Psi_4 + X_2 . \tag{5.5.23}$$

From the point of view of (5.5.1), the adjoint of so(8) decomposes as $\mathbf{28} \to \mathbf{21} \oplus \mathbf{7}$; the second summand is $\text{spin}(7)^\perp = \mathbf{7}$. Now

$$(\text{adjoint})^\perp \otimes (\text{fundamental}) = \mathbf{7} \otimes \mathbf{8} = \mathbf{8} \oplus \mathbf{48} , \tag{5.5.24}$$

which agrees with (5.5.23).

The spinorial point of view is a little more complicated in this case; we cannot quite use something similar to (5.5.17). The reason is that a Spin(7)-structure is defined by a Majorana–Weyl spinor $\eta = \eta_{\text{MW}}$, which in particular has fixed chirality (say positive). Spinors of the form $\gamma_m \eta$ would be a basis for the space of negative chirality, but $D_m \eta$ has positive chirality. For a positive-chirality basis, we need to use two gamma matrices, using $\gamma_{mn}\eta$. Naively, there are 28 of these, but in fact 21 vanish because they span the stabilizer of η, the adjoint of Spin(7); the remaining seven form a basis for positive-chirality spinors, together with η itself. This leads to writing

$$D_m \eta = \text{i} q_{mnp} \gamma^{np} \eta . \tag{5.5.25}$$

(Once again we have excluded a term $p_m \eta$ by imposing η to be of unit norm.) This expression is greatly redundant; the np indices of q_{mnp} should be understood to be in the **7**, and together with the index m, this reproduces (5.5.24).

5.5.3 Nijenhuis tensor as intrinsic torsion

We now look at a different sort of example. Recall that an almost complex structure I without a metric defines a $GL(d/2, \mathbb{C})$-structure (Section 5.1.1). In Section 5.3.1, we discussed the geometry that results from imposing a certain integrability condition; we will now see how that fits in the present discussion of intrinsic torsion.

First of all, let us look for a $\tilde{\nabla}$ that preserves I. We parameterize a connection as $\tilde{\Gamma}^m_{np} = \Gamma^m_{np} + \delta\Gamma^m_{np}$, where as usual Γ^m_{np} are the coefficients of the Levi-Civita connection. We impose

$$0 = \tilde{\nabla}_m I_n{}^p = \nabla_m I_n{}^p - \delta\Gamma^q_{mn} I_q{}^p + I_n{}^q \delta\Gamma^p_{mq} . \tag{5.5.26a}$$

We can free ourselves from a few indices by considering $\delta\Gamma^p_{mn}$ as a matrix $\delta\Gamma_m$ with components ${}^p{}_n$. From this point of view, for example, we can rewrite $\delta\Gamma^q_{mn} I_q{}^p = (\delta\Gamma_m)^q{}_n I_q{}^p = (\delta\Gamma^t_m)_n{}^q I_q{}^p = (\delta\Gamma^t_m I)_n{}^p$. Then (5.5.26a) reads

$$0 = \nabla_m I + [I, \delta\Gamma^t_m] . \tag{5.5.26b}$$

Proceeding in the same index-free symbolic fashion, we recall from (5.3.22) the holomorphic projector $\Pi = \frac{1}{2}(1 - \text{i}I)$, which satisfies $\Pi I = I\Pi = \text{i}\Pi$. Since $I^2 = -1$, we also have $(\nabla_m I)I + I(\nabla_m I) = 0$, which upon multiplication by Π gives

$$\Pi \nabla_m I \Pi = 0 \,. \tag{5.5.27}$$

Its complex conjugate $\bar{\Pi}\nabla_m I\bar{\Pi} = 0$; together these say that $\nabla_m I_n{}^p$ is only nonzero when one of the indices n, p is holomorphic and the other is antiholomorphic. In other words,

$$\nabla_m I = \Pi \nabla_m I \bar{\Pi} + \bar{\Pi} \nabla_m I \Pi \,. \tag{5.5.28}$$

It is now easy to verify that

$$\begin{aligned} \delta\Gamma^t_m &= \frac{\mathrm{i}}{2}\left(\Pi \nabla_m I \bar{\Pi} - \bar{\Pi} \nabla_m I \Pi\right) = -\Pi \nabla_m \Pi \bar{\Pi} - \bar{\Pi} \nabla_m \bar{\Pi} \Pi \\ &= \Pi \nabla_m \bar{\Pi} + \bar{\Pi} \nabla_m \Pi \end{aligned} \tag{5.5.29a}$$

satisfies (5.5.26). (In the second line, we used $\nabla_m(\bar{\Pi}\Pi) = 0$, $\Pi^2 = \Pi$.) It would be possible to also add terms $\Pi\delta\Gamma^{(++)t}_m\Pi + \bar{\Pi}\delta\Gamma^{(--)t}_m\bar{\Pi}$, which would cancel out in (5.5.26); these are nonintrinsic, and are analogous to the terms n^{ab}_m in (5.5.14). We will set them to zero in what follows. Reintroducing explicit indices, (5.5.29a) reads

$$\delta\Gamma^p_{mn} = \Pi_n{}^r \nabla_m \bar{\Pi}_r{}^p + \bar{\Pi}_n{}^r \nabla_m \Pi_r{}^p \,. \tag{5.5.29b}$$

$\tilde{\Gamma}$ has torsion, because this $\delta\Gamma$ is not symmetric under exchange of the lower indices. Projecting them holomorphically and antisymmetrizing them, we obtain

$$\Pi_{[m}{}^r \Pi_{n]}{}^s \delta\Gamma^p_{rs} = \Pi_{[m}{}^r \Pi_{n]}{}^s \nabla_r \bar{\Pi}_s{}^p = -\Pi_{[m|}{}^r \nabla_r \Pi_{|n]}{}^s \bar{\Pi}_s{}^p \overset{(5.3.24)}{=} -\frac{1}{2} N_{mn}{}^q \bar{\Pi}_q{}^p \,. \tag{5.5.30}$$

So the Nijenhuis tensor (5.3.25) can be interpreted as a piece of intrinsic torsion. This is not entirely surprising, in view of our earlier result (5.5.8) for SU(3)-structures.

Exercise 5.5.1 Repeat the argument of Section 5.5.1 to SU(2)-structures in $d = 4$. Show in particular that the intrinsic torsion consists of three one-forms. Compute the analogue of (5.5.11).

Exercise 5.5.2 Now consider instead SU($d/2$)-structures in even $d \geq 4$. What changes in the W_i?

Exercise 5.5.3 Upon using (3.4.81), show that the metric (5.1.51) reduces to the metric defined by an SU(3)-structure in Section 5.1.2.

Exercise 5.5.4 Check (5.2.37) by using (3.4.81) and (5.2.15). It is convenient to work in flat indices and take $\eta = e^7$; (5.2.34) then gives a set of constraints on α_{abc} and α_{ab7}.

Exercise 5.5.5 Show that if $W_2 = W_3 = W_4 = W_5$, then W_1 is constant.

Exercise 5.5.6 Under a rescaling $\eta_+ \to \eta'_+ = \mathrm{e}^{\alpha}\eta_+$, show that the only intrinsic torsion components that transform are

$$W_4 \to W'_4 = W_4 + 2\mathrm{dRe}\alpha \,, \qquad W'_5 \to W_5 + 2\partial\alpha \,. \tag{5.5.31}$$

Work out the transformation law for the p_m, q_{mn}, and check that the result is compatible with (5.5.12).

Exercise 5.5.7 Do the same exercise as the previous one, but now for a Weyl transformation $g_{mn} \to \mathrm{e}^{2\lambda}g_{mn}$: the W_i will now transform as

$$W_4 \to W'_4 = W_4 + 2\mathrm{d}\lambda \,, \qquad W_5 \to W'_5 = W_5 + 3\partial\lambda \,. \tag{5.5.32}$$

It can be useful to recall (4.3.69).

Exercise 5.5.8 Work out the relation between the X_i in (5.5.19) and the W_i in (5.5.6) if the G_2- and SU(3)-structures are related as in (3.4.81).

5.6 Calibrations

The forms associated to a spinor are special in that they define a G-structure. We will now see that in a sense they can be used to measure volumes.

5.6.1 Forms and volumes

Consider a tangent distribution, a k-dimensional subspace D of the tangent space T_pM, the fiber at a point p of the tangent bundle (Section 4.1.7). The restriction of the metric g evaluated at p to this subspace defines a quadratic form $g|_D$. Consider now a k-form ω_k; evaluating it at p, we can take its component $\omega|_D$ along the subspace D. In a coordinate system such that $x^1, \ldots, x^k$ are along D, $g|_D$ is the upper $k \times k$ block of g, and $\omega|_D = \omega_{1\ldots k}$ (both evaluated at p). A form ω is an *almost calibration* if

$$\omega|_D \le \sqrt{\det g|_D} \tag{5.6.1}$$

for every p and every D. Both quantities in (5.6.1) have some dependence on the choice of coordinates, but their ratio does not, so the inequality is meaningful.

If D is tangent space to a k-dimensional submanifold S_k to which p belongs, then (5.6.1) can be reformulated in terms of pull-backs of the inclusion map $i\colon S_k \to M$, which we denote by $|_{S_k}$:

$$\omega|_{S_k} \equiv i^*\omega\,, \qquad g|_{S_k} \equiv i^*g\,. \tag{5.6.2}$$

(If ω_k is not already a k-form but a polyform ω, we define $\omega|_{S_k}$ to be the pull-back of its k-form part $i^*(\omega_k)$.) Then (5.6.1) can be rewritten as

$$\omega|_{S_k} \le \mathrm{vol}_{S_k}\,, \tag{5.6.3}$$

where vol_{S_k} is the volume form of the pull-back metric $g|_{S_k}$.

We call S_k *calibrated* if the equality in (5.6.3) is saturated:

$$\omega|_{S_k} = \mathrm{vol}_{S_k}\,, \tag{5.6.4}$$

at every point $p \in S_k$. For such a manifold, the volume is given by the integral of the pull-back of ω:

$$\mathrm{Vol}(S_k) = \int_{S_k} \mathrm{vol}_{S_k} = \int_{S_k} \omega\,. \tag{5.6.5}$$

A *calibration* is an almost calibration that is closed:

$$d\omega = 0\,. \tag{5.6.6}$$

In this case, the lower bound (5.6.3) becomes more useful. Consider a calibrated S_k, and any other submanifold S'_k that is homologous to it: namely, there exists N_{k+1} such that $\partial N_{k+1} = S'_k - S_k$. Then

$$\begin{aligned}\mathrm{Vol}(S'_k) &= \int_{S'_k} \mathrm{vol}_{S'_k} \ge \int_{S'_k} \omega = \int_{S_k} \omega + \int_{\partial N_{k+1}} \omega \\ &\overset{(4.1.131)}{=} \int_{S_k} \omega + \int_{N_{k+1}} d\omega \overset{(5.6.6)}{=} \int_{S_k} \omega\,.\end{aligned} \tag{5.6.7}$$

5.6.2 Calibrations from spinors

To see how (almost) calibrations might arise, consider an SU(3)-structure defined by a spinor η_+ on a $d = 6$ Riemannian manifold, which for simplicity we take to have unit norm, $||\eta_+||^2 = 1$. Most of the following results hold more generally for SU($d/2$)-structures.

In the adapted coordinate system described previously, consider the product of all gamma matrices in directions parallel to D_p:

$$\gamma_{\parallel} \equiv \frac{1}{\sqrt{\det g|_{S_k}}} \gamma_{1\ldots k} \, . \tag{5.6.8}$$

This is inspired by (1.3.52), but in this section we take it as a mathematical definition; we will use the two definitions together in Section 9.2. One can check that this is unitary: $\gamma_{\parallel}\gamma_{\parallel}^{\dagger} = 1$. (The computation is slightly more complicated than usual because the lower indices are curved.) From the point of view of a submanifold S_k, (5.6.8) can be viewed as follows. The inclusion map $i\colon S_k \to M$ can be used to pull back forms from M to S_k, but not vice versa (Section 4.1.4). In particular, the volume form vol_{S_k} cannot be pushed forward to M. However, we can raise its indices to turn it into a multivector $\widehat{\mathrm{vol}_{S_k}}$, a section of $\Lambda^k T S_k$ that we can call *covolume*. This can now be pushed forward with i, to obtain a section of $\Lambda^k TM$. Choosing then a set of coordinates $\sigma^1, \ldots, \sigma^k$ on S_k, and parameterizing the inclusion map $i\colon S_k \to M$ as $x^m(\sigma^1, \ldots, \sigma^k)$, (5.6.8) can be written more intrinsically as

$$\gamma_{\parallel} = \frac{1}{\sqrt{\det g|_{S_k}}} \partial_{\sigma_1} x^{m_1} \ldots \partial_{\sigma_k} x^{m_k} \gamma_{m_1\ldots m_k} = (i_* \widehat{\mathrm{vol}_{S_k}})_{/} \, . \tag{5.6.9}$$

In the last step, we have used the push-forward map and the slash formalism from (3.1.15).

Even calibrations

We now assume k is even. Since $\gamma_{\parallel}$ is a bispinor, we can expand it in the basis (3.4.94), keeping only even terms:

$$\gamma_{\parallel} = U_{00}\eta_+ \otimes \eta_+^{\dagger} + U_{20}^m \eta_+ \otimes \eta_-^{\dagger}\gamma_m + \cdots \, . \tag{5.6.10}$$

(For example, a hypothetical term $U_{10}^m \eta_+ \otimes \eta_+^{\dagger}\gamma_m$ doesn't appear, since it corresponds to an odd form.) We have used real indices, at the cost of some redundancy, recalling the comment preceding (3.4.87); we can assume $U_{20}^m \Pi_m{}^n = U_{20}^n$ without loss of generality.

Using now cyclicity of the trace, (3.4.20), (3.2.34), and (3.4.88), we get

$$\begin{aligned} U_{00} &= \eta_+^{\dagger}\gamma_{\parallel}\eta_+ = \mathrm{Tr}(\gamma_{\parallel}\Phi_+) = \frac{1}{\sqrt{\det g|_{S_k}}}(\Phi_+)_{k\ldots 1} \, , \\ U_{20}^m &= \frac{1}{2}\eta_+^{\dagger}\gamma_{\parallel}\gamma^m\eta_- = \frac{1}{2}\mathrm{Tr}(\gamma_{\parallel}\gamma^m\bar{\Phi}_-) = \frac{1}{2\sqrt{\det g|_{S_k}}}(\gamma^m\bar{\Phi}_-)_{k\ldots 1} \, . \end{aligned} \tag{5.6.11}$$

Multiplying by a $(-1)^{\lfloor k/2 \rfloor} d\sigma^1 \wedge \ldots \wedge d\sigma^k$, these are proportional respectively to

$$U_{00} \propto \bar{\Phi}_+|_{S_k} \, , \qquad U_{20}^m \propto ((\mathrm{d}x^m - g^{mn}\iota_n)\bar{\Phi}_-)|_{S_k} \, . \tag{5.6.12}$$

Now

$$1 = \eta_+^\dagger \gamma_\| \gamma_\|^\dagger \eta_+ \overset{(5.6.10)}{=} (U_{00}\eta_+^\dagger + U^i_{20}\eta_-^\dagger \gamma_i)(\bar{U}_{00}\eta_+ + \bar{U}^{\bar{i}}_{20}\gamma_{\bar{i}}\eta_-) \overset{(3.4.88)}{=} |U_{00}|^2 + 2U^i_{20}\Pi_{i\bar{j}}\bar{U}^{\bar{j}}_{20}\,. \tag{5.6.13}$$

In the first step, we used the fact that $\gamma_\|$ is unitary. In the second step, the terms with $(\dots)$ in (5.6.10) canceled out. (5.6.13) implies in particular that $|U_{00}|^2 = \mathrm{Re}U^2_{00} + \mathrm{Im}U^2_{00} \leq 1$. Recalling (5.6.1) and (5.6.11), we have obtained that both $\mathrm{Re}\Phi_+$ and $\mathrm{Im}\Phi_+$ are almost calibrations: they satisfy (5.6.1) (and hence (5.6.3)) for even k. We can unify the two by saying that $\mathrm{Re}(e^{i\theta}\Phi_+)$ is an almost calibration for any θ.

From the definition (5.6.4), we then see that S_k is calibrated by this $\mathrm{Re}(e^{i\theta}\Phi_+)$ if $U^i_{20} = 0$, and U_{00} has a constant phase. Writing (5.6.13) as $\mathrm{Re}(e^{i\theta}U_{00})^2 + \mathrm{Im}(e^{i\theta}U_{00})^2 + \frac{1}{2}U^i_{20}\Pi_{i\bar{j}}\bar{U}^{\bar{j}}_{20} = 1$, we have an alternative formulation of (5.6.4):

$$\mathrm{Re}(e^{i\theta}\Phi_+)|_{S_k} = \mathrm{vol}_{S_k} \quad \Leftrightarrow \quad \begin{cases} \mathrm{Im}(e^{i\theta}\Phi_+)|_{S_k} = 0\,, \\ ((dx^m \wedge - g^{mn}\iota_n)\Phi_-)|_{S_k} = 0\,. \end{cases} \tag{5.6.14}$$

Let us see what this means more concretely, recalling from (3.4.20) that $\Phi_+ = e^{-iJ} = 1 - iJ - \frac{1}{2}J^2 + \frac{i}{6}J^3$ and $\Phi_- = \Omega$. For $k = 2$, taking $\theta = \pi/2$, J is a calibration for two-dimensional submanifolds S_2. The latter is calibrated if it obeys $J|_{S_2} = \mathrm{vol}_{S_2}$, or equivalently if $(dx^m \wedge + g^{mn}\iota_n)\Omega)|_{S_2} = 0$; only the second term survives, so the condition is $(\iota_n\Omega)|_{S_2} = 0$. That says that the pull-back of $(2,0)$-forms to S_2 should vanish. Summarizing,

$$J|_{S_2} = \mathrm{vol}_{S_2} \quad \Leftrightarrow \quad \Lambda^{2,0}|_{S_2} = 0\,. \tag{5.6.15}$$

We call such S_2 *almost* (or *pseudo-*) *holomorphic*. In terms of the almost complex structure, the condition can also be written as

$$I^t T_{0,1}S_2 = T_{0,1}S_2\ : \tag{5.6.16}$$

I behaves as a complex structure on S_2 itself.[8] For $k = 4$, taking $\theta = 0$ we see that $-\frac{1}{2}J^2$ is a calibration for submanifolds S_4. According to (5.6.14), S_4 is calibrated if $(-\frac{1}{2}J^2)|_{S_4} = \mathrm{vol}_{S_4}$, or equivalently if $(dx^m \wedge \Omega)|_{S_4} = 0$. This says that the pull-back of $(3,1)$-forms to S_4 should vanish and can be reformulated in terms of I as in (5.6.16). Again, such S_4 are said to be almost holomorphic. In general for k even, we can write

$$\pm \frac{1}{(k/2)!} J^{k/2}|_{S_k} = \mathrm{vol}_{S_k} \quad \Leftrightarrow \quad I^t T_{0,1}S_k = T_{0,1}S_k\,. \tag{5.6.17}$$

This is a version of *Wirtinger's theorem.*

Equation (5.6.14) is valid even if there is no integrability condition on I or J. If the almost complex structure I associated to Ω is integrable, the $k = 2$ and 4 submanifolds calibrated by J and $-\frac{1}{2}J^2$ are simply called "holomorphic" and can be described as zeros of systems of holomorphic functions.

[8] I was introduced in (5.1.11) with index structure $I_m{}^n$, so as to act naturally on one-forms; the action on vectors is then $(I^t v)^m = I_n{}^m v^n = (\iota_v I)^m$.

Odd calibrations

The story is similar if k is odd. Now the expansion is

$$\gamma_{\|} = U_{30}\eta_{+} \otimes \eta_{-}^{\dagger} + U_{10}^{m}\eta_{+} \otimes \eta_{+}^{\dagger}\gamma_{m} + \cdots \tag{5.6.18}$$

The relevant terms in the expansion of $\gamma_{\|}$ are

$$U_{30} \propto \bar{\Phi}_{-}|_{S_k}\,, \qquad U_{10}^{m} \propto ((\mathrm{d}x^m \wedge - g^{mn}\iota_n)\bar{\Phi}_{+})|_{S_k} \tag{5.6.19}$$

and similar to (5.6.13) we get $|U_{30}|^2 + 2U_{10}^{i}\bar{\Pi}_{i\bar{j}}\bar{U}_{10}^{\bar{j}} = 1$. So everything works as in the k = even case, but with $\Phi_{+} \leftrightarrow \Phi_{-}$. In particular, $\mathrm{Re}(\mathrm{e}^{\mathrm{i}\theta}\Phi_{-})$ is a calibration, and (5.6.4) can be written as

$$\mathrm{Re}(\mathrm{e}^{\mathrm{i}\theta}\Phi_{-})|_{S_k} = \mathrm{vol}(S_k) \quad \Leftrightarrow \quad \begin{cases} \mathrm{Im}(\mathrm{e}^{\mathrm{i}\theta}\Phi_{-})|_{S_k} = 0\,, \\ ((\mathrm{d}x^m \wedge - g^{mn}\iota_n)\Phi_{+})|_{S_k} = 0\,. \end{cases} \tag{5.6.20}$$

The only interesting case is now $k = 3$, where we see that $\mathrm{Re}(\mathrm{e}^{\mathrm{i}\theta}\Omega)$ is a calibration for three-dimensional submanifolds S_3, for any θ. The right-hand side of (5.6.20) gives an alternative condition for calibrated S_3:

$$\mathrm{Re}(\mathrm{e}^{\mathrm{i}\theta}\Omega)|_{S_3} = \mathrm{vol}_{S_3} \quad \Leftrightarrow \quad \begin{cases} \mathrm{Im}(\mathrm{e}^{\mathrm{i}\theta}\Omega)|_{S_3} = 0\,, \\ \mathrm{d}x^m \wedge J|_{S_3} = 0\,. \end{cases} \tag{5.6.21}$$

Such S_3 are called *almost special Lagrangian (sLag)* submanifolds. This is in part because an S_3 where the pull-back of J vanishes is called *almost Lagrangian.* If J is closed and S_6 is symplectic, we can drop the "almost" and call these calibrated submanifolds simply "special Lagrangian" and "Lagrangian" respectively.

Exercise 5.6.1 Adapt the arguments of Section 5.6.2 to $d = 7$, and use them to show that a form ϕ defining a G_2-structure is a calibration.

5.7 Special holonomy

We now focus on cases where the G-structure is covariantly constant. The covariant derivative ∇_m is the local counterpart of parallel transport (see Section 4.1.2). Parallel transport along a closed curve γ rotates a vector, in a way related to the Riemann tensor in (4.1.14). Since curves can be naturally composed by concatenation, these rotations form a group, called *holonomy group* Hol. Curves that can be continuously deformed to the trivial curve $(\gamma(t) = p \ \forall t \in [0, 1])$ form a subgroup called the *restricted holonomy* subgroup Hol_0.

Via (4.1.21), we can let these rotations act on any tensor or spinor, not just on vectors. If a tensor T is invariant under Hol, then it is in particular invariant under small curves; so T is in fact covariantly constant, $\nabla_m T = 0$. Conversely, if T is covariantly constant, the action of parallel transport (and hence of Hol) on it is trivial. So

$$\nabla_m T = 0 \quad \Leftrightarrow \quad T \text{ singlet of Hol}\,. \tag{5.7.1}$$

It is perhaps useful to contrast this new notion of holonomy with that of G-structure, which instead only signals that on M_d we have a *globally defined* tensor or spinor with stabilizer G, rather than a covariantly constant one. A G-structure only reduces the structure group of G, rather than the holonomy group.

We are using the Levi-Civita connection, for which $\nabla_m g_{np} = 0$; (5.7.1) then tells us that the metric is one such singlet, and (5.7.1) implies

$$\text{Hol} \subset \text{SO}(d) . \tag{5.7.2}$$

At the infinitesimal level, the parallel transport R_{pqmn} in (4.1.14) is an element of Hol near the identity, or in other words of the Lie algebra of Hol; by (4.1.25b) it is antisymmetric, and hence an element of so(d), in agreement with (5.7.2).

If a manifold M_d has further covariantly constant T, (5.7.1) implies Hol $\subset$ Stab(T). In this case, we say that M_d has *special holonomy*. We usually say that M_d is a (proper) G-holonomy manifold if

$$\text{Hol} = G \tag{5.7.3}$$

exactly, and not a subgroup of G.

The possible G that can occur as holonomy groups have been classified; this is known as *Berger's list* [150]. We present it in Table 5.1 for the Riemannian case.

We have given the relevant dimension in parentheses in the two cases where it is not obvious from the notation. All the dimensions in the table are the *minimal* ones; for example, if M_4 has holonomy SU(2), so will $M_4 \times \mathbb{R}^k$. Some further comments:

- In non-Euclidean signature, the list has to be modified by using different real forms of the complexification of the groups in Table 5.1; this would include, for example, $\text{SO}(d-1,1)$.
- A classification is also available for the allowed holonomy groups of connections that are not compatible with any metric. This results in a much longer list that has only recently been completed [151, 152].
- In some examples, reduced holonomy implies Ricci-flatness. These are the cases where the holonomy singlet is a spinor

$$D_m \eta = 0 \tag{5.7.4}$$

rather than by a tensor as in (5.7.1). Indeed

Table 5.1. Possible holonomy groups for Riemannian manifolds.

G	name	Ricci-flat?
$\text{U}(d/2)$	Kähler	no
$\text{SU}(d/2)$	Calabi–Yau	yes
G_2 ($d = 7$)		yes
Spin(7) ($d = 8$)		yes
$\text{Sp}(d/4)$	Hyper-Kähler	yes
$\text{Sp}(d/4) \cdot \text{Sp}(1)$	Quaternionic-Kähler	no
$\text{SO}(d)$		no

$$0 \overset{(5.7.4)}{=} 4\gamma^m[D_m, D_n]\eta \overset{(4.3.33)}{=} \gamma^m R^{ab}{}_{mn}\gamma_{ab}\eta$$
$$\overset{(3.1.22)}{=} R^{ab}{}_{mn}(\gamma^m{}_{ab} + 2\delta^m_a\gamma_b)\eta \overset{(4.1.25c)}{=} 2R_{nm}\gamma^m\eta = 0\,. \qquad (5.7.5)$$

Now we observe that the $\gamma^n\eta$ are all independent over the real numbers; for example, in Section 3.4.7, we saw that they are part of the spinor basis associated to η for $G = \mathrm{SU}(3)$ and G_2. (For SU(3), (3.4.87) gives a vanishing linear combination of the $\gamma^n\eta$, but with *complex* coefficients.) So we conclude

$$D_m\eta = 0 \quad \Rightarrow \quad R_{mn} = 0\,. \qquad (5.7.6)$$

- The case $\mathrm{U}(d/2)$ can also be formulated in terms of a spinor η, but one that is not covariantly constant; so the conclusion (5.7.6) does not hold in that case. We will analyze this *Kähler* case extensively in the next chapter.
- As we will see, the $\mathrm{Sp}(d/4)$ and $\mathrm{Sp}(d/4) \cdot \mathrm{Sp}(1) \equiv \mathrm{Sp}(d/4) \times \mathrm{Sp}(1)/\mathbb{Z}_2$ examples are modeled on the corresponding structures discussed in Section 5.1.4.

6 Kähler geometry

6.1 Kähler manifolds

6.1.1 Covariantly constant U(3)-structure

In Section 5.1.1, we saw that an $\mathrm{U}(d/2)$-structure is defined by an almost complex structure I and a Hermitian metric g. The latter is invariant under the Levi-Civita connection by definition; so the holonomy is $\mathrm{U}(d/2)$ if and only if

$$\nabla_m I_n{}^p = 0\,. \tag{6.1.1}$$

This implies that I is integrable. To see this, recall that in a Lie derivative (and hence in a Lie bracket) we can replace ordinary partial derivatives by covariant ones, as noticed in (4.1.60) and (4.1.61), and the definition of the Nijenhuis tensor in terms of Lie brackets in (5.3.24). If we define as usual J by lowering the p index of I as in (5.1.16), we also see from (6.1.1) that $\nabla_m J_{np} = 0$, and in particular that J is symplectic. So a Kähler manifold has both a complex structure I and a symplectic structure J, which are compatible in the sense defined after (5.1.39).

Intrinsic torsion

Alternatively, as also pointed out in Section 5.1.1, we can describe an $\mathrm{U}(d/2)$-structure by trading off the almost complex structure I for a $(d/2,0)$-form Ω, or more generally a twisted form, a section of $\Lambda^{d/2,0}T^*\otimes K^{-1}$. Now J and Ω should satisfy the compatibility conditions (5.1.40). Using this point of view, it is instructive to analyze the $d=6$ case using the language of intrinsic torsion, such as (5.5.6).

Raising the index p in (5.5.10b), we see that (6.1.1) imposes $q_{mn}=0$ and $\mathrm{Re}p_m = 0$; from the definition (5.5.2), then

$$(D_m - \mathrm{i}\,\mathrm{Im}p_m)\eta_+ = 0\,. \tag{6.1.2}$$

In the language of the W_i defined in (5.5.6), from (5.5.12) we see that

$$W_1 = W_2 = W_3 = W_4 = 0\,, \qquad \bar{W}_5 = 2p_{0,1}\,. \tag{6.1.3}$$

Conversely, $\mathrm{Im}p = \mathrm{Im}\bar{W}_5$. Recalling (5.5.6):

$$\mathrm{d}J = 0\,, \qquad \mathrm{d}\Omega = \bar{W}_5\wedge\Omega \qquad \text{(Kähler)}. \tag{6.1.4}$$

From (5.5.8) and (5.5.9), we again conclude that a Kähler manifold has compatible complex and symplectic structures. Moreover, we now see that this is in fact an alternative definition.

Curvature

As anticipated in the previous subsection, η_+ is not covariantly constant, and the Ricci tensor is not necessarily zero. A generalization of the computation in (5.7.5) using (3.4.87) gives

$$\rho = 2\mathrm{d}\,\mathrm{Im}p = 2\mathrm{d}\,\mathrm{Im}\bar{W}_5 \,, \qquad R_{mn} = \rho_{pn} I^p{}_m \,. \tag{6.1.5}$$

The antisymmetric tensor ρ_{pn} is sometimes called *Ricci form*; (6.1.5) says it is the curvature of a connection on K, in agreement with our earlier remark that follows (5.3.18). (This also tells us that η_+ is really a section of $\Sigma \otimes K^{-1/2}$.) By (4.2.57) and (5.1.33), the cohomology class $\frac{1}{2\pi}[\rho] = c_1(M)$.

The *$\partial\bar{\partial}$-lemma*, which holds on Kähler manifolds (and some other complex manifolds), says that an exact form that is annihilated by $\mathrm{d}^c = \mathrm{i}(\bar{\partial} - \partial)$ can be written as $\mathrm{dd}^c = 2\mathrm{i}\partial\bar{\partial}$ of some other form:

$$\mathrm{d}^c(\mathrm{d}\alpha) = 0 \quad \Rightarrow \quad \mathrm{d}\alpha = \mathrm{dd}^c\beta = 2\mathrm{i}\partial\bar{\partial}\beta \,. \tag{6.1.6}$$

An exact $(1, 1)$-form is automatically annihilated by d^c; applying this to (6.1.5), we can write it in terms of $\partial\bar{\partial}$. We can also derive this more directly. Recall (5.3.3):

$$J_{i\bar{\jmath}} = \mathrm{i}g_{i\bar{\jmath}} \,. \tag{6.1.7}$$

The Levi-Civita connection is (5.3.4) in the integrable case. But the $\partial_{[\bar{k}} g_{\bar{l}]j}$ appearing there is in fact proportional to $\mathrm{d}J_{1,2} = 0$. So the only surviving component is

$$\Gamma^i_{jk} = g^{i\bar{l}} \partial_j g_{k\bar{l}} \tag{6.1.8}$$

and its complex conjugate. The usual formulas for the Riemann (4.1.13) and Ricci (4.1.27) tensors now boil down to

$$\rho = -\mathrm{i}\partial\bar{\partial} \log \sqrt{g} \,, \tag{6.1.9}$$

where as usual $g = \det_{(mn)} g_{mn}$.[1] Comparison with (6.1.5) gives us

$$\mathrm{Im}p = \mathrm{Im}\bar{W}_5 = -\frac{1}{4}\mathrm{d}^c \log \sqrt{g} \,, \tag{6.1.10}$$

up to an exact form.

Another application of (6.1.6) is to J itself; since it is closed, *locally* it is exact, and so it can be written as

$$J = \mathrm{i}\partial\bar{\partial}K = \frac{1}{2}\mathrm{dd}^c K \tag{6.1.11}$$

for some function K, called *Kähler potential*. By (6.1.7),

$$g_{i\bar{\jmath}} = \partial_i \partial_{\bar{\jmath}} K \,. \tag{6.1.12}$$

As well as being only locally defined, it is not unique, since

$$K \to K + f + \bar{f} \,, \qquad \bar{\partial} f = 0 \tag{6.1.13}$$

leaves it invariant.

[1] In some references, this formula is given in terms of the determinant of the $g_{i\bar{\jmath}}$ block alone: $\det_{(mn)} g_{mn} = (\det_{(i\bar{\jmath})} g_{i\bar{\jmath}})^2$.

Summary of definitions

While some of the analysis in this section was specialized to $d = 6$, our conclusions actually hold in any d. So, summarizing, a Kähler structure can be defined as follows:

(1) A pair (I, J) of a complex and a symplectic structure that are compatible (recalling (5.1.39))
(2) A U(d/2)-structure (I, J) such that I and J are covariantly constant
(3) A U(d/2)-structure (I, J) such that (6.1.4) hold

6.1.2 Hodge diamond

When we introduced Dolbeault cohomology in (5.3.41), we mentioned that it had no clear relation with the de Rham cohomology (4.1.132). We will now see that on Kähler manifolds the latter is a direct sum of the former.

Hodge identity

We begin by simplifying (5.4.1). We could have computed it using $\mathrm{d} = \mathrm{d}x^m \wedge \nabla_m$ rather than $\mathrm{d} = \mathrm{d}x^m \wedge \partial_m$; so we can replace $\partial_m \to \nabla_m$ in its right-hand side. (This is similar to the equivalence of (4.1.60) and (4.1.61).) In the Kähler case, $\nabla J = 0$, so on the right-hand side we are left with only $J^{mn}\iota_m\nabla_n$; with (5.1.16), this becomes

$$[\mathrm{d}, J\cdot] = I_m{}^n \iota^m \nabla_n \,. \tag{6.1.14}$$

We can apply the same logic to (5.3.35); again $\nabla I = 0$ for the Kähler case, and we conclude

$$\mathrm{d}^{\mathrm{c}} = -I_m{}^n \mathrm{d}x^m \wedge \nabla_n \,. \tag{6.1.15}$$

Using (4.1.110) and comparing to (6.1.14) gives

$$[\mathrm{d}, J\cdot] = (\mathrm{d}^{\mathrm{c}})^\dagger \,. \tag{6.1.16}$$

This is called *Hodge identity*; let us stress that we have used the Kähler condition several times in deriving it.

Hodge diamond

We can also rewrite it in terms of the Dolbeault differential ∂, using the decompositions (5.3.30) and (5.3.31) of d and d^{c}. The adjoint of ∂ goes in the opposite direction as (5.3.28); so $\partial^\dagger : \Lambda^{p,q} \to \Lambda^{p-1,q}$. Applying (6.1.16) to a (p, q)-form and projecting the result on $\Lambda^{p-1,q}$ and $\Lambda^{p,q-1}$, we get

$$[\partial, J\cdot] = \mathrm{i}\bar\partial^\dagger \,, \qquad [\bar\partial, J\cdot] = -\mathrm{i}\partial^\dagger \,. \tag{6.1.17}$$

Using these, the definitions of the Laplacian and the decompositions (5.3.30) and (5.3.31), we get $\Delta = \Delta_\partial + \Delta_{\bar\partial}$, where $\Delta_\partial \equiv \partial\partial^\dagger + \partial^\dagger\partial$; so

$$\Delta_\partial = \Delta_{\bar\partial} \,, \qquad \Delta = 2\Delta_\partial \,. \tag{6.1.18}$$

Now, every de Rham cohomology class has a harmonic representative A, which satisfies $\Delta A = 0$, as we saw around (4.1.140). By (6.1.18), Δ of a (p, q)-form is again a (p, q)-form. So if we decompose $A = \sum_{p,q} A_{p,q}$, each of the $A_{p,q}$ will be separately harmonic. Equation (6.1.18) also says that $\Delta_\partial A_{p,q} = 0$. An argument similar to (4.1.116) now implies that $\partial A_{p,q} = \partial^\dagger A_{p,q} = 0$. So we have that every de Rham cohomology class can be decomposed as a sum of forms in the Dolbeault cohomology. In other words:

$$H^p = H^{p,0} \oplus H^{p-1,1} \oplus \ldots \oplus H^{0,p} = \oplus_{k=0}^{p} H^{p,k-p} \, . \qquad (6.1.19)$$

This is often also called Hodge decomposition for Kähler manifolds (not to be confused with (4.1.140), which is true in general). It refines (5.3.46). With a similar argument, (6.1.18) also implies

$$h^{p,q} = h^{q,p} \, . \qquad (6.1.20)$$

Finally, we can also use (5.3.48) for $E = \mathcal{O}$:

$$h^{p,q} = h^{d/2-q,d/2-p} \, . \qquad (6.1.21)$$

It makes sense to organize the Dolbeault cohomology groups in a *Hodge diamond*. In $d = 6$:

$$\begin{array}{ccccccc} & & & 1 & & & \\ & & h^{1,0} & & h^{1,0} & & \\ & h^{2,0} & & h^{1,1} & & h^{2,0} & \\ h^{3,0} & & h^{2,1} & & h^{2,1} & & h^{3,0} \\ & h^{2,0} & & h^{1,1} & & h^{2,0} & \\ & & h^{1,0} & & h^{1,0} & & \\ & & & 1 & & & \end{array} \, , \qquad \begin{array}{ccc} \nwarrow & & \nearrow \\ \bar\partial^\dagger & & \partial^\dagger \\ \partial & & \bar\partial \\ \swarrow & & \searrow \end{array} \, . \qquad (6.1.22)$$

This is similar to (5.2.15), but now the entries are cohomology dimensions rather than spaces of forms. We have used (6.1.20), (6.1.21), and the fact that $h^0 = h^{0,0} = 1$, $h^d = h^{d/2,d/2} = 1$.

Lefschetz decomposition

A further type of decomposition is obtained by looking at the action of the Lefschetz operators $J\wedge$ and $J\cdot$ from Section 5.2.1. On Kähler manifolds,

$$[J\wedge, \Delta] = 0 \, , \qquad [J\cdot, \Delta] = 0 \, . \qquad (6.1.23)$$

It follows that $J\wedge$ maps cohomology to cohomology, and not simply forms to forms as in Section 5.2.1. Thus we can define primitive cohomology classes, extending the notion of primitive forms in (5.2.10). Moreover, from (5.2.9) it follows that the cohomology of Kähler manifolds decomposes in $\mathrm{sl}(2,\mathbb{R})$ representations. This is called *Lefschetz decomposition*. Visually, $J\wedge$ and $J\cdot$ act vertically in the Hodge diamond (6.1.22).

Exercise 6.1.1 Derive (6.1.5) from (6.1.2), using (5.7.5) and (3.4.87).

Exercise 6.1.2 Check (6.1.9) for the metric (4.1.47), $\mathrm{d}s^2 = 4\frac{\mathrm{d}z\mathrm{d}\bar z}{(1+|z|^2)^2}$, using (4.5.2).

Exercise 6.1.3 Still for S^2, find an explicit solution to (6.1.2); compare the result to (5.3.13). (Hint: in the patch $U_\mathrm{N} \cap U_\mathrm{S}$, η can be taken to be constant.)

6.2 Complex projective space

Part of what makes Kähler manifolds special is how easy it is to produce examples.

A trivial one is $\mathbb{C}^N$. The complex structure is obtained by taking the complex vielbein $e^a = \mathrm{d}z^a$, in the language of Section 5.3.1, and the symplectic structure by writing $J = \frac{\mathrm{i}}{2} h^a \wedge \overline{h^a}$ as in (5.1.35).

The simplest compact example is *complex projective space*, introduced in (4.4.49). A more common definition is

$$\mathbb{CP}^N \equiv \frac{\mathbb{C}^{N+1} - \{\underline{0}\}}{\mathbb{C}^*} . \tag{6.2.1}$$

Here $\underline{0} \equiv \{0, \ldots, 0\}$ is the origin, and $\lambda \in \mathbb{C}^* = \mathbb{C} - \{0\}$ acts as

$$\lambda \cdot (Z^0, \ldots, Z^N) = (\lambda Z^0, \ldots, \lambda Z^N) . \tag{6.2.2}$$

In other words, $\mathbb{CP}^N$ is the space of complex lines in $\mathbb{C}^{N+1}$. Equivalence of (4.4.49) and (6.2.1) follows by the general principle (2.4.25).

We can cover $\mathbb{CP}^N$ with charts $U_I = \{Z^I \neq 0\}$, $A = 0, \ldots, N\}$, with coordinates

$$\left\{ z^1_{(A)} \equiv \frac{Z^0}{Z^i}, \ldots, z^{i-1}_{(A)} \equiv \frac{Z^{i-1}}{Z^i}, z^i_{(A)} \equiv \frac{Z^{i+1}}{Z^i}, \ldots, z^{d-1}_{(A)} \equiv \frac{Z^d}{Z^i} \right\} , \tag{6.2.3}$$

invariant under the $\mathbb{C}^*$-action.

6.2.1 Bundles and projections

One can also view (6.2.1) as the base of a bundle. The most direct way is to view $\mathbb{C}^{N+1} - \{\underline{0}\}$ as the total space and $\mathbb{C}^*$ as the fiber, following (4.4.31). Another way is to consider the sphere $S^{2N+1} \subset \mathbb{C}^{N+1}$, which is left invariant by the action in (6.2.1) if we take $\lambda = \mathrm{e}^{\mathrm{i}\varphi} \in \mathrm{U}(1) \cong S^1$. So from this point of view, we can write the sphere as a bundle over $\mathbb{CP}^N$. In the notation of (4.2.2):

$$\begin{array}{ccc} S^1 & \hookrightarrow & S^{2N+1} \\ & & \downarrow \\ & & \mathbb{CP}^N . \end{array} \tag{6.2.4}$$

Another relevant bundle is

$$\mathbb{R}_+ \hookrightarrow (\mathbb{C}^{N+1} - \{\underline{0}\}) \to S^{2N+1} . \tag{6.2.5}$$

In a sense, (6.2.4) and (6.2.5) are a way of performing the quotient (6.2.1) in stages, viewing $\mathbb{C}^*$ as $\mathbb{R}_+ \times \mathrm{U}(1)$. (Yet another bundle results from (4.4.31) and the coset structure (4.4.49).)

To build some intuition, consider the case $N = 1$. Comparing (4.4.49) with (4.4.44) shows

$$S^2 \cong \mathbb{CP}^1 . \tag{6.2.6}$$

Equation (6.2.4) is then a generalization of (4.2.54), so it is often also called Hopf fibration. (A different generalization was considered in (4.4.64) and (4.4.65).) There are two charts. Already $U_0 = \{Z^0 \neq 0\}$ covers almost all of $\mathbb{CP}^1$: the only locus missing is $\{Z^0 = 0\}$, which by the identification in (6.2.1) is a single point. $U_1 = \{Z^1 \neq 0\}$ misses instead the point $\{Z^1 = 0\}$. This corresponds to our discussion following (4.1.41); we can identify $z_{(0)} = z^1/z_0$ in the current notation (6.2.3) with z_N in (4.1.43), and $z_{(1)}$ with z_S. Alternatively, the single point missing to U_1 is the point at infinity in the Riemann sphere (4.1.45).

Projected derivatives

To define forms on $\mathbb{CP}^N$, we will make repeated use of the idea (4.2.9), according to which forms that are basic (horizontal and invariant) under a projection really represent forms on the base of a bundle. We will apply this first to define forms on S^{2N+1}, then on $\mathbb{CP}^N$ itself, using (6.2.4) and (6.2.5).

On $\mathbb{C}^{N+1}$, we define the usual radial coordinate by $r^2 \equiv |Z^1|^2 + \ldots + |Z^{N+1}|^2 = Z^I \bar{Z}_I$ (where $\overline{Z^I} \equiv \bar{Z}_I$). In this language, the Euler vector field (4.3.51) and its one-form dual read

$$r\partial_r = 2\text{Re}\left(Z^I \partial_I\right), \qquad r\text{d}r = \text{Re}\left(\bar{Z}_I \text{d}Z^I\right). \tag{6.2.7}$$

The complex structure of $\mathbb{C}^N$ is defined by $I_I{}^J = \text{i}\delta_I^J$, and produces another vector and another one-form:

$$\begin{aligned} \xi &\equiv I^t r\partial_r = \text{i}(Z^I \partial_I - \bar{Z}_I \bar{\partial}^I) = -2\text{Im}\left(Z^I \partial_I\right), \\ r^2\eta &\equiv \iota_{r\partial_r} J = \frac{\text{i}}{2}(Z^I \text{d}\bar{Z}_I - \bar{Z}_I \text{d}Z^I) = \frac{1}{r^2}\text{Im}\left(\bar{Z}_I \text{d}Z^I\right). \end{aligned} \tag{6.2.8}$$

The definition ensures $\iota_\xi \eta = 1$. The vectors ξ and $r\partial_r$ are vertical for (6.2.4) and (6.2.5) respectively.

It is useful to define a projected derivative:

$$\begin{aligned} DZ^I &\equiv \text{d}Z^I - \frac{1}{r^2} Z^I \bar{Z}_J \text{d}Z^J = \mathcal{P}^I{}_J \text{d}Z^I \\ &= \text{d}Z^I - Z^I\left(\frac{\text{d}r}{r} + \text{i}\eta\right). \end{aligned} \tag{6.2.9}$$

(The projector $\mathcal{P}^I{}_J \equiv \delta^I_J - \frac{1}{r^2} Z^I \bar{Z}_J$ is a complex analogue of the first fundamental form (4.1.85).) The operator (6.2.9) has the property that it is annihilated by both $r\partial_r$ and ξ:

$$\iota_{r\partial_r} DZ^I = 0, \qquad \iota_\xi DZ^I = 0. \tag{6.2.10}$$

By (6.2.9) and (6.2.10), the functions Z^I and the form DZ^I have weight 1 under $L_{r\partial_r}$ and i under L_ξ.

Forms on S^{2N+1}

We then expect the function $\frac{Z^I}{r}$ to be invariant under $r\partial_r$. Indeed,

$$\text{d}\left(\frac{Z^I}{r}\right) = \frac{1}{r}(DZ^I + \text{i}Z^I\eta). \tag{6.2.11}$$

Since the dr term has dropped out, from (4.1.79) it follows that $L_{r\partial_r}\frac{Z^I}{r} = 0$. By (4.2.9), a function that is invariant under vertical vectors on a bundle is called basic and can be thought of as a function on the base. Applying this to the bundle (6.2.5), we see that $\frac{Z^I}{r}$ represents a function on S^{2N+1}.

The one-form $\frac{DZ^I}{r}$ satisfies

$$\mathrm{d}\left(\frac{DZ^I}{r}\right) = \frac{\mathrm{i}}{r}(\eta \wedge DZ^I - Z^I \mathrm{d}\eta) . \tag{6.2.12}$$

A direct computation from (6.2.8) and (6.2.10) gives

$$\mathrm{d}\eta = \frac{\mathrm{i}}{r^2} DZ^I \wedge D\bar{Z}_I \equiv 2J_{\mathrm{FS}} , \tag{6.2.13}$$

which in particular is annihilated by $\iota_{r\partial_r}$. It follows that $\frac{DZ^I}{r}$ is horizontal and invariant, so again by (4.2.9) and (6.2.5), it is a form on S^{2N+1}.

Kähler structure on $\mathbb{CP}^N$

We now define a Kähler structure (I, J) on $\mathbb{CP}^N$. The complex structure I can be defined by saying that the z^i are complex coordinates; in other words, the $(1,0)$-forms are linear combinations of the $\mathrm{d}z^i$ in each chart, and this defines I.

To define J, we use again the principle (4.2.9), this time by using the bundle (6.2.4). Both $\frac{Z^I}{r}$ and $\frac{DZ^I}{r}$ are horizontal, but they have weight i under L_ξ; so they are not invariant and in particular not basic. But they can be combined to obtain basic forms. One example is J_{FS} in (6.2.13): the weights of DZ^I/r and $D\bar{Z}_I/r$ cancel, and $L_\xi J_{\mathrm{FS}} = 0$. So (6.2.13) is really a form on $\mathbb{CP}^N$; it is called the *Fubini–Study* two-form. It is easy to check that it is closed, using (6.2.12). Thus we will take J_{FS} to be our symplectic structure on $\mathbb{CP}^N$.

While (6.2.13) might seem to also indicate that J_{FS} is exact, η is not horizontal and thus not basic, and so cannot be considered a form in $\mathbb{CP}^N$; so J_{FS} is not exact on $\mathbb{CP}^N$. On the other hand, η is basic with respect to $r\partial_r$, so J_{FS} is exact in S^{2N+1}, consistent with (4.1.134).

We can express J_{FS} in a chart by picking one of the coordinate systems in (6.2.1), say the one on U_0, $z^i \equiv z^i_{(0)}$. Since the form is invariant under the $\mathbb{C}^*$ action, we can use it to fix $Z^0 = 1$, and then $z^i = Z^i$ for $a = 1, \ldots, d$. By (6.2.13) and (6.2.9),

$$J_{\mathrm{FS}} = \frac{\mathrm{i}}{2}\left(r^{-2}\mathrm{d}z^i \wedge \mathrm{d}\bar{z}^{\bar{i}} - r^{-4}\bar{z}^{\bar{i}} z^j \mathrm{d}z^i \wedge \mathrm{d}\bar{z}^{\bar{j}}\right) , \qquad r^2 = 1 + z^i \bar{z}^{\bar{i}} = Z^I \bar{Z}_I . \tag{6.2.14}$$

A possible choice of Kähler potential (6.1.11) is

$$K = \frac{1}{2}\log(1 + z^i \bar{z}^{\bar{i}}) = \frac{1}{2}\log(Z^I \bar{Z}_I) . \tag{6.2.15}$$

We can now check if the I and J we have described are compatible. Since we have chosen the complex structure I so that the $(1,0)$-forms are linear combinations of the $\mathrm{d}z^i$, we see that g in (5.1.39) is obtained by dropping an i and the $\wedge$ symbol (so that the product among the differentials becomes a symmetric one), just as we saw in (5.1.35) and (5.1.37). So we get the *Fubini–Study metric*:

$$g_{\mathrm{FS}} = \frac{1}{r^2} DZ^I D\bar{Z}_I = r^{-2} dz^i \mathrm{d}\bar{z}^{\bar{i}} - r^{-4}\bar{z}^{\bar{i}} z^j \mathrm{d}z^i \mathrm{d}\bar{z}^{\bar{j}} . \tag{6.2.16}$$

Since this is symmetric and positive definite, I and J are compatible; we know already that they are respectively integrable and closed, so together they define a Kähler structure.

For $N = 1$, the form η becomes simply $\mathrm{Im}\left(\frac{\bar{z}\mathrm{d}z}{1+|z|^2}\right)$. This z is in fact z_{N}, since we are working in the patch where $Z^0 = 1$, $Z^1 = z$. In polar coordinates,

$$\eta = \frac{1}{2}(1 - \cos\theta)\mathrm{d}\phi = -A_{\mathrm{N}}\,, \tag{6.2.17}$$

recalling (4.2.53). Equation (6.2.16) becomes $r^{-4}\mathrm{d}z\mathrm{d}\bar{z}$, $r^2 = 1 + |z|^2$, which is 1/4 of the round S^2 metric (4.1.47). The Kähler form (6.2.14) becomes

$$J_{\mathbb{CP}^1} = \frac{\mathrm{i}}{2}r^{-4}\mathrm{d}z \wedge \mathrm{d}\bar{z} = \frac{1}{4}\mathrm{vol}_{S^2} = \frac{1}{4}\sin\theta\mathrm{d}\theta \wedge \mathrm{d}\phi\,. \tag{6.2.18}$$

So the identification (6.2.6) is not only topological: the coset metrics and the Kähler structures are also identified.

6.2.2 Cohomology

J_{FS} and its powers J^k_{FS} are the only nontrivial elements in de Rham cohomology:

$$h^k(\mathbb{CP}^N) = \begin{cases} 1 & k \in \{0, 2, \ldots, 2N\} \\ 0 & \text{otherwise.} \end{cases} \tag{6.2.19}$$

This can be established with the methods of Section 4.4.2. In particular, since line bundles are characterized by their c_1 (4.2.70), there is only one line bundle with $c_1 = n$, which by analogy with S^2 we call $O(n)$. The U(1)-bundle (6.2.4) is related to $O(-1)$ by structure group reduction, as in Section 4.2.4.

Dual to (6.2.19), there is also a single generator of homology in each even degree. A representative for the generator of $H_2(\mathbb{CP}^N)$ is the copy of $\mathbb{CP}^1$ defined by $N - 1$ linear equations, say $\{Z^2 = \cdots = Z^N = 0\}$. On this locus, J_{FS} reduces to (6.2.18), which integrates to $\frac{1}{4} \cdot 4\pi = \pi$. So the generator of the integer-valued cohomology is

$$\frac{1}{\pi}J_{\mathrm{FS}} \in H^2(\mathbb{CP}^N, \mathbb{Z})\,. \tag{6.2.20}$$

This is the Poincaré dual (4.1.136) of the generator of $H_{2N-2}(\mathbb{CP}^N)$, which is a hypersurface cut by a linear equation f_1, also called a *hyperplane*. An integer multiple of this, $\frac{d}{\pi}J_{\mathrm{FS}}$, is the Poincaré dual of the locus cut in $\mathbb{CP}^N$ by an equation of degree d. To see this, we can imagine a process by which such an equation degenerates to the dth power of a linear one, f_1^d. We will study such loci at length in Section 6.3.

Since the forms J^k_{FS} are (k, k), we can refine (6.2.19) to a statement about Dolbeault cohomology:

$$h^{i,j}(\mathbb{CP}^N) = \begin{cases} 1 & i = j \in \{0, 1, \ldots, N/2\} \\ 0 & \text{otherwise.} \end{cases} \tag{6.2.21}$$

So the diamond (6.1.22) would have all ones in the central vertical column, and all zeros in any other entry. This agrees with (6.1.19) and (6.1.20) for Kähler manifolds.

Twisted holomorphic volume form

We will now see to what extent we can introduce a holomorphic volume form Ω. Since the forms DZ^I are $(1, 0)$, we can try assembling $d/2 = N$ of them:

$$\Omega = \frac{1}{N!\,r^{N+1}}\epsilon_{I_0\ldots I_N}Z^{I_0}DZ^{I_1} \wedge \ldots \wedge DZ^{I_N}\,. \tag{6.2.22}$$

This is annihilated by $\iota_{r\partial_r}$ and has weight zero under $L_{r\partial_r}$, so it is a form on S^{2N+1}. Under ξ, it is again vertical, $\iota_\xi \Omega = 0$, but it has weight $(N+1)\mathrm{i}$:

$$L_\xi \Omega = (N+1)\mathrm{i}\Omega\,; \tag{6.2.23}$$

so it is not a form on $\mathbb{CP}^N$. This tells us that $c_1 \neq 0$: Ω does not exist as a genuine $(N,0)$-form on $\mathbb{CP}^N$.

Recall from (5.1.34) that, if $c_1 \neq 0$, Ω can be interpreted as a twisted form. Let us see this explicitly by again working in the patch U_0 and taking

$$Z^0 = \mathrm{e}^{\mathrm{i}\psi}\,, \qquad Z^i = \mathrm{e}^{\mathrm{i}\psi} z^i\,. \tag{6.2.24}$$

We have used invariance under $r\partial_r$ to fix the norm of the Z^I. Since ξ acts by changing the phases of the Z^I simultaneously, it acts only on ψ and not on the z^i; in other words,

$$\xi = \partial_\psi\,. \tag{6.2.25}$$

Using (6.2.24) in (6.2.22), we then have

$$\Omega = \mathrm{e}^{\mathrm{i}(N+1)\psi}\Omega_0\,, \qquad \Omega_0 \equiv \frac{1}{r^{N+1}}\mathrm{d}z^1 \wedge \ldots \wedge \mathrm{d}z^N\,. \tag{6.2.26}$$

Now Ω_0 has weight zero under L_ξ, but it does not glue as a form with the expressions on the other patches, but rather as a bundle-valued form. This is in the spirit of the remark around (4.2.10). For example, for $N = 1$, $\Omega = \frac{1}{2}\mathrm{e}^{2\mathrm{i}\psi}h$, where h is the holomorphic vielbein defined in (5.3.11) and (5.3.13), which, as we showed there, is a section of $K \otimes U_{-2}$.

First Chern class

Since (6.2.4) has $c_1 = -1$, (6.2.26) suggests that $c_1 = N+1$; but let us see this more explicitly. By (6.2.11) and (6.2.12):

$$\mathrm{d}\Omega = (N+1)\mathrm{i}\eta \wedge \Omega \tag{6.2.27}$$

(from which (6.2.23) can be rederived). Now Ω_0 in (6.2.26) obeys $\mathrm{d}\Omega_0 = (N+1)\mathrm{i}(\eta - \mathrm{d}\psi) \wedge \Omega_0$. The one-form $\eta - \mathrm{d}\psi$ is horizontal, so we can write it as

$$\eta = \mathrm{d}\psi + \frac{2}{N+1}\mathrm{Im}\bar{W}_5 \tag{6.2.28}$$

with a notation chosen to match (6.1.4), recalling that $(\bar{W}_5)_{1,0} = 0$. By (6.2.13), $\mathrm{dIm}\bar{W}_5 = (N+1)J_{\mathrm{FS}}$, and by (6.1.5)[2],

$$\rho = 2(N+1)J_{\mathrm{FS}}\,. \tag{6.2.29}$$

So now from (6.2.20) $\frac{1}{2\pi}\int_{\mathbb{CP}^1} \rho = (N+1)$, and by (4.2.60):

$$c_1(\mathbb{CP}^N) = N+1\,. \tag{6.2.30}$$

This should be read with some care. c_1 is not a number, but rather a class in H^2. Here on $\mathbb{CP}^N$ there is only one generator of this group, the class of J_{FS}; (6.2.30) means that c_1 is $(N+1)$ times this generator. A more precise way of writing (6.2.30) is

[2] The definition of Ricci form (6.1.5) also tells us that $R_{mn} = 2(N+1)g_{mn}$: this says that $\mathbb{CP}^N$ is an Einstein manifold, as defined in (4.5.2).

$$K_{\mathbb{CP}^N} \cong \mathcal{O}(-N-1). \tag{6.2.31}$$

For $N = 1$, (6.2.27) gives $\mathrm{d}h = -2\mathrm{i}A_\mathrm{N} \wedge h$, where again h is the holomorphic vielbein in (5.3.11) and (5.3.13), and A_N was given in (4.2.53). The result for c_1 in (4.2.57) agrees with the statement (5.3.10) that $T_{1,0} = \mathcal{O}(2)$ for S^2.

Equation (6.2.30) also follows from the *Euler quotient*

$$T_{1,0}(\mathbb{CP}^N) = ((N+1)\mathcal{O}(1))/\mathcal{O}, \tag{6.2.32}$$

where $(N+1)\mathcal{O}(1)$ is the direct sum of $N+1$ copies of $\mathcal{O}(1)$. This can be proven by observing that the components of a vector field $Z^I \partial_B$ have degree one under ξ, and that the particular vector field $Z^I \partial_I$ acts trivially on $\mathbb{CP}^N$.

With the method of long exact sequences mentioned in (5.3.50), (6.2.32) also implies

$$H^{0,1}(\mathbb{CP}^N, T_{1,0}) = 0. \tag{6.2.33}$$

Recalling (5.3.60), this means that $\mathbb{CP}^N$ has no complex deformations.

This concludes our discussion of $\mathbb{CP}^N$, arguably the prototype of compact Kähler manifolds.

Exercise 6.2.1 Show (6.2.12); using (4.1.79), compute explicitly the charges of $\frac{DZ^I}{r}$ under $L_{r\partial_r}$ and L_ξ.

Exercise 6.2.2 Recover the Fubini–Study metric as an example of coset metric (4.4.40), using (4.4.49).

Exercise 6.2.3 Use (6.2.20), Poincaré duality (4.1.136) and Wirtinger's theorem (5.6.17) to compute

$$\mathrm{Vol}(\mathbb{CP}^N) = \frac{\pi^N}{N!} \tag{6.2.34}$$

with respect to the Fubini–Study metric. Confirm this by using (6.2.4) and the formula for $\mathrm{Vol}(S^d)$ given after (1.3.59).

Exercise 6.2.4 Write S^7 as a join $S^3 * S^3$, (4.5.7). Now write each of the S^3 as a Hopf fibration, (4.2.51), with two sets of coordinates $(\psi_a, \theta_a, \phi_a)$, $a = 1, 2$. Now write the S^7 metric in the bundle metric form (4.2.35), with $\psi \equiv \psi_1 + \psi_2$. You should get a base metric

$$\mathrm{d}s^2 = \mathrm{d}x^2 + \frac{1}{4}\left(\cos^2 x \mathrm{d}s^2_{S^2_1} + \sin^2 x \mathrm{d}s^2_{S^2_2} + \sin^2 x \cos^2 x(\mathrm{d}a - A_1 + A_2)\right). \tag{6.2.35}$$

Recalling (6.2.4), this is the Fubini–Study metric on $\mathbb{CP}^3$, written as a foliation whose leaves are the total space of an S^1-bundle over $S^2 \times S^2$.

6.3 Algebraic geometry

A powerful procedure to obtain Kähler manifolds is to consider zero loci of holomorphic or meromorphic functions. This is one of the main topics of interest of algebraic geometry.[3] Unfortunately, we will not do justice to this important (and prestigious) field of mathematics. Classic introductions are [119, 153] and the perhaps more accessible [154].

6.3.1 Hypersurfaces

A *hypersurface* is simply the locus cut by a single holomorphic function. For example, in $\mathbb{C}^N$ we can consider

$$Z(f, \mathbb{C}^N) \equiv \{f = 0\} \subset \mathbb{C}^N, \tag{6.3.1}$$

where f is holomorphic. If f is a polynomial, we say this manifold is *algebraic*. A natural complex structure I is defined by declaring the pull-backs $i^* dZ^i$ to be $(1,0)$-forms, where $i\colon Z(f, \mathbb{C}^N) \to \mathbb{C}^N$ is the inclusion map. The two-form J can be similarly defined as pull-back $i^* J$ of (5.1.35). This is again a Kähler structure.

However, such a locus $Z(f, \mathbb{C}^N)$ is always noncompact. In order to define a compact Kähler manifold in this fashion, we can apply a similar procedure to $\mathbb{CP}^N$, which is compact itself. By Liouville's theorem (see Section 5.3.1), there are no holomorphic functions on a compact space, but we can use meromorphic functions, which are allowed to have poles. As we mentioned in Section 4.2.1 for the case of S^2, an alternative and equivalent point of view is to use holomorphic sections of nontrivial line bundles. On $\mathbb{CP}^N$, we saw after (6.2.19) that line bundles $O(d)$ are classified by a single number d. Concretely, their sections can be thought of as degree-d functions f_d of the Z^I. Indeed, such f_d have weight d under ξ, so they are functions on the total space of a bundle and not on $\mathbb{CP}^N$. (For $N = 1$, specifying the f_d to both charts, we indeed recover (4.2.47.)

So we end up with the submanifolds

$$Z(f_d, \mathbb{CP}^N) = \{f_d = 0\} \subset \mathbb{CP}^N, \tag{6.3.2}$$

which we also call hypersurfaces, and are again Kähler. Being cut by a single complex equation, $Z(f_d, \mathbb{CP}^N)$ has complex dimension $N - 1$. From now on in this subsection, we will restrict to compact examples. We will also just write $Z(f_d) = Z(f_d, \mathbb{CP}^N)$ when the ambient space $\mathbb{CP}^N$ should be clear from context.

We now sketch how to compute some features of these hypersurfaces. The cohomology is partially determined by the *Lefschetz hyperplane theorem*. Given an embedding $M_{N_1} \subset M_{N_2}$, with N_1, N_2 denoting the dimensions:

$$h^i(M_{N_1}) = h^i(M_{N_2}) \quad 0 \le i \le N_2 - 2. \tag{6.3.3}$$

So the cohomology in low degrees of the smaller M_{N_1} is partially inherited from that of M_{N_2}. For us, the latter will often be $\mathbb{CP}^N$.

To compute c_1 of (6.3.2), consider the holomorphic normal bundle $N_{1,0}$, defined as in (4.2.8) but with holomorphic tangent bundles $T_{1,0}$. Again we first consider a more general situation, with $Z = Z(f) \subset P$ a hypersurface (complex codimension one) in a manifold P. Just like in the previous example $Z(f_d, \mathbb{CP}^N)$, the function f

[3] In that context, one often prefers the name "variety" to "manifold." Implicit in the name is the fact that the natural maps in an algebraic context are rational rather than holomorphic. One then identifies varieties if they are *birational* rather than biholomorphic. Here we will overlook these important subtleties.

whose zeros cut $Z \subset P$ will not really be a function, but a section of a bundle, which we will call $O(Z)$. The *adjunction formula* now says

$$N_{1,0} = O(Z)|_Z . \tag{6.3.4}$$

Taking the determinant bundle (see discussion in (4.2.7)) on both sides of definition (4.2.8), we obtain

$$K_Z = K_P|_Z \otimes N_{1,0} \overset{(6.3.4)}{=} (K_P \otimes O(Z))|_Z . \tag{6.3.5}$$

We now apply these to $Z = Z(f_d) \subset P = \mathbb{CP}^N$. (6.3.4) and (6.3.5) give respectively

$$N_{1,0} = O(d)|_{Z(f_d)} , \qquad K_{Z(f_d)} = (K_{\mathbb{CP}^N} \otimes O(d))|_{Z(f_d)} . \tag{6.3.6}$$

By (6.2.31), it now follows that

$$c_1(Z(f_d)) = (N + 1 - d)[J_{\rm FS}]|_{Z(f_d)} . \tag{6.3.7}$$

Recall that $[J_{\rm FS}]$ is the generator of $H^2(\mathbb{CP}^N)$; we have denoted by $[J_{\rm FS}]|_{Z(f_d)}$ its pull-back to $Z(f_d)$.

A manifold is called *Fano* if K^{-1} is a positive line bundle. From the definition that precedes (5.3.49), we see that $c_1(K)$ should have a positive $(1, 1)$-representative; after (6.1.5), we saw that a representative of $c_1(K)$ is the Ricci form ρ. Equation (6.2.29) shows that $\mathbb{CP}^N$ is Fano; (6.3.7) shows that $Z(f_d)$ is also Fano for $d < N+1$.

6.3.2 Riemann surfaces and plane curves

As an example, let us study hypersurfaces $Z(f_d, \mathbb{CP}^2)$; they have complex dimension one, so they should be Riemann surfaces (Section 5.3.5), which in algebraic geometry are often called *curves* because one tends to count complex dimensions. From this point of view, $\mathbb{CP}^2$ is called a projective *plane* (complex dimension two), and $Z(f_d, \mathbb{CP}^2)$ are called *plane curves*. The c_1 is given in (6.3.7), but a subtlety appears: even though $[J_{\rm FS}]$ was the generator of $H^2(\mathbb{CP}^2)$, its pull-back to $Z(f_d, \mathbb{CP}^2)$ is not. This is clear from the point of view of the Poincaré duals: $\mathrm{PD}(J_{\rm FS})$ is the generator of $H_2(M, \mathbb{Z})$, and can be identified with a linear hyperplane $H = \{a_I Z^I = 0\}$; its intersection $H \cap Z(f_d) = \{a_I Z^I = 0,\ f_d\}$ is a degree-d system, and as such it consists of d points, which is d times the generator of $H_0(Z(f_d))$. Using again Poincaré duality, this time in $Z(f_d)$, we see that $[J_{\rm FS}]|_{Z(f_d)}$ is d times the generator of $H^2(Z(f_d))$. So for $N = 2$, (6.3.7) is to be read as

$$c_1(Z(f_d, \mathbb{CP}^2)) = (3 - d)d . \tag{6.3.8}$$

Comparing this with (5.3.72), we get

$$g = \frac{1}{2}(d - 1)(d - 2) . \tag{6.3.9}$$

In particular, for $d = 1$ and 2 we have an S^2, while for $d = 3$ we have a T^2.

Tori

Let us now focus on the T^2 case, $d = 3$. The space of cubic polynomials is ten dimensional; using the nine linear coordinate changes of GL(3, $\mathbb{C}$), we can set $f_3 = 0$ to a canonical form called *Weierstrass equation*:

$$Z_0 Z_2^2 = Z_1^3 + f Z_0^2 Z_1 + g Z_0^3 \,. \tag{6.3.10}$$

In the chart where $Z_0 \neq 0$, defining $x \equiv Z_1/Z_0$, $y \equiv Z_2/Z_0$, we can write this as

$$y^2 = x^3 + f x + g \,. \tag{6.3.11}$$

In this chart, we miss a single "point at infinity" $\{Z_0 = Z_1 = 0\}$. Thus this realization of a T^2 has a preferred, or marked, point. In algebraic geometry, this T^2 is often called an *elliptic curve* to emphasize this point. (The term originates from the theory of elliptic integrals.)

The choice of parameters f, g in (6.3.11) implicitly defines a choice of complex structure, induced by the ambient $\mathbb{CP}^2$. So there should be a bijective map from the modular parameter τ we introduced in (5.3.64), which was a coordinate on the complex structure moduli space, to the pair (f, g). Explicitly, this is given by

$$f = -4^{1/3} 60 \sum_{(m,n)\neq(0,0)} (m + n\tau)^{-4} \,, \qquad g = -560 \sum_{(m,n)\neq(0,0)} (m + n\tau)^{-6} \,. \tag{6.3.12}$$

The inverse map can be described by introducing the function[4]

$$j(\tau) \equiv 4 \frac{(12 f)^3}{4 f^3 + 27 g^2} \,. \tag{6.3.13}$$

This is invariant under Möbius transformations (5.3.65), so the value of j identifies τ uniquely.

An intuitive way to see that (6.3.11) describes a T^2 is to solve $y = \sqrt{x^3 + f x + g}$. This is a function with four branching points (including the point $x = \infty$), two cuts, and two Riemann sheets. Each copy of $\mathbb{C}$, with the point at infinity, is an S^2; gluing two S^2 along the two cuts, one obtains a T^2.

To be more precise, this is the generic picture. There are special values of the parameters where the cuts touch, or where a cut shrinks. In such cases, the torus degenerates, for example to a sphere with two points identified. This happens when two roots of $x^3 + f x + g = 0$ coincide, or in other words when the *discriminant*

$$\Delta = 27 g^2 + 4 f^3 \tag{6.3.14}$$

vanishes, $\Delta = 0$.

6.3.3 Higher codimension

To go beyond hypersurfaces, we consider systems of equations

$$Z(\{f_a\}, \mathbb{CP}^N) \equiv \{f_1 = 0, \ldots, f_k = 0\} \subset \mathbb{CP}^N \,. \tag{6.3.15}$$

[4] We used normalizations that are more a bit more standard in the physics literature, but that differ a bit from more common ones, as in, e.g., [155, chap. 23]: the relation is $f = -4^{1/3} g_2$, $g = -4 g_3$, $\Delta = -\Delta_{\text{there}}/16$. The function $\eta = (2\pi)^{-1/2}(-\Delta/16)^{1/24}$ is known as the *Dedekind eta function*; it can also be written as $\eta = \mathrm{e}^{\pi i \tau/12} \prod_{n=1}^{\infty}(1 - \mathrm{e}^{2n\pi i \tau})$. It also plays a role in discussions of the string spectrum, in contexts where τ is the modular parameter for a $g = 1$ world-sheet.

A more intrinsic way of describing such a locus is by introducing the *ideal* I of all functions that vanish on it: it has the property that if $f_a \in I$, and g_a are any functions, then $\sum_a g_a f_a \in I$. A manifold associated to an ideal I is denoted by $Z(I)$.

One would naively expect $\dim_{\mathbb{C}} Z(\{f_a\}, \mathbb{CP}^N) = N - k$. This is not always true; when it is, we call $Z(\{f_a\}, \mathbb{CP}^N)$ a *complete intersection manifold*. A classic counterexample is the image of the (generalized) *Veronese embedding*:[5]

$$\begin{array}{ccc} \mathbb{CP}^1 & \to & \mathbb{CP}^3 \\ (W_0, W_1) & \mapsto & (Z_0 = W_0^3,\ Z_1 = W_0^2 W_1,\ Z_2 = W_0 W_1^2,\ Z_3 = W_1^3)\,. \end{array} \tag{6.3.16}$$

The image of (6.3.16) is cut by three equations:

$$\mathbb{CP}^1 \cong Z(\{f_a\}, \mathbb{CP}^3)\,, \qquad \begin{array}{l} f_1 = Z_0 Z_2 - Z_1^2\,, \\ f_2 = Z_1 Z_3 - Z_2^2\,, \\ f_3 = Z_0 Z_3 - Z_1 Z_2\,. \end{array} \tag{6.3.17}$$

All three f_a are necessary: if one tries to omit one of the three, say f_3, then $\{f_1 = f_2 = 0\}$ contains the image of (6.3.16) but also the spurious line $\{Z_1 = Z_2 = 0\}$. The task of the third equation $f_3 = 0$ is to get rid of this line, and not to lower the dimension of the space. So the complex dimension of (6.3.17) is one, and not $3 - 3 = 0$, as naive dimension counting would suggest.

On the other hand, consider the *Segre embedding*:[6]

$$\begin{array}{ccc} \mathbb{CP}^1 \times \mathbb{CP}^1 & \to & \mathbb{CP}^3 \\ (W_0, W_1, W_0', W_1') & \mapsto & (Z_0 = W_0 W_0',\ Z_1 = W_0 W_1',\ Z_2 = W_1 W_0',\ Z_3 = W_1 W_1')\,. \end{array} \tag{6.3.18}$$

Here the image is cut by the single equation

$$\mathbb{CP}^1 \times \mathbb{CP}^1 \cong Z(f, \mathbb{CP}^3)\,, \qquad f = Z_0 Z_3 - Z_1 Z_2\,. \tag{6.3.19}$$

So in this case, the naive formula $\dim Z(f, \mathbb{CP}^3) = 3 - 1 = 2$ does work: this is a complete intersection.

In fact, a single equation always describes a locus of codimension one; non-complete intersections only appear for higher codimensions. We implicitly used this in our earlier discussion of hypersurfaces. To be more precise, this is only true on *smooth* manifolds; this is an important caveat, given that algebraic varieties can quite easily fail to be smooth, as we will see soon. To deal with cases with singularities, one uses the notion of *divisor* as a more precise generalization of hypersurface. One distinguishes between Cartier divisors, which consist of zeros of a single equation, and Weil divisors, which are loci of codimension one. It is of course allowed to use the word "divisor" even in the smooth case, where as we said Cartier's and Weil's versions coincide.

In general, submanifolds of $\mathbb{CP}^N$ are called *projective*. As is probably clear by now, this is a very large class of manifolds; in fact, algebraic geometry often gives a procedure to embed an abstract complex manifold in a projective space $\mathbb{CP}^N$. If on

[5] The original Veronese embedding is actually a similar map $\mathbb{CP}^2 \to \mathbb{CP}^5$, but the idea is easy to generalize to other dimensions.

[6] A real analogue of (6.3.18) shows that a hyperboloid in three dimensions contains two families of lines. This can be visualized, for example, by making a cube spin around a diagonal that connects two opposite vertices: some of the edges of the cube will then describe a piece of hyperboloid.

(a compact) M we find a positive line bundle $\mathcal{L}$, then *Kodaira's embedding theorem* says that for large enough n, given the global sections s_a of $\mathcal{L}^n$, the map

$$\begin{array}{ccc} M & \to & \mathbb{CP}^N \\ x & \mapsto & s_a(x) \end{array} \tag{6.3.20}$$

is an embedding. So M is now a submanifold of $\mathbb{CP}^N$ for some large enough N. *Chow's theorem* now guarantees that M is also algebraic: it can be described as a $Z(\{f_a\}, \mathbb{CP}^N)$, where the f_a are polynomials in $\mathbb{CP}^N$.

6.3.4 Singularities

So far, we have treated the $Z(f)$ or $Z(\{f_a\})$ as if they were manifolds. However, sometimes they have loci where they don't look locally like $\mathbb{R}^d$. At such points, the metric develops a curvature singularity, meaning that curvature invariants such as (4.1.31) diverge. A simple criterion to identify them is the system

$$f_a = \partial_{z_i} f_a = 0\,, \tag{6.3.21}$$

where z_i are local coordinates in the ambient space. Any locus where this system has a solution is called *singular*.

For example, if there is only one f, we are looking for critical points for f. More concretely, if we consider for simplicity the origin $z = \underline{0}$, f is singular if and only if its Taylor expansion doesn't have a linear term, but starts with terms of degree two or higher. This makes sense: if the Taylor expansion contains a linear term, for example $f = z_1 + \tilde{f}(z_2, \ldots, z_N)$, then on $Z(f)$ we can eliminate $z_1 = -\tilde{f}$ and use $z_2, \ldots, z_N$ as local coordinates; so around the origin, $Z(f)$ looks like $\mathbb{C}^{N-1}$.

Another possible way of defining a singularity is via a discrete quotient, also called *orbifold singularities*. The most basic example is perhaps

$$\mathbb{C}/\mathbb{Z}_n, \tag{6.3.22}$$

where we identify $w \sim e^{i\pi/n} w$. A possible fundamental region for this identification is a sector $\{re^{i\theta}, \theta \in [0, 2\pi/n)\}$. The topology induced by the quotient is such that this sector looks like a two-dimensional cone, whose vertex is the singularity. In higher dimensions, perhaps the most basic example is $\mathbb{C}^2/\mathbb{Z}_2$, with the $\mathbb{Z}_2$ identification $(w_1, w_2) \sim (-w_1, -w_2)$.

Orbifold singularities can sometimes be also realized in terms of algebraic equations. For the two-dimensional example we just saw:

$$\begin{array}{ccc} \mathbb{C}^2/\mathbb{Z}_2 & \cong & Z(z_1 z_3 - z_2^2, \mathbb{C}^3) \\ (w_1, w_2) \mapsto & (z_1 = w_1^2, & z_2 = w_1 w_2,\; z_2 = w_2^2). \end{array} \tag{6.3.23}$$

One can see that this map is one-to-one. The zero locus of the quadric $z_1 z_3 - z_2^2$, which by (6.3.21) is singular at the origin (because it has no linear terms). Any quadric of the form $M_{ij} z_i z_j = 0$, where M is nondegenerate, such as $z_1^2 + z_2^2 + z_3^2$, can be brought to this form by a coordinate change.

Going still up in dimension, a $d = 6$ analogue of (6.3.23) is the quadric in $\mathbb{C}^4$:

$$z_1^2 + z_2^2 + z_3^2 + z_4^2 = 0 \quad \text{(conifold)}\,. \tag{6.3.24}$$

This is known as the *conifold* and will play an important role later on. Unlike the quadric in $\mathbb{C}^3$ (6.3.23), this is not isomorphic to an orbifold singularity.

6.3.5 Blowup

Given a manifold M with a singular locus S, a *resolution* is a smooth $\hat{M}$ with a map $\pi: \hat{M} \to M$ that away from S is biholomorphic, Section 5.3.1 (or birational; see footnote at the beginning of Section 6.3). So $M - S$ and $\hat{M} - \phi^{-1}(S)$ are equivalent as complex manifolds.

The simplest resolution is called a *blowup*. We will first illustrate it in absence of singularities, and then apply them to the singular case. The origin of the idea is this. Suppose we have a function $f: \mathbb{R}^N \to \mathbb{R}$ that is discontinuous at the origin, but whose directional limits are well defined. For such a function, it might be natural to replace the origin with a sphere S^{N-1}, whose points would represent all the possible directions one can take to approach the origin in $\mathbb{R}^N$. This would look a bit like "inflating" the origin, or "blowing it up." To do something similar in $\mathbb{C}^N$, we would replace the origin with the space of possible complex lines, which as we know by definition is $\mathbb{CP}^{N-1}$.

To formalize this, one introduces the space

$$\widehat{\mathbb{C}^N} = \{(z, s) |\; z \in \mathbb{C}^N, \; s \in \mathbb{CP}^{N-1}, \; z_i s_j = z_j s_i\} . \tag{6.3.25}$$

Consider the map

$$\begin{aligned} \pi: \widehat{\mathbb{C}^N} &\to \mathbb{C}^N \\ (z, s) &\mapsto z . \end{aligned} \tag{6.3.26}$$

If we pick any point in $\mathbb{C}^N$ that is not the origin, $z \neq \underline{0}$, the condition $z_i s_j = z_j s_i$ in (6.3.25) determines a unique $s \in \mathbb{CP}^{N-1}$; so away from the origin, the map is one-to-one, and in fact it is biholomorphic. On the other hand, for $z = \underline{0}$, s is completely undetermined; it can be any point in $\mathbb{CP}^{N-1}$. So $\widehat{\mathbb{C}^N}$ is a space where the origin has been replaced by a copy of $\mathbb{CP}^{N-1}$, which is called *exceptional locus*. This is the simplest example of blowup.

Since π does change volumes, it can be viewed as a deformation of the Kähler structure; recall Wirtinger's theorem (5.6.17), which relates the volume of holomorphic cycles to the pull-back of a power of J.

To see how this helps with singularities, consider now a single equation in $\mathbb{C}^N$. For example, take the quadric (6.3.23). If we blow up the origin of $\mathbb{C}^3$, we replace it with a copy of $\mathbb{CP}^2$. The equation $z_1 z_3 - z_2^2$ on the z_i implies a similar equation,

$$s_1 s_3 - s_2^2 = 0, \tag{6.3.27}$$

in this $\mathbb{CP}^2$. This is a plane curve, and by (6.3.9) it describes a genus-zero Riemann surface, or in other words an $S^2 \cong \mathbb{CP}^1$. So we have learned that the blowup of $\mathbb{C}^2/\mathbb{Z}_2$ at the origin is a new complex manifold $\widehat{\mathbb{C}^2/\mathbb{Z}_2}$, which has an exceptional curve $\mathbb{CP}^1$ replacing the singularity. In this particular case, it turns out that this manifold is the total space of the bundle $O(-2) \to \mathbb{CP}^1$:

$$\widehat{\mathbb{C}^2/\mathbb{Z}_2} \cong O_{\mathbb{CP}^1}(-2) . \tag{6.3.28}$$

The exceptional curve $\mathbb{CP}^1$ is the base of the bundle.

For a singularity defined by a single homogeneous hypersurface $f(z_1, \ldots, z_N)$, the story is rather similar: the exceptional locus will be the locus

$$\{f(s_1, \ldots, s_N) = 0\} \subset \mathbb{CP}^N , \tag{6.3.29}$$

generalizing (6.3.27). As another example, if we blow up the conifold singularity (6.3.24), the exceptional locus will be described by a single quadric in $\mathbb{CP}^3$. All quadrics $M_{ab}s^a s^b$, where M is nondegenerate, are equivalent up to coordinate change; so we can appeal to (6.3.19), where we argued that a quadric in $\mathbb{CP}^3$ is $\mathbb{CP}^1 \times \mathbb{CP}^1$. The total space can be described as a line bundle,

$$O(-1,-1) \to \mathbb{CP}^1 \times \mathbb{CP}^1 , \tag{6.3.30}$$

with Chern classes $(-1,-1)$. We will see in Section 6.4.3 that there is a more economical resolution, which replaces the singularity with only one copy of $\mathbb{CP}^1$.

6.3.6 Milnor number

In general, the f or f_a defining an algebraic singularity need not be homogeneous. A simple generalization is $Z(z_1 z_3 - z_2^n, \mathbb{C}^3)$. As in (6.3.23), this is equivalent to an orbifold: this time $\mathbb{C}^2/\mathbb{Z}_n$, with the $\mathbb{Z}_n$ acting as $(w_1, w_2) \sim (\omega_n w_1, \omega_n^{-1} w_2)$, with $\omega_n \equiv e^{2\pi i/n}$. The identification is

$$\begin{aligned} \mathbb{C}^2/\mathbb{Z}_n \quad &\cong \quad Z(z_1 z_3 - z_2^n, \mathbb{C}^3) \\ (w_1, w_2) &\mapsto (z_1 = w_1^n, z_2 = w_1 w_2, z_2 = w_2^n) . \end{aligned} \tag{6.3.31}$$

In this case, a blowup at the origin produces in $\mathbb{CP}^2$ an equation that is still singular; one has to blow up again several more times. In the end, the exceptional locus is a sequence of $n-1$ copies of $\mathbb{CP}^1$ intersecting each other.

We expect that only the first terms in the Taylor expansion of the f_a are actually important for the singularity's properties. In the quadric case, we would expect to be able to discard all the terms of degree three and higher. In the (6.3.31) case, things are less clear, because we don't want to discard z_2^n, but we would still expect to discard z_2^{n+1}.

Fortunately there is a whole theory of how to classify algebraic singularities; for great introductions, see [156, 157]. For example, for the hypersurface case (a single f), one defines a *Jacobi ring*:

$$J_f \equiv \mathbb{C}[z_1, \ldots, z_N]/\{\partial_i f\} . \tag{6.3.32}$$

Its dimension

$$\mu \equiv \dim J_f \tag{6.3.33}$$

is called the *Milnor number*. This means that we consider polynomial functions in $\mathbb{C}^N$, but we set to zero all the $\partial_i f$. For a quadratic f, we would set all linear functions to zero, and hence (6.3.32) is just $\mathbb{C}$, of dimension $\mu = 1$. For $f = z_1 z_3 - z_2^n$, J_f is generated by all monomials of degree up to $n-1$, so $\mu = 1$. *Tougeron's theorem* says that all terms of degree larger than μ can be reabsorbed by a local change of coordinates that leaves the origin unaffected.

If we instead deform f by adding polynomials of lower degree, the singularity does change, and generically it can become nonsingular. This process is a complex structure deformation, and so it is distinct from the blowup of the previous section (or other resolutions we will see later), which are biholomorphic away from the singular point and can be viewed as Kähler structure deformations. (The $d = 4$ case is a little special, in that complex and Kähler structures are more entangled, as we will see in Section 7.2.4.)

Since deforming by f itself does not change the singularity, deformations are parameterized not quite by (6.3.32) but by elements of

$$\mathbb{C}[z_1, \ldots, z_N]/\{f, \partial_i f\}, \tag{6.3.34}$$

whose dimension τ is called *Tjurina number*. Clearly $\tau \leq \mu$ in general, although they often coincide (because f might be a linear combination of the $\partial_i f$, as when f is homogeneous). Taking a basis g_a of (6.3.34), a deformation is then

$$f \to f' = f + \sum_a \alpha_a g_a\,. \tag{6.3.35}$$

For a generic choice of $\alpha_a \in \mathbb{C}$, the resulting space $Z(f')$ is nonsingular; the singular locus is replaced by a set consisting of μ copies of the sphere $S^{d/2}$, glued at one point: $(S_1 \cup \ldots \cup S_\mu)/\{x_1 \sim \ldots \sim x_\mu\}$, $x_i \in S_i$. (This is also called a *bouquet* of spheres.) So the Milnor number μ has a topological meaning.

For example, for the conifold singularity (6.3.24), the relations in (6.3.32) set to zero all $\partial_i f = 2x_i$, and J_f is generated by the constant function alone, so it has dimension one. The deformation (6.3.35) is

$$z_1^2 + z_2^2 + z_3^2 + z_4^2 + \alpha = 0 \quad \text{(deformed conifold)}\,. \tag{6.3.36}$$

This is called the *deformed conifold.* Since $\mu = 1$, the singularity is replaced here by a single S^3.

6.3.7 Stability and Kähler quotients

Yet another technique to produce Kähler manifolds is the *Kähler quotient*, a generalization of the definition (6.2.1) of $\mathbb{CP}^N$. We will first describe it in terms of symplectic geometry, and then in more algebraic-geometric terms.

Symplectic quotient of complex manifolds

The symplectic quotient (5.4.14) (Section 5.4.2) produces a symplectic manifold $\mu^{-1}(c)/G$ from one M of larger dimension, on which a group G acts by symplectomorphisms: its generators ξ preserve the symplectic form, $L_\xi J = 0$. For each generator ξ, we lose two directions in the quotient: one because we restrict to the level set of μ, the other because of equivalence by the G action. If M is Kähler and the action also preserves the complex structure, $L_\xi I = 0$, one can show that

$$\frac{\mu_\xi^{-1}(c)}{G} \tag{6.3.37}$$

is also Kähler. This refined procedure is called Kähler quotient.

For $\mathbb{CP}^N$, $G = \mathrm{U}(1)$, with a single generator ξ, (6.2.8). To find a moment map, we compute

$$\iota_\xi J_{\mathbb{C}^{N+1}} = \mathrm{i}\left(Z^I \iota_{\partial_I} - \bar{Z}_I \iota_{\bar{\partial}^I}\right) \frac{\mathrm{i}}{2} \mathrm{d}Z^J \wedge \mathrm{d}\bar{Z}_J = -r\mathrm{d}r \tag{6.3.38}$$

so that we can satisfy (5.4.11) by taking

$$\mu_\xi = -\frac{1}{2}r^2 = -\frac{1}{2}Z^I\bar{Z}_I\,. \tag{6.3.39}$$

A level set for this map, for example $\mu_\xi^{-1}(-1/2)$, is just a sphere $S^{2N+1} \subset \mathbb{C}^{N+1}$. Now (6.4.25) gives

$$\mathbb{CP}^N \cong S^{2N+1}/S^1\,, \tag{6.3.40}$$

which reproduces (6.2.4).

Geometric invariant theory

Another quotient procedure is inspired by complex (rather than symplectic) geometry. We will give here only a rough introduction to this vast subject; see [158] for foundational work and [159, 160] for reviews.

Intuitively, one would like to be able to quotient a complex manifold M by a Lie group. To obtain another complex manifold, we want the group to be the complexification $G_\mathbb{C} \equiv G \otimes \mathbb{C}$ of a compact real Lie group G. (At the infinitesimal level, the generators of G are vector fields ξ_a, and the remaining generators of $G_\mathbb{C}$ are of the form $I^t\xi_a$.) Taking a straightforward set-theoretic quotient usually does not produce a manifold; rather, one usually has to exclude a locus F from M:

$$\frac{M-F}{G_\mathbb{C}}\,, \tag{6.3.41}$$

as we did in (6.2.1). This is called a *geometric invariant theory (GIT) quotient*. The points in the excluded locus F are called *unstable*.

One way to understand this locus is in terms of invariant functions. The idea comes from a wider program, whose aim is to reformulate geometry in terms of rings of functions, and in particular complex geometry in terms of rings of holomorphic functions. In this framework, one would like to define the quotient (6.3.41) so that its ring of functions is the space of G-invariant functions over M. This does not happen if we directly define $M/G_\mathbb{C}$; one needs to exclude any points over which all invariant functions vanish. Another problem is that often there are too few holomorphic functions; on a compact manifold, for example, there are none, by Liouville's theorem (Section 5.3.1).

So one considers a line bundle $\mathcal{L}$ over M, and sections of $\mathcal{L}$ that are invariant under an action over both the base and fiber. A point p is said to be unstable if there is no invariant section s such that $s(p) \neq 0$. A point that is not unstable is called *semistable*; if, moreover, its stabilizer is finite, p is said to be *stable*.

The simplest example is $\mathbb{CP}^N$ itself. If we want to quotient $\mathbb{C}^{N+1}$ by $G_\mathbb{C} = \mathbb{C}^*$ acting by the rescaling (6.2.2), we can take $\mathcal{L}$ to be trivial, and lift the action to the coordinate W on the fiber as $\lambda \cdot W = \lambda^{-w}W$ for some weight $w > 0$. Now the invariant functions are the homogeneous polynomials, namely those such that

$$p(\lambda Z^0, \dots, \lambda Z^N) = \lambda^w p(Z^0, \dots, Z^N)\,. \tag{6.3.42}$$

These all vanish at the origin $\underline{0}$. By definition, such a point is said to be unstable, and we have to exclude it from M before taking the quotient; this results in (6.2.1).

In the compact case, we have seen in (6.3.20) that many Kähler manifolds M can be embedded in $\mathbb{CP}^N$ using sections of (powers of) a positive line bundle $\mathcal{L}^n$, which then can be viewed as the pull-back of $O(-1)$ to M. In practice, this means that we have to promote the action on M to a linear action on $\mathbb{C}^{N+1}$; this is called a *linearization* of the original action on M.

In Section 6.4, we will see many more examples of (6.3.41), with $G_{\mathbb{C}}$ Abelian.

Hilbert–Mumford numerical criterion

A perhaps more practical way to understand the unstable locus F comes from the study of the orbits of one-parameter subgroups of $G_{\mathbb{C}}$. In other words, one considers the action of one generator ξ at a time. To decide if a point x is (un)stable, consider the limit point $x_0 = \lim_{\lambda\to 0}\lambda \cdot x$ of its action on M. Since x_0 is a fixed point, the action will only be along the fiber of $\mathcal{L}$ over it; calling again W the fiber coordinate, this action will be of the form $\lambda \cdot W = \lambda^{\rho} W$ for some $\rho = \rho(\xi, x)$. Now the *Hilbert–Mumford numerical criterion* [158] is

$$\begin{aligned} &\exists\, \xi \mid \rho(x,\xi) > 0 && \Leftrightarrow && x \text{ is unstable}\,; \\ &\forall\, \xi \;\; \rho(x,\xi) \le 0 && \Leftrightarrow && x \text{ is semistable}\,; \\ &\forall\, \xi \;\; \rho(x,\xi) < 0 && \Leftrightarrow && x \text{ is stable}\,. \end{aligned} \tag{6.3.43}$$

For example, for $\mathbb{CP}^N$, let us consider again the lift $\lambda \cdot W = \lambda^{-w} W$ we took preceding (6.3.42), with $w > 0$. For a point $x = (Z^0, \ldots, Z^N) \neq \underline{0}$ in $\mathbb{C}^{N+1}$, under the rescaling (6.2.2) the limit point is the origin $\underline{0}$; then $\rho(x,\xi) = -w < 0$, and x is stable. On the other hand, for the origin $\underline{0}$ we can also consider the action λ^{-1}, with generator $-\xi$, for which it is still a fixed point, and in this case $\rho(\underline{0}, -\xi) = w > 0$, so the point $\underline{0}$ is unstable, as we concluded previously.

Equivalence with symplectic quotient

The notions of symplectic and GIT quotient are related by the *Kempf–Ness theorem*:

$$\frac{M - F}{G_{\mathbb{C}}} \cong \frac{\mu_\xi^{-1}(c)}{G}\,. \tag{6.3.44}$$

In the case of $\mathbb{CP}^N$, this is the equality of (6.2.1) with (6.3.40), or with the base of (6.2.4). While in complex quotient we lose $2\dim(G)$ directions in one go, in the symplectic quotient we first lose $\dim(G)$ directions by restricting to the level set of μ, and then $\dim(G)$ more when we take the quotient by G.

Here is a rough idea of the proof. As we have seen, to define stability we have lifted the action of $G_{\mathbb{C}}$ to one on the total space of a line bundle $\mathcal{L}$. We consider a point $\tilde{x}$ in $\mathcal{L}$ that projects to a stable x. As the orbit goes near a fixed point $x_0 = \lim_{\lambda\to 0}\lambda \cdot x$, the positivity of $\rho(x,\xi)$ indicates that it will have the fiber $\mathcal{L}_{x_0}$ over it as a vertical asymptote. So the graph of the orbit will be a convex function; the "vertical distance" from the zero section of $\mathcal{L}$ is a norm on the fiber, and will have a minimum at some other point x' – see Figure 6.1.

If we choose the norm on the fiber appropriately, such a minimum will also be a zero for the moment map μ. Take a Kähler potential (6.1.11) that is invariant, $L_\xi K = \iota_\xi \mathrm{d}K = 0$, and recall we have assumed also $L_\xi I = 0$:

$$\begin{aligned} \mathrm{d}\mu_\xi = \iota_\xi J &\overset{(6.1.11)}{=} \frac{1}{2}\iota_\xi \mathrm{d}\mathrm{d}^c K \overset{(5.3.32)}{=} -\frac{1}{2}\iota_\xi \mathrm{d}I \cdot \mathrm{d}K \overset{(4.1.79)}{=} \frac{1}{2}\mathrm{d}\iota_\xi I \cdot \mathrm{d}K \\ &\overset{(3.2.54)}{=} \frac{1}{2}\mathrm{d}(I \cdot \iota_\xi - \iota_{I^t\xi})\mathrm{d}K = -\frac{1}{2}\mathrm{d}L_{I^t\xi} K\,. \end{aligned} \tag{6.3.45}$$

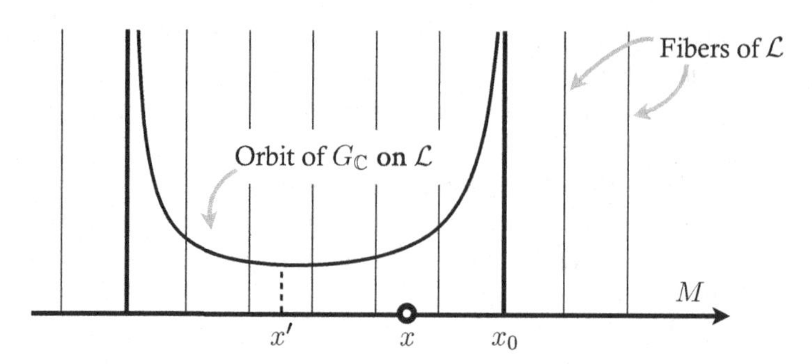

Figure 6.1 Convexity of fiber norm function for a stable point.

So up to a constant, we can take

$$\mu_\xi = -\frac{1}{2} L_{I^t\xi} K \,. \tag{6.3.46}$$

Hence the critical points of K under the $G_\mathbb{C}$ action are the zeros of μ_ξ.

With one generator, we would be done, but if G is nonabelian, one has to work a little harder. First one shows that (6.3.46) is equivariant, as defined in (5.4.13). (This requires us to set to zero the constant c in (6.3.44).) Using this property, moving in the zero level set $\mu_\xi^{-1}(0)$, one can find a locus where $\mu_{\xi'} = 0$ for a second generator ξ', and so on until one finds a point x' such that $\mu_{\xi_a} = 0$ for all generators of $G_\mathbb{C}$. With a similar computation, one can show that the critical points are actually minima.

So if we take the norm on the fiber to be K, we have shown that every stable point is in the same $G_\mathbb{C}$ orbit of a zero of the moment map μ_ξ. x' is not unique: the action of the real compact G leaves K invariant, and so there is a whole G orbit of critical points. This leads to (6.3.44).

Exercise 6.3.1 Find a map $\mathbb{CP}^2 \to \mathbb{CP}^5$ similar to (6.3.16) (as mentioned in footnote 5). Is this a complete intersection?

Exercise 6.3.2 Check that a Riemann surface in Weierstrass form (6.3.11) has a singularity if and only if $\Delta = 0$.

Exercise 6.3.3 Consider $\{s_1 s_2 = 0\} \subset \mathbb{C}^2$. It is a singularity that consists of two copies of $\mathbb{C}$ that intersect at one point. Blow it up using (6.3.29); show that now the two $\mathbb{C}$ do not intersect any more, but that they each intersect the exceptional $\mathbb{CP}^1$ in one point.

Exercise 6.3.4 Compute the Milnor number for a singularity $\{\sum_{i=1}^4 z_i^{w_i} = 0\}$.

Exercise 6.3.5 Check (6.3.46) for $\mathbb{CP}^N$.

6.4 Toric geometry

6.4.1 Real and algebraic tori

From algebraic tori

A *toric manifold* is a manifold of dimension $d = 2N$ that contains

$$(\mathbb{C}^*)^N \tag{6.4.1}$$

as a dense open subset. It comes with a natural Kähler metric; given its connections to algebraic geometry, often the name "toric variety" is preferred. The name originally comes from the fact that (6.4.1) is often called *algebraic torus*, but the definition does also imply that there is an action of an ordinary torus $T^N = (S^1)^N$, defined by taking $\mathrm{U}(1) \subset \mathbb{C}^*$ subgroups.

Toric manifolds differ in how sets of smaller dimension are added to (6.4.1). For example, in $d = 2$, $N = 1$, we can add to $\mathbb{C}^*$ a single point $z = 0$, to make it $\mathbb{C}$, or two points, $z = \{0, \infty\}$, to make it the Riemann sphere (4.1.45). As we learned in (6.2.6), this is also $\mathbb{CP}^1$, and is our first example of a compact toric manifold.

Going up in dimension, in $d = 4$, $N = 2$, we have $\mathbb{CP}^1 \times \mathbb{CP}^1$, but also $\mathbb{CP}^2$. Indeed, the $U_0 = \{Z^0 \neq 0\} \cong \mathbb{C}$ chart of $\mathbb{CP}^2$ contains (6.4.1) for $d = 4$ as a dense open subset; the remaining locus is $\{Z^0 = 0\}$, which is a copy of $\mathbb{CP}^1$.

It is useful to have a pictorial way of keeping track of what loci we have added to the algebraic torus to build our toric manifold M. The *toric polyhedron* $P(M)$ is built so that a k-dimensional face represents a copy of $(\mathbb{C}^*)^k$, with $k = 0$ being interpreted as a single point. The angles of the faces of the polyhedron encode the holomorphic transition functions for these copies of $\mathbb{C}^*$.

Before we give a formal combinatorial definition, here are some simple examples:

- $P(\mathbb{CP}^1)$ is a segment: the interior of the segment corresponds to $\mathbb{C}^*$, while the two endpoints correspond to the points $z = 0, \infty$.
- $P(\mathbb{C})$ is a half-line: there is now only one point, corresponding to $z = 0$.
- $P(\mathbb{CP}^1 \times \mathbb{CP}^1)$ is the cartesian product of two segments: a square.
- $P(\mathbb{C}^2)$ is the cartesian product of two half-lines: a quadrant in $\mathbb{R}^2$.
- $P(\mathbb{CP}^2)$ is a triangle (Figure 6.2, left). The interior corresponds to the points where all three coordinates $Z^I \neq 0$. The three sides correspond to loci where only one of the coordinates is zero. The three vertices correspond to the three points where two of the coordinates are zero; using the $\mathbb{C}^*$ quotient in the definition (6.2.1), these can be put in the form $(1, 0, 0)$, $(0, 1, 0)$, $(0, 0, 1)$.
- $P(\mathbb{CP}^3)$ is a tetrahedron. Continuing in higher dimensions, $P(\mathbb{CP}^N)$ is its N-dimensional analogue, sometimes called the *N-simplex*.

Now let us describe how the angles of P describe the transition functions. We will illustrate the procedure in the $\mathbb{CP}^2$ example. To each vertex, we have associated a point; we now also associate to them the charts $\cong \mathbb{C}^2$ of which they are the origin. So, for example, $(0, 0, 1)$ is the origin of the chart $U_0 = \{Z^0 \neq 0\}$, where we have local coordinates (according to (6.2.3))

$$z^1_{(0)} = \frac{Z^1}{Z^0}, \qquad z^2_{(0)} = \frac{Z^2}{Z^0}. \tag{6.4.2}$$

Then $(0, 1, 0)$ is the origin of $U_1 = \{Z^1 \neq 0\}$, with coordinates

$$z^1_{(1)} = \frac{Z^0}{Z^1}, \qquad z^2_{(1)} = \frac{Z^2}{Z^1}. \tag{6.4.3}$$

The two are related by transition functions:

$$z^1_{(1)} = (z^1_{(0)})^{-1}, \qquad z^2_{(1)} = z^2_{(0)}(z^1_{(0)})^{-1}. \tag{6.4.4}$$

This is encoded in the polygon P as follows. From the vertex corresponding to $(0, 0, 1)$, draw two vectors $v^1_{(0)} = \binom{0}{1}$, $v^2_{(0)} = \binom{1}{0}$ along the sides of the triangle,

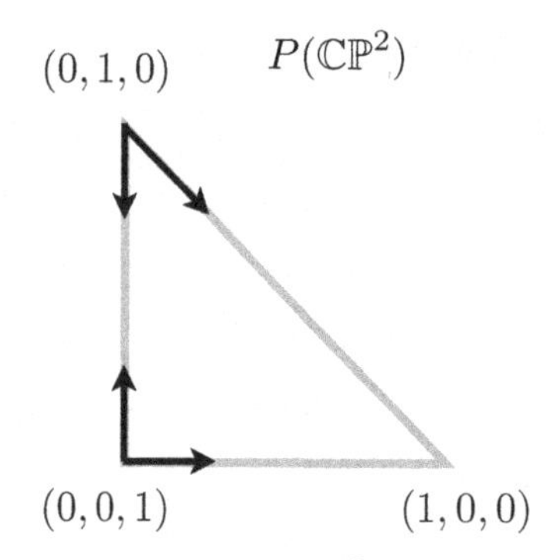

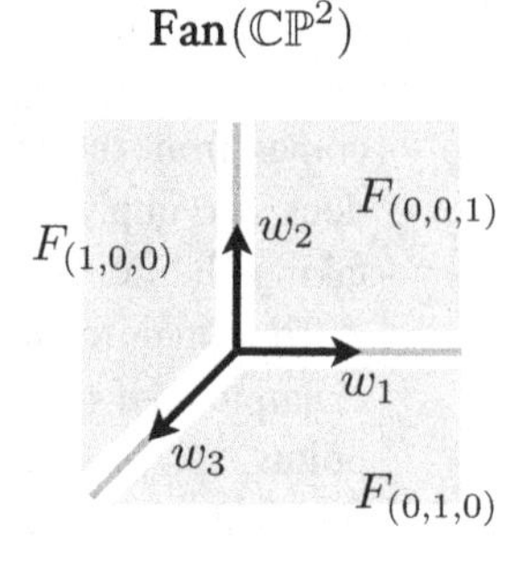

Figure 6.2 Toric polyhedron P and fan for $\mathbb{CP}^2$.

and with integer coefficients. Likewise, draw two vectors $v^1_{(1)} = \binom{0}{-1}$, $v^2_{(1)} = \binom{1}{-1}$ from the vertex $(0, 1, 0)$; see Figure 6.2. The two sets are related by linear relations:

$$v^1_{(1)} = -v^1_{(0)}\,, \qquad v^2_{(1)} = v^2_{(0)} - v^1_{(0)}\,. \tag{6.4.5}$$

From this, one can read off (6.4.4) by formally associating $v^i_{(0)} \to \log z^i_{(0)}$, $v^i_{(1)} \to \log z^i_{(1)}$.

In general, for any vertex p of the polyhedron there is a basis of v^i_p; they have integer coefficients. The linear relations between the v^i relative to two vertices will encode the transition functions of the toric manifold, generalizing (6.4.5). Notice also that (6.4.5) only really uses linear relations among the v^i_p and not the v^i_p themselves. So acting on all these data with an $\mathrm{SL}(N,\mathbb{Z})$-transformation will not affect the geometry.

From torus T^N actions

We now reinterpret toric manifolds from the point of view of the action of an ordinary torus $T^N = (S^1)^N$. This leads to a more intuitive picture.

The algebraic torus (6.4.1) can be foliated in copies of T^N by writing its coordinates as $r_i e^{i\theta_i}$. Varying $\theta_i \to \theta_i + \theta_{0i}$ defines a group action of T^N on $(\mathbb{C}^*)^N$, and so on a dense open set of any toric manifold. This can also be thought of as an obvious fiber bundle $T^N \hookrightarrow (\mathbb{C}^*)^N \to \mathbb{R}^N_+$, where $\mathbb{R}_+ \equiv \{x \in \mathbb{R}, x > 0\}$. This action can be extended to the lower-dimensional loci, by making it degenerate there. The base becomes none other than the polyhedron P:

$$\begin{array}{ccc} T^N & \hookrightarrow & M_N \\ & & \downarrow \\ & & P(M_N)\,. \end{array} \tag{6.4.6}$$

The simplest example is $\mathbb{CP}^1$: the action $\theta \to \theta + \theta_0$ on $\mathbb{C}^*$ degenerates on the two points $z = 0, \infty$. In this case, (6.4.6) is an S^1 fibration $S^1 \hookrightarrow \mathbb{CP}^1 \to P(\mathbb{CP}^1) = I$. This is shown in Figure 6.3 on the left.

For $\mathbb{CP}^2$, the T^2 action degenerates on the vertices of the triangle P so that the fiber there is only a point, and on the sides so that the fiber there is an S^1. This circle is obtained from the generic T^2 by shrinking it along one of its homology cycles. This is encoded in P: if a side has orthogonal vector $\binom{p}{q}$, the S^1 is obtained by shrinking

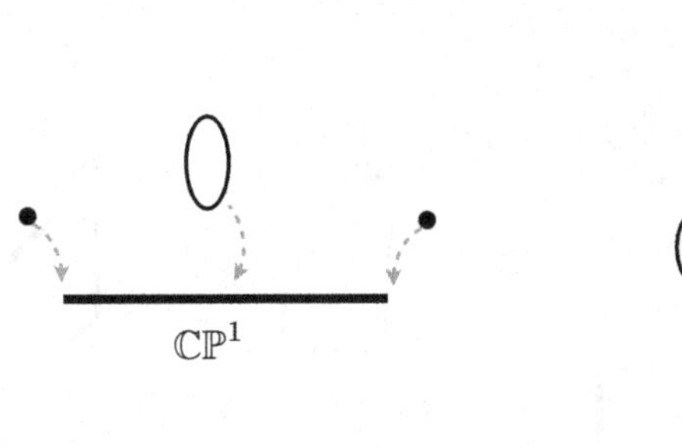

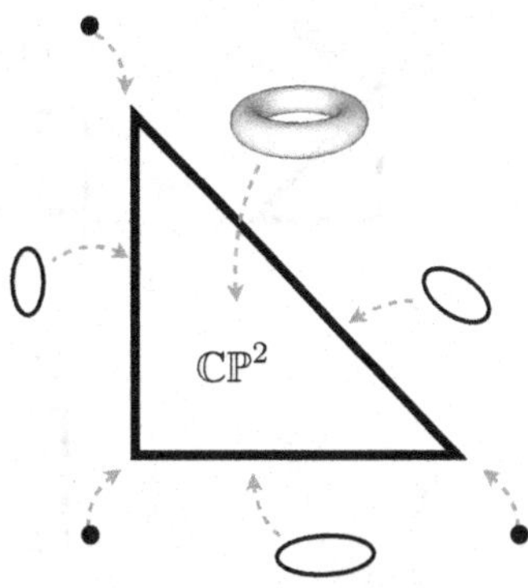

Figure 6.3 S^1-fibration for $\mathbb{CP}^1$, and T^2-fibration for $\mathbb{CP}^2$.

the T^2 along its cycle $(p, q) \in \mathbb{Z}_2 = H_1(T^2, \mathbb{Z})$, a circle winding p times along one direction, q along the other. This is illustrated in Figure 6.3 on the right.

In general, in higher dimensions, on the $(N-1)$-dimensional faces f_I of P the T^N degenerates to a T^{N-1}, and the vectors w_I orthogonal to the faces dictate which one-cycles shrink.[7] On k-dimensional faces of P, the T^N degenerates to T^{N-k}.

Toric fan

The vectors w_I orthogonal to the $(N-1)$-dimensional faces f_I generate half-lines called *rays*. The rays are sometimes drawn in a separate diagram; see Figure 6.2 on the right for the $\mathbb{CP}^2$ case.

In general, the w_I have integer coefficients, and we assume them to be the generators over $\mathbb{Z}$ of their rays: in other words, one cannot divide a w_I by an overall number and find another vector that has integer coefficients. Again, the geometry of the associated manifold remains invariant if we act on the w_I with an $\mathrm{SL}(N, \mathbb{Z})$ transformation.

To a vertex p, one can also associate an N-dimensional region

$$F_p = \{w \mid w \cdot v_p^i \geq 0\}. \tag{6.4.7}$$

These are real N-dimensional, and they are conical, in the sense that they are left invariant by an overall rescaling. One associates with a similar procedure conical regions of real dimension k to any $(N-k)$-dimensional face of P. All these regions, including the rays, are often just called *cones*, and the set of all cones is called *toric fan*. We denote a cone by

$$\langle w_{I_1}, \ldots, w_{I_p} \rangle \equiv \left\{ \sum_i c_i w_{I_i}, c_i \in \mathbb{R}_+ \right\}. \tag{6.4.8}$$

We say that (6.4.8) is generated by the $w_{I_1}, \ldots, w_{I_p}$, or that these vectors are its generators.

For $\mathbb{CP}^2$, Figure 6.2 on the right shows the fan. The rays are generated by the three vectors shown, w_1, w_2, and w_3. There are three two-dimensional cones, shown in shades of gray; they are $F_{(1,0,0)} = \langle w_2, w_3 \rangle$, $F_{(0,1,0)} = \langle w_2, w_3 \rangle$, adn $F_{(0,0,1)} = \langle w_1, w_2 \rangle$.

[7] In this section, the index I will run over the number of $(N-1)$-dimensional faces, which we will soon see to be the same as the number of homogeneous coordinates. This extends our notation in Section 6.2, where $I \in \{0, \ldots, N\}$; indeed, there are $N+1$ faces in $P(\mathbb{CP}^N)$.

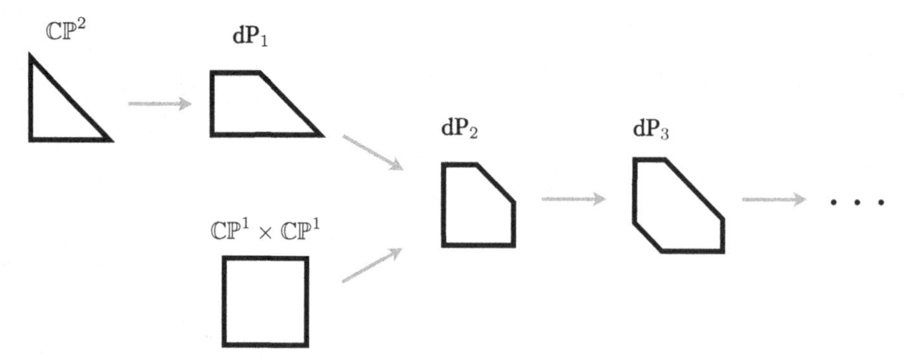

Figure 6.4 Toric polyhedra P for del Pezzo surfaces.

Del Pezzo surfaces

So far, the toric manifolds we have seen are only projective spaces, or products thereof. One way to generate more interesting examples is by using the blowup procedure introduced in (6.3.25). This is important for resolving singularities, but can also be performed at a regular point, which is how we introduced it in (6.3.25). Consider, for example, $\mathbb{CP}^2$, and blow up at $(0, 0, 1)$. Since we are in two complex dimensions, the corresponding vertex should be replaced by a copy of $\mathbb{CP}^1$. So P has now four sides. We can now blow up again at another vertex, obtaining a space whose P has five sides, and so on.

The manifolds dP_n one obtains by blowing up $\mathbb{CP}^2$ at $n \geq 8$ points are all Fano manifolds (their K^{-1} is positive; see Section 6.3.1). In $d = 4$, the only other Fanos are $\mathbb{CP}^2$ and $\mathbb{CP}^1 \times \mathbb{CP}^1$; together, all these are called *del Pezzo surfaces.*[8] We show the corresponding P in Figure 6.4; the arrows denote blowup procedures.

For $n \geq 3$, the points where we blow up can always be assumed to be among the vertices of P, by using linear transformations of the Z^I. For $n \geq 4$, this is not generically possible; the positions of $n - 3$ of the points become complex moduli of the dP_n. $\mathbb{CP}^2$ itself has no complex moduli, as remarked in (6.2.33), so these are the only moduli for del Pezzo manifolds. For $n \geq 4$, the generic point in the complex moduli space is nontoric.

Cohomology

The cohomology of toric manifolds can be computed from the polyhedron. We will sketch here what one can say about $H^2(M)$, or dually codimension-one subspaces $H_{d-2}(M)$, $d = 2N$.

We associated loci of complex codimension one, namely hypersurfaces, to $(N-1)$-dimensional faces of P; let us call them D_I. They generate H^2, but there are relations that can be read off from the vectors w_I orthogonal to the faces of P as follows:

$$\sum_I (w \cdot w_I) D_I = 0 \,, \tag{6.4.9}$$

where w is any vector. There are thus N independent relations. The intersection properties of these D_I can be just read off visually from the toric polyhedron, although we will not describe this in detail. The canonical bundle is given by

[8] In algebraic geometry, a manifold with complex dimension two is often called a surface, just like a manifold with complex dimension one is called a curve.

$$K = -\sum_I D_I \,. \tag{6.4.10}$$

As an example, for $\mathbb{CP}^2$ these vectors were the three columns of (6.4.13). So there are three generators, and $N = 2$ relations (6.4.9); taking $w = \binom{1}{0}$ and $\binom{0}{1}$ these are $D_0 = D_2$, $D_1 = D_2$. So all D_i are equivalent in homology; the single generator is dual to a hyperplane, which as we saw following (6.2.19) is the single generator of $H_{d-2}(\mathbb{CP}^2)$. Equation (6.4.10) gives $K = -3D$, in agreement with (6.2.31).

As another example consider dP_1, whose rays are the columns of (6.4.15). Again, (6.4.9) give two relations:

$$D_0 - D_2 = 0 \,, \qquad D_1 - D_2 - D_3 = 0 \,. \tag{6.4.11}$$

We can solve these constraints by defining $D_0 = D_2 = F$, and $D_1 = B$, so that $D_3 = B - F$. The meaning of these letters will become clear shortly. Here we see $K = -2(B + F)$, in agreement with our previous statement that dP_1 is Fano.

6.4.2 Toric Kähler quotients

Complex quotient

Yet another point of view on toric manifolds is to consider them as GIT quotients: (6.3.41) here gives[9]

$$\frac{\mathbb{C}^{N+k} - F}{(\mathbb{C}^*)^k} \,. \tag{6.4.12}$$

The data required in (6.4.12) are how $(\mathbb{C}^*)^k$ acts on $\mathbb{C}^{N+k}$, summarized by a *charge matrix* C, and the excluded locus F. As we discussed in Section 6.3.7, the latter is needed to obtain a manifold with a reasonable topology. Recall that the prototype is the definition (6.2.1) of $\mathbb{CP}^N$, where $k = 1$, the charge matrix is simply $C = (1, 1, \ldots, 1)$, and the excluded locus is just the origin, $F = \{\underline{0}\}$.

There is an algorithm to compute these data from the polyhedron P. One associates a complex coordinate Z^I to each face, and so to each vector w_I orthogonal to it. Intuitively, the Ith face can be thought of as the locus $\{Z^I = 0\}$.

- The charge matrix C: collect the w_I as columns in a matrix W; find a basis for $\ker(W)$; then construct C as a matrix that has this basis as its rows.
- The excluded locus F: $\{Z^{I_1} = \cdots = Z^{I_p} = 0\}$ belongs to F if the corresponding subset $\{f_{I_1}, \ldots, f_{I_p}\}$ of $(N-1)$-dimensional faces has zero intersection, or alternatively, if the set of vectors $\{w_{I_1}, \ldots, w_{I_p}\}$ is not a subset of the generators of a cone in the fan.

As a first example, consider $\mathbb{CP}^2$: the orthogonal vectors to the faces are $w_1 = \binom{1}{0}$, $w_2 = \binom{0}{1}$, and $w_3 = \binom{-1}{-1}$, as illustrated in Figure 6.2. So

$$W_{\mathbb{CP}^2} = \begin{pmatrix} 1 & 0 & -1 \\ 0 & 1 & -1 \end{pmatrix} . \tag{6.4.13}$$

[9] More precisely, this is true for smooth manifolds; in presence of singularities, the group we quotient by might also include some discrete group Γ besides a $(\mathbb{C}^*)^k$. We will see this in the next subsection.

$\ker(W)$ is one dimensional, and is generated by a single vector $(1, 1, 1)$; so

$$C_{\mathbb{CP}^2} = (1\ 1\ 1)\,. \tag{6.4.14}$$

To compute the excluded locus, we can use either the polyhedron or the fan in Figure 6.2. All pairs of faces intersect, but of course the set of all three faces has zero intersection. In terms of the fan, all pairs of w_I span a two-dimensional cone; but the set of all three w_I generates all of $\mathbb{R}^2$, which is not a cone. So the excluded locus is $\{Z^0 = Z^1 = Z^2 = 0\}$. All this is in agreement with our definition (6.2.1).

For a slightly more complicated example, consider dP_1, whose rays are generated by those of $\mathbb{CP}^2$, plus an additional one, $w_4 = \binom{0}{-1}$ (orthogonal to the upper horizontal side in the $P(\mathrm{dP}_1)$ in Figure 6.4). So

$$W_{\mathrm{dP}_1} = \begin{pmatrix} 1 & 0 & -1 & 0 \\ 0 & 1 & -1 & -1 \end{pmatrix}. \tag{6.4.15}$$

Its kernel is two dimensional, and two possible generators are the rows of

$$C_{\mathrm{dP}_1} = \begin{pmatrix} 1 & 1 & 1 & 0 \\ 0 & 1 & 0 & 1 \end{pmatrix}. \tag{6.4.16}$$

Explicitly, this means that there are identifications

$$(Z^0, Z^1, Z^2, Z^3) \sim (\lambda Z^0, \lambda Z^1, \lambda Z^2, Z^3) \sim (Z^0, \mu Z^1, Z^2, \mu Z^3) \tag{6.4.17}$$

with $\lambda, \mu \in \mathbb{C}^*$. Looking again at $P(\mathrm{dP}_1)$ in Figure 6.4, we see that neighboring faces meet, but opposite faces don't. In terms of the fan, all $\langle w_0, w_1\rangle$, $\langle w_1, w_2\rangle$, $\langle w_2, w_3\rangle$, and $\langle w_3, w_0\rangle$ are two-dimensional cones in the fan; but $\{w_1, w_3\}$ and $\{w_0, w_2\}$ do not generate cones in the fan. So the excluded loci are

$$F = \{Z^0 = Z^2 = 0\} \cup \{Z^1 = Z^3 = 0\}\,. \tag{6.4.18}$$

In this example, it is intuitively clear why this locus needs to be excluded: if we focus on the coordinates Z_1 and Z_3, leaving the others constant, then we see from (6.4.17) that they describe a copy of $\mathbb{CP}^1$, so they cannot possibly be zero simultaneously. The previously described general rule makes sure that this doesn't happen for any copies of $\mathbb{CP}^N$ inside the toric space.

Fibrations of $\mathbb{CP}^1$ over itself

Inspection of (6.4.17) also suggests defining the projection map

$$\begin{array}{ccc} \mathrm{dP}_1 & \to & \mathbb{CP}^1 \\ (Z^0, Z^1, Z^2, Z^3) & \mapsto & (Z^0, Z^2)\,. \end{array} \tag{6.4.19}$$

This is well defined, in the sense that it maps equivalent points to equivalent points: the identifications (6.4.17) are mapped to $(Z^0, Z^2) \sim (\lambda Z^0, \lambda Z^1)$ (which defines $\mathbb{CP}^1$) and to the trivial map respectively. The counterimage of a point in $\mathbb{CP}^1$ is the copy of $\mathbb{CP}^1$ discussed earlier, obtained by varying Z^1 and Z^3. So we can write

$$\begin{array}{ccc} \mathbb{CP}^1 & \hookrightarrow & \mathrm{dP}_1 \\ & & \downarrow \\ & & \mathbb{CP}^1\,. \end{array} \tag{6.4.20}$$

The letters B and F discussed after (6.4.11) were chosen to evoke the words "base" and "fiber" respectively.

We can describe this even more explicitly. Were it not for the second identification in (6.4.17), we could treat Z^1 and Z^3 as describing the fibers of line bundles over $\mathbb{CP}^1$. Indeed, in the chart $U_{\rm N} = \{Z^0 \neq 0\}$, we can parameterize a point as $(1, w^1_{\rm N}, z_{\rm N}, w^3_{\rm N})$; in the chart $U_{\rm S} = \{Z^2 \neq 0\}$, we can parameterize a point as $(z_{\rm S}, w^1_{\rm S}, 1, w^3_{\rm S})$. Using the first in (6.4.17), we obtain

$$w^1_{\rm S} = z_{\rm N}^{-1} w^1_{\rm N}\,, \qquad w^3_{\rm S} = w^3_{\rm N}\,. \tag{6.4.21}$$

Comparing with the definition of $\mathcal{O}(n)$ in (4.2.45), we see that w^1 and w^3 behave as sections of $\mathcal{O}(1)$ and $\mathcal{O}$ respectively. Their direct sum would be a rank-two vector bundle; reinstating the second identification in (6.4.17) turns it into a bundle with fiber $\mathbb{CP}^1$, which we can write as

$$\mathrm{dP}_1 \cong \mathbb{P}(\mathcal{O}(1) \oplus \mathcal{O})\,. \tag{6.4.22}$$

An interesting generalization is the *Hirzebruch surface*:

$$\mathbb{F}_n \cong \mathbb{P}(\mathcal{O}(n) \oplus \mathcal{O})\,. \tag{6.4.23}$$

For $n \neq 1$, this is not a del Pezzo. The charge matrix is now

$$C_{\mathbb{F}_n} = \begin{pmatrix} 1 & n & 1 & 0 \\ 0 & 1 & 0 & 1 \end{pmatrix}. \tag{6.4.24}$$

While these are all different complex manifolds, many are equivalent topologically: by the discussion leading to (4.2.71), S^2-bundles over S^2 are classified by elements in $\pi_1(\mathrm{SO}(3)) = \mathbb{Z}_2$. So topologically, all $\mathbb{F}_n$ with n even are homeomorphic to $S^2 \times S^2$, while all $\mathbb{F}_n$ with n odd are homeomorphic to each other.

Kähler quotient

Using the Kempf–Ness theorem (6.3.44), we can realize the quotient (6.4.12) also as a symplectic quotient. We take $G = \mathrm{U}(1)^N$ in (5.4.14). There are then k generators ξ_α, $\alpha = 1, \ldots, k$, one for each U(1), and N moment maps μ_α; the restriction to the level set $\mu^{-1}(c_\alpha)$, then fixes the $\mathbb{R}_+^N$ part of $(\mathbb{C}^*)^N$. This gives

$$\frac{\mathbb{C}^{N+k} - F}{(\mathbb{C}^*)^k} \cong \frac{\mu_\xi^{-1}(c_\alpha)}{\mathrm{U}(1)^k}\,. \tag{6.4.25}$$

As we commented before (6.3.44), the idea is to write $\mathbb{C}^* \cong \mathbb{R}_+ \times \mathrm{U}(1)$, and to perform the quotient in two steps. The choice of the c_α is related to the choice of F, and hence in turn to the fan and toric polyhedron. (Since G is abelian, the constant value c_α is not fixed to zero by equivariance.)

In this section, we will not discuss explicit Kähler metrics; we could do so using a formalism similar to that in Section 6.2 for $\mathbb{CP}^N$. The locus $\mu^{-1}(c_\alpha)$ is invariant under $\mathrm{U}(1)^{N+k}$, so starting from the trivial Kähler metric, we end up on the quotient (6.4.25) with a metric that has (at least) a torus

$$\mathrm{U}(1)^N \tag{6.4.26}$$

of isometries, acting on the fiber of (6.4.6).

We already saw an example of this formalism for $\mathbb{CP}^N$: we computed the moment map in (6.3.39). It is easy to generalize that to an arbitrary charge matrix C. Each of its rows encodes the action of a U(1) generator:

$$\xi_\alpha = -2\mathrm{Im} \sum_I C_{\alpha I} Z^I \partial_I \, , \tag{6.4.27}$$

generalizing (6.2.8). The moment map is then

$$\mu_{\xi_\alpha} = -\frac{1}{2} \sum_I C_{\alpha I} Z^I \bar{Z}_I \, , \tag{6.4.28}$$

generalizing (6.3.39).

In the $\mathbb{CP}^N$ case, there were no choices in either F or the c_α in (6.4.25); things are subtler in general, as we will see in the next subsection.

6.4.3 Toric singularities

So far, we have considered smooth manifolds, but toric geometry is also very useful for singularities.

There is a simple criterion for a toric space to be smooth. At the vertex of the toric polyhedron, the v^i_p of Section 6.4.1 (see (6.4.5) and Figure 6.2) should be related by an SL(2, $\mathbb{Z}$) transformation to a canonical basis $(v^i)_j = \delta^i_j$. A similar condition should hold around k-dimensional faces. In terms of the fan, this condition says that a cone $\langle w_{I_1}, \ldots, w_{I_p} \rangle$, in the sense of (6.4.8), is related to a subset of a canonical basis by an SL(n, $\mathbb{Z}$) transformation. This is possible only if the matrix W whose columns are the $\{w_{I_1}, \ldots, w_{I_p}\}$ has

$$|\mathrm{det} W| = 1 \, . \tag{6.4.29}$$

As anticipated in footnote 9 at the beginning of Section 6.4.2, in presence of singularities we might need to include some discrete group Γ in the denominator of (6.4.12). Given the matrix W collecting the ray generators w_I, (6.4.12) should read now

$$\frac{\mathbb{C}^{N+k} - F}{G_W} \, , \tag{6.4.30}$$

where $G_W \subset (\mathbb{C}^*)^{N+k}$ consists of the transformations $Z^I \to \lambda_I Z^I$, $\lambda_I \in \mathbb{C}^*$ that leave invariant the functions

$$\Pi_I (Z^I)^{W_{iA}} \, . \tag{6.4.31}$$

In the nonsingular case, this recipe agrees with our earlier one involving ker(W).

Toric singularities can also be described as zeros of polynomials as in Section 6.3.3, but in general they are not complete intersections; so the number of equations f_a is not expected to be equal to the k in (6.4.12) or (6.4.30). A singularity is toric if and only if its f_a are of the form

$$\mathrm{monomial}_1 = \mathrm{monomial}_2 \, . \tag{6.4.32}$$

Resolving $\mathbb{C}^2/\mathbb{Z}_n$

As a first example, we consider the $\mathbb{C}^2/\mathbb{Z}_2$ singularity discussed around (6.3.23). The ray generators in this case consists of the two columns of

$$W_{\mathbb{C}^2/\mathbb{Z}_2} = \begin{pmatrix} 0 & 2 \\ 1 & 1 \end{pmatrix} . \tag{6.4.33}$$

According to (6.4.31), we should look for the group transformations $(Z^1, Z^2) \to (\lambda_1 Z^1, \lambda_2 Z^2)$ such that $(Z^2)^2$ and $Z^1 Z^2$ are invariant. This imposes $\lambda_2 = \pm 1$, $\lambda_1 = \lambda_2^{-1}$. So $G_W \cong \mathbb{Z}_2$, acting as $(Z^1, Z^2) \to (\pm Z^1, \pm Z^2)$. This agrees with our earlier definition of this orbifold in (6.3.23).

The fan only has a single other cone $\langle w_1, w_2 \rangle$. Since $\det W = -2$, (6.4.29) is not met, as should be the case for a singularity. The toric polyhedron P now is noncompact, and also looks like a cone.

Just as for the del Pezzo surfaces in Section 6.4.1, a blowup is described by adding a vector $w_3 = \binom{1}{1}$ between the previous two. The two-dimensional cones of this new fan are $\langle w_1, w_3 \rangle$ and $\langle w_3, w_2 \rangle$. Each of these satisfies (6.4.29), so the corresponding manifold is now nonsingular.

Collecting all the w_I in a matrix

$$W_{\widetilde{\mathbb{C}^2/\mathbb{Z}_2}} = \begin{pmatrix} 0 & 2 & 1 \\ 1 & 1 & 1 \end{pmatrix} , \tag{6.4.34}$$

we can compute the corresponding charge matrix of this resolved space by computing its kernel as we did in the previous subsection:

$$C_{\widetilde{\mathbb{C}^2/\mathbb{Z}_2}} = \begin{pmatrix} 1 & 1 & -2 \end{pmatrix} . \tag{6.4.35}$$

Since $\{w_1, w_2\}$ do not generate a cone in the resolved fan, the locus $\{Z_1 = Z_2 = 0\}$ is excluded. Applying the logic around (6.4.21), we see that (6.4.35) describes the bundle $O(-2)$ over $\mathbb{CP}^1$, as we previously claimed in (6.3.28). The toric polyhedron P is still noncompact, but instead of being a cone it has a segment at its tip, representing the exceptional locus $\mathbb{CP}^1$.

The generalization to $\mathbb{C}^2/\mathbb{Z}_n$ is easy. The original singularity has

$$W_{\mathbb{C}^2/\mathbb{Z}_n} = \begin{pmatrix} 0 & n \\ 1 & 1 \end{pmatrix} . \tag{6.4.36}$$

To obtain a resolution $\widetilde{\mathbb{C}^2/\mathbb{Z}_n}$, we need to introduce $n-1$ new vectors $w_3 = \binom{1}{1}, \ldots, w_{n+1} = \binom{n-1}{1}$. In the polyhedron P, these new vectors are orthogonal to $n-1$ segments, representing $n-1$ copies of $\mathbb{CP}^1$, as we claimed after (6.3.31). We show both the (noncompact) toric polyhedron P and the fan in Figure 6.5, in the case $n = 3$.

The orbifold $\mathbb{C}^3/\mathbb{Z}_3$

A more complicated orbifold singularity is obtained with

$$W_{\mathbb{C}^3/\mathbb{Z}_3} = \begin{pmatrix} 1 & 0 & -1 \\ 0 & 1 & -1 \\ 1 & 1 & 1 \end{pmatrix} . \tag{6.4.37}$$

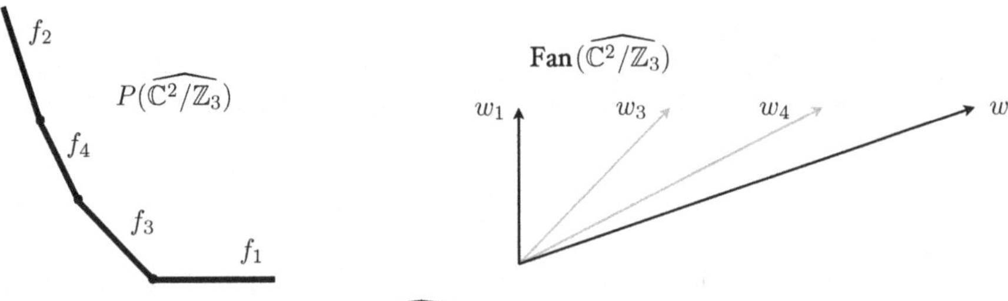

Figure 6.5 Noncompact toric polyhedron and fan for $\widehat{\mathbb{C}^2/\mathbb{Z}_3}$.

Applying (6.4.31) to $(Z^1, Z^2, Z^3) \to (\lambda_1 Z^1, \lambda_2 Z^2, \lambda_3 Z^3)$ now imposes $\lambda_1 = \lambda_2 = \lambda_3$ and $\lambda_1\lambda_2\lambda_3 = 1$; so we have

$$\mathbb{C}^3/\mathbb{Z}_3 \tag{6.4.38}$$

with $\mathbb{Z}_3$ acting as $(Z^1, Z^2, Z^3) \to (\omega_3 Z^1, \omega_3 Z^2, \omega_3 Z^3)$, and $\omega_3 \equiv e^{2\pi i/3}$.

To realize this singularity in terms of zeros of equations, we can use the map $\mathbb{C}^3/\mathbb{Z}_3 \to \mathbb{C}^{10}$ that maps the Z^A to the cubic monomials $c_{ABC} Z^A Z^B Z^C$. This map is well defined: points in $\mathbb{C}^3$ that are equivalent under $\mathbb{Z}_3$ are mapped all to the same point in $\mathbb{C}^{10}$. This is not a complete intersection, for reasons similar to the ones discussed for the Veronese embedding (6.3.16).

Conifold

The conifold singularity (6.3.24) is described by the charge matrix

$$C_{\text{cfd}} = \begin{pmatrix} 1 & 1 & -1 & -1 \end{pmatrix}. \tag{6.4.39}$$

To see this, we can use the embedding

$$\begin{aligned} \text{conifold} \quad &\to \quad \mathbb{C}^4 \\ (Z^1, Z^2, Z^3, Z^4) &\mapsto (X^1 = Z^1 Z^3,\ X^2 = Z^1 Z^4,\ X^3 = Z^2 Z^3,\ X^4 = Z^2 Z^4). \end{aligned} \tag{6.4.40}$$

(This is very similar to the Segre embedding (6.3.18), but the spaces being mapped have one less $\mathbb{C}^*$ action.) The image is cut by a single quadric

$$X^1 X^4 - X^2 X^3 = 0 \tag{6.4.41}$$

in $\mathbb{C}^4$.[10] All nondegenerate quadrics are equivalent; indeed, (6.4.41) can be mapped to (6.3.24) by taking $X^1 = x^1 + ix^4$, $X^4 = x^1 - ix^4$, $X^2 = x^2 + ix^3$, $X^3 = x^2 - ix^3$.

The ray generators are the columns of

$$W_{\text{cfd}} = \begin{pmatrix} 0 & 1 & 1 & 0 \\ 0 & 1 & 0 & 1 \\ 1 & 1 & 1 & 1 \end{pmatrix}. \tag{6.4.42}$$

The other cones in the fan are the two-dimensional $\langle w_1, w_3 \rangle$, $\langle w_2, w_3 \rangle$, $\langle w_2, w_4 \rangle$, $\langle w_1, w_4 \rangle$, and the three-dimensional $\langle w_1, w_2, w_3, w_4 \rangle$. There is no excluded locus, so we conclude that the conifold is simply

$$\mathbb{C}^4/\mathbb{C}^*. \tag{6.4.43}$$

[10] One should not confuse this $\mathbb{C}^4$ with the $\mathbb{C}^4$ parameterized by the Z^I, which in this case is the numerator of (6.4.30).

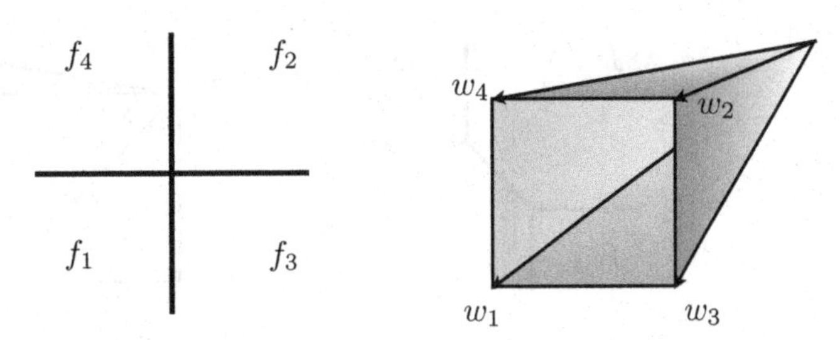

Figure 6.6 Toric polyhedron (viewed from above) and fan for the conifold.

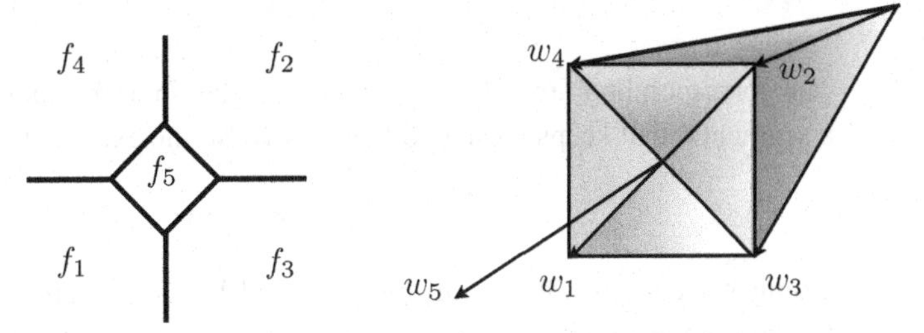

Figure 6.7 Toric polyhedron and fan for the blowup of the conifold. With respect to Figure 6.6, there is a new face in P and correspondingly an extra ray w_5; there are also additional two-dimensional cones.

Perhaps confusingly, both the fan and the toric polyhedron P look like an infinite cone with square base. We show both of them in Figure 6.6. On the left, we show the toric polyhedron P viewed from above, while on the right we show the fan in a more visually explicit three-dimensional fashion.

Let us now try to resolve the singularity. Our first attempt is to blow it up. We guess we should use a vector that goes "between" the w_A already present. A natural choice is

$$W_{\widehat{\text{cfd}}} = \begin{pmatrix} 0 & 1 & 1 & 0 & 1 \\ 0 & 1 & 0 & 1 & 1 \\ 1 & 1 & 1 & 1 & 2 \end{pmatrix}. \tag{6.4.44}$$

Polyhedron and fan are shown in Figure 6.7. The cones in the fan are such that the excluded locus is $F = \{Z^1 = Z^2 = 0\} \cup \{Z^3 = Z^4 = 0\}$. Now the charge matrix is

$$C_{\widehat{\text{cfd}}} = \begin{pmatrix} 1 & 1 & 0 & 0 & -1 \\ 0 & 0 & 1 & 1 & -1 \end{pmatrix}. \tag{6.4.45}$$

(There are no discrete components in C_W this time.) By the logic around (6.4.21), this describes the bundle $O(-1,-1)$ over $\mathbb{CP}^1 \times \mathbb{CP}^1$, as claimed in (6.3.30).

As we anticipated in that equation, there is another way to resolve this particular singularity. The idea is to change the fan without actually introducing any new w_A. For example, we can add a two-dimensional cone:

$$\langle w_3, w_4 \rangle . \tag{6.4.46}$$

As a consequence, there are now two three-dimensional cones, $\langle w_1, w_3, w_4 \rangle$ and $\langle w_2, w_3, w_4 \rangle$. The polyhedron and fan are shown in Figure 6.8.

Looking at the corresponding minors of (6.4.42), we see that both satisfy (6.4.29), so now the space is smooth. $\{w_1, w_2\}$ are not a subset of generators of any cone.

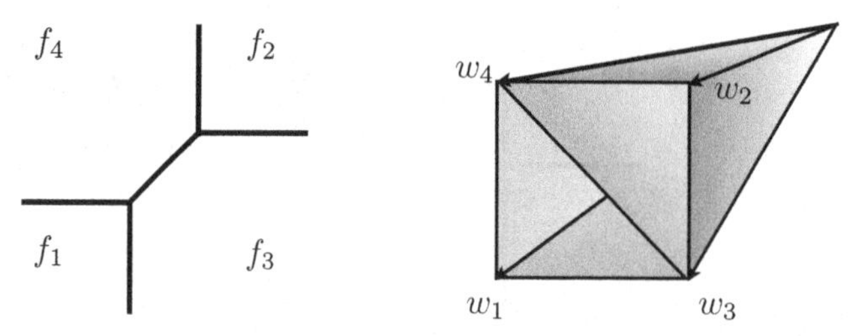

Figure 6.8 Toric polyhedron and fan for the small resolution (6.4.48) of the conifold. The rays of the fan are the same as in the singular conifold in Figure 6.6, but there is an additional two-dimensional cone $\langle w_3, w_4 \rangle$.

The corresponding faces f_1, f_2 do not meet; the tip of the cone has been replaced by a segment I that keeps these two faces away. So the excluded locus is

$$\{Z^1 = Z^2 = 0\}. \tag{6.4.47}$$

Going back to (6.4.39), with this excluded locus we can interpret Z^1 and Z^2 as the coordinates on a $\mathbb{CP}^1$, which corresponds to the segment I. Along the lines in (6.4.21), the remaining two coordinates Z^3, Z^4 can be interpreted as parameterizing the fiber of a bundle:

$$\mathcal{O}(-1) \oplus \mathcal{O}(-1) \to \mathbb{CP}^1 \quad \text{(resolved conifold)}. \tag{6.4.48}$$

This is called a *small resolution* because we have replaced the singularity with a single copy of $\mathbb{CP}^1$ rather than with a $\mathbb{CP}^1 \times \mathbb{CP}^1$ as with a blowup. Alternatively, (6.4.48) is called a *resolved conifold.*

Had we introduced the cone $\langle w_1, w_2 \rangle$ rather than $\langle w_3, w_4 \rangle$, the discussion would have been very similar; now Z^3 and Z^4 would play the role of the coordinates on $\mathbb{CP}^1$, and Z^1, Z^2 the role of coordinates on the fiber of (6.4.48). Replacing one small resolution with the other is called a *flop*.

The difference between the conifold and its small resolution can also be understood from the point of view of Kähler quotients, using the equivalence (6.4.25). There is only one $\mathbb{C}^*$; according to (6.4.28), $\mu = -\frac{1}{2}(|Z^1|^2 + |Z^2|^2 - |Z^3|^2 - |Z^4|^2)$. We only have to fix a single constant c. The original conifold singularity corresponds to taking it to be zero:

$$|Z^1|^2 + |Z^2|^2 - |Z^3|^2 - |Z^4|^2 = 0. \tag{6.4.49}$$

On the other hand, the small resolution is obtained if we choose $c = -\frac{1}{2}v < 0$:

$$|Z^1|^2 + |Z^2|^2 - |Z^3|^2 - |Z^4|^2 = v. \tag{6.4.50}$$

This has no solution where $Z^1 = Z^2 = 0$, which is the excluded locus (6.4.47).

Exercise 6.4.1 Draw the polyhedron P for the Hirzebruch surfaces (6.4.24).

Exercise 6.4.2 Write the charge matrix and draw the toric polyhedron for a $\mathbb{CP}^1$-fibration over $\mathbb{CP}^2$. Find any examples that are Fano.

Exercise 6.4.3 Blow up (6.4.38). What is the exceptional locus? Is $\widehat{\mathbb{C}^3/\mathbb{Z}_3}$ the total space of a bundle?

6.5 Complex and symplectic beyond Kähler

All the examples of complex and symplectic manifolds we have seen so far are also Kähler. The reader might perhaps conjecture at this point that this is always so. In this short section, we will see some counterexamples to this conjecture.

6.5.1 Complex non-Kähler

We first note that in a compact Kähler manifold, the form J is nontrivial in de Rham cohomology H^2. Indeed, $\mathrm{d}J = 0$, and from (5.2.31) we see $\mathrm{d}^\dagger J = 0$, so it is harmonic and nontrivial in cohomology (Section 4.1.10). Alternatively, we can use (5.5.10b) and (6.1.3) to deduce $\nabla_m J_{np} = 0$ and hence harmonicity.

Consider now the *Calabi–Eckmann* manifold:

$$M_{\mathrm{CE}} \equiv S^1 \times S^{2N+1} . \qquad (6.5.1)$$

From (4.1.134) and from the Künneth theorem, we see that M_{CE} has no even cohomology, and in particular $h^2 = 0$. By the previous remark, it cannot be Kähler. But it does admits a complex structure. Write S^{2N+1} as a Hopf fibration (6.2.4), and call y the coordinate on the S^1: then I can be defined via its $(2,0)$-form

$$\Omega_{\mathrm{CE}} = (\mathrm{d}y + \mathrm{i}\mathrm{d}\psi) \wedge \Omega , \qquad (6.5.2)$$

where Ω is (6.2.22). From (6.2.27), we then see that $\mathrm{d}\Omega = -(N+1)\mathrm{i}\eta \wedge \Omega$ for (6.5.2) as well. Recalling one of our criteria for integrability, (5.3.18), we see that M_{CE} is indeed complex.

For $N = 1$, there is a related construction, called a *Hopf surface*, as a quotient of $\mathbb{C}^2$ by a discrete complex rescaling. One can also generalize the idea to $S^{2N+1} \times S^{2M+1}$.

6.5.2 Symplectic non-Kähler

One can find several examples of symplectic non-Kähler manifolds among nilmanifolds, which were introduced in Section 4.4.2 as compact right quotients G/Γ, with G a nilpotent group and Γ a discrete subgroup. An example is the *Thurston four-manifold*

$$\mathrm{Nil}_3 \times S^1 , \qquad (6.5.3)$$

where the three-dimensional nilmanifold Nil_3 was discussed around (4.4.57). This is tantamount to adding to (4.4.58), a fourth one-form λ^4 such that $\mathrm{d}\lambda^4 = 0$. In the notation introduced following (4.4.58), this would now read $(0, 0, 12, 0)$. The two-form

$$J = \lambda^1 \wedge \lambda^3 + \lambda^2 \wedge \lambda^4 . \qquad (6.5.4)$$

is clearly nondegenerate ($J^2 \neq 0$), and from (4.4.58) it is immediate to check that $\mathrm{d}J = 0$. So this is a symplectic manifold. But it cannot be Kähler. Indeed, by (4.4.56) we can compute the de Rham cohomology algebraically. The only non-closed one-form is λ^3; the remaining three one-forms cannot be exact, because the only left-invariant zero-form is 1. So $h^1 = 3$. But (6.1.19) and (6.1.20) for a Kähler manifold would give $h^1 = 2h^{1,0}$, an even number. There are many examples like this one in higher dimensions.

A more sophisticated class of counterexamples can be obtained with the so-called *symplectic sum* construction [161]. Here we take two M_i of dimension d and a V of dimension $d-2$, all symplectic; V is a submanifold of both M_i. There is then a procedure that "glues" the M_i along V and produces a new symplectic manifold, called $M_1 \# M_2$. This gives many examples of symplectic manifolds that are not Kähler. In dimension $d = 4$, one can obtain examples that have fundamental group π_1 equal to any possible group that has a finite presentation, namely that can be written in terms of a finite number of generators and relations. So in particular, h^1 can be any number, not just 3, as in the example of (6.5.3).

7 Ricci-flatness

7.1 Calabi–Yau geometry

As we saw in Section 5.7, Calabi–Yau manifolds are manifolds with $SU(d/2)$ holonomy. They are a particular case of Kähler manifolds, which have $U(d/2)$ holonomy.

From Table 5.1, we see that Calabi–Yau manifolds are Ricci-flat. This is because their holonomy is reduced by a spinor η, whose stabilizer is $\mathrm{Stab}(\eta) = SU(d/2)$; and we have seen in the discussion around (5.7.4) that a covariantly constant spinor implies Ricci-flatness. This can also be seen by specializing (6.1.2)–(6.1.4). Since now $D_m \eta = 0$, now the intrinsic torsion vanishes identically:

$$W_1 = W_2 = W_3 = W_4 = W_5 = 0\,, \tag{7.1.1}$$

and

$$dJ = 0\,, \qquad d\Omega = 0 \qquad \text{(Calabi–Yau)}. \tag{7.1.2}$$

From (6.1.5), we now recover $R_{mn} = 0$.

The remark that follows (6.1.5) also shows that the canonical bundle K is now topologically trivial:

$$c_1(M) = 0\,. \tag{7.1.3}$$

The discussion after (5.3.52) shows that K is also *holomorphically* trivial, $K \cong O$. If M is simply connected, $\pi_1(M) = 0$, by the Hurewicz theorem (Section 4.1.10) we have $h^1(M) = 0$; because of (6.2.19) and (6.2.21), this is equivalent to $h^{0,1}(M) = 0$; and we mentioned after (5.3.52) that in this case $K \cong O$ is equivalent to (7.1.3). Even if the space isn't simply connected but $\pi_1(M)$ is finite, we can consider the universal covering (4.1.122), $\hat{M}$, which will have $\pi_1(\hat{M}) = 0$.[1]

Given our definition, nonsimply connected Calabi–Yaus also exist. The most notable example is the torus T^d (with d even); in this case, the holonomy group is trivial. Some versions of the definition of a Calabi–Yau require that the holonomy be *equal* to $SU(d/2)$ and not just contained in it. We will call these *proper* Calabi–Yau manifolds, and in this section we will have primarily have these in mind.

[1] If $\pi_1(M) \neq 0$, a Kähler manifold can very well have $K \ncong O$ and $c_1 = 0$; by the preceding discussion, it is a discrete quotient of a Calabi–Yau. A notable example in $d = 4$ is the *Enriques surface*, where $K^2 \cong O$ but $K \ncong O$.

Summarizing so far, we have seen two definitions of a Calabi–Yau manifold:

(1) A manifold with SU($d/2$) holonomy, which can be defined by a covariantly constant pure spinor, $D_m\eta = 0$;
(2) An SU($d/2$)-structure (J, Ω) such that $\mathrm{d}J = \mathrm{d}\Omega = 0$. Recall from Section 5.1.2 that the metric is determined by (J, Ω).

7.1.1 Yau's theorem

Recall from Section 5.1.2 that an SU($d/2$)-structure consists of a real two-form J and of a complex decomposable $d/2$-form Ω that satisfy (5.1.40). Thus J describes a symplectic structure (Section 5.2.1), while Ω describes a complex structure with $K \cong O$ (Section 5.3.1). The condition (5.1.40a) is equivalent to the compatibility condition (5.1.39) between J and the complex structure I associated to Ω, which we already had for a Kähler manifold. Thus, altogether we can satisfy (5.1.40a) and (7.1.2) by taking a Kähler manifold with $K \cong O$.

But we still haven't imposed the other compatibility condition (5.1.40c). The symplectic and complex conditions already tell us that $J^{d/2}$ and $\Omega \wedge \bar{\Omega}$ do not vanish anywhere, but we don't know yet that the proportionality coefficient is the one required by (5.1.40c). With a random Kähler metric, with two-form J_0, this coefficient could even be a function:

$$\frac{(-1)^{\lfloor d/4 \rfloor}}{(d/2)!} J_0^{d/2} = \mathrm{e}^f \left(\frac{\mathrm{i}}{2}\right)^{d/2} \Omega \wedge \bar{\Omega} . \tag{7.1.4}$$

Such a Kähler metric does not satisfy all the conditions for a manifold of SU($d/2$) holonomy, and in particular would not be Ricci-flat.

A minimal modification consists in keeping Ω, while changing J_0 by an exact piece, $J_0 \to J = J_0 + \mathrm{d}j$. (This sounds like it might be a coordinate transformation, but it is not, because we leave Ω fixed.) By the $\partial\bar{\partial}$-lemma (6.1.6), this can be written as

$$J_0 \to J = J_0 + \mathrm{i}\partial\bar{\partial}\mu . \tag{7.1.5}$$

Now if we can find a function μ such that

$$J^{d/2} = (J_0 + \mathrm{i}\partial\bar{\partial}\mu)^{d/2} = \mathrm{e}^{-f} J_0^{d/2} , \tag{7.1.6}$$

then we have solved the missing compatibility condition (5.1.40c). This is a non-linear partial differential equation (PDE) called a *Monge–Ampère equation*, and as such is rather hard to solve; however, Yau proved that, if M is *compact*, for any f there exists a μ such that (7.1.6) is satisfied. So the condition (5.1.40c) can always be solved with the trick (7.1.5), if we have already solved the remaining conditions by taking a Kähler manifold with $K \cong O$.

To summarize, a compact simply connected Kähler manifold with $c_1 = 0$ is a Calabi–Yau.

7.1.2 Cohomology and moduli spaces

Cohomology

The Hodge diamond for a Calabi–Yau manifold is even simpler than the one for Kähler manifolds (6.1.22). By the Dolbeault theorem (5.3.45), (5.3.48), and $K \cong O$,

$$h^{0,j}(M) = h^{d/2,d/2-j}(M) = h^{0,d/2-j}(M,K) = h^{0,d/2-j}(M)\,. \tag{7.1.7}$$

If M is simply-connected and $d = 6$, we end up with

$$\begin{array}{ccccccc} & & & 1 & & & \\ & & 0 & & 0 & & \\ & 0 & & h^{1,1} & & 0 & \\ 1 & & h^{2,1} & & h^{2,1} & & 1 \\ & 0 & & h^{1,1} & & 0 & \\ & & 0 & & 0 & & \\ & & & 1 & & & \end{array} \,. \tag{7.1.8}$$

In particular, we see

$$h_2 = h^{1,1}\,, \qquad h_3 = 2h^{2,1} + 2 \tag{7.1.9}$$

for the Betti numbers.

To (7.1.8), we can further apply the Lefschetz decomposition, which says that cohomology should decompose in $\mathrm{sl}(2,\mathbb{R})$ representations (Section 6.1.2). Starting from $H^{0,0}$, we can build such a representation acting with $J\wedge$; this will contain an element in $H^{1,1}$ (the class of J), one in $H^{2,2}$ (the class of $-\frac{1}{2}J^2$), and $H^{3,3}$. Others are formed by the primitive forms $H_0^{1,1}$ and $H_0^{2,2}$. In other columns, we don't see possible actions of $J\wedge$ of $J\cdot$: for example, above and below $H^{2,1}$ there are vanishing cohomology groups. Indeed, in this degree all cohomology is primitive: a form $J\cdot\alpha_{2,1}$ would belong to $H^{1,0}$, which is zero.

For $d > 6$, there are more undetermined numbers in the Hodge diamond. On the other hand, for $d = 4$ we will see that the whole diamond is fixed. This case is peculiar enough to deserve a separate discussion, which we defer to Section 7.2.4.

No Killing vectors on simply connected Ricci-flat manifolds

An immediate consequence of the Hodge diamond (7.1.8) is the *absence of Killing vectors*. From the defining equation $\nabla_{(m}K_{n)} = 0$ in (4.1.65), contracting the two indices we see $\nabla^m K_m = -\mathrm{d}^\dagger K = 0$. Now

$$\begin{aligned}(\Delta K)_m &= (\mathrm{d}^\dagger \mathrm{d}K)_m = \nabla^n \nabla_{[m}K_{n]} \overset{(4.1.65)}{=} \nabla_n \nabla_m K^n \\ &\overset{(4.1.13)}{=} \nabla_m \nabla_n K^n + R_{mn}K^n = 0\,.\end{aligned} \tag{7.1.10}$$

So, on a Ricci-flat manifold, if we lower the index of a Killing vector, we obtain a harmonic one-form K_m. These are in one-to-one correspondence with elements of the de Rham cohomology (Section 4.1.10), so if such a form existed it would signal $h_1 \neq 0$, which would contradict (7.1.8). We conclude that proper Calabi–Yau manifolds have no Killing vectors, and so no continuous group of isometries. More generally, this argument applies to any Ricci-flat manifolds with $h_1 = 0$. There can still be *discrete* isometries, and these will in fact have an important role in our discussion of mirror symmetry in Section 9.1.4.

Infinitesimal deformations

A quick way to analyze deformations is to use (5.1.39), which shows how the metric is determined by J and I already for a $U(d/2)$-structure. Deforming it gives $\delta g_{mn} = (\delta J_{mp}) I_n{}^p + J_{mp}\delta I_n{}^p$; going to holomorphic indices and using (5.3.2), (5.3.3) results in

$$\delta g_{i\bar{j}} = -\mathrm{i}\delta J_{i\bar{j}}\,, \qquad \delta g_{\bar{i}\bar{j}} = -\mathrm{i} g_{\bar{i}k}\mu_{\bar{j}}{}^k\,, \tag{7.1.11}$$

where μ is the Beltrami differential discussed (Section 5.3.4), and the components δg_{ij} are determined by complex conjugation. We have characterized these earlier in terms of cohomology, for δJ in (5.4.6) and for μ in (5.3.55), where we now take $K \cong O$. So we conclude that the deformations are given by

$$H^{1,1}(M) \oplus H^{d/2-1,1}_{\mathbb{C}}(M)\,. \tag{7.1.12}$$

The two summands are called Kähler and complex moduli, respectively. The subscript $_{\mathbb{C}}$ in (7.1.12) is meant to stress that the latter form a complex vector space. The dimensions of (7.1.12) are the two undetermined numbers in (7.1.8).

This was a little too slick: there is a compatibility condition among J and I, which consists in imposing that g_{mn} is symmetric and positive definite. In the language of J and Ω, compatibility becomes (5.1.40). Deforming that gives

$$\delta J\wedge\Omega + J\wedge\delta\Omega = 0\,, \qquad \frac{(-1)^{\lfloor d/4\rfloor}}{(d/2-1)!} J^{d/2-1}\delta J = \left(\frac{\mathrm{i}}{2}\right)^{d/2} (\delta\Omega\wedge\bar{\Omega} + \Omega\wedge\delta\bar{\Omega})\,. \tag{7.1.13}$$

To see what these mean, we can use our group-theory decompositions in Section 5.2.2; we will do so in $d = 6$, but the results are similar for any d. Looking at the diamond in (5.2.15), we can rewrite (7.1.13) as

$$\delta J_3 \propto \delta\Omega_3\,, \qquad \delta J_1 \propto \delta\Omega_1\,. \tag{7.1.14}$$

The component $\delta J_{2,0} = \delta J_3$ is thus related to the nonprimitive part of the $(2,1)$-part of $\delta\Omega$. On the other hand, the component $\delta\Omega_{3,0} = \delta\Omega_1$ is related to the non-primitive part of $\delta J_{1,1}$. So these two equations seem to mix complex and symplectic deformations. It is natural to resolve the ambiguity by declaring $\delta J_{2,0}$ to be part of the complex deformations, and $\delta\Omega_{3,0}$ to be part of the symplectic deformations. On the other hand, the primitive components δJ_8 and $\delta\Omega_{\bar{6}}$ are unconstrained. All this leads again to (7.1.12).

Kähler moduli space

So far we have discussed infinitesimal deformations. Let us now look at the full space of Ricci-flat metrics on a given Calabi–Yau metric.

We first consider the moduli spaces of symplectic structures J compatible to a given complex structure I. The closure condition $\mathrm{d}J = 0$ is linear, and so the space of two-forms might look like a linear space. But we have to take into account non-degeneracy, and again the compatibility condition (5.1.39): $g_{mn} = J_{mp} I_n{}^p$ should be symmetric and positive definite. The symmetric requirement was taken into account in the discussion (7.1.13), while positive definiteness did not appear, because it is never broken by infinitesimal deformations.

To see how to describe this positivity, we give a more geometrical interpretation to Kähler moduli using the theory of calibrations. We have seen in Section 5.6.1 that J and $\mathrm{Re}(e^{i\theta}\Omega)$ are almost calibrations for (J, Ω) any $SU(d/2)$-structure. For a Calabi–Yau, these forms are closed, and hence become calibrations. According to the general argument in (5.6.7), the volume of a two-dimensional subspace S_2,

$$\mathrm{Vol}(S_2) = \int_{S_2} J\,, \tag{7.1.15}$$

will be the smallest in its homology class. So if we demand (7.1.15) to be positive, the volume of any other S_2 in that class will be automatically positive too. By (5.6.15), a particular case of Wirtinger's theorem, a calibrated S_2 is holomorphic: it is described locally by a system of holomorphic equations.

The subspace of two-cycles in $H_2(M, \mathbb{Z})$ with a holomorphic representative is called *Mori cone*: it is a cone in the sense that it remains invariant under rescalings by integers. It can be obtained by intersecting hypersurfaces (elements of $H_{d-2}(M, \mathbb{Z})$; in $d = 6$, four-cycles). The *Lefschetz theorem on* $(1,1)$*-forms* says that

$$\forall\, \omega \in H^{1,1}(M) \cap H^2(M, \mathbb{Z}) \quad \exists\, \mathcal{L} \text{ such that } c_1(\mathcal{L}) = \omega\,. \tag{7.1.16}$$

In our case, from (7.1.8), we know that

$$H^2(M) = H^{1,1}(M) \qquad (d \geq 6)\,. \tag{7.1.17}$$

So the Poincaré dual of any element $S_{d-2} \in H_{d-2}(M, \mathbb{Z})$ has an associated line bundle $\mathcal{L}$, and by (4.2.63) the zero locus of one of its sections will be a holomorphic representative of S_{d-2}. The intersections of holomorphic subspaces are also holomorphic, and so we realize holomorphic curves this way.

Now we can impose (7.1.15) to be positive for any S_2 in the Mori cone. The set of $J \in H^2(M)$ with this property is called *Kähler cone*. This is the moduli space

$$\mathcal{M}_{\mathrm{K}} \tag{7.1.18}$$

of symplectic structures J for a given complex structure I. Again it is a cone because a multiple of a J in (7.1.18) is still in it. It has some boundaries, which correspond to values of J where one of the curves has zero volume.

A consequence of this discussion is that every Calabi–Yau for $d \geq 6$ is algebraic. To see this, it is enough to pick an element ℓ in the Kähler cone (7.1.18) which is integral, namely $\ell \in H^2(M, \mathbb{Z})$. This element can then be thought of as the first Chern class c_1 of a line bundle,

$$\ell = c_1(\mathcal{L})\,. \tag{7.1.19}$$

Since it is in the Kähler cone, by definition it is positive; by Kodaira's embedding theorem and Chow's theorem (Section 6.3.3), it gives an embedding of our Calabi–Yau $M \hookrightarrow \mathbb{CP}^N$ for large enough N.

Complex moduli space

We now discuss the moduli space of complex structures I. In general, this is described by the Kodaira–Spencer equation (5.3.63a). For the Calabi–Yau case, a theorem by Tian and Todorov shows that every infinitesimal deformation can be promoted to a full deformation: there are no obstructions, in the language of Section 5.3.4. So the moduli space

$$\mathcal{M}_{\rm c} \tag{7.1.20}$$

of complex structures is a complex manifold of the expected dimension $H^{2,1}(M)$. On some locus $\mathcal{M}_{\rm sing} \subset \mathcal{M}_{\rm c}$, however, M will have some singular points (6.3.21).

Similar to (7.1.15), a three-cycle S_3 is calibrated if it is special Lagrangian, (5.6.21), for some choice of θ. Again its volume

$$\mathrm{Vol}(S_3) = \int_{S_3} \mathrm{Re}(\mathrm{e}^{\mathrm{i}\theta}\Omega) \tag{7.1.21}$$

will be the smallest in its homology class. Thanks to θ, there is no positivity issue here.

7.1.3 Examples

Hypersurfaces

We now want to show how to find examples. We have seen in (6.2.30) that the simple example $\mathbb{CP}^N$ of a Kähler manifold has $c_1 = N + 1$, so this is not a Calabi–Yau. We saw in Section 6.3.1 that we can decrease c_1 already by taking a hypersurface $Z(f_a, \mathbb{CP}^N)$, where f_d is a homogeneous polynomial of degree d. Equation (6.3.7) implies that we can obtain $c_1 = 0$ by choosing

$$d = N + 1\,. \tag{7.1.22}$$

A concrete Ω for $Z(f_a, \mathbb{CP}^N)$ is given by the form such that

$$\Omega \wedge D f_d = \Omega_{\mathbb{CP}^N}\,, \tag{7.1.23}$$

where $\Omega_{\mathbb{CP}^N}$ is the one defined in (6.2.22), and D is the projected exterior differential in (6.2.9). This Ω is called a *Poincaré residue*. By (6.2.26), and the fact that Df_d has charge d, Ω is a form on $Z(f_a, \mathbb{CP}^N)$ (rather than a twisted form) precisely when $d = N + 1$. On the patch (6.2.24), where $\Omega_{\mathbb{CP}^N}$ is given by (6.2.26), f becomes a function of the z^i, and

$$\Omega = \frac{\mathrm{d}z^1 \wedge \ldots \wedge \mathrm{d}z^N}{\mathrm{d}f} = \frac{\mathrm{d}z^2 \wedge \ldots \wedge \mathrm{d}z^N}{\partial_1 f} = \ldots = \frac{\mathrm{d}z^1 \wedge \ldots \wedge \mathrm{d}z^{N-1}}{\partial_N f}\,. \tag{7.1.24}$$

The first expression is a bit symbolic; the N following ones are all valid and equal.

Applying (6.3.3) to $Z(f_a, \mathbb{CP}^N)$ and using (6.2.19), we see that it is simply connected for $N > 2$. By our comment following (7.1.3), this is also enough to guarantee that $K \cong O$.

Quintic

Thus we have concluded that $Z(f_a, \mathbb{CP}^N)$ is a Calabi–Yau if $d = N + 1$. We are especially interested in examples of real dimension six, or complex dimension $3 = N - 1$; so $N = 4$ and $d = 5$, and thus the space is called the *quintic*:

$$Q = \{p_{I_1 \dots I_5} Z^{I_1} \dots Z^{I_5} = 0\} \subset \mathbb{CP}^4 . \tag{7.1.25}$$

Let us compute its moduli and cohomology. By (6.3.3) and (6.2.21),

$$h^{1,1}(Q) = h^{1,1}(\mathbb{CP}^4) = 1 . \tag{7.1.26}$$

So the moduli space $\mathcal{M}_K$ is one dimensional. There is only one class in $H_2(M, \mathbb{Z})$; imposing positivity, we see that $\mathcal{M}_{\mathcal{K}}$ is a half-line $\mathbb{R}_{\geq 0}$.

The story is more complicated for the complex moduli space $\mathcal{M}_c$. Around (7.1.19), we saw that every Calabi–Yau with $d \geq 6$ is algebraic. So it is enough to count the space of possible $p_{I_1 \dots I_5}$ in (7.1.25): there are no deformations that take us away from this form. The space of degree d polynomials in n variables has dimension

$$\binom{d+n-1}{d} ; \tag{7.1.27}$$

so for a degree-five polynomial in five variables, we have $\binom{9}{5} = 126$ parameters. But there are many redundancies: polynomials that are related by linear transformations $Z^I \to M^{IJ} Z^J$ describe the same geometry. The space of matrices M^{IJ} is GL(5), of dimension 25. Thus the number of complex moduli is $126 - 25 = 101$, and

$$h^{2,1}(Q) = 101 . \tag{7.1.28}$$

$\mathcal{M}_c$ contains some singular locus. For example, the particular quintic

$$Z_0^5 + Z_1^5 + Z_2^5 + Z_3^5 + Z_4^5 - 5 Z_0 Z_1 Z_2 Z_3 Z_4 = 0 \tag{7.1.29}$$

is singular in the 125 points $(1, \omega_5^i, \omega_5^j, \omega_5^k, \omega_5^{-i-j-k})$, where $\omega_5 = e^{2\pi i/5}$.

Complete intersections

Equation (7.1.25) is arguably the simplest example of CY_3. A possible generalization is the locus $Z(\{f_a\})$ cut by a system of k equations of degrees $d_1, \dots, d_k$. If we assume this to be a complete intersection (Section 6.3.3), then most of our previous logic goes through: we can consider the submanifolds obtained by adding one f_a at a time, and apply the Lefschetz hyperplane theorem (6.3.3) to each step. The generalization of (6.3.7) is

$$c_1(Z(\{f_a\})) = N + 1 - \sum_a d_a . \tag{7.1.30}$$

Imposing $c_1 = 0$ and the condition $N - k = 3$ on the dimension, there are finitely many new possibilities [162]. For example, one is $d_1 = d_2 = 3$, so two cubics in $\mathbb{CP}^5$; this turns out to have $h^{2,1} = 73$, $\chi = -144$.

Even more generally, we can start considering systems in products such as $\mathbb{CP}^{N_1} \times \dots \times \mathbb{CP}^{N_l}$, or in toric manifolds. These techniques produce a huge number of examples; it is currently not even clear whether the total number of different topologies that admit Calabi–Yau metrics is finite. Typically at least one of the Hodge numbers $h^{1,1}$, $h^{2,1}$ is large, as in our quintic example, but this can be moderated by taking quotients by finite groups, even achieving small numbers such as $h^{1,1} = h^{2,1} = 1$ [163–166]. Topologically, recall from Wall's theorem in Section 4.2.5 that any value of h^2 and (even) h^3 is realized by a six-manifold M_6. Empirically, so far not all of the $h^{1,1}$, $h^{2,1}$ are realized by Calabi–Yaus. For those pairs where a Calabi–Yau does exist, there are many possibilities for the triple intersection form D_{abc}.

7.1.4 Topology change

The moduli spaces of Calabi–Yau manifolds that are topologically different are connected by two types of processes. Both are particularly natural from the point of view of string theory, as we will see later.

Flops

The first is a compact version of the flop we mentioned in Section 6.4.3. This happens when we go to a boundary of the Kähler cone (7.1.18), which as we saw corresponds to a J such that a curve shrinks to zero volume. Around such a curve, the geometry will look like the small resolution of the conifold discussed around (6.4.48). The distance from the wall can be parameterized by the parameter v in (6.4.50), which goes to zero when the $\mathbb{CP}^1$ in the small resolution shrinks. Now we can choose to continue the process to negative v; in other words, to replace the conifold singularity with the other small resolution, the one where a $\mathbb{CP}^1$ appears parameterized by Z^3, Z^4. In the conifold, the two possibilities yield the same space (6.4.48) parameterized differently, but in a compact Calabi–Yau they are "glued" in a different way to the rest of the space. In this process, $h^{1,1}$ and $h^{2,1}$ remain the same, but the triple intersection D_{abc} changes.

Extremal transitions

A second type of topology change is more drastic, and changes the Hodge numbers as well. We again reach a boundary of the Kähler cone, but we then deform the resulting conifold singularity; locally this will become like (6.3.36). The Hodge numbers are changed because the two-cycle present in the small resolution is replaced by an S^3, as remarked after (6.3.36). The reverse process is also possible: we mentioned earlier that at some locus $\mathcal{M}_{\text{sing}} \subset \mathcal{M}_{\text{c}}$ of the complex structure moduli space, a Calabi–Yau can develop singularities. On a generic point on $\mathcal{M}_{\text{sing}}$, the Calabi–Yau has conifold singularities (6.3.24). In such a case, one can resolve them and obtain a new Calabi–Yau where some three cycles have been replaced by two cycles.

A famous example is the quintic [167]

$$Z_3 g_4 + Z_4 h_4 = 0 \,, \tag{7.1.31}$$

where g_4 and h_4 are quartic polynomials of all coordinates. This defines a subspace of $\mathcal{M}_{\text{c}}$ of dimension 86, and hence of codimension $101 - 86 = 15$. For generic g_4 and h_4, this has 16 conifold singularities, located at $\{Z_3 = Z_4 = 0,\ g_4 = h_4 = 0\}$, so it is part of $\mathcal{M}_{\text{sing}}$. ($\{Z_3 = Z_4 = 0\}$ is a copy of $\mathbb{CP}^2 \subset \mathbb{CP}^4$, and $\{g_4 = h_4 = 0\}$ give two quartic equations in it, with $4 \times 4 = 16$ solutions.) On this locus, we have shrunk some copies of S^3, just like in going from (6.3.36) to (6.3.24). However, all these copies A_a of S^3 lie on the $\mathbb{CP}^2$, and as a consequence their sum is a boundary of a four-dimensional subspace S_4, $\partial S_4 = \sum_a A_a$; so only 15 of these three-cycles are independent in homology. We also lose 15 more three-cycles B^a, dual to the A^a in the sense of (4.1.128). All in all, b_3 decreases by 30 and $h_{2,1}$ by 15. After the transition, B_4 has no boundary any more, and so it is a new four-cycle; $h_{2,2} = h_{1,1}$ increases by one. Summing up, $\Delta h_{2,1} = -15$, $\Delta h_{1,1} = 1$, and we have transformed the quintic to a new Calabi–Yau threefold with

$$h_{2,1} = 86 \,, \qquad h_{1,1} = 2 \,. \tag{7.1.32}$$

There are many more examples of such *extremal* transitions. *Reid's fantasy* is the conjecture that all Calabi–Yau's can in fact be connected in this way.

Exercise 7.1.1 Show that each of the 125 singular points (7.1.29) is a conifold singularity (6.3.24). (Hint: choose a coordinate patch in $\mathbb{CP}^4$, and do a Taylor expansion of (7.1.29).)

Exercise 7.1.2 Find a few more solutions to (7.1.30). Compute the Hodge numbers for each.

7.2 Other Ricci-flat geometries

In Table 5.1, there are some other cases where special holonomy implies Ricci-flatness, namely Sp(d/4), G_2, and Spin(7). We will discuss the first two in this section (with a brief mention of Sp(d/4) · Sp(1) too), while we will leave Spin(7) for Section 7.5.3, where we focus on the conical case.

7.2.1 K3 manifolds

We start in this section with the simplest example of Sp(d/4) holonomy, namely $d = 4$. The group $\mathrm{Sp}(d/4) = \mathrm{Sp}(1) \cong \mathrm{SU}(2)$, so this is in fact a particular case of SU(d/2) holonomy; so we could have discussed this as a particular case of Calabi–Yau manifold, but we will see that there are some peculiarities that justify a separate treatment. For a dedicated review, see [168].

As we saw in Section 5.1.4 and earlier, the (J, Ω) that define the SU(2)-structure can also be thought of as a triplet

$$J_a = \{J,\ \mathrm{Re}\Omega,\ \mathrm{Im}\Omega\}\,, \tag{7.2.1}$$

with algebraic compatibility conditions $J_a \wedge J_b = -2\delta_{ab}\mathrm{vol}_4$ and differential equations $\mathrm{d}J_a = 0$.

There are only two compact manifolds with SU(2) holonomy: the torus T^4, and a simply connected manifold called *K3*.

Hodge diamond and cohomology

We can obtain a manifold with SU(2) holonomy with the techniques of Section 7.1. Indeed, Yau's theorem tells us that such a metric exists on a Kähler manifold with $c_1 = 0$. (An explicit expression of the metric was recently found in [169, 170].) Using (6.3.7) with $d = 4$ leads to a quartic:

$$Z(f_4, \mathbb{CP}^3) = \{f_4 = p_{I_1\ldots I_4} Z^{I_1} \ldots Z^{I_4} = 0\}\,. \tag{7.2.2}$$

It turns out that any other compact simply connected manifold with SU(2) holonomy is diffeomorphic to this one. So we can use the realization (7.2.2) to study topological features such as the Hodge diamond. However, we will soon see that the complex and Kähler structures are not all equivalent to each other; in this respect, (7.2.2) does not cover all possibilities.

We next compute the Euler characteristic χ. By the splitting principle (4.2.67) and the Euler quotient (6.2.32), the total Chern class (4.2.66) is

$$c(TM_4) = \frac{(1+x)^4}{1+4x}\,, \tag{7.2.3}$$

where $x = c_1(\mathcal{O}(1))$; expanding this, we find $c_2 = 6x^2$. Then by the Gauss–Bonnet theorem (4.2.72):

$$\chi(M_4) = \int_{M_4} c_2 = \int_{M_4} 6x^2 = 24. \tag{7.2.4}$$

In the last step, we used (4.2.63) to say that x is the Poincaré dual of a hyperplane, namely the zero locus $Z(f_1)$ of a degree $d = 1$ equation in $\mathbb{CP}^3$. Then $6x^2 = 6x \wedge x$ is the Poincaré dual of the zero locus $Z(f_6, f_1)$ of two equations, of degrees six and one; this uses the logic that follows (6.2.20). Finally the integral $\int_{M_4} 6x^2$ is the intersection between this zero locus and M_4, which is a system of three equations, $Z(f_6, f_1, f_4)$, with total degree 24, whose solutions are 24 points.

The Calabi–Yau conditions (7.1.7) on (6.1.22) leave only $h^{1,1}$ undetermined, with the four corners equal to 1 and the rest 0; so the Euler characteristic $\chi = \sum_i (-1)^i b_i = 4 + h^{1,1}$, and thus $h^{1,1} = 20$. The Hodge diamond is then

$$\begin{matrix} & & 1 & & \\ & 0 & & 0 & \\ 1 & & 20 & & 1 \\ & 0 & & 0 & \\ & & 1 & & \end{matrix}\ . \tag{7.2.5}$$

We can also apply the Lefschetz decomposition: $H^{1,1}$ decomposes further as the part proportional to J, and the primitive part $H_0^{1,1}$. This suggests a way of computing the signature of the intersection pairing

$$\int_{M_4} \alpha_2 \wedge \beta_2 \tag{7.2.6}$$

in H^2 (Section 4.1.10). As discussed around (5.2.30) (and even earlier in flat space in Section 3.3.2), J and Ω are anti-self-dual, while primitive (1, 1)-forms are self-dual. Using the fact that (4.1.106b) is symmetric, it is easy to see that the two subspaces are orthogonal; moreover, $\int \alpha_\pm \wedge \alpha_\pm = \pm \int *\alpha_\pm \wedge \alpha_\pm = \pm||\alpha_\pm||^2$ where $\alpha_\pm$ is an (anti-)self-dual form. So the signature of the intersection pairing on H^2 is

$$(3, 19)\,. \tag{7.2.7}$$

K3 moduli spaces

The moduli space of K3's is subtler than the one we saw for higher-dimensional Calabi–Yaus. The first peculiarity is that $h^{1,1}(M)$ and $h^{d/2-1,1}$ coincide: so symplectic and complex deformations belong to the same space, which by (7.2.5) has dimension 20. It can be described more geometrically by noticing that $H^{2,0} \subset H^2$ is a subspace of dimension 2; every complex structure corresponds to the choice of such a two-subspace. This is a Grassmannian, but not quite the one in (4.4.51) because the natural metric in H^2 is not Euclidean: we just saw it has signature (3, 19). So (4.4.51) in this case should be replaced by

$$\mathcal{M}_c = \frac{O(3,19)}{O(2)\times O(1,19)\times \Gamma_{3,19}}, \tag{7.2.8}$$

with a discrete component $\Gamma_{3,19}$ we won't describe here. By counting group dimensions, $\dim\mathcal{M}_c = 2 \cdot 20$, namely complex dimension 20, as promised.

Let us try to describe (7.2.8) starting from (7.2.2), along the lines of (7.1.28) for the quintic. This time we need to count quartic polynomials in four variables, which by (7.1.27) gives $\binom{4+3}{4} = 35$, minus the dimension of GL(4); this gives $35 - 16 = 19$, which is less than the expected 20.

In other words, not all K3s can be put in the form (7.2.2). If we try with more complicated constructions, such as loci $Z(\{f_a\})$ cut by systems of equations, we always fall short of 20 (see, for example, [119, chap. 4.5]). The reason for this mismatch is that *not all K3s are algebraic*: indeed, the discussion around (7.1.19) shows that all Calabi–Yaus are algebraic, but assuming $d \geq 6$. What fails in $d = 4$ is that it is no longer true that $H^2 = H^{1,1}$. This is captured by the dimension of the *Picard lattice*

$$H^2(M,\mathbb{Z}) \cap H^{1,1}(M). \tag{7.2.9}$$

For a generic complex structure, the dimension of (7.2.9) is zero. The space of K3s with complex structure induced by (7.2.2) is a complex codimension-one subspace of $\mathcal{M}_c$, on which (7.2.9) has dimension one. More generally, demanding that (7.2.9) should have dimension p specifies a subspace of $\mathcal{M}_c$ of dimension $20 - p$.

The moduli space $\mathcal{M}_K$ of symplectic structures J is again of dimension 20, this time real. Together with the 20 complex dimensions of $\mathcal{M}_c$, it might seem that overall the moduli space $\mathcal{M}_{HK}$ of hyper-Kähler metrics should have dimension $20+40 = 60$. However, there are redundancies: rotating the three J_a (7.2.1) by an SO(3) does not actually change the metric. This gives $60 - 3 = 57$. We still have the freedom to rescale all three J_a by a number, rescaling Vol(K3). So the final dimension is 58.

Similar to (7.2.8), a more intrinsic way of describing $\mathcal{M}_{HK}$ is by thinking of the J_a as spanning a three-subspace inside $H^2 = \mathbb{R}^{3,19}$. Ignoring the overall rescaling for the J_a, this is again a Grassmannian:

$$\mathcal{M}_{HK} = \frac{O(3,19)}{O(3)\times O(19)\times \Gamma_{3,19}}. \tag{7.2.10}$$

Counting group dimensions, we see $\dim\mathcal{M}_{HK} = 3 \cdot 19 = 57$, as anticipated.

Let us now give some additional constructions of K3 manifolds, to complement (7.2.2), which as we just saw is not the most general case.

Orbifold resolution

The first starts from an orbifold

$$T^4/\mathbb{Z}_2, \tag{7.2.11}$$

where we take T^4 to be the "square" torus, obtained by quotienting $\mathbb{R}^4_{x_i}$ by the same translation $x_i \sim x_i + 1$ for all four of its coordinates, and $\mathbb{Z}_2$ acts by reversing the sign of all coordinates. The origin is a fixed point of this $\mathbb{Z}_2$; so is any point whose coordinates are either 0 or 1/2. Each of these $2^4 = 16$ points is locally a $\mathbb{C}^2/\mathbb{Z}_2$ singularity. Blowing them up, as we learned in Section 6.3.5, produces 16 exceptional S^2s. This produces a smooth manifold, which is again a K3.

Elliptic fibrations

Another notable construction is via a fibration whose generic fiber is T^2:

$$\begin{array}{ccc} T^2 & \hookrightarrow & \mathrm{K3} \\ & & \downarrow \\ & & \mathbb{CP}^1 . \end{array} \tag{7.2.12}$$

We require the fibers to be complex in one of the complex structures of the K3, but we allow a finite number of degenerate fibers. We can describe such a K3 by taking the Weierstrass form (6.3.11) of a torus, but now making τ depend on the coordinates of the $\mathbb{CP}^1$ base. As we noticed following (6.3.11), the T^2 in this realization has a marked point at infinity, and it is called an elliptic curve to emphasize this. This becomes a global section of our T^2 fibration, which is then called an *elliptic fibration*.

One can generalize this idea to elliptic fibrations $T^2 \hookrightarrow E \to B$ over any B: we take f and g in (6.3.11) to depend holomorphically on B. By Liouville's theorem, there are no holomorphic functions on a compact space. But we can take them to be sections of a bundle; making y and x also sections, we can try to make both sides transform in the same way, so that the entire equation (6.3.11) makes sense. In other words, we can embed the whole elliptic fibration in a larger fibration $\mathbb{CP}^2 \hookrightarrow M \to B$. This can be achieved by picking a line bundle $\mathcal{L}$, and taking

$$\begin{array}{ccccc} & x & y & f & g \\ \text{sections of} & \mathcal{L}^2 & \mathcal{L}^3 & \mathcal{L}^4 & \mathcal{L}^6 . \end{array} \tag{7.2.13}$$

Degenerate fibers appear over points where the discriminant (6.3.14) vanishes; it is a section of $\mathcal{L}^{12}$. The canonical bundle can be computed using the more general version (6.3.5) of the adjunction formula, applied to $E \subset M$. This gives $c_1(E) = c_1(B) + c_1(L)(3 + 2 - 6)$, because of the degrees in (7.2.13). So the condition that $K_E \cong O$ gives is the following:

$$\mathcal{L} \cong K_B^{-1} . \tag{7.2.14}$$

Returning to our case (7.2.12), where $B = \mathbb{CP}^1$, (7.2.14) gives $\mathcal{L} = O(2)$. The discriminant is now a section of $\mathcal{L}^{12} = O(24)$, so it will vanish at 24 points on $\mathbb{CP}^1$ (counted with multiplicity). The positions of these points are not independent, since they are determined by f and g. Since f is a section of $\mathcal{L}^4 \cong O(8)$, it is a degree 8 polynomial in one patch of $\mathbb{CP}^1$, so it has nine free parameters; similarly g, section of $\mathcal{L}^6 \cong O(12)$, has 13 parameters, for a total of 22 (all complex). We should now also notice that both a rescaling $(f, g) \to (\alpha^2 f, \alpha^3 g)$ (one parameter) and a Möbius transformation on x (three parameters) do not change the complex structure and should be subtracted. This brings us to a total of 18 complex moduli, corresponding to a Picard number of $p = 2$, the two algebraic curves being the base and the fiber. A modification of the logic behind (7.2.8) gives a moduli space:

$$\mathcal{M}_{\mathrm{c,ell}} = \frac{\mathrm{O}(2,18)}{\mathrm{O}(2) \times \mathrm{O}(18) \times \Gamma_{2,18}} . \tag{7.2.15}$$

7.2.2 Noncompact spaces with SU(2) holonomy

We will now see an explicit construction of noncompact manifolds with SU(2) holonomy. These are sometimes colloquially referred to as "noncompact K3," even though strictly speaking the term "K3" refers to the compact manifold we discussed in the previous subsection.

The general class

We consider the forms

$$J_a = \mathrm{d}y_a \wedge (\mathrm{d}\psi + \sigma) + \frac{1}{2}\epsilon_{abc} V \mathrm{d}y_b \wedge \mathrm{d}y_c \,. \tag{7.2.16}$$

Here ψ is a coordinate on an S^1; V and A are respectively a function and one-form on $\mathbb{R}^3$, with coordinates y_a, $a = 1, 2, 3$. Equation (7.2.16) satisfies the algebraic conditions (3.3.17) for an SU(2)-structure, with $J_a \equiv \{J, \mathrm{Re}\Omega, \mathrm{Im}\Omega\}$ and $\mathrm{vol} = \mathrm{d}y_1 \wedge \mathrm{d}y_2 \wedge \mathrm{d}y_3 \wedge (\mathrm{d}\psi + \sigma)$. The metric can be determined with the procedure described in Section 5.1.2:

$$\mathrm{d}s^2_{\mathrm{GH}} = V \mathrm{d}s^2_{\mathbb{R}^3} + V^{-1}(\mathrm{d}\psi + \sigma)^2 \,. \tag{7.2.17}$$

$\mathrm{d}s^2_{\mathbb{R}^3}$ refers to the Euclidean $\mathbb{R}^3$ metric. This is called *Gibbons–Hawking* metric [171].

The Calabi–Yau condition now imposes $\mathrm{d}J = \mathrm{d}\Omega = 0$, or in other words $\mathrm{d}J_a = 0$; it is straightforward to check that it reduces to $\partial_1 V = \partial_{[2}\sigma_{3]}$, etc.; or in other words,

$$\mathrm{d}\sigma = *\mathrm{d}V \,, \tag{7.2.18}$$

where the $*$ is taken with respect to the Euclidean metric of $\mathbb{R}^3$. By taking d of (7.2.18), we obtain $\Delta V = 0$. Given a harmonic V, we can solve for σ (since we are on $\mathbb{R}^3$) and obtain a noncompact space with SU(2) holonomy.

Coordinate singularities

One can generalize the construction to include functions whose Laplacian has delta-function singularities. Suppose we take, for example,

$$V = \frac{1}{r}, \tag{7.2.19}$$

with $r = \sqrt{y_a y_a}$ the radial coordinate in $\mathbb{R}^3$. This is not quite harmonic, but rather satisfies[2]

$$-\nabla^2 \frac{1}{r} = 4\pi \delta(y_1)\delta(y_2)\delta(y_3) \,. \tag{7.2.20}$$

With the choice (7.2.19), in (7.2.17) now the metric of the $\mathbb{R}^3$ diverges, while the size of the ψ goes to zero. This might look like a singularity. And yet it is a smooth geometry, as we now show. From (7.2.18), we get $\mathrm{d}\sigma = -*\frac{\mathrm{d}r}{r^2} = -\mathrm{vol}_{S^2}$, so $\sigma = \cos\theta \mathrm{d}\phi$. After defining $\rho \equiv \sqrt{r}$, (7.2.17) then becomes the following:

[2] A possible proof, famous from electrodynamics, is obtained by integrating over a round ball B using Stokes (4.1.131): with coordinates y^m in $\mathbb{R}^3$, $-\int_B \mathrm{d}^3 y \nabla^2 \frac{1}{r} = \int_B \partial_m (y^m/r^3) = \int_{\partial B} (\vec{y}\cdot \mathrm{d}\vec{a})/r^3 = 4\pi$.

$$
\begin{aligned}
\frac{1}{r}\mathrm{d}s^2_{\mathbb{R}^3} &+ r(\mathrm{d}\psi+\sigma)^2 \\
&= \frac{1}{r}\mathrm{d}r^2 + r\mathrm{d}s^2_{S^2} + r(\mathrm{d}\psi+\sigma)^2 = 4\mathrm{d}\rho^2 + \rho^2\left((\mathrm{d}\psi+\sigma)^2 + \mathrm{d}s^2_{S^2}\right) \\
&\overset{(4.2.51)}{=} 4\left(\mathrm{d}\rho^2 + \rho^2\mathrm{d}s^2_{S^3}\right) = 4\mathrm{d}s^2_{\mathbb{R}^4}\,.
\end{aligned} \tag{7.2.21}
$$

(We have redefined ψ with a factor of 2 relative to (4.2.51), so our ψ in this section has periodicity $\Delta\psi = 4\pi$.) We conclude that (7.2.17) with $V = 1/r$ is just flat space, and in particular is smooth.

Equation (7.2.21) can be read as a fibration $S^1 \hookrightarrow \mathbb{R}^4 \to \mathbb{R}^3$; it is not quite a bundle, because the S^1 fiber degenerates at the origin $r = 0$. It can be read as an extension of the Hopf fibration (4.2.54) to include radial coordinates. The projection can also be seen as a map:

$$
\begin{array}{ccc}
\mathbb{H} \cong \mathbb{C}^2 \cong \mathbb{R}^4 & \to & \mathbb{R}^3 \\
q \equiv \begin{pmatrix} w_1 \\ w_2 \end{pmatrix} & \mapsto & y_a \equiv q^\dagger \sigma_a q
\end{array} \tag{7.2.22}
$$

so that

$$
y_3 = |w_1|^2 - |w_2|^2\,, \qquad y_1 + \mathrm{i}y_2 = 2w_1\bar{w}_2\,. \tag{7.2.23}
$$

In a more explicitly quaternionic formalism, recalling (2.1.23), we can write the map as

$$
q = q_4 + \sum_{a=1}^{3} q_a i_a \mapsto y_a = \frac{1}{2}\mathrm{Re}(q^* i_a q i_a) \quad (\text{no sum})\,, \tag{7.2.24}
$$

where * reverses sign to all three imaginary units. With both definitions, we can see

$$
r^2 = y_a y_a = \left(|w_1|^2 + |w_2|^2\right)^2 = \left(\sum_{m=1}^{4} q_m^2\right)^2 \equiv \rho^4\,. \tag{7.2.25}
$$

This explains the peculiar redefinition $\rho = \sqrt{r}$ of radial variables in (7.2.21).

Multicenter solutions

We just saw in (7.2.21) that a function whose Laplacian has a delta function, rather than being harmonic, does not lead to a singularity in the metric (7.2.17). So in general we can allow functions V proportional to the potential of k point-like electric charges:

$$
V = V_0 + \sum_{\alpha=1}^{k} \frac{1}{|y - y_\alpha|}\,, \qquad |y - y_\alpha|^2 = \sum_{a=1}^{3}(y_a - y_{a\,\alpha})^2\,. \tag{7.2.26}
$$

Near each point $y_a = y_{a\,\alpha}$, (7.2.21) shows that the geometry is smooth. We call these k points in $\mathbb{R}^3$ *nuts*.

The topology of our four-dimensional manifold GH_k is not $\mathbb{R}^3 \times S^1$ as one might naively think from (7.2.17), because at each nut the S^1 shrinks. For example, in (7.2.21), we saw that $\mathrm{GH}_1 \cong \mathbb{R}^4$. In presence of more nuts, nontrivial elements in $H_2(\mathrm{GH}_k, \mathbb{Z})$ are generated. To see this, consider a segment in $\mathbb{R}^3$ with endpoints on two nuts. The S^1_ψ fiber shrinks at the endpoints; so on this segment, (7.2.17) looks like the left side of Figure 6.3, which reconstructs an S^2, which is not the boundary of anything. There are $k-1$ such independent cycles $A_\alpha \cong S^2$, so

$$h_2(\mathrm{GH}_k, \mathbb{Z}) = k - 1\,. \qquad (7.2.27)$$

There are no cycles in h_1 or h_3. Since the manifold is noncompact, we also have the option of considering relative homology. This is a noncompact space, not one with a boundary, but we can compactify it by adding a set of points at infinity; now $\mathbb{R}^3$ is replaced by a ball B_3. In relative homology, there is a new set of two-cycles. This is obtained by considering lines that begin and end on ∂B_3 and go between some of the nuts. Adding the fiber S^1_ψ now produces cylinders $B^\alpha \cong S^1 \times \mathbb{R}$; again there are $k-1$ independent cycles of this type. The old A_α are still included in relative homology, but in fact the two sets are related by[3]

$$A_\alpha \cong C_{\alpha\beta} B^\beta \qquad (7.2.28)$$

where $C_{\alpha\beta}$ is the Cartan matrix of $\mathrm{su}(k)$. So in fact,

$$h_2(\mathrm{GH}_k, \partial\mathrm{GH}_k, \mathbb{Z}) = k - 1\,. \qquad (7.2.29)$$

A similar discussion holds in cohomology. There are two sets of forms [173], dual to the A_α and B^α in the sense of Poincaré duality (4.1.145).

Singularities

We saw already in many instances that shrinking a cycle leads to singularities. For example, a blowup (Section 6.3.5) replaces a singularity with a cycle, and the inverse process creates a singularity from a smooth space. We saw the conifold singularity as arising from shrinking a two-cycle in Section 6.4.3, and from shrinking a three-cycle in Section 6.3.6.

For the present Gibbons–Hawking geometries, this can happen if we make two or more nuts coincide, which shrinks the A_α arising from a segment joining them. Indeed, near such a "multiple" nut the solution will now look like $V \sim \frac{n}{r}$. We can repeat (7.2.21) almost identically with this n; defining this time $\rho = \sqrt{nr}$, the result only changes in that $\psi \to \psi/n$. If $n = k$, so that all nuts have coalesced, we can take the periodicity $\Delta\psi = 4\pi n$, and the result is again topologically $\mathbb{R}^4$. But if $n < k$ and there are nuts with multiplicity one, then the periodicity is already fixed to be $\Delta\psi = 4\pi$, and near the multiple nut we have to conclude that the topology is

$$\mathbb{R}^4/\mathbb{Z}_n\,. \qquad (7.2.30)$$

The way $\mathbb{Z}_n$ acts on $\mathbb{R}^4$ is precisely the same as in (6.3.31), as can be seen by writing $\psi \to \psi + \delta\psi$ as an action on complex coordinates:

$$(w_1, w_2) \to \left(\mathrm{e}^{\mathrm{i}\delta\psi} w_1, \mathrm{e}^{-\mathrm{i}\delta\psi} w_2\right), \qquad (7.2.31)$$

compatibly with (7.2.22) and (4.2.50).

If we take $V_0 = 0$ in (7.2.26), at $r \to \infty$ the function $V \sim \frac{k}{r}$; so the metric (7.2.17) at large distances looks like the metric on $\mathbb{R}^4/\mathbb{Z}_n$. At smaller distances, if the n nuts are not coincident, the metric is nonsingular, since near every nut (7.2.21) holds. So we have obtained a metric on a resolution:

$$\widehat{\mathbb{C}^2/\mathbb{Z}_n}\,. \qquad (7.2.32)$$

[3] For an explicit explanation of this point, see [172, app. B].

This has in fact the same topology as the toric resolution described after (6.4.36) (see also Figure 6.5 for $n = 3$). The case $n = 2$ can also be written in other notable coordinate systems, and is known as *Eguchi–Hanson* metric.

On the other hand, if $V_0 > 0$ the Gibbons–Hawking metric (7.2.17) is asymptotic to $\mathbb{R}^3 \times S^1$ at $r \to \infty$. In this case, we obtain the Euclidean counterpart of a class of black hole solutions called Taub–Newman–Unti–Tamburino (*Taub–NUT*). This partially explains the name “nut” we introduced earlier.

In string theory, $\mathbb{R}^6 \times \text{GH}$ is a solution, since $R_{MN} = 0$; the $V_0 > 0$ solutions are T-dual to NS5-branes, and in this context the nuts are called *KK5-monopoles* (Exercise 7.2.5) [174]. In M-theory, we can define a KK6 as $\mathbb{R}^6 \times \text{GH}$.

7.2.3 D6 and O6 lift

We pause here our purely geometric discussion to discuss an application of the previous subsection to D-brane physics. Long ago, in Section 1.4.1, we pointed out that the lift of D6-branes and O6-planes should be a purely geometrical solution, since C_1 lifts to M-theory as part of the metric. These turn out to involve metrics of SU(2) holonomy.

D6 lift

Our starting point is (1.3.59), which we reproduce here for $p = 6$:

$$\begin{aligned} ds^2 &= h^{-1/2} ds^2_{\mathbb{R}^7} + h^{1/2} ds^2_{\mathbb{R}^3} , \qquad e^{\phi} = g_s h^{-\frac{3}{4}} , \\ F_2 &= \frac{l_s}{4} \epsilon_{ijk} y^i dy^j \wedge dy^k , \end{aligned} \tag{7.2.33}$$

where $y^m = x^{6+m}$, $r^2 = y^m y^m$, and $h = 1 + \frac{r_0}{r}$, with

$$r_0 = \frac{g_s l_s}{2} . \tag{7.2.34}$$

Now we lift to eleven dimensions with (1.4.5a). We see that $A_3 = 0$, and

$$ds^2_{11} = g_s^{-2/3} ds^2_{\mathbb{R}^7} + g_s^{-2/3} h ds^2_{\mathbb{R}^3} + g_s^{4/3} h^{-1} (dx^{10} - C_1)^2 . \tag{7.2.35}$$

This is a direct product of flat $\mathbb{R}^7$ and a four-dimensional metric, which we recognize to be of the form (7.2.17) upon rescaling the y_i.

So we can apply all the geometric intuition from the previous subsection. We already observed after (4.2.37) that the M-theory circle is fibered over the IIA spacetime. Here we see that D6s are loci where the circle shrinks, and the bundle becomes a fibration; in other words, *D6-branes are M-theory's nuts*. The discussion around (7.2.21) shows that for a single D6, the lifted metric is completely smooth; (7.2.32) shows that the lift of N D6s has an $\mathbb{R}^4/\mathbb{Z}_N$ singularity.

Taking h in (7.2.33) to be a different harmonic function of the $\mathbb{R}^3$ one can also obtain metrics where the D6s are parallel but separated. The lift of these corresponds to GH metrics with (7.2.26).

O6 lift

The Op_- supergravity solution is obtained from that for Dp-branes, (1.3.59), by replacing the function h with (1.4.55). As we observed there, the resulting metric has a hole $r < 2^{p-5} r_0$, where it becomes purely imaginary, but really it gets strongly coupled and curved well before one gets there. In particular, for O6$_-$-planes we need to use (7.2.33) with

$$h = 1 - 2\frac{r_0}{r}, \tag{7.2.36}$$

and the hole is $r < 2r_0$.

Lifting this O6 metric directly as in (7.2.35) would not make much sense in the strongly coupled region. Fortunately, there exists a solution to $d = 11$ supergravity that is a strong candidate for the lift:

$$\begin{aligned} ds^2_{11} &= g_s^{-2/3} ds^2_{\mathbb{R}^7} + \frac{g_s^{4/3} l_s^2}{4} ds^2_{\mathrm{AH}}, \\ ds^2_{\mathrm{AH}} &= \frac{b^2}{r^2} dr^2 + a^2 \lambda_1^2 + b^2 \lambda_2^2 + c^2 \lambda_3^2, \end{aligned} \tag{7.2.37}$$

so again a direct product as in (7.2.35). ds^2_{AH} is a smooth four-dimensional hyper-Kähler metric called *Atiyah–Hitchin metric* [175]. λ^a are the left-invariant forms in (4.4.27) on S^3; from this, we see that the solution has SO(3) isometry. a, b, c are functions of r/r_0,[4] with range

$$r \in [\pi r_0, \infty] \tag{7.2.38}$$

and with asymptotic behaviors:

$$\begin{aligned} a &\sim \frac{r}{r_0}\sqrt{h} - p_a e^{-r/r_0} + O(e^{-2r/r_0}), & c &\sim \frac{2}{\sqrt{h}} + O(e^{-2r/r_0}), \\ b &\sim \frac{r}{r_0}\sqrt{h} - p_b e^{-r/r_0} + O(e^{-2r/r_0}), & r &\to \infty; \end{aligned} \tag{7.2.39a}$$

$$\begin{aligned} a &\sim \frac{2}{r_0}(r - \pi r_0) + O(r - \pi r_0)^2, & c &\sim \pi - \frac{1}{2r_0} r - \pi r_0 + O(r - \pi r_0)^2, \\ b &\sim \pi + \frac{1}{2r_0} r - \pi r_0 + O(r - \pi r_0)^2, & r &\to \pi r_0, \end{aligned} \tag{7.2.39b}$$

where p_a, p_b are polynomials in r/r_0.

We can reduce (7.2.37) to IIA along the angle called ψ in (4.4.27) using (1.4.5a), essentially following (7.2.35) backward:

$$ds^2_{10} = \frac{c}{2} ds^2_{\mathbb{R}^7} + \frac{g_s^2 l_s^2 c}{8} \left(\frac{b^2}{r^2} dr^2 + a^2 \lambda_1^2 + b^2 \lambda_2^2 \right) \tag{7.2.40a}$$

$$\overset{r\to\infty}{\sim} \frac{1}{\sqrt{h}} ds^2_{\mathbb{R}^7} + \sqrt{h}\left(dr^2 + r^2 ds^2_{S^2}\right) + O(e^{-r/r_0}). \tag{7.2.40b}$$

The dilaton is given by

$$e^{\phi} = 2^{-3/2} g_s c^{3/2} \overset{r\to\infty}{\sim} g_s h^{-3/4} + O(e^{-r/r_0}). \tag{7.2.41}$$

So at large distances, the solution is asymptotic to the O6 solution obtained from (7.2.33) and (7.2.36). But there are corrections; from (1.3.25) and (1.3.28), $T_{\mathrm{D0}} = \frac{1}{2\pi l_s g_s} = \frac{1}{4\pi r_0}$, so it is natural to interpret these as coming from D0 instantons wrapping the M-theory circle.

[4] For more details, see [175; 176, sec. 3; 177, sec. 3.1]. We are using here the coordinates in the second reference.

As r gets smaller, these corrections change the naive O6 solution completely; (7.2.40a) no longer solves the IIA supergravity equations of motion, but since it came from a smooth solution in $d = 11$, it should be a solution of the equations of motion of *full IIA string theory*. To see how the solution manages to be smooth near $r = \pi r_0$, change coordinates so that $(\lambda_1, \lambda_2, \lambda_3) = (-\tilde{\lambda}_3, \tilde{\lambda}_2, \tilde{\lambda}_1)$, and use (7.2.39b):

$$\mathrm{d}s^2_{\mathrm{AH}} \sim \frac{1}{r_0^2}(\mathrm{d}r^2 + 4(r - \pi r_0)^2 D\tilde{\psi}^2) + \pi^2 \mathrm{d}s^2_{S^2} \,. \tag{7.2.42}$$

If we shift $r \to r + \pi r_0$ and take the periodicity $\Delta\tilde{\psi} = \pi$, we recognize this as the metric of an $\mathbb{R}^2$ (the first two terms) fibered over an S^2.

So M-theory resolves the singularity of the O6. The naive hole $r \leq r_0$ is replaced by a smooth end of spacetime at $r = \pi r_0$. We will see in Section 9.4.3 another example where we can resolve the supergravity hole of an O-plane.

7.2.4 Hyper-Kähler manifolds

As we see from Table 5.1, a *hyper-Kähler* manifold is defined as having holonomy $\mathrm{Sp}(d/4)$. We already treated the case $d = 4$ in the previous two subsections, where $\mathrm{Sp}(1) \cong \mathrm{SU}(2)$; the differential conditions for a hyper-Kähler manifold and for a Calabi–Yau, (7.2.44) and (7.1.2), coincide. Of the higher-dimensional hyper-Kähler manifolds, obviously only those for $d = 8$ can be part of spacetime for string theory. (Spaces with even higher dimensions do actually play a role as field theory spaces for low-energy effective actions.)

We discussed $\mathrm{Sp}(d/4)$-structures in Section 5.1.4. For the corresponding special holonomy, we just impose that the tensors defining the structure be covariantly constant:

$$\nabla I_a = 0 \,. \tag{7.2.43}$$

We did not develop the theory of intrinsic torsion for this case, but by analogy with the previous cases (in particular with Calabi–Yau geometry) it is by now not surprising that (7.2.43) is equivalent to

$$\mathrm{d}J_a = 0 \qquad \text{(hyper-Kähler)}. \tag{7.2.44}$$

(This is consistent with the fact that the (J_a, Ω_a) structures discussed after (5.1.61) define $\mathrm{SU}(d/2)$ holonomy.)

As we remarked already while discussing $\mathrm{Sp}(d/4)$-structures in Section 5.1.4, the spinorial point of view is clumsier: the structure is associated with the $(d/4 + 1)$ spinors (5.1.70), which in the $\mathrm{Sp}(d/4)$-holonomy case all become covariantly constant. The cases relevant for us are mostly $d = 4$, where there are two such spinors, and $d = 8$, where there are three; see (5.1.71) for a local model.

Higher dimensions

Finding compact examples in higher dimensions is rather nontrivial. One is

$$(\mathrm{K3})^n / S_n \,, \tag{7.2.45}$$

where S_n is the group of permutations and resolving its singularities [178–180].

A way to obtain noncompact examples is via *hyper-Kähler quotients*, similar to the Kähler quotients described in Section 6.4.2. Now for every vector field that leaves the J_a invariant, instead of a moment map we have a hypermoment map, defined as a triple of functions μ^a_ξ such that

$$\iota_\xi J_a = \mathrm{d}\mu^a_\xi \,. \tag{7.2.46}$$

An example is already provided by our $\mathbb{R}^4 \to \mathbb{R}^3$ projection (7.2.22), where the y_a are the components of a hypermoment map for the vector field ∂_ψ, as can be readily seen from (7.2.16).

With this point of view, one can obtain a quaternionic analogue of the Kähler quotients of Section 6.4.2. In particular, we can take an action of $\mathrm{U}(1)^k$ on $\mathbb{H}^{N+k} \cong \mathbb{R}^{4(N+k)}$ by a charge matrix C, fix the μ_a to certain values $c_{a\alpha}$, $\alpha = 1, \ldots, k$, and quotient it by the $\mathrm{U}(1)^k$:

$$\mu_a^{-1}(c_{a\alpha})/\mathrm{U}(1)^k \,. \tag{7.2.47}$$

This is similar to the right-hand side of (6.4.25), and it defines the class of *hypertoric manifolds*. Explicitly, (7.2.47) reads

$$\frac{\{\sum_I C_{\alpha I} 2 Z^I \bar{W}_I = c_{1\alpha} + \mathrm{i} c_{2\alpha} \,,\ \sum_I C_{\alpha I}(|Z^I|^2 - |W_I|^2) = c_{3\alpha}\}}{\{(Z^I, W_I) \sim (\mathrm{e}^{\mathrm{i}C_{I\alpha}} Z^I, \mathrm{e}^{\mathrm{i}C_{I\alpha}} W_I)\}} \tag{7.2.48}$$

in terms of the charge matrix $C_{I\alpha}$, and using (7.2.22) as a model. Once again one can introduce a matrix W, defined as the kernel of the charge matrix C as in Section 6.4.2, and introduce a hypertoric fan whose rays are its column vectors w_I.

Remarkably, there is an explicit higher-dimensional metric on the hyper-Kähler quotients (7.2.47), generalizing (7.2.17) [181, 182]. We introduce coordinates y^I_a, ψ^I, for $I = 1, \ldots, d/4$, and

$$J_a = \mathrm{d}y^I_a \wedge (\mathrm{d}\psi_I + \sigma_I) + \frac{1}{2}\epsilon_{abc} V_{IJ} \mathrm{d}y^I_b \wedge \mathrm{d}y^J_c \tag{7.2.49}$$

extending (7.2.16). One can check that

$$V_{IJ} = V^0_{IJ} + \sum_{\alpha=1}^{k} \frac{w^\alpha_I w^\alpha_J}{|w^\alpha_K (y^K - y^K_\alpha)|} \,, \qquad |w^\alpha_K (y^K - y^K_\alpha)|^2 \equiv \sum_{a=1}^{3} (w^\alpha_K (y^K_a - y^K_{a\alpha}))^2$$

$$\mathrm{d}\sigma_I = -\frac{1}{2}\epsilon_{abc} \sum_{\alpha=1}^{k} \frac{w^\alpha_L (y^L_a - y^L_{a\alpha})}{|w^\alpha_K (y^K - y^K_\alpha)|^3} w^\alpha_I w^\alpha_J w^\alpha_K \mathrm{d}y^J_b \wedge \mathrm{d}y^K_c \tag{7.2.50}$$

solves $\mathrm{d}J_a = 0$. This is directly inspired by (7.2.26); the w_I are the column vectors of the matrix W mentioned earlier. The corresponding metric is

$$\mathrm{d}s^2 = V_{IJ} \mathrm{d}y^I_a \wedge \mathrm{d}y^J_a + (V^{-1})^{IJ} (\mathrm{d}\psi_I + \sigma_I)(\mathrm{d}\psi_J + \sigma_J) \,. \tag{7.2.51}$$

We will see some examples in Section 7.5.1.

Quaternionic Kähler

We mention here a type of geometry which is *not* Ricci-flat – it would not belong to this chapter, but it arises as a natural generalization of hyper-Kähler geometry.

In Section 5.1.4, we also saw a G-structure defined by the sphere of almost complex structures $x_a I_a$ rather than by the individual I_a; the stabilizer was $G = \mathrm{Sp}(d/4) \cdot \mathrm{Sp}(1)$. To obtain a manifold with this holonomy, we don't require the I_a to be individually covariantly constant, but rather the covariant derivative ∇_m to close on the space of I_a: namely,

$$\nabla_m(I_a)^n{}_p = -\epsilon_{abc}A_{bm}(I_c)^n{}_p \qquad (7.2.52)$$

for some SU(2) connection A. Such a manifold has $\mathrm{Sp}(d/4)\cdot\mathrm{Sp}(1)$ holonomy, and is called *quaternionic-Kähler*. Correspondingly, (7.2.44) is modified to

$$\mathrm{d}J_a + \epsilon_{abc}A_b \wedge J_c = 0\,. \qquad (7.2.53)$$

In general, none of the J_a is closed, so in spite of the name a quaternionic-Kähler manifold is in general *not Kähler*.

Acting on (7.2.53) with d one also gets $\epsilon_{abc}F_bJ_c = 0$, where $F_b = \mathrm{d}A_b + \frac{1}{2}\epsilon_{bcd}A_cA_d$ is the usual nonabelian field-strength (4.2.21). So

$$F_a = \lambda J_a \qquad (7.2.54)$$

for some λ (which is fixed to be constant by (4.2.24)). Since A is related to the Levi-Civita connection ∇_m, (7.2.54) implies constraints on the Riemann tensor. In particular, one sees that the Ricci tensor is proportional to the metric (so that the space is Einstein), and that Riemann vanishes when acting on the space of two-forms spanned by the J_a.

For $d = 4$, actually

$$\mathrm{sp}(1)\oplus\mathrm{sp}(1)\cong\mathrm{so}(4)\,, \qquad (7.2.55)$$

which is the holonomy algebra defined by a metric alone. Indeed, (7.2.53) becomes automatic (see Exercise 5.5.1). So the quaternionic-Kähler condition would end up being trivial in $d = 4$. Many authors instead treat $d = 4$ separately, and define it by (7.2.54). This can be again rephrased in terms of the Riemann tensor, recalling (4.1.30) and Exercise 4.3.11. The upshot is that a quaternionic-Kähler four-manifold is one that is Einstein (4.5.2) and such that the Weyl tensor is self-dual, $W^- = 0$. The only such manifolds that are compact and have no singularities are [183]

$$S^4\,, \qquad \mathbb{CP}^2 \qquad (7.2.56)$$

for positive curvature; and T^4, K3, or one of its orbifolds for zero curvature (which takes us back to the hyper-Kähler case). If one allows for orbifold singularities or noncompactness, there are many more examples; see, for example, [184, sec. 12.5]. We see from the two examples (7.2.56) that indeed quaternionic-Kähler geometry is not Ricci flat, as we anticipated.

7.2.5 G_2-holonomy manifolds

We recall that a G_2-structure can be defined by a single Majorana spinor η in $d = 7$, or by its three- and four-form bilinears ϕ, $*\phi$. We discussed the theory of intrinsic torsion for this case following (5.5.17). The case with G_2 holonomy corresponds to $D_m\eta = 0$, so we set $q_{mn} = 0$. From (5.5.21), we see that then the $X_i = 0$ too, which implies

$$\mathrm{d}\phi = 0\,, \qquad \mathrm{d}*\phi = 0\,. \qquad (7.2.57)$$

So this gives an alternative definition of G_2 holonomy.

It is easy to find manifolds with holonomy that is *contained* in G_2: for example, $\mathrm{CY}_6 \times S^1$ or $\mathrm{K3} \times T^3$ would have holonomies SU(3) and SU(2) respectively. The real challenge is to find examples where $\mathrm{Hol} = G_2$, and not one of its subgroups; according to our terminology back in (5.7.3), this is what we want to call a (proper) *G_2-holonomy manifold*, or directly *G_2-manifold*.

Finding explicit examples is harder than for the other holonomy groups we have discussed so far, basically because we no longer get help from complex geometry. Noncompact examples can be found by solving (7.2.57) explicitly in some cases with many symmetries. The first were found by considering metrics on the total spaces of some bundles [185, 186]:

$$\begin{array}{ccc} \mathbb{R}^4 \hookrightarrow S & \qquad & \mathbb{R}^3 \hookrightarrow \Lambda^2_+ \\ \downarrow & & \downarrow \\ S^3\,; & & B_4\,. \end{array} \tag{7.2.58}$$

The first is the spinor bundle over S^3; the second is the bundle of self-dual two-forms over a quaternionic-Kähler four-manifold B_4. In both these cases, the metrics depend on the radial coordinates on the fibers. We will see some more details in Section 7.5.2, where we will focus on limits of these metrics where the space becomes conical.[5]

The compact case is a lot more challenging and less explicit. We now review the two main strategies.

Resolving orbifolds

The first is to start from orbifolds, and then resolve their singularities; these are called *Joyce manifolds* [188, chap. 11–12]. A famous example is a

$$T^7/\mathbb{Z}_2^3 \tag{7.2.59}$$

with the torus defined by coordinates $x_i \sim x_i + 1$, and the three $\mathbb{Z}_2$ acting as

$$\begin{aligned} (x_1, \ldots, x_7) &\overset{\alpha}{\sim} (x_1, -x_2, -x_3, x_4, -x_5, -x_6, x_7) \\ &\overset{\beta}{\sim} \left(-x_1, x_2, \frac{1}{2} - x_3, -x_4, x_5, -x_6, x_7\right) \\ &\overset{\gamma}{\sim} \left(x_1, x_2, x_3, -x_4, \frac{1}{2} - x_5, \frac{1}{2} - x_6, -x_7\right). \end{aligned} \tag{7.2.60}$$

All three actions leave invariant ϕ and $*\phi$ in (5.1.50), with $e_i = \mathrm{d}x_i$. Quotienting by only one of these would produce a $T^3 \times T^4/\mathbb{Z}_2$, where the $T^4/\mathbb{Z}_2$ is exactly of the type in (7.2.11); resolving would then produce $T^3 \times \mathrm{K3}$, with holonomy SU(2). This seems promising, but as we said earlier, we actually want manifolds with holonomy in G_2, not a subgroup; using three copies of $\mathbb{Z}_2$ ensures that.

As we saw already in (7.2.11), a sign reversal of four coordinates has 16 fixed loci; for us, each of these loci is now a T^3. One might think that the three actions would now give rise to 48 fixed loci, which might moreover intersect; at such intersections, the resolution would presumably become a lot more complicated than a blowup.

[5] A more recent large class of noncompact examples is [187], which have an interpretation in terms of D6-branes on a Calabi–Yau (see Section 10.4.1).

However, in fact the 16 fixed loci of α are acted upon by β and γ: there are no points that are fixed by more than one action, and the number of fixed loci of α is only $16/4 = 4$. The same is true for the fixed loci of β and γ. Summing up, (7.2.60) is chosen carefully so that the $\mathbb{Z}_2$ singularities don't intersect; there are 12 of them, each being a T^3. Because of these, the resolution of each of these loci is basically just a blowup; this produces a smooth G_2-holonomy manifold

$$\widetilde{T^7/\mathbb{Z}_2^3}\,. \tag{7.2.61}$$

We can also compute the Betti numbers h_k. The cohomology of T^7 is spanned by the forms $\mathrm{d}x_i$ and their wedge products. As we argued earlier, blowing up replaces the 12 singularities with copies of $S^2 \times T^3$.

- H^1: none of the $\mathrm{d}x_i$ is invariant under (7.2.60). The S^2s don't contribute any new one-forms, so $h_1(\widetilde{T^7/\mathbb{Z}_2^3}) = 0$.
- H^2: again, none of the $\mathrm{d}x_i \wedge \mathrm{d}x_j$ is invariant under (7.2.60). Each of the $S^2 \times T^3$ contributes a two-form, namely the Poincaré dual of the S^2; so $h_2(\widetilde{T^7/\mathbb{Z}_2^3}) = 12$.
- H^3: this time there are seven three-forms $\mathrm{d}x_i \wedge \mathrm{d}x_j \wedge \mathrm{d}x_k$, which are invariant under (7.2.60): they are exactly the seven summands of (5.1.50). Moreover, now each of the $S^2 \times T^3$ introduces three three-forms, namely the Poincaré dual of the S^2 wedged with the three generators of the T^3. So $h_3(\widetilde{T^7/\mathbb{Z}_2^3}) = 7 + 3 \cdot 12 = 43$.

A generalization of this idea was recently proven in [189]. It applies to any M_7 with Hol $\subset G_2$ with a $\mathbb{Z}_2$ involution that preserves the G_2 form: $R\colon M_7 \to M_7$, $R^2 = 1$, $R^*\phi = \phi$. If the fixed locus L has a nowhere-vanishing harmonic one-form α_1, then one can again resolve the singularities of $M_7/\mathbb{Z}_2$. This could be applied, for example, to $M_7 = \mathrm{CY}_6 \times S^1$ or to any G_2-manifold produced with other methods; the hard part is to show that α_1 exists on L.

Twisted connected sums

A second strategy to obtain compact G_2-holonomy manifolds has emerged more recently; it goes under the name of *twisted connected sums*.

A Calabi–Yau Y is said to be *asymptotically cylindrical* if outside a compact set the metric is asymptotic to that of $\mathrm{K3} \times \mathbb{C}^*$, defined by the forms

$$J \sim j + \frac{\mathrm{i}}{2}\frac{\mathrm{d}z \wedge \mathrm{d}\bar{z}}{z\bar{z}}\,, \qquad \Omega \sim \omega \wedge \frac{\mathrm{d}z}{z}\,, \tag{7.2.62}$$

adapted from (3.4.46); (j, ω) define the K3, and z is the coordinate on $\mathbb{C}^*$. The asymptotic agreement should be such that the differences and all their gradients vanish faster than some (fixed) inverse power of $|z|$.

With such a Y, the product $Y \times S^1$ asymptotes to

$$\mathrm{K3} \times (\mathbb{R}_+)_t \times S^1_\theta \times S^1_x\,; \tag{7.2.63}$$

we have decomposed $z = \mathrm{e}^{t+\mathrm{i}\theta}$, and x is the coordinate on the extra S^1. The holonomy is $\mathrm{SU}(3) \subset G_2$; by (5.1.52), (5.1.53), and (7.2.62),

$$\phi \sim \mathrm{Re}\omega \wedge \mathrm{d}\theta + \mathrm{Im}\omega \wedge \mathrm{d}t + j \wedge \mathrm{d}x + \mathrm{d}t \wedge \mathrm{d}\theta \wedge \mathrm{d}x\,. \tag{7.2.64}$$

Now consider two asymptotically cylindrical Calabi–Yau manifolds Y_+, Y_-. We would like to "glue" $Y_+ \times S^1$ to $Y_- \times S^1$ so that the holonomy is properly G_2 and not in SU(3). This is accomplished by swapping the two S^1 in (7.2.63): in other words, we would glue the two copies so that

$$\theta_+ = x_- , \qquad \theta_- = x_+ . \tag{7.2.65}$$

We should also glue $t_+ = t_0 - t_-$ for some large constant t_0. Looking at (7.2.64), we see that it is possible to identify the ϕ of $Y_\pm \times S^1$ if there is a diffeomorphism $r: Y_+ \to Y_-$ such that

$$r^* j_- = \mathrm{Re}\omega_+ , \qquad r^*\mathrm{Re}\omega_- = j_+ , \qquad r^*\mathrm{Im}\omega_- = -\mathrm{Im}\omega_+ , \tag{7.2.66}$$

where $(j_\pm, \omega_\pm)$ are the hyper-Kähler structures on the two copies of the K3. *Kovalev's theorem* [190] says that if there are two asymptotically cylindrical Calabi–Yau manifolds $Y_\pm$ with a diffeomorphism (7.2.66), then $Y_\pm \times S^1$ can indeed be "glued" by identifying their asymptotically cylindrical region as in (7.2.65) and (7.2.66), and that one can deform the resulting space to a smooth G_2-holonomy manifold.

While the conditions we have imposed might seem complicated to meet, there are in fact various constructions of asymptotically cylindrical Calabi–Yau manifolds using algebraic-geometric techniques. As of this writing, there are already millions of G_2-holonomy manifolds obtained with this method [191].

Exercise 7.2.1 Use the Hirzebruch signature theorem (4.2.74) to check (7.2.7).

Exercise 7.2.2 Use the same logic in (7.2.4) to compute the Euler characteristic of the quintic (7.1.25); check that the result agrees with Section 7.1.3.

Exercise 7.2.3 Find other ways to realize an algebraic K3, along the lines of (7.1.30). Count the complex moduli; check that you always get < 20.

Exercise 7.2.4 Check that (7.2.49) and (7.2.50) solve the hyper-Kähler conditions $\mathrm{d}J_a = 0$.

Exercise 7.2.5 When $V_0 \neq 0$, T-dualize (7.2.17) and (7.2.26) along ψ. Show that the resulting solution is a smeared version (Section 4.2.6) of the NS5-brane solution (1.3.65). (String corrections to the usual Buscher rules unsmear the NS5-brane [128].)

7.3 Killing spinors

7.3.1 Cones

Given a manifold L_d with a line element $\mathrm{d}s^2_{L_d}$, its *cone* $C(L_d)$ is a $(d+1)$ manifold with topology

$$\frac{\mathbb{R}_+ \times L_d}{(0,p) \sim (0,p')} \tag{7.3.1}$$

(in notation similar to (4.5.8)), and line element (see Figure 7.1)

$$\mathrm{d}s^2_{C(L_d)} = \mathrm{d}r^2 + r^2 \mathrm{d}s^2_{L_d} . \tag{7.3.2}$$

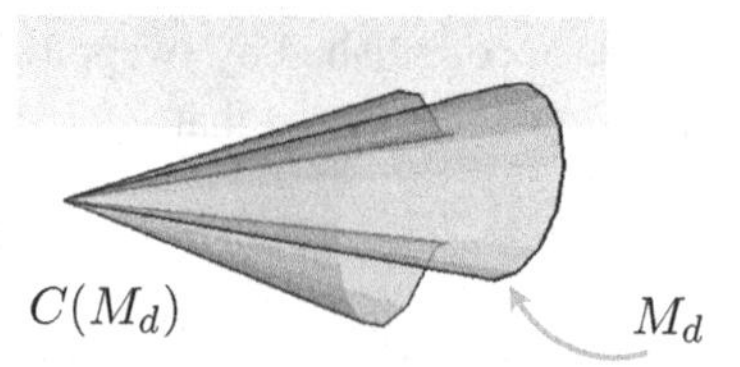

Figure 7.1 A cartoon of a conical manifold $C(L_d)$.

It has a conformal transformation $r \to Rr$, called *homothety*; its generator $r\partial_r$ is then a conformal Killing vector, sometimes called *Euler vector field*, which appeared already in flat space in Sections 4.3.4 and 6.2.1. More specifically,

$$(L_{r\partial_r} g^{C(L_d)})_{mn} = 2g^{C(L_d)}_{mn} \,. \tag{7.3.3}$$

The manifold L_d is called the *link* or the *base* of the cone.

In the rest of this chapter, we will study conical manifolds $C(L_d)$ with special holonomy. We will focus on cases that are defined by a covariantly constant spinor η, and hence that are also Ricci-flat. For all these manifolds, lowering the index of $r\partial_r$ with (7.3.2), we obtain rdr, which is closed; so (4.3.39) gives

$$L_{r\partial_r} \eta = 0 \,. \tag{7.3.4}$$

As a consequence, bilinears $B_{m_1 \ldots m_k} \equiv \eta^\dagger \gamma_{m_1 \ldots m_k} \eta$ only transform under $r\partial_r$ because of the transformation of the vielbein; just like in the flat space example of (4.3.53), from (4.3.50) we obtain $(L_{r\partial_r} B)_{m_1 \ldots m_k} = kB_{m_1 \ldots m_k}$.

7.3.2 From covariantly constant to Killing spinors

The condition that a spinor η is covariantly constant can be reformulated in terms of L_d [192].

Covariantly constant spinors on a cone

When d is odd, the gamma matrices in $d + 1$ dimensions can be expressed in terms of those in d dimensions by adapting (2.2.22) to Euclidean signature and generalizing the three-dimensional part there to any odd d (as partially illustrated in the rest of Section 2.2.3). The spin connection can be taken from (4.3.63) in Exercise 4.3.3. The spinorial covariant derivative (4.3.20) on $C(L_d)$ becomes

$$D^{C(L_d)}_m = D_m + \frac{\mathrm{i}}{2} \sigma_3 \otimes \gamma_m \,, \qquad D^{C(L_d)}_r = \partial_r \,, \tag{7.3.5}$$

where D_m, γ_m are on L_d. For all the cases we have seen in even dimension, special holonomy is defined by a chiral spinor. The chirality matrix in the basis we have used is $\pm\sigma_3 \otimes 1$, with the sign depending on the c chosen in (2.1.17). We may then assume $\eta_{C(L_d)} = \binom{\eta}{0}$ or $\binom{0}{\eta}$. So

$$D^{C(L_d)} \eta_{C(L_d)} = 0 \,, \qquad \eta_{C(L_d)} \text{ chirality } \pm 1 \quad \Rightarrow \quad D_m \eta = \pm \frac{\mathrm{i}}{2} \gamma_m \eta \quad (d \text{ odd}). \tag{7.3.6}$$

When d is even, there is a chiral γ, and gamma matrices in $d + 1$ dimensions can be obtained by simply declaring that $\gamma_{d+1} = \gamma$ (adapting (2.2.14) to Euclidean signature). The spinors on $C(L_d)$ and L_d can be taken to be equal. The analogue of (7.3.6) now reads

$$D^{C(L_d)}\eta_{C(L_d)} = 0 \quad \Rightarrow \quad D_m\eta = -\frac{1}{2}\gamma_m\gamma\eta \quad (d \text{ even}). \tag{7.3.7}$$

In any d, a spinor satisfying

$$D_m\eta = \frac{\mu}{2}\gamma_m\eta \tag{7.3.8}$$

is called a *Killing spinor*. Rescaling the metric as $g_{mn} \to \lambda^2 g_{mn}$ rescales the factor in (7.3.8) as $\mu \to \mu/\lambda$. Thus, as long as μ is purely imaginary, it can always be brought to $\mu = \pm\mathrm{i}$ as in (7.3.6). When d is even, there is also alternative equation

$$D_m\eta = \mathrm{i}\frac{\mu}{2}\gamma_m\gamma\eta\,, \tag{7.3.9}$$

inspired by (7.3.7). Solutions of (7.3.8) and (7.3.9) are in one-to-one correspondence: this is because, given a basis γ_m of gamma matrices,

$$\gamma'_m = \mathrm{i}\gamma_m\gamma \tag{7.3.10}$$

also satisfy the Clifford algebra (2.1.3).

Killing spinors and vectors

One reason for the name "Killing spinor" is that the bilinear

$$K_m = \eta^\dagger\gamma_m\eta \tag{7.3.11}$$

is a Killing vector when μ is purely imaginary, as one can check by symmetrizing the two indices in (4.3.35). (For a more general complex μ, K_m is still a *conformal* Killing vector (4.1.68).)

Moreover, the Lie–Kosmann action of a Killing vector on a Killing spinor produces another Killing spinor. Let us show this explicitly, in the spirit of Section 4.1.6. We first compute

$$\begin{aligned}[L_K, \gamma_m] &\overset{(4.3.39)}{=} \left[K^n D_n + \frac{1}{4}\partial_n K_p \gamma^{np}, \gamma_m\right] \\ &\overset{(4.3.73),(2.1.13)}{=} K^n\Gamma^p_{mn}\gamma_p - \partial_{[m}K_{p]}\gamma^p\,,\end{aligned} \tag{7.3.12}$$

$$\begin{aligned}[D_m, L_K] &\overset{(4.3.39)}{=} \left[D_m, K^n D_n + \frac{1}{4}\partial_n K_p \gamma^{np}\right] \\ &\overset{(4.3.33),(4.3.74)}{=} \partial_m K^n D_n + \frac{1}{4}K^n R^{ab}{}_{mn}\gamma_{ab} + \frac{1}{4}\nabla_m\partial_n K_p\gamma^{np} \\ &\overset{(4.1.67)}{=} \partial_m K^n D_n\,.\end{aligned} \tag{7.3.13}$$

The latter is similar to (4.3.65). Now assume η satisfies (7.3.8):

$$\begin{aligned}\frac{2}{\mu}D_m(L_K\eta) &\overset{(7.3.13)}{=} \frac{2}{\mu}L_K(D_m\eta) + \frac{2}{\mu}\partial_m K^n D_n\eta \overset{(7.3.8)}{=} L_K(\gamma_m\eta) + \partial_m K^n\gamma_n\eta \\ &\overset{(7.3.12)}{=} \gamma_m L_K\eta + (\nabla_m K_p - \nabla_{[m}K_{p]})\gamma^p\eta \overset{(4.1.65)}{=} \gamma_m L_K\eta\,.\end{aligned} \tag{7.3.14}$$

In other words, $L_K\eta$ *is itself a Killing spinor.*

The finite analogue of this result is that the action of an isometry on a Killing spinor produces another Killing spinor. In other words, the space of Killing spinors is a representation of the isometry group. Using this and (7.3.11), one can build a superalgebra whose bosonic part is the Lie algebra of the isometry group; see, for example, [193, 194].

The Einstein condition

Just like a covariantly constant spinor implies Ricci-flatness, so a Killing spinor implies a condition on the Ricci tensor:

$$\begin{aligned} R_{mn}\gamma^m\eta &\overset{(5.7.5)}{=} 4\gamma^m D_{[m}D_{n]}\eta = 2\mu\gamma^m D_{[m}(\gamma_{n]}\eta) \overset{(4.3.73)}{=} \mu^2\gamma^m\gamma_{[n}D_{m]}\eta \\ &= \mu^2\gamma^m\gamma_{nm}\eta \overset{(3.1.17)}{=} (1-d)\mu^2\gamma_n\eta \,. \end{aligned} \tag{7.3.15}$$

Since the $\gamma^m\eta$ are independent (Section 3.4.7), it follows that

$$R_{mn} = (1-d)\mu^2 g_{mn}\,, \tag{7.3.16}$$

so L_d is Einstein manifold, as defined in (4.5.2). This also shows that only two cases are allowed:

- μ is real, in which case the Ricci tensor is negative definite.
- μ is purely imaginary, in which case the Ricci tensor is positive definite.

In particular, we saw that for the link L_d of a Ricci-flat cone $C(L_d)$, μ is purely imaginary, and thus the curvature is positive.

The Einstein property (7.3.16), with positive curvature, is also implied directly by Ricci-flatness of the cone, as one can see from (4.3.64) by taking $a = r$. In principle, however, an Einstein manifold need not have Killing spinors, just like a Ricci-flat manifold need not have covariantly constant spinors.

Later in this section, we will reformulate the existence of a Killing spinor in terms of forms, as we have often done in this book. For some later applications, however, it is sometimes useful to deal with the Killing spinors directly, especially in presence of symmetries. In this spirit, we will now solve (7.3.8) on a sphere.

7.3.3 Killing spinors on spheres

The cone $C(S^d)$ is of course just $\mathbb{R}^{d+1}$; the metric (7.3.2) is simply the flat metric in polar coordinates. In this case, the holonomy group is trivial, so by the previous subsection we should be able to find Killing spinors. Rather than tackling the general d right away, we will start with S^2.

Killing spinors on S^2

The metric is (4.1.46). We choose the vielbein

$$e^1 = \mathrm{d}\theta\,, \qquad e^2 = \sin\theta\mathrm{d}\phi\,. \tag{7.3.17}$$

The spin connection reads $\omega^{12} = -\omega^{21} = -\cos\theta d\phi$; so (4.3.20) gives $D_\theta = \partial_\theta$, $D_\phi = \partial_\phi - \frac{1}{2}\cos\theta\gamma_{12}$. The Killing spinor equation (7.3.8) becomes, for $\mu = 1$:

$$\partial_\theta \eta = \frac{i}{2}\gamma_1 \eta \tag{7.3.18a}$$

$$\partial_\phi \eta - \frac{1}{2}\cos\theta \gamma_{12}\eta = \frac{i}{2}\sin\theta\gamma_2\eta\,. \tag{7.3.18b}$$

First we solve (7.3.18a) with $\eta = \exp\left[\frac{i}{2}\theta\gamma_1\right]\eta_1$, where $\partial_\theta\eta_1 = 0$. Now (7.3.18b) can be rewritten as

$$0 = \left(\partial_\phi - \frac{1}{2}e^{i\theta\gamma_1}\gamma_{12}\right)\eta = e^{\frac{i}{2}\theta\gamma_1}\left(\partial_\phi - \frac{1}{2}\gamma_{12}\right)\eta_1\,. \tag{7.3.19}$$

(The first step is a mere rewriting of (7.3.18b), as one sees expanding the exponential; the second uses $\gamma_{12}e^{\frac{i}{2}\theta\gamma_1} = e^{-\frac{i}{2}\theta\gamma_1}\gamma_{12}$, which follows from $\gamma_{12}\gamma_1 = -\gamma_1\gamma_{12}$.) Equation (7.3.19) in turn can be solved by taking $\eta^1 = \exp\left[\frac{1}{2}\phi\gamma_{12}\right]\eta_0$, for η_0 now a constant. So in conclusion,

$$\eta = \exp\left[\frac{i}{2}\theta\gamma_1\right]\exp\left[\frac{1}{2}\phi\gamma_{12}\right]\eta_0 \qquad (S^2)\,. \tag{7.3.20}$$

This is the solution to (7.3.8) on S^2 with $\mu = 1$; it depends on the choice of a constant spinor η_0 in two dimensions, so it depends on two parameters. These correspond to the two constant spinors on $\mathbb{R}^3$. Solutions for $\mu = -1$ are obtained by conjugation $\eta \to \eta^c$.

As shown in (7.3.14), the action of an isometry on a Killing spinor produces another Killing spinor. Indeed, at the infinitesimal level, one can show

$$L_{\ell_a}\begin{pmatrix}\eta_\pm\\ \eta^c_\mp\end{pmatrix} = \frac{i}{2}\sigma^a_{\alpha\beta}\begin{pmatrix}\eta_\pm\\ \eta^c_\mp\end{pmatrix}, \tag{7.3.21}$$

where η_+ are obtained from $\eta_0 = \binom{1}{0}$ and $\binom{0}{1}$ respectively. The ℓ_a are also bilinears of the Killing spinors as in (7.3.11). This can be used to build a superalgebra extension of the so(3) isometry algebra.

Killing spinors on S^d

Going up in dimension, one can write a metric on S^d by recursion as in (4.5.4). A natural vielbein is

$$e^d = d\theta_d\,, \qquad e^{d-1} = \sin\theta_d d\theta_{d-1}\,, \qquad \ldots\,, \qquad e^1 = \Pi^d_{a=2}\sin\theta_i d\theta_1\,. \tag{7.3.22}$$

In this frame, we can extend the computation that led to (7.3.20) recursively. This leads to the following [195]:

$$\eta = \exp\left[\frac{i}{2}\theta_d\gamma_d\right]\exp\left[-\frac{1}{2}\theta_{d-1}\gamma_{d-1,d}\right]\ldots\exp\left[-\frac{1}{2}\theta_1\gamma_{12}\right]\eta_0 \quad (S^d)\,. \tag{7.3.23}$$

Alternative frames for S^3

It is important to stress that the preceding expressions for the Killing spinors are relative to the vielbein we have chosen. If one wants to work in a different frame, one has to change the spinor using (4.3.7) and (4.3.8).

For example, let us look at S^3, where several other frames exist that are more convenient than (7.3.22). We can take the vielbein to coincide with the left-invariant forms (4.4.5):

$$e^a = \frac{1}{2}\lambda^a . \tag{7.3.24}$$

(The factor corresponds to a unit radius, as we see following (4.4.27).) With this choice, by (4.3.55) and (4.4.6b) the anholonomy coefficients $F^a{}_{bc} = -f^a_{bc} = -2\epsilon^a{}_{bc}$. Since we are in three dimensions $\gamma_{ab} = \mathrm{i}\epsilon_{abc}\gamma^c$, by (3.2.29) (or by using the basis $\gamma_a = \sigma_a$ of Pauli matrices). Then the $\frac{1}{4}\omega_m^{ab}\gamma_{ab}$ in the spinorial covariant derivative simplifies to $-\frac{\mathrm{i}}{2}\gamma_m$, which when brought to the right-hand side matches (7.3.8) for $\mu = 1$. So with this vielbein there are Killing spinors that are just constant:

$$\eta = \eta_0 . \tag{7.3.25}$$

(The Killing spinors with $\mu = -1$ are constant in the vielbein $e^a = \rho^a$ instead.)

Another convenient vielbein is adapted to the expression of the metric in Hopf coordinates (4.2.51):

$$e^1 = \frac{1}{2}\mathrm{d}\theta , \qquad e^2 = \frac{1}{2}\sin\theta \mathrm{d}\phi , \qquad e^3 = \mathrm{d}\psi + A . \tag{7.3.26}$$

To obtain the Killing spinor in these coordinates, we observe from (4.4.27) that the λ^a are related to (7.3.26) by a simple rotation matrix (and a rescaling by 1/2). We know that the Killing spinor is constant with $e^a = \lambda^a$, so we can use (4.3.8) to find

$$\eta = \exp\left[\psi\gamma_{12}\right]\eta_0 . \tag{7.3.27}$$

In the same vein, one can in principle obtain any of these expressions by starting from spinors on $\mathbb{R}^{d+1} = C(S^d)$. These are constant in the vielbein $e^a = \delta^a_m \mathrm{d}x^m$, but become more complicated once we change to a vielbein that separates the radial direction from the ones tangent to the sphere.

Exercise 7.3.1 Check that (7.3.11) is a (conformal) Killing vector if μ is real (complex).

Exercise 7.3.2 Using (4.3.73) and (7.3.12), show that if K is a Killing vector, the spinorial Lie derivative L_K commutes with the Dirac operator (4.3.21).

Exercise 7.3.3 Compute what happens to the Killing spinor (7.3.20) when we exchange $e^1 \to e^2$.

Exercise 7.3.4 Given two cones $C(M_1)$, $C(M_2)$, write their product as a cone over a third manifold M_{12}. Show that it is of the join form (4.5.7).

7.4 Sasaki–Einstein manifolds

7.4.1 Cones and $\mathrm{SU}(d)$-structures

We now consider the case where $C(L_d)$ has holonomy Hol $= \mathrm{SU}((d+1)/2)$, and hence is called a *conical Calabi–Yau*; by definition, L_d is said to be a *Sasaki–Einstein* manifold. Indeed, we know already from (7.3.16) that L_d is Einstein. (L_d is called *Sasaki* if $C(L_d)$ is only Kähler.) The cases most relevant for us are $d = 5$, where

$C(L_5)$ has holonomy SU(3), and $d = 7$, where $C(L_7)$ has holonomy SU(4). To fix ideas, we will consider here mostly the $d = 5$ case, with occasional comments for general d.

Rather than work with spinors, we will use the formulation (7.1.2) of the Calabi–Yau condition, $\mathrm{d}J = \mathrm{d}\Omega = 0$. Recall that (J, Ω) define an SU(3)-structure, which we discussed in Section 5.1.2. Importantly, Yau's theorem only applies to compact manifolds, so it cannot be used in the conical case. So we will have to use other techniques to find examples.

Structure on the link

By (7.3.4) and the discussion that follows it,

$$L_{r\partial_r} J = 2J\,, \qquad L_{r\partial_r}\Omega = 3\Omega\,. \tag{7.4.1}$$

(In general d, the weight of Ω becomes $(d+1)/2$.) Acting with the complex structure on $r\partial_r$ immediately gives a second vector field:

$$\xi = I^t\, r\partial_r\,. \tag{7.4.2a}$$

(Recall footnote 8 in Chapter 5 following (5.6.16): explicitly this reads $\xi^m = rI_r{}^m$.) Lowering the index of ξ with the metric, we obtain a one-form $r^2\eta_m = g_{mn}\xi^n = rg_{mn}I_r{}^n = rJ_{rm}$. In other words,

$$r^2\eta = \iota_{r\partial_r} J\,. \tag{7.4.2b}$$

We can also write $r^2\eta_m = rg_{mn}I_r{}^n = -rI_m{}^n g_{nr} = -rI_m{}^r$, which gives the alternative characterization

$$r^2\eta = rI\cdot \mathrm{d}r = r\mathrm{d}^{\mathrm{c}} r \tag{7.4.2c}$$

in notation introduced in (5.1.24) and (5.3.32). Equation (7.4.2) extends (6.2.8) for $\mathbb{CP}^N$. The definitions are such that $\iota_\xi\eta = 1$, and hence

$$L_\xi \eta = 0\,. \tag{7.4.3}$$

Moreover, $\iota_\xi \mathrm{d}r = rI_r{}^r = 0$, so ξ is tangent to the level sets of r, or in other words to L_5.

It follows from the definition (7.4.2c) of η that the one-form $v = \mathrm{d}r + \mathrm{i}r\eta$ is a $(1,0)$-form. We can use it to decompose the SU(3)-structure (J,Ω) on the cone to an SU(2)-structure, as we mentioned around (5.1.45). The details were given in flat space in Section 3.4.3; in particular, (3.4.46) shows how to define projected j and ω. This gives

$$J = r\mathrm{d}r\wedge\eta + r^2 j\,, \qquad \Omega = r^2(\mathrm{d}r + \mathrm{i}r\eta)\wedge\omega\,. \tag{7.4.4}$$

The powers of r are chosen so that j and ω do not rescale under homothety,

$$L_{r\partial_r} j = L_{r\partial_r}\omega = 0\,. \tag{7.4.5}$$

By the usual principle (4.2.9), they become forms on L_5. Moreover, η, j, and ω satisfy the algebraic constraints of an SU(2)-structure in $d = 5$:[6]

$$\eta \cdot j = \eta \cdot \omega = 0\,, \qquad \omega^2 = 0\,, \qquad j \wedge \omega = 0\,, \qquad \omega \wedge \bar{\omega} = 2j^2 \neq 0\,. \quad (7.4.6)$$

In general d, we would have

$$J = r\mathrm{d}r \wedge \eta + r^2 j\,, \qquad \Omega = r^{(d-1)/2}(\mathrm{d}r + \mathrm{i}r\eta) \wedge \omega \quad (7.4.7)$$

together with (5.1.40) with $d \to d - 1$, and again $\eta \cdot j = \eta \cdot \omega = 0$.

Roughly speaking, the idea of the decomposition, (7.4.4) and (7.4.7), is that we take a holomorphic vielbein whose elements are

$$\mathrm{d}r + \mathrm{i}\, r\eta\,, \qquad r h^a \quad a = 1, \ldots, (d-1)/2\,, \quad (7.4.8)$$

where $\{h^a\}$ are a holomorphic vielbein for j and ω; recall (5.1.35) and (5.1.22). This reproduces (7.4.4), including the powers of r.

Differential conditions

We now impose the Calabi–Yau condition $\mathrm{d}J = \mathrm{d}\Omega = 0$. For this, it is convenient to split the coordinates of $C(L_d)$ into r and the rest, obtaining

$$\mathrm{d}_{C(L_d)} = \mathrm{d}r\partial_r + \mathrm{d}, \quad (7.4.9)$$

where now $\mathrm{d} = \mathrm{d}_{L_d}$ is the exterior differential along L_d. Using this on (7.4.4), the Calabi–Yau condition becomes

$$\mathrm{d}\eta = 2j\,, \quad (7.4.10a)$$

$$\mathrm{d}\omega = 3\mathrm{i}\, \eta \wedge \omega\,. \quad (7.4.10b)$$

We can take (7.4.6) and (7.4.10) as a possible alternative definition of Sasaki–Einstein manifold in $d = 5$. In general d, (7.4.10b) is replaced by

$$\mathrm{d}\omega = \frac{d+1}{2}\mathrm{i}\, \eta \wedge \omega. \quad (7.4.10c)$$

Equation (7.4.10) should not be completely unfamiliar: we saw them already as (6.2.13) and (6.2.27) in our discussion of $\mathbb{CP}^N$. Indeed, those equations were valid on S^{2N+1}, which is a Sasaki–Einstein because $C(S^{2N+1}) = \mathbb{R}^{2N+2}$; we discussed its Killing spinors in the previous subsection. Recall that the idea of the formalism in Section 6.2.1 was to work on S^{2N+1}, and then to recognize that some forms are basic (namely horizontal, and invariant under vertical vector fields) and thus represent forms on the base $\mathbb{CP}^N$ of the bundle (6.2.4).

Equation (7.4.10) has an interpretation as defining a *transversely holomorphic foliation (THF)*, which we will not cover here. It also implies

$$\mathrm{d}(\eta \wedge \omega) = 0\,. \quad (7.4.11)$$

[6] We did not see the algebraic constraints (7.4.6) yet, but we did discuss the very similar case of SU(3)-structures in $d = 7$: see (3.4.75) in flat space, with some additional comments regarding the corresponding G-structure in Section 5.1.2. Our (η, j, ω) here play the role of (v, J, Ω) there.

If on an odd-dimensional manifold we have a one-form η and a decomposable ω such that $\eta\wedge\omega\wedge\bar{\omega} \neq 0$, (7.4.11) defines a so-called *Cauchy–Riemann (CR) structure*. This is the odd-dimensional analogue of a complex structure; its theory parallels in many ways the discussion in Section 5.3.1 for complex geometry. It arises naturally on the base L_d of a conical complex manifold $C(L_d)$, so it is a lot more general than Sasaki–Einstein. CR-structures also exist on spaces that are not only not bases of cones, but not even boundaries of complex manifolds; an obstruction exists, involving *Kohn–Rossi cohomology*, the CR analogue of Dolbeault cohomology [196].

It follows from (7.4.6), (7.4.10), and (4.1.79) that

$$L_\xi j = 0\,, \qquad L_\xi \omega = 3\mathrm{i}\,\omega\,. \tag{7.4.12}$$

The metric is determined by these forms (and by η, which also satisfies (7.4.3)) in such a way that rescaling ω by a phase does not affect it; so it follows that $(L_\xi g)_{mn} = 0$, namely ξ is a Killing vector. (Compact Calabi–Yau's cannot have isometries by (7.1.10), but in the noncompact case there is no such an obstruction.)

7.4.2 Orbifolds

The simplest examples are obtained by taking orbifolds.

For $d = 5$, we want a Calabi–Yau cone $C(L_5)$ of the type $\mathbb{C}^3/\Gamma$, so that $L_5 = S^5/\Gamma$. $\mathbb{C}^3$ is flat, so it is Calabi–Yau in several ways: there are four independent constant spinors. All we have to do, then, is make sure that at least one of these, η_+, is invariant under Γ. Since the stabilizer of η_+ is SU(3), this means

$$\Gamma \subset \mathrm{SU}(3)\,. \tag{7.4.13}$$

(If there are several invariant spinors, Γ is in their common stabilizer.)

The fixed loci of the Γ action are singularities (Section 6.3.4). Usually the origin of $\mathbb{C}^3$ is the only fixed point of Γ; it is then said to be *isolated*. If on the contrary the singular locus is extended along some directions, it will intersect the link L_5, which will then have some singularities itself.

Cyclic groups

Consider first $\Gamma = \mathbb{Z}_n$. A well-known theorem in group theory says that all irreducible representations of an abelian finite group are one dimensional; so up to change of basis, we can assume it to act on each coordinate separately:

$$(z_1, z_2, z_3) \sim (\omega_n^{a_1} z_1, \omega_n^{a_2} z_2, \omega_n^{a_3} z_3)\,; \qquad \omega_n = \mathrm{e}^{2\pi\mathrm{i}/n}\,. \tag{7.4.14}$$

Checking when this satisfies (7.4.13) is easiest if we look at the bilinears (J, Ω), using their flat-space expressions (3.4.17) and (3.4.12):

$$J = \frac{\mathrm{i}}{2}\mathrm{d}z^i \wedge \mathrm{d}\bar{z}^{\bar{i}} \to J\,, \qquad \Omega = \mathrm{d}z^1 \wedge \mathrm{d}z^2 \wedge \mathrm{d}z^3 \to \mathrm{e}^{2\pi\mathrm{i}(a_1+a_2+a_3)/n}\Omega\,. \tag{7.4.15}$$

So (7.4.14) satisfies (7.4.13) if

$$\frac{a_1 + a_2 + a_3}{n} \in \mathbb{Z}\,. \tag{7.4.16}$$

The computation from the spinorial point of view is similar to that in Section 2.2.1 for the stabilizer of a pure spinor, and can be carried out using (2.1.9) and (2.4.28) with the spinor $\eta_+ = |+ + +\rangle$.

The same discussion applies to general even dimension, where (7.4.13) and (7.4.16) become

$$\Gamma \subset \mathrm{SU}(d/2) \quad \Leftrightarrow \quad \frac{1}{n}\sum_{i=1}^{d/2} a_i \in \mathbb{Z}. \tag{7.4.17}$$

Let us specialize the discussion even further to some particular choices of a_i. Of the orbifold examples we have seen earlier, in $d = 2$ (6.3.22) is immediately ruled out. On the other hand, in $d = 4$ (6.3.23) and its generalization $\mathbb{C}^2/\mathbb{Z}_n$ in (6.3.31) do satisfy (7.4.17): they have $a_1 = n$, $a_2 = -n$. So $\Gamma \subset \mathrm{SU}(2)$. In $d = 6$, an example where (7.4.14) is satisfied is $\mathbb{C}^3/\mathbb{Z}_3$, discussed around (6.4.38); here all $a_i = 1$.

The $\mathbb{C}^2/\mathbb{Z}_n$ case can be made relevant for $d = 6$ by taking a further product with $\mathbb{C}$:

$$\mathbb{C}^2/\mathbb{Z}_n \times \mathbb{C}. \tag{7.4.18}$$

The $\mathbb{C}$ factor does not change Γ, so this is an example where there are two invariant spinors. The spinor $|++\rangle$ in the first four directions can be tensored with both spinors in the $\mathbb{C}$ factor, yielding $|+++\rangle$ and $|++-\rangle$. The invariant forms are a first SU(3)-structure as in (7.4.15), and a second obtained by conjugating the third coordinate, $z_3 \to \bar{z}_3$. Together these define also an SU(2)-structure that has the expression (3.4.50). In this case, the singularity is not isolated; this implies that $L_5/\mathbb{Z}_n$ has itself two orbifold singularities.

Of course, $\Gamma = \mathbb{Z}_n$ is not the only possibility. One can consider products, such as in

$$\mathbb{C}^3/\mathbb{Z}_n \times \mathbb{Z}_n \tag{7.4.19}$$

with the first $\mathbb{Z}_n$ acting as before: $(z_1, z_2, z_3) \sim (\omega_n z_1, \omega_n^{-1} z_2, z_3)$, and the second acting as $(z_1, z_2, z_3) \sim (z_1, \omega_n z_2, \omega_n^{-1} z_3)$. This is described by the single equation

$$Z_1 Z_2 Z_3 = Z_4^n. \tag{7.4.20}$$

According to the general criterion (6.4.32), this is also toric.

McKay correspondence

Even more generally, one can consider nonabelian discrete groups. In $d = 4$, the possible subgroups of SU(2) belong to three sequences:

- $\Gamma = \mathbb{Z}_n$, which we already analyzed.
- $\Gamma = \mathrm{BD}_n$, the *binary dihedral* group. This is a 2 : 1 cover under $\mathrm{SU}(2) \to \mathrm{SO}(3)$ of the *dihedral* group D_n, the symmetry group of a polygon.
- $\Gamma =$ BT, BO, BD, the 2 : 1 covers under $\mathrm{SU}(2) \to \mathrm{SO}(3)$ of the symmetry groups of the tetrahedron, octahedron, or dodecahedron.

Generalizing our comments following (6.3.31), we can resolve these singularities by a sequence of blowups, resulting in a space that has several two-spheres intersecting each other. Curiously, if one associates a node to each sphere and a link to each intersection, one obtains the extended Dynkin diagram of $\mathrm{SU}(n+1)$, $\mathrm{SO}(2n)$, and E_n respectively. This is part of a set of coincidences called the *McKay correspondence*. Based on this, we define

$$\begin{aligned} &\Gamma_{\mathrm{SU}(n)} \equiv \mathbb{Z}_n\,, \qquad \Gamma_{\mathrm{SO}(2n)} = \mathrm{BD}_n\,; \\ &\Gamma_{\mathrm{BT}} = E_6\,, \qquad \Gamma_{\mathrm{BO}} = E_7\,, \qquad \Gamma_{\mathrm{BD}} = E_8\,. \end{aligned} \tag{7.4.21}$$

M-theory and ADE singularities

This has an application to M-theory on $\mathbb{R}^7 \times \mathbb{C}^2/\Gamma$. We learned after (7.2.35), applying a result in (7.2.32), that the lift of N D6-branes contains a GH space, which in the limit where the D6 coincide has a $\mathbb{C}^2/\mathbb{Z}_N$ singularity. From the IIA perspective, we know that there should be a super-YM effective theory on the D6-world-volume, with gauge group $G = \mathrm{U}(N)$ (Section 1.3.6).

Let us recover this gauge group from the M-theory perspective. We can resolve the singularity by distancing the D6, as in (7.2.32). The gauge fields in the Cartan subalgebra $\mathrm{u}(1)^{\mathrm{N}} \subset \mathrm{su}(N)$ originate from integrating the RR three-form potential on the N two-spheres of the resolution:

$$a_\alpha = \int_{A_\alpha} C_3\,. \tag{7.4.22}$$

This notation means that we are taking the components of C_3 with one leg along $\mathbb{R}^7$ and two along the internal GH space; of this internal two-form we are taking a pull-back to the two-cycle A_α of Section 7.2.2. The gauge fields outside the Cartan subalgebra are more interesting. In the IIA realization, they come from strings that begin and end on different D6-branes. When we lift to M-theory, these become M2-branes that wrap the compact two-cycles A_α of the GH space. In $\mathbb{R}^7$, these look like particles with mass:

$$T_{\mathrm{M2}}\mathrm{vol}(A_\alpha)\,. \tag{7.4.23}$$

If we make the D6 coincide again, the strings become light and the nonabelian $\mathrm{SU}(N)$ gauge symmetry emerges. Correspondingly, in the singular limit the A_α shrink and (7.4.23) goes to zero, generating massless particles whose modes make the gauge field in $\mathbb{R}^7$ nonabelian.

We can now apply this lesson to the other ADE groups:

$$\text{M-theory on } \mathbb{R}^7 \times \mathbb{C}^2/\Gamma_G \quad \to \quad G \text{ gauge symmetry on } \mathbb{R}^7\,, \tag{7.4.24}$$

where Γ_G was defined in (7.4.21). In the $\Gamma = \mathrm{BD}_N$ case, there is a IIA realization in terms of D6s on an O6; in the $\Gamma = E_n$ case, the geometry has no small circle, and the phenomenon is purely M-theoretical.

7.4.3 Regularity

Orbit closure

The $L_5 = S^5$ example might suggest to consider ξ as the vertical vector field of a fiber bundle

$$\begin{array}{ccc} S^1 & \hookrightarrow & L_5 \\ & & \downarrow \\ & & B_4\,, \end{array} \tag{7.4.25}$$

generalizing (6.2.4) over $\mathbb{CP}^N$. The advantage of this would be that the base of such a bundle would be even-dimensional, hopefully getting us back to more familiar concepts.

The problem is that it is not guaranteed that the one-dimensional group of isometries generated by ξ is U(1). It could be another one-dimensional Lie algebra: $\mathbb{R}$. It might seem strange that this could be realized inside a compact L_5, but in fact even just a torus $T^2 = \mathrm{U}(1)_x \times \mathrm{U}(1)_y$ offers an example. Consider the vector field

$$\partial_x + a\partial_y \,. \tag{7.4.26}$$

Recall that an orbit of a group action is the set of points related to one another by the action of a group element. If $a = \frac{p}{q}$ is rational, then the orbits of the vector field (7.4.26) close after p turns along $\mathrm{U}(1)_x$ and q turns along $\mathrm{U}(1)_y$, but if a is irrational, the orbits form copies of $\mathbb{R}$ inside the T^2.

We say that the Sasaki–Einstein L_5 is *regular* if the orbits of ξ are all copies of S^1 with the same size, *quasiregular* if they are copies of S^1 but not all of the same size, and *irregular* if they are copies of $\mathbb{R}$. The latter case might look exotic, but in fact most known examples are of this type. The way this happens is exactly as in (7.4.26): the isometry group contains a T^k, and ξ is an irrational combination of its generators.

Regular L_5 do exist, but the base B_4 of the bundle (7.4.25) is severely restricted. To see this, we follow the case (6.2.4) as a blueprint. First we notice that the form j is basic for this bundle: it is horizontal by (7.4.6) and invariant by (7.4.12). So by (4.2.9), it represents a form on B_4. Equation (7.4.10a) implies that it is closed on B_4:

$$\mathrm{d}_4 j = 0 \,. \tag{7.4.27a}$$

η is not horizontal, so it does *not* descend to a form on B_4, and (7.4.10a) does not imply that j is exact on B_4. (Compare to our comment preceding (6.2.14), on J_{FS} not being exact despite (6.2.13).)

To see what (7.4.10b) implies, we can follow the logic in (6.2.27). Introduce a coordinate ψ such that $\xi = 3\partial_\psi$. In terms of this, $\eta = \frac{1}{3}(\mathrm{d}\psi + w)$ for w a one-form on B_4. Inspired by (6.2.26), we also write

$$\omega_0 = \mathrm{e}^{\mathrm{i}\psi}\omega \,, \qquad \mathrm{d}_4\omega_0 = \mathrm{i}w \wedge \omega_0 \,, \tag{7.4.27b}$$

where ω_0 is a bundle-valued form on B_4. The generalization of (6.1.4) to $d = 4$ tells us that B_4 is Kähler. The metric is also determined by the SU(2)-structure, adapting the comments in (5.1.49) to $d = 5$. Recall that η is taken to be orthogonal to j, ω; so $\mathrm{d}s_5^2 = \eta^2 + \mathrm{d}s_4^2$, where $\mathrm{d}s_4^2$ is determined by j and ω as a anti-Hermitian metric in the usual way for an SU(2)-structure (Section 5.1.2 and in particular (5.1.39)). So

$$\mathrm{d}s^2_{L_5} = \mathrm{d}s^2_{B_4} + \frac{1}{9}(\mathrm{d}\psi + w)^2 \,. \tag{7.4.27c}$$

(These formulas are still valid in nonregular cases, but only make sense locally; we will see examples in Section 7.4.5.) By (6.1.5), we can also determine the Ricci form, which gives $\rho = \mathrm{d}w = 6j$. So in conclusion, B_4 is a *Kähler–Einstein manifold* with positive curvature.

The cohomology class of ρ represents $c_1(B_4)$, as noted after (6.1.5). Since U(1)-bundles are classified by their c_1, we might be tempted to conclude that (7.4.25) is the U(1)-bundle associated to the determinant of the tangent bundle $\det T_{1,0} = K^{-1}$. However, the first Chern class of a U(1)-bundle (4.2.60) actually depends on the periodicity we choose on the fiber coordinate. If the periods of $c_1(B_4)$ on all two-cycles are all divisible by an integer i, then we can take the periodicity to be $2\pi i$ rather than the usual 2π, and c_1 of (7.4.25) will be still in $H^2(B_4, \mathbb{Z})$. This integer $i = i(B_4)$ is called the *index* of the Kähler–Einstein B_4. For example, for $B_4 = \mathbb{CP}^2$ there is only one cycle, and by (6.2.29) $i = 3$. We can take 3ψ in (6.2.26) (which corresponds to ψ in (7.4.27b)) to have periodicity 2π, but also 6π; these two lead to $L_5 = S^5/\mathbb{Z}_3$ and S^5 respectively.

From this perspective, we can also think of the radial coordinate r as complexifying the fiber coordinate ψ of (7.4.25), reconstructing the total space of

$$K^{-1}_{B_4} \, . \tag{7.4.28}$$

This cannot be literally true, because the total space of a line bundle over a smooth space is nonsingular. Rather, (7.4.28) is a blowup of $C(L_5)$. Yet another point of view is that $C(L_5)$ as a "complex cone" over B_4, where the real homothety is now promoted to a complex one generated by $r\partial_r + \mathrm{i}\xi$; then B_4 is a complex quotient of $C(L_5)$, similar to those of Section 6.4.2. Finally we can also consider B_4 a symplectic quotient, along the lines of Sections 5.4.2 and 6.4.2. The radial coordinate r of the cone plays the role of the moment map, while ξ is the symplectomorphism.

In the quasiregular case, most of the preceding discussion applies, but the base has orbifold singularities.

Kähler–Einstein bases

In four dimensions, there are only a few Kähler–Einstein manifolds with positive curvature [197]:

- $B_4 = \mathbb{CP}^2 = \mathrm{SU}(3)/\mathrm{U}(2)$. Since $K^{-1} = \mathcal{O}(3)$ (from (6.2.31)), and there is only one two-cycle, the index here is $i(\mathbb{CP}^2) = 3$, and there is the possibility of taking the periodicity of the fiber coordinate ψ to be either $\Delta\psi = 2\pi$ or $\Delta\psi = 3 \cdot 2\pi$. These yield $L_5 = S^5/\mathbb{Z}_3$ or $L_5 = S^5$ respectively.
- $B_4 = \mathbb{CP}^1 \times \mathbb{CP}^1 = (\mathrm{SU}(2)/\mathrm{U}(1))^2$. Here $K^{-1} = \mathcal{O}(2,2)$, a line bundle which has $c_1 = 2$ over both copies of $\mathbb{CP}^1$; the index is $i(\mathbb{CP}^1 \times \mathbb{CP}^1) = 2$, and so once again we have two choices:

$$T_{1,1} = (\mathrm{SU}(2) \times \mathrm{SU}(2))/\mathrm{U}(1) \tag{7.4.29}$$

  and $T_{2,2}/\mathbb{Z}_2$. These names come about because they are the total space of $\mathcal{O}(2,2)$ or $\mathcal{O}(1,1)$ respectively.[7] According to the discussion in Section 4.4.2, the metric is not uniquely determined by writing (7.4.29): in particular, the size of the S^1-fiber is not fixed. We will discuss $T_{1,1}$ in more detail in Section 7.4.4.
- $B_4 = \mathrm{dP}_k$, the del Pezzo manifolds of Section 6.4.1, for $3 \le k \le 8$.

[7] Another possible way to understand the label is this. The notation in (7.4.29) is incomplete in this case because there are several ways to embed U(1) in $\mathrm{SU}(2)^2$. These are parameterized by a linear combination of the Cartan subalgebras of the two copies of su(2). The resulting spaces are known as $T_{p,q}$, but they are not Sasaki–Einstein except for the two cases we mentioned.

Table 7.1. Homogeneous Sasaki–Einstein manifolds in $d = 7$.

Kähler–Einstein base B_6	Sasaki–Einstein L_7	index
$\mathbb{CP}^3 \cong \mathrm{SU}(4)/\mathrm{U}(3)$	S^7	4
$\mathbb{CP}^2 \times \mathbb{CP}^1 \cong (\mathrm{SU}(3) \times \mathrm{SU}(2))/\mathrm{U}(2) \times \mathrm{U}(1)$	$M_{3,2} \equiv \mathrm{SU}(3) \times \mathrm{SU}(2))/\mathrm{U}(2)$	1
$(\mathbb{CP}^1)^3 \cong (\mathrm{SU}(2)/\mathrm{U}(1))^3$	$Q_{1,1,1} \equiv \mathrm{SU}(2)^3/\mathrm{U}(1)^2$	2
$\mathbb{F}(1,2;3) \cong \mathrm{SU}(3)/\mathrm{U}(1)^2$	$N_{1,1} \cong \mathrm{SU}(3)/\mathrm{U}(1)^2$	1
$Z(\text{quadric}, \mathbb{CP}^4) \cong \mathrm{SO}(5)/(\mathrm{SO}(3) \times \mathrm{SO}(2))$	$V_{5,2} \equiv \mathrm{SO}(5)/\mathrm{SO}(3)$	3

The first two cases are cosets, while the others are not. Notice the absence from this list of dP_1 and dP_2. The complex cones over these spaces are in fact conical Calabi–Yau manifolds, but over an irregular Sasaki–Einstein L_5, which is not an S^1-bundle over them.

Homogeneous Kähler–Einstein in $d = 6$

In higher dimensions, it is still true that the base B_{d-1} of the bundle $\mathrm{U}(1) \hookrightarrow L_d \to B_{d-1}$ is Kähler–Einstein with positive curvature, but there are many more examples. The homogeneous ones are the easiest to find: by this we mean that we look for solutions to the analogue of the conditions (7.4.27) for higher d among invariant forms, in the sense of Section 4.4.2. The differential equations are then reduced to algebraic equations thanks to (4.4.6). We show the complete list of such homogeneous Kähler–Einstein B_6 in Table 7.1 [198–202]. (There are no irregular homogeneous Sasaki–Einstein L_7.) The indices in the notation are related to footnote 7. Notation for the last two examples is as in (4.4.53) and (6.3.2).

Nonhomogeneous Kähler–Einstein manifolds

More generally, finding Kähler–Einstein manifolds is an important geometrical problem. It can be reformulated in terms of an analogue of the Monge–Ampère equation (7.1.6) for the Calabi–Yau case.

For some simple examples, consider the zero locus $Z(f_k) \subset \mathbb{CP}^N$ of a homogeneous polynomial of degree k for $k < N + 1$. We observed after (6.3.7) that it is Fano, and thus has positive curvature. For $d = 3$, we take $N = 4$, and this says that polynomials f_k of degree $k = 2$, 3, or 4 in $\mathbb{CP}^4$ are Fano. It turns out [203, 204] that these are all Kähler–Einstein:

$$Z(f_k, \mathbb{CP}^4) \ : \ \text{positive Kähler–Einstein for } k < 5 \, . \tag{7.4.30}$$

The $k = 2$ case is homogeneous and appears in Table 7.1, but the $k = 3, 4$ examples are not. (Recall that $Z(f_5, \mathbb{CP}^4)$ is the famous quintic Calabi–Yau we studied in Section 7.1.3.)

More recently, the technique called *K-stability* has started producing many more Kähler–Einstein manifolds with positive curvature. This also has a more direct application to Sasaki–Einstein manifolds directly, regular or not; so we will discuss it in this more general setting in Section 7.4.7.

7.4.4 The conifold and $T_{1,1}$

We now return to $d = 5$ and focus on a particularly important example, the conifold (6.3.24). This brings together many techniques and observations we have seen so far.

We have already noticed in Section 7.1.4 that this singularity can appear in a compact Calabi–Yau, and that in fact it is generic on $\mathcal{M}_{\rm sing} \subset \mathcal{M}_{\rm c}$. This already suggests that the singularity itself should have a Calabi–Yau metric. Moreover, it has a natural homothety $v = Z^I \partial_{Z^I}$ inherited from $\mathbb{C}^4$. So we expect the conifold to support a conical Calabi–Yau metric, and hence to lead to a Sasaki–Einstein L_5.

To see what space L_5 is, we intersect the conifold with a level set for the radius function r defined by the homothety via $v = r\partial_r$; in particular, we can take $\{|Z^1|^2 + |Z^2|^2 + |Z^3|^2 + |Z^4|^2 = 1\}$. So L_5 is given by the intersection of this sphere in $\mathbb{C}^4$ with the conifold. If we take a further quotient by the vector $\xi = \mathrm{i}(Z^I\partial_I - \bar{Z}_I\bar{\partial}^I)$, the total effect will be to add another $\mathbb{C}^*$ action to the definition of the conifold as a GIT quotient in Section 6.4.3. The original $\mathbb{C}^*$ action was (6.4.39); the new one acts with charges $(1, 1, 1, 1)$. By taking sums and differences, in total we see that we have the same charge matrix that would describe $\mathbb{CP}^1 \times \mathbb{CP}^1$; reassuringly, this is a Kähler–Einstein manifold with positive curvature. So we conclude

$$\begin{array}{ccc} S^1 & \hookrightarrow & T_{1,1} \\ & & \downarrow \\ & & S^2 \times S^2 \, . \end{array} \tag{7.4.31}$$

This also agrees with (7.4.29).

To confirm this conclusion, here is the solution to (7.4.6) and (7.4.10c):

$$\eta = \frac{1}{3}\left(\mathrm{d}\psi - 2A_1 - 2A_2\right), \qquad j = \frac{1}{6}(\mathrm{vol}_{S^2_1} + \mathrm{vol}_{S^2_2}), \qquad \omega = \frac{1}{6}\mathrm{e}^{\mathrm{i}\psi} h_1 \wedge h_2 \, . \tag{7.4.32}$$

On each of the two copies of S^2, the A_a, $\mathrm{vol}_{S^2_a}$, and h_a are a connection for the bundle U_1, the volume form and the holomorphic vielbein. These were given as $A_a = \frac{1}{2}\cos\theta_a \mathrm{d}\phi_a$, $\mathrm{vol}_{S^2_a} = \sin\theta_a \mathrm{d}\theta_a \wedge \mathrm{d}\phi_a$, and $h_a = \mathrm{d}\theta_a + \mathrm{i}\sin\theta_a \mathrm{d}\phi_a$ in spherical coordinates, and discussed more carefully in other coordinate systems in (4.2.53) and (5.3.11). The periodicity of ψ is taken to be $\Delta\psi = 4\pi$, as appropriate for $T_{1,1}$ (see discussion around (7.4.29)). Using this and (4.2.60), we can check that indeed the $c_1 = 1$ over both copies of S^2. Equation (7.4.32) also determine the metric, as in (7.4.27c):

$$\mathrm{d}s^2_{T_{1,1}} = \frac{1}{9}(\mathrm{d}\psi - 2A_1 - 2A_2)^2 + \frac{1}{6}\mathrm{d}s^2_{S^2_1} + \frac{1}{6}\mathrm{d}s^2_{S^2_2} \, . \tag{7.4.33}$$

Another notable feature of this example is its coset structure, which we anticipated in (7.4.29). There are two ways to see this. The first is that $\mathbb{CP}^1 \cong \mathrm{SU}(2)/\mathrm{U}(1)$; the total space of a U(1)-bundle over $(\mathrm{SU}(2)/\mathrm{U}(1))^2$ can be obtained by "forgetting" a U(1) in the denominator, following the pattern $\mathrm{U}(1) \hookrightarrow G/H \to G/(H \times \mathrm{U}(1))$. This leads to the coset expression in (7.4.29). The second way is to notice that the original conifold equation (6.3.24) has an SO(4) symmetry that permutes the four coordinates. This action is transitive; according to the general principle (2.4.25),

L_5 will be a coset G/H, with H being the stabilizer of a point. The latter can be computed at any point; for example, the point $(Z^1, Z^2, Z^3, Z^4) = (\frac{1}{\sqrt{2}}, \frac{i}{\sqrt{2}}, 0, 0)$ is left invariant by rotations in the Z^3–Z^4 plane, which are an SO(2). This tells us $L_5 \cong \mathrm{SO}(4)/\mathrm{SO}(2)$, which is the same as (7.4.29).

The coset structure also determines the isometry group. According to (4.4.43):

$$\mathrm{ISO}(T_{1,1}) = \mathrm{SU}(2) \times \mathrm{SU}(2) \times \mathrm{U}(1)\,. \tag{7.4.34}$$

The two SU(2) factors can be viewed as the isometries of the $S^2 \times S^2$ base, lifted to the full space, while the U(1) comes from the isometries of the S^1 fiber. (See comments that follow (4.2.35).) Notice that (7.4.34) contains a $\mathrm{U}(1)^3$, which via the discussion in (6.4.26) signals that $C(T_{1,1})$ is toric, as we of course already know from Section 6.4.3.

As an alternative way to understand the topology of this space, decompose $Z^I = x_I + \mathrm{i}y_I$. Now the intersection of the conifold equation (6.3.24) and of the sphere $\{|Z^1|^2 + |Z^2|^2 + |Z^3|^2 + |Z^4|^2 = 1\}$ is given by

$$\sum_I x_I^2 = \sum_I y_I^2 = \frac{1}{2}\,, \qquad \sum_I x_I y_I = 0\,. \tag{7.4.35}$$

The vector with components x_I thus belongs to an S^3; moreover, for every choice of x_I, the vector with components y_I is orthogonal to it, and hence belongs to an S^2. So we have an S^2-bundle over S^3. By the logic around (4.2.71), such bundles are classified by $\pi_2(\mathrm{SO}(3)) = 0$, so they are all trivial, and in fact topologically

$$T_{1,1} \cong S^2 \times S^3\,. \tag{7.4.36}$$

The conifold has already occurred in many guises, and this is perhaps a good place to list them all:

- It is the zero locus $Z(\text{quadric}, \mathbb{C}^4)$ of the quadric equation (6.3.24) in $\mathbb{C}^4$.
- It is a complex quotient $\mathbb{C}^4/\mathbb{C}^*$ of the type discussed in Section 6.4.2, with charge matrix (6.4.39).
- It is a cone $C(T_{1,1})$, where we can see the regular Sasaki–Einstein $T_{1,1}$ as
  - An homogeneous space as in (7.4.29)
  - An S^1-bundle over $\mathbb{CP}^1 \times \mathbb{CP}^1 = S^2 \times S^2$, (7.4.31)
  - Topologically, $S^2 \times S^3$

7.4.5 Toric examples

Toric Calabi–Yau manifolds

We have just seen that the conifold is toric; we will now consider toric conical Calabi–Yau's more broadly.

There is a necessary condition for a noncompact toric manifold (conical or not) to be Calabi–Yau: the vectors w_I generating the rays of the fan (Section 6.4.1) should all lie on a hyperplane (a plane of codimension one). In other words, there should be a k such that

$$k \cdot w_I = 1 \tag{7.4.37}$$

for every I. For example:

- For $\mathbb{C}^2/\mathbb{Z}_n$ and its resolutions $\widetilde{\mathbb{C}^2/\mathbb{Z}_n}$, we see from (6.4.36) and visually from Figure 6.5 that the condition is met: all w_I have the second component $w_I^2 = 1$, and so they all lie on the plane $\{w^2 = 1\}$. In other words, (7.4.37) is satisfied with $k = \binom{0}{1}$. (We have indeed shown $\mathbb{C}^2/\mathbb{Z}_n$ to be Calabi–Yau around (7.4.18).)
- For the conifold, the ray generators were the columns of (6.4.42); their third entry $w_I^3 = 1$ for all of them, and hence they all end on the plane $\{w^3 = 1\}$. So (7.4.37) is satisfied with $k = (0, 0, 1)^t$. This can also be seen in Figure 6.6.
- For the small resolution (6.4.48) of the conifold, (7.4.37) is still satisfied, since the w_I have not changed. This can also be seen in Figure 6.8. So the small resolution is also Calabi–Yau; again we might have expected this from our discussion of compact examples in Section 7.1.4, where small resolutions appeared prominently.
- On the other hand, for the blowup of the conifold, we see from the columns of (6.4.44) and visually from Figure 6.7 that (7.4.37) is not met: there is no hyperplane on which all vectors end. So this is not a Calabi–Yau.

The *Futaki–Ono–Wang theorem* [205] says that toric conical manifolds satisfying (7.4.37) are indeed Calabi–Yau.

The $Y_{p,q}$ manifolds

Explicit metrics also exist; they were actually found before the existence theorem. The most famous set of explicit Sasaki–Einstein metrics is called $Y_{p,q}$ [206]. The topology is again $S^2 \times S^3$, the same as for $T_{1,1}$. We give the solutions first, and will comment later on how they were obtained. The SU(2)-structure forms read

$$\begin{aligned} \eta &= \frac{1}{3}(1-y)(\mathrm{d}\psi - 2A) - 2y\mathrm{d}\alpha\,, \\ j &= \frac{1}{6}(1-y)\mathrm{vol}_{S^2} + \frac{1}{6}\,(6\mathrm{d}\alpha + \mathrm{d}\psi - 2A)\wedge \mathrm{d}y\,, \\ \omega &= \sqrt{\frac{1-y}{6}}h_{S^2}\wedge\left(\frac{\mathrm{d}y}{\sqrt{U}} - \frac{\mathrm{i}}{6}\sqrt{U}(6\mathrm{d}\alpha + \mathrm{d}\psi - 2A)\right)\mathrm{e}^{\mathrm{i}\psi}\,, \end{aligned} \tag{7.4.38}$$

where $U \equiv 2\frac{a-3y^2+2y^3}{1-y}$. The notation is similar to (7.4.32): in spherical coordinates on the round S^2, $A = \frac{1}{2}\cos\theta\mathrm{d}\phi$ is the connection for the bundle U_1, $\mathrm{vol}_{S^2} = \sin\theta\mathrm{d}\theta \wedge \mathrm{d}\phi$, and $h_{S^2} = \mathrm{d}\theta + \mathrm{i}\sin\theta\mathrm{d}\phi$ is the S^2 holomorphic vielbein, discussed also in other coordinate patches in (4.2.53) and (5.3.12). As for $T_{1,1}$, the metric is now determined in the usual way from an SU(2)-structure as $\mathrm{d}s_5^2 = \eta^2 + \mathrm{d}s_4^2$. To determine the four-dimensional piece $\mathrm{d}s_4^2$, it is useful to think of (7.4.38) as being written in terms of a holomorphic vielbein: $j = \frac{\mathrm{i}}{2}(h^1\wedge\bar{h}^{\bar{1}} + h^2\wedge\bar{h}^{\bar{2}})$, $\omega = h^1\wedge h^2$, where

$$h^1 = \sqrt{\frac{1-y}{6}}h_{S^2}\,, \qquad h^2 = \frac{\mathrm{d}y}{\sqrt{U}} - \frac{\mathrm{i}}{6}\sqrt{U}(6\mathrm{d}\alpha + \mathrm{d}\psi - 2A)\,. \tag{7.4.39}$$

After some manipulations, the metric is then

$$\begin{aligned} \mathrm{d}s^2_{Y_{p,q}} &= \eta^2 + h^1\bar{h}^{\bar{1}} + h^2\bar{h}^{\bar{2}} \qquad\qquad (7.4.40) \\ &= \frac{1}{6}(1-y)\mathrm{d}s^2_{S^2} + \frac{1}{U}\mathrm{d}y^2 + \frac{U}{9w}(\mathrm{d}\psi - 2A)^2 + w(\mathrm{d}\alpha + f(\mathrm{d}\psi - 2A))^2, \end{aligned}$$

where $w = 2\frac{a-y^2}{1-y}$ and $f = \frac{a-2y+y^2}{6(a-y^2)}$. From the coefficient of the round S^2, we see that the range of y should be inside the interval $[0, 1]$.

Regularity of $Y^{p,q}$

Having written the relevant formulas for this geometry, let us now make sense of them.

The y–ψ part of the metric describes a sphere. To see this, let us first consider the simplified metric

$$ds^2_{y\psi} = \frac{1}{U}dy^2 + \frac{U}{9w}d\psi^2\,. \tag{7.4.41}$$

Take ψ to be periodic. The length of this S^1_ψ is $\sqrt{U/9w}$, so in particular it shrinks at the zeros $y = y_\pm$ of U. So we have a S^1-fibration over the interval $I_y \equiv [y_-, y_+]$, that shrinks at two points. Even if the metric is not round, topologically this is an S^2, as we see from the left side of Figure 6.3. For positive definiteness of the metric, we have to make sure that $I_y \subset [0, 1]$, and that w is positive in I_y.

One might now worry that (7.4.41) is singular near the zeros of U: after all, $g_{yy} \to \infty$ there. But this is a coordinate singularity, which is eliminated by a coordinate change. Near $y = y_\pm$, expanding the functions of y at first order and defining $r = \sqrt{|y - y_\pm|}$,

$$ds^2_{y\psi} \sim \frac{1}{3y_\pm}(dr^2 + r^2 d\psi^2), \tag{7.4.42}$$

which is the flat two-dimensional metric in polar coordinates; if we take $\Delta\psi = 2\pi$, this is nonsingular. So with these choices (7.4.41) represents a topological S^2. It is not round; it only has a single isometry ∂_ψ rather than the usual three.

Equation (7.4.41) appears in (7.4.40) with $d\psi \to d\psi - 2A$. For the Hopf metric (4.2.51), this covariant derivative signaled that S^1_ψ was fibered over the S^2. Here ψ is part of an S^2, and so it is the S^2 that is fibered over $S^2_{\theta\phi}$. In fact, (7.4.40) (forgetting for now the $(d\alpha + \ldots)^2$ term) is an example of the general bundle metric (4.2.35), with the fiber metric $g^{\rm F}_{ij}$ being (7.4.41), and the Dy^i in (4.2.27) being gauged with respect to only one isometry $T^a = \partial_\psi$. So the y–ψ part of (7.4.40) represents an S^2 fibered over $S^2_{\theta\phi}$. By our comment following (6.4.24), we know that topologically there are only two possibilities for such a fibration; our case turns out to be the trivial one, $S^2 \times S^2$.

Finally, we look at α. We can take it to be periodically identified; then we have an S^1_α fibered over the $S^2 \times S^2$ we have discussed so far. For this to fully make sense, however, we should make sure the c_1 of this bundle is quantized as in (4.2.60). Since the base of this S^1-bundle is $S^2 \times S^2$, there are only two periods to consider, leading to two integrality conditions; one imposes that $y_+ - y_-$ should be rational, while the other fixes the periodicity of α:

$$y_+ - y_- = \frac{3q}{2p}\,, \qquad \Delta\alpha = \frac{2\pi q}{3q^2 - 2p^2 + p\sqrt{4p^2 - 3q^2}}\,. \tag{7.4.43}$$

The first condition also fixes the parameter a in terms of p and q; it turns out $0 < a < 1$. This funny-looking value for the periodicity will soon have profound consequences.

So to summarize, we have that the $S^2_{y\psi}$ is fibered over the $S^2_{\theta\phi}$, but in a topologically trivial way; and the S^1_α is fibered over the resulting $S^2_{\theta\phi} \times S^2_{y\psi}$.

Isometries of $Y^{p,q}$

The Reeb vector field can be obtained by lowering the index of η with the following metric. The result is as follows:

$$\xi = 3\partial_\psi - \frac{1}{2}\partial_\alpha \,. \tag{7.4.44}$$

Recall that the periodicity of ψ is $\Delta\psi = 2\pi$, while $\Delta\alpha$ is given by (7.4.43). So, while the coefficients of (7.4.44) are rational, the ratio $\frac{\Delta\alpha}{\Delta\psi}$ in general is not. So we are in the situation discussed around (7.4.26); in general, the orbits of ξ do not close, and the Sasaki–Einstein $Y_{p,q}$ are irregular.

Put another way, one might try to redefine $\psi' = \psi$, $\alpha' = -(\alpha + \psi)/6$, so that $\xi = 3\partial_{\psi'}$, with respect to this new ψ' the metric (7.4.40) is in form (7.4.27c), with $ds^2_{B_4} = h^1\bar{h}^{\bar{1}} + h^2\bar{h}^{\bar{2}}$; however, the latter is now no longer regular.

The metric (7.4.40) has other isometries. There is an SU(2) factor coming from the $S^2_{\theta\phi}$. It has to be lifted to the total space, as explained around (4.2.43); the lift can be seen by rewriting the metric in terms of the left-invariant forms λ^a in (4.4.27) (with $\psi \to -\psi$), recalling $ds^2_{S^2} = (\lambda^1)^2 + (\lambda^2)^2$. There are also two abelian isometries, generated by ∂_ψ and ∂_α. So in conclusion, the isometry group is

$$\mathrm{SU}(2) \times \mathrm{U}(1)^2 \,. \tag{7.4.45}$$

In particular, there is an $\mathrm{U}(1)^3$ subgroup, signaling that the cone $C(Y_{p,q})$ is toric (compare the discussion around (6.4.26)). The charge matrix (Section 6.4.2) is [207]

$$C = (p, p, -p + q, -p - q) \,. \tag{7.4.46}$$

Even though the isometry group (7.4.45) is relatively large, it does not act transitively: two points with different values of y are not mapped to each other by an isometry. In other words, the generic orbits of the isometry group are the four-dimensional $S^2_{\theta\phi} \times S^1_\psi \times S^1_\alpha$. So the space is not homogeneous. The codimension of the orbits of the isometry group is in general called *cohomogeneity*; so the $Y_{p,q}$ spaces have cohomogeneity one.

This is actually how these solutions were obtained: looking for a solution with isometry (7.4.45), our building blocks should be the two-forms vol_{S^2}, h_{S^2}, and the one-forms $d\psi - 2A$, $d\alpha$, dy. If we write a general Ansatz for η, j, ω in terms of these, the coefficients f_i will have to be all functions of y, the coordinate on which no isometry is acting. Imposing (7.4.6) and (7.4.10) on this Ansatz produces a system of ordinary differential equations on the $f_i(y)$, which was then solved analytically [206]. In general, a cohomogeneity-k Ansatz leads to differential equations depending on k coordinates; this will play a role later in our search for string theory vacua.

One can show that the $Y_{p,q}$ are the only Sasaki–Einstein five-manifolds with cohomogeneity one [208]. Remarkably, there also exist explicit Sasaki–Einstein metrics $L_{p,q,r}$ with cohomogeneity two [209].

Higher dimensions

In higher dimensions, a similar metric exists [210, 211], called $Y_{p,k}(\text{KE})$. In $d = 7$, it reads[8]

$$ds^2_{Y_{p,k}(\text{KE}_4)} = (1-y)ds^2_{\text{KE}_4} + \frac{3}{4U}dy^2 + \frac{U}{3w}(d\psi - A)^2 + w(d\alpha + f(d\psi - A))^2\,, \quad (7.4.47)$$

where

$$U = \frac{(1-y)^3(1+3y) - a}{(1-y)^2}\,, \quad w = \frac{(1-y)^2(1+2y) - a}{3(1-y)^2}\,, \quad f = \frac{(1-y)^3 - a}{(1-y)^2(1+2y) - a}\,. \quad (7.4.48)$$

Regularity requires $0 \le a \le 1$. The metric KE_4 is a $d = 4$ positive-curvature Kähler–Einstein that replaces the S^2 in (7.4.40), and $dA = J_{\text{KE}_4}$. When $\text{KE}_4 = S^2 \times S^2$, there is a further generalization, which now depends on three integers (see [212, sec. 4.5], [213], [214, sec. 3]). For a general KE_4, (7.4.47) is no longer cohomogeneity-one, but the equations still only depend on the coordinate y. Finally we note that for $a = 1$ and KE_4 one recovers the homogeneous $M_{3,2}$ case from Table 7.1.

7.4.6 Volume minimization

General results

We will now see some remarkable general results concerning the volume of a Sasaki–Einstein [215–217].

(1) For a Kähler conical metric on $C(L_5)$ with a fixed complex structure, the volume only depends on the vector ξ defined by (7.4.2a):

$$\text{Vol}(L_5) \equiv V(\xi)\,. \quad (7.4.49)$$

(2) For a Calabi–Yau conical metric $C(L_5)$, the volume $V(\xi)$ is minimum among the Kähler conical metrics; in particular,

$$\delta V(\xi)|_{\text{SE}} = 0\,. \quad (7.4.50)$$

This was originally motivated by holography; for us it will be also be useful in Section 7.4.7.

In these results, we are expanding our scope to consider not only Calabi–Yau metrics on $C(L_5)$, but also Kähler ones. Recall from (6.1.4) that a Kähler metric is defined by $dJ = 0$, $d\Omega = \bar{W}_5 \wedge \Omega$. In the conical case, we can also set $W_5 = 0$, but at the cost of not achieving the volume normalization constraint (5.1.40b), so that we only obtain

$$-\frac{1}{6}J^3 = -\frac{i}{8}e^{\varphi}\Omega \wedge \bar{\Omega} \quad (7.4.51)$$

for some real function φ. This is the same situation as in (7.1.4) during our discussion of Yau's theorem. So all our definitions in Section 7.4.1 still apply, and in particular

[8] To make this similar to (7.4.40), we took $\lambda = 8$, $\Lambda = 8$, $a = -\kappa$ in [210] and changed coordinates as $4\rho^2 = 1 - y$.

the equations (7.4.10) are still satisfied for a conical Kähler; the problem of finding a Calabi–Yau metric is reduced to achieving $\varphi = 0$ in (7.4.51).

Now suppose we fix a complex structure on $C(L_5)$ and the vector ξ. The Kähler metric can still vary because of the choice of the homothety $r\partial_r$, or in other words, of what function on $C(L_5)$ we choose to call r. Varying this will also make η vary: from (7.4.2c), we see that

$$\delta\eta = \mathrm{d}^c \delta r \,. \tag{7.4.52}$$

This in turn modifies the rest of the decomposition (7.4.4), and a priori we might imagine it to vary the volume $\mathrm{Vol}(L_5)$. This variation receives two contributions: one because the locus $\{r = 1\} \subset C(L_5)$ changes, and the other because the forms η, j, and ω do. Both these can be shown to vanish by an integration by parts argument. So in fact $\mathrm{Vol}(L_5)$ does not depend on the choice of r; this is (7.4.49).

Suppose now we have a conical Calabi–Yau metric that is part of a family of conical Kähler metrics. If there is only one isometry, then this must be ξ; by the result (7.4.49), the deformations of the Kähler structure do not change the volume. It is more common, however, that there are several isometries for the metrics in the family. Let us call

$$T = \mathrm{U}(1)^r \tag{7.4.53}$$

the *isometry torus*, the largest possible torus subgroup of commuting isometries inside the isometry group. For example, for a toric manifold we commented in (6.4.26) that $r = N = d/2$, the complex dimension.

Now ξ varies in the isometry torus (7.4.53), and $V(\xi)$ is a function on this space. The second nontrivial result is that $V(\xi)$ attains its minimum when ξ is the Reeb vector of the conical Calabi–Yau metric. To show this, first one uses (6.1.9) and (7.4.51) to relate the Ricci scalar $R_{C(L_5)}$ to $\Delta_{C(L_5)}\varphi$, the Laplacian of φ on $C(L_5)$. From (7.4.12), one sees φ does not depend on r, so this is in fact $\frac{1}{r^2}\Delta_{L_5} f$. We thus conclude

$$\int_{r\le 1} \mathrm{vol}_{C(L_5)} R_{C(L_5)} = \int_{L_5} \Delta_{L_5}\varphi = 0 \,. \tag{7.4.54}$$

Then using (4.3.64), we compute $R_{C(L_5)} = \frac{1}{r^2}(R_{L_5} - 20)$. Using this in (7.4.54), we see that the volume is proportional to the integral of R_{L_5} and hence to the Einstein–Hilbert action $\int_{L_5} \mathrm{vol}_{L_5}(R_5 - 12)$, whose extrema are Einstein metrics $R_{mn} = 4g_{mn}$; by (4.3.64) again, these are such that $C(L_5)$ is Ricci-flat. This leads to (7.4.50). One can further show that this extremum is in fact a minimum by explicitly computing the derivatives.

Hypersurfaces

To make the discussion more concrete, let us see what happens if we specialize to cones that are hypersurfaces; in the notation of Section 6.3.1,

$$Z(f, \mathbb{C}^4) = \{f = 0\} \subset \mathbb{C}^4 \,. \tag{7.4.55}$$

If we keep the complex structure induced by $\mathbb{C}^4$, a trial Reeb vector can be written as

$$\xi = \mathrm{i} \sum_I d_I (X^I \partial_{X^I} - \bar{X}_I \partial_{\bar{X}_I}) \,. \tag{7.4.56}$$

If f transforms homogeneously under it, $L_\xi f = \mathrm{i} d_f f$, then it leaves the locus $Z(f)$ invariant, or in other words ξ is tangent to $Z(f)$. The holomorphic volume form Ω may be taken to be induced by that of $\mathbb{C}^4$ via the Poincaré residue map (7.1.24). If we impose that the degree of Ω is 3 as in (7.4.12), we obtain

$$d_I = b\alpha_I\,, \qquad b = \frac{3}{-\alpha_f + \sum_I \alpha_I}\,. \tag{7.4.57}$$

From our general discussion, the volume is only a function of ξ, and hence of the α_I. This can be evaluated with the techniques mentioned earlier, and gives

$$V(\xi) = \frac{\pi^3 \alpha_f}{b^3 \Pi_I \alpha_I} = \frac{\pi^3 \alpha_f}{27 \Pi_I \alpha_I}\left(-\alpha_f + \sum_I \alpha_I\right)^3, \tag{7.4.58}$$

where we also defined $\alpha_f \equiv \frac{d_f}{b}$, similar to $\alpha_I = \frac{d_I}{b}$. This is invariant under simultaneous rescaling of the (α_I, α_f), so we need not worry about normalizations. (This is the advantage of working with these variables rather than directly with the (d_I, d_f).) So now to determine the Reeb vector and the volume, all we have to do is minimize (7.4.58).

We specialize our discussion further and focus on $T_{1,1}$. The cone $C(T_{1,1})$ is the conifold, which is toric, so the isometry torus is $\mathrm{U}(1)^3$. In this case, it is more convenient to use the equation $X^1X^4 - X^2X^3 = 0$, from (6.4.41). The three possible $\mathbb{C}^*$ actions on the X^I are the rows of the charge matrix:[9]

$$\left(\begin{array}{cccc|c} 1 & 1 & 1 & 1 & 2 \\ 1 & 0 & 0 & -1 & 0 \\ 0 & 1 & -1 & 0 & 0 \end{array}\right). \tag{7.4.59}$$

In each row, the first four entries are the α_I, and the last is α_f. In other words, we can parameterize the possible vectors as

$$\alpha_1 = x_1 + x_2\,, \qquad \alpha_2 = x_1 + x_3\,, \qquad \alpha_3 = x_1 - x_3\,, \qquad \alpha_4 = x_1 - x_2\,, \qquad \alpha_f = 2x_1\,. \tag{7.4.60}$$

Applying this to (7.4.58) gives

$$V(\xi) = \frac{16 x_1^4 \pi^3}{27(x_1^2 - x_2^2)(x_1^2 - x_3^2)}\,. \tag{7.4.61}$$

The nontrivial minimum of this occurs for $x_2 = x_3 = 0$, and yields

$$\mathrm{Vol}(T_{1,1}) = \min V(\xi) = \frac{16}{27}\pi^3\,. \tag{7.4.62}$$

This can also be easily computed from (7.4.33), recalling that $\Delta\psi = 4\pi$ there and that $\mathrm{Vol}(S^2) = 4\pi$. Here, however, we managed to obtain it without using the metric at all.

7.4.7 K-stability

We have seen in the previous subsection that the Reeb vector field of a conical Calabi–Yau metric can be found by minimizing the volume $V(\xi)$ in the isometry

[9] This is not to be confused with the charge matrix (6.4.39), which acts on the coordinates Z^I of the $\mathbb{C}^4$ of the complex quotient $\mathbb{C}^4/\mathbb{C}^*$.

torus. We will now see that in certain situations this result can be generalized to an existence theorem for conical Calabi–Yau metrics.

Origin of the idea

The relevant concept is called *K-stability* and was originally applied to Kähler–Einstein manifolds M_{2k}. The idea originates from the concept of stability in Section 6.3.7 and the Kempf–Ness theorem (6.3.44).

We want to define an infinite-dimensional quotient $(\mathcal{M}_{\mathrm{K}} - F)/\mathcal{G}$, with $\mathcal{M}_{\mathrm{K}}$ being the space of Kähler structures, and $\mathcal{G}$ being the group of diffeomorphisms that keep the Kähler form J invariant (or symplectomorphisms). Putting aside many subtleties, the moment map for this action is the integral $\int_{M_{2k}} \mathrm{vol}(R - R_0)$, where R is the scalar curvature and R_0 its average. The zero locus of this moment map consists of Kähler manifolds with constant scalar curvature (cscK). In particular, if M_{2k} is Fano, namely if $c_1(M_{2k})$ is positive (Section 6.3.1), then we can take J to be in a cohomology class proportional to c_1. Since a representative for this is the Ricci form (Section 6.1.1), we can say that $\rho - \Lambda J$ is exact; the proportionality constant Λ can be determined by (5.2.5) to be $\Lambda = \frac{2}{d} R_0$. By the $\partial\bar{\partial}$-lemma (6.1.6), we can then write

$$\rho - \Lambda J = \mathrm{i}\partial\bar{\partial}\varphi \tag{7.4.63}$$

for some function φ. Acting with the contraction $J\cdot$, the left-hand side gives zero by construction; the right-hand side gives $\Delta\varphi$. If M_{2k} is compact, it now follows that φ is constant, and $\rho = \Lambda J$; in other words, M_{2k} is Kähler–Einstein.

So if M_{2k} is compact and Fano, the zero locus of the moment map $\mu^{-1}(0) \subset \mathcal{M}_{\mathrm{K}}$ is the space of Kähler–Einstein metrics. According to (6.3.44), the orbit of a point under $G_{\mathbb{C}}$ intersects $\mu^{-1}(0)$ if and only if the point is stable. As we saw in (6.3.43) and Figure 6.1, the Hilbert–Mumford criterion instructs us to compute a certain weight on a limit point of the action of $G_{\mathbb{C}}$. In our case, this should lead to a stability criterion to tell us which Kähler manifolds admit a Kähler–Einstein metric; it should involve a computation not on the original manifold, but on a limit of the action of $\mathcal{G}_{\mathbb{C}}$, the complexified group of symplectomorphisms. We will now see that this is a *degeneration* of the original manifold, and that the weight is called *Futaki invariant*. This conjecture [218, 219] was proven in [220].

Application to Sasaki–Einstein manifolds

We will now describe the K-stability test concretely, in a version adapted directly to Sasaki–Einstein manifolds. As in the previous subsection, we will focus on hypersurfaces $Z(f, \mathbb{C}^4)$, but the idea can be applied more generally.

We need to minimize the volume not just over the space of conical Kähler manifolds over $C(L_5)$, but also on some degenerations induced by the so-called *test configurations*. Given the embedding $C(L_5) \subset \mathbb{C}^4$, this is a one-dimensional subgroup

$$\lambda : \mathbb{C}^* \to \mathrm{U}(4) \tag{7.4.64}$$

that commutes with the isometry torus (7.4.53). The action of λ on $C(L_5)$ takes it to a manifold $C(L_5)_\lambda$. These are all biholomorphic to each other, but the limit for $\lambda \to 0$, the degeneration $C(L_5)_0$, is usually different. We will also require the degenerations to be *normal*: the ring of functions R_0 over it is an integrally closed domain, or in other words there are no solutions f to an algebraic equation $f^n + \lambda_{n-1} f^{n-1} + \cdots + f_0 = 0$, where $\lambda_i \in R_0$.[10]

Now, by definition the action of λ leaves the degeneration $C(L_5)_0$ invariant; so this new manifold will have one more U(1) symmetry than the original $C(L_5)$. We should now perform volume minimization by taking into account this new U(1) as well. If the minimum we had prior to taking this new direction into account is now no longer a minimum, we say that $C(L_5)$ is *destabilized* by λ. If $C(L_5)$ is not destabilized by any test configuration, we say that it is *K-stable*. A theorem now guarantees [221] that $C(L_5)$ admits a conical Calabi–Yau metric.

In practice, stability with respect to a test configuration is established by computing the derivative with respect to the new parameter. We saw earlier the formula (7.4.58) for the volume of a hypersurface. Let us suppose we already minimized this with respect to the U(1) isometries of the original $C(L_5)$. Let us denote by t_I the weights of the action of λ:

$$\lambda \cdot X^I = \lambda^{t_I} X^I \,, \tag{7.4.65}$$

and by t_f the degree of the degenerated $f_0 = f|_{\lambda \to 0}$ under it. λ doesn't destabilize $C(L_5)$ if the variation with respect to the new charges is positive:

$$\mathrm{Fut}(\lambda) = \partial_\epsilon (V(\xi + \epsilon\lambda))|_{\epsilon=0} > 0 \,. \tag{7.4.66}$$

This is called *Futaki invariant*. The derivative should be performed by imposing that Ω has still charge 3i as in (7.4.12). Explicitly, this is

$$\mathrm{Fut}(\lambda) = \left(-t_f + \sum_I t_I + \frac{1}{3}\left(\sum_I \alpha_I - \alpha_p \right)\left(\frac{t_f}{\alpha_f} - \sum_I \frac{t_I}{\alpha_I} \right) \right) \mathrm{Vol}(L_5) \,. \tag{7.4.67}$$

Examples

As an example, let us consider the *Brieskorn–Pham* singularity

$$\mathrm{BP}(p,q) \equiv Z(f, \mathbb{C}^4) \,, \qquad f = X^1 X^2 + (X^2)^p + (X^4)^q \,. \tag{7.4.68}$$

It is a generalization of the conifold: if $p = q = 2$, it can be taken to either (6.3.24) or (6.4.41) by a change of coordinates. We only find two $\mathbb{C}^*$ symmetries of (7.4.68), because it is not toric. (Indeed, we cannot put it in the form (6.4.32).) In the same notation as in (7.4.59), these actions are the rows of the charge matrix:

$$\left(\begin{array}{cccc|c} pq & pq & p & q & 2pq \\ 1 & -1 & 0 & 0 & 0 \end{array} \right) . \tag{7.4.69}$$

Applying volume minimization as in (7.4.62), we get a putative minimum:

$$\min V(\xi) = \frac{4(p+q)^3}{27 p^2 q^2} \pi^3 \,. \tag{7.4.70}$$

[10] A classic example of nonnormal space is $x^2 - y^3 = 0$. The function $f = \frac{y^2}{x}$ satisfies $f^2 - x = 0$: indeed, $(y^2/x)^2 - y = \frac{y}{x^2}(y^3 - x^2) = 0$. Another nonnormal space is $xy = 0$. The singularities we have considered so far, such as the conifold (6.3.24), are normal. Intuitively, a nonnormal space is obtained by a normal one by gluing or pinching it in some way.

However, in this case we don't know yet whether the singularity admits a conical Calabi–Yau metric, since it is not toric and the Futaki–Ono–Wang theorem does not apply. So we should perform the K-stability test.

It is easy to find test configurations. For example, consider [221]

$$\lambda_1 \cdot (X^1, X^2, X^3, X^4) = (X^1, X^2, \lambda_1 X^3, X^4), \\ \lambda_2 \cdot (X^1, X^2, X^3, X^4) = (X^1, X^2, X^3, \lambda_2 X^4). \tag{7.4.71}$$

Under the first, (7.4.68) becomes $C(L_5)_{\lambda_1} = \{X^1X^2 + \lambda_1^p (X^2)^p + (X^4)^q = 0\}$; so the degeneration is $C(L_5)_0 = \{X^1X^2 + (X^4)^q = 0\} \cong \mathbb{C}^2/\mathbb{Z}_q \times \mathbb{C}$. To compute the Futaki invariant, we can use (7.4.67) with $t_1 = t_2 = t_4 = t_f = 0$, $t_3 = 1$. For the second $\mathbb{C}^*$-action λ_2, computations are similar, with $X^3 \leftrightarrow X^4$, $t_3 \leftrightarrow t_4$. All in all,

$$\mathrm{Fut}(\lambda_1) = \frac{2q-p}{3q}\mathrm{Vol}(L_5), \qquad \mathrm{Fut}(\lambda_2) = \frac{2p-q}{3p}\mathrm{Vol}(L_5). \tag{7.4.72}$$

Imposing that both are positive, we get

$$\frac{1}{2} < \frac{p}{q} < 2. \tag{7.4.73}$$

It is harder to show that these are the only test configurations. A hypothetical rescaling of the first coordinate, $\lambda \cdot (X^1, X^2, X^3, X^4) \equiv (\lambda X^1, X^2, X^3, X^4)$, would lead to a degeneration $C(L_5)_0 = \{(X^2)^p + (X^4)^q = 0\}$, which is nonnormal (see footnote 10 after (7.4.64)); a rescaling of the second coordinate is excluded for the same reason. One might also embed (7.4.68) in a higher-dimensional $\mathbb{C}^N$ as the zero-locus of a system, and there in principle more $\mathbb{C}^*$-actions might appear. Fortunately, for cases in which the isometry torus (7.4.53) is $\mathrm{U}(1)^2$, such as this one, a combinatorial tool is available that can compute all possible test configurations [222]. (This is in part thanks to the fact that a degeneration has one more U(1) isometry than the original equation, so that it is toric.) In this case, this confirms that there are no more test configurations to consider, and so all Brieskorn–Pham singularities (7.4.68) that satisfy (7.4.73) admit a conical Calabi–Yau metric [221]. This leads to infinitely many new examples of Sasaki–Einstein manifolds. It turns out that these all have the topology of S^5.

Beyond this example, K-stability can be used to find many more Sasaki–Einstein metrics, especially for cases with isometry torus $\mathrm{U}(1)^2$. This has probably not been explored yet to its full potential.

Exercise 7.4.1 Apply (7.4.4) to $\mathbb{C}^2 = C(S^3)$. For example, you can use the coordinates in (4.2.50), with $J = \frac{\mathrm{i}}{2}\mathrm{d}w_a \wedge \mathrm{d}\overline{w_a}$. In this case, you should find

$$\eta = \mathrm{d}\psi + A, \qquad j = -\frac{1}{4}\mathrm{vol}_{S^2}, \qquad \omega = \frac{1}{2}\mathrm{e}^{2\mathrm{i}\psi}\bar{h}, \tag{7.4.74}$$

in the notation of (4.2.51) and (5.3.13).

Exercise 7.4.2 Find the SU(3)-structure for (7.4.47), similar to the SU(2)-structure (7.4.38) for the $Y_{p,q}$ in $d = 5$.

Exercise 7.4.3 Analyze the regularity of (7.4.47).

Exercise 7.4.4 Check that (7.4.32) satisfies (7.4.10).

7.5 Other Ricci-flat cones

7.5.1 3-Sasaki manifolds

Hyper-Kähler cones $C(L_{4k+3})$. These can be defined in any dimension $d = 4k + 3$, but the cases relevant for spacetime geometry are $d = 3$ and $d = 7$.

Definition

For $d = 3$, the hyper-Kähler cone has dimension four, so it is a conical K3. The only such manifolds are the orbifolds $\mathbb{C}^2/\Gamma$ we already discussed in Section 7.4.2.

For $d = 7$ (and higher), the story is more interesting. A hyper-Kähler manifold is Kähler with respect to three complex structures I_a; so we have a particular case of Sasaki–Einstein geometry, where we can now define three Reeb vectors $\xi_a \equiv I^t_a r\partial_r$, and their duals $\eta_a \equiv \iota_{r\partial_r} J_a$. By (5.1.56),

$$\xi_a \cdot \eta_b = \delta_{ab}\,; \qquad I_1 \cdot (\eta_2 - \mathrm{i}\eta_3) = \mathrm{i}(\eta_2 - \mathrm{i}\eta_3)\,. \tag{7.5.1}$$

So $\eta_2 - \mathrm{i}\eta_3$ is a $(1,0)$-form with respect to I_1. Similar equations for the actions of I_2 and I_3 also hold.

The decomposition (7.4.4) was obtained by subtracting the $(1,0)$-vector $\mathrm{d}r + \mathrm{i}r\eta$ and its conjugate from J; here it would read $J_a = r\mathrm{d}r \wedge \eta_a + r^2 j_a$. Now it can be refined by further projections; for example, from j_1 we can subtract the $(1,0)$-vector $\eta_2 - \mathrm{i}\eta_3$ and its conjugate. We are then led to defining the $\tilde{j}_a$ such that

$$J_a = r\mathrm{d}r \wedge \eta_a - \frac{1}{2}r^2\epsilon_{abc}\eta_b \wedge \eta_c + r^2\tilde{j}_a \tag{7.5.2}$$

with the property

$$\iota_{\xi_a}\tilde{j}_b = 0\,. \tag{7.5.3}$$

The $\tilde{j}_a$ now define an $\mathrm{Sp}\left(\frac{d+1}{4} - 1\right)$-structure. (So for $d = 7$, they define an $\mathrm{Sp}(1) \cong \mathrm{SU}(2)$-structure.) Imposing $\mathrm{d}J_a = 0$:

$$\mathrm{d}\eta_a = -\epsilon_{abc}\eta_b \wedge \eta_c + 2\tilde{j}_a\,, \tag{7.5.4a}$$
$$\mathrm{d}\tilde{j}_a = \epsilon_{abc}\tilde{j}_b \wedge \eta_c\,. \tag{7.5.4b}$$

If on a manifold L_d there is an $\mathrm{Sp}\left(\frac{d+1}{4} - 1\right)$-structure that satisfies (7.5.4), it is said to be *3-Sasaki*.

Fibration over quaternionic-Kähler

Equation (7.5.4a) signals a nonabelian Lie algebra. This can be seen by adapting the computation of anholonomy coefficients in Section 4.3.6 to the ξ_a and η_b, even if they are not a basis. In particular, similar to (4.3.56) we obtain

$$\iota_{[\xi_a,\xi_b]}\eta_c = -\iota_{\xi_b}\iota_{\xi_a}\mathrm{d}\eta_c \overset{(7.5.4a),(7.5.3)}{=} \epsilon_{abc}\,. \tag{7.5.5}$$

Moreover, from (7.5.4b) $\iota_{[\xi_a,\xi_b]}\tilde{j}^c = 0$. All this determines

$$[\xi_a, \xi_b] = 2\epsilon_{abc}\xi_c\,, \tag{7.5.6}$$

proportional to the su(2) Lie algebra. Equations (7.5.4) also imply the Lie derivatives

$$L_{\xi_a}\eta_b = 2\epsilon_{abc}\eta_c\,, \qquad L_{\xi_a}\tilde{\jmath}_b = 2\epsilon_{abc}\tilde{\jmath}_c\,. \tag{7.5.7}$$

The orbits of the SU(2) should be homogeneous spaces, of the form SU(2)/H; H could only be U(1) or the trivial subgroup, and the orbits could be either S^2 or S^3 (up to further quotients by discrete subgroups). But actually S^2 orbits are not allowed. Indeed, the volume form determined by the J_a is $\eta^1 \wedge \eta^2 \wedge \eta^3 \wedge \mathrm{vol}_{d-3}$ (where vol_{d-3} is determined by the $\tilde{\jmath}^a$). So the η^a are everywhere independent, and so are their dual vector fields ξ_a. So we conclude that the orbits are always S^3 or a quotient thereof.

So considering the orbits of the ξ_a gives us a bundle

$$S^3/\Gamma \hookrightarrow M_d \to B_{d-3} \tag{7.5.8}$$

similar to (7.4.25). There is no 3-Sasaki analogue of irregular Sasaki–Einstein manifolds (Section 7.4.3). The crucial difference is that u(1) can be the Lie algebra of the Lie groups U(1) and $\mathbb{R}$, while su(2) is the Lie algebra only of SU(2) or SO(3) $\cong$ SU(2)/$\mathbb{Z}_2$, both compact. There is, however, a notion of quasiregular 3-Sasaki, where the orbits are not all S^3.

We can obtain more information on B_{d-3} by adapting the logic leading to (7.4.27b) to a nonabelian setting. We introduce coordinates on the S^3-fiber, and realize the ξ_a as left-invariant vector fields r_a as in (4.4.28). Then their duals have to be of the form $\eta_a = \frac{1}{2}(\lambda_a + \hat{A}_a)$, with λ^a the left-invariant one-forms in (4.4.27) and $\hat{A}_a$ some horizontal forms. However, from (7.5.7) we see that the η_a transform in the adjoint under the ξ_a, so the $\hat{A}_a$ are not invariant but should transform as well. It is convenient to introduce $\eta = \eta_a T^a$, $\tilde{\jmath} = \tilde{\jmath}_a T^a$; then (7.5.4) read

$$\mathrm{d}\eta = -[\eta, \eta] + 2\tilde{\jmath}\,, \qquad \mathrm{d}\tilde{\jmath} = 2[\tilde{\jmath}, \eta]\,, \tag{7.5.9}$$

in a compact notation similar to (4.2.21). Equation (7.5.7) now can be solved as follows:

$$\eta = \frac{1}{2}(g^{-1}\mathrm{d}g + g^{-1}Ag) = g^{-1}(\mathrm{d}gg^{-1} + A)g\,, \qquad \tilde{\jmath} = g^{-1}jg\,, \tag{7.5.10}$$

where $j = j_a T^a$, $A = A_a T^a$, and now the A_a are independent of the S^3 coordinates. Plugging this into (7.5.9):

$$F = 4j\,, \qquad \mathrm{d}j + [A, j] = 0\,, \tag{7.5.11}$$

where F is the nonabelian curvature of A, (4.2.21). The second condition is the nonabelian Bianchi identity (4.2.24). If we convert it back in index notation, we obtain

$$F_a = 4j_a\,, \tag{7.5.12a}$$

$$\mathrm{d}j_a = \epsilon_{abc} j_b \wedge A_c\,. \tag{7.5.12b}$$

Equation (7.5.12b) is in fact (7.2.53); so we see that B_{d-3} is quaternionic-Kähler. (7.5.12a) fixes the constant in (7.2.54). We saw there that (7.5.12b) is trivial for a four-manifold, but it is precisely (7.5.12a) that takes its place; so, even for $d = 7$, the base B_4 is quaternionic-Kähler.

The metric is

$$\mathrm{d}s^2_{L_d} = \eta^a\eta^a + \mathrm{d}s^2_{B_{d-3}} = -2\mathrm{Tr}(\mathrm{d}gg^{-1} + A)^2 + \mathrm{d}s^2_{B_{d-3}}\,, \tag{7.5.13}$$

which is of the bundle type (4.2.35), with respect to the ℓ_a. (Compare with Exercise 4.4.1.) This is compatible with (4.2.41): the S^3 fiber has the $\mathrm{SO}(3) \times \mathrm{SO}(3)$ isometry group, but the isometries that have been fibered are not isometries of the total space.

Hypertoric examples

A large class of examples can be obtained by specializing the construction of hypertoric manifolds in (7.2.47). In order to obtain conical manifolds, one needs to take the constants in the hypermoment map to zero, $c_{a\alpha} = 0$. Here are two notable examples in $d = 7$:

- If we take $w_1 = \binom{1}{0}$, $w_2 = \binom{0}{1}$, $w_3 = \binom{-1}{-1}$, the hypertoric manifold (7.2.47)–(7.2.51) is the homogeneous space

$$N_{1,1} \cong \mathrm{SU}(3)/\mathrm{U}(1)\,. \tag{7.5.14}$$

  The indices indicate that the U(1) is diagonally embedded in the Cartan of SU(3). This is the only homogeneous 3-Sasaki in $d = 7$.
- If we take the charge matrix $C = (k_1, k_2, k_3)$, (7.2.47)–(7.2.51) is not homogeneous; it is a *biquotient*

$$\mathrm{U}(1)\backslash\mathrm{U}(3)/\mathrm{U}(1)\,, \tag{7.5.15}$$

  where the left subgroup is embedded as $\mathrm{diag}(\mathrm{e}^{\mathrm{i}|k_1|\theta}, \mathrm{e}^{\mathrm{i}|k_2|\theta}, \mathrm{e}^{\mathrm{i}|k_3|\theta})$, and the right subgroup as $\mathrm{diag}(1, 1, \mathrm{e}^{\mathrm{i}\phi})$. This S_{k_1,k_2,k_3} is also called *Eschenburg space*.

More examples can be obtained by considering more complicated charge matrices.

In all these cases, the quaternionic-Kähler base B_4 has orbifold singularities. Conversely, one can use the quaternionic-Kähler orbifolds we mentioned at the end of Section 7.2.4 to obtain other 3-Sasaki manifolds.

7.5.2 Nearly Kähler manifolds

We now consider cones $C(L_6)$ with G_2 holonomy.

Definition

Because of the discussion in (7.3.4), the form ϕ defining the G_2-structure and its dual $\tilde\phi = *_7\phi$ rescale under the homothety with weights three and four:

$$L_{r\partial_r}\phi = 3\phi\,, \qquad L_{r\partial_r}\tilde\phi = 4\tilde\phi\,. \tag{7.5.16}$$

Unlike in the Sasaki–Einstein case, now we cannot define a vector field on L_6 from the homothety $r\partial_r$; the closest we can get is to define a two-form $J \equiv \frac{1}{r^3}\iota_{r\partial_r}\phi$. The analogue of the decomposition (7.4.4) is now modeled on (5.1.52):

$$\phi = r^3 \mathrm{Im}\Omega + r^2 \mathrm{d}r \wedge J\,, \qquad \tilde\phi = -\frac{r^4}{2}J^2 - r^3 \mathrm{d}r \wedge \mathrm{Re}\Omega\,. \tag{7.5.17}$$

The powers of r are chosen so that $L_{r\partial_r} J = L_{r\partial_r}\Omega$. Moreover, just as in (5.1.52), the decomposition (7.5.17) implies that (J, Ω) satisfy the by now familiar algebraic constraints (5.1.40) of an SU(3)-structure on L_6.

We again decompose the exterior differential as in (7.4.9) and apply this to (7.5.17), imposing $\mathrm{d}\phi = \mathrm{d} * \phi = 0$ (recall (7.2.57)). This gives

$$\mathrm{d}J = 3\mathrm{Im}\Omega\,, \qquad \mathrm{dRe}\Omega = 2J^2\,. \tag{7.5.18}$$

A manifold L_6 with an SU(3)-structure that obeys this equation is said to be *nearly Kähler*. This condition is equivalent to the cone $C(L_6)$ having G_2 holonomy.

In terms of intrinsic torsion (5.5.6), we see that this corresponds to

$$W_1 = 2\,, \qquad W_2 = W_3 = W_4 = W_5 = 0\,. \tag{7.5.19}$$

So in a sense, this is very close to being Calabi–Yau (where of course all $W_i = 0$), which justifies the name. $W_2 = W_3 = W_4 = W_4 = 0$ imply that W_1 is constant (Exercise 5.5.5). Any constant W_1 can then be reduced to the value $W_1 = 2$ by rescaling the volume of L_6 and redefining $\Omega \to \mathrm{e}^{\mathrm{i}\theta}\Omega$. So another possible definition of nearly Kähler is as an SU(3)-structure whose only nonzero intrinsic torsion component is W_1.

Homogeneous spaces

There are four homogeneous examples:

- S^6. This is somewhat trivial, since of course the holonomy group of $C(S^6) = \mathbb{R}^7$ is the identity, which is $\subset G_2$. The nearly Kähler structure (J, Ω) is still quite interesting.[11]
- $S^3 \times S^3$. Here we give the explicit SU(3)-structure in terms of right-invariant forms in (4.4.27), which satisfy $\mathrm{d}\rho^a = \frac{1}{2}\epsilon^a{}_{bc}\rho^b \wedge \rho^c$, (4.4.6b). We call ρ_1^a and ρ_2^a the sets on the first and second copy of S^3. Then [223]

$$J = \frac{1}{6\sqrt{3}}\rho_1^a \wedge \rho_2^a\,, \qquad \Omega = \frac{1}{27}(\rho_1^1 + \omega_3\rho_2^1)\wedge(\rho_1^2 + \omega_3\rho_2^2)\wedge(\rho_1^3 + \omega_3\rho_2^3)\,, \tag{7.5.20}$$

  where as usual $\omega_3 = \mathrm{e}^{2\pi\mathrm{i}/3}$. This example is related to a limit of the bundle metric on the spinor bundle on S^3, which we mentioned in (7.2.58). We can view that metric as a resolution of $C(S^3 \times S^3)$.
- $\mathbb{CP}^3 \cong \mathrm{Sp}(2)/\mathrm{Sp}(1) \times \mathrm{U}(1)$. This is a particular case of the squashed $\mathbb{CP}^N$ coset in (4.4.50). It is topologically $\mathbb{CP}^3$, but the metric is not the Fubini–Study (6.2.16). Indeed, the isotropy representation (4.4.34) in this case is not irreducible, and so according to the discussion leading to (4.4.41) there is a family of metrics that depends on a parameter, which gets fixed by the nearly Kähler condition (7.5.18), or indeed even just the Einstein condition [136].
- $\mathbb{F}(1, 2; 3) \cong \mathrm{SU}(3)/\mathrm{U}(1) \times \mathrm{U}(1)$. This time, there is a two-parameter family of metrics compatible with the coset in the sense of (4.4.41). The nearly Kähler metric is not the same as the Kähler–Einstein metric we saw in Table 7.1.

Cohomogeneity one

The natural next step is to look for examples of cohomogeneity one, as in Section 7.4.5; recall that this means that the orbits of the isometry group have codimension one, so for nearly Kähler manifolds, they should have dimension five. In such a case, equations (7.5.18) reduce to a system of ordinary differential equations

[11] It is not known currently whether S^6 admits an *integrable* complex structure.

(ODEs) in a single variable. It was shown in [224] that there exist cohomogeneity-one nearly Kähler metrics on

$$S^6\,, \qquad S^3 \times S^3\,. \tag{7.5.21}$$

Of course, these are distinct from the homogeneous nearly Kähler metrics we mentioned earlier on the same two spaces.

Twistor bundles

A final set of examples is obtained by considering the so-called *twistor* S^2-bundles over a quaternionic-Kähler space B_4. As we mentioned in Section 7.2.4, only two compact nonsingular examples of the latter exist, but with orbifold singularities there are more [184, sec. 12.5]. In any case, this gives us an alternative understanding of two homogeneous nearly Kähler examples presented earlier, which will be useful later.

The S^2 bundle is obtained as the sphere bundle inside the bundle $\Lambda^2_+ T^* B_4$ of self-dual two-forms (recall our discussion around (4.2.5)). Indeed, this has fiber $\mathbb{R}^3$, and the sphere inside it is an S^2. These examples are related to the second non-compact G_2-manifold we mentioned in (7.2.58), which can be viewed as a resolution of the conical G_2 spaces we are studying in this section.

We can describe the nearly Kähler SU(3)-structure in two ways, both inspired by the S^3-bundle on B_4 we studied in Section 7.5.1. The first is obtained [225, sec. 5] by considering the 3-Sasaki total space of that bundle as a particular case of Sasaki–Einstein manifold. We then reduce along the vector field dual to $\eta \equiv \eta_1$, say; this is a Hopf reduction (4.2.51) of the S^3 to an S^2, turning the 3-Sasaki into our twistor bundles. Now we define the SU(3)-structure as

$$J = -2\tilde{j}_1 - \eta_2 \wedge \eta_3\,, \qquad \Omega = 2(\eta_2 - \mathrm{i}\eta_3) \wedge (\tilde{j}_2 + \mathrm{i}\tilde{j}_3)\,, \tag{7.5.22}$$

with the same forms as in Section 7.5.1. It is easy to see that they satisfy (5.1.40), and using (7.5.4) that they satisfy (7.5.18).

An alternative description makes the S^2 more explicit. We introduce coordinates y_a, $a = 1, 2, 3$ on the S^2, such that $y_a y_a = 1$, and its covariant derivative $Dy_a = \mathrm{d}y_a + \epsilon_{abc} A_b y_c$, or more succinctly $\mathrm{d}y + [A, y] = 0$. The connection is the same we used in (7.5.12b), since those j_a were after all self-dual forms on B_4; we can view the twistor bundle as being associated to the S^3 bundle of Section 7.5.1. It is useful to recall from (4.2.18) and (4.2.24) that $D^2 y_a = \epsilon_{abc} F_b y_c$ and $\mathrm{d}F_a + \epsilon_{abc} A_b F_c = 0$. Next we introduce

$$j_{\mathrm{F}} \equiv \frac{1}{2}\epsilon_{abc} y_a D y_b \wedge D y_c\,, \qquad j_{\mathrm{B}} \equiv y^a j_a\,; \tag{7.5.23a}$$

$$\psi = D y_a \wedge j_a\,, \qquad \tilde{\psi} = \epsilon_{abc} y_b D y_c \wedge j_a\,. \tag{7.5.23b}$$

As the names imply, j_{F} and j_{B} are to be interpreted as being along the fiber and base, respectively; so in particular $j_{\mathrm{F}}^2 = 0 = j_{\mathrm{B}}^3$. These forms obey

$$\mathrm{d}j_{\mathrm{F}} = 4\psi\,, \qquad \mathrm{d}j_{\mathrm{B}} = \psi\,, \qquad \mathrm{d}\tilde{\psi} = 2 j_{\mathrm{F}} \wedge j_{\mathrm{B}} + 8 j_{\mathrm{B}}^2\,, \tag{7.5.24}$$

where we also used (7.5.12a). Now the SU(3)-structure is[12]

[12] To check that Ω is decomposable, we can work at $y_a = (0, 0, 1) \in S^2$, since all points on the S^2 are related by rotations. Then $\Omega \propto (Dy^1 + \mathrm{i}Dy^2) \wedge J^1 + (Dy^2 - \mathrm{i}Dy^1) \wedge J^2 = (Dy^1 + \mathrm{i}Dy^2) \wedge (J^1 - \mathrm{i}J^2)$. Now $J^1 - \mathrm{i}J^2$ is decomposable by (5.1.62).

$$J = 2j_{\mathrm{F}} + \frac{1}{4}j_{\mathrm{B}}\,, \qquad \Omega = \tilde{\psi} + \mathrm{i}\psi\,. \tag{7.5.25}$$

Using (7.5.24), one now sees that these define a nearly Kähler structure (7.5.18).

Two of the preceding homogeneous examples can be viewed as twistor bundles:

$$\begin{array}{ccc} S^2 \hookrightarrow \mathbb{CP}^3 & \qquad & S^2 \hookrightarrow \mathbb{F}(1,2;3) \\ \downarrow & & \downarrow \\ S^4\,. & & \mathbb{CP}^2\,. \end{array} \tag{7.5.26}$$

For $\mathbb{CP}^3$, the projection can be understood as

$$\begin{array}{ccc} \mathbb{CP}^3 & \to & \mathbb{HP}^1 \\ (Z^1, Z^2, Z^3, Z^4) & \mapsto & (Z^1 + \mathrm{j}Z^2, Z^3 + \mathrm{j}Z^4)\,. \end{array} \tag{7.5.27}$$

Similar to $\mathbb{CP}^N$, the quaternionic plane is defined as $\mathbb{HP}^1 \equiv (\mathbb{H}^2 - \{0\})/\mathbb{H}^*$, with the equivalence relation $(q_1, q_2) \sim (q_1 q, q_2 q)$; it is topologically $\mathbb{HP}^1 \cong S^4$.

Twistor bundles also admit a *complex* structure $\tilde{I}$, which is integrable but has $c_1 \neq 0$. For example, in the $\mathbb{CP}^3$ case this is the ordinary complex structure we studied in Section 6.2. On the other hand, Ω in (7.5.25) defines an almost complex structure I that is nonintegrable (since $W_1 \neq 0$) but whose $c_1 = 0$. The two are related roughly speaking by complex conjugation in the fiber directions: in block-diagonal form,

$$\tilde{I} = \begin{pmatrix} I_{\mathrm{F}} & 0 \\ 0 & I_{\mathrm{B}} \end{pmatrix}, \qquad I = \begin{pmatrix} -I_{\mathrm{F}} & 0 \\ 0 & I_{\mathrm{B}} \end{pmatrix}. \tag{7.5.28}$$

This sign reversal is also visible in the opposite sign in the two factors of Ω in (7.5.22).

7.5.3 Weak G_2

We have not discussed Spin(7) holonomy at all so far, either compact or noncompact. So we begin with a general discussion.

Spin(7)-holonomy manifolds

We saw in (5.1.54) that Spin(7)-structures can be defined in terms of a real four-form Ψ. The theory of Spin(7) intrinsic torsion was sketched in (5.5.23) and (5.5.25); the latter was a bit clumsy, but the important point is that the two points of view contained the same representations, so $D_m \eta = 0$ is equivalent to

$$\mathrm{d}\Psi_4 = 0\,. \tag{7.5.29}$$

Hence this can be taken as an alternative definition of Spin(7) holonomy.

Finding examples presents the same challenges we discussed for G_2 holonomy: algebraic geometry is of much less help than for the Calabi–Yau case. Once again, the first examples were noncompact, on the total space of the spinor bundle over S^4 [185, 186]. The strategy of resolving singularities discussed for G_2-manifolds around (7.2.60) can be applied to produce compact Spin(7)-holonomy manifolds as well. The twisted connected sum idea in (7.2.62) is also helpful [226, 227].

Spin(7) cones

In the conical case, we proceed similar to earlier subsections. Equation (7.3.4) implies $L_{r\partial_r}\Psi_4 = 4\Psi_4$. Adapting (3.4.84) to our case, we have

$$\Psi_4 = r^3\phi \wedge \mathrm{d}r + r^4\tilde{\phi}\,. \tag{7.5.30}$$

Decomposing the exterior differential as in (7.4.9) gives now

$$\mathrm{d}\phi = -4 * \phi\,, \tag{7.5.31}$$

where the $*$ is now meant to be along L_7. Such a space is said to have *weak G_2 holonomy*. Comparing with (5.5.19), we see that it corresponds to $X_1 = -1$, $X_2 = X_3 = X_4$ in terms of intrinsic torsion. Only the singlet survives; in this respect, this case is similar to nearly Kähler geometry in $d = 6$.

Once again, one can find examples among homogeneous spaces:

- $S^7 \cong \mathrm{Sp}(2)/\mathrm{Sp}(1)$. This is a particular case of the squashed sphere metrics in (4.4.48). There is a single free parameter in the metrics induced by (4.4.41), and it is fixed by the condition (7.5.31).
- $N_{p,q} \cong \mathrm{SU}(3)/\mathrm{U}(1)$. These generalize (7.5.14). Here the integers p and q parameterize the embedding of U(1) into the Cartan of SU(3).
- SO(5)/SO(3).

More examples can be found by squashing a general seven-dimensional 3-Sasaki [228]. Namely, we rescale (7.5.13) with two factors R and s:

$$\mathrm{d}s^2_{\hat{L}_7} = R^2\left(s\eta^a\eta^a + \mathrm{d}s^2_{B_4}\right)\,. \tag{7.5.32}$$

We can obtain an Ansatz for the G_2-structure by recalling its relation (5.1.53) to an SU(2)-structure:

$$\begin{aligned}\phi &= R^3\left(s^3\eta_1 \wedge \eta_2 \wedge \eta_3 + s\,\eta_a \wedge \tilde{j}_a\right),\\ *\phi &= R^4\left(\mathrm{vol}_4 - \frac{s^2}{2}\epsilon_{abc}\eta_a \wedge \eta_b \wedge \tilde{j}_c\right).\end{aligned} \tag{7.5.33}$$

From (3.3.17), we read $\tilde{j}_a \wedge \tilde{j}_b = -2\delta_{ab}\mathrm{vol}_4$. Now we can compute $\mathrm{d}\phi$ using (7.5.4); imposing (7.5.31), we obtain

$$s = \frac{1}{\sqrt{5}}\,, \qquad R = \frac{3}{\sqrt{5}}\,. \tag{7.5.34}$$

So there is a weak G_2 manifold for any 3-Sasaki. The aforementioned squashed metric on S^7 is a particular case of this construction.

Exercise 7.5.1 Using coordinates y^m such that $y^m y^m = 1$, write an explicit expression for the nearly Kähler structure on S^6.

Exercise 7.5.2 Check (7.5.34).

Exercise 7.5.3 Use (5.5.2) and (5.5.12) to check that the spinor η_+ defining a nearly Kähler structure is Killing.

8 Vacua and reductions

After the technical and mathematical build-up of Chapters 2–7, in this chapter we give an introduction to the compactification problem, which will be our focus for the rest of the book.

We begin in Section 2.4.2 by considering pure general relativity, with no additional matter fields, in $d = 5$ and $d = 10$. In this case, vacuum solutions are only of the type $\text{Mink}_4\times$ a Ricci-flat manifold. In Section 8.2, we then start looking for vacuum solutions in string theory. Here we have to refine the notion of vacuum to take supersymmetry into account. The general problem is quite complicated, so we warm up by considering the case where all form fields are set to zero. This will set the stage for a more thorough search in later chapters. In Section 8.3, we give a general qualitative introduction to effective $d = 4$ theories for string theory vacua, highlighting some of the challenges we face in writing down such theories.

Finally, in Section 8.4 we review $d = 4$ supergravity theories. Despite our warnings about the difficulties in finding effective theories, this technique will still be quite useful in later chapters.

8.1 Pure gravity

In this section, we will look at the problem of compactifications in pure general relativity, which is considerably easier than string theory.

8.1.1 Five-dimensional pure gravity

We start by considering a single extra dimension, with topology S^1.

Vacuum solutions

The action is the usual

$$S_{\text{GR5}} = -\frac{1}{2\kappa_5^2}\int \mathrm{d}^4x\,\mathrm{d}\psi\,\sqrt{-g_5}R_5 \tag{8.1.1}$$

for general relativity, but now for a five-dimensional metric g^5_{MN}, $M, N = 0, \ldots, 4$; we called $\psi \equiv x^4$ the S^1 coordinate. The equations of motion are

$$R_{MN} = 0\,. \tag{8.1.2}$$

In the Introduction, we defined a solution to be a vacuum if the macroscopic spacetime appears to have zero stress-energy tensor, possibly except a cosmological constant, and we concluded that it is a maximally symmetric space MS_4; we studied these in Section 4.5, where we mentioned that locally only three geometries are possible, which all satisfy (4.5.1) and (4.5.2). So

$$R_{\mu\nu} = \Lambda g_{\mu\nu} \,. \tag{8.1.3}$$

As also mentioned in the Introduction, the $g_{\mu\nu}$ of a four-dimensional observer might differ from the components $g^5_{\mu\nu}$ of the five-dimensional metric by a function $A = A(\psi)$ of the internal coordinate called warping:

$$g^5_{\mu\nu} = \mathrm{e}^{2A(\psi)} g_{\mu\nu} \,. \tag{8.1.4}$$

Indeed, this does not break any spacetime symmetries. The mixed components $g^5_{4\mu} = g^5_{\psi\mu}$ look like a vector in four dimensions, so they break some of the symmetries of MS_4, and should not be allowed in a vacuum solution:

$$g^5_{\psi\mu} = 0 \,. \tag{8.1.5}$$

Finally we have the single component $g^5_{\psi\psi} = g^5_{44}$. All metrics on S^1 are equivalent, or in other words diffeomorphic: under a coordinate change $\psi = \psi(\tilde{\psi})$, every one-dimensional metric can be turned into a constant:

$$g^5_{\psi\psi} = \mathrm{e}^{2\sigma_0} \,. \tag{8.1.6}$$

We can also set $\sigma_0 = 1$ if we are willing to change $L \to L\mathrm{e}^{\sigma_0}$. With these assumptions, the $\psi\mu$ components of the equations of motion (8.1.2) vanish identically; the $\mu\nu$ and $\psi\psi$ components read (Exercise 4.3.8)

$$\mathrm{e}^{-2A}\Lambda = \partial^2_\psi A + 4(\partial_\psi A)^2 \,, \qquad 0 = \partial^2_\psi A + (\partial_\psi A)^2 \,. \tag{8.1.7}$$

The only solution is the trivial

$$\Lambda = 0 \,, \qquad A = A_0, \tag{8.1.8}$$

where now A_0 is a constant. (If the ψ direction were noncompact, we would have a solution $\mathrm{e}^A = \sqrt{\Lambda/3}\,\psi$, ultimately arising from the embedding of dS_4 in Mink_5 in (4.5.18).) So all vacuum solutions have the somewhat trivial form $\mathrm{d}s^2_5 = g^5_{MN}\mathrm{d}x^M\mathrm{d}x^N = \mathrm{d}s^2_{\mathbb{R}^4} + \mathrm{d}\psi^2$.

Dimensional reduction for gravity

For the S^1 not to have been detected so far, we would want it to be small enough that our current experiments (and our senses) would somehow average over it.[1] A crude way of implementing this average is to make the metric independent of ψ:

$$\partial_\psi g_{MN} = 0 \,. \tag{8.1.9}$$

This is called *dimensional reduction*, and we would expect it to be appropriate at distances larger than the S^1. We then wonder what physics would be observed in four dimensions. For this, we want to examine the action for fluctuations of the

[1] Another possibility would be that some or all matter fields are localized on a brane. The extra dimension would still be detectable by gravitational experiments at small distances. With an action more general than (8.1.1), a nonconstant A might be allowed, and the observed m_{Pl} might depend on x^5; the so-called Randall–Sundrum models [229, 230] used this fact to suggest a solution to the hierarchy problem.

metric around the vacuum (8.1.8). We expect the components $g_{\mu\nu}$, $g_{\mu 4}$, g_{44} (μ, $\nu = 0, \dots, 3$) to behave as a four-dimensional metric, vector field, and scalar, respectively. It is more convenient to organize them as in the bundle metric (4.2.37) and in the M-theory reduction (1.4.5a) to IIA:

$$ds_5^2 = ds_4^2 + e^{2\sigma}(d\psi + C_1)^2 . \tag{8.1.10}$$

So $g^5_{\mu\nu} = e^{-\sigma} g_{\mu\nu} + e^{2\sigma} C_\mu C_\nu$, $g^5_{\mu\psi} = e^{2\sigma} C_\mu$, and $g^5_{\psi\psi} = e^{2\sigma}$. Equation (8.1.10) is just a parameterization of the general five-dimensional metric; for now, we have made no assumptions.

We can now evaluate the Ricci scalar R_5 and the action (8.1.1) with the parameterization (8.1.10) and the hypothesis (8.1.9). Using (4.3.68):

$$S_{\rm GR5} = \frac{1}{2\kappa_4^2} \int d^4x \sqrt{-g_4} \left(e^\sigma R_4 - \frac{1}{4} e^{3\sigma} F_{\mu\nu} F^{\mu\nu} \right) , \tag{8.1.11}$$

with $\kappa_4^2 = L_0 \kappa_5^2$, and $F_{\mu\nu} = \partial_\mu C_\nu - \partial_\nu C_\mu$. (Indices are raised now with the metric $g_{\mu\nu}$.) Here it looks as though the scalar σ has no kinetic term. This is an illusion similar to the wrong-sign kinetic term for the dilaton in the string frame metric, which got resolved in the Einstein frame as in (1.2.23). Indeed if we define $g^{\rm E}_{\mu\nu} = e^{-\sigma/2} g_{\mu\nu}$ we obtain, using (4.3.70) and discarding total derivative terms:[2]

$$S_{\rm E,\,GR5} = \frac{1}{2\kappa_4^2} \int d^4x \sqrt{-g_{\rm E}} \left(R_{\rm E} - \frac{1}{4} e^{3\sigma} F_{\mu\nu} F^{\mu\nu} - \frac{3}{2} \partial_\mu \sigma \partial^\mu \sigma \right) . \tag{8.1.12}$$

So now we have obtained the usual GR action in four dimensions, together with that for a vector and a scalar field.

The vector field is a gauge field: it comes with a gauge invariance $C_\mu \sim C_\mu + \partial_\mu \lambda$. This is a relief: a massless vector without a gauge invariance would also propagate longitudinal degrees of freedom, which have the wrong sign in the kinetic energy. The gravity gauge transformations are the action of diffeomorphisms (4.1.64), $g^5_{MN} \to g^5_{MN} + 2\nabla_{(M}\xi_{N)}$. Roughly speaking, the vectors ξ_μ become the gauge transformations for four-dimensional gravity, while the extra component ξ_4 becomes the desired gauge transformation for C_μ.

Thus already in this simple model, the one originally considered by Kaluza and Klein [12, 13], we see that extra dimensions provide a possibility to unify geometrically two forces, in this case gravity and electromagnetism. Notice, however, the presence of the scalar σ. It has zero potential, $V(\sigma) = 0$, and in particular zero mass. So it would give to a long-distance "fifth force," so far not observed. The reason the potential vanishes is that (8.1.10) is a vacuum solution for any constant value of σ, as we can see from (8.1.6) and (8.1.8). In other words, the S^1 can have any size. This free parameter in the geometry becomes a line of vacua in the four-dimensional effective theory.

We will see that many compactifications suffer from this problem, sometimes generating dozens or hundreds of massless scalars. In general, massless scalars generated by dimensional reduction are called *moduli*, especially those with a geometric origin, such as our σ, borrowing a word from differential and algebraic geometry that we saw in earlier chapters, first in Section 5.3.4.

[2] It can be useful to notice that $\int d^4x \sqrt{-g} e^f (\nabla^2 f + |df|^2) = \int d^4x \sqrt{-g} \nabla^2 e^f = 0$.

KK tower

The approximation we have made in (8.1.9) will break down at scales comparable to Le^{σ}, the size of the fifth dimension: at that point, we should no longer "average" over $\psi = x^4$. We already mentioned in (0.2.1) that the field dependence on ψ gives rise to a "tower" of four-dimensional fields. First we quickly go over the case of a single scalar σ on $\mathbb{R}^4 \times S^1$, with flat metric. Its equation of motion is $\nabla^2_{(5)}\sigma = \partial_M \partial^M \sigma = 0$. We decompose $\partial_M \partial^M = \partial_\mu \partial^\mu + (\partial_\psi)^2$. Since σ should be periodic in ψ, we can expand it in a Fourier series:

$$\sigma(x^0, \dots, x^4) = \sum_k \sigma_k(x^0, \dots, x^3) e^{\frac{2\pi i k}{L}\psi} . \tag{8.1.13}$$

Now the equation of motion gives

$$0 = \partial_M \partial^M \sigma \quad \Rightarrow \quad \left(\partial_\mu \partial^\mu - \frac{4\pi^2 k^2}{L^2} \right) \sigma_k = 0 . \tag{8.1.14}$$

So each σ_k is a scalar field in four dimensions, with mass (0.2.1), $m_k = \frac{2\pi k}{L}$.

After (0.2.1), we also mentioned the result for pure gravity, which we now describe in slightly more detail. The fields σ, C_μ, and $g_{\mu\nu}$ in (8.1.10) can all be expanded in Fourier modes similar to (8.1.13). However, recall from (8.1.6) that with a coordinate change we can always set the internal component $g_{\psi\psi}$ to a constant. So in (8.1.10) we can make σ not depend on ψ. This indicates that the higher KK modes σ_k, $k \neq 0$ might not be physical after all. The reason for this is a Brout–Englert–Higgs mechanism: the higher KK modes $g^k_{\mu\nu}$, $k > 0$ are massive spin-2 fields, and so they have more propagating degrees of freedom than a massless spin-2 field; these are exactly the KK modes for A^k_μ, σ^k, which are "eaten." The only uneaten fields are A^0_μ and σ^0. So the KK spectrum for gravity on S^1 is

$$g^k_{\mu\nu} ; \qquad A^0_\mu , \qquad \sigma^0 . \tag{8.1.15}$$

With respect to our earlier, cruder dimensional reduction, we now have found a tower of spin-2 fields.

The model in this section is peculiar in that the metric of the extra dimension depends on only a single parameter. Models with more than one extra dimension are significantly more complicated, in part because there are infinitely many inequivalent internal metrics. Let us now turn to this.

8.1.2 Ten-dimensional pure gravity

We again consider pure gravity, but to fix ideas we take the number of extra dimensions to be six, as in string theory. We take M_6 compact, and we call y^m, $m = 1, \dots, 6$ its local coordinates. The possible topologies are classified by Wall's theorem (Section 4.2.5), but in the following we will be more interested in the choice of metric on M_6.

Vacuum solutions

To find vacuum solutions, we still need to take $g^{10}_{\mu\nu} = e^{2A} g_{\mu\nu}$, now with $A = A(y^m)$, and $g^{10}_{\mu m} = 0$, as in (8.1.4) and (8.1.5). It is now far from being the case that all internal metrics are equivalent, as in (8.1.6) with a single internal coordinate. The equations of motion are still $R_{MN} = 0$, and now decompose as

$$e^{-2A}\Lambda = \nabla_6^2 A + 4\partial_m A \partial^m A\,, \qquad (8.1.16a)$$

$$R^6_{mn} = 4(\nabla^6_m \partial_n A + \partial_m A \partial_n A)\,, \qquad (8.1.16b)$$

where the label 6 denotes use of the internal metric. Equation (8.1.16a) is of the form (0.5.3) in the introduction, with $T_{MN} = 0$ because we are now in pure gravity; integrating over M_6 and using Stokes's theorem (4.1.131) as we did there, we conclude that the cosmological constant vanishes. Going back to (8.1.16a) and integrating it over M_6 again, we then find $0 = \int_{M_6} d^6 y \sqrt{g_6} \partial_m A \partial^m A$; since the integrand is always positive, it follows that A is constant. Finally (8.1.16b) tells us that the Ricci tensor vanishes as well. So:

$$\Lambda = 0\,, \qquad A = A_0\,, \qquad R_{mn} = 0\,, \qquad (8.1.17)$$

similar to (8.1.8). Again, the conclusion (8.1.17) can be avoided if the internal space is noncompact (see [231] for analytic examples in other dimensions), although such solutions of course cannot be called compactifications.

Reduction

Once again we would like to have an effective action that describes physics for a four-dimensional observer. This is superficially similar to the five-dimensional case, but with complications that give us a flavor of bigger problems to come.

- The internal components g_{mn} give rise to scalars in $d = 4$. There are infinitely many internal metrics, and infinitely many scalars; unlike in Section 8.1.1, they will not all be "eaten" by massive spin-2 fields. It is natural to keep only the lightest scalars around a vacuum. Ricci-flat manifolds come in finite-dimensional families $\mathcal{M}_{\rm Rf}$. The four-dimensional theory should then have $k = \dim \mathcal{M}_{\rm Rf}$ massless scalars σ^i, whose expectation values lead to vacua for the effective theory. This is the closest analogue to the "dimensional reduction" procedure of Section 8.1.1: we look at physics at length scales well above the size of M_6, where only massless degrees of freedom are relevant.
- The mixed components $g_{\mu m}$ vanish in a vacuum solution, but in a $d = 4$ effective theory, we keep their fluctuations as we did in (8.1.10). Since they have a single index m, they are vector fields on M_6; imposing them to be constant would be coordinate dependent. If we want to obtain massless vectors in $d = 4$, we should also have an associated gauge invariance; in Section 8.1.1, these were provided by translations along the S^1, generated by the vector field ∂_y. So in the present case, we should look for internal vector fields that generate symmetries of the internal metric: Killing vector fields K^a. This leads to

$$g_{\mu m} = \sum_a A^a_\mu(x) K^a_m(y)\,. \qquad (8.1.18)$$

- Finally we have the spacetime components of the metric. Here there are no internal indices, and so it makes sense to impose that $g_{\mu\nu}$ is constant.

All this would lead to taking the $d = 10$ metric to have the bundle metric form (4.2.35). Intriguingly, in such a reduction the group of isometries of the internal space becomes the gauge group of the $d = 4$ theory. However, (7.1.10) shows that Ricci-flat spaces have isometries only if they are not simply connected. Proper Calabi–Yaus have no vectors; this would leave only the torus T^6 and $T^2 \times$ K3. So in the vast majority of cases, there are no massless vectors in the reduction. So we take simply

$$\mathrm{d}s^2_{10} = \mathrm{d}s^2_4 + \mathrm{d}s^2_6(\sigma^i)\,, \tag{8.1.19}$$

with a Ricci-flat internal metric $g_{mn}(\sigma^i)$. The dynamical fields are the $d = 4$ metric, and the $k = \dim \mathcal{M}_{\mathrm{Rf}}$ scalars σ^i. Evaluating the $d = 10$ Einstein–Hilbert action on (8.1.19) gives

$$\int \mathrm{d}^4x\sqrt{-g_4}\mathrm{Vol}(M_6)(R_4 - G_{ij}\partial_\mu\sigma^i\partial^\mu\sigma^j)\,, \tag{8.1.20}$$

similar to (8.1.12);

$$G_{ij} = \frac{1}{\mathrm{Vol}(M_6)}\int_{M_6} \mathrm{d}^6y\sqrt{g_6}\partial_{\sigma_i}g_{mn}\partial_{\sigma_j}g_{pq}g^{mp}g^{nq} \tag{8.1.21}$$

is a metric on the space of Ricci-flat deformations.

As in the five-dimensional case, at higher energies this low-energy action approximation will gradually break down, and it will be necessary to include some massive fields. The potential of such a theory would be

$$V_{d=4} = -\int_{M_6} \sqrt{g_6}R_6\,, \tag{8.1.22}$$

obtained evaluating the $d = 10$ action on configurations where all the scalars are constant along $d = 4$ spacetime. Equation (8.1.22) did not appear in (8.1.20) because the internal g_{mn} was Ricci-flat there. To see that (8.1.22) is sensible, we can extremize it with respect to an scalar describing a deformation δg_{mn} of the internal space; this is a familiar computation from general relativity, and it gives $R_{mn} = 0$, which is indeed the vacuum in (8.1.17). At the quadratic level, (8.1.22) gives masses to the scalars we have not included in (8.1.20); as in $d = 5$, we call the set of these masses KK spectrum, even though it will not follow (0.2.1).

On any M_6, we can make (8.1.22) arbitrarily negative. For example, starting from any metric, rescaling it by an overall function as in (4.1.158) and (4.3.70), and using footnote 2 preceding (8.1.12), we have the following:

$$\int_{M_6} \sqrt{g_6}R_6 \to \int_{M_6} \sqrt{g_6}(R_6 + 20|\mathrm{d}\lambda|^2)\,. \tag{8.1.23}$$

This might make (8.1.22) look unbounded from below. This is solved [232] once one takes into account the constraints of general relativity.

KK tower for a free scalar

As we did for $d = 5$, we first consider the toy model of a free scalar. The Laplace operator ∇^2 (4.1.114) decomposes as $\partial_\mu\partial^\mu + \nabla^2_6$. Instead of using a Fourier series, we should now decompose $\sigma = \sum_k \sigma_k(x)s_k(y)$, and

$$\nabla^2_6 s_k = \lambda_k s_k\,. \tag{8.1.24}$$

The equation of motion now gives

$$0 = g^{MN}\nabla_M\nabla_N\sigma \quad\Rightarrow\quad \left(\partial_\mu\partial^\mu + \lambda_k\right)s_k = 0\,. \tag{8.1.25}$$

One eigenvalue λ_0 is always zero, corresponding to $\sigma_0 = 1$. If M_6 is compact, specializing (4.1.116) to functions, we see that the remaining λ_k are negative, as in (0.2.1). Explicit expressions for the λ_k can be obtained for metrics that have many symmetries, such as the homogeneous spaces G/H of Section 4.4.2. More generally, various mathematical results exist about the first nontrivial mass λ_1, relating it to the curvature or to the *diameter* $d(M_6)$ (the largest distance among two points); see [233]. For example, if the Ricci curvature of M_6 is nonnegative, then [234]

$$\lambda_1 \geq \frac{\pi^2}{4d^2(M_6)} . \tag{8.1.26}$$

So a larger size corresponds to lower masses, as in (0.2.1).

KK spectrum for gravity

We now turn to the KK spectrum for gravity, around any vacuum (8.1.16); see [235, sec. 5]. The computation involves different operators for the various components of the metric:

- The spin-2 spectrum consists of a massless graviton $g^0_{\mu\nu}$, and of massive $g^k_{\mu\nu}$; since these are internal scalars, the masses are given by the eigenvalues $\lambda_k \neq 0$ of the scalar Laplacian (8.1.24).
- The spin-1 spectrum has a massless vector A^a_μ for each internal Killing vector, as we concluded in (8.1.18), and a tower of massive vectors A^I_μ with masses $\lambda_I \neq 0$, one for every non-Killing co-closed eigenform α_I of the Laplacian (4.1.115) on one-forms: $\Delta\alpha_I = \lambda_I\alpha_I$, $\nabla_m\alpha^m_I = 0$, $\nabla_{(m}\alpha^I_{n)} \neq 0$.
- The scalar spectrum has two types of scalars $\lambda_a \neq 0$, $\hat{\lambda}_a$: one for every eigenfunction f_a of the scalar Laplacian $\nabla^2 f_a = \lambda_a f_a$ such that $\partial_m f_a$ is *not* a conformal Killing vector (4.1.68); and one for every symmetric tensor $h^{\rm TT}_{a\,mn}$ that is transverse traceless (TT) ($\nabla^m h^{\rm TT}_{a\,mn} = 0$, $h^{\rm TT}_{a\,m}{}^m = 0$), which is an eigentensor of the *Lichnerowicz operator*: $\Delta^{\rm L} h^{\rm TT}_{a\,mn} = \hat{\lambda}_a h^{\rm TT}_{a\,mn}$, where

$$\Delta^{\rm L} h_{mn} \equiv -\nabla^2 h_{mn} - 2R_{mpnq}h^{pq} + 2R_{(m}{}^p h_{n)p} \, ; \tag{8.1.27}$$

  see Exercise 8.1.3. In particular, the modes $\hat{\lambda}_a = 0$ correspond to Ricci-flat moduli.

This applies to an internal space of any dimension, and in particular to the S^1 of the previous subsection. In that case, the only co-closed one-form is constant, and so also Killing; there are no internal traceless tensors, and every vector is a conformal Killing vector. So there are no massive KK modes except in spin-2, reproducing (8.1.15).

Truncating massive modes is not consistent

Our $d = 4$ action (8.1.20) contains only massless modes; at low energies, ignoring the massive KK modes is physically justified. From a formal point of view, however, there is no guarantee that a solution of the effective action that only includes massless fields corresponds to a solution in $d = 10$.

A toy model is provided by a massless scalar L and a heavy scalar H, and action

$$S_{\rm LH} = \frac{1}{2}\int \sqrt{-g}{\rm d}^4x \left(-\partial_\mu L\partial^\mu L - \partial_\mu H\partial^\mu H + M^2H^2 + L^2H\right) . \qquad (8.1.28)$$

The equations of motion are

$$\nabla^2 L = LH , \qquad (\nabla^2 - M^2)H = \frac{1}{2}L^2 . \qquad (8.1.29)$$

Setting $H = 0$ in the action (8.1.28), we obtain $S_{\rm L} = -1/2\int {\rm d}^4x\partial_\mu L\partial^\mu L$, the action for a massless scalar, with equation of motion $\nabla^2 L = 0$. But setting $H = 0$ in (8.1.29) gives $\nabla^2 L = 0$ *and* $L^2 = 0$, which have no solution. So setting $H = 0$, or *truncating* it, is not consistent: the action obtained by doing so forgets an important equation of motion.

A *consistent truncation* is an action $S_{\rm tr}$ obtained from another S by setting to imposing some constraints on the fields, such that every solution of $S_{\rm tr}$ is also a solution of S; in other words, such that the phenomenon we saw after (8.1.29) does not happen. In particular, this definition can be applied to a reduction, where we set to zero most KK modes. Consistency here would mean that every solution to the effective action in $d = 4$ is in fact also a solution in $d = 10$.

Unfortunately, the reduction of gravity on a Ricci-flat space is believed *not to be consistent* [236]. The aim of the KK spectrum computation we described earlier was finding the masses; so cubic interactions of the type L^2H as in (8.1.28) were not visible there. However, the action in $d = 10$ is nonlinear, and there is no symmetry principle protecting against such as couplings. In this particular example, this is a rather formal point: at low energies, we can perform the path integral over the KK modes, integrating them out. This will change the effective theory of the massless modes by some operators of dimension greater than four, irrelevant at low energies, giving a more physical justification to the truncation. Moreover, for classical solutions we often have a good idea of a $d = 10$ origin of a classical solution. For example, classical black hole solutions in $d = 4$ may not be guaranteed to have a $d = 10$ origin because of the inconsistency of the reduction, but we usually know that they originate from black holes or p-brane solutions in $d = 10$, especially in the supersymmetric case.

On the other hand, consistency becomes an important issue for reductions to vacua with negative cosmological constant, which we will start discussing in Section 8.3.

Exercise 8.1.1 Using (4.3.66b) and integrating by parts, show (8.1.20).

Exercise 8.1.2 Count the number of massless fields in (8.1.20) for $M_6 = T^6$, K3$\times T^2$, or a Calabi–Yau.

Exercise 8.1.3 Deforming a metric $g_{mn} \to g_{mn} + h_{mn}$, with small h_{mn}, show using (4.1.21) that

$$R_{mn} \to R_{mn} + \frac{1}{2}\Delta^{\rm L} h_{mn} + \nabla_{(m}\nabla^p h_{n)p} - \frac{1}{2}\nabla_m\nabla_n h_p{}^p , \qquad (8.1.30)$$

where $\Delta^{\rm L}$ is the Lichnerowicz operator (8.1.27). So in particular if $h_{mn} = h^{\rm TT}_{mn}$ the last two terms vanish, and we see that TT zero modes of $\Delta^{\rm L}$ are moduli.

Exercise 8.1.4 On a Calabi–Yau M_6, show that both types of moduli (7.1.11) are zero modes of $\Delta^{\rm L}$, as expected from the previous exercise.

8.2 String theory vacua with no flux

We now consider the vacuum problem more specifically in string theory. The most important new element is supersymmetry. Rather than considering the supersymmetry transformations in their full generality right away, we will introduce them gradually. In this section, we set all the form fields to zero.

For most of this section, we will consider type II supergravity. After briefly discussing the supersymmetry transformations in Section 8.2.1, we refine in Sections 8.2.2 and 8.2.3 what we mean by vacuum compactification, building on our work in Section 8.1.2. We then conclude in Section 8.2.4 that with our simple ingredients, we need the internal space to be Ricci-flat. In Section 8.2.5, we briefly examine how our discussion changes for the heterotic string.

8.2.1 Supersymmetry

If we set all form fields to zero, in both type II theories the supersymmetry transformations are

$$\begin{aligned} \delta\psi^a_M &= D_M\epsilon^a\,, \qquad & \delta e^A_M &= \bar\epsilon^a\Gamma^A\psi_{aM}\,; \\ \delta\lambda^a &= \partial_M\phi\gamma^M\epsilon^a\,, \qquad & \delta\phi &= \frac{1}{2}\bar\epsilon^a\lambda_a\,. \end{aligned} \tag{8.2.1}$$

The gravitinos ψ^a_M are vector-spinors, sections of the bundle $(S\otimes T)M$ (Section 2.3.1). The two Majorana–Weyl spinors ϵ^a, $a=1,2$, are the infinitesimal parameters of the transformation, spinorial analogues of the vector field v_m appearing in the infinitesimal transformation (4.1.64) of the metric under diffeomorphisms. The chirality of all spinors is summarized in Table 8.1.

The equations of motion in this setup are simply

$$R_{MN}+2\nabla_M\partial_N\phi=0\,,\qquad R=4e^{\phi}\nabla^2e^{\phi}\,. \tag{8.2.2}$$

Supersymmetric solutions are defined as those for which $\delta\psi^a_M=\delta\lambda^a=0$; so

$$D_M\epsilon^a=0\,, \tag{8.2.3a}$$

$$\partial_M\phi\gamma^M\epsilon^a=0\,. \tag{8.2.3b}$$

By our discussion in Section 5.7, (8.2.3a) implies that spacetime has reduced holonomy. We presented in that section the classification of such spaces, but only in the Riemannian case, mentioning that the list is longer for Lorentzian signature.

But we will now focus on vacuum solutions, where (8.2.3a) implies reduced holonomy on the internal space, as we will now see.

Table 8.1. Spinor chirality.

	ϵ^1	ϵ^2	ψ^1_M	ψ^2_M	λ^1	λ^2
IIA	+	−	+	−	−	+
IIB	+	+	+	+	−	−

8.2.2 Vacua and spinors

In the Introduction and again in Section 8.1, we defined a vacuum solution to be one where macroscopic spacetime is a maximally symmetric space MS_4, and where the expectation values of all fields preserve its symmetries. We concluded that the off-diagonal components should vanish, but that the external components are allowed a simple dependence on the internal coordinates y. We summarize this as

$$ds^2_{10} = e^{2A(y)} ds^2_{MS_4} + ds^2_{M_6} . \tag{8.2.4}$$

The only other field we are considering in this section is ϕ; for it not to break the symmetries of MS_4, it should only be a function of the internal coordinates, $\phi = \phi(y)$.

We will now see what the same ideas imply on fermions.

Gravitinos

The space of spinors on $\mathbb{R}^{1,9}$ is a tensor product of spinors on $\mathbb{R}^{1,3}$ and on $\mathbb{R}^6$; recall (2.2.20). In curved space, we can apply this decomposition to the gamma matrices with flat indices (Section 4.3.1). We already did so in Section 7.3.3 when we discussed spinors on conical manifolds. For the metric (8.2.4), it is natural to take a vielbein

$$\begin{aligned} e^\alpha &= e^A e^\alpha_{(4)} , & \alpha &= 0, \ldots, 3 ; \\ e^{a+3} &= e^a_{(6)} , & a &= 1, \ldots, 6 . \end{aligned} \tag{8.2.5}$$

α and a are the flat indices relative to MS_4 and M_6 respectively (just like μ and m are their curved indices). In curved indices, (2.2.20) then becomes

$$\Gamma_\mu = e^A \gamma_\mu \otimes 1 , \qquad \Gamma_{m+3} = \gamma_{(4)} \otimes \gamma_m , \tag{8.2.6}$$

where now γ_μ and γ_m generate the Clifford algebras on MS_4 and M_6 respectively, and $\gamma_{(4)} = i\gamma^{0123}$ is the chiral matrix in MS_4. (As remarked earlier, the notation γ_5, customary in QFT, seems potentially confusing when we work with additional dimensions.) We will usually have in mind the real basis we mentioned after (2.2.20). The chirality matrix for (8.2.6) reads

$$\Gamma = \gamma_{(4)} \otimes \gamma_{(6)} . \tag{8.2.7}$$

(Each chirality matrix is the product of all gamma matrices with *flat* indices, so that they square to the identity.)

We now want to impose invariance under the symmetries of MS_4. Consider first the internal components ψ^a_m ($a = 1, 2$). Fix a point $P \in M_{10}$, and let p be its projection to MS_4. Since the space of spinors at a point decomposes as a tensor product, by definition we can expand

$$\psi^a_m(P) = \sum_{IJ} c_{IJ} \psi^{(4)a}_I \otimes \psi^{(6)a}_{Jm} \tag{8.2.8}$$

with $\{\psi_I^{(4)a}\}$ and $\{\psi_{Jm}^{(6)a}\}$ bases for spinors and vector-spinors on MS_4 and M_6 respectively, and c_{IJ} some coefficients. Among the isometries of MS_4, those in

$$\text{Stab}(p) = \text{SO}(1,3)\,, \tag{8.2.9}$$

leave p invariant. (This stabilizer is clear for Mink_4, and follows from (4.5.19), (4.5.29) for $\Lambda \neq 0$.) Already under (8.2.9), there is no invariant spinor on MS_4: the space of chiral spinors is an irreducible Lorentz representation, and in particular does not contain a singlet. (In our analysis of orbits in Section 2.4, such a spinor would have shown up as a one-dimensional orbit.) We conclude that ψ_m^a vanish in P, but since the isometry group acts transitively, all points are equivalent and in fact $\psi_m^a = 0$ everywhere.

The same argument also tells us that the dilatinos have to vanish, $\lambda^a = 0$. As for the gravitino spacetime components ψ_μ^a, we can adapt the argument by using the fact that the spinorial⊗fundamental representation of the Lorentz group contains no singlet. All in all, we are led to take

$$\psi_M^a = \lambda^a = 0\,. \tag{8.2.10}$$

This has the added benefit of making δe_M^A and $\delta\phi$ in (8.2.1) vanish automatically. So (8.2.3) is all we need to impose for unbroken supersymmetry. This will still hold when we give expectation values to the other string fields.

One often assumes (8.2.10) even beyond the vacuum case. Supersymmetric solutions with nonzero fermionic fields, however, do exist: given a solution where $\psi_M^a = 0$, acting with one of the ϵ such that $\delta\psi_M^a \neq 0$ will produce a new solution with nonzero fermions, since supersymmetry maps solutions into solutions.

8.2.3 Supercharges

We now turn to the spinorial parameter ϵ^a of the supersymmetry transformations (8.2.1); we are especially interested in those such that satisfy (8.2.3). By an abuse of language, these are sometimes called supercharges, since they appear in the conserved charges $Q(\epsilon^a)$.

Factorization

The argument that made us set (8.2.10) does not apply to ϵ. If we again select $P \in M_{10}$, $p \in MS_4$, we can decompose ϵ as a tensor product as in (8.2.8), $\epsilon(P) = \sum_{IJ} c_{IJ}\zeta_I^{(4)a} \otimes \eta_{Jm}^{(6)a}$. Let us rewrite this as

$$\epsilon^a = \sum_J \zeta_J^{(4)a} \otimes \eta_J^{(6)a}\,, \tag{8.2.11}$$

where now $\zeta_a^{(4)}$ and $\eta_J^{(6)a}$ are a set of four- and six-dimensional spinors rather than a basis. In order not to break the isometries $\text{ISO}(MS_4)$, we have to pick a *set* of $\zeta_J^{(4)}$, not just a single one, which is invariant under $\text{ISO}(MS_4)$. This set is a vector space, since the equation $D_M\epsilon = 0$ is linear. So in (8.2.11), we need to choose a vector space spanned by the $\zeta_J^{(4)a}$, that contains an orbit under $\text{Stab}(p)$ (8.2.9), the Lorentz group. An example is the space $S_\pm$ of chiral spinors of either chirality. There are other possibilities, such as the space of spinors of the form $\zeta_+ + \alpha\zeta_-$ for some number α. All these possibilities can be summarized by refining (8.2.11) as

$$\epsilon^1 = \sum_J (\zeta^1_{+J} \otimes \eta^1_{+J} + \zeta^1_{-J} \otimes \eta^1_{-J}) ,$$
$$\epsilon^2 = \sum_J (\zeta^2_{+J} \otimes \eta^2_{\mp J} + \zeta^2_{-J} \otimes \eta^2_{\pm J}) , \qquad {}^{\text{IIA}}_{\text{IIB}} \tag{8.2.12}$$

with the understanding that the $\zeta^a_{\pm J}$ are free to roam over the full spaces of chiral spinors $S_\pm$, possibly with some Lorentz-covariant constraints among them. We will see some examples of such constraints soon. We define the amount of supersymmetry $\mathcal{N}$ as the number of independent $\zeta^a_{\pm J}$ in (8.2.12). With respect to (8.2.11), we have now dropped the superscript for ease of reading, and we just call $\zeta \equiv \zeta^{(4)}$, $\eta \equiv \eta^{(6)}$. The chirality of the $\eta^{(6)a}_J$ has been taken to agree with the ten-dimensional chirality prescribed for the ϵ^a by Table 8.1, recalling also (8.2.7).

The ϵ^a should obey the Majorana condition. From (2.3.10), we see that conjugation $(\ )^{\text{c}}$ changes chirality both in $d = 4$ Lorentzian and $d = 6$ Euclidean. So $(\epsilon^a)^{\text{c}} = \epsilon^a$ imposes $(\zeta^a_{\pm I})^{\text{c}} = m^{a\pm}_{IJ} \zeta^a_{\mp J}$, with the inverse relation on the η^a_I; conjugating these equations demands $\bar{m}^{a+} m^{a-} = 1$ (no sum over a). But by redefining $\zeta^{a\pm}_I \to c^{a\pm}_{IJ} \zeta^{a\pm}_J$, the matrix $m^{a+} \to (c^{a-})^{-1} m^{a+} c^{a+}$, so we can always set $m^{a\pm} = 1$. In conclusion,

$$(\zeta^a_{\pm I})^{\text{c}} = \zeta^a_{\mp I} , \qquad (\eta^a_{\pm I})^{\text{c}} = \eta^a_{\mp I} . \tag{8.2.13}$$

In other words, the spinors $\zeta^a_I \equiv \zeta^a_+ + \zeta^a_-$, $\eta^a_I \equiv \eta^a_{+I} + \eta^a_{-I}$ should both be Majorana.

Killing spinors

We are not quite done: we should require that (8.2.12) is invariant also under the isometries that move p. The η^a_I are simply scalars as far as MS_4 is concerned, so they should again be constant. The ζ^a_I should form a representation of the isometry group of MS_4. For Mink_4, a natural choice is simply the space of constant spinors. For the other MS_4, we saw in (7.3.14) (at the infinitesimal level) that one such representation is offered by the space of *Killing spinors* defined in (7.3.8). The $\zeta^a_{\pm I}$ are chiral, but we can project on the two chiralities; this leads us to the assumptions

$$D^{(4)}_\mu \zeta^a_{\pm I} = \frac{\mu}{2} \gamma_\mu \zeta^a_{\mp I} , \tag{8.2.14a}$$
$$\partial_m \zeta^a_I = 0 . \tag{8.2.14b}$$

An advantage of Killing spinors is that they automatically define a superalgebra, recalling a comment following (7.3.8). This uses the fact that the Killing vectors are bilinears of the Killing spinors. In this book, we will take (8.2.14) as a *definition* of supercharge for a supersymmetric vacuum. More general choices might conceivably make physical sense. We will briefly consider some of these generalizations at the end of this subsection.

By conjugating (8.2.14a) and using (8.2.13), we see that μ is *real*.[3] Moreover, by (7.3.16),

$$R^{(4)}_{\mu\nu} = \Lambda g^{(4)}_{\mu\nu} , \qquad \Lambda = -3\mu^2 \leq 0 . \tag{8.2.15}$$

[3] It would be possible to consider the alternative equation (7.3.9); even more generally, one might mix this with our (8.2.14a). This would lead to $D^{(4)}_\mu \zeta^a_{+I} = \frac{\mu}{2} \gamma_\mu \zeta^a_{-I}$, $D^{(4)}_\mu \zeta^a_{-I} = \frac{\bar{\mu}}{2} \gamma_\mu \zeta^a_{+I}$, and now μ would be allowed to be complex; (8.2.15) would be modified as $\Lambda = -3|\mu|^2$. However, this seemingly more general possibility can be mapped back to (8.2.14a) by $\zeta^a_{\pm I} \to \text{e}^{\pm i\alpha} \zeta^a_{\pm I}$.

Equation (4.5.2) now can only be satisfied with $k < 0$, corresponding to $\mathrm{MS}_4 = \mathrm{AdS}_4$, or $k = 0$, $\mathrm{MS}_4 = \mathrm{Mink}_4$. So there are *no supersymmetric de Sitter compactifications*.[4]

For AdS_4, the value of k was set to -1 in presenting the metric in Section 4.5.4; this gives $\mu = \pm 1$. We can also use $g_{\mu\nu} = L^2_{\mathrm{AdS}} g^{\mathrm{AdS}_4}_{\mu\nu}$, where $g^{\mathrm{AdS}_4}_{\mu\nu}$ is the AdS_4 metric we presented in various forms in Section 4.5.4; then

$$\mu = \pm\frac{1}{L_{\mathrm{AdS}}} . \tag{8.2.16}$$

Usually we can reabsorb this L_{AdS} by rescaling the warping function e^{2A} in (8.2.4).

We still need to demonstrate that (8.2.14a) can in fact be solved on AdS_4. In the Poincaré coordinates (4.5.37), x^α, $\alpha = 0, 1, 2$, and $\tilde{r} \equiv -\log z = \log r$, one can use the result of Exercise 4.3.3 with $a = \mathrm{e}^{\tilde{r}}$. This solves $D_\mu \zeta = \frac{1}{2}\zeta$ with

$$\zeta = \mathrm{e}^{\tilde{r}/2}\zeta_0 , \qquad \gamma_{\tilde{r}}\zeta_0 = \zeta_0 ; \tag{8.2.17a}$$

$$\zeta = \left(\mathrm{e}^{-\tilde{r}/2} + \mathrm{e}^{\tilde{r}/2}(x^\alpha\gamma_\alpha)\right)\zeta_0' , \qquad \gamma_{\tilde{r}}\zeta_0' = \zeta_0' , \tag{8.2.17b}$$

where both ζ_0 and ζ_0' are constant spinors. Splitting $\zeta = \zeta_+ + \zeta_-$ produces a solution to (8.2.14a).

Possible generalizations

We will now consider briefly possible generalizations to (8.2.14a).

A first idea might be to consider

$$D^{(4)}_\mu \zeta^a_{\pm I} = \frac{1}{2}\mu_{IJ}\gamma_\mu \zeta^a_{\mp J} . \tag{8.2.18}$$

The integrability argument in (8.2.15) now implies $\mu_{IK}\bar{\mu}_{KJ} = -\frac{1}{3}\Lambda\delta_{IJ}$. For $I = 1, 2$, starting from this condition, one can show that with some redefinitions (8.2.18) reduces to (8.2.14a) [237, sec. 2.1] (see also a more explicit argument for AdS_3 in [238, app. B]).

Another idea would be considering the spectrum of the Dirac operator $D^{(4)} = \gamma^\mu D^{(4)}_\mu$ on MS_4. Since $[D, L_K] = 0$, by Exercise 7.3.2, the eigenspace of every eigenvalue of D is a representation for the L_K, and hence for the isometry group $\mathrm{ISO}(\mathrm{MS}_4)$. The ζ^a_I could be taken to be any of these.

Yet another possibility is that the ζ^a_I might depend on the internal coordinates y. For an example [239], consider a solution

$$\mathrm{AdS}_5 \times M_5 , \tag{8.2.19}$$

with a direct product metric. In Poincaré coordinates (4.5.37), we can rewrite (8.2.19) formally as a warped product $\mathrm{Mink}_4 \times M_6$:

$$\mathrm{d}s^2_{\mathrm{AdS}_5} + \mathrm{d}s^2_{M_5} = r^2\mathrm{d}s^2_{\mathrm{Mink}_4} + \frac{1}{r^2}(\mathrm{d}r^2 + r^2\mathrm{d}s^2_{M_5}) . \tag{8.2.20}$$

(This trick was, for example, used in [240] to classify AdS compactifications from $d = 11$.) The right-hand side is indeed of the type (8.2.4), with a warping

[4] There are superalgebras whose bosonic part contains a summand $\mathrm{iso}(\mathrm{dS}_d) = \mathrm{so}(1, d)$, but for these one can show $\sum_\alpha \{Q_\alpha, Q^{\alpha\dagger}\} = 0$, which implies the presence of ghosts [241, 242].

$e^{2A} = r^2$ and a six-dimensional metric proportional to the cone $C(M_5)$ (7.3.2). The "internal space" is noncompact, but other than this, (8.2.20) does formally satisfy the definition of a vacuum solution. Suppose now M_5 is such that (8.2.19) is supersymmetric; we will see in Section 11.2.4 that there are many examples, for example where M_5 is a Sasaki–Einstein manifold. The supercharges are a tensor product similar to (8.2.12), except for the fact that there is no chirality in $d = 5$. Along AdS_5, they are Killing spinors, again given by (8.2.17), only now with $\alpha = 0, \ldots, 3$. Trying to rewrite these as spinors on Mink_4, we see that they are *not* of the factorized form (8.2.14b): they have a dependence on $\tilde{r} = \log r$, now one of the internal coordinates. The second type, (8.2.17b), are not even constant along Mink_4. Even more strikingly, in (4.5.44) we even have a foliation of AdS_{d+1} in copies of dS_d rather than Mink_4; so by dropping (8.2.14b), one would seem to get supersymmetric de Sitter solutions.

However, supercharges of this type will obey the superalgebra of a supersymmetric theory in four dimensions. For example, in the preceding example, the supercharges of the type (8.2.17b) with $\alpha = 0, \ldots, 3$ are related to the additional fermionic generators of the superconformal algebra (often called S_α, beyond the usual Q_α). This suggests that Mink_4 solutions where such exotic supercharges are present might always in the end be of the type (8.2.20). But I am not aware of a proof of this fact, and this issue remains open.

In conclusion, in this section we have given a *definition* of supersymmetric compactifications that seems to capture the physics requirements; in the rest of the book, we will use this definition, but it might be interesting to explore generalizations in the future.

8.2.4 Ricci-flat internal space

We now apply the Ansatz (8.2.12), (8.2.13), and (8.2.14a) we developed to the gravitino equation (8.2.3a). The spin connection corresponding to (8.2.5) is obtained from (4.3.28):

$$\omega^{\alpha}{}_{\beta} = \omega^{\alpha}_{(4)\beta}\,, \qquad \omega^{\alpha}{}_{b} = e^{\alpha} E^{m}_{b}\partial_m A\,, \qquad \omega^{a}{}_{b} = \omega^{a}_{(6)b}\,. \tag{8.2.21}$$

The spinorial covariant derivative (4.3.20) is then, recalling also (8.2.6),

$$D_\mu = D^{(4)}_\mu + \frac{1}{2}\mathrm{e}^{A}\partial_m A \gamma_\mu \gamma_{(4)} \otimes \gamma^m\,, \qquad D_m = D^{(6)}_m\,. \tag{8.2.22}$$

For the time being, we put no constraints on the ζ^a_{+J} (a possibility we mentioned before (8.2.12)). From the $M = m$ components, we obtain

$$\begin{aligned}
&\sum_J \left(\zeta^1_{+J} \otimes D_m \eta^1_{+J} + \zeta^1_{-J} \otimes D_m \eta^1_{-J}\right) = 0\,,\\
&\sum_J \left(\zeta^2_{+J} \otimes D_m \eta^2_{\mp J} + \zeta^2_{-J} \otimes D_m \eta^2_{\pm J}\right) = 0\,;
\end{aligned} \tag{8.2.23a}$$

from the $M = \mu$ components,

$$\sum_J \left(\gamma_\mu \zeta^1_{+J} \otimes (\mu e^{-A}\eta^1_{-J} + \mathrm{d}A \cdot \eta^1_{+J}) + \gamma_\mu \zeta^1_{-J} \otimes (\mu e^{-A}\eta^1_{+J} - \mathrm{d}A \cdot \eta^1_{-J})\right) = 0$$
$$\sum_J \left(\gamma_\mu \zeta^2_{+J} \otimes (\mu e^{-A}\eta^2_{\pm J} + \mathrm{d}A \cdot \eta^2_{\mp J}) + \gamma_\mu \zeta^2_{-J} \otimes (\mu e^{-A}\eta^2_{\mp J} - \mathrm{d}A \cdot \eta^2_{\pm J})\right) = 0 \tag{8.2.23b}$$

Here $\mathrm{d}A\cdot = \partial_m A \gamma^m = \not{\mathrm{d}}A$, following the old definition (2.4.13). The $\zeta^a_{\pm J}$ are all independent by assumption, so the six-dimensional spinors they multiply must all vanish separately. The parenthesis multiplied by $\gamma_\mu \zeta^a_{-J}$ are the conjugates of those multiplied by $\gamma_\mu \zeta^a_{+J}$, as follows from the Majorana property of the whole equation. Had we not assumed the Killing spinor condition (8.2.14a), the terms in (8.2.23b) would not have combined with a common γ_μ. In this case, this is not going to make much difference in the end, but in the more complicated cases of Chapter 10 it will be essential.

So from (8.2.23) we extract the equations

$$D_m \eta^a_{+I} = 0 \tag{8.2.24a}$$
$$\mu e^{-A}\, \eta^1_{+I} - \mathrm{d}A \cdot \eta^1_{-I} = 0\,, \tag{8.2.24b}$$
$$\mu e^{-A}\, \eta^2_{+I} \pm \mathrm{d}A \cdot \eta^2_{-I} = 0\,. \tag{8.2.24c}$$

Equation (8.2.24a) shows that M_6 has reduced holonomy. Even if there is only one such spinor, we know from Section 7.1 that this is a Calabi–Yau manifold, with holonomy SU(3). For (8.2.24b), recall from (3.4.86) that η^1_{+1} and $\gamma^m \eta^1_{-1}$ are linearly independent, so their coefficients have to vanish separately:[5]

$$\mu = 0\,, \qquad \mathrm{d}A = 0\,. \tag{8.2.25}$$

So $\mathrm{MS}_4 = \mathrm{Mink}_4$, and the warping function A is a constant, which can be reabsorbed by a rescaling of the Minkowski coordinates x^μ. Summing up [243]:

$$\mathrm{d}s^2_{10} = \mathrm{d}s^2_{\mathrm{Mink}_4} + \mathrm{d}s^2_{\text{Calabi–Yau}}\,. \tag{8.2.26}$$

We only looked at one value of J in (8.2.24b), but now that equation and (8.2.24c) are both satisfied for all J.

Finally we should also look at (8.2.3b). Following the previous steps, we obtain $\partial_m \phi \gamma^m \eta^1_{+J} = 0$, which implies

$$\mathrm{d}\phi = 0\,. \tag{8.2.27}$$

We have solved the supersymmetry equations. Given that all fluxes are zero and we just found (8.2.27), the equations of motion from (1.2.17) are those of pure gravity, which we analyzed in (8.1.17). The first two are (8.2.25), while Ricci-flatness follows from (5.7.6). We will show in Chapter 10 that in general the equations of motion are implied by supersymmetry and the Bianchi identities.

It is perhaps a bit disappointing that all our work only got us (8.2.26) and (8.2.27) as a solution. This will change when we allow expectation values for fluxes.

[5] One may object that $\gamma^m \eta^1_{-1}$ vanishes for $m = \bar{\imath}$, so we only conclude $\partial_i A = 0$; but A is real, which implies also $\partial_{\bar{\imath}} A = 0$ and thus also $\mathrm{d}A = 0$.

Amount of supersymmetry

The holonomy group depends on how many covariantly constant η^a_+ we have:

- If there is a single $\eta^1_{1+} = \eta_+$, then the holonomy group is SU(3) and not smaller. One might think we then have obtained a solution with $\mathcal{N} = 1$, in the terminology introduced after (8.2.12): it seems we are only involving one $\zeta^1_{1+} \equiv \zeta^1_+$. But we can introduce a second $\zeta^2_{1+} \equiv \zeta^2_+$ for free:

$$\begin{aligned} \epsilon^1 &= \zeta^1_+ \otimes \eta_+ + \zeta^1_- \otimes \eta_- , \\ \epsilon^2 &= \zeta^2_+ \otimes \eta_\mp + \zeta^2_- \otimes \eta_\pm . \end{aligned} \qquad \text{(CY vacua)} . \tag{8.2.28}$$

  This is (8.2.12) with $J = 1$ in both sums and $\eta^1 = \eta^2$. So the solution $\text{Mink}_4 \times \text{CY}_6$ with constant dilaton has $\mathcal{N} = 2$ supersymmetry. This illustrates that fixing the amount of supersymmetry is not so easy as one might think: a solution might have more supersymmetry than one imposed. We will see later how to obtain $\mathcal{N} = 1$ vacua.
- If there are two linearly independent internal spinors η^1_{1+}, η^1_{2+}, then the amount of supersymmetry is double the naive amount, namely $\mathcal{N} = 4$. The holonomy group is further reduced. From the discussion around (5.1.45), the two spinors in $d = 6$ define an SU(2)-structure; when they are covariantly constant, they reduce the holonomy group to SU(2). The complex vector $g^{-1}v$, with v in (5.1.45), is covariantly constant; its real and imaginary parts commute, and by the Frobenius theorem (4.1.95), we have a foliation with two-dimensional leaves. On each leaf, the holonomy group is trivial because we have a basis of covariantly constant vectors. By the dual Frobenius theorem (4.1.97), the one-forms $\text{Re}v$, $\text{Im}v$ define four-dimensional leaves. Together all these facts imply that

$$M_6 = T^2 \times \text{K3} . \tag{8.2.29}$$

- If there are more than two independent internal spinors, then the holonomy group is reduced to the identity. Indeed, if we start from the previous case and add another independent spinor, it will define a further covariantly constant spinor on K3, which will reduce its holonomy from SU(2), and there is then no nontrivial possibility contemplated in Table 5.1. So

$$M_6 = T^6 , \tag{8.2.30}$$

  and in fact there are four independent chiral covariantly constant spinors, with $\mathcal{N} = 8$ supersymmetry.

8.2.5 Heterotic Ricci-flat vacua

The equations of motion for the heterotic string without fluxes are the same as in (8.2.2). We have a single supersymmetry parameter ϵ, of positive chirality. We can decompose it as in (8.2.12):

$$\epsilon = \sum_J (\zeta_{+J} \otimes \eta_{+J} + \zeta_{-J} \otimes \eta_{-J}) . \tag{8.2.31}$$

We also have a single gravitino ψ_M and dilatino λ, whose supersymmetry transformations are just as in (8.2.1), ignoring the a index there. Imposing these conditions, we arrive again at the solution (8.2.26).

The three-form field-strength H and the nonabelian gauge field F_{MN} are related by (1.2.48), whose exterior differential gives

$$dH = \frac{l_s^2}{4}(\mathrm{Tr}(R \wedge R) - \mathrm{Tr}(F \wedge F)) . \tag{8.2.32}$$

By (4.2.65), the right-hand side is proportional to the difference of $C_2(E)$ for the gauge bundle E and the tangent bundle T respectively. If we set $H = 0$ as we did in type II, we actually need $F \neq 0$. So we need to make an exception for this gauge field and allow it to be nonzero.

As usual, we need to demand that F_{MN} should not break the maximal symmetry of $\mathrm{MS}_4 = \mathrm{Mink}_4$. There are no one- or two-forms that satisfy this condition; the only forms that preserve maximal symmetry are the zero-form 1 and the four-form vol_4. So

$$F_{\mu\nu} = F_{\mu n} = 0 , \tag{8.2.33}$$

while a purely internal F_{mn} is allowed.

Now an obvious solution to (8.2.32) presents itself: we take $E = TM$, so that the two Chern class densities are equal. Actually, the two have different structure groups: the heterotic gauge group G_{het} (1.1.42) for E, and SU(3) for T. But the latter can be taken to be a subgroup of G_{het}. Concretely, this means that we take the internal gauge field equal to the spin connection of M_6:

$$A = \omega . \tag{8.2.34}$$

This is called *standard embedding*; it breaks G_{het} to

$$\mathrm{SO}(26) \times \mathrm{U}(1) \qquad \text{or} \qquad E_6 \times E_8 . \tag{8.2.35}$$

Next we consider the gaugino supersymmetry transformation:

$$\delta\chi = F_{MN}\Gamma^{MN}\epsilon . \tag{8.2.36}$$

In the language of Section 2.4.4, $\delta\chi = 0$ says that the two-form $F \in \mathrm{Stab}(\epsilon)$. In $d = 10$, we know already that this is (2.4.49). But using (8.2.31) and (8.2.33), we obtain $f_{mn}\gamma^{mn}\eta = 0$. This now says that $f \in \mathrm{Stab}(\eta) = \mathrm{SU}(3)$; recalling (2.4.33), we get $J^{mn} f_{mn} = 0$, $\Omega^{mnp} f_{mn} = 0$, or

$$J \cdot f = 0 , \qquad f_{0,2} = 0 , \tag{8.2.37}$$

with the usual pointwise inner product $\cdot$ in (4.1.106b). Equations (9.2.34) are known as *anti-Hermitian Yang–Mills (HYM)* equations [244, 245]. With the standard embedding (8.2.34), the gauge curvature is the Riemann tensor, and $J \cdot f$ reconstructs the Ricci form (6.1.5), which vanishes for a Calabi–Yau. We will study (8.2.37) in more detail in Section 9.2 with stability techniques, similar to the K-stability of Section 7.4.7. This can be used to find other solutions beyond (8.2.34), to break G_{het} to smaller subgroups than (8.2.35).

Exercise 8.2.1 Work out the analogue of (8.2.28) for a compactification of type II on K3. Recall from Section 7.2.1 that a K3 has a single covariantly constant η_+, and from Section 2.3.1 that the Majorana condition cannot be imposed in $d = 4$ Lorentzian dimensions, but that we can find instead symplectic Majorana pairs (2.3.8). In $d = 6$, $\mathcal{N} = (p, q)$ supersymmetry means that there are p (q) such pairs of positive (negative) chirality. You should then find

$$\mathcal{N} = (1,1) \text{ in IIA}, \qquad \mathcal{N} = (2,0) \text{ in IIB} \quad (\text{K3}). \tag{8.2.38}$$

Exercise 8.2.2 Similar to the previous exercise, find the spinor decomposition for an $\mathcal{N} = 1$ compactification of M-theory on a G_2-holonomy manifold. What holonomy groups do you need for $\mathcal{N} > 1$?

8.3 Introduction to effective theories

After finding a vacuum solution, one would like to write down an effective theory that describes the physics one would observe in it, so that we don't have to always work with the $d = 10$ theory for phenomena at energies much lower than the compactification scale.

In Section 8.2, we found the simple string theory vacua $\text{Mink}_4 \times \text{CY}_6$. The effective theory for string theory around these vacua is a supersymmetric analogue of (8.1.20); we will study it extensively in Section 9.1.2. Finding more general string theory vacua will require more work, and will be the main topic of the rest of the book. Perhaps we might think of saving some time by reversing the procedure, using effective theories to *find* new vacua. This section is a general introduction to the ideas behind this strategy. As anticipated in the Introduction, in most of the rest of the book we will not use this method; rather, we will look for solutions directly using the $d = 10$ equations of motion and supersymmetry constraints.

8.3.1 Effective theories and nonlinear reductions

Superficially, an effective theory might look easy to define: we choose an internal space M_6 and we look for a theory that describes low-energy physics. The space of metrics $\mathcal{M}$ on M_6 is infinite dimensional; this would lead to a $d = 4$ theory with infinitely many scalars. In an effective theory, we should keep the lightest ones, and discard those above a certain mass scale. But this will depend greatly on the region of $\mathcal{M}$ we are exploring; and the concept of mass is only really defined around vacua.

Space of internal metrics

So the idea is to identify a "valley" in the four-dimensional potential V: a finite-dimensional subspace

$$\mathcal{M}_{\text{fin}} \tag{8.3.1}$$

in the space of all metrics, along which V varies relatively mildly, while it has a very steep dependence on all directions orthogonal to $\mathcal{M}_{\text{fin}}$. We would integrate out the scalars corresponding to these orthogonal directions, and obtain an effective potential $V_{\text{eff}}(\sigma^i)$, with σ^i the coordinates parameterizing $\mathcal{M}_{\text{fin}}$.

An example is provided by (8.1.20), where we took $\mathcal{M}_{\text{fin}} = \mathcal{M}_{\text{Rf}}$, the space of Ricci-flat internal metrics, and $V_{\text{eff}} = 0$. In that case, we already knew that such metrics lead to vacuum solutions. The directions outside $\mathcal{M}_{\text{fin}}$ are the higher KK modes of the compactification, and at low energies we can ignore them. In string theory, we can proceed in the same way: taking $\mathcal{M}_{\text{fin}} = \mathcal{M}_{\text{CY}}$ leads us to a supersymmetric analogue of (8.1.20), to be discussed in Sections 9.1.1 and 9.1.2.

In string theory, we can also include the form fields, as we will see in Section 9.1.5. The contributions of these other fields now make the potential V_{eff} nonzero. Sometimes this destroys the vacuum, but sometimes some vacua do survive; we say then that some of the moduli have been *fixed*. If we manage to do so in a regime where V_{eff} is still much smaller than the dependence of the full potential V on the directions outside $\mathcal{M}_{\text{fin}}$, then we can predict the existence of vacua for the higher-dimensional theory. In Section 8.3.2, we will make this more concrete by giving rough estimates on how various terms depend on the size of the internal space and on the string coupling.

This approach is very useful, but it depends in part on having modified a space of previously established vacua $\text{Mink}_4 \times \text{CY}_6$. One would like to be able to identify other examples of finite-dimensional spaces of metrics $\mathcal{M}_{\text{fin}}$ that lead to interesting effective theories, beyond the Calabi–Yau case.

Additional issues for AdS solutions

Finding this $\mathcal{M}_{\text{fin}}$ is particularly challenging for AdS vacua, because of additional complications:

- The size of the internal space is usually fixed, and in the vast majority of cases it is of the same order as the external curvature radius L_{AdS}. In other words, recalling also (8.1.26), the smallest KK masses are of the same order of the scale of the cosmological constant:

$$L_{\text{AdS}}^{-1} = \sqrt{-\Lambda/3} \sim m_{\text{KK}} \,. \tag{8.3.2}$$

  One says that there is no *scale separation* between the two. As a consequence, we cannot physically justify discarding the higher KK modes.
- The KK spectrum around an AdS solution usually has tachyons. As we will discuss in Section 11.1, these do not always lead to instabilities: if the mass is not too negative, the corresponding wavelength will be very large and form a standing wave that will not grow over time.
- Related to the previous two points, the KK spectrum may depend quite nontrivially on the metric. Some scalars that were very heavy around one vacuum might become very light at another vacuum with the same M_6 — a phenomenon called "space invaders" in [246].

AdS solutions are unrealistic because of the negative cosmological constant, and because of the first two points. (Observations give $L_{\text{dS}}^{-1} \sim \text{meV}$.) They are still useful as an intermediate step toward more interesting ones, and often for their holographic applications; indeed, we will dedicate the whole Chapter 11 to them. So, while we cannot use an effective field theory point of view, we would still like to have for them

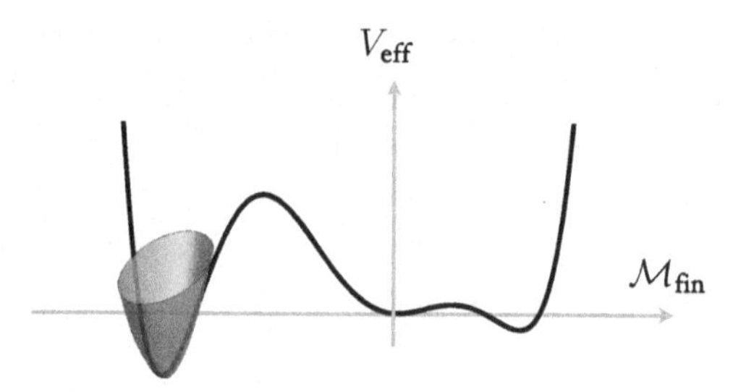

Figure 8.1 Nonlinear reduction and KK spectrum.

a lower-dimensional description of some kind. We will still select a family $\mathcal{M}_{\rm fin}$ of metrics on the internal space, with coordinates σ^i, and plug it in the $d = 10$ action to obtain a $d = 4$ action, with a potential $V_{\rm red}(\sigma^i)$. This more general *nonlinear reduction* will not describe a "valley" in the space of all metrics as in the effective theory case, but in favorable conditions it can still capture some $d = 10$ physics, usually at the classical level.

For example, its vacua sometimes still correspond to vacuum solutions of the higher-dimensional theory. More impressively, sometimes the reduced theory is a consistent truncation, so that *all* its solutions have a higher-dimensional lift. This gives a different type of justification: even if the $d = 4$ action is not an effective action, it is a useful device to obtain solutions in $d = 10$. In recent years, more and more nonlinear reductions to AdS vacua have been shown to be consistent truncations, relying on a large amount of supersymmetry, on the presence of a coset structure on the internal space, or on more sophisticated *exceptional geometry* techniques. We will discuss this further in Chapter 11. On the other hand, not even the simpler question of what choice of $\mathcal{M}_{\rm fin}$ results in a supersymmetric $d = 4$ theory seems to have a general answer. We will discuss this in Section 8.3.3.

Figure 8.1 offers a pictorial summary of the difference between a nonlinear reduction and a KK spectrum. The former is a finite-dimensional slice $\mathcal{M}_{\rm fin}$ in the infinite space of all metrics, on which we evaluate the theory's potential. Once we find a promising vacuum, we might have reasons to compute the potential around it even in the directions not tangent to $\mathcal{M}_{\rm fin}$, but we then do so only at quadratic level, thus computing a KK spectrum, the infinite (discrete) set of all masses.

8.3.2 The effective potential

Following the logic of the previous subsection, we now want to see what terms contribute to the effective potential $V_{\rm eff}$ in $d = 4$, and estimate their dependence on the overall size and the dilaton. These two $d = 4$ scalars are particularly important because they also determine the importance of the string theory corrections to supergravity. We will use our results to reach some rough conclusions about vacuum solutions; in later chapters, we will largely abandon this approach and work directly in $d = 10$ or 11, but several references explore this point of view further [247–249].

Normalizations

The reduction of the kinetic term of the graviton is similar to (8.1.20); the main difference is the presence of an overall factor $e^{-2\phi}$. Usually one defines a four-dimensional dilaton

$$e^{-2\varphi} = e^{-2\phi}\mathrm{Vol}(M_6) \tag{8.3.3}$$

so that the $\mathrm{Vol}(M_6)$ factor is reabsorbed. Then the $e^{-2\varphi}$ multiplying R_4 is reabsorbed with a frame change similar to that in (1.2.22):

$$g^{4\mathrm{E}}_{\mu\nu} = e^{-2\varphi} g^4_{\mu\nu} . \tag{8.3.4}$$

This also changes the potential, because

$$\int d^4x\sqrt{-g_4}\,\tilde{V}(\sigma^i) = \int d^4x\sqrt{-g_{4\mathrm{E}}}\,e^{4\varphi}\,\tilde{V}(\sigma^i)\,; \tag{8.3.5}$$

so the actual potential is $e^{4\varphi}\tilde{V}(\sigma^i)$.

Potential scaling

We now discuss how several terms scale with

$$r \equiv \mathrm{Vol}(M_6)^{1/6}\,, \qquad g_{s4} \equiv e^{\varphi} \sim g_s r^{-3} . \tag{8.3.6}$$

r is interpreted as the overall scale of M_6, measured in string units so as to be dimensionless. In other words,

$$g_{mn} = r^2 g^0_{mn} . \tag{8.3.7}$$

(It would also be possible to keep g_{mn} fixed and rescale the coordinate periodicity, as, for example, for $M_6 = T^6$ by taking $y^m \sim y^m + 2\pi r$, but we will not follow this strategy.)

- A first contribution to V_{eff} comes from the internal curvature, similar to (8.1.22) in pure gravity; the only change is the presence of the dilaton. $\sqrt{g_6}R$ scales as $r^6 r^{-2} = r^4$; from (8.3.3), $e^{-2\phi} = g_s^{-2} \sim g_{s4}^{-2} r^{-6}$. Finally, we have the overall factor g^4_{s4} from (8.3.5). So

$$V_R = -e^{4\varphi}\int_{M_6} e^{-2\phi} R \sim -g^2_{s4} r^{-2} . \tag{8.3.8a}$$

- The kinetic term for the internal NSNS three-form also contributes to the potential, if we take it to be purely internal. $|H|^2$ contains three inverse metrics and thus scales as r^{-6}, the measure as $\sqrt{g_6} \sim r^6$. We again have the factor $e^{-2\phi}$ and the overall factor (8.3.5). All in all,

$$V_{H^2} = e^{4\varphi}\int_{M_6} e^{-2\phi}|H|^2 \sim g^2_{s4} r^{-6} . \tag{8.3.8b}$$

  H is constrained by flux quantization (1.3.50b), but this is unaffected by r, since we are not rescaling the coordinates but rather the metric, (8.3.7).
- The internal components of F_k give a similar contribution, the main difference being the absence of the dilaton:

$$V_{F^2} = e^{4\varphi}\int_{M_6} |F_k|^2 \sim g^4_{s4} r^{6-2k} . \tag{8.3.8c}$$

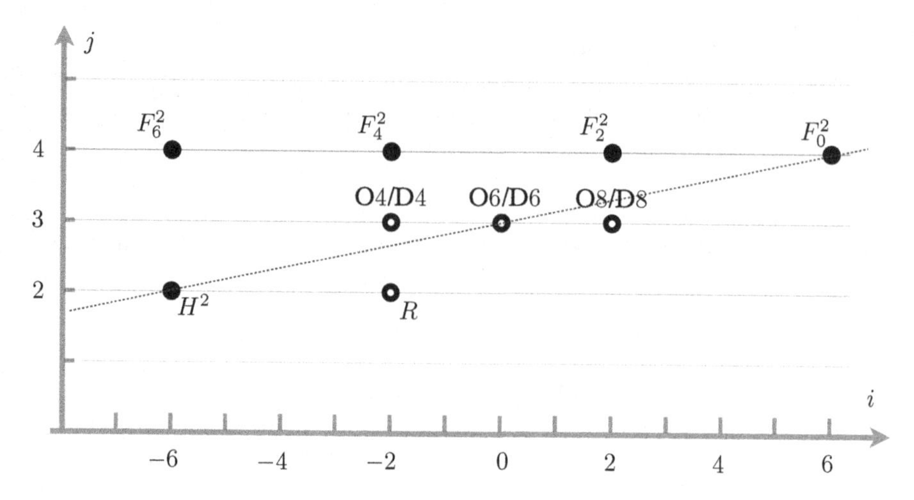

Figure 8.2 Various terms $r^i g_{s4}^j$ in V_{eff} for IIA string theory.

- Dp-branes and Op-planes also give contributions. Here we have a factor $\mathrm{e}^{-\phi}$, and the internal integral is over $p-3$ internal directions:

$$V_{\mathrm{D}p} = \mathrm{e}^{4\varphi} \int_{S_{p-3}} \sqrt{g|_{S_{p-3}}} \mathrm{e}^{-\phi} \sim g_{s4}^3 r^{p-6} . \tag{8.3.8d}$$

  The computation for an Op_--plane is identical, but opposite sign due to the negative tension (1.4.49).
- Finally, we can consider string corrections. For example, in (1.2.24) we need four inverse metrics; $R^m{}_{npq}$ does not scale under (8.3.7). This gives

$$V_{R^4} \sim \mathrm{e}^{4\varphi} \int_{M_6} \mathrm{e}^{-2\phi} (\text{Riemann})^4 \sim g_{s4}^2 r^{-8} . \tag{8.3.8e}$$

We summarize this behavior in Figure 8.2, for IIA. The black dots correspond to terms whose coefficients are positive; the white dots, to terms whose coefficients can have either sign. The dotted line has a role in an argument against de Sitter solutions that we will see in (8.3.11).

We will now try to combine (8.3.8) together to get some intuition about vacuum solutions. However, general relativity is a nonlinear theory, and it is not clear that we can just sum these terms. For this to be justified, presumably all the integrands have to be small everywhere, so that the equations of motion are a linear perturbation around a well-understood solution such as a Ricci-flat compactification. This is not hard to imagine for (8.3.8a)–(8.3.8c), but the D-brane and O-plane back-reactions are certainly not small everywhere, as we saw in the discussions of their gravity solutions in Sections 1.3.7 and 1.4.4. We can imagine, however, that if these sources are away from each other, the regions where their back reaction is strong don't overlap, and summing their contributions should still be justified. The issue is discussed further in [250].

This idea, however, cannot be implemented for low codimension: D8- and D7-branes, and O8- and O7-planes. Their gravitational effects don't die off as one goes away from them, but even grow, sometimes creating critical distances beyond which the solutions don't make sense; see (1.3.63)–(1.3.64) and (1.4.56)–(1.4.57).

De Sitter vacua and effective theory

Keeping in mind the previous warnings, let us try to draw some conclusion about the existence of de Sitter vacua, for which $V_{\text{eff}} > 0$. Consider first the dependence on g_{s4}, which in (8.3.8a)–(8.3.8e) is of the form [249]

$$V_{\text{eff}} = a g_{s4}^2 - b g_{s4}^3 + c g_{s4}^4 , \tag{8.3.9}$$

for some a, b, c that are functions of r and of the other scalars. c comes from (8.3.8c), which are integrals of positive quantities as in (4.1.107); so $V > 0$ for large g_{s4}. If $a < 0$, there will be a negative minimum, which will correspond to an AdS vacuum; so we better take $a > 0$. But then if we take $b < 0$, all coefficients in (8.3.9) are positive, and V_{eff} has no minimum at all; so we should take $b > 0$ as well. From (8.3.8d), it then follows that Op_--planes are necessary for de Sitter vacua, in agreement with the no-go argument (0.5.1)–(0.5.4).

If we impose $\partial_{g_{s4}} V = 0$, we obtain a second-order equation; from its discriminant, we see that $9b^2 > 32ac$ in order for a minimum $g_{s4\,\text{m}}$ to exist. Substituting the g_{s4}^4 term with that in the equation $g_{s4}\partial_{g_{s4}} V = 0$, we obtain $V_{\text{eff}}|_{\text{min}} = g_{s4\,\text{m}}(2a - b g_{s4\,\text{m}})/4$, whose positivity gives $4ac > b^2$. So

$$1 < \frac{4ac}{b^2} < \frac{9}{8} \tag{8.3.10}$$

for the potential to have a de Sitter vacuum. This tight window suggests that such vacua are hard to find even with O-planes.

We can refine this result by including the dependence on r. For example, consider IIA supergravity with M_6 a Calabi–Yau, with all possible fluxes but only an O6-plane as source. Then

$$V_{\text{eff}} \sim -g_{s4}^3 + h^2 g_{s4}^2 r^{-6} + \sum_{k=0}^{4} f_{2k} g_{s4}^4 r^{6-2k} . \tag{8.3.11}$$

Since this is only an estimate, we have not bothered to keep the coefficient of the first term, which is the O6 one, stressing only its sign. The other coefficient are symbolic, reminding us of their origin in (8.3.8b) and (8.3.8c). Now [251]

$$0 = \left(-r\partial_r + 6 g_{s4}\partial_{g_{s4}}\right) V_{\text{eff}} = 18 V_{\text{eff}} + 4 \sum_{k=1}^{4} k f_{2k} g_{s4}^4 r^{6-2k} > 18 V_{\text{eff}} . \tag{8.3.12}$$

So we see that with these ingredients we cannot have a dS vacuum, even if we have an O-plane. It is useful to understand this visually from Figure 8.2. The terms on the dotted line are such that $-r\partial_r + 6 g_{s4}\partial_{g_{s4}} r^i g_{s4} = 18 r^i g_{s4}$; all the remaining terms are above that line, and have a positive coefficient. The picture shows that a V_R term (8.3.8a) with a positive coefficient would invalidate the argument, if the coefficient is taken to be positive (with $R_6 < 0$). But if $V_R \neq 0$ and $F_0 = 0$, one can prove another no-go using a different oblique line in Figure 8.2, going through R, the O6, and the F_2 terms. The operator to use in (8.3.12) would now be $-r\partial_r + 2 g_{s4}\partial_{g_{s4}}$. For similar arguments, see [252].

One might think of excluding dS vacua with an O8 with a similar strategy, by taking an oblique line going through the R, O8, and F_0^2 in Figure 8.2, corresponding to an operator

$$- r\partial_r + 4 g_{s4}\partial_{g_{s4}} . \tag{8.3.13}$$

However, as we warned previously, an O8 is one of the objects whose internal back-reaction is not localized; it is then not clear that our linearized approximation, where we just sum the terms (8.3.8), is justified. We will see arguments for and against this conclusion in Sections 10.3.1 and 12.1.

AdS vacua

We now turn to AdS vacua, which we instead expect to be plentiful. In the previous subsection, we discussed how one might still be able to use $\mathcal{M}_{\rm fin} = \mathcal{M}_{\rm CY}$ for an effective theory, even in presence of fluxes. So we take M_6 to be Calabi–Yau. The argument in (8.3.12) still applies; we can also read it as predicting that $V_{\rm eff} < 0$ when at least one of F_2, F_4, F_6 is nonzero. We choose $F_4 \neq 0$, and set $F_2 = F_6 = 0$. To get a vacuum, we also need a negative term; so we introduce an O6, as in the setup of (8.3.12). So we still have $V_{\rm eff}$ as in (8.3.11), but now with $f_2 = f_6 = 0$. This setup was considered in [253].

Taking sums and differences of $\partial_r V_{\rm eff} = \partial_{g_{s4}} V_{\rm eff} = 0$, we get a second-order equation for $g_{s4} r^6$ and a linear one for r^{-8}. This gives

$$g_{s4} r^6 = \frac{3 + \sqrt{9 + 640 f_0^2 h^2}}{32 f_0^2}\,, \qquad r^{-8} = \frac{9}{5 f_4^2}\left((2 g_{s4} r^6)^{-1} - f_0^2\right), \tag{8.3.14}$$

in terms of the parameters in (8.3.11); so we have fixed the two scalars r, g_{s4}, which were previously massless. The numerical factors in (8.3.14) are not very significant, since $V_{\rm eff}$ is an estimate anyway. If we keep all parameters of order one, both the string coupling and the volume are of order one too, and the supergravity approximation does not apply, in keeping with the Dine–Seiberg argument in the Introduction. As anticipated there, the solution is to take some of the flux quanta to be large. We choose to take the $n_{4i} = \frac{1}{(2\pi)^4} \int_{S_{4i}} dC_3$ of order $N \gg 1$, while keeping those of F_0 and H of order one. In this limit,

$$r \sim N^{1/4}\,, \qquad g_s = g_{s4} r^3 \sim N^{-3/4}\,. \tag{8.3.15}$$

The effective potential on the minimum is $\Lambda = V_{\rm eff}|_{\rm min} \sim -N^{-9/2}$, with a negative coefficient as predicted by (8.3.12). The AdS curvature radius (8.2.16) is $L_{\rm AdS} = \sqrt{-3/\Lambda} \sim N^{9/4}$, much larger than the internal radius $r \sim N^{1/4}$; so this model evades the problem (8.3.2), achieving scale separation. It is argued in [254] (although with some possible loopholes) that AdS vacua with scale separation in type II supergravity always need O-planes, just as Minkowski ones do.

Additional fields

This sort of computation is too rough for us to really claim to have found a vacuum. We have only considered $V_{\rm eff}$ as a function of r and g_{s4}, but it depends on many more light scalars, such as the many moduli in $\mathcal{M}_{\rm CY}$ besides the volume. For (8.3.14), this more sophisticated analysis was carried out in [253], taking also into account $\mathcal{N} = 1$ supersymmetry.

We also have to make sure we are in a regime where the effective theory is appropriate, by estimating the potential for the infinitely many scalars that parameterize metrics not in the finite-dimensional space $\mathcal{M}_{\rm fin}$. For (8.3.15), in such directions the nonzero curvature would give $V_R \sim g_{s4}^2 r^{-2} \sim N^{-7/2}$, which for large N is larger than $V_{\rm eff}|_{\rm min} \sim -N^{-9/2}$, as required.

Even when we think we have found a valid vacuum with effective field theory, the effective theory cannot predict all the minute details of the higher-dimensional fields. The large-mass scalars we have ignored will cause small adjustments in the position of the vacuum, while not destroying it. In the Calabi–Yau case, $V_{\rm eff}$ will fix some of the moduli; the dependence of the full potential higher KK modes means that those get small expectation values, too. Thus the internal metric will be Calabi–Yau only approximately, due to the back-reaction of the fluxes, O-planes, and D-branes. For the vacua in (8.3.14), a fully satisfactory $d = 10$ realization of these vacua has been elusive because of the back-reaction of the O6-planes; this has generated some debate. We will discuss this class further in Section 11.3. In Section 9.5, we will see another important set of vacua with $\mathcal{N} = 1$-supersymmetric effective description, where the Calabi–Yau metric gets multiplied by an overall function, in a way that can be instead fully analyzed.

When we are dealing with a nonlinear reduction rather than an effective theory, recall from Section 8.3.1 that the potential in the directions not included in $\mathcal{M}_{\rm fin}$ is of comparable order to that along it. There is now no guarantee that minimizing in the excluded directions does not destroy the vacuum; a potential might have a minimum along one direction but not be a minimum at all along another, as for $V = x^2 + e^y$, which on $\{y = 0\}$ has a minimum at the origin, but on the full $\mathbb{R}^2$ has none. A consistent truncation is a situation where one can prove that this does not happen. Even so, it happens relatively often that a vacuum that appears stable in the truncated theory is unstable in the full theory. We will see some examples in Chapter 11.

Even if $d = 4$ effective theories or consistent truncations are useful ways to find vacuum solutions, there is no reason that *all* vacua should admit such a description. Hence in this book we will devote a lot of attention to methods to find vacuum solutions working directly in $d = 10$ or $d = 11$, especially in Chapter 10.

8.3.3 Reduction and supersymmetry

In this subsection, we consider the problem of what choice $\mathcal{M}_{\rm fin}$ of finite-dimensional space of metrics on M_6 leads to a supersymmetric action in $d = 4$. By itself, this cannot replace the requirements of having either an effective theory or a consistent truncation; but it is already an interesting and challenging problem, for which no general solution exists.

Besides the choice of $\mathcal{M}_{\rm fin}$, to reduce the form fields, we need a finite-dimensional space $\Omega_{\rm fin}$ of forms on M_6. The internal action involves the exterior differential d and the Hodge star $*$. For type II supergravity (1.2.15) and (1.2.36), the Chern–Simons terms involve d, and the kinetic terms can be rewritten using (4.1.107) as $\int_{M_6} *F_k \wedge F_k$, $\int_{M_6} *H \wedge H$. We should then demand that a basis of $\Omega_{\rm fin}$ consists of two-forms ω_i, three-forms α^A, β_A and four-forms $\tilde\omega_i$ that are closed under d and $*$ [255, 256]. (We ignore one- and five-forms for simplicity.) Imposing d^2, this reads

$$\begin{aligned} &d\omega_i = m_i^A \alpha_A + e_{iA}\beta^A\,, \qquad d\alpha_A = -e_{iA}\tilde\omega^i\,, \qquad d\beta^A = m_i^A\tilde\omega^i\,, \qquad d\tilde\omega^i = 0\,; \\ &*\alpha_A = A_{AB}\alpha^B + B_A{}^B\beta_B\,, \qquad *\beta^A = C^A{}_B\alpha^B + D^{AB}\beta_B\,, \qquad *\omega_i = A_{ij}\tilde\omega^j\,. \end{aligned} \tag{8.3.16}$$

The idea is now that we would expand the form fields on this basis and obtain lower-dimensional fields such as scalars. Since $*$ depends on the metric, the set Ω_{fin} is related to our earlier choice of metrics $\mathcal{M}_{\text{fin}}$. We might take it to be the space of harmonic forms (4.1.139). Or on a coset, we might take the space of left-invariant forms (4.4.38). Another possible idea is to consider the Laplacian on M_6: by its definition (4.1.112),

$$[\Delta, \mathrm{d}] = [\Delta, *] = 0\,, \tag{8.3.17}$$

so its eigenspaces are closed under d and $*$, and hence satisfy (8.3.16).

The scalars coming from both $\mathcal{M}_{\text{fin}}$ and Ω_{fin} are needed to obtain complete representations of the supersymmetry algebra. If we reduce on a manifold with SU(3) structure, it is natural to expand J and Ω as a linear combination of the forms in (8.3.16):

$$J = v^i \omega_i\,, \qquad \Omega = X^A \alpha_A + \mathcal{F}_A \beta^A\,, \tag{8.3.18}$$

for some coefficients v^i, X^A, $\mathcal{F}_A$, which would become scalar fields in $d = 4$. Since (J, Ω) determine the metric, this gives a further constraint relating $\mathcal{M}_{\text{fin}}$ to Ω_{fin}. To see why this complicates the problem, suppose we start from a particular SU(3)-structure (J^0, Ω^0) with corresponding metric g^0, and we choose a basis of forms such that (8.3.18) and (8.3.16) hold for some choice of coefficients. Now when we modify the coefficients v^i, X^A, $\mathcal{F}_A$, the SU(3)-structure and the metric will change, and with it the Hodge $*$; so it is no longer clear that (8.3.16) still holds.

So far there are few examples with a satisfactory solution to this problem. When is $\mathcal{M}_{\text{fin}}$ = Calabi–Yau manifolds, Ω_{fin} = harmonic forms, which will be studied at length in Section 9.1. Another class is $\mathcal{M}_{\text{fin}}$ = coset spaces, Ω_{fin} = left-invariant forms, of which we will see examples in Section 10.3.6 and Chapter 11.

Exercise 8.3.1 Along the lines of Section 8.3.2, find other AdS vacua of the effective $d = 4$ potential where the internal space has positive curvature rather than being Calabi–Yau, and without D-branes or O-planes.

Exercise 8.3.2 Repeat the analysis for a reduction to $d = 5$, again with an internal M_5 of positive curvature and without sources, with only $F_5 \neq 0$.

Exercise 8.3.3 Repeat the analysis for a reduction of M-theory to $d = 4$.

8.4 Four-dimensional supergravity

We now introduce some rudiments of $d = 4$ supergravity. We will review the structure of the most common $\mathcal{N} = 1$ and $\mathcal{N} = 2$ supergravity, both gauged and not, in Sections 8.4.1 and 8.4.2; we will then comment on their solutions in Section 8.4.3, focusing on vacua. Far more complete introductions to these topics exist [257, 258].

8.4.1 $\mathcal{N} = 1$ supergravity

We saw in Chapter 1 that supergravity theories in high dimensions are very constrained: in $d = 11$, there is only one model, while in $d = 10$ there are a few, corresponding the low-energy limits of perturbative strings. In $d = 4$, there is a lot more freedom.

The fields in a relativistic model collect creators of states that produce representations of Poincaré symmetry. For a supersymmetric relativistic theory, fields produce representations of the $\mathcal{N} = 1$ superalgebra, called *supermultiplets*. These can be presented as collections of ordinary bosonic and fermionic fields. The $\mathcal{N} = 1$ multiplets are as follows:

- The gravity multiplet: it consists of the metric field $g_{\mu\nu}$ and a gravitino ψ_μ.
- Vector multiplets: each with a vector A_μ and a gaugino λ.
- Chiral multiplets: a complex scalar z, a chiralino χ.

Each of the spinorial fields can be taken to be either Weyl or Majorana. (The two are isomorphic, as we learned at the end of Section 2.3.1.) For example, we can take λ to be Majorana, and ψ_μ, χ to be Weyl, say of positive chirality.

Action

The bosonic action can be written as

$$S_{\mathcal{N}=1\,\text{sugra}} = \int \mathrm{d}^4x\sqrt{-g}\Big(\frac{1}{2\kappa^2}R_4 - g_{i\bar{j}}D_\mu z^i D^\mu z^{\bar{j}} - V_{\mathcal{N}=1\,\text{sugra}} \\ - \frac{1}{4}\mathrm{Re} f_{ab}F^a \cdot F^b + \frac{1}{4}\mathrm{Im} f_{ab}F^a \cdot *F^b\Big). \tag{8.4.1}$$

In this section, the Newton constant is $\kappa \equiv \kappa_4$, not to be confused with the one in $d = 10$ from Chapter 1. The index i goes over the number of chiral multiplets; $g_{i\bar{j}}$ is a metric over the space $\mathcal{M}_{\text{ch}}$ of the scalars z^i. Supersymmetry demands $\mathcal{M}_{\text{ch}}$ to be a the Kähler manifold such that

$$\frac{1}{2\pi}J \in H^2(M,\mathbb{Z}). \tag{8.4.2}$$

Recalling (7.1.16), this also guarantees the existence of a line bundle $\mathcal{L}_{\text{KH}}$ with $c_1(\mathcal{L}_{\text{KH}}) = [J]$, which is called the *Kähler–Hodge* condition. Recall also that for a Kähler manifold we can define locally a Kähler potential (6.1.11), $J = \mathrm{i}\partial\bar{\partial}K$. A connection for the line bundle $\mathcal{L}_{\text{KH}}$ is then given by

$$\mathcal{A}_{\text{KH}} = \frac{\mathrm{i}}{2}(\bar{\partial} - \partial)K, \tag{8.4.3}$$

since then $\mathrm{d}\mathcal{A}_{\text{HK}} = J$, the curvature on $\mathcal{L}_{\text{KH}}$. For this connection, (6.1.13) is then a gauge transformation. The term $g_{i\bar{j}}\partial_\mu z^i \partial_\nu z^{\bar{j}}$ in (8.4.1) is the pull-back of an embedding $x^\mu \mapsto z^i$ of spacetime into $\mathcal{M}_{\text{ch}}$. From this point of view, the kinetic term can be viewed as a sigma model, much like the string action (1.1.4).

The covariant derivative

$$D_\mu z^i \equiv \partial_\mu z^i - A^a_\mu K_a \tag{8.4.4}$$

is of the form (4.2.27), which we introduced in the context of fiber bundles. K_a are Killing vectors of the Kähler metric; they preserve the Kähler structure. For example, for free scalars the $g_{i\bar{j}}$ is the flat Euclidean metric, and the K_a can be either translations or rotations, leading to a covariant derivative of Higgs or Stückelberg type, (4.2.15) or (4.2.28). So the scalars are in general charged under the vector fields; in other words, some of the symmetries of $\mathcal{M}_{\rm ch}$ are gauged. For this reason, one says that (8.4.1) is a *gauged supergravity* when $K_a \neq 0$, and an *ungauged* one when $K_a = 0$.

The complex matrix f_{ab} in (8.4.1) may also depend on z; we demand its real part to be positive definite. If the gauge group G is nonabelian, it should be a quadratic invariant in the adjoint representation; in each simple factor of G, one can take it to be the Killing form.[6] Finally, the potential is

$$V_{\mathcal{N}=1\ \rm sugra} = e^{\kappa^2 K}\left(g^{i\bar{j}}\, D_i W D_{\bar{j}}\bar{W} - 3\kappa^2 |W|^2\right) + \frac{1}{2}({\rm Re} f)^{-1\,ab}\mu_a \mu_b\,. \tag{8.4.5}$$

K is the Kähler potential of the metric $g_{i\bar{j}}$, (6.1.11); $W = W(z^i)$ is the *superpotential*, a holomorphic function of the z^i, in that it does not depend on the $\bar{z}^i$; and

$$D_i W \equiv \partial_{z^i} W + \kappa^2 (\partial_{z^i} K) W\,. \tag{8.4.6}$$

Finally, μ_a are the components of a moment map that is equivariant. When G is nonabelian, this fixes it as in (6.3.46), which here reads

$$\mu_a = \frac{1}{2{\rm i}}(\xi^i \partial_i K - \xi^{\bar{\imath}} \partial_{\bar{\imath}} K)\,. \tag{8.4.7}$$

On the other hand, when G is abelian there is a freedom to add constants:

$$\mu_a \to \mu_a - c_a\,. \tag{8.4.8}$$

The c_a are real constants called *Fayet–Iliopoulos parameters*. When the scalars are free and the K_a are rotations, the moment map is of the form (6.4.28), which we studied in the context of toric manifolds.

The fermionic completion of (8.4.1) can be found, for example, in [258, chap. 18] (as well as in the classic [257]). One aspect we highlight here is that the mass terms for the fermions depend on the superpotential; for example, for the chiralinos we have

$$-\frac{1}{2}\bar{\chi}^i \chi^j e^{\kappa^2 K} D_i D_j W\,. \tag{8.4.9}$$

These are called *Yukawa couplings* because of their similarity to the terms $\bar{\psi}\psi\phi$ in the Standard Model.

Supersymmetry

The supersymmetry transformations of the fermions are

$$\delta\psi_{+\mu} = \left(D_\mu - \frac{{\rm i}}{2}\mathcal{A}_\mu\right)\zeta_+ + \frac{1}{2}\kappa^2 e^{\kappa^2 K/2} W \gamma_\mu \zeta_-\,, \tag{8.4.10a}$$

$$\delta\lambda^a = \left(\frac{1}{2}F + \frac{{\rm i}}{2}({\rm Re} f)^{-1\,ab}\mu_b\right)\zeta\,, \tag{8.4.10b}$$

$$\delta\chi^i = dz^i \cdot \zeta_- - e^{\kappa^2 K/2} g^{i\bar{j}} D_{\bar{j}}\bar{W}\zeta_+\,. \tag{8.4.10c}$$

[6] By *Levi's theorem*, a Lie algebra can be decomposed as the direct sum of a semisimple summand $\mathfrak{g}_{\rm ss}$ and a solvable one $\mathfrak{g}_{\rm sol}$. The Killing form is non-zero on $\mathfrak{g}_{\rm ss}$, and negative definite if it corresponds to a compact group; it vanishes on $\mathfrak{g}_{\rm sol}$.

The first term in the gravitino transformation (8.4.44a) has the same form as its ten-dimensional counterpart in (8.2.1). The transformation parameter $\zeta = \zeta_+ + \zeta_-$ is a Majorana spinor, and $\zeta_\pm$ its two chiral projections. If one prefers working with a Majorana gravitino, the transformation of the negative chirality is $\delta\psi_{-\mu} = (\delta\psi_{+\mu})^c$. We also defined the composite connection

$$\mathcal{A}_\mu = \frac{i}{2}\kappa^2(\partial_\mu z^i \partial_{z^i} K - \partial_\mu \bar{z}^{\bar{\imath}} \partial_{\bar{z}^{\bar{\imath}}} K) - \kappa^2 A_{a\,\mu}\mu^a \,. \tag{8.4.11}$$

Note that the first two terms are proportional to the pull-back of (8.4.3) under the sigma model map $x^\mu \mapsto z^i$.

Generalizations

One possible extension is to introduce *constrained superfields*; these are especially convenient for models with spontaneous supersymmetry-breaking. A famous example is that of a chiral superfield $S = s + \theta^\alpha \chi_\alpha + \theta^2 F$, with a constraint $S^2 = 0$, whose bosonic component gives $s = \frac{1}{4F}\chi^\alpha \chi_\alpha$. This upsets the usual balance among bosons and fermions; one can even obtain in this way a model with no boson and only a fermion, which can then be thought of as a Goldstino, the fermionic analogue of a Nambu–Goldstone boson, signaling supersymmetry-breaking. For a sample of the literature, see [259–263]. In more recent years, this formalism has seen a revival, for example, to models with higher derivatives [264, 265] and with (nonsupersymmetric) de Sitter vacua [266, 267].

8.4.2 $\mathcal{N} = 2$ supergravity

An $\mathcal{N} = 2$ supergravity theory contains the following supermultiplets:

- The $\mathcal{N} = 2$ *gravity multiplet* contains the metric field $g_{\mu\nu}$, *two* gravitinos ψ^a_μ, $a = 1, 2$, and a vector field A^0_μ (often called *graviphoton*). We can think of it as a gravity and a vector multiplet for $\mathcal{N} = 1$ supersymmetry.
- The $\mathcal{N} = 2$ *vector multiplet* contains a vector field A_μ, a gaugino λ^a, and a complex scalar z; it has same content as an $\mathcal{N} = 1$ vector plus a chiral multiplet.
- The *hypermultiplet* contains four real scalars or two complex scalars q^a, and a hyperino κ; it has same content as two $\mathcal{N} = 1$ chiral multiplets.

Again, there are several possibilities for the spinors; this time, we will take the gravitinos of positive chirality, the gauginos of negative, and the hyperino to be Dirac (no restriction). There also exist variants of these multiplets where one or more of the scalars are replaced by a two-form potential, by a dualization procedure we will discuss shortly, resulting in *tensor multiplets*.

For $\mathcal{N} = 1$ supergravity (8.4.1), the field space for the scalars was a Kähler–Hodge manifold. For $\mathcal{N} = 2$, it is a product:

$$\mathcal{M}_{\rm qK} \times \mathcal{M}_{\rm sK} \,. \tag{8.4.12}$$

$\mathcal{M}_{\rm qK}$ is a quaternionic-Kähler manifold, from Section 7.2.4. As remarked there, this is in general not Kähler; so, perhaps surprisingly, an $\mathcal{N} = 2$ supergravity model is not necessarily an $\mathcal{N} = 1$ one. The second factor $\mathcal{M}_{sK}$ is a *special Kähler* manifold; in this case, the terminology is appropriate, as it is a particular type of Kähler manifold.

Special Kähler manifolds

A special Kähler manifold of complex dimension n is a Kähler–Hodge manifold whose Kähler potential reads

$$K = -\log[\mathrm{i}(\bar{v}|v)]\,, \tag{8.4.13}$$

where v is a holomorphic section of $\mathcal{L}_{\mathrm{KH}} \otimes S$, with S a holomorphic vector bundle with structure group $\mathrm{Sp}(2n+2,\mathbb{R})$, and such that

$$(v|\partial v) = 0\,. \tag{8.4.14}$$

Here $(\,|\,)$ is the pairing defining the $\mathrm{Sp}(2n+2,\mathbb{R})$-structure on $\mathcal{S}$. It is similar to the almost symplectic form in our discussion of $\mathrm{Sp}(d,\mathbb{R})$-structures at the end of Section 5.1.1.

Equation (8.4.13) is the constraint on K imposed by $\mathcal{N} = 2$ supersymmetry. There are several alternative formulations; for reviews, see [268, 269]. A slightly different type of geometry, called *rigid* special Kähler, is appropriate for the vector multiplets of $\mathcal{N} = 2$ theories without gravity; the one in (8.4.13) is then sometimes also called *local* special Kähler, emphasizing that in this case supersymmetry is a local symmetry (its parameters are point dependent). (Mathematicians studying these spaces have a different set of names; see [270], where a reformulation of these conditions is proposed.)

The peculiar rank $2n+2$ of the bundle $\mathcal{S}$ arises from the need to describe together the $n+1$ vector fields A^I (the graviphoton and the $n = n_{\mathrm{v}}$ vectors in the vector multiplets), as well as the $n+1$ *magnetic duals* $\tilde{A}_I$, defined by $*F^I = *\mathrm{d}A^I = \mathrm{d}\tilde{A}^I$ (when the gauge group is abelian). As we mentioned after (1.3.12), in $\mathcal{N} = 2$ it is possible to trade a gauge field for its magnetic dual without breaking supersymmetry, and this is realized on $\mathcal{S}$ by an action $v \to Mv$, with

$$M \in \mathrm{Sp}(2n+2,\mathbb{Z})\,. \tag{8.4.15}$$

One often also defines a normalized section

$$\mathcal{V} \equiv \mathrm{e}^{K/2} v\,, \qquad \mathrm{i}(\bar{\mathcal{V}}|\mathcal{V}) = 1\,. \tag{8.4.16}$$

Holomorphicity of v, $\bar{\partial} v = 0$, becomes then

$$\partial_{\bar{i}}\mathcal{V} - \frac{1}{2}\partial_{\bar{i}} K\mathcal{V} = D_{\bar{i}}\mathcal{V} = 0\,, \tag{8.4.17}$$

in terms of local complex coordinates z^i, $i = 1, \ldots, n$; $D = \mathrm{d} + \mathrm{i}\mathcal{A}_{\mathrm{KH}}$ is the covariant derivative with respect to the Kähler–Hodge connection in (8.4.3). Since $\partial_i \mathcal{V} = \mathrm{e}^{K/2}(\partial_i v + 1/2\partial_i K v)$,

$$D_i\mathcal{V} = \partial_i\mathcal{V} + \frac{1}{2}\partial_i K\mathcal{V} = \mathrm{e}^{K/2}(\partial_i v + \partial_i K v) = \mathrm{e}^{K/2}\left(\partial_i v - \frac{(\bar{v}|\partial_i v)}{(\bar{v}|v)} v\right) \tag{8.4.18}$$

In particular,

$$(\bar{\mathcal{V}}|D_i\mathcal{V}) = 0\,. \tag{8.4.19}$$

Now the metric can be written as

$$g_{i\bar{j}} = \partial_i\partial_{\bar{j}} K \overset{(8.4.13)}{=} -\frac{(\partial_{\bar{j}}\bar{v}|\partial_i v)}{(\bar{v}|v)} + \frac{(\partial_{\bar{j}}\bar{v}|v)(\bar{v}|\partial_i v)}{(\bar{v}|v)^2} \overset{(8.4.18)}{=} -\mathrm{i}(D_{\bar{j}}\bar{\mathcal{V}}|D_i\mathcal{V})\,. \tag{8.4.20}$$

This is also similar to (6.2.16) for the Fubini–Study metric on $\mathbb{CP}^N$, which is also written in terms of a projected derivative of $N + 1$ objects X^I.

It is useful to spell out the conditions more concretely; if the pairing looks like (5.1.18), we can divide v into its upper and lower components, both $n + 1$-dimensional. Introducing an index $I = 0, \ldots, n$:

$$v = \begin{pmatrix} X^I \\ \mathcal{F}_I \end{pmatrix}, \qquad \mathcal{V} = \begin{pmatrix} L^I \\ M_I \end{pmatrix}. \tag{8.4.21}$$

So the constraint (8.4.14) reads

$$X^I \partial_i \mathcal{F}_I - \mathcal{F}_I \partial_i X^I = 0 . \tag{8.4.22}$$

Equation (8.4.13) reads

$$K = -\log\left[\mathrm{i}(\bar{X}^I \mathcal{F}_I - X^I \bar{\mathcal{F}}_I)\right] . \tag{8.4.23}$$

Each component of v is a section of $\mathcal{L}_{\mathrm{KH}}$, but the ratios X^i/X^0 are functions. Generically, we expect these to be functionally independent, in the sense that

$$\det\left[\partial_{z^j}(X^i/X^0)\right] \neq 0 . \tag{8.4.24}$$

If so, we can simplify matters by taking directly $z^i = X^i/X^0$. The ratios $\mathcal{F}_I/X^0$ are also functions, and they should depend on the local coordinates; so we can write

$$\mathcal{F}_I = X^0 \hat{\mathcal{F}}_I(z^i) . \tag{8.4.25}$$

But $X^i = X^0 z^i$, so we can write $X^I = X^0 \partial_{X^0} X^I$. Then we also have $X^I \partial_{X^0} \mathcal{F}_I = X^I \hat{\mathcal{F}}_I = X^0 \partial_{X^0} X^I \hat{\mathcal{F}}_I = \mathcal{F}_I \partial_{X^0} X^I$. This is similar to (8.4.22); we can collect them together introducing derivatives $\partial_I \equiv \partial_{X^I}$. The idea is that the z^i are the coordinates, but we sometimes have to include factors of X^0 to obtain sections of $\mathcal{L}_{\mathrm{KH}}$, as we did in (8.4.25). In other words, the X^I can be thought of as homogeneous coordinates, similar to those on $\mathbb{CP}^N$ (Section 6.2). All in all, we can promote (8.4.22) to

$$X^I \partial_J \mathcal{F}_I = \mathcal{F}_J , \tag{8.4.26}$$

or in other words,

$$\mathcal{F}_I = \partial_I \mathcal{F} , \qquad \mathcal{F} \equiv \frac{1}{2} X^I \mathcal{F}_I . \tag{8.4.27}$$

The function $\mathcal{F}$ is called *prepotential*. Rewriting (8.4.27) as $X^I \partial_I \mathcal{F} = 2\mathcal{F}$ tells us that it is homogeneous of degree two. Taking a ∂_I of (8.4.23), we also find

$$L^I K_I = -1 , \tag{8.4.28}$$

again with $K_I \equiv \partial_I K$.

While (8.4.24) (and hence (8.4.27)) is generically true, it is still an assumption. However, the examples where it does not hold are usually generated by starting with a case where it does, and exchanging one of the F_I with one of the X^I. In what follows, we will always assume that a prepotential $\mathcal{F}$ exists.

It is also useful to introduce the *period matrix* $\mathcal{N}_{IJ}$, defined by

$$M_I = \mathcal{N}_{IJ} L^J , \qquad D_{\bar{i}} M_I = \bar{\mathcal{N}}_{IJ} D_{\bar{i}} L^J . \tag{8.4.29}$$

To see that such an $\mathcal{N}$ exists, we can define $M_{0I} \equiv \bar{M}_I$, $M_{jI} \equiv D_j M_I$, and a similar $L_I{}^J$; then (8.4.29) becomes a matrix equation, solved by $\mathcal{N}_{KJ} = (L_K{}^I)^{-1} M_{IJ}$. When a prepotential exists, there is a more explicit solution:

$$\mathcal{N}_{IJ} = \bar{\mathcal{F}}_{IJ} + 2\mathrm{i}\frac{\mathrm{Im}\mathcal{F}_{IK}\mathrm{Im}\mathcal{F}_{JL}L^K L^L}{\mathrm{Im}\mathcal{F}_{KL}L^K L^L} , \tag{8.4.30}$$

where $\mathcal{F}_{IJ} \equiv \partial_I \mathcal{F}_J = \partial_I \partial_J \mathcal{F}$. The first property in (8.4.29) is immediate to check; the second requires showing $\mathcal{F}_{IJ} \nabla_i L^J = \nabla_i M_I$ from the definitions, and then noting

$$\mathrm{Im}\mathcal{F}_{JK} L^J D_{\bar{i}} \bar{L}^K \overset{(8.4.26)}{=} \frac{1}{2\mathrm{i}}(M_K D_{\bar{i}} \bar{L}^K - \bar{\mathcal{F}}_{JK} L^J D_{\bar{i}} \bar{L}^K) = \frac{1}{2\mathrm{i}}(\mathcal{V}|D_{\bar{i}}\bar{\mathcal{V}}) \overset{(8.4.19)}{=} 0 . \tag{8.4.31}$$

To see why (8.4.30) is useful, observe that (8.4.29) simplifies (8.4.20) to

$$g_{i\bar{j}} = -2\mathrm{Im}\mathcal{N}_{IJ} D_i L^I D_{\bar{j}} \bar{L}^J . \tag{8.4.32}$$

We will see in Section 9.1.2 that Calabi–Yau moduli spaces are examples of special Kähler manifolds.

Dualizing two-forms

As we mentioned earlier, a two-form potential in $d = 4$ is physically equivalent to a scalar. This duality is very similar to the world-sheet T-duality in (1.4.16)–(1.4.21). Consider the action

$$S_{B_2,\alpha_1} = -\frac{1}{2}\int \mathrm{d}^4 x(\alpha_1^2 + 2\mathrm{d}B_2 \wedge \alpha_1) \tag{8.4.33a}$$

$$\overset{(4.1.156b)}{=} -\frac{1}{2}\int \mathrm{d}^4 x \left(|\alpha_1 + *\mathrm{d}B_2|^2 + |\mathrm{d}B_2|^2\right) , \tag{8.4.33b}$$

in the notation of (4.1.106b) and (4.1.107b). In (8.4.33a), we can redefine $\tilde{\alpha}_1 \equiv \alpha_1 + *\mathrm{d}B_2$; so this action is equivalent to the natural free action for a two-form:

$$S_{B_2} = -\frac{1}{2}\int \mathrm{d}^4 x |\mathrm{d}B_2|^2 . \tag{8.4.34}$$

On the other hand, if we vary the original action (8.4.33a) with respect to B_2, we find the equation of motion $\mathrm{d}\alpha_1$, locally solved by $\alpha_1 = \mathrm{d}a$; inserting this back in (8.4.33a) gives

$$S_a = -\frac{1}{2}\int \mathrm{d}^4 x |\mathrm{d}a|^2 , \tag{8.4.35}$$

the standard action for a free scalar a. So we conclude that (8.4.34) and (8.4.35) are equivalent: a two-form in four dimensions is equivalent to a scalar.

Gauging data

As for $\mathcal{N} = 1$, it is possible for an $\mathcal{N} = 2$ supergravity to be gauged: in other words, for some of its scalars to be charged under the vector fields. To introduce a covariant derivative (8.4.4), we again need Killing vectors on the field space; this is now the product (8.4.12), so we have separate Killing vectors on each factor. Along $\mathcal{M}_{\mathrm{sK}}$, we can apply the usual definitions and get a moment map μ. For $\mathcal{M}_{\mathrm{qK}}$, we need a slight generalization of the hypermoment map (7.2.46) that we introduced for hyper-Kähler manifolds.

Recall from Section 7.2.4 that quaternionic-Kähler manifold are defined by imposing that the set of J^α is preserved by parallel transport, rather than each being individually preserved; this leads to (7.2.53). So the action of a Killing vector ξ should not necessarily annihilate the J^α, but can mix them:

$$L_\xi J^\alpha = \epsilon^{\alpha\beta\gamma} w^\beta J^\gamma , \tag{8.4.36}$$

for some functions w_β; this generalizes (7.2.46). Now we can modify the definition of moment map (5.4.11) with SU(2) covariant derivatives. Using (4.1.79) and (4.2.15),

$$D(\iota_\xi J^\alpha) = \mathrm{d}(\iota_\xi J^\alpha) + \epsilon^{\alpha\beta\gamma} A^\beta \iota_\xi J^\gamma = \epsilon^{\alpha\beta\gamma}(w^\beta + \iota_\xi A^\beta) J^\gamma . \tag{8.4.37}$$

The square of D gives

$$D^2 \mu^\alpha \overset{(4.2.18)}{=} \epsilon^{\alpha\beta\gamma} F^\beta \mu^\gamma \overset{(7.2.54)}{=} \lambda \epsilon^{\alpha\beta\gamma} J^\beta \mu^\gamma . \tag{8.4.38}$$

So (8.4.37) can be solved by

$$\iota_\xi J^\alpha = D\mu^\alpha_\xi = \mathrm{d}\mu^\alpha_\xi + \epsilon^{\alpha\beta\gamma} A^\beta \mu^\gamma_\xi , \tag{8.4.39}$$

with $\mu^\alpha_\xi = -\lambda^{-1}(w^\alpha + \iota_\xi A^\alpha)$. This defines a hypermoment map for the quaternionic-Kähler case.[7] The analogue of the equivariance condition (5.4.13) is

$$L_\xi \mu^\alpha_{\xi'} + \epsilon^{\alpha\beta\gamma} w^\beta \mu^\gamma_{\xi'} = \mu^\alpha_{[\xi,\xi']} \quad \Rightarrow \quad \mu^\alpha_{[\xi,\xi']} = \iota_\xi \iota_{\xi'} J^\alpha - \lambda \epsilon^{\alpha\beta\gamma} \mu^\beta_\xi \mu^\gamma_{\xi'} . \tag{8.4.40}$$

Action

In Section 8.4.1, we kept κ explicit, to show how to decouple gravity and describe $\mathcal{N} = 1$ super-YM as well; in this subsection, this would become clumsier, and we choose to work in natural units. Earlier we introduced indices $i = 1, \ldots, n_\mathrm{v}$, $I = 0, \ldots, n_\mathrm{v}$, where n_v is the number of vector multiplets; now we also introduce $u = 1, \ldots, n_\mathrm{h}$, where n_h is the number of hypermultiplets. Recall that I has one more entry to collect together both the graviphoton and the vector fields in the vector multiplets. The bosonic action is then

$$S_{\mathcal{N}=2\ \mathrm{sugra}} = \int \mathrm{d}^4 x \sqrt{-g} \Big(\frac{1}{2} R_4 - g_{i\bar{\jmath}} D_\mu z^i D^\mu z^{\bar{\jmath}} - \frac{1}{2} h_{uv} D_\mu q^u D^\mu q^v - V_{\mathcal{N}=2\ \mathrm{sugra}} + \frac{1}{2} \mathrm{Im} \mathcal{N}_{IJ} F^I \cdot F^J - \frac{1}{2} \mathrm{Re} \mathcal{N}_{IJ} F^I \cdot * F^J \Big) . \tag{8.4.41}$$

(The fermionic part can be found in [258, chap. 21] or [269, sec. 8].) Recall that $g_{i\bar{\jmath}}$ is the metric on $\mathcal{M}_{sK}$; g_{uv} is the metric on $\mathcal{M}_{qK}$. The covariant derivatives are defined as in (8.4.4) and its earlier avatar (4.2.27), which now reads

$$D_\mu z^i = \partial_\mu z^i - A^I_\mu K^i_I , \qquad D_\mu q^u = \partial_\mu q^u - A^I_\mu K^u_I . \tag{8.4.42}$$

The period matrix $\mathcal{N}_{IJ}$ was defined in (8.4.30) and (8.4.29). Finally,

[7] The μ^α_I are sometimes also called *Killing prepotentials*. We will avoid this name, as we already introduced another completely different "prepotential" in (8.4.27), and the name *hypermoment map* is also well attested.

$$V_{\mathcal{N}=2\,\text{sugra}} = (g^{i\bar{j}} D_i L^I D_{\bar{j}} \bar{L}^J - 3\bar{L}^I L^J)\mu_I^\alpha \mu_J^\alpha + \bar{L}^I L^J (g_{i\bar{j}} K_I^i K_J^{\bar{j}} + 2g_{uv} K_I^u K_J^v), \tag{8.4.43}$$

where $\mu_{\alpha I} \equiv \mu_{\alpha\, K_I}$ is an equivariant hypermoment map (8.4.39) relative to the vector K_I describing the gauging (8.4.42) on the vector multiplet moduli space. One can also simplify this a bit using Exercise 8.4.4. One can make (8.4.43) more similar to (8.4.5) by rewriting the Killing vectors in terms of the moment maps μ_I, μ_I^α. Morally speaking, when we break to $\mathcal{N} = 1$ the hypermoment map generates the D-term and the F-term: μ^3 assembles with the moment map μ on $\mathcal{M}_{\text{sK}}$ to become the $\mathcal{N} = 1$ moment map, and $\mu^1 + \mathrm{i}\mu^2$ becomes the $\mathcal{N} = 1$ F-term.

Supersymmetry

The supersymmetry transformations read

$$\delta\psi_\mu^A = \left(D_\mu - \frac{\mathrm{i}}{2}\mathcal{A}_\mu\right)\zeta^A - \mathrm{i}\left(\partial_\mu q^u A_u^\alpha + \frac{1}{2}\mu_I^\alpha A_\mu^I\right)\sigma_\alpha{}^A{}_B \zeta^B \tag{8.4.44a}$$

$$- \frac{1}{2} T^+ \gamma^\mu \epsilon^{AB} \zeta_B + \frac{\mathrm{i}}{2}(\bar{L}^I \mu_{I\alpha})\gamma_\mu \sigma^{\alpha AB}\zeta_B ,$$

$$\delta\lambda_A^i = Dz^i \zeta_A - \frac{1}{2}\mathbb{G}^{i+}\epsilon_{AB}\zeta^B - \mathrm{i}D^i \bar{L}^I(\mu_I \epsilon_{AB} + \mu_{I\alpha}\sigma_{AB}^\alpha)\zeta^B , \tag{8.4.44b}$$

$$\delta\kappa^a = \mathrm{i}U_u^{aA}\left(\frac{1}{2}Dq^u \cdot \zeta_A - (\bar{L}^I K_I^u)\epsilon_{AB}\zeta^B\right) . \tag{8.4.44c}$$

The ζ_A, $A = 1, 2$ are the supersymmetry parameters, and the ζ^A their conjugates, respectively, of positive and negative chirality. In general, the upper (lower) index I denotes positive (negative) chirality.[8] These indices are raised and lowered with $\epsilon = \left(\begin{smallmatrix} 0 & 1 \\ -1 & 0 \end{smallmatrix}\right)$. $\sigma_\alpha{}^A{}_B$ are the Pauli matrices, and $\sigma_{AB}^\alpha = \sigma_A^{\alpha C}\epsilon_{CB}$ are symmetric, as one can check.

In (8.4.44a), $\mathcal{A}_\mu$ is the same composite connection as in (8.4.11), only now with the index $\alpha \to I$; A_u^α are the components of the quaternionic-Kähler connection in (7.2.53) and (8.4.39). The two-form $T^+ \equiv \frac{1}{2}(T - \mathrm{i} * T)$, which by (3.1.11) or (3.2.27) obeys $*T^+ = \mathrm{i}T^+$, is the imaginary-self-dual part of the *graviphoton* field-strength:

$$T_{\mu\nu}^+ \equiv L^I \mathrm{Im}\mathcal{N}_{IJ} F_{\mu\nu}^{J+} . \tag{8.4.45}$$

It is the particular combination of the F^I which is the partner of graviton and gravitinos.

The remaining n_{v} vector field-strengths in (8.4.44b) are

$$G_{\mu\nu}^{i+} \equiv g^{i\bar{j}} D_{\bar{j}} \bar{L}^I \mathrm{Im}\mathcal{N}_{IJ} F_{\mu\nu}^{J+} . \tag{8.4.46}$$

In general, for the action of a two-form on a chiral spinor,

$$F\zeta_\pm = \pm F\gamma\zeta_\pm = \pm\gamma F\zeta_\pm \overset{(3.2.27)}{=} \mp\mathrm{i}{*F}\zeta_\pm , \tag{8.4.47}$$

[8] In type II supergravity in $d = 10$, we called a the index on the supersymmetry parameters, which was always up irrespective of chirality (see, for example, (8.2.1)). Here we use a different index to avoid a clash with the index a on hypermultiplets.

using $c = \mathrm{i}$, consistent with (2.1.20). In other words, only the part of F with the appropriate chirality remains. So for both T and G^i, the imaginary-self-dual projector in (8.4.44) can be dropped.

Finally, in (8.4.44c), U_{aAu}, $\alpha = 1, \ldots, 2n_{\mathrm{h}}$, are a vielbein on $\mathcal{M}_{\mathrm{hm}}$: a pair of indices αI is a flat $4n_{\mathrm{h}}$-dimensional index. This index splitting is related to the quaternionic-Kähler structure: the forms J_a defining it can be written as

$$J_\alpha = \sigma_{\alpha A}{}^{B} U^{aAu} U_{aBv} \mathrm{d}q^u \wedge \mathrm{d}q^v \,. \tag{8.4.48}$$

Generalizations

Because of the dualization procedure (8.4.33)–(8.4.35), we have not included tensor multiplets, although some gaugings would be more natural in this framework (see, for example, [271]). There is also the possibility of including Chern–Simons-like couplings of the form $A^2 \mathrm{d}A$ and A^4; these are discussed, for example, in [258, chap. 21]. The modern framework for finding supergravity models is the *embedding tensor formalism* [272–274], which starts from the gauging data (how scalars transforms under vectors, as in (8.4.4)) and then works out the supersymmetry transformations a posteriori, without relying on multiplet structure. For more thorough general discussions of $\mathcal{N} = 2$ supergravity, see [258, 269, 274, 275].

8.4.3 Supersymmetric solutions

$\mathcal{N} = 1$ vacuum solutions

We begin with $\mathcal{N} = 1$ theories. To find vacuum solutions that preserve supersymmetry, following the logic in Section 8.2.2 we take spacetime to have maximally symmetry, and we impose that no other fields break it; this implies that all the scalars are constant, and that $F^a_{\mu\nu} = 0$. Then (8.4.10) reduces to the conditions

$$D_i W = \partial_{z^i} W + \kappa^2 (\partial_{z^i} K) W = 0 \,, \qquad \mu_a = 0 \,; \tag{8.4.49a}$$

$$D_\mu \zeta_+ = -\frac{1}{2} \kappa^2 \mathrm{e}^{\kappa^2 K/2} W \gamma_\mu \zeta_- \,. \tag{8.4.49b}$$

These are called *F-term* and *D-term* equations respectively. The names of these two equations come from the traditional names of the auxiliary scalars in the chiral and vector multiplets respectively, which when integrated out contribute to (8.4.5). The G action is a symmetry, so $L_\xi W = 0$; from the F-term, it follows that $\xi^i \partial_i K W = 0$, which from (6.3.46) and (8.4.7) implies $\mu_\xi W = 0$. So if $W \neq 0$, the D-term condition is redundant.

Already with (8.4.49a), going back to (8.4.5) we see that the vacuum energy is

$$V_0 = \Lambda = -3\kappa^2 |W|^2 \mathrm{e}^{\kappa^2 K} \leq 0 \,; \tag{8.4.50}$$

so spacetime is either AdS_4 or Mink_4. Alternatively, (8.4.49b) and its conjugate are the Killing spinor equations (8.2.14a) with $\mu = -\kappa \mathrm{e}^{\kappa^2 K/2} W$. The cosmological constant from (8.2.15) confirms (8.4.50). (This explains the factor of -3 in (8.4.5).) We expect most supersymmetric vacua to be AdS_4, since for Minkowski we need to impose the extra condition $W = 0$.

These arguments do not forbid the presence of supersymmetry-breaking vacua with $\Lambda > 0$.

Decoupling gravity in $\mathcal{N} = 1$

Taking the $\kappa \to 0$ limit in (8.4.1) gives the action of an $\mathcal{N} = 1$ gauge theory. The gravitational term decouples, and the potential (8.4.5) simplifies to

$$V_{\mathcal{N}=1\,\text{SYM}} = g^{i\bar{j}}\partial_{z^i} W \partial_{\bar{z}^{\bar{j}}} \bar{W} + \frac{1}{2}(\text{Re}f)^{-1\,ab}\mu_a\mu_b\,. \tag{8.4.51}$$

The conditions (8.4.49) for supersymmetric vacua now read

$$\partial_{z^i} W = 0\,, \qquad \mu_a = 0\,. \tag{8.4.52}$$

Alternatively, in a supersymmetric theory without gravity, the vacua that preserve supersymmetry have $V = 0$. From the positive definitess of g and $\text{Re}f$, we again conclude (8.4.52).

The moduli space of vacua is obtained by first considering the locus of zeros $Z(\partial_{z^i} W, \mathcal{M}_{\text{ch}})$, and then performing a Kähler quotient as in Section 6.3.7 (possibly after shifting (8.4.8) in the abelian case). Indeed, the D-term in (8.4.52) restricts to a level set of the moment map for the action of the K_a, that represent the gauging; vacua that differ by an action of the K_a should be identified. This is of the form $\mu^{-1}(c)/G$, as in (6.3.37). For example, if the z^i are free and $W = 0$, if the gauging is by a rotation, with a Higgs-type covariant derivative, then the moduli space of vacua becomes toric (Section 6.4).

$\mathcal{N} = 2$ vacuum solutions

For $\mathcal{N} = 2$, imposing that all fields respect the symmetries of $\mathcal{M}S_4$ makes $\delta\psi^A_\mu = 0$ resemble (8.2.18). As we mentioned there, this reduces to the usual Killing spinor Ansatz (8.2.14a). Substituting this in $\delta\psi^A_\mu = \delta\lambda^{iA} = 0$:

$$\bar{L}^I \mu_{I\alpha}\sigma^\alpha_{AB}\zeta^B = -\mathrm{i}\sqrt{-\frac{\Lambda}{3}}\zeta_A\,, \qquad D_{\bar{j}}\bar{L}^I(\mu_I\epsilon_{AB} + \mu_{I\alpha}\sigma^\alpha_{AB})\zeta^B = 0\,. \tag{8.4.53}$$

These two can be reassembled together. One of the properties of the moment map μ_I on a special Kähler manifold (as opposed to a manifold that is just Kähler) is that $L^I\mu_I = 0$ [258, sec. 21.1.2]. So the left-hand sides of the two (8.4.53) contain the same parenthesis, multiplied by $\bar{L}^I$ and $D_{\bar{j}}\bar{L}^I$ respectively. When a prepotential exists, as we have assumed, these are independent, and span a basis of $\mathbb{C}^{n_v+1}$. Using this and (8.4.28), (8.4.53) becomes [276]

$$(\mu_I\epsilon_{AB} + \mu_{I\alpha}\sigma^\alpha_{AB})\zeta^B = \mathrm{i}K_I\sqrt{-\frac{\Lambda}{3}}\mathrm{e}^{-K/2}\zeta_B\,. \tag{8.4.54}$$

Finally, from the hyperinos (8.4.44c):

$$\bar{L}^I K^u_I \epsilon_{AB}\zeta^B = 0\,. \tag{8.4.55}$$

For $\Lambda = 0$, since $v_a\sigma^a$ has eigenvalues $\pm|v|/2$, (8.4.54) implies $\mu_I = \mu_{I\alpha} = 0$, so the gauging vanishes. The conditions (8.4.54) and (8.4.55) are then satisfied by all spinors; the vacuum preserves $\mathcal{N} = 2$, and there is no supersymmetry-breaking [277]. One way out is to look at cases where the prepotential does not exist, which we used to get to (8.4.54). These models are obtained by exchanging some of the X^I with some of the $\mathcal{F}_I$ in a model with prepotential; recalling the remarks before (8.4.15), we can achieve the same results by dualizing one of the vector fields A^I to its magnetic dual $\tilde{A}_I$.

Such a *magnetic gauging* is not included in the gauging (8.4.42); it requires using the embedding tensor formalism, mentioned at the end of Section 8.4.2. We should now introduce a new index α running over the vectors we want to gauge, which now might include some of the $\tilde{A}_I$. Alternatively, we can introduce new Killing vectors $\tilde{K}^{Iu}$ and their hypermoment maps $\tilde{\mu}^I_\alpha$, and change (8.4.44) with the rule

$$\begin{aligned} L^I \mu_{\alpha I} &\to L^I \mu_{I\alpha} + M_I \tilde{\mu}^I_\alpha , \\ D^i \bar{L}^I \mu_{I\alpha} &\to D^i \bar{L}^I \mu_{I\alpha} + D^i \bar{M}_I \tilde{\mu}^I_\alpha , \\ \bar{L}^I K^u_I &\to \bar{L}^I K^u_I + \bar{M}_I \tilde{K}^{Iu} . \end{aligned} \tag{8.4.56}$$

This is enough to invalidate the previous argument against $\mathcal{N} = 1$ supersymmetry-breaking. (For a general analysis, see, for example, [276, 278].) Moreover, we will see in Section 10.3.6 that the vacuum equations that result from applying (8.4.56) to (8.4.54) and (8.4.56) are in a precise (albeit a bit formal) correspondence with the equations that describe the most general $\mathcal{N} = 1$ compactification of supergravity.

Beyond vacua

It is also very interesting to look for supergravity solutions beyond the vacua, just like for general relativity [279]. Those that are preserved by some supercharges have been classified to some extent. One method is to use the spinor bilinears and the G-structures they define (Chapters 3 and 5); we will use this extensively to study string vacua. Another, called *spinorial geometry* [280, 281], uses the group theory of the possible orbits for the supersymmetry parameters ζ (Sections 2.4, 3.3, and 3.4).

For simple theories, one can obtain a classification as a list of possible solutions; for example, this was achieved for $\mathcal{N} = 2$ ungauged [282, 283] and gauged [284] supergravity. For more complicated theories, one can at least reduce to lists of cases, each requiring a list of conditions that is simpler or at least more geometrical in character than the original supersymmetry equations. For $\mathcal{N} = 2$ theories with an arbitrary number of vector multiplets, spinorial geometry reformulates supersymmetry in terms of a system of partial differential equations [285]. For the class of theories in Section 8.4.2, with an arbitrary number of vector and hypermultiplets, one can at least reformulate the supersymmetry conditions in terms of G-structures and spinor bilinears [286, 287]. If the two ζ^A are different, they define an identity structure (Exercise 5.1.6), or in other words a vielbein. With the methods of Section 3.3.6, this can be parameterized as

$$\zeta_A \otimes \overline{\zeta}^B = (1 + \mathrm{i}*)(k\, \delta^B_A + v^\alpha \sigma_\alpha{}^B{}_A) , \qquad \zeta_A \otimes \overline{\zeta}_B = \mu \epsilon_{AB}(1 + \mathrm{i}\,\mathrm{vol}_4) + \sigma^\alpha_{AB}\omega_\alpha , \tag{8.4.57}$$

where μ is a function, the one-forms k, v^α are a vielbein; and the ω_α are two-forms. The gravitino equations $\delta\psi^A_\mu = 0$ are equivalent to

$$\begin{aligned} &\mathrm{d}\mu - \mathrm{i}\mathcal{A}\mu = -\bar{L}^I \mu_{I\alpha} v^\alpha + \iota_k T^+ , \qquad \mathrm{d}k = 2\mathrm{Re}(\mathrm{i}\bar{L}^I \mu_{I\alpha}\bar{\omega}^\alpha + \bar{\mu} T^+) , \\ &\mathrm{d}v_\alpha = \epsilon_{\alpha\beta\gamma}(v^\beta \wedge (2\mathrm{d}q^u A^\gamma_u + A^I \mu^\gamma_I) + 2\mathrm{Re}(L^I \mu^\beta_I \omega^\gamma)) . \end{aligned} \tag{8.4.58a}$$

The $\delta\lambda^i_A = 0 = \delta\kappa^a$ equations become

$$\begin{aligned} &\mathrm{i}\bar{\mu} Dt^a + \iota_k G^{i+} - \mathrm{i}D^i \bar{L}^I (\mu_I k + \mu_{I\alpha} v^\alpha) = 0 , \\ &\mathrm{i}\iota_k Dq^u + \mathrm{i}J^{\alpha\, u}{}_v \iota_{v_\alpha} Dq^v + 2\bar{L}^I K^u_I \mu = 0 . \end{aligned} \tag{8.4.58b}$$

For $\mathcal{N} = 1$ supergravity, an analysis of solutions using spinorial geometry is given in [288].

In higher dimensions, G-structures are quite successful for $d = 5$ [289] and $d = 6$ [290] minimal supergravity. In $d = 10$ and 11, some classification theorems can be obtained for large numbers of supercharges n. For $n = 32$, one can show that there are only Mink_{10}, $\text{AdS}_5 \times S^5$, and a certain "plane wave" geometry [291]. In M-theory and type II, one can show that $n = 31 \Rightarrow n = 32$ [292–294]. Moreover, all solutions for $n > 16$ are cosets [295].

The broader problem of obtaining all supersymmetric solutions seems too hard. In Chapter 10, we will see how to reformulate it in terms of forms as in (8.4.58); however, in this book we will apply that result only to vacuum solutions $\text{MS}_4 \times M_6$. Another class where classifications exist is that of near-horizon geometries, where, for example, one can show that there can only be an even number of supercharges [296]. For a recent review on the classification of supersymmetric solutions beyond vacua, see [281].

Exercise 8.4.1 Using Chapters 6 and 7, find examples of Kähler–Hodge manifolds.

Exercise 8.4.2 Find $\mathcal{F}$ such that $g_{i\bar{j}}$ is the Poincaré metric (5.3.69), after a suitable coordinate change.

Exercise 8.4.3 Work out the metric $g_{i\bar{j}}$ for $\mathcal{F} = (X^1)^3/X^0$. Using a symplectic matrix (8.4.15), dualize it to a model with $\mathcal{F} \propto \sqrt{X^0(X^1)^3}$.

Exercise 8.4.4 Show that the inverse of (8.4.32) can be written as

$$g^{i\bar{j}} D_i L^I D_{\bar{j}} \bar{L}^J = -\frac{1}{2}(\text{Im}\mathcal{N})^{-1IJ} - \bar{L}^I L^J \,. \tag{8.4.59}$$

(Hint: compute $C \text{Im}\mathcal{N} C^{-1}$, where

$$C = (L^I \; D_{\bar{j}} \bar{L}^I) \tag{8.4.60}$$

is an $(n_\text{v} + 1) \times (n_\text{v} + 1)$-matrix. (8.4.16) and (8.4.19) should also be useful.)

9 Minkowski compactifications

The most notable Minkowski vacua are those where M_6 is a Calabi–Yau; we study this case in detail in Section 9.1, adding D-branes in Section 9.2. In Sections 9.3–9.5, we then study some notable classes of Minkowski vacua with internal form field-strengths. Here the internal geometry is sometimes still related to a Calabi–Yau in certain ways. A more general analysis will be deferred to the next chapter, where we will introduce a more systematic formalism that allows us to also describe AdS_4 vacua.

9.1 Calabi–Yau compactifications

In Section 8.2, we have concluded that in absence of fluxes the only supersymmetric vacua are $\mathrm{Mink}_4 \times \mathrm{CY}_6$, with constant dilaton ϕ and warping function $A = 0$. In this section, we explore the physics of these vacua. We will assume the Calabi–Yau to be proper: to have holonomy SU(3) and not smaller.

In Sections 9.1.1 and 9.1.2, we study their $d = 4$, $\mathcal{N} = 2$ effective Lagrangian. In Section 9.1.3, we describe the world-sheet description of Calabi–Yau compactifications, and in Section 9.1.4 their important mirror symmetry property. Finally, in Section 9.1.5 we consider how the effective action changes when introducing fluxes, in the spirit of the discussion of Section 8.3. This anticipates more dramatic changes to be considered in later sections.

9.1.1 Multiplet structure

Since $\mathrm{Mink}_4 \times \mathrm{CY}_6$ vacua have $\mathcal{N} = 2$, we expect the effective action to be described by the formalism in Section 8.4.2. In this subsection, we work out its multiplets from counting arguments. As in Section 8.4.2, we will use indices from the middle of the Latin alphabet for the vector multiplets, $i = 1, \ldots, n_\mathrm{v}$, $I = 0, \ldots, n_\mathrm{v}$ for the vector multiplets, and from its beginning, $a = 1, \ldots, n_\mathrm{h}$, for the hypermultiplets. As we will see, some scalars with a geometrical origin belong to one set in IIA and to the other in IIB; for example, we will see Kähler and complex moduli change type of index from one theory to the other.

Multiplets for IIA reduction

First we look at the scalars. Some of these should come from the moduli of the Calabi–Yau, as in Section 8.1.2. From Section 7.1.2, we get the following:

- $h^{1,1}$ real massless scalars v^i, coordinates on $\mathcal{M}_\mathrm{K}$
- $h^{2,1}$ complex massless scalars z^a, coordinates on $\mathcal{M}_\mathrm{c}$

The v^i can be thought of as the coefficients of the expansion

$$J = v^i \omega_i \,, \tag{9.1.1}$$

where $\{\omega_i\}$ is a basis in $H^2(M_6, \mathbb{Z})$, which has dimension $h^{1,1}$. We will take the elements of this and other cohomology bases to be harmonic (4.1.139).

Other massless scalars come from reducing the other IIA bosonic fields. One is the dilaton ϕ. The form field-strengths should be zero, but the potentials can be nonzero as long as they are closed. For example, we can expand

$$B = b^i \omega_i \,. \tag{9.1.2}$$

The b^i should be constant on M_6, so that $\mathrm{d}_6 B = 0$, but they can fluctuate as functions of Mink$_4$; so they are additional scalar fields. Because of the large gauge transformation (1.3.50a), these scalars are periodic: if the ω_i are normalized so as to have integer periods on all two-cycles, then $b^i \sim b^i + (2\pi l_s)^2$. The spacetime components $B_{\mu\nu}$ can also be dualized to a scalar a, as we have seen in (8.4.33)–(8.4.35).

We can expand the other form potentials with the same logic. C_1 doesn't lead to new scalars, because $h^1(M_6) = 0$, but

$$(C_3)_{\rm int} = \zeta^A \alpha_A + \tilde{\zeta}_A \beta^A \,, \tag{9.1.3}$$

where $\{\alpha_A, \beta^A\}$ is a basis for $h^{2,1}(M_6)$, usually taken to be Poincaré dual to the A^A and B_A cycles in odd dimensions described in (4.1.128):

$$\int_{A^A} \alpha_B = \delta^A_B \,, \qquad \int_{B_A} \beta^B = \delta^B_A \,, \qquad \int_{A^A} \alpha^B = 0 = \int_{B_A} \beta^B \,. \tag{9.1.4}$$

Because of (4.1.137),

$$\int_{M_6} \alpha_A \wedge \alpha_B = 0 = \int_{M_6} \beta^A \wedge \beta^B \,, \qquad \int_{M_6} \alpha_A \wedge \beta^B = \delta^B_A \,. \tag{9.1.5}$$

The ζ^A, $\tilde{\zeta}_A$ in (9.1.3) give $b_3 = 2h^{2,1} + 2$ new scalar fields.

So all in all, we have obtained

$$v^i \,, \qquad b^i \,; \tag{9.1.6a}$$

$$z^a \,, \qquad \zeta^a \,, \qquad \tilde{\zeta}_a \,; \tag{9.1.6b}$$

$$\phi \,, \qquad a \,, \qquad \zeta^0 \,, \qquad \tilde{\zeta}_0 \,. \tag{9.1.6c}$$

This is a total of $(2h^{1,1}) + 4(h^{2,1} + 1)$ scalars. It is natural to conjecture that they assemble in

$$h^{1,1} \text{ vector multiplets}\,, \qquad h^{2,1} \text{ hypermultiplets} \qquad \text{(IIA)}\,. \tag{9.1.7}$$

More precisely, (9.1.6a) would be assembled in $h^{1,1}$ complex scalars:

$$t^i \equiv b^i + \mathrm{i} v^i \,; \tag{9.1.8}$$

(9.1.6b) would be the two complex scalars of $h^{2,1}$ hypermultiplets; and (9.1.6c) would form together the two complex scalars of an additional *universal* hypermultiplet.

To confirm our guess, we also look at what vector fields are generated by reduction. The metric might generate a vector $g_{\mu m}$. We have argued in Sections 8.1 and 8.2.2 that such vector fields should have zero vacuum expectation value, but perhaps they could still give rise to fluctuations, to be included in the effective action. This would mean expanding $g_{\mu m} \stackrel{?}{=} A_\mu \alpha_m$. However, α_m would be a one-form in the internal M_6, and we know from Section 7.1.2 that there are no one-form cohomology in a Calabi–Yau. A similar logic rules out possible vector fields from $B_{\mu m}$.

Next we consider the RR potentials. An obvious vector is the spacetime part C_μ of C_1. Other vector fields are generated by C_3, taking one index along Mink$_4$, and the remaining two on M_6; we already used this strategy in (7.4.22). In the dual form language:

$$(C_3)_{\mu np} \mathrm{d}y^n \wedge \mathrm{d}y^p = A^i_\mu \omega_i \,. \tag{9.1.9}$$

Thus we have obtained $h^{1,1} + 1$ vector fields. If one of these vector fields is the graviphoton, the remaining $h^{1,1}$ match our conjecture (9.1.7).

Multiplets for IIB reduction

We now apply more briefly the same logic to IIB. We again begin with scalars. This time we call v^a, $a = 1, \ldots, h^{1,1}$ those coming from coordinates on $\mathcal{M}_\mathrm{K}$, and z^i, $i = 1, \ldots, h^{2,1}$ those coming from $\mathcal{M}_\mathrm{c}$. Just as in IIA we have the dilaton ϕ and the dual a produced by the dualization procedure (8.4.35). B can be expanded just as in (9.1.2), but now we call the resulting scalars b^a.

More scalars come from the RR forms. C_0 is already a scalar. C_2 with legs along Mink$_4$ produces a scalar $\tilde{C}^0$ by the same dualization procedure as in (8.4.35). Purely internal C_2 and C_4 can be expanded similar to (9.1.2) and (9.1.3):

$$(C_2)_\mathrm{int} = \zeta^a \omega_a \,, \qquad (C_4)_\mathrm{int} = \tilde{\zeta}_a * \omega^a \,, \tag{9.1.10}$$

where ω_a is again the basis for $H^{1,1}$, and $\tilde{\omega}^a$ is a basis for $H^{2,2}$. The two are dual, in the sense that the intersection form (4.1.137) is the identity:

$$\int_{M_6} \omega_i \wedge \tilde{\omega}^j = \delta_i^j \,. \tag{9.1.11}$$

We could also try producing more spacetime two-forms by expanding $\frac{1}{2}(C_4)_{\mu\nu pq} \mathrm{d}y^p \wedge \mathrm{d}y^q = \hat{\zeta}^a_{\mu\nu} \omega_a$; these two-forms could then be dualized into scalars. However, because of the self-duality of F_5, these scalars are in fact nothing but the $\tilde{c}_a$ we already introduced in (9.1.10).

So in total, we have

$$z^i \,; \tag{9.1.12a}$$

$$v^a \,, \qquad b^a \,, \qquad \zeta^a \,, \qquad \tilde{\zeta}_a \,; \tag{9.1.12b}$$

$$\phi \,, \qquad a \,, \qquad C_0 \,, \qquad \tilde{C}^0 \,. \tag{9.1.12c}$$

This time, we conjecture

$$h^{2,1} \text{ vector multiplets}, \qquad h^{1,1} \text{ hypermultiplets} \qquad \text{(IIB)}. \tag{9.1.13}$$

Equation (9.1.12a) would be the $h^{2,1}$ complex scalars of the vector multiplets; in (9.1.12b), $t^a = b^a + \mathrm{i}v^a$ and $\zeta^a + \mathrm{i}\tilde{\zeta}_a$ would be the two complex scalars of $h^{1,1}$ hypermultiplets; and (9.1.12c) would form together the two complex scalars of another hypermultiplet, which again we call universal.

Again we confirm our conjecture by counting vector fields. These are all produced by expanding

$$\frac{1}{6}(C_4)_{\mu npq}\mathrm{d}y^n \wedge \mathrm{d}y^p \wedge \mathrm{d}y^q = \tilde{A}_{\mu I}\alpha^I + A^I_\mu \beta_I \,. \tag{9.1.14}$$

Self-duality of F_5 implies that the $\tilde{A}_{\mu I}$ are not really independent; they are the magnetic duals of the A^I_μ, as in our comment preceding (8.4.15). So we only have to count the A^I_μ as independent, and we have $h^{1,1} + 1$ vector fields. This agrees with the conjecture (9.1.13), since one of them will be the graviphoton A^0_μ.

9.1.2 Effective four-dimensional actions

We now turn to the $d = 4$ action, which should be of the form (8.4.41). In particular, we expect the space of scalars to be of the form (8.4.12).

We begin with vector multiplet scalars. In IIA, these are (9.1.6a): coordinates on the moduli space of symplectic structures, together with a choice of B field. As we remarked following (9.1.2), the moduli b^i are periodic, so they describe a fibration of a torus $T^{h^{1,1}}$ over $\mathcal{M}_{\mathrm{K}}(M_6)$. This is now a complex manifold

$$\mathcal{M}^{\mathbb{C}}_{\mathrm{K}}(M_6) \tag{9.1.15}$$

of dimension $h^{1,1}$. In IIB, the vector multiplet scalars are (9.1.12a), which are more simply coordinates on $\mathcal{M}_{\mathrm{c}}(M_6)$. We will now show why these two spaces are special Kähler, as predicted by (8.4.12).

Special Kähler structure on complex moduli

We start with $\mathcal{M}_{\mathrm{c}}$. The bundle $\mathcal{S}$ here has fiber $H^3(M_6)$, which indeed has dimension $2h^{2,1} + 2$. The symplectic pairing $(\,|\,)$ is just the integrated wedge product (4.1.137), dual to the intersection pairing in homology. We take the section to be

$$v = \Omega \,. \tag{9.1.16}$$

By definition, $\Omega \in H^{3,0}$, while

$$\partial_{z^i}\Omega = \mu_{(i)} \cdot \Omega + K_i \Omega \in H^{2,1} \oplus H^{3,0} \,, \tag{9.1.17}$$

where the first component is the action of the ith Beltrami differential $\mu_{(i)} \in H^{0,1}(M_6, T_{1,0})$, and the second is the variation of the volume; see discussions in Sections 5.3.4 and 7.1.2. The coefficient K_i is constant on M_6. So Ω satisfies (8.4.14). One should be careful about the index i: in this subsection, it parameterizes coordinates in $\mathcal{M}_{\mathrm{c}}$, not on M_6.

To write out the section more explicitly as in (8.4.23), we can use an integral basis $\{\alpha^I, \beta^I\}$ in (9.1.4), and expand

$$\Omega = X^I \alpha_I + \mathcal{F}_I \beta^I \,. \tag{9.1.18}$$

Recalling (9.1.4), these can be written as *periods*, namely as the integrals

$$X^I = \int_{A^I} \Omega \,, \qquad \mathcal{F}_I = \int_{B_I} \Omega \,. \tag{9.1.19}$$

The X^I are then functionally independent, because Ω is a calibration (Section 5.6); if a combination of its periods was zero, the volume of a three-cycle would be identically zero all over moduli space. So, according to the discussion after (8.4.24), a prepotential exists, defined as in (8.4.27) as $\mathcal{F} = \frac{1}{2} X^I \mathcal{F}_I$.

The Kähler potential reads

$$\mathrm{e}^{-K} \overset{(8.4.23),(9.1.18)}{=} \mathrm{i} \int_{M_6} \bar{\Omega} \wedge \Omega \overset{(5.1.40)}{=} 8V \,. \tag{9.1.20}$$

Finally, we compute the metric. $\nabla_i \mathcal{V} = \mathrm{e}^{K/2} (\partial_{z^i} \Omega)_{2,1}$, since the subtracted term in (8.4.18) is exactly such that $(\bar{\mathcal{V}} | \partial_i \mathcal{V}) = 0$. So

$$g_{i\bar{j}} \overset{(8.4.20),(9.1.17)}{=} \frac{\mathrm{i}}{8V} \int_{M_6} (\bar{\mu}_{(\bar{j})} \cdot \bar{\Omega}) \wedge (\mu_{(i)} \cdot \Omega) \,. \tag{9.1.21}$$

Special Kähler structure on Kähler moduli

We now turn to the complexified Kähler moduli space $\mathcal{M}_{\mathrm{K}}^{\mathbb{C}}$. This time we take the fiber of the bundle $\mathcal{S}$ to be

$$H^{\mathrm{even}}(M_6) = H^0(M_6) \oplus H^2(M_6) \oplus H^4(M_6) \oplus H^6(M_6) \,, \tag{9.1.22}$$

which has indeed dimension $2+2n = 2+2h^{1,1}$. The symplectic pairing is the integral of the Chevalley–Mukai pairing (3.2.40):

$$(\alpha|\beta) = \int_{M_6} (\alpha \wedge \lambda(\beta))_6 = \int_{M_6} (\alpha_6 \beta_0 - \alpha_4 \wedge \beta_2 + \alpha_2 \wedge \beta_4 - \alpha_0 \beta_6) \,. \tag{9.1.23}$$

For the section ν, we can draw inspiration from Ω being a pure form, as defined in Section 3.2.5. Another prominent pure form we saw there is $\mathrm{e}^{-\mathrm{i}J}$, which has already played an important role alongside Ω in (3.4.20). To include the moduli of the B field as well, we recall (9.1.1), (9.1.2), and (9.1.8). So we are led to

$$\nu = X^0 \mathrm{e}^{B+\mathrm{i}J} = X^0 \mathrm{e}^{t^i \omega_i} \,. \tag{9.1.24}$$

The normalization X^0 will be useful soon. Equation (9.1.24) is indeed in (9.1.22), and it satisfies (8.4.14):

$$\frac{1}{(X^0)^2} (\nu | \partial_{t^i} \nu) = (\mathrm{e}^{B+\mathrm{i}J} | \omega_i \wedge \mathrm{e}^{B+\mathrm{i}J}) \overset{(3.2.62)}{=} (\mathrm{e}^{\mathrm{i}J} | \omega_i \wedge \mathrm{e}^{\mathrm{i}J}) \overset{(9.1.23)}{=} \int_{M_6} (\omega_i)_6 = 0 \,. \tag{9.1.25}$$

(Again, here the index i parameterizes coordinates on $\mathcal{M}_{\mathrm{K}}^{\mathbb{C}}$, not on M_6.) The final integral vanishes because ω_i has no six-form part.

The fact that both (9.1.16) and (9.1.24) are pure is responsible for many similarities between $\mathcal{M}_{\mathrm{K}}^{\mathbb{C}}$ and $\mathcal{M}_{\mathrm{c}}$, which will culminate with mirror symmetry (Section 9.1.4). It would be possible to give a unified treatment of the two cases [256], but at this point we have not developed the necessary mathematical machinery; we will introduce it in the next chapter and briefly return to this point then.

The Kähler potential defined by v is now

$$\frac{1}{|X^0|^2}\mathrm{e}^{-K} = \mathrm{i}(\overline{\mathrm{e}^{B+\mathrm{i}J}}|\mathrm{e}^{B+\mathrm{i}J}) \overset{(3.2.62)}{=} \mathrm{i}(\mathrm{e}^{-\mathrm{i}J}|\mathrm{e}^{\mathrm{i}J}) \\ = \mathrm{i}\int_{M_6}\mathrm{e}^{-2\mathrm{i}J} = 8\int_{M_6}\left(-\frac{J^3}{6}\right) \overset{(5.1.40b)}{=} 8\mathrm{Vol}(M_6) \equiv 8V\,. \tag{9.1.26}$$

We can evaluate this volume more explicitly expanding $J = v^i\omega_i$ as in (9.1.1) and using the triple intersection form (4.1.138):

$$V = -\frac{1}{6}v^iv^jv^k\int_{M_6}\omega_i\wedge\omega_j\wedge\omega_k = -\frac{1}{6}D_{ijk}v^iv^jv^k\,. \tag{9.1.27}$$

The metric is then

$$g_{i\bar{j}} = \partial_i\partial_{\bar{j}}K = -\frac{\partial_i\partial_{\bar{j}}V}{V}+\frac{\partial_iV\partial_{\bar{j}}V}{V^2} = \frac{1}{V}\int_{M_6}\omega_i\wedge\omega_j\wedge J+\frac{1}{4V^2}\int_{M_6}\omega_i\wedge J^2\int_{M_6}\omega_j\wedge J^2\,, \tag{9.1.28}$$

which can also be further evaluated in terms of the D_{ijk}. Another alternative expression is obtained from the identity (5.2.25) for the action of the Hodge $*$ on $(1,1)$-forms. This is valid for any SU(3)-structure, but when M_6 is Kähler, J is also covariantly constant; if $\alpha_{1,1}$ is harmonic, then $J\cdot\alpha_{1,1}$ is harmonic itself, and hence a constant. So it is equal to its average over M_6. Moreover, its integral can be rewritten by using $(J\cdot\alpha_{1,1})\mathrm{vol}_6 = -\frac{1}{2}J^2\wedge\alpha_{1,1}$. (The factor in this formula can be fixed using (5.1.40b) and (5.2.5).) All in all,

$$J\cdot\alpha_{1,1} = \frac{1}{V}\int_{M_6}(J\cdot\alpha_{1,1})\mathrm{vol}_6 = -\frac{1}{2V}\int_{M_6}\alpha_{1,1}\wedge J^2\,. \tag{9.1.29}$$

Using this and (5.2.25), the metric (9.1.28) on $\mathcal{M}_{\mathrm{K}}^{\mathbb{C}}$ can be rewritten as

$$g_{i\bar{j}} = \frac{1}{V}\int_{M_6}*\omega_i\wedge\omega_j\,. \tag{9.1.30}$$

This time not only does the prepotential exist, but we can find it explicitly. We choose X^I to be the zero- and two-form part of $\mathrm{e}^{B+\mathrm{i}J}$. In particular,

$$X^i = X^0t^i = X^0(b^i+\mathrm{i}v^i)\,. \tag{9.1.31}$$

Then

$$\mathcal{F} \overset{(8.4.27)}{=} \frac{1}{2}X^I\mathcal{F}_I = \frac{1}{2}(X^0)^2\big(1+B+\mathrm{i}J\,|\,(B+\mathrm{i}J)^2/2+(B+\mathrm{i}J)^3/6\big) \\ \overset{(9.1.23)}{=} \frac{1}{6}(X^0)^2\int_{M_6}(B+\mathrm{i}J)^3 \overset{(4.1.138)}{=} \frac{1}{6}(X^0)^2D_{ijk}t^it^jt^k = \frac{D_{ijk}X^iX^jX^k}{6X^0}\,. \tag{9.1.32}$$

As expected by (8.4.27) and the observation that follows it, this is of overall degree two in the X^I. We can also check (8.4.23) directly:

$$\mathrm{i}(X^I\bar{\mathcal{F}}_I - \bar{X}^I\mathcal{F}_I) = -\frac{\mathrm{i}X^0}{6(\bar{X}^0)^2}D_{ijk}\bar{X}^i\bar{X}^j\bar{X}^k + \frac{\mathrm{i}}{2\bar{X}^0}D_{ijk}X^i\bar{X}^j\bar{X}^k + \mathrm{c.\,c.} \\ = \frac{1}{6\mathrm{i}}|X^0|^2D_{ijk}(t^i-\bar{t}^i)(t^j-\bar{t}^j)(t^k-\bar{t}^k) = 8|X^0|^2V = \mathrm{e}^{-K}\,. \tag{9.1.33}$$

When discussing both Kähler and complex moduli, to avoid confusion we will call the prepotentials $\mathcal{F}_{\mathrm{c}}$ and $\mathcal{F}_{\mathrm{K}}$. We anticipate that $\mathcal{F}_{\mathrm{K}}$ is valid in supergravity, but receives l_s corrections in full string theory; these will play a role in Section 9.1.4.

Quaternionic-Kähler manifolds via c-map
Hypermultiplet scalars should be coordinates on a quaternionic-Kähler manifold. In fact, given a special Kähler manifold $\mathcal{M}_{\rm sK}$ of complex dimension n, one can find a quaternionic-Kähler $\mathcal{M}_{\rm qK}$ by fibering a copy of $\mathbb{C}^* \times \mathbb{C}^{n+1}$ over it, with metric

$$\begin{aligned} ds^2_{qK} &= ds^2_{\rm sK} + d\varphi^2 + \frac{1}{4}e^{4\varphi}(da + \tilde{\xi}_A d\xi^A - \xi^A d\tilde{\xi}_A)^2 \\ &- \frac{1}{2}e^{2\varphi}{\rm Im}\mathcal{N}^{-1\,AB}(d\tilde{\xi}_B + \mathcal{N}_{AC}d\xi^C)(d\tilde{\xi}_B + \bar{\mathcal{N}}_{BD}d\xi^D)\,, \end{aligned} \tag{9.1.34}$$

where $\mathcal{N}_{AB}$ is defined as usual in (8.4.30), with $\mathcal{F}$ the prepotential of $\mathcal{M}_{\rm sK}$. (We switched to using indices from the beginning of the alphabet, because now this special Kähler manifold is part of the hypermultiplet moduli space $\mathcal{M}_{\rm qK}$.) This fibration is called *c-map* [297]. To see that (9.1.34) is indeed the metric over a quaternionic-Kähler manifold, one can exhibit a vielbein $U^{\alpha A u}$; this can be found in [298], and in the present notation, for example, in [278, (3.61)]. This map exists because after dimensional reduction to $d = 3$ a vector can be dualized to a scalar, via a procedure similar to that in (8.4.33)–(8.4.35) for two-forms in $d = 4$; using this, a vector multiplet becomes a hypermultiplet. Another reason is mirror symmetry, to be discussed in Section 9.1.4.

The metric (9.1.34) has some isometries along the fiber directions, irrespective of what $\mathcal{M}_{\rm sK}$ is:

$$\partial_a\,, \qquad D_A \equiv \partial_{\xi^A} + \tilde{\xi}_A \partial_a\,, \qquad \tilde{D}^A \equiv \partial_{\tilde{\xi}_A} - \xi^A \partial_a\,. \tag{9.1.35}$$

These will become important later for gaugings induced by fluxes.

Reduction
We finally get to the reduction procedure. It is lengthy but in principle straightforward: following the ideas in Section 8.3.1, we evaluate the ten-dimensional action (1.2.15) and (1.2.36) on the Ansatz for the fields described in Section 9.1.1. So we expand the form fields in harmonic forms on M_6, and we take the internal metric to be Calabi–Yau. In the following we only highlight the main steps; for more details, see [299], [300, app. D].

Reducing the Einstein–Hilbert term $\int_{M_{10}} e^{-2\phi}R$ in (1.2.17) is similar to (8.1.20), but now we recall from (7.1.11) that the space of Ricci-flat deformations factorizes into the space of Kähler and complex moduli. So

$$\int d^4x\sqrt{-g_4}e^{-2\phi}{\rm Vol}(M_6)\left(R_4 - G_{ij}\partial_\mu v^i \partial^\mu v^j - G_{a\bar{b}}\partial_\mu z^a \partial^\mu z^{\bar{b}}\right)\,. \tag{9.1.36}$$

where the metrics G_{ij} and $G_{a\bar{b}}$ are the ones on $\mathcal{M}_{\rm K}$ and $\mathcal{M}_{\rm c}$ respectively. As discussed in (8.3.4), it is convenient to switch to the four-dimensional Einstein frame, and to introduce the four-dimensional dilaton as in (8.3.3); this will play the role of φ in the metric (9.1.34).

Next we assemble the various flux terms we postulated in Section 9.1.1. For example, in IIA we collect (9.1.2), (9.1.3), (9.1.9), and the external component of B to write

$$B = B^{\rm ext} + b^i\omega_i\,, \qquad C_3 = A^i \wedge \omega_i + \zeta^A\alpha_A + \tilde{\zeta}_A\beta^A\,. \tag{9.1.37}$$

C_1 is purely external. We then plug this in the action (1.2.15). To evaluate the kinetic terms for the fluxes, we can use (4.1.106) to write them as $\int_{M_6} *F_k \wedge F_k$. Now we need to evaluate the $*$ on the internal forms. Take, for example, the three-forms α_I, β^I in (9.1.4): since they are harmonic, $*\alpha_I$, $*\beta^I$ are harmonic as well, so there have to exist matrices A, B, C, D as in (8.3.16):

$$* \alpha_I = A_{IJ}\alpha^J + B_I{}^J\beta_J\,, \qquad *\beta^I = C^I{}_J\alpha_J + D^{IJ}\beta_J\,. \tag{9.1.38}$$

To determine these matrices, we notice that the variation $\partial_I\Omega = \alpha_I + \mathcal{F}_{IJ}\beta^J$ belongs to $H^{2,1} \oplus H^{3,0}$. Equation (5.2.22) tells us that $*\alpha^0_{2,1} = \mathrm{i}\alpha^0_{2,1}$, $*\alpha_{3,0} = -\mathrm{i}\alpha_{3,0}$; these are valid more generally for a manifold with an SU(3)-structure. On a Calabi–Yau, a (2, 1)-form is automatically primitive. So

$$* \,\partial_I\Omega = \mathrm{i}\partial_I\Omega - 2\mathrm{i}\frac{(\bar\Omega|\partial_I\Omega)}{(\bar\Omega|\Omega)}\Omega\,. \tag{9.1.39}$$

The idea is that we have first acted as if $\partial_I\Omega$ was all in $H^{2,1}$, and then corrected by subtracting $-2\mathrm{i}$ times its projection on $H^{3,0}$. Using (9.1.18) now gives four real constraints that determine the matrices completely:

$$\begin{aligned}\begin{pmatrix} B & A \\ D & C\end{pmatrix} &= \begin{pmatrix} \mathrm{Im}\mathcal{N} + \mathrm{Re}\mathcal{N}(\mathrm{Im}\mathcal{N})^{-1}\mathrm{Re}\mathcal{N} & -\mathrm{Re}\mathcal{N}(\mathrm{Im}\mathcal{N})^{-1} \\ (\mathrm{Im}\mathcal{N})^{-1}\mathrm{Re}\mathcal{N} & -(\mathrm{Im}\mathcal{N})^{-1}\end{pmatrix} \\ &= \begin{pmatrix} 1 & \mathrm{Re}\mathcal{N} \\ 0 & 1\end{pmatrix}\begin{pmatrix} (\mathrm{Im}\mathcal{N})^{-1} & 0 \\ 0 & -\mathrm{Im}\mathcal{N}\end{pmatrix}\begin{pmatrix} 1 & 0 \\ -\mathrm{Re}\mathcal{N} & 1\end{pmatrix},\end{aligned} \tag{9.1.40}$$

where $\mathcal{N}$ is the matrix we defined in (8.4.30), from the prepotential $\mathcal{F} = \mathcal{F}_{\mathrm{c}}$ on the complex structure moduli space $\mathcal{M}_{\mathrm{c}}$.

The action of the $*$ on even forms can be evaluated in a similar way by using again (5.2.25), as we did to arrive at (9.1.30). The final result can be put in a similar form to (9.1.40), if we also include the zero-form 1 and the six-form vol_6.

With this preliminary work, the reduced action is an $\mathcal{N} = 2$ supergravity as in (8.4.41), with

$$K^i_I = K^u_I = 0 \quad \Rightarrow \quad \mu_I = \mu^a_I = 0\,, \qquad V_{\mathcal{N}=2\ \mathrm{sugra}} = 0\,; \tag{9.1.41a}$$

$$\mathcal{F}_{\mathrm{sK}} = \begin{cases} \mathcal{F}_{\mathrm{K}} & (\mathrm{IIA}), \\ \mathcal{F}_{\mathrm{c}} & (\mathrm{IIB});\end{cases} \qquad \mathcal{F}_{\mathrm{qK}} = \begin{cases} \mathcal{F}_{\mathrm{c}} & (\mathrm{IIA}), \\ \mathcal{F}_{\mathrm{K}} & (\mathrm{IIB}).\end{cases} \tag{9.1.41b}$$

The metric $g_{i\bar{\jmath}}$ on $\mathcal{M}$ is associated to the prepotential $\mathcal{F}_{\mathrm{sK}}$, and the metric g_{uv} is associated to the prepotential $\mathcal{F}_{\mathrm{qK}}$ via the c-map (9.1.34). From the point of view of the reduction, the off-diagonal terms in (9.1.34) originate from the Chern–Simons terms in (1.2.15) and (1.2.36).

The reason the potential $V_{\mathcal{N}=2\ \mathrm{sugra}}$ vanishes is that each value of the scalar fields should correspond to a vacuum solution in $d = 10$. Had $V_{\mathcal{N}=2\ \mathrm{sugra}}$ been nonzero along some directions in field space, those would have corresponded to massive modes, which we have not included in the reduction.[1]

[1] It might be possible to contemplate a potential that has vanishing second-derivative but is nonzero. This would correspond to an infinitesimal deformation that cannot be promoted to a finite one, but as observed before (7.1.20), this cannot happen.

The heterotic case

The discussion is similar for the heterotic string. Since it has half as many supercharges already in $d = 10$, after reduction we obtain an $\mathcal{N} = 1$ effective theory. We did not discuss the multiplet structure in Section 9.1.1: mostly we just need to ignore all the fields there that originate from the RR fields and interpret the remaining ones in terms of $\mathcal{N} = 1$ chiral multiplets. All the vectors disappear; all the scalars in (9.1.6) and (9.1.12) now coincide; the field theory space they span is now

$$\frac{\mathrm{SU}(1,1)}{\mathrm{U}(1)} \times \mathcal{M}_K^{\mathbb{C}} \times \mathcal{M}_c\,. \tag{9.1.42}$$

The first factor is a coset, spanned by the fields ϕ and a; the other two are our familiar special Kähler manifolds associated with M_6.

There is, however, a new sector, coming from the gauge sector in the heterotic action (1.2.44). We only sketch its discussion; for more details, see, for example, [1, 7, 301]. The gauge field A can have its index along M_6 or Mink$_4$. These two generate respectively scalars and vectors in $d = 4$. The gauge group is now partially broken by having selected an internal gauge field (Section 8.2.5). For the $E_8{\times}E_8$ case, the standard embedding (8.2.34) breaks one of the E_8 to E_6; the adjoint decomposes as a sum of $\mathrm{SU}(3) \times E_6$ representations:

$$\mathbf{248} \to (\mathbf{3},\mathbf{27}) \oplus (\bar{\mathbf{3}},\overline{\mathbf{27}}) \oplus (\mathbf{1},\mathbf{78}) \oplus (\mathbf{8},\mathbf{1})\,. \tag{9.1.43}$$

So, for example, the scalars originating from the $(\mathbf{3},\mathbf{27})$ are one-forms valued in $T_{1,0}$, A^a_m, with an extra index in the $\mathbf{27}$ (the fundamental) of E_6. The antiholomorphic part is of the type $A^i_{\bar{j}}$, and its zero modes are classified by the bundle-valued Dolbeault cohomology $H^{0,1}(M_6, T_{1,0})$, defined in (5.3.44); as noted after (5.3.60), this is isomorphic to $H^{2,1}(M_6)$. The $(\bar{\mathbf{3}},\overline{\mathbf{27}})$ gives rise to an $A^{\bar{i}}_{\bar{j}}$; its zero modes will be given by $H^{0,1}(M_6, T^*_{1,0}) \cong H^{1,1}(M_6)$. A similar argument gives $H^1(M_6, T_{1,0} {\otimes} T^*_{1,0})$ for the $(\mathbf{8},\mathbf{1})$. We have not computed this latter cohomology; this can be done with the method (5.3.50).

The superpotential on these new bundle scalars is nonzero; its F-term is the condition $F_{0,2} = 0$ in (8.2.37). The cubic term for the $(\mathbf{3},\mathbf{27})$ scalars has a geometrical interpretation as

$$\int_{M_6} \bar{\Omega} \wedge \mu_{(i)} \cdot \mu_{(j)} \cdot \mu_{(k)} \cdot \Omega\,, \tag{9.1.44}$$

where the $\mu_{(i)}$ are Beltrami differentials, in $H^{0,1}(M_6, T_{1,0})$. This is proportional to $\partial_i\partial_j\partial_k\mathcal{F}_c$. There is also a cubic term for $(\bar{\mathbf{3}},\overline{\mathbf{27}})$, proportional to $\partial_i\partial_j\partial_k\mathcal{F}_K$; this will get corrected away from the supergravity limit, just as $\mathcal{F}_K$ itself. These cubic terms are called *Yukawa couplings* because they contribute to the $\bar{\chi}\chi z$ terms in (8.4.9).

Physical interpretation of the conifold transition

The effective theory we have studied so far breaks down on some loci of the moduli space. For example, let us work in IIB. When M_6 acquires a singularity, the metric on $\mathcal{M}_c$ becomes singular too. If M_6 has a conifold singularity at a locus $\{z^1 = 0\} \subset \mathcal{M}_c$, a three-cycle $A_1 \cong S^3$ has shrunk (Sections 6.3.6 and 7.1.4). Under a path in $\mathcal{M}_c$ that circles around $\{z^1 = 0\}$, the dual cycle B_1 undergoes the *Picard–Lefschetz* transformation:

$$B_1 \to B_1 + A^1 \tag{9.1.45}$$

in homology; all other cycles are unchanged. Integrating Ω and recalling (9.1.19), (9.1.45) implies that $\mathcal{F}_1 \to \mathcal{F}_1 + X^1$; so near $\{z^1 = 0\}$, we can write $\mathcal{F}_1 \sim \frac{1}{2\pi i} z^1 \log z^1 + \cdots$. This in turn implies that the curvature of the metric on $\mathcal{M}_c$ diverges as $z^1 \to 0$.

A singularity in an effective theory often signals that a field has been integrated out even though its mass was lower than the theory's cutoff; this happens for example in the $\mathcal{N} = 2$ SW effective theory discussed around (1.3.12). Indeed, a D3-brane wrapping the shrinking cycle A^1 has mass

$$T_{\rm D3} \mathrm{vol}(A^1) , \tag{9.1.46}$$

which is light near $\{z^1 = 0\}$, and should have been included in the effective theory [302]. Since a brane is charged under RR fields, the new light fields should form a hypermultiplet $\mathcal{B}$ charged under the gauge fields (9.1.14).

As an example, consider the quintic Q. The effective theory has originally $n_{\rm v} = h^{2,1}(Q) = 101$ and $n_{\rm h} = h^{1,1}(Q) = 1$. The locus defined by (7.1.31) is a subspace of codimension 15 in $\mathcal{M}_c(Q)$, where $M_6 = Q$ has 16 conifold singularities; so we have to include 16 new hypermultiplets $\mathcal{B}_I$. As noticed after (7.1.31), one linear combination of the shrinking three-cycles A^I is trivial in homology. This implies that the 16 $\mathcal{B}_I$ are charged under only 15 vector multiplets. Writing an action that includes these $\mathcal{B}_I$, one finds a new branch of the moduli space where a Brout–Englert–Higgs (BEH) effect takes place, and the 15 vectors "eat" 15 of the 16 $\mathcal{B}_I$. In this new branch, we thus have $n_{\rm v} = 101 - 15 - 86$, and $n_{\rm h} = 1 + 1$. Remarkably, these are exactly the Hodge numbers (7.1.32) of a new Calabi–Yau threefold obtained by an extremal transition. It is natural to identify the new branch obtained by the BEH effect with the moduli space of this new Calabi–Yau.[2] So once again paying attention to branes that become light has produced a string duality.

9.1.3 World-sheet approach

We have seen in (1.1.29) that the world-sheet action can always be made supersymmetric on the world-sheet, by promoting the coordinates $x^M(\sigma)$ to superfields. This $\mathcal{N} = (1, 1)$ supersymmetry is present even when the background breaks supersymmetry. For Calabi–Yau vacua, where supersymmetry is instead preserved, there are further supercharges in the world-sheet. We will focus on the sigma model that describes the internal coordinates $x^m(\sigma)$, with the understanding that it needs to be supplemented with other terms that describe the external $x^\mu(\sigma)$ (as well as ghosts if one chooses to quantize using the BRST method).

From $\mathcal{N} = (1, 1)$ superfields

We will start by using $\mathcal{N} = (1, 1)$ superfields and impose the presence of additional world-sheet supersymmetry; then we will recast the result in terms of $\mathcal{N} = (2, 2)$ superfields. For more details, on both approaches, see, for example, [28].

[2] With these methods one can also infer the existence of Minkowski solutions that are not Calabi–Yau and have fluxes [303, 304]. These are peculiar in that they don't require the presence of O-planes to overcome the no-go arguments in the introduction, so they presumably they are intrinsically strongly curved; for this reason they unfortunately will not appear in our later analysis.

We start with the internal part of the NSR action (1.1.29):

$$S^{1,1}_{\rm intNSR} = -\frac{1}{2}T_{\rm F1}\int_\Sigma {\rm d}^2\sigma {\rm d}^2\theta (g+B)_{mn}(X)D_+X^m D_-X^n\,. \qquad (9.1.47)$$

This is automatically $\mathcal{N} = (1,1)$-supersymmetric, since it is written in terms of superfields. The supersymmetry transformation is of the form $\delta^1 X^m = -{\rm i}\varepsilon^\alpha Q_\alpha X^m$, with the algebra

$$[\delta^{(1)}_\varepsilon, \delta^{(1)}_{\varepsilon'}] = -{\rm i}\bar\varepsilon^\alpha \gamma^\mu_{\alpha\beta}\varepsilon'^\beta\partial_\mu\,. \qquad (9.1.48)$$

To find new supercharges, we can combine of the superfields at our disposal. In the world-sheet, scalars are dimensionless, and spinorial derivatives have dimension length$^{-1/2}$. This leaves us with only one possibility:

$$\delta^{(2)}X^m = \varepsilon^+ D_+X^n I_{+n}{}^m(X) + \varepsilon^- D_-X^n I_{-n}{}^m(X)\,. \qquad (9.1.49)$$

In this section, we simplify this by taking

$$I_+ = I_- \quad\Rightarrow\quad \delta^{(2)}X^m = \varepsilon^\alpha D_\alpha X^n I_n{}^m(X)\,. \qquad (9.1.50)$$

We will study the more general case $I_+ \neq I_-$ in Section 10.5.5. The transformations (9.1.50) should complete (9.1.48) into its $\mathcal{N} = 2$ extension, including a δ^I_J term on the right-hand side. Imposing this superalgebra on (9.1.49) gives

$$I_m{}^p I_p{}^n = -\delta^n_m\,, \qquad N_{mn}{}^p = 0\,, \qquad (9.1.51)$$

where N is the Nijenhuis tensor (5.3.25); in other words, I is a complex structure. Invariance of the action (9.1.47) imposes

$$I_m{}^p g_{pn} + g_{mp}I_n{}^p = 0\,, \qquad \nabla_m I_n{}^p = 0\,, \qquad H = {\rm d}B = 0\,. \qquad (9.1.52)$$

The first says that I is Hermitian with respect to g, (5.1.16); the second, that I reduces the holonomy to its stabilizer U(3). So

$$I_+ = I_-\ : \qquad (9.1.47) \text{ has } \mathcal{N} = (2,2) \quad\Leftrightarrow\quad M_6 \text{ is Kähler}\,. \qquad (9.1.53)$$

From $\mathcal{N} = (2,2)$ superfields

A model of the type (9.1.53) can also be realized in terms of $\mathcal{N} = (2,2)$ superfields. We define superderivatives

$$\mathbb{D}_\pm = \partial_{\theta^\pm} + {\rm i}\theta^\pm\partial_\pm\,, \qquad (9.1.54)$$

similar to (1.1.28) for $\mathcal{N} = (1,1)$, but now θ_α are complex rather than real. We can also decompose $\mathbb{D}_\alpha = D^1_\alpha + {\rm i}D^2_\alpha$, where each D_α is of the type (1.1.28). By definition, a chiral superfield Z satisfies

$$\bar{\mathbb{D}}_\pm Z = 0\,. \qquad (9.1.55)$$

Other types of $\mathcal{N} = (2,2)$ superfields will appear in Section 9.3.3; they are not needed for the present case (9.1.53). We will need three Z^i, whose bosonic components are the local complex coordinates z^i on M_6. A (2,2)-model can then be written as

$$S^{(2,2)}_{\text{intNSR}} = T_{\text{F1}} \int \mathrm{d}^2\sigma \mathrm{d}^2\theta \mathrm{d}^2\bar{\theta}\, K(Z, \bar{Z}) \,, \tag{9.1.56}$$

where K is the Kähler potential (6.1.11) of M_6.

Superconformal invariance

So far we have not imposed conformal invariance; indeed, in (9.1.53) M_6 is only required to be Kähler, not necessarily a Calabi–Yau. As we saw already for the bosonic string, conformal invariance is essential for string theory's consistency, so we should impose that it holds at the quantum level.

The beta functions should reproduce the theory's equations of motion (recall (1.1.14) and (1.1.17)). If the fluxes vanish and the dilaton is constant, *at leading order in l_s* the equations of motion are those of pure gravity, leading to $R_{mn} = 0$; together with the Kähler condition, this makes M_6 a Calabi–Yau. However, even when the only active field is gravity, we saw in (1.2.24) that there are corrections to the Einstein–Hilbert term; the equations of motion are now of the type $R_{MN} + l_s^6 (R^4)_{MN}$. So the beta function equations no longer require Ricci-flatness, and in particular the Calabi–Yau metric does not satisfy them.

Fortunately, one can show that a metric solving the corrected equations always exists on a Calabi–Yau [305]. Working in terms of $\mathcal{N} = (2,2)$ chiral superfields, we can reformulate the analysis directly in terms of the beta function β_K for K in (9.1.56). At leading order the beta functions for $g_{i\bar{j}}$ would require $R_{i\bar{j}} = 0$; from (6.1.9), we see that $R_{i\bar{j}} = -\partial_i \partial_{\bar{j}} \log \sqrt{g}$. So the leading order beta function for K is

$$\beta^{(0)}(K) = -\log \sqrt{g} = -\frac{1}{2} \mathrm{Tr} \log g \,. \tag{9.1.57}$$

At all orders, $\beta(K) = \sum_p l_s^p \beta^{(p)}(K) = \beta^{(0)}(K) + l_s^6 \beta^{(6)}(K) + \ldots$. Calling K^0 the Kähler potential for the Ricci-flat metric $g^{(0)}_{i\bar{j}}$, we want to show that by deforming $g_{i\bar{j}} = \sum_p l_s^p g^{(p)}_{i\bar{j}} = \partial_i \partial_{\bar{j}} \sum_p l_s^p K^{(p)}$ we can obtain

$$\beta(K) = -\frac{1}{2} \mathrm{Tr} \log g^0 \,. \tag{9.1.58}$$

Indeed, if we can arrange for this, then by taking $\partial_i \partial_{\bar{j}}$ on both sides we see that the beta function for the full metric $g_{i\bar{j}}$ is zero (since g^0 is the Ricci-flat metric). Expanding the left-hand side of (9.1.58) in powers of l_s, we get $\sum_{p,q} l_s^p \beta^{(p)}(\sum l_s^q K^{(q)})$. Now

$$\begin{aligned} \beta^{(0)}(K + \delta K) - \beta^{(0)}(K) &\sim -\mathrm{Tr} \log(1 + g^{-1} \delta g) \\ &\sim -\mathrm{Tr}(g^{-1} \delta g) = -\nabla^2 \delta K \,. \end{aligned} \tag{9.1.59}$$

So at each order l_s^N, (9.1.58) reads $\nabla^2_{(0)} K^{(N)}$ plus a lengthy expression in terms of the $K^{(p)}$ with $p < N$. From the Hodge decomposition (4.1.140), we know that this can be solved for $K^{(N)}$ whenever the right-hand side has zero harmonic component. The latter condition can be shown to hold even if the $\beta^{(p)}$ are not known at all orders, basically because it appears with many derivatives [305]. So (9.1.58) can be solved, and a metric with a zero beta function can be found on any Calabi–Yau.

Had it not been for this argument, a $\text{Mink}_4 \times \text{CY}_6$ would have been a solution of pure gravity and type II supergravity, but not of string theory. This might look puzzling: by taking the volume large and the dilaton small, we can make the stringy corrections as small as possible. But in this case, the leading term is already zero, so the corrections are indeed dangerous. In terms of the effective action, the danger is that they give rise to a nonzero potential $V_{\rm corr}$, when the supergravity approximation gave $V = 0$; a priori, $V_{\rm corr}$ might have had no vacua. The previous argument tells us that fortunately this does not happen.

For string theory, it is also important that the value of the total Lagrangian has a vanishing central charge. (Recall that a nonzero c signals a breakdown of the invariance for nonflat world-sheet metric $h_{\alpha\beta}$.) In the discussion that follows (1.1.33), we saw that the total contribution to the central charge c from the ghosts was -15. In our case, the Mink_4 part of the action contains four bosons and four fermions, which gives a contribution $4(1 + 1/2)$. So the contribution from the internal sigma model (9.1.56) should have $c = 9$.

Superconformal algebra and spacetime supersymmetry

Once our world-sheet model is conformal, the algebra of world-sheet Lorentz symmetries gets enhanced to the infinite-dimensional Virasoro algebra of conformal transformations (Section 1.1.1). Correspondingly, the $\mathcal{N} = (2,2)$ supersymmetry algebra gets enhanced to an infinite-dimensional $\mathcal{N} = (2,2)$ *superconformal* algebra, with generators

$$L_m\,, \qquad J_m\,, \qquad G^{\pm}_{r=m\pm a}\,. \tag{9.1.60}$$

The $m \in \zeta$; the parameter a can in principle be any real number in $[0,1)$, but the cases of interest are $a = 0$ and $a = 1/2$. These are called respectively Ramond and Neveu–Schwarz versions of the algebra, because they are relevant to the flat space quantization in the R and NS sectors respectively. The L_m are the generators of the Virasoro algebra (1.1.19); the remaining (anti)-commutation relations read

$$[L_m, J_n] = -nJ_{m+n}\,, \qquad [J_m, J_n] = \frac{c}{3}m\delta_{m+n,0}\,; \tag{9.1.61a}$$

$$[L_m, G^{\pm}_r] = \left(\frac{m}{2} - r\right)G^{\pm}_{m+r}\,, \qquad [J_m, G^{\pm}_r] = \pm G^{\pm}_{m+r}\,; \tag{9.1.61b}$$

$$\{G^+_r, G^-_s\} = 2L_{r+s} + (r-s)J_{r+s} + \frac{c}{3}\left(r^2 - \frac{1}{4}\right)\delta_{r+s,0}\,. \tag{9.1.61c}$$

The $[L_m, \cdot]$ express the fact that $J(z) = \sum_n J_n z^{-n-1}$, $G\pm(z) = \sum_r G^{\pm}_r z^{-r-3/2}$ (in the notation of (1.1.20)) are *conformal primary* operator: namely, they transform as

$$\phi(z) \to |\partial_z z'|^w \phi(z') \tag{9.1.62}$$

under a conformal transformation. (This definition is extended to any d by replacing $|\,|$ with the determinant; compare footnote 5 before (4.1.102).) Of course, not all operators transform this way: not even $\partial_z\phi$ does. The $[J, J]$ are an example of *Kac–Moody* algebra, and the $[J, \cdot]$ can be interpreted in terms of its representation theory. Finally, the G^{α}_r, $\alpha = 1, 2$ are the infinite-dimensional extension of the supercharges Q^1, Q^2 in (9.1.48) and (9.1.49). All this was for left-moving degrees of freedom; there is another copy of the algebra for right-movers, with generators $\tilde{L}_m$, $\tilde{J}_m$, $\tilde{G}^{\alpha}_r$.

The algebras (9.1.61) obtained by taking different a are all isomorphic under

$$L^{a=0}_n = U(L^{a=1/2}_n)U^{\dagger}\,, \tag{9.1.63}$$

and similar for the other generators, where U is the *spectral flow* operator:

$$U = \exp\left[-\frac{\mathrm{i}}{2}\sqrt{\frac{c}{3}}\varphi\right], \qquad J \equiv \sum_n J_n z^{-n-1} \equiv \mathrm{i}\sqrt{\frac{c}{3}}\partial_z \varphi . \tag{9.1.64}$$

This isomorphism is a manifestation of spacetime supersymmetry: indeed, the spectral flow for left-movers maps the NSNS to the RNS sector and the RNS to RR, thus exchanging bosons and fermions. The presence of a separate spectral flow for the left- and right-movers signals that in the spacetime, we have $\mathcal{N} = 2$ supersymmetry. Finally, to implement the GSO projection (Section 1.1.2), the U(1) charges under J_0 should be odd integers in the full model containing (9.1.56) and the free $d = 4$ spacetime part.

Once we find a $\mathcal{N} = (2, 2)$ superconformal model with $c = 9$ and this odd charge property, we can use it to define a string theory model with a Mink_4 spacetime. We still expect it to have a four-dimensional effective $\mathcal{N} = 2$ supergravity. The number of vector and hypermultiplets can be expressed in terms of representations of the $\mathcal{N} = (2, 2)$-superconformal algebra. The relevant states $|\phi\rangle$ are those that are primary, namely that are created by a conformal primary (9.1.62); in terms of modes, $L_n|\phi\rangle = 0$ for $n > 0$, $L_0|\phi\rangle = h|\phi\rangle$, so they are "highest weight" states. A *superconformal* primary is also annihilated by all the other positive modes J_n, $G^{\pm}_r$, $n > 0$, $r > 0$, and $J_0|\phi\rangle = q|\phi\rangle$. Now, massless fields in the NS sector are associated to *chiral primary* states, namely superconformal primary states that are also chiral, in the sense that $G^+_{-1/2}|\phi\rangle = 0$. It follows that

$$\langle\phi|\{G^-_{1/2}, G^+_{-1/2}\}|\phi\rangle \overset{(9.1.61)}{=} \langle\phi|(2L_0 - J_0)|\phi\rangle = 2h - q . \tag{9.1.65}$$

Conversely, a state with $h = q/2$ is a chiral primary, as one can show with a version of the Hodge decomposition (4.1.140), [306]. An *anti*chiral primary obeys $G^-_{-1/2}|\phi\rangle = 0$, and can be similarly characterized by $h = -q/2$. Massless fields in the R sector are associated to *Ramond ground states*, namely states that are annihilated by $G^{\pm}_0$. The (anti)chiral primary states in the NS sector and Ramond ground states in the R sector are related by the spectral flow, consistent with its interpretation as spacetime supersymmetry.

For example, for a sigma model with target space a Calabi–Yau M_6, NS states that are chiral primary for both the left- and right-movers are associated to $H^{1,1}(M_6)$, while those that are chiral for left-movers and antichiral for right-movers are associated to $H^{2,1}(M_6)$:

$$\begin{aligned} \{|\phi\rangle \text{ such that } G^+_{-1/2}|\phi\rangle = \tilde{G}^+_{-1/2}|\phi\rangle = 0\} &\cong H^{1,1}(M_6) ; \\ \{|\phi\rangle \text{ such that } G^+_{-1/2}|\phi\rangle = \tilde{G}^-_{-1/2}|\phi\rangle = 0\} &\cong H^{2,1}(M_6) . \end{aligned} \tag{9.1.66}$$

This reproduces the geometric scalars in Section 9.1.1. Notice how similar the two cohomologies look from the point of view of the world-sheet; this will be important in Section 9.1.4.

Gepner and Landau–Ginzburg models

In our discussion so far, we started from the nonlinear sigma model (9.1.56). Then we concluded that the internal part of a Mink_4 compactification with no flux should be an $\mathcal{N} = 2$ superconformal field theory (SCFT) model with $c = 9$. But is it really necessary that this SCFT should be a nonlinear sigma model? Perhaps we could dispose with the geometrical interpretation of the internal space. After all, physically we only really need the Mink_4 directions.

We can thus try considering generalizations of (9.1.56). For example, we can include a world-sheet superpotential:

$$S_{\text{LG}}^{(2,2)} = T_{\text{F1}} \left[\int \mathrm{d}^2\sigma \mathrm{d}^2\theta \mathrm{d}^2\bar{\theta}\, K(Z, \bar{Z}) + \int \mathrm{d}^2\sigma \mathrm{d}^2\theta\, W(Z) \right] . \tag{9.1.67}$$

W is holomorphic: it only depends on the chiral superfields Z^i and not their conjugates $\bar{Z}^{\bar{\imath}}$. Now the Z^i now no longer have a geometrical interpretation, and so the index i no longer ranges from 1 to 3. Equation (9.1.67) is called *Landau–Ginzburg (LG) model* (in analogy to the effective theory for superconductivity). The new W term does not change under renormalization group (RG) flows to low energy, other than by field rescaling. In particular, if we take W to be a homogeneous polynomial of the Z^i, (6.3.42), then the effect of the RG flow can be completely reabsorbed. More generally, this is true for *quasihomogeneous* polynomials, namely

$$W(\lambda_1 Z^1, \ldots, \lambda_k Z^k) = \lambda^w W(\lambda_1 Z^1, \ldots, \lambda_k Z^k) . \tag{9.1.68}$$

The Kähler potential K does change, but the model cannot become trivial at low energies because of the W term. So when (9.1.68) holds, (9.1.67) is expected to be conformal [307].

The simplest case is when only one Z is present; the model with

$$W = Z^{k+2} \tag{9.1.69}$$

is called a *minimal* $\mathcal{N} = (2, 2)$*-model* MM_k, with

$$c = \frac{3k}{k+2} . \tag{9.1.70}$$

(A similar definition exists for conformal models without supersymmetry.) To achieve $c = 9$, we can take a tensor product of r minimal models MM_{k_i} by summing their actions, making sure $\sum_{i=1}^{r} \frac{3k_i}{k_i+2} = 9$. If we take all k_i equal, $rk = 3(k + 2)$; this leads to

$$r = 5 , \qquad k_i = k = 3 . \tag{9.1.71}$$

This tensor product can also be realized by having r superfields Z_i, each with its superpotential, and summing the individual actions; in other words, by having a superpotential

$$W = \sum_{i=1}^{r} Z_i^{k_i+2} . \tag{9.1.72}$$

We mentioned after (9.1.64) that the U(1)-charges should be odd integers for the GSO projection. The minimal model does not satisfy this, but this can be fixed by quotienting by a discrete symmetry. For example, the model in (9.1.71) has states with U(1)-charges that are multiples of 1/5; the integrality constraint is enforced with a $\mathbb{Z}_5$ quotient. In conclusion, we may denote our model symbolically as

$$\left(W = \sum_{i=1}^{5} Z_i^5 \right) / \mathbb{Z}_5 . \tag{9.1.73}$$

Gauged linear sigma models

The spectrum for the Gepner model (9.1.73) turns out to be equal to the one for the quintic. More generally, for every Gepner model one can identify a corresponding Calabi–Yau that has the same spectrum.

To see why this happens, we now consider a model that is not conformal itself, but that flows in the IR either to a sigma model with a Calabi–Yau target space, or to a LG model, thus interpolating between the two [308]. The model is obtained by *gauging* an ordinary sigma model. For a bosonic sigma model $\int d^2\sigma\sqrt{-h}g_{mn}\partial_\alpha x^m \partial^\alpha x^n$, with fields $x^m : \Sigma \to M$, if M is a Killing vector K^m, then we can introduce a world-sheet vector field A and a covariant derivative $D_\alpha x^m \equiv \partial_\alpha x^m - A_\alpha K^m$, similar to the gauging (8.4.4) in $d = 4$ supergravity. Adding a kinetic term for A, we arrive at the gauged sigma model action:

$$S_{\text{gauged}} = -\frac{1}{2}\int d^2\sigma\sqrt{-h}\left(g_{mn}D_\alpha x^m D^\alpha x^n + \frac{1}{e^2}|F|^2\right). \tag{9.1.74}$$

The generalization involving more than one K^m is straightforward. Physically, (9.1.74) describes a string in the space $M/\mathrm{U}(1)$, where $\mathrm{U}(1)$ is generated by K^m.

Since e has dimension of mass, at low energies it becomes large, making the gauge field A strongly coupled and its kinetic term F^2 disappear. We can then integrate out A_m from (9.1.74), obtaining a new sigma model with metric

$$g^{M/\mathrm{U}(1)}_{mn} = g_{mn} - \frac{K_m K_n}{K^2}. \tag{9.1.75}$$

This is indeed an alternative expression for the metric on the base of (4.2.35). The flow of (9.1.74) to a sigma model with metric (9.1.75) is "fast": the gauge field becomes strongly coupled for dimensional reasons. At lower energies still, if (9.1.75) is not Ricci-flat, the model will undergo a "slow" logarithmic RG flow with a beta-function R_{mn}, regulated by world-sheet quantum effects.

We can also introduce a coupling to B as in (1.1.10); now the gauging also requires that $L_K H = 0$. The metric (9.1.75) is also modified. Another ingredient one can introduce is a world-sheet potential V.

For an $\mathcal{N} = (1,1)$-supersymmetric version, a gauge field is part of a superfield Γ_α, with a super covariant derivative $D_\alpha X^m = \partial_\alpha X^m - \Gamma_\alpha X^m$; see, for example, [309, sec. 2.3]. (Much of supersymmetry in $d = 2$ is similar to $d = 3$, for which one may consult [310, chap. 2].) For $\mathcal{N} = (2,2)$, the gauge field is contained in a superfield V, together with a gaugino, an auxiliary field D similar to the one in $d = 4$, and another scalar σ. (One can obtain this by dimensional reduction from $d = 4$.) The field-strength F is contained in the superfield $\Sigma \equiv \bar{\mathbb{D}}_+\mathbb{D}_- V$. Gauging is now implemented by $Z^I \to \mathrm{e}^{C_I V} Z^I$ on the chiral superfields. For our purposes, it will be enough to consider a gauged *linear* sigma model (GLSM), which means that the metric on M is Euclidean; in other words, prior to gauging we consider (9.1.67) with $K = \sum_I \bar{Z}_I Z^I$. Moreover, we will take a Killing vector[3]

$$\mathrm{i}\sum_{I=1}^{N} C_I(Z^I\partial_{Z^I} - \bar{Z}_I\partial_{\bar{Z}_I}). \tag{9.1.76}$$

[3] Here we keep a notation similar to (6.4.27), for reasons that will soon be apparent. For this reason, we will keep for this model an upper-case letter also for the bosonic components of the chiral superfields Z^I.

The $\mathcal{N} = (2,2)$-supersymmetric version of (9.1.74) then reads as follows [308]:

$$S^{(2,2)}_{\rm GLSM} = \int \mathrm{d}^2\sigma \left[\mathrm{d}^2\theta \mathrm{d}^2\bar{\theta} \left(-\frac{1}{4e^2}\bar{\Sigma}\Sigma + \sum_{I=1}^{N} \bar{Z}_I e^{2C_I V} Z^I \right) + \mathrm{d}^2\theta W + c\mathrm{Re}(\mathrm{d}^2\tilde{\theta}\Sigma) \right]. \tag{9.1.77}$$

Here $\mathrm{d}^2\tilde{\theta} = \mathrm{d}\theta^+\mathrm{d}\bar{\theta}^-$; this strange measure arises because Σ is a *twisted chiral* superfield rather than a chiral one, a concept we will further explore in Section 9.3.3. In general, a term that is integrated with this measure is called a *twisted superpotential.* This last term of (9.1.77) implements the shift (8.4.8) in this model, which is allowed for abelian gauge groups, as we saw there. The world-sheet potential is

$$V^{(2,2)}_{\rm GLSM} = \frac{e^2}{2}\left(\sum_I C_I |Z^I|^2 - c \right)^2 + \sum_i |\partial_{z^I} W|^2 + 2|\sigma|^2 \sum_I |z^I|^2 ; \tag{9.1.78}$$

for $c \neq 0$, the vacua are

$$\partial_{z^I} W = 0 , \qquad \sum_I C_I |z^I|^2 = c . \tag{9.1.79}$$

This is similar to (8.4.52) in $d = 4$.

At low energies, (9.1.77) first flows to a sigma model whose target space is the moduli space of vacua (9.1.79) quotiented by (9.1.76). As in our discussion after (9.1.75), there will then be a slow RG flow with beta function R_{mn}. For example, with $W = 0$, $c > 0$, and all the $C_I = 1$, the fast flow will be $\mathbb{CP}^{N-1}$; in this case, there is no Ricci-flat metric, and so the model will eventually not flow to an SCFT. If instead we take $W = 0$, $c = 0$, $C_I = (1, 1, -1, -1)$, by (6.4.39), the model will flow to the conifold, which does admit a Ricci-flat metric (Section 7.4.4). But for $W = 0$, the target space has isometries and hence cannot be a *compact* Calabi–Yau (Section 7.1.2).

So for compact examples, we need $W \neq 0$. We can take $N = 6$ chiral superfields Z^I, with charges and superpotential [308]

$$C_I = (1, 1, 1, 1, 1, -5) , \qquad W = Z^6 f_5(Z^1, \ldots, Z^5) , \tag{9.1.80}$$

where f_5 is a quintic polynomial whose only singularity is at the origin: there are no solutions to (6.3.21) except $Z^1 = \cdots = Z^5 = 0$. The D-term equation is $\sum_i |Z^i|^5 - 5|Z^6|^2 = c$. The F-terms give $f_5 = 0$, $Z^6 \partial_i f_5 = 0$, $i = 1, \ldots, 5$. The space of solutions (9.1.79) depends crucially on c.

- First take $c > 0$. Assuming $Z^6 \neq 0$, the F-terms give $f_5 = \partial_i f_5 = 0$, which have the only solution $Z^1 = \cdots = Z^5 = 0$. But the D-term then becomes $-5|Z^6|^2 = c$, which cannot be satisfied. So we conclude $Z^6 = 0$. The D-term is now $\sum_i |Z^i|^5 = c$, which together with the $\mathrm{U}(1)$ quotient describes a $\mathbb{CP}^4$; the F-term is $f_5 = 0$, which describes our familiar quintic $Q \subset \mathbb{CP}^4$, which as we know admits a Ricci-flat metric. By our discussion around (9.1.58), there is on this space a solution to the beta function equations, and the slow RG flow will converge to this metric at low energies.
- Now consider $c < 0$. This time, if we take $Z^6 = 0$, the D-term has no solution. So we conclude $Z^6 \neq 0$, and by the argument for the previous case we conclude $Z^1 = \cdots = Z^5 = 0$. At low energies, the gauge field as usual disappears, and the potential (9.1.78) gives a mass to the scalars Z^6 and σ. The remaining fields

$Z^1, \dots, Z^5$ are now described by an LG theory (9.1.67), with a superpotential $W = f_5$, which breaks the U(1) action to $\mathbb{Z}_5$. This theory is a generalization of the Gepner model (9.1.73).

So the model (9.1.77) and (9.1.80) interpolates between the sigma model for the quintic Calabi–Yau at $c > 0$, and an LG model similar to (9.1.73) at $c < 0$. There is a phase transition at $c = 0$. This parameter c is in correspondence with the single Kähler modulus for the quintic, where indeed $h^{1,1} = 1$. One can obtain similar descriptions for more complicated Calabi–Yau manifolds by considering more than one gauge field and more complicated superpotentials. Besides the purely geometrical phases described by sigma models and the LG models, in general there are also hybrid phases, and many more phase transitions.

9.1.4 Mirror symmetry

While in Sections 5.3 and 5.4 we have tried to highlight some of the similarities between complex and symplectic geometry, the two might still seem very different. But on a Calabi–Yau, the two play a very similar role: in (9.1.7) and (9.1.13) for the multiplet structure, (9.1.41b) for the effective theories, and in (9.1.66) from the world-sheet point of view.

These similarities suggested [306, 311] that perhaps there might be a *mirror symmetry* mapping a Calabi–Yau M_6 to another Calabi–Yau $\tilde{M}_6$ such that

$$h^{1,1}(\tilde{M}_6) = h^{2,1}(M_6) , \qquad h^{2,1}(\tilde{M}_6) = h^{1,1}(M_6) . \tag{9.1.81}$$

Around the same time, this symmetry manifested itself [163] while producing explicit examples of Calabi–Yau manifolds, using techniques such as those in Section 7.1.3. The name "mirror" is because the Hodge diamond (7.1.8) of $\tilde{M}_6$ looks like a reflection of the one of M_6 around a $\pi/4$-oblique axis. Another reason for the name is that the symmetry (9.1.81) becomes apparent if one plots all the known values of $(h_{1,1}, h_{2,1})$ for Calabi–Yaus.

A stronger version of (9.1.81) is

$$\text{IIA on } M_6 \ \cong \text{IIB on } \tilde{M}_6 . \tag{9.1.82}$$

In this section, we will see evidence that this is actually correct. For more details, see [312, 313].

Greene–Plesser procedure

A simple procedure to find examples of mirror pairs is the *Greene–Plesser orbifold* [314]. For example, for the quintic Q defined in (7.1.25), one defines

$$\tilde{Q} = \{Z_0^5 + Z_1^5 + Z_2^5 + Z_3^5 + Z_4^5 - 5\psi Z_0 Z_1 Z_2 Z_3 Z_4 = 0\}/\mathbb{Z}_5^3 , \tag{9.1.83}$$

where the generators of the three copies of $\mathbb{Z}_5$ are given by

$$\begin{aligned}(Z_0, Z_1, Z_2, Z_3, Z_4) &\sim (Z_0, \omega_5 Z_1, Z_2, Z_3, \omega_5^4 Z_4)\\ &\sim (Z_0, Z_1, \omega_5 Z_2, Z_3, \omega_5^4 Z_4)\\ &\sim (Z_0, Z_1, Z_2, \omega_5 Z_3, \omega_5^4 Z_4) .\end{aligned} \tag{9.1.84}$$

The two quintic monomials in (9.1.83) are the only ones left invariant under (9.1.84); so there is only one modulus ψ left, and $h^{2,1}(\tilde{Q}) = 1$. We can also compute $\chi(\tilde{Q})$ by using (4.2.72). At first, one might think $\chi(\tilde{Q}) \stackrel{?}{=} \chi(Q)/5^3$, because of the $\mathbb{Z}_5^3$ identification. But we have to take into account that there are ten curves invariant under one of the $\mathbb{Z}_5$; and ten points invariant under a $\mathbb{Z}_5^2$. The ten curves are

$$\{Z_i^5 + Z_j^5 + Z_k^5 = 0\}, \tag{9.1.85}$$

with i, j, k all different; these are in a linear subspace $\cong \mathbb{CP}^2$ and thus by (6.3.9) have genus $g = 6$, and $\chi = 2 - 2g = -10$. Using these facts and additivity of χ, one finds $\chi(\tilde{Q}) = 200$ [315, sec. 2].

So $h^{1,1}(\tilde{Q}) = h^{2,1}(\tilde{Q}) + \chi(\tilde{Q})/2 = 101$. On the other hand, we computed the cohomology of the quintic in (7.1.26) and (7.1.28). So (9.1.81) is satisfied.

World-sheet argument

An argument for mirror symmetry in models such as (9.1.83) was given in [99]. The idea is to dualize the gauged linear sigma model (9.1.77), using methods similar to T-duality; this is related to ideas in Section 9.2.4.

Compact Calabi–Yau manifolds have no isometries, by (7.1.10); applying T-duality directly is difficult. So the first step is to dualize (9.1.77) with $W = 0$. The procedure is similar to that in (1.4.17)–(1.4.16), but now directly at the level of (2, 2) superfields. The result is that N chiral superfields Z_I coupled to a single vector multiplet with charges C_I are dual to a model with *twisted* chiral superfields Y_I, which satisfy $\bar{\mathbb{D}}_+ Y_I = \mathbb{D}_- Y_I$ rather than (9.1.55), and a term

$$\int d^2\tilde{\theta}\tilde{W}_0 , \qquad \tilde{W}_0 = \Sigma\left(\sum_I C_I Y_I - t\right) \tag{9.1.86}$$

with t a constant. The measure $d^2\tilde{\theta}$ was defined after (9.1.77). The second step is to follow the IR flow of this model; it is argued that the twisted superpotential $\tilde{W}$ gets corrected to $\tilde{W} = \tilde{W}_0 + \sum_I e^{-Y_I}$. If we integrate out Σ, we find a constraint $\sum_I C_I Y_I = t$; since it is linear, we can solve it. For example, for (9.1.80), $Y_6 = (Y_1 + \cdots + Y_5)/5$; after rescaling,

$$\tilde{W} = \sum_I e^{-Y_I} = e^{-5Y_1} + \cdots + e^{-5Y_5} + e^{-t+Y_1+\cdots+Y_5} . \tag{9.1.87}$$

The last step is to reintroduce the superpotential, which for the quintic was in (9.1.80). It is argued in [99] that the effect of this is to change $e^{-Y_I} \to X_I$. With this change, $\tilde{W}$ reproduces the equation of the mirror quintic (9.1.83).

Moduli spaces

The two special Kähler manifolds in Section 9.1.2 were $\mathcal{M}_c$, and the complexified Kähler moduli space $\mathcal{M}_c^{\mathbb{C}}$. (9.1.81) suggests that

$$\mathcal{M}_c(M_6) \stackrel{?}{\cong} \mathcal{M}_K^{\mathbb{C}}(\tilde{M}_6) , \qquad \mathcal{M}_K^{\mathbb{C}}(M_6) \stackrel{?}{\cong} \mathcal{M}_c(\tilde{M}_6) . \tag{9.1.88}$$

In the example where $M_6 = Q$ is the quintic and $\tilde{M}_6 = \tilde{Q}$ is (9.1.83), $\mathcal{M}_K(Q)$ has dimension $h^{1,1}(Q) = 1$. We can picture it as a segment, with one endpoint representing the large volume limit and the other endpoint representing the LG model

(9.1.73). Adding the single B field modulus turns this into a topological S^2. As we saw at the end of Section 9.1.3, there are a geometric and an LG phase, separated by a phase transition at the equator of the S^2. On the mirror side, $\mathcal{M}_c(\tilde{Q})$ has the single holomorphic coordinate ψ in (9.1.83); this describes a copy of $\mathbb{C}$, to which we should add a point at infinity $\psi \to \infty$, representing the (singular) $\{Z_0 Z_1 Z_2 Z_3 Z_4 = 0\}/\mathbb{Z}_5^3$. So this is also a copy of $\mathbb{CP}^1 \cong S^2$. So in this case, we indeed conclude

$$\mathcal{M}_K^{\mathbb{C}}(Q) \cong \mathcal{M}_c(\tilde{Q}), \tag{9.1.89}$$

at least topologically; a comparison of the geometries will come from the study of their prepotentials, which we will undertake soon. There is another point on $\mathcal{M}_c$ where $\tilde{Q}$ is singular: $\psi = 1$, as we know from (7.1.29). The 125 points mentioned there are all identified by the $\mathbb{Z}_3^5$ to be a single one, which is a conifold singularity by Exercise 7.1.1.

In more complicated examples with $h^{1,1} > 1$, some puzzles appear. $\mathcal{M}_K^{\mathbb{C}}$ has some special loci that don't seem to be present in $\mathcal{M}_c(\tilde{M}_6)$: all the walls of the Kähler cone, the loci where a curve shrinks to zero size, creating a conifold singularity (Section 7.1.2). As we have seen in Section 7.1.4, once we arrive at such a wall, we can flop it to create a new Calabi–Yau M_6'. Its topology differs from the original M_6, but $h^{1,1}$ remains the same, since a new curve replaces the one that shrunk to zero size. Now we have two $\mathcal{M}_K(M_6)$, $\mathcal{M}_K(M_6')$ that have a wall in common; it is natural to glue them together and define the *enlarged Kähler moduli space*:

$$\mathcal{M}_{eK}\,. \tag{9.1.90}$$

In fact, there are many walls, so the full extended moduli space will in general consist of the Kähler moduli spaces of many Calabi–Yaus with different topologies. Taking also into account the B field moduli, we arrive at the statement

$$\mathcal{M}_{eK}^{\mathbb{C}}(M_6) \cong \mathcal{M}_c(\tilde{M}_6)\,. \tag{9.1.91}$$

This now does work; for more details, see [313, sec. 7.3–7.5].

Prepotentials

We saw in Section 9.1.2 that the effective four-dimensional theory is governed by two prepotential functions, $\mathcal{F}_{vm}$ and $\mathcal{F}_{hm}$, for the vector multiplets and hypermultiplets respectively. If (9.1.82) is to hold, these two prepotentials should also be exchanged by mirror symmetry:

$$\mathcal{F}_{vm}(M_6) = \mathcal{F}_{hm}(\tilde{M}_6)\,, \qquad \mathcal{F}_{hm}(M_6) = \mathcal{F}_{vm}(\tilde{M}_6)\,. \tag{9.1.92}$$

It is best to start by computing $\mathcal{F}_c$, the prepotential on the complex structure moduli space $\mathcal{M}_c$. This can be computed reliably in supergravity, since l_s corrections should depend on Kähler moduli. From (9.1.19), we see that the X^I, $\mathcal{F}_I$ that we need to define the prepotential are periods of Ω. This might look difficult, because after all we don't know the metric of M_6 explicitly. However, recall from Section 7.1.1 that the Ricci-flat metric on M_6 is obtained by modifying the two-form as in (7.1.5); in the process, the complex structure and the total volume form are not touched, and we can use their expression as Poincaré residue in (7.1.24).

To simplify the problem further, we can select a model with few complex moduli as the mirror quintic (9.1.83), whose $h^{2,1} = 1$ [315], already discussed around

(9.1.89). There are only $2 + 2h^{2,1} = 4$ periods to compute, depending on the single parameter ψ. These can be found using a Taylor expansion in ψ or by solving a differential equation called *Picard–Fuchs equation*: they are all annihilated by the differential operator

$$(\psi\partial_\psi)^5 - \psi^{-5}(5\psi\partial_\psi - 1)(5\psi\partial_\psi - 2)(5\psi\partial_\psi - 3)(5\psi\partial_\psi - 4)\,. \tag{9.1.93}$$

A procedure to find the correct differential operator in general is in [316].

Knowing four solutions to (9.1.93), which can be written in terms of generalized hypergeometric functions, we still choose which linear combinations we assign to the four entries $(X^I, \mathcal{F}_I)$. This is done in part by checking how they behave around the point $\psi = 1$ discussed following (9.1.89). At that conifold point, a single cycle $A^1 \cong S^3$ shrinks; (9.1.45) implies a property on the periods of Ω. This helps finding the $(X^I, \mathcal{F}_I)$ from solutions of (9.1.93), up to an overall symplectic transformation (8.4.15), and then $\mathcal{F}_c(\tilde{Q}) = \frac{1}{2}X^I\mathcal{F}_I$ as usual from (8.4.27).

Now to check (9.1.92), we should compare this to $\mathcal{F}_K(Q)$. At large $t = b + iv$, the curvature of Q is small, and this can be computed reliably in supergravity with (9.1.32). There is a single two-cycle, which using a logic similar to that presented after (7.2.4) has triple intersection $D_{111} = 5$; so

$$\mathcal{F}_K = \frac{5}{6}(X^0)^2 t^3 \tag{9.1.94}$$

in the supergravity approximation. This matches with what one gets from the $\mathcal{F}_c$ in the $\psi \to \infty$ limit, if the symplectic basis $\{A^I, B_I\}$ is appropriately chosen. By comparing $X^1/X^0 = t^1 = t$ on the Kähler side with $X^1(\psi)/X^0(\psi)$ on the complex side, we get a *mirror map* relating t to ψ. At small t, the curvature of Q is large, and (9.1.94) can no longer be trusted; but the mirror map provides a prediction for the stringy corrections to $\mathcal{F}_K(t)$. The result is that there are quadratic and linear terms in t, a constant $\mathcal{F}_0$, and exponential corrections. The quadratic and linear terms are real, and as such actually disappear from the metric, as we see from (8.4.30) and (8.4.32). The constant is imaginary and does contribute, as it appears multiplied by the usual $(X^0)^2$; it originates from the stringy corrections (1.2.24). Finally, the exponential terms can be interpreted as world-sheet instantons, briefly discussed around (1.3.18).

Alternatively, the triple derivatives of $\mathcal{F}$ have a physical interpretation in the heterotic theory as the Yukawa couplings (9.1.44). Setting $X^0 = 1$, the result predicted by mirror symmetry is [315, (5.12)]

$$\partial_t^3\mathcal{F}_K = 5 + 2875\,e^{it/(4\pi^2 l_s^2)} + 4876875\,e^{2it/(4\pi^2 l_s^2)} + \cdots\,. \tag{9.1.95}$$

The leading constant comes from (9.1.94). The exponents shown, and those in the following terms, are all proportional to $dit = d\int(-v^1 + ib^1)\omega_1 = \int_{\Sigma_d}(-J + iB)$. Here Σ_d is a two-dimensional subspace defined by a system of equations of degree d, using the argument that follows (6.2.20). Taking into account the possibility of an instanton $\Sigma \to \Sigma_d$ that is a k-to-one map, one can reconstruct from (9.1.95) the number of Σ_d of degree d in the quintic. For low d, this agrees with results obtained with techniques from algebraic geometry; in particular, for $d = 1$ the number is the 2,875 in (9.1.95), in agreement with [317]. But for large d, it becomes a lot quicker than the classical mathematical techniques. This computation has been greatly generalized, culminating in mathematical theorems for complete intersections in toric manifolds [318, 319].[4]

Besides giving a nice mathematical application, this result gives a powerful indication that string theory is a well-defined quantum theory, in spite of all the challenges in defining it that we mentioned in Chapter 1.

9.1.5 Flux and gauging

So far we have considered the form fields only as fluctuations, such as in (9.1.2), (9.1.3), taking them to vanish on the vacuum. Indeed, from the type II action (1.2.15) and (1.2.36), we expect that they contribute a nonzero stress-energy tensor T_{MN}, in particular no longer demanding M_6 to be Ricci-flat. Still, as we discussed in Section 8.3, we can use the moduli space of Calabi–Yau manifolds as the basis for an effective field theory, hoping that some vacua are retained. As we discussed there, the form fields might be accommodated by some modification of the internal geometry.

In this section, we will include fluxes (form fields without sources) in the $\mathcal{N}=2$ formalism we have used for the Calabi–Yau effective action in Section 9.1.2. Unfortunately, we will not find any vacua this way. But the idea does work if we first break to $\mathcal{N}=1$ supersymmetry. This is ultimately because we need O-planes for Minkowski vacua with fluxes and, as we mentioned in Section 8.3.2, probably also for AdS vacua with scale separation. We will give an important example of flux vacua on a Calabi–Yau with an $\mathcal{N}=1$ effective description in Section 9.5.3.

We will mostly work in IIA. The field-strengths are obtained by adding purely internal components to the exterior differential of (9.1.37):

$$H = H^{\rm int} + d_4 b^i \wedge \omega_i + *d_4 a\,, \qquad F_2 = d_4 C_1 + F_2^{\rm int}\,,$$
$$F_4 = F_4^{\rm int} - H \wedge C_1 + d_4 A^i \wedge \omega_i + d_4 \zeta^A \wedge \alpha_A + d_4 \tilde{\zeta}_A \wedge \beta^A + d_4 c_3\,. \tag{9.1.96a}$$

(We have set $F_0 = 0$ for simplicity, to avoid the terms in (1.2.27) and (1.2.29) with B.) It is natural to take the internal fluxes to be harmonic, so that they cannot be fluctuations, because they cannot be continuously deformed to zero for topological reasons. So we can expand them in cohomology:

$$H^{\rm int} = h^A \alpha_A + \tilde{h}_A \beta^A\,, \qquad F_2^{\rm int} = f_2^i \omega_i\,, \qquad F_4^{\rm int} = f_{4i} \tilde{\omega}^i\,. \tag{9.1.96b}$$

The coefficients h^A, $\tilde{h}_A$, f_{4i} are quantized because of (1.3.47). The last term in F_4 in (9.1.96a) can be dualized to a constant:

$$* d_4 c_3 \equiv f_6\,. \tag{9.1.96c}$$

This is similar to the parameter playing a role in the dynamical neutralization of the cosmological constant in [321]. From the point of view of string theory, it can be considered a purely internal component of F_6, as defined in (1.3.40).

Gauging data

If we rewrite the CS terms in (1.2.15) as

$$\int C_3 \wedge H \wedge F_4 \tag{9.1.97}$$

[4] Another way of computing the Kähler potential is via localization of the GLSM on a sphere [320].

by integrating by parts, using (9.1.11), (9.1.37), and (9.1.96), it contributes a term

$$f_{4i}\int *\mathrm{d}_4 a \wedge A^i = f_{4i}\int \mathrm{d}^4x\sqrt{-g_4}\,\partial_\mu a\, A^{i\mu} \tag{9.1.98}$$

in the four-dimensional action. When one collects all terms in the reduction, this ends up being part of a Stückelberg-type term $D_\mu a D^\mu a$, with $D_\mu a \equiv \partial_\mu a + f_{4i}A^i_\mu$ (recall (1.2.26)). This is allowed by the general structure (8.4.42) of gauging for an $\mathcal{N}=2$ theory. So fluxes also cause some scalars to become gauged under some vectors.

Keeping track of all the new terms is laborious, so we just give the results. The Killing vectors that are activated in (8.4.42) are the K^u_I, representing the action of the Ith vector multiplet on the hypermultiplets. These are a linear combination of the Killing vectors (9.1.35), and the coefficients are related to the flux coefficients in (9.1.96):

$$K_0 = 2f_6\partial_a - h^A D_A - \tilde{h}_A\tilde{D}^A\,, \qquad K_i = 2f_{4i}\partial_a\,. \tag{9.1.99}$$

If the fluxes $F_2^{\rm int}$ and F_0 do not vanish, we have the option of performing a symplectic transformation (discussed before (8.4.15)) until the vector of RR flux quanta

$$(f_6\; f_{4i}\; f_2^i\; f_0)^t \tag{9.1.100}$$

becomes of the form $(f_6'\; f_{4i}'\; 0\; 0)$; then (9.1.99) applies with $f_6 \to f_6'$, $f_{4i}\to f_{4i}'$. Alternatively, one can introduce magnetic gaugings as well. These were discussed around (8.4.56) as a way to circumvent the no-go argument against supersymmetry-breaking in [277]. In that formalism we also introduce the magnetic dual vectors $\tilde{A}_I$, and corresponding Killing vectors

$$\tilde{K}^0 = 2f_0\partial_a\,, \qquad \tilde{K}^i = 2f_2^i\partial_a\,. \tag{9.1.101}$$

We observe a certain regularity in how the RR quanta (9.1.100) appear in (9.1.99) and (9.1.101). In particular, it is easy to guess the result for IIB: the vector (9.1.100) is replaced by

$$(f_{3,0}\; f_{2,1\,i}\; f^i_{1,2}\; f_{0,3})\,. \tag{9.1.102}$$

It is also natural to wonder if there is any way to also activate components in K_i, $\tilde{K}^i$, $\tilde{K}^0$ along D_A, $\tilde{D}^A$. Formally, this can be achieved by modifying the internal geometry so that it is no longer Calabi–Yau; the internal forms are then no longer harmonic, but should be taken to satisfy (8.3.16). We will discuss this further in Section 10.3.6.

Potential

Equation (9.1.96b) gives rise to a nonzero potential $V_{\mathcal{N}=2\text{ sugra}}$, because the internal part of the action is no longer just $\int_{M_6}\sqrt{g_6}R_6$, but also receives contributions from the $\int_{M_6}\sqrt{g_6}|F_k|^2$ and $\int_{M_6}\sqrt{g_6}|H|^2$. Knowing the gauging data, we can also compute the potential from the general expression (8.4.43). It is helpful to notice that for the Killing vectors (9.1.35), the hypermoment maps is of the form

$$\mu^\alpha_I = A^\alpha_u K^u_I \tag{9.1.103}$$

[322, 323]; the quaternionic-Kähler connection was found in [298]. For example, in the case $H=0$, one gets the following [300]:

$$V_{\mathcal{N}=2\text{ sugra}} = -(f_I + \mathcal{N}_{IK}\tilde{f}^K)(\mathrm{Im}\mathcal{N})^{-1\,IJ}(f_J + \bar{\mathcal{N}}_{JL}\tilde{f}^L), \tag{9.1.104}$$

where we have collected $f_I = (f_6\, f_{4I})$, $\tilde{f}^I = (f_2^i\, f_0)$. Using the fact that $\mathrm{Im}\mathcal{N}_{IJ}$ is negative definite, we see that this has no vacua unless the fluxes are all zero.

We postpone a more detailed analysis of the vacuum conditions until Section 10.3.6.

Exercise 9.1.1 Repeat the analysis of Section 9.1.1 for $\mathrm{Mink}_6 \times \mathrm{K3}$ vacua.

Exercise 9.1.2 Repeat the analysis of Section 9.1.1 for the M-theory solutions $\mathrm{Mink}_4 \times M_7$, where M_7 has G_2-holonomy (Exercise 8.2.2).

Exercise 9.1.3 Show that (9.1.35) are indeed Killing vectors of (9.1.34). (It is best to first compute the Lie derivatives of the one-forms appearing in it, using (4.1.79).)

Exercise 9.1.4 Check from (8.4.23) and (9.1.45) and the formulas that follow them that $g_{1\bar{1}} \sim \log|z^1|^2$ near a conifold point in $\mathcal{M}_c$.

9.2 D-brane geometry

In this section, we will study some properties of D-branes and O-planes, mostly in the probe approximation.

We saw back in Section 1.3.5 that D-branes preserve supersymmetry if (1.3.51) holds:

$$\epsilon^1 = \Gamma_{\parallel}\epsilon^2 . \tag{9.2.1}$$

The generalization of (1.3.52) for $\mathcal{F} = 2\pi l_s^2 f + B|_{\mathrm{D}p} \neq 0$ reads (see, for example, [324])

$$\begin{aligned}\Gamma_{\parallel} &= \frac{1}{\sqrt{-\det(g|_{\mathrm{D}p} + \mathcal{F})}} \sum_l \frac{(-1)^l}{l!\,(p+1-2l)!\,2^l} \epsilon_{(0)}^{a_1\ldots a_{p+1}} \mathcal{F}_{a_1 a_2} \cdots \mathcal{F}_{a_{2l-1}a_{2l}} \Gamma_{a_{2l+1}\ldots a_{p+1}} \\ &= \frac{\sqrt{-\det(g|_{\mathrm{D}p})}}{\sqrt{-\det(g|_{\mathrm{D}p} + \mathcal{F})}} (\mathrm{e}^{-\mathcal{F}})_{\llcorner} \Gamma_{\parallel\, \mathcal{F}=0} .\end{aligned} \tag{9.2.2}$$

In the second line, we have basically used (3.1.21) repeatedly on the world-volume: only the contraction terms survive. The denominator is the square root in the DBI term in (1.3.24). In part thanks to this factor, $\Gamma_{\parallel}$ is unitary, as one can show using (3.2.56).

For $\mathcal{F} = 0$, we interpreted $\Gamma_{\parallel}$ in Section 5.6.2 as the push-forward of the *covolume* $\widehat{\mathrm{vol}}_{\mathrm{D}p}$, a section of $\Lambda^{p+1}T$ on the Dp-brane. In the same spirit, we can interpret (9.2.2) as the push-forward of the contraction $\iota_{\mathrm{e}^{\mathcal{F}}} \widehat{\mathrm{vol}}_{\mathrm{D}p}$.

We will now analyze solutions to (9.2.1) in various situations. We will focus on flat space and Calabi–Yau vacuum solutions. Equation (9.2.2) is valid more generally; roughly speaking, the algebraic equations in this section will be applicable to any manifold, while the differential ones are more specific to the Calabi–Yau case. We will return to this issue in the next chapter.

9.2.1 Branes in flat space

We begin with branes in flat ten-dimensional space $\mathbb{R}^{10}$, with constant dilaton and all other fields set to zero. At first, we also take $\mathcal{F} = 0$.

It might seem that (9.2.1) simply determines ϵ^1 in terms of ϵ^2; this would mean that a D-brane is supersymmetric along any subspace. However, the ϵ^a we have to consider in (9.2.1) are the supercharges of the background: in (8.2.3), we saw that when all field-strengths are zero and the dilaton is constant, that means that $D_M \epsilon^a = 0$. In flat space, the ϵ^a should then be *constant*. If we try to place a Dp-brane along an arbitrary subspace, the $\Gamma^a = \partial_M x^a \Gamma^M$ in (9.2.2) will not be constant, and (9.2.1) will in general have no solution. If on the other hand we consider a flat subspace $\mathbb{R}^{p+1} \subset \mathbb{R}^{10}$, the factors $\partial_M x^a$ are constant, and (9.2.1) indeed always has a solution, determining the constant spinor ϵ^1 in terms of ϵ^2. The latter is constant but otherwise unconstrained, so there are 16 supercharges.

Next, if we have two D-branes, we have to solve (9.2.1) for each of them. If they are parallel, the $\partial_M x^a$ is the same for both, and so (9.2.1) is in fact the same equation for both branes; so we still have 16 supercharges. For a Dp – anti-Dp pair, we saw in (1.3.58) that supersymmetry is completely broken. More generally, we will have

$$\epsilon^1 = \Gamma^1_{\parallel}\epsilon^2 = \Gamma^2_{\parallel}\epsilon^2 \quad \Rightarrow \quad (\Gamma^1_{\parallel})^{-1}\Gamma^2_{\parallel}\epsilon^2 = \epsilon^2 , \tag{9.2.3}$$

which now gives a constraint on ϵ^2.

Orthogonal subspaces

We now consider a Dp and a Dq, which are all either extended or localized along every direction. We assume without loss of generality that the Dp is extended along directions $0 \ldots p$ and localized along directions $(p+1) \ldots 9$. Take the Dq extended along directions $M_0 \ldots M_q$, and localized along the others. Then $\Gamma^1_{\parallel} = \Gamma_{0\ldots p}$, $\Gamma^2_{\parallel} = \Gamma_{M_0 \ldots M_q}$, and the constraint (9.2.3) on ϵ^2 is

$$(\Gamma^1_{\parallel})^{-1}\Gamma^2_{\parallel}\epsilon^2 = \Gamma_{N_1 \ldots N_k}\epsilon^2 = \epsilon^2 . \tag{9.2.4}$$

The N_a are now the directions that are parallel to one brane and transverse to another. These are often called the *Neumann–Dirichlet (ND) directions*, since the open strings connecting them will have N boundary condition at one end and D at the other. If $q > p$ and the Dq is extended along directions $0 \ldots q$, the ND directions are $(p+1) \ldots q$, and (9.2.4) reads $\Gamma_{(p+1)\ldots q}\epsilon^2 = -(-1)^{\lfloor p/2 \rfloor}\epsilon^2$. Recall that T-duality turns an N direction into D, and vice versa; so by T-dualizing, we can always reach such a situation. (This means compactifying one or more directions, T-dualizing along them, and decompactifying them again.) Since p and q are both even in IIA and both odd in IIB, $p - q$ is even, and so is the number of ND directions #(ND).

Now we observe that $\Gamma^2_{N_1 \ldots N_k} = (-1)^{\lfloor k/2 \rfloor} 1$. So for k = #(ND) = 2 and 6, the eigenvalues of $\Gamma_{N_1 \ldots N_k}$ are $\pm$i, and (9.2.4) has no solution. If k = #(ND) = 0, 4, 8, the eigenvalues are ± 1, and (9.2.4) can be solved. So supersymmetry is broken unless

$$\frac{1}{4}\#(\mathrm{ND}) \in \mathbb{Z} . \tag{9.2.5}$$

Since $\mathrm{Tr}\Gamma_{N_1\ldots N_k} = 0$, the eigenvalues ± 1 appear with equal multiplicity, so the space of solutions of (9.2.4) has dimension eight; in other words, there are now eight supercharges.

We can also derive (9.2.5) more explicitly in the basis (2.2.13). By a rotation, we can always arrange so that the ND directions appear together on the rows of that equation; for example, 26, 37, and so on. For $k = \#(\mathrm{ND}) = 2$, $\Gamma_{N_1\ldots N_k} = 1 \otimes \mathrm{i}\sigma_3 \otimes 1 \otimes 1 \otimes 1$, whose eigenvalues are $\pm\mathrm{i}$, incompatible with (9.2.4). For $k = 4$, $\Gamma_{N_1\ldots N_k} = 1 \otimes \mathrm{i}\sigma_3 \otimes \mathrm{i}\sigma_3 \otimes 1 \otimes 1$, and (9.2.4) has solutions, such as $\epsilon^2 = |\pm++++\rangle \pm |\pm----\rangle$ (similar to (2.4.46)).

Branes at angles

Next we consider two Dp-branes (same dimension), still along flat subspaces $\mathbb{R}^{p+1}$, but no longer orthogonal. For example, take two D6-branes on the subspaces:

$$\begin{aligned} \mathrm{D6}_1 &= \{x^7 = x^8 = x^9 = 0\}\,, \qquad (9.2.6)\\ \mathrm{D6}_2 &= \{\cos\theta_1 x^7 + \sin\theta_1 x^4 = \cos\theta_2 x^8 + \sin\theta_2 x^5 = \cos\theta_3 x^9 + \sin\theta_2 x^6 = 0\}\,. \end{aligned}$$

The particular cases where the θ_i are 0 or $\pm\pi/2$ are covered by the analysis for (9.2.4). For example for $\theta_1 = \pi/2$, $\theta_2 = \theta_3 = 0$, we see that the only ND directions are 47, so $\#(\mathrm{ND}) = 2$ and according to (9.2.5), supersymmetry is broken. On the other hand, if $\theta_1 = \pi/2$, $\theta_2 = -\pi/2$, we have $\#(\mathrm{ND}) = 4$, and supersymmetry is preserved.

To see how to deal with the case with general θ_i, we go back to (9.2.3), which here reads

$$\begin{aligned} \epsilon^2 &= \Gamma_{0123456}^{-1}\Gamma_{0123}(\cos\theta_1\Gamma_4 + \sin\theta_1\Gamma_7)(\cos\theta_2\Gamma_5 + \sin\theta_2\Gamma_8)(\cos\theta_3\Gamma_6 + \sin\theta_3\Gamma_9)\epsilon^2\\ &= (\cos\theta_1 1 + \sin\theta_1\Gamma_{47})(\cos\theta_2 1 + \sin\theta_2\Gamma_{58})(\cos\theta_3\Gamma_6 1 + \sin\theta_3\Gamma_{69})\epsilon^2\,. \qquad (9.2.7) \end{aligned}$$

Since $\Gamma_{47}^2 = -1$, Γ_{47} has eigenvalues $\pm\mathrm{i}$, so $\cos\theta_1 1 + \sin\theta_1\Gamma_{47}$ has eigenvalues $\mathrm{e}^{\pm\mathrm{i}\theta_1}$. Altogether, the matrix in (9.2.7) has eigenvalues $\mathrm{e}^{\mathrm{i}(\sum_{i=1}^3 s_i\theta_i)}$, where $s_i \in \{+1, -1\}$. Taking, for example, $s_i = 1$ leads to imposing

$$\frac{1}{2\pi}\sum_i \theta_i \in \mathbb{Z}\,. \qquad (9.2.8)$$

Indeed, this time the 16 supercharges preserved by the $\mathrm{D6}_1$ are divided in eight groups of four supercharges, each characterized by a choice of the s_i; so when (9.2.8) holds, (9.2.7) is satisfied by four supercharges. The same holds for any other choice of the s_i. If $\theta_3 = 0$, then the eigenvalues are $\mathrm{e}^{\mathrm{i}(\sum_{i=1}^2 s_i\theta_i)}$ and only two s_i matter; this time we have eight supercharges left. The particular cases (9.2.6) are reproduced by this. Another cross-check is provided by $\theta_1 = \pi$, $\theta_2 = \theta_3 = 0$, where (9.2.8) is not satisfied: this is indeed a D6–anti-D6 configuration.

We can interpret (9.2.8) as saying that

$$(\Gamma_{\parallel}^1)^{-1}\Gamma_{\parallel}^2 \in \mathrm{SU}(3)\,. \qquad (9.2.9)$$

This is the condition for four supercharges; if more particularly $(\Gamma_{\parallel}^1)^{-1}\Gamma_{\parallel}^2 \in \mathrm{SU}(2)$, our discussion following (9.2.8) gives eight supercharges. This is reminiscent of our discussion in (7.4.17) for orbifolds.

In the presence of $\mathcal{F} \neq 0$, the conclusions of this subsection will change. We will revisit this issue in the next subsection, in the more geometrical language of Section 5.6.

Gravity solutions

So far in this subsection we have treated the branes as probes. Indeed, when there are few of them and g_s is small, the typical radius r_0 of the solutions goes to zero, and this back-reaction can be neglected (Section 1.3.7). Outside of this regime, one might want to have a description of the gravity solution they generate.

Unfortunately, this is a challenging problem. Consider the simplest situation with Dp- and Dq-branes along orthogonal subspaces, studied around (9.2.4). The natural guess would be the *harmonic function rule* [325–327]:

$$ds^2 \overset{?}{=} h_p^{-1/2} h_q^{-1/2} ds^2_{\|\,\|} + h_p^{1/2} h_q^{-1/2} ds^2_{\perp\|} + h_p^{-1/2} h_q^{1/2} ds^2_{\|\perp} + h_p^{1/2} h_q^{1/2} ds^2_{\perp\perp} . \tag{9.2.10}$$

Here $\|\,\|$ denotes the directions of spacetime parallel to both species of branes, and so on. Unfortunately, the equations of motion imply $\partial_m h_p \partial_n h_q = 0$, where m is one of the $\perp\|$ directions and n is one of the $\|\perp$ directions. One way to satisfy this is to smear one of the two sets of branes, taking them to be a uniform distribution of parallel planes (Section 4.2.6). If, for example, we smear the Dp-branes, then h_p becomes constant and the problem is overcome. A technically similar but perhaps physically more interesting situation is when the Dq are subspaces inside the Dp: then the $\perp\|$ directions don't exist. Recalling (9.2.5), supersymmetry implies that $p - q = 4$ or 8; so this strategy works, for example, for a distribution of D3-branes inside D7-branes. Among the other components of the Einstein equations of motion, we find

$$\nabla^2_{\perp\perp} h_p = 0 , \qquad \nabla^2_{\perp\perp} h_q + h_p \nabla^2_{\|\perp} h_q = 0 . \tag{9.2.11}$$

The first can be solved by a harmonic function as in (1.3.59) and (1.3.61); the second is more complicated, but it can be solved explicitly if we work in a limit near the Dp-branes, where (1.3.61) simplifies to $h_p = N(r_0/r)^{7-p}$. Some solutions were found in this way in [328–332], some of which will play a role later; but the general problem is still open.

9.2.2 Branes on Calabi–Yau manifolds

We now consider branes that fill a four-dimensional Minkowski spacetime in a Calabi–Yau vacuum:

$$\text{D}p = \text{Mink}_4 \times S_{p-3} \subset \text{Mink}_4 \times M_6 . \tag{9.2.12}$$

Calibration interpretation

We can use the decomposition (8.2.6) to rewrite (9.2.1). The matrix $\Gamma_\|$ decomposes as $\text{i}1 \otimes \gamma_\|$ in IIA, and as $\text{i}\gamma \otimes \gamma_\|$ in IIB. For the ϵ^a, we assume the $\mathcal{N} = 2$ spinorial decomposition (8.2.28) for a Calabi–Yau; it would be easy to adapt the results to the more general (8.2.12). With these decompositions, (9.2.1) becomes

$$\zeta^1_+ \otimes \eta_+ + \zeta^1_- \otimes \eta_- = \text{i}(\zeta^2_+ \otimes \gamma_\| \eta_\mp \pm \zeta^2_- \otimes \gamma_\| \eta_\pm) . \tag{9.2.13}$$

By chirality, the first terms on each side have to be equal. It follows that

$$\zeta_+^1 = \alpha \zeta_+^2 , \qquad \eta_+ = \mathrm{i}\alpha^{-1}\gamma_{\|}\eta_{\mp} . \tag{9.2.14}$$

(Had we started from (8.2.12), we would have a matrix α^{IJ} instead of this single α.) Since the two ζ_+^a of the Calabi–Yau background are now constrained to be proportional, we are breaking to $\mathcal{N} = 1$, as expected in presence of a D-brane.

To analyze the internal part of (9.2.14), we can recall the analysis in Section 5.6.2. As remarked after (5.6.8), $\gamma_{\|}$ is unitary, so $|\alpha| = 1$; we can expand as in (5.6.10) or (5.6.18), and we end up with the calibration conditions (5.6.14) and (5.6.20), with θ determined by α. Since M_6 is a Calabi–Yau, the $\Phi_\pm$ in those equations are closed, and they are calibrations. On a more general vacuum, we could only call them *almost* calibrations, and we could not use the argument (5.6.7) about volume minimization.

So we obtain that a space-filling D-brane (9.2.12) preserves some supersymmetry if and only if S_{p-3} is calibrated. Since (9.2.14) determines one ζ in terms of the other, only four supercharges are preserved.

- In IIA, the calibration condition becomes (5.6.21), which requires $p = 6$; the internal three-cycle S_3 is calibrated by

$$\mathrm{Re}(\mathrm{e}^{\mathrm{i}\theta}\Omega) . \tag{9.2.15}$$

We called such an S_3 an almost special Lagrangian cycle, but when (9.2.15) is closed, we drop the "almost" and call it simply *special Lagrangian (sLag)*, or sometimes also *A-branes*.

- In IIB, we need $p = 3 + 2k$; by (5.6.17), the internal S_{2k}-cycle is calibrated by

$$(-1)^{\lfloor k/2 \rfloor} \frac{1}{k!} J^k . \tag{9.2.16}$$

When J is closed these are called *holomorphic* cycles, or sometimes *B-branes*. From (5.6.14) and $\Phi_+ = \mathrm{e}^{-\mathrm{i}J} = 1 - \mathrm{i}J - \frac{1}{2}J^2 + \frac{\mathrm{i}}{6}J^3$ it follows that

$$\theta = \begin{cases} 0 & k = 0, 4 ; \\ \pi/2 & k = 2, 6 . \end{cases} \tag{9.2.17}$$

Recall that θ is related to the phase of α in (9.2.14); so it characterizes the supercharges preserved by the brane. So for two D-branes to preserve some supersymmetry together, they should have the same θ. This agrees with (9.2.5).

Transverse scalars

The scalars x^i on a D-brane parameterize transverse fluctuations (Section 1.3.2). When the brane wraps a subspace S_k of a curved manifold, there is an important subtlety [333]: the transverse fluctuations of S_k are in fact sections of the normal bundle we defined in (4.2.8).

For B-branes, wrapping an even cycle S_{2k}, such sections should be also *holomorphic*, so that the deformation is still holomorphic. So they should be sections of the holomorphic normal bundle $N_{1,0}$. We can then use the adjunction formula (6.3.4), (6.3.5) to view the transverse scalars as forms on S_{2k}.

For example, consider a four-cycle $S_4 \subset M_6$, wrapped by a D7. There is a single holomorphic $z = x^8 + \mathrm{i}x^9$. We can apply the first identity in (6.3.5), replacing $P \to M_6$ and $Z \to S_4$; since M_6 is a Calabi–Yau, $K_{M_6} = \mathcal{O}$, and

$$K_{S_4} \cong N_{1,0}\,. \tag{9.2.18}$$

z can be seen also as the local coefficient of (2, 0)-form $Z = z\mathrm{d}z^1 \wedge \mathrm{d}z^2$ on S_4. The procedure is similar for other holomorphic branes.

This phenomenon can be regarded as an example of a field-theory procedure called *twisting*. A QFT model is usually defined in flat space. The simplest option to define it on curved space, called *minimal coupling to gravity*, is to promote all partial derivatives to covariant derivatives. If the flat space model is supersymmetric, however, its minimally coupled version will no longer be supersymmetric. The twist is an alternative coupling to gravity that instead preserves supersymmetry [334]: fields are promoted to curved space as sections of appropriate bundles. A more general idea is to couple the model to *super*gravity, without making it dynamical [335]. This gives rise to a wider range of possibilities, which can be analyzed with the methods developed in this book (see, for example, [336–338] for $d = 4$ models). These more general couplings can to some extent also be realized on D-brane world-volumes [339].

For A-branes, wrapping sLag cycles $S_3 \subset M_6$, transverse scalars are sections of the (real) normal bundle NS_3. On a sLag the two-form J restricts to zero, (5.6.21); this implies that its contraction $\iota_n J$ by a normal vector n gives a section of T^*S_3. Working with an adapted vielbein (5.1.36a) and (5.1.36b), one obtains on S_3

$$\iota_n \mathrm{Im}(\mathrm{e}^{\mathrm{i}\theta}\Omega) = *_3 \iota_n J\,. \tag{9.2.19}$$

Imposing that the sLag condition is preserved by the deformation gives

$$0 = L_n J = \mathrm{d}(\iota_n J) = 0\,, \qquad 0 = L_n \mathrm{Im}(\mathrm{e}^{\mathrm{i}\theta}\Omega) = \mathrm{d}\iota_n \mathrm{Im}(\mathrm{e}^{\mathrm{i}\theta}\Omega) = \mathrm{d}(*_3 \iota_n J)\,. \tag{9.2.20}$$

So $\iota_n J$ is a harmonic form on S_3, and thus infinitesimal deformations are parameterized by

$$H^1(S_3, \mathbb{R})\,. \tag{9.2.21}$$

McLean's theorem says that any such infinitesimal deformations are unobstructed: they can always be promoted to finite ones [340].

Flat calibrated cycles

It is worth recasting some previous results in the language of calibrations.

Let us take (9.2.6) with two D6s at angles. The "internal" space is $M_6 = \mathbb{R}^6$; it has an SU(3)-structure defined by (3.4.21), where $z^1 = x^4 + \mathrm{i}x^7$, $z^2 = x^5 + \mathrm{i}x^8$, $z^3 = x^6 + \mathrm{i}x^9$. (Since $\mathbb{R}^6$ is flat, there are other SU(3)-structures we could pick, each related to a choice of internal η.) The two subspaces in $\mathbb{R}^6$ defined by (9.2.6) can be rewritten as

$$S_{3,1} \equiv \{\bar z^{\bar\imath} = z^i\}\,, \qquad S_{3,2} \equiv \{\bar z^{\bar\imath} = \mathrm{e}^{2\mathrm{i}\theta_i} z^i\}\,. \tag{9.2.22}$$

So the pull-back $J = \mathrm{i}/2(\mathrm{d}z^i \wedge \mathrm{d}\bar z^{\bar\imath})$ to both $S_{3,a}$ vanishes (and both are Lagrangian). The pull-back of $\mathrm{Im}(\mathrm{e}^{\mathrm{i}\theta}\mathrm{d}z^1 \wedge \mathrm{d}z^2 \wedge \mathrm{d}z^3)|_{S_{3,2}}$ is proportional to

$$(\mathrm{e}^{\mathrm{i}\theta}\mathrm{d}z^1 \wedge \mathrm{d}z^2 \wedge \mathrm{d}z^3 - \mathrm{e}^{-\mathrm{i}\theta}\mathrm{e}^{2\mathrm{i}\sum_i \theta_i}\mathrm{d}z^1 \wedge \mathrm{d}z^2 \wedge \mathrm{d}z^3)|_{S_{3,2}}\,; \tag{9.2.23}$$

so it vanishes if

$$\theta = \sum_i \theta_i\,, \tag{9.2.24}$$

modulo integer multiples of 2π. For $S_{3,1}$, the computation is the same but with $\theta_i \to 0$; so $\theta = 0$. But recall, as we saw after (9.2.17), that for the D-branes to preserve the same supercharges, they should have the same value of θ. This gives us back (9.2.8).

The role of world-volume flux

The world-volume $\mathcal{F} \neq 0$ modifies the supersymmetry requirement for space-filling branes in interesting ways. We will take it to be purely internal. Equation (9.2.2) then instructs us to modify

$$\gamma_{\|} \to \frac{\sqrt{-\det(g|_{\mathrm{D}p})}}{\sqrt{-\det(g|_{\mathrm{D}p} + \mathcal{F})}} (\mathrm{e}^{\mathcal{F}})_{/} \gamma_{\|} \,. \tag{9.2.25}$$

For example, consider B-branes with $\mathcal{F}$ along the internal subspace S_k. The computation leading to (5.6.14) can still be carried out: the left-hand version of (5.6.14) now becomes

$$\mathrm{Re}(\mathrm{e}^{\mathrm{i}\theta}\Phi_+)|_{S_k} = \sqrt{\det(g|_{\mathrm{D}p} + \mathcal{F})}\mathrm{d}\sigma^1 \wedge \ldots \wedge \mathrm{d}\sigma^k \,, \tag{9.2.26a}$$

with $\sigma^1, \ldots, \sigma^{k=p-3}$ being the internal world-volume coordinates. The right-hand version of (5.6.14) becomes [341]

$$\mathrm{Im}\left(\mathrm{e}^{\mathrm{i}\theta}\mathrm{e}^{-\mathcal{F}\wedge}\Phi_+\right)|_{S_k} = 0 \,, \tag{9.2.26b}$$

$$((\mathrm{d}x^m \wedge - g^{mn}\iota_n)\,\mathrm{e}^{-\mathcal{F}\wedge}\Phi_-)|_{S_k} = 0 \,. \tag{9.2.26c}$$

Recall that on a Calabi–Yau $\Phi_+ = \mathrm{e}^{-\mathrm{i}J}$, $\Phi_- = \Omega$. The computation for A-branes is similar, but in the end it imposes that the world-volume flux vanishes, $\mathcal{F} = 0$.

We will explore the effect of world-volume flux on B-branes at length in this subsection. While the S_k will still be holomorphic, the presence of $\mathcal{F}$ has important consequences.

Effect on flat branes

Again for $M_6 = \mathbb{R}^6$, we now consider a D9 and a D3, the latter filling the external spacetime $\mathbb{R}^4$ and localized, for example, at the origin of $\mathbb{R}^6$ [342]. The number of ND directions is six, so for $\mathcal{F} = 0$ according to (9.2.5) or (9.2.17), these two branes together should break all supersymmetry.

For the D3, (9.2.26) is satisfied with $\theta = 0$. For the D9, (9.2.26c) gives $B \wedge \Omega = 0$, so B should be $(1,1)$. Equation (9.2.26b) reads

$$\mathrm{Im}\left(\mathrm{e}^{\mathrm{i}\theta}(\mathcal{F} + \mathrm{i}J)^3\right) = 0 \,. \tag{9.2.27}$$

To be even more specific, let us take $f = 0$ and $B = -\cot\beta_1 \mathrm{d}x^4 \wedge \mathrm{d}x^7 - \cot\beta_2 \mathrm{d}x^5 \wedge \mathrm{d}x^8 - \cot\beta_3 \mathrm{d}x^6 \wedge \mathrm{d}x^9$; recall that $J = \mathrm{d}x^4 \wedge \mathrm{d}x^7 + \mathrm{d}x^5 \wedge \mathrm{d}x^8 + \mathrm{d}x^6 \wedge \mathrm{d}x^9$. Now (9.2.27) reduces to

$$\theta = \sum_i \beta_i \tag{9.2.28}$$

modulo integer multiples of 2π. For the D9 to preserve the same supersymmetry as the D3, we need $\theta = 0$. $B = 0$ corresponds to $\beta_i = \pi/2$, which does not satisfy

(9.2.28), in agreement with (9.2.5). But with $B \neq 0$ it is possible to make the D9 and D3 mutually supersymmetric. The case $\beta_i = 0$ would correspond to $B \to \infty$ and is not acceptable. But there are still many solutions to (9.2.28), such as

$$\beta_i = \frac{2\pi}{3}\,. \tag{9.2.29}$$

The similarity between (9.2.28) and (9.2.24) is not accidental: the two configurations are related by T-duality along the directions 789. Indeed, from IIB the D3 gains three directions and is mapped to the $D6_1$ in (9.2.6). The D9 loses three directions and is also mapped to a D6, which B makes oblique, as the world-volume flux f did in our D2–D0 examples in Figure 1.1. The map is such that

$$\beta_i \overset{T_{789}}{\longleftrightarrow} \theta_i\,. \tag{9.2.30}$$

Hermitian Yang–Mills equations

We now study (9.2.26) on B-branes more generally.

First we look at (9.2.26c), which for $\mathcal{F} = 0$ gave in (5.6.17) the condition that the cycle S_k is holomorphic. For $k = 2$ we find $\iota_n \Omega = 0$, which is again that condition. For $k = 4$, we find

$$dx^m \wedge \Omega - g^{mn} \iota_n(\mathcal{F} \wedge \Omega) = 0\,. \tag{9.2.31}$$

Taking $m = i$, a holomorphic index, we have $(\iota_{\bar{j}} \mathcal{F}) \wedge \Omega = 0$, implying $\mathcal{F}_{\bar{j}\bar{k}} = 0$. Taking $m = \bar{\imath}$ we are then left with $dz^{\bar{\imath}} \wedge \Omega = 0$, which is again holomorphicity of $S_4 \subset M_6$. Finally, for $k = 6$ we have $dx^m \wedge \mathcal{F} \wedge \Omega = 0$, which reduces to $\mathcal{F}_{\bar{j}\bar{k}} = 0$ again. Summarizing, (9.2.26c) implies that S_k is holomorphic and $\mathcal{F}_{0,2} = 0$. Since $\mathcal{F}$ is real, this implies $\mathcal{F}_{2,0} = 0$ as well.

Next we consider (9.2.26b). For $k = 2$ it is $\mathcal{F} = -\cot\theta J$. For $k = 4$, $(\mathcal{F}^2 - J^2) = -2\cot\theta \mathcal{F} \wedge J$. Finally, for $k = 6$, from (9.2.27):

$$\mathcal{F}^3 - 3\mathcal{F} \wedge J^2 = -\cot\theta(3\mathcal{F}^2 \wedge J - J^3)\,. \tag{9.2.32}$$

Except for $k = 2$, these are nonlinear partial differential equations for the world-volume potential a of the flux $f = da$; so it is not easy to find solutions, or even to prove their existence. To make progress, we can take the large-volume limit, making J large while leaving f fixed. For $B = 0$, (9.2.26b) becomes

$$f \wedge \frac{J^{k/2-1}}{(k/2-1)!} = -s\,\frac{J^{k/2}}{(k/2)!}\,, \qquad s \equiv -\tan\left(\theta - \frac{k}{4}\pi\right)\,. \tag{9.2.33}$$

In this limit, $\theta \sim k\pi/4$. If we decompose $f = f^0_{1,1} + f_0 J$, (9.2.33) constrains the nonprimitive coefficient as $f_0 = s/k$. Taking the $*$ and using Section 5.2.2,

$$J \cdot f = -\frac{s}{2} 1\,, \qquad f_{0,2} = 0\,, \tag{9.2.34}$$

with the usual pointwise inner product $\cdot$ in (4.1.106b). These are the Hermitian Yang–Mills (HYM) equations we have already seen in the heterotic case in (8.2.37), now with an extra parameter s.

Here we derived (9.2.34) for f abelian, but in the large volume limit they are believed to also hold for a stack of N D-branes, where the world-volume flux f

becomes $u(N)$ valued. The non-abelian version of the brane action is less understood (Section 1.3.6), but in the large volume limit we keep only the leading terms in $f = da$. At that level, we do know that the nonabelian action is SU(N) super-Yang–Mills; its supersymmetry transformations lead to (9.2.34) again, much as in (8.2.36). In this context, the parameter s comes about from an invariance of the form $\delta\chi = \epsilon$ [343]; in the abelian case, this becomes the spinorial translation (1.3.54).

Non-abelian scalars

On a stack of N D-branes, the transverse scalars become nonabelian too, and not just the gauge field. Their presence should modify (9.2.34).

In flat space, on a D9 (9.2.34) reads

$$f_{1\bar{1}} + f_{2\bar{2}} + f_{3\bar{3}} = -\frac{i}{4}s1 , \qquad f_{\bar{1}\bar{2}} = f_{\bar{1}\bar{3}} = f_{\bar{2}\bar{3}} = 0 . \tag{9.2.35}$$

Recall that maximally supersymmetric super-YM is obtained by dimensional reduction from $d = 10$ (Section 1.3.6). Consider, for example, a D7. There are two transverse scalars, which can be assembled into a single complex $z \equiv x^6 + ix^9$. Reducing (9.2.35) along z gives the *Hitchin equations* [344]:

$$f_{1\bar{1}} + f_{2\bar{2}} + [z, z^\dagger] = -\frac{i}{4}s1 , \qquad f_{\bar{1}\bar{2}} = D_{\bar{1}}\bar{z} = D_{\bar{2}}\bar{z} = 0 . \tag{9.2.36}$$

To make (9.2.36) a globally valid expression, recall that transverse scalars are sections of (9.2.18). So z can be seen also as the local coefficient of (2, 0)-form $Z = z dz^1 \wedge dz^2$ on S_4, and $[z, \bar{z}]$ is a (2, 2)-form. Then (9.2.36) can be written globally as

$$2iJ \wedge f + [Z, Z^\dagger] = -\frac{i}{4}s1\mathrm{vol}_4 , \qquad f_{0,2} = 0 , \qquad \bar{\partial}_E Z = 0 , \tag{9.2.37}$$

where $\bar{\partial}_E$ is the twisted Dolbeault differential (5.3.42). The procedure is similar for other holomorphic branes: for example, for D5-branes we have two complex transverse scalars Z_i on a two-cycle, and (9.2.37) has $\sum_i [Z_i, Z_i^\dagger]$.

Equation (9.2.34) is related to other famous problems in mathematical physics. For $k = 4$ and $s = 0$, it says that f is orthogonal to an SU(2)-structure (j, ω) on the world-volume; recalling (5.2.30), we conclude $f = *f$, the self-duality equation (1.3.17), which also describes instantons in YM theories. This also generates other interesting equations via dimensional reduction (Exercise 9.2.2)

Four-dimensional effective action

Introducing space-filling D-branes to a Calabi–Yau compactification adds a new sector to the effective theory. As we have seen, we can preserve half of the supersymmetry of the background, if S_{p-3} wraps either a sLag or a holomorphic cycle. So the new sector only has $\mathcal{N} = 1$ supersymmetry. The additional terms in the Lagrangian can be obtained by reducing the Dp action (1.3.24) on the internal cycle S_{p-3}. The world-volume gauge field a_a gives rise to a spacetime gauge field a_μ in Mink$_4$, but also to scalar fields when the index a along S_{p-3}. More scalars are generated by the usual brane scalars x^i.

All these fields have a KK tower, but we are especially interested in the massless fields. There is a single massless vector; the number of massless scalars depends on S_{p-3} and on the gauge bundle E on it. Just like for the closed string sector, the massless scalars are in correspondence with the moduli of the D-brane. In general, the F-terms are a complex equation, depending holomorphically on the z_I, while the D-terms are real (Section 8.4.3). Indeed, all the conditions for brane supersymmetry consist of a holomorphic and a real equation, starting from our geometrical (5.6.14) and (5.6.20), and more recently in (9.2.26), in the HYM equations (9.2.34), and in the Hitchin equations (9.2.37). So in all these cases we can identify the holomorphic equation with the F-terms of the effective $d = 4$ potential, and the real equation with the D-terms.

To determine the number of massless scalars, we can count the dimension of the space of infinitesimal deformations.

- Space-filling branes in IIA wrap A-branes, special Lagrangian cycles S_3. Their infinitesimal deformations are given by (9.2.21). As mentioned following (9.2.26), the world-volume flux vanishes; the gauge bundle on S_3 is flat. Deformations of a flat bundle are given by δa such that $d\delta a = 0$; this again belongs to $H^1(S_3)$. The two assemble in

$$h^1(S_3) \tag{9.2.38}$$

chiral multiplets in four dimensions.
- The relevant data for B-branes are a pair (S_k, E), $k = p - 3$, of a cycle and a gauge bundle over it. The F-term conditions demand that the cycle S_{p-3} is holomorphic. Its deformations are given by the transverse scalars, which are global holomorphic sections of the normal bundle; recalling (5.3.47), they are parameterized by

$$H^{0,0}(S_k, N_{1,0})\,. \tag{9.2.39}$$

For the gauge field, the F-terms in this section were always of the form $f_{0,2} = 0$. Deforming this gives $\bar\partial_E \delta a_{0,1}$, where again ∂_E is the twisted Dolbeault differential (5.3.42). $a_{0,1}$ is in the adjoint of the structure group of E, so the relevant cohomology is

$$H^{0,1}(S_k, \mathrm{End}(E))\,, \qquad \mathrm{End}(E) \equiv E \otimes E^*\,. \tag{9.2.40}$$

The D-term equations are significantly harder; we consider them in the next subsection with the help of the stability techniques from Section 6.3.7. We will show that they are still satisfied after infinitesimal deformations, so they do not alter the conclusions (9.2.39) and (9.2.40).

9.2.3 D-brane stability

Existence theorems and stability

The equation $f_{0,2} = 0$ in (9.2.34) is equivalent to the existence of a twisted Dolbeault differential for the gauge bundle E on S_k, by (5.3.42). As we commented there, this is equivalent to the transition functions being holomorphic. Unsurprisingly, bundles produced by algebraic-geometric techniques do have this property.

$J \cdot f = -\frac{s}{2}1$ is a real equation. To fix ideas take $k = 6$, relevant for a D9-brane. If we trace and integrate it in its form (9.2.33) we find

$$s = \frac{\deg(E)}{N\mathrm{Vol}(M_6)}\,, \tag{9.2.41}$$

called the *slope* of E, where the *degree* $\deg(E)$ is[5]

$$\deg(E) \equiv \int J^2 \wedge \mathrm{Tr} f \overset{(4.1.138)}{=} 2\pi\, D_{ijl} v^i v^j (c_1(E))^l \,. \tag{9.2.42}$$

(For $k \neq 6$, the degree is $\int J^{k/2-1} \wedge \mathrm{Tr} f$.)

It is not easy to find explicit solutions for $J \cdot f = -\frac{s}{2} 1$, other than in a few simple cases such as $M_6 = \mathbb{R}^6$. Fortunately, there is an existence theorem, thanks to the stability techniques of Section 6.3.7. This is a bit similar to the K-stability techniques of Section 7.4.7 (and was an inspiration for it). We want to apply the Kempf–Ness theorem (6.3.44) to a quotient

$$\frac{\mathcal{A} - F}{\mathcal{G}_{\mathbb{C}}} \,, \tag{9.2.43}$$

where $\mathcal{A}$ is the infinite-dimensional space of connections on M_6, and $\mathcal{G}$ is the group of gauge transformations, whose Lie algebra is generated by functions λ on M_6. A vector field on $\mathcal{A}$ is an infinitesimal variation of the connection δa. The expression

$$\iota_{\delta a_1} \iota_{\delta a_2} \mathbb{J} \equiv \int_{M_6} J^2 \wedge \mathrm{Tr}(\delta a_1 \wedge \delta a_2) \tag{9.2.44}$$

defines a symplectic form $\mathbb{J}$ on $\mathcal{A}$. In this context, it is better to write the usual definition (5.4.11) of the moment map as $d\mu_\xi = \iota_{X_\xi} \omega$, distinguishing the element of the Lie algebra $\xi \in \mathfrak{g}$ from the vector field X_ξ that represents it. The exterior differential we need in this relation is the one on $\mathcal{A}$, which we call δ in order not to confuse it with the one on M_6. The moment map is

$$\mu = J^2 \wedge f \,. \tag{9.2.45}$$

Indeed, the variation $\delta\mu$ with respect to the connection is

$$\iota_{\delta a} \delta\mu = J^2 \wedge \delta f \overset{(4.2.21)}{=} J^2 \wedge D\delta a \,. \tag{9.2.46}$$

Now the gauge algebra of $\mathcal{G}$ is given by functions λ, so

$$\begin{aligned} \iota_{\delta a} \delta\mu_\lambda &= \int_{M_6} \mathrm{Tr}(\iota_{\delta a} \delta\mu\lambda) = \int_{M_6} J^2 \wedge \mathrm{Tr}(D\delta a \lambda) \\ &= \int_{M_6} J^2 \wedge \mathrm{Tr}(\delta a \wedge D\lambda) \overset{(9.2.44)}{=} \iota_{\delta a} \iota_{D\lambda} \mathbb{J} \,. \end{aligned} \tag{9.2.47}$$

Interpreting $D\lambda$ as the vector field X_ξ representing the Lie algebra of gauge transformations on $\mathcal{A}$, we see that (9.2.45) is indeed a moment map. For $s = 0$, this gives (9.2.33) an interpretation in terms of symplectic quotients. It is possible to modify the argument to include $s \neq 0$.

Now the Kempf–Ness theorem tells us that a connection can be made to satisfy the HYM equation (9.2.34) if and only if it satisfies an appropriate stability condition. This requires computing a certain weight on a limit of the action of the complexified gauge group. In our case, this involves a degeneration of the bundle to a subbundle. We define the bundle E to be *slope-stable* if for every subbundle $E' \hookrightarrow E$ (Section 4.2.1):

[5] In (4.2.57), there is an extra i because the F there is taken to be *anti*-Hermitian.

$$s(E') < s(E)\,. \tag{9.2.48}$$

The application of the Kempf–Ness theorem now becomes the *Donaldson–Uhlenbeck–Yau (DUY)* theorem: (9.2.34) admits a solution if and only if E is slope-stable [244, 245]. Since (9.2.48) is a set of inequalities, it should remain valid after infinitesimal deformations, as we promised at the end of Section 9.2.2. (One sometimes says that stability is an *open condition*: it defines an open set.)

Let us show directly that (9.2.48) is necessary. Consider the bundle $V \equiv (E')^* \times E$; its sections are maps $E' \hookrightarrow E$, so if F has a section σ, then E' is a subbundle of E. The curvature f^V of a connection on V is the difference of curvatures on E and E'; then

$$||\partial_V \sigma||^2 = \int \mathrm{d}^6 y g^{i\bar{j}} (\bar{\partial}^V_{\bar{j}} \bar{\sigma})(\partial^V_i \sigma) = -\int \mathrm{d}^6 y g^{i\bar{j}} \bar{\sigma} \nabla^V_{\bar{j}} \partial^V_i \sigma \overset{(4.2.18)}{=} -\mathrm{i} \int \mathrm{d}^6 y g^{i\bar{j}} \bar{\sigma} f^V_{i\bar{j}} \sigma$$
$$\overset{(5.3.3)}{=} -\frac{1}{2} \int \mathrm{d}^6 y J \cdot f^V |\sigma|^2 \overset{(9.2.34)}{=} \frac{1}{4}(s(E) - s(E'))||\sigma||^2\,. \tag{9.2.49}$$

Since the norms are positive, (9.2.48) follows.

Beyond the large-volume limit

Since the slope (9.2.41) depends on J, so does the stability condition (9.2.48): it is satisfied in an open set in $\mathcal{M}_{\mathrm{K}}$, bounded by *walls*. But we derived (9.2.34) in the limit $f \ll J$; beyond this regime, we don't know the nonabelian version of the equations (9.2.33), and also the supergravity approximation is expected to break down.

For the closed string sector, mirror symmetry and the world-sheet approach allow sometimes to go beyond the large volume limit. In this spirit, we will now try to guess the right stability condition for B-branes that extends (9.2.48) to all of $\mathcal{M}_{\mathrm{K}}$, even if we don't know the supersymmetry equations that deform (9.2.34) [345]. We need to decide what replaces the notion of slope, and whether the notion of subbundle is still appropriate. We will analyze these two issues in turn. This will get us to the idea of Π-*stability* [345, 346]; for reviews, see [347–349].

Central charge

Going back to (9.2.33), we see that the slope s originates from the angle θ in the supersymmetry conditions (9.2.26). This suggests that the proper generalization of (9.2.48) should involve an inequality on θ directly.

This can be motivated by using branes that are particles in Mink_4 rather than space-filling. In that case, the parameter θ characterizes the $\mathcal{N} = 1$ preserved subalgebra of the $\mathcal{N} = 2$ supersymmetry of the background, via the central charge:

$$Z = |Z| \mathrm{e}^{-\mathrm{i}\theta}. \tag{9.2.50}$$

This is defined by the way it appears in the $\mathcal{N} = 2$ algebra (1.3.11). As we saw for $\mathcal{N} = 2$ super-YM following (1.3.10), the BPS bound is $M \geq |Z|$. The value of Z for a state can be computed explicitly by looking at the currents that integrate to the fermionic generators Q^I [350]; this gives $Z = a(q_{\mathrm{m}} + \mathrm{i}q_{\mathrm{e}})$, which for monopoles reproduces the BPS bound (1.3.9). The generalization of Dirac quantization (1.3.7) gives $Z = n + m\tau$. For more general $\mathcal{N} = 2$ theories, we get

$$Z = n_I X^I - m^I \mathcal{F}_I \,. \tag{9.2.51}$$

For the particles obtained from wrapped D-branes, recall that the symplectic section $\binom{X^I}{\mathcal{F}_I}$ is obtained by expanding $\Phi_+ = \mathrm{e}^{-\mathrm{i}J}$ or $\Phi_- = \Omega$ over a basis of forms (Section 9.1.2). The vector $\binom{n_I}{m^I}$ comes from an even or odd form $\alpha_\pm \in H(M_6, \mathbb{Z})$. So the lift of (9.2.51) is $Z = \int_{M_6} \alpha_\pm \wedge \Phi_\pm$. Since $\alpha_\pm$ is integer valued, we can take it to be the Poincaré dual of some subspace S_k; then (4.1.135) gives $Z = \int_{S_k} \Phi_\pm$, with k even or odd. For sLags, this gives

$$Z = \int_{S_3} \Omega \,. \tag{9.2.52}$$

Recalling from (9.2.15) that a sLag is calibrated by $\mathrm{Re}(\mathrm{e}^{\mathrm{i}\theta}\Omega)$, this gives $Z = \mathrm{e}^{-\mathrm{i}\theta}\mathrm{Vol}(S_3)$, confirming (9.2.50). For even cycles, the same logic takes us to $Z = \int_{S_k} (\mathrm{e}^{\mathcal{F}-\mathrm{i}J})_k$. This confirms (9.2.50), with the θ in (9.2.26).

The coupling of the particles to the gauge fields in $d = 4$ originates from the coupling of the D-brane to the RR fields. Some information about these is obtained beyond the large-volume limit by requiring the absence of anomalies on brane intersections [67, 351, 352]:

$$\int_{\mathrm{D}p} C\mathrm{e}^B \wedge \mathrm{Tr}\left(\mathrm{e}^{2\pi l_s^2 f}\right) \wedge \frac{\hat{A}(T\mathrm{D}p)\mathrm{e}^{\frac{\delta}{2}}}{\sqrt{\hat{A}(TM_{10}|_{\mathrm{D}p})}} \,. \tag{9.2.53}$$

Here $C = \sum_p C_p$ is the formal sum of potential of all degrees. $\hat{A}(TM)$ was defined after (4.3.22); in practice, its density is a polynomial in the Riemann curvature of TM. The exponential in the flux also has a characteristic class interpretation, in terms of the Chern character $\mathrm{ch}(E)$ in (4.2.68). Finally, δ is only nonzero when the world-volume is not a spin manifold (Section 4.3.2). In the abelian case, (9.2.53) reconstructs $\mathrm{e}^{\mathcal{F}}$, which gives us back the WZ term in (1.3.24) when expanded in degrees. In particular, for a D9 the two $\hat{A}$ simplify, and the relation that was discussed after (4.3.22) gives us $\hat{A} = \mathrm{Td}$. If we have two D9s with different gauge bundles, the symplectic pairing of their electric and magnetic charges gives

$$n_{1I} m_2^I - n_{2I} m_1^I = \int_{M_6} \mathrm{ch}(E_1)\mathrm{ch}(E_2^*)\mathrm{Td}(TM_6) \overset{(5.3.51)}{=} \chi(E_1 \otimes E_2^*) \,. \tag{9.2.54}$$

Two contributions $\sqrt{\hat{A}} = \sqrt{\mathrm{Td}}$ have combined. The left-hand side had to be an integer by a generalization of the Dirac quantization (1.3.7). This explains why the square root appears in (9.2.53).

Formally, in (9.2.51) and the discussion that follows it, we were obtaining Z by replacing the gauge field with the symplectic section $\binom{X^I}{\mathcal{F}_I}$ or its lift $\Phi_\pm$. So

$$Z = \int_{M_6} \mathrm{e}^{B-\mathrm{i}J} \wedge \mathrm{Tr}\left(\mathrm{e}^{2\pi l_s^2 f}\right) \wedge \sqrt{\mathrm{Td}(TM_6)} \,. \tag{9.2.55}$$

This is valid for a D9, where the gauge bundle has support all over M_6. But we will see soon that there is a way to write any brane as a formal difference of D9s and anti-D9s; the Z of Dp-branes with $p < 9$ can be computed as a sum of terms of the type (9.2.55). This allows us to apply (9.2.54) more broadly.

Coherent sheaves and derived category*

So far we considered B-branes as pairs $\mathcal{B} = (S_k, E)$ of a cycle and a gauge bundle. Before we take into account the D-term equations and stability issues, this notion requires only notions of complex geometry, which are insensitive to Kähler moduli.

Still, recall that for some regions at small volume in $\mathcal{M}_K$ are better described by LG models, which don't have a geometrical nature. D-branes for these models can still be defined [353], but in terms of a completely algebraic concept called *quiver representation*, that no longer refers directly to the Calabi–Yau geometry. (This is because of the discrete quotient mentioned after (9.1.71).) When we extrapolate one of these Gepner D-branes back to the large-volume region of $\mathcal{M}_K$, its charges don't match those of a pair $\mathcal{B} = (S_k, E)$, but rather of a formal sum with signs

$$\sum_i (-1)^a \mathcal{B}_a \,. \tag{9.2.56}$$

The presence of an alternating sign is reminiscent of the brane–antibrane system discussed following (1.3.58). There is a generalization of cohomology called *K-theory*, defined by taking formal differences of vector bundles. Here, however, we need to be able to take differences of bundles over different subspaces, so we need a different formalism.

The first step is to describe each $\mathcal{B} = (S_k, E)$ as an object on M_6 itself. This is also useful to describe systems of branes $\mathcal{B}_a = (S_{k\,a}, E_a)$ on different cycles. If these all wrap the same cycle $S_{k\,a} = S_k$, it is natural to generalize (9.2.40) to $\oplus_{a,b} H^{0,1}(S_k, E_a \otimes E_b^*)$; in a sense, this is already covered by (9.2.40), because we could consider a single $\mathcal{B}$ with gauge bundle $\oplus_a E_a$. When the $S_{k\,a}$ are all different, we need a new type of cohomology to describe the fields arising from strings connecting the $\mathcal{B}_a$.

We can achieve this by extending the concept of vector bundle, so that the fiber can be nonzero on a subspace and zero everywhere else. Actually, it is better to avoid the concept of fiber altogether. A *presheaf* is a map that associates a ring to every open set:

$$U \mapsto \mathcal{S}(U) \,, \tag{9.2.57}$$

with a restriction homomorphism $\rho_{UV} : \mathcal{S}(U) \to \mathcal{S}(V)$ for any open subset $V \subset U$, which is associative: $\rho_{UW}\rho_{WV} = \rho_{UV}$ if $W \subset V \subset U$. It is called a *sheaf* if

$$\frac{\mathcal{S}(U_1) \times \mathcal{S}(U_2)}{\mathcal{S}(U_1 \cup U_2)} \cong \mathcal{S}(U_1 \cap U_2) \,. \tag{9.2.58}$$

The isomorphism $\cong$ is the map $(s_1, s_2) \mapsto \rho_{U_1, U_1 \cap U_2}(s_1) - \rho_{U_2, U_1 \cap U_2}(s_2)$. This boils down to the intuitive properties that if two sections s_i over two U_i agree over the intersection, there is a section over the union, and that if a section is zero over the union, it is zero over each U_i.

To every vector bundle E, we can naturally associate the sheaf $\mathcal{E}$, whose $\mathcal{E}(U)$ are the sections of E over U. In particular, to the trivial line bundle we associate the sheaf $\mathcal{O}$, whose $\mathcal{O}(U)$ is the ring of holomorphic functions over U. A sheaf C is called *coherent* if every point has a neighborhood U such that

$$C(U) \cong \frac{\mathcal{O}(U)^n}{\mathcal{O}(U)^m} \,. \tag{9.2.59}$$

The sheaf associated to a vector bundle is coherent, because locally its sections are simply n-uples of functions. A more interesting example is the sheaf $\mathcal{O}_S$, where S is a holomorphic submanifold, defined by $\mathcal{O}_S(U) \equiv \mathcal{O}(U \cap S)$. This is empty over open sets that don't intersect S. To see that it is coherent, write locally $S = \{z^1 = \cdots = z^m = 0\}$. Now $\mathcal{O}(U \cap S)$ is the space of holomorphic functions over U, quotiented by any function of the form $\sum_{i=1}^m z_i g_i$, since such functions vanish on S. So $\mathcal{O}(U \cap S) \cong \mathcal{O}(U)/(\mathcal{O}(U)^m)$, and according to (9.2.59) $\mathcal{O}_S$ is coherent. More generally, to a bundle E over a holomorphic submanifold S, we can associate a coherent sheaf $\mathcal{E}_S$ defined by $\mathcal{E}_S(U) \cong \mathcal{E}(U \cap S)$.

This formalism allows to describe bundles over subspaces, as we wanted. With some more effort, one can define groups $\mathrm{Ext}^i(\mathcal{E}_1, \mathcal{E}_2)$, which reduce to bundle cohomology $H^{0,i}(E_1 \otimes E_2^*)$ when $\mathcal{E}_i$ are associated to vector bundles E_i; in particular $i = 1$ generalizes (9.2.40).

Finally, we can give a meaning to (9.2.56). A map of sheaves $\mathcal{B}_1 \to \mathcal{B}_2$ is given by a morphism of rings $\mathcal{B}_1(U) \to \mathcal{B}_2(U)$ for every U. A *complex* is a sequence of maps

$$\mathcal{B}_\bullet \equiv \quad \cdots \to \mathcal{B}_{k-1} \overset{m_{k-1}}{\to} \mathcal{B}_k \overset{m_k}{\to} \mathcal{B}_{k+1} \to \cdots \tag{9.2.60}$$

such that $m_k m_{k-1} = 0$ for every k. The *cohomology* of the complex is defined on each U as the complex $H^\bullet\mathcal{B}(U) \equiv \mathrm{kerd}_k/\mathrm{Imd}_{k-1}$; this is similar to the definition of de Rham cohomology.[6] Given a second complex $\mathcal{B}'_\bullet$, a map $\phi : \mathcal{B}'_\bullet \to \mathcal{B}_\bullet$ is a sequence of maps

$$\phi_k : \mathcal{B}'_k \to \mathcal{B}_k\,, \qquad \text{such that} \quad \phi_k m_{k-1} = m'_k \phi_{k-1}\,. \tag{9.2.61}$$

Such a map induces a map $H\phi\colon H^\bullet\mathcal{B}' \to H^\bullet\mathcal{B}$. The *derived category* $\mathcal{D}(M_6)$ of M_6 is the set of all complexes of coherent sheaves, where we consider two complexes equivalent if there is a map ϕ between them such that $H\phi$ is an isomorphism.

Physically, a complex is as a system of branes and antibranes. For example,

$$(\mathcal{B}_0 = \mathcal{L}^{-1}) \overset{T}{\to} (\mathcal{B}_1 = \mathcal{O}) \tag{9.2.62a}$$

is interpreted as a brane–antibrane pair, with T being the tachyon we expect in such a pair (Section 1.3.6). The map ϕ from (9.2.62a) to the complex with the single entry

$$\mathcal{B}_1 = \mathcal{O}_S\,, \tag{9.2.62b}$$

defined by $\phi_1 : \mathcal{O} \to \mathcal{O}_S$ being the restriction of a function to S, is a quasi-isomorphism. Equation (9.2.62b) represents a D7, and its equivalence to (9.2.62a) is interpreted as the tachyon condensation of a brane–antibrane pair [354].

This suggests that B-branes on M_6 are described mathematically by a derived category of coherent sheaves $\mathcal{D}(M_6)$, at least at the level of F-term equations. To take into account D-terms, we need to define a notion of subcomplex, extending that of subbundle in (9.2.48).

We might want to define the map $\phi\colon \mathcal{B}_\bullet \to \mathcal{B}'_\bullet$, defined as in (9.2.61), to be injective if ϕ_k is injective. But this condition is not preserved by the identification defining $\mathcal{D}(M_6)$: if we replace $\mathcal{B}''_\bullet$ with an equivalent complex, the new map $\mathcal{B}_\bullet \to \mathcal{B}'_\bullet$ is not guaranteed to be injective. The correct notion is called *triangle*: a sequence of maps

$$\mathcal{B}'_\bullet \to \mathcal{B}_\bullet \to \mathcal{B}''_\bullet \to \mathcal{B}'_{\bullet+1}\,, \tag{9.2.63}$$

[6] The definition (9.2.60) can also be used to define complexes of rings, of groups, etc. So for the complex $\Lambda^\bullet$ of forms, whose kth entry is the space of k-forms Λ^k and whose maps are d, the cohomology is the de Rham cohomology (4.1.132).

where the last complex is equal to the first, but with the position of each sheaf shifted by one. It is said to be *distinguished* if $\mathcal{B}'_\bullet$ is obtained with a certain *cone construction* from $\mathcal{B}_\bullet$ and $\mathcal{B}''_\bullet$, which physically means that it can be interpreted as a bound state of them. For example, for the simple case where $\mathcal{B}'$ and $\mathcal{B}$ represent two vector bundles E', E on the same holomorphic submanifold, E' is a subbundle of E if and only if $\mathcal{B}'$ and $\mathcal{B}$ belong to a triangle (9.2.63).

Π-stability

Our digression on derived categories was relatively outside the main line of development of this book, so let us summarize it. The upshot is that, to go away from the large-volume limit, we often need to consider not just a single brane $\mathcal{B} = (S_k, E)$, but more complicated configurations of branes and antibranes. The notion of a derived category, while developed by algebraic geometers for different reasons, formalizes this idea.

We can now go back to our original aim of defining a stability condition that extends slope stability (9.2.48). We are led to conjecturing that the D-term of a B-brane is equivalent to this property: a B-brane $\mathcal{B}$ is said to be Π-*stable* if and only if [345]

$$\theta(\mathcal{B}') < \theta(\mathcal{B}) \tag{9.2.64}$$

for any $\mathcal{B}'$ that belongs with $\mathcal{B}$ to a distinguished triangle (9.2.63), the analogue of subbundle for the derived category. The angle θ is defined here by $Z = |Z|e^{-i\theta}$, (9.2.50); the expression for the central charge Z is given in (9.2.55) for D9-branes, which wrap the whole M_6. For a Dp-brane $\mathcal{B} = (S_{p-3}, E)$ with $p < 9$, one can derive a more complicated expression from (9.2.53), but it is usually more practical to use the derived category construction to rewrite it as a system $\sum_i (-1)^i \mathcal{B}_i$ of D9 and anti-D9, and then write $Z(\mathcal{B}) \equiv \sum_i (-1)^i Z(\mathcal{B}_i)$.

Equation (9.2.64) might look ill defined, because θ was originally defined as a phase. But in fact Z has zeros in some loci of the complexified Kähler moduli space $\mathcal{M}_K^{\mathbb{C}}$. From (9.2.55), we see that Z depends only on $B - iJ$, so it is holomorphic in the moduli $t^i = v^i - ib^i$, and it undergoes monodromy around these zeros, under which $\theta \to \theta + 2\pi$. To avoid the presence of cuts, we can just declare that under this monodromy the complex $\mathcal{B}_\bullet \to \mathcal{B}_{2+\bullet}$, shifting the position of each brane in (9.2.60) by two spots to the right. (Shifting the position by one spot would turn branes into antibranes.) With this specification, (9.2.64) is no longer ambiguous.

Some checks of the Π-stability conjecture come from B-branes near a Gepner point, which have an alternative description in terms of a quiver. The D-term equations are algebraic equations; they are again equivalent to a stability condition called θ-stability [355]. It can be checked that Π-stability reduces to this θ-stability near a Gepner point [345]. Other checks come from mirror symmetry, as we will see in the next subsection.

Assuming Π-stability to be correct, an interesting physical picture emerges. Quite by coincidence, the word "stability" turns out to describe physical stability under possible decay channels. Consider again a distinguished triangle (9.2.63):

- Suppose (9.2.64) is satisfied. A world-sheet analysis [346] shows that the map $\mathcal{B}''_\bullet \to \mathcal{B}'_{1+\bullet}$ represents a tachyon. Its condensation can be read as a process where they form a bound state $\mathcal{B}_\bullet$. This is a generalization of the brane–antibrane tachyon condensation in flat space.
- Suppose instead (9.2.64) is not satisfied. Then the map $\mathcal{B}''_\bullet \to \mathcal{B}'_{1+\bullet}$ represents a field with a positive mass; there is no tachyon condensation, and the two objects do not form a bound state. Conversely, $\mathcal{B}$ decays to the two separate branes $\mathcal{B}''_\bullet$ and $\mathcal{B}'_{1+\bullet}$.

For example, even a B-brane that is a point in M_6 can become Π-unstable when the Calabi–Yau is small [356].

While our presentation in this section focused on space-filling branes, a similar picture holds for branes that are particles in Mink_4. Π-stability is then reinterpreted as a condition for such particles to create a bound state in spacetime, where their gravitational attractions and gauge field repulsions balance exactly. This has been demonstrated in exquisite detail [357].

9.2.4 D-branes and mirror symmetry

Given the impressive results of mirror symmetry on the closed string moduli in Section 9.1.4, it is natural to explore its consequences on the physics of D-branes too.

A-branes and B-branes

We have seen that A-branes wrap special Lagrangian cycles and are calibrated by $\text{Re}(e^{i\theta}\Omega)$; B-branes wrap holomorphic cycles and are calibrated by $\text{Re}(e^{i\theta}e^{-iJ})$. The two pure forms $\Phi_- = \Omega$ and $\Phi_+ = e^{-iJ}$ get exchanged under mirror symmetry (Section 9.1.4). This leads us to conclude

$$\text{A-branes} \quad \leftrightarrow \quad \text{B-branes}. \tag{9.2.65}$$

So the stability ideas for B-branes in Section 9.2.3 should have an A-brane counterpart.

To identify the D- and F-term conditions for A-branes, we go back to our geometrical study in (5.6.14) and (5.6.20). In both cases, the F-term condition is the second condition on the right; recalling also (5.6.21), for A-branes this says that the pull-back of J vanishes, or in other words that the brane is a Lagrangian submanifold (not yet special). The D-term is the condition that

$$\text{Im}(e^{i\theta}\Omega)|_{M_3} = 0\,, \tag{9.2.66}$$

which should be related to a stability condition on the space of Lagrangian submanifolds, again with θ playing the role of the slope in (9.2.48).

As an example, we again take the intersecting D6-branes of (9.2.6), now viewed as A-branes in the Calabi–Yau $M_6 = \mathbb{R}^6$; they are mutually BPS if (9.2.8) holds. It is also possible to find a sLag that asymptotes to the two $\mathbb{R}^3$ planes, depending on a parameter [358]. Now if we consider the case where $\sum_i \theta_i \neq 0$, the *angle theorem* [358] says that the union of the two original $\mathbb{R}^3$ is volume minimizing if and only if

$$\sum_i \theta_i < 0\,. \tag{9.2.67}$$

This exactly parallels the picture we found in Section 9.2.3 for Π-stability. When (9.2.67) is satisfied, the two D6 cannot form a bound state. If, on the other hand, $\sum_i \theta_i > 0$, the two original D6s are not volume minimizing, so they should decay to another brane that asymptotes to them, which we can think of as a bound state. A more general result is found in [359], varying the complex structure of a Calabi–Yau rather than the θ_i. There is also a mirror of the derived category for B-branes, called Fukaya category [360, 361].

In the example with two flat D6s, the string spectrum can be analyzed exactly [362]. This confirms the BPS condition (9.2.8), but more interestingly confirms the absence of a tachyon when (9.2.67) holds, and its presence when it does not. (For $\theta_1 = \pi$, $\theta_2 = \theta_3 = 0$, which is a D6–anti-D6 pair, the tachyon was discussed after (1.3.58).)

It is also useful to view the situation from the point of view of the four-dimensional effective action [363]. There are (9.2.21) chiral multiplets for each A-brane, but there is also another chiral for each intersection point. For B-branes, there are (9.2.39) plus (9.2.40) chiral multiplets, but we mentioned in Section 9.2.3 that in presence of multiple branes there are also extra chiral multiplets, counted by the Ext^1 groups of the associated coherent sheaves.

Let us consider a single chiral multiplet σ due to such an intersection. Since we have solved the F-terms (by taking Lagrangian cycles for A-branes, and holomorphic cycles for B-branes), the first term in the $\mathcal{N} = 1$ potential (8.4.51) vanishes, and schematically we get

$$V \sim (|\sigma|^2 - \sigma_0)^2 . \tag{9.2.68}$$

The sign of this c is correlated with the sign of $\Delta\theta$ for the two intersecting branes, for both A- and B-branes:

- When $\sigma_0 > 0$, we see that σ is a tachyon near $\langle\sigma\rangle = 0$; making it condense simply means letting it relax to a vev $\langle\sigma\rangle = \sqrt{\sigma_0}$. This indicates the formation of a bound state.
- When $\sigma_0 < 0$, σ is not a tachyon, and its true vacuum is at $\langle\sigma\rangle = 0$. Since $V \neq 0$ at this vacuum, supersymmetry is spontaneously broken.

Mirror symmetry as T-duality

Given that A- and B-branes have to be exchanged by mirror symmetry, it is natural to look for the mirror of a B-brane that exists on any Calabi–Yau: one that wraps a point. Any point is a holomorphic submanifold, so the moduli space is M_6 itself, with complex dimension three. So there are three chiral multiplets in the $d = 4$ effective action. Mirror symmetry should leave the effective action unchanged; so the A-brane mirror $\tilde{S}_3$ should also have three chiral multiplets, which by (9.2.38) implies $h^1(\tilde{S}_3) = 3$. All this suggests that $\tilde{M}_6$ is a T^3-fibration over M_3. Reversing the role of M_6 and of its mirror $\tilde{M}_6$, we see that M_6 should be T^3-fibered as well. In other words, *any* Calabi–Yau is T^3-fibered:

$$\begin{array}{ccc} T^3 & \hookrightarrow & \tilde{M}_6 \\ & & \downarrow \\ & & M_3 \, . \end{array} \tag{9.2.69}$$

There is another duality that acts on torus fibrations: T-duality. It acts on D-branes by exchanging N and D directions. T-dualizing along all three directions of a T^3 maps a brane localized to a point to a point wrapping the whole T^3. This is the same action we just found for mirror symmetry. This motivates the *SYZ conjecture*: mirror symmetry is nothing but T-duality along the T^3 fiber of (9.2.69) [364].

Another B-brane that we can always define, on any M_6, is the D9 with trivial gauge bundle, $(S, E) = (M_6, \mathcal{O})$. There are no transverse deformations, and no moduli from (9.2.40) either, since $h^{0,1}(M_6, \mathcal{O}) = h^{0,1}(M_6) = 0$. The mirror $\tilde{S}'_3$ should have $h^1(\tilde{S}'_3) = 0$, suggesting $S'_3 \cong S^3$. If mirror symmetry is really T-duality, it should map a D9 to the base of (9.2.69):

$$M_3 \cong S^3 . \tag{9.2.70}$$

We showed in (7.1.10) that a compact Calabi–Yau manifold has no isometries, and the formulas we have seen in Section 4.2.6 for T-duality rely on the existence of an isometries along the torus. As we briefly mentioned there, however, those formulas are an approximation; the sigma model duality does not always automatically produce a world-sheet CFT on the dual side, and one might need flowing to the IR, as demonstrated for example in [128] (see Exercise 7.2.5). In such a flow, there are world-sheet instanton effects to take into account, as in the Hori–Vafa approach to mirror symmetry (Section 9.1.4).

At the *topological* level, we saw in (5.1.3) that the existence of even a single vector field without zeros implies that $\chi = 0$; on a torus bundle, we would have three such independent vector fields. But we know that, for example, $\chi(Q) = 200$ for the quintic Q, and in fact most Calabi–Yaus have $\chi \neq 0$. The apparent contradiction is avoided by admitting degenerations of the T^3 fiber. There is a topological model of such a fibration, with two types of degenerating fibers; see, for example, [365] and [366] for a review of mathematical progress on the conjecture.

Exercise 9.2.1 Find all solutions to (9.2.4) in the basis (2.2.13) for a D1 along 01 and a D4 along $0 \ldots 4$.

Exercise 9.2.2 Consider a D6-brane stack along a subspace $\mathbb{R}^7 \subset \mathbb{R}^{10}$. Allowing the transverse scalars X^a, $a = 1, 2, 3$ to depend only on one world-volume coordinate $x \equiv x^6$ and only one world-volume gauge field $a \equiv a_6$, use dimensional reduction from (9.2.34) to show that the BPS equations become the *Nahm equations*

$$D_x X^a - \frac{1}{2}\epsilon^{abc}[X^b, X^c] = 0 . \tag{9.2.71}$$

Exercise 9.2.3 Consider now a space-filling D6, and find the BPS equations on the world-volume of the internal three-cycle S_3.

Exercise 9.2.4 Work out the analogue of (9.2.14) for a brane that along Mink_4 only extends in direction 0.

Exercise 9.2.5 How is (9.2.38) generalized when there are several A-branes?

9.3 NSNS flux

We now start introducing flux in our analysis of vacuum solutions. We already tried this unsuccessfully in Section 9.1.5 at the level of the effective $d = 4$ theory, but in that approach we were keeping the internal space a Calabi–Yau. We will now work directly in $d = 10$, adding an internal NSNS three-form, without a priori assumptions on M_6.

9.3.1 Supersymmetry

Supersymmetry transformations in $d = 10$

The NSNS three-form H modifies the supersymmetry transformations (8.2.1) in a rather mild way:

$$\begin{aligned} \delta\psi^1_M &= \left(D_M - \frac{1}{4}H_M\right)\epsilon^1\,, & \delta\lambda^1 &= \left(-\frac{1}{2}H + \mathrm{d}\phi\right)_{/}\epsilon^1\,; \\ \delta\psi^2_M &= \left(D_M + \frac{1}{4}H_M\right)\epsilon^2\,, & \delta\lambda^2 &= \left(\frac{1}{2}H + \mathrm{d}\phi\right)_{/}\epsilon^2\,. \end{aligned} \tag{9.3.1}$$

Recall that the symbol $(\,)_{/}$ denotes the Clifford map (3.1.13); in particular, $\mathrm{d}\phi = \partial_M\phi\gamma^M$. The NSNS field-strength also appears via $H_M \equiv (\iota_M H)_{/} = \frac{1}{2}H_{MNP}\Gamma^{NP}$; in the following we will drop the slash symbol, as we often did in Chapter 3. The transformations of e^A_M and ϕ in (8.2.1) are not modified by H (nor by the RR fields). In any case, from now on we are only going to focus on the fermion transformations, for the reasons we explained in Section 8.2.2. The contribution to (9.3.1) of the RR form fields will be partially introduced in later sections, and in full in Section 10.1, using the democratic formalism. We will also see there that preserved supersymmetry plus the Bianchi identities implies the equations of motion.

Vacuum solutions

For vacuum solutions, H needs to be purely internal, as we already saw for the heterotic string in Section 8.2.5. The analysis in Section 8.2.4 does not change much: the spinor Ansatz (8.2.12) gives

$$\left(D_m - \frac{1}{4}H_m\right)\eta^1_{+I} = 0\,, \qquad \left(D_m + \frac{1}{4}H_m\right)\eta^2_{+I} = 0\,; \tag{9.3.2a}$$

$$\mu e^{-A}\eta^1_{+I} = \mathrm{d}A\cdot\eta^1_{-I}\,, \qquad \mu e^{-A}\eta^2_{+I} = \mp\mathrm{d}A\cdot\eta^2_{-I}\,; \tag{9.3.2b}$$

$$\left(\mathrm{d}\phi - \frac{1}{2}H\right)\eta^1_{+I} = 0\,, \qquad \left(\mathrm{d}\phi + \frac{1}{2}H\right)\eta^2_{+I} = 0\,. \tag{9.3.2c}$$

These are respectively $\delta\psi^a_m = 0$, $\delta\psi^a_\mu = 0$, and $\delta\lambda^a = 0$. The internal gravitino (9.3.2a) generalizes (8.2.24a), which had $H = 0$. As it was the case there, the equations for the η^1_I and the η^2_I are decoupled, and they differ from one another only by some signs. The external gravitino (9.3.2b) is identical to (8.2.24b) and (8.2.24c), and so it still implies $\mu = 0$, $\mathrm{d}A = 0$ as in (8.2.25). So we are again restricted to Minkowski unwarped compactifications.

We are now going to focus on the case of minimal supersymmetry, keeping only one spinor: $\eta^1_{+I} \equiv \delta^1_I \eta_+$, $\eta^2_{+I} = 0$, or in other words

$$\epsilon^1 = \zeta_+ \otimes \eta_+ + \zeta_- \otimes \eta_- , \qquad \epsilon^2 = 0 ; \qquad (\mathcal{N} = 1 \text{ NS vacua}) . \tag{9.3.3}$$

With this, (9.3.2) become

$$\left(D_m - \frac{1}{4} H_m \right) \eta_+ = 0 ; \tag{9.3.4a}$$

$$\left(\mathrm{d}\phi - \frac{1}{2} H \right) \eta_+ = 0 . \tag{9.3.4b}$$

In the Calabi–Yau case, this attempt at getting $\mathcal{N} = 1$ supersymmetry was foiled by the fact that the equations for η^1 and η^2 were identical, as explained following (8.2.28). Here, however, this is no longer the case: even if we just look at (9.3.2a), if we solve the equation for η^1 with $H \neq 0$, we cannot also solve the second with $\eta^2 \stackrel{?}{=} \eta^1$. If that were the case, sum and difference would give $D_m \eta^1_+ \stackrel{?}{=} 0 \stackrel{?}{=} H_m \eta^1_+$, implying that M_6 is Calabi–Yau and that $H = 0$ (using (3.4.92a)).

9.3.2 Analysis in terms of SU(3)-structures

We now analyze (9.3.4) using G-structures; historically, this was the first such application to string theory [367, 368]. η_+ defines an SU(3)-structure, which as we know can also be defined in terms of forms (J, Ω).

Torsionful connection

As a first step, we compute their covariant derivatives. This is quite similar to (5.5.10):

$$\mathrm{i}\nabla_m J_{np} \stackrel{(5.5.10a)}{=} (D_m \eta^\dagger_+) \gamma_{np} \eta_+ + \eta^\dagger_+ \gamma_{np} D_m \eta_+ \stackrel{(9.3.2a)}{=} \frac{1}{8} H_{mqr} \eta^\dagger_+ [\gamma_{np}, \gamma^{qr}] \eta_+$$
$$\stackrel{(2.1.6),(5.1.16)}{=} \mathrm{i} H_{mr[n} I_{p]}{}^r . \tag{9.3.5}$$

Computing $\nabla\Omega$ would be similar. But for later use, we prefer completing the analysis using the more advanced methods of Sections 3.1–3.2. We recall the definition

$$\Phi_\pm = \eta_+ \otimes \eta^\dagger_\pm , \tag{9.3.6}$$

as usual omitting the slash relating bispinors and forms. Then

$$\nabla_m \Phi_\pm \stackrel{(4.3.46)}{=} D_m \eta_+ \otimes \eta^\dagger_\pm + \eta_+ \otimes D_m \eta^\dagger_\pm$$
$$\stackrel{(9.3.4a)}{=} \frac{1}{8} H_{mnp} [\gamma^{np}, \Phi_\pm] = \frac{1}{2} H_{mnp} \mathrm{d}x^n \wedge \iota^p \Phi_\pm . \tag{9.3.7}$$

Recalling the definition (4.3.20) of the spinorial covariant derivative, we see that we can interpret (9.3.4a) as $D^H_m \eta = 0$, where $D^H_m = \partial_m + \frac{1}{4} \omega^{H\,ab}_m$, $\omega^{H\,ab}_m \equiv \omega^{ab}_m - \frac{1}{2} H_m{}^a b$. We then want to modify (4.3.27) so that it is valid, $\partial_m e^a_n - \Gamma^{H\,p}_{mn} e^a_p + \omega^{H\,ab}_m e_{bn} = 0$; this gives

$$\Gamma^{H\,p}_{mn} \equiv \Gamma^p_{mn} + \frac{1}{2} H^p{}_{mn} . \tag{9.3.8}$$

This connection still preserves the metric, but it is no longer torsionless. It is useful in the analysis of stringy corrections [37]. However, (9.3.8) will actually not play a major role for us, basically because H also appears in (9.3.4b) with a different role. We will find an apparently more natural conceptual home for H in Chapter 10, where we will see that the two appearances of H in (9.3.4) conspire to give a wedge operator $H\wedge$.

Bispinors

With $H = 0$, back in (7.1.2) we reformulated the condition $D_m\eta = 0$ in terms of the closure of (J, Ω); that helped establish Yau's theorem. So it might also be useful to reformulate (9.3.2) in terms of $(\mathrm{d}J, \mathrm{d}\Omega)$. Antisymmetrizing the free indices of (9.3.5) and recalling the definition (5.1.24), we get

$$\mathrm{d}J = -I \cdot H \,. \tag{9.3.9}$$

Multiplying (9.3.7) by a further $\mathrm{d}x^m\wedge$, we get

$$\mathrm{d}\Phi_\pm = \frac{1}{2}H_{mnp}\mathrm{d}x^m \wedge \mathrm{d}x^n \wedge \iota^p\Phi_\pm = (\iota_m H)\wedge \iota^m\Phi_\pm \,. \tag{9.3.10}$$

Recalling $\Phi_+ = \mathrm{e}^{-\mathrm{i}J}$, (9.3.7) and (9.3.10) imply (9.3.5) and (9.3.9) respectively. The zero-form part $\mathrm{d}(\eta_+^\dagger\eta_+) = 0$, which could have been found easily more directly, and can be solved without loss of generality by taking η_+ to have norm one.

We now rewrite also the dilatino equation (9.3.2c) in terms of forms. This would be possible with using bilinears and explicit indices as in (9.3.5), but once again we prefer working with slightly more abstract methods; it will be a valuable warm-up for the more challenging computations of Chapter 10. We can write

$$\begin{aligned} 2\mathrm{d}\phi\wedge\Phi_+ &\overset{(3.2.9\mathrm{a})}{=} \left(\mathrm{d}\phi\cdot\Phi_\pm \pm \Phi_\pm\cdot\mathrm{d}\phi\right) \overset{(9.3.4\mathrm{b})}{=} \frac{1}{2}(H\Phi_\pm \mp \Phi_\pm H) \\ &\overset{(3.2.55)}{=} \frac{1}{12}H_{mnp}\left((\mathrm{d}x\wedge +\iota)^3 - (\mathrm{d}x\wedge -\iota)^3\right)^{mnp}\Phi_\pm \\ &= \frac{1}{6}H_{mnp}(3\mathrm{d}x^m\wedge\mathrm{d}x^n\wedge\iota^p + \iota^{mnp})\Phi_\pm \overset{(3.2.64)}{=} ((\iota_m H)\wedge\iota^m + \iota_H)\,\Phi_\pm \,. \end{aligned} \tag{9.3.11}$$

(In the third step, we also used the dagger of (9.3.2c), and $H^\dagger = \frac{1}{6}H_{mnp}\gamma^{pnm} = -H$.) Putting this together with (9.3.10) and using $\mathrm{d} - 2\mathrm{d}\phi\wedge = \mathrm{e}^{2\phi}\mathrm{d}(\mathrm{e}^{-2\phi}\,\cdot)$ gives

$$\mathrm{e}^{2\phi}\mathrm{d}(\mathrm{e}^{-2\phi}\Phi_\pm) = -\iota_H\Phi_\pm \,. \tag{9.3.12}$$

Since η_+ has norm one, we can directly use (3.4.20), namely $8\Phi_+ = \mathrm{e}^{-\mathrm{i}J}$, $8\Phi_- = \Omega$. Separating (9.3.12) by form degrees, one finds the following [369]:

$$\mathrm{e}^{2\phi}\mathrm{d}(\mathrm{e}^{-2\phi}J) = -\iota_H\mathrm{vol}_6 \overset{(3.2.65)}{=} -*H \,, \tag{9.3.13a}$$

$$\mathrm{d}(\mathrm{e}^{-2\phi}\Omega) = 0 \,, \tag{9.3.13b}$$

$$\mathrm{e}^{2\phi}\mathrm{d}(\mathrm{e}^{-2\phi}J^2) = 0 \,; \tag{9.3.13c}$$

as well as

$$\iota_H \Omega = 0\,, \qquad \mathrm{d}\phi = -\frac{1}{4}\iota_H J^2 \overset{(5.2.16),(3.2.31)}{=} -\frac{1}{2} J \cdot * H\,. \tag{9.3.14}$$

Equivalence to supersymmetry

We derived (9.3.13) from (9.3.4); we now show that the opposite implication holds, using the intrinsic torsion formalism from Section 5.5.1. We already used it to show the equivalence of spinor and form equations, most notably in the Calabi–Yau case (7.1.2).

In the spirit of the definitions (5.5.6) of the intrinsic torsion components W_i, it proves useful to decompose H in SU(3) representations:

$$H = h\Omega + h_0^{2,1} + h^{1,0} \wedge J + h_0^{1,2} + h^{0,1} \wedge J + \bar{h}\bar{\Omega}\,. \tag{9.3.15}$$

(Recall that the $(\,)_0$ label denotes primitivity.) Using also (5.2.22) for the action of the Hodge $*$, (9.3.13) gives

$$\begin{aligned} &W_1 = h = 0\,, \qquad W_2 = 0\,, \qquad W_3 = -\mathrm{i}(h_0^{2,1} - h_0^{1,2})\,, \\ &W_4 = \mathrm{d}\phi = -\mathrm{i}(h^{1,0} - h^{0,1})\,, \qquad W_5 = 2\mathrm{d}\phi\,. \end{aligned} \tag{9.3.16}$$

Equations (9.3.13a) and (9.3.13b) were enough to determine the W_i, and one more equation. In the definition (5.5.6), W_1 appears twice; comparing these two in (9.3.13a) and (9.3.13b) gives $h = 0$, which is nothing but $\iota_H \Omega = 0$. Equation (9.3.13c) is not needed to determine the W_i: it gives an alternative expression for W_4, which when compared with the one from $\mathrm{d}J$ gives $\mathrm{d}\phi = -\mathrm{i}(h^{1,0} - h^{0,1})$. This is the second equation in (9.3.14). So we have just found that (9.3.13) implies (9.3.14).

We now go back to the supersymmetry equations (9.3.4) using SU(3) representations. Recall that $D_m \eta$ determines the W_i, by (5.5.2) and (5.5.12). Expanding (9.3.4b) in the basis associated to η_+ and using (3.4.93), we find (9.3.14). Since we showed earlier that (9.3.13) give the W_i and imply (9.3.14), we conclude that they also imply (9.3.4).

There are several other ways of rewriting (9.3.13). For example, (9.3.13b) implies that M_6 is complex, by (5.3.18). It can now be convenient to divide (9.3.9) in (p,q)-forms, recalling (5.1.30) and the decomposition of d in terms of the Dolbeault differential: $0 = (\mathrm{d}J)_{3,0} = -3\mathrm{i}H_{3,0}$, which reproduces $\iota_H \Omega = 0$ from (9.3.14), and $\partial J = (\mathrm{d}J)_{2,1} = -\mathrm{i}H_{2,1}$. Subtracting the complex conjugate, we arrive at

$$\mathrm{d}^c J \overset{(5.3.31)}{=} \mathrm{i}(\bar{\partial} - \partial)J = -H\,. \tag{9.3.17}$$

Replacing (9.3.13a) with (9.3.17), we again obtain a system equivalent to supersymmetry. Yet another possibility is to swap (9.3.13c) for the second equation in (9.3.14).

Extended supersymmetry

So far, we discussed the $\mathcal{N} = 1$ case (9.3.3). Generalizing our treatment to $\mathcal{N} > 1$ is straightforward: as we noted at the beginning of our analysis, the equations for each internal spinor are decoupled. So we can take, for example, $\eta_I^1 \equiv \delta_I^1 \eta^1$, $\eta_I^2 \equiv \delta_I^1 \eta^2$, which leads to

$$\begin{aligned} \epsilon^1 &= \zeta_+^1 \otimes \eta_+^1 + \zeta_-^1 \otimes \eta_-^1 , \\ \epsilon^2 &= \zeta_+^2 \otimes \eta_\mp^2 + \zeta_-^2 \otimes \eta_\pm^2 . \end{aligned} \qquad (\mathcal{N} = 2 \text{ NS vacua}) . \tag{9.3.18}$$

In (9.3.2a), the equations for $\eta_\mp^2$ are the same as for η_+^1, with the single change $H \to -H$. So taking $\eta_\mp^2 \neq 0$ will lead to a second SU(3)-structure, related to the same metric as the first and obeying equations (9.3.13) with

$$H \to -H . \tag{9.3.19}$$

Alternatively to (9.3.18), we can take $\eta_1^1 \equiv \eta^1$, $\eta_2^1 \equiv \eta^2$, $\eta_{I>2}^1 = 0$, $\eta_I^2 = 0$. Again this leads to two SU(3)-structures, this time both satisfying (9.3.13) with the same H. More generally, we can consider p SU(3)-structures satisfying (9.3.13) with H, and q with $H \to -H$.

No compact solutions in type II

The upshot so far is that for $\mathcal{N} = 1$ supersymmetry M_6 needs to be a complex manifold, but not symplectic ($\mathrm{d}J \neq 0$), and that it needs to obey the system (9.3.13), where (9.3.13a) can be replaced by (9.3.17). However, imposing the Bianchi identities leads to a nasty surprise: in type II supergravity, there are no solutions with $H \neq 0$ and M_6 compact and smooth. There is a very simple argument [369]:

$$0 \overset{(4.1.106)}{\leq} \int_{M_6} \mathrm{e}^{-2\phi} * H \wedge H \overset{(9.3.13a)}{=} - \int_{M_6} \mathrm{d}(\mathrm{e}^{-2\phi} J) \wedge H = \int_{M_6} \mathrm{e}^{-2\phi} J \wedge \mathrm{d}H , \tag{9.3.20}$$

the last step being integration by parts using (4.1.131). In type II supergravity, the Bianchi identity is $\mathrm{d}H = 0$, so the right-hand side is zero. Since by (4.1.106a) $*H \wedge H = \mathrm{vol}_6 |H|^2$, which is nonnegative pointwise, the only possibility is that $H = 0$ identically.

We could try introducing an NS5-brane source as in (1.3.68), $\mathrm{d}H = -2\kappa^2 \tau_{\mathrm{NS5}} \delta_{\mathrm{NS5}}$. In order not to break spacetime symmetries, this NS5 would be extended along Mink_4 and wrap an internal subspace S_2. Plugging this in (9.3.20), the right-hand side is no longer zero, but $\int_{M_6} J \wedge \delta_4 = \int_{S_2} J$. For a supersymmetric solution, the cycle should be calibrated (Section 9.2), so $\int_{S_2} J = \mathrm{Vol}(S_2) > 0$. This is again in tension with (9.3.20), and so an NS5 does not help. We instead need an object whose contribution to $\mathrm{d}H$ has the opposite sign. Such an object is called *ONS5*-plane: it is the S-dual of an O5-plane, defined by quotienting IIB string theory by the operator $(-1)^{F_L} R_5$, where R_4 is a reflection of four coordinates, similar to (1.4.45). For more details, see, for example, [370, 371]. These objects, however, add a layer of mystery over the already less understood NS5, and for this reason are less thoroughly explored than other O-planes. All this is a manifestation of the no-go argument mentioned in the introduction: without sources, a Minkowski compactification allows no nonzero form fields.

A better explored (and perhaps more natural) way out is to consider heterotic string theory. The previous discussion about supersymmetry is unchanged; (9.3.13) or its variant with (9.3.17) are still valid. But the NSNS Bianchi identity is now (8.2.32); so in this case (9.3.20) is not an obstruction. Indeed, solutions to the combined system (8.2.32) and (9.3.13) can be found; see, for example, [372–376] for a sample of work on this. An issue to evade is that the R^2 term in the heterotic Bianchi (8.2.32) is a stringy correction, so when we solve an equation of motion involving it we are likely to end up with some order one curvature, which takes us away from the supergravity limit. In other words, it is difficult to use one stringy correction without involving them all; this is another manifestation of the Dine–Seiberg argument we saw in the Introduction.

Linear dilaton background

A famous example of a solution to (9.3.13) is $M_6 = \mathbb{R}^3 \times S^3$,

$$J = \mathrm{d}x \wedge \mathrm{d}y + N\left(\mathrm{d}\rho \wedge D\psi - \frac{1}{4}\mathrm{vol}_{S^2}\right), \quad \Omega = \frac{N}{2}\mathrm{e}^{2\mathrm{i}\psi}(\mathrm{d}x + \mathrm{i}\mathrm{d}y) \wedge (\mathrm{d}\rho + \mathrm{i}D\psi) \wedge \bar{h}_{S^2}, \tag{9.3.21a}$$

with x, y, ρ being the coordinates on the $\mathbb{R}^3$, and the S^3 written as a Hopf fibration as in (4.2.51) and (5.3.13). The fields are

$$\mathrm{d}s^2_6 = \mathrm{d}x^2 + \mathrm{d}y^2 + N(\mathrm{d}\rho^2 + \mathrm{d}s^2_{S^3}), \qquad \mathrm{e}^{\phi} = \sqrt{N}\mathrm{e}^{-\rho}, \qquad H = 2N\mathrm{vol}_{S^3}. \tag{9.3.21b}$$

This is called *linear dilaton background*, for obvious reasons. (We mentioned it following (1.1.18) for the bosonic string.) It can be obtained from a near-horizon limit of the NS5 solution (1.3.65), as detailed in Exercise 9.3.2.

While we can make the x- and y-direction compact, we cannot do that with the ρ coordinate, in agreement with the general argument (9.3.20).

9.3.3 World-sheet

Since we are still taking the RR fields to vanish, it is possible to use the NSR formalism to describe the internal space. As in the Calabi–Yau case, we can start from an $\mathcal{N} = (1,1)$ action in (9.1.47), and then impose the presence of extra supercharges. This time, if we are interested in spacetime $\mathcal{N} = 1$ supersymmetry, we don't want to impose $\mathcal{N} = (2,2)$ in the world-sheet, but rather $\mathcal{N} = (2,1)$. The $\mathcal{N} = 2$ superconformal algebra will then only exist for the left-movers, and there will be a spectral flow (9.1.64) only for them, resulting in a map NSNS → RNS and NSR → RR, corresponding to a spacetime supersymmetry where only $\eta^1 \neq 0$.

The enhancement of (9.1.47) to $\mathcal{N} = (2,1)$ can be achieved by imposing the presence of a second world-sheet supercharge $\delta^{(2)}$ of the form (9.1.49), but now only with

$$I_+ \equiv I, \qquad I_- = 0 \quad \Rightarrow \quad \delta^{(2)}X^m = \varepsilon^+ D_+ X^n I_n{}^m(X). \tag{9.3.22}$$

Closure of the algebra again imposes (9.1.51), and in particular that I is integrable. Invariance of the action (9.1.47) gives

$$I_m{}^p g_{pn} + g_{mp} I_n{}^p = 0, \qquad \nabla^H_m I_n{}^p = 0, \tag{9.3.23}$$

in terms of the connection (9.3.8). This modifies (9.1.52). The first condition is the usual Hermiticity (5.1.16), while the second is just (9.3.5).

So the differential conditions imposed by $\mathcal{N} = (2,1)$ supersymmetry are integrability of I and $\nabla^H I = 0$. With the metric g, I defines an U(3)-structure, not an SU(3)-structure; so $\nabla^H I$ then determines $W_1 \ldots W_4$, but not W_5. Integrability of I is equivalent to $W_1 = W_2 = 0$, by (5.5.8). So we get two equations for W_1, one of which implies $h = 0$. This gives all the equations in (9.3.16), except the ones involving W_5

and ϕ. We can also replace $\nabla^H I$ by (9.3.17), since $W_2 = 0$ is already contained in the integrability of I. So $\mathcal{N} = (2,1)$ can be summarized as giving a U(3)-structure (g, I) such that

$$I \text{ is integrable}, \qquad \mathrm{d}^c J = -H\,. \tag{9.3.24}$$

The reason the world-sheet result (9.3.24) is not equivalent to (9.3.13) is that we have not imposed conformal invariance yet. This is similar to the $H = 0$ case, where imposing $\mathcal{N} = (2,2)$ supersymmetry corresponded to M_6 being Kähler, but not yet Calabi–Yau. The beta function equations at leading order are the same as (1.1.14) in the bosonic string, since we know that its Lagrangian is the same as that of the common sector (1.2.17) in the superstring.

As an example of a solution to (9.3.24), we can consider again (9.3.21). Since the dilaton does not appear, there is now no reason not to make ρ compact (as well as x, y) to obtain a solution with topology $T^3 \times S^3$. This model has in fact $\mathcal{N} = (4,4)$ supersymmetry; the world-sheet action *without* the dilaton term $\phi R_{(2)}$ is conformally invariant [377].

$\mathcal{N} = (2,2)$ supersymmetry

We can generalize this analysis to $\mathcal{N} = (p,q)$ supersymmetry. This corresponds to spacetime η^1_{+I}, $I = 1, \ldots, p$ and η^2_{+I}, $I = 1, \ldots, q$ respectively. In particular, for $\mathcal{N} = (2,2)$ we obtain two complex structures I^1, I^2, such that [378]

$$\mathrm{d}^c_1 J_1 = -\mathrm{d}^c_2 J_2 = -H\,, \tag{9.3.25}$$

where as usual $\mathrm{d}^c_a = \mathrm{i}(\bar\partial_a - \partial_a)$, with ∂_a the Dolbeault differential associated to I_a. An M_6 with this property is called *bi-Hermitian*. This generalizes the analysis for $H = 0$ in Section 9.1.3, where $\mathcal{N} = (2,2)$ world-sheet supersymmetry led to M_6 being Kähler.

Writing these more general models with $\mathcal{N} = (2,2)$ superspace requires using superfields other than the chiral ones we introduced in (9.1.55):

- The models with $[I^1, I^2] = 0$ ("quite lovely," in the words of [308]) involve both chiral and *twisted chiral* superfields, mentioned in Section 9.1.4 and defined by [378]

$$\bar{\mathbb{D}}_+ \chi = \mathbb{D}_- \chi = 0\,. \tag{9.3.26}$$

  These are generated by T-duality from chiral superfields [379].
- When $[I^1, I^2] \neq 0$, we also need to involve *semichiral* superfields, which involve only one condition, such as [380]

$$\bar{\mathbb{D}}_+ X_{\mathrm{L}} = 0 \tag{9.3.27}$$

  or $\bar{\mathbb{D}}_- X_{\mathrm{R}} = 0$. Each of these multiplets now describes four real directions in M_6, rather than two as is the case for chiral and twisted chiral multiplets.

To be more precise, note as in [28] that $[I^1, I^2] = (I^1 + I^2)(I^1 - I^2)$, so $\ker([I^1, I^2]) = \ker(I^1 + I^2) \oplus \ker(I^1 - I^2)$. One can then decompose the tangent space at each point as

$$\ker(I^1 + I^2) \oplus \ker(I^1 - I^2) \oplus \mathrm{im}([I^1, I^2])\,. \tag{9.3.28}$$

These three summands are the directions spanned by the chiral, twisted chiral, and semichiral multiplets, respectively. To write the action, we need to define the analogue of the Kähler potential in (9.1.56). This is relatively easy without semichirals [378, sec. 8], but in general it requires reformulating bi-Hermitian geometry in terms of pure forms [28], as we will see in Section 10.5.5.

Exercise 9.3.1 Write the metric (1.3.65) as $\text{Mink}_4 \times M_6$. Define now

$$J = dx^1 \wedge dx^4 + h(dx^2 \wedge dx^5 + dx^3 \wedge dx^6)\,,$$
$$\Omega = h(dx^1 + i dx^4) \wedge (dx^2 + i dx^5) \wedge (dx^3 + i dx^6)\,. \tag{9.3.29}$$

Show that the metric associated to this SU(3)-structure by the procedure in Section 5.1.2 is the internal metric on M_6. Show also that (9.3.29) solves (9.3.13) with H as in (9.3.29).

Exercise 9.3.2 Rewrite (9.3.29) by going to polar coordinates in the 2, 3, 5, and 6 directions, taking the $r \to 0$ limit, and defining $\rho \equiv \log r$. Show that (9.3.29) reproduces (9.3.21a). (Hint: you might find Exercise 7.4.1 useful.)

9.4 F-theory

The next flux we consider is the one-form field-strength F_1 in IIB string theory. Its peculiarity is that it is mixed with the string coupling e^{ϕ} by the $SL(2,\mathbb{Z})$ symmetry (1.2.38); the two form together the axio-dilaton $\tau_{\text{IIB}} = C_0 + i e^{-\phi}$.

We observed after (1.4.44) that the possible complex structures on a torus T^2 are classified by a so-called modular parameter $\tau = \tau_{T^2}$, subject to an $SL(2,\mathbb{Z})$ identification: two τ_{T^2} represent the same complex structure if and only if they are related by a Möbius transformation:

$$\tau \to m \cdot \tau \equiv \frac{a\tau + b}{c\tau + d}\,, \qquad m = \begin{pmatrix} a & b \\ c & d \end{pmatrix} \in SL(2,\mathbb{Z})\,. \tag{9.4.1}$$

Recall that in Section 1.4.3 we concluded that M-theory on a T^2 with modular parameter τ_{T^2} is dual to IIB with axio-dilaton τ_{IIB}, and that the $SL(2,\mathbb{Z})$ acting on the modular parameter is identified the IIB symmetry (1.2.38).

F-theory is a method to study IIB solutions where τ depends on spacetime [381–383]. In most nontrivial solutions, τ undergoes *monodromy*: it is not single-valued as a function $M_{10} \to \mathbb{C}$, but only as a function to the moduli space (5.3.66) of τ,

$$M_{10} \to \mathcal{M}_\tau = H_\tau / SL(2,\mathbb{Z})\,. \tag{9.4.2}$$

(A fundamental region for the equivalence relation defining M_τ was given in (5.3.66) and Figure 5.1.) This can happen as one goes around one-cycles of M_{10}, or around codimension-two loci, called *seven-branes*. This is an allowed solution because $SL(2,\mathbb{Z})$ is a symmetry of IIB string theory. In a sense, we are using this symmetry to glue different regions of spacetime, much like geometrical transition functions allow us to glue together different pieces of space in a nontrivial manifold.

More specifically, F-theory studies such solutions by considering an auxiliary T^2-fibration over M_{10}, whose τ_{T^2} is identified with τ_{IIB}, as in the M-theory/IIB duality.

(More general solutions where one does not use this method are sometimes called *S-folds*.) In most situations, one can promote this method to a duality with M-theory. Also, usually τ is taken to be a holomorphic function of spacetime, for reasons we will see soon. F-theory is a vast subject; more complete reviews can be found, for example, in [247, 384].

9.4.1 Supersymmetry

The supersymmetry transformations with only F_1, dilaton and metric read in $d = 10$

$$\begin{aligned} \delta\psi^1_M &= D_M\epsilon^1 + \frac{e^\phi}{8}F_1\Gamma_M\epsilon^2 , \qquad \delta\lambda^1 = \mathrm{d}\phi\,\epsilon^1 - \mathrm{e}^\phi F_1\epsilon^2 ; \\ \delta\psi^2_M &= D_M\epsilon^2 - \frac{e^\phi}{8}F_1\Gamma_M\epsilon^1 , \qquad \delta\lambda^2 = \mathrm{d}\phi\,\epsilon^2 + \mathrm{e}^\phi F_1\epsilon^1 . \end{aligned} \tag{9.4.3}$$

Again we have dropped the slash on $\mathrm{d}\phi$ and F_1. As usual, both of these one-forms have to be purely internal for a vacuum solution.

For vacuum solutions, the Ansatz for the ϵ^a should be a particular case of (8.2.12). Unlike in Section 9.3, the F_1 term now mixes ϵ^1 and ϵ^2, so setting one of the two to zero does not look particularly promising. A source for F_1 is a D7-brane, which is half-BPS. So we expect only $\mathcal{N} = 1$ in four dimensions; so we would like to use only one of the ζ^a_{+I} of (8.2.12). For the internal spinors, we can take inspiration from (9.2.14), which leads to

$$\begin{aligned} \epsilon^1 &= \zeta_+ \otimes \eta_+ + \zeta_- \otimes \eta_- , \\ \epsilon^2 &= \zeta_+ \otimes \alpha\eta_+ + \zeta_- \otimes \bar\alpha\eta_- , \end{aligned} \tag{9.4.4}$$

with $\alpha \in \mathbb{C}$. For a D7, $\gamma_\parallel$ is the product of four internal gamma matrices, and this has eigenvalues ± 1. Looking at (9.2.14), we would then guess $\alpha = \pm\mathrm{i}$. For now, however, we leave α arbitrary.

We will discuss the choice of factorization for vacua with RR fluxes in a lot more detail in Section 10.3.2, justifying the choice (9.4.4) better. Later in that chapter, we will also reanalyze supersymmetry for F-theory in a more general framework, and the derivation of the BPS equations will also be smoother.

We now insert (9.4.4) in (9.4.3) and obtain

$$D_m\eta_+ = -\frac{\mathrm{e}^\phi}{8}\alpha F_1\gamma_m\eta_+ , \qquad D_m\eta_+ = \frac{\mathrm{e}^\phi}{8}\alpha^{-1}F_1\gamma_m\eta_+ ; \tag{9.4.5a}$$

$$\mu\mathrm{e}^{-A}\eta_- = \left(-\mathrm{d}A + \frac{\mathrm{e}^\phi}{4}\alpha F_1\right)\eta_+ , \qquad \mu\bar\alpha\mathrm{e}^{-A}\eta_- = -\left(\alpha\mathrm{d}A + \frac{\mathrm{e}^\phi}{4}F_1\right)\eta_+ ; \tag{9.4.5b}$$

$$(\mathrm{d}\phi - \mathrm{e}^\phi\alpha F_1)\eta_+ = 0 , \qquad (\mathrm{d}\phi + \mathrm{e}^\phi\alpha^{-1}F_1)\eta_+ = 0 . \tag{9.4.5c}$$

Comparing the first column with the second, say in (9.4.5b), gives $\alpha = -\alpha^{-1}$; this agrees with our guess following (9.4.4). We will pick the solution

$$\alpha = -\mathrm{i} . \tag{9.4.6}$$

Now (9.4.5b) and (9.4.5c) can be analyzed similarly to (8.2.24), and lead to

$$\mu = 0 , \qquad (\mathrm{d}\phi - \mathrm{i}\mathrm{e}^\phi F_1)_{1,0} = 0 , \qquad \phi = \phi_0 + 4A . \tag{9.4.7}$$

(The subscript $_{1,0}$ is because the one-form in parenthesis is not real, so footnote 5 before (8.2.25) does not apply.) Recalling $F_1 = \mathrm{d}C_0$ and the definition (1.2.39) of τ, we obtain

$$\bar{\partial}\tau = 0\,. \tag{9.4.8}$$

Solving this implies that C_0 exists, and so the only nontrivial Bianchi identity $\mathrm{d}F_1 = 0$ is satisfied.

The gravitino equation

We are left with the internal gravitino, (9.4.5a). We could work out the corresponding $\mathrm{d}J$ and $\mathrm{d}\Omega$ as we did in Section 9.3.2. Since it is a relatively easy equation, there are faster alternatives. We can compare $f\gamma_m = f_m + f_n\gamma^n{}_m$ to (5.5.2). Using (3.4.92a) and (5.5.5a), we recognize that only p_m and $\hat{q}_m$ are nonzero. From (5.5.12) and (5.5.31), after a computation we see that the only nonzero intrinsic torsion components are

$$W_4 = \frac{1}{2}\mathrm{d}\phi\,, \qquad \bar{W}_5 = \frac{1}{4}\bar{\partial}\phi\,. \tag{9.4.9}$$

If we go to the Einstein frame (1.2.22), we have $J_\mathrm{E} = \mathrm{e}^{-\phi/2}J$, $\Omega_\mathrm{E} = \mathrm{e}^{-3\phi/4}\Omega$, and by (5.5.32):[7]

$$W_4^\mathrm{E} = 0\,, \qquad \bar{W}_5^\mathrm{E} = -\frac{1}{2}\bar{\partial}\phi\,, \tag{9.4.10}$$

To solve (9.4.10), notice that $\hat{\Omega} \equiv \mathrm{e}^{\phi/2}\Omega_\mathrm{E}$ is closed, $\mathrm{d}\hat{\Omega} = 0$. One might naively think we have obtained a Calabi–Yau with this redefinition, but

$$-\frac{1}{6}J_\mathrm{E}^3 = -\frac{\mathrm{i}}{8}\Omega_\mathrm{E}\wedge\bar{\Omega}_\mathrm{E} = -\frac{\mathrm{i}}{8}\mathrm{e}^{-\phi}\hat{\Omega}\wedge\bar{\hat{\Omega}}\,. \tag{9.4.11}$$

So the forms $(J_\mathrm{E}, \hat{\Omega})$ are both closed, but now they satisfy an unusual normalization, rather than (5.1.40b). But Yau's theorem is still useful. Suppose we have found a pair (J_0, Ω_0) that are closed and nondegenerate and satisfy $J_0 \wedge \Omega_0 = 0$. Then comparing their associated volume forms J_0^3 and $\Omega_0 \wedge \bar{\Omega}_0$, we will in general find (7.1.4) (for $d = 6$) for some f. Yau's theorem tells us that by modifying J_0 by an exact piece as in (7.1.5), we can change the function f to anything we want. In the Calabi–Yau case, we wanted to set $f = 0$; but now we can also set $f = \mathrm{e}^{-\phi}$, solving (9.4.11).

The procedure (7.1.5) does not change the cohomology class; so there is a topological obstruction. In the Calabi–Yau case, this requires $c_1(M) = 0$; in the present case, we now argue that (9.4.12) requires $c_1(M_6)$ to be positive. Recall from (6.1.3) that an SU(3)-structure where the only nonzero component is W_5 is a Kähler manifold. Taking the exterior differential and recalling (6.1.9) and (6.1.10), we obtain

$$\rho_\mathrm{E} = \mathrm{i}\partial\bar{\partial}\phi \quad \Rightarrow \quad \partial\bar{\partial}\log(\mathrm{e}^{\phi}\sqrt{g_\mathrm{E}}) = \partial\bar{\partial}\log(\mathrm{e}^{-\phi/2}\sqrt{g}) = 0\,. \tag{9.4.12}$$

(If we solve this and $h^1(M_6) = 0$, then (9.4.10) is solved too.) The cohomology class of the Ricci form ρ_E is $2\pi c_1(M_6)$. We will see in the next subsections that we need to

[7] Yet another alternative would be to transform (9.4.5a) to the Einstein frame directly using (4.3.69); the right-hand side of (9.4.5a) now becomes $\mathrm{i}\mathrm{e}^{\phi}[\partial_i\tau\gamma^i{}_m + F_m]\eta_+$, and using (3.4.92a) the equation becomes of the form (6.1.2).

take $e^{-\phi}$ to be not quite a function, but a section of a bundle $\mathcal{L}$; generalizing (4.2.61) to higher dimensions, $-i\partial\bar{\partial}\phi = i\partial\bar{\partial}\log(e^{-\phi})$ is a representative for $c_1(\mathcal{L})$. So at the level of cohomology, (9.4.12) can be interpreted as the equality

$$c_1(M_6) = c_1(\mathcal{L}) . \tag{9.4.13}$$

In particular, (5.3.49) implies that $c_1(M_6)$ is positive; at the end of Section 6.3.1 we defined this to be a *Fano* manifold. (The case $c_1(M_6) = 0$ is allowed, but it takes us back to constant dilaton and M_6 Calabi–Yau.)

Finally recall that the equations of motion are automatically satisfied once supersymmetry is preserved and the Bianchi identities hold.

Summary

We have concluded in this subsection that vacua with only F_1 and the dilaton should be of the form

$$ds^2_{10} = e^{\phi/2} ds^2_{\text{Mink}_4} + e^{\phi/2} ds^2_{\text{Kähler}} , \tag{9.4.14}$$

and that the axio-dilaton τ should be holomorphic. The only torsion class of the Kähler metric is fixed in terms of the dilaton by (9.4.10), but this can be achieved on any Kähler manifold with positive $c_1(M)$, a Fano manifold.

9.4.2 Axio-dilaton monodromy

The D7 solution and its duals

Before we consider compactifications, let us see why considering an axio-dilaton with monodromy might be a good idea. We already know one solution with monodromy: the back-reaction of a D7-brane. From (1.3.59a) and (1.3.63), we see $e^{\phi} = -\frac{2\pi}{N}\log(r/r_0)$. From (1.3.59b), rewritten as (4.1.154), the field-strength $F_1 = \frac{1}{2\pi r^2}(x^8 dx^9 - x^9 dx^8) = \frac{1}{2\pi} d\theta$, in polar coordinates (r, θ). So

$$\tau = \frac{N}{2\pi i}\log\left(\frac{z}{r_0}\right) , \qquad z \equiv r e^{i\theta} \qquad \text{(D7)} . \tag{9.4.15}$$

This is holomorphic, as predicted by (9.4.8), and undergoes monodromy

$$M = t^N : \tau \to \tau + N \tag{9.4.16}$$

around the locus $\{z = 0\}$; compare the discussion around (1.4.36), where we see that this is a large gauge transformation. So in this case, the seven-brane is simply a D7 stack.

We can act on (9.4.15) with a Möbius transformation to generate solutions with more interesting monodromies. We saw in Section 1.4.3 that an F1 is mapped in general to a (p, q)-string, a bound state of p F1s and q D1s. The element $m \in \mathrm{SL}(2, \mathbb{Z})$ doing this should map $\binom{1}{0} \to \binom{p}{q}$, so it has the form

$$m = \begin{pmatrix} p & r \\ q & s \end{pmatrix} . \tag{9.4.17}$$

The two integers r, s are determined by imposing that $m \in \mathrm{SL}(2, \mathbb{Z})$; a solution always exists when p and q are coprime, by *Bézout's theorem* in number theory.

By analogy, we call the solution with monodromy (9.4.18) a *(p, q)-seven-brane* for $N = 1$, or a stack of such objects for $N > 1$. Acting on (9.4.15) with (9.4.1), we get a new solution, where τ is still holomorphic and now has monodromy

$$M_{(p,q)} = mt^N m^{-1} = \begin{pmatrix} 1 - pq & p^2 \\ -q^2 & 1 + pq \end{pmatrix} . \tag{9.4.18}$$

Notice that r and s in (9.4.17) are not unique, since we can shift $(r, s) \to a(q, -p)$, but this ambiguity is canceled in (9.4.18).

To find the metric, it is convenient to work in the *Einstein* frame, which by (1.2.38) is invariant. From (1.2.22) and (1.3.59a):

$$\sqrt{g_s} ds^2_{\rm E} = ds^2_{\mathbb{R}^8} + h dz d\bar{z} , \tag{9.4.19}$$

where $h = -\frac{g_s N}{2\pi} \log(|z|/r_0)$; so this is the form of the $d = 10$ Einstein-frame metric for all (p, q)-seven-brane solutions. Going back to the string-frame metric, we find $\phi = 4A$, in agreement with (9.4.7), and an internal metric

$$ds^2_6 = e^{-\phi/2}(ds^2_4 + h dz d\bar{z}) , \tag{9.4.20}$$

where ds^2_4 is flat. We will usually compactify this to be a T^4.

For all these metrics, (9.4.12) reduces to $\partial_z \partial_{\bar{z}} \log(e^\phi h) = 0$. In the D7 case, this is satisfied because $e^\phi h = g_s$. More generally, for the (p, q)-seven-branes, h is still equal to that of the D7, while from (1.2.51) we find $e^\phi = |q\tau_{\rm D7} + q'|^2 e^{\phi_{\rm D7}}$, so $e^\phi h = |q\tau_{\rm D7} + q'|^2$, which satisfies (9.4.12).

For the (p, q)-seven-brane solutions, the dilaton is in general *not small.* For example, consider the $(0, 1)$-seven-brane, the S-dual of the D7 (the dual under the particular element s, (1.4.30)). Here, according to (9.4.18), the monodromy is sts^{-1}, which acts on τ as

$$\tau \mapsto -\frac{1}{-(1/\tau) + 1} = \frac{\tau}{1 - \tau} . \tag{9.4.21}$$

A small e^ϕ (large ${\rm Im}\tau = e^{-\phi}$) is mapped to $O(1)$ under this transformation. So this solution is beyond the supergravity approximation. However, we obtained it by applying a duality to a D7 solution, so we should trust it just as much.

The D7 solution itself is not everywhere under control: it is only sensible in a disk $r \le r_0$ from the source, outside which the string coupling $e^\phi = ({\rm Im}\tau)^{-1}$ and h in (9.4.15) and (9.4.19) become negative. Using (1.2.51), the same is true also of the (p, q)-seven-brane solution.

So we would now like to find some solutions where we don't have this limitation. Intuitively, this might come about if the back-reactions of several (p, q)-seven-branes interact with one another, so that we never get to the critical radius for any of them. We will see that this is indeed possible, and that it allows us to evade the no-go theorem about Minkowski compactifications with fluxes, and obtain a compact internal space.

Solutions from the j-function

To find more general solutions, we first solve for τ. Taking it to be holomorphic everywhere would not require any sources, but is impossible on a compact manifold by Liouville's theorem (Section 5.3.1). To include sources, we need to allow τ to have branch cuts, across which it jumps according to a Möbius transformation.

A concrete way to satisfy this condition [385] is to consider the function

$$j(\tau) = 4\frac{(12f)^3}{\Delta}, \qquad \Delta \equiv 4f^3 + 27g^2. \tag{9.4.22}$$

As we saw in Section 6.3.2, this is invariant under Möbius transformations (9.4.1):

$$j(\tau_1) = j(\tau_2) \quad \Leftrightarrow \quad \tau_1 = M \cdot \tau_2, \qquad M \in \mathrm{SL}(2,\mathbb{Z}). \tag{9.4.23}$$

It is a one-to-one map of the moduli space $H_\tau/\mathrm{SL}(2,\mathbb{Z})$ of modular parameters to $\mathbb{C}$. So if we give a function $j(z)$, where z collectively denotes the complex coordinates on M_6, we implicitly determine a function $\tau(z)$. In general, this has branch cuts, but since j is invariant, the two values of τ across such a branch cut are related by (9.4.1), as required.

The dilaton $\mathrm{e}^{\phi} = (\mathrm{Im}\tau)^{-1}$ is now positive by construction, and so there is no critical radius. The branch points are those where τ is one of the three special values in (5.3.67), where

$$j(\tau = \mathrm{i}) = 12^3, \qquad j(\tau = \mathrm{e}^{\pi \mathrm{i}/3}) = 0, \qquad \lim_{\mathrm{Im}\tau\to\infty} j(\tau) = \infty. \tag{9.4.24}$$

The last of these may be refined as[8]

$$j(\tau) \sim \mathrm{e}^{-2\pi \mathrm{i}\tau} + \cdots, \qquad \mathrm{Im}\tau \to \infty. \tag{9.4.25}$$

In particular, $\mathrm{Im}\tau$ is large if and only if $|j(\tau)|$ is. So around a pole z_0 of j:

$$j(z) \sim \frac{j_0}{(z-z_0)^N} \quad \Rightarrow \quad \tau(z) \sim \frac{N}{2\pi \mathrm{i}}\log(z-z_0), \tag{9.4.26}$$

which from (9.4.15) we recognize as being a stack of N D7s. To be more precise, j only determines τ up to $\mathrm{SL}(2,\mathbb{Z})$; so this could also be a stack of (p,q)-seven-branes, with monodromy (9.4.18). Indeed, all these solutions are dual, so we should not be able to tell them apart. But once we fix the duality frame around one such pole, the (p,q) numbers for all the other poles are fixed.

The other two cases in (9.4.24), $\tau = \mathrm{i}$ and $\mathrm{e}^{\pi \mathrm{i}/3}$, also correspond to branch points and introduce monodromies. In the fundamental region (5.3.66), the $\mathrm{Re}\tau > 0$ and $\mathrm{Re}\tau < 0$ segments of the unit circle are related by $\tau \to -1/\tau$, so this is the monodromy. Around $\tau = \mathrm{e}^{\pi \mathrm{i}/3}$, we have $\tau \to -1 - 1/\tau$. In the usual notation of (1.4.30) and (1.4.36):

$$M_{\tau=\mathrm{i}} = s, \qquad M_{\tau=\mathrm{e}^{\pi \mathrm{i}/3}} = t^{-1}s, \tag{9.4.27}$$

up to conjugation and an overall sign (which disappears in the Möbius transformation (9.4.1)). $s^2 = -1$, $(t^{-1}s)^3 = 1$ are consistent with i and $\mathrm{e}^{\mathrm{i}\pi/3}$ being $\mathbb{Z}_2$ and $\mathbb{Z}_3$ orbifold points, respectively (recall the discussion that follows (5.3.67)). The monodromy t around $\tau = \infty$ is not unipotent, consistent with the path "around $z = \infty$" being horizontal in Figure 5.1.

With this method, we can easily produce solutions for τ. Moreover, there is no particular problem in obtaining compact solutions: τ evades Liouville's theorem

[8] One can expand in powers of $\mathrm{e}^{2\pi \mathrm{i}\tau}$; the subleading terms, $+744 + 196684\mathrm{e}^{2\pi \mathrm{i}\tau} + \cdots$, have integer coefficients, which happen to be the dimensions of representations of the *monster group*, the largest exceptional finite simple group.

because it is multivalued, and j evades it because it is allowed to have poles, so it is in general meromorphic rather than holomorphic.

Metric

We now turn to (9.4.10). We first consider the situation where the Einstein-frame metric is nontrivial only in one complex direction, as in the Ansatz (9.4.20); to obtain a compactification, we can make ds_4^2 a T^4. As we mentioned there, the equation of motion (9.4.12) for this Ansatz reduces to $\partial_z \partial_{\bar{z}} \log(e^{\phi} h) = 0$. One might first think to solve this by taking $h = e^{-\phi}$, but the metric needs to be invariant under monodromy, while the dilaton transforms as (1.2.51). So we need a function of τ that transforms under monodromy by picking up a factor $|c\tau + d|^2$. We can also use that $\partial\bar{\partial}|f|^2 = 0$ if f is holomorphic. Fortunately, the function Δ in (9.4.22) has the property

$$|\Delta(M \cdot \tau)| = |c\tau + d|^{12} |\Delta(\tau)|. \tag{9.4.28}$$

So $h = e^{-\phi}|\Delta|^{1/6}$ is modular invariant, and is still annihilated by $\partial\bar{\partial}$. Near the positions z_I of the seven-branes (9.4.26), Δ has a zero, but we can cancel it with an extra prefactor:

$$h = e^{-\phi} |\Delta(\tau(z))|^{1/6} \prod_I |z - z_I|^{-1/6}. \tag{9.4.29}$$

This is now a solution for the simple Ansatz (9.4.20) [385].

Near a single seven-brane, the two rational powers cancel out, and we simply have $h \sim e^{-\phi}$, reproducing the D7 solution (9.4.19). If we go away from the seven-brane to a region where τ is nearly constant,

$$h \sim |z - z_0|^{-1/12}. \tag{9.4.30}$$

The nontrivial term in (9.4.19) is $|z - z_0|^{-1/12} dz d\bar{z}$, which upon defining $w = (z - z_0)^{1-1/24}$ turns into the usual flat $\mathbb{R}^2$ expression $dw d\bar{w}$, up to an overall constant. But if we turn once around the source, $z - z_0 \to (z - z_0)e^{2\pi i}$, the new variable does not return to itself:

$$w \to e^{\left(2 - \frac{1}{12}\right)\pi i} w. \tag{9.4.31}$$

Far from the seven-brane, we recover flat space in the variable w, but we need to identify the polar angle 0 not with 2π, but with $(2 - 1/12)\pi$. We say there is a *deficit angle* equal to $\pi/12$. Intuitively, if we have 24 seven-branes, then the sum of the deficit angles is 2π, and space becomes compact, namely an S^2. This can be interpreted as the gravitational back-reaction of the seven-branes, which we had not taken into account by just finding a holomorphic τ with the right monodromies. Alternatively, we could reach the same conclusion by integrating (9.4.12).

So we have seen that solving (9.4.10) imposes a global condition on compact solutions.

Properties of the solutions

We have seen that it is possible to obtain a compact internal M_6, *without O-planes*. It is natural to ask why this evades the general no-go against Minkowski compactifications with fluxes. The answer is that the solutions of this section use $SL(2, \mathbb{Z})$ transition functions to glue different regions. Already the monodromy around a single (p, q)-seven-brane is an example of such a transition function. The solution for a single such object can be dualized back to that of the D7, whose monodromy is a more conventional large gauge transformation. But in a compact solution, the (p, q) numbers of each seven-brane are different and cannot be all dualized simultaneously to conventional D7s.

Since these solutions are outside the supergravity limit, should we trust them? It is difficult to give a completely rigorous argument in general, but there are various regimes where things look under control. For example, it is easy to construct solutions that are weakly coupled almost everywhere; we will see an explicit example in Section 9.4.3. More generally, by making M_6 very large we can make τ and the metric vary slowly almost everywhere. As we pointed out before (for example, after (1.4.6)), when fields vary slowly the two-derivative action should be valid, because higher derivatives are suppressed by the slow rate of variations; and the two-derivative action is uniquely fixed by supersymmetry to be the familiar (1.2.36). So even if the dilaton is large, the supergravity action should apply in most of M_6. The regions where this is not true are those near the seven-branes; but for (p, q)-seven-branes, those where j has a pole should be trusted as much as the usual D7, because they are locally dual to it. More general monodromies can be often obtained as limits of these.

Supersymmetry demands that the moduli space of solutions be a complex manifold (as in Section 8.4.1); if we believe that these solutions exist in some limits, we expect them to continue existing for other values of the complex moduli as well, just like the zero set of a holomorphic function cannot be a segment. We will soon show that for most F-theory solutions there is a duality to M-theory that makes the case much stronger.

9.4.3 O7 resolution

Recall from Section 1.4.4 that, like most O-plane solutions, the O7 has a "hole" region where it stops making sense. The metric and dilaton are the same as for the D7 solution, but with h now given by (1.4.56). This essentially amounts to taking $N = -4$ in the results for D7, so, for example, (9.4.15) becomes $\tau_{O7} = \frac{4i}{2\pi} \log(z/r_0)$, which loses meaning for $|z| < r_0$.

We will now see evidence that in fact the O7 is a supergravity approximation to a pair of (p, q)-seven-branes [386, 387]. Consider

$$j = \frac{64(4z^2 + 3\epsilon(z^2 - z_0^2))^3}{\epsilon^2(z^2 - z_0^2)^2(z^2 + \epsilon(z^2 - z_0^2))}, \tag{9.4.32}$$

where $\epsilon \ll 1$ is a small parameter. (This corresponds to taking $f = \frac{9}{4}\epsilon(z_0^2 - z^2) - 3z^2$, $g = z(\frac{9}{4}\epsilon(z_0^2 - z^2) - 2z^2)$ in (9.4.22), at first order in ϵ.) At large distance, we have

$$|z| \to \infty \qquad j \to \frac{4096}{\epsilon^2}; \tag{9.4.33}$$

this is large, and so the string coupling is small, $e^\phi = (\mathrm{Im}\tau)^{-1} \ll 1$; by (9.4.26), $e^\phi \sim -1/\log \epsilon$. There are poles at $z = \pm z_0$ and at $z \sim \pm\sqrt{\epsilon} z_0$. By the argument in (9.4.26), all of these are (p, q)-seven-branes, but we can be more precise. The poles at $z = \pm z_0$ can be reached from infinity without ever having to make j small; so the monodromy around them is that of ordinary D7s. Since the pole is double at both $z = \pm z_0$, these are two stacks of two D7s each, so $M_{z=\pm z_0} = T^2$.

The monodromy around the poles at $z \sim \pm\sqrt{\epsilon} z_0$ is less clear: to get there we have to go through regions where j is small, and the dilaton is large. The monodromy around each point has to be of the form (9.4.18), but we don't know for which values of (p, q). The monodromy M_{C_R} around a large circle of radius $R > |z_0|$ and the monodromy M_{C_r} around a circle of radius $\sqrt{\epsilon}|z_0| < r < |z_0|$ must obey

$$M_{C_R} = M_{z=z_0} M_{z=-z_0} M_{C_r} = T^4 M_{C_r} \,. \tag{9.4.34}$$

This is because we can continuously deform the path C_r, plus two paths around $z = \pm z_0$, to the path C_R; and the monodromy around $z = \pm z_0$ was just argued to be T^2 for each. In (9.4.33), we see that at large distances the axio-dilaton is constant, including the phase: so $M_{C_R} \cdot \tau = \tau$, which by (9.4.1) implies $M_{C_R} = \pm 1$. So M_{C_r} is either T^{-4} or $-T^{-4}$. Now we can deform C_r to two paths around $z \sim \sqrt{\epsilon} z_0$: each of these has to be of the form (9.4.18). This gives $M_{(p_+,q_+)} M_{(p_-,q_-)} = -T^{-4}$, which in turn implies ([387]; Exercise 9.4.2)

$$M_{z\sim\sqrt{\epsilon}z_0} = M_{(p,1)} \,, \qquad M_{z\sim\sqrt{\epsilon}z_0} = M_{(2+p,1)} \,, \qquad M_{C_R} = -1 \,. \tag{9.4.35}$$

This argument alone does not determine p. It is possible to fix this ambiguity by explicitly following τ as it varies on a path in the z plane; we do not give the details, but this gives $p = -1$. All in all, we have

$$z = \pm z_0 : \; 2\text{ D7 (each)} \,, \qquad z \sim \mp\sqrt{\epsilon} z_0 : \; (\pm 1, 1)\text{-seven-brane.} \tag{9.4.36}$$

Now, if we take $\epsilon \to 0$, the two $(\pm 1, 1)$-seven-branes coalesce into a single object at $z = 0$. What is it? We saw earlier that the total monodromy around it plus the four D7 is the matrix -1, which doesn't act on τ. In particular, there is no monodromy for C_0, and so $\oint_{C_R} F_1 = 0$ on the path C_R that encircles all branes. In other words, the total system has zero D7 charge. So our object at the origin must have D7-charge -4. By (1.4.49), this is the appropriate charge for an *O7-plane*, or more precisely an O7_-. This is confirmed by (1.2.38), where we see that $m = -1$ flips the sign of both C_2 and B, which by (1.4.46) agrees with the action of an O7.

So the O7 "hole" we observed in Section 1.4.4 is really an artifact of supergravity; there is a strong-coupling solution that looks similar at larger distances, but has no hole:

$$\text{O7}_- \;\to\; (1, 1)\text{-seven-brane} + (-1, 1)\text{-seven-brane} \,. \tag{9.4.37}$$

In a way, this is similar to the resolution of the O6 in M-theory (Section 7.2.3).

We have included the two D7 in (9.4.36) so that τ is constant at infinity, and the full system has zero RR fields; now that we know (9.4.37), we could just as well write a version of (9.4.32) where only the $(\pm 1, 1)$-seven-branes are present. Either with D7s or without, recall that it is possible to write the metric explicitly using (9.4.29). This replaces (1.4.56) for the case without D7s; for the case with D7, the supergravity solution would have been $h = \frac{g_s}{2\pi}(4 \log |z| - \sum_I \log |z - z_I|)$, by the generalization of (1.3.62) to $p = 7$.

Finally, let us see how to make this solution compact. Within the Ansatz (9.4.20), where everything only depends on one complex coordinate, we concluded around (9.4.31) that we need 24 seven-branes to close the spacetime and obtain an S^2. The total number of seven-branes in (9.4.36) is six, so we can obtain a compact solution by putting together four copies of (9.4.32). This leads to an S^2 with four O7-planes and 16 D7s, which can also be understood as

$$T^2/\mathbb{Z}_2 \tag{9.4.38}$$

after an orientifold involution ΩR_7. This is the same as the definition of an O7 in (1.4.45), but now with R_7: $x^{8,9} \to -x^{8,9}$ on T^2 rather than $\mathbb{R}^2$. Because of the periodic identification on T^2, $x^8 \sim x^8 + 2\pi R_8$, $x^9 \sim x^9 + 2\pi R_9$, there are now four fixed loci rather than just one:

$$(x^8, x^9) = (0,0)\,,\ (\pi R_8, 0)\,,\ (0, \pi R_9),\ (\pi R_8, \pi R_9)\,. \tag{9.4.39}$$

These should be identified with the positions of the four O7-planes. While this orientifold can be defined perturbatively, the F-theory method gives us a way to resolve the hole of the O7s. It also reveals a larger space of solutions, given by the positions of the seven-branes and by the background value of the axio-dilaton. We will return to this in Section 9.4.4.

9.4.4 The associated elliptic fibration

The axio-dilaton τ_{IIB} has a formal similarity with the modular parameter τ_{T^2} parameterizing complex structures on a torus; they are both identified under Möbius transformations. This creates an opportunity to find solutions for τ_{IIB} using a more geometrical method.

Geometrical auxiliary fibration

In Section 9.4.2, we found a τ_{IIB} with Möbius monodromies by using the function j in (9.4.22). One can use the same strategy to find a torus fibration over M_6,

$$\begin{array}{ccc} T^2 & \hookrightarrow & M_8 \\ & & \downarrow \\ & & M_6\,. \end{array} \tag{9.4.40}$$

We discussed such fibrations briefly following (7.2.12): we described the T^2 with a Weierstrass model (6.3.11), with an f and g that vary over the base manifold, which now is M_6:

$$y^2 = x^3 + f(z)x + g(z)\,, \tag{9.4.41}$$

where again z denotes collectively the coordinates on M_6. There are loci in M_6 where the fibers degenerate and are no longer tori; these are the zeros of $\Delta = 4f^2 + 27g^3$, which from (7.2.13) and (7.2.14) is a section of $K_{M_6}^{-12}$. The model (9.4.41) has a global section, defined by taking for any z the point at infinity on the torus (recall the comment after (6.3.11)); it is called elliptic fibration to emphasize this.

Whenever we have such an elliptic fibration, the modular parameter τ_{T^2} of the fiber can be found by computing the j invariant as in (9.4.22); since f and

g are holomorphic, τ will also be. Moreover, its monodromies will be Möbius transformations $\tau \to M \cdot \tau$, because the complex structure of the torus should return to itself after we follow a path in M_6. So τ_{T^2} of a torus fibration gives us a solution for our initial IIB problem. Conversely, given a solution found giving $j(z)$ as in Section 9.4.2, one can associate to it an elliptic fibration: from the $\tau(z)$, one computes $f(z) = f(\tau(z))$ and $g(z) = g(\tau(z))$ using their expressions in (9.4.22). The sign of M acquires a geometrical meaning as well: it acts on the space

$$H_1(T^2, \mathbb{Z}) = \mathbb{Z}^2 \tag{9.4.42}$$

of one-cycles of the fiber. So, for example, a monodromy $M = -1$ leaves τ invariant, but it reverses the orientation of both one-cycles in the T^2 fiber.

At first it might be unclear what we gain from this geometrical correspondence. One immediate advantage is that it provides a simple way to satisfy the global condition demanded by gravity, (9.4.12) or (9.4.13). Comparing the transformation laws (1.2.51) and (9.4.28) of e^{ϕ} and Δ, (9.4.13) becomes the equality of bundles

$$\mathcal{L} \cong K_{M_6}^{-1}, \tag{9.4.43}$$

where Δ is a section of $\mathcal{L}^{12}$. This is exactly the condition (7.2.14), that makes the total space of the fibration a Calabi–Yau. So finding IIB solutions with varying axio-dilaton is equivalent to finding elliptic fibrations which are Calabi–Yau, a well-studied problem in algebraic geometry.

For example, in the Ansatz (9.4.20), where both the metric and the dilaton only depend on one complex direction, we have $M_6 = T^4 \times M_2$; our problem is reduced to finding a fibration over a Riemann surface whose total space is Calabi–Yau, or in other words a K3. The condition (9.4.43) reduces to $\mathcal{L} = K_{M_2}^{-1}$. A line bundle has a nontrivial global section only if it is positive (Section 5.3.3); by (5.3.72), this implies that only $M_2 = \mathbb{CP}^1$ is allowed. In this case, we saw in Section 7.2.1 that K3 fibrations over $\mathbb{CP}^1$ exist and have 24 degenerate fibers. These correspond to the 24 seven-branes we found in the previous subsection by using deficit angles.

In higher dimensions, this method quickly becomes more practical than using the j-invariant alone.

Weak-coupling limit in general

We now revisit the example in Section 9.4.3 from the point of view of elliptic fibrations. The fundamental objects are now f and g, which determine Δ and j through (9.4.22). Recall from (7.2.13) that f and g are sections of $\mathcal{L}^4 = K_{M_6}^{-4}$ and $\mathcal{L}^6 = K_{M_6}^{-6}$ respectively.

One important feature of (9.4.32) was that the string coupling was small and almost constant, when $\epsilon \to 0$. In general, this happens in a limit when j is constant and large; (9.4.22) instructs us to take

$$\frac{f^3}{g^2} = \text{const} \sim -\frac{27}{4}. \tag{9.4.44}$$

In this case, τ is constant and $g_s \ll 1$. The most general deformation near this is [387]

$$f \sim -3h^2 + \epsilon\eta, \qquad g \sim -2h^3 + \epsilon h\eta - \frac{1}{2}\epsilon^2\chi, \tag{9.4.45}$$

where again $\epsilon \ll 1$ and h, η, and χ are sections of $K_{M_6}^{-2}$, $K_{M_6}^{-4}$, $K_{M_6}^{-6}$ respectively. Our example (9.4.32) can be recovered by taking $h = z$, $\eta = \frac{9}{4}(z_0^2 - z^2)$, $\chi = 0$.

Similar to that example, at leading order $\Delta \sim -9\epsilon^2 h^2(\eta^2 - h\chi)$; the locus $\Delta = 0$ consists of various sources, which are revealed to be an O7-plane at $h = 0$, and a D7 at $\eta^2 = h\chi$. (The latter is in general a singular subspace of M_6, which generates subtleties analyzed in [388].) We expect that in this limit M_6 is the $\mathbb{Z}_2$ quotient induced by the O7 of a doubled manifold. To see this, we can introduce a new coordinate ξ and the manifold [387]

$$\tilde{M}_6 = \{\xi^2 = h\}. \tag{9.4.46}$$

This is a submanifold of the total space $\hat{M}$ of the bundle $\mathcal{L} = K_{M_6}^{-1}$. When $h \neq 0$, there are two solutions of this equation, which collapse to one on $\{h = 0\}$. In other words, this is a $\mathbb{Z}_2$ branched covering of M_6, with a branch cut at $h = 0$. The first Chern class can be computed applying the adjunction formula (6.3.5) to $\hat{M}$. Similar to the computation that precedes (7.2.14), this gives $c_1(M_6) + c_1(\mathcal{L})(1-2) = c_1(M_6) - c_1(\mathcal{L}) = 0$. So in the weak-coupling limit $\epsilon \to 0$, $\tilde{M}_6$ is a Calabi–Yau, of which M_6 is an O7 quotient. Indeed, the O7-plane sits at $h = 0$.

Going at the next order in ϵ as we did for (9.4.32), the O7 again splits in seven-branes, but in general for $\chi \neq 0$ the locus $\Delta = 0$ no longer factorizes.

We can also revisit our compact examples in Section 9.4.3 in light of this general discussion. We took there $M_6 = T^4 \times M_2$, and we found $M_2 = \mathbb{CP}^1$. Recalling $K_{\mathbb{CP}^1} \cong \mathcal{O}(-2)$, h, η, and χ are sections of $\mathcal{O}(4)$, $\mathcal{O}(8)$, and $\mathcal{O}(12)$ respectively. At leading order in ϵ, the O7 is at $h = 0$; a section of $\mathcal{O}(4)$ has four zeros, so there are four O7-planes, as we concluded in Section 9.4.3. The Calabi–Yau (9.4.46) is now $T^4\times$ a $\mathbb{Z}_2$ covering of $\mathbb{CP}^1$ with four branching points, which below (6.3.13) we argued to be a T^2. The original M_2 should be an O7 quotient of this torus; this agrees with (9.4.38).

Duality with M-theory

We have seen that the total space M_8 of the elliptic fibration (9.4.40) is required by the gravitino condition (9.4.12) to be a Calabi–Yau manifold of dimension $d = 8$. From the IIB point of view, this is an auxiliary space of no direct physical significance, but given the role of Calabi–Yau's in string compactifications, it is natural to wonder if this is a coincidence.

The explanation comes from M-theory. If we switch off the four-form flux G_4, the bosonic equations of motion of eleven-dimensional supergravity become $R_{MN} = 0$. So $\mathbb{R}^3 \times M_8$ is a solution if M_8 is Ricci-flat, and is also supersymmetric if M_8 is Calabi–Yau. Now take M_8 to be elliptically fibered as in (9.4.40). We may consider the limit where the size of the T^2 goes to zero. We might think we end up with a nine-dimensional theory, but just as in Section 1.4.3, the M2s wrapping the shrinking T^2 provides a tower of states, which can be interpreted as a KK tower for a new S^1; the resulting ten-dimensional theory is IIB on

$$\mathbb{R}^3 \times S^1 \times M_6, \tag{9.4.47}$$

where the size of the S^1 is again given by (1.4.42). In the limit where the T^2 is totally shrunk, the S^1 expands to infinity and we recover $\mathbb{R}^4 \times M_6$. In Section 1.4.3, we also argued that τ_{T^2} becomes τ_{IIB} under this duality. In our present case, τ_{T^2} in (9.4.40) depends on M_6, and so we end up with a varying axio-dilaton in IIB. To sum up,

$$\text{M-theory on } M_8 \quad \cong \quad \text{IIB on } M_6 \quad \equiv \quad \text{F-theory on } M_8\,, \tag{9.4.48}$$

where M_8 is a Calabi–Yau and M_6 is its base as in (9.4.40). The last relation is just a definition of what "F-theory on an elliptically fibered Calabi–Yau" means.

There are situations where one cannot use this logic, at least not directly; for example, when one wants to deal with an external AdS spacetime, it is not clear how to recover it as a limit of some different spacetime $M_3 \times S^1$ by expanding the S^1 to infinite size. Sometimes, however, one can still use the elliptic fibration to find solutions. For AdS solutions, this was sketched in [389] and realized in various ways, for example, in [238, 390, 391]; black hole solutions were studied in [392].

Gauge algebra

Another advantage of the auxiliary fibration (9.4.40) is that it provides an easy and intuitive way to understand the physics of seven-branes. So far, we have only focused on (p,q)-seven-branes, defined by poles of j, but we mentioned in (9.4.27) that the special points $\tau = \mathrm{i}$ and $\mathrm{e}^{\pi \mathrm{i}/3}$ also give rise to monodromy.

Equation (9.4.48) implies that the seven-branes in IIB map to degenerate fibers of the elliptic fibration (9.4.40). As a cross-check of this, we can think of the duality as first a T-duality to IIA, and then a lift to M-theory. Then D7-branes T-dualize to D6, which then lift to loci where the M-theory fibration degenerates, as we saw in Section 7.2.3. Over a D7 or to a (p,q)-seven-brane, the fiber has the topology of a sphere with two points identified. We can picture this as arising from a T^2 after shrinking an S^1 to a single point. This is called an I_1 fiber. More generally on a stack of N D7-branes (or (p,q)-seven-branes of the same type), the T^2 degenerates to a necklace of S^2s, called an I_N fiber. We expect this to give rise to a $\mathrm{U}(N)$ gauge group, as would any stack of D-branes.

From the M-theory point of view, this gauge group comes about because of a logic very similar to that in (7.4.24). Integrating the M-theory three-form potential A_3 on the N copies of S^2 of the I_N fiber provides N abelian gauge fields. Moreover, M2-branes wrapping the S^2s become massless in the limit where we shrink the fiber, and provide additional gauge fields. Because the necklace is the extended Dynkin diagram of the Lie algebra $\mathrm{su}(N)$, these additional gauge fields combine with the abelian ones to give an $\mathrm{U}(N)$ gauge group.

Kodaira and Néron classified the generic degenerate fibers of a Calabi–Yau elliptic fibration and obtained the list in Table 9.1 [393, 394]. (See, for example, [395, sec. V.7].) They all consist of a collection of S^2s; in most cases, they intersect according to the extended Dynkin diagram of an ADE group, as in the I_N case. One can then reconstruct the gauge algebra using the previous M-theory logic.[9] In the first three columns, "ord" denotes the order of vanishing as a function of a complex transverse coordinate; a zero means that the function does not vanish at all. Notice

[9] The presence of $\mathrm{u}(1)$ summands is a lot subtler to establish, and requires using the so-called *Mordell–Weil group*, whose elements are the sections of the fibration; the global structure of the gauge group is even subtler to obtain. For more details, see, for example, [384].

Table 9.1. Types of seven-branes in F-theory.

ord(f)	ord(g)	ord(Δ)	Name	τ	Monodromy	Gauge algebra
0	0	n	I_n	$i\infty$	t^n	su(n) [sp($\lfloor n/2 \rfloor$)]
≥ 1	1	2	II	$e^{\pi i/3}$	$-t^{-1}s$	–
1	≥ 2	3	III	i	$-s$	su(2)
≥ 2	2	4	IV	$e^{\pi i/3}$	$-st^{-1}$	su(3) [su(2)]
≥ 2	≥ 3	6	I_0^*	any	-1	so(8) [so(7), g_2]
2	3	$n+6 \geq 7$	I_n^*	$i\infty$	$-t^n$	so($2n+8$) [so($2n+7$)]
≥ 3	4	8	IV*	$e^{\pi i/3}$	$t^{-1}s$	e_6 [f_4]
3	≥ 5	9	III*	i	s	e_7
≥ 4	5	10	II*	$e^{\pi i/3}$	st^{-1}	e_8

that all degenerate fibers happen where the discriminant Δ does vanish; indeed, $\Delta \neq 0$ corresponds to a smooth T^2 (Section 6.3.2).

For the simplest cases in Table 9.1, while the fiber is degenerate, the total space M_8 of the elliptic fibration is smooth. Technically, this can happen because when we apply (6.3.21) to the coordinates x and y, we obtain a singularity, but when we apply (6.3.21) to z as well, we have an extra equation, which is not satisfied (Exercise 9.4.4). So in this case, the singularity of the fiber is an artifact of the way we are "slicing" the total space M_8. The same happens for a fiber of type II. For all other fibers, the total space M_8 is singular itself. Still, in all cases of Table 9.1, the singularity can be resolved (Section 6.3.5) so that the resulting nonsingular $\hat{M}_8$ is a Calabi–Yau.

A few comments on the individual entries:

- We have already commented on the I_n fiber; the first gauge algebra in the last column on the table is the expected su(n) gauge algebras. The other possibility sp($n/2$) may arise because of a phenomenon called *Tate monodromy* that will be explained shortly. Also in the other entries in this column, the algebras in brackets are due to this.
- The I_n^* fiber has a similar behavior to the I_n, but the monodromy matrix has an extra minus sign, and the S^2s intersect as in the extended Dynkin diagram of so($2n+8$), which is then the gauge algebra. Both features signal that it should be interpreted as an $O7_-$-plane with $n+4$ D7-branes on top. Indeed, the sign in the monodromy acts by reversing the sign of C_2 and B, from (1.2.38); we already encountered this sign in the example in Section 9.4.3. The gauge algebra is consistent with our general finding for an Op_--plane following (1.4.52).
- One might wonder why we didn't interpret I_n^* as an $O7_+$ with $n-4$ D7s on top, which according to (1.4.49) has the same RR charge. This is indeed possible; we don't show it in Table 9.1, but it leads to a gauge algebra sp($n-4$). This has a smaller dimension than so($2n+8$), so these singularities can be deformed in fewer ways, and for this reason they are said to be *frozen* and denoted by $\widehat{I_n^*}$. Some of the later discussion in this section, notably about matter fields, also changes, but we are not going to discuss this further; see [110, 396, 397].

- An I_0^* fiber is special in that an O7$_-$-plane with four D7s has zero RR charge. So there should still be a sign reversal on C_2 and B, but $\oint F_1 = 0$ on a path around it, and so no action on τ. Its value on the fiber can then be arbitrary, unlike in all other cases. We find this fiber in the $\epsilon = 0$ case in the weak-coupling limit (9.4.44), at the zeros of $\Delta \propto f^3$.
- The IV*, III*, II* fibers have S^2 intersecting as in the extended Dynkin diagrams of e_6, e_7, e_8, which are then the gauge algebras. We see from the table that τ on these fibers corresponds to strong string coupling. This is consistent with the fact that on ordinary D-branes do not yield exceptional gauge groups.
- Finally, the S^2 in the II, III, and IV fibers do not intersect according to a Dynkin diagram. In the II case, there is a single S^2 with a so-called *cusp* singularity, $y^2 = x^3$. In the III case, there are two S^2 intersecting once but tangentially. In the IV case, there are three S^2 intersecting together at a single point. In the last two cases, we still have local $\mathbb{C}^2/\mathbb{Z}_2$ and $\mathbb{C}^2/\mathbb{Z}_3$ singularities, giving rise to gauge algebras su(2) and su(3) respectively.

It is also possible to reconstruct the gauge algebra directly in a IIB language; it requires introducing multipronged open strings, whose existence can be argued using string dualities [398].

Tate monodromy

While Table 9.1 classifies the possible fibers at a single point, globally it might happen that the S^2s of a fiber get exchanged as we follow a loop *on the seven-brane.* This is called *Tate monodromy*. It is, of course, distinct from the monodromy we studied so far, which describes how the nondegenerate fibers T^2 are transformed as one goes *around* the seven-brane. In our elliptic fibration (9.4.41), there is a global section, and so at least one S^2 is preserved by Tate monodromy. The others, however, may be exchanged by it. If so, we say that the seven-brane is called *non-split*, and we mark it with a superscript $(\,)^{\mathrm{ns}}$.

This is what generates the gauge algebras in brackets in Table 9.1. While the local physics on the seven-brane remains the same, from the point of view of the effective $d = 4$ physics the gauge fields generated by two exchanged S^2s get identified. They both acquire a dependence on one direction of M_6, and keeping only the lowest mode results in the identification. One can work out a refinement of Table 9.1 where a local geometry is given for each seven-brane and each possible Tate monodromy [399, table 2]; it is expressed in terms of a generalization of the Weierstrass model, where lower powers of x and y are also kept.

Visually, Tate monodromy gives an action on the Dynkin diagram that one can use to infer the resulting $d = 4$ gauge algebra. For example, the usual I_n has n copies of S^2 intersecting in a necklace; if n is even and under Tate monodromy, the ith S^2 is exchanged with the $(n - i)$th, then the resulting $\mathrm{I}_{2n}^{\mathrm{ns}}$ seven-brane gives rise to an sp(n) gauge algebra. This happens, for example, whenever an I seven-brane intersects one of type I*; this particular intersection can be realized perturbatively as a D7 intersecting an O7, and the sp(n) can then be motivated with perturbative computations in the spirit of Section 1.4.4 (see, for example, [397, sec. 2] for more details). For another example, consider an I_0^* curve; Tate monodromy here may permute three of the S^2, thus identifying them. This results in the g_2 gauge algebra.

Matter fields

The fibers classified in Table 9.1 are the generic ones. Namely, they occur at loci of complex codimension one; this is why they describe seven-branes. In codimension two, we might have more complicated degenerations, for example when two seven-branes intersect. At such intersection points, matter fields also appear. This is no surprise for us: we have already seen this phenomenon at the intersection of ordinary D-branes, for example D6s in Sections 9.2.3 and 9.2.4. This can be used to compute the matter spectrum appearing at the intersection of curves that have a perturbative origin, namely I_n and I_n^*. A more general method consists in realizing an intersection of two seven-branes with gauge algebras $\mathfrak{g}_1$ and $\mathfrak{g}_2$ as a complex deformation of a single seven-brane with algebra $\mathfrak{g} \supset \mathfrak{g}_1 \oplus \mathfrak{g}_2$ [400]. A list has been compiled of the matter fields to which a gauge algebra has to couple on each seven-brane, depending on its geometry [401, table 2]; this is also summarized in a slightly different form in [402, table 1]. These rules get modified in presence of frozen $\widehat{\mathrm{I}_n^*}$ curves [397, table 3.1].

Conformal matter

We mentioned earlier that the Kodaira–Néron classification of Table 9.1 allows M_8 to be singular, but only if a Calabi–Yau resolution $\hat{M}_8$ exists. Often, the singularity at an intersection of two seven-branes does not admit such a resolution. This happens when

$$\mathrm{ord}(f) \geq 4\,, \qquad \mathrm{ord}(g) \geq 6\,, \qquad \mathrm{ord}(\Delta) \geq 12\,. \tag{9.4.49}$$

In such a case, one can improve the singularity by performing a blow-up (Section 6.3.5) in the base M_6. This separates the two seven-branes that now intersect the newly created exceptional locus (Exercise 6.3.3). The latter might now itself be a seven-brane. By iterating this procedure if necessary, one eventually obtains a new M_6' whose degenerate fibers are all in Table 9.1 [403]. As an example, we may consider

$$y^2 = x^3 + z^5 w^5\,. \tag{9.4.50}$$

Using Table 9.1, we see that this corresponds to two II* seven-branes, at $\{z = 0\}$ and $\{w = 0\}$, with two e_8 gauge algebras. The singularity at $\{z = w = 0\}$ has $\mathrm{ord}(g) = 10$, $\mathrm{ord}(\Delta) = 20$, which satisfies (9.4.49) and hence does not appear in Table 9.1. In this case, the blow-up in M_6 has to be repeated until we have created eleven new exceptional curves, with new gauge algebras

$$\mathrm{su}(2) \oplus g_2 \oplus f_4 \oplus g_2 \oplus \mathrm{su}(2) \tag{9.4.51}$$

besides the original $e_8 \oplus e_8$.

The physics interpretation is that the original singularity (9.4.49) corresponds to superconformal theory called *conformal matter*, and that the blow-ups correspond to spontaneous breaking of conformal symmetry [404]. In other words, we interpret (9.4.49) as a limit of singularities that are in Table 9.1. This was used to produce many superconformal theories using F-theory, culminating in a classification of such theories in $d = 6$ [405, 406]; for a review, see [407]. A duality to M-theory (different from (9.4.48)) relates conformal matter models to a nonperturbative phenomenon where an M5-brane on an $\mathbb{C}^2/\Gamma_{\mathrm{ADE}}$ singularity can fractionate [404].

9.4.5 Duality with heterotic theory

The appearance of E_8 in F-theory and the duality (9.4.48) gives a way to realize an E_8 gauge group in M-theory. We already saw in Section 1.4.5 another way to realize an E_8 in M-theory: including a boundary. It is natural to suspect that a relation between the two exists.

A more direct line of attack is to see if there is a relation between the E_8 gauge groups realized by F-theory and those in the heterotic string. The simplest type of F-theory solutions are those of the form (9.4.19), where spacetime in IIB is $\mathbb{R}^8 \times \mathbb{CP}^1$, which we called "F-theory on K3" in the previous subsection. (A general discussion of the dualities involving K3s is [168].) While in Section 9.4.1 we set up our analysis for minimal supersymmetry, the presence of four extra flat directions enhances that, since along those directions the internal η_+ can be taken to be any constant spinor. This leads to 16 supercharges, the same as heterotic in flat space. So the only possible relation would seem to be [381]

$$\text{F-theory on K3} \quad \cong \quad \text{heterotic on } T^2 \,, \tag{9.4.52}$$

where the K3 is elliptically fibered.

Scalars and vectors

To check (9.4.52), we can compare the effective theories on both sides. We could also compactify (9.4.52) further to $d = 4$ on a T^4, but let us work directly in $d = 8$. We have not reviewed the relevant supergravity machinery in that dimension, but let us at least compare scalars and vectors.

The moduli space of heterotic T^d compactifications is a discrete quotient of a Grassmannian (4.4.51):

$$\mathcal{M}_{\text{het},T^d} = \frac{\mathrm{O}(d, 16+d)}{\mathrm{O}(d) \times \mathrm{O}(16+d) \times \Gamma_{d,16+d}} \,. \tag{9.4.53}$$

This can be obtained with a logic similar to (1.4.65) for heterotic S^1 compactifications; indeed, the latter gave 17 moduli, and (9.4.53) has dimension $d(16+d)$, which for $d = 1$ gives 17. (9.4.53) can also be obtained by modifying (4.2.83) to include the extra right-moving bosons on the heterotic world-sheet.

The complex structure moduli space for an elliptically–fibered K3 is (7.2.15), which matches (9.4.53) for $d=2$, in agreement with (9.4.52); so there are $2(16+2) = 18$ scalars.

Supersymmetry in $d = 8$ is restrictive enough that the number of vector fields is now guaranteed to work, but let us go through it anyway. On the F-theory side, the generic solution only has single poles, or in other words (p, q)-seven-branes. Of these, we can arrange only for at most 18 of them to be D7s; this should be the generic case, with the positions of the D7s corresponding to the scalar expectation values. Two more gauge fields come the RR fields, C_1 and $\int_{S^2} C_3$, for a total of 20. For a T^d heterotic compactification, the generic gauge group is $\mathrm{U}(1)^{16}$ from the Cartan subalgebra of the heterotic gauge group, and $2d$ more vectors from $g_{\mu m}$, $B_{\mu m}$, as in (1.4.66), for a total of

$$16 + 2d \,. \tag{9.4.54}$$

So with $d = 2$, we get 20 vector fields again.

The gauge group can enhance and become nonabelian. On the F-theory side, this can happen from choices of elliptic fibrations that have special fibers. For example, we can achieve two E_8 by making two II^* fibers appear. On the heterotic side, the moduli come from the choice of lattice for the periodic world-sheet bosons; the gauge group can become nonabelian for special choices of that lattice, as happened in (1.4.66) for S^1 compactifications (in the context of heterotic T-duality). One can check that these phenomena appear on the same loci on the two sides. This is a much more impressive check of (9.4.52).

M-theory

Let us now compactify (9.4.52) on a further S^1. By (9.4.48), on the F-theory side we obtain M-theory on K3. So we find [72, 408]

$$\text{M-theory on K3} \quad \cong \quad \text{heterotic on } T^3 \,. \tag{9.4.55}$$

M-theory is defined on any K3, not just on elliptically fibered ones. So we should consider the full moduli space $\mathcal{M}_{\mathrm{HK}}$ of K3 metrics (7.2.10). (The other bosonic field, A_3, does not contribute any scalars, because $h_3(\mathrm{K3}) = 0$.) This matches with (9.4.53) for $d = 3$, confirming (9.4.55).

We can also compare the vector fields in $d = 7$. On the M-theory side, we have $\int_{S_{2i}} A_3$, of which there are $h_2(\mathrm{K3}) = 22$. This matches (9.4.54) for $d = 3$.

IIA

Compactifying on a further S^1, M-theory reduces to IIA; so we arrive at [174]

$$\text{IIA on K3} \quad \cong \quad \text{heterotic on } T^4 \,. \tag{9.4.56}$$

(We are presenting this last, but historically it was in fact the first string duality [174].)

We again compare the effective theories on both sides, this time in a little more detail. In Section 8.4, we did not review $d = 6$ supergravity, so let us quickly give the basics. We have $\mathcal{N} = (1, 1)$ supersymmetry, as seen in (8.2.38). The following are the relevant multiplets:

- The gravity multiplet, whose bosonic fields are the metric, four graviphotons A^a_μ, a two-form potential $b_{\mu\nu}$, and a scalar σ
- The vector multiplet, with a vector field A_μ and four scalars ϕ^a

Locally, the space spanned by the vector multiplets scalars is constrained by supersymmetry to be again a Grassmannian $\mathrm{Gr}(4, n_{\mathrm{v}})$, (4.4.51). See, for example, [409] or [410, sec. 2] for more details.

As in M-theory, we have the geometrical moduli from $\mathcal{M}_{\mathrm{HK}}$, which are $3 \cdot 19 = 57$, plus one for the overall volume modulus. But now there are also $h_2(\mathrm{K3}) = 22$ massless scalars from B, and one from the dilaton, for a total of $57 + 1 + 22 + 1 = 81$.

The vectors all come from the RR fields: C_1; $\int_{S_{2i}} C_3$, of which there are $h_2(\mathrm{K3}) = 22$; and one from dualizing $*\mathrm{d}_6 C_3 = \mathrm{d}_6 a$ along Mink$_6$, adapting the by now familiar procedure in (8.4.33)–(8.4.35). Finally, we have a two-form, which is just the Mink$_6$ part of B. All these fields can be arranged in one gravity multiplet and $n_\mathrm{v} = 20$ vector multiplets. So the moduli space of scalars is

$$\mathcal{M}_{\mathrm{IIA,K3}} = \mathrm{Gr}(4, 24) = \frac{\mathrm{O}(4, 20)}{\mathrm{O}(4) \times \mathrm{O}(20)} \tag{9.4.57}$$

up to discrete a quotient, matching (9.4.53) for $d = 4$. The total number of vectors, 24, also matches (9.4.54) for $d = 4$. In this case too, one can compare the special loci where the gauge group enhances on the two sides of the duality, finding agreement [72]; again, see [168] for a review.

Exercise 9.4.1 For the Ansatz we considered in this section, the IIB pseudo-action (1.2.36) simplifies (up to an overall constant) to

$$\int d^{10}x\sqrt{-g}\left(\mathrm{e}^{-2\phi}(R + 4|\mathrm{d}\phi|^2) - \frac{1}{2}|F_1|^2\right) = \int d^{10}x\sqrt{-g_\mathrm{E}}\left(R_\mathrm{E} - \frac{\mathrm{d}\tau \cdot \mathrm{d}\bar{\tau}}{2(\mathrm{Im}\tau)^2}\right). \tag{9.4.58}$$

Compute the equations of motion in your preferred frame; show that they are implied by (9.4.8) and (9.4.10).

Exercise 9.4.2 Reproduce (9.4.35). (Hint: start from $T^4 M_{(p_+,q_+)} = M^{-1}_{(p_-,q_-)}$, and use (9.4.18).)

Exercise 9.4.3 Check that all the seven-branes in Section 9.4.3 correspond to I_n fibers, using the first line in Table 9.1.

Exercise 9.4.4 Consider the Weierstrass model (9.4.41) with

$$f = -3 + z\,, \qquad g = 2 + z\,. \tag{9.4.59}$$

Check that the fiber at $z = 0$ is of I_1 type. Applying the definition (6.3.21) to the coordinates x and y, check that it has a singularity. By applying (6.3.21) to x, y, and z, check that the total space of the elliptic fibration is smooth around this point.

Exercise 9.4.5 Find an elliptically fibered K3 with two singularities of type II* (so that the gauge group contains $E_8 \times E_8$).

9.5 The conformal Kähler class

Our last class is one of the most famous we will see in the whole book [411–414]. In IIB, one can introduce both three-form fluxes H, F_3 together in a certain way, so that the Calabi–Yau condition is modified only by an overall function. As we will see, this is morally because part of the solution can be thought of as the back-reaction of D3-branes. The construction works only for some choices of the complex structure; a welcome development, which decreases the often large number of massless scalars in a fluxless Calabi–Yau compactification. This class is also the starting point for a perhaps even more famous proposal of dS vacua, which we will review in Section 12.1.

9.5.1 Supersymmetry

We will dedicate most of our discussion to the supersymmetric case; later we will also describe an extension of this class where supersymmetry is broken. After anticipating pieces of the supersymmetry transformations for various particular cases, for IIB supergravity we now give them in full, with all fields switched on. We will revisit them in Section 10.1.3 in a slightly different form, extending them to IIA.

$$\delta\psi_M = D_M\epsilon + \frac{1}{8}(-2H_M + \mathrm{i}e^{\phi}F_3\Gamma_M)\epsilon^c - \mathrm{i}\frac{e^{\phi}}{16}(2F_1 + F_5)\Gamma_M\epsilon = 0\,, \tag{9.5.1a}$$

$$\delta\lambda = -\frac{1}{2}(H + \mathrm{i}e^{\phi}F_3)\epsilon^c + (\mathrm{d}\phi + \mathrm{i}e^{\phi}F_1)\epsilon\,, \tag{9.5.1b}$$

where

$$\psi_M \equiv \psi^1_M + \mathrm{i}\psi^2_M\,, \qquad \lambda \equiv \lambda^1 + \mathrm{i}\lambda^2\,, \qquad \epsilon \equiv \epsilon^1 + \mathrm{i}\epsilon^2\,. \tag{9.5.2}$$

As usual, $(\,)^c$ denotes conjugation, as defined in (2.3.4). Equation (9.5.1) generalizes both (9.3.1) and (9.4.3); as in the latter, we have dropped all the slash signs from all forms. As we mentioned earlier and we will show in the next chapter, when (9.5.1) are zero and the Bianchi identities are satisfied, the equations of motion follow.

Spinor Ansatz

The F-theory solutions of the previous subsection were partially inspired by D7s; now we want to draw inspiration from D3s. To see why these are particularly promising, recall from (1.3.59) that a D-brane back-reaction treats the transverse directions differently from the parallel ones. If we consider a D3 extended along spacetime and pointlike in M_6, all the internal directions will be multiplied by the same function. So the Calabi–Yau metric should be modified only by multiplication by an overall function.

The D3 also suggests an Ansatz for $\epsilon = \epsilon^1 + \mathrm{i}\epsilon^2$, via (9.2.14). The internal $\gamma_{\parallel}$ for a D3 is the identity, so $\alpha = \pm\mathrm{i}$; the sign is immaterial, and this leads us back to (9.4.4) and (9.4.6) for F-theory, which upon rescaling ζ we rewrite as

$$\epsilon = \epsilon^1 + \mathrm{i}\epsilon^2 = \zeta_+ \otimes \eta_+\,. \tag{9.5.3}$$

Since we are only using a single ζ_+ in $d = 4$, solutions of this type have $\mathcal{N} = 1$ in four dimensions. It was to be expected that a D3 would lead to the same spinor Ansatz as a D7, given that the two together have #(ND) $= 4$ (recall (9.2.5)). This also suggests that we might be able to combine the new D3 effect with the F-theory class. This is the last time we have to guess a spinor Ansatz; in the next chapter, we will carry out a systematic analysis of (9.5.1) and of its IIA counterpart.

Vacuum conditions

As usual, for a vacuum we need F_1, F_3 and H to have only internal components. For F_5, this cannot be quite true, because self-duality $F_5 = *F_5$ relates the internal and external components. This is not a problem, however, because the components F_{0123m}, with m internal, do not break any of the symmetries of the external spacetime. These components are completely determined by the internal part f_5: we can write

$$F_5 = f_5 + e^{4A}\mathrm{vol}_4 \wedge *_6 f_5\,. \tag{9.5.4}$$

The internal and external components give equal contributions to (9.5.1). The one-form $*_6 f_5 \equiv *f_5$ is easier to manage than the five-form; using (3.2.27) (with $c = -\mathrm{i}$ as usual), we obtain $\not{f}_5\gamma = \mathrm{i}\not{*f}_5$.

Plugging (9.5.3) in (9.5.1) now gives

$$D_m\eta_+ = \frac{\mathrm{e}^\phi}{8}(\mathrm{i}F_1 + *f_5)\gamma_m\eta_+\,, \qquad H_m\eta_+ = -\frac{\mathrm{i}}{2}\mathrm{e}^\phi F_3\gamma_m\eta_+\,; \tag{9.5.5a}$$

$$\left(\mathrm{d}A + \frac{1}{4}\mathrm{e}^\phi(\mathrm{i}F_1 - *f_5)\right)\eta_+ = 0\,, \qquad \mu\mathrm{e}^{-A}\,\eta_- = -\frac{\mathrm{i}}{4}\mathrm{e}^\phi F_3\eta_+\,; \tag{9.5.5b}$$

$$(\mathrm{d}\phi + \mathrm{i}\mathrm{e}^\phi F_1)\eta_+ = 0\,, \qquad (H - \mathrm{i}\mathrm{e}^\phi F_3)\eta_+ = 0\,. \tag{9.5.5c}$$

These three rows come from $\delta\psi_m = 0$, $\delta\psi_\mu = 0$ and $\delta\lambda = 0$ respectively. The system reduces to (9.4.5) for $f_5 = f_3 = 0$.

9.5.2 Geometry and fluxes

The first in (9.5.5c) coincides with (9.4.5c) and (9.4.6); we know already that this leads to (9.4.8), the holomorphicity of τ. The first in (9.5.5b) is very similar: it tells us $(\mathrm{d}A + \mathrm{e}^\phi(\mathrm{i}F_1 - *f_5)/4)_{1,0} = 0$, and using again (9.4.8), we obtain the following:

$$*\,f_5 = \mathrm{e}^{-\phi}\mathrm{d}(4A - \phi)\,. \tag{9.5.6}$$

We now turn to the equations involving three-form fluxes, which are decoupled from the rest. Recalling the definition $G_3 = F_3 - \mathrm{i}\mathrm{e}^{-\phi}H$ from (1.2.41), the second in (9.5.5c) is $\bar{G}_3\eta_+ = 0$; with (3.4.93) we see $\bar{G}_{0,3} = 0$, $J \cdot G_3 = 0$. For the second in (9.5.5a), rewrite $H_m = \frac{1}{2}\{\gamma_m, H\}$, which is just (3.2.9b), and take a linear combination with γ_m times the second in (9.5.5c). This gives $(\gamma_m G_3 + 2G_3\gamma_m)\eta_+ = 0$. If we multiply this further by γ^m and use (3.2.23), we obtain $G_3\eta_+ = 0$. So we also have $G_3\gamma_m\eta_+ = 0$, and again using (3.2.9b) we obtain $G_m\eta_+ = 0$. With (3.4.92a), this gives $G_{m\bar{j}\bar{k}} = 0$, or in other words $G_{0,3} = G_{1,2} = 0$. All in all,

$$G_3 = G^0_{2,1}\;: \tag{9.5.7}$$

G_3 is (2, 1) and *primitive* (Section 5.2.2). In particular, from (5.2.19) it follows that

$$*\,G_3 = \mathrm{i}G_3 \quad\Rightarrow\quad *F_3 = \mathrm{e}^{-\phi}H\,. \tag{9.5.8}$$

Using again (3.4.93), the second in (9.5.5b) simply gives $\mu = 0$. So once again with the Ansatz in this section, we cannot introduce a cosmological constant.

We are left with the first in (9.5.5a). Its analysis is similar to (9.4.5a); the only nonzero intrinsic torsion coefficients are

$$W_4 = \mathrm{d}(-2A + \phi)\,, \qquad W_5 = \partial(-3A + \phi)\,. \tag{9.5.9}$$

Again we can eliminate W_4 using (5.5.32), but this time we use a different rescaling: we define $\mathrm{d}\tilde{s}^2_6 \equiv \mathrm{e}^{2A-\phi}\mathrm{d}s^2_6$, so that the $d = 10$ string frame metric reads

$$\mathrm{d}s^2_{10} = \mathrm{e}^{2A}\mathrm{d}s^2_{\mathrm{Mink}_4} + \mathrm{e}^{\phi-2A}\mathrm{d}\tilde{s}^2_6\,. \tag{9.5.10}$$

Now the rescaled internal metric has $\tilde{W}_4 = 0$, and

$$\tilde{W}_5 = -\frac{1}{2}\partial\phi\,, \tag{9.5.11}$$

as in (9.4.10). So again the internal space is Kähler; the rest of the analysis in Section 9.4.1 also applies. Now M_6 can be Fano, but it can also be Calabi–Yau, the latter case being still rather nontrivial, as we will see.

Bianchi identities

We now have to impose the Bianchi identities (1.2.10) on the form field-strengths. The general no-go argument against Minkowski vacua with flux tells us we should include O-planes, and use (1.3.43). Given our motivation at the beginning of this section, we will postulate that these are O3-planes, perhaps with additional D3-branes.

The Bianchi identity for F_1 is satisfied because τ is holomorphic, just like in the F-theory case. For F_3, we would have $\mathrm{d}F_3 = H \wedge F_1$. Given a G_3 that satisfies (9.5.7), this is tantamount to imposing $\mathrm{d}G_3 = -\mathrm{d}\tau \wedge H$, as we can see from the first definition (1.2.41). For example, one can exhibit several solutions where both sides of this equation are closed. This is particularly easy when $C_0 = 0$ and $\phi = \phi_0$ is constant; then it is enough to take G_3 to be closed. We will comment further on this case later, and in that instance we will also investigate flux quantization.

Finally, we come to the F_5 Bianchi identity. In terms of the internal f_5 and exterior differential, this reads

$$\mathrm{d}f_5 - H \wedge F_3 = \delta_3 \equiv 2\kappa^2\left(\sum_I \tau_{\mathrm{D3}}\delta_{\mathrm{D3}_I} + \tau_{\mathrm{O3}}\delta_{\mathrm{O3}}\right), \tag{9.5.12}$$

where I runs over the number of D3-branes; recall from (1.3.42) that it is a six-form. A solution for f_5 exists if and only if $H \wedge F_3 + 2\kappa^2\delta$ is a trivial six-form. Since $h^6(M_6) = 1$ by compactness, it is enough that its integral over M_6 vanishes. This gives

$$2\kappa^2\tau_{\mathrm{D3}}\left(n_{\mathrm{D3}} - \frac{1}{4}n_{\mathrm{O3}}\right) = -\int_{M_6} H \wedge F_3 \overset{(9.5.8),(4.1.106)}{\leq} 0\,, \tag{9.5.13}$$

confirming the need for O3-planes.

Having shown that f_5 solving (9.5.12) exists, we determine A by reading (9.5.6) backward. One can obtain solutions where A has large variations over M_6; a famous example was worked out for applications to holography [415], with M_6 the deformed conifold (6.3.36). This is a possible local prototype for compact models with complex moduli near a conifold locus; these have very similar features to the Randall–Sundrum models [229, 230] mentioned in footnote 1 after (8.1.8).

Alternatively, we can substitute in (9.5.12) the expression (9.5.6), and take its Hodge dual. This is particularly convenient in the Einstein frame; using $*_{\mathrm{E}}\alpha_k = e^{(k-3)\phi/2}\alpha_k$ (compare Exercise 4.1.17), $f_5 = -*_{\mathrm{E}}\,\mathrm{d}(4A - \phi)$, and we can write (9.5.12) in terms of the Laplacian Δ_{E} (4.1.112). By our discussion of the Hodge decomposition around (4.1.140), we can then determine A by inverting this operator, provided (9.5.13) holds.

Breaking supersymmetry

In solving the equations of motion, one does not really need (9.5.7) but only its consequence (9.5.8) [414]:

$$*G_3 = \mathrm{i}G_3\,. \tag{9.5.14}$$

While (9.5.14) is implied by $*F_3 = \mathrm{e}^{-\phi}H$, the opposite is not true. Indeed, as summarized in (5.2.22), on an SU(3)-structure manifold we have other imaginary-self-dual forms: a nonprimitive $(1,2)$-form and a $(0,3)$-form:

$$*\,\bar{\Omega} = \mathrm{i}\bar{\Omega}\,, \qquad *(\alpha_{0,1}\wedge J) = \mathrm{i}\alpha_{0,1}\wedge J\,. \tag{9.5.15}$$

It is possible to see using the formalism in Section 10.3.6 that these correspond respectively to F-term and D-term supersymmetry-breaking: in a $d = 4$ effective description, they fail to satisfy either the first or the second equation in (8.4.52). Harmonic $(0,1)$-forms are not present, but Ω always is.

The conformal Calabi–Yau case

A case of particular note is when the dilaton is constant:

$$\mathrm{e}^{\phi} = \mathrm{e}^{\phi_0} \equiv g_s\,. \tag{9.5.16}$$

Since $\mathrm{e}^{-\phi} = \mathrm{Im}\tau$ and τ should be holomorphic, C_0 is a constant, which we will take to be zero, and $F_1 = 0$. From (9.5.11), $\tilde{W}_5 = 0$, and so the rescaled metric $\mathrm{d}\tilde{s}^2_6$ is Calabi–Yau. So we call this the *conformal Calabi–Yau* class.

As we anticipated, the $d = 10$ string-frame metric (9.5.10) now becomes like that for a D3-brane or O3-plane (recall (1.3.59)) with $\mathrm{e}^{-4A} = h$; the Mink_4 and M_6 metrics play the role of the parallel and transverse directions respectively. This similarity becomes more precise for $F_3 = H = 0$. Away from the sources, the Bianchi identity (9.5.12) and (9.5.6) become $\Delta A = 0$; by (4.1.114), this becomes $\tilde{\Delta}\mathrm{e}^{-4A} = 0$ with respect to the rescaled metric $\mathrm{d}\tilde{s}^2$. So now the metric is

$$h^{-1/2}\mathrm{d}s^2_{\mathrm{Mink}_4} + h^{1/2}\mathrm{d}s^2_{M_6} \tag{9.5.17}$$

with h harmonic on M_6. For $M_6 = \mathbb{R}^6$, this is just (1.3.59) for $p = 3$. (Remember that both there and in the present class the Hodge dual $*_{10}$ of this internal component is also switched on; compare (9.5.4).) So for $F_3 = H = 0$, the conformal Calabi–Yau class is interpreted as the back-reaction of D3-branes and O3-planes on $\mathbb{R}^4 \times M_6$.

Let us reintroduce the three-form field-strengths. We can pick G_3 to be a harmonic form $\in H^{2,1}(M_6,\mathbb{C})$. For proper Calabi–Yaus (excluding T^6 and $T^2\times$ K3), such forms are automatically primitive, as discussed after (7.1.8); so (9.5.7) is satisfied. Now from (1.2.41), we determine $H = g_s\mathrm{Im}G_3$, $F_3 = \mathrm{d}C_2 = \mathrm{Re}G_3$.

Flux quantization (1.3.47) and (1.3.50b) imposes that the integrals of $\frac{1}{4\pi^2 l_s^2}H$, $\frac{1}{4\pi^2 l_s^2}F_3$ on any three-cycles should be integer. It is enough to impose it on a basis $\{A^A, B_A\}$; it follows that

$$H = 4\pi^2 l_s^2(n^A\alpha_A + \tilde{n}_A\beta^A)\,, \qquad F_3 = 4\pi^2 l_s^2(m^A\alpha_A + \tilde{m}_A\beta^A)\,, \tag{9.5.18}$$

in terms of the basis $\{\alpha_A, \beta^A\}$ in $H^3(M_6,\mathbb{Z})$, dual in the sense (9.1.4).

Recalling (9.5.8), we are quantizing the periods both of a three-form H and of its dual $*H$. This does not work with an arbitrary Calabi–Yau metric, but only for some particular choice of the moduli. For a baby example, we can consider T^2 and the complex $(1,0)$-form $\mathrm{d}z = \mathrm{d}x + \mathrm{i}\mathrm{d}y$. Take A, B the two sides of the parallelogram in Figure 1.2; then $\int_A \mathrm{d}z = 1$, $\int_B \mathrm{d}z = \tau$. Imposing that both of these be integer quantizes τ. In other words, if we start with H and F_3 that satisfy (9.5.18), then $G_3 = F_3 - \mathrm{i}H/g_s$ is $(2,1)$ only for some choice of the complex moduli. We conclude that *some of the complex moduli are fixed*, or *stabilized*, and the number of massless scalars is reduced. In Section 12.1, we will see that quantum effects fix the Kähler moduli as well.

Finally, we impose flux quantization for the five-form. $\mathrm{d}C_4$ is not a form in $H^5(M_6,\mathbb{Z})$, because of the delta in (9.5.12). But (1.3.47) still applies: we should integrate $\mathrm{d}C_4$ on any compact five-dimensional subspace S_5. A priori the result can be nonzero either because S_5 is topologically nontrivial or because it surrounds a source (Section 1.3.4). If M_6 is a proper Calabi–Yau, there are no nontrivial five-cycles. For S_5 that surrounds a source, flux quantization should be automatically satisfied because of the values of the D-brane tensions $\tau_{\mathrm{D}p}$. Integrating (9.5.12) over M_6 and using Stokes's theorem (4.1.131) as well as (9.5.13) and (9.1.5), we obtain

$$n_{\mathrm{D3}} - \frac{1}{4} n_{\mathrm{O3}} = \tilde{n}_A m^A - n^A \tilde{m}_A \,, \tag{9.5.19}$$

the factors of $2\pi l_s$ conspiring to simplify.

F-theory with flux, and its M-theory dual

We know from Section 9.4.1 that (9.5.11) admits more general solutions, where M_6 is not Calabi–Yau but a Fano manifold. For

$$G_3 = 0\,, \qquad \phi = \frac{A}{4} + \phi_0 \;\Rightarrow\; f_5 = 0\,, \tag{9.5.20}$$

the class in this section reduces to F-theory. It is then natural to conjecture that the more general case $G_3 \neq 0$, $\phi \neq A/4$ also has a similar interpretation.

To establish this, we could proceed along the lines of Section 9.4.2, making sure that the new ingredients transform as needed under $\mathrm{SL}(2,\mathbb{Z})$. By (1.2.38), f_5 is invariant. In (1.2.42), we found $G_3 \to G_3/(c\tau + d)$; comparing with the transformation law (9.4.28) of Δ, which is a section of $K_{M_6}^{-12}$, tells us that G_3 is a section of $\Omega^3 \otimes K_{M_6}$. The auxiliary elliptic fibration of Section 9.4.4 makes it easier to keep track of these transformation properties and can be used to dualize to M-theory as in (9.4.48).

We can use that duality on the new extension of F-theory beyond (9.5.20). The new fluxes we have introduced will all map to components of G_4, M-theory's only form field-strength. Starting from F-theory, we can first T-dualize IIB to IIA along the spacetime direction x^3, and then lift to M-theory, introducing a new coordinate x^{10}. F_3 is purely internal, so it has no leg along x^3 and T-dualizes to F_4 in IIA; then it lifts to internal G_4 in M-theory. H_3 T-dualizes to itself, and then it lifts to G_4. We can summarize these two as

$$G_4^{\mathrm{int}} = \mathrm{d}\sigma^1 \wedge H + \mathrm{d}\sigma^2 \wedge \mathrm{d}C_2 = \mathrm{e}^{\phi}\mathrm{Im}(\mathrm{d}z \wedge \bar{G}_3)\,, \tag{9.5.21}$$

where $\sigma^1 \equiv x^{10}$ and $\sigma^2 \equiv \tilde{x}^4$ are the two coordinates on the T^2, and $z = \sigma^1 + \tau\sigma^2$. The internal and external components of F_5 in (9.5.4) get T-dualized to F_6 and F_4 respectively, which are Hodge dual to each other as in (1.3.40); then F_4 lifts to

$$G_4^{\text{ext}} = \text{vol}_3 \wedge g\,, \tag{9.5.22}$$

with g a one-form on the base M_6 of the elliptic fibration M_8. The D3-brane sources are T-dualized to D2-branes, and then lifted to M2-branes. Finally, the F-theory seven-branes are mapped to degenerate fibers of the $M_8 \to M_6$ elliptic fibration, as we know from Section 9.4.4.

It is possible to analyze directly the conditions imposed by supersymmetry on $\text{Mink}_3 \times M_8$. With an Ansatz similar to (9.5.3), one obtains [411] that the metric of M_8 is a Calabi–Yau, up to an overall function e^{-A}, and that the internal field-strength should be a primitive $(2,2)$-form:

$$G^{\text{int}} = (G^{\text{int}})^0_{2,2}\,. \tag{9.5.23}$$

Moreover, there is an external component of the form (9.5.22), with $g = 3\text{d}A$. (See [416, 417] for three-dimensional M-theory compactifications with a more general spinor decomposition.) If we apply these results to an elliptically fibered M_8 and internal flux of the form (9.5.21), we recover the results obtained previously from F-theory [412]. For example, the condition (9.5.23) becomes (9.5.7). Historically, this class first arose in this way.

String corrections

String corrections to supergravity modify (9.5.19). In general, taking into account only the leading correction to supergravity can lead to misleading results, but the effect on a topological obstruction should be robust.

The M-theory action has a correction of the form $\int A_3 \wedge X_8$, where $X_8 = (P_1^2 - 4P_2)/192$ is a quartic polynomial in the Riemann tensor, P_i being the Pontryagin densities of Section 4.2.5. This term can be inferred by anomalies [35, 36]; we mentioned it in IIA in the discussion that follows (1.2.25). It contributes to the G_4 equation of motion, which now reads

$$\text{d} * G_4 + \frac{1}{2} G_4 \wedge G_4 = \delta_8 + \frac{1}{4\pi^2 l_{\text{P}11}} X_8\,, \tag{9.5.24}$$

where δ_8 is now the M2-brane source, an eight-form. Under duality with IIB, the first three terms in this equation map to (9.5.12). We can integrate (9.5.24) on M_8 to find global obstructions, as in (9.5.19). Since $c_1(M_8) = 0$, X_8 simplifies to become C_4, the density for the fourth Chern class, which by the Gauss–Bonnet theorem (4.2.72) integrates to the Euler characteristic $\chi(M_8)$. Translating this result back to IIB, (9.5.19) is modified to

$$n_{\text{D3}} - \frac{1}{24}\chi(M_8) = \tilde{n}_A m^A - n^A \tilde{m}_A\,. \tag{9.5.25}$$

In IIB, the new term on the left-hand side is the D3-brane charge of the F-theory D7-branes. This charge is induced by the corrected WZ term (9.2.53), much in the same way as for the ordinary WZ term, as discussed after (1.3.37).

9.5.3 Effective description

Finally, we briefly discuss a $d = 4$ effective action description of these vacua. This will give a supersymmetric realization of the ideas discussed in Section 8.3.1. We focus on the case where M_6 is Calabi–Yau. One might think that the effective action remains $\mathcal{N} = 2$ in this case, as it was in the case without fluxes in Section 9.1, and that perhaps supersymmetry is broken to $\mathcal{N} = 1$ spontaneously. But the analysis in Section 9.1.5 suggested that introducing fluxes in a Calabi–Yau reduction with no further modification is unlikely to lead to $\mathcal{N} = 1$ vacua. Moreover, in (9.5.13) we saw that the presence of O3-planes is necessary; an O3 projection breaks supersymmetry by half already prior to reducing. For both these reasons, we expect the effective action to only have $\mathcal{N} = 1$ supersymmetry.

Some features of the reduction are inherited from Sections 9.1.2 and 9.1.5; this was sketched in [412, 414], carried out in greater detail and generality in [418], and is reviewed in [419, sec. 5]. The orientifold projects away some of the fields according to (1.4.46): among the scalars in (9.1.12), $b^a = \zeta^a = 0$, and $a = \tilde{C}_0 = 0$. The remaining fields are now collected in chiral multiplets: z^i, $z^0 \equiv \tau = C_0 + \mathrm{i}\mathrm{e}^{-\phi}$, and $T_a \equiv \frac{1}{2} D_{abc} v^b v^c + \tilde{\zeta}_a$. The Kähler potential and superpotential are

$$K = -2\log\left(-\frac{1}{6}\int_{M_6} J^3\right) - \log\left(\mathrm{i}\int_{M_6} \bar{\Omega}\wedge\Omega\right) - \log(\mathrm{i}(\bar{\tau}-\tau))\,; \qquad W = \int_{M_6} \Omega\wedge G\,. \tag{9.5.26}$$

K is roughly speaking inherited from (9.1.20) and (9.1.26); the τ term can be inferred from (9.1.34). (The nonconstant warping function A introduces some subtleties that are believed to change K [420–422]). W is called the *Gukov–Vafa–Witten* superpotential. It was motivated in [423] with indirect methods, studying BPS domain walls; it can also be found by breaking to $\mathcal{N} = 1$ the results in Section 9.1.5. We will describe this computation in greater generality in Section 10.3.6.

The metric (9.1.28) on the Kähler moduli satisfies

$$g^{a\bar{b}}\partial_a K \partial_{\bar{b}} K = 3\,; \tag{9.5.27}$$

for example, when $h^{1,1} = 1$, $\partial_T K = -3/(T + \bar{T})$, and $g_{T\bar{T}} = 3/(T + \bar{T})^2$. So the contribution of the T_a to the first term of (8.4.5) cancels the term $-3|W|^2$, and

$$V_{\mathcal{N}=1\ \mathrm{sugra}} = \mathrm{e}^K \left(g^{I\bar{J}} D_I W D_{\bar{J}} \bar{W}\right)\,, \tag{9.5.28}$$

where now I runs over i and τ. So now $V_{\mathcal{N}=1\ \mathrm{sugra}} \geq 0$; the scalar expectation values in the locus

$$\{V_{\mathcal{N}=1\ \mathrm{sugra}} = 0\} = \{D_I W = 0\}\,, \tag{9.5.29}$$

where all the F-terms along the directions I vanish, correspond to vacua. But recall from (8.4.49a) that supersymmetry is preserved only if *all* the F-terms are zero. So if we are on (9.5.29) but $D_a W \neq 0$, then supersymmetry is broken, while the cosmological constant is still zero because of (8.4.50). This mechanism to break supersymmetry in $d = 4$ models is called *no-scale* supersymmetry-breaking [424, 425]. It corresponds to the supersymmetry-breaking mechanism in (9.5.14). Applying to $\mathcal{V} = \Omega$ the general result (8.4.18), $D_i W = \int_{M_6} (\mu_{(i)} \cdot \Omega) \wedge G$, where $\mu_{(i)}$ is a Beltrami differential. The variation $D_\tau W \propto \int_{M_6} \Omega \wedge \bar{G}$. Setting these two to zero, we get $G_{1,2} = G_{3,0} = 0$; since, as we mentioned, primitivity is automatic on a proper Calabi–Yau, we conclude from (5.2.22) that G is imaginary-self-dual, (9.5.14). But W does not depend on the Kähler moduli, so $D_a W = \partial_{T^a} K W \propto W \propto G_{0,3}$. This is exactly the supersymmetry-breaking parameter we found in (9.5.15).

Exercise 9.5.1 Check that (9.5.1) reduces to (9.3.1) and (9.4.3) in the appropriate cases.

Exercise 9.5.2 Show that the form $\bar{G}_3 = \mathrm{d}z^1 \wedge \mathrm{d}z^2 \wedge \mathrm{d}z^3$ defines an $\mathcal{N} = 3$ Minkowski solution. (Hint: be careful how you choose the complex structure associated to each internal η_{I+} [426].)

10 The vacuum problem in general

In the previous chapter, we have seen several classes of Minkowski compactifications, which we obtained by making increasingly general assumptions on the supersymmetry parameters. In this chapter, we are finally going to reveal the supersymmetry transformations in all their glory, and learn how to tame their apparent complication by using geometrical ideas.

10.1 The democratic formalism

We start in this section by reviewing the democratic formalism for type II theories [427].

10.1.1 Doubling the RR fluxes

The RR fields of type II supergravity in Chapter 1 are F_0, F_2, and F_4 in IIA, and F_1, F_3, and F_5 in IIB, the latter with a self-duality constraint

$$F_5 = *F_5 \,. \tag{10.1.1a}$$

Several aspects of type II are simplified by including the magnetic duals (1.2.50) of the RR fields:

$$F_6 \equiv *F_4 \,, \qquad F_8 \equiv - * F_2 \,, \qquad F_{10} \equiv *F_0 \tag{10.1.1b}$$

in IIA, and

$$F_7 \equiv - * F_3 \,, \qquad F_9 = *F_1 \tag{10.1.1c}$$

in IIB. Recall also from (1.2.34b) that $F_5 = *F_5$ is a twisted field-strength (1.2.9). It is possible to solve the constraints (10.1.1b) and (10.1.1c) by working only with the forms of lowest degrees, or with those of highest degrees. In the *democratic formalism*, we instead give equal representation to all forms.[1] They can be collected efficiently by defining an RR polyform

$$F = \begin{cases} F_0 + F_2 + F_4 + F_6 + F_8 + F_{10} & \text{(IIA)} \\ F_1 + F_3 + F_5 + F_7 + F_9 & \text{(IIB)} \,. \end{cases} \tag{10.1.2}$$

[1] Relative to [427], in IIA we have changed the signs of B, F_2, F_6, F_{10}. This makes the supersymmetry transformations more similar to those of IIB, and in particular it will make many of the equations in Section 10.2 identical in the two theories.

The duality relations (10.1.1) can then be summarized as[2]

$$F = *\lambda F \overset{(3.2.59)}{=} \mp \lambda * F . \quad \overset{\text{IIA}}{\text{IIB}} \tag{10.1.3}$$

λ is the familiar sign reversal (3.2.28), which already appeared in close relation with Hodge duality in Chapter 3. The Bianchi identity (1.3.43) then becomes

$$\mathrm{d}_H F \equiv (\mathrm{d} - H\wedge)F = \delta . \tag{10.1.4}$$

The RR source is obtained by varying the WZ term in (1.3.24):

$$\delta = \sum_I (-1)^{p_I} \delta_I \mathrm{e}^{\mathcal{F}} \wedge \lambda \mathrm{d}x^1 \wedge \ldots \wedge \mathrm{d}x^{9-p_I} , \qquad \delta_I = 2\kappa^2 \tau_I \delta(x^1) \ldots \delta(x^{9-p_I}) . \tag{10.1.5}$$

This is essentially a sum over terms of the type (1.3.42). As in that equation, the x^i, $i = 1, \ldots, 9 - p_I$ are transverse coordinates. The index I runs over the number of sources, with p_I space dimensions; each can be a Dp_I or Op_I, with charge densities τ_I given by (1.3.25) and (1.4.49). Since the O-planes are nondynamical, they have no f, and (1.4.46) makes $B = 0$ too; so $\mathcal{F} = 0$ on them. The signs in (10.1.5) are a convention on what we call a Dp-brane and what an anti-Dp-brane; we have chosen it so as to agree with our supersymmetry conventions that are introduced later.

There is no need to introduce a dual for H, whose Bianchi identity is

$$\mathrm{d}H = 2\kappa^2 \tau_{\mathrm{NS5}} \delta_{\mathrm{NS5}} , \tag{10.1.6}$$

and whose equation of motion is

$$\mathrm{d}(\mathrm{e}^{-2\phi} * H) + \frac{1}{2} \sum_k *F_k \wedge F_{k-2} = 2\kappa^2 T_{\mathrm{F}_1} \delta_{\mathrm{F1}} , \tag{10.1.7a}$$

with source terms δ_{NS5}, δ_{F1} localized on NS5-branes and fundamental strings F1. From now on in this chapter we will set these sources to zero, while keeping the RR source (10.1.5) non-zero; nothing essential is lost with this simplification. For compactifications, on which we will focus soon, δ_{F1} would be a rather exotic source anyway: for it not to break the maximal symmetry of spacetime, it would have to be a distribution of world-sheet instantons. (In any case, solutions are typically found first by solving the equations outside the sources, and then recognizing the latter by the comparing their singularities to those in the brane solutions (1.3.59).) Setting thus $\delta_{\mathrm{F1}} = 0$, we can rewrite (10.1.7a) by using the pairing (3.2.42) and self-duality (10.1.3) as

$$\mathrm{d}(\mathrm{e}^{-2\phi} * H) = \frac{1}{2}(F, F)_8 . \tag{10.1.7b}$$

[2] From now on, we will write many formulas that are valid for IIA with the upper sign, and for IIB with the lower sign. We will often but not always include the symbol $\overset{\text{IIA}}{\text{IIB}}$ as a reminder.

Away from the sources for H, the differential operator d_H in (10.1.4) squares to zero:

$$d_H^2 = \frac{1}{2}\{d - H\wedge, d - H\wedge\} = -\{d, H\wedge\} \overset{(4.1.77)}{=} -(dH)\wedge \\ = 0\,. \tag{10.1.8}$$

So one can introduce an analogue of de Rham cohomology using d_H rather than d. More relevant for us, since (10.1.4) says that F is d_H-closed, locally it should be d_H exact. This would suggest writing

$$F \overset{?}{=} d_H C, \tag{10.1.9}$$

where C would be a polyform

$$C = \begin{cases} C_1 + C_3 + C_5 + C_7 + C_9 & \text{(IIA)} \\ C_0 + C_2 + C_4 + C_6 + C_8 & \text{(IIB)}\,. \end{cases} \tag{10.1.10}$$

Equation (10.1.9) is almost correct; it is fine in IIB, but it cannot be quite right for the *massive* IIA case, namely when $F_0 \neq 0$. Indeed, in degree zero it would imply $F_0 \overset{?}{=} dC_{-1}$, which doesn't make sense. To repair it, begin by observing that locally $H = dB$; then by the Hadamard lemma

$$e^{B\wedge} d e^{-B\wedge} \overset{(2.1.15)}{=} d - [d, B\wedge] = d - (dB\wedge) = d - H\wedge\,. \tag{10.1.11}$$

Now (10.1.4) is equivalent to

$$d(e^{-B\wedge} F) = 0\,. \tag{10.1.12}$$

$e^{-B\wedge}F$ can be locally written as d of a polyform, which we can call $e^{-B\wedge}C$, except for the zero-form part. This leads to $e^{-B\wedge}F = d(e^{-B\wedge}C) + F_0$, or in other words,

$$F = e^{B\wedge}\left(d(e^{-B\wedge}C) + F_0\right) \overset{(10.1.11)}{=} d_H C + F_0\, e^B\,. \tag{10.1.13}$$

This finally corrects (10.1.9). It reproduces (1.2.27) and (1.2.29), extending them to the new higher-degree field-strengths.

The gauge transformations can also be summarized conveniently in this polyform language. The RR gauge transformations become

$$\delta C = d_H \lambda, \tag{10.1.14}$$

while the NSNS gauge transformation is

$$\delta B = d\hat{\lambda}_1\,, \qquad \delta C = -F_0 \hat{\lambda}_1 \wedge e^B\,, \tag{10.1.15}$$

extending (1.2.30).

10.1.2 Action and equations of motion

In IIB, we already saw that one usually gives up deriving the self-duality constraint $F_5 = *F_5$ from S_{IIB}, which for this reason is called a pseudo-action; our new (10.1.3) extends this issue to IIA as well. For reasons explained in Section 8.2.2, we will not include fermions in the following discussion.

Pseudo-action

For both type II theories, the pseudo-action is

$$S_0 = \frac{1}{2\kappa^2}\int_{M_{10}} \mathrm{d}^{10}x\sqrt{-g}\left(\mathrm{e}^{-2\phi}\left(R + 4|\mathrm{d}\phi|^2 - \frac{1}{2}|H|^2\right) - \frac{1}{4}\sum_k |F_k|^2\right). \quad (10.1.16)$$

Here we used the old definition (4.1.107a) for the square of forms; if we like, we can also use (4.1.106a), which is still valid in Lorentzian signature (Exercise 4.1.11) to rewrite $\int_{M_{10}} \sqrt{-g}\alpha^2 = \int_{M_{10}} *\alpha \wedge \alpha$. As we warned, (10.1.16) is a pseudo-action, so it is not appropriate to use the constraint (10.1.3) before varying. Indeed, if we did so, we would get

$$\sum_k |F_k|^2 \overset{(4.1.156b)}{=} 0\,. \quad (10.1.17)$$

In terms of the polyform potential C, the D-brane action (1.3.24) can be written

$$S_{\mathrm{D}p} = \tau_{\mathrm{D}p}\int_{\mathrm{D}p}\left(-\mathrm{d}^p\sigma \mathrm{e}^{-\phi}\sqrt{-\det(g|_{\mathrm{D}p} + \mathcal{F})} \mp C \wedge \mathrm{e}^{-\mathcal{F}}\right) \quad {}^{\mathrm{IIA}}_{\mathrm{IIB}}\,, \quad (10.1.18)$$

with the Dp-brane tension (1.3.25), $\tau_{\mathrm{D}p} = \frac{1}{(2\pi)^p l_s^{p+1}}$. The contribution from O-planes differs only in that $\tau_{\mathrm{O}p} = -2^{5-p}\tau_{\mathrm{D}p}$, and $\mathcal{F} = 0$. In massive IIA, the WZ term should really be understood as a form wz such that $\mathrm{d}wz = F \wedge \mathrm{e}^{-\mathcal{F}}$, as discussed before (1.3.39).

Equations of motion

Varying (10.1.16) gives (10.1.4) and (10.1.7), as well as the equations of motion for the metric and dilaton:

$$R_{MN} + 2\nabla_M\nabla_N\phi - \frac{1}{2}H_M \cdot H_N \quad (10.1.19a)$$
$$= \frac{1}{4}\mathrm{e}^{2\phi}F_M \cdot F_N - \frac{1}{4}\mathrm{e}^{\phi}\sum_I j^I_{MN}\,;$$
$$2R - |H|^2 - 8\mathrm{e}^{\phi}\nabla^2\mathrm{e}^{-\phi} = \mathrm{e}^{\phi}\sum_I j^I\,, \quad (10.1.19b)$$

where $H_M = \iota_M H$, $F_M \equiv \iota_M F$; the bispinor H_M already appeared after (9.3.1). Following (4.1.106b), the contraction $F_M \cdot F_N$ is a sum of the individual degrees,

$$\sum_k (F_k)_M \cdot (F_k)_N = \sum_k \frac{1}{(k-1)!}F_{MM_2\ldots M_k}F_N{}^{M_2\ldots M_k}\,. \quad (10.1.20)$$

The contribution of each k-form to the stress-energy tensor would be proportional to

$$(F_k)_M \cdot (F_k)_N - \frac{1}{2}g_{MN}|F_k|^2\,, \quad (10.1.21)$$

but the sum over all k of the second term drops out from (10.1.19a) because of (10.1.17). If we prefer restricting the sum over the original $k = 0, 2, 4$ in IIA or

$k = 1, 3, 5$ in IIB, we can use the fact that (10.1.21) is invariant under $F_k \to *F_k$, as is clear from (3.2.27) and (3.2.67).

The localized sources in (10.1.19) are a sum of the Dp and Op contributions:

$$j^I_{MN} \equiv (g_{MN} - 2\Pi^I_{MN})j^I\,, \qquad j^I \equiv \frac{\sqrt{-E|_I}}{\sqrt{-g}}\delta_I\,; \qquad \Pi^I_{MN} \equiv (E^{-1})^{ab}\partial_a x^P \partial_b x^Q g_{MP} g_{NQ}\,. \tag{10.1.22}$$

δ_I was given in (10.1.5), and as in that notation I can be a Dp-brane or an Op-plane; $|_I$ denotes pull-back to the Ith object. As usual, $a = 0, 1, \ldots, p-1$ are world-volume indices, $E_{MN} = g_{MN} + \mathcal{F}_{MN}$, and $E_{ab} = E_{MN}\partial_a x^M \partial_b x^N$ is its pull-back, (4.1.51).

Supergravity versus string theory

Equations (10.1.7), (10.1.4), and (10.1.19) are the equations of motion and Bianchi identities of type II *supergravity*. To claim a string theory vacuum, we also need to make sure of the following:

- That flux quantization, (1.3.47) and (1.3.50b), is satisfied
- That we are in the regime of applicability of the supergravity approximation

The second point in particular requires us to make sure that the string coupling and curvature (in string units) are small. As with any approximation, we cannot quote a particular threshold that guarantees with absolute certainty that string corrections will be under control; for many supergravity solutions, however, one can make both parameters as good as one wishes, a situation called *parametric control*. This is usually achieved by noting that

$$g_{MN} \to e^{2c} g_{MN}\,, \qquad \phi \to \phi - c\,, \qquad H \to e^{2c} H\,, \qquad F_k \to e^{kc} F_k \tag{10.1.23}$$

is a symmetry of the equations of motion away from sources. Here c is a real constant, which often becomes related to an integer upon imposing flux quantization.

More generally, flux quantization often discretizes any free parameters present in a supergravity solution. If this does not happen and the solution has any moduli c_i left, they can pose a danger to the supergravity approximation even if g_s and the curvature are small. This is because if the supergravity action variation with respect to c_i is identically zero, a nonzero variation in these directions of any string correction becomes potentially fatal, no matter how small. From the point of view of an effective four-dimensional theory, the c_i would manifest themselves as flat directions in the effective potential; string corrections can give contributions to this $V_{\rm eff}(c_i)$, whose zeros are then difficult to establish. This problem can manifest itself even in supersymmetric $\mathcal{N} = 1$ compactifications; for example, instanton corrections might generate a superpotential W whose corresponding potential does not vanish.

Metric equation of motion

The most subtle equation of motion to derive is (10.1.19a), for which we now give some details. Recalling the standard identity

$$\delta \log \det g = \delta \mathrm{Tr} \log g = \mathrm{Tr}(g^{-1}\delta g) = g^{MN}\delta g_{MN}\,, \tag{10.1.24}$$

and computing explicitly δR_{MN} from the definition (4.1.27), the variation of the usual Einstein–Hilbert (EH) action gives

$$\delta(\sqrt{-g}R) = \sqrt{-g}\left[\delta g^{MN}\left(R_{MN} - \frac{1}{2}g_{MN}R\right) + \nabla_M V^M\right], \tag{10.1.25}$$

where

$$V^M = \delta\Gamma^M{}_N{}^N - \delta\Gamma^N{}_N{}^M = -2g^{M[P}g^{Q]N}\nabla_P\delta g_{QN}\,. \tag{10.1.26}$$

The term $\nabla_M V^M$ in (10.1.25) is a total derivative and is usually discarded, except when in presence of boundaries (where it is important in the discussion of the Gibbons–Hawking–York boundary term). But in (10.1.16), the EH term appears multiplied by $e^{-2\phi}$, so $\nabla_M V^M$ becomes relevant. Integrating it by parts gives

$$\int \delta(\sqrt{-g}e^{-2\phi}R) = \int \sqrt{-g}\,\delta g^{MN}\Big(e^{-2\phi}\Big(R_{MN} - \frac{1}{2}g_{MN}R\Big) + (-\nabla_M\nabla_N + g_{MN}\nabla^2)e^{-2\phi}\Big)\,. \tag{10.1.27}$$

Now one simply needs to compute the contributions of the stress–energy tensors. In particular, the one from the dilaton gives

$$\frac{\delta(\sqrt{-g}\,e^{-2\phi}(d\phi)^2)}{\sqrt{-g}\delta g^{MN}} = e^{-2\phi}\left(\partial_M\phi\partial_N\phi - \frac{1}{2}g_{MN}(d\phi)^2\right). \tag{10.1.28}$$

This combines with the terms already involving ϕ in (10.1.27). In fact, there are even more dramatic cancellations if one combines the equation of motion of g_{MN} with that of ϕ, (10.1.19b); considering

$$\frac{\delta S}{\delta g^{MN}} + \frac{1}{4}g_{MN}\frac{\delta S}{\delta\phi} = 0 \tag{10.1.29}$$

produces (10.1.19a).

Massive IIA cannot be strongly coupled

As a simple consequence of the equations of motion in the democratic formalism, we now derive a general result about IIA with $F_0 \neq 0$, which we promised back in Section 1.4.1. The two-derivative action of type II is fixed by supersymmetry, so if ϕ varies slowly, the supergravity approximation can be valid even if $g_s = e^\phi$ is small; in the massless $F_0 = 0$ case, we identified the $g_s \gg 1$ limit as M-theory (Section 1.4.1). We now show that the same limit cannot be taken if $F_0 \neq 0$ classically: if we take $g_s \gg 1$, the curvature also gets large, so we lose control of the supergravity limit anyway [87].

Multiply (10.1.19a) by $E_0^M E_0^N$, where E is the inverse vielbein and $A = 0$ is a flat index:

$$e^{-2\phi}E_0^M E_0^N\left(R_{MN} + 2\nabla_M\nabla_N\phi - \frac{1}{2}H_M\cdot H_N\right) = \frac{1}{4}|\iota_{E_0}F|^2 = \frac{1}{4}\sum_k\frac{(F_{0A_1\dots A_k})^2}{(k-1)!}\,. \tag{10.1.30}$$

(We have ignored the source terms in (10.1.19a) because we can just pick a generic point away from them; we will not integrate over the internal space.) The polyform $\iota_{E_0}F = \sum_k \frac{1}{(k-1)!}F_{0A_1\dots A_k}e^{A_1}\wedge\dots\wedge e^{A_k}$ does not contain e^0, so each term on the right-hand side of (10.1.30) is separately positive. In particular, the $k = 0$ term is F_0^2, which by (1.3.49) is $(n_0/2\pi l_s)^2$. So the whole right-hand side is $> l_s^{-2}$, up to order one factors. On the left-hand side of (10.1.30), the supergravity approximation requires curvature to be small in string units; so $R_{MN} \ll l_s^{-2}$. The other two-derivative NSNS terms on the left-hand side of (10.1.30) should also be $\ll l_s^{-2}$ for the same reason. So we conclude

$$e^{\phi} \ll 1\,, \tag{10.1.31}$$

as promised. The Ricci tensor is of order r_{curv}^{-2}, where r_{curv} is the local curvature radius; for a *generic* solution, all the terms in the equations of motion are expected to be of the same order. This leads to the estimate

$$e^{\phi} \lesssim \frac{l_s}{r_{\text{curv}}}\,. \tag{10.1.32}$$

This argument is eluded if $F_0 = 0$ because all the $k > 0$ terms on the right-hand side of (10.1.30) can be "diluted": flux quantization (1.3.47) constrains their integrals, but the fields themselves can be made small by taking the volume to be large. Indeed, we will see several massless IIA solutions where the dilaton can be made large but several fluxes are present.

10.1.3 Supersymmetry

We finally get to the supersymmetry transformations:

$$\delta\psi^1_M = \left(D_M - \frac{1}{4}H_M\right)\epsilon^1 + \frac{e^{\phi}}{16}F\Gamma_M\epsilon^2\,, \tag{10.1.33a}$$

$$\delta\psi^2_M = \left(D_M + \frac{1}{4}H_M\right)\epsilon^2 \pm \frac{e^{\phi}}{16}\lambda(F)\Gamma_M\epsilon^1\,; \quad {}^{\text{IIA}}_{\text{IIB}} \tag{10.1.33b}$$

$$\Gamma^M\delta\psi^1_M - \delta\lambda^1 = \left(D - \frac{1}{4}H - d\phi\right)\epsilon^1\,, \tag{10.1.33c}$$

$$\Gamma^M\delta\psi^2_M - \delta\lambda^2 = \left(D + \frac{1}{4}H - d\phi\right)\epsilon^2\,. \tag{10.1.33d}$$

Let us again review the notation. The upper sign is for IIA, the lower for IIB. (The IIB case was already discussed in (9.5.1) in the more conventional nondemocratic formalism.) D_M is the spinorial covariant derivative, (4.3.20); $D \equiv \Gamma^M D_M$ is the Dirac operator (4.3.21). The NSNS terms are as in (9.3.1); in particular, recall $H_M = \iota_M H$. The λ operator is the degree-dependent sign reversal defined in (3.2.28). In all the forms appearing, namely H_M, F, $\lambda(F)$ and $d\phi$, for readability we have left implicit the Clifford map $\alpha \mapsto \not{\alpha} \equiv \alpha_/$ (Section 3.1). So, for example, $d\phi = \partial_M\phi\Gamma^M$.

The combination in (10.1.33c) and (10.1.33d) of gravitino and dilatino transformation is convenient because the RR terms cancel out. The original dilatino transformations are

$$\delta\lambda^1 = \left(-\frac{1}{2}H + d\phi\right)\epsilon^1 + \frac{e^{\phi}}{16}\Gamma^M F\Gamma_M\epsilon^2\,, \tag{10.1.34a}$$

$$\delta\lambda^2 = \left(\frac{1}{2}H + d\phi\right)\epsilon^2 \pm \frac{e^{\phi}}{16}\Gamma^M\lambda(F)\Gamma_M\epsilon^1\,. \quad {}^{\text{IIA}}_{\text{IIB}} \tag{10.1.34b}$$

In (10.1.33), terms involving the fermionic fields ψ^a_M, λ^a have been omitted. Moreover, we have not given the supersymmetry transformations of bosons. This is because in this book we are primarily interested in vacuum solutions, for which fermions are set to zero (Section 8.2.2).

10.1.4 Equations of motion from supersymmetry

We will now see that imposing $\delta\psi^a_M = 0$, $\delta\lambda = 0$ and the Bianchi equations $\mathrm{d}H = 0$, $\mathrm{d}_H F = 0$ implies almost all the equations of motion. This can be shown in more traditional, nondemocratic formulations (see [428, sec. 2] and [429, app. B] for the metric and dilaton equations in IIA and IIB respectively, and [430, 431] for H), or in the democratic formalism [432, sec. 11, app. E].

For the metric and B, we first introduce the operators

$$\begin{aligned} \mathcal{D}_M &= D_M \otimes 1_2 - \frac{1}{4} H_M \otimes \sigma_3 + \frac{\mathrm{e}^\phi}{16} \begin{pmatrix} 0 & F\Gamma_M \\ \pm\lambda(F)\Gamma_M & 0 \end{pmatrix}, \\ \mathcal{O} &= \frac{1}{2} H \otimes \sigma_3 - \mathrm{d}\phi \otimes 1_2 - \frac{\mathrm{e}^\phi}{16} \begin{pmatrix} 0 & \Gamma^M F\Gamma_M \\ \pm\Gamma^M \lambda(F)\Gamma_M & 0 \end{pmatrix}. \end{aligned} \tag{10.1.35}$$

They act on a doublet of spinors; the supersymmetry conditions from (10.1.33a), (10.1.33b) and (10.1.34) read $\mathcal{D}_M \binom{\epsilon^1}{\epsilon^2} = \mathcal{O} \binom{\epsilon^1}{\epsilon^2} = 0$. The key step is now to show

$$\begin{aligned} &\Gamma^M [\mathcal{D}_M, \mathcal{D}_N] - \left[D_N \otimes 1 - \frac{1}{4} H_N \otimes \sigma_3, \mathcal{O} \right] + \frac{\mathrm{e}^\phi}{16} \mathcal{O} \begin{pmatrix} 0 & F\Gamma_N \\ \pm\lambda(F)\Gamma_N & 0 \end{pmatrix} = \\ &\quad \frac{1}{2} E(g)_{NP} \Gamma^P \otimes 1 + \left(\frac{1}{4} E(B)_{NP} \Gamma^P + \iota_N \mathrm{d}H \right) \otimes \sigma_3 + \frac{\mathrm{e}^\phi}{8} \begin{pmatrix} 0 & \mathrm{d}_H F\Gamma_N \\ \pm\lambda(\mathrm{d}_H F)\Gamma_N & 0 \end{pmatrix}, \end{aligned} \tag{10.1.36}$$

where $E(g)_{MN} = R_{MN} + \ldots = 0$, $E(B)_{MN} = \nabla^P H_{PMN} + \ldots = 0$ are the equations of motion (10.1.19a) and (10.1.7b) respectively. Most of the steps to show this are applications of techniques we know from Chapters 2–5. The Ricci tensor is obtained by adapting (5.7.5), which is in fact the inspiration for the first term in (10.1.36). Several commutators with D_M are evaluated with the help of (4.3.46), such as $[D_M, F] = \slashed{\nabla_M F}$ and $[D_M, H_N] = \frac{1}{2} \nabla_M H_N{}^{PQ} \Gamma_{PQ}$. One also uses several gamma matrix commutators, evaluated using the techniques of Section 3.2.2 or even Section 2.1.4. In particular, the former is most appropriate for the off-diagonal terms $H \wedge F$; we will see a similar computation in (10.2.11). By far the most challenging terms are the RR contributions to (10.1.7b) and (10.1.19a), which are reproduced with the help of the identities

$$\begin{aligned} \Gamma^M F\Gamma_N \lambda(F)\Gamma_M &= \pm 16 [F_N \cdot F_P + [*(F \wedge \lambda F)_8]_{NP}] \Gamma_P (1 + \Gamma), \\ \Gamma^M \lambda(F)\Gamma_N F\Gamma_M &= \pm 16 [F_N \cdot F_P - [*(F \wedge \lambda F)_8]_{NP}] \Gamma_P (1 \mp \Gamma). \end{aligned} \tag{10.1.37}$$

To see how these come about, consider $R \equiv F v \lambda(F) = F v F^\dagger$, $\tilde{R} \equiv \lambda(F) v F = F^\dagger v F$, where v is a real one-form. These bispinors are odd and Hermitian by construction. So their Fierz decomposition (3.3.26) contains only the terms with $k = 1, 5$ and 9. Moreover, $\Gamma F = F$ by (3.2.27) and (10.1.3); so $\Gamma R = R$, $\Gamma \tilde{R} = \pm \tilde{R}$, and the $k = 9$ and $k = 1$ terms are related by duality. Now by (3.2.23) we have $\Gamma^M R \Gamma_M = 8(1 + \Gamma) R_1$, $\Gamma^M \tilde{R} \Gamma_M = 8(1 \pm \Gamma) \tilde{R}_1$; the whole computation is similar to (3.4.59). Finally we compute the one-form part: (3.3.26) gives $(R_1)_M = \mathrm{Tr}(R_1 \Gamma_M)/32$ and similar for $(\tilde{R}_1)_M$. Taking $v = \mathrm{d}x_N$, using (3.2.67), (3.2.68), and $(\mathrm{d}x^M \wedge \mathrm{d}x^N \wedge F, F) = -(\iota^M F) \cdot (\mathrm{d}x^N F)$ results in (10.1.37).

A similar identity concerns the operator $\mathcal{D} = D \otimes 1 - \frac{1}{4}H \otimes \sigma_3 - \mathrm{d}\phi \otimes 1$ related to the modified dilatino (10.1.33c) and (10.1.33d):

$$\left(\mathcal{D} + \frac{e^\phi}{16}\begin{pmatrix} 0 & F\Gamma_N \\ \pm\lambda(F)\Gamma_N & 0 \end{pmatrix}\right)\mathcal{D} - \left((D_M - 2\partial_M\phi)\otimes 1 - \frac{1}{4}H_M \otimes \sigma_3\right)\mathcal{D}^M$$
$$= -\frac{1}{8}E(\phi) + \frac{1}{4}\mathrm{d}H \otimes \sigma_3 + \frac{1}{8}\begin{pmatrix} 0 & \mathrm{d}_H F \\ \pm\lambda(\mathrm{d}_H F) & 0 \end{pmatrix}, \tag{10.1.38}$$

where $E(\phi) = 0$ is (10.1.19b). Here it is useful to recall the Bismut identity (4.3.75). (The role of this identity for restricting string corrections was emphasized in [433].)

We now let (10.1.37) and (10.1.38) act on the supercharges $\binom{\epsilon^1}{\epsilon^2}$. The left-hand side is easily seen to be zero; on the right-hand side, we impose the Bianchi identities $\mathrm{d}H = \mathrm{d}_H F = 0$ and we obtain $(2E(g) - E(B))_{NP}\Gamma^P\epsilon^1 = (2E(g) + E(B))_{NP}\Gamma^P\epsilon^2 = 0$, and $E(\phi) = 0$. We cannot quite conclude that all the components of the equations of motion hold, because the ϵ^a are Majorana–Weyl (MW) spinors, and each has a real annihilator V_a, (2.4.47), and (3.4.64); in other words,

$$(2E(g) - E(B))_{NP}V_1^P\,, \qquad (2E(g) + E(B))_{NP}V_2^P \tag{10.1.39}$$

are the only equations of motion that are *not* implied by supersymmetry.

For a four-dimensional vacuum solution, the equations of motion in the MS_4 spacetime directions are mixed by its symmetries. In particular, the light-cone vectors V_a^P are transformed to other lightlike vectors. So the missing equations of motion (10.1.39) are equal to others that are implied by supersymmetry.

Exercise 10.1.1 Compute $\overline{\delta\psi_M^a}$ and $\overline{\partial\lambda^a}$ from (10.1.33), both in IIA and IIB. Pay attention to the role of Γ_0 in the definition of $\overline{\epsilon}$, and to the parity of F.

Exercise 10.1.2 Check that (10.1.33a), (10.1.33b), and (10.1.34) reduce to (9.5.1) and (9.5.2). (Recall (3.2.23).)

Exercise 10.1.3 Check (10.1.19) for the D-brane solutions (1.3.59). For (10.1.19b), it might be useful to recall (4.1.114).

Exercise 10.1.4 Check if (10.1.31) holds on the D8-brane solution (1.3.59) and (1.4.57).

Exercise 10.1.5 The strong energy condition is

$$\left(T_{MN} - \frac{1}{D-2}g_{MN}T_P{}^P\right)n^M n^N \geq 0 \tag{10.1.40}$$

for any nonspacelike n^M; in (0.5.3), we saw it for $n = \partial_0$. Show that (10.1.40) holds for (10.1.21) for $k \geq 1$, but not for $k = 0$ (the Romans mass).

10.2 Ten-dimensional bilinear equations

We saw in Chapter 3 that spinors are deeply related to differential forms. In various instances, we saw how spinorial equations can be rewritten as form equations, for example in Sections 5.5 and 9.3.

We now show some notable form equations that follow from imposing

$$\delta\psi_M^a = 0 = \delta\lambda^a \tag{10.2.1}$$

in (10.1.33).

10.2.1 Bilinears of two spinors

For a single Majorana–Weyl spinor ϵ in $d = 10$ Lorentzian dimensions, all possible bilinears are collected in the bispinor (3.4.57). In the flat-space discussion that follows (3.4.57), we also found some algebraic relations among these forms.[3] We saw in (3.4.64) that ϵ is not pure: its annihilator is one-dimensional, being spanned by V itself.

Applying (3.4.57) to both ϵ^a:

$$\begin{aligned} \epsilon^1 \otimes \overline{\epsilon^1} &= \frac{1}{32}(V_1 + \Omega_1 + *V_1)_{/} \,, \\ \epsilon^2 \otimes \overline{\epsilon^2} &= \frac{1}{32}(V_2 + \Omega_2 \mp *V_2)_{/} \,. \quad {\scriptstyle \mathrm{IIA} \atop \mathrm{IIB}} \end{aligned} \tag{10.2.2}$$

The $V_a = \overline{\epsilon^a}\Gamma_M \epsilon^a dx^M$ are one-forms, and the Ω_a are five-forms. The sign of the last term is correlated with the chirality of ϵ^2.

Both of the V_a are lightlike, by (3.4.62). Moreover, they are "future directed": their zero components

$$(V_a)^0 = \overline{\epsilon^a}\Gamma^0\epsilon^a = \epsilon^{a\,\dagger}\Gamma_0\Gamma^0\epsilon^a = \epsilon^{a\,\dagger}\epsilon^a \geq 0 \,. \tag{10.2.3}$$

It is convenient to work with their sum and difference:

$$K_M \equiv \frac{1}{2}(V^1 + V^2)_M \,, \qquad \tilde{K}_M \equiv \frac{1}{2}(V^1 - V^2)_M \,, \tag{10.2.4}$$

which by construction satisfy

$$K^2 \geq 0 \,, \qquad K \cdot \tilde{K} = 0 \,. \tag{10.2.5}$$

Recalling the chiralities of the ϵ^a from Table 8.1, the bilinears $\overline{\epsilon^2}\Gamma_{M_1 \ldots M_k}\epsilon^1$ survive if k is even in IIA, and if k is odd in IIB. Collecting them in the bispinor (Section 3.3.3)

$$\Psi = \epsilon^1 \otimes \overline{\epsilon^2} \,, \tag{10.2.6}$$

we have that Ψ is an even form in IIA, and is odd in IIB.

Since the ϵ^a are not pure, the logic (3.4.4) that we applied repeatedly in six dimensions does not apply, and Ψ is not a pure form. Its annihilator is two dimensional, and is spanned by the left action of V^1 and the right action of V^2:

$$V^1 \cdot \Psi = \Psi \cdot V^2 = 0 \,. \tag{10.2.7}$$

The bilinears $\epsilon^2 \otimes \overline{\epsilon^1}$ are not independent, by (3.4.56):

$$\epsilon^2 \otimes \overline{\epsilon^1} = \mp \lambda \Psi \qquad {\scriptstyle \mathrm{IIA} \atop \mathrm{IIB}} \,. \tag{10.2.8}$$

The forms we have found are in loose correspondence with the extended objects:

- Equation (10.2.2) contains one-forms V_a and five-forms Ω_a, matching the NSNS objects of the theory, F1 strings and NS5-branes.
- The polyform Ψ in (10.2.6) is even in IIA and odd in IIB, corresponding to Dp-branes.

[3] In (3.4.65), we also computed the stabilizer of ϵ. In curved space, this becomes the statement that ϵ defines a $\mathrm{Spin}(7) \ltimes \mathbb{R}^8$-structure, but this will not be very important for us.

This is because the extended BPS objects of the theory are related to the central charges in the superalgebra; in computing the supercharge anticommutator $\{Q(\epsilon^a), Q(\epsilon^a)\}$, the bilinears of the ϵ^a will appear. This will play a role later on.

Having identified all possible bilinears, we now proceed to compute some differential equations for them.

10.2.2 Bispinor equation

We begin by computing the exterior differential of the polyform Ψ. Here we will be careful to include all the slash symbols we have often omitted for readability.

We begin with

$$
\begin{aligned}
2(\mathrm{d}\Psi)_{/} &\overset{(4.1.73)}{=} 2(\mathrm{d}x^M \wedge \nabla_M \Psi)_{/} \overset{(3.2.8a)}{=} \Gamma^m \not{\nabla}_m \not{\Psi} \pm \not{\nabla}_M \not{\Psi} \Gamma^M \\
&\overset{(4.3.46),(10.2.6)}{=} \Gamma^M (D_M(\epsilon^1 \otimes \overline{\epsilon^2})) \pm (D_M(\epsilon^1 \otimes \overline{\epsilon^2}))\Gamma^M \\
&\overset{(2.3.13)}{=} (D\epsilon^1) \otimes \overline{\epsilon^2} + \Gamma^M \epsilon^1 \otimes \overline{D_M \epsilon^2} \pm D_M \epsilon^1 \otimes \overline{\epsilon^2}\Gamma^M \mp \epsilon^1 \otimes \overline{D\epsilon^2} \\
&\overset{(10.1.33)=0}{=} \left(\frac{1}{4}H + \mathrm{d}\phi\right)_{/} \epsilon^1 \otimes \overline{\epsilon^2} + \Gamma^M \epsilon^1 \otimes \left(\frac{1}{4}\overline{\epsilon^2} H_M + \frac{\mathrm{e}^\phi}{16}\overline{\epsilon^1}\Gamma_M F\right) \\
&\pm \left(\frac{1}{4}H_M \epsilon^1 - \frac{\mathrm{e}^\phi}{16} F \Gamma_M \epsilon^2\right) \otimes \overline{\epsilon^2}\,\Gamma^M \pm \epsilon^1 \otimes \overline{\epsilon^2} \left(\frac{1}{4}H + \mathrm{d}\phi\right)_{/} . \quad {}^{\text{IIA}}_{\text{IIB}}
\end{aligned}
\tag{10.2.9}
$$

The result might look messy, but there is hidden beauty. The dilaton terms reconstruct a wedge:

$$
\mathrm{d}\phi \cdot \Psi \pm \Psi \cdot \mathrm{d}\phi \overset{(3.2.8)}{=} 2\mathrm{d}\phi \wedge \Psi . \tag{10.2.10}
$$

Next, all the H terms in (10.2.9) can be collected as an operator acting on $\epsilon^1 \otimes \overline{\epsilon^2}$. We learned in Section 3.2.1 how to evaluate such expressions quickly; see (3.2.15), (3.2.18), and (3.2.55). With a similar notation:

$$
\begin{aligned}
&H \not{\alpha}_\pm \pm \not{\alpha}_\pm H + \Gamma^M \not{\alpha}_\pm H_M \pm H_M \not{\alpha}_\pm \Gamma^M \\
&\quad = H_{MNP} \left[\frac{1}{6}\left(\overset{\rightarrow}{\gamma}{}^{MNP} \pm \overset{\leftarrow}{\gamma}{}^{MNP}\right) + \frac{1}{2}\overset{\rightarrow}{\gamma}{}^{M}\overset{\leftarrow}{\gamma}{}^{NP} \pm \frac{1}{2}\overset{\rightarrow}{\gamma}{}^{NP}\overset{\leftarrow}{\gamma}{}^{M}\right] \not{\alpha}_\pm \\
&\quad = H_{MNP} \Big[\Big(\frac{1}{6}(\mathrm{d}x \wedge + \iota)^3 + \frac{1}{6}(\mathrm{d}x \wedge - \iota)^3 \\
&\qquad + \frac{1}{2}(\mathrm{d}x \wedge + \iota)(\mathrm{d}x \wedge - \iota)^2 + \frac{1}{2}(\mathrm{d}x \wedge - \iota)(\mathrm{d}x \wedge + \iota)^2\Big)^{MNP} \alpha_\pm\Big]_{/} \\
&\quad = H_{MNP} \Big[\Big(\frac{1}{3}\mathrm{d}x^M \wedge \mathrm{d}x^N \wedge \mathrm{d}x^P \wedge + \mathrm{d}x^M \wedge \iota^N \iota^P \\
&\qquad + \mathrm{d}x^M \wedge (\mathrm{d}x^N \wedge \mathrm{d}x^P - \iota^N \iota^P)\Big)\alpha_\pm\Big]_{/} \\
&\quad = \frac{4}{3} H_{MNP} \left[\mathrm{d}x^M \wedge \mathrm{d}x^N \wedge \mathrm{d}x^P \wedge \alpha_\pm\right]_{/} \\
&\quad = 8(H \wedge \alpha_\pm)_{/} .
\end{aligned}
\tag{10.2.11}
$$

So all the terms in (10.2.9) reconstruct $H \wedge \Psi$.

Finally we look at the RR terms. Generalizing (3.4.59):

$$
\Gamma^M \epsilon_\pm \otimes \overline{\epsilon_\pm} \Gamma_M = -\frac{1}{4}(1 \mp \gamma)V \tag{10.2.12}
$$

The self-duality constraint (10.1.3) in bispinor language reads

$$F \overset{(10.1.3),(3.2.27)}{=} \Gamma F = \pm F\Gamma\,. \tag{10.2.13}$$

(We have already fixed $c = 1$ when we took (8.2.7).) So the RR terms in (10.2.9) give, omitting a $\mathrm{e}^{\phi}/16$ prefactor:

$$\begin{aligned}
&\Gamma^M \epsilon^1 \otimes \overline{\epsilon^1}\,\Gamma_M F \mp F\Gamma^M \epsilon^1 \otimes \overline{\epsilon^1}\,\Gamma_M = -\frac{1}{4}\left(V_1(1+\Gamma)F \mp F(1\pm\Gamma)V_2\right)\\
&= -\frac{1}{2}\left(V_1 F \mp F V_2\right) = -\frac{1}{2}\left(K\cdot F \mp F\cdot K + \tilde{K}\cdot F \pm F\cdot \tilde{K}\right)\\
&= -(\iota_K F + \tilde{K}\wedge F)_{/}\,.
\end{aligned} \tag{10.2.14}$$

Putting together (10.2.9)–(10.2.14), the slash symbol $_/$ is now overall; so we obtain an equation that is purely in terms of forms:

$$(\mathrm{d} - H\wedge + \mathrm{d}\phi\wedge)\Psi = -\frac{\mathrm{e}^{\phi}}{32}(\iota_K + \tilde{K}\wedge)F\,, \tag{10.2.15}$$

or in other words [434]

$$\mathrm{d}_H(\mathrm{e}^{-\phi}\Psi) = -\frac{1}{32}(\iota_K + \tilde{K}\wedge)F\,. \tag{10.2.16}$$

As anticipated at the end of Section 10.2.1, this Ψ will be related to calibration forms for D-branes. It will also be our main tool to reformulate supersymmetry in Section 10.3.

10.2.3 Canonical isometry

We now show that the vector field K is a symmetry of a supersymmetric solution.

Killing vector

First we show that K is a Killing vector [193]. We begin by computing the covariant derivative

$$\begin{aligned}
\nabla_M V^1_N &= \nabla_M(\overline{\epsilon^1}\Gamma_N\epsilon^1) \overset{(4.3.35)}{=} \overline{D_M\epsilon^1}\Gamma_N\epsilon^1 + \overline{\epsilon^1}\Gamma_N D_M\epsilon^1\\
&\overset{(10.1.33a)=0}{=} \frac{1}{4}\overline{\epsilon^1}[\Gamma_N, H_M]\epsilon^1 + \frac{\mathrm{e}^{\phi}}{16}\left(\pm\overline{\epsilon^2}\Gamma_M\lambda(F)\Gamma_N\epsilon^1 - \overline{\epsilon^1}\Gamma_N F\Gamma_M\epsilon^2\right)\\
&\overset{(2.1.13),(3.4.56)}{=} \frac{1}{2}V_1^P H_{MNP} - \frac{\mathrm{e}^{\phi}}{8}\overline{\epsilon^1}\Gamma_N F\Gamma_M\epsilon^2\,.
\end{aligned} \tag{10.2.17a}$$

The result for $\nabla_M V^2_N$ is similar:

$$\nabla_M V^2_N = -\frac{1}{2}V_2^P H_{MNP} + \frac{\mathrm{e}^{\phi}}{8}\overline{\epsilon^1}\Gamma_M F\Gamma_N\epsilon^2\,. \tag{10.2.17b}$$

Taking the sum of (10.2.17) and symmetrizing indices, the RR terms cancel out:

$$\nabla_{(M}K_{N)} = 0\,. \tag{10.2.18}$$

So any supersymmetric solution, even with only one supercharge, has an isometry. This generalizes the statement after (7.3.11) for Killing spinors.[4] Another way to make the RR terms drop out is to take the difference of (10.2.17), and to antisymmetrize the indices: this leads to

$$\mathrm{d}\tilde{K} = \iota_K H\,. \tag{10.2.19}$$

So it is natural to view K as a vector field, and $\tilde{K}$ as a one-form.

Invariance of other fields

Acting on (10.2.19) with d:

$$0 = \mathrm{d}\iota_K H \overset{(4.1.79)}{=} L_K H\,. \tag{10.2.20}$$

The invariance of the dilaton under K follows most easily from the original dilatino transformations (10.1.34). If we multiply (10.1.34a) from the left by $\overline{\epsilon^1}$ and (10.1.34b) by $\overline{\epsilon^2}$, and sum the resulting two equations, both the H and RR terms cancel because of (3.4.56). This results in

$$K^M \partial_M \phi = L_K \phi = 0\,. \tag{10.2.21}$$

For the RR fields, first we evaluate

$$\begin{aligned}\{\mathrm{d}_H, \tilde{K}\wedge + \iota_K\} &= \{\mathrm{d}, \tilde{K}\wedge\} - \{H\wedge, \iota_K\} + \{\mathrm{d}, \iota_K\} \\ &\overset{(4.1.77),\,(3.2.3a),\,(4.1.79)}{=} (\mathrm{d}\tilde{K} - \iota_K H)\wedge + L_K \overset{(10.2.19)}{=} L_K\,.\end{aligned} \tag{10.2.22}$$

Now we act with d_H on (10.2.16). The left-hand side gives zero, because $\mathrm{d}_H^2 = 0$:

$$\begin{aligned}0 = \mathrm{d}_H((\tilde{K}\wedge + \iota_K)F) &= \{\mathrm{d}_H, (\tilde{K}\wedge + \iota_K)\}F - (\tilde{K}\wedge + \iota_K)\mathrm{d}_H F \\ &\overset{(10.2.22)}{=} L_K F - (\tilde{K}\wedge + \iota_K)\mathrm{d}_H F\,.\end{aligned} \tag{10.2.23}$$

If we recall the Bianchi identities (10.1.4), away from sources we conclude

$$L_K F = 0\,. \tag{10.2.24}$$

Conversely, if in a situation we know already for some reason that $L_K F = 0$, from (10.2.23), we can conclude that some components of the Bianchi identities (10.1.4) are automatically satisfied.

Invariance of supercharge

We now show that K leaves even the supercharge invariant. First we observe

$$(\tilde{K}\wedge + \iota_K)^2 = \frac{1}{2}\{\tilde{K}\wedge + \iota_K, \tilde{K}\wedge + \iota_K\} \overset{(3.2.3)}{=} \iota_K \tilde{K} \overset{(10.2.4)}{=} \frac{1}{4}(V_1^2 - V_2^2) = 0\,. \tag{10.2.25}$$

Now it is natural to act with $(\tilde{K}\wedge + \iota_K)$ on (10.2.16):

$$\begin{aligned}0 &\overset{(10.2.25)}{=} (\tilde{K}\wedge + \iota_K)\mathrm{d}_H(e^{-\phi}\Psi) \overset{(10.2.22)}{=} L_K(e^{-\phi}\Psi) - \mathrm{d}_H(e^{-\phi}(\tilde{K}\wedge + \iota_K)\Psi) \\ &\overset{(10.2.7)}{=} L_K(e^{-\phi}\Psi).\end{aligned} \tag{10.2.26}$$

[4] For this reason, some people also call the supersymmetry equations (10.1.33) "Killing spinor equations," but we will not do that in this book.

So we conclude also

$$L_K\Psi = 0. \tag{10.2.27}$$

From (10.2.19), we can also obtain $L_K V_1 = L_K V_2 = 0$. By the definition (10.2.6) of Ψ and our Leibniz identities for bispinors (4.3.48) and (4.3.50), together all these imply $L_K\epsilon^a = 0$.

10.2.4 Five-form

We now look at the five-forms Ω_a in (10.2.2). So we also consider the bispinors $\Psi^{ab} \equiv \epsilon^a \otimes \overline{\epsilon^b}$. Similar to (10.2.9), we compute

$$\begin{aligned} 2(\mathrm{d}\Psi^{11})_{/} &= \left[\Gamma^M, D_M(\epsilon_1 \otimes \overline{\epsilon_1})\right] \\ &= \left[\Gamma^M, \left(\frac{1}{4}H_M\epsilon^1 - \frac{e^\phi}{16}F\Gamma_M\epsilon^2\right)\otimes\overline{\epsilon^1} + \epsilon^1 \otimes \left(-\frac{1}{4}\overline{\epsilon^1}H_M \mp \frac{e^\phi}{16}\overline{\epsilon^2}\Gamma_M\lambda(F)\right)\right] \\ &\overset{(3.2.9a),(3.2.19)}{=} H_{MNP}\mathrm{d}x^M \wedge \mathrm{d}x^N \wedge \iota^P\epsilon^1 \otimes \overline{\epsilon^1} \\ &\quad - \frac{e^\phi}{16}\left[\Gamma^M, F\Gamma_M\epsilon^2 \otimes \overline{\epsilon^1} \mp \epsilon^1 \otimes \overline{\epsilon^2}\Gamma_M\lambda(F)\right]. \end{aligned} \tag{10.2.28}$$

Now a computation similar to (10.2.11) gives

$$H_{MNP}\left[\left(\mathrm{d}x^M \wedge \mathrm{d}x^N \wedge \iota^P + \frac{1}{3}\iota^M\iota^N\iota^P\right)\alpha_\pm\right]_{/} = \not{H}\not{\alpha}_\pm \mp \not{\alpha}_\pm\not{H}. \tag{10.2.29}$$

Using also the original dilatino equations (10.1.34), repeating the computation for Ψ^{22}, and recalling (3.2.64):

$$\begin{aligned} [(\mathrm{d} + \iota_H - 2\mathrm{d}\phi\wedge)\Psi^{11}]_{/} &= \mp\frac{e^\phi}{16}\{\Gamma^M, F\Gamma_M\lambda(\Psi) - \Psi\Gamma_M\lambda(F)\}, \\ [(\mathrm{d} - \iota_H - 2\mathrm{d}\phi\wedge)\Psi^{22}]_{/} &= \pm\frac{e^\phi}{16}\{\Gamma^M, \lambda(F)\Gamma_M\Psi - \lambda(\Psi)\Gamma_M F\}. \end{aligned} \tag{10.2.30}$$

We now focus on the particular five-form:

$$\Omega \equiv \frac{1}{2}(\Omega_1 \pm \Omega_2). \quad {}^{\text{IIA}}_{\text{IIB}} \tag{10.2.31}$$

So we take the six-form parts of (10.2.30); we sum them in IIA, while we subtract them in IIB. The H contraction can be rewritten using (3.2.63). For the RR, we are out of tricks, and as far as I know a lengthier degree-by-degree algebraic manipulation is needed; but the final result is elegant [431]:

$$e^{2\phi}\mathrm{d}(e^{-2\phi}\Omega) = -\iota_K * H + 32e^\phi(\Psi, F)_6, \tag{10.2.32}$$

where as usual $(\alpha, \beta)_6 = (\alpha \wedge \lambda(\beta))_6$, recalling (3.2.42). Related to the comment at the end of Section 10.2.1, this equation can be shown to be related to NS5-brane calibrations [431].

Multiplying (10.2.32) by $e^{-2\phi}$ and acting with d:

$$\frac{1}{32}\mathrm{d}(\iota_K e^{-2\phi} * H) = \mathrm{d}(e^{-\phi}\Psi, F)_6 \overset{(3.2.58)}{=} (\mathrm{d}(e^{-\phi}\Psi), F)_7 + (e^{-\phi}\Psi, \mathrm{d}F)_7$$
$$\overset{(10.1.4),(3.2.61)}{=} (\mathrm{d}_H(e^{-\phi}\Psi), F)_7 \overset{(10.2.16)}{=} -\frac{1}{32}((\iota_K + \tilde{K}\wedge)F, F)_7$$
$$\overset{(3.2.61),(3.2.66)}{=} -\frac{1}{64}\iota_K(F, F)_8 \,. \tag{10.2.33}$$

Since $L_K = \{\mathrm{d}, \iota_K\}$ by (4.1.79), and K is a symmetry of the solution as shown in Section 10.2.3, it follows that

$$\iota_K\left(\mathrm{d}(e^{-2\phi} * H) - \frac{1}{2}(F, F)_8\right) = 0 \,. \tag{10.2.34}$$

We recognize in the parenthesis the equation of motion (10.1.7b) for the NSNS flux. This shows again (after Section 10.1.4) that the H equation of motion is almost totally implied by supersymmetry, and completely for vacua [431]. (A similar, earlier argument for vacua is in [430].)

10.2.5 Equivalence to supersymmetry*

We have derived several interesting equations from supersymmetry: (10.2.16), (10.2.18), (10.2.19), and (10.2.32). We now ask if they also imply the supersymmetry equations $\delta\psi^a_M = 0 = \delta\lambda^a$, with (10.1.33). Unfortunately, this is not the case, as we will now discuss.

To compare a spinorial equation to one involving form exterior differentials, we can use the method of *intrinsic torsion*, which parameterizes deviation from special holonomy. We used it in Section 7.1 to show that $\mathrm{d}J = 0 = \mathrm{d}\Omega$ are equivalent to $D_m\eta_+ = 0$, and don't just follow from it. So we need an appropriate definition of intrinsic torsion. We have not developed much the theory of G-structures in $d = 10$ Lorentzian dimensions. The stabilizer of a single chiral spinor is (3.4.65); with two spinors it is more complicated, as we saw in (3.4.67) and in Exercise 3.4.9. The proper way to avoid dealing with all these cases would be to work with the generalized tangent bundle $T \oplus T^*$, a strategy we will develop in Section 10.5 for vacua. Here we can take a little shortcut, and present a definition of intrinsic torsion that is just enough for our aims. In (5.5.2) for SU(3)-structures, the idea was to expand $\nabla\eta_+$ in the basis associated to η_+. So we do the same for each ϵ, using the basis given in (3.4.98) (with $\eta_{\mathrm{MW}} \to \epsilon$):

$$\left(D_M - \frac{1}{4}H_M\right)\epsilon^1 \equiv Q^1_{MNP}\Gamma^{NP}\epsilon^1 \,, \qquad \left(D_M + \frac{1}{4}H_M\right)\epsilon^2 \equiv Q^2_{MNP}\Gamma^{NP}\epsilon^2 \,. \tag{10.2.35}$$

To expand (10.1.33c) and (10.1.33d), we also need to expand $D\epsilon$, in a basis of opposite chirality to that of ϵ, such as (3.4.99). An additional complication is that there should be a Γ_{-_1} for ϵ^1, and a different Γ_{-_2} for ϵ^2. The idea is now to write the supersymmetry equations using these Q, T coefficients, and to also expand the form equations on the basis obtained by taking tensor products of two copies of (3.4.98) and (3.4.99) (similar to what we did in (3.4.95)). The computation is lengthy and technical; the result is that unfortunately not all the supersymmetry equations are reproduced [434]:

- Equation (10.2.16) has no information on the components $Q^1_{+_1NP}$, $Q^2_{+_2NP}$.
- Equations (10.2.18) and (10.2.19) together give $Q^1_{MN_{-1}} = -Q^2_{NM_{-2}}$, so they contain some of the components missing in (10.2.16) but not all.
- The content of (10.2.32) depends on whether K is lightlike or timelike.

It is natural to wonder whether we can complete these equations to a system that is equivalent to supersymmetry, written in terms of natural operations on forms. The definition of "natural" sounds subjective, but one would tend to give preference to the exterior differential d over the covariant derivative ∇_m. Here are some proposals:

- If we assume that K is *time-like*, $K^2 < 0$, the system consisting of (10.2.16), (10.2.32), (10.2.21), and the two additional equations

$$e^{2\phi}d(e^{-2\phi}\tilde{\Omega}) = *(\tilde{K} \wedge H) \mp 16e^{\phi}(\iota_M \Psi, \iota^M F)_6 , \tag{10.2.36a}$$

$$d * \tilde{K} = -4e^{\phi}(\Psi, (10 - \deg)F) , \quad {}^{\text{IIA}}_{\text{IIB}} \tag{10.2.36b}$$

  is equivalent to preserved supersymmetry [431]. Here $\tilde{\Omega} = \frac{1}{2}(\Omega_1 \mp \Omega_2)$ (compare (10.2.31)), and as usual $\deg\alpha_k = k\alpha_k$. Equation (10.2.36a) seems likely to have an interpretation in terms of KK5-monopoles (Section 7.2.2), which would complement the interpretations of (10.2.16) and (10.2.32) as calibrations for D-branes and NS5-branes. However, the assumption $K^2 < 0$ excludes many vacuum solutions, such as $\mathcal{N} = 1$ in $d = 4$.
- Without any assumptions on K, the system consisting of (10.2.16), (10.2.18), (10.2.19), and the two *pairing equations*

$$\begin{aligned} &\left(\psi_- \otimes \overline{\epsilon^2} \cdot \hat{V}_2, \pm d_H(e^{-\phi}\Psi \cdot \hat{V}_2) + \sigma_1\Psi - F\right) = 0 , \\ &\left(\hat{V}_1 \cdot \epsilon^1 \otimes \overline{\psi_\pm}, d_H(e^{-\phi}\hat{V}_1 \cdot \Psi) - \sigma_2\Psi - F\right) = 0 , \end{aligned} \qquad \forall \psi_\pm \tag{10.2.37}$$

  is equivalent to supersymmetry [434]. Here $\hat{V}_a$ are two light-like vectors such that $2\hat{V}_a \cdot V_a = 1$, their left- and right-actions are $\hat{V}_a \cdot = \hat{V}_a \wedge + \iota_{\hat{V}_a}$, $\cdot \hat{V}_a = (\hat{V}_a \wedge - \iota_{\hat{V}_a})$ (by (3.1.21), (3.1.22)), and $\sigma_a \equiv \frac{1}{2}e^{\phi}d^{\dagger}(e^{-2\phi}\hat{V}_a)$. The pairing has the usual definition (3.2.40). The advantage of this system is that it is valid both if K is light-like and time-like; in particular, it is applicable to vacuum solutions. The disadvantage is that it introduces the somewhat spurious vectors $\hat{V}_a$, and that it involves pairing with the bispinors $\psi_- \otimes \overline{\epsilon^2}$, $\epsilon^1 \otimes \overline{\psi_\pm}$, which have no obvious characterization purely in terms of forms. In this sense, this system is not completely natural.
- If we allow the use of covariant derivatives, we can complete the system (10.2.16), (10.2.18), and (10.2.19) by using equations on $\nabla_M \Psi$ [435, (2.31)].

I suspect there is a better formulation still awaiting discovery.

Exercise 10.2.1 Compute the stabilizer of two MW spinors ϵ^a in $d = 10$ in the generalized Lorentz group $O(10, 10)$ [434].

Exercise 10.2.2 Show (10.2.36a), using steps as in Section 10.2.4.

10.3 The pure spinor formalism

We now turn to the problem of vacuum solutions. In Section 10.3.1, we extend the discussion of Sections 8.2.2 and 8.2.3 to fluxes, and we apply it to $\mathcal{N} = 1$ vacua to obtain a set of supersymmetry equations on the internal parameters $\eta^a_\pm$. Those equations look complicated, but we will show in Section 10.3.3 that they become elegant in terms of the pure forms defined by $\Phi_\pm = \eta^1_+ \otimes \eta^{2\dagger}_\pm$. Later in this chapter, we will see that the forms $\Phi_\pm$ encode all the data of the metric, B, dilaton, and of the $\eta^a_\pm$. So one can completely reformulate the problem of supersymmetric vacua in terms of forms.

10.3.1 Vacuum solutions

We already discussed in Sections 8.2.2 and 8.2.3 some of the requirements a vacuum should satisfy. We concluded that the metric should factorize, as in (8.2.4), that the fermionic fields should be set to zero, and that the supersymmetry parameters should be split as in (8.2.12).

Fluxes

Along MS_4, all form fields can only be proportional either to the zero-form 1 or to the volume form vol_4; in other words, along MS_4 they can either have no legs or all legs. So H is purely internal. The RR polyform should decompose as $F = f_{\text{int}} + \text{e}^{4A}\text{vol}_4 \wedge f_{\text{ext}}$. To impose the self-duality constraint (10.1.3), we use

$$*_{10}\, \alpha = \text{e}^{4A}\text{vol}_4 \wedge *\alpha\,, \qquad *_{10}(\text{e}^{4A}\text{vol}_4 \wedge \alpha) = -\alpha\,, \tag{10.3.1}$$

where α is a form on M_6; from now on, $* \equiv *6$, and more generally unlabeled quantities are relative to the internal metric. This gives $f_{\text{ext}} = *\lambda f_{\text{int}}$, $f_{\text{int}} = -*\lambda f_{\text{ext}}$. Calling $f \equiv f_{\text{int}}$,

$$F = f + \text{e}^{4A}\text{vol}_4 \wedge *\lambda f\,. \tag{10.3.2}$$

We saw an example of this in (9.5.4).

Equation (10.3.2) completes the vacuum Ansatz in Section 8.2.2. The Bianchi identities (10.1.4) become

$$(\text{d} - H\wedge)f = \delta_{\text{int}}\,, \tag{10.3.3a}$$

$$(\text{d} - H\wedge)(\text{e}^{4A} * \lambda f) \overset{(3.2.58),(3.2.59)}{=} \mp\lambda(\text{d} + H\wedge)(\text{e}^{4A} * f) = \delta_{\text{ext}}\,, \tag{10.3.3b}$$

where we decomposed the sources in (10.1.4) as $\delta = \delta_{\text{int}} + \text{vol}_4 \wedge \delta_{\text{ext}}$.

Equations of motion

Using (4.1.161), we find that on a function α_0 of the internal coordinates

$$\begin{aligned} &\nabla^{10}_m\nabla^{10}_\nu\alpha_0 = \text{e}^{2A}g^4_{\mu\nu}\text{d}A\cdot\text{d}\alpha_0\,,\\ &\nabla^{10}_\mu\nabla^{10}_m\alpha_0 = \nabla^{10}_m\nabla^{10}_\mu\alpha_0 = 0\,, \qquad \nabla^{10}_m\nabla^{10}_n\alpha_0 = \nabla_m\nabla_n\alpha_0\,. \end{aligned} \tag{10.3.4}$$

In particular

$$\nabla^2_{10}\alpha_0 = \nabla^2_4\alpha_0 + 4\mathrm{d}A\cdot\mathrm{d}\alpha_0\,. \tag{10.3.5}$$

The Ricci tensor was also given in (4.3.71). We again stress that the unlabeled ∇_m and the form contractions are now relative to the internal metric g^6_{mn}.

Specializing the equations of motion (10.1.19a) and (10.1.19b), now we get

$$\mathrm{e}^{-2A}\Lambda - \nabla^2 A - 4|\mathrm{d}A|^2 + 2\mathrm{d}A\cdot\mathrm{d}\phi = -\frac{1}{4}(\mathrm{e}^{2\phi}|f|^2 + \mathrm{e}^{\phi}\sum_I j^I)\,; \tag{10.3.6a}$$

$$R_{mn} - 4(\nabla_m\partial_n A + \partial_m A\partial_n A) + 2\nabla_m\nabla_n\phi - \frac{1}{2}\iota_m H\cdot\iota_n H \tag{10.3.6b}$$

$$= \frac{1}{4}\mathrm{e}^{2\phi}(2\iota_m f\cdot\iota_n f - g_{mn}|f|^2) + \frac{1}{4}\mathrm{e}^{\phi}\sum_I j^I_{mn}\,;$$

$$2\nabla^2\phi - 4|\mathrm{d}\phi|^2 + 8\mathrm{d}A\cdot\mathrm{d}\phi + |H|^2 = \frac{1}{2}\mathrm{e}^{2\phi}\sum_k(5-k)|f_k|^2 + \frac{1}{2}\mathrm{e}^{\phi}\sum_I(3-p_I)j^I\,. \tag{10.3.6c}$$

In the last term of (10.3.6c), recall that p_I is the space dimension of the Ith source. The derivatives on the left-hand side of (10.3.6a) and (10.3.6c) can be rewritten in various ways; for example:

$$\begin{aligned}\nabla^2 A + 4|\mathrm{d}A|^2 - 2\mathrm{d}A\cdot\mathrm{d}\phi &= -\mathrm{e}^{-4A+2\phi}\nabla_m(\mathrm{e}^{4A-2\phi}g^{mn}\partial_n A)\,,\\ \nabla^2\phi - 2|\mathrm{d}\phi|^2 + 4\mathrm{d}A\cdot\mathrm{d}\phi &= \mathrm{e}^{-4A+2\phi}\nabla_m(\mathrm{e}^{4A-2\phi}g^{mn}\partial_n\phi)\,.\end{aligned} \tag{10.3.7}$$

No-go results for de Sitter and Minkowski

We saw in the introduction that de Sitter vacua (and Minkowski vacua with fluxes) in the supergravity regime necessarily need to include Op-planes. The argument we gave there had the advantage of being quite general, being applicable to any gravitational theory with an Einstein–Hilbert term coupled to matter fields obeying the strong energy condition. As we saw in Exercise 10.1.5 and we mentioned in the Introduction, in type II supergravity this holds for all contributions to the stress energy tensor except for the Romans mass F_0, whose $T_{MN} = -\frac{1}{8\kappa^2}g_{MN}F_0^2$. In principle, this would leave open the possibility that de Sitter vacua exist with $F_0 \neq 0$ and without O-planes. But we will now give a different derivation of the no-go argument, which rules out this possibility [18].

For IIA, we consider the linear combination $\mathrm{e}^{4A-2\phi}$(10(10.3.6a)+2 (10.3.6c)), chosen so that the F_0 term cancels out. Using also (10.3.7):

$$\begin{aligned}&10\mathrm{e}^{2(A-\phi)}\Lambda + \nabla_m(\mathrm{e}^{4A+2\phi}g^{mn}\partial_n(-10A+2\phi)) + \mathrm{e}^{4A-2\phi}|H|^2\\ &\quad= \frac{\mathrm{e}^{4A}}{2}\left(-\sum_k k|f_k|^2 + \mathrm{e}^{-\phi}\sum_I(p_I-8)\delta_I\right).\end{aligned} \tag{10.3.8}$$

Integrating both sides on a compact M_6, the ∇_m total derivative vanishes; the f_k^2 terms on the right-hand side are all negative, and so are the sources, if they are all D-branes. So we find that *without O-planes, there are no de Sitter (or fluxless Minkowski) vacua*. Unlike in (0.5.4), we did not assume $F_0 = 0$.

As noted in [436, A.7] and [437], in the linear combination (10.3.8) the $p = 8$ sources cancel out too; so it seems that an O8-plane cannot help in finding dS_4 solutions. This argument is a higher-dimensional version of the one in (8.3.13) using the effective potential. However, this is still not rigorous, because it uses the supergravity equations of motion in a region where the string coupling and curvature are both large, as mentioned following (1.4.57). In fact a proposal for a dS_4 solution with O8-planes exists [438]. We will return on the issue in Section 12.1.

It is also possible to consider other linear combinations in the same spirit. For example, $e^{4A-2\phi}$(6(10.3.6a)+2(10.3.6c)) gives an argument that if $F_0 = 0$ an O6 alone is not enough to produce a dS solution. This extends a similar argument in Section 8.3.2, again with imperfect rigor. In any case, all this does suggest that $F_0 \neq 0$ is important for de Sitter vacua in IIA.

In IIB, the argument in the Introduction has no loophole, but we can show it again along the lines in (10.3.8). This time we consider $e^{4A-2\phi}$(8(10.3.6a)+2(10.3.6c)), obtaining [414]:

$$\begin{aligned} 8e^{2(A-\phi)}\Lambda + \nabla_m(e^{4A+2\phi}g^{mn}\partial_n(-8A+2\phi)) + e^{4A-2\phi}|H|^2 \\ = \frac{e^{4A}}{2}\left(\sum_k (1-k)|f_k|^2 + e^{-\phi}\sum_I (p_I - 7)\delta_I\right). \end{aligned} \tag{10.3.9}$$

(This is also the equation for $g^{\mathrm{E}}_{\mu\nu}$ in the Einstein frame.)

10.3.2 Supersymmetry and factorization

Minimal supersymmetry: a first attempt

The Ansatz for the supercharges ϵ^a is (8.2.12), (8.2.13), and (8.2.14a). This was conceived as generally appropriate for any amount of supersymmetry. However, to simplify the problem it looks wise to start first with the minimal possible amount, namely a single nonzero ζ in those equations, or in other words $\mathcal{N} = 1$ supersymmetry (Chapter 9).

The simplest possibility might appear to take $\zeta^{a=1}_{I=1} \equiv \zeta \neq 0$, setting all the others to zero. According to (8.2.14a), this single spinor satisfies the Killing spinor equation

$$D^4_\mu \zeta_\pm = \frac{\mu}{2}\gamma_\mu \zeta_\mp . \tag{10.3.10}$$

Equation (8.2.12) would now read as (9.3.3):

$$\epsilon^1 \stackrel{?}{=} \zeta_+ \otimes \eta_+ + \zeta_- \otimes \eta_- , \qquad \epsilon^2 \stackrel{?}{=} 0 . \tag{10.3.11}$$

Let us plug this Ansatz in the internal gravitino equations $\delta\psi^a_m = 0$, with (10.1.33a) and (10.1.33b). For concreteness, we focus on IIB. We get

$$\begin{aligned} \zeta_+ \otimes \left(D_m - \frac{1}{4}H_m\right)\eta_+ + \zeta_- \otimes \left(D_m - \frac{1}{4}H_m\right)\eta_- = 0 , \\ \zeta_+ \otimes \lambda(f)\gamma_m\eta_- + \zeta_- \otimes \lambda(f)\gamma_m\eta_+ = 0 . \end{aligned} \tag{10.3.12}$$

The two chiralities $\zeta_\pm$ belong to different spaces, so their coefficients have to vanish independently; but they are each other's conjugate. So we get two equations:

$$\left(D_m - \frac{1}{4}H_m\right)\eta_+ = 0\,, \qquad (10.3.13a)$$

$$\lambda(f)\gamma_m\eta_+ = 0\,. \qquad (10.3.13b)$$

This already doesn't look promising: we already saw (10.3.13a) in (9.3.2a), for compactifications without RR fields. For the external gravitino equations $\delta\psi^a_\mu = 0$, the analysis is similar: this time there is an overall $\gamma_\mu\zeta_\pm$, once we use (10.3.10). This leads to

$$\mu e^{-A}\eta^1_- + dA\eta^1_+ = 0 \qquad (10.3.13c)$$

$$\lambda(f)\eta^1_+ = 0\,. \qquad (10.3.13d)$$

We could also work out the modified dilatino (10.1.33c), (10.1.33d), but (10.3.13) is enough to show we are not on the right track. We can rewrite (10.3.13b) and (10.3.13d) as $\lambda(f)\gamma_i\eta_+ = 0$, $\lambda(f)\gamma_{\bar\imath}\eta_- = 0$, $\lambda(f)\eta_\pm = 0$. Then we can project each of these on the basis (3.4.86). For example, multiplying (10.3.13d) from the left by η_- we get

$$0 = \eta_-^\dagger\lambda(f)\eta_+ = \mathrm{Tr}(\lambda(f)\eta_+\otimes\eta_-^\dagger) \overset{(3.2.38),\,(3.4.20b)}{=} 8(*\lambda(f)\wedge\Phi_-)\,. \qquad (10.3.14)$$

So $\lambda(f)$ is orthogonal to Φ_- with respect to the inner product (4.1.106b). Further projections reveal that $\lambda(f)$ is orthogonal to all odd forms in the basis (3.4.95); so we conclude $f = 0$. Moreover, (10.3.13c) already appeared in (8.2.24b), where it was found to lead to $\mu = 0$ and $dA = 0$.

In conclusion, the Ansatz (10.3.11) leads back to the case without RR fields, analyzed in Section 9.3.

Minimal supersymmetry

Since (10.3.11) did not work as intended, we next try to switch on two spinors in (8.2.12): namely, we take only $I = 1$ there, but with both $\eta^a_{1+} \equiv \eta^a_+$. The internal gravitino conditions $\delta\psi^a_m = 0$ from (10.1.33a) now give

$$\zeta^1_+\otimes\left(D_m - \frac{1}{4}H_m\right)\eta^1_+ + \zeta^2_+\otimes\frac{e^\phi}{8}f\,\gamma_m\eta^2_+ + \text{c.c.} = 0\,. \qquad (10.3.15)$$

If the ζ^a_+ are independent, we fall back in the previous case (10.3.13). But we now spot an opportunity: if we impose the constraint $\zeta^1_+ = \zeta^2_+ \equiv \zeta_+$, the two terms in (10.3.15) can combine and give rise to less constraining internal equations. With this idea, (8.2.12) becomes

$$\begin{aligned}\epsilon^1 &= \zeta_+\otimes\eta^1_+ + \zeta_-\otimes\eta^1_-\,,\\ \epsilon^2 &= \zeta_+\otimes\eta^2_\mp + \zeta_-\otimes\eta^2_\pm\,. \quad {\scriptstyle \mathrm{IIA}/\mathrm{IIB}}\end{aligned} \qquad (\mathcal{N} = 1 \text{ RR vacua})\,. \qquad (10.3.16)$$

We have finally made use of the possibility, first mentioned after (8.2.12), of introducing a constraint among the ζ^a_I appearing there. One might want to consider a more general constraint in (8.2.12), such as $\zeta^1_{1+} = \alpha\zeta^2_{1+}$ for some α, but this can be reabsorbed in the definition of the internal spinors η^a. Notice that at this point we have not imposed any constraints on the latter.

Factorizing now the supersymmetry equations (10.1.33) with (10.3.16) leads to the following set of purely six-dimensional spinor equations:

$$\left(D_m - \frac{1}{4}H_m\right)\eta^1_+ + \frac{e^\phi}{8} f\,\gamma_m \eta^2_\mp = 0\,, \tag{10.3.17a}$$

$$\left(D_m + \frac{1}{4}H_m\right)\eta^2_\mp \pm \frac{e^\phi}{8}\lambda(f)\,\gamma_m \eta^1_+ = 0\,; \tag{10.3.17b}$$

$$\mu e^{-A}\eta^1_- + \mathrm{d}A\,\eta^1_+ \pm \frac{e^\phi}{4} f\,\eta^2_\mp = 0\,, \tag{10.3.17c}$$

$$\mu e^{-A}\eta^2_\pm + \mathrm{d}A\,\eta^2_\mp + \frac{e^\phi}{4}\lambda(f)\,\eta^1_+ = 0\,; \tag{10.3.17d}$$

$$2\mu e^{-A}\eta^1_- + D\eta^1_+ + \left(\mathrm{d}(2A - \phi) - \frac{1}{4}H\right)\eta^1_+ = 0\,, \tag{10.3.17e}$$

$$2\mu e^{-A}\eta^2_\pm + D\eta^2_\mp + \left(\mathrm{d}(2A - \phi) + \frac{1}{4}H\right)\eta^2_\mp = 0\,. \quad {}^{\text{IIA}}_{\text{IIB}} \tag{10.3.17f}$$

Equations (10.3.17a) and (10.3.17b) are the gravitino equations $\delta\psi^a_m = 0$, in the internal directions; (10.3.17c) and (10.3.17d) are $\delta\psi^a_\mu = 0$, in the external directions; and (10.3.17e) and (10.3.17f) are the modified dilatino equations $\Gamma^M\delta\psi^a_M - \delta\lambda^a = 0$.

Equations (10.3.17) look scary, but if we do solve them, we have obtained an $\mathcal{N} = 1$ vacuum solution (thanks to Section 10.1.4). A different worry is that at this point (10.3.17) appear uninspiring and devoid of any geometrical meaning. This is quite unlike our analysis of Section 8.2, where we ended up with the condition of special holonomy. The main aim of this chapter will be to show that they do in fact have a geometrical meaning.

10.3.3 The pure spinor equations

In Sections 9.3 and 9.5, we have seen that the analysis of the supersymmetry equations becomes more transparent in terms of bilinears. We reexpressed the data of the internal supersymmetry parameter η in terms of its associated forms (J, Ω), using ideas from Sections 3.4.2, 5.1.2, and 5.5.1. We now would like to perform a similar analysis for the supersymmetry equations (10.3.17), describing the most general $\mathcal{N} = 1$ vacuum with RR flux [439].

The presence of two internal spinors η^a is a significant complication, because there are now additional bilinear forms, which we saw in Section 3.4.3 when the η^a are orthogonal; more generally, their inner product $(\eta^{1\dagger})\eta^2$, a function on M_6, also plays a role. One can express the supersymmetry equations (10.3.17) in terms of all these data, but the result is not particularly elegant or inspiring.

Compatible pairs of pure forms

Fortunately, we will see now that (10.3.17) simplify enormously if we rewrite them directly in terms of the bispinors

$$\Phi_+ \equiv \eta^1_+ \otimes \eta^{2\dagger}_+\,, \tag{10.3.18a}$$

$$\Phi_- \equiv \eta^1_+ \otimes \eta^{2\dagger}_-\,, \tag{10.3.18b}$$

which we also saw in Section 3.4. Perhaps surprisingly, the other bispinors $\eta^1_+ \otimes \eta^{1\,\dagger}_\pm$, $\eta^2_+ \otimes \eta^{2\,\dagger}_\pm$ will not play much of a role.

We know already from the general principle (3.4.4) that the $\Phi_\pm$ are pure forms. Two more properties that follow from (10.3.18) are the following (Exercise 3.4.8):

$$(\bar\Phi_+, \Phi_+) = (\bar\Phi_-, \Phi_-) = \frac{\mathrm{i}}{8} N^2 \equiv \frac{\mathrm{i}}{8} ||\eta^1_+||^2 ||\eta^2_+||^2 \qquad (10.3.19a)$$

$$(\Gamma_A \Phi_+, \Phi_-) = (\Gamma_A \bar\Phi_+, \Phi_-) = 0 \,. \qquad (10.3.19b)$$

Recall that Γ_A is any wedge or contraction, as in (3.2.4), and that $(\,,\,)$ is the Chevalley–Mukai pairing (3.2.40). In Section 10.5.3, we will prove that the $\Phi_\pm$ can be completely characterized as a pair of pure forms that obey (10.3.19); such a pair is called *compatible*.

In Section 10.3.4, we will give a more down-to-earth characterization of compatible pure pairs in terms of SU(3)- and SU(2)-structures. In Section 10.5.3, we will show that a compatible pair of pure forms $\Phi_\pm$ is enough to determine the metric and the spinorial parameters $\eta^a_\pm$.

Supersymmetric vacua and pure forms

In type II supergravity, the existence of four-dimensional vacuum solution with nonzero RR fields and $\mathcal{N} = 1$ supersymmetry is equivalent to the system

$$\mathrm{d}_H(\mathrm{e}^{A-\phi}\mathrm{Re}\Phi_\mp) = \mp\frac{c_-}{16} f \,, \qquad (10.3.20a)$$

$$\mathrm{d}_H(\mathrm{e}^{2A-\phi}\Phi_\pm) = \pm 2\mu \mathrm{e}^{A-\phi}\mathrm{Re}\Phi_\mp \,, \qquad {}^{\mathrm{IIA}}_{\mathrm{IIB}} \qquad (10.3.20b)$$

$$\mathrm{d}_H(\mathrm{e}^{3A-\phi}\mathrm{Im}\Phi_\mp) = \pm 3\mu \mathrm{e}^{2A-\phi}\mathrm{Im}\Phi_\pm \mp \frac{c_+}{16}\mathrm{e}^{4A} * \lambda f \,, \qquad (10.3.20c)$$

where $\mu \equiv \sqrt{-\Lambda/3}$;[5] $c_+ \neq 0$, c_- are two real constants, related to the normalization in (10.3.19a) by

$$||\eta^1_+||^2 \pm ||\eta^2_+||^2 = c_\pm \mathrm{e}^{\pm A} \quad \Rightarrow \quad 4N^2 = c_+^2 \mathrm{e}^{2A} - c_-^2 \mathrm{e}^{-2A} \,; \qquad (10.3.21)$$

and the internal Bianchi identities

$$\mathrm{d}H = 0 \,, \qquad (10.3.22a)$$

$$\mathrm{d}_H f = \delta \,, \qquad (10.3.22b)$$

should also be satisfied.

Recall also that H is the NSNS flux, which is purely internal. f is the internal RR field-strength, collected in a polyform; it determines the external flux too, via (10.3.2). The function A is the warping, introduced in (8.2.4). The sign-flip operator λ was defined in (3.2.28). Finally, δ is the physical source defined in (10.1.5); it describes either D-branes or O-planes.

We will show that (10.3.20) follow from supersymmetry in Section 10.3.7, and that they are equivalent to it in Section 10.3.8. The Bianchi identities are $\mathrm{d}H = 0$, $\mathrm{d}_H F = -\delta$ (Section 10.1.1). Recalling (10.3.2), the latter is equivalent to (10.3.22b) and

[5] The equations are sometimes written with a complex μ, such that $\Lambda = -3|\mu|^2$. The phase of μ, however, can always be reabsorbed by using $\eta^1_+ \to \mathrm{e}^{\mathrm{i}\alpha}\eta^1_+$, $\eta^2_\pm \to \mathrm{e}^{-\mathrm{i}\alpha/2}\eta^2_\pm$, $\mu \to \mathrm{e}^{\mathrm{i}\alpha}\mu$, in the usual notation where the upper (lower) sign is for IIA (IIB). See also footnote 3 that precedes (8.2.15).

$$d_H(e^{4A} * \lambda f) = 0 . \tag{10.3.23}$$

But this is implied by (10.3.20), as one can see by acting on (10.3.20c) with d_H, using d_H (showed in (10.1.8)), and using the imaginary part of (10.3.20b). So the Bianchi identities are just (10.3.22); the equations of motion then all follow (Section 10.1.4).

We specified the presence of RR fields before (10.3.20), because $\mathcal{N} = 1$ solutions with $f = 0$ are described by the system (9.3.13). This is not a particular case of (10.3.20), because it arises from $\eta^2 = 0$ as in (9.3.3), and with this choice (10.3.18) both vanish. The case $H = 0$ in (10.3.20) corresponds to both $\eta^a \neq 0$, which was discussed around (9.3.19) and argued to lead to $\mathcal{N} = 2$. We will discuss it again in Section 10.5.5.

The Minkowski case

For $\Lambda = 0$, (10.3.20) simplify to

$$d_H(e^{A-\phi}\mathrm{Re}\Phi_\mp) = \mp\frac{c_-}{16} f , \tag{10.3.24a}$$
$$d_H(e^{2A-\phi}\Phi_\pm) = 0 , \tag{10.3.24b}$$
$$d_H(e^{3A-\phi}\mathrm{Im}\Phi_\mp) = \mp\frac{c_+}{16} e^{4A} * \lambda f . \tag{10.3.24c}$$

We will see in Section 10.5 that (10.3.24b) in particular has an interpretation: it implies that M_6 is a *generalized complex manifold.*

The AdS case

For $\Lambda < 0$, by acting on (10.3.20b) with d_H we obtain $d_H(e^{A-\phi}\mathrm{Re}\Phi_\mp) = 0$. This is (10.3.20a), but with $c_- = 0$. So in this case

$$\Lambda < 0 \quad \Rightarrow \quad c_- = 0 , \quad \text{(10.3.20a) is redundant} . \tag{10.3.25}$$

and the normalization (10.3.21) becomes $N = \frac{1}{2}c_+ e^A$. Equations (10.3.20) become invariant under

$$\mu \to \mu\, e^{A_0} , \qquad A \to A + A_0 , \tag{10.3.26}$$

where A_0 is a constant. This is as it should be: the warping function appears in the metric multiplying the AdS_4 part, and so shifting A by a constant only redefines the curvature radius.

10.3.4 From spinors to pure forms

In Section 10.5.3, we will give a conceptual understanding of $\Phi_\pm$. For now, in this section we give a more concrete characterization of the forms $\Phi_\pm$, building on Section 3.4. We will drop the unit-norm assumption $||\eta^a_+||^2 = 1$ we made there. We divide our discussion into three cases: the η^a are parallel, orthogonal, or generic. The structure group they define on M_6 may be SU(3) or SU(2). In Section 10.5, we will see, however, their *generalized structure*, namely the structure group they define on the generalized tangent bundle $T \oplus T^*$, is $SU(3) \times SU(3)$ in all three cases.

SU(3)-structure

We first consider the case where the two spinors are proportional. We know from Section 5.1.2 that a single chiral spinor defines an SU(3)-structure. $\eta^2_+ \propto \eta^1_+$ implies that $|\eta^{2\dagger}_+ \eta^1_+| = ||\eta^2_+|| \, ||\eta^1_+||$, but the proportionality factor can include a phase:

$$\eta^{2\dagger}_+ \eta^1_+ \equiv N e^{i\theta} . \tag{10.3.27}$$

In this case, $\Phi_\pm$ can be read off from (3.4.20):

$$\Phi_+ = \frac{N}{8} e^{i\theta} e^{-iJ} , \tag{10.3.28a}$$

$$\Phi_- = \frac{N}{8} \Omega . \tag{10.3.28b}$$

Recall from (10.3.19a) that $N = ||\eta^1_+|| \, ||\eta^2_+||$. We should have included a factor $e^{i\theta}$ also in Φ_-, but we chose to reabsorb it in Ω. Here J and Ω are the forms of an SU(3)-structure, so, as discussed in Section 5.1.2, Ω has to be decomposable, both have to be nondegenerate, and they have to satisfy the compatibility conditions (5.1.40):

$$J \wedge \Omega = 0 , \qquad \mathrm{vol}_6 = -\frac{1}{6} J^3 = -\frac{i}{8} \Omega \wedge \bar{\Omega} . \tag{10.3.29}$$

As expected, both (10.3.28) are of the form (3.2.53); according to the definition given after that equation, they have type 0 and type 3 respectively.

Orthogonal SU(2)-structure

The next case we consider is when the two spinors are orthogonal:

$$\eta^{2\dagger}_+ \eta^1_+ = 0 . \tag{10.3.30}$$

Here we take the bispinors from (3.4.44):

$$\Phi_+ = \frac{N}{8} e^{\frac{1}{2} v \wedge \bar{v}} \wedge \omega , \tag{10.3.31a}$$

$$\Phi_- = -\frac{N}{8} v \wedge e^{-ij} . \tag{10.3.31b}$$

The forms v, j, ω define an SU(2)-structure:

$$\begin{aligned} \iota_v j = \iota_{\bar{v}} j = 0 , \qquad & \iota_v \omega = \iota_{\bar{v}} \omega = 0 , \\ j \wedge \omega = \omega \wedge \omega = 0 , \qquad & \omega \wedge \bar{\omega} - 2 j^2 . \end{aligned} \tag{10.3.32}$$

The normalization condition (10.3.19a) also implies the nondegeneracy condition $v \wedge \bar{v} \wedge \omega \wedge \bar{\omega} = 2 v \wedge \bar{v} \wedge j^2 \neq 0$ everywhere. After we associate a metric to the $\Phi_\pm$, as we will learn in Section 10.5.3, v will automatically have square norm $|v|^2 = 2$ as in Section 3.4.3.

Again (10.3.31b) are of the form (3.2.53); they have type 2 and type 1.

Generic case: dynamic $\mathrm{SU}(2)$-structure, or $\mathrm{SU}(3)\times\mathrm{SU}(3)$-structure

Finally we tackle the generic case, in the usual sense that it covers all possibilities except a set of zero measure, which are (10.3.28) and (10.3.31). It is called "dynamic SU(2)" because now the inner product $\eta_+^{2\dagger}\eta_+^1$ is nonzero and can be a function on M_6. Another common expression is "SU(3) × SU(3)-structure," after the generalized structure (Section 10.5.3), although the previous cases would also deserve this name.

The inner product is always such that $|\eta_+^{2\dagger}\eta_+^1| \leq ||\eta_+^1||||\eta_+^2|| = N$, so we can parameterize it as

$$\eta_+^{2\dagger}\eta_+^1 \equiv N\cos\psi e^{i\theta}\,. \tag{10.3.33}$$

The two spinors

$$\eta_+ = \frac{1}{2\cos(\psi/2)}\left(e^{-i\theta}\frac{\eta_+^1}{||\eta_+^1||} + \frac{\eta_+^2}{||\eta_+^2||}\right), \qquad \eta_+' = \frac{1}{2\sin(\psi/2)}\left(e^{-i\theta}\frac{\eta_+^1}{||\eta_+^1||} - \frac{\eta_+^2}{||\eta_+^2||}\right) \tag{10.3.34}$$

are orthogonal, and have norm one; so according to Section 3.4.3, we can write $\eta_+' = \frac{1}{2}v\cdot\eta_-$, with v being the one-form bilinear defined by the two. This leads to the parameterization

$$\begin{aligned}\eta_+^1 &= ||\eta_+^1||e^{i\theta}\left(\cos(\psi/2)\,\eta_+ + \frac{1}{2}\sin(\psi/2)\,v\cdot\eta_-\right),\\ \eta_+^2 &= ||\eta_+^2||\left(\cos(\psi/2)\,\eta_+ - \frac{1}{2}\sin(\psi/2)\,v\cdot\eta_-\right).\end{aligned} \tag{10.3.35}$$

Now η_+ defines an SU(3)-structure, and together with v it defines an SU(2)-structure. The computation of $\Phi_\pm$ follows the same strategy as in (3.4.37) and (3.4.39), now decorated by the parameter ψ. (One also needs the results of Exercises 3.4.3 and 3.4.4.) The result reads as follows [440, 441]:

$$\begin{aligned}\Phi_+ &= \frac{N}{8}e^{i\theta}e^{\frac{1}{2}v\wedge\bar v}\wedge\left(\cos^2(\psi/2)\,e^{-ij} - \sin^2(\psi/2)\,e^{ij} - \frac{1}{2}\sin\psi\,(\omega+\bar\omega)\right),\\ \Phi_- &= \frac{N}{8}e^{i\theta}v\wedge\left(\cos^2(\psi/2)\,\omega - \sin^2(\psi/2)\,\bar\omega + \frac{1}{2}\sin\psi\,\left(e^{-ij}+e^{ij}\right)\right).\end{aligned} \tag{10.3.36}$$

Just as in (10.3.28b), the factor $e^{i\theta}$ would have been present in Φ_- as well, but we have chosen to reabsorb it in the definition of v.

From the expression (10.3.36), it is not immediately clear that $\Phi_\pm$ are pure forms, as predicted by the general argument (3.4.4); it does become apparent by rewriting them as

$$\Phi_+ = \frac{N}{8}e^{i\theta}\cos\psi\exp\left[\frac{1}{2}v\wedge\bar v - \frac{i}{\cos\psi}j - \tan\psi\,\mathrm{Re}\omega\right], \tag{10.3.37a}$$

$$\Phi_- = \frac{N}{8}\sin\psi\, v\wedge\exp\left[\frac{1}{\sin\psi}(\cos\psi\,\mathrm{Re}\omega + i\mathrm{Im}\omega)\right], \tag{10.3.37b}$$

which are both of the form (3.2.53), of type 0 and 1. The SU(3)-structure forms (J^1, Ω^1), (J^2, Ω^2) associated to η^1_+ and η^2_+ are

$$\Omega^a = v \wedge \omega^a \,, \qquad J^a = j^a + \frac{\mathrm{i}}{2} v \wedge \bar{v} \,, \tag{10.3.38}$$

where

$$\begin{pmatrix} j^{1,2} \\ \mathrm{Im}\omega^{1,2} \end{pmatrix} = \begin{pmatrix} \cos\psi & \mp\sin\psi \\ \pm\sin\psi & \cos\psi \end{pmatrix} \begin{pmatrix} j \\ \mathrm{Im}\omega \end{pmatrix}, \qquad \mathrm{Re}\omega^{1,2} = \mathrm{Re}\omega \,. \tag{10.3.39}$$

10.3.5 Calibration interpretation

In this book, the pure forms $\Phi_\pm$ have already played a role, as *calibrations* (Sections 5.6 and 9.2). This leads to a remarkable interpretation for the pure spinor equations (10.3.20) [442]. For simplicity, we will limit ourselves to the Minkowski case (10.3.24), taking $c_+ = 2$, $c_- = 0$. The AdS case is similar [443].

Spacetime-filling branes

For Calabi–Yau compactifications, the supersymmetry parameters were (8.2.28), and the BPS condition for D-branes extended along all four directions of spacetime led to (9.2.14). For the $\mathcal{N} = 1$ compactifications of this section, the supersymmetry parameters are (10.3.16). So we have to change (9.2.14) in two ways: the external spinors are now equal, $\zeta^1_\pm = \zeta^2_\pm$, so $\alpha = 1$; and the internal spinors are now different, $\eta^1_\pm \neq \eta^2_\pm$. Equation (9.2.14) then becomes

$$\eta^1_+ = \mathrm{i}\gamma_{||}\eta^2_\mp \,. \tag{10.3.40}$$

We can again translate this in forms as in (5.6.11), modifying that analysis to allow the two spinors to be different. This gives $\mathrm{Tr}(\Phi_\mp \gamma^\dagger_{||}) = \mathrm{i}||\eta_+||^2$, and yields

$$8\mathrm{e}^{-A}\mathrm{Im}\Phi_\mp \tag{10.3.41}$$

as an almost calibration for the brane: its pull-back to a brane Dp is $\leq$ vol(Dp), as in (5.6.3).

Had (10.3.41) been closed, we could have called it a calibration (without the "almost"): its calibrated cycles would have minimized the volume in its homology class. However, from a physics point of view, we don't care about the volume anyway, but rather about the potential energy $V_{\mathrm{D}p}$ of the brane, inferred from its action (10.1.18) by considering a time-independent configuration. In absence of fluxes, $V_{\mathrm{D}p}$ is indeed proportional to the volume, as discussed after (1.3.29); but more generally we have to minimize

$$\int \left(\mathrm{e}^{4A-\phi} \sqrt{\det(g|_{\mathrm{D}p} + \mathcal{F})}\, \mathrm{d}\sigma^1 \wedge \ldots \wedge \mathrm{d}\sigma^k \pm \tilde{c} \wedge \mathrm{e}^{-\mathcal{F}} \right), \qquad \mathrm{d}_H \tilde{c} = \mathrm{e}^{4A} * \lambda f \,. \tag{10.3.42}$$

We need to find a form whose pull-back is less than or equal to this quantity, rather than the volume. The e^{4A} is because the brane we are considering is spacetime filling, and simply modifies (10.3.41) as an overall factor. As in (9.2.26a), the effect of the world-volume flux $\mathcal{F}$ further changes (10.3.41) to $\mathrm{e}^{3A-\phi}\mathrm{e}^{-\mathcal{F}} \wedge \mathrm{Im}\Phi_\mp$. Finally, the WZ contribution is already the pull-back of a form, so we can just add it to (10.3.41). All this leads to

$$e^{-\mathcal{F}} \wedge (8e^{3A-\phi}\text{Im}\Phi_{\mp} \pm \tilde{c}) \quad \text{(calibration for spacetime-filling D-branes)}. \quad (10.3.43)$$

This *generalized calibration* [444] has the property that the subspaces it calibrates minimize not the volume, but the physically relevant $V_{\text{D}p}$. Its closure is equivalent to the pure spinor equation (10.3.24c).

Domain walls

Next we consider D-branes that are extended along an $\text{Mink}_3 \subset \text{Mink}_4$ subspace. From a four-dimensional point of view, it represents a domain wall. Choosing it, for example, to be along directions 012, we have $\Gamma_{\parallel} = \gamma_{012} \otimes \gamma_{\parallel}$ for IIA, $\Gamma_{\parallel} = \gamma_{012}\gamma_{(4)} \otimes \gamma_{\parallel}$ in IIB. So the calibration condition (9.2.1), $\epsilon^1 = \Gamma_{\parallel}\epsilon^2$ gives

$$\gamma_{012}\zeta_+ = \alpha\zeta_- , \qquad \gamma_{\parallel}\eta^2_{\pm} = \alpha^{-1}\eta^1_+ , \qquad (10.3.44)$$

where α is a constant, similar to (9.2.14). Now the logic in Section 5.6.2 tells us

$$8e^{2A-\phi}e^{-\mathcal{F}} \wedge \text{Re}(\alpha^{-1}\Phi_{\pm}) \quad \text{(calibration for domain-wall D-branes)}. \quad (10.3.45)$$

This is closed for all α, due to (10.3.24b). There is no WZ contribution, because in our decomposition (10.3.2), we don't have any components $e^0 \wedge e^1 \wedge e^2 \wedge$ internal.

Strings

We now turn to D-branes extended along a $\text{Mink}_2 \subset \text{Mink}_4$, a string in four dimensions. We put it along directions 01; then we have $\Gamma_{\parallel} = \gamma_{01}\gamma_{(4)} \otimes \gamma_{\parallel}$ for IIA, $\Gamma_{\parallel} = \gamma_{01} \otimes \gamma_{\parallel}$ in IIB. This leads to

$$\gamma_{01}\zeta_+ = \zeta_+ , \qquad \gamma_{\parallel}\eta^2_{\mp} = \eta^1_+ . \qquad (10.3.46)$$

Unlike in (10.3.44), we don't have a parameter α here because ζ_+ is an eigenspinor of γ_{01}, which squares to +1 and thus only has ± 1 eigenvalues. Equation (10.3.46) now gives

$$8e^{A-\phi}e^{-\mathcal{F}} \wedge \text{Re}\Phi_{\mp} \quad \text{(calibration for string D-branes)}. \quad (10.3.47)$$

For $c_- = 0$, this is closed on a supersymmetric solution, this time because of (10.3.24a).

Particles

Finally, suppose we want to consider D-branes that are only extended along time in Mink_4, and thus look like particles. Here the four-dimensional part of (9.2.1) becomes

$$\gamma_0\zeta_+ = \zeta_- . \qquad (10.3.48)$$

This is inconsistent. Taking the conjugate gives $\gamma_0\zeta_- = \zeta_+$; but $(\gamma_0)^2 = -1$, so we also obtain $\gamma_0\zeta_+ = -\zeta_-$, in contradiction with (10.3.48). This result is familiar from the study of supersymmetric field theories, where it is well known that BPS particles can only be defined for $\mathcal{N} = 2$ theories, while in this section we are considering $\mathcal{N} = 1$ compactifications.

Interpretation

In conclusion, we have seen that there are three types of possible BPS branes, extended along two, three, or four dimensions in Mink_4. They are calibrated respectively by (10.3.43), (10.3.44), and (10.3.46). The condition that these calibrations should be closed precisely reproduces the pure spinor equations (10.3.24) [442].

Given our previous discussion in Section 5.6, it is perhaps reasonable that the conditions for a brane to be BPS should admit a reformulation in terms of calibrations. It is more surprising that this is *equivalent* to supersymmetry of the background solution. This coincidence will be partially explained in Section 10.3.6, where we will see that these calibrations also play a role in the effective four-dimensional theory describing the fluctuations around a vacuum solution.

Calibrations are forms whose integrals on a brane measure the minimal energy it can attain. In this sense, they are a sort of "geometrical flux." The pure spinor equations relate it to the more customary H and F.

Extensions of the pure spinor equations also exist to Mink_d and AdS_d vacuum solutions in $d \neq 4$, some of which we will see in Section 10.6.1. They can also be given a calibration interpretation [445]. This might suggest that supersymmetry is related to calibrations even in general, beyond vacuum solutions; this conjecture is still open, and is related to the discussion in Section 10.2.5.

O-planes

Calibrations also describe O-planes. Recall that these are defined as the fixed loci of reflections R_p of $9 - p$ coordinates, where the orientifold itself is defined by quotienting type II strings by (1.4.45). First we need to determine the action of R_p on the pure forms $\Phi_\pm$. One way to do this is to extrapolate it from the world-sheet in models without RR fields, where the NSR model for superstrings is available (for example, [446]). Another method is consistency with the action (1.4.46) of R_p on the fields, and the supersymmetry transformations (10.1.33); this gives $\epsilon^2 = R_p^* \epsilon^1$, where R_p^* now includes the transverse gamma matrices as in (2.1.12), and the action on the pure forms can be inferred from it. For us, the quickest method is to use (1.4.46) directly on the pure spinor equations; the result is

$$R_p^* \Phi_\pm = -(-1)^{\lfloor \frac{p}{2} \rfloor} \lambda(\Phi_\pm) , \qquad R_p^* \Phi_\mp = (-1)^{\lfloor \frac{p-1}{2} \rfloor} \lambda(\bar{\Phi}_\mp) \quad {}^{\text{IIA}}_{\text{IIB}} , \tag{10.3.49}$$

and that $c_\pm \to \pm c_\pm$. The latter also follows from the orientifold action exchanging the two ϵ^a. In particular, we also have that $||\eta^1||^2 - ||\eta^2||^2$ is odd under R_p; from (10.3.21), this is equal to $c_- \mathrm{e}^{-A}$, which is even, so we conclude $c_- = 0$.

From (10.3.49), we see that $R_p^*(X \cdot \Phi_\pm)_{p-3} = -(X \cdot \Phi_\pm)_{p-3}$, $R_p^*(\text{Re}\Phi_\mp)_{p-3} = -(\text{Re}\Phi_\mp)_{p-3}$, so the pull-back of these forms vanishes on the internal $(p-3)$-cycle wrapped by the Op-plane. In other words, the Op-plane is automatically calibrated. On the other hand,

$$R_p^* (\text{Im}\Phi_\mp)_{p-3} = (\text{Im}\Phi_\mp)_{p-3} , \tag{10.3.50}$$

compatibly with it being the calibration for the Op, (10.3.43).

Another no-go argument

We can now give an alternative derivation of the no-go of the introduction for supersymmetric Minkowski solutions, this time directly inspired by (9.3.20):

$$\mp \int_{M_6} e^A(\delta, \mathrm{Im}\Phi_\mp) \overset{(10.3.22b),(3.2.61)}{=} \mp \int_{M_6} (f, d_H(e^A \mathrm{Im}\Phi_\mp)) \overset{(10.3.24c)}{=} \frac{c_+}{16} \int_{M_6} e^{4A}(f, *\lambda(f)) \overset{(3.2.59)}{=} -\frac{c_+}{16} \int_{M_6} e^{4A} * f \wedge f \overset{(4.1.106)}{\leq} 0 . \tag{10.3.51}$$

By the explicit expression (10.1.5), for a Dp the left-hand side of (10.3.51) is

$$-2\kappa^2 \tau_{\mathrm{D}p} \int_{M_6} (e^{\mathcal{F}} \wedge \lambda \delta_{\mathrm{D}p}, e^A \mathrm{Im}\Phi_\mp) = 2\kappa^2 \tau_{\mathrm{D}p} \int_{\mathrm{D}p} e^A e^{-\mathcal{F}} \wedge \mathrm{Im}\Phi_\mp , \tag{10.3.52}$$

which is positive because $e^{-\mathcal{F}} \wedge \mathrm{Im}\Phi_\mp$ is an almost calibration for space-filling branes, (10.3.41). This contradicts (10.3.51). With Op-planes, the sign of $\tau_{\mathrm{O}p}$ is negative and the contradiction is avoided.

10.3.6 Correspondence with $d = 4$ supergravity

We will now show that the pure spinor equations have natural interpretations in terms of $d = 4$ supergravity.

$\mathcal{N} = 1$ supergravity

First we rewrite type II supergravity as an $\mathcal{N} = 1$ theory in $d = 4$ [447] (see also [278, 448]). Here we are not really attempting any truncation: we will keep all the infinitely many fields that are generated by the KK reduction. This avoids all the issues we described in Section 8.3. We work with $c_- = 0$, for simplicity and because it is necessary anyway for AdS vacua, (10.3.25).

The superpotential (9.5.26) was originally motivated by domain walls; it is naturally generalized to

$$\pi \int_{M_6} (e^{2A-\phi}\Phi_\pm, F) \quad {}^{\mathrm{IIA}}_{\mathrm{IIB}} , \tag{10.3.53}$$

since $\Phi_\pm$ is the calibration for domain walls (Section 10.3.5). A superpotential should be holomorphic in the chiral multiplet scalars, which in our treatment should be somehow assembled from our forms on M_6. For example, $Z \equiv e^{2A-\phi}\bar{\Phi}_\pm$ is a complex form that upon reduction contains infinitely many complex scalars z^i. In IIB, for the SU(3)-structure case, $\Phi_- = \Omega$, so in this case these scalars are the complex moduli of M_6. But F is a real form, so by itself it cannot be the sum of complex scalars. To complete it, one can compute the action of BPS instantons wrapping an internal cycle $S_k \subset M_6$ [447]:

$$S_{\mathrm{E}} = 2\pi \int_{S_k} e^{-\mathcal{F}} \wedge (e^{-A-\phi}\mathrm{Im}\Phi_\pm - iC) . \tag{10.3.54}$$

The first term comes about because $\mathrm{Im}\Phi_\pm$ is the calibration for space-filling branes; the i in the second term comes from the Wick rotation to Euclidean signature, needed for an instanton action. The nonperturbative superpotential is a holomorphic function of the action of BPS instantons, so the complex form

$$T \equiv e^{-A-\phi}\mathrm{Im}\Phi_\pm - iC \tag{10.3.55}$$

is a good candidate to generate more $d = 4$ chiral multiplet scalars. One concludes that (10.3.53) should really be

$$W = \pm\pi \int_{M_6} (e^{2A-\phi}\Phi_\pm, d_H T) . \qquad (10.3.56)$$

In the Calabi–Yau case, we have identified the Kähler potential of both $\mathcal{M}_K^{\mathbb{C}}$ and $\mathcal{M}_c$ to be the volume, (9.1.20) and (9.1.26). In the present more general setup,

$$K = -3 \log \int_{M_6} \sqrt{g_6} e^{2A-2\phi} . \qquad (10.3.57)$$

As we mentioned in Section 9.5, the presence of a nontrivial warping function A in fact creates subtleties in deriving the effective action [420–422], but we will not need to understand these for our comparison of the supersymmetry vacuum equations.

Comparison with the pure spinor equations

Among the equations (8.4.49a) for a supersymmetric vacuum in $\mathcal{N} = 1$ supergravity, the D-term equation is obtained by restricting to a level set for a moment map. From (6.3.46) and (8.4.7), we see that this is related to variations of K. The isometries ξ are the gauge symmetries (10.1.14) of the RR potentials, $\delta C = d_H \lambda$. In (6.3.46), we need the vector fields $I^t \xi$; because of (10.3.55), we identify these with the variations

$$\delta(e^{-A-\phi}\text{Im}\Phi_\pm) = d_H \lambda . \qquad (10.3.58)$$

Using then (10.3.19a):

$$\delta K \propto \int_{M_6} (e^{A-\phi}\text{Re}\Phi_\pm, \delta(e^{-A-\phi}\text{Im}\Phi_\pm)) \overset{(3.2.61)}{=} - \int_{M_6} (\lambda, d_H(e^{A-\phi}\text{Re}\Phi_\pm)) . \qquad (10.3.59)$$

Since this has to vanish for any λ, we recover (10.3.20a) for $c_- = 0$; so that equation is interpreted as a D-term. The fact that (10.3.20a) is redundant for $\Lambda < 0$ (as noted in (10.3.25)) agrees with our remarks following (8.4.49).

The other vacuum equation in (8.4.49a) is the F-term, which can be obtained by varying (10.3.56). Doing this carefully requires the formalism in Section 10.5.3 [447, sec. 4.2]. The result is that

$$D_T W = 0 \quad \Leftrightarrow \quad (10.3.20\text{b}) , \qquad D_Z W = 0 \quad \Leftrightarrow \quad (10.3.20\text{c}) . \qquad (10.3.60)$$

So the last two equations in (10.3.18) have to be considered F-terms.

This interpretation of the pure spinor equations as D- and F-terms could be the starting point of a stability analysis. The D-terms would be interpreted as a moment map under an appropriate infinite-dimensional group action, and we could try to show that they are equivalent to a stability condition using the Kempf–Ness theorem (6.3.44), along the lines of K-stability (Section 7.4.7) and of the DUY theorem (9.2.48). This would reduce the system to (10.3.20b) and (10.3.20c). This point of view was used recently in [449], in the framework of exceptional geometry (Section 10.6.4) to analyze moduli in terms of cohomology.

$\mathcal{N} = 2$ supergravity

We now consider the point of view of $\mathcal{N} = 2$ supergravity [256, 278, 450]. To avoid the aforementioned subtleties with the warped reduction, we will set $A = 0$. For definiteness, we work in IIA, but it is easy to adapt the results to IIB.

The first step is to identify the field spaces $\mathcal{M}_{\text{sK}}$ and $\mathcal{M}_{\text{qK}}$. As in the Calabi–Yau case, the latter is obtained as a fibration (9.1.34) over a second special Kähler space. The two are associated with the moduli spaces of deformations of $v = \Omega$ and $v = e^{B+iJ}$ as described in Section 9.1.2. As noted there, both are pure forms $\Phi_\pm$ for an SU(3)-structure. The Kähler potentials were (9.1.20) and (9.1.26):

$$e^{-K_\pm} = i \int_{M_6} (\bar{\Phi}_\pm, \Phi_\pm) . \tag{10.3.61}$$

Unlike in Section 9.1.2, these are now to be read as Kähler potentials for the infinite-dimensional moduli spaces of all pure spinors, not only the closed v we had for a Calabi–Yau M_6. For this generalization, it is useful to note that the integrand of (10.3.61) is the square root of a quartic function called *Hitchin invariant*, which we will discuss in Section 10.5.2.

The next step is to discuss the gauging. We work in the formalism introduced in (8.4.56), which allows magnetic gaugings. The hypermoment map for the vectors and their magnetic duals appears there in the combination

$$\mu^\alpha \equiv X^I \mu_I^\alpha + \mathcal{F}_I \tilde{\mu}^{\alpha I} . \tag{10.3.62}$$

Comparing the supersymmetry transformations in $d = 10$ and $d = 4$, (10.1.33) and (8.4.44), leads to the identification

$$\begin{aligned} \mu^1 + i\mu^2 &= 2e^{\frac{1}{2}K_- + \varphi} \int_{M_6} (\Phi_+, d_H \Phi_-) , \\ \mu^1 - i\mu^2 &= 2e^{\frac{1}{2}K_- + \varphi} \int_{M_6} (\Phi_+, d_H \bar{\Phi}_-) , \\ \mu^3 &= -\frac{1}{\sqrt{2}} e^{2\varphi} \int_{M_6} (\Phi_+, f) . \end{aligned} \tag{10.3.63}$$

(φ is the four-dimensional dilaton (8.3.3).) As we observed after (8.4.43), morally two of the μ_I^α can be assembled into an F-term, and correspondingly two of the μ^α can be assembled into a superpotential W. This depends on the way we break $\mathcal{N} = 2$ to $\mathcal{N} = 1$. For example, (10.3.56) corresponds to $\mu^2 + i\mu^3$.

The corresponding Killing vectors can be described as a generalization to (9.1.99) and (9.1.101):

$$K_I = 2 f_I \partial_a + M_I^A D_A + E_{IA} \tilde{D}^A , \qquad \tilde{K}^I = 2 \tilde{f}^I \partial_a , \tag{10.3.64}$$

linear combinations of the Killing vectors (9.1.35) of $\mathcal{M}_{\text{qK}}$. The flux quanta are $f_I = (f_6\, f_{4I})$, $\tilde{f}^I = (f_2^i\, f_0)$ as in Section 9.1.5. The other parameters are an extension of (8.3.16):

$$d_H \omega_I = M_I^A \alpha_A + E_{IA} \beta^A , \quad d_H \alpha_A = -E_{IA} \tilde{\omega}^I , \quad d_H \beta^A = M_I^A \tilde{\omega}^I , \quad d\tilde{\omega}^I = 0 . \tag{10.3.65}$$

We have included the effect of d_H also on $\omega_0 = 1$ and $\tilde{\omega}^0 = \text{vol}_6$. For $H = 0$, (10.3.65) reduces to (8.3.16). For $H \neq 0$, we see that $M_0^A = -h^A$, $E_{0A} = -\tilde{h}_A$; in the Calabi–Yau case, all forms are closed, and $H \wedge \omega_i = 0$ because it would be a harmonic five-form. This reproduces (9.1.99).

A further generalization can be obtained on T-folds (Section 4.2.6). Here d_H is replaced by a differential that maps a k-form also to forms of *lower* degree; as a result, (10.3.64) now also has D_A and $\tilde{D}^A$ entries in $\tilde{K}^I$. Such gaugings were studied, for example, in [131, 278, 450]. These more general theories should be considered with care, for the reasons explained in Section 4.2.6.

Vacuum equations

The vacuum equations are (8.4.53) and (8.4.55), with the modification (8.4.56). For an $\mathcal{N} = 1$ vacuum, we want to choose a particular combination $\bar{a}\zeta^1 + b\zeta^2$. A projection of (8.4.55) has a similar form as the first in (8.4.53), and together they give [278, (4.23)]

$$\begin{aligned} \frac{a}{2}(\mathrm{i}\mu^1 + \mu^2) &= \mathrm{i}\bar{b}\mu^3 = \bar{a}\sqrt{-\Lambda/3}\,\mathrm{e}^{\varphi - K_+/2}\,, \\ \frac{\bar{b}}{2}(-\mathrm{i}\mu^1 + \mu^2) &= \mathrm{i}a\mu^3 = b\sqrt{-\Lambda/3}\,\mathrm{e}^{\varphi - K_+/2}\,. \end{aligned} \tag{10.3.66}$$

This implies $(|a|^2 - |b|^2)\Lambda = 0$. If $\Lambda \neq 0$, we conclude $|a|^2 = |b|^2$. Since in $d = 10$ the supersymmetry parameters factorize as in (10.3.16), this can be interpreted as

$$||\eta^1||^2 = ||\eta^2||^2\,. \tag{10.3.67}$$

This agrees with (10.3.25). When $\Lambda = 0$, if $a \neq 0$, $b \neq 0$ we conclude that $\mu^\alpha = 0$; so with this assumption we fall back to the absence of $\mathcal{N} = 1$ compactifications, which first manifested itself with simpler assumptions in Section 8.4.3. The $d = 10$ interpretation is that Minkowski compactifications require the presence of an O-plane, which is by now familiar. We know that an orientifold imposes the conditions (1.4.46) on the fields; it also acts on the supersymmetry parameters, breaking supersymmetry by a half. So an effective theory with an orientifold involution is expected to have $\mathcal{N} = 1$ supersymmetry, and this matches the $d = 4$ vacuum analysis.

Beyond (10.3.66), one can show using (10.3.63) that the other supersymmetry conditions match the pure spinor equations; roughly speaking, the hyperino condition reproduces (10.3.20b), and the gaugino gives (10.3.20c), both mixed with the gravitino [278, sec. 4.2].

Truncation

So far we just rewrote type II supergravity as a $d = 4$ theory with infinitely many fields. If we want to translate this formal analysis to a usable $d = 4$ effective action, we face the hard task of truncation, namely choosing the finite basis $\Omega_{\rm fin}$ of forms (Section 8.3.3).

Putting aside the Calabi–Yau case, there are various cases where this issue can be tackled, to various degrees of satisfaction. The following is a partial list, including also reductions from M-theory, which face similar challenges as in type II:

- When M_6 is a coset, one can take the basis of forms in (10.3.65) to be left-invariant, (4.4.38). This was done in [451] for IIA, and in [452] for M-theory. Thanks to the group action, the resulting $\mathcal{N} = 2$ effective theories are consistent truncations.
- One can take the forms in (10.3.65) to be those defining a G-structure on M_6. For $G = \mathrm{SU}(3)$, there is only one $\omega_1 = J$, one $\alpha_1 = \mathrm{Re}\Omega$, $\beta^1 = \mathrm{Im}\Omega$, and $\tilde{\omega}^1 = -J^2/2$. The Hodge $*$ closes on these automatically, but closure of d as in (8.3.16) demands the nearly Kähler condition (7.5.18). This Ansatz for the forms was used in [453] for type IIA; this theory has $\mathcal{N} = 1$ vacua with $\Lambda < 0$, which

we will see in Section 11.3.2. This class partially overlaps with the previous one, since most nearly Kähler manifolds are also cosets (Section 7.5.2).

- A similar strategy in M-theory is to use an SU(3)-structure on M_7, requiring forms (w, J, Ω), (5.1.48). One way to impose closure of the forms is the Sasaki–Einstein condition (7.4.10) [454–456]. Its $\mathcal{N} = 1$ vacua $AdS_4 \times SE_7$ are famous and will be treated in Section 11.3.1.
- Another way to obtain a set of forms closed under d is by using fibrations. For example, on an S^1-fibration with the bundle metric (4.2.37), we can consider the one-form $D\psi$, along with a basis of forms on the base B_6. This was used in M-theory for S^1-fibrations over a Calabi–Yau's [457].
- Consider a Calabi–Yau M_6 with $H \neq 0$. Both the structure of (10.3.64) and the Strominger–Yau–Zaslow (SYZ) picture discussed in Section 9.2.4 suggest that mirror symmetry maps this to a manifold that is no longer a Calabi–Yau, where $H = 0$ but the action of d on the forms is nonzero. Looking at (10.3.64) and exchanging the role of the I and A indices leads to the conjecture that the mirror $\tilde{M}_6$ is a *half-flat* manifold, defined by $\mathrm{dRe}\Omega = 0 = \mathrm{d}J^2$ [458], and that only one even and one odd cycles get lost [459]. (The action of mirror symmetry on SU(3) intrinsic torsion was studied via T-duality on T^3 in [127].)
- For Sasaki–Einstein manifolds, a natural set of forms larger than the (η, J, Ω) defining the SU(3)-structure is that defining the so-called *Kohn–Rossi cohomology*. This plays a role in the computation of the KK spectrum [460], and it was proposed that it also leads to a supersymmetric reduction to an $\mathcal{N} = 2$ supergravity [461], at least solving the issues mentioned around (8.3.18).

10.3.7 Derivation of pure spinor equations*

We will now derive (10.3.20) in two different ways. This will be involved (especially with the second strategy), but remember that we are dealing with the most general case; once this is out of the way, we never have to look at such computations again.

Derivation from ten dimension

The quickest way to derive the equations is by using (10.2.16), (10.2.18), and (10.2.19) [434, sec. 4.1].

We start by evaluating K and $\tilde{K}$ in (10.2.4); $V^a_M = \overline{\epsilon^a}\Gamma_M\epsilon^a$, from (10.2.2). The internal components vanish because $(\eta^a_\pm)^\dagger\gamma_m\eta^a_\mp = 0$, from (3.4.19). The external components read

$$V^a_\mu = \mathrm{e}^A\left(\overline{\zeta_+}\gamma_\mu\zeta_+||\eta^a_+||^2 + \overline{\zeta_-}\gamma_\mu\zeta_-||\eta^a_-||^2\right) = 2\mathrm{e}^A v_\mu||\eta^a_+||^2\,, \tag{10.3.68}$$

where v_μ is the four-dimensional bilinear (3.3.45). We found in (10.2.18) that K^M is Killing. Now, raising the index, we find

$$K^\mu = \mathrm{e}^{-A}v^\mu\left(||\eta^1_+||^2 + ||\eta^2_+||^2\right)\,. \tag{10.3.69}$$

But v^μ is already a Killing vector: if $MS_4 = Mink_4$ it is constant, and if $MS_4 = AdS_4$ it is a bilinear of Killing spinors (Exercise 7.3.1). So the function it multiplies must be a constant c_+:

$$||\eta^1_+||^2 + ||\eta^2_+||^2 = c_+\mathrm{e}^A\,. \tag{10.3.70a}$$

Moreover, in (10.2.19) the right-hand side vanishes, since K is only along MS_4 and H is purely internal. So

$$||\eta_+^1||^2 - ||\eta_+^2||^2 = c_- e^{-A} \tag{10.3.70b}$$

for another constant c_-. (These two equations are also easy to derive from (10.3.17) directly.) In conclusion,

$$K = c_+ v\,, \qquad \tilde{K} = c_- v\,. \tag{10.3.71}$$

The bispinors (10.3.18) are related to the ten-dimensional Ψ defined in (10.2.6). To see the precise relation, we apply the Fierz decomposition (3.3.28):

$$\begin{aligned}\Psi &= \frac{1}{32}\sum_p \frac{1}{p!}\overline{\epsilon^2}\Gamma_{M_p\ldots M_1}\epsilon^1 dx^{M_1}\wedge\ldots\wedge dx^{M_p}\\ &= \frac{1}{32}\sum_{j,k}\frac{1}{k!\,j!}\overline{\epsilon^2}\Gamma_{m_j\ldots m_1\mu_k\ldots\mu_1}\epsilon^1 dx^{\mu_1}\wedge\ldots\wedge dx^{\mu_k}\wedge dy^{m_1}\wedge\ldots dy^{m_j}\,.\end{aligned} \tag{10.3.72}$$

In the second line, we have divided the ten-dimensional coordinates in spacetime, x^μ, and internal, y^m. Now we can factorize the gamma matrices according to (8.2.6), and the spinors according to (10.3.16). This splits the sum (10.3.72) according to $d = 4$ and $d = 6$ chirality, creating four types of terms:

$$\begin{aligned}\Psi = \frac{1}{32}\sum_{j,k}\frac{e^{kA}}{k!\,j!}\Big[&\overline{\zeta_\mp}\gamma_{(4)}^j\gamma_{\mu_k\ldots\mu_1}\zeta_+\eta_+^{2\dagger}\gamma_{m_j\ldots m_1}\eta_+^1 + \overline{\zeta_\mp}\gamma_{(4)}^j\gamma_{\mu_k\ldots\mu_1}\zeta_-\eta_+^{2\dagger}\gamma_{m_j\ldots m_1}\eta_-^1\\ &+\overline{\zeta_\pm}\gamma_{(4)}^j\gamma_{\mu_k\ldots\mu_1}\zeta_+\eta_-^{2\dagger}\gamma_{m_j\ldots m_1}\eta_+^1 + \overline{\zeta_\pm}\gamma_{(4)}^j\gamma_{\mu_k\ldots\mu_1}\zeta_-\eta_-^{2\dagger}\gamma_{m_j\ldots m_1}\eta_-^1\Big]\\ &\times dx^{\mu_1}\wedge\ldots\wedge dx^{\mu_k}\wedge dy^{m_1}\wedge\ldots dy^{m_j}\,, \quad {}^{\text{IIA}}_{\text{IIB}}\end{aligned} \tag{10.3.73}$$

$\gamma_{(4)}$ being the chiral matrix in four dimensions. The four-dimensional coefficients reconstruct the bispinors $\zeta_+\otimes\overline{\zeta_\pm}$, which we can read off from (3.3.53). We also have to take care of the factors of e^{kA} in (10.3.73), which ultimately come from (8.2.6). In the internal directions, we recognize the forms $\Phi_\pm$ defined in (10.3.18). All in all,

$$\begin{aligned}\Psi &= \mp(e^A v + i e^{3A} *_4 v)\wedge\Phi_\mp + e^{2A}\omega\wedge\Phi_\pm + \text{complex conj.}\\ &= \mp 2e^A v\wedge\text{Re}\Phi_\mp + 2e^{2A}\text{Re}(\omega\wedge\Phi_\pm) \pm 2e^{3A} * v\wedge\text{Im}\Phi_\mp\,.\end{aligned} \tag{10.3.74}$$

To evaluate the exterior derivative on (10.3.74), we split $d = dx^\mu\wedge\partial_\mu + dy^m\partial_m = d_4 + d_6$. The action of d_4 is conceptually similar to (10.2.9), but much simpler. Taking some liberty to confuse bispinors and forms:

$$\begin{aligned}2d(\zeta_+\otimes\overline{\zeta_+}) &\overset{(3.2.8a),(4.3.46)}{=} [\gamma^\mu, D_\mu(\zeta_+\otimes\overline{\zeta_+})] \overset{(10.3.10)}{=} \frac{\mu}{2}[\gamma^\mu, \gamma_\mu\zeta_-\otimes\overline{\zeta_+} - \zeta_+\otimes\overline{\zeta_-}\gamma_\mu]\\ &= \mu(4\text{Re}(\zeta_+\otimes\overline{\zeta_-}) - \gamma^\mu\text{Re}(\zeta_+\otimes\overline{\zeta_-})\gamma_\mu) \overset{(3.2.23)}{=} 4\mu\text{Re}(\zeta_+\otimes\overline{\zeta_-})\,.\end{aligned} \tag{10.3.75}$$

The computation for $d_4\zeta_+\otimes\overline{\zeta_-}$ is similar. With (3.3.53), we conclude

$$d_4 v = 2\mu\text{Re}\omega\,, \qquad d_4\omega = 3i\mu * v\,. \tag{10.3.76}$$

Now we evaluate (10.2.16), using (10.3.76) on the left-hand side and (10.3.2) and (10.3.71) on the right-hand side. We also use $\iota_v\text{vol} = -* v$ (easy to check in a basis, or using Section 3.2). This gives

$$v\wedge(10.3.20\text{a}) + \text{Re}(\omega\wedge(10.3.20\text{b})) + *v\wedge(10.3.20\text{c}) = 0\,, \tag{10.3.77}$$

which completes the derivation of (10.3.20).

We have used (10.2.16), (10.2.18), and (10.2.19). In Section 10.2.5, we have seen that these three equations are not equivalent to supersymmetry; one possible way to complement them is (10.2.37). For the present case of four-dimensional vacua, those equations are redundant [434]. This shows that (10.3.20) are not only implied, but also equivalent to supersymmetry. We will give an alternative argument for this in Section 10.3.8.

Derivation from factorized supersymmetry

It is also possible to derive (10.3.20) more directly from (10.3.17) [140, app. A]. This mimics (10.2.9), and we will be brief.

The first equation we derive is (10.3.20b):

$$
\begin{aligned}
2\mathrm{d}\Phi_\pm &\overset{(3.2.8a),(4.3.46)}{=} \gamma^m D_m(\eta^1_+ \otimes \eta^{2\,\dagger}_\pm) \pm D_m(\eta^1_+ \otimes \eta^{2\,\dagger}_\pm)\gamma^m \qquad (10.3.78)\\
&= (D\eta^1_+) \otimes \eta^{2\,\dagger}_\pm + \gamma^m \eta^1_+ \otimes D_m \eta^{2\,\dagger}_\pm \pm (D_m\eta^1_+) \otimes \eta^{2\,\dagger}_\pm \gamma^m \pm \eta^1_+ \otimes (D\eta^2_\pm)^\dagger\\
&\overset{(10.3.17)}{=} \left(-2\mu \mathrm{e}^{-A}\eta^1_- - \mathrm{d}(2A-\phi)\eta^1_+ + \frac{1}{4}H\eta^1_+\right) \otimes \eta^{2\,\dagger}_\pm\\
&+ \gamma^m\eta^1_+ \otimes \left(\frac{1}{4}\eta^{2\,\dagger}_\pm H_m + \frac{\mathrm{e}^\phi}{8}\eta^{1\,\dagger}_- \gamma_m f\right) \pm \left(\frac{1}{4}H_m\eta^1_+ - \frac{\mathrm{e}^\phi}{8} f\gamma_m \eta^2_\mp\right) \otimes \eta^{2\,\dagger}_\pm \gamma^m\\
&\pm \eta^1_+ \otimes \left(2\mu \mathrm{e}^{-A}\eta^{2\,\dagger}_\mp - \eta^{2\,\dagger}_\pm \mathrm{d}(2A-\phi) + \frac{1}{4}\eta^{2\,\dagger}_\pm H\right) \qquad {}^{\text{IIA}}_{\text{IIB}}\\
&\overset{(3.2.23),\,(3.2.8a),\,(10.2.11)}{=} -2\mathrm{d}(2A-\phi)\wedge\Phi_\pm + 2H\wedge\Phi_\pm - 2\mu \mathrm{e}^{-A}(\eta^1_- \otimes \eta^{2\,\dagger}_\pm \mp \eta^1_+ \otimes \eta^{2\,\dagger}_\mp)\,.
\end{aligned}
$$

In the last step, (3.2.23) and (3.4.20) together imply $\gamma_m \eta^a_+ \otimes \eta^{a\,\dagger}_- \gamma^m = 0$. Now for the last parenthesis we have to recall that $\overline{\Phi_-} = -\not{\bar\Phi}_-$, as in (3.4.29). Taking this into account, we obtain (10.3.20b).

Next we consider (10.3.20a) and (10.3.20c) together. We skip some of the steps that are similar to (10.3.78):

$$
\begin{aligned}
2\mathrm{d}\Phi_\mp &= (D\eta^1_+) \otimes \eta^{2\,\dagger}_\mp + \gamma^m\eta^1_+ \otimes D_m\eta^{2\,\dagger}_\mp \mp (D_m\eta^1_+) \otimes \eta^{2\,\dagger}_\mp \gamma^m \mp \eta^1_+ \otimes (D\eta^2_\mp)^\dagger\\
&\overset{(10.3.17)}{=} \left(-2\mu \mathrm{e}^{-A}\eta^1_- - \mathrm{d}(2A-\phi)\eta^1_+ + \frac{1}{4}H\eta^1_+\right) \otimes \eta^{2\,\dagger}_\mp \qquad (10.3.79)\\
&+ \gamma^m\eta^1_+ \otimes \left(\frac{1}{4}\eta^{2\,\dagger}_\mp H_m \mp \frac{\mathrm{e}^\phi}{8}\eta^{1\,\dagger}_+ \gamma_m f\right) \mp \left(\frac{1}{4}H_m\eta^1_+ - \frac{\mathrm{e}^\phi}{8} f\gamma_m \eta^2_\mp\right) \otimes \eta^{2\,\dagger}_\mp \gamma^m\\
&\mp \eta^1_+ \otimes \left(-2\mu \mathrm{e}^{-A}\eta^{2\,\dagger}_\pm - \eta^{2\,\dagger}_\mp \mathrm{d}(2A-\phi) + \frac{1}{4}\eta^{2\,\dagger}_\mp H\right) \qquad {}^{\text{IIA}}_{\text{IIB}}\\
&\overset{(3.2.8a),(10.2.11),(3.4.30)}{=} -2\mathrm{d}(2A-\phi)\wedge\Phi_\mp + 2H\wedge\Phi_\mp \pm 4\mathrm{i}\mu \mathrm{e}^{-A}\mathrm{Im}\Phi_\pm\\
&\mp \frac{\mathrm{e}^\phi}{8}\left[\left(||\eta^1_+||^2(1-\gamma) - 2\eta^1_- \otimes \eta^{1\,\dagger}_-\right) f - f\left(||\eta^2_\mp||^2(1\pm\gamma) - 2\eta^2_\pm \otimes \eta^{2\,\dagger}_\pm\right)\right]\,.
\end{aligned}
$$

The last line can be rewritten using (10.3.17c), (10.3.17d), and (10.3.70), and $\gamma f = -\mathrm{i} * \lambda f$ (from (3.2.27)). This finally gives

$$
\begin{aligned}
\mathrm{d}_H(\mathrm{e}^{2A-\phi}\Phi_\mp) &= \mp\frac{1}{16}(c_-\mathrm{e}^A f + \mathrm{i}c_+\mathrm{e}^{3A} * \lambda f)\\
&+ \mathrm{e}^{2A-\phi}\mathrm{d}A\wedge\bar\Phi_\mp \pm 3\mathrm{i}\mu \mathrm{e}^{A-\phi}\mathrm{Im}\Phi_\pm\,,
\end{aligned}
\qquad (10.3.80)
$$

whose real and imaginary parts are (10.3.20a) and (10.3.20c) respectively.

10.3.8 Equivalence to supersymmetry*

We have shown in two different ways that the pure spinor equations (10.3.20) follow from supersymmetry. We will now show that they are actually equivalent to them: (10.3.20) implies the original spinorial system (10.3.17).

As in Section 10.2.5, we need an appropriate notion of intrinsic torsion. For a single SU(3)-structure, (5.5.2) was obtained by expanding $\nabla\eta_+$ in the natural basis (3.4.86) associated to η_+ itself. But now we have two $\eta_+^{1,2}$. When they coincide, we can use the theory of SU(3) intrinsic torsion; when they don't, we have seen in Section 10.3.4 that they define an SU(2)-structure in $d = 6$. It would be annoying to have to consider these cases separately. We will define a notion of intrinsic torsion by expanding $\nabla\eta_+^1$ in the basis associated to η_+^1, and $\nabla\eta_+^2$ in the basis associated to η_+^2, as if the two SU(3)-structures defined by $\eta_+^{1,2}$ ignored each other. It is convenient to include in the definition the H_m terms in (10.3.17):

$$\left(D_m - \frac{1}{4}H_m\right)\eta_+^1 = (\mathrm{i}Q_m^1 + \partial_m \log||\eta_+^1||)\eta_+^1 + \mathrm{i}Q_{mn}^1\gamma^n\eta_-^1$$
$$\left(D_m + \frac{1}{4}H_m\right)\eta_+^2 = (\mathrm{i}Q_m^2 + \partial_m \log||\eta_+^2||)\eta_+^2 + \mathrm{i}Q_{mn}^2\gamma^n\eta_-^2 . \qquad (10.3.81a)$$

We would also need coefficients for the action of H on $\eta_+^{1,2}$, but it is more convenient to define

$$\left(D - \frac{1}{4}H\right)\eta_+^1 = T_m^1\gamma^m\eta_+^1 + T^1\eta_-^1 , \qquad \left(D + \frac{1}{4}H\right)\eta_+^2 = T_m^2\gamma^m\eta_+^2 + T^2\eta_-^2 . \qquad (10.3.81b)$$

Next we need a basis for forms. We have seen one in (3.4.94), associated to a single spinor; in view of (10.3.20), we would like to have a version of that diamond associated to two spinors, where the $\Phi_\pm = \eta_+^1 \otimes \eta_\pm^{2\dagger}$ and their conjugates sit at the four corners. A subtlety arises here: in (3.4.94) we were using (anti)holomorphic indices adapted to η_+, but here we have two $\eta_\pm^{1,2}$. We then need to define *two* sets of (anti)holomorphic indices, which we denote by $i_1, i_2, \bar{\imath}_1, \bar{\imath}_2$.[6] The definition is the natural extension of (2.4.5) and (3.4.87):

$$\gamma_{\bar{\imath}_a}\eta_+^{1,2} = \frac{1}{2}(\delta + \mathrm{i}I^{1,2})_{\bar{\imath}_a}{}^n\gamma_n\eta_+^{1,2} = 0 . \qquad (10.3.82)$$

With this definition, we can now write the natural generalization of (3.4.94):

$$\begin{array}{ccccccc} & & & \Phi_+ & & & \\ & & \Phi_+\gamma_{\bar{\imath}_2} & & \gamma_{i_1}\Phi_+ & & \\ & \Phi_-\gamma_{i_2} & & \gamma_{i_1}\Phi_+\gamma_{\bar{\jmath}_2} & & \gamma_{\bar{\imath}_1}\bar{\Phi}_- & \\ \Phi_- & & \gamma_{i_1}\Phi_-\gamma_{j_2} & & \gamma_{\bar{\imath}_1}\bar{\Phi}_-\gamma_{\bar{\jmath}_2} & & \bar{\Phi}_- \\ & \gamma_{i_1}\Phi_- & & \gamma_{\bar{\imath}_1}\bar{\Phi}_+\gamma_{j_2} & & \bar{\Phi}_-\gamma_{\bar{\imath}_2} & \\ & & \gamma_{\bar{\imath}_1}\bar{\Phi}_+ & & \bar{\Phi}_+\gamma_{i_2} & & \\ & & & \bar{\Phi}_+ & & & \end{array} . \qquad (10.3.83)$$

The orthogonality property (3.4.96) still holds.

[6] There is an abuse of language here: the almost complex structures $I^{1,2}$ associated to the $\eta_+^{1,2}$ are in general not integrable. A holomorphic index should just be understood as the presence of a holomorphic projector $\frac{1}{2}(\delta - \mathrm{i}I)$.

Now the sufficiency proof works as follows. From now on, we specialize to IIB; very little changes for IIA. First we expand the RR flux F in the basis (10.3.83):

$$\begin{aligned} F = R^{10}_{i_2}\,\Phi_+\gamma^{i_2} + R^{01}_{\bar{i}_1}\,\gamma^{\bar{i}_1}\Phi_+ + \\ R^{30}\,\Phi_- + R^{21}_{\bar{i}_1\bar{j}_2}\gamma^{\bar{i}_1}\,\Phi_-\,\gamma^{\bar{j}_2} + R^{12}_{i_1 j_2}\gamma^{i_1}\,\bar{\Phi}_-\gamma^{j_2} + R^{03}\,\bar{\Phi}_- \\ + R^{32}_{i_1}\gamma^{i_1}\,\bar{\Phi}_+ + R^{23}_{\bar{i}_2}\,\bar{\Phi}_+\gamma^{\bar{i}_2}\,. \end{aligned} \tag{10.3.84}$$

The action of F on the $\eta^{1,2}$ can be evaluated along the lines of Section 3.4.7, and in particular using (3.4.88); for example,

$$F\gamma_m\eta^2_+ = 2R^{10}_{i_2}\Pi^{i_2}_{2\ m}\eta^1_+ + 2R^{12}_{i_1 j_2}\Pi^{j_2}_{1\ m}\gamma^{i_1}\eta^1_-\,. \tag{10.3.85}$$

With this and (10.3.81a) and (10.3.81b), we obtain from (10.3.17) a set of spinorial equations, linear combinations of the basis (3.4.86), either relative to η^1 or to η^2. So the coefficients have to vanish independently, leading to

$$\begin{aligned} &Q^1_{\bar{i}_2 j_1} = 0\,, \qquad \mathrm{i}Q^1_{i_2 j_1} = \frac{\mathrm{e}^{\phi}}{4}||\eta^2_+||^2 R^{12}_{j_1 i_2}\,,\\ &\mathrm{i}Q^1_{\bar{i}_2} + \partial_{\bar{i}_2}\log||\eta^1_+|| = 0 = \mathrm{i}Q^1_{i_2} + \partial_{i_2}\log||\eta^1_+|| + \frac{\mathrm{e}^{\phi}}{4}||\eta^2_+||^2 R^{10}_{i_2}\,,\\ &\mu\,\mathrm{e}^{-A} + \frac{\mathrm{e}^{\phi}}{4}||\eta^2_+||^2 R^{03} = 0\,, \qquad \frac{\mathrm{e}^{\phi}}{4}||\eta^2_+||^2 R^{01}_{\bar{i}_1} = \partial_{\bar{i}_1}A\,,\\ &2\mu\,\mathrm{e}^{-A} + T^1 = 0\,, \qquad T^1_{\bar{i}_1} + \partial_{\bar{i}_1}(2A-\phi) = 0, \end{aligned} \tag{10.3.86}$$

along with a second set obtained by $1 \leftrightarrow 2$, $R^{ij} \to R^{3-j,3-i}$.

On the other hand, we can also reformulate (10.3.20) using the intrinsic torsion (10.3.81a) and (10.3.81b). The computation is similar to (10.3.78) and (10.3.79), and gives

$$\begin{aligned} \mathrm{d}_H\Phi_+ &= (-\mathrm{i}Q^2_{\bar{i}_1} + T^1_{\bar{i}_1} + \partial_{\bar{i}_1}\log||\eta^2_+||)\gamma^{\bar{i}_1}\Phi_+ + (\mathrm{i}Q^1_{i_2} + T^2_{i_2} + \partial_{i_2}\log||\eta^1_+||)\Phi_+\gamma^{i_2}\\ &\quad + T^1\,\bar{\Phi}_- + \bar{T}^2\,\Phi_- - \mathrm{i}Q^2_{\bar{i}_1\bar{i}_2}\gamma^{\bar{i}_1}\Phi_-\gamma^{\bar{i}_2} - \mathrm{i}Q^1_{i_2 i_1}\gamma^{i_1}\bar{\Phi}_-\gamma^{i_2}\,;\\ \mathrm{d}_H\Phi_- &= (\mathrm{i}Q^2_{\bar{i}_1} + T^1_{\bar{i}_1} + \partial_{\bar{i}_1}\log||\eta^2_+||)\gamma^{\bar{i}_1}\Phi_- - (\mathrm{i}Q^1_{\bar{i}_2} + T^2_{\bar{i}_2} + \partial_{\bar{i}_2}\log||\eta^1_+||)\Phi_-\gamma^{\bar{i}_2}\\ &\quad + T^1\,\bar{\Phi}_+ + T^2\,\Phi_+ - \mathrm{i}Q^2_{\bar{i}_1 i_2}\gamma^{\bar{i}_1}\Phi_+\gamma^{i_2} - \mathrm{i}Q^1_{\bar{i}_2 i_1}\gamma^{i_1}\bar{\Phi}_+\gamma^{\bar{i}_2}\,. \end{aligned} \tag{10.3.87}$$

These can be understood as the expansion of $\mathrm{d}_H\Phi_\pm$ on (10.3.83). Each of the parentheses can be thought of as a generalization of the intrinsic torsion W_i in (5.5.6) for SU(3)-structures; (10.3.87) is then similar to (5.5.12).

Comparing (10.3.87) to the expansion (10.3.84) of F in the same basis, we get a set of equations that is in fact a linear combination of (10.3.86). So (10.3.17) and (10.3.20) are indeed equivalent.

Exercise 10.3.1 Check (10.3.1). (It is enough to do so on $\alpha = e^4 \wedge \ldots \wedge e^k$, where e^A is a vielbein on M_{10}.)

Exercise 10.3.2 Check that (10.3.17) reduce to (9.3.2), (9.4.5), and (9.5.5).

Exercise 10.3.3 Check using (3.2.67) that the f contribution in (10.3.6b) changes sign under $f \to *f$.

Exercise 10.3.4 Check that the conformal Kähler class solution (Section 9.5) solves (10.3.6). (The result of the previous exercise might be useful.)

Exercise 10.3.5 Derive (10.3.70) directly from (10.3.17).

Exercise 10.3.6 Check (10.3.39) by computing $\eta^1_+ \otimes \eta^{1\dagger}_+$ and $\eta^2_+ \otimes \eta^{2\dagger}_+$, similar to how we computed (10.3.36).

Exercise 10.3.7 Check that the limit of (10.3.37) for $\psi \to 0$ reproduces (10.3.28), and for $\psi \to \pi/2$ reproduces (10.3.31).

Exercise 10.3.8 Derive (10.3.49) from (10.3.20) and (1.4.46). (Notice that $R^*_p F = -(-1)^{\lfloor p/2 \rfloor} \lambda(F)$.)

Exercise 10.3.9 Set $f = 0$ in (10.3.20). Show that this implies A is constant. (Hint: you can use that $\mathrm{d}\Phi_\pm$ in (10.3.87) has no component along $\bar{\Phi}_\pm$, and that the coefficient of $\mathrm{d}\Phi_\pm$ along $\Phi_\mp$ are related. Going back to (9.3.2) is considered cheating.)

10.4 Minkowski examples

In this section, we will specialize the pure spinor equations (10.3.20) to various Ansätze for the pure forms $\Phi_\pm$. We will focus on Minkowski solutions, leaving AdS_4 to the next chapter. We will recover some classes we know already, and derive some new ones. Because we need O-planes, in this section we will take $c_- = 0$, as we concluded after (10.3.49). We can also set $c_+ = 2$ without loss of generality.

We will say that a solution is "of Dp type" if a space-filling Dp probe is BPS on it. This can be useful to determine what Op-plane it can admit, which in turn is important to overcome the no-go arguments in the Introduction and in Section 10.3.5. Some Minkowski solutions can be interpreted as the back-reaction of space-filling Dp-branes on a preexisting $\mathrm{Mink}_3 \times M_6$; in this case, a probe Dp of the same type should be BPS. For example, for the conformal Kähler class in Section 9.5, the solutions with $G_3 = 0$ arise from the backreaction of D3-branes on a Calabi–Yau. When $G_3 \neq 0$, this is not literally true, but we still say the class is of D3 type.

In Section 10.4.1, we will consider several classes where the pair $\Phi_\pm$ is of SU(3)-structure type (10.3.28), and in Section 10.4.2 a class where it is of SU(2)-structure type (10.3.31). This is not meant to provide an exhaustive list of solutions, but only to present some notable cases. There are many other local solutions we will not cover here. Some are motivated by holography; for example, we know from (8.2.20) that AdS_5 vacua can be seen as Mink_4 solutions with a noncompact M_6, but solutions representing RG-flows are also of this type [462, 463].

With compact M_6, there are far fewer known types of solutions. All currently known solutions are uncomfortably close to Calabi–Yau geometry: either they represent the brane back-reaction on a $\mathrm{Mink}_4 \times \mathrm{CY}_6$, or they are related to such a solution by dualities (such as those in [464]). This comes a bit as a surprise, given that the pure spinor equations (10.3.20) don't seem to suggest a particular role for Calabi–Yau manifolds; rather, we will see in Section 10.5 that they naturally select a more general type of geometry. This scarcity of examples might be in part caused by technical difficulties, such as the no-go arguments. One interesting set of possible counterexamples was obtained in [465–468], where several analytic solutions were found, which were argued to become compact upon gluing by S- and T-dualities (as with the T-folds of Section 4.2.6). It would be important to analyze these solutions further, for example understanding better the physics interpretation of their degeneration loci.

In a different direction, a general classification was undertaken [469–471] of Mink_4 solutions with an S^2 factor of internal space, in IIA, IIB, and M-theory. In particular, in type IIA and IIB it was shown that all solutions can be derived from two "master" classes and sequences of dualities.

10.4.1 Minkowski solutions from SU(3)-structures

The simplest type of pure spinor pair is (10.3.28), associated to an SU(3)-structure. Since we are taking $c_- = 0$, $c_+ = 2$, the norm $N = \mathrm{e}^A$, and (10.3.28) becomes

$$\Phi_+ = \frac{1}{8}\mathrm{e}^{A+\mathrm{i}\theta}\mathrm{e}^{-\mathrm{i}J}\,, \qquad \Phi_- = \frac{1}{8}\mathrm{e}^{A}\Omega\,. \tag{10.4.1}$$

As usual, recall that the SU(3)-structure (J, Ω) satisfies the algebraic conditions (5.1.40).

A first general result now follows from (10.3.24b):

$$\begin{aligned} &\mathrm{IIA}:\ \mathrm{d}J = 0 && \Rightarrow \quad M_6 \text{ is symplectic}\,,\\ &\mathrm{IIB}:\ \mathrm{d}(\mathrm{e}^{3A-\phi}\Omega) = 0 && \Rightarrow \quad M_6 \text{ is complex}\,, \end{aligned} \tag{10.4.2}$$

recalling respectively the definition of symplectic manifold in Section 5.4.1, and the criterion (5.3.18) for integrability of the almost complex structure associated to a complex, decomposable three-form Ω.

Solutions of D3/D7 type: M_6 Kähler

We begin by considering IIB. First we want to consider solutions with O3-planes. These are expected to reproduce the important conformal Kähler class of Section 9.5 [411–414].

From (10.3.50), we see that $(\mathrm{Im}\Phi_+)_0 = \mathrm{e}^A \sin\theta$ should be even under R_p; in other words,

$$R_p^*\theta = \pi - \theta\,. \tag{10.4.3}$$

On an O3 we then have $\theta = \pi/2$. The same conclusion also holds for an O7, because (10.3.49) remains unchanged under $p \to p + 4$.

We choose to satisfy (10.4.3) by making the *Ansatz* that

$$\theta = \frac{\pi}{2} \tag{10.4.4}$$

everywhere. We now use (10.4.1) in the pure spinor equations (10.3.24). We look at each form degree separately. For example, $8\mathrm{Re}\Phi_+ = \mathrm{e}^A(-J + \mathrm{vol}_6)$, so (10.3.24a) has a three-form part $\mathrm{d}(\mathrm{e}^{2A-\phi}J) = 0$ and a five-form part $H \wedge J = 0$. We continue in this way for (10.3.24b) and (10.3.24c):

$$\begin{aligned} &\mathrm{d}(\mathrm{e}^{3A-\phi}\Omega) = 0\,, \qquad \mathrm{d}(\mathrm{e}^{2A-\phi}J) = 0\,, \qquad H\wedge\Omega = H\wedge\tilde{J} = 0\,,\\ &*f_5 = \mathrm{e}^{-\phi}\mathrm{d}(4A-\phi)\,, \qquad *f_3 = e^{-\phi}H\,, \qquad *f_1 = -\frac{1}{2}\mathrm{e}^{-\phi}\mathrm{d}\phi\wedge J^2\,. \end{aligned} \tag{10.4.5}$$

The first two of (10.4.5) determine the geometry; in terms of the intrinsic torsion (5.5.6), we see that $W_1 = W_2 = W_3 = 0$, $W_4 = \mathrm{d}(-2A + \phi)$, $W_5 = \bar\partial(-3A + \phi)$.

This reproduces (9.5.9), with a lot less effort. Of course, this is thanks to the long general work in Section 10.3, which now pays off. As we discussed in (9.5.10), these torsion classes correspond to a conformal Kähler metric.

Likewise, the equations for f_5, f_3 in (10.4.5) reproduce (9.5.6) and (9.5.8). Using (3.1.11) and (5.2.21), the f_1 equation becomes (9.4.8), the holomorphic property of the axio-dilaton. We will see in Section 10.5.6 an expression for the RR flux alternative to (10.3.24c), where the Hodge $*$ will not appear, and $\bar{\partial}\tau = 0$ will be reproduced even more directly.

Because of (10.4.4), (5.6.14) and the discussion after it show that spacetime-filling D3-branes are BPS at every point, and D7-branes are BPS on every holomorphic four-cycle $S_4 \subset M_6$. So the conformal Kähler class is of D3/D7-type. As a simple check, the D3-brane and D7-brane solutions from (1.3.59) can be thought of as solutions to (10.4.5) (Exercises 10.4.2 and 10.4.4). In Sections 9.4 and 9.5, we saw how to construct several more interesting examples; this is indeed the most abundant known class of Minkowski solutions.

Solutions of D5-type: M_6 complex

Remaining in IIB, we now look at solutions with O5-planes. Now (10.3.50) implies that $(\mathrm{Im}\Phi_+)_0 = \mathrm{e}^A \cos\theta$ should be even under R_5, so (10.4.3) is replaced by $R_5^*\theta = -\theta$. On the O5-plane, extended in M_6 along a two-dimensional locus, we have $\theta = 0, \pi$. Again we strengthen this conclusion by making an Ansatz:

$$\theta = 0\,. \tag{10.4.6}$$

Now (10.4.1) and (10.3.24), separated by form degree, give

$$\begin{aligned} &A = \frac{\phi}{2} + \text{const}\,, \qquad \mathrm{d}(\mathrm{e}^{-\phi/2}\Omega) = 0\,, \qquad \mathrm{d}J^2 = 0\,, \qquad H = 0\,; \\ &f_5 = f_1 = 0\,, \qquad *f_3 = \mathrm{e}^{-2\phi} * \mathrm{d}(\mathrm{e}^{\phi} J)\,. \end{aligned} \tag{10.4.7}$$

Equations (5.6.14) and (10.4.6) guarantee that D5-branes wrapping any holomorphic two-dimensional subspace $S_2 \subset M_6$ will be BPS. The Ω equation in (10.4.7) is the general result (10.4.2), that M_6 is complex. Since $\mathrm{d}J \neq 0$ in general, it is not Kähler any more.

Equation (10.4.7) is formally rather similar to (9.3.13). Recalling from (9.3.29) that the NS5 is a solution of the latter, we are led to conclude that the two systems are S-dual to one another. This can indeed be checked directly from the S-duality transformations (1.2.38) (Exercise 10.4.1).

It is not easy to find compact examples of this type. Some formal solutions exist on the nilmanifold defined by the structure constants [140]

$$(0,\, 0,\, 0,\, 12,\, 23,\, 14 - 35)\,, \tag{10.4.8}$$

in the notation discussed following (4.4.58), and on the solmanifold with structure constants (25, −15, 45, −35, 0, 0). They are formal in that they have been shown to exist only with a smeared source for the O5. We have discussed this smearing, for example in Section 4.2.6 and after (9.2.10); it makes sense for distributions of large numbers of dynamical objects, such as D-branes, but not for O-planes. So these solutions don't make physical sense; it remains to be seen if the smearing assumption can be removed. Similar solutions have been obtained in [472] for SU(2)-structure and in [473] for solutions with other Op-planes.

Solutions of D6-type: M_6 symplectic

We now consider the IIA case. Here (10.3.49) says that

$$R_6^*\theta = \theta\,, \qquad R_6^*J = -J\,, \qquad R_6^*\mathrm{Re}\Omega = -\mathrm{Re}\Omega\,, \qquad R_6^*\mathrm{Im}\Omega = \mathrm{Im}\Omega\,. \quad (10.4.9)$$

Now the one-form part of (10.3.24b) immediately gives that $3A - \phi$ and θ are constant; moreover, in the rest of the equations, the latter drops out, in contrast with the important role it played in (10.4.5) and (10.4.7). The full list of conditions from (10.3.24) is

$$\begin{aligned} &A = \frac{\phi}{3} + \text{const}\,, \qquad \mathrm{d}J = H = 0\,, \qquad \mathrm{d}(\mathrm{e}^{-A}\mathrm{Re}\Omega) = 0\,, \\ &f_6 = f_4 = f_0 = 0\,, \qquad \mathrm{d}(\mathrm{e}^{4A-\phi}\mathrm{Im}\Omega) = \mathrm{e}^{4A} * f_2\,. \end{aligned} \quad (10.4.10)$$

The J equation is (10.4.2): M_6 is symplectic.

In solutions of this type, D6-brane probes wrapping a subspace $S_3 \subset M_6$ are BPS by construction, calibrated by $\mathrm{Im}\Omega$. We recall from (5.6.20) that such S_3 are called almost special Lagrangian. The back-reaction of a D6 stack itself on flat space gives a solution in (10.4.10). To see this, take the internal space M_6 topologically $\mathbb{R}^6$, with a vielbein $e^{4,5,6} = h^{-1/4}\mathrm{d}x^{4,5,6}$, $e^{7,8,9} = h^{1/4}\mathrm{d}x^{7,8,9}$, with $h = 1 + Ng_s/(2\pi r)$, $r^2 = \sum_{i=7}^{9}(x^i)^2$. Now

$$\begin{aligned} &\mathrm{e}^{\phi} = g_s h^{-3/4} = g_s \mathrm{e}^{3A}\,, \qquad f_2 = \frac{N}{2\pi r^3}(x^7\mathrm{d}x^8 \wedge \mathrm{d}x^9 + x^8\mathrm{d}x^9 \wedge \mathrm{d}x^7 + x^9\mathrm{d}x^7 \wedge \mathrm{d}x^8)\,; \\ &J = e^{47} + e^{58} + e^{69}\,, \qquad \Omega = \mathrm{i}(e^4 + \mathrm{i}e^7) \wedge (e^5 + \mathrm{i}e^8) \wedge (e^6 + \mathrm{i}e^9) \end{aligned} \quad (10.4.11)$$

is a solution of (10.4.10). The corresponding metric reproduces (1.3.59) for $p = 6$.

Since $F_0 = 0$, we may lift a solution of (10.4.10) to M-theory. The only form field present is F_2, so the resulting solution should be purely geometrical. Using (1.4.5a) and $\phi = 3A + \text{const}$, we see that $\mathrm{d}s^2_{11} = \mathrm{d}s^2_4 + \mathrm{d}s^2_7$, so the product is unwarped; and

$$\mathrm{d}s^2_7 = \mathrm{e}^{-2A}\mathrm{d}s^2_6 + \mathrm{e}^{4A}(\mathrm{d}\psi - C_1)^2\,, \quad (10.4.12)$$

A being the IIA warping and ψ the M-theory circle. We know that this solution has $\mathcal{N} = 1$; the logic in Section 8.2 tells us that the metric (10.4.12) should admit a covariantly constant spinor, and at least have G_2 holonomy [474, 475].

We can prove this expectation correct by introducing the G_2-structure forms

$$\phi_3 = e^{-3A}\mathrm{Im}\Omega + (\mathrm{d}\psi - C_1) \wedge J\,, \qquad *\phi_3 = -\frac{1}{2}\mathrm{e}^{-4A}J^2 + \mathrm{e}^{-A}\mathrm{Re}\Omega \wedge (\mathrm{d}\psi - C_1)\,, \quad (10.4.13)$$

which can be obtained from (5.1.52) upon rescaling $J \to \mathrm{e}^{-2A}J$, $\Omega \to \mathrm{e}^{-3A}\Omega$ as suggested by (10.4.12). The equations in (10.4.10) imply that $\mathrm{d}\phi_3 = \mathrm{d} * \phi_3 = 0$, and thus they indeed make (10.4.12) a space of G_2 holonomy.

For M_6 to be compact, there should be O6-planes. If we integrate the Bianchi identity $\mathrm{d}f_2 = \delta$ on a compact M_6 and recall (1.4.49) and (10.1.5), we see

$$n_{\mathrm{D}6} = 2n_{\mathrm{O}6}\ : \quad (10.4.14)$$

the net D6-charge should be zero. Very close to a source, a solution to (10.4.10) will look like the D6 or O6 solution in flat space. Near a stack of N D6-branes, the metric will lift to a $\mathbb{R}^3 \times \mathbb{R}^4/\mathbb{Z}_N$ orbifold singularity (Section 7.2.3). Near an O6-plane, the solution is expected to be affected significantly by string corrections, its true lift being $\mathbb{R}^3\times$ the Atiyah–Hitchin solution (7.2.37).

A solution of this type is expected to exist when M_6 is a Calabi–Yau. Many *noncompact* examples of this type were constructed in [187], mentioned in footnote 5 after (7.2.58). When M_6 is compact, one can define an involution R_6 that acts like (10.4.9) by conjugating the complex coordinates Z_I in the ambient space [476]; we may then add D6-branes wrapping any special Lagrangian three-cycles, until (10.4.14) is satisfied. Beyond the probe approximation, the D6s and O6-planes will distort the Calabi–Yau metric and turn it into a solution of (10.4.10). This is expected to lead to G_2-holonomy manifolds, although it has not been shown mathematically, which is why we did not present this construction in Section 7.2.5.

10.4.2 An intersecting-branes class

We will now consider an SU(2)-structure class in IIA. This was initially found in [330], generalizing the earlier [332, sec. 3], without using G-structure or pure forms, but inspired by an Ansatz for intersecting branes.

Derivation from pure spinor equations

We take $M_6 = T^2 \times M_4$, with

$$\Phi_+ = \frac{1}{2}(1 - \mathrm{i}e^{2A}\mathrm{d}x^4 \wedge \mathrm{d}x^5) \wedge \phi^1 , \qquad \Phi_- = \frac{1}{2}e^A(\mathrm{d}x^4 + \mathrm{i}\mathrm{d}x^5) \wedge \phi^2 . \quad (10.4.15a)$$

$$\phi_1 = \frac{\mathrm{i}}{4}v_1 \wedge e^{\frac{1}{2}v_2\wedge\bar{v}_2} , \qquad \phi_2 = \frac{1}{4}v_2 \wedge e^{\frac{1}{2}v_1\wedge\bar{v}_1} . \quad (10.4.15b)$$

These are of the form (10.3.31), associated to an SU(2)-structure. By comparing with the pure forms for two orthogonal spinors in flat $\mathbb{R}^6$, we see that

$$\{\mathrm{Re}v_1, \mathrm{Re}v_2, \mathrm{Im}v_1, \mathrm{Im}v_2\} \quad (10.4.16)$$

is a vielbein in M_4. (One can also make the T^2 noncompact; the class then becomes a solution of the Mink_6 system in Section 10.6.1.)

From (10.3.24a)–(10.3.24b) we first obtain the two-form equations:

$$\mathrm{d}(e^{2A-\phi}\mathrm{Im}v_1) = \mathrm{d}(e^{4A-\phi}\mathrm{Re}v_1) = \mathrm{d}(e^{4A-\phi}v_2) = 0 . \quad (10.4.17)$$

The first two are of the form (4.1.97) for $k = d - 1$, so each defines a codimension-one foliation, and by (4.1.98) there exist local coordinates x, y^3 such that $\mathrm{Im}v_1 = e^{-2A+\phi}\mathrm{d}x$, $\mathrm{Re}v_1 = e^{-4A+\phi}\mathrm{d}y^3$. The third equation in (10.4.17) is also of the form (4.1.97), for $k = d - 2$, so it defines a codimension-two foliation. But it is also a complex one-form equation, and this implies the existence of local coordinates y^1, y^2 such that $v_2 = e^{-4A+\phi}(\mathrm{d}y^1 + \mathrm{i}\mathrm{d}y^2)$.[7] So we have determined the vielbein, and hence the following metric:[8]

$$\mathrm{d}s^2_{10} = h_6^{-1/2}\mathrm{d}s^2_{\mathrm{Mink}_4\times T^2} + h_5(h_6^{-1/2}\mathrm{d}x^2 + h_6^{1/2}\mathrm{d}s^2_{\mathbb{R}^3}) , \quad (10.4.18)$$

[7] This is a special case of the *complex Frobenius theorem* [477]; for a short proof that applies to our case, see [478, app. C]. In general, the theorem concerns a rank-k subbundle $\mathcal{E} \subset (T^*M_d)_\mathbb{C}$ in a dimension-d manifold M_d. Suppose $\mathcal{E}$ has a basis ω_a such that (4.1.97) holds, and suppose the same is true for $\mathcal{E}' \equiv \mathcal{E} \cap \bar{\mathcal{E}}$, of rank k'. Then there exist local coordinates such that $\mathcal{E}$ is spanned by $\mathrm{d}y^a + \mathrm{i}\mathrm{d}y^{a+l}$, $a = 1, \ldots, l \equiv k - k'$ and $\mathrm{d}y^p$, $p = d - k' + 1, \ldots, d$.

[8] h_6 and h_5 are called S and K respectively in [330].

where $h_6 \equiv e^{-4A}$ and $h_5 \equiv e^{-6A+2\phi}$. Comparing with (1.3.59), (1.3.65), and (9.2.10), we see that this is of the form one would expect for an NS5–D6 system, associated to harmonic functions h_6 and h_5. Because of (1.3.69) and the possible presence of F_0, it is also natural to imagine that the Ansatz includes D8-branes.

Continuing with the analysis of (10.3.24a)–(10.3.24b) with (10.4.15), we find that the higher form degrees simply determine H, and don't place any further restriction on the geometry. Finally (10.3.24c) determines f. This gives us

$$\begin{aligned} H &= -\partial_x(h_5 h_6)\mathrm{d}y^1 \wedge \mathrm{d}y^2 \wedge \mathrm{d}y^3 + \frac{1}{2}\epsilon_{ijk}\partial_i h_5 \mathrm{d}y^j \wedge \mathrm{d}y^k \wedge \mathrm{d}x\,, \\ F_0 &= -\frac{\partial_x h_6}{h_5}\,, \qquad F_2 = -\frac{1}{2}\epsilon_{ijk}\partial_i h_6 \mathrm{d}y^j \wedge \mathrm{d}y^k\,, \qquad F_4 = 0\,. \end{aligned} \tag{10.4.19}$$

(We can use the capital F for these RR field-strengths, because all of them only have internal components, $F_k = f_k$.) As usual, we now have to impose the Bianchi identities on these fluxes. When $F_0 = 0$, (10.4.19) implies $\partial_x h_6 = 0$, and $\mathrm{d}F_2 - HF_0 = 0$, $\mathrm{d}H = 0$ give respectively

$$\nabla_3^2 h_6 = 0\,, \qquad \nabla_3^2 h_5 + h_6 \partial_x^2 h_5 = 0 \qquad (F_0 = 0), \tag{10.4.20}$$

where $\nabla_3^2 = \partial_i\partial_i$ is the Laplacian on $\mathbb{R}^3$; this is reminiscent of (9.2.11). When $F_0 \neq 0$, we can eliminate $h_5 = -\partial_x h_6/F_0$ from the equations; both Bianchi identities are satisfied if

$$\nabla_3^2 h_6 + \frac{1}{2}\partial_x^2 h_6^2 = 0 \qquad (F_0 \neq 0)\,. \tag{10.4.21}$$

This second case is particularly remarkable in that the whole vacuum solution problem has been reduced to a single partial differential equation.

Examples

For noncompact M_4, as expected we find some solutions with brane interpretations. For $F_0 = 0$, $h_5 = 1$, $h_6 = 1 + r_0/y$, $y^2 \equiv y^i y^i$, describes a D6, and $h_6 = 1$, $h_5 = 1 + r_0^2(y^2+x^2)^{-2}$ describes an NS5, as can be checked from (1.3.65). A more complicated solution is

$$h_5 = 1 + c(x^2 + 4r_0 y)^{-3/2}\,, \qquad h_6 = \frac{r_0}{y}\,, \qquad F_0 = 0\,. \tag{10.4.22}$$

This should describe an NS5 in the near-horizon region of a D6, in the spirit of the discussion around (9.2.11). It can be obtained upon dimensional reduction from the M5 solution (Exercise 10.4.6).

For $F_0 \neq 0$, we can take

$$h_6 = -F_0 g_s^2 x + S_3\,, \qquad \nabla_3^2 S_3 + F_0^2 g_s^4 = 0\,, \tag{10.4.23}$$

where $g_s \equiv e^{-3A+\phi}$ is constant, and S_3 only depends on the y^i. If we take it to depend only on $y = \sqrt{y^i y^i}$, we obtain noncompact solutions that might describe an NS5–D6 system far from a D8 [330, 332], but that have a critical distance; gluing several such solutions seems to be better behaved [330]. More relevant for us, we can compactify $\mathbb{R}^3 \to T^3$ in (10.4.18); the partial differential equation in (10.4.23) can be solved. The x direction can also be compactified with O8-planes. This describes a compact solution [479, 480]:

$$\begin{aligned} \mathrm{d}s^2 &= S_3^{1/2}(\mathrm{d}s^2_{\mathbb{R}^4} + \mathrm{d}s^2_{T^2} + g_s^2 \mathrm{d}z^2) + g_s^2 S_3^{-1/2} \mathrm{d}s^2_{T^3} \\ F_2 &= -\frac{1}{2}\epsilon_{ijk}\partial_i S_3 \mathrm{d}y^j \wedge \mathrm{d}y^k\,, \qquad H = F_0 g_s^4 \mathrm{vol}_{T^3}\,. \end{aligned} \tag{10.4.24}$$

For more details, see, for example, [480, sec. 4.1].

Finally, we will see in Section 11.2.1 that with an appropriate change of coordinates the class of this subsection contains all type IIA AdS$_7$ solutions.

Exercise 10.4.1 Start from an unwarped direct product $M_4 \times M_6$, and apply (1.2.38) and (1.4.30). Conclude that $\phi \to -\phi'$, $J \to J' = \mathrm{e}^{-\phi} J$, $\Omega \to \Omega' = \mathrm{e}^{-3/2\phi}\Omega$. Use this to check that (9.3.13) is S-dual to (10.4.7).

Exercise 10.4.2 Take M_6 to be topologically $\mathbb{R}^6$, with the vielbein

$$e^{4,5,7,8} = h^{-1/4}\mathrm{d}x^{4,5,7,8}\,, \qquad e^{6,9} = h^{1/4}\mathrm{d}x^{6,9}\,, \tag{10.4.25}$$

with $h = h(x^6, x^9)$. Show that

$$\begin{aligned} &\mathrm{e}^A = h^{-1/4}\,, \qquad \mathrm{e}^\phi = g_s h^{-1}\,, \qquad H = f_3 = f_5 = 0\,, \\ &J = e^{47} + e^{58} + e^{69}\,, \qquad \Omega = (e^4 + \mathrm{i}e^7) \wedge (e^5 + \mathrm{i}e^8) \wedge (e^6 + \mathrm{i}e^9) \end{aligned} \tag{10.4.26}$$

is a solution of (10.4.5); determine f_1. Show that for (1.3.63) this is the D7 solution from (1.3.59).

Exercise 10.4.3 Again for $M_6 = \mathbb{R}^6$ and getting some inspiration from (10.4.26), find an SU(3)-structure solution of (10.4.7).

Exercise 10.4.4 Find a solution to (10.4.5) representing a D3-brane on a Calabi–Yau.

Exercise 10.4.5 Show that (10.4.13) are indeed closed and co-closed. This involves showing that

$$f_2 \wedge J = \mathrm{d}(\mathrm{e}^{-3A}\mathrm{Im}\Omega)\,, \tag{10.4.27}$$

which can be derived from (10.4.10) using (5.2.24) and (5.2.27). (We will learn how to obtain it more directly in Section 10.5.6.)

Exercise 10.4.6 Consider the M5 solution (1.4.70), and write the transverse $\mathbb{R}^5$ as

$$\mathrm{d}s^2 = \mathrm{d}x^2 + \frac{1}{4r}\mathrm{d}s^2_{\mathbb{R}^3} + \frac{r}{4}(\mathrm{d}\psi + \sigma)^2 \tag{10.4.28}$$

following (7.2.21). Reduce along ψ (recalling (1.4.5a)), and check that you obtain a solution of the form (10.4.22).

10.5 Generalized complex geometry

We will now give an interpretation to the results of Section 10.3 in terms of *generalized complex geometry*, introduced in [481, 482]. (Supergravity-oriented reviews are [483, 484]). In particular, we will see the following:

- On the generalized tangent bundle $T \oplus T^*$, the $\Phi_\pm$ define an SU(3)×SU(3)-structure in all three cases of Section 10.3.4. This is in contrast to the ordinary G-structure defined by the internal supersymmetry parameters $\eta_+^{1,2}$, which together can give $G = \mathrm{SU}(3)$ or SU(2).

- The data of the two $\Phi_\pm$ are enough to reconstruct both the internal supersymmetry parameters $\eta_+^{1,2}$ *and the metric*:

$$(\Phi_+, \Phi_-) \quad \longleftrightarrow \quad (g_{mn}, \eta_+^1, \eta_+^2) \tag{10.5.1}$$

  This generalizes the observation in Section 5.1.2 that the metric can be reconstructed from an SU(3)-structure.[9]
- One of the pure spinor equations, (10.3.20b), has a geometrical interpretation as an integrability condition, encompassing both complex and symplectic geometry (Sections 5.3 and 5.4).
- There is an alternative version of the pure spinor equations without the Hodge star operator, thus eliminating the last explicit appearance of the metric in them. We will achieve this at the mild cost of introducing a generalization of the Dolbeault differential (5.3.28) of complex geometry.

10.5.1 Pure forms and generalized almost complex structures

We begin in this subsection by exploiting a particular property of the $\Phi_\pm$: they are *pure forms*. Recall that a pure spinor is one which is annihilated by half of the gamma matrices, and that forms are spinors for the "doubled" Clifford algebra $\mathrm{Cl}(d, d)$ in (3.2.3) (Sections 2.4.2 and 3.2). Putting the two ideas together, a pure form is one that is pure as a $\mathrm{Cl}(d, d)$ spinor (Section 3.2.5). In this subsection only, we work in arbitrary even dimension d; we will later specialize to $d = 6$.

Definition

We have seen in Section 5.1, and earlier in Sections 2.4 and 3.4, that a pure (chiral) spinor η_+ defines an almost complex structure I:

$$\eta_+ \to I\,. \tag{10.5.2}$$

Concretely, this is because the $d/2$ matrices that annihilate η_+ can be written as (3.4.87), $(\delta_m^n + \mathrm{i} I_m{}^n)\,\gamma_n \eta_+ = 0$. In other words,

$$\alpha \cdot \eta_+ = \alpha_m \gamma^m \eta_+ = 0 \quad \Leftrightarrow \quad I \cdot \alpha = \mathrm{i}\alpha\,. \tag{10.5.3}$$

(In flat space, or when I is integrable, we can write this in holomorphic indices as $\gamma^i \eta_+ = 0$ or $\gamma_{\bar{\imath}} \eta_+ = 0$.) So the $(1,0)$-forms (or the $(0,1)$-vectors) of I annihilate η_+; this defines the map (10.5.2).

We would like to apply the same idea to a pure form Φ. The generators of the doubled Clifford algebra (3.2.3) are vectors and one-forms; so the corresponding almost complex structure should act on the space of such objects. We define a *generalized almost complex structure* (GACS) $\mathcal{J}$ as a (real) map[10]

$$\mathcal{J} : T \oplus T^* \to T \oplus T^* \text{ such that } \mathcal{J}^2 = -1_{2d}\,, \tag{10.5.4}$$

[9] For an even more general perspective on how spinors can be reconstructed for forms, see [485, 486].
[10] As stressed in [487], topologically nontrivial B actually turns $T \oplus T^*$ into a bundle E such that $E/T^* \cong T$. We will ignore this subtlety in our treatment.

where 1_{2d} is simply the identity map $T \oplus T^* \to T \oplus T^*$. This becomes perhaps a little clearer if we write it out in blocks:

$$\mathcal{J} = \begin{pmatrix} K & P \\ L & I \end{pmatrix}. \tag{10.5.5}$$

Here K represents the part of the map $T \to T$, so its index structure would be $K^m{}_n$; then $P\colon T^* \to T$ has index structure P^{mn}, and so on. Then the condition in (10.5.4) reads explicitly

$$\begin{aligned} K^m{}_n K^n{}_p + P^{mn} L_{np} &= -\delta^m_p\,, & K^m{}_n P^{np} + P^{mn} I_n{}^p &= 0\,, \\ L_{mn} K^n{}_p + I_m{}^n L_{np} &= 0\,, & L_{mn} P^{np} + I_m{}^n I_n{}^p &= -\delta^p_m\,. \end{aligned} \tag{10.5.6}$$

Just as in (10.5.2), we want to associate a generalized almost complex structure to a pure form:

$$\Phi \to \mathcal{J}\,. \tag{10.5.7}$$

This map exists for any *nondegenerate* pure form Φ, namely one such that[11]

$$(\Phi, \bar{\Phi}) \neq 0\,, \tag{10.5.8}$$

with the usual Chevalley–Mukai pairing (3.2.40), $(\alpha, \beta) = (\alpha \wedge \lambda(\beta))_d/\mathrm{vol}_d$. The correspondence is defined just as in (10.5.3):

$$X \cdot \Phi = 0 \quad \Longleftrightarrow \quad \mathcal{J} X = \mathrm{i} X\,. \tag{10.5.9a}$$

Here X is now a section of $T \oplus T^*$, so a formal pair $\binom{v^n}{\alpha_n}$; and

$$X \cdot \equiv v^n \iota_n + \alpha_n \mathrm{d}x^n \wedge = \iota_v + \alpha \wedge \tag{10.5.9b}$$

is its natural action on forms. We denote the i-eigenspace of $\mathcal{J}$ by $L_{\mathcal{J}}$, so another way of writing (10.5.9a) is

$$L_{\mathcal{J}} = \mathrm{Ann}(\Phi)\,. \tag{10.5.9c}$$

Since Φ is pure, this has dimension d, half the number of gamma matrices for $\mathrm{Cl}(d, d)$.

The almost complex structure I associated to a spinor η_+ also has the Hermitian property (5.1.16), $I_m{}^p g_{pn} + g_{mp} I_n{}^p = 0$. The same property holds for the $\mathcal{J}$ associated to a pure form Φ, with the metric being (3.2.6), $\mathcal{I} = \begin{pmatrix} 0 & 1_d \\ 1_d & 0 \end{pmatrix}$, the metric appearing in the Clifford algebra $\mathrm{Cl}(d, d)$:

$$\mathcal{J}\mathcal{I} + \mathcal{I}\mathcal{J}^t = 0\,. \tag{10.5.10}$$

In other words, $\mathcal{J}\mathcal{I}$ is antisymmetric. This can be spelled out in terms of the blocks (10.5.5):

$$L_{mn} = -L_{nm}\,, \qquad P^{mn} = -P^{nm}\,, \qquad K^m{}_n = -I_n{}^m\,. \tag{10.5.11}$$

So L and P are antisymmetric, while $K = -I^t$. Unless otherwise stated, all our GACS $\mathcal{J}$ will satisfy (10.5.10).

[11] For ordinary spinors η in Euclidean signature, the analogue $\eta^\dagger \eta \neq 0$ of this condition is automatically satisfied.

Conversely, if we have a GACS $\mathcal{J}$ (with the property (10.5.10)), at every point we can associate to it a pure Φ. Indeed, (10.5.10) implies that $\mathcal{J}$ commutes with its transpose, so it is diagonalizable; since $\mathcal{J}^2 = -1$ the eigenvalues are $\pm$i. Reality then demands the i-eigenspace $L_{\mathcal{J}}$ to be $d/2$-dimensional. Moreover, taking the inverse of (10.5.10) we get $\mathcal{I}\mathcal{J} + \mathcal{J}^t\mathcal{I} = 0$, and multiplying this by two elements $X_{1,2}$ of $L_{\mathcal{J}}$, we find

$$X_1^t \mathcal{I} X_2 = 0 \,. \tag{10.5.12}$$

So according to the definition in Section 2.4.2 $L_{\mathcal{J}}$ is *isotropic*: the metric $\mathcal{I}$ is zero when restricted to it. Since it also has maximal dimension, according to the discussion in that subsection there exists a pure Φ such that (10.5.9c) holds. According to the discussion in Section 2.4.2, this Φ is unique up to rescaling. This is true pointwise; we will see soon that global issues might obstruct the existence of Φ.

Sometimes it is convenient to use the notation in (3.2.4), with "doubled" indices $A = 1, \ldots, 2d$. Then (10.5.5) and (10.5.10) read respectively

$$\mathcal{J}_A{}^C \mathcal{J}_C{}^B = -\delta_A^B \tag{10.5.13a}$$

$$0 = \mathcal{J}_A{}^C \mathcal{I}_{CB} + \mathcal{I}_{AC}\mathcal{J}_B{}^C \equiv \mathcal{J}_{AB} + \mathcal{J}_{BA} \,. \tag{10.5.13b}$$

From now on, we will use $\mathcal{I}_{AB}$ to raise indices. This is rather natural, but it does give rise to a little numerical issue: (3.2.5) has the usual form for a Clifford algebra, but with metric $\frac{1}{2}\mathcal{I}_{AB}$. So, for example, (2.1.33) reads

$$\Gamma_A\Gamma_B = \Gamma_{AB} + \frac{1}{2}\mathcal{I}_{AB} \,. \tag{10.5.14}$$

In general, when importing formulas from the world of ordinary Cl(d) spinors, one should map

$$\gamma_\mu \to \sqrt{2}\Gamma_A \,. \tag{10.5.15}$$

Examples

It is time to give some examples. Rather than trying to find solutions to (10.5.4) and (10.5.10) from scratch, we will find them by using the map (10.5.9a) and the discussion of pure forms in Section 3.2.5.

The simplest pure form we saw there is $\Phi = 1$; its annihilator was given in (3.2.43). But this is degenerate: it does not satisfy (10.5.8). Indeed, (10.5.9c) would tell us that $\mathcal{J}$ has to be i on T, so it cannot be real as prescribed.

The next-easiest example of pure spinor is a decomposable form (3.2.44). The top holomorphic form Ω of a complex structure is decomposable, and is nondegenerate: according to (5.1.22), it can be written as

$$\Phi = \Omega = h^1 \wedge \ldots \wedge h^{d/2} \,, \tag{10.5.16}$$

and it is nondegenerate because $\Omega \wedge \bar{\Omega}$ is proportional to the volume form of the metric for which h^i are a holomorphic vielbein. The annihilator was given in (3.2.45); for our case (10.5.16), it consists of all (1, 0)-forms and all (0, 1)-vectors, so

$$\mathrm{Ann}(\Omega) = L_{\mathcal{J}_\Omega} = T_{0,1} \oplus T^*_{1,0} \,. \tag{10.5.17}$$

The corresponding GACS is then

$$\mathcal{J}_\Omega = \mathcal{J}_I \equiv \begin{pmatrix} -I^t & 0 \\ 0 & I \end{pmatrix}, \qquad (10.5.18)$$

where I is the almost complex structure associated to Ω, which indeed is defined so that $I \cdot h^a = \mathrm{i}h^a$ (Section 5.1).

Another notable example of pure form in Section 3.2.5 was the exponential (3.2.51), $\Phi = \mathrm{e}^{-\beta}$. Nondegeneracy gives

$$(\mathrm{e}^{-\beta}, \mathrm{e}^{-\bar\beta}) = (\mathrm{e}^{-\beta} \wedge \lambda \mathrm{e}^{-\bar\beta})_d = (\mathrm{e}^{-\beta} \wedge \mathrm{e}^{\bar\beta})_d = (\mathrm{e}^{-\beta+\bar\beta})_d = \frac{1}{d!}(-2\mathrm{i}\mathrm{Im}\beta)^{d/2} \neq 0\,. \qquad (10.5.19)$$

The easiest way to satisfy this is by taking β to be purely imaginary and nondegenerate. This leads to our old friend $\mathrm{e}^{-\mathrm{i}J}$. The annihilator was found in (3.2.50):

$$\mathrm{Ann}(\mathrm{e}^{-\mathrm{i}J}) = L_{\mathcal{J}_{\mathrm{e}^{-\mathrm{i}J}}} = \{\iota_m + \mathrm{i}J_{mn}\mathrm{d}x^n \wedge\} \overset{(10.5.9b)}{=} \left\{ \begin{pmatrix} \delta^n_m \\ \mathrm{i}J_{mn} \end{pmatrix} \cdot \right\}. \qquad (10.5.20)$$

We notice a formal similarity with the way one defines complex coordinates (2.2.5), whose almost complex structure was given in (2.4.3). This helps in finding $\mathcal{J}$, of which $\mathrm{Ann}(\mathrm{e}^{-\mathrm{i}J})$ is the i-eigenspace:

$$\mathcal{J}_{\mathrm{e}^{-\mathrm{i}J}} = \mathcal{J}_J \equiv \begin{pmatrix} 0 & J^{-1} \\ -J & 0 \end{pmatrix}. \qquad (10.5.21)$$

It is immediate to check that it squares to minus the identity.

The two examples (10.5.18) and (10.5.21) are associated respectively to an almost complex structure I and to an almost symplectic structure J. These were the original motivating examples for the definition of generalized almost complex structure. They will serve as running examples throughout this section.

The b-transform (3.2.14)

$$\Phi \to \Phi^b \equiv \mathrm{e}^{b\wedge}\Phi \qquad (10.5.22)$$

preserves purity, as noted after (3.2.53). To see how $\mathcal{J}$ changes under (3.2.14), consider an element $X = \binom{v}{\alpha}$ in the annihilator, $X \cdot \Phi = 0$. Then the element

$$X^b \cdot \equiv \mathrm{e}^{b\wedge} X \cdot \mathrm{e}^{-b\wedge} = \mathrm{e}^{b\wedge}(\iota_v + \alpha\wedge)\mathrm{e}^{-b\wedge} \overset{(2.1.15),(3.2.3)}{=} \iota_v + (\alpha - \iota_v b)\wedge \qquad (10.5.23)$$

is in the annihilator of Φ^b. (The exponential on the right-hand side of (2.1.15) truncates at the first step.) This is also

$$X^b = \begin{pmatrix} v \\ \alpha - \iota_v b \end{pmatrix} = \begin{pmatrix} 1 & 0 \\ -b & 1 \end{pmatrix} \begin{pmatrix} v \\ \alpha \end{pmatrix} \equiv \mathcal{B}X\,. \qquad (10.5.24)$$

Since $\mathcal{J}X = \mathrm{i}X$ by (10.5.9a), we also have $\mathcal{J}^b X^b = \mathrm{i}X^b$, with

$$\mathcal{J}^b \equiv \mathcal{B}\mathcal{J}\mathcal{B}^{-1} = \begin{pmatrix} 1 & 0 \\ b & 1 \end{pmatrix} \mathcal{J} \begin{pmatrix} 1 & 0 \\ -b & 1 \end{pmatrix}. \qquad (10.5.25)$$

In Section 10.3.4 we saw additional examples of $\Phi_\pm$. Those for an SU(3)-structure, (10.3.28), are of the form (10.5.18) and (10.5.21) in $d = 6$. For (10.3.31), the expressions are more complicated but still conceptually clear: we decompose T and T^* in two-dimensional and four-dimensional subspaces, and then write $\mathcal{J}_{\Phi_\pm}$ with four blocks instead of just two.

We can also derive the annihilators for all the $\Phi_\pm$ in Section 10.3.4 directly from their bispinor expression $\eta^1_+ \otimes \eta^{2\dagger}_\pm$. For example, from (10.3.82) and its conjugate, we see

$$\gamma_{\bar{\imath}_1} \eta^1_+ \otimes \eta^{2\dagger}_+ = 0 = \eta^1_+ \otimes \eta^{2\dagger}_+ \gamma_{i_2} , \tag{10.5.26}$$

and similar for $\eta^1_+ \otimes \eta^{2\dagger}_-$. This gives

$$L_{\Phi_+} = \text{Span}\left(\overrightarrow{\gamma}_{\bar{\imath}_1}, \overleftarrow{\gamma}_{i_2}\right) , \qquad L_{\Phi_-} = \text{Span}\left(\overrightarrow{\gamma}_{\bar{\imath}_1}, \overleftarrow{\gamma}_{\bar{\imath}_2}\right) , \tag{10.5.27}$$

in the language of Section 3.1.3. One can translate (10.5.27) into more customary wedges and contractions using (3.1.21) and (3.1.24).

$\mathrm{U}(d/2, d/2)$-structures

A G-structure was defined in Section 5.1 as a reduction from $\mathrm{GL}(d, \mathbb{R})$ to the group G in which the transition functions of the frame bundle FM, or equivalently of the tangent bundle T, take value. Recall that already the presence of a metric reduces to $\mathrm{O}(d)$. In d = even, a pure spinor η_+ reduces further to $\mathrm{SU}(d/2)$, while the almost complex structure I associated to η_+ reduces to the slightly larger $\mathrm{U}(d/2)$ (Sections 5.1.1 and 5.1.2). It is possible to extend the idea to reductions of any bundle. We define a *generalized G-structure* to be a reduction on the generalized tangent bundle $T \oplus T^*$.

The structure group of $T \oplus T^*$ is $\mathrm{GL}(2d, \mathbb{R})$; but the metric $\mathcal{I} = \begin{pmatrix} 0 & 1_d \\ 1_d & 0 \end{pmatrix}$ always exists, and always reduces the structure group, just as an ordinary metric does on T. Since the signature of $\mathcal{I}$ is (d, d), this reduction is to

$$\mathrm{gStab}(\mathcal{I}) = \mathrm{O}(d, d) . \tag{10.5.28}$$

The "g" is for "generalized": this is a stabilizer not in the ordinary $\mathrm{GL}(d, \mathbb{R})$, but in $\mathrm{GL}(2d, \mathbb{R})$. The group (10.5.28) occurred also in Section 4.2.6. The reason for this coincidence is that T-duality can be viewed as exchanging one direction of T with one of T^*. The occurrence of $\mathrm{O}(d, d)$, so peculiar to string theory, is one of the signs that generalized complex geometry is a natural formalism for it.

Next we consider a pure and nondegenerate Φ in even d. Its stabilizer is computed in the same way as (2.4.35) for ordinary spinors, but this time because of the split signature

$$\mathrm{gStab}(\Phi) = \mathrm{SU}(d/2, d/2) . \tag{10.5.29}$$

Finally the GACS $\mathcal{J}$ reduces to

$$\mathrm{gStab}(\mathcal{J}) = \mathrm{U}(d/2, d/2) . \tag{10.5.30}$$

The slight difference between (10.5.29) and (10.5.30) is completely analogous to the one between $\mathrm{Stab}(\eta_+)$ and $\mathrm{Stab}(I)$, discussed in Sections 5.1.1 and 5.1.2. Just like in those cases, while Φ determines a GACS $\mathcal{J}$, the opposite is true locally, but globally Φ is a section of a *generalized canonical* line bundle, which then needs to be trivial in order for Φ to exist.

Clifford levels

By taking the complex conjugate of (10.5.9c), we see $\bar{L}_{\mathcal{J}} = \mathrm{Ann}(\bar{\Phi})$. As for ordinary pure spinors, the space of gamma matrices is a direct sum of $L_{\mathcal{J}} = \mathrm{Ann}(\Phi)$ and of

$\bar{L}_{\mathcal{J}}$, which can be thought of as creators; the situation is similar to that illustrated in the diagram (2.4.18) for ordinary (pure) spinors in $d = 6$. Since $\bar{L}_{\mathcal{J}}$ has dimension d, by acting with it on Φ it takes d steps to reach $\bar{\Phi}$. So the space of all forms decomposes in $(d+1)$ *Clifford levels* U_k similar to those in (2.4.18):

$$\Lambda = U^{\mathcal{J}}_{d/2} \oplus U^{\mathcal{J}}_{d/2-1} \oplus \ldots \oplus U^{\mathcal{J}}_{-d/2+1} \oplus U^{\mathcal{J}}_{-d/2} \,. \tag{10.5.31}$$

(We will sometimes omit to write $\mathcal{J}$.) Here $U^{\mathcal{J}}_{d/2} = \mathrm{Span}(\Phi)$, the space of forms proportional to Φ, and $U^{\mathcal{J}}_{-d/2} = \mathrm{Span}(\bar{\Phi})$. The reason for the labels will become clear soon. They are not directly related to form degree: an element of U_k will in general be a polyform, just as Φ itself often is.

For example, for $\Phi = \Omega$ the creators are given by (10.5.17), and

$$U^{\mathcal{J}_\Omega}_k = \oplus_j \Lambda^{k+j,j} \,. \tag{10.5.32}$$

In $d = 6$, $U_3 = \Lambda^{3,0}$, $U_2 = \Lambda^{2,0} \oplus \Lambda^{3,1}$, and so on.

For the pure forms $\Phi_\pm$ of Section 10.3.4, the annihilators are in (10.5.27). In the generalized Hodge diamond (10.3.83), we see that acting with $\bar{L}_{\Phi_+}$ makes one go one row lower in (10.3.83); so the U_k in (10.5.31) for Φ_+ are just the rows of (10.3.83). For Φ_-, the U_k are the columns of (10.3.83).

Action on forms

The action on forms of an almost complex structure I and an almost symplectic structure J played an important role in Section 5.1. For example, the action of I gave a possible definition of (p,q)-forms in (5.1.30), and the action of J and its inverse allowed us to define the Lefschetz operators (5.2.2a)–(5.2.2b).

There is a natural generalization of both concepts: we can let each block in (10.5.5) act as in (3.2.13), with lower indices acting by wedge and upper indices acting by contraction:

$$\begin{aligned}\mathcal{J}\cdot &\equiv \frac{1}{2}\mathcal{J}_{AB}\Gamma^{AB} \\ &\overset{(10.5.11)}{=} \frac{1}{2}\left(L_{mn}\mathrm{d}x^m \wedge \mathrm{d}x^n + P^{mn}\iota_m\iota_n + I_m{}^n(\mathrm{d}x^m \wedge \iota_n - \iota_n \mathrm{d}x^m \wedge)\right) \\ &\overset{(3.2.3a)}{=} L\wedge + \iota_P + I\cdot \,.\end{aligned} \tag{10.5.33}$$

In the two examples (10.5.18) and (10.5.21):

$$\mathcal{J}_\Omega \cdot = I\cdot \,, \qquad \mathcal{J}_{e^{-iJ}}\cdot = -J\wedge + J\cdot \,. \tag{10.5.34}$$

The eigenspaces of $\mathcal{J}\cdot$ are nothing but the U_k in (10.5.31). To see this, observe first that

$$\begin{aligned}X \in L_{\mathcal{J}} \;\Rightarrow\; [\mathcal{J}\cdot, X\cdot] &= \frac{1}{2}[\mathcal{J}_{AB}\Gamma^{AB}, X^C\Gamma_C] \\ &\overset{(3.2.5),(2.1.13),(10.5.15)}{=} \mathcal{J}_{AB}\mathrm{I}^{\cdot A}X^B = \mathrm{i}X_A\mathrm{I}^{\cdot A} = \mathrm{i}X\cdot \,,\end{aligned} \tag{10.5.35}$$

recalling also the notation (10.5.9b). Applying this to Φ:

$$X \in L_{\mathcal{J}} \quad \Rightarrow \quad 0 = \mathcal{J}\cdot(X\cdot\Phi) = X\cdot(\mathcal{J}\cdot\Phi + \mathrm{i}\Phi) = X\cdot(\mathcal{J}\cdot\Phi) \,. \tag{10.5.36}$$

So $X\cdot$ annihilates also $\mathcal{J}\cdot\Phi$. By the discussion after (10.5.12), $\mathcal{J}\cdot\Phi$ must then be proportional to Φ itself, or in other words Φ is an eigenvector of $\mathcal{J}\cdot$. (10.5.35) also shows that $X\cdot$ is a raising operator: if $\mathcal{J}\cdot\alpha = \mathrm{i}k\alpha$, then $\mathcal{J}\cdot(X\cdot\alpha) = \mathrm{i}(k+1)\alpha$. By complex conjugation, an element of $\bar{L}_{\mathcal{J}}$ is then a lowering operator. So all the U_k are eigenspaces. Finally, the lowest level $U_{d/2}$ is obtained by acting d times with lowering operators; since this is spanned by $\bar{\Phi}$, we conclude

$$\mathcal{J}\cdot\Phi = \frac{d}{2}\mathrm{i}\Phi \tag{10.5.37}$$

and more generally

$$\mathcal{J}\cdot u_k = k\,\mathrm{i}\,u_k\,, \qquad u_k \in U_k\,, \tag{10.5.38}$$

finally justifying the labels for the U_k in (10.5.31). An analogue of this holds for ordinary pure spinors: for example, in $d = 6$ (3.4.92) implies $J\eta_+ = -3\mathrm{i}\eta_+$.

As another application of $\mathcal{J}\cdot$, we now show an orthogonality property for the U_k. First remark that (3.2.61) can be written as

$$(\Gamma^A\alpha,\beta) = -(\alpha,\Gamma^A\beta)\,; \tag{10.5.39}$$

in other words, Γ^A is anti-self-adjoint with respect to the Chevalley–Mukai inner product. From this it follows that $(\Gamma^{AB}\alpha,\beta) = (\alpha,\Gamma^{BA}\beta)$, and so in particular

$$(\mathcal{J}\cdot\alpha,\beta) = -(\alpha,\mathcal{J}\cdot\beta)\,. \tag{10.5.40}$$

Now the argument is familiar from other, formally similar contexts: consider elements $u_k \in U_k$, $u_l \in U_l$:

$$\begin{aligned} 0 &\overset{(10.5.40)}{=} (\mathcal{J}\cdot u_k, u_l) + (u_k, \mathcal{J}\cdot u_l) \overset{(10.5.38)}{=} (k+l)\mathrm{i}(u_k,u_l) \\ &\Rightarrow \quad (u_k, u_l') = 0 \quad \text{unless } k = -l\,. \end{aligned} \tag{10.5.41}$$

This leads to another definition of purity, inspired by (3.4.9) for ordinary spinors:

$$(\Phi,\Gamma_{A_1\dots A_k}\Phi) = 0 \qquad \text{unless } k = d \qquad (\Phi \text{ pure})\,. \tag{10.5.42}$$

The idea is to use (10.5.41): by taking $k = d/2$ there, we see that Φ is orthogonal to all forms except $\bar{\Phi} \in U_{-d/2}$, but $\Gamma_{A_1\dots A_k}\Phi$ can belong to $U_{-d/2}$ only if $k = d$.

10.5.2 The six-dimensional case

We now specialize to $d = 6$. This is the case that will be most important for us physically, but pure forms in this dimension also have some special mathematical properties. This goes back to Table 2.3, where we saw that the space of pure spinors in $d = 12$ has dimension 32, which happens to be the same as the space Λ^+ or even forms (or the space Λ^- of odd forms) in $d = 6$, recalling (3.1.12). So in $d = 6$, a pure form Φ has the same number of degrees of freedom as its real part $\mathrm{Re}\Phi$.

Indeed, we will see in this subsection [256, 488]:[12]

- A real form ρ is the real part of a pure spinor, $\rho = \mathrm{Re}\Phi$, if and only if a certain quartic function $q(\rho)$, called *Hitchin invariant*, is positive.
- If $q > 0$, then an explicit formula exists for Φ and for the associated $\mathcal{J}$.

[12] This is a bit similar to the result of Exercise 5.1.5 in $d = 7$.

This is partially inspired by the case of ordinary $\mathrm{Cl}(d)$ pure spinors η_+. In that case, one way to define the almost complex structure I associated to η_+ is to consider the two-form bilinear (3.4.5), $J_{mn} = -i\eta_+^\dagger \gamma_{mn}\eta_+$, and then raise one index with the metric, $I_m{}^n = J_{mp}g^{pn}$. This suggests that $\mathcal{J}$ should be related to

$$\begin{aligned} Q_{AB} &\equiv \frac{1}{2}(\bar{\Phi}, \Gamma_{AB}\Phi) =^{(10.5.39),(3.2.41)} \frac{1}{2}(\Phi, \Gamma_{AB}\bar{\Phi}) \\ &\overset{(10.5.42)}{=} (\mathrm{Re}\Phi, \Gamma_{AB}\mathrm{Re}\Phi)\,. \end{aligned} \tag{10.5.43}$$

Indeed, we mentioned in (3.2.40) that the Chevalley–Mukai inner product is the natural product for $\mathrm{Cl}(d,d)$, in the sense that it is $\mathrm{O}(d,d)$-invariant. To see if (10.5.43) is the correct guess, we have to use Fierz identities, just like for ordinary $\mathrm{Cl}(d)$ spinors.

A $\mathrm{Cl}(d,d)$ Fierz identity

We now introduce a new class of such operators on the space of forms Λ. For β and β' two (poly)forms,

$$\beta\,(\bar{\beta}', \cdot) \tag{10.5.44}$$

is the operator that acts on a (poly)form as $\alpha \to \beta(\bar{\beta}', \alpha)$. So, in (10.5.44), $\cdot$ denotes the slot in which the form is applied. This is similar to the operator

$$\psi \to \eta((\eta')^\dagger \psi) \tag{10.5.45}$$

acting on spinors, which is in fact the familiar $\eta \otimes (\eta')^\dagger$. Both (10.5.44) and (10.5.45) are in turn similar to the projector $|\psi_1\rangle \otimes \langle\psi_2|$ in quantum mechanics, as we already remarked in (3.3.24). In this spirit, we could write the operator (10.5.44) also as $|\beta\rangle\langle\beta'|$.

Thus in a sense, (10.5.44) is a *bispinor for* $\mathrm{Cl}(d,d)$. This can be confusing: the examples of pure forms we have seen in (10.5.18) and (10.5.21) were already obtained as bispinors $\eta^1_+ \otimes \eta^{2\,\dagger}_\pm$ of ordinary $\mathrm{Cl}(d)$ spinors. A $\mathrm{Cl}(d)$ bispinor is a form; a $\mathrm{Cl}(d,d)$ bispinor such as (10.5.44) is an operator on forms.

So we can apply Fierz identities such as (3.3.28) or (3.3.29) to (10.5.44), but now using the $\mathrm{Cl}(d,d)$ gamma matrices (so that one has to remember also the factor (10.5.15)). In $d = 6$, for $\beta = \Phi$, $\beta' = \bar{\Phi}$, with Φ pure:

$$\begin{aligned} \Phi(\Phi, \cdot) &= \frac{1}{64}\sum_{k=1}^{12} \frac{2^k}{k!}(\Phi, \Gamma_{A_1\ldots A_k}\Phi)\Gamma^{A_k\ldots A_1} \\ &\overset{(10.5.42)}{=} \frac{1}{6!}(\Phi, \Gamma_{A_1\ldots A_6}\Phi)\Gamma^{A_6\ldots A_1}\,. \end{aligned} \tag{10.5.46}$$

Formal manipulations aside, (10.5.46) is simply expressing that $\Phi(\Phi, \cdot)$ gives zero on all forms except on $\bar{\Phi}$, on which it returns $\Phi(\Phi, \bar{\Phi})$. So it maps $U_{-d/2} = U_{-3} \to U_{d/2} = U_3$, and gives zero on all the other U_k. Such an operator can only consist of an (antisymmetrized) product of six Γ^A.

We also need to import two formulas from ordinary $\mathrm{Cl}(d)$ spinors, (3.2.23) and (3.2.57). We apply them to (10.5.46) recalling (10.5.15):

$$\Gamma^A \Phi(\Phi, \Gamma_A \cdot) = 0 \tag{10.5.47a}$$

$$\Gamma^{AB} \Phi(\Phi, \Gamma_{AB} \cdot) = 3\Phi(\Phi, \cdot)\,. \tag{10.5.47b}$$

Explicit formula for $\mathcal{J}$

Now consider the matrix

$$2\Pi_A{}^B \equiv \delta_A^B + \frac{4}{(\Phi,\bar\Phi)} Q_A{}^B \overset{(10.5.43)}{=} \frac{1}{(\Phi,\bar\Phi)}\left(\Phi, (\delta_A^B + 2\Gamma_A{}^B)\bar\Phi\right) \overset{(10.5.14)}{=} \frac{2}{(\Phi,\bar\Phi)}\left(\Phi, \Gamma_A\Gamma^B\bar\Phi\right). \tag{10.5.48}$$

This behaves like a holomorphic projector, similar to (5.3.22):

$$\Pi_A{}^B\bar\Pi_B{}^C = -\frac{1}{(\Phi,\bar\Phi)^2}(\Phi,\Gamma_A\Gamma^B\bar\Phi)(\bar\Phi,\Gamma_B\Gamma^C\Phi) \overset{(10.5.47a)}{=} 0. \tag{10.5.49}$$

In $d = 6$, the pairing is antisymmetric, so

$$(\Phi,\bar\Phi) \overset{(3.2.41)}{=} -2\mathrm{i}(\mathrm{Re}\Phi, \mathrm{Im}\Phi) \tag{10.5.50}$$

is purely imaginary, and Π is of the form $(1 - \mathrm{i}\mathcal{J})/2$; it then follows from (10.5.49) that $\mathcal{J}^2 = -1$. By (10.5.48):

$$\mathcal{J}_A{}^B = -\frac{2}{(\mathrm{Re}\Phi, \mathrm{Im}\Phi)} Q_A{}^B = \frac{1}{\sqrt{-\mathrm{Tr}(Q^2)/12}} Q_A{}^B. \tag{10.5.51}$$

The Hermitian condition (10.5.13b) also holds, because Q_{AB} is automatically antisymmetric by its definition (10.5.43). So (10.5.51) is a generalized almost complex structure. Moreover, it is the one associated to Φ according to the definition (10.5.9c): we can check

$$\mathcal{J}\cdot\Phi = \frac{4\mathrm{i}}{(\Phi,\bar\Phi)}\frac{1}{2}Q_{AB}\Gamma^{AB}\Phi \overset{(10.5.43)}{=} \frac{4\mathrm{i}}{(\Phi,\bar\Phi)}\frac{1}{4}\Gamma^{AB}\Phi(\Phi,\Gamma_{AB}\bar\Phi) = \overset{(10.5.47b)}{=} 3\mathrm{i}\Phi, \tag{10.5.52}$$

in agreement with (10.5.37) with $d = 6$. So if we have a pure form Φ, we now have an explicit formula for $\mathcal{J}$.

Q_{AB} in (10.5.43) also had an expression involving $\mathrm{Re}\Phi$ alone. So if we use the last expression in (10.5.51) for the normalization factor, we have a formula for $\mathcal{J}$ purely in terms of $\mathrm{Re}\Phi$. Taking the real part of (10.5.52) gives a formula for $\mathrm{Im}\Phi$:

$$\mathrm{Im}\Phi = -\frac{1}{3}\mathcal{J}\cdot\mathrm{Re}\Phi = -\frac{1}{6\sqrt{-\mathrm{Tr}(Q^2)/12}} Q_{AB}\Gamma^{AB}\mathrm{Re}\Phi. \tag{10.5.53}$$

This, however, makes sense only if

$$q(\mathrm{Re}\Phi) \equiv -\frac{1}{12}\mathrm{Tr}(Q^2) > 0. \tag{10.5.54}$$

This q is called the *Hitchin invariant*. Of course, (10.5.54) has to hold everywhere on the manifold.

To summarize: a real form ρ (even or odd) can be the real part of a pure form if and only if $q(\rho) = -\mathrm{Tr}(Q^2(\rho)) > 0$, with $Q_{AB}(\rho) = (\rho, \Gamma_{AB}\rho)$. If this test is passed, then indeed $\rho = \mathrm{Re}\Phi$, and $\mathrm{Im}\Phi$ is determined explicitly by (10.5.43) and (10.5.53); the corresponding $\mathcal{J}$ is given explicitly by (10.5.51).

Examples: three-forms

Let $\rho = \rho_3$ be a three-form. From the general form (3.2.53) of a pure spinor, we know that if ρ_3 is the real part of a pure form, then $\Phi = \Omega$, for Ω a nondegenerate decomposable three-form, which would then be the holomorphic three-form for some almost complex structure I.

In this case, two blocks of Q_{AB} will be automatically zero, namely those where the Γ^A are either both wedges or both contractions: $(\rho_3, \mathrm{d}x^m \wedge \mathrm{d}x^n \wedge \rho_3) = 0 = (\rho_3, \iota_m \iota_n \rho_3)$. The remaining nonzero blocks are related to each other:

$$k_m{}^n \equiv (\rho_3, \mathrm{d}x^n \wedge \iota_m \rho_3) = -(\rho_3, \iota_m \mathrm{d}x^n \wedge \rho_3)\,, \tag{10.5.55}$$

as we expect already from the last of (10.5.11). Then the Hitchin invariant reads $q(\rho_3) = -\mathrm{Tr}(k^2)/6$; so we associate to $\rho_3 = \mathrm{Re}\Omega$ the almost complex structure and three-form

$$I_m{}^n = \frac{1}{\sqrt{-\mathrm{Tr}(k^2)/6}} k_m{}^n\,, \qquad \mathrm{Im}\Omega = -\frac{1}{3} I \cdot \mathrm{Re}\Omega\,, \tag{10.5.56}$$

the latter consistent with (5.1.30).

As a first concrete example, we consider

$$\rho_3 = \mathrm{d}x^1 \wedge \mathrm{d}x^2 \wedge \mathrm{d}x^3 + \mathrm{d}x^4 \wedge \mathrm{d}x^5 \wedge \mathrm{d}x^6 \tag{10.5.57}$$

in $M_6 = \mathbb{R}^6$. Applying the definition (10.5.55), we find $k = 2\mathrm{diag}(1, 1, 1, -1, -1, -1)$, and so $q(\rho_3) = -\mathrm{Tr}(k^2)/6 = -1 < 0$. So we conclude that there is no Ω such that $\rho_3 = \mathrm{Re}\Omega$.

If we start instead with a more familiar example:

$$\rho_3 = \mathrm{d}x^1 \wedge \mathrm{d}x^2 \wedge \mathrm{d}x^3 - \mathrm{d}x^4 \wedge \mathrm{d}x^5 \wedge \mathrm{d}x^3 - \mathrm{d}x^4 \wedge \mathrm{d}x^2 \wedge \mathrm{d}x^6 - \mathrm{d}x^1 \wedge \mathrm{d}x^5 \wedge \mathrm{d}x^6\,, \tag{10.5.58}$$

then we obtain $k_1{}^4 = k_2{}^5 = k_3{}^6 = 2$, $k_4{}^1 = k_5{}^2 = k_6{}^3 = -2$, where up to normalization we already recognize an almost complex structure I; indeed, $q > 0$, and (10.5.53) produces $\Omega = (\mathrm{d}x^1 + \mathrm{i}\mathrm{d}x^4) \wedge (\mathrm{d}x^2 + \mathrm{i}\mathrm{d}x^5) \wedge (\mathrm{d}x^3 + \mathrm{i}\mathrm{d}x^6)$.

Deformations*

We now study how deforming $\mathrm{Re}\Phi \to \mathrm{Re}\Phi + \delta\mathrm{Re}\Phi$ changes Φ [461, app. A].

For this we need a slight generalization of (10.5.46), involving $\mathrm{Re}\Phi(\mathrm{Re}\Phi, \cdot)$. The first line works in the same way, but in the second we cannot use (10.5.42), since $\mathrm{Re}\Phi$ is not pure. Still most $(\mathrm{Re}\Phi, \Gamma_{A_1 \ldots A_K} \mathrm{Re}\Phi)$ vanish; those with odd k vanish because $\mathrm{Re}\Phi$ is even or odd, and $k = 0, 4, 8, 12$ because $\Gamma^t_{A_1 \ldots A_k} = (-1)^{\lfloor k/2 \rfloor} \Gamma_{A_1 \ldots A_k}$ and antisymmetry of the pairing, (3.2.41). Moreover, the $k = 10$ bilinear is related to $k = 2$. All this implies

$$\mathrm{Re}\Phi(\mathrm{Re}\Phi, \cdot) \overset{(10.5.43)}{=} -\frac{1}{32} Q_{AB} \Gamma^{AB} (1 + \Gamma) + \frac{1}{6!} (\mathrm{Re}\Phi, \Gamma_{A_1 \ldots A_6} \mathrm{Re}\Phi) \Gamma^{A_6 \ldots A_1}\,, \tag{10.5.59}$$

where Γ is the chirality matrix for Cl(6, 6), which just gives $\Gamma \alpha_\pm = \pm \alpha_\pm$ on even or odd forms. Using (3.2.57) and (10.5.15) gives

$$\Gamma_{AB} \mathrm{Re}\Phi(\mathrm{Re}\Phi, \Gamma^{AB} \cdot) = \frac{1}{2} Q_{AB} \Gamma^{AB} (1 + \Gamma) + 3\Phi(\Phi, \cdot)\,. \tag{10.5.60}$$

Now $\delta Q_{AB} = 2(\mathrm{Re}\Phi, \Gamma_{AB}\delta\mathrm{Re}\Phi)$, and from $(\mathrm{Re}\Phi, \mathrm{Im}\Phi) = \sqrt{-\mathrm{Tr}(Q^2)/3}$ we also have $\delta(\mathrm{Re}\Phi, \mathrm{Im}\Phi) = 2(\delta\mathrm{Re}\Phi, \mathrm{Im}\Phi)$. Using these and (10.5.60), one finds

$$\delta\mathrm{Im}\Phi = -\mathcal{J}\cdot\delta\mathrm{Re}\Phi - \frac{2}{(\mathrm{Re}\Phi, \mathrm{Im}\Phi)}((\delta\mathrm{Re}\Phi, \mathrm{Re}\Phi)\mathrm{Re}\Phi + (\delta\mathrm{Re}\Phi, \mathrm{Im}\Phi)\mathrm{Im}\Phi)\,. \tag{10.5.61}$$

10.5.3 Compatible pairs

Compatible generalized almost complex structures

Two generalized almost complex structures $\mathcal{J}_1$, $\mathcal{J}_2$ are called *compatible* if they commute

$$[\mathcal{J}_1, \mathcal{J}_2] = 0\,, \tag{10.5.62a}$$

and if

$$\mathcal{G} = \mathcal{I}\mathcal{J}_1\mathcal{J}_2 \tag{10.5.62b}$$

is a positive-definite metric on $T \oplus T^*$. We will call it a *generalized metric*.

To see why this condition is interesting, we can immediately consider an example. Take $\mathcal{J}_\Omega$, $\mathcal{J}_{e^{-iJ}}$ as in (10.5.18) and (10.5.21). Their commutator only contains two non-zero blocks, and imposing that they vanish gives the single condition

$$JI^t + IJ = 0\,, \tag{10.5.63a}$$

which is nothing but the Hermitian property (5.1.39). Moreover,

$$\mathcal{G} = \begin{pmatrix} JI^t & 0 \\ 0 & J^{-1}I \end{pmatrix} = \begin{pmatrix} g & 0 \\ 0 & g^{-1} \end{pmatrix}, \tag{10.5.63b}$$

using again (5.1.39). So compatibility of two GACS is a generalization of the old compatibility condition between an almost complex structure and an almost symplectic structure.

Since $\mathcal{J}_1$ and $\mathcal{J}_2$ commute, we can refine the definition (10.5.9a) and introduce the eigenspaces

$$\begin{aligned} L_{\pm\pm} &\equiv \{X | \mathcal{J}_1 X = \pm \mathrm{i} X\,,\ \mathcal{J}_2 X = \pm \mathrm{i} X\}\,, \\ L_{\pm\mp} &\equiv \{X | \mathcal{J}_1 X = \pm \mathrm{i} X\,,\ \mathcal{J}_2 X = \mp \mathrm{i} X\}\,. \end{aligned} \tag{10.5.64}$$

So, for example, $L_{++} = L_{\mathcal{J}_1} \cap L_{\mathcal{J}_2}$, $L_{+-} = L_{\mathcal{J}_1} \cap \bar{L}_{\mathcal{J}_2}$, and so on. Conversely, if there is a splitting

$$L_{\mathcal{J}_1} = L_{++} \cup L_{+-}\,, \qquad L_{\mathcal{J}_2} = L_{++} \cup L_{-+}\,, \tag{10.5.65}$$

then we know that $\mathcal{J}_1$ and $\mathcal{J}_2$ commute. All (10.5.64) have dimension $d/2$ each, which shows that d must be even.

By (2.1.6) and (10.5.62a), the actions of the $\mathcal{J}_{1,2}$ on forms defined by (10.5.33) also commute:

$$[\mathcal{J}_1\cdot, \mathcal{J}_2\cdot] = 0\,. \tag{10.5.66}$$

So they can be block-diagonalized simultaneously, and we can refine (10.5.31) and (10.5.38) to split the space of all forms as a sum over common eigenspaces of both:

$$\Lambda = \oplus_{p,q} U_{p,q}\,, \qquad \begin{aligned} \mathcal{J}_1 \cdot U_{p,q} &= p\mathrm{i} U_{p,q} \\ \mathcal{J}_2 \cdot U_{p,q} &= q\mathrm{i} U_{p,q}\,, \end{aligned} \tag{10.5.67}$$

In $d = 6$, this decomposition is the pure diamond (10.3.83), as we will see shortly.

The presence of two $\mathcal{J}_{1,2}$ now reduces the structure group further from (10.5.30). Indeed,

$$(\mathcal{I}\mathcal{G})^2 = (\mathcal{J}_1\mathcal{J}_2)^2 = 1\,, \tag{10.5.68}$$

since the $\mathcal{J}_{1,2}$ square to -1 and commute. So $\frac{1}{2}(1\pm\mathcal{I}\mathcal{G})$ are two orthogonal projectors, and reduce the structure group of $T \oplus T^*$ to $\mathrm{O}(d) \times \mathrm{O}(d)$. Overall,

$$\mathrm{gStab}(\mathcal{J}_1, \mathcal{J}_2) = \mathrm{U}(d/2, d/2) \cap (\mathrm{O}(d) \times \mathrm{O}(d)) = \mathrm{U}(d/2) \times \mathrm{U}(d/2)\,. \tag{10.5.69}$$

Equation (10.5.62b) also has the Hermitian property

$$\mathcal{G}\mathcal{I} = \mathcal{I}\mathcal{G}^t\,, \tag{10.5.70}$$

similar to (5.1.39). If we write in blocks $\mathcal{J}_1\mathcal{J}_2 = \left(\begin{smallmatrix} M_1 & M_2 \\ M_3 & M_4 \end{smallmatrix}\right)$, (10.5.70) imposes $M_1 = M_4^t$, $M_2^t = M_2$, $M_3^t = M_3$. Moreover, the fact that $\mathcal{G} = \mathcal{I}\mathcal{J}_1\mathcal{J}_2$ is positive definite requires M_2 and M_3 to be positive definite too; we define $M_2 \equiv g^{-1}$, which is going to be the inverse of a Riemannian metric, and $M_1 \equiv -g^{-1}b$. Now imposing that $(\mathcal{J}_1\mathcal{J}_2)^2 = 1$ gives $b^t = -b$, and determines M_3, M_4. All this gives an explicit parameterization for the generalized metric:

$$\mathcal{G} = \mathcal{I}\mathcal{J}_1\mathcal{J}_2 = \begin{pmatrix} g - bg^{-1}b & bg^{-1} \\ -g^{-1}b & g^{-1} \end{pmatrix} = \begin{pmatrix} 1 & b \\ 0 & 1 \end{pmatrix}\begin{pmatrix} g & 0 \\ 0 & g^{-1} \end{pmatrix}\begin{pmatrix} 1 & 0 \\ -b & 1 \end{pmatrix}. \tag{10.5.71}$$

Via this formula, a compatible pair $\mathcal{J}_{1,2}$ determines a metric g and a two-form b.

We can now also block-diagonalize $\mathcal{I}\mathcal{G}$:

$$\mathcal{I}\mathcal{G} = \mathcal{J}_1\mathcal{J}_2 = \mathcal{E}\begin{pmatrix} 1 & 0 \\ 0 & -1 \end{pmatrix}\mathcal{E}^{-1}\,,$$
$$\mathcal{E} \equiv \begin{pmatrix} 1 & 1 \\ g+b & -g+b \end{pmatrix} = \begin{pmatrix} 1 & 0 \\ b & 1 \end{pmatrix}\begin{pmatrix} 1 & 1 \\ g & -g \end{pmatrix}. \tag{10.5.72}$$

This has some consequence on the $\mathcal{J}_{1,2}$ too. Since they both commute with $\mathcal{J}_1\mathcal{J}_2$, $\hat{\mathcal{J}}_{1,2} \equiv \mathcal{E}^{-1}\mathcal{J}_{1,2}\mathcal{E}$ should both commute with $\left(\begin{smallmatrix} 1 & 0 \\ 0 & -1 \end{smallmatrix}\right)$; this requires them to be block-diagonal. Since $\hat{\mathcal{J}}_{1,2}^2 = -1$, their blocks are almost complex structures. Moreover, imposing that they commute and that their product gives $\left(\begin{smallmatrix} 1 & 0 \\ 0 & -1 \end{smallmatrix}\right)$ fixes them so that

$$\mathcal{J}_1 = \mathcal{E}\begin{pmatrix} -I_1^t & 0 \\ 0 & I_2^t \end{pmatrix}\mathcal{E}^{-1}\,, \qquad \mathcal{J}_2 = \mathcal{E}\begin{pmatrix} -I_1^t & 0 \\ 0 & -I_2^t \end{pmatrix}\mathcal{E}^{-1}\,, \tag{10.5.73}$$

with $I_{1,2}$ two almost complex structures. (Since in our conventions I has indices $I_m{}^n$, the transpose ^{t} is needed for the index structure to work out.)

T-duality

The matrix $\mathcal{G}$ also appeared in Section 4.2.6; this is related to the occurrence of $\mathrm{O}(d,d)$ in both T-duality and (10.5.28).[13] The generalized complex structures transform as [140]

[13] Equation (10.5.71) is also similar to (9.1.40); this is not accidental, but that matrix acts on the space of forms over a manifold, while (10.5.71) acts on $T \oplus T^*$.

$$\mathcal{J}_a \to (O^t)^{-1}\mathcal{J}_a O^t\,. \tag{10.5.74}$$

Indeed, one can show that this is compatible with (4.2.77), using $\mathcal{G} = \mathcal{I}\mathcal{J}_1\mathcal{J}_2 = -\mathcal{J}_1^t\mathcal{I}\mathcal{J}_2$ and $O \in \mathrm{O}(d,d)$. The corresponding action on the Φ_a was studied in [127, 140, 487]. For a single T-duality along an abelian symmetry v, it is given by the following [487]:

$$\Phi \to (\iota_v + \zeta\wedge)\Phi\,, \tag{10.5.75}$$

where $L_v B = \mathrm{d}\zeta$. This action in general changes the type of Φ, as defined after (3.2.53). It is checked in [487] that this action maps a solution of the pure spinor equations (10.3.20) to another solution.

Compatible pure forms

Two pure forms Φ_1, Φ_2 are said to be compatible, or a *compatible pure pair*, if their associated $\mathcal{J}_{\Phi_1}$, $\mathcal{J}_{\Phi_2}$ are compatible in the sense of (10.5.62), and if they have equal norm:

$$(\Phi_1, \bar{\Phi}_1) = (\Phi_2, \bar{\Phi}_2)\,. \tag{10.5.76}$$

This requirement is not necessary from a mathematical point of view, but it is convenient for subsequent applications.

Now L_{++} in (10.5.64) is the common annihilator, while L_{-+} is the annihilator of $\bar{\Phi}_1$ and Φ_2, and so on. If we act with L_{-+} on Φ_1, we will keep generating new forms until we use all of its $d/2$ elements; the last form is still annihilated by L_{++}, and so it must be Φ_2. So [256, 489]

$$\Phi_2 \in U_0^{\mathcal{J}_1} = U_{0,d/2} \quad \Rightarrow \quad \mathcal{J}_1 \cdot \Phi_2 = 0\,. \tag{10.5.77}$$

Similar to this, $\Phi_1 \in U_0^{\mathcal{J}_2} = U_{d/2,0}$. It follows that in $d = 4k$, the $\Phi_{1,2}$ have the same form parity (they are both even or both odd); for $d = 4k+2$, they have opposite parity. In $d = 6$, a pair will be of the form $\Phi_{1,2} = \Phi_\pm$.

In fact, the property (10.5.77) is pointwise *equivalent* to $[\mathcal{J}_1, \mathcal{J}_2] = 0$. To see this, consider $X_2 \in L_{\mathcal{J}_2}$; in other words, $X_2 \cdot \Phi_2 = 0$. Adapting (10.5.35), we see $[\mathcal{J}_1\cdot, X_2\cdot] = (\mathcal{J}_1 X_2)\cdot$; so

$$(\mathcal{J}_1 X_2)\cdot\Phi_2 = [\mathcal{J}_1\cdot, X_2\cdot]\Phi_2 \overset{(10.5.77)}{=} 0\,. \tag{10.5.78}$$

This means that $\mathcal{J}_1$ maps $L_{\mathcal{J}_2} \to L_{\mathcal{J}_2}$, and by reality $\bar{L}_{\mathcal{J}_2} \to \bar{L}_{\mathcal{J}_2}$; so it is block-diagonal. $\mathcal{J}_1$ is diagonalizable, so each block is also diagonalizable: there exists a basis $\{X_{2a}\}$ of $L_{\mathcal{J}_2}$ such that $\mathcal{J}_1 X_{2a} = \mathrm{i}\lambda_a X_{2a}$, with $\lambda_a \in \{+1,-1\}$. By definition, we also know $\mathcal{J}_2 X_{2a} = \mathrm{i}X_{2a}$. Together with the conjugates $\bar{X}_{2a}$, we have obtained a basis of $T \oplus T^*$ on which both $\mathcal{J}_1$ and $\mathcal{J}_2$ are diagonal. So (10.5.66) holds.

From a global point of view, a pure form Φ reduces the structure group a bit more than the associated $\mathcal{J}$, as in (10.5.29) and (10.5.30). Similar to this, a compatible pair reduces to

$$\mathrm{gStab}(\Phi_1, \Phi_2) = \mathrm{SU}(d/2) \times \mathrm{SU}(d/2) \tag{10.5.79}$$

rather than (10.5.69). This is the reason we called "SU(3) × SU(3)-structure" the generic pair of compatible pure forms (10.3.37). As we mentioned in that subsection, every compatible pure pair defines such a structure on $T \oplus T^*$, but in the special cases (10.3.28) and (10.3.31), they also define an SU(3)- and SU(2)-structure already on TM, and so it is customary to use those names to denote those pure pairs.

The six-dimensional case

Let us specialize the preceding discussion to $d = 6$ a bit further. Since we reach Φ_2 from Φ_1 by acting with $d/2 = 3$ gamma matrices, Φ_1 and Φ_2 have opposite parity: one is even, the other is odd. The $U_{p,q}$ can be visualized as a diamond:

$$\begin{array}{ccccccc} & & & U_{3,0} & & & \\ & & U_{2,1} & & U_{2,-1} & & \\ & U_{1,2} & & U_{1,0} & & U_{1,-2} & \\ U_{0,3} & & U_{0,1} & & U_{0,-1} & & U_{0,-3} \\ & U_{-1,2} & & U_{-1,0} & & U_{-1,-2} & \\ & & U_{-2,1} & & U_{-2,-1} & & \\ & & & U_{-3,0} & & & \end{array} . \tag{10.5.80}$$

Recall that the action of $\mathcal{J}_{1,2}$ is given in (10.5.67). Applying the orthogonality property (10.5.41) to both actions, we obtain

$$(u_{j,k}, u_{j',k'}) = 0 \quad \text{unless } j = -j' \, , \; k = -k' \, . \tag{10.5.81}$$

To ensure (10.5.77), it is enough to exclude that Φ_2 has any component on $U^{\mathcal{J}_1}_{\pm 2}$, the Clifford levels obtained by acting with one Γ^A on Φ_1 or $\bar{\Phi}_1$. So we obtain

$$(\Gamma_A \Phi_1, \Phi_2) = (\Gamma_A \bar{\Phi}_1, \Phi_2) = 0 \, . \tag{10.5.82}$$

In $d = 6$, this is equivalent to the condition that the $\mathcal{J}_{1,2}$ commute. (The other condition for compatibility, namely that the metric is positive definite, still needs to be imposed separately.) In particular, for $\Gamma_A = \mathrm{d}x^m \wedge$ this gives

$$(\Phi_1, \Phi_2)_5 = (\bar{\Phi}_1, \Phi_2)_5 = 0 \, . \tag{10.5.83}$$

The pair $\Phi_1 = \Omega$, $\Phi_2 = \mathrm{e}^{-\mathrm{i}J}$ is compatible, recalling (10.5.63). These two pure forms occurred together as the bispinors of a pure spinor η_+ in (3.4.6) and (3.4.11), in any even d. We can generalize this example by taking two different pure spinors η^a_+:

$$\Phi_1 = \eta^1_+ \otimes \eta^{2\,\dagger}_+ \, , \qquad \Phi_2 = \eta^1_+ \otimes (\eta^{2\,c}_+)^\dagger \, . \tag{10.5.84}$$

They are pure forms because of the general principle (3.4.4). To see that they are compatible, we compute

$$\begin{aligned} L_{++} &= \mathrm{Ann}(\Phi_1) \cap \mathrm{Ann}(\Phi_2) = \mathrm{Span}\left(\overrightarrow{\gamma}_{\bar{i}_1}\right) , \\ L_{+-} &= \mathrm{Ann}(\Phi_1) \cap \mathrm{Ann}(\bar{\Phi}_2) = \mathrm{Span}\left(\overleftarrow{\gamma}_{i_2}\right) , \\ L_{-+} &= \mathrm{Ann}(\bar{\Phi}_1) \cap \mathrm{Ann}(\Phi_2) = \mathrm{Span}\left(\overleftarrow{\gamma}_{\bar{i}_2}\right) , \\ L_{--} &= \mathrm{Ann}(\bar{\Phi}_1) \cap \mathrm{Ann}(\bar{\Phi}_2) = \mathrm{Span}\left(\overrightarrow{\gamma}_{i_1}\right) , \end{aligned} \tag{10.5.85}$$

with the same logic that led to (10.5.27). The diamond (10.5.80) is now nothing but (10.3.83). The orthogonality property (10.5.81) becomes the old (3.4.96).

In fact, it is now easy to show that *the most general compatible pure pair is a b-transform of* (10.5.84). First we go back to our general expression (10.5.73) for a compatible pair $\mathcal{J}_{1,2}$. This gives, for example,

$$L_{++} = \left\{ \mathcal{E} \begin{pmatrix} v^1_{0,1} \\ 0 \end{pmatrix} \right\} = \left\{ \begin{pmatrix} v^1_{0,1} \\ (g+b)v^1_{0,1} \end{pmatrix} \right\}, \qquad (10.5.86)$$

where $v^1_{0,1}$ is a $(0,1)$-vector with respect to I^1. With a similar notation, and using the same trick as (10.5.23), the action on forms of $L_{\pm\pm}$ is

$$\begin{aligned} L_{++} &= \{\iota_{v^1_{0,1}} + (g+b)v^1_{0,1}\wedge\} = \{e^{b\wedge}(\iota_{v^1_{0,1}} + (gv^1_{0,1})\wedge)e^{-b\wedge}\}, \\ L_{+-} &= \{\iota_{v^2_{1,0}} + (-g+b)v^2_{1,0}\wedge\} = \{e^{b\wedge}(\iota_{v^2_{1,0}} - (gv^2_{1,0})\wedge)e^{-b\wedge}\}, \\ L_{-+} &= \{\iota_{v^2_{0,1}} + (-g+b)v^2_{0,1}\wedge\} = \{e^{b\wedge}(\iota_{v^2_{0,1}} - (gv^2_{0,1})\wedge)e^{-b\wedge}\}, \\ L_{--} &= \{\iota_{v^1_{1,0}} + (g+b)v^1_{1,0}\wedge\} = \{e^{b\wedge}(\iota_{v^1_{1,0}} + (gv^1_{1,0})\wedge)e^{-b\wedge}\}. \end{aligned} \qquad (10.5.87)$$

For $b = 0$, using (3.1.21) and (3.1.24), these are exactly (10.5.85). Imposing also the equal norm condition (10.5.76), the $\Phi_{1,2}$ are determined to be of the form (10.5.84). More generally for $b \neq 0$, we conclude that they are of the form

$$\Phi^b_{1,2} \equiv N e^{b\wedge}\Phi_{1,2}, \qquad (10.5.88)$$

where N is a normalization function, and again $\Phi_{1,2}$ are as in (10.5.84).

Compatible pure forms and metric

It is worth taking stock of what we have obtained, focusing on $d = 6$. The compatible pair is of the form $\Phi_{1,2} = \Phi_\pm$. Adapting the result of (10.5.88), and with a slight abuse of language, we saw that the most general compatible pair is of the form

$$\Phi^b_\pm \equiv e^{b\wedge}\eta^1_+ \otimes \eta^{2\dagger}_\pm. \qquad (10.5.89)$$

The $\Phi_\pm$ were originally introduced in Section 10.3 for practical reasons: they were bilinears on which the supersymmetry equations simplified, and which were enough to encode supersymmetry completely. We have now found that they are also the most general compatible pair of pure forms, which in $d = 6$ can be formulated as (10.5.82). An important consequence is that they determine the metric and a two-form b, via (10.5.71). Moreover, in (10.5.88) we also have a normalization function N, which in the application to supersymmetry we know to be related to the warping function e^A. To summarize, we can now refine (10.5.1):

$$(\Phi^b_+, \Phi^b_-) \quad \longleftrightarrow \quad (g_{mn}, b_{mn}, A, \eta^1_+, \eta^2_+). \qquad (10.5.90)$$

Once again the arrow means that the two sets of data are in one-to-one correspondence.

10.5.4 Integrability

The discussion so far has been *purely algebraic*: nowhere did we involve derivatives of pure forms Φ or of generalized almost complex structures $\mathcal{J}$. We will now study what happens when a pure form Φ is closed or d_H-closed; this is relevant to (10.3.24b).

The discussion will be modeled on two examples we already know:

- If $\Phi = \Omega$, the top holomorphic form of an almost complex structure I, the condition $d\Omega = 0$ is a particular case of (5.3.17b), which implies that I is integrable. So in this case we are discussing complex geometry (Section 5.3).
- If $\Phi = e^{-iJ}$, closure is simply $dJ = 0$, which defines symplectic geometry (Section 5.2.1).

Courant bracket

Complex geometry is a slightly deeper source of inspiration, and we will follow it closely. We saw that integrability can be reformulated in many ways, summarized at the end of Section 5.3.2. One of those is that the holomorphic tangent bundle $T_{1,0}$ should be closed under the Lie bracket, (5.3.21). The proof that this is implied by $d\Omega = 0$ was given in (5.3.27); it relied crucially on the property (4.1.81) of the Lie bracket of vector fields:

$$[\{\iota_v, d\}, \iota_w] = \iota_{[v,w]} \,. \tag{10.5.91}$$

The guiding idea of this section has been to use the generalized tangent bundle $T \oplus T^*$ rather than T; so to generalize the ideas of complex geometry, we need a bracket that extends $[v, w]$ to $T \oplus T^*$. In other words, we need a bracket $[X, Y]$ whose two entries are both pairs of a vector and a one-form:

$$X = \begin{pmatrix} v \\ \alpha \end{pmatrix}, \qquad Y = \begin{pmatrix} w \\ \beta \end{pmatrix}. \tag{10.5.92}$$

Ideally, this would have the usual properties for the bracket of a Lie algebra: linearity in each entry, antisymmetry under exchange $X \leftrightarrow Y$, and the Jacobi identity. Unfortunately, a bracket with all these properties does not exist on $T \oplus T^*$, but we will now see that we can get close.

The natural extension of (10.5.91) might seem to be

$$(X \circ Y)\cdot \equiv [\{X\cdot, d\}, Y\cdot] \,. \tag{10.5.93}$$

This is called *Dorfman bracket*. It is not antisymmetric:

$$[\{X\cdot, d\}, Y\cdot] \overset{(2.1.32c)}{=} -[\{X\cdot, Y\cdot\}, d] + [X\cdot, \{Y\cdot, d\}] = d(X^t \mathcal{I} Y) \wedge -Y \circ X \,. \tag{10.5.94}$$

We can also check this by evaluating it explicitly on (10.5.92):

$$\begin{aligned} [\{X\cdot, d\}, Y\cdot] &\overset{(10.5.91),(4.1.79),(4.1.77)}{=} \iota_{[v,w]} + [L_v, \beta\wedge] + [d\alpha\wedge, \iota_w + \beta\wedge] \\ &= \iota_{[v,w]} + \left(L_v\beta - \iota_w d\alpha\right) \wedge \,. \end{aligned} \tag{10.5.95}$$

We can antisymmetrize this by hand by defining the *Courant bracket* as

$$[X, Y]_C \equiv \frac{1}{2}(X \circ Y - Y \circ X) = X \circ Y - \frac{1}{2} d(X^t \mathcal{I} Y) \,, \tag{10.5.96a}$$

or in other words

$$[X,Y]_C \cdot = \frac{1}{2}([\{X\cdot,\mathrm{d}\},Y\cdot] - [\{Y\cdot,\mathrm{d}\},X\cdot]) \tag{10.5.96b}$$

$$\overset{(10.5.95)}{=} \begin{pmatrix} [v,w] \\ L_v\beta - L_w\alpha + \frac{1}{2}\mathrm{d}(\iota_w\alpha - \iota_v\beta) \end{pmatrix} \cdot \, . \tag{10.5.96c}$$

$[X,Y]_C$ is antisymmetric by construction; we now check if it obeys the Jacobi identity. After remarking that

$$[\{X\cdot,\mathrm{d}\},\mathrm{d}] = X\cdot\mathrm{d}^2 + \mathrm{d}X\cdot\mathrm{d} - \mathrm{d}X\cdot\mathrm{d} - \mathrm{d}^2X\cdot \overset{(4.1.75)}{=} 0\,, \tag{10.5.97}$$

we find an encouraging Leibniz identity for $\circ$:

$$\begin{aligned} ((X\circ Y)\circ Z)\cdot &= [\{[\{X\cdot,\mathrm{d}\},Y\cdot],\mathrm{d}\},Z\cdot] = [[\{X\cdot,\mathrm{d}\},\{Y\cdot,\mathrm{d}\}],Z\cdot] \\ &\overset{(2.1.32a)}{=} [[\{X\cdot,\mathrm{d}\},Z\cdot],\{Y\cdot,\mathrm{d}\}] + [\{X\cdot,\mathrm{d}\},[\{Y\cdot,\mathrm{d}\},Z\cdot]] \\ &= ((X\circ Z)\circ Y)\cdot + (X\circ(Y\circ Z))\cdot\,. \end{aligned} \tag{10.5.98}$$

A lengthy but straightforward computation now gives [482, prop. 3.16]

$$\begin{aligned} &[[X,Y]_C,Z]_C + [[Y,Z]_C,X]_C + [[Z,X]_C,Y]_C \\ &\quad = \frac{1}{3}\mathrm{d}\left([X,Y]_C^t\mathcal{I}Z + [Y,Z]_C^t\mathcal{I}X + [Z,X]_C^t\mathcal{I}Y\right)\,. \end{aligned} \tag{10.5.99}$$

So the Courant bracket does not obey the Jacobi identity. But we will soon see that in our applications the right-hand side of (10.5.99) vanishes, basically because we will restrict it to isotropic subspaces of $T\oplus T^*$, where $\mathcal{I}$ is zero.

Brackets obtained by antisymmetrizing an expression such as $[\{X\cdot,\mathrm{d}\},Y\cdot]$ as in (10.5.94), with d replaced by any operator squaring to zero, are called *derived brackets* [490].

Generalized complex structures

In (5.3.21), we defined a complex structure to be one for which $T_{1,0}$ is closed under the Lie bracket. The analogue of $T_{1,0}$ for a generalized almost complex structure $\mathcal{J}$ is its i-eigenspace $L_{\mathcal{J}}$, and the analogue of the Lie bracket is the Courant bracket (10.5.96). This leads to defining $\mathcal{J}$ to be *integrable*, or a *generalized complex structure (GCS)*, if

$$[L_{\mathcal{J}},L_{\mathcal{J}}]_C \subset L_{\mathcal{J}}\,. \tag{10.5.100}$$

This is the direct analogue of (5.3.21). A manifold with such a $\mathcal{J}$ is said to be a generalized complex manifold.

We can also rewrite (10.5.100) slightly more concretely:

$$\bar{\Pi}[\Pi X,\Pi Y]_C = 0 \quad \forall X,Y\,; \qquad \Pi = \frac{1}{2}(1 - \mathrm{i}\mathcal{J})\,, \tag{10.5.101}$$

similar to (5.3.24) for ordinary complex structures. We could even write this out explicitly using blocks, as we did in (10.5.6) for the condition $\mathcal{J}^2 = -1$. This gives lengthy and uninspiring relations, except for the upper-right block, which has to satisfy

$$p^{q[m}\partial_q p^{np]} = 0\,. \tag{10.5.102}$$

We recognize this from (5.4.9) as the definition of a Poisson tensor. So any generalized complex structure $\mathcal{J}$ defines a Poisson tensor (which may be zero) [491]. There is also a more direct proof, using the Frobenius theorem [441, sec. 4.1].

Once (10.5.100) holds, we can restrict the Courant bracket to $L_{\mathcal{J}}$. Since it is an isotropic subspace (see the discussion following (10.5.12)), as anticipated, the right-hand side of (10.5.99) vanishes on it, and the Jacobi identity holds. So sections of $L_{\mathcal{J}}$ form a Lie algebra under the Courant bracket.

Complex and symplectic examples

For $\mathcal{J}_I$ associated to an almost complex structure I, (10.5.100) says that (10.5.17) should be closed under the Courant bracket. Thus we take two general elements of $L_{\mathcal{J}_I}$, $X = \binom{v_{0,1}}{\alpha_{1,0}}$, $Y = \binom{w_{0,1}}{\beta_{1,0}}$; recalling (10.5.96c), we should impose

$$[X,Y]_{\mathrm{C}} = \begin{pmatrix} [v_{0,1}, w_{0,1}] \\ \iota_{v_{0,1}}\mathrm{d}\beta_{1,0} - \iota_{w_{0,1}}\mathrm{d}\alpha_{1,0} \end{pmatrix} \in T_{0,1} \oplus T^*_{1,0}\,. \tag{10.5.103}$$

The first entry is in $T_{0,1}$ if the usual integrability (5.3.21) holds. So $\mathcal{J}_I$ is a generalized complex structure only if I is a complex structure – as expected, given that this was our inspiration for (10.5.100). The second entry is in fact also equivalent to integrability of I: $\iota_{v_{0,1}}\mathrm{d}\beta_{1,0}$ has no $(0,1)$-component if and only if (5.3.15) holds, which was one of the several definitions of complex structure in Section 5.3.2.

Next we check what integrability means for $\mathcal{J}_J$. By (10.5.20), now we need to take $X = \binom{v}{-\mathrm{i}\iota_v J}$, $Y = \binom{w}{-\mathrm{i}\iota_w J}$. Equation (10.5.96c) gives

$$[X,Y]_{\mathrm{C}} \overset{(4.1.79),(4.1.81)}{=} \begin{pmatrix} [v,w] \\ -\mathrm{i}\iota_{[v,w]}J + \mathrm{i}\iota_v\iota_w \mathrm{d}J \end{pmatrix}. \tag{10.5.104}$$

So for the right-hand side to be in $L_{\mathcal{J}_{e^{-\mathrm{i}J}}}$, we see we have to require

$$\mathrm{d}J = 0\,. \tag{10.5.105}$$

Given that J is nondegenerate, this is the definition of symplectic geometry.

Integrability and closure

As is the case for complex geometry, the definition (10.5.100), while conceptually pleasing, is not the most practical way to check integrability. If $\mathcal{J}$ comes from a pure form Φ, integrability is more easily formulated by looking at Φ directly. Consider an integrable $\mathcal{J}$, and $X, Y \in L_{\mathcal{J}} = \mathrm{Ann}(\Phi)$; then

$$0 \overset{(10.5.100)}{=} [X,Y]_{\mathrm{C}} \cdot \Phi \overset{(10.5.96b)}{=} \frac{1}{2}(X\cdot Y \cdot - Y \cdot X \cdot)\mathrm{d}\Phi\,. \tag{10.5.106}$$

So $\mathrm{d}\Phi$ should vanish when we act on it with two annihilators; in the terminology of (10.5.31),

$$\mathrm{d}\Phi \in U^1_{\mathcal{J}}\,, \tag{10.5.107}$$

or in other words,

$$d\Phi = W \cdot \Phi \qquad (10.5.108)$$

for some $W \in T \oplus T^*$ (which can be taken to be in $\bar{L}_{\mathcal{J}}$). Conversely, if (10.5.108) holds, we can follow the derivation backward, to deduce that $\mathcal{J}_\Phi$ is integrable. Equation (10.5.108) is a direct generalization of (5.3.18), to which it reduces if $\Phi = \Omega$. The equivalence of (10.5.108) and (10.5.100) generalizes the equivalence of (5.3.18) and (5.3.27).

For the same reasons discussed before (5.3.17a), and again after (5.5.6), even without integrability $d\Phi$ can only belong to the Clifford levels $U^{\mathcal{J}}_{d/2-1}$ and $U^{\mathcal{J}}_{d/2-3}$. The right-hand side of (10.5.108) is by definition in $U^{\mathcal{J}}_{d/2-1}$, so (10.5.108) is equivalent to

$$(d\Phi)|_{U^{\mathcal{J}}_{d/2-1}} = 0\,. \qquad (10.5.109)$$

We could regard the left-hand side of (10.5.109) as a generalization of the Nijenhuis tensor.

If we consider now our other canonical example $\Phi = e^{-iJ}$, an element $W \in \bar{L}_{\mathcal{J}}$ would be of the form $\iota_v + i\iota_v J\wedge$; but then the one-form part of (10.5.108) imposes $v = 0$. So in this case (10.5.108) reduces to (10.5.105), as expected.

Generalized Dolbeault differential

One more integrability criterion for complex geometry in Section 5.3.2 was $\bar{\partial}^2 = 0$, the fact that the Dolbeault differential should square to zero. This holds if I is integrable, but in (5.3.40) we proved that it is equivalent to it.

Mimicking (5.3.32) and (5.3.34), we now define the *generalized Dolbeault differential* $\partial_{\mathcal{J}}$ by

$$d^{\mathcal{J}} \equiv [d, \mathcal{J}\cdot]\,, \qquad \partial_{\mathcal{J}} \equiv \frac{1}{2}(d + i\,d^{\mathcal{J}})\,, \qquad (10.5.110)$$

where the action $\mathcal{J}\cdot$ was defined in (10.5.33). Now integrability of $\mathcal{I}$ can be reformulated along the lines of (5.3.36) as [492]

$$[[d, \mathcal{J}\cdot], \mathcal{J}\cdot] = -d\,, \qquad (10.5.111)$$

which in turn is equivalent to

$$\partial^2_{\mathcal{J}} = 0\,. \qquad (10.5.112)$$

In our two canonical examples, (10.5.110) can be computed using (10.5.34):

$$d^{\mathcal{J}_I} = [d, I\cdot] \overset{(5.3.32)}{=} d^c \qquad (10.5.113a)$$

$$d^{\mathcal{J}_J} = [d, -J\wedge + J\cdot] \overset{(10.5.105)}{=} [d, J\cdot] \overset{(5.4.1)}{=} d_J\,. \qquad (10.5.113b)$$

The second case is perhaps a bit less familiar; we discussed it in Section 5.4.1, where we showed that it indeed plays the same role as d^c in complex geometry (see in particular (5.4.5)). We also evaluated it for the Kähler case in (6.1.16).

It is also possible to develop a theory of generalized complex deformations following Section 5.3.4, with generalizations of Beltrami differentials $\delta\mathcal{J}$ and of the Kodaira–Spencer equation; see, for example, [489, sec. 3.4].

Twisted integrability

The nice properties of the Courant bracket hinge on d being a differential, $d^2 = 0$. Another prominent differential we encountered in this chapter is $d_H = d - H\wedge$ (see (10.1.8)). This leads to the definition of the *twisted* version of the Courant bracket, inspired by (10.5.96b):

$$[X,Y]_H \cdot \equiv \frac{1}{2}\left([\{X\cdot, d_H\}, Y\cdot] - [\{Y\cdot, d_H\}, X\cdot]\right) . \tag{10.5.114}$$

This is a relatively small modification of the usual Courant bracket:

$$[X,Y]_H = [X,Y] - \iota_v \iota_w H , \tag{10.5.115}$$

in the usual notation (10.5.92). This leads to a notion of twisted integrability for a $\mathcal{J}$, which is equivalent to

$$d_H \Phi = W \cdot \Phi , \tag{10.5.116}$$

by the same logic leading to (10.5.108). There is also a twisted version of (10.5.110):

$$d_H^{\mathcal{J}} \equiv [d_H, \mathcal{J}\cdot] , \qquad \partial_H^{\mathcal{J}} \equiv \frac{1}{2}(d + i\, d_H^{\mathcal{J}}) . \tag{10.5.117}$$

If we write locally $H = dB$, then

$$d_{dB}^{\mathcal{J}} \overset{(10.1.11)}{=} [e^{B\wedge} d e^{-B\wedge}, \mathcal{J}\cdot] \overset{(10.5.25)}{=} e^{B\wedge}[d, \mathcal{J}^{-B}] e^{-B\wedge} . \tag{10.5.118}$$

Noncomplex, nonsymplectic examples

So far, the examples of generalized complex structures we have seen are of the form $\mathcal{J}_I$ or $\mathcal{J}_J$. It is easy to give examples that are neither, by considering pure spinors that are neither of type 0 (such as e^{-iJ}) nor of type $d/2$ (such as Ω). In $d = 6$, we can consider Φ_+ or Φ_- as in (10.3.31): if these are closed, they define generalized complex structures that are neither complex nor symplectic, but a mix of the two types.

This leads to asking if there are manifolds that admit a generalized complex structure but not a symplectic or ordinary complex one. In $d = 6$, there are 34 distinct nilmanifolds. Restricting attention to *left-invariant* Φ's (Sections 4.4.2 and 4.4.4), one finds that all these nilmanifolds admit a generalized complex structure, while five admit neither a complex nor a symplectic structure [139]. This is similar to the symplectic non-Kähler manifold in (6.5.3).

10.5.5 Generalized Kähler geometry

We can now put together the ideas of Sections 10.5.3 and 10.5.4: namely, we consider

$$(\mathcal{J}_1, \mathcal{J}_2) \text{ compatible and both integrable} . \tag{10.5.119}$$

This is called a *generalized Kähler* structure, and a manifold that admits it a generalized Kähler manifold.

$\mathcal{N} = 2$ supersymmetry, world-sheet $\mathcal{N} = (2,2)$, bihermitian geometry

Generalized Kähler geometry has a natural physics interpretation. Setting to zero the RR fields, $f = 0$, the conditions for spacetime supersymmetry are (9.3.2a) and (9.3.2c) (a particular case of (10.3.17)). We showed there that $\mu = \mathrm{d}A = 0$. With these assumptions, the pure spinor equations (10.3.24) reduce to

$$\mathrm{d}_H(\mathrm{e}^{-\phi}\Phi_\pm) = 0 \,. \tag{10.5.120}$$

(See also Exercise 10.3.9.) In the case where $\eta^1_+ \equiv \eta_+ \neq 0$, $\eta^2_+ = 0$, on which we focused in Section 9.3.1, the $\Phi_\pm$ are identically zero, and (10.5.120) is valid, but trivially so. When both $\eta^a_+ \neq 0$, we have (9.3.18) and $\mathcal{N} = 2$ spacetime supersymmetry. This led to two solutions (J_a, Ω_a) to (9.3.13), with H and $-H$. We now see that this is equivalent to the more compact (10.5.120). (It is possible, but a bit involved, to show this equivalence directly by using the explicit expressions (10.3.37) and (10.3.39).)

We also saw in Section 9.3.3 that spacetime $\mathcal{N} = 2$ implies world-sheet $\mathcal{N} = (2,2)$ supersymmetry, but that the opposite is not quite true; the latter leads to the weaker *bihermitian* equations (9.3.25), which involve neither the dilaton nor the Ω_a. (Recall that this is because world-sheet supersymmetry does not imply conformal invariance.) From the point of view of generalized complex geometry, it is easy to guess what this weaker geometry should be: both equations in (10.5.120) should be modified as in (10.5.116):

$$\mathrm{d}_H(\mathrm{e}^{-\phi}\Phi_\pm) = W_\pm \cdot \Phi_\pm \tag{10.5.121}$$

for some $W_\pm$, sections of $T \oplus T^*$. In other words, the corresponding $\mathcal{J}_\pm$ should both be integrable with respect to the twisted Courant bracket (10.5.114). This is precisely the generalized Kähler condition (10.5.119), for $\mathcal{J}_{1,2} = \mathcal{J}_\pm$.

Indeed, we now show that *bihermitian geometry* (9.3.25) *is equivalent to generalized Kähler geometry* [482, prop. 6.17]. Recalling (10.5.65), for $L_{\mathcal{J}_a}$ to be closed under the twisted Courant bracket, we need all four $L_{\pm\pm}$ to be closed too. For example, for L_{++} we can use the expression (10.5.86); its integrability with respect to (10.5.115) is then a mix of (10.5.103) and (10.5.104):

$$[X,Y]_{\mathrm{H}} = \begin{pmatrix} [v^1_{0,1}, w^1_{0,1}] \\ -\mathrm{i}\iota_{[v^1_{0,1}, w^1_{0,1}]} J_1 + \iota_{v^1_{0,1}} \iota_{w^1_{0,1}}(-H + \mathrm{i}\mathrm{d}J_1) \end{pmatrix}. \tag{10.5.122}$$

The first entry implies that I_1 is integrable; the second that $\iota_{v^1_{0,1}} \iota_{w^1_{0,1}}(H - \mathrm{i}\mathrm{d}J_1) = 0$. This means $H_{0,3} = (H - \mathrm{i}\mathrm{d}J_1)_{1,2} = 0$, the latter of which becomes $H = -\mathrm{d}^c_1 J_1$. Integrability of the other $L_{\pm\pm}$ completes (9.3.25).

Generalized Kähler potential

As an application, we now sketch the analogue of Kähler potential for generalized Kähler manifolds [28]. As we mentioned in Section 9.3.3, this is the function of chiral, twisted chiral, and semichiral multiplets that appear in (9.1.56) for the most general case of $\mathcal{N} = (2,2)$ model, where the $I^{1,2}$ do not commute. For simplicity, we focus on the four directions described by a semichiral multiplet, which is the hardest part. We know from (9.3.28) and the remark that follows it that this is a subspace spanned by $\mathrm{im}([I_1, I_2])$.

The upper-right block of a generalized complex $\mathcal{J}$ is a Poisson tensor P, (10.5.102). Using (10.5.73), we compute the Poisson tensors associated to $\mathcal{J}_{1,2}$ to be

$$P_1 = (-I_1 + I_2)^t g^{-1}\,, \qquad P_2 = (-I_1 - I_2)^t g^{-1}\,. \tag{10.5.123}$$

The tensor $\sigma \equiv g^{-1}[I_1, I_2]$ is also Poisson [493]. We are working along $\mathrm{im}([I^1, I^2])$, so σ is also invertible and defines a symplectic structure σ^{-1} (Section 5.4.1). Since $(\Pi_1)_p{}^m(\bar{\Pi}_1)_q{}^n\sigma^{pq} = (\Pi_1)_p{}^m(\bar{\Pi}_1)_q{}^n\sigma^{pq} = 0$, σ is $(2,0) + (0,2)$ for both I_a. Its $(2,0)$ part $\tilde{\sigma}_a \equiv (\Pi_1)_p{}^m(\Pi_1)_q{}^n\sigma^{pq}$ with respect to either complex structure is then a holomorphic Poisson tensor. By the complex analogue of the Darboux theorem (5.4.8), we can define complex coordinates q, p such that $\tilde{\sigma}_1^{-1} = \mathrm{d}q \wedge \mathrm{d}p$, and Q, P such that $\tilde{\sigma}_2^{-1} = \mathrm{d}Q \wedge \mathrm{d}P$. These are redundant, and we can take, say, q and P to be independent. As in classical mechanics, we can then define a canonical transformation between the two sets of coordinates: a function K exists such that

$$p = \partial_q K\,, \qquad Q = \partial_P K\,. \tag{10.5.124}$$

This K is the function in terms of which the world-sheet action can be written as $\int \mathrm{d}^2\theta \mathrm{d}^2\bar{\theta} K$ [28]; it generalizes the Kähler potential, which played that role in (9.1.56), the case with only chiral multiplets. We can also write this in terms of pure spinors [441, (5.23)]:

$$\begin{aligned} \Phi_+ &= \exp[8\,\partial\bar{\partial}K]\,, \\ \Phi_- &= \mathrm{d}w \wedge \exp\left[8\,\mathrm{d}q \wedge \mathrm{d}(\partial_q K) + 8\,\mathrm{d}(\partial_P K) \wedge \mathrm{d}P\right], \end{aligned} \tag{10.5.125}$$

where $\partial \equiv \mathrm{d}q\partial_q + \mathrm{d}P\partial_P + \mathrm{d}w\partial_w$.

Generalized intrinsic torsion*

While this is not really needed for us, it is amusing to try and define an analogue of intrinsic torsion for the $SU(3) \times SU(3)$ structures defined by a compatible pair $\Phi_{1,2}$. The idea is to expand $\mathrm{d}\Phi_\pm$ in the diamond (10.5.80). To fix ideas, let us say $\Phi_+ \in U_{3,0}$, $\Phi_- \in U_{0,3}$. As observed before (10.5.109), $\mathrm{d}\Phi$ only has nonzero projections to the Clifford levels $U^{\mathcal{J}}_{d/2-1} = U^{\mathcal{J}}_0$ and $U^{\mathcal{J}}_{d/2-3} = U^{\mathcal{J}}_0$; integrability sets the latter to zero. So $\mathrm{d}\Phi_+$ only has coefficients in the second and fourth row, while $\mathrm{d}\Phi_-$ only in the second and fourth column. Up to b-transform, the most general pair is of the form $\eta^1_+ \otimes \eta^{2\dagger}_\pm$; the diamond (10.5.80) is (10.3.83) (Section 10.5.3). So $U_{2,1}$ is generated by elements of the form $\Phi_+\gamma_{\bar{\imath}_2}$, and so on. We then introduce coefficients $W^{p,q}$ for the basis expansion of $\mathrm{d}\Phi_\pm$:

$$\begin{aligned} \mathrm{d}_H\Phi_+ &= W^{2,1}_{i_2}\Phi_+\gamma^{i_2} + W^{2,-1}_{\bar{\imath}_1}\gamma^{\bar{\imath}_1}\Phi_+ \\ &\quad + W^{0,3}\Phi_- + W^{0,1}_{\bar{\imath}_1\bar{\imath}_2}\gamma^{\bar{\imath}_1}\Phi_-\gamma^{\bar{\imath}_2} + W^{0,-1}_{i_1\bar{\imath}_2}\gamma^{i_1}\bar{\Phi}_-\gamma^{i_2} + W^{0,-3}\bar{\Phi}_-\,; \\ \mathrm{d}_H\Phi_- &= W^{1,2}_{\bar{\imath}_2}\Phi_-\gamma^{\bar{\imath}_2} + W^{-1,2}_{\bar{\imath}_1}\gamma^{\bar{\imath}_1}\Phi_- \\ &\quad + W^{3,0}\Phi_+ + W^{1,0}_{\bar{\imath}_1 i_2}\gamma^{\bar{\imath}_1}\Phi_+\gamma^{i_2} + W^{-1,0}_{i_1\bar{\imath}_2}\gamma^{i_1}\bar{\Phi}_+\gamma^{\bar{\imath}_2} + W^{-3,0}\bar{\Phi}_+\,. \end{aligned} \tag{10.5.126}$$

This is the natural extension of (5.5.6) for SU(3)-structures. Moreover, comparing these $W^{i,j}$ with the Q's and T's in (10.3.87) gives the extension of (5.5.12) for SU(3)-structures. In (10.3.87), we seem to have some extra information, namely the equalities $W^{0,-3} = -W^{-3,0}$, $W^{0,3} = -\bar{W}^{3,0}$. This is a generalization of the property in (5.5.6), that the two singlets in $\mathrm{d}J$ and $\mathrm{d}\Omega$ are proportional because $\mathrm{d}(J \wedge \Omega) = 0$. We can reobtain it directly from the perspective of (10.5.126):

$$
\begin{aligned}
(\mathrm{d}_H \Phi_\mp, \Phi_\pm)\,\mathrm{vol} &\overset{(3.2.40)}{=} (\mathrm{d}_H \Phi_\mp \wedge \lambda(\Phi_\pm))_6 \\
&= (\mathrm{d}(\Phi_\mp \wedge \lambda(\Phi_\pm)) \pm \Phi_\pm \wedge \mathrm{d}\lambda(\Phi_\pm) \pm \Phi_\mp \wedge H \wedge \lambda(\Phi_\pm))_6 \\
&\overset{(10.5.83),(3.2.58)}{=} (\Phi_\mp \wedge \lambda(\mathrm{d}_H \Phi_\pm))_6 = \mathrm{vol}(\Phi_\mp, \mathrm{d}_H \Phi_\pm)\,.
\end{aligned}
\tag{10.5.127}
$$

10.5.6 Dolbeault form of pure spinor equations

Our final application of generalized complex geometry will be an alternative formulation [489] of the pure spinor equations (10.3.20).

Hodge $*$ and generalized almost complex structures

The Hodge $*$ is the only place where the metric appears directly in (10.3.20); so we would like to reexpress it in terms of generalized complex quantities. First we recall

$$
*\,\lambda \Phi_\pm \overset{(3.2.27)}{=} \mathrm{i}\gamma \Phi_\pm = \mathrm{i}\Phi_\pm\,. \tag{10.5.128}
$$

(We already used this, for example way back in (3.4.22) and (3.4.24) for an SU(3)-structure.) Moreover, the left gamma action $\overrightarrow{\gamma}_m$ anticommutes with the left action by the chiral matrix, while $\overleftarrow{\gamma}_m$ anticommutes with it. Comparing with (10.3.83), this leads to the following eigenvalues on the $U_{j,k}$ of (10.5.80):

$$
\begin{array}{ccccccc}
 & & & \mathrm{i} & & & \\
 & & \mathrm{i} & & -\mathrm{i} & & \\
 & \mathrm{i} & & -\mathrm{i} & & \mathrm{i} & \\
\mathrm{i} & & -\mathrm{i} & & \mathrm{i} & & -\mathrm{i} \\
 & -\mathrm{i} & & \mathrm{i} & & -\mathrm{i} & \\
 & & \mathrm{i} & & -\mathrm{i} & & \\
 & & & -\mathrm{i} & & &
\end{array}\,. \tag{10.5.129}
$$

Recall that in (10.5.80), the $U_{j,k}$ are such that $\mathcal{J}_1 \cdot U_{j,k} = \mathrm{i}jU_{j,k}$, $\mathcal{J}_2 \cdot U_{j,k} = \mathrm{i}kU_{j,k}$. By comparing this with (10.5.129), we obtain the curious formula

$$
*\,\lambda = \mathrm{e}^{-\frac{\pi}{2}(\mathcal{J}_+ + \mathcal{J}_-)\cdot}\,. \tag{10.5.130}
$$

This is simple, but has the drawback of involving exponentials. An alternative is to expand the form in the diamond; for an even form f_+, we have

$$
f_+ = f_{3,0} + f_{1,2} + f_{1,0} + f_{1,-2} + f_{-1,2} + f_{-1,0} + f_{-1,-2} + f_{-3,0}\,. \qquad f_{j,k} \in U_{j,k}\,. \tag{10.5.131}
$$

The action of the Hodge operator can then be written as

$$
*\,\lambda f_+ = -\mathcal{J}_+ \cdot f_+ + 2\mathrm{i}(f_{1,2} + f_{1,-2} - f_{-1,2} - f_{-1,-2}) + 4\mathrm{i}(f_{3,0} - f_{-3,0})\,. \tag{10.5.132}
$$

We also note

$$
-\mathcal{J}_+^2 \cdot f_+ = f_+ + 8(f_{3,0} + f_{-3,0})\,. \tag{10.5.133}
$$

Similar formulas for odd forms f_- are obtained by replacing $f_{j,k} \to f_{k,j}$.

Rewriting the flux pure spinor equation

Once again we assume in the rest of this subsection that $c_- = 0$, and then $c_+ = 2$. (We know from (10.3.25) that this is the case anyway for $\Lambda < 0$.) We will project the pure spinor equations on several entries of the diamond (10.3.83), using the orthogonality property (10.5.81). We begin with (10.3.20b), projected on $\Phi_\pm$:

$$(d_H\Phi_\pm, \Phi_\mp) = \pm\mu e^{-A}(\bar{\Phi}_\mp, \Phi_\mp)\,. \tag{10.5.134}$$

Next we project on $\bar{\Phi}_\pm$ (10.3.20a) and (10.3.20c), or more directly their combination (10.3.80); we recall from (10.5.126) that $d_H\Phi_\pm$ has no component along it. We also use (10.5.127) and (10.5.134):

$$f_{-3,0} = -4\mu e^{-2A-\phi}\bar{\Phi}_+ \ \text{(IIA)}\,, \qquad f_{0,-3} = -4\mu e^{-2A-\phi}\bar{\Phi}_- \ \text{(IIB)}\,. \tag{10.5.135a}$$

Finally, we project (10.3.80) on $\gamma^m\bar{\Phi}_\mp$ and $\bar{\Phi}_\mp\gamma^m$:

$$f_{-1,2} - f_{1,2} = -8e^{-A-\phi}dA\wedge\bar{\Phi}_- \ \text{(IIA)}\,, \qquad f_{2,-1} - f_{2,1} = 8e^{-A-\phi}dA\wedge\bar{\Phi}_+ \ \text{(IIB)}\,. \tag{10.5.135b}$$

Now we can evaluate (10.5.132):

$$*\lambda f \overset{(10.5.132),(10.5.135)}{=} -\mathcal{J}_\pm f \mp 32e^{-A-\phi}dA\wedge \text{Im}\Phi_\mp + 32\mu e^{-2A-\phi}\text{Im}\Phi_\pm \ {}^{\text{IIA}}_{\text{IIB}}\,. \tag{10.5.136}$$

Using this in (10.3.20c):

$$\mp\, d_H(e^{-A-\phi}\text{Im}\Phi_\mp) = \mu e^{-2A-\phi}\text{Im}\Phi_\pm - \frac{1}{8}\mathcal{J}_\pm\cdot f_\pm\,. \tag{10.5.137}$$

We have succeeded in eliminating the $*$. It is perhaps preferable to write the equation as a direct expression for the RR flux; this can be achieved by using (10.5.133) and its odd-form counterpart. It is also useful to recall (10.5.52):

$$\frac{1}{8}f = \mp\mathcal{J}_\pm\cdot d_H(e^{-A-\phi}\text{Im}\Phi_\mp) + 5\mu e^{-2A-\phi}\text{Re}\Phi_\mp\,. \tag{10.5.138}$$

We stress once again that this is for $c_- = 0$, but it is valid for both AdS and Minkowski.

Alternative expression for the Minkowski case

In the Minkowski case ($\mu = 0$), $\mathcal{J}_\pm$ is integrable by (10.3.20b); the term in (10.5.138) with the $\mathcal{J}_\pm$ action can then also be written as

$$\mathcal{J}_\pm\cdot d_H(e^{-A-\phi}\text{Im}\Phi_\mp) \overset{(10.5.117),(10.5.77)}{=} -d_H^{\mathcal{J}_\pm}(e^{-A-\phi}\text{Im}\Phi_\mp)\,. \tag{10.5.139}$$

So in this case, (10.5.138) becomes

$$\frac{1}{8}f = \pm d_H^{\mathcal{J}_\pm}(e^{-A-\phi}\text{Im}\Phi_\mp)\,. \tag{10.5.140}$$

We can use (10.5.139) for AdS_4 case too, but in that case $d_H^{\mathcal{J}_\pm}$ does not square to zero. An alternative differential is $e^{-\frac{\pi}{2}\mathcal{J}_\pm}de^{\frac{\pi}{2}\mathcal{J}_\pm}$; using this with (10.5.130) leads to another equivalent rewriting of (10.5.138) [489]. In practice, however, (10.5.138) has proven more useful.

One advantage of (10.5.138) (or (10.5.140)) is that it behaves well under a b-transform:

$$f \to e^{b\wedge} f\,, \qquad H \to H + db\,, \qquad \Phi \to e^{b\wedge}\Phi\,, \qquad \mathcal{J}\cdot \to e^{b\wedge}\mathcal{J}\cdot e^{-b\wedge}\,, \tag{10.5.141}$$

the last coming from (10.5.25). This works less well in the original (10.3.20c) and (10.3.24c), because $e^{b\wedge}$ does not commute with $*\lambda$.

Exercise 10.5.1 Write down the $\mathcal{J}_{\Phi_\pm}$ associated to the pure spinors in (10.3.31).

Exercise 10.5.2 Check (10.5.37) for the two cases in (10.5.34). (Sections 5.1.2 and 5.2.1 might be helpful.)

Exercise 10.5.3 Let α_4 be a four-form, and suppose we want to write it as $\alpha_4 = J \wedge J$. Use the methods of this section to determine a condition under which this is possible. When it is, use (10.5.53) to find an explicit formula for J in terms of α_4.

Exercise 10.5.4 On T^6, show that the pure forms $(dx^1 + idx^4) \wedge (dx^2 + idx^5) \wedge (dx^3 + idx^6)$, $\exp[-i(dx^1 \wedge dx^4 + dx^2 \wedge dx^5 + dx^3 \wedge dx^6)]$ are exchanged under T-duality along directions 4, 5, and 6.

Exercise 10.5.5 Check that (10.5.60) is correct when it acts on $\mathrm{Re}\Phi$, $\mathrm{Im}\Phi$.

Exercise 10.5.6 Show (10.5.61).

Exercise 10.5.7 Show that the analogue of (10.5.82) in $d = 4$ is

$$(\Phi_1, \Phi_2) = (\Phi_1, \bar{\Phi}_2) = 0\,. \tag{10.5.142}$$

Exercise 10.5.8 Use the previous exercise to explain (9.4.57) geometrically, along the lines of (7.2.8), (7.2.10), and (7.2.15). (Hint: consider $H^{\text{even}}(\text{K3})$.)

Exercise 10.5.9 Show that the Courant bracket gets twisted under b-transform (3.2.14):

$$[e^{b\wedge}X, e^{b\wedge}Y]_C = e^{b\wedge}[X,Y]_{H=db}\,. \tag{10.5.143}$$

(Hint: use (10.5.96b).) This implies that if $\mathcal{J}$ is integrable, then its b-transform is twisted integrable; notice how this follows more quickly from (10.5.116).

Exercise 10.5.10 Apply (10.5.140) to the three cases in Section 10.4.1. In particular, show (10.4.27).

Exercise 10.5.11 For $\Lambda = \mu = c_- = 0$, apply (10.3.24a), (10.3.24b), and (10.5.138) to the $SU(3) \times SU(3)$-structure case (10.3.37). (The results are in [494, sec. 5.3]; to get shorter expressions, it might be helpful to define two-forms J_ψ, ω_ψ as in (11.3.24).)

10.6 Some extensions

10.6.1 Other dimensions

It is relatively easy to derive equations similar to (10.3.20) for $\mathrm{Mink}_d \times M_{10-d}$ or $\mathrm{AdS}_d \times M_{10-d}$ compactifications for any d. For example, the first method in Section 10.3.7, where we specialized the ten-dimensional bispinor equation (10.2.16), often works very similar to the $d = 4$ case. Sometimes it is quicker to exploit relations between different d, as we will now see. We will not cover all possible values of d, but illustrate the main ideas by looking at $d = 6$ and $d = 3$. (A systematic description of all Mink$_d$ is given in [445].)

Mink$_6$ **compactifications**

The factorization Ansatz for the ten-dimensional supersymmetry parameters works a little differently for Mink$_6$ than for Mink$_4$. This is because of (2.3.10): conjugation preserves chirality in both $d = 6$ Lorentzian and $d = 4$ Euclidean dimensions. So the conjugate of $\zeta_{(6)+} \otimes \eta_{(4)+}$, where $\zeta_{(6)+}$ is a Mink$_6$ and $\eta_{(4)+}$ an M_4 spinor, is of the form $\zeta^c_{(6)+} \otimes \eta^c_{(4)+}$, both factors being still of positive chirality. So for both ϵ^1 and ϵ^2, in principle we could choose to use only positive-chirality spinors, or only negative-chirality ones, unlike for our familiar Mink$_4 \times M_6$ factorization (8.2.12). For vacuum solutions with Ramond–Ramond (RR) field-strengths, the Mink$_6$ spinors should be equal: the RR contribution to the supersymmetry equations mixes ϵ^1 and ϵ^2 (Section 10.3.2). Dropping dimension labels, for minimal supersymmetry this leads to

$$\begin{aligned} \epsilon_1 &= \zeta_+ \otimes \eta^1_+ + \zeta^c_+ \otimes \eta^{1\,c}_+ \\ \epsilon_2 &= \zeta_+ \otimes \eta^2_\mp + \zeta^c_+ \otimes \eta^{2\,c}_\mp \end{aligned} \quad {}^{\text{IIA}}_{\text{IIB}} \quad (\mathcal{N} = 1 \text{ Mink}_6 \text{ RR vacua}), \tag{10.6.1}$$

to be contrasted with (10.3.16).

To derive the supersymmetry conditions, we can now use the first method of Section 10.3.7. Alternatively, notice that a Mink$_4 \times M_6$ whose internal metric is

$$ds^2_{M_6} = e^{2A}((dx^4)^2 + (dx^5)^2) + ds^2_{M_4}, \tag{10.6.2}$$

is also a warped Mink$_6$ vacuum. We should also require that the field-strengths H, f are now all along M_4 only, with no 4, 5 components. As for the supercharges, starting from (10.3.16) we can take $\eta^1_+ = \eta_{(2)+} \otimes \eta^1_{(4)+}$, $\eta^2_\mp = \eta_{(2)+} \otimes \eta^2_{(4)\mp}$ and obtain a particular case of (10.6.1). It follows that

$$\Phi_\pm = \phi_{(2)-} \wedge \phi^2_\mp, \qquad \Phi_\mp = \phi_{(2)+} \wedge \phi^1_\mp, \quad {}^{\text{IIA}}_{\text{IIB}} \tag{10.6.3}$$

where $\phi_{(2)}$, $\phi^{1,2}$ are pure forms in $\mathbb{R}^2$, M_4 respectively, and the $\pm$ as always denotes form degree. On $\mathbb{R}^2$, the $\eta_{(2)\pm}$ are constant; so we can go back all the way to the flat-space expressions (3.3.33), (3.3.35), $2\phi_{(2)+} = 1 - ie^4 \wedge e^5$, and $2\phi_{(2)-} = e^4 + ie^5$, with the only difference that now the vielbein one-forms $e^{4,5}$ on $\mathbb{R}^2$ are of the form $e^A dx^{4,5}$ to agree with (10.6.2). Using all this in (10.3.20), one obtains, for $c_- = 0$, $c_+ = 2$, [445]

$$d_H\left(e^{A-\phi}\text{Re}\phi^1_\mp\right) = 0, \tag{10.6.4a}$$

$$d_H\left(e^{3A-\phi}\text{Im}\phi^1_\mp\right) = 0, \quad {}^{\text{IIA}}_{\text{IIB}} \tag{10.6.4b}$$

$$d_H\left(e^{3A-\phi}\phi^2_\mp\right) = 0, \tag{10.6.4c}$$

$$d_H\left(e^{5A-\phi}\text{Re}\phi^1_\mp\right) = \pm\frac{1}{4}e^{6A} * \lambda f. \tag{10.6.4d}$$

The normalization is $(\bar\phi_+, \phi_+) = (\bar\phi_-, \phi_-) = e^{2A}/4$. The two last equations, (10.6.4b) and (10.6.4c), can be written in terms of a triplet $(\text{Re}\phi^2_\mp, \text{Im}\phi^2_\mp, \text{Im}\phi^1_\mp)$: this is related to the fact that $d = 6$ $\mathcal{N} = 1$ supersymmetry reduces to $d = 4$ $\mathcal{N} = 2$. The flux equation (10.6.4d) can be combined with (10.6.4a) to become

$$e^{\phi} f = \pm 16 * \lambda(\mathrm{d}A \wedge \mathrm{Re}\phi^1_{\mp}) . \qquad (10.6.5)$$

For example, the intersecting-brane class (10.4.15b) is a solution of (10.6.4).

Mink$_3$, AdS$_3$ compactifications

In this book, we have mostly focused on $d = 4$ compactification; we motivated our foray in $d = 6$ by its relation to $d = 4$. For $d = 3$, we don't have this justification; we discuss it briefly to illustrates how (10.2.16) is not always enough for supersymmetry.

The supersymmetry parameters are decomposed in this case as $\epsilon^1 = v_+ \otimes \zeta \otimes \eta^1$, $\epsilon^2 = v_{\mp} \otimes \zeta \otimes \eta^2$, in the basis (2.2.26). Each of the $d = 7$ spinors is Majorana and defines a separate G_2-structure. These two can define either a G_2- or an SU(3)-structure, similar to Section 10.3.4. But working with $T \oplus T^*$, we describe both cases as a $G_2 \times G_2$-structure [495], defined by the polyforms

$$\eta_1 \otimes \eta_2^\dagger \equiv (\psi_+)_{/} = \mathrm{i}(\psi_-)_{/} , \qquad \psi_- = - * \lambda \psi_+ , \qquad (10.6.6)$$

recalling (3.2.29). Supersymmetry is equivalent to the following system [495–498]:[14]

$$\mathrm{d}_H(\mathrm{e}^{A-\phi}\psi_{\mp}) = 0 , \quad \mathrm{d}_H(\mathrm{e}^{2A-\phi}\psi_{\pm}) \mp 2\mu \mathrm{e}^{A-\phi}\psi_{\mp} = \frac{1}{8}\mathrm{e}^{3A} * \lambda F , \quad (\psi_{\mp}, F) = \pm\frac{\mu}{2}\mathrm{e}^{-\phi} , \qquad (10.6.7)$$

with the normalization $8(\psi_+, \psi_-) = \mathrm{e}^{2A}$. (We set to zero a constant c_- in the right-hand side of the first equation, similar to the c_- in (10.3.20a).) The first two equations in (10.6.7) come from (10.2.16) and are similar in spirit to (10.3.20). The last comes from the pairing equation (10.2.37); without it, the system (10.6.7) would not be equivalent to supersymmetry.

10.6.2 Extended supersymmetry

We now turn to four-dimensional vacua with $\mathcal{N} > 1$ supersymmetry.

For the spinor Ansatz, we again start from the general expression (8.2.12) and introduce relations among the ζ^a_I. For $\mathcal{N} = 2$, we already saw two possible Ansätze in the case with vanishing RR fields: one is (9.3.18), another was mentioned after (9.3.19). But neither of these gives $\mathcal{N} = 2$ when $F \neq 0$, again because the RR fields force us to relate the ζ^a_+ to one another (Section 10.3.2), leading us back to (10.3.16) and $\mathcal{N} = 1$. The way out is to include $2\mathcal{N}$ internal spinors:

$$\begin{aligned} \epsilon^1 &= \sum_I (\zeta_{+I} \otimes \eta^1_{+I} + \zeta_{-I} \otimes \eta^1_{-I}) , \\ \epsilon^2 &= \sum_I (\zeta_{+I} \otimes \eta^2_{\mp I} + \zeta_{-I} \otimes \eta^2_{\pm I}) , \quad {}^{\mathrm{IIA}}_{\mathrm{IIB}} \end{aligned} \qquad (\mathcal{N} > 1 \text{ RR vacua}) , \qquad (10.6.8)$$

which is just (8.2.12) with $\zeta^1_I = \zeta^2_I$.

Plugging (10.6.8) in the supersymmetry equations, we find (10.3.17) with an extra spectator index I on the η^a. In other words, the system for $I = 1$ is decoupled from the one with $I = 2$, and so on, except for the physical fields being the same. So we can just apply the good old pure spinor equations (10.3.20) separately to each $\Phi^{II}_\pm \equiv \eta^1_{+I} \otimes \eta^{2\dagger}_{\pm I}$ (no sum over I). In other words, we should find $\mathcal{N}$ different solutions to the pure spinor equations (10.3.20), all with the same physical fields (the metric,

[14] For similar results in $d = 2$, see [499–501].

$B, \phi, f)$. One practical way to find such $\mathcal{N} > 1$ solutions is to make an Ansatz for the pure forms that is acted upon by an internal isometry. For example, for $\mathcal{N} = 2$ we may look for a family $\Phi_\pm(\alpha)$, such that the parameter α is rotated by an internal U(1). This is called an *R-symmetry* and will be discussed further in the next chapter; for an example of this strategy, see [87, app. A].

An alternative method, which is better for classification purposes, is to obtain pure spinor equations for all the $\Phi_\pm^{IJ} \equiv \eta^1_{+I} \otimes \eta^{2\dagger}_{\pm J}$. This can be done from the $d = 10$ equations as in the first part of Section 10.3.7. In particular, for AdS_4, (10.2.16) implies the following equations [478, 502]:

$$d_H \left(e^{2A-\phi}\Phi_\pm^{(IJ)}\right) \mp 2e^{A-\phi}\mathrm{Re}\Phi_\mp^{(IJ)} = 0\,, \tag{10.6.9a}$$

$$d_H \left(e^{3A-\phi}\mathrm{Re}\Phi_\mp^{[IJ]}\right) = 0\,, \tag{10.6.9b}$$

$$d_H \left(e^{A-\phi}\mathrm{Im}\Phi_\mp^{[IJ]}\right) \mp e^{-\phi}\mathrm{Im}\Phi_\pm^{[IJ]} = \pm\frac{1}{8}e^{A} f^{IJ} F\,; \tag{10.6.9c}$$

and

$$d_H \left(e^{3A-\phi}\mathrm{Im}\Phi_\mp^{(IJ)}\right) \mp 3e^{2A-\phi}\mathrm{Im}\Phi_\pm^{(IJ)} = \mp\frac{1}{16}c^{IJ}e^{4A} * \lambda(F)\,, \tag{10.6.10a}$$

$$d_H \left(e^{-\phi}\Phi_\pm^{[IJ]}\right) = -\frac{1}{16}(\bar{\bar{\xi}} \wedge + \iota_\xi)^{IJ} F\,, \tag{10.6.10b}$$

$$d_H \left(e^{4A-\phi}\Phi_\pm^{[IJ]}\right) \mp 4e^{3A-\phi}\mathrm{Re}\Phi_\mp^{[IJ]} = \frac{\mathrm{i}}{16}(\bar{\bar{\xi}} \wedge + \iota_\xi)^{IJ}e^{4A} * \lambda(F)\,, \tag{10.6.10c}$$

where $c^{IJ} \equiv 2e^{-A}\eta^{1\dagger}_{(I+}\eta^1_{J)+} = 2e^{-A}\eta^{2\dagger}_{+(I}\eta^2_{+J)}$ are constant, and $f^{IJ} \equiv -\mathrm{i}\eta^{1\dagger}_{+[I}\eta^1_{+J]} = \mp\mathrm{i}\eta^{2\dagger}_{+[I}\eta^2_{+J]}$; we also introduced

$$\xi^{IJ}_m \equiv \eta^{1\dagger}_{+[I}\gamma_m\eta^1_{-J]} + \eta^{2\dagger}_{\mp[I}\gamma_m\eta^2_{\pm J]}\,, \qquad \bar{\bar{\xi}}^{IJ}_m \equiv \eta^{1\dagger}_{+[I}\gamma_m\eta^1_{-J]} - \eta^{2\dagger}_{\mp[I}\gamma_m\eta^2_{\pm J]}\,. \tag{10.6.11}$$

Equation (10.2.18) implies that the ξ_{IJ} are internal Killing vectors.

This system looks intimidating, but it is highly redundant. For example, we may only keep the diagonal $I = J$ equations, which reduce to the pure spinor equations (10.3.20) for $c_- = 0$. This is the strategy we outlined earlier: imposing each $\mathcal{N} = 1$ subalgebra separately. Alternatively, one may show that all the (10.6.10) follow from the (10.6.9); for example, (10.6.10b) is obtained from (10.6.9c) upon acting with $(\bar{\bar{\xi}} \wedge + \iota_\xi)^{IJ}$. But keeping the full redundant system proves useful for classification purposes: some information that in a truncated system would appear at high form degree is more immediately garnered by the full system at low form degrees. We will see some examples of results obtained with these equations in Section 11.3.

10.6.3 M-theory

While M-theory reduces to massless IIA, the presence of the eleventh direction introduces several differences.

Equations of motion and supersymmetry

We saw the action for eleven-dimensional supergravity in (1.4.4); the equations of motion read, away from sources,

$$R_{MN} = \frac{1}{2}\iota_M G_4 \cdot \iota_N G_4 - \frac{1}{6}g_{MN}|G_4|^2\,, \qquad d * G_4 + \frac{1}{2}G_4 \wedge G_4 = 0\,. \tag{10.6.12}$$

(We already saw the second in Exercise 4.1.13 and in (9.5.24).) As we know, there are 32 supercharges, and hence a single supersymmetry parameter ϵ, a Majorana spinor. Setting to zero the supersymmetry transformations of the gravitino gives

$$\delta\psi_M = \nabla_M \epsilon + \frac{1}{12}(\Gamma_M G_4 - 3\iota_M G_4)\epsilon = 0\,. \qquad (10.6.13)$$

As hopefully familiar by now, we have omitted the slash on the forms G_4 and $\iota_M G_4$. An analysis similar to that in Section 10.1.4 shows that most of the Einstein equations follow from preserved supersymmetry, the Bianchi identity and the flux equations of motion [503].

Since there is only one supersymmetry parameter, it defines an ordinary G-structure, and working on $T \oplus T^*$ at this stage does not seem natural. There are two orbits for ϵ, which work a bit like $d = 7$ in Table 2.5, in that

$$K_M \equiv \bar{\epsilon}\Gamma_M \epsilon \qquad (10.6.14)$$

can be either time-like or space-like; but now

$$\text{Stab}(\epsilon) = \begin{cases} \text{Spin}(7) \ltimes \mathbb{R}^9 & \text{if } K^2 = 0\,, \\ \text{SU}(10) & \text{if } K^2 < 0\,. \end{cases} \qquad (10.6.15)$$

Spin(7) appears for the same reasons as (3.4.65) in $d = 10$; the two are of course related by dimensional reduction.

The time-like case

One can now analyze the two cases (10.6.15) separately. We present here the time-like case [503], which was historically one of the first examples of G-structure analysis of supersymmetry; the light-like case is considerably more involved [504].

The time-like case can be understood by formally reducing along the time direction, as we saw in $d = 5$ in (2.4.57). At the level of the metric, this means

$$ds^2 = -V^2(dt + \omega)^2 + V^{-1} ds^2_{M_{10}}\,, \qquad (10.6.16)$$

reminiscent of the reduction to IIA (1.4.5a) along a space-like direction.

The SU(5)-structure defined by ϵ can be described by K, by a two-form J, and by a five-form Ω, much like the case (5.1.48) of an SU(3)-structure in $d = 7$. (That discussion was in Euclidean signature, but little changes.) Supersymmetry demands that the geometry of the space slice M_{10} satisfies

$$\text{Re}W_5 = \frac{3}{4}\frac{dV}{V}\,, \qquad (10.6.17)$$

following the same notation for SU(3)-structures in Section 5.5.1, and that the flux

$$\begin{aligned} G_4 = &- d[(dt + \omega) \wedge J] + *(V^{-3/2}(\text{Re}W_5 - W_4) \wedge \text{Re}\Omega) \\ &+ \Lambda + \frac{1}{2} * d(V^{-3/2}\text{Re}\Omega) - \frac{1}{2} * (J \wedge d(V^{-3/2}\text{Re}\Omega)) \wedge J\,. \end{aligned} \qquad (10.6.18)$$

Λ is a (2, 2)-form such that $J \cdot \Lambda = 2(d\omega)_{1,1}$.

Vacuum solutions

We now turn to supersymmetric vacuum solutions. We will describe various results on AdS vacua in Chapter 11. The Mink_4 case suffers from the usual no-go argument, and in this case we don't have the analogue of O-planes to circumvent it. The compactification decomposition is

$$\epsilon = \zeta_+ \otimes \eta + \zeta_- \otimes \eta^c \,, \tag{10.6.19}$$

where as usual $\zeta_- = \zeta_+^c$. A Majorana η would define a G_2-structure, but there exist no solutions of this type. So η defines an SU(3)-structure, described as in (5.1.48) by forms w, J, Ω. If the real and imaginary parts of η are taken to be orthogonal and with equal norms and for $\Lambda = 0$, the resulting equations are relatively nice [505, 506]:

$$\begin{aligned} \mathrm{d}(\mathrm{e}^{2A} v) &= 0 \,, \qquad & \mathrm{d}(\mathrm{e}^{4A} J) &= -\mathrm{e}^{4A} * G_4 \,, \\ \mathrm{d}(\mathrm{e}^{3A}\Omega) &= 0 \,, \qquad & \mathrm{d}(\mathrm{e}^{2A} J^2) &= -2\mathrm{e}^{2A} v \wedge G \,. \end{aligned} \tag{10.6.20}$$

In general, they become a lot more involved [507]. A more satisfactory description is obtained within exceptional geometry, to which we now turn.

10.6.4 Exceptional geometry

We commented after (10.5.28) and (10.5.72) that the reason for the success of generalized complex geometry in supergravity is that the structure group on $T \oplus T^*$ is the continuous version $\mathrm{O}(d,d)$ of the T-duality group $\mathrm{O}(d,d,\mathbb{Z})$. This often combines with S-duality to generate a larger symmetry group called *U-duality*. For example, E_7 plays a role in $d = 4$ effective supergravity from type II [508] or M-theory [509]; it is discretized in string theory, just like $\mathrm{SL}(2,\mathbb{R})$ (1.2.38) in IIB. More generally, compactifying on T^d leads to E_{d+1}. M-theory on T^d has a similar U-duality E_d.[15]

This suggests there might be an E_d-covariant description of type II supergravity vacua, alternative to that of Section 10.3; and a description of M-theory vacua, hopefully more universally valid than (10.6.20). In this subsection, we will give the outline of this formalism; we focus on four-dimensional vacua, where the relevant group is E_7.

The central bundle in generalized complex geometry is $T \oplus T^*$: it has $\mathrm{O}(d,d)$ structure group, and is roughly associated in T-duality to momentum and winding states. The analogue for exceptional geometry is as follows [510, 511]:

$$E \equiv T \oplus T^* \oplus \Lambda^{\pm} T^* \oplus \Lambda^5 T^* \oplus (T^* \otimes \Lambda^6 T^*) \quad (\tfrac{\text{IIA}}{\text{IIB}} \text{ on } \mathrm{MS}_4 \times M_6) \tag{10.6.21a}$$

$$E \equiv T \oplus \Lambda^2 T^* \oplus \Lambda^5 T^* \oplus (T^* \otimes \Lambda^7 T^*) \quad (\text{M-theory on } \mathrm{MS}_4 \times M_7)\,. \tag{10.6.21b}$$

This has rank 56 and E_7 structure group, acting in its fundamental representation. (The same bundles also give representations of E_d for M_d in type II and M_{d+1} in M-theory respectively.) In (10.6.21a), the summands are associated to quantum numbers on which U-duality acts: momentum and winding modes, NS5 states, and modes of KK5-monopoles (Section 7.2.2). The interpretation of (10.6.21b) is similar, in terms of momentum, M2, M5, and KK6-monopole state.

[15] The complex Lie group $\mathfrak{e}_d$ is the complexification of many real Lie groups; in this subsection, we are considering the maximally noncompact one, also called *split*, often denoted by $E_{d(d)}$. For $d < 6$, $E_2 \equiv \mathrm{SL}(2,\mathbb{R}) \times \mathbb{R}$, $E_3 \equiv \mathrm{SL}(3,\mathbb{R}) \times \mathrm{SL}(3,\mathbb{R})$, $E_4 \equiv \mathrm{SL}(5,\mathbb{R})$, $E_5 \equiv \mathrm{Spin}(5,5)$.

On $T \oplus T^*$, the generalized metric $\mathcal{G}$ in (10.5.72) defines an $\mathrm{O}(d) \times \mathrm{O}(d)$-structure, and the pure spinor pairs further reduce the structure group to $\mathrm{SU}(d/2) \times \mathrm{SU}(d/2)$; recall that these are called "generalized structures." On the bundle (10.6.21), there is a metric that reduces to the maximal compact subgroup SU(8), which we may similarly call an *exceptional structure* [511, 512]. This acts on internal spinors in the fundamental representation, so the internal spinor η in (10.6.19) reduces further to SU(7). We can sum this up as

$$\begin{array}{ccccc} \mathrm{SO}(6,6) & \supset & \mathrm{SO}(6) \times \mathrm{SO}(6) & \supset & \mathrm{SU}(3) \times \mathrm{SU}(3) \\ \cap & & \cap & & \cap \\ E_7 & \supset & \mathrm{SU}(8) & \supset & \mathrm{SU}(7)\,. \end{array} \tag{10.6.22}$$

The conditions for supersymmetric vacua are in the E_7 representations [513–515]

$$\mathbf{56} \oplus \mathbf{512}\,. \tag{10.6.23}$$

In type II, they involve a differential that now also involves the RR flux (and not just NSNS as in d_H), acting on bispinors built out of the $\Phi_\pm$. For a finer analysis, we can analyze them as SU(7) representations, just like, for example, we used SU(3)-structures in ordinary geometry in Section 9.3.2. The only nonzero equations are those in the SU(7) representations [516]:

$$\mathbf{1} \oplus \mathbf{7} \oplus \mathbf{21} \oplus \mathbf{35}\,. \tag{10.6.24}$$

The equations can be interpreted as intrinsic torsion components. We first need to define an appropriate bracket [,] on (10.6.21). This can be found by imposing properties similar to (10.5.143) under b-transform for the Courant bracket [511]; as in Section 10.5.4, it is an antisymmetrization of a Dorfman bracket $X \circ Y$ with a Jacobi property. It can also be recovered using the language of derived brackets (Section 10.5.4) [517]. (A deformation is needed in massive IIA [518].) Related to this, it can be written as

$$(X \circ Y)^\Lambda = X^\Sigma \partial_\Sigma Y^\Lambda - (\partial \times_{\mathrm{ad}} X)^\Lambda{}_\Sigma Y^\Sigma\,, \tag{10.6.25}$$

where now Λ and Σ are indices on E, and $\partial_\Lambda = (\partial_m, 0, \ldots, 0)$; the symbol $\times_{\mathrm{ad}}$ denotes projection on the adjoint representation **133**. Equation (10.6.25) is an analogue on E of a Lie derivative, recalling (4.1.60). Given now a connection D on E, we define $\circ_D$ as obtained from (10.6.25) by replacing $\partial \to D$, and the torsion tensor as $T_E(X,Y) \equiv X \circ Y - X \circ_D Y$; this is similar to the alternative definition of torsion given after (4.1.61). Ordinary torsion is a tensor T^p_{mn}; the exceptional torsion T_E is valued in the E_7 representations (10.6.23).

We can further define the *intrinsic* torsion for the SU(7)-structure defined on E by the spinors. Recall from Section 5.5.1 that this is the part of torsion one cannot eliminate by varying among the connections that preserves the G-structure. For example, in (5.5.14) it is spanned by the p_m and q_{mn} along the complement $\mathrm{su}(3)^\perp$ of the adjoint, and not by the n_m^{ab} along the adjoint. This measures the distance from special holonomy. With the same logic, we can ask what part of T_E is intrinsic. In ordinary geometry, we could just use (5.5.1), but we just saw that T_E only takes values in the representations (10.6.23), so we should repeat the analysis. The result is that the SU(7) intrinsic torsion is exactly in the representation (10.6.24). We conclude that supersymmetry is equivalent to E having special holonomy [516]:

$$\mathcal{N} = 1 \text{ Mink}_4 \text{ vacua} \quad \Leftrightarrow \quad \text{Hol}(E) \subset \text{SU}(7)\,, \tag{10.6.26}$$

both for type II and M-theory. (The detailed proof also relies on [519–521].) Much as for ordinary G-structures in (5.7.5), this exceptional special holonomy implies Ricci flatness on E, which is nothing but the equations of motion; this provides a more elegant derivation than that in Section 10.1.4 [516, app. B].

This remarkable result was further extended in various directions. For AdS, the only change is that the singlet in (10.6.24) should now be allowed to be a nonzero constant [522]. For extended supersymmetry, the holonomy group is $\text{SU}(\mathcal{N}-8)$ [523, 524]; in the $\mathcal{N} = 2$ case, SU(6) can be realized as the intersection of a $\text{Spin}^*(12)$ and an $E_{6(2)}$, which can be interpreted in terms of four-dimensional supergravity as being associated to hypermultiplets and vector multiplets respectively [523]. Similar results also hold for Mink_d and AdS_d with $d \geq 4$ [523–525]. The methods of exceptional geometry also have interesting applications to reductions and more specifically to consistent truncations (Section 11.1.2).

Exercise 10.6.1 Recover the intersecting-brane class of Section 10.4.2 from the Mink_6 system (10.6.4). In particular, use (10.6.5) to compute the RR flux in (10.4.19).

Exercise 10.6.2 Work out how the two cases in (10.6.15) reproduce the various cases in Exercise 3.4.9 upon dimensional reduction.

11 AdS vacua

As we argued in the Introduction, most vacua have a negative cosmological constant, $\Lambda < 0$, and a macroscopic spacetime $\mathrm{MS}_4 = \mathrm{AdS}_4$. While AdS solutions are of course not realistic, they are useful as a first step toward more realistic ones, and for holographic applications. We will begin in Section 11.1 with a general discussion. We will then present of catalogue of known AdS vacua, with special emphasis on the supersymmetric case. We will include higher dimensions as well, for holography but also because they generate many solutions in $d = 4$, as we will see.

11.1 Overview

11.1.1 Near-horizon limits

The first type of AdS solution one encounters in string theory is the result of a near-horizon limit.

Extremal charged black holes

Let us recall how this concept arises in general relativity. The *extremal* charged (Reissner–Nordström) black hole in $d = 4$ has metric

$$\begin{aligned} \mathrm{d}s^2_{\mathrm{ERN}} &= -\left(1 - \frac{m}{\rho}\right)^2 \mathrm{d}t^2 + \left(1 - \frac{m}{\rho}\right)^{-2} \mathrm{d}\rho^2 + \rho^2 \mathrm{d}s^2_{S^2} \\ &\overset{r=\rho-m}{=} -\left(1 + \frac{m}{r}\right)^{-2} \mathrm{d}t^2 + \left(1 + \frac{m}{r}\right)^2 \mathrm{d}s^2_{\mathbb{R}^3} \end{aligned} , \tag{11.1.1}$$

where $m \equiv G_{\mathrm{N}} M$, and r is the radial coordinate in $\mathbb{R}^3$; the coordinate system in the second line only covers the region $\rho \geq m$. The horizon is at $\rho = m$, $r = 0$; to focus on the region around it, we can take $r \ll m$; the m/r dominates over the 1 in the parentheses. After rescaling $t \to m^2 t$, we obtain $\mathrm{AdS}_2 \times S^2$:

$$\mathrm{d}s^2 = m^2 \left(-r^2 \mathrm{d}t^2 + \frac{\mathrm{d}r^2}{r^2}\right) + m^2 \mathrm{d}s^2_{S^2} \overset{(4.5.37)}{=} m^2 \left(\mathrm{d}s^2_{\mathrm{AdS}_2} + \mathrm{d}s^2_{S^2}\right) . \tag{11.1.2}$$

Branes in flat space

The second line of (11.1.1) is quite similar to a brane solution: $1 + m/r$ is a harmonic function in $\mathbb{R}^3$, although it appears with different powers than in (1.3.59). This suggests that something like (11.1.2) might work for some brane solutions. For the

D3-brane solution from (1.3.59), the near-horizon region is $r \ll L \equiv r_0 N^{1/4} = (4\pi g_s N)^{1/4}$; here we can neglect the 1 in (1.3.61):[1]

$$\begin{aligned} ds^2_{D3} &= \left(1+\frac{L^4}{r^4}\right)^{-1/2} ds^2_{\mathrm{Mink}_4} + \left(1+\frac{L^4}{r^4}\right)^{1/2} (dr^2 + r^2 ds^2_{S^5}) \\ &\to L^2 \left(ds^2_{\mathrm{AdS}_5} + ds^2_{S^5}\right) , \end{aligned} \tag{11.1.3}$$

after rescaling the coordinates of Mink_4 by $x^\mu \to r_0^2 x^\mu$. (This extends our observation in (8.2.20).) So the near-horizon limit of a D3 stack is $\mathrm{AdS}_5 \times S^5$. Recall from (1.3.59) that the dilaton ϕ is constant. The remaining field is F_5, which by (4.1.154) is the volume form of S^5; since it is self-dual, we also have to include its dual along AdS_5, and we obtain

$$F_5 = 16\pi N \left(\mathrm{vol}_{\mathrm{AdS}_5} + \mathrm{vol}_{S^5}\right) . \tag{11.1.4}$$

Before the near-horizon limit, the F_5 was sourced by the D3-branes, satisfying the Bianchi identity $dF_5 = \delta_{D3}$. After the limit, $dF_5 = 0$: *the D3-branes have dissolved* under the limit. The locus where the D3 were sitting and the transverse S^5 was shrinking is now no longer present in the geometry; the integer N, which was measuring the number of D3-branes, is now a flux quantum. (Recall also the discussion following (1.3.47).)

If we try to use the same procedure on Dp-branes with $p \neq 3$, it does not quite work: we get a solution of the type $f(r)(ds^2_{\mathrm{AdS}_{p+2}} + ds^2_{8-p})$. This is not an AdS compactification according to our definition in Chapter 8, not even of the warped product type, because $f(r)$ depends on one of the coordinates of the external spacetime AdS_{p+2}, which breaks its isometry group to that of Mink_{p+1}, $\mathrm{ISO}(1, p+1)$. A near-horizon limit of the NS5-brane also fails (Exercise 9.3.2). The near-horizon limit does produce AdS solutions in M-theory:

$$\mathrm{M2} \to \mathrm{AdS}_4 \times S^7 , \qquad \mathrm{M5} \to \mathrm{AdS}_7 \times S^4 . \tag{11.1.5}$$

The field-strength G_4 is proportional to $\mathrm{vol}_{\mathrm{AdS}_4}$ and vol_{S^4}, respectively.

A solution where the field-strength is proportional to the volume forms of either AdS, or the internal space (as in (11.1.4) and (11.1.5)) is said to be of *Freund–Rubin* type [526]. These are the simplest AdS solutions; in the rest of this chapter, we will see that they are far from being typical.

More general brane systems

Indeed, a near-horizon limit often produces AdS solutions that are not of Freund–Rubin type. One example we can exhibit already now is the solution (10.4.18) and (10.4.22), describing an NS5 on top of a D6-brane: setting $c = 1$, in the limit $x^2 + 4r_0 y \ll 1$, with the coordinate change $x = 4r^2 \cos\theta$, $y = 4r^4 \sin^2\theta / r_0$, it produces a warped product

$$ds^2 = \frac{2}{r_0^2} \sin\theta\, ds^2_{\mathrm{AdS}_7} + \frac{\sin\theta}{2r_0} \left(d\theta^2 + \frac{1}{4}\sin^2\theta\, ds^2_{S^2}\right) . \tag{11.1.6}$$

[1] In this chapter, we will often work in string or Planck units, setting l_s or $l_{\mathrm{P}11}$ to 1.

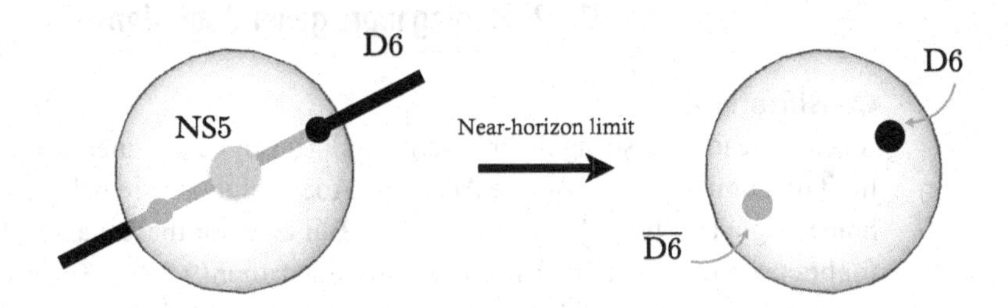

Figure 11.1 Near-horizon limit on a brane system.

This particular example is actually dual to the M5 solution (Exercise 10.4.6), but it does illustrate some general points:

- An appropriate near-horizon limit on an intersection of branes can yield an AdS solution even if the individual ingredients do not.
- Equation (11.1.6) has singularities at $\theta = 0$ and $\pi/2$, which can be recognized as D6 backreaction (Exercise 11.1.1). The near-horizon limit focuses on the region near the NS5-branes. A topological S^3 surrounding the NS5 does not intersect them, but it does intersect the D6; so when this S^3 becomes the internal space M_3 in (11.1.6), the NS5-branes are no longer seen, but the D6-branes have remained. See Figure 11.1.

Branes probing geometries

A further generalization is obtained by considering branes in spaces that are curved even prior to their back-reaction.

For example, the limit (11.1.3) can be greatly generalized by placing D3-branes not in flat space, but at the tip of a Calabi–Yau cone $C(\text{SE}_5)$ [527]; SE_5 is a Sasaki–Einstein manifold (Section 7.4). The limit (11.1.3) still works after replacing $S^5 \to \text{SE}_5$ in (11.1.3), and produces a large class of new solutions $\text{AdS}_5 \times \text{SE}_5$. We expect them to be supersymmetric, since $\text{Mink}_4 \times C(M_5)$ preserves eight supercharges (Section 9.1).

A similar trick works also for the other Freund–Rubin solutions (11.1.5); we will see in Sections 11.2.1 and 11.3.1 that the S^4 can only be generalized by a discrete quotient, while S^7 can be replaced by an SE_7, a 3-Sasaki, or a weak-G_2 manifold (Sections 7.5.1 and 7.5.3).

Another possibility is to wrap a brane on a cycle inside a supersymmetric solution. For example, one can take an M5 on the subspace [18]:

$$\mathbb{R}^4 \times \Sigma \subset \mathbb{R}^5 \times \text{CY}_6 \,. \tag{11.1.7}$$

The idea is that in the near-horizon limit, the transverse radial direction to the M5 assembles with the $\mathbb{R}^4$ to create an AdS_5; the angular directions create once again a sphere, but fibered over S^2. In other words, the internal space is in this case of the type $S^4 \hookrightarrow M_6 \to \Sigma$. If $\Sigma \subset \text{CY}_6$ is calibrated, supersymmetry should be preserved. We will see a few such examples; an early comprehensive review is [528].

11.1.2 Finding more general solutions

Classifications

It is not clear if *all* string theory AdS solutions can be obtained from a near-horizon limit of a brane system. All the AdS_7 and AdS_6 solutions are believed to have such a brane origin (Sections 11.2.1 and 11.2.2). But even for these classes, finding metrics for brane systems is rather hard, as we discussed after (9.2.10). So only a tiny fraction of vacua has really been found by performing a near-horizon limit as in (11.1.6).

Rather, AdS solutions have been found using the methods introduced earlier in the book. G-structures are especially effective when we have no idea what the internal space might look like; they relate the existence of vacua to the geometrical problems studied in Chapters 5–7. When we already know that the internal space has a large symmetry group, working with the original spinorial equations $\delta\psi^a_M = \delta\lambda^a = 0$ is competitive, with the possible advantage that the differential equations to be solved are still linear in the spinors, whereas G-structure consider bilinears.

For AdS_7 in M-theory, we will see that a direct spinorial analysis is faster, but the far more populated IIA case was first classified using the pure spinor equations. For AdS_6, the direct spinorial classification was historically faster. For AdS_5 and AdS_4 most results come from G-structures, but for large supersymmetry we will again see an exception in Section 11.3.5.

The meaning of "classification" in these applications varies wildly. The most ambitious meaning is a complete list of all solutions; this is the case for AdS_7 and AdS_6. At the other extreme, it can also just mean that the problem is reduced to a set of partial differential equations, even if they are difficult to solve; this is the case for $\mathcal{N} = 1$ AdS_4 solutions, especially in M-theory and IIB supergravity.

Recall that if the generic orbit of the symmetry group has codimension k, the solution is said to be of cohomogeneity k; the conditions of preserved supersymmetry or the equations of motion reduce to a system of partial differential equations of k variables.[2] For example, if the space is homogeneous, the fluxes are invariant, and A and ϕ are constant, then the orbit of the symmetry group is the internal space itself, and the system becomes algebraic. For cohomogeneity one, we end up with ordinary differential equations, as noted, for example, for the $Y_{p,q}$ solutions in Section 7.4.5.

Once we obtain a solution to supersymmetry and the equations of motion, we need to impose flux quantization and to check that the curvature and dilaton are small, so that using the supergravity approximation is justified. These steps are important but usually not problematic for AdS solutions.

R-symmetry

The isometry group is often partially constrained by the amount of supersymmetry. This is because a superalgebra that contains the isometry algebra $\mathrm{iso}(AdS_d) = \mathrm{so}(d-1, 2)$ with the required amount of supercharges might also contain an additional bosonic subalgebra, called R-symmetry.[3] which then has to be realized by an action on the internal directions. In detail:

[2] A general method to determine how many local solutions exist for such a system is that of exterior differential systems [529].

[3] In the more familiar superalgebra that extends the Poincaré group, R-symmetry is optional and can be broken; in superconformal theories or for AdS superalgebras, the R-symmetry generators also appear on the right-hand side of the commutation relations of other generators, and so they cannot be discarded.

- For AdS_7, the superalgebra is called $\mathrm{osp}(6,2|\mathcal{N})$. The notation is meant to suggest that $\mathrm{o}(6,2) \times \mathrm{sp}(\mathcal{N})$ is the bosonic subalgebra; the second factor $\mathrm{sp}(\mathcal{N})_R$ is the bosonic R-symmetry algebra.[4]
- For AdS_6, the only possible superalgebra is called $f^*(4)$, with R-symmetry $SU(2)_R$.
- For AdS_5, we have $\mathrm{su}(2,2|\mathcal{N}/2)$, with R-symmetry $\mathrm{u}(\mathcal{N})_R$; for $\mathcal{N} = 8$, we have to quotient this by a $\mathrm{u}(1)$.
- For AdS_4, $\mathrm{osp}(\mathcal{N}|4)$, with R-symmetry $\mathrm{so}(\mathcal{N})$.

(For more details, see the recent discussion [530].) For example, $AdS_7 \times S^4$ has $\mathrm{sp}(2)_R = \mathrm{so}(5)_R$, $AdS_5 \times S^5$ has $\mathrm{su}(4)_R \cong \mathrm{so}(6)_R$, and $AdS_4 \times S^7$ has $\mathrm{so}(8)_R$; in all three cases, R-symmetry acts by rotating the internal sphere.

Solution-generating techniques

There is a long tradition of solution-generating procedures in general relativity (see, for example, [279, chap. 10]). In string theory, we know already several dualities, but even more interesting are procedures that map solutions to new solutions which are *not* physically equivalent. A famous example is *nonabelian T-duality* [531–533]. It acts on a space with a nonabelian isometry group (hence the name); for example, on a solution with an S^3 factor, it acts on the metric as

$$ds_7^2 + e^{2f} ds_{S^3}^2 \to ds_7^2 + e^{-2f} dr^2 + \frac{r^2 e^{2f}}{r^2 + e^{4f}} ds_{S^2}^2 . \tag{11.1.8}$$

So the map breaks the initial SO(4) isometry to SO(3). The new solution is usually noncompact; often this can be interpreted as a brane system with infinitely many branes [534]. So the map is not perfect for our purposes, but it can provide important proofs of concept by populating previously unexplored classes, an important example being AdS_6 in IIB, where this method gave the first genuinely new solution [535].

Consistent truncations

Yet another method to find solutions is that of consistent truncations, which we mentioned back in Section 8.3.1. The vast majority of known AdS vacua have KK scale and cosmological constant of the same order of magnitude, (8.3.2); for all these, we cannot imagine a useful effective field theory description. There are a couple proposed exceptions [22, 253]; we will give an approximate solution for one of these in Section 11.3.2. We mentioned, however, that for many more solutions there is a nonlinear reduction, a lower-dimensional action S_d obtained by considering an appropriate slice $\mathcal{M}_{\mathrm{fin}} \neq \mathcal{M}_{\mathrm{CY}}$ in the space of all metrics; sometimes this is a consistent truncation, in the sense that all the solutions of S_d correspond of string or M-theory in $d = 10$ or 11.

Consistent truncations have been found for the Freund–Rubin solutions (11.1.3), (11.1.5), from IIB/M-theory. These supergravity models are maximally supersymmetric as those solutions are, and showing consistency is a hard task whose final details have been worked out only recently (on which more later). The reduced theory is a *gauged* supergravity: intuitively, the internal curvature and the flux contribute to

[4] In $d = 5$ and 7, there is some possible naming confusion about what we call $\mathcal{N}$, because of the absence of Majorana spinors; we count a single Majorana pair (2.3.8) as $\mathcal{N} = 1$.

the gauging parameters (Section 10.3.6). For example, reducing M-theory on a T^7 gives ungauged $\mathcal{N} = 8$ supergravity, while reducing it on S^7 gives a gauged version.

Several consistent truncations, especially for maximal supersymmetry, are now understood in the framework of generalized and exceptional structures, namely using G-structures on $T \oplus T^*$ or on (10.6.21). Reductions on compact group manifolds are consistent [536], thanks to their tangent bundle being trivial (Exercise 4.2.2). Similar to this, reductions are also consistent if $T \oplus T^*$ is trivial [537].[5] Similar methods work for exceptional geometry [518, 541, 542]. In this language, the truncation becomes a simple-looking separation of variables. The formalism was recently extended to half-maximal supersymmetry [543, 544].

For solutions with less supersymmetry, there often exist nonlinear reductions that are far less complex and whose consistency is correspondingly easier to show. It has been conjectured that at least one consistent truncation exists *for every* AdS vacuum [454]. A (small) price to pay for this simplicity is that the space of metrics $\mathcal{M}_{\text{fin}}$ has small dimension, and sometimes many different solutions end up having a consistent truncation to the same lower-dimensional theory.

Consistent truncations often help finding new AdS vacua. The potential of the d-dimensional theory usually has more vacua than the original AdS_d solution it was built to replicate. Those other vacua have fewer symmetries than the original, and in particular often break supersymmetry. Another possibility is that the d-dimensional theory might have a solution:

$$\text{AdS}_{d'} \times M_{d-d'} \,. \tag{11.1.9}$$

Lifting this to type II or M-theory then yields an AdS'_d solution whose internal space consists of a fibration over $M_{d-d'}$. These solutions are often of the type (11.1.7).

Breaking supersymmetry

As in the rest of the book, we will mostly focus on *supersymmetric* AdS vacua. It is for these that geometrical methods are most helpful, and where we can hope to find classifications. Without supersymmetry, the equations of motion are in general an uninspiring set of partial differential equations. Here is a partial list of examples, to illustrate several possible methods:

- Freund–Rubin solutions only require the internal space to be Einstein. A list of homogeneous Einstein manifolds is given in [246, table 6] for AdS_4 M-theory vacua; for example, there are generalizations $M_{p,q}$, $Q_{p,q,r}$ of the Sasaki–Einstein's in Table 7.1. (In IIA with Romans mass, F_0 is related to F_4 by the equations of motion [40].)
- The forms in a G-structure often obey quadratic relations relating them to the metric; for example, for an SU(3)-structure, (5.1.39) and (3.4.16) give

$$J_{mp} J_{nq} g^{pq} = \iota_m J \cdot \iota_n J = g_{mn} \,, \qquad \iota_m \Omega \cdot \iota_n \bar{\Omega} = 4\Pi_{mn} \,. \tag{11.1.10}$$

For a G_2-structure, from (2.2.29), (5.1.51), or simply from Exercise 2.2.3, we get $\iota_m \phi \cdot \iota_n \phi = 3 g_{mn}$. This suggests an Ansatz where the fluxes are proportional to the G-structure forms. The first solution of this type was obtained by deforming

[5] This was also discussed [538, 539] in the slightly different language of double field theory [540], a formalism originally introduced to describe T-folds (Section 4.2.6).

$\mathrm{AdS}_4 \times S^7$ by an internal four-form $F_4 \propto \tilde{\phi}$ [545], where $\tilde{\phi} = *\phi$, in terms of the G_2-structure making S^7 a weak G_2 manifold (Section 7.5.3); of course, this works also for other weak G_2 [246, sec. 10.2]. Other examples use Sasaki–Einstein manifolds in M-theory [546, 547] and IIB [548]; and Kähler–Einstein [40, 549, 550] and nearly Kähler [432, sec. 11.4] in massive IIA.

- The equations of motion are algebraic homogeneous spaces; in cohomogeneity one, they reduce to ordinary differential equations. This strategy was used, for example, to obtain AdS_8 solutions in [551].
- Another common strategy is to change the sign of some fluxes, exploiting the fact that they appear quadratically in the equations of motion, an idea called *skew-whiffing* [246, 552].
- A more ambitious strategy has been proposed [432] to deform the pure spinor methods of Chapter 10 to obtain nonsupersymmetric solutions (see also [480]).
- Finally, we will see that a large number of nonsupersymmetric solutions are obtained from consistent truncations originally built around supersymmetric ones.

Nonsupersymmetric AdS vacua are not much used in holography because it is hard to produce dual pairs, unless one can show some kind of relation to supersymmetric ones [550, 553]. For these reasons, we will mention nonsupersymmetric AdS vacua much less systematically than their supersymmetric counterparts.

11.1.3 Stability

Tachyons

An extreme manifestation of how nonlinear reductions are not effective theories is the presence of tachyons, fields with negative masses.

Recall how a tachyon behaves in flat space: while it was initially introduced in the hope it would describe particles that are faster than light, it actually signals an instability. Indeed, the Klein–Gordon equation $\partial_\mu \partial^\mu \phi = m^2 \phi$ is solved by a plane wave $\mathrm{e}^{\mathrm{i}(\omega t + k_i x^i)}$, and so $\omega^2 - k_i k_i = m^2$; for $m^2 < 0$, one might have thought that $|k| > \omega$, but in fact the correct interpretation is that $\omega = \sqrt{k_i k_i + m^2}$ can become imaginary, which makes the plane wave turn into an exponentially growing function. (If the potential is globally not quadratic, the field may eventually settle in a new vacuum, a process called condensation.)

In AdS, things are slightly different [554]. In global coordinates, the analogue of plane waves can be found by separation of variables. In global coordinates (4.5.32), a basis of solutions is of the type $\mathrm{e}^{\mathrm{i}\omega\tau} Y_l f(\rho)$, where Y_l is an eigenfunction of the Laplacian on S^{d-2} (a "spherical harmonic"), whose spectrum is $l(l+d-3)$, $l \in \mathbb{Z}_{\geq 0}$. For m^2 negative but very small, the Compton wavelength m^{-1} will be very large; it can become so large as to create a standing wave that reflects off the boundary, thus eliminating the instability (recall Figure 4.6). The frequency of this wave is given by $\pm\omega L_{\mathrm{AdS}} = \lambda + l + 2n$, where n is the number of nodes of the standing wave in the ρ direction, and λ satisfies

$$\lambda(\lambda - d + 1) = m^2 L^2_{\mathrm{AdS}_d} \, . \qquad (11.1.11)$$

We see that λ (and thus ω) can be real if and only if the mass satisfies the *Breitenlohner–Freedman (BF) bound*

$$m^2 L^2_{\mathrm{AdS}_d} \geq -\frac{(d-1)^2}{4} . \tag{11.1.12}$$

As expected, this mechanism works for small negative masses, such that m^{-1} is as large as AdS itself. Similar bounds exist for nonscalar fields; see, for example, [555].

Around supersymmetric vacua, all fields have masses above the bound [554], and thus this potential instability is avoided. Nonsupersymmetric vacua don't enjoy such a protection and usually have tachyons below the bound. Among those discussed in Section 11.1.2, some known exceptions are the skew-whiffed solutions [552]; the massive IIA solutions based on a Kähler–Einstein metric [40, 549, 550], which can be argued to be stable for small F_0 by a small-deformation argument [550, sec. 4.1]; the manifolds $M_{p,q}$, $Q_{p,q,r}$ for a certain range of p/q [556, 557] (see also [246, sec. 6]); and some of the vacua of an S^6 reduction in Section 11.3.4 [558].

Tunnel effects

Even in absence of tachyons below the BF bound, a vacuum might be unstable because of tunnel effects to other vacua. Unlike in quantum mechanics, we cannot expect the field to simultaneously move everywhere to a new vacuum, as this would require infinite energy; rather, a bubble of a different vacuum might appear (a bit like a bubble of vapor in boiling water) and then expand. This is a possible concern for any value of the cosmological constant, but for AdS such a bubble would take finite time to swallow the entire Universe, basically because of Exercise 4.5.6.

For a single particle in a potential, the probability of such a tunnel effect can be computed with path integral methods [559]. The leading contribution is $e^{-S_{\mathrm{inst}}}$, where the S_{inst} is the classical action of a *bounce*, a path that goes from the original vacuum toward the new one, and comes back, extremizing the Euclidean action. This estimates the probability that the particle tunnels on the other side of the potential; its ensuing relaxation toward the new vacuum can then be described by classical mechanics.

For vacuum decay, the bounce is a path in configuration space: a bubble of new vacuum is created, and then annihilates, as in the gray line in Figure 11.2. The action of such a configuration gives the probability that the bubble is created; its subsequent expansion is then classical, and is shown as a hyperbola in Figure 11.2. In string theory, the surface of the bubble itself is often the Euclidean version of a brane, and

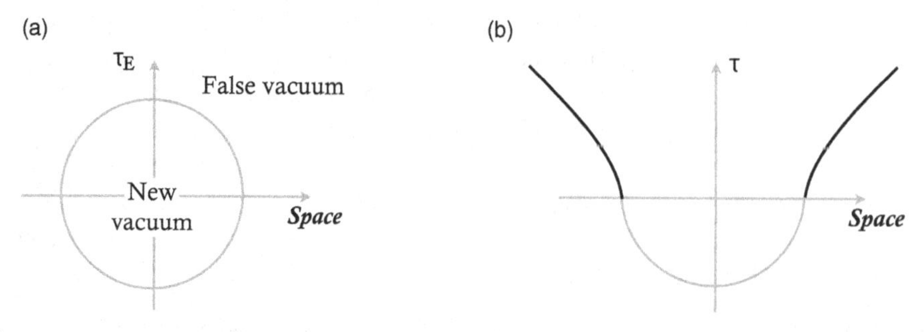

Figure 11.2 (a) A bounce in Euclidean signature. (b) In black the bubble's classical trajectory in Lorentzian signature.

we can treat it as a probe. Consider, for example, a D$(d-2)$-brane that is point-like in the internal space and extended along $d-1$ directions in AdS$_d$. As we see by comparing (4.5.10) and (4.5.29), Euclidean AdS is hyperbolic space; for the bounce to extremize the Euclidean action S_E, it has to be highly symmetric [560], and we thus take it to be a sphere S^d at $\{\rho = \rho_0\}$ in the coordinates (4.5.14). (This is not immediately apparent in the Wick-rotated $\tau \to \tau_E$ global AdS coordinates (4.5.32) [561].) The brane's DBI term gives Vol(S^d). The WZ term is the coupling to C_{d-1}, whose field-strength $F_d = dC_{d-1}$ should be proportional to the volume form:

$$F_d = f \mathrm{vol}_{\mathrm{EAdS}_d} = f \sinh^{d-1} \rho \, d\rho \wedge \mathrm{vol}_{S^{d-1}}, \tag{11.1.13}$$

as usual in order to preserve the maximal symmetry of the AdS vacuum. From (1.3.24), the Euclidean action is then

$$S_E = \mathrm{Vol}(S^{d-1})(-T \sinh^{d-1}(\rho_0) + Qc(\rho_0)) . \tag{11.1.14}$$

$T \equiv \tau_{D(d-2)} e^{(d-1)A-\phi}$ and $Q \equiv \tau_{D(d-2)} f$ are the effective tension and charge of the brane, and $\partial_\rho c = \sinh^{d-1} \rho$. We still have to extremize S_E with respect to the remaining parameter ρ_0: this gives[6]

$$\tanh \rho_0 = \frac{(d-1)T}{Q} . \tag{11.1.15}$$

Intuitively, the gravitational term T tends to shrink the bubble, because the gravitational potential (4.5.35) attracts toward the center; the WZ charge Q tends to expand it, much like [321] in the Schwinger effect, the creation of a particle–antiparticle pair by a strong electric field. Now we have two cases:

- If $(d-1)T < Q$, (11.1.15) has a solution; the new vacuum has a finite probability of being created, and when that happens it expands as in Figure 11.2a. Integrating the Bianchi identity for F_d, we find that its flux quantum is smaller inside the bubble; the AdS radius L_{AdS} decreases, as in (11.1.3), and the vacuum energy $\Lambda = -3/L^2_{\mathrm{AdS}}$ becomes more negative.
- If $(d-1)T \geq Q$, there is no solution; the original vacuum is stable against decay to the candidate new vacuum, even if the candidate new vacuum has a lower energy.

Outside string theory, a more traditional type of bubble is a domain wall $\sigma^i(\rho)$ where the scalars interpolate between an old vacuum at $\rho \to \infty$ and a new one at $\rho = 0$; in practice, one often makes a *thin-wall* approximation [560], where the transition is sharp. In this case, Figure 11.2 is still appropriate; (11.1.14) becomes $-T\mathrm{Vol}(S^{d-1}) + \mathrm{Vol}(B_d(\rho))\Delta V$, where T is now the tension of the thin wall, which one may estimate from the scalar action; ΔV is the difference in energy between the two vacua; and $B_d(\rho)$ is a ball of radius ρ.[7] At large r, the volumes of the sphere and of the ball scale in the same way, unlike in flat space: $\mathrm{Vol}(S^{d-1}) \sim (d-1)\mathrm{Vol}(B_d) \sim e^{(d-1)\rho}$, so if $(d-1)T < \Delta V$ the decay happens, while if $(d-1)T > \Delta V$ there is no

[6] If A, ϕ or f are functions of the internal directions y^m, some interesting dynamics arises where one has to extremize with respect to those as well [562, 563].

[7] This second point of view is related to that in (11.1.14): the second term there is also related to the difference in vacuum energy inside and outside the bubble by the equation of motion (10.3.6a) (for $d = 4$). A D-brane can also be seen as a domain wall in string field theory, as we discussed in Section 9.2.3.

extremum of S_E, and once again the original vacuum is stable against decay to the candidate new one [560].

Supersymmetric AdS vacua are protected against both types of vacuum decay [564, 565]. For example, BPS brane bubbles have $(d-1)T = Q$ (Exercise 11.3.4), and the BPS bound is $(d-1)T \geq Q$, similar to (1.3.9) in QFT.[8] In $d = 4$ this bound can be obtained along the lines of Section 10.3.5: Q is related to the integral of $*\lambda f$ on the internal cycle C, which by (10.3.20c) is related to $\mathrm{Im}\Phi_\pm$, which as a calibration gives a bound on $\mathrm{Vol}(C)$. For nonsupersymmetric vacua, there is no such protection, and indeed several examples have been found where this tunnel effect does happen [550, 561–563]. As we saw previously, the flux quanta tend to decrease in this process; when they become small, curvature ($\sim L_{\mathrm{AdS}}^{-2}$) becomes small, and one ends up beyond the domain of applicability of supergravity, where neither the existence of the solutions nor their stability is easy to establish. Moreover, in string theory there are many possible brane bubbles, and the space of scalars is in principle infinite dimensional, with a large number of vacua; excluding decay without some general argument is difficult. We will come back on this in Section 12.2.

Bubbles of nothing

A final mode of decay is known as *bubble of nothing* [566]. The original example was in $\mathrm{Mink}_4 \times S^1$: at a fixed time, the bubble is an S^2 on which the KK S^1 shrinks. Space ends on this S^2, and there is literally nothing inside. This can happen smoothly, much as for the S^1 at the endpoints of the segment Figure 6.3a.

Evaluating the probability of nucleation for such a bubble is similar to the previous procedure for tunneling to other vacua: in this case, one goes to Euclidean-signature $\mathbb{R}^4 \times S^1$ and looks for a solution where the S^1 shrinks on an $S^3 \subset \mathbb{R}^4$. One such solution is the following [566]:

$$ds_5^2 = \left(1 - \frac{r_0^2}{r^2}\right)^{-1} dr^2 + r^2 ds_{S^3}^2 + \left(1 - \frac{r_0^2}{r^2}\right) d\psi^2, \qquad (11.1.16)$$

where r is the radial direction in $\mathbb{R}^4$, and ψ is the KK S^1. (This is an analytic continuation of the Schwarzschild solution in Mink_5.) Once again, after being nucleated the bubble expands rapidly; Figure 11.2 again applies, but this time instead of the candidate new vacuum, one should imagine that there is no spacetime at all.

With one extra dimension, this decay mode can be avoided by having the usual antiperiodic boundary conditions for fermions along the S^1. More generally, a bubble of nothing is sometimes impossible for topological reasons: there might simply not exist any space B such that $\partial B = M$, where M is the internal space. Or sometimes one may prove *positive energy theorems* [567–570], protecting against all decay modes, which show that the ADM mass is bounded from below, and that the original vacuum realizes the bound. These typically rely on the *dominant energy condition* (see again [19, 4.3;8.2] and [20, chap. 9]) and on the existence of spinors that are asymptotically covariantly constant [568]; see [571] for a recent discussion.

[8] Strictly speaking, it is not the brane bubble that is BPS. A D$(d-2)$-brane domain wall sitting at $\{r = r_0\}$ in Poincaré coordinates (4.5.37) has action $\propto e^{(d-1)r}(-T + Q/(d-1))$; if it is BPS at a certain internal point p, then $(d-1)T = Q$, and so the D$(d-2)$-brane bubble at the same p saturates the bound for stability [562, sec. 5.1].

In the AdS case, a bubble of nothing can be thought of as an extreme limit of the previous decay [572], where the curvature $L_{\text{AdS}} \to 0$. For explicit AdS examples, see [573, 574].

11.1.4 Holography

According to the *holographic principle* [575, 576], the number of states for a theory of quantum gravity in a region of space V is proportional to the volume of the boundary ∂V, not of V itself. This is suggested by the fact that the object of largest entropy one can create in V is a black hole. Its most explicit realization is the AdS/CFT correspondence [577–579], which predicts that all the data of the interior of AdS_d can be encoded on the boundary $\mathbb{R} \times S^{d-2}$ (recall Figure 4.6). The isometry group (4.5.30) is isomorphic to the conformal group (4.1.69) of Mink_{d-1}; this and other observations suggest that the data at the boundary describe a conformal field theory (CFT) model, without gravity:

$$\text{Quantum gravity on AdS}_d \quad \cong \quad \text{CFT in } d \text{ dimensions}. \tag{11.1.17}$$

For us, the quantum gravity theory will be string theory. The spacetime on which the CFT_d lives is specified by the choice of radial coordinate: if $ds^2_{\text{AdS}_d} \sim \frac{dr^2}{r^2} + r^2 ds^2_{M_{d-1}}$ at large r, the CFT_{d-1} lives on M_{d-1}. So if we work in Poincaré coordinates (4.5.37), the theory on the boundary lives on Mink_{d-1} (the wedge in Figure 4.6); if we work in global coordinates (4.5.32), the theory on the boundary lives on the full $\mathbb{R} \times S^{d-2}$. Both these M_{d-1} are conformally flat, and so locally conformal to each other.

The first example of this correspondence was suggested by the identification of the brane metric (1.3.59) with the back-reaction of a Dp stack, taking the $l_s \to 0$ limit in the $p = 3$ case [577].

- From the point of view of the D3-branes, the gravitational constant (1.2.16) $\kappa^2 \to 0$; the closed string modes become free, and moreover, the corrections we considered in Section 9.2.3 disappear. Thus in the limit we have free gravitons in flat space, decoupled from $\mathcal{N} = 4$ super-YM on the $d = 4$ world-volume (Section 1.3.6).
- From the point of view of the gravity solution, the region $r \leq r_0 N^{1/4} = 2\sqrt{\pi}(g_s N)^{1/4}$, where the back-reaction is appreciable and becomes very small; modes starting from the asymptotical Mink_{10} find it harder and harder to enter in the near-horizon region. Thus in the limit we have free gravitons in flat space, decoupled from the near-horizon region, which is $\text{AdS}_5 \times S^5$ (11.1.3).

Comparing the two points of view, we obtain the famous identification [577]

$$\text{AdS}_5 \times S^5 \quad \cong \quad \mathcal{N} = 4 \text{ super-YM in } d = 4. \tag{11.1.18}$$

This is an example of (11.1.17), since $\mathcal{N} = 4$ super-YM is indeed a CFT.

For Dp-branes, $p \neq 3$, we mentioned before (11.1.5) that the near-horizon limit does not produce AdS solutions; in $d = p + 1 \neq 4$, the YM coupling (1.3.33) is dimensionful, and the world-volume effective theory cannot be conformal. For M2- and M5-branes, (11.1.5) predicts a conformal world-volume theory. As we mentioned in Section 1.4.1, for M2s this was found in [74–77], while for M5-branes a Lagrangian description is not known yet, but holography is one of the sources of indirect information.

We also saw earlier that more complicated brane systems provide many more AdS solutions, of which we gave (11.1.6) as a first example; the brane interactions provide additional matter fields (as in Section 9.2.4), and this modifies the dynamics of super-YM so that it flows to a CFT at low energies, or on the contrary has a CFT as a UV completion, at high energies. For solutions that have no clear brane origin, there is no general method to obtain the CFT.

In the rest of this chapter, we will mention what is known about the CFTs dual to each AdS solution we encounter, but without a detailed field theory discussion. Holography also has various generalizations beyond AdS that we will not consider; for example, to RG flows connecting CFTs, more broadly to nonconformal theories, or to nonrelativistic theories.

Exercise 11.1.1 Consider the region near $\theta = 0$ or $\pi/2$ in (11.1.6); check that after the coordinate change $\theta \sim r^{2/3}$ the singularity has the same behavior as the $r \to 0$ in the D6 solution from (1.3.59).

Exercise 11.1.2 Using the equations of motion (10.3.6), find solutions $\mathrm{AdS}_4 \times \mathrm{KE}_6$ in massive IIA.

11.2 Higher dimensions

We begin with a discussion of $d > 4$ solutions. These are important for holography, but they are also useful to illustrate some issues that appear in $d = 4$, while being easier to classify. Finally, they generate $d = 4$ solutions as in (11.1.9).

11.2.1 AdS_7

M-theory

The M-theory case is easy enough to be analyzed directly using the supersymmetry transformations (10.6.13), without bilinears or bispinors.

As usual, the field-strength should be purely internal, and thus $G_4 = g\mathrm{vol}_{S^4}$ for some function g. The gamma matrices factorize as in (8.2.6) and more broadly in Section 2.2.3: $\Gamma_\mu = e^A \gamma_\mu \otimes \gamma^{(4)}$, $\Gamma_m = 1 \otimes \gamma_m$. The supersymmetry parameter factorizes as $\epsilon = \zeta \otimes \eta + \zeta^c \otimes \eta^c$. The internal components of (10.6.13) are easily evaluated: restoring slashes, $\not{G} = g(1 \otimes \gamma^{(4)})$, $\not{\iota}_m \not{G} = \frac{g}{6} \epsilon_{mnpq} \Gamma^{npq} = g(1 \otimes \gamma_m \gamma^{(4)})$, using also (3.2.26). Then the internal components imply $D_m \eta + \frac{g}{6} \gamma_m \gamma^{(4)} \eta = 0$, which is the variant (7.3.9) of the Killing spinor equation.

By Section 7.3, the cone $C(M_4)$ should then admit a covariantly constant spinor, and thus have restricted holonomy. There is no $d = 5$ example in the classification of Section 5.7, but we can take products; even so, the only possibility is $C(M_4) = \mathbb{R} \times C_4$, with C_4 a conical space of SU(2) holonomy. We saw in Section 7.5.1 that conical hyper-Kähler manifolds are a bundle with fiber S^3 or a quotient thereof, so we conclude that $C_4 = C(S^3/\Gamma)$, one of the orbifolds of Section 7.4.2. Putting it all together, $C(M_4) = \mathbb{R} \times C(S^3/\Gamma)$, or in other words,

$$M_4 = S^4/\Gamma\,. \tag{11.2.1}$$

The discrete group acts only on four of the embedding coordinates x^I, $S^4 = \{\sum_{I=1}(x^I)^2 = 1\}$. For example, for $\Gamma = \mathbb{Z}_k$, $z_1 \equiv x^2 + \mathrm{i}x^3$, $z_2 \equiv x^4 + \mathrm{i}x^5$, and Γ acts as $(x^1, z_1, z_2) \mapsto (x^1, \mathrm{e}^{2\pi \mathrm{i}/k} z_1, \mathrm{e}^{-2\pi \mathrm{i}/k} z_2)$. In particular, there are two orbifold singularities on M_4, at $(\pm 1, 0, 0, 0, 0)$.

Using (7.3.10) and (7.3.16), we now determine the radius of the S^4 in terms of g. The spacetime components of (10.6.13) are evaluated using (8.2.21); they set the warping to a constant and determine the AdS_7 radius. The end result is as follows [526]:

$$\mathrm{d}s^2 = r_0^2 \left(4\mathrm{d}s^2_{\mathrm{AdS}_7} + \mathrm{d}s^2_{S^4/\Gamma}\right), \qquad G = 3r_0^3 \mathrm{vol}_{S^4/\Gamma}, \tag{11.2.2}$$

where $r_0^3 = \pi N l_{\mathrm{P}11}^3$. Our argument shows that this is the most general solution, but as anticipated in (11.1.5) it can also be obtained from the M5 (1.4.70): in the near-horizon limit $r \to 0$, the harmonic function $h \sim (r_0/r)^{-3}$, and

$$\mathrm{d}s^2_{\mathrm{M5}} \sim \frac{r}{r_0}\mathrm{d}s^2_{\mathbb{R}^6} + r_0^2\left(\frac{\mathrm{d}r^2}{r^2} + \mathrm{d}s^2_{S^4}\right), \tag{11.2.3}$$

which becomes (11.2.2) upon rescaling the $\mathbb{R}^6$ coordinates by $2r_0$, defining $r = r_0\rho^2$, and comparing with Poincaré coordinates.

Without the Γ quotient in (11.2.2), there are *two* independent Killing spinors η_I on S^4 (coming as usual from the constant ones on $\mathbb{R}^5$). These can be multiplied by two independent ζ_I, in the spirit of (8.2.12); so there are 32 supercharges. The M5-brane solution only preserves 16, being half-BPS as are all other branes; the near-horizon limit restores the missing supersymmetry. If we do introduce Γ, however, we break to 16 supercharges again. All this is also in agreement with our discussion of R-symmetry in Section 11.1.2.

Type II classification procedure

We now look for supersymmetric solutions in type II, using the pure spinor methods of Chapter 11. One possible strategy would be to start from the $\mathrm{d}\Phi$ equation (10.2.16), as we did for $d = 4$. We can use a gamma matrix decomposition similar to (2.2.26), and a decomposition for the supersymmetry parameters:

$$\epsilon^1 = (\zeta \otimes \chi_1 + \zeta^c \otimes \chi_1^c) \otimes v_+, \qquad \epsilon^2 = (\zeta \otimes \chi_2 \mp \zeta^c \otimes \chi_2^c) \otimes v_\pm, \quad {}^{\mathrm{IIA}}_{\mathrm{IIB}} \tag{11.2.4}$$

where $v_\pm$ are chosen so that ϵ^a are Majorana and have the correct chirality; as usual, ζ is a Killing spinor. To use (10.2.16), we would need the bispinors of ζ; from Table 2.5, we see that there are two possible orbits. But this strategy would unnecessarily duplicate an earlier computation: just like in (8.2.20), $\mathrm{AdS}_7 \times M_3$ solutions are a particular case of $\mathrm{Mink}_6 \times M_4$:

$$\mathrm{e}^{2A}\mathrm{d}s^2_{\mathrm{AdS}_7} + \mathrm{d}s^2_{M_3} = \mathrm{e}^{2A}r^2\mathrm{d}s^2_{\mathrm{Mink}_6} + \left(\mathrm{e}^{2A}\frac{1}{r^2}\mathrm{d}r^2 + \mathrm{d}s^2_{M_3}\right). \tag{11.2.5}$$

So we can use the Mink_6 system in (10.6.4) [445]. The M_4 pure forms have to be chosen so that the metric determined by the map (10.5.90) is as in (11.2.5). Rewriting (11.2.4) in terms of a 6 + 4 gamma matrix decomposition, the $d = 4$ spinors read $\eta^1 = z^{-1/2}\chi_1 \otimes v_+$, $\eta^2 = z^{-1/2}\chi_2 \otimes v_\mp$, and the pure forms of (10.6.4) are $\Phi_1 = \eta^1 \otimes \eta^{2\dagger}$, $\Phi_2 = \eta^1 \otimes \eta^{2c\,\dagger}$. These involve the M_3 bispinors $\chi_1 \otimes \chi_2^\dagger$, $\chi_1 \otimes \overline{\chi_2}$, which can be parameterized along the lines of Section 3.4.6. Since $\mathrm{Stab}(\chi) = \{1\}$ in $d = 3$, we have an identity structure (5.1.5), or in other words a vielbein. So we expect that the equations will determine the metric.

In IIB, already the zero-form part of (10.6.4) does not have a solution, so we conclude that there are *no AdS_7 solutions in IIB supergravity* [580].[9]

Local solutions in IIA

In IIA, as expected the pure spinor equations determine the metric, and in fact all remaining fields [580]. In appropriate coordinates, the BPS equations reduce to the following single ordinary differential equation [97]:

$$\dddot{\alpha} = -162\pi^3 F_0\,, \qquad \dot{(\,)} \equiv \partial_z(\,)\,. \tag{11.2.6}$$

All fields in the solution are expressed in terms of this single function. The metric and dilaton are

$$\frac{1}{\pi}ds^2 = 8\sqrt{2}\sqrt{-\frac{\alpha}{\ddot{\alpha}}}ds^2_{AdS_7} + \sqrt{2}\sqrt{-\frac{\ddot{\alpha}}{\alpha}}\left(dz^2 + \frac{\alpha^2}{\dot{\alpha}^2 - 2\alpha\ddot{\alpha}}ds^2_{S^2}\right)\,; \tag{11.2.7a}$$

$$e^{\phi} = 162\cdot 2^{1/4}\pi^{5/2}\frac{(-\alpha/\ddot{\alpha})^{3/4}}{\sqrt{\dot{\alpha}^2 - 2\alpha\ddot{\alpha}}}\,. \tag{11.2.7b}$$

The only form fields are F_0, F_2, and $H = dB$, where

$$B = \pi\left(-z + \frac{\alpha\dot{\alpha}}{\dot{\alpha}^2 - 2\alpha\ddot{\alpha}}\right)\text{vol}_{S^2}\,, \qquad F_2 = \left(\frac{\ddot{\alpha}}{162\pi^2} + \frac{\pi F_0 \alpha\dot{\alpha}}{\dot{\alpha}^2 - 2\alpha\ddot{\alpha}}\right)\text{vol}_{S^2}\,. \tag{11.2.7c}$$

The fields are sensible provided $\alpha > 0$, $\ddot{\alpha} < 0$. The Bianchi identities are automatically satisfied; as we saw in Section 10.6.2, this sometimes happens for a large amount of supercharges. All the equations of motion are then automatically satisfied, as follows from the general arguments in Sections 10.1 and 10.2.

The appearance of a round S^2 was expected from our discussion of R-symmetry in Section 11.1.2: the superalgebra containing $\text{iso}(AdS_7) = \text{so}(2,6)$ also contains an additional su(2). So $\text{iso}(M_3) \supset \text{su}(2)$. The orbits of the isometry group should then be a coset of SU(2) or SO(3), and up to a discrete quotient this should contain an S^2.[10]

Solutions with D8-branes

To make M_4 compact, we might try to make the z periodic, but this cannot work, because $\dot{\alpha}$ is monotonous (since $\ddot{\alpha} < 0$), and appears as a coefficient in the metric. So we make z belong to an interval, say $I_z = [0, z_0]$; to avoid a boundary in M_3, we make the S^2 shrink at the endpoints, which gives $M_3 \cong S^3$ topologically. From (11.2.7a), this requires $\alpha \to 0$; making e^{2A} a nonzero constant requires $\ddot{\alpha} \to 0$ as well. Thus smoothness requires

$$\alpha = \ddot{\alpha} = 0 \tag{11.2.8}$$

[9] As we mentioned in Section 9.4.4, in F-theory one can engineer many CFT_6 [404–406]. These theories might conceivably have AdS_7 duals in F-theory, but they would be intrinsically strongly coupled and thus beyond the reach of supergravity.

[10] One could use this expectation to shorten the derivation of the solutions, declaring the χ_a right away to be two spinors on S^2 that transform as a doublet under $SO(3) = ISO(S^2)$. Notice, however, that a priori there are other such doublets. The pure spinor equations naturally pick the χ_a to be the Killing spinors (7.3.21).

at both endpoints $z = 0, z_0$. This gives four conditions on the four coefficients of the cubic $\alpha(z)$, leading to $\alpha(z) = 0$ as the only solution. We conclude that there is no completely smooth AdS_7 solution in IIA.

Next we look for solutions with sources; we start with *D8-branes*. From the zero-form part $\mathrm{d}F_0 = \delta$ of the Bianchi identity (10.1.4), this is a locus across which F_0 jumps. Suppose at various loci $z = z_a$ there are stacks of $n_{\mathrm{D8}a}$ D8-branes; between two such stacks, $\alpha(z)$ has a different cubic expression. Now there are more free coefficients, and (11.2.8) is easy to arrange at both endpoints $z = 0, z_0$. At each $z = z_a$, we demand that the metric be continuous, but not smooth: the Riemann tensor will contain delta functions. This is as it should be: it reproduces the δ term due to the D8-branes in the metric equation of motion (10.1.19a), with the correct coefficient.

Further constraints come flux quantization. Defining $B \equiv \beta(z)\mathrm{vol}(S^2)$ and integrating by parts:

$$N \overset{(1.3.50b)}{=} \frac{1}{4\pi^2}\int_{M_3} H = \frac{\mathrm{Vol}(S^2)}{4\pi^2}(\beta(z_0) - \beta(0)) \overset{(11.2.7c)}{=} -z_0 \,. \tag{11.2.9}$$

So the interval has integral length. The F_0 and F_2 quantizations (1.3.47) give, using (11.2.6):

$$-\frac{\dddot{\alpha}}{81\pi^2} = n_{2a} + n_{0a} z \qquad z \in [z_a, z_{a+1}] \,. \tag{11.2.10}$$

Another way to phrase this is $-\ddot{\alpha}(i)/81\pi^2 \in \mathbb{Z}$, $\forall i \in \mathbb{Z}$. In particular, continuity of $\ddot{\alpha}$ at $z = z_a$ gives $z_a = -\Delta n_{2a}/\Delta n_{0a}$. Equations (10.1.4) and (10.1.5) imply that the jump $\Delta n_{2a} \equiv n_{2a} - n_{2\,a-1} = n_{\mathrm{D8}a}\mu_a$, where $\mu_a = \frac{1}{2\pi}\int_{S^2} f$ is the D6-charge of the D8-branes at $z = z_a$. So the D8s are really D8/D6 bound states, with quantized positions:

$$z_a = \mu_a \in \mathbb{Z} \,. \tag{11.2.11}$$

It might naively seem that H has a delta at the endpoint $z = N$: by Stokes's theorem, the integral of H on a small disk D_ϵ around this point equals the integral of B on its boundary $\{z = N - \epsilon\}$, which gives $4\pi^2 N$ as $\epsilon \to 0$. This is a familiar issue from Sections 1.3.1 and 1.3.4; it is solved by patching together two local potentials. Since z only appears explicitly in B, its large gauge transformation (1.3.50a) is equivalent to a translation by an integer $z \to z - N$; so, by (11.2.10), n_2 is not gauge invariant. The D6-charges of the D8-branes also transform, $\mu \to \mu - N$.

Integrating (11.2.10) twice and imposing α to be continuous and differentiable, we find infinitely many solutions, with arbitrary numbers of D8-branes. There are no moduli, in agreement with general group-theoretic arguments [581, 582]. (The discrete parameters are in correspondence with the data $\{N, Y_{\mathrm{L}}, Y_{\mathrm{R}}\}$, with N in (11.2.9) and $Y_{\mathrm{L,R}}$ two Young tableaux with the same number of boxes k; for more details about this, see [97, sec. 2.2].) As a simple example:

$$\alpha = \frac{27}{2}\pi^2 n \begin{cases} z(3\mu(N-\mu) - z^2) & z \in [0, \mu]\,, \\ \mu(3z(N-z) - \mu^2) & z \in [\mu, N-\mu]\,, \\ (N-z)(3\mu(N-\mu) - (N-z)^2) & z \in [N-\mu, N]\,. \end{cases} \tag{11.2.12}$$

From (11.2.6), $F_0 = \frac{1}{2\pi}\{n, 0, -n\}$ in the three subintervals; so there are n D8s at $z = \mu, N - \mu$. According to the discussion that follows (11.2.10), they each have

D6-charge μ and $N - \mu$ respectively. If we perform the large gauge transformation for B in the $n_0 = 0$ region, then the second stack has charge $-\mu$.

Other sources

We can explore alternative boundary conditions, beyond (11.2.8). For example, we can still demand that the S^2 should shrink, which gives $\alpha \to 0$, but now give up on smoothness. Around $z = 0$, writing $\alpha = \alpha_1 z - \alpha_2 z^2 + \alpha_3 z^3$, (11.2.7a) gives

$$\mathrm{d}s^2 \sim \sqrt{\frac{\alpha_1 z}{2\alpha_2}}\mathrm{d}s^2_{\mathrm{AdS}_7} + \sqrt{\frac{2\alpha_2}{\alpha_1 z}}(\mathrm{d}z^2 + z^2\mathrm{d}s^2_{S^2})\,. \tag{11.2.13}$$

This local $z \to 0$ behavior is singular, but it has a physical interpretation: it is of the form (1.3.59) with $h \sim 1/z$, which is (1.3.61) for $p = 6$, $r = z$. (See also Exercise 11.1.1.) The dilaton and F_2 are also consistent with this D6 interpretation. The number of D6-branes is equal to the flux quantum n_2 computed near $z = 0$.

The easiest example of solution with such D6-brane singularities is

$$\alpha = \frac{81}{2}\pi^2 k z(N - z)\,. \tag{11.2.14}$$

There are k D6-branes at $z = 0$, and k anti-D6-branes at $z = N$; there is no tachyon because they are far from each other. (For $k = 1$, this is the same solution as (11.1.6).) Equation (11.2.6) gives $F_0 = 0$ everywhere, so we can lift this solution to M-theory, where it becomes $\mathrm{AdS}_7 \times S^4/\mathbb{Z}_k$, (11.2.2). Restricted to each copy of S^2 at fixed z, the M-theory fibration is of Hopf type, (4.2.51). The D6s lift to the two $\mathbb{Z}_k$ singularities at $\{x^1 = \pm 1, x^2 = x^3 = x^4 = x^5 = 0\}$, as in Section 7.2.3. So our identification of (11.2.13) with a D6 singularity is sensible.

Another simple example is

$$\alpha = \frac{27}{2}\pi^2 n_0 z(N^2 - z^2)\,, \tag{11.2.15}$$

which has $F_0 = n_0/2\pi \neq 0$ everywhere, and $k = n_0 N$ anti-D6-branes at $z = N$. Now M-theory cannot help; there is an additional diverging (but integrable) H, consistent with earlier analysis in [583, 584], and in particular of [436, (4.15)], predating the global AdS_7 solutions. It is natural, however, to regard this as the result of placing a D6 in an F_0 background; H and F_2 are linked after all by a Stückelberg mechanism (Section 1.2.2).

Other physical singularities are obtained by setting to zero some other Taylor coefficients of α around $z = 0$; we collect the possibilities in Table 11.1. (Some are absent because they violate the conditions for positivity of the metric: for example,

Table 11.1. AdS_7 boundary conditions.

$\alpha(0)$	$\dot\alpha(0)$	$\ddot\alpha(0)$	Interpretation
0			D6
	0		O8 (e^ϕ finite)
		0	O6
0		0	regular
	0	0	O8 ($e^\phi \to \infty$)

$\alpha(0) = \dot{\alpha}(0) = 0$ gives $\frac{\alpha}{\ddot{\alpha}} \sim z^2/2$, which cannot be used in (11.2.7a).) The two types of O8-planes were discussed after (1.4.57). Our final example

$$\ddot{\alpha} = \frac{27}{2}\pi^2 n_0 (N^3 - z^3) \tag{11.2.16}$$

has an O8-plane with diverging dilaton at $z = 0$, and again k anti-D6-branes at $z = N$. Here the topology is a half-S^3, because of the orientifold projection.

Intersecting brane origin

The solutions (11.2.7) are related to the intersecting-branes class of Mink$_6$ solutions from Section 10.4.2 [469, 585]. From (11.2.5), the Mink$_6$ warping function $e^{A_6} = e^{2A} r^2$; the A in Section 10.4.2 is this A_6, and in particular $h_6 = e^{-4A_6} = r^{-4} e^{-4A}$. With the coordinate change,

$$x = \frac{8}{81\pi}\dot{\alpha}\, r^2 , \qquad y = \frac{64}{81}\alpha\, r^4 , \tag{11.2.17}$$

(10.4.18) turns into (11.2.7a), and so do the other fields. (Equation (11.1.6) is a particular case of (11.2.17).) Equation (10.4.21) becomes $\partial_z^4 \alpha = 0$, and the expression for F_0 in (10.4.19) becomes (11.2.6).

The solutions with D8-branes are believed to be the result of a near-horizon limit on a brane system with NS5s, D6s, and D8s shown in Figure 11.3. Naively, the D6s end on the D8s, but in fact the two fuse into funnel-shaped objects carrying both charges [586, sec. 3.5]. They intersect a stack of NS5-branes, which dissolve in the limit, leaving behind $\int H = 4\pi^2 N$, much as the D3s in AdS$_5\times S^5$. The angular sphere around them, in the spirit of Figure 11.1, becomes M_3; it intersects the funnels, which become the D8/D6 bound states at $z_a = \mu_a$.

Holographic duals

The holographic dual of AdS$_7 \times S^4$ is the CFT on a stack of N M5-branes, with $\mathcal{N} = (2,0)$ supersymmetry, for which no Lagrangian description is known (Section 1.4.1). The quotient by Γ as in (11.2.2) breaks to $\mathcal{N} = (1,0)$. In both of these cases, we at least know the equations of motion when the M5s are separated; this theory is nonrenormalizable, and is thought to be UV completed by the CFT. For the AdS$_7$ vacua in massive IIA, a similar picture is suggested by the intersecting-brane

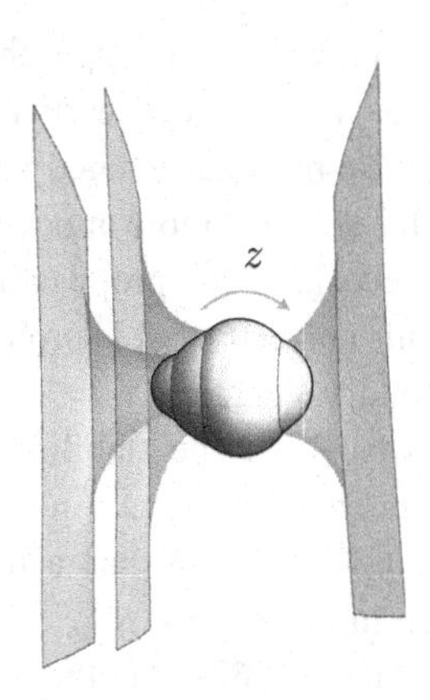

Figure 11.3 The space M_3 and its brane origin.

origin (11.2.17). The field theories described by these brane intersections are known explicitly [587, 588]; for a summary, see [97, sec. 2.1]. They are nonrenormalizable, and are believed to be completed at high energies by $d = 6$ SCFTs. Checks of this duality were made in [97, 589]. The solutions with O6s, D6s, and O8s also pass these tests [590, 591].

Consistent truncations

For every AdS_7 solution, there exists a consistent truncation to a $d = 7$ supergravity. For M-theory, this is [592], the so-called *maximal* gauged supergravity, with $\mathcal{N} = 2$ supersymmetry [593]. The space $\mathcal{M}_{\rm fin}$ of internal metrics consists of ellipsoids whose topology is still S^4; the corresponding space of scalars is

$$\mathrm{SL}(5,\mathbb{R})/\mathrm{SO}(5)\,. \tag{11.2.18}$$

There are also nonabelian vector fields with gauge group SO(5), five three-form fields, four gravitinos, and four spin-1/2 fields.

For the IIA solutions, the consistent truncation is to a *minimal* supergravity [594], whose fields are the metric $g_{\mu\nu}$; an SU(2) gauge field A^i_μ; a two-form potential $b_{\mu\nu}$; a scalar X; gravitini ψ^A_μ and dilatini λ^A, both SU(2) doublets. The uplift of the bosonic fields is [544, 595]

$$\begin{aligned}
&\frac{1}{\pi\sqrt{2}}\mathrm{d}s^2 = \frac{120X^{-1/2}}{V_0}\sqrt{-\frac{\alpha}{\ddot\alpha}}ds_7^2 + X^{5/2}\sqrt{-\frac{\ddot\alpha}{\alpha}}\left(\mathrm{d}z^2 + \frac{\alpha^2}{\dot\alpha^2 - 2X^5\alpha\ddot\alpha}Ds^2_{S^2}\right),\\
&V_0 = X^{-8}(8X^{10} + 8X^5 - 1)\,, \qquad e^{\phi} = 162\cdot 2^{1/4}\pi^{5/2}X^{5/4}\frac{(-\alpha/\ddot\alpha)^{3/4}}{\sqrt{\dot\alpha^2 - 2X^5\alpha\ddot\alpha}}\,,\\
&B = \pi\left(-z + \frac{\alpha\dot\alpha}{\dot\alpha^2 - 2X^5\alpha\ddot\alpha}\right)\mathrm{vol}_{S^2}\,, \qquad F_2 = \left(\frac{\ddot\alpha}{162\pi^2} + \frac{\pi F_0\alpha\dot\alpha}{\dot\alpha^2 - 2X^5\alpha\ddot\alpha}\right)\mathrm{vol}_{S^2}\,.
\end{aligned} \tag{11.2.19}$$

V_0 is the potential for the scalar X in the $d = 7$ theory. $Ds^2 \equiv Dy^iDy^i$, $Dy_i \equiv \mathrm{d}y_i + \epsilon_{ijk}A_jy_k$; so (11.2.19) is a bundle metric in the sense of (4.2.35). The internal metric is still topologically S^3. The original AdS_7 solution (11.2.7) is recovered for $X = 1$, $A_i = 0$, $\mathrm{d}s_7^2 = \mathrm{d}s^2_{\mathrm{AdS}_7}$.

It might be surprising that a single theory describes all the infinitely many AdS_7 vacuum solutions we described earlier. It means that all vacua have a "universal sector" captured by minimal gauged supergravity. It should be possible to reduce from $d = 10$ by keeping more modes; the differences among the various vacua should then be revealed. One proposal is given in [596]; this reproduces vacua and BPS domain walls correctly, but is probably not a consistent truncation.

V_0 has a second vacuum at $X - 2^{-1/5}$; this lifts to a nonsupersymmetric AdS_7 solution of massive IIA. The stability and potential holographic interpretation of this solution was also considered in [562, 597, 598]. In the consistent truncation (11.2.19), the single scalar X has a tachyon here, but it is above the BF bound (11.1.12), which in this case is $m^2_{\rm BF} = -9/L^2_{\rm AdS}$. With more scalars, however, most solutions are unstable [562]. Another application of (11.2.19) is its $\mathrm{AdS}_d \times M_{7-d}$ solutions, which will be useful later along the lines of (11.1.9).

11.2.2 AdS_6

No AdS_6 in M-theory

We begin by showing that there are no supersymmetric AdS_6 solutions in M-theory [599, app. B]. The analysis is similar to that in Section 11.2.1, but a bit more complicated, and we will be less detailed.

The gamma matrix decomposition this time is $\Gamma_\mu = e^A \gamma_\mu \otimes 1$, $\Gamma_m = \gamma^{(6)} \otimes \gamma_m$. The spinor decomposition is

$$\epsilon = \zeta_+ \otimes \eta_+ + \zeta_- \otimes \eta_- + \zeta_+^c \otimes \eta_+^c + \zeta_-^c \otimes \eta_-^c \,, \tag{11.2.20}$$

in similar notation to (10.6.1). The two $d = 6$ spinors $\zeta_\pm$ are needed for the Killing spinor equations (8.2.14a), while the last two conjugate terms are needed to make ϵ Majorana. The field-strength needs to be purely internal; it is convenient to dualize it to a one-form $g \equiv *_5 G_4$. The supersymmetry conditions (10.6.13) then become

$$\left(D_m \pm \frac{1}{12} g_m \mp \frac{1}{6} \gamma_m{}^n g_n\right) \eta_\pm \,, \qquad \frac{\mu}{2} e^{-A} \eta_\mp \pm \frac{1}{2} dA \eta_\pm + \frac{1}{2} g \eta_\pm = 0 \,. \tag{11.2.21}$$

As usual, we omitted the slash symbols on the one-forms g and dA. We have used (3.2.29) (with $c = 1$) and (3.2.9b). Using the first in (11.2.21), we compute $g = \mp 6 d \log ||\eta_\pm|| \equiv d\lambda$. But now the first in (11.2.21) can be brought to the form $D_m \eta_\pm = 0$, upon rescaling the spinors $\eta_\pm \to e^{\mp \lambda/12} \eta_\pm$ and the metric $g_{mn} \to e^{\pm \frac{2}{3}\lambda} g_{mn}$, and using (4.3.69). So the internal space has restricted holonomy; in $d = 5$, this can be T^5 or $S^1 \times K3$. Moreover, for both $e^{\pm \frac{2}{3}\lambda} g_{mn}$ to have this property, λ should be constant, which in turn requires $g = 0$ and makes the whole system collapse to the flat-space case, by an analysis similar to Section 8.2.4.

Only one solution in massive IIA

In (9.2.11), we described a general strategy to find gravity solutions for Dq-branes inside Dp-branes [329]. This also works with Op-planes. For the particular case of D4-branes inside O8-planes with n_{D8} D8-branes on top, a near-horizon limit leads to the solution [600]

$$ds^2 = 3\pi \sqrt{\frac{2N}{n_0}} (\cos\theta)^{-1/3} \left(ds^2_{AdS_6} + \frac{4}{9} (d\theta^2 + \sin^2\theta ds^2_{S^3}) \right) \,, \tag{11.2.22}$$

$$e^\phi = 2^{3/4} \sqrt{\frac{\pi}{3}} N^{-1/4} n_0^{-3/4} (\cos\theta)^{-5/6} \,, \qquad G_4 = \frac{80}{9} N\pi (\cos\theta)^{1/3} \sin^3\theta d\theta \wedge \mathrm{vol}_{S^3} \,.$$

The internal metric is that of an S^4 in the usual join form (4.5.4). There is a singularity at $\theta = 0$, but it has a physical meaning: it is due to the back-reaction of the O8–D8 system, as in the AdS_7 solution (11.2.16). (One may also compare with the result of Exercise 1.3.3.) As usual, $n_0 = 2\pi F_0 \in \mathbb{Z}$; it is odd under the O8, and it jumps by one unit across a D8, so $n_0 = 8 - n_{D8}$. Finally, N is the flux quantum defined by (1.4.69). It originates from the N D4-branes on the O8–D8 prior to the near-horizon limit. It was shown in [601] that this solution is unique up to discrete quotients of the S^3, by direct analysis of the supersymmetry equations.

Given its brane origin, the holographic dual of this solution is believed [602] to be one of the theories in [603, 604]. As in Section 11.2.1, these are nonrenormalizable super-YM theories with an SCFT completion at high energies. A simple generalization of this duality where $S^3 \to S^3/\Gamma$ was considered in [605]. There is a consistent truncation [606] to a six-dimensional supergravity with $F(4)$ exceptional superalgebra [607].

IIB

We presented the pure spinor equations for Mink_6 solutions in Section 10.6.1. Its generalization to AdS_6 is not as straightforward as one might think. In the factorization (10.6.1), we only have positive-chirality spinors on Mink_6, but the Killing spinor equation (8.2.14a) needs both chiralities. This forces us to add to (10.6.1) a second copy of spinors with negative chiralities. This complicates the bispinor analysis, and results in a much uglier system than (10.6.4) [599, (2.8)]. The internal space is of the form

$$S^2 \hookrightarrow M_4 \to M_2\,, \tag{11.2.23}$$

with a local geometry completely determined in terms of two partial differential equations for two functions on M_2, which is, however, hard to solve (see also [608]). The round S^2 in (11.2.23) is expected from the R-symmetry algebra $\text{su}(2)_\text{R}$ (Section 11.1.2). Imposing its existence at the outset leads to a simpler system with a single Toda-like equation [469], which was later mapped to a Laplace equation [470].

Analytic solutions were found with a direct spinor analysis in terms of two holomorphic functions $\mathcal{A}_a$, $a = 1, 2$ [609]. The metric in Einstein frame is

$$\mathrm{d}s^2_\text{E} = \frac{|\partial_z G|\sqrt{1-R^2}}{\kappa\sqrt{R}}\left(\mathrm{d}s^2_{\text{AdS}_6} + \frac{4\kappa^4 R}{|\partial_z G|^2(1-R)^2}\mathrm{d}z\mathrm{d}\bar{z} + \frac{1}{9}\left(\frac{1-R}{1+R}\right)^2\mathrm{d}s^2_{S^2}\right), \tag{11.2.24a}$$

where

$$\begin{aligned} \kappa^2 &\equiv -|\partial_z\mathcal{A}_1|^2 + |\partial_z\mathcal{A}_2|^2\,, \qquad & G &\equiv |\mathcal{A}_1|^2 - |\mathcal{A}_2|^2 + 2\text{Re}\mathcal{B}\,, \\ \partial_z\mathcal{B} &\equiv \mathcal{A}_1\partial_z\mathcal{A}_2 - \mathcal{A}_2\partial_z\mathcal{A}_1\,, \qquad & R + R^{-1} &= 2 + \frac{6\kappa^2 G}{|\partial_z G|^2}\,. \end{aligned} \tag{11.2.24b}$$

The other fields are

$$\tau = \mathrm{i}\frac{\partial_z(\mathcal{A}_1+\mathcal{A}_2)\partial_{\bar{z}}G - R\partial_{\bar{z}}(\bar{\mathcal{A}}_1+\bar{\mathcal{A}}_2)\partial_z G}{\partial_z(-\mathcal{A}_1+\mathcal{A}_2)\partial_{\bar{z}}G - R\partial_{\bar{z}}(\bar{\mathcal{A}}_1-\bar{\mathcal{A}}_2)\partial_z G}\,, \tag{11.2.24c}$$

$$B_2 + \mathrm{i}C_2 = \frac{2}{3}\mathrm{i}\left(\left(\frac{1-R}{1+R}\right)^2\frac{\partial_z\mathcal{A}_1\partial_{\bar{z}}G + \partial_{\bar{z}}\bar{\mathcal{A}}_2\partial_z G}{3\kappa^2} - \bar{\mathcal{A}}_2 - \mathcal{A}_1\right)\text{vol}_{S^2}\,.$$

Solutions with five-branes

The functions κ^2 and G in (11.2.24b) need to be positive for (11.2.24a) to have the correct signature. One shows from the definition that $\partial_z\partial_{\bar{z}}G = -\kappa^2$; integrating this over M_2, we find a contradiction unless M_2 has a boundary ∂M_2. For M_4 to be compact, we need the S^2 to shrink at ∂M_2, so $R \to 1$; imposing the other coefficients in (11.2.24a) to stay finite, we find

$$\kappa^2 \to 0\,, \qquad G \to 0 \text{ on } \partial M_2\,. \tag{11.2.25}$$

We need to find $\mathcal{A}_a$ such that this holds, with $\kappa^2, G > 0$ on M_2. A strategy [610] is to consider the function $\lambda = \partial_z \mathcal{A}_1 / \partial_z \mathcal{A}_2$; this is meromorphic, but $\kappa^2 > 0$ implies $|\lambda| < 1$, so λ also cannot have poles in M_2, but only zeros s_I. From (11.2.25) it follows that $\lambda = 1$ on ∂M_2. When M_2 is a disk,[11] we can work on a double copy, the Riemann sphere $\mathbb{C} \cup \{z = \infty\}$, we take $M_2 = H_z = \{z \in \mathbb{C},\ z > 0\}$. Then a λ solving these constraints is obtained by taking

$$\lambda = \frac{\partial_z \mathcal{A}_1}{\partial_z \mathcal{A}_2} = \lambda_0^2 \prod_I \frac{z - s_I}{z - \bar{s}_I} \tag{11.2.26}$$

with λ_0 a constant phase. The $\partial_z \mathcal{A}_a$ themselves will have poles x_i on the Riemann sphere, but for a regular solution they can only be on the real axis ∂M_2. Also, regularity at infinity requires them to have a double zero at $z = \infty$, so the number L of x_i is two less than the number of s_I. This leads to

$$\mathcal{A}_a = \mathcal{A}_a^0 + \sum_{i=1}^{L} Z_a^i \log(z - x_i)\,, \tag{11.2.27}$$

where $Z_1^i = \omega_0 \lambda_0 \Pi_I (x_i - s_I) \Pi_{j \neq i} (x_i - x_j)^{-1}$, $Z_2^i = \omega_0 \bar{\lambda}_0 \Pi_I (x_i - \bar{s}_I) \Pi_{j \neq i} (x_i - x_j)^{-1}$, with $\mathcal{A}_a^0$, ω_0 constant. Demanding $G = 0$ as in (11.2.25) imposes constraints on these constants and on the positions of the zeros and poles.

This determines the solution completely as a supergravity solution. The geometry is regular everywhere except at the p_i on the boundary; there is no solution without such punctures, in agreement with [613]. Near such a point, defining $r e^{i\theta} = z - x_i$, the local behavior is $R \to 1 - \sqrt{12} |\log r|^{-1/2} \sin\theta$, $\kappa^2 \sim \kappa_0^2 \frac{\sin\theta}{r}$, $G \sim 2\kappa_0^2 r |\log r| \sin\theta$; (11.2.24a) becomes the local behavior of a (p_i, q_i)-five-brane, with

$$p_i + \mathrm{i} q_i = \frac{4}{3} Z_1^i\,. \tag{11.2.28}$$

For example, if $q_i = 0$, we approximately get the linear dilaton background (9.3.21b). (The transverse S^3 is reproduced as $\mathrm{d}\theta^2 + \sin^2\theta \mathrm{d}s^2_{S^2}$.) So these singularities have a physical interpretation, just like, for example, the D6 in Section 11.2.1. Demanding that the p_i and q_i be integers is a further constraint, which fixes all remaining free parameters; infinitely many solutions remain, but without free moduli. This is similar to AdS_7, and in agreement with a group-theoretical argument [582].

Other sources

The next natural source to consider is a (p, q)-seven-brane [614]. This has codimension two, so we don't expect it to be at the boundary ∂M_2, where the S^2 shrinks, but in the interior. (This is a bit similar to the way the D8s in the AdS_7 vacua were located at values of z inside the interval of definition.) The appropriate monodromy matrix is (9.4.18), acting as usual with (1.2.38). The action on the local AdS_6 solution (11.2.24) can be summarized as

$$\begin{pmatrix} \partial_z \mathcal{A}_1 \\ \partial_z \mathcal{A}_2 \end{pmatrix} \to \begin{pmatrix} 1 + \frac{\mathrm{i}}{2}(p^2 + q^2) & -\frac{\mathrm{i}}{2}(p - \mathrm{i}q)^2 \\ \frac{\mathrm{i}}{2}(p + \mathrm{i}q)^2 & 1 - \frac{\mathrm{i}}{2}(p^2 + q^2) \end{pmatrix} \begin{pmatrix} \partial_z \mathcal{A}_1 \\ \partial_z \mathcal{A}_2 \end{pmatrix}. \tag{11.2.29}$$

[11] The only case where M_2 has another topology so far is the T-dual of (11.2.22), worked out in [535], where M_2 is an annulus [611]; this, however, has smeared NS5-branes and D7/O7 on the two boundaries. For the nonabelian T-dual [612], M_2 is again a disk [611].

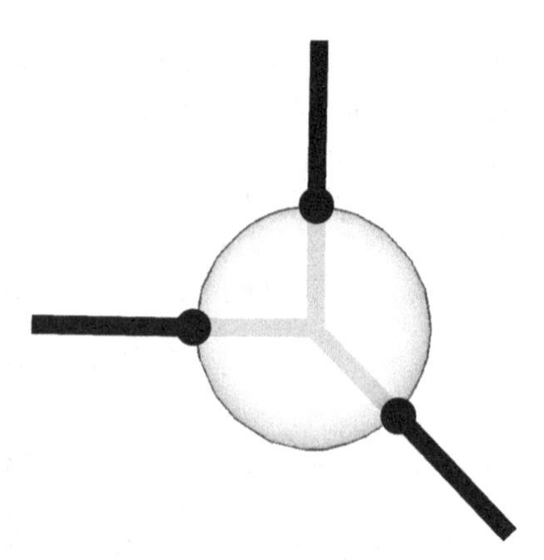

Figure 11.4 The sphere M_4, the punctures at the boundary of M_2, and its brane origin.

So we now need to find holomorphic functions $\mathcal{A}_a$ which again have positive κ^2 and G and satisfy (11.2.25), but also such that there is monodromy (11.2.29) around one or more points inside M_2. This can be achieved by starting with a "seed" solution $\partial_z \mathcal{A}_a^0$ without seven-branes, and adding a piece with a branch cut: for example, to introduce D7-branes at the points z_i in the upper half-plane, this reads

$$\partial_z \mathcal{A}_a = \partial_z \mathcal{A}_a^0 + \partial_z (\mathcal{A}_1^0 - \mathcal{A}_2^0) \frac{1}{4\pi} \sum_i \log \frac{z - z_i}{z - \bar{z}_i} . \qquad (11.2.30)$$

Around each z_i this realizes (11.2.29) for $p = 1, q = 0$. A variant of this idea, starting from solutions with a $\mathbb{Z}_2$ symmetry, realizes O7-planes [615].

Holography

The AdS_6 solutions with (p, q)-five-branes are thought to arise as a near-horizon limit from intersecting (p, q)-five-branes in flat space, as illustrated in Figure 11.4, similar to Figure 11.1. The sphere is M_4, projected to the disk M_2. The (p, q)-five-branes intersect at the origin, where they engineer a CFT_5. Their intersections with the sphere produce the punctures at the boundary of the disk.

The AdS_6 solutions with seven-branes would then be obtained by letting the five-branes end on seven-branes; this would produce funnel-shaped bound states, similar to those in Figure 11.3, that would make the puncture migrate to the interior of the disk. The CFTs engineered by these brane systems are again a high-energy completion of nonrenormalizable gauge theories. The combinatorial possibilities are even richer than those in $d = 6$. Various aspects of this holographic duality were studied, for example, in [616–618].

Consistent truncation

Similar to AdS_7, for every AdS_6 solution in IIB there is a consistent truncation [544, 619] to the same $F(4)$ supergravity [607] we mentioned after (11.2.22). Again we can interpret this as a "universal sector" common to all AdS_6 vacua. It should be possible to include more scalars and have more vacua in the same theory, along the lines of [596] for AdS_7, allowing us to see the vacua with seven-branes as the result of a partial symmetry breaking of those without.

11.2.3 AdS_5: M-theory and IIA

$\mathcal{N} = 1$ in M-theory

A classification of AdS_5 solutions with $\mathcal{N} = 1$ in M-theory can be obtained using G-structures [620]. It can be derived from scratch, or using $Mink_4$ results, similar to (11.2.5). Already for $Mink_4$ we needed an SU(3)-structure in (10.6.19); the AdS_5 analogue of (11.2.5) introduces a further one-form, so the structure becomes SU(2). This leads to a local expression for the internal ds^2_6 that depends on a Kähler metric in $d = 4$. The system simplifies if one demands that M_6 be complex: everything depends then on a single coordinate y. The differential equations can be solved explicitly, leading to several analytic solutions: for example,

$$ds^2_{11} = e^{2A} ds^2_{AdS_5} + \frac{1-x^2}{6e^{4A}} ds^2_{KE_4} + \frac{(1-x^2)^2}{p\, e^{4A}} dx^2 + \frac{L^2 p}{9(1-x^2)^2} D\psi^2 , \qquad (11.2.31)$$

where $e^{6A} = \frac{-12x(1-x^2)^2}{L^2 p'}$, $p = 1 - 6x^2 - 2cx^3 - 3x^4$; $D\psi \equiv d\psi - A$, $dA = 2J_{KE_4}$, with KE_4 a Kähler–Einstein four-manifold. The flux G is also analytically known [620, (4.3)].[12] There are also generalizations where KE_4 is replaced by a product of two Riemann surfaces. When one of the two is a T^2, this can be dualized to IIB, yielding $AdS_5 \times Y_{p,q}$, with $Y_{p,q}$ the Sasaki–Einstein we discussed at length in Section 7.4.5, which is how those metrics were first discovered.

$\mathcal{N} = 2$ in M-theory

A general classification of the $\mathcal{N} = 2$ AdS_5 solution was obtained [621] by imposing the R-symmetry $su(2)_R$ and partially exploiting the previous $\mathcal{N} = 1$ classification. The result is

$$ds^2_{11} = (\pi/2)^{2/3} e^{2A} \Big[4ds^2_{AdS_5} + y^2 e^{-6A} ds^2_{S^2} + \frac{4D\psi^2}{1 - y\partial_y D} - \frac{\partial_y D}{y} (dy^2 + e^D dx^i dx^i) \Big] , \qquad (11.2.32a)$$

$$G_4 = \pi \text{vol}_{S^2} \wedge \Big(D\psi \wedge d(y^3 e^{-6A}) + y(1 - y^2 e^{-4A}) dv - \frac{1}{2} \partial_y e^D dx^1 \wedge dx^2 \Big) , \qquad (11.2.32b)$$

the only assumption being $G_{x^1 x^2 y \psi} = 0$. Here $e^{6A} = \frac{y(y\partial_y D - 1)}{\partial_y D}$; $D\psi \equiv d\psi + v$, where $v = \frac{1}{2} *_2 d_2 D$ is along the two flat dimensions x^i; and $D = D(x^1, x^2, y)$ should satisfy the *Toda equation*

$$\partial_{x^i} \partial_{x^i} D + \partial^2_y e^D = 0 , \qquad (11.2.33)$$

away from sources (since it originates from the G_4 Bianchi identity).

[12] From now on, we will give the expressions for the form fields (and dilaton) only occasionally, when they are particularly nice or when we are describing a very general class.

Solutions from M5-branes wrapping cycles

The easiest solution to (11.2.33) is obtained by separation of variables:

$$e^D = d(y)e^{2f(x^1,x^2)} . \tag{11.2.34}$$

This works if $\partial_y^2 d = 2k$, and if f satisfies the Liouville equation (4.5.43), so that it is the conformal factor for a constant-curvature metric. By translation, we may take $d = k(y^2 - N^2)$. If $k = 1$, the admitted range for y will be $[N, \infty]$, which leads to a noncompact geometry; if $k = 0$, the solution is singular; so only $k = -1$ remains, where the range $y \in [-N, N]$ is compact. (Flux quantization imposes that the parameter $N \in \mathbb{Z}$.) In (11.2.32), $e^D dx^i dx^i$ then reconstructs hyperbolic space, which as usual we may quotient to produce a Riemann surface Σ_g with $g \geq 2$ (Section 5.3.5). All in all, (11.2.32) becomes the *Maldacena–Nuñez solution* [18, sec. 4.1]

$$ds^2_{11} = L^2(1+\cos^2\theta)^{1/3}\left(4ds^2_{\text{AdS}_5} + 8ds^2_{\Sigma_g} + 2d\theta^2 + \frac{2\cos^2\theta}{1+\cos^2\theta}ds^2_{S^2} + \frac{4\sin^2\theta}{1+\cos^2\theta}D\psi^2\right). \tag{11.2.35}$$

The coefficient of $ds^2_{S^2}$ goes to zero at $\theta = 0$ and stays finite at $\theta = \pi/2$, and the one of $d\psi$ has the opposite behavior. By (4.5.8), the four-dimensional space described by the directions θ, ψ, and the S^2 has the topology of an S^4. The covariant derivative $D\psi$ signals that ψ is fibered over the Riemann surface Σ_g. The internal isometry algebra is $\text{u}(2)_\text{R}$, as expected for an $\mathcal{N} = 2$ vacuum (Section 11.1.2), made of an su(2) acting on the S^2 and a u(1) acting on ψ.

This topology suggests an interpretation for (11.2.35) as the near-horizon around an M5, along the lines of (11.1.7). The M5 wraps a subspace $\Sigma_g \subset \text{CY}_6 = \mathbb{R}^2 \times$ K3. The normal bundle to such a subspace is trivial along the $\mathbb{R}^2$ dimensions, and isomorphic to K_{Σ_g} inside the K3; the logic is similar to (9.2.18). Already in (11.1.7) we have a transverse $\mathbb{R}$, so overall the normal bundle to the M5 is

$$N \cong \mathbb{R}^3 \oplus K_{\Sigma_g} . \tag{11.2.36}$$

When we take the near-horizon limit, the radial direction inside the fiber should assemble with the $\mathbb{R}^4$ parallel to the M5 to produce an AdS_5. The angular directions in the fiber give a topological S^4; the angle ψ inside K_{Σ_g} is nontrivially fibered on Σ_g, while the angular $S^2 \subset \mathbb{R}^3$ is unfibered. All this agrees with the structure of (11.2.35).

The solution (11.2.35) was originally obtained as an

$$\text{AdS}_5 \times \Sigma_g \tag{11.2.37}$$

vacuum solution in the consistent truncation of M-theory on S^4 to gauged supergravity in $d=7$, described around (11.2.18) [18, sec. 4.1]. As mentioned in Section 11.2.1, this theory is quite complicated, but there are several further truncations that are more manageable, such as one with two gauge fields and two scalars. The solution (11.2.37) has a field-strength for one of the gauge fields, which in the uplift becomes the connection for ψ in (11.2.35). We can also reproduce (11.2.35) by finding a solution (11.2.37) in *minimal* $d = 7$ gauged supergravity, lift it to IIA using (11.2.19), and then lift it to IIA.

It is possible to find similar $\mathcal{N} = 1$ solutions. The aim is now to describe an M5 as in (11.1.7), but with a proper CY_6; the normal bundle now only has one trivial direction, the other four forming the fiber of $O(p) \oplus O(q)$. The adjunction formula logic that led to (9.2.18) and (11.2.36) now gives $p + q = 2g - 2$. Again in the near-horizon limit we expect an internal space with topology $S^4 \hookrightarrow M_6 \to \Sigma_g$, but with two angles ψ_a that are nontrivially fibered rather than just one as in (11.2.35). This solution was first found for $p = q$ [18, sec. 4.2], and in [622] for general p, q, again using a clever truncation of maximal gauged supergravity in $d = 7$.

An important generalization arises when we add M5-branes that extend along AdS_5 and the circle S^1_ψ. One could check that these are BPS by a probe analysis along the lines of Section 9.2, but it is also interesting to work out the back-reaction they create when many of them are added. This can be done explicitly in a neighborhood of the point $p \in \Sigma$ where the M5s are located, called *puncture* [623]. The resulting solutions are still $\mathcal{N} = 2$ and thus still fall in the classification (11.2.35), but now the right-hand side of (11.2.33) has a delta function localized at p, due to the extra M5 source. The relevant solution is no longer obtained by separation of variables as in (11.2.34). The general global solution with several p on a Riemann surface with no isometries is not known; some aspects can be captured by solving (11.2.33) with a source in gauged supergravity [624]. For the $\mathcal{N} = 1$ case, a richer array of possible solutions should be available, so far only partially explored [625].

IIA

A classification similar to (11.2.32) can be achieved in IIA [590, 626]; we will focus on the $F_0 \neq 0$, since the $F_0 = 0$ case can be lifted to M-theory. The metric for the general solution reads[13]

$$ds^2_{10} = \left(-\frac{3 + 2s_1\partial_{s_1}D_1}{9\partial_{s_2}D_2 + 6s_1\partial D}\right)^{1/2}\left(ds^2_{AdS_5} + \frac{3\partial_{s_2}D_2 + 2s_1\partial D}{48s_1}\left(s_1^{3/2}e^{D_1}dx^i dx^i\right.\right.$$
$$\left.\left. - \frac{8D\psi^2}{\partial D(3 + 2s_1\partial_{s_1}D_1)} - \frac{\partial_{s_i}D_j ds^i ds^j}{s_1\partial D}\right)\right). \quad (11.2.38)$$

Here the internal coordinates are ψ, and s_i, x^i, both with $i = 1, 2$; we defined $\partial D \equiv \det_{(ij)}\partial_{s_i}D_j$; $D\psi \equiv d\psi - \frac{1}{2}\epsilon_{ij}\partial_{x_i}D_1 dx^j$; and finally, D_i depend on all except ψ, and satisfy

$$\partial_{x^i}\partial_{x^i}D_1 = \partial_{s_1}(s_1^{5/2}e^{D_1}\partial D) + \frac{F_0 e^{D_1}}{48\sqrt{2}}(3 + 2s_1\partial_{s_1}D_1), \quad (11.2.39)$$
$$\partial_{x^i}\partial_{x^i}(\partial_{s_2}D_2) = \partial^2_{s_2}(s_1^{5/2}e^{D_1}\partial D) + \frac{F_0\sqrt{s_1}e^{D_1}}{72\sqrt{2}}\partial_{s_1}\left(s_1^{3/2}e^{D_1}(3s_1^{-1}\partial_{s_2}D_2 + \partial D)\right).$$

In spite of the complexity of these equations, several solutions can be obtained by separation of variables. A large class is in one-to-one correspondence with the AdS_7 solutions in massive IIA of Section 11.2.1 [626]. These can also be obtained similarly to (11.2.35), by finding a vacuum solution of the form (11.2.37) in minimal $d = 7$ gauged supergravity:

$$ds^2_7 = \frac{19}{30}\left(ds^2_{AdS_5} + \frac{4}{5}ds^2_{\Sigma_2}\right), \qquad X^5 = \frac{3}{4}, \quad (11.2.40)$$

[13] With respect to [590], notation has been changed as $s \to s_1$, $u \to s_2$, $\tilde{D}_s \to D_1$, $D_u \to D_2$.

and lifting it to massive IIA with (11.2.19).[14] More general solutions also include D4-brane sources; these are similar to the punctures described earlier, but remarkably they can be described globally [590]. Finally, one can obtain $\mathcal{N} = 2$ solutions by reducing the M-theory solutions with punctures [627, 628].

Holography

When an AdS_5 vacuum originates from wrapped M5-branes as in (11.1.7), the dual CFT is obtained by putting the M5 CFT on a spacetime $\mathbb{R}^4 \times \Sigma$, and performing a certain twist on Σ [18], for reasons similar to those explained after (9.2.18). This breaks conformal invariance, but at long distances, Σ appears small, and we should have an effective four-dimensional description. The existence of the AdS_5 vacuum suggests that this $d = 4$ theory is itself a CFT.

The $\mathcal{N} = 2$ CFT dual to (11.2.35) has a famous description in terms of a *trinion* T_N theory [623, 629]. This is a non-Lagrangian theory (for $N > 2$) with $\mathrm{SU}(N)^3$ flavor symmetry and $\sim N^3$ degrees of freedom. We write the Riemann surface Σ_g as a union of $2g - 2$ copies of a sphere with three holes (using footnote 12, which precedes (4.2.72) in Chapter 4). We associate a copy of T_N to each sphere; the gluing of two holes corresponds to gauging two copies of $\mathrm{SU}(N)$ with an $\mathcal{N} = 2$ super-YM theory. The number of gauge groups needed is $3g - 3$, and their values of τ_{YM} correspond to the complex moduli of Σ_g (5.3.73). The solutions with punctures [623] are dual to theories where some of the $\mathrm{SU}(N)$ symmetries are left ungauged. For the $\mathcal{N} = 1$ analogues of (11.2.35), some of the gauge groups are copies of $\mathcal{N} = 1$ super-YM instead [622]. Such a description has not been developed for the massive IIA solutions (11.2.40); presumably one needs a generalization of the T_N theory.

The holographic dual to the $\mathcal{N} = 1$ class in M-theory does not seem to have a similar brane origin, except for $M_4 = S^2 \times \Sigma_g$ [630]. The case that dualizes to $\mathrm{AdS}_5 \times Y_{p,q}$ will be discussed in the next section.

11.2.4 AdS_5 in IIB

Sasaki–Einstein solutions

In IIB, we know already from Section 11.1.1 that we should be able to generalize (11.1.3) and (11.1.4) by replacing the S^5 with a Sasaki–Einstein manifold (see also Exercise 10.4.4). The spinor decomposition is similar to (11.2.4), in that it requires an auxiliary spinor v_+. besides the external and internal ones. With a Freund–Rubin Ansatz $F = f(\mathrm{vol}_{\mathrm{AdS}_5} + \mathrm{vol}_{M_5})$, the internal gravitino equations are of the form $\nabla_m \chi_a + \epsilon_{ab} \frac{f}{8 g_s} \gamma_m \chi_b = 0$; by taking $\chi_2 = \pm i \chi_1$, they reduce to the usual Killing spinor equations (7.3.8), which as we know admit solutions on a Sasaki–Einstein manifold. This eventually gives

$$\mathrm{d}s^2 = L^2 \left(\mathrm{d}s^2_{\mathrm{AdS}_5} + \mathrm{d}s^2_{\mathrm{SE}_5} \right) , \qquad F_5 = 4 L^4 g_s^{-1} \left(\mathrm{vol}_{\mathrm{AdS}_5} + \mathrm{vol}_{\mathrm{SE}_5} \right) , \qquad (11.2.41)$$

where $L^4 = 4\pi^4 g_s N / \mathrm{Vol}(\mathrm{SE}_5)$. Supersymmetry is $\mathcal{N} = 2$; the $\mathrm{u}(1)_{\mathrm{R}}$ is the Reeb vector ξ.[15] This is a vast class, because of the abundance of Sasaki–Einstein manifolds (Section 7.4).

[14] In this case, historically the consistent truncation was first found inspired by the AdS_5 solutions and by similar AdS_4 solutions we will see in Section 11.3.3.

[15] An interesting variant is a quotient of $\mathrm{AdS}_5 \times S^5$ by an element of $\mathrm{SL}(2, \mathbb{Z})$; this can realize $\mathcal{N} = 3$ supersymmetry [391].

Holography

From the point of view of supergravity, (11.2.41) are all of the same Freund–Rubin type. Still, the solutions (11.2.41) have been an important testing ground for developing the AdS/CFT correspondence; many techniques have been developed to find their holographic duals. Orbifolds can be understood using D3-brane physics [631]. The dual of $\mathrm{AdS}_5 \times T_{1,1}$ was another particularly important early example [527, 632]: it consists of two $\mathrm{SU}(N)$ gauge groups, with four bifundamental hypermultiplets and a quartic superpotential. For the toric case, there is a combinatorial method using dimer models [633, 634]. For more general SE_5, a procedure has been proposed in [635] using a mathematical device called noncommutative crepant resolutions, partly based on the derived category ideas presented in Section 9.2.3. This, however, does not seem to cover all possible Sasaki–Einstein manifolds [636].

These CFT duals have $\mathcal{N} = 1$ supersymmetry. (Being superconformal, they have eight rather than four supercharges, which map to the $\mathcal{N} = 2$ of (11.2.41).) Their abelian superpotential W_{ab} vanishes. For example, $\mathcal{N} = 4$ super-YM (the dual to $\mathrm{AdS}_5 \times S^5$) has $W = \mathrm{Tr}(Z_1[Z_2, Z_3])$; the associated potential V (via the rigid limit of (8.4.5)) is just the bosonic potential we saw for a D-brane stack on (1.3.57). In the abelian case, $W_{\mathrm{ab}} = 0$ because of cyclicity of the trace. This reflects the fact that an individual D3-brane can be removed from a stack without breaking supersymmetry, if it is kept parallel. More generally, if from a D3-stack at a singularity $C(\mathrm{SE}_5)$ we remove a single D3 and move it somewhere else along the cone, supersymmetry is unbroken. The dual to this is the statement that $W_{\mathrm{ab}} = 0$.

This property is not necessary for a CFT, and one expects to be able to relax it. Indeed, for $\mathcal{N} = 4$ super-YM, [637] showed that the superpotential can be deformed to

$$W = W_{\mathcal{N}=4} + \mathrm{Tr}\left(\alpha \sum_a Z_a^3 + \beta Z_1\{Z_2, Z_3\}\right) \tag{11.2.42}$$

while keeping the model superconformal, albeit with only $\mathcal{N} = 1$. Now $W_{\mathrm{ab}} \neq 0$. Similar deformations can be found for models dual to other $\mathrm{AdS}_5 \times \mathrm{SE}_5$ [638]. All this suggests the existence of AdS deformations of (11.2.41), where D3-branes are no longer BPS everywhere.

Solution-generating techniques

One such vacuum has been obtained using a solution-generating technique [639]. This can be applied whenever there is a $\mathrm{U}(1)^2$ of symmetries. First one T-dualizes along one of the $\mathrm{U}(1)$; then on the dual $\tilde{T}^2$, with coordinates x, $\tilde{y}$, one changes coordinates $x \to x + \gamma\tilde{y}$; then one T-dualizes back. A shortcut is to act with a Möbius transformation

$$\rho \to \frac{\rho}{\gamma\rho + 1} \,. \tag{11.2.43}$$

on the ρ of Exercise 4.2.8. As we saw there, the two $\mathrm{SL}(2,\mathbb{Z})$ actions in a T^2 compactification are exchanged by T-duality along a single direction. So for $\gamma \in \mathbb{Z}$, (11.2.43) is the T-dual of a modular transformation; for more general γ, this is not an equivalent solution but still a solution of the equations of motion. To apply this trick to $\mathrm{AdS}_5 \times S^5$, one rewrites the spherical metric as the "multiple join" (4.5.42), which makes a $\mathrm{U}(1)^3$ isometry manifest; the diagonal $\sum_{i=1}^3 \partial_{\phi_i}$ is the R-symmetry and is best left untouched in order not to break supersymmetry completely, so one applies

(11.2.43) to the remaining $U(1)^2$. The resulting solution is still regular [639]. This was proposed to be dual to the β deformation in (11.2.42).

A simple test of this proposed duality is that in these new solutions a D3-brane is no longer BPS everywhere on M_5. For this, it is convenient to write the AdS_5 vacua as $Mink_4$ solutions, as in (11.2.5), and to apply the pure spinor equations (10.3.20), now with $e^{2A} \to r^2 e^{2A}$ and an M_6 metric of the form

$$ds^2_{M_6} = e^{2A}\frac{dr^2}{r^2} + ds^2_{M_5}\,. \tag{11.2.44}$$

We can now use the formalism of Section 9.2.2. The conditions (9.2.26b)–(9.2.26c) for brane supersymmetry were interpreted there as an F-term and a D-term respectively. Specializing to $k = 0$, the one-form part $(\Phi_-)_1 = 0$ is the F-term equation. Because of (10.3.24a), Φ_- is closed, so locally $\Phi_- = dw$ defines a function w that is naturally identified with the abelian superpotential [640]. Indeed, $W_{ab} \neq 0$ in the β-deformation of (11.2.42). The solution (11.2.43) arises as a D3 back-reaction on a preexisting space that is generalized-Kähler [441], similar to the origin of (11.2.41) from a conical Calabi–Yau.

The deformations (11.2.42) with $\alpha \neq 0$ have not been found yet, except for an approximate solution in a small-flux approximation [641].

Classification

A general characterization of AdS_5 solutions has been obtained using G-structures [429] and then recast in the language of pure forms and generalized complex structures [642]. The two internal spinors χ_a have to be taken linearly independent, to avoid falling back to (11.2.41). One spinor in $d = 5$ defines an SU(2)-structure, as we saw, for example, in Section 7.4; two spinors define an identity structure (Exercise 5.1.8), because the intersection of two SU(2) subgroups in SO(5) is the identity. A more concrete way of saying this is that the five one-forms $\chi_a^\dagger \gamma_m \chi_b$ and $(\chi_1^c)^\dagger \gamma_m \chi_2$ are independent, and define a vielbein. So a general solution different from $AdS_5 \times SE_5$ has an identity structure. Its internal metric is then completely determined, subject to solving a system of partial differential equations for this vielbein. The case with $F_5 = 0$ was analyzed separately in [643]; many local solutions were obtained using nonabelian T-duality [644], but as usual for that technique they are noncompact or singular (Section 11.1.2).

The equations are most suggestive when written in terms of generalized complex structures [642], again using (11.2.44). The presence of the radial direction plays a similar role as for conical manifolds. We can define $\xi \equiv \mathcal{J}_- r\partial_r$, $\eta = \mathcal{J}_- d\log r$, with $\mathcal{J}_-$ the generalized complex structure associated to Φ_-; these are similar to the definitions of ξ and η for Sasaki–Einstein manifolds in (7.4.2), and as in that case they induce a fibration $M_5 \to M_4$. (This may in fact make sense only locally, as we know from the existence of irregular Sasaki–Einstein manifolds.) The pure forms can now be decomposed as

$$\begin{aligned} e^{2A-\phi}e^{-B}\wedge\Phi_+ &= r(1 + ie^{2A}d\log r\cdot\eta)\cdot e^{-b_2}\wedge\tilde{\Phi}_+\,,\\ e^{2A-\phi}e^{-B}\wedge\Phi_+ &= re^{-3i\psi}(d\log r - i\eta)\cdot e^{-b_2}\wedge\tilde{\Phi}_-\,, \end{aligned} \tag{11.2.45}$$

which generalizes (7.4.4). Here $\tilde{\Phi}_\pm$ are pure forms on M_4, $\cdot$ is the action of a section of $T \oplus T^*$ as in (10.5.9b), and b_2 is a closed two-form. Now the pure spinor equations give nice-looking geometrical conditions on M_4:

$$\begin{aligned} d\tilde{\Phi}_- &= -3i\tilde{\eta}\cdot\tilde{\Phi}_-\,, & d(e^{A}\tilde{\eta}\cdot \mathrm{Im}\tilde{\Phi}_+) &= -2e^{-A}\mathrm{Re}\tilde{\Phi}_+\,, \\ d(e^{A}\mathrm{Im}\tilde{\Phi}_+) &= 0\,, & \mathcal{J}_-\cdot d(e^{-A}\tilde{\eta}\cdot \mathrm{Re}\tilde{\Phi}_+) &= 0\,. \end{aligned} \tag{11.2.46}$$

$\tilde{\eta} \equiv \eta - d\psi + \iota_\eta b_2$, with ι_η denoting contraction by the vector part of η. The first equation is a generalization of (7.4.27b); by (10.5.108), it implies that the $\tilde{\mathcal{J}}_-$ associated to $\tilde{\Phi}_-$ is a generalized complex structure. A more surprising consequence uncovered in [642] is that the cone over M_5 is a symplectic manifold. These intriguing geometrical conditions can be thought of as a generalized complex analogue of Sasaki–Einstein geometry, and deserve to be studied further.

Consistent truncations

$AdS_5 \times S^5$ has a nonlinear reduction to a $d = 5$ gauged supergravity theory that realizes all of its maximal $\mathcal{N} = 8$ supersymmetry [593, 645]. This was shown to be a consistent truncation in stages, working with larger and larger subsectors [646–648], with some details getting clarified only recently [537, 649]. These subsectors have brought to light several vacua over the years, culminating with partially computer-aided searches that identified 32 vacua [650, 651]. So far, only two are supersymmetric: the original $\mathcal{N} = 8$ vacuum, and an $\mathcal{N} = 2$ vacuum [646]. Its uplift has a metric [652]:

$$\begin{aligned} ds^2_{10} &= L^2\sqrt{1+x^2}\Big(ds^2_{AdS_5} + \frac{2}{3}\frac{dx^2}{1-x^2} + \frac{1-x^2}{3(1+x^2)}ds^2_{S^2} \\ &\qquad + \frac{8x^2(1-x^2)}{3(1+x^2)^2}D\psi^2 + \frac{4}{9}\Big(d\alpha + \frac{1-x^2}{1+x^2}D\psi\Big)^2\Big)\,; \\ G_3 &= g e^{2i(\alpha+\psi)} h_{S^2}\wedge\left(\frac{2ix(1-x^2)}{1+x^2}d\alpha\wedge D\psi - d\alpha\wedge dx + 2\frac{1-x^2}{(1+x^2)^2}dx\wedge D\psi\right) \end{aligned} \tag{11.2.47}$$

in our usual notation as in (1.2.41), (4.2.51), and (5.3.12); moreover, $g = (2/3)^{3/2}L^2$. We also have $F_5 = 8L^2(\mathrm{vol}_{AdS_5} + \mathrm{vol}_{S^5})$, and $F_1 = 0$; the dilaton is constant. (This solution is holographically dual to a fixed point obtained by deforming $\mathcal{N} = 4$ super-YM by a mass operator [637].)

In agreement with a conjecture in Section 11.1.2, it can be proven that any AdS_5 solution has a consistent truncation to $d = 5$ minimal gauged supergravity, a theory whose bosonic fields are the metric and a gauge field [454].

11.2.5 Summary

In $AdS_{6,7}$, we claimed a complete classification, which we summarize in Table 11.2. The notation

$$F \curvearrowright B \quad \text{is shorthand for} \quad F \hookrightarrow M \to B\,: \tag{11.2.48}$$

Table 11.2. Supersymmetric $AdS_{6,7}$ solutions.

	M-theory	IIA	IIB
AdS_7	S^4/Γ	∞: $S^2 \curvearrowright I$	$\emptyset$
AdS_6	$\emptyset$	one	∞: $S^2 \curvearrowright D$

Table 11.3. Some notable examples of supersymmetric AdS_5 solutions.

M-theory	IIA	IIB
Top. $S^4 \curvearrowright \Sigma_g$ (11.2.35)	Top. $S^3 \curvearrowright \Sigma_g$ (11.2.40)	Sasaki–Einstein$_5$
+ punctures	+ punctures	Def. SE_5 (11.2.43)
Top. $S^2 \curvearrowright KE_4$ (11.2.31), $\Sigma_g \times \Sigma_{g'}$		Top. S^5 (11.2.47)

the internal space M is obtained by fibering F over B. In AdS_5, our understanding is less complete: in M-theory for $\mathcal{N} = 2$ and IIA $\mathcal{N} = 1$, we have a local expression for the solution, (11.2.32) and (11.2.38), with respectively one and two partial differential equations to solve; in other cases, we have several notable classes, which we list in Table 11.3. "Top." and "Def." stand for topological and deformed, respectively, the difference being that in the latter case there is a family connecting to the undeformed case.

Exercise 11.2.1 Consider the $AdS_7 \times S^4$ solution (11.2.2). Write $ds^2_{S^4} = da^2 + \sin^2 a ds^2_{S^3}$ (as in (4.5.4) for $d = 4$). Now write the S^3 metric as in (4.2.51), and reduce to IIA along the ψ direction, recalling (1.4.5). Find a coordinate change that reduces this to the solution obtained using (11.2.14) in (11.2.7).

Exercise 11.2.2 Write an AdS_7 solution whose only source is a single D8-brane.

Exercise 11.2.3 Write an analogue of (11.2.15) with D6/anti-D6 branes at both $z = 0$ and $z = N$.

Exercise 11.2.4 In the solution (11.2.15), how is it possible for D6-branes to be present without any other source? Why doesn't the argument in (10.4.14) apply?

Exercise 11.2.5 Check that (11.2.35) is smooth. (The crucial points are $\theta = 0$ and $\pi/2$.)

Exercise 11.2.6 Check that (11.2.47) is smooth. (Hint: positivity of the metric requires $y \in [0, 1]$; expand around these two points.) Check that it has the topology of an S^5. (Use the ideas in Section 4.5.1.)

11.3 AdS_4

11.3.1 M-theory

Cones with special holonomy

In M-theory, the most famous solution is $AdS_4 \times S^7$, the near-horizon limit of a stack of N M2-branes, anticipated in (11.1.5); as we mentioned there, it can also be generalized to other manifolds with Killing spinors. The quickest way to see this is directly from the supersymmetry transformations (10.6.13), with a decomposition spinor Ansatz $\epsilon = \zeta_I \otimes \eta_I$. If we take the internal η_I to be Killing spinors, the computation is very similar to that leading to (11.2.2), now with the role of external and internal dimensions exchanged. This leads to

$$ds^2_{11} = r_0^2\left(\frac{1}{4}ds^2_{\text{AdS}_4} + ds^2_{M_7}\right), \qquad G_4 = \frac{3}{8}r_0^3\text{vol}_{\text{AdS}_4}, \tag{11.3.1}$$

where $r_0^6 = \frac{(2\pi)^6 N l_{\text{P}11}^6}{6\text{Vol}(M_7)}$. The number of allowed independent ζ_I is the same as the number of independent η_I, which in turn equals the number of covariantly constant spinors on $C(M_7)$ (Section 7.3). Supersymmetry fixes the sign in the Killing spinor equation, which is related by (7.3.6) to chirality on $C(M_7)$. We have several cases [653]:

- For $M_7 = S^7$, there are eight Killing spinors η_I, which come as in Section 7.3.3 from the constant spinors on $\mathbb{R}^8 = C(S^7)$. Just like for AdS$_7 \times S^4$ and AdS$_5 \times S^5$, supersymmetry gets restored in the near-horizon limit to the maximal amount, 32 supercharges.
- If $C(M_7)$ has Sp(2) holonomy, it is hyper-Kähler, and by (5.1.71) it has three covariantly constant spinors; M_7 is a 3-Sasaki and has three Killing η_I. So AdS$_4 \times M_7$ has $\mathcal{N} = 3$ supersymmetry.
- If $C(M_7)$ has SU(4) holonomy, it is a Calabi–Yau fourfold, with one Weyl (or two MW) covariantly constant spinor η. M_7 is a Sasaki–Einstein, with two η_I. So AdS$_4 \times M_7$ has $\mathcal{N} = 2$ supersymmetry.
- If $C(M_7)$ has Spin(7) holonomy, it has only one Killing spinor. M_7 is a weak-G_2 manifold; AdS$_4 \times M_7$ has only $\mathcal{N} = 1$ supersymmetry.[16]

Supersymmetry $3 < \mathcal{N} < 8$ may be obtained on quotients $M_7 = S^7/\Gamma$, along the lines of Section 7.4.2. A famous example is the quotient where $\Gamma = \mathbb{Z}_k$ acts by

$$z^i \to e^{2\pi i/k} z^i. \tag{11.3.2}$$

We can compute the amount of supersymmetry as in (7.4.15), by associating $|++++\rangle \to dz^1 \wedge dz^2 \wedge dz^3 \wedge dz^4$, $|-+++\rangle \to d\bar{z}^1 \wedge dz^2 \wedge dz^3 \wedge dz^4$, and so on. The invariant spinors are those that have an equal number of + and −; there are six of these, so we obtain $\mathcal{N} = 6$ supersymmetry. When k is large, the S^1 direction spanned by the vector $\xi = 2\text{Re}(z^i \partial_{z^i})$ (the continuous version of (11.3.2)) is made very small by the quotient, and we are effectively reducing to IIA. According to (6.2.4) and (6.2.16), this reduces to $\mathbb{CP}^3$ with its Fubini–Study metric, resulting in a famous solution [655–657]:

$$\text{AdS}_4 \times \mathbb{CP}^3. \tag{11.3.3}$$

In the next two subsections, we will see various other solutions with $\mathbb{CP}^3$ internal space, but with different metrics and amounts of supersymmetry.

Holography

The holographic dual to (11.3.2), including its IIA limit (11.3.3), was found in [77] and is known as *ABJM model*; it is particularly important, as it gives an explicit Lagrangian for the CFT describing M2s at a singularity, and in particular the CFT for an M2 stack, for $k = 1$. It is a Chern–Simons theory with gauge group $\text{U}(N)\times\text{U}(N)$,

[16] With singularities, there are other possibilities. If M_6 is nearly Kähler, $\mathbb{R} \times C(M_6)$ has special holonomy; we can write it as a cone $C(M_7)$, where $ds^2_{M_7} = d\theta^2 + \sin^2\theta ds^2_{M_7}$ (Exercise 7.3.4) is called *sine-cone*. Now AdS$_4 \times M_7$ is again a solution (11.3.1), with two conical singularities $C(M_6)$ at $\theta = 0, \pi/2$, which should be allowed in string theory [654].

coupled to four bifundamental chiral multiplets, with the same superpotential used in the dual to $AdS_5 \times T_{1,1}$ (Section 11.2.4); the geometrical origin of this coincidence is that $\mathbb{CP}^3$ can be written as a $T_{1,1}$-fibration over an interval (Exercise 6.2.4).

For the more general (11.3.1), the duals are again Chern–Simons matter theories with bifundamentals, summarized by a quiver diagram. The formal relation to a $d = 4$ field theory persists, [211, 658–660]; the duals were studied for M_7 a 3-Sasaki [658, 661, 662] and a Sasaki–Einstein [211, 659].

Consistent truncations

Similar to $AdS_7 \times S^4$ and $AdS_5 \times S^5$, $AdS_4 \times S^7$ also has a consistent truncation [663] to a maximally supersymmetric gauged $d = 4$ supergravity [664]; the formula for how to lift the G_4 field-strength has been clarified only recently [665]. Several supersymmetric vacua are known:

- $\mathcal{N} = 2$ supersymmetry, $SU(3) \times U(1)_R$ isometry [666], uplifted in [667]
- $\mathcal{N} = 1$: one with isometry G_2 [666]; one with isometry $U(1)^2$ [668]; one with isometry $SO(3)$ [669, 670]

A computer search has also produced hundreds of nonsupersymmetric vacua [669].

A far simpler consistent truncation exists [454] from any vacuum $AdS_4 \times SE_7$ to minimal gauged supergravity in $d = 4$, with bosonic fields $g_{\mu\nu}$, A_μ and equations of motion:

$$R_{\mu\nu} = -3g_{\mu\nu} + \frac{1}{2}F_{\mu\rho}F_\nu{}^\rho - \frac{1}{8}g_{\mu\nu}F_{\rho\sigma}F^{\rho\sigma}\,, \qquad \mathrm{d} *_4 F_2 = 0\,. \tag{11.3.4}$$

The uplift is provided by

$$r_0^{-2}\mathrm{d}s^2_{11} = \frac{1}{4}\mathrm{d}s^2_4 + \left(\mathrm{d}\psi + \frac{1}{4}(A - w)\right)^2 + \mathrm{d}s^2_6\,, \qquad r_0^{-3}G_4 = \frac{3}{8}\mathrm{vol}_4 - \frac{1}{4}*_4F_2\wedge J\,, \tag{11.3.5}$$

where the two-form J and the Reeb vector $\xi = \partial_\psi$ are defined as in Section 7.4.1, and $\mathrm{d}s^2_6$ is defined by writing the Sasaki–Einstein metric as $\mathrm{d}s^2_{SE_7} = (\mathrm{d}\psi - \frac{1}{4}w)^2 + \mathrm{d}s^2_6$, as in (7.4.27c) (generalized to $d = 7$). As usual, consistency means that plugging (11.3.5) in the M-theory equations of motion (10.6.12), we get (11.3.4). For $\mathrm{d}s^2_4 = \mathrm{d}s^2_{AdS_4}$, $A_\mu = 0$, we recover (11.3.1) with $M_7 = SE_7$. A slightly more complicated extension also includes a volume mode [455].

Classifications

G-structures can be used to analyze the most general AdS_4 compactification in the usual way [507, 671]. The problem is reduced to a set of partial differential equations on an SU(3)-structure that generalize (10.6.20). For $\mathcal{N} = 2$ supersymmetry, the situation improves; a class of solutions was identified [237, 667, 672] by imposing an Ansatz that a certain spinor bilinear is a Killing vector. This yields the solutions

$$\frac{\mathrm{d}s^2_{11}}{L^{2/3}(1+2x)^{2/3}} - \mathrm{d}s^2_{AdS_4} + \frac{3(1-x)}{2(1+2x)}\mathrm{d}s^2_{KE_4} + \frac{1}{8x(1-x)}\Big(\mathrm{d}x^2 + \frac{4x^2(1-x)^2}{1+8x^2}D\psi^2 + \frac{x(1-x)(1+8x^2)}{2(1+2x)^2}\Big(\mathrm{d}\alpha + 2\frac{1-x}{1+8x^2}D\psi\Big)^2\Big). \tag{11.3.6}$$

Here $D\psi \equiv \mathrm{d}\psi - A$ with $\mathrm{d}A = j$, and (j, ω) define the SU(2)-structure on the Kähler–Einstein KE_4. For $KE_4 = \mathbb{CP}^2$, these become the aforementioned $\mathcal{N} = 2$ vacua of $\mathcal{N} = 8$ gauged supergravity [666, 667]; for other KE_4, it has a conical singularity CY at $x = 1$. Equation (11.3.6) bears some resemblance to the Sasaki–Einstein metrics (7.4.47), and it is found more or less in the same way, by reducing the system to a system of ordinary differential equations that then are solved analytically. Additional numerical solutions were uncovered in [237, 672]. The holographic dual CFTs are deformations by relevant operators of M2-branes on a $\mathbb{R} \times C(\mathrm{SE}_5)$ singularity.

Another known class of $\mathcal{N} = 2$ solutions is obtained [673] as an

$$\mathrm{AdS}_4 \times H_3/\Gamma \tag{11.3.7}$$

solution of maximal $d = 7$ gauged supergravity, with H_3 the hyperbolic space of Section 4.5.2, in a three-dimensional analogue of the quotient (5.3.68) for Riemann surfaces of $g \geq 2$. This can be lifted to $d = 11$, similarly to how (11.2.37) lifts to (11.2.35). As in that case, the holographic duals are obtained by compactifying the M5 CFT on H_3/Γ [674].

11.3.2 IIA: $\mathcal{N} = 1$ with SU(3)-structures

In IIA, we will focus on the $F_0 \neq 0$ case, since solutions with $F_0 = 0$ can be lifted to M-theory, which we already discussed.

Recall from Section 10.3.4 that there are three types of compatible pairs $\Phi_\pm$:

- SU(3)-structure (10.3.28). This is by far the simplest, and there are several explicit solutions. We will see that in this case supersymmetry cannot be enhanced to $\mathcal{N} > 1$, and that sources are admitted only if smeared.
- Orthogonal SU(2)-structure (10.3.31). The one-form part of (10.3.20b) receives no contribution from the left-hand side, and so we end up requiring $\mathrm{Re}v = 0$, which is impossible by the nondegeneracy condition mentioned following (10.3.32) [675, 676]. So this case is excluded.
- The generic case (10.3.37), defining an $\mathrm{SU}(3) \times \mathrm{SU}(3)$ structure on $T \oplus T^*$. Most solutions should be of this type, but in practice at present few examples with $\mathcal{N} = 1$ are known (Section 11.3.3). The situation improves drastically with extended supersymmetry (Section 11.3.4).

SU(3)-structure

For SU(3)-structures, $\Phi_+ = \mathrm{e}^{A+\mathrm{i}\theta}\mathrm{e}^{-\mathrm{i}J}$, $\Phi_- = \mathrm{e}^{A}\Omega$. We plug these in (10.3.20) and divide the equations by form degrees. The one-form part of (10.3.20a) set θ and $\mathrm{e}^{3A-\phi}$ to constants. For the RR fields, it is more convenient to use the alternative Dolbeault formulation (10.5.138), using from (10.5.34) that $\mathcal{J}\cdot = -J\wedge + J\cdot$. The zero-form is $F_0 = 5\mu\cos\theta \mathrm{e}^{-A}$. Since we assumed $F_0 \neq 0$, the Bianchi $\mathrm{d}F_0 = 0$ implies that A and ϕ are separately constant. We can then set $A = 0$ without loss of generality, by using the invariance (10.3.26); and we define $g_s = \mathrm{e}^{\phi}$ as usual. The results can be summarized as

$$\mathrm{d}J = -2\tilde{m}\mathrm{Re}\Omega\,, \qquad \mathrm{d}\Omega = \mathrm{i}\left(W_2^- \wedge J + \frac{4}{3}\tilde{m}J^2\right), \qquad H = -2m\mathrm{Re}\Omega\,; \tag{11.3.8}$$
$$g_s f_0 = 5m\,, \qquad g_s f_2 = W_2^- + \frac{1}{3}\tilde{m}J\,, \qquad g_s f_4 = \frac{3}{2}mJ^2\,, \qquad g_s f_6 = -\frac{1}{2}\tilde{m}J^3\,.$$

(This was found before the pure spinor equations [428].) We defined $m = \mu \cos\theta$, $\tilde{m} = \mu \sin\theta$. The primitive two-form $W_2^- \equiv \mathrm{Im} W_2$ is one of the intrinsic torsion coefficients defined back in (5.5.6). We also see from there that $W^1 = -(4i/3)\tilde{m}$, $W^2 = iW_-^2$, $W_3 = W_4 = W_5 = 0$.

We already commented on the Bianchi identity for $F_0 = f_0$; those of f_4, f_6 are also automatically satisfied. We are left with $\mathrm{d}f_2 - Hf_0 = \delta$: by (11.3.8),

$$\mathrm{d}W_2^- = \frac{2}{3}(\tilde{m}^2 - 15m^2)\mathrm{Re}\Omega - g_s\delta\,. \tag{11.3.9}$$

This is a purely geometrical condition on the SU(3)-structure.

If there are no sources, $\delta = 0$, we can derive a bound on the coefficient in (11.3.9):

$$\frac{2}{3}(\tilde{m}^2 - 15m^2)||\mathrm{Re}\Omega||^2 \overset{(4.1.107)}{=} \int_{M_6} *\mathrm{Re}\Omega \wedge \mathrm{d}W_2^- = \int_{M_6} \mathrm{Im}\Omega \wedge \mathrm{d}W_2^- \tag{11.3.10}$$
$$= \int_{M_6} \mathrm{dIm}\Omega \wedge W_2^- \overset{(11.3.8)}{=} \int_{M_6} J \wedge W_2^- \wedge W_2^- \overset{(5.2.24)}{=} ||W_2^-||^2\,.$$

So it follows that [428]

$$\tilde{m}^2 \geq 15m^2 \quad \Rightarrow \quad \tan\theta \leq \frac{1}{\sqrt{15}}\,. \tag{11.3.11}$$

No extended supersymmetry

Solutions of (11.3.8) with $F_0 \neq 0$ have exactly $\mathcal{N} = 1$ supersymmetry and not larger. Indeed, for $\mathcal{N} = 2$ we would need a second pair $\Phi_+ = e^{A+i\theta}e^{-iJ'}$, $\Phi_- = e^A\Omega'$ that also solves (11.3.8), but with the same physical fields. Writing $H = -\frac{2}{5}g_s f_0 \mathrm{Re}\Omega$, we see that $\mathrm{Re}\Omega' = \mathrm{Re}\Omega$. But we know from (10.5.56) that a decomposable three-form is determined by its real part; so in fact $\Omega' = \Omega$. Likewise, (11.3.8) gives $f_4 = \frac{3}{10}f_0 J^2$, which shows that $(J')^2 = J^2$. The methods of Section 10.5.2 now also imply that $J' = J$ (Exercise 10.5.3). So the hypothetical second pair coincides with the first, and the solution has $\mathcal{N} = 1$ supersymmetry.

Nearly-Kähler solutions

The easiest way to solve (11.3.9) is to take $\delta = 0$, and [677]

$$W_2^- = 0\,, \qquad \tilde{m} = \pm\sqrt{15}m\,. \tag{11.3.12}$$

The only remaining nonvanishing intrinsic torsion coefficient is W_1^-. We recall from Section 7.5.2 that this defines *nearly Kähler* geometry; in (7.5.19), we had $W^1 = 2$, but as we commented there, that value can be changed by rescaling J and Ω.

We gave a list of known nearly Kähler manifolds in Section 7.5.2. A summary is that there are four homogeneous examples (S^6, $S^3 \times S^3$, $\mathbb{CP}^3$, and the flag manifold $\mathbb{F}(1, 2; 3)$, defined in (4.4.53)); two nonhomogeneous ones (with topology S^6 and $S^3 \times S^3$ again); and finally twistor bundles $S^2 \hookrightarrow M_6 \to B_4$, with B_4 quaternionic-Kähler four-manifolds, which have orbifold singularities except for (7.5.26).

The particular case $\mathrm{AdS}_4 \times S^3 \times S^3$, whose SU(3)-structure we saw in (7.5.20), was rediscovered from the point of view of effective actions [678–680]; see the discussion in [681]. It is especially amenable to that approach because $S^3 \times S^3$ is a group. An effective description covering all nearly Kähler spaces was given in [453] (Section 10.3.6).

Given that a nearly Kähler manifold M_6 is defined such that the cone $C(M_6)$ has G_2 holonomy, one would imagine that (11.3.12) should originate as a near-horizon limit of some brane system on this cone. The presence of all fluxes in (11.3.8) means that this is not a Freund–Rubin solution, however; so such a potential origin cannot be as trivial as in (11.1.3). (A D2 origin has been suggested in a limit where the dilaton vanishes asymptotically [682].)

Homogeneous/twistor vacua

Vacua with $W_2 \neq 0$ have also been found [223, 683, 684]. They can be understood in terms of twistor bundles, or in terms of cosets.

The twistor construction [684] is a generalization of (7.5.25):

$$J = R^2\left(\frac{2}{\sigma}j_{\rm F} + \frac{1}{4}j_{\rm B}\right), \qquad \Omega = \frac{R^3}{\sigma}(\psi - {\rm i}\tilde{\psi}). \tag{11.3.13}$$

Here $j_{\rm F}$ and $j_{\rm B}$ are almost symplectic forms on the S^2 fiber and the B_4 base of the twistor bundle:

$$S^2 \hookrightarrow M_6 \to B_4. \tag{11.3.14}$$

By (7.5.24), the nonzero intrinsic torsion is

$$W_1^- = \frac{4}{3}\tilde{m} = \frac{2}{3R}(\sigma + 2), \qquad W_2^- = \frac{2}{3}R(\sigma - 1)\left(\frac{4}{\sigma}j_{\rm B} - j_{\rm F}\right). \tag{11.3.15}$$

Taking a further exterior differential, we obtain that (11.3.9) is satisfied:

$$dW_2^- = \frac{8}{3R^2}(\sigma - 1)^2{\rm Re}\Omega, \qquad m = -\frac{1}{2R}\sqrt{(2-\sigma)(\sigma - 2/5)} \tag{11.3.16}$$

The latter implies $\sigma \in [2/5, 2]$. Flux quantization discretizes the parameters of the solution, and leaves no free moduli. Curvature and string coupling can be made parametrically small [684].

For $\sigma = 1$, (11.3.13) becomes (7.5.25) and we reproduce the nearly Kähler vacua (11.3.12), as can also be seen from the expression of W_2^- in (11.3.15). From (11.3.8) and (11.3.16), we see that

$$F_0 = 0 \quad \Leftrightarrow \quad \sigma = 2 \text{ or } 2/5. \tag{11.3.17}$$

For these two particular values, we can lift the solutions to M-theory. The extra S^1 is Hopf-fibered as in (4.2.51) over the S^2 in (11.3.14). We obtain $S^3 \hookrightarrow M_7 \to B_4$, up to a $\mathbb{Z}_k$ acting on the S^3, where k is interpreted as the flux quantum of $(dC)_2$ in IIA.

- For $\sigma = 2$ this is a 3-Sasaki (7.5.13), which according to Section 11.3.1 has in fact $\mathcal{N} = 3$; notice that for $F_0 = 0$ the argument against extended supersymmetry that we saw earlier in this subsection does not apply. For $M_6 = \mathbb{CP}^3$, the metric is the Fubini–Study (6.2.16). The lift is the AdS$_4 \times S^7/\mathbb{Z}_k$ solution we considered in (11.3.1) and (11.3.2), with supersymmetry further enhanced to $\mathcal{N} = 6$.[17]
- For $\sigma = 2/5$, we reconstruct the weak G_2 (7.5.32) and (7.5.34). This has $\mathcal{N} = 1$ supersymmetry. For $M_6 = \mathbb{CP}^3$, M_7 is the squashed S^7 we mentioned in Section 7.5.3.

[17] The $\mathcal{N} = 6$ solution also admits various other deformations involving F_0. One breaks supersymmetry completely [40, 549] and is obtained by keeping the metric the Fubini–Study one but changing the fluxes; it is an example of a general strategy mentioned in Section 11.1.2. There are also $\mathcal{N} = 2$ and $\mathcal{N} = 3$ deformations, to be discussed in Section 11.3.4.

The aforementioned 3-Sasaki manifolds M_7 also give an alternative way of writing the SU(3)-structure (11.3.13), generalizing (7.5.22) [225, (5.17)]:

$$J = R^2\left(-\frac{2}{\sigma}\tilde{j}_1 + \eta_2 \wedge \eta_3\right), \qquad \Omega = \frac{2\mathrm{i}R^3}{\sigma}(\eta_2 - \mathrm{i}\eta_3) \wedge (\tilde{j}_2 + \mathrm{i}\tilde{j}_3), \qquad (11.3.18)$$

in terms of the forms in (7.5.4a). While this specific structure came from a 3-Sasaki, the trick of reversing the complex structure along one of the directions, discussed in (7.5.28), can be greatly generalized to toric manifolds [685]; it would be interesting to see if this can lead to other families of vacua.

As we mentioned, most compact quaternionic-Kähler manifolds B_4 have orbifold singularities, the exceptions being S^4 and $\mathbb{CP}^2$, for which the twistor bundles (7.5.26) are respectively $\mathbb{CP}^3$ and the flag manifold $\mathbb{F}(1,2;3)$. For these two examples, the solutions can also be written as cosets [223, 683]. For $S^2 \hookrightarrow \mathbb{CP}^3 \to S^4$, the relevant coset structure is $\mathbb{CP}^3 \cong \mathrm{Sp}(2)/(\mathrm{Sp}(1)\times\mathrm{U}(1))$, from (4.4.50). For $S^2 \hookrightarrow \mathbb{F}(1,2;3) \to \mathbb{CP}^2$, the coset is $\mathbb{F}(1,2;3) \cong \mathrm{SU}(3)/\mathrm{U}(1)^2$ from (4.4.53). In this second case, the coset structure covers a more general solution, with one more parameter than in the twistor description (11.3.13). With the structure constants of $\mathfrak{g} = \mathrm{su}(3)$ as in (4.4.62) and $\mathfrak{h} = \mathrm{u}(1) \oplus \mathrm{u}(1) = \mathrm{Span}\{\lambda^7, \lambda^8\}$, the SU(3)-structure is defined by

$$J = a\lambda^{12} - b\lambda^{34} + c\lambda^{56}, \qquad \Omega = \sqrt{abc}(\lambda^1 + \mathrm{i}\lambda^2)\wedge(\lambda^3 - \mathrm{i}\lambda^4)\wedge(\lambda^5 + \mathrm{i}\lambda^6). \quad (11.3.19)$$

These satisfy (11.3.8) and (11.3.9) with

$$m^2 = \frac{-5(a^2 + b^2 + c^2) + 6(ab + ac + bc)}{80abc}, \qquad \tilde{m} = +\frac{1}{4}\frac{a + b + c}{\sqrt{abc}}. \qquad (11.3.20)$$

The parameters are discretized by flux quantization, again leaving no moduli. For $a = b$, this solution becomes (11.3.13), calling $c = R^2/4$ and $a = b = R^2/4\sigma$. Reality of m imposes an inequality that describes the interior of an ellipse in the space of b/a and c/a. The boundary of the ellipse corresponds to massless solutions that lift in M-theory to the weak-G_2 spaces $N_{p,q} \cong \mathrm{SU}(3)/\mathrm{U}(1)$ in Section 7.5.3. The special case $a = b = R^2/8$, $c = R^2/4$ lifts to the 3-Sasaki $N_{1,1}$, and falls in the $\mathcal{N} = 3$ cases we discussed previously.

The holographic duals of these vacua can be found by changing the CS levels in the duals to the $F_0 = 0$ theories discussed in Section 11.3.1 (in particular, ABJM for $\mathrm{AdS}_4 \times \mathbb{CP}^3$) and requiring that the action preserve $\mathcal{N} = 1$ supersymmetry [550]. Finally, an effective gauged supergravity was given in [451].

Scale separation with smeared sources

The solutions we have seen so far don't have any sources. More generally, for $F_0 \neq 0$ we concluded in (11.3.8) that the dilaton ϕ and warping function A are constant; but near a source we would expect nontrivial $\mathrm{e}^A \sim h^{-1/4}$, $\mathrm{e}^\phi \sim g_s h^{(3-p)/4}$ from (1.3.59). So it looks very unlikely that a solution with localized sources can solve the SU(3)-structure system (11.3.8).

One possible way out is to consider smeared sources. For example, let us set

$$\theta = 0, \qquad F_2 = 0. \qquad (11.3.21)$$

This implies $\tilde{m} = 0$, $m = \mu$, and violates our bound (11.3.11) for sourceless solutions maximally. Now (11.3.8) becomes [686]

$$\begin{aligned} &\mathrm{d}J = 0\,, \qquad \mathrm{d}\Omega = 0\,, \qquad H = -2\mu\mathrm{Re}\Omega\,; \\ &g_s f_0 = 5\mu\,, \qquad f_2 = 0\,, \qquad g_s f_4 = \frac{3}{2}\mu J^2\,, \qquad f_6 = 0\,. \end{aligned} \tag{11.3.22}$$

So we see that M_6 is a Calabi–Yau. The Bianchi identity for F_2 now becomes

$$2\mu F_0\,\mathrm{Re}\Omega = \delta\,, \tag{11.3.23}$$

showing that the source is indeed smeared and not localized. It can be checked that it is an O6 by the test (10.3.52).

The ingredients in the solution (11.3.22) are the same as those in the vacuum (8.3.14), which was itself a rough way to understand the $\mathcal{N} = 1$ solution of [253] with scale separation: an internal Calabi–Yau, F_0, F_4, H, and an O6-plane. The parameters of the solution scale in the same way. Indeed, let us take F_0 of order one, the flux quanta of F_4 of order N, and $\mathrm{Vol}(M_6) \sim r^6$. If we integrate (11.3.23) on a three-cycle, it should give an order one number; this says $\mu \sim r^{-3}$. From the expression for F_0 in (11.3.22), we then see $g_s \sim \mu$, and from that for F_4 we see $r^4 \sim N$. This is in agreement with (8.3.15), as promised. All this suggests that (11.3.22) should be identified with the vacuum (8.3.14), and more importantly with the $\mathcal{N} = 1$ solution of [253].[18]

However, as we discussed following (10.4.8), smearing is physically reasonable for D-branes, not for O-planes. Still, (11.3.22) might signal the existence of a more complicated unsmeared solution. We will see a possible step in that direction in the next subsection.

11.3.3 IIA: $\mathcal{N} = 1$ with $\mathrm{SU}(3) \times \mathrm{SU}(3)$-structures

The generic pure spinor pair is the so-called $\mathrm{SU}(3)\times\mathrm{SU}(3)$-structure, with $\Phi_\pm$ given by (10.3.37). Here it is convenient to distinguish two cases, according to θ, the phase of the zero-form $(\Phi_+)_0 = \eta_+^{2\dagger}\eta_+^1$.

$\mathrm{SU}(3) \times \mathrm{SU}(3)$**-structure:** $\theta \neq 0$
We define[19]

$$J_\psi \equiv \frac{1}{\cos\psi} j + \frac{\mathrm{i}}{2} v \wedge \bar{v}\,, \qquad \mathrm{i}\omega_\psi \equiv \frac{1}{\sin\psi}\left(\frac{1}{\cos\psi}\mathrm{Re}\omega + \mathrm{i}\mathrm{Im}\omega\right)\,, \tag{11.3.24}$$

so that (10.3.37) reads

$$\Phi_+ = N\mathrm{e}^{\mathrm{i}\theta}\cos\psi\,\mathrm{e}^{-\mathrm{i}J_\psi - \tan\psi\mathrm{Re}\omega}\,, \qquad \Phi_- = N\sin\psi\, v \wedge \mathrm{e}^{\mathrm{i}\omega_\psi - \tan\psi\mathrm{Re}\omega}\,. \tag{11.3.25}$$

We first plug these pure forms in (10.3.20b) [494]. (Recall from (10.3.25) that (10.3.20a) is redundant in the AdS case.) The one- and three-form equations give some purely geometrical conditions:

[18] Several other solutions with smeared O6-planes were also found on cosets, again using invariant forms [676].
[19] In [494, 687], the one-form v is rescaled by a $\tan\psi$ relative to here.

$$d(e^{3A-\phi}\cos\psi\,\sin\theta)=0\,,\qquad \mathrm{Re}v=-\frac{e^A}{2\mu}\cot\psi\frac{d\theta}{\sin\theta} \tag{11.3.26a}$$

$$d\left(\frac{1}{\sin\theta}J_\psi\right)=-2\mu e^{-A}\tan\psi\,\mathrm{Im}(v\wedge\omega_\psi), \tag{11.3.26b}$$

as well as the NSNS three-form:

$$H=dB\,,\qquad B=-\cot(\theta)J_\psi-\tan\psi\mathrm{Re}\omega\,. \tag{11.3.27}$$

To determine the RR flux, it is convenient to use the generalized Dolbeault equation (taking $c_+=2$); we can then use the b-transform (10.5.141) to set $H=0$ in (10.5.138). We compute this way:

$$F=e^{B\wedge}\hat{F}\,, \tag{11.3.28}$$

with

$$\begin{aligned}
F_0&=J_\psi\cdot d(e^{-\phi}\sin\psi\mathrm{Im}v)+5\mu e^{-A-\phi}\cos\psi\cos\theta\,,\\
\hat{F}_2&=F_0\cot\theta J_\psi+J_\psi\cdot d\,\mathrm{Re}(\sin\psi e^{-\phi}v\wedge\omega_\psi)\\
&\quad+\mu\frac{\sin^2\psi}{\cos\psi}e^{-A-\phi}\left[(2+5\cot^2\psi)\sin\theta J_\psi+2\sin\theta\mathrm{Re}v\wedge\mathrm{Im}v-2\cos\theta\mathrm{Im}\omega_\psi\right],\\
\hat{F}_4&=F_0\frac{J_\psi^2}{2\sin^2\theta}-d\left[\sin\psi\,e^{-\phi}(J_\psi\wedge\mathrm{Im}v-\cot\theta\mathrm{Re}(v\wedge\omega_\psi))\right],\\
\hat{F}_6&=-\frac{1}{\cos^2\psi}\mathrm{vol}_6\left(F_0\frac{\cos\theta}{\sin^3\theta}+3\frac{\mu\cos\psi e^{-A-\phi}}{\sin\theta}\right).
\end{aligned} \tag{11.3.29}$$

The Bianchi identities are now $d\hat{F}=0$. One can show that $\hat{F}_4$ is in fact automatically closed [494]; this implies the Bianchi identity $dF_4-H\wedge F_2=0$.

SU(3) × SU(3)-structure: $\theta=0$
For $\theta=0$, the equations are different:

$$\mathrm{Re}v=\frac{e^A}{2\mu}\cot\psi\,d\log(\cos\psi e^{3A-\phi}) \tag{11.3.30a}$$

$$d(e^{3A-\phi}\cos\psi J_\psi)=0\,. \tag{11.3.30b}$$

This time H is not exact:

$$H=\hat{H}-d(\tan\psi\mathrm{Re}\omega)\,,\qquad \hat{H}=-2\mu e^{-A}\mathrm{Re}(iv_1\wedge\omega_\psi)\,. \tag{11.3.31}$$

We cannot eliminate it completely from the system using a b-transform. So we write $F=e^{\tan\psi\mathrm{Im}\omega}\wedge\hat{F}$, with

$$\begin{aligned}
F_0&=J_\psi\cdot d(\sin\psi e^{-\phi}\mathrm{Im}v)+5\mu\cos\psi e^{-A-\phi}\,,\\
\hat{F}_2&=J_\psi\cdot d\,\mathrm{Im}(i\sin\psi e^{-\phi}v\wedge\omega_\psi)-2\mu\frac{\sin^2\psi}{\cos\psi}e^{-A-\phi}\mathrm{Im}\omega_\psi\,,\\
\hat{F}_4&=J_\psi^2\left[\frac{1}{2}F_0-\mu\cos\psi e^{-A-\phi}\right]-J_\psi\wedge d\,\mathrm{Im}(\sin\psi e^{-\phi}v)\,,\\
\hat{F}_6&=0\,.
\end{aligned} \tag{11.3.32}$$

In this case, the Bianchi identities become $d_{\hat{H}}\hat{F}=0$.

Fibrations over maximally symmetric three-manifolds

Unfortunately, not many examples of AdS$_4$ $\mathcal{N} = 1$ vacua with SU(3) × SU(3) structure are known.

A first class of known solutions [688] has the internal topology

$$S^3 \hookrightarrow M_6 \to \Sigma_3 \,, \tag{11.3.33}$$

where Σ_3 is a compact three-manifold that is locally maximally symmetric. For example, this can be S^3, or a compact hyperbolic manifold as in (11.3.7). The metric is of the form

$$\mathrm{d}s^2_{M_6} = \mathrm{d}x^2 + f^2 D s^2_{S^2} + g^2 \mathrm{d}s^2_{\Sigma_3} \,, \tag{11.3.34}$$

where f and g are functions of x, and $Ds^2_{S^2} = Dy^a Dy^a$ is the round S^2 metric fibered on Σ_3 as defined following (11.2.19), so a slice at constant x is of bundle metric form (4.2.35). The coordinate x parameterizes an interval, and f goes to zero at its two endpoints, thus creating the topological S^3. The SU(2)-structure is

$$\begin{aligned} v &= \mathrm{d}x + g y^a e^a \,, \qquad j = fg\, \epsilon^{abc} e^a \wedge y^b D y^c \,, \\ \omega &= -fg\, e^a \wedge Dy^a + \mathrm{i}\left(f^2 \mathrm{Vol}_{S^2} - \frac{1}{2} g^2 y^a *_3 e^a \right), \end{aligned} \tag{11.3.35}$$

where e^a is a vielbein on Σ_3, and $\mathrm{Vol}_{S^2} = \frac{1}{2}\epsilon^{abc} y^a Dy^b \wedge Dy^c$ is the S^2 volume form, fibered over Σ_3. The pure $\Phi_\pm$ are then of the form (10.3.37), for $\theta = 0$; using (11.3.30a)–(11.3.32), the Bianchi identities reduce to a system of ordinary differential equations in x. Numerical solutions were found in [688], but in particular the subset where $g e^{-A}$ is constant admits analytic solutions of the form (11.2.19) with

$$\mathrm{d}s^2_7 = \frac{19}{40}\left(\mathrm{d}s^2_{\mathrm{AdS}_4} + \frac{4}{5} \mathrm{d}s^2_{\Sigma_3} \right), \qquad X^5 = \frac{5}{8}, \tag{11.3.36}$$

where Σ_3 has scalar curvature -6. This particular set of solutions is thus in one-to-one correspondence with the massive IIA AdS$_7$ solutions of Section 11.2.1. Their holographic duals are the $d = 6$ superconformal theories mentioned in that section, compactified on Σ_3. Thus (11.3.36) is in a sense an analogue of (11.3.7) in massive IIA.

Scale separation from O6-planes?

The next set of examples is a recent approximate solution [687, 689], meant to be a version of (11.3.22) but with localized sources. While a solution with some of the expected features was found for $\theta \neq 0$ [494], global considerations seem to favor the $\theta = 0$ branch. Indeed, we now show that any compactification with an O6 wrapping an internal nontrivial three-cycle should have $f_6 = 0$. The source δ_3 is purely internal; if $f_6 \neq 0$, (10.3.3b) gives

$$\mathrm{d}(\mathrm{e}^{4A} * f_4) + H \wedge \mathrm{e}^{4A} * f_6 = 0 \,, \qquad \mathrm{d}(\mathrm{e}^{4A} * f_6) = 0 \,. \tag{11.3.37}$$

So $\mathrm{e}^{4A} * f_6$ is a nonzero constant, and it follows that H is exact. But then (10.3.3a) gives $\mathrm{d}f_2 - H f_0 = \delta_3$, which implies that δ_3 is exact, which cannot be Poincaré dual to a nontrivial cycle, contradicting the hypothesis.

So we take $\theta = 0$, which in (11.3.32) has $f_6 = 0$ automatically. We assume $\mu = \sqrt{-\Lambda/3} \sim g_s \ll 1$ and the expected scaling (8.3.15), and expand in g_s. With the Ansatz

$$\psi = g_s\psi_1 + O(g_s^3), \qquad e^{\phi} = g_s e^{3A_0+g_s^2\phi_2+O(g_s^4)}, \qquad e^{A} = e^{A_0+g_s^2A_2+O(g_s^4)}, \tag{11.3.38}$$

the equations (11.3.30)–(11.3.32) simplify. (j,ω) are determined by an auxiliary SU(3)-structure (J,Ω). The form of $F_2 \sim g_s^{-1}J\cdot \mathrm{dIm}\Omega$ leads to guessing a solution for $\mathrm{Im}\Omega$, which in turn determines $\mathrm{Re}\Omega$ by the Hitchin formula (10.5.61). In the end, the leading-order solution is

$$J_\psi \sim J_{\mathrm{CY}}, \qquad \omega = -\frac{\mathrm{i}}{2}\bar{v}\cdot\Omega, \qquad e^{A_0} \sim 1 - g_s f, \tag{11.3.39}$$
$$\mathrm{Im}\Omega \sim (1+g_s f)\mathrm{Im}\Omega_{\mathrm{CY}} - g_s *_{\mathrm{CY}} K, \qquad \mathrm{Re}\Omega \sim (1-g_s f)\mathrm{Re}\Omega_{\mathrm{CY}} + g_s K,$$
$$v \sim \frac{e^{A_0}}{2\psi_1}\partial_{\mathrm{CY}}\tilde{f}, \qquad \Delta_{\mathrm{CY}}\tilde{f} = 8g_s F_0 f, \qquad \Delta_{\mathrm{CY}}K = 2g_s F_0 \mathrm{Re}\Omega_{\mathrm{CY}}.$$

ψ_1 is determined by $J\cdot v\wedge\bar{v} = -2\mathrm{i}$. The next-order warping and dilaton are constrained by $3A_2 - \phi_2 = \frac{1}{2}\psi_1^2 - \frac{1}{5}F_0\tilde{f}$. If this approximate solution could somehow be made exact, it would prove the existence of the vacua with scale separation found in [253] by effective $\mathcal{N}=1$ field theory. For a skeptical viewpoint, see [690].[20]

Scale separation from quantum effects?

Another set of AdS vacua with scale separation is again suggested by an effective $d=4$ supergravity [22], but this time by considering non-perturbative corrections to the GKV superpotential (9.5.26). It is the starting point for the KKLT proposal of dS vacua, and for this reason we defer its treatment to Section 12.1. The lift to $d=10$ was argued to involve an SU(3)×SU(3)-structure [447]. The equations solved by the $\Phi_\pm$ are no longer the pure spinor equations (10.3.20), because of the lift to $d=10$ of the instanton corrections.

As noticed in Section 10.1.2, most Minkowski $\mathcal{N}=1$ supergravity solutions will also have nonperturbative corrections; see [691] for a review. This would either destabilize the vacuum or make it AdS. The case in (12.1.1) is more explicitly calculable because of the presence of W_0, constant in the Kähler moduli.

Examples from a consistent truncation

In the next section, we will review a consistent truncation that uplifts to $\mathrm{AdS}_4\times S^6$ topology in massive IIA. As well as solutions with $\mathcal{N}>1$ of interest there, it also produces several $\mathcal{N}=1$ solutions: with isometry group SU(3) [692], U(1) [693], and no isometries [694]. The uplift of the first was in particular discussed in detail in [695]. It has cohomogeneity one; the isometry group orbits are copies of a squashed S^5, an S^1-fibration over $\mathbb{CP}^2$ (Section 4.4.3). The pure $\Phi_\pm$ for this solution have not been obtained, but they are thought to be of SU(3) × SU(3)-structure type with the "singlet" Ansatz in [549]. The latter also produces other solutions with smeared sources.

11.3.4 IIA: extended supersymmetry

$\mathcal{N}=2$ supersymmetry

For $\mathcal{N}=2$ solutions, it is convenient to start from the system of Section 10.6.2. There are now four possible internal spinors η_I^a. Generically, they are linearly independent

[20] An alternative idea to obtain AdS_5 and AdS_4 solutions with scale separation using seven-branes was suggested in [389].

and define an identity structure; this case has not been fully analyzed yet. (For $F_0 = 0$, it should be related to the solution (11.3.6).) A nongeneric case is when the four η^a_I are linear combinations of only two different spinors; this defines an SU(2)-structure on M_6. So far, all known solutions are of this type; they were obtained over the years perturbatively in F_0 [225], numerically [87, 214, 696, 697], and then analytically [95, 502, 698].

There are two types of solutions [502]. In the first, M_6 is the product of a hyper-Kähler manifold with a noncompact metric, so we will not consider it further. The second contains a Kähler–Einstein KE$_4$ and is much more flexible. As an Ansatz, one may demand that M_6 be complex, which makes everything depend on a single coordinate y; the analysis is remarkably similar to that in Section 11.2.3 for $\mathcal{N} = 1$ AdS$_5$ solutions. Such vacua can be fully classified: the metric reads as follows [502]

$$ds^2_{10} = L^2\sqrt{\frac{x^2q' - 4xq}{q'}}\left(ds^2_{\text{AdS}_4} - \frac{q'}{4xq}dx^2 + \frac{q}{4q - xq'}D\psi^2 + \frac{q'}{3q' - xq''}ds^2_{\text{KE}_4}\right), \tag{11.3.40}$$

where as usual $D\psi \equiv d\psi - A$, $dA = 2J_{\text{KE}_4}$; x belongs to an interval, $(\,)' \equiv \partial_x(\,)$; and generically

$$q = x^6 + \frac{3}{4}\frac{1 - a - 4\beta^2}{\beta}x^4 + \sqrt{8\frac{1-a}{\beta}}x^3 + 3x^2 - \beta\,, \tag{11.3.41}$$

with a, β constant. The dilaton and forms are also analytically known [502, (3.25a), (4.9c), (4.28)]; all fluxes allowed by symmetries are switched on. (With respect to [502], we have introduced $a = 1 - 12s = 1 - 8\beta\gamma^2$, for ease of comparison with the massless case.)

The regularity analysis is similar to that for AdS$_7$ solutions (Section 11.2.1) and for the $Y_{p,q}$ manifolds (Section 7.4.5). The interpretation of the solution depends on the local behavior of q at the endpoints of the x interval:

- A simple zero of q is a regular point for (11.3.40).
- A triple zero of q gives a local conical Calabi–Yau singularity $C(\text{SE}_5)$, with SE$_5$ as in (7.4.27c).
- An inflection point at the origin ($q'(0) = q''(0) = 0$) means that (11.3.40) has an O8-singularity.

The O8 possibility is only allowed on a separate branch where the metric is still (11.3.40), but q has a different sextic expression. With (11.3.41), the physically allowed solutions are such that

$$0 \le a \le 1\,, \qquad 0 \le \beta < \beta_0(a)\,, \tag{11.3.42}$$

where $\beta_0(a)$ is the smallest positive solution of $144\beta^6 + 36(3a - 11)\beta^4 + 9(3a^2 + 14a - 1)\beta^2 + a(a - 1)^2 = 0$. The domain (11.3.42) has an approximately triangular shape:

- The first side $\{\beta = 0, 0 \le a \le 1\}$ is subtle because of the appearance of β in the denominator of (11.3.41). The limit $\beta \to 0$ can be taken with care; q degenerates to a quartic polynomial, and yields solutions with $F_0 = 0$, which upon M-theory lift become AdS$_4 \times Y_{p,k}(\text{KE}_4)$ from (7.4.47). Here $0 \le a \le 1$ reproduces the same requirement discussed after (7.4.48).

- On the side $\{\beta = \beta_0(a),\ 0 \le a \le 1\}$, the solution has a single conical Calabi–Yau $C(\mathrm{SE}_5)$ singularity. For $\mathrm{KE}_4 = \mathbb{CP}^2$, this is a $\mathbb{C}^3/\mathbb{Z}_3$ singularity, and can be avoided by changing the periodicity of ψ. Especially notable is the point $\{\beta = 1,\ a = 1\}$, which corresponds to the first analytic solution ever found [95], obtained with a consistent truncation to be explained soon.
- The third side $\{a = 1,\ 0 \le \beta \le 1\}$ corresponds to the first ever numerical solutions, for $\mathrm{KE}_4 = \mathbb{CP}^2$ [696]. (It was later pointed out that they can be generalized to any KE_4 [697].)
- Finally, for parameters in its interior of (11.3.42), where the inequalities are strict, the solutions are fully regular.

One can also replace the KE_4 term in (11.3.40) by [502]

$$\frac{\kappa_1 q'}{x u_1} \mathrm{d}s^2_{\Sigma_1} + \frac{\kappa_2 q'}{x u_2} \mathrm{d}s^2_{\Sigma_2}, \tag{11.3.43}$$

where Σ_a are two Riemann surfaces of scalar curvature κ_a; u_a are two quartic polynomials, q is still sextic but has a slightly different expression from (11.3.41). These solutions had also been previously found numerically [87, 214]. Here the parameter space has pyramidal shape; at its boundary we find orbifold and conifold singularities. When $\Sigma_a = S^2$, a certain choice of parameters gives an $F_0 \neq 0$ deformation of $\mathrm{AdS}_4 \times \mathbb{CP}^3$, originally found perturbatively in F_0 [225].

Consistent truncations

$\mathcal{N} = 8$ supergravity in four dimensions, mentioned earlier in Section 11.3.1, has a family of deformations [32], found with the embedding tensor formalism. Most of these are not thought to have an uplift to $d = 11$, unlike their undeformed counterpart. An exception is the theory with gauge group

$$\mathrm{ISO}(7) = \mathrm{SO}(7) \ltimes \mathbb{R}^7, \tag{11.3.44}$$

where the $\mathbb{R}^7$ part of the gauge group is gauged. The gauging is *dyonic*, in that some scalars are charged under some vector fields and their magnetic duals. This has an uplift to massive IIA in $d = 10$ [95, 692]; the internal space is S^6. Unlike in the usual maximal supergravity, there is no vacuum where all of the $\mathcal{N} = 8$ supersymmetry of the Lagrangian is realized. But there are many other vacua:

- One solution has $\mathcal{N} = 3$ supersymmetry, and $\mathrm{SO}(3) \times \mathrm{SO}(3)_\mathrm{R}$ isometries [692, 699–701]. The isometry orbits are S^2-fibrations over S^3: the metric is

$$\mathrm{d}s^2_{10} = L^2\sqrt{1+x^2}\left(\mathrm{d}s^2_{\mathrm{AdS}_4} + \frac{2\mathrm{d}x^2}{1-x^2} + \frac{4x^2(1+x^2)}{2+3x^2+3x^4} Ds^2_{S^2} + 4\frac{1-x^2}{1+x^2}\mathrm{d}s^2_{S^3}\right), \tag{11.3.45}$$

where as usual $Ds^2_{S^2} = Dy^a Dy^a$, with $Dy^a = \mathrm{d}y^a + \frac{1}{2}\frac{1-x^2}{1+x^2}\rho^a$, ρ^a being the right-invariant forms in (4.4.27). This is likely to admit generalizations; a massive $\mathcal{N} = 3$ deformation of (11.3.3) was found [702, sec. 5.4] at first order in F_0; it is are likely to be connected to (11.3.45) in a larger web of solutions, perhaps similar to (11.3.40).
- One solution has $\mathcal{N} = 2$ supersymmetry [95]. It was first found with the consistent truncation, but as we mentioned it can be recovered from (11.3.40): with $a = 0$, $\beta = 1$, $q = (x^2 - 1)^3$, and

$$ds_{10}^2 = L^2\sqrt{2+x^2}\left(ds^2_{\text{AdS}_4} + \frac{3}{2}\frac{dx^2}{1-x^2} + \frac{9}{2}\frac{1-x^2}{2+x^2}D\psi^2 + 3\frac{1-x^2}{1+x^2}ds^2_{\mathbb{CP}^2}\right). \quad (11.3.46)$$

Here we rescaled ψ by a factor of 3 to avoid the $\mathbb{C}^2/\mathbb{Z}_3$ singularity noted in the discussion that follows (11.3.42).

- Several have $\mathcal{N} = 1$ supersymmetry. One is the nearly Kähler solution $\text{AdS}_4 \times S^6$ of Section 11.3.2; others are mentioned at the end of Section 11.3.3.
- Finally, there are a hundreds of nonsupersymmetric solutions [694].

$\mathcal{N} \geq 2$

The only known $\mathcal{N} = 3$ solutions are (11.3.45) and the conjectural ones described after it. For $\mathcal{N} = 4$, the R-symmetry algebra is $\text{so}(4)_\text{R}$. We might imagine its orbits to be copies of S^3, but if we take the internal supersymmetry parameters η_I to be Killing spinors on it, they transform as a vector under its so(4) rather than as in the spinor representation as would be expected. One would rather expect its orbits to be $S^2 \times S^2$, with the η_I transforming as a doublet along each sphere by (7.3.21). There is a single example with an $S^3 \times S^2$ fibered over an interval: this is a reduction of $\text{AdS}_4 \times S^7/\mathbb{Z}_k$ along a particular S^1, and can be understood as the near-horizon limit on a D2–D6 brane system [703, sec. 5].

Holography

The holographic duals of the vacua in this subsection are again Chern–Simons theories coupled to chiral multiplets. The main difference with the duals to (11.3.1) is that the sum of Chern–Simons levels k_a for the gauge group factors is no longer zero, but [550, 704]

$$\sum_a k_a = n_0 = 2\pi F_0 . \quad (11.3.47)$$

For example, the aforementioned $F_0 \neq 0$ deformation of $\text{AdS}_4 \times \mathbb{CP}^3$ has the same quiver diagram as the ABJM model [87, 550]. A new possibility is having a single gauge group: the dual to (11.3.45) is $\mathcal{N} = 3$ Chern–Simons coupled to two chiral multiplets Z_i with $W = \text{Tr}[Z_1, Z_2]^2$ [705]; the dual to (11.3.46) is $\mathcal{N} = 2$ Chern–Simons coupled to three Z_i with $W = \text{Tr}(Z_1[Z_2, Z_3])$ [95]. (The generalization of the latter with $\mathbb{CP}^2 \to \text{KE}_4$ is again a quiver [706].)

11.3.5 IIB

So far, G-structures have been far less successful for AdS_4 in IIB; the largest class has been obtained by direct analysis of the spinorial BPS equations.

Classifications

If we start from the pure spinor equations, we notice immediately that the SU(3)-structure case does not work [707]: the zero-form part of (10.3.20b) only receives a contribution on the right-hand side, which implies $\mu = 0$. With $\Phi_\pm$ associated to SU(2)-structures, there are no obvious obstructions, but so far this method has only produced vacua with smeared sources [549, 708, 709].

With $\mathcal{N} = 2$ supersymmetry, one can again apply the methods of Section 10.6.2. Following by now familiar steps, the local form of the metric is completely determined in terms of two functions of five coordinates, satisfying a system of four partial differential equations [478, (6.1)–(6.4)]. Local solutions can be found with relative ease, but so far no compact internal M_6 has been found with this method.

Even putting aside G-structure and pure spinor methods, only few solutions are known with $\mathcal{N} < 4$. A single $\mathcal{N} = 1$ solution has recently been found numerically with topology $S^1 \times S^5$ [710]. Some more $\mathcal{N} = 1, 2, 3$ solutions can be obtained if we glue the internal space to itself using S-duality transition functions; we will describe this *S-fold* idea later in this chapter. A class of $\mathcal{N} = 2$ solutions can be obtained by combining the AdS_4 solutions of $F(4)$ supergravity [711] with the consistent truncation from IIB we mentioned in Section 11.2.2 [544, 619].

The $\mathcal{N} = 4$ class

On the other hand, a large $\mathcal{N} = 4$ class exists on spaces M_6 of the following form [712]:

$$S^2 \times S^2 \hookrightarrow M_6 \to M_2 \,. \tag{11.3.48}$$

The local analysis is somewhat similar to the one for AdS_6 solutions in IIB (Section 11.2.2), which it predates. Everything is determined by two real harmonic functions h_1, h_2 on the strip $\mathrm{Im}z \in [0, \pi/2]$, such that $h_1 = \partial_\perp h_2 = 0$ at $\mathrm{Im}z = 0$, and $h_2 = \partial_\perp h_1 = 0$ at $\mathrm{Im}z = \pi/2$. As usual, the h_i can be related to holomorphic functions: $h_1 = 2\mathrm{Im}\mathcal{A}_1$, $h_2 = 2\mathrm{Re}\mathcal{A}_2$. The metric is

$$\begin{aligned} \mathrm{d}s^2_{10} &= 2\sqrt{\frac{N_2}{W}}\cdot\left(\mathrm{d}s^2_{\mathrm{AdS}_4} + 2\frac{W}{h_1 h_2}\mathrm{d}z\mathrm{d}\bar{z} + \frac{h_1^2 W}{N_1}\mathrm{d}s^2_{S_1^2} + \frac{h_2^2 W}{N_2}\mathrm{d}s^2_{S_2^2}\right), \\ W &= \partial_z\partial_{\bar{z}}(h_1 h_2)\,, \qquad N_1 = 2h_1 h_2|\partial_z h_1|^2 - h_1^2 W\,, \qquad N_2 = 2h_1 h_2|\partial_z h_2|^2 - h_2^2 W\,. \end{aligned} \tag{11.3.49}$$

The string coupling is $\mathrm{e}^\phi = (N_2/N_1)^{1/2}$; $F_1 = 0$, and

$$\begin{aligned} B &= (-4h_1^2 h_2 N_1^{-1}\mathrm{Im}(\partial_z h_1 \partial_{\bar{z}} h_2) + 2h_2^D)\mathrm{vol}_{S_1^2}\,, \\ C_2 &= (-4h_1 h_2^2 N_2^{-1}\mathrm{Im}(\partial_z h_1 \partial_{\bar{z}} h_2) - 2h_1^D)\mathrm{vol}_{S_2^2}\,, \\ F_5 &= -4(1 + *)\mathrm{vol}_{\mathrm{AdS}_4} \wedge \mathrm{d}\mathrm{Re}\left(6\mathcal{B} - 6\mathcal{A}_1\bar{\mathcal{A}}_2 + 2\mathrm{i}h_1 h_2 W^{-1}\partial_z h_1 \partial_{\bar{z}} h_2\right), \end{aligned} \tag{11.3.50}$$

where $\mathcal{B}$ is defined as in (11.2.24b), $h_1^D = 2\mathrm{Re}\mathcal{A}_1$, $h_2^D = -2\mathrm{Im}\mathcal{A}_2$. More general solutions can be obtained by acting with $\mathrm{SL}(2, \mathbb{R})$.

The infinite strip suggests that the solutions should be noncompact. This is interpreted as the effect of the back-reaction of D3-branes, which puts them at infinite distance (Section 1.3.7). But for the class of functions [713]

$$h_1 = -2\mathrm{Re}\sum_a \gamma_a \log\tanh\frac{\pi\mathrm{i} + 2\delta_a - 2z}{4}\,, \qquad h_2 = -2\mathrm{Re}\sum_a \hat{\gamma}_a \log\tanh\frac{z - \hat{\delta}_a}{2}\,, \tag{11.3.51}$$

with $\gamma_a, \hat{\gamma}_a, \delta, \hat{\delta}_a \in \mathbb{R}$, the limits $\mathrm{Re}z \to \pm\infty$ in the strip are at finite distance and can be shown to be regular points with an appropriate change of coordinates. So M_2 is more properly thought of as a disk. From (11.3.49), we see that at the lower boundary of the strip $\{\mathrm{Im}z = 0\}$, where $h_1 = 0$, S_1^2 shrinks, while at the upper boundary $\{\mathrm{Im}z = \pi/2\}$, $h_2 = 0$ and the S_2^2 shrinks. On a vertical segment $\{\mathrm{Re}z = x\,, \mathrm{Im}z \in [0, \pi/2]\}$, the topology is that of a join $S^2 * S^2 \cong S^5$, from (4.5.8), (4.5.9). Since this S^5 shrinks in the two limits $\mathrm{Re}z \to \pm\infty$, the topology of M_6 is S^6.

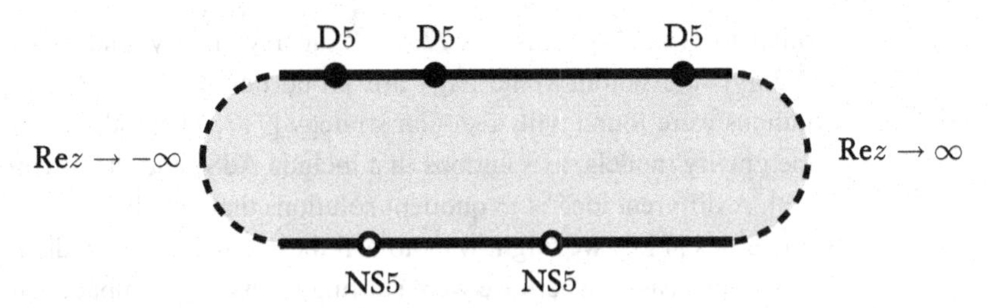

Figure 11.5 The strip M_2 for $\mathcal{N} = 4$ IIB vacua.

There are singularities in $\{z = \hat{\delta}_a\}$ and $\{z = \delta_a + i\pi/2\}$, at the lower and upper strip boundary. A local analysis, similar in spirit to that in (11.2.13) for D6-branes, shows that these have a physical interpretation as NS5-branes and D5-branes respectively. These points are shown on Figure 11.5, with the points at infinity on the strip brought to finite distance.

There is also a variant of this construction where the strip is glued with itself and has the topology of an annulus [714]; here the idea is to introduce infinitely many sources in (11.3.51), repeating periodically with $z \to z + z_0$, and then to quotient the strip by such a translation, $z \sim z + z_0$.

S-folds

So far we have described supergravity solutions. In string theory, we are also allowed to use transition functions in the S-duality group $\mathrm{SL}(2,\mathbb{Z})$. We used this possibility already in F-theory, but more general vacua where the axion is not holomorphic are usually called *S-folds*. This possibility should not be taken too lightly, because in such a solution there is usually a region where the string coupling is large. This can still be under control if all the fields are varying slowly, since that would suppress string corrections, and the two-derivative IIB action is fixed by supersymmetry. But this does mean that an alleged S-fold solution should be scrutinized carefully. The discussion is similar to that for T-folds (Section 4.2.6).

A common strategy is to look for a supergravity solution of the type

$$\mathrm{d}s^2_{\mathrm{E}} = \mathrm{e}^{2A}\mathrm{d}s^2_{\mathrm{AdS}_4} + \mathrm{e}^{\lambda}\mathrm{d}x^2 + \mathrm{d}s^2_{M_5}\,, \qquad \phi = -2x + f\,, \qquad C_2 = \mathrm{e}^{x}c_2\,, \qquad B_2 = \mathrm{e}^{-x}b_2\,, \tag{11.3.52}$$

where nothing depends on x except where explicitly noted; E denotes the Einstein frame. Moreover, $C_0 = 0$ and F_5 has no dependence on x. Such a solution is non-compact, but is invariant under a shift $x \to x + T$ together with an $\mathrm{SL}(2,\mathbb{R})$ transformation $m_0 \equiv \begin{pmatrix} \mathrm{e}^{-T} & 0 \\ 0 & \mathrm{e}^{T} \end{pmatrix}$, recalling (1.2.38). For the special values $2\mathrm{e}^T = n + \sqrt{n^2 - 4}$, $n \in \mathbb{Z}$, $m_0 = C s t^n C^{-1}$. So if we apply the $\mathrm{SL}(2,\mathbb{R})$ matrix C to (11.3.52), it will become more complicated, but now it will be invariant under $x \to x + T$ and the $\mathrm{SL}(2,\mathbb{Z})$ transformation st^n. Quotienting under this symmetry, we obtain a compact solution. Even if $C \in \mathrm{SL}(2,\mathbb{R})$ is a symmetry of supergravity and not of string theory, we can take the fundamental region for (11.3.52) to be such that ϕ is arbitrarily small. Indeed, an explicit computation [715, sec. 2.3] shows that for the final solution the supergravity action is much larger than any corrections.

An example of (11.3.52) was first obtained in [716] as a vacuum in a $d = 4$ supergravity model with dyonic gauging and a semidirect product gauge group, similar to (11.3.44). This has $\mathcal{N} = 4$ supersymmetry, and is a particular case of (11.3.49): the holomorphic $\mathcal{A}_{1,2}$ are proportional to $e^{\pm z}$. More general $\mathcal{N} \leq 2$ solutions were found with a similar strategy [717, 718]; others were found in $d = 5$ supergravity models, as solutions that include AdS_4 and the direction x in (11.3.52) [719]. A different idea is to quotient solutions that are already compact by $SL(2,\mathbb{Z})$ symmetries [715]; we might want to call these *S-quotients*, to distinguish them from the previous cases, where we were starting from a noncompact one.

Holography

The $\mathcal{N} = 4$ solutions (11.3.49) are thought to arise as near-horizon limits of a D3–D5–NS5 system; the CFT duals suggested by this are the so-called *Hanany–Witten theories*, a chain of gauge groups of SU type with bifundamental hypermultiplets [720]. (An equivalent interpretation is as the $SCFT_3$ living on a superconformal interface in $d = 4$ $\mathcal{N} = 4$ super-YM [721].) In the variant where M_2 has the topology of an annulus, the chain becomes a necklace. The dual to the $\mathcal{N} = 4$ S-folds is also a necklace, but it involves [715] a non-Lagrangian CFT called $T(SU(N))$ [721]. Some properties of these theories are studied in [722, 723]. Superconformal interfaces are expected to exist in many other $SCFT_4$; this suggests that many more AdS_4 solutions with $\mathcal{N} < 4$ should exist.

11.3.6 Summary

For AdS_4, in general there are no claims of complete classifications as for AdS_7 and AdS_6, except for $\mathcal{N} = 4$ in IIB, (11.3.49). However, we have seen several classes of examples; we collect some prominent ones in Table 11.4, dividing them by amount of supersymmetry. Notation is as in Tables 11.2 and 11.3. In some cases, especially in IIB, the only known solutions are obtained by S-folds or S-quotients, where we quotient by an element of the S-duality group $SL(2,\mathbb{Z})$ a noncompact or compact solution respectively.

Table 11.4. Some notable examples of supersymmetric AdS_4 solutions.

	M-theory	IIA	IIB
$\mathcal{N} \geq 4$	S^7/Γ	?	$S^2 \times S^2 \curvearrowright$ disk, ann. (11.3.49)
$\mathcal{N} = 3$	3-Sasaki	Top. S^6 (11.3.45)	S-quotient [715]
$\mathcal{N} = 2$	SE_7 Top. S^7 (11.3.6) Top. $S^4 \curvearrowright H_3/\Gamma$ (11.3.7)	Top. $S^2 \curvearrowright KE_4$ (11.3.40) $\mathbb{CP}^3$ Top. S^6 (11.3.46)	S-folds (11.3.52)
$\mathcal{N} = 1$	Weak G_2 Top. S^7 [666, 668, 669]	Nearly Kähler (11.3.12) Coset/twistor (11.3.19) Top. $S^3 \curvearrowright H_3/\Gamma$ (11.3.36) Approx. CY + O6 (11.3.39) Top. S^6 [692–694]	Top. $S^5 \times S^1$ [710] S-folds (11.3.52)

Exercise 11.3.1 Show that (11.2.2) and (11.3.1) satisfy the flux quantization (1.4.69).

Exercise 11.3.2 Find (11.3.1) from (1.4.71).

Exercise 11.3.3 In a supersymmetric AdS$_4$ vacuum, consider a domain wall D2-brane, at $\{r = r_0\}$ in Poincaré coordinates (4.5.37). Using (10.3.45), write down the conditions for it to be BPS, in both the SU(3)- and SU(3) $\times$ SU(3)-structure cases.

Exercise 11.3.4 In the SU(3)-structure case, show that if the D2-brane in the previous exercise is BPS at a point in M_6, then a D2-brane bubble at the same internal point has $3T = Q$ in the language of (11.1.14), so that it does not lead to an instability.

12 Frontiers

In Section 12.1, we finally discuss dS vacua. As shown by no-go results discussed in the Introduction and in Section 10.3.1, these need either string theory corrections beyond supergravity or O-planes. There are several proposals, but it is safe to say that none of these has reached the same level of rigor of $\Lambda \leq 0$ vacua. In Section 12.2, we give a short review of the swampland program, which aims at characterizing the QFT models that can be coupled to quantum gravity, for example via string theory. I offer some short parting words in Section 12.3.

12.1 De Sitter vacua

12.1.1 From string corrections

To overcome the no-go theorem in the Introduction and in Section 10.3.1, one can introduce O-planes or consider string theory corrections to supergravity. In this section, we consider models that use both ingredients. In heterotic theory, which does not have O-planes, world-sheet arguments exclude de Sitter solutions [724]; this is exact in l_s, but does not include effects nonperturbative in g_s.

KKLT: $\Lambda < 0$

The proposal that has attracted most attention is the *KKLT model* [22]. It is obtained by a two-step modification of the conformal Kähler class (Section 9.5). First, the Kähler moduli are fixed by quantum corrections, producing an AdS vacuum. Second, anti-D3-branes modify the vacuum to $\Lambda > 0$.

The first step is to consider string theory correction to the effective potential of Section 9.5.3, whose notation we follow. The superpotential W is corrected only nonperturbatively. One example is brane instantons, which was one of our motivations for introducing D-branes in Section 1.3. Euclidean D3-branes wrapping a four-cycle $S_4 \subset M_6$ [725] give contributions of the form $Ae^{2\pi i T}$, where T is the modulus Poincaré dual to S_4; recall that $\mathrm{Re}T \sim r^4$. Alternatively, in presence of a stack of N D7-branes wrapping S_4, the gauge theory dynamics on their worldvolume becomes $\mathcal{N} = 1$ super-YM at low energies, which is known to have similar corrections $e^{2\pi i T/N}$. In a model with a single Kähler modulus $T = \sigma + i$, the leading approximation to the superpotential is

$$W \sim W_0 + Ae^{-2\pi T}\,, \tag{12.1.1}$$

where W_0 is the supergravity contribution (9.5.26), and A comes from a fluctuations integral (Section 1.3.1). Suppose we stabilized the complex moduli as described after

(9.5.18); then W_0 is a function of T alone. By (8.4.49a), a supersymmetric vacuum is given by $D_T W = 0$. As discussed prior to (9.5.28), $\partial_T K = -3/(T + \bar{T})$, and taking for simplicity $\zeta = 0$, $T = \frac{1}{2}v^2 \equiv \sigma \sim r^4$, we get

$$W_0 = -A e^{-2\pi\sigma_0}\left(1 + \frac{4}{3}\pi\sigma_0\right) \tag{12.1.2}$$

on the vacuum $\sigma = \sigma_0$. With generic numbers, we fall in the Dine–Seiberg argument in the Introduction. On the vacuum, the classical W_0 and the nonperturbative exponential are equally important, but then further string theory corrections should likely be considered: those due to multiple instantons, but also corrections to the Kähler potential K, which can also come from perturbative terms such as (1.2.24). This can be partially avoided by making the parentheses in (12.1.2) large; so σ_0 large, and W_0 small. In other words, we should make the classical and nonperturbative terms compete by making the former very small, rather than the latter large. W_0 is not completely arbitrary, since it results from stabilizing the complex moduli z^i, but by fine-tuning the choice of the flux quanta one can achieve very small and negative values, as studied statistically in [726]. This is key to evading the Dine–Seiberg problem. This strategy can work on Calabi–Yau manifolds with $h^{1,1} > 1$ [727, 728].

The cosmological constant is $\Lambda = -3e^{-K}|W|^2 = -2\pi^2 A^2 e^{-4\pi\sigma_0}/3\sigma_0 < 0$, so (12.1.2) is an AdS vacuum; $\sqrt{-\Lambda/3} \ll m_{\rm KK} \sim \sigma_0^{-1/2}$, so (8.3.2) is evaded, and there is scale separation. As briefly mentioned in Section 11.3.3, their lift to $d = 10$ is not obvious because of the instanton effects, and was argued in [447] to involve $SU(3) \times SU(3)$-structures.

KKLT: $\Lambda > 0$

The second step is to introduce anti-D3-branes. This is achieved by choosing the flux quanta on the right-hand side of (9.5.25) so that it requires $n_{\rm D3} < 0$. The effect of this is to break supersymmetry; now the $d = 4$ effective potential $V_{d=4}$ is no longer of the $\mathcal{N} = 1$ form $V_{\mathcal{N}=1}$ in (8.4.5), but has an extra term

$$\delta V = 2T_{\rm D3}\sigma^{-3}e^{-4A(y_{\overline{\rm D3}})}\,, \tag{12.1.3}$$

where $y_{\overline{\rm D3}}$ is the anti-D3 position. The factor of 2 comes from two contributions: one from the anti-D3 itself, another from the flux contributions. These would cancel for a D3, getting us back to the $V_{\mathcal{N}=1}$ associated to (12.1.1). Finally, the factor of $\sigma^{-3} \sim r^{-12} \sim \text{Vol}(M_6)^{-2}$ comes essentially from the rescaling (8.3.5). Adding this to the potential $V_{\mathcal{N}=1}$ obtained from (12.1.1) destroys the AdS vacuum if $e^{-4A(y_0)}$ is too large. But statistical studies of flux vacua indeed indicate that for a large fraction of flux quanta, the complex moduli z^i are stabilized near a conifold locus in $\mathcal{M}_c$ [726]. There is then a region in M_6 where the metric is approximately that of the resolved conifold, and the solution is similar to that in [415]. The anti-D3 position y_0 is then in this region, where A is large; so the coefficient $e^{-4A(y_0)}$ can be made small. The AdS vacuum is then not destroyed, but "uplifted" to $\Lambda > 0$.

The vacuum has no tachyons, but it is unstable under tunnel effect: in other words, it is metastable. This is probably inevitable in any theory of quantum gravity [729]. The decay time to Minkowski space was estimated in [22] using the methods we saw in Section 11.1.3 for AdS. If there are multiple anti-D3s, they tend to first polarize in an NS5 [730]; this is the S-dual of a brane-polarization phenomenon observed elsewhere in string theory [731, 732]. This NS5 can disappear by collapsing to zero size at the other pole of the S^3, but it has to first get larger to do so. For this reason, the vacuum is metastable; the decay can only happen via an instanton.

Similar constructions

Each stage of the KKLT strategy can be generalized. At the stage of finding the AdS vacuum, one can work outside the small-W_0 regime. The perturbative corrections to K then start competing with (12.1.1); taking this into account, the effective potential has another, nonsupersymmetric AdS vacuum where the internal volume is very large (but not necessarily all cycles are) [733, 734]. This can yield better control over the quantum corrections.

Variants have also been suggested for the uplifting. The term (12.1.3) breaks supersymmetry explicitly; mechanisms have been suggested where terms similar to (12.1.3) would be part of an $\mathcal{N} = 1$ potential, so that supersymmetry would be broken spontaneously. One proposal is to stabilize already the complex moduli at a nonsupersymmetric vacuum, allowing G_3 to no longer be imaginary-self-dual [735]. Another general strategy is to break supersymmetry using the D-term contributions μ_a to $V_{\mathcal{N}=1}$ (8.4.5), originally motivated by switching on a world-volume flux on the D7-branes; this appears to be challenging [736–739]. There is also a mechanism called Kähler uplift, where corrections to K would be sufficient for the uplift [740].

There are *many* other proposals that don't quite follow the KKLT two-step procedure, but rely on quantum effects in various corners of string theory. Examples include brane instantons in IIA [741], and membrane instantons for M-theory on G_2-holonomy manifolds (Section 7.2.5) [742] or in heterotic M-theory (Section 1.4.5) [743].

Criticisms

All these strategies rely on one or more particular string correction, while trying to argue that all others can be neglected, by estimating them. Moreover, the relevant corrections themselves are often introduced in a four-dimensional effective approach by extrapolating from known cases, sometimes by making assumptions about their coefficients. While these arguments are often carefully laid out, a sufficiently hard-nosed physicist may still find room for skepticism.

A more radical worry is that perhaps using effective field theory in string theory might be misleading [744, 745]. According to this view, while the method is extremely well established in quantum field theory, it would be inappropriate in quantum gravity. In particular, the perturbative view of string theory (Section 1.1) relies on having a conformal model describing the world-sheet. So a vacuum would need to first have a leading-order perturbative definition before we could even start describing how it is corrected in l_s and g_s. This would mean one cannot combine supergravity and string corrections to obtain a vacuum that did not exist in supergravity alone. Given our relatively scarce experience with quantum gravity, it is good to consider all possibilities, but my personal opinion is that effective field theory methods are valid in quantum gravity and string theory, and that we should not give excessive importance to the perturbative world-sheet description.

For the KKLT model, this worry would imply that we first have to establish that the supersymmetry-breaking vacua of (9.5.14) survive higher-derivative string corrections before we can even consider computing the nonperturbative term in (12.1.1). A justification of this procedure from the effective field theory point of view is given in [746].

The KKLT construction has also attracted more specific criticism. One long-standing debate concerns the actual metastability of the anti-D3 potential. Its analysis was carried out in the probe approximation [730], and this raised concerns that taking back-reaction into account would make the local minimum in the potential disappear. Solving the equations of motion directly is too complicated, but it becomes doable after smearing the anti-D3 along the S^3 of the resolved conifold; this smeared solution has a diverging flux density, which was suggested to signal that the anti-D3-branes are directly unstable rather than metastable [747–750]. (The D6 singularity in (11.2.15), first found in [436, (4.15)], was also considered a possible toy model for this problem [583, 584].) One issue with this approach is that it tries to draw conclusions based on supergravity, which breaks down near the antibranes. Using the brane effective action beyond the probe approximation appears problematic, because it involves the pull-back to the world-volume of the closed string fields, which diverge; this can be solved by treating the brane as an effective field theory, matching it to string theory in the UV [751, 752]. The result is that the divergence does not seem to indicate an instability for the anti-D3. Meanwhile, it has been argued that the singularity is due to the smearing [753]. Also, approximate solutions obtained with the blackfold formalism [754] confirm [755] the anti-D3 potential of [730].

The potential should be studied, however, as a function of all moduli, and in yet another line of attack it was suggested [756] that the complex modulus can be destabilized by the anti-D3-brane, making the resolved conifold become singular.

Ten-dimensional description

The aforementioned fears about the use of $d = 4$ effective actions should be at least partially assuaged by a ten-dimensional description. If the nonperturbative effects in (12.1.1) are replaced by those on the D7, they can be described by adding a fermion bilinear on the D7 world-volume action, modeling gaugino condensation due to the world-volume super-YM dynamics [757, 758]. This generates an additional term in the pure spinor equation (10.3.20b), proportional to Ae^{-aT} [447, (6.3)]; one can use this formalism to compute the back-reaction of the nonperturbative effects on the geometry [758, 759], eventually confirming (12.1.2) from a ten-dimensional point of view.

Inclusion of anti-D3-branes initially seemed not to provide the desired uplift to $\Lambda > 0$ [760, 761], but a more recent analysis yields a match with the expected features of the KKLT dS vacua [762].

12.1.2 In supergravity

Given the subtleties introduced by stringy corrections, one may want to find dS solutions using supergravity alone, relying on O-planes to overcome the no-go theorem in the Introduction.

Four-dimensional approach

The simplest method is perhaps working with an effective $d = 4$ potential as in Section 8.3.2. As discussed there, a general problem with that approach is that when M_6 is not Ricci-flat, it is hard to pick a finite-dimensional space $\mathcal{M}_{\rm fin}$ of metrics for the reduction; in other words, it is not clear if the scalars we are including are really the lightest and separated from the others by a hierarchy. The danger is that we are not describing a "valley" in the space of all metrics, but a random slice; if so, minimizing the potential along this slice does not guarantee minimization with respect to the other scalars. Ideally one should find a $d = 10$ lift of the proposed vacua. This is most likely to be possible when the space $\mathcal{M}_{\rm fin}$ has some geometrical relevance, such as for cosets, which sometimes lead to consistent truncations.

Early careful examples of this $d = 4$ strategy are [249, 763], on products or Riemann surfaces or of nilmanifolds (Section 4.4.4); the latter uses some extra ingredients beyond those of Section 8.3.2. One is KK5-monopoles (Section 7.2.2); they are a geometrical space rather than a D-brane, but the idea is that they still give a localized contribution to the potential, which can be summed to the others in an appropriate regime. Another new ingredient is form potentials in the torsion part of cohomology (Section 4.1.10); these, however, cause issues [764]. Similar examples are [252, 765], and [766] on cosets with SU(3)-structures (discussed in more detail later in this chapter). A common issue with these vacua is that they have tachyons.

Solutions in ten dimensions with smeared O-planes

At the end of Section 8.2.3, we saw that when M_6 is noncompact, there even exist supersymmetric $\mathrm{dS}_4 \times M_6$ solutions of type II supergravity. Additional noncompact examples are analytically known [231, 767].

When M_6 is compact, the need to include O-planes complicates matters significantly. Since G-structures are useful for the study of supersymmetric compactifications, one can try to use them for dS vacua as well [768–770]. They are no longer directly invoked by the supersymmetric parameters, but they are still convenient in that the metric is determined by forms, which can then be used to write an Ansatz for the fluxes as well. Concretely, we can draw inspiration from (11.3.8) for $\mathcal{N} = 1$ SU(3)-structure AdS_4 vacua, but now changing the coefficients of J and W_2 in the fluxes [768]. With a smeared O6-plane as in (11.3.23), the equations of motion reduce to a quadratic algebraic equation on W_2, which unfortunately has no known solution. Looking at broader SU(3)-structures, another Ansatz is as follows [769]:

$$\mathrm{Re} W_1 = W_2 = W_4 = W_5 = 0\,, \tag{12.1.4}$$

with the fluxes and the smeared source $\delta_{\rm O6}$ taken to be linear combinations of J, Ω and the remaining torsion form W_3. The equations of motion reduce to the condition $dW_3 \propto J^2$ and to three algebraic equations on W_3. These were found to admit a solution on $S^3 \times S^3$.[1]

Another approach is to use an internal space that is a group manifold (Section 4.4.1), taking the internal field-strengths to be invariant forms. The equations of motion reduce to algebraic equations on the structure constants and on the flux parameters, but some numerical solutions can be found with O5-planes [771].

[1] A potentially more general formalism, using pure spinor pairs, was set up in [144] and studied for solmanifolds, but solutions were obtained there only by generalizing the source terms.

As we discussed earlier, a smeared source is reasonable for a distribution of sources such as D-branes, but not for O-planes, which should be located at the fixed loci of an orientifold involution. So these solutions are still not completely physical. Smeared solutions can have localized counterparts, as for T-duality of D-brane solutions (Section 4.2.6), but sometimes also not [584].

Solutions in ten dimensions with cohomogeneity one

To avoid this smearing issue, one can draw inspiration from the various solutions with localized O-planes in Chapter 11. Some of the examples we saw, such as (11.2.16) and its AdS_4 descendant via (11.3.36), have cohomogeneity one: there are enough symmetries to reduce the equations of motion (and supersymmetry) to ordinary differential equations depending on a single coordinate z.

It is natural to apply the same strategy to dS vacua. The simplest Ansatz is as follows [438]:

$$ds^2_{10} = e^{2A} ds^2_{dS_4} + e^{-2A}(dz^2 + e^{2\lambda} ds^2_{M_5}), \tag{12.1.5}$$

with A, λ and the dilaton ϕ only dependent on $z \in I$, and M_5 a compact quotient of compact hyperbolic space. There do exist numerical solutions with two singularities at the two endpoints of I, which at leading order in z look like $O8_\pm$-planes; the $O8_-$ is of the type $h_0 = 0$ pointed out after (1.4.57), which has a strongly coupled region, where supergravity breaks down. The solutions (1.4.57) with $h_0 = 0$ are believed to exist in full string theory in spite of this because of their world-sheet definition; this might give hope that the solutions (12.1.5) also exist in string theory. The supergravity equations of motion in this region have an ambiguity because the O8 delta function is multiplied by a divergent function. A pessimistic interpretation would exclude these solutions [437], but the ambiguity is related to analytic conditions for the field fluctuations [772, sec. 4] and cannot be resolved in supergravity alone. So unfortunately, determining the validity of these simple solutions depends on applying the full string theory action in the near-$O8_-$ region. There also exist similar solutions with $O6_-$-planes [772, sec. 3], but the hole region cannot be resolved by M-theory as in Section 7.2.3, because of the nonzero F_0.

This illustrates a general issue. For AdS solutions, in Chapter 11 we identified O-plane singularities by comparing the leading-order behavior of the fields to those in flat space from Section 1.4.4; this could at least be cross-checked using holography. In the de Sitter case, we need more robust criteria to recognize an O-plane singularity before we can claim with certainty to have found a dS solution in supergravity.

Alternatives

Given the difficulties we have encountered in standard type II supergravity, several alternatives have been explored.

One is to consider string theory in $d > 10$. In this case, flat space is not a solution; but this is irrelevant if we are interested in dS compactifications, and indeed we have reviewed many dS schemes for massive IIA, for which flat $\mathbb{R}^{10}$ is not a solution either. As discussed following (1.1.18) for the bosonic string, finding a solution now requires being able to control the string corrections to the effective action, balancing them with the effective potential $e^{-2\phi}$ in (1.1.17); but this is not much different from the challenges we encountered in Section 12.1.1 anyway. A very early proposal for dS solutions of this type in superstrings is [773]; a recent discussion in the bosonic string is in [774].

Another possibility is to consider T-folds (Section 4.2.6). As for the S-folds of Section 11.3.5, an alleged solution of this type should be scrutinized with care, because there are usually regions where some directions are very small, which might take us outside the supergravity approximation. Some dS proposals of this type are considered in [775, 776].

A possibility that currently looks more exotic is to consider condensates of fermionic bilinears. Such effects are well known in gauge theories, and we mentioned it briefly in Section 12.1.1 on D7 world-volumes. *If* similar effects happen in supergravity, it has been argued that they would lead to dS vacua [777, 778].

Finally, several alternatives to dS vacua have been considered, where dark energy is produced in some other way. We will discuss these in the next section.

12.2 Swampland conjectures

The swampland program takes inspiration from string theory, but has a broader ambition. Its aim is to characterize which QFT models can be coupled to a theory of quantum gravity. The more obstructions we find in this program, the more we can claim quantum gravity (and string theory) is useful.

Here we will focus on the aspects more relevant to the topics of this book; for extensive reviews, see [779, 780].

No global symmetries

One of the oldest conjectures regarding quantum gravity is that it should have no global symmetries G. Suppose we take particles charged under G, and throw them in a black hole. The black hole will evaporate via Hawking radiation; if it leaves nothing behind, G is broken. This conclusion can be avoided by postulating that the evaporation stops when the semiclassical Hawking computation breaks down. This might reasonably happen when the black hole has mass $\sim m_{\text{Pl}}$; at this stage, its Compton wavelength coincides with its Schwarzschild radius, and we have entered in a quantum gravity regime, where the black hole has effectively become a particle. However, this would mean that the spectrum of the theory contains a "remnant" particle for each possible charge under G, of mass $\sim m_{\text{Pl}}$. For $G = \text{U}(1)$, this implies an infinite number of remnants, which creates various problems in quantum field theory [781]. For recent discussions, see [782–784]; the last reference gives an independent argument using holography.

Weak gravity conjecture

A gauge symmetry with very small coupling g is very similar to a continuous global symmetry. So one would expect a version of the problem to arise already for a weakly coupled gauge symmetry. To quantify this for $G = \text{U}(1)$, recall that extremal black holes (11.1.1) have $gq = G_{\text{N}}^{1/2} m = m/m_{\text{Pl}}$; for $gq < m/m_{\text{Pl}}$ we have two distinct horizons, while for $gq > m/m_{\mathrm{Pl}}$ we have a naked singularity, which the cosmic censorship conjecture postulates not to occur in reality. Since remnants should have $m \sim m_{\mathrm{Pl}}$, there could conceivably be $\sim 1/g$ of them; this is now a finite number, but it can be made arbitrarily large, which would still lead to QFT problems.

But if there is at least one particle such that

$$m \leq gq\,, \tag{12.2.1}$$

a black hole can emit it by Hawking radiation, thereby losing more charge than mass, and lowering its charge-to-mass ratio q/m. So a black hole can lose all its charge while it evaporates, which would allow it to decay, leaving nothing behind, without violating charge conservation; so there would be no remnants after all. If only particles violating (12.2.1) exist, then q/m can only grow during evaporation and the remnants would exist. So we are led to conjecture that in quantum gravity, there must exist at least one particle satisfying (12.2.1).

Equation (12.2.1) is called the *weak gravity conjecture (WGC)* [785], because on this particle gravity is weaker than the force associated to the gauge symmetry. For supersymmetric solutions, the BPS bound $m \geq gq$ means that there should be at least one particle saturating it, $m = gq$. For nonsupersymmetric solutions, generically we expect the particle satisfying (12.2.1) to have $m < gq$, "violating" the BPS bound.

The conjecture was tested in various ways in string theory, including in heterotic [785] and type II [786]; see [780, sec. 4] Several refinements and generalizations have been considered, for example to theories with multiple abelian gauge groups [787], to theories with scalars [788], and to p-form potentials with $p > 1$ [789].

AdS instability

The p-form version of the WGC has a particularly important implication. The objects charged under it are branes; in the spirit of (12.2.1), we are led to conjecture that there should exist at least one brane saturating the BPS bound for supersymmetric solutions, and violating it for nonsupersymmetric ones. In particular, for a $(d-1)$-form potential, in the language of (11.1.15), the BPS bound reads $(d-1)T \geq Q$, so we are led to conjecture that there is at least one D$(d-2)$ satisfying

$$(d-1)T \leq Q\,. \tag{12.2.2}$$

Moreover, in a nonsupersymmetric case, one would generically expect the D$(d-2)$-brane satisfying this to have $(d-1)T < Q$. This is exactly the case that causes an instability (Section 11.1.3). For example, for $d = 4$ this predicts that nonsupersymmetric vacua should be unstable toward nucleating D2-brane bubbles, coupling to C_3.

One may object that the relevant potential might just be absent in the solution. In $d = 4$, there might exist solutions with $C_3 = 0$ in the spacetime, or in other words where the internal $f_6 = 0$. The argument around (11.3.37) implies that this should be the case for any solution with an O6 wrapping a topologically nontrivial cycle; the supersymmetric approximate solution (11.3.39) is like that. But even in such solutions, there will be another internal flux f_{6-k}, and the preceding argument would then apply to a D$(2+k)$-brane bubbles wrapping an internal k-cycle.

Perhaps more importantly, the argument relies on a genericity assumption, which may have some outliers.[2] In any case, in view of these considerations it was conjectured in [790, 791] that all nonsupersymmetric AdS vacua are unstable. As we mentioned in Section 11.1.3, few examples are known without tachyons anyway.

Distance conjecture

Other conjectures have been inspired by peculiar features of string theory. A common thread of Section 1.4 was the emergence of light states in various limits; this is how we derived all our dualities. For T-duality, shrinking an S^1 leads to a tower of light states (0.2.1) becoming light. Abstracting from this and other examples suggests the *swampland distance conjecture* [792]: if a theory coupled to quantum gravity has a moduli space $\mathcal{M}$, there is a point $q \in \mathcal{M}$ at infinite distance, and a tower of states with mass proportional to

$$m \sim \mathrm{e}^{-\alpha d} \tag{12.2.3}$$

for some $\alpha > 0$, as a function of distance d. The latter is computed with the metric g_{ij} appearing in the kinetic terms $g_{ij}\partial_\mu \sigma^i \partial^\mu \sigma^j$.

For T-duality, the moduli space is $\mathcal{M} = \mathbb{R}_+$, parameterizing the size L of the S^1. From (8.1.10) and (8.1.12), one finds that $\sigma = \log(L/L_0)$ is the canonically normalized scalar, so the line element on $\mathcal{M}$ is $\mathrm{d}s^2 = \mathrm{d}\sigma^2$. There are two points at infinite distance: one is when $\sigma \to \infty$, where (0.2.1) with $L = L_0 \mathrm{e}^{\sigma}$ gives

$$m = 2\pi k \mathrm{e}^{-\sigma} \tag{12.2.4}$$

for the smallest mass in the KK tower. The other is $\sigma \to -\infty$, where the S^1 shrinks, and we have a tower of modes coming from strings winding l times, of mass $m = 2\pi l T_{\mathrm{F1}} \mathrm{e}^{\sigma}$. The two limits are of course T-dual.

Another example is a singular locus in the complex structure moduli space $\mathcal{M}_{\mathrm{c}}$ of a Calabi–Yau. The conifold point is at finite distance (Exercise 9.1.4), but there can be other loci at infinite distance; it can be checked that there is indeed a tower of D3-branes getting massless with an exponential rate, consistent with (12.2.3) [786].

It has even been proposed that the point of view should be reversed, and that the infinite distance loci in moduli space are in fact caused by integrating out the asymptotically massless towers of states. (This is a bit similar to how the singularity at the finite-distance conifold point is reproduced by integrating out a single massless D3; see Section 9.1.2.) This suggests the *emergence proposal* [786, 793, 794]: that perhaps *all* kinetic terms might be generated by integrating out some fields at some high-energy scale, and that the fields we observe are actually composite.

A particular limit in string compactifications is $\Lambda \to 0$. This is not quite covered by the swampland distance conjecture (12.2.3) because usually we cannot take it while following a modulus. Λ is fixed in terms of the flux integers; taking those to infinity does achieve $\Lambda \to 0$, but it is a discrete rather than continuous limit. If we ignore this and proceed anyway, we can use the distance (8.1.21) to find a logarithmic distance

[2] An AdS_8 solution was found [551] with an O8-plane; the only flux is F_0, and the D8 brane bubble that would couple to it is forbidden by consistency with the O8. However, this solution presents the same subtleties of the dS solutions with O8s considered in Section 12.1.2.

on the space of Λ; (12.2.3) then leads to conjecture the appearance of a tower of states with mass proportional to the following [795]:

$$m \sim |\Lambda|^{\alpha} \tag{12.2.5}$$

for some $\alpha > 0$.

Recall also (8.3.2): for the vast majority of vacua, there is no scale separation between the Λ and $m_{\rm KK}$, or in other words the scale of the internal M_6 is of the same order as the AdS curvature radius. This might suggest a stronger version of (12.2.5), namely that $\alpha = 1/2$ for all supersymmetric vacua [795]. This would mean in particular that there is something wrong with the few approximate solutions with scale separation that we have seen earlier: the KKLT AdS solution (12.1.2), and the vacua we saw in (8.3.14) and (11.3.39) [253]. Recall from our earlier discussions that the latter are motivated by an effective $d = 4$ approach [253] and were later given an uplift to $d = 10$ with smeared O6-planes [686] and an approximate one with localized O6s [687, 689]. (From (8.3.15) and the scaling of Λ that follows it, we find $m \sim \Lambda^{1/18}$ in this case.)

Inconsistency of de Sitter?

Among the swampland conjectures we have seen so far, some were motivated by general quantum gravity arguments, such as the WGC conjecture (12.2.1). Others were motivated by extrapolation from large sets of examples, such as the distance conjecture (12.2.3). The spirit of this second procedure is that if something happens many times, perhaps it does so for a deep reason. The danger is, of course, that we might miss important sporadic counterexamples to general trends.

With this in mind, a more speculative conjecture is that there are no dS vacua in quantum gravity [796, 797]. This is motivated by the difficulty to prove the existence of dS vacua beyond any doubt (Section 12.1). Looking at the no-go arguments in this book even suggests a general property of the $d = 4$ potential V that might forbid dS: for example, once we canonically normalize g_{s4} and r, (8.3.12) becomes a bound $|\nabla V|/V > \sqrt{54/13}M_{\rm Pl}$ [251]. Once again extrapolating from this and other examples leads to a conjecture that $|\nabla V|/V > cM_{\rm Pl}$ for some order-one c [797]. (A different argument had been suggested earlier in [798].)

This would rule out even unstable dS vacua, such as the one at the top of the Higgs potential in the Standard Model [799]. But an argument from the distance conjecture revealed a way out [800]. A bound on $|\nabla V|$ was found in regions near the boundary of moduli space $\mathcal{M}$, by comparing the entropy of the light fields in the tower of states (12.2.3) with the entropy associated to the cosmological horizon in a space that is dS (discussed after (4.5.27)) or near-dS. The argument, however, is not applicable when there are tachyons, and more specifically when the Hessian matrix in $\mathcal{M}$ satisfies $\nabla_i \nabla_j V \leq -c' V M_{\rm Pl}$. This would allow unstable dS, while excluding stable dS; see also [801]. This entropy argument is valid near the boundary of $\mathcal{M}$, so it is an extension of the Dine–Seiberg argument in the Introduction.

This conjecture, if correct, would seem to imply that string theory is in conflict with observations. But strictly speaking, dark energy does not automatically imply a dS vacuum. An old alternative is called quintessence [802, 803], although this seems even harder to obtain than dS vacua: rather than a vacuum, it requires rolling down a very flat potential (for a recent assessment, see [804]). Other recent alternatives include a proposal where gravity is localized to a four-dimensional shell inside an AdS_5 [805] and one for a FLRW cosmological solution (4.5.20) in F-theory [806].

12.3 Epilogue

We have now come to the end of the book.

A summary

We have seen several geometrical techniques that help understanding string theory compactifications. Because of the connection between spinors and forms (Chapters 2 and 3), supersymmetry is intimately connected to the geometry of forms, and notably to G-structures (Chapter 5). This has been particularly useful for the vacuum problem, where it has led, for example, to the pure spinor equations (Chapter 10) and to several classification results for AdS solutions. We have mentioned broader applications, for example to nonvacuum solutions (Section 8.4.3) and to reductions (Section 10.3.6).

Concretely, the results of these methods have been most impressive for AdS vacua (Chapter 11). Not only there is now a rich catalogue of exact solutions, but in some cases, such as for supersymmetric AdS_7 and AdS_6, that list is believed to be a complete classification. In lower dimensions and especially without supersymmetry, the picture is less clear, but a lot more progress is likely in the near future.

For Minkowski vacua, several important classes are known (Chapter 9). They are currently all based on Calabi–Yau and Kähler geometry, even though the general analysis of supersymmetry with pure spinor methods seem to suggest the existence of larger classes. It remains to be seen if this is a selection effect due to our present technical limitations, or if it has a deeper reason.

All these geometrical methods have been developed mostly to construct and classify supersymmetric solutions, and so far they have not been very efficient at solving what is perhaps the most important problem in string compactifications, the existence of de Sitter vacua. This is in part because those methods can and should be refined further, but mostly because of physics issues that geometry alone cannot help with: our very incomplete knowledge of string theory corrections to supergravity, and of O-plane back-reaction.

In absence of exact results, a long list of approximate dS vacua have been proposed, mostly using effective field theory methods (Section 12.1). On the other hand, general arguments in quantum gravity, together with some extrapolation from the more solid exact solutions, has led to conjecturing that there are no dS vacua at all (Section 12.2). These opposite conclusions might almost look embarrassing, but such temporary disagreements are not uncommon in science, and should in fact be regarded as healthy. I think it is fair to say that neither side of the argument has achieved full rigor so far. Many dS proposals are too vague and should be more fully worked out; many rely on optimistic assumptions such as fine-tunings, or that some string corrections can be ignored while others are taken into account, or that smeared sources can be localized. On the other hand, while some of the swampland conjectures are based on general and solid arguments, some are based on generalizations from known exact examples, which might be premature. My personal expectation is that metastable dS vacua do exist in string theory, and that in the end some models such as KKLT will survive in some form. However, this is a gut feeling, and I don't regard the issue as settled.

Outlook

While great progress has been obtained in AdS vacua, much of it in recent years, there is probably still a lot to do. I suspect a lot of interesting structure remains in the analysis of AdS_4 solutions with extended supersymmetry, and a classification is conceivable in the near future for $\mathcal{N} \geq 3$. For $\mathcal{N} = 1$ solutions, a full classification might be too much to hope, but at least the list of known solutions can be expanded in various directions; for example, there probably exist many more $SU(3) \times SU(3)$-structure solutions than the few in Section 11.3.3.

It would be important to break free from the need to find explicit solutions. Calabi–Yau metrics are difficult to obtain explicitly, but Yau's theorem reduces their existence to some properties that are simple to check using algebraic geometry. More recently, the Kähler–Einstein problem was reduced to a so-called stability condition (Section 7.4.7). One might be able to reduce the D-term equations to stability conditions along similar lines, simplifying the system for Minkowski vacua, as briefly commented in Section 10.3.6.

Some of the ingredients for finding new solutions are in dire need of better understanding. Quantum corrections are usually treated in a four-dimensional effective approach (Section 12.1.1); it would also be important to understand them directly in $d = 10$ and 11. As we stressed multiple times, the back-reaction of sources cannot be described by supergravity alone, because the string coupling and/or the curvature diverge near them. If we want to include sources in supergravity solutions, we need to recognize which singularities to allow. So far, this has been done by comparing with flat-space behavior, but it is a priori not too clear why this procedure is correct, and up to what order in distance the comparison should be carried out. This issue has created confusion both for the anti-D3-branes in the KKLT construction and for proposed dS supergravity solutions with O8- and O6-planes (Section 12.1). Since O-planes are one of the main antidotes against dS no-go arguments, this issue is particularly urgent.

Among the proposed possible alternatives to O-planes, there are also T-folds and S-folds, solutions that are made compact by using transition functions that are symmetries of string theory but not supergravity. These have been studied formally quite a bit; many are simply dual of supergravity solutions and should be trusted, but those that are not inevitably involve small cycles or large string coupling, and might lie beyond the supergravity approximation. A relatively new development is that some cases can be checked using holography (Section 11.3.5); finding more examples of this sort would help build confidence in this technique.

G-structures and other geometrical methods can be useful beyond the vacuum problem (Section 8.4.3). It is possible to write systems similar to the pure spinor equations to lift more general $d = 4$ solutions, such as black holes or their horizons, but a probably more promising avenue is to expand the current techniques to associate a consistent truncation to a vacuum. These once relied heavily on internal symmetries but have now been related to generalized complex and exceptional geometry (Section 11.1.2).

The geometrical methods of this book were mostly developed to look for supersymmetric solutions; for nonsupersymmetric ones, the motivations to use G-structures or pure spinors are less clear. Nevertheless, at least *some* classes of solutions are still captured by these methods (Section 11.1.2). For example, one might hope that the pure spinor equations might be replaced by a system involving second-order operators instead of d_H. So far, however, the most practical procedure to find nonsupersymmetric solutions has been to use consistent truncations. I hope some of these methods will eventually be useful for de Sitter vacua as well.

The vacuum selection problem is central to string theory. I hope I have managed to convince the reader that it is also beautiful. Even should string theory not be our last step in our quest for the correct theory of reality, the geometrical ideas behind it will be one of its most enduring aspects.

References

[1] M. B. Green, J. H. Schwarz, and E. Witten, *Superstring Theory*, vol. 1, 2, Cambridge University Press, 1988.

[2] J. Polchinski, *String Theory*. Cambridge Monographs on Mathematical Physics. Cambridge University Press, 1998.

[3] B. Zwiebach, *A First Course in String Theory*. Cambridge University Press, 2004.

[4] K. Becker, M. Becker, and J. H. Schwarz, *String Theory and M-Theory: A Modern Introduction*. Cambridge University Press, 2006.

[5] E. Kiritsis, *String Theory in a Nutshell*, vol. 21. Princeton University Press, 2019.

[6] L. Ibáñez and A. Uranga, *String Theory and Particle Physics: An Introduction to String Phenomenology*. Cambridge University Press, 2012.

[7] R. Blumenhagen, D. Lüst, and S. Theisen, *Basic Concepts of String Theory*. Springer Science & Business Media, 2012.

[8] P. West, *Introduction to Strings and Branes*. Cambridge University Press, 2012.

[9] V. Schomerus, *A Primer on String Theory*. Cambridge University Press, 2017.

[10] M. Graña and H. Triendl, *String Theory Compactifications*. Springer, 2017.

[11] M. Polyakov, "Gauge fields and strings," in *Contemporary Concepts in Physics*, vol. 3. Harwood Academic Publishers, 1987.

[12] T. Kaluza, "Zum unitätsproblem der physik," *International Journal of Modern Physics* **1921** (1921), 1803.08616.

[13] O. Klein, "Quantentheorie und fünfdimensionale relativitätstheorie," *Zeitschrift für Physik* **37** (1926), no. 12, 895–906.

[14] J. F. Plebanski, "On the separation of Einsteinian substructures," *Journal of Mathematical Physics* **18** (1977) 2511–2520.

[15] R. Capovilla, J. Dell, T. Jacobson, and L. Mason, "Self-dual 2-forms and gravity," *Classical and Quantum Gravity* **8** (1991) 41–57.

[16] G. W. Gibbons, "Aspects of supergravity theories," Lectures at San Feliu de Guixols, Spain, June 1984.

[17] B. de Wit, D. J. Smit, and N. D. Hari Dass, "Residual supersymmetry of compactified $d = 10$ supergravity," *Nuclear Physics* **B283** (1987) 165.

[18] J. M. Maldacena and C. Núñez, "Supergravity description of field theories on curved manifolds and a no-go theorem," *International Journal of Modern Physics* **A16** (2001) 822–855, `hep-th/0007018`.

[19] S. Hawking and G. Ellis, *The Large Scale Structure of Space-Time*, Cambridge University Press, 1975.

[20] R. M. Wald, *General Relativity*. University of Chicago Press, 2010.

[21] M. Dine and N. Seiberg, "Is the superstring weakly coupled?," *Physics Letters* **162B** (1985) 299–302.

[22] S. Kachru, R. Kallosh, A. Linde, and S. P. Trivedi, "De Sitter vacua in string theory," *Physical Review* **D68** (2003) 046005, hep-th/0301240.

[23] L. Susskind, "The anthropic landscape of string theory," (2003), hep-th/0302219.

[24] C. Vafa, "The string landscape and the swampland," (2005), hep-th/0509212.

[25] M. H. Goroff and A. Sagnotti, "The ultraviolet behavior of Einstein gravity," *Nuclear Physics* **B266** (1986) 709–736.

[26] A. Sen and B. Zwiebach, "A proof of local background independence of classical closed string field theory," *Nuclear Physics* **B414** (1994) 649–714, hep-th/9307088.

[27] S. V. Ketov, *Quantum Nonlinear Sigma Models: From Quantum Field Theory to Supersymmetry, Conformal Field Theory, Black Holes and Strings.* Springer, 2000.

[28] U. Lindström, M. Roček, R. von Unge, and M. Zabzine, "Generalized Kaehler manifolds and off-shell supersymmetry," *Communications in Mathematical Physics* **269** (2007) 833–849, hep-th/0512164.

[29] N. Berkovits, "Super-Poincaré covariant quantization of the superstring," *Journal of High Energy Physics* **04** (2000) 018, hep-th/0001035.

[30] N. Berkovits, "ICTP lectures on covariant quantization of the superstring," *ICTP Lecture Notes Series* **13** (2003) 57–107, hep-th/0209059.

[31] D. J. Gross, J. A. Harvey, E. J. Martinec, and R. Rohm, "The heterotic string," *Physical Review Letters* **54** (1985) 502–505.

[32] G. Dall'Agata, G. Inverso, and M. Trigiante, "Evidence for a family of SO(8) gauged supergravity theories," *Physical Review Letters* **109** (2012) 201301, 1209.0760.

[33] J. H. Schwarz, "Superstring theory," *Physics Reports* **89** (1982) 223–322.

[34] M. T. Grisaru, A. E. M. van de Ven, and D. Zanon, "Four loop beta function for the $\mathcal{N} = 1$ and $\mathcal{N} = 2$ supersymmetric nonlinear sigma model in two dimensions," *Physics Letters* **B173** (1986) 423–428.

[35] M. J. Duff, J. T. Liu, and R. Minasian, "Eleven-dimensional origin of string-string duality: a one loop test," *Nuclear Physics* **B452** (1995) 261–282, hep-th/9506126.

[36] D. Freed, J. A. Harvey, R. Minasian, and G. W. Moore, "Gravitational anomaly cancellation for M theory five-branes," *Advances in Theoretical and Mathematical Physics* **2** (1998) 601–618, hep-th/9803205.

[37] J. T. Liu and R. Minasian, "Higher-derivative couplings in string theory: dualities and the B-field," *Nuclear Physics* **B874** (2013) 413–470, 1301.3137.

[38] K. Peeters, P. Vanhove, and A. Westerberg, "Supersymmetric higher derivative actions in ten-dimensions and eleven-dimensions, the associated superalgebras and their formulation in superspace," *Classical and Quantum Gravity* **18** (2001) 843–890, hep-th/0010167.

[39] G. Policastro and D. Tsimpis, "R^4, purified," *Classical and Quantum Gravity* **23** (2006) 4753–4780, hep-th/0603165.

[40] L. J. Romans, "Massive N=2a supergravity in ten dimensions," *Physics Letters* **B169** (1986) 374.

[41] P. S. Howe, N. Lambert, and P. C. West, "A new massive type IIA supergravity from compactification," *Physics Letters B* **416** (1998) 303–308, hep-th/9707139.

[42] D. Tsimpis, "Massive IIA supergravities," *Journal of High Energy Physics* **10** (2005) 057, hep-th/0508214.

[43] E. Stückelberg, "Die wechselwirkungskräfte in der elektrodynamik und in der feldtheorie der kräfte," *Helvetica physica acta* **11** (1938) 225.

[44] W. Nahm, "Supersymmetries and their representations," *Nuclear Physics* **B135** (1978) 149.

[45] M. B. Green and J. H. Schwarz, "Extended supergravity in ten-dimensions," *Physics Letters* **122B** (1983) 143–147.

[46] J. H. Schwarz and P. C. West, "Symmetries and transformations of chiral $\mathcal{N} = 2$ $D = 10$ supergravity," *Physics Letters* **126B** (1983) 301–304.

[47] M. B. Green and J. H. Schwarz, "Anomaly cancellation in supersymmetric D=10 gauge theory and superstring theory," *Physics Letters* **149B** (1984) 117–122.

[48] A. Sen, "Covariant action for type IIB supergravity," *Journal of High Energy Physics* **07** (2016) 017, 1511.08220.

[49] A. Sen, "Self-dual forms: action, Hamiltonian and compactification," *Journal of Physics* **A53** (2020), no. 8, 084002, 1903.12196.

[50] M. B. Green and M. Gutperle, "Effects of D instantons," *Nuclear Physics* **B498** (1997) 195–227, hep-th/9701093.

[51] L. A. Dickey, *Soliton Equations and Hamiltonian Systems*. World Scientific, 2003.

[52] N. Seiberg and E. Witten, "Electric–magnetic duality, monopole condensation, and confinement in $\mathcal{N} = 2$ supersymmetric Yang–Mills theory," *Nuclear Physics* **B426** (1994) 19–52, hep-th/9407087.

[53] S. Coleman, *Aspects of Symmetry: Selected Erice Lectures*. Cambridge University Press, 1988.

[54] M. Marino, *Instantons and Large N: An Introduction to Non-Perturbative Methods in Quantum Field Theory*. Cambridge University Press, 2015.

[55] H. Kleinert, *Path Integrals in Quantum Mechanics, Statistics, Polymer Physics, and Financial Markets*. World Scientific, 2009.

[56] N. Dorey, T. J. Hollowood, V. V. Khoze, and M. P. Mattis, "The calculus of many instantons," *Physics Reports* **371** (2002) 231–459, hep-th/0206063.

[57] N. A. Nekrasov, "Seiberg–Witten prepotential from instanton counting," *Advances in Theoretical and Mathematical Physics* **7** (2003), no. 5, 831–864, hep-th/0206161.

[58] A. Lichnerowicz, *Problèmes globaux en mécanique relativiste*, vol. 833. Hermann & cie, 1939.

[59] Y. Choquet-Bruhat, *Introduction to General Relativity, Black Holes, and Cosmology*. Oxford University Press, 2014.

[60] J. Polchinski, "Dirichlet–Branes and Ramond–Ramond charges," *Physical Review Letters* **75** (1995) 4724–4727, hep-th/9510017.

[61] E. Bergshoeff and M. De Roo, "D-branes and T-duality," *Physics Letters* **B380** (1996) 265–272, hep-th/9603123.

[62] M. B. Green, C. M. Hull, and P. K. Townsend, "D-brane Wess–Zumino actions, T-duality and the cosmological constant," *Physics Letters* **B382** (1996) 65–72, hep-th/9604119.

[63] A. Sen, "NonBPS states and Branes in string theory," in *Supersymmetry in the Theories of Fields, Strings and Branes. Proceedings, Advanced School, Santiago de Compostela, Spain, July 26–31, 1999*, pp. 187–234. 1999. hep-th/ 9904207. [,45(1999)].

[64] A. Sen, "Tachyon condensation on the brane anti-brane system," *Journal of High Energy Physics* **08** (1998) 012, hep-th/9805170.

[65] A. Sen and B. Zwiebach, "Tachyon condensation in string field theory," *Journal of High Energy Physics* **03** (2000) 002, hep-th/9912249.

[66] E. Witten, "D-branes and K-theory," *Journal of High Energy Physics* **12** (1998) 019, hep-th/9810188.

[67] R. Minasian and G. W. Moore, "K-theory and Ramond–Ramond charge," *Journal of High Energy Physics* **11** (1997) 002, hep-th/9710230.

[68] O. Aharony, S. S. Gubser, J. M. Maldacena, H. Ooguri, and Y. Oz, "Large N field theories, string theory and gravity," *Physics Reports* **323** (2000) 183–386, hep-th/9905111.

[69] T. Ortín, *Gravity and Strings*. Cambridge University Press, 2004.

[70] G. W. Gibbons, G. T. Horowitz, and P. K. Townsend, "Higher dimensional resolution of dilatonic black hole singularities," *Classical and Quantum Gravity* **12** (1995) 297–318, hep-th/9410073.

[71] E. Cremmer, B. Julia, and J. Scherk, "Supergravity theory in 11 dimensions," *Physics Letters* **B76** (1978) 409–412.

[72] E. Witten, "String theory dynamics in various dimensions," *Nuclear Physics* **B443** (1995) 85–126, hep-th/9503124.

[73] E. Bergshoeff, E. Sezgin, and P. Townsend, "Supermembranes and eleven-dimensional supergravity," *Physics Letters* **B189** (1987) 75–78.

[74] J. Bagger and N. Lambert, "Modeling multiple M2's," *Physical Review* **D75** (2007) 045020, hep-th/0611108.

[75] J. Bagger and N. Lambert, "Gauge symmetry and supersymmetry of multiple M2-Branes," *Physical Review* **D77** (2008) 065008, 0711.0955.

[76] A. Gustavsson, "Algebraic structures on parallel M2-branes," *Nuclear Physics B* **811** (2009) 66–76 0709.1260.

[77] O. Aharony, O. Bergman, D. L. Jafferis, and J. Maldacena, "$\mathcal{N} = 6$ superconformal Chern–Simons-matter theories, M2-branes and their gravity duals," *Journal of High Energy Physics* **10** (2008) 091, 0806.1218.

[78] P. Pasti, D. P. Sorokin, and M. Tonin, "Covariant action for a $D = 11$ five-brane with the chiral field," *Physics Letters* **B398** (1997) 41–46, hep-th/ 9701037.

[79] J. A. Harvey, R. Minasian, and G. W. Moore, "Nonabelian tensor multiplet anomalies," *Journal of High Energy Physics* **09** (1998) 004, hep-th/ 9808060.

[80] M. Henningson and K. Skenderis, "The holographic Weyl anomaly," *Journal of High Energy Physics* **07** (1998) 023, hep-th/9806087.

[81] K. A. Intriligator, "Anomaly matching and a Hopf–Wess–Zumino term in 6d, $\mathcal{N} = (2,0)$ field theories," *Nuclear Physics* **B581** (2000) 257–273, hep-th/0001205.

[82] T. Maxfield and S. Sethi, "The conformal anomaly of M5-Branes," *Journal of High Energy Physics* **06** (2012) 075, 1204.2002.

[83] C. Córdova, T. T. Dumitrescu, and X. Yin, "Higher derivative terms, toroidal compactification, and Weyl anomalies in six-dimensional (2, 0) theories," *Journal of High Energy Physics* **10** (2019) 128, 1505.03850.

[84] C. Beem, M. Lemos, L. Rastelli, and B. C. van Rees, "The (2, 0) superconformal bootstrap," *Physical Review* **D93** (2016), no. 2, 025016, 1507.05637.

[85] N. Lambert, C. Papageorgakis, and M. Schmidt-Sommerfeld, "Deconstructing (2,0) proposals," *Physical Review* **D88** (2013), no. 2, 026007, 1212.3337.

[86] C. Saemann and L. Schmidt, "Towards an M5-Brane model I: a 6d superconformal field theory," *Journal of Mathematical Physics* **59** (2018), no. 4, 043502, 1712.06623.

[87] O. Aharony, D. Jafferis, A. Tomasiello, and A. Zaffaroni, "Massive type IIA string theory cannot be strongly coupled," *Journal of High Energy Physics* **1011** (2010) 047, 1007.2451.

[88] E. Bergshoeff, Y. Lozano, and T. Ortín, "Massive branes," *Nuclear Physics* **B518** (1998) 363–423, hep-th/9712115.

[89] A. Sagnotti and T. N. Tomaras, "Properties of eleven-dimensional supergravity," 1982.

[90] K. Bautier, S. Deser, M. Henneaux, and D. Seminara, "No cosmological $D=11$ supergravity," *Physics Letters* **B406** (1997) 49–53, hep-th/9704131.

[91] S. Deser, "Uniqueness of $D = 11$ supergravity," 1997, hep-th/9712064.

[92] A. G. Tumanov and P. West, "E_{11}, Romans theory and higher level duality relations," *International Journal of Modern Physics* **A32** (2017), no. 05, 1750023, 1611.03369.

[93] C. M. Hull, "Massive string theories from M-theory and F-theory," *Journal of High Energy Physics* **11** (1998) 027, hep-th/9811021.

[94] N. Chaemjumrus and C. M. Hull, "Degenerations of K3, orientifolds and exotic Branes," *Journal of High Energy Physics* **10** (2019) 198, 1907.04040.

[95] A. Guarino, D. L. Jafferis, and O. Varela, "String theory origin of dyonic $\mathcal{N}=8$ supergravity and its Chern–Simons duals," *Physical Review Letters* **115** (2015), no. 9, 091601, 1504.08009.

[96] D. L. Jafferis and S. S. Pufu, "Exact results for five-dimensional superconformal field theories with gravity duals," *Journal of High Energy Physics* **05** (2014) 032, 1207.4359.

[97] S. Cremonesi and A. Tomasiello, "6d holographic anomaly match as a continuum limit," *Journal of High Energy Physics* **05** (2016) 031, 1512.02225.

[98] T. H. Buscher, "A symmetry of the string background field equations," *Physics Letters* **B194** (1987) 59–62.

[99] K. Hori and C. Vafa, "Mirror symmetry," 2000, hep-th/0002222.

[100] C. M. Hull, "Gravitational duality, branes and charges," *Nuclear Physics* **B509** (1998) 216–251, hep-th/9705162.

[101] E. Witten, "Bound states of strings and p-branes," *Nuclear Physics* **B460** (1996) 335–350, hep-th/9510135.

[102] C. Montonen and D. I. Olive, "Magnetic Monopoles as Gauge Particles?" *Physics Letters* **B72** (1977) 117.

[103] J. H. Schwarz, "An $\mathrm{SL}(2,\mathbb{Z})$ multiplet of type IIB superstrings," *Physics Letters* **B360** (1995) 13–18, hep-th/9508143. [Erratum: Phys. Lett.B364,252(1995)].

[104] P. S. Aspinwall, "Some relationships between dualities in string theory," *Nuclear Physics B – Proceedings Supplements* **46** (1996) 30–38, hep-th/9508154.

[105] A. Sagnotti, "Open strings and their symmetry groups," in *NATO Advanced Summer Institute on Nonperturbative Quantum Field Theory (Cargese Summer Institute) Cargese, France, July 16–30, 1987*, pp. 521–528. 1987. hep-th/0208020.

[106] P. Horava, "Strings on world sheet orbifolds," *Nuclear Physics* **B327** (1989) 461–484.

[107] J. Dai, R. G. Leigh, and J. Polchinski, "New connections between string theories," *Modern Physics Letters* **A4** (1989) 2073–2083.

[108] C. Angelantonj and A. Sagnotti, "Open strings," *Physics Reports* **371** (2002) 1–150, hep-th/0204089.

[109] A. Dabholkar, "Lectures on orientifolds and duality," in *High-Energy Physics and Cosmology. Proceedings, Summer School, Trieste, Italy, June 2–July 4, 1997*, pp. 128–191. 1997. hep-th/9804208.

[110] E. Witten, "Toroidal compactification without vector structure," *Journal of High Energy Physics* **02** (1998) 006, hep-th/9712028.

[111] O. Bergman, E. G. Gimon, and S. Sugimoto, "Orientifolds, RR torsion, and K theory," *Journal of High Energy Physics* **05** (2001) 047, hep-th/0103183.

[112] P. Horava and E. Witten, "Heterotic and type I string dynamics from eleven dimensions," *Nuclear Physics* **B460** (1996) 506–524, hep-th/9510209.

[113] J. Polchinski and E. Witten, "Evidence for heterotic – type I string duality," *Nuclear Physics* **B460** (1996) 525–540, hep-th/9510169.

[114] K. S. Narain, "New heterotic string theories in uncompactified dimensions < 10," *Physics Letters* **169B** (1986) 41–46.

[115] K. S. Narain, M. H. Sarmadi, and E. Witten, "A note on toroidal compactification of heterotic string theory," *Nuclear Physics* **B279** (1987) 369–379.

[116] H. Lawson and M. Michelsohn, *Spin Geometry*, vol. 38. Princeton University Press 1989.

[117] A. Bilal, J.-P. Derendinger, and K. Sfetsos, "(Weak) G_2 holonomy from selfduality, flux and supersymmetry," *Nuclear Physics* **B628** (2002) 112–132, hep-th/0111274.

[118] S. Cappell, D. DeTurck, H. Gluck, and E. Y. Miller, "Cohomology of harmonic forms on riemannian manifolds with boundary," in *Forum Mathematicum*, vol. 18, pp. 923–931. 2006.

[119] P. Griffiths and J. Harris, *Principles of Algebraic Geometry*. Wiley-Interscience [John Wiley & Sons], 1978.

[120] C. T. C. Wall, "Classification problems in differential topology. V," *Inventiones Mathematicae* **1** (1966), no. 4, 355–374.

[121] A. Giveon, E. Rabinovici, and G. Veneziano, "Duality in string background space," *Nuclear Physics* **B322** (1989) 167–184.

[122] K. A. Meissner and G. Veneziano, "Symmetries of cosmological superstring vacua," *Physics Letters* **B267** (1991) 33–36.

[123] A. Sen, "$O(d) \times O(d)$ symmetry of the space of cosmological solutions in string theory, scale factor duality and two-dimensional black holes," *Physics Letters* **B271** (1991) 295–300.

[124] A. Giveon and M. Rocek, "Generalized duality in curved string backgrounds," *Nuclear Physics* **B380** (1992) 128–146, hep-th/9112070.

[125] A. Giveon, M. Porrati, and E. Rabinovici, "Target space duality in string theory," *Physics Reports* **244** (1994) 77–202, hep-th/9401139.

[126] S. Hassan, "Supersymmetry and the systematics of T-duality rotations in type II superstring theories," *Nuclear Physics B – Proceedings Supplements* **102** (2001) 77–82, hep-th/0103149.

[127] S. Fidanza, R. Minasian, and A. Tomasiello, "Mirror symmetric SU(3)-structure manifolds with NS fluxes," *Communications in Mathematical Physics* **254** (2005) 401–423, hep-th/0311122.

[128] D. Tong, "NS5-branes, T-duality and world sheet instantons," *Journal of High Energy Physics* **0207** (2002) 013, hep-th/0204186.

[129] P. Bouwknegt, J. Evslin, and V. Mathai, "On the topology and H flux of T dual manifolds," *Physical Review Letters* **92** (2004) 181601, hep-th/0312052.

[130] S. Kachru, M. B. Schulz, P. K. Tripathy, and S. P. Trivedi, "New supersymmetric string compactifications," *Journal of High Energy Physics* **03** (2003) 061, hep-th/0211182.

[131] J. Shelton, W. Taylor, and B. Wecht, "Nongeometric flux compactifications," *Journal of High Energy Physics* **10** (2005) 085, hep-th/0508133.

[132] C. M. Hull, "A geometry for non-geometric string backgrounds," *Journal of High Energy Physics* **10** (2005) 065, hep-th/0406102.

[133] M. F. Atiyah and I. M. Singer, "The index of elliptic operators on compact manifolds," *Bulletin of the American Mathematical Society* **69** (1963), no. 3, 422–433.

[134] L. Alvarez-Gaumé and E. Witten, "Gravitational anomalies," *Nuclear Physics* **B234** (1984) 269.

[135] Y. Kosmann, "Dérivées de Lie des spineurs," *Annali di matematica pura ed applicata* **91** (1971), no. 1, 317–395.

[136] W. Ziller, "Homogeneous Einstein metrics on spheres and projective spaces," *Mathematische Annalen* **D72** (1982) 351–358.

[137] L. Magnin, "Sur les algèbres de Lie nilpotentes de dimension ≤ 7," *Journal of Geometry and Physics* **3** (1986), no. 1, 119–144.

[138] V. V. Morozov, "Classification of nilpotent Lie algebras of sixth order," *Izvestiya Vysshikh Uchebnykh Zavedenii. Matematika* (1958), no. 4, 161–171.

[139] G. R. Cavalcanti and M. Gualtieri, "Generalized complex structures on nilmanifolds," *Journal of Symplectic Geometry* **2** (2004), no. 3, 393–410. math/0404451 (2004).

[140] M. Graña, R. Minasian, M. Petrini, and A. Tomasiello, "A scan for new $\mathcal{N} = 1$ vacua on twisted tori," *Journal of High Energy Physics* **05** (2007) 031, hep-th/0609124.

[141] L. Auslander, "An exposition of the structure of solvmanifolds. Part I: Algebraic theory," *Bulletin of the American Mathematical Society* **79** (1973), no. 2, 227–261.

[142] M. Saito, "Sous-groups discrets des groups résolubles," *American Journal of Mathematics* **83** (1961), no. 2, 369–392.

[143] C. Bock, "On low-dimensional solvmanifolds," *Asian Journal of Mathematics* **20** (2016), no. 2, 199–262, 0903.2926.

[144] D. Andriot, E. Goi, R. Minasian, and M. Petrini, "Supersymmetry breaking branes on solvmanifolds and de Sitter vacua in string theory," *Journal of High Energy Physics* **05** (2011) 028. 1003.3774.

[145] S. Weinberg, *Gravitation and Cosmology: Principles and Applications of the General Theory of Relativity*. Wiley, 1972.

[146] J. Marsden and A. Weinstein, "Reduction of symplectic manifolds with symmetry," *Reports on Mathematical Physics* **5** (1974), no. 1, 121–130.

[147] A. Gray and L. Hervella, "The sixteen classes of almost Hermitian manifolds and their linear invariant," *Annali di Matematica Pura ed Applicata (IV)* **123** (1980) 35.

[148] S. Chiossi and S. Salamon, "The intrinsic torsion of SU(3) and G_2 structures," *International Conference on Differential Geometry held in honor of the 60th Birthday of A.M. Naveira*. math/0202282.

[149] M. Fernàndez and A. Gray, "Riemannian manifolds with structure group g_2," *Annali di matematica pura ed applicate* **132** (1982), no. 1, 19–45.

[150] M. Berger, "Les espaces symétriques noncompacts," *Annales scientifiques de l'École Normale Supérieure* **74** (1957) 85–177.

[151] Q.-S. Chi, S. A. Merkulov, and L. J. Schwachhöfer, "On the existence of infinite series of exotic holonomies," *Inventiones mathematicae* **126** (1996), no. 2, 391–411.

[152] R. Bryant, "Recent advances in the theory of holonomy," *Asterisquesociete Mathematique de France* **266** (2000) 351–374, hep-th/0202208.

[153] R. Hartshorne, *Algebraic Geometry*, vol. 52. Springer, 2013.

[154] J. Harris, *Algebraic Geometry: A First Course*, vol. 133. Springer, 2013.

[155] F. W. Olver, *NIST Handbook of Mathematical Functions*. Cambridge University Press, 2010.

[156] Arnold, S. Gusein-Zade, and A. Varchenko, *Singularities of Differentiable Maps*, vol. 2. Birkhäuser Boston, 2012.

[157] V. I. Arnold, V. V. Goryunov, O. Lyashko, and V. Vasilev, *Singularity Theory I*, vol. 6. Springer Science & Business Media, 2012.

[158] D. Mumford, J. Fogarty, and F. Kirwan, *Geometric Invariant Theory*, vol. 34. Springer, 1994.

[159] R. P. Thomas, "Notes on GIT and symplectic reduction for bundles and varieties," *Surveys in Differential Geometry* **10** (2005), no. 1, 221273, math/0512411.

[160] I. Dolgachev, *Lectures on Invariant Theory*. Cambridge University Press, 2003.

[161] R. E. Gompf, "A new construction of symplectic manifolds," *Annals of Mathematics* (1995) 527–595.

[162] P. Candelas, A. M. Dale, C. A. Lutken, and R. Schimmrigk, "Complete intersection Calabi–Yau manifolds," *Nuclear Physics* **B298** (1988) 493.

[163] P. Candelas, M. Lynker, and R. Schimmrigk, "Calabi–Yau manifolds in weighted $\mathbb{P}^4$," *Nuclear Physics* **B341** (1990) 383–402.

[164] M. Kreuzer and H. Skarke, "Complete classification of reflexive polyhedra in four dimensions," *Advances in Theoretical and Mathematical Physics* **4** (2002) 1209–1230, hep-th/0002240.

[165] Y.-H. He, "Calabi–Yau spaces in the string landscape," *Oxford Research Encyclopedia of Physics*, Oxford University Press, 2020 2006.16623.

[166] L. B. Anderson, F. Apruzzi, X. Gao, J. Gray, and S.-J. Lee, "A new construction of Calabi–Yau manifolds: generalized CICYs," *Nuclear Physics* **B906** (2016) 441–496, 1507.03235.

[167] B. R. Greene, D. R. Morrison, and A. Strominger, "Black hole condensation and the unification of string vacua," *Nuclear Physics* **B451** (1995) 109–120, hep-th/9504145.

[168] P. S. Aspinwall, "K3 surfaces and string duality," in *Theoretical Advanced Study Institute in Elementary Particle Physics (TASI 96)*, World Scientific, pp. 421–540. 1996. hep-th/9611137.

[169] S. Kachru, A. Tripathy, and M. Zimet, "K3 metrics from little string theory," 2018. 1810.10540.

[170] S. Kachru, A. Tripathy, and M. Zimet, "K3 metrics," 2020. 2006.02435.

[171] G. W. Gibbons and S. W. Hawking, "Gravitational multi-instantons," *Physics Letters* **78B** (1978) 430.

[172] S. Katmadas and A. Tomasiello, "Gauged supergravities from M-theory reductions," *Journal of High Energy Physics* **04** (2018) 048, 1712.06608.

[173] G. Franchetti, "Harmonic forms on ALF gravitational instantons," *Journal of High Energy Physics* **12** (2014) 075, 1410.2864.

[174] C. M. Hull and P. K. Townsend, "Unity of superstring dualities," *Nuclear Physics* **B438** (1995) 109–137, hep-th/9410167.

[175] M. Atiyah and N. J. Hitchin, "Low-energy scattering of nonabelian monopoles," *Physics Letters* **A107** (1985) 21–25.

[176] G. Gibbons and N. Manton, "Classical and quantum dynamics of BPS monopoles," *Nuclear Physics* **B274** (1986) 183.

[177] A. Hanany and B. Pioline, "(Anti-)instantons and the Atiyah–Hitchin manifold," *Journal of High Energy Physics* **0007** (2000) 001, hep-th/0005160.

[178] A. Fujiki, "On primitive symplectic compact Kahler V-manifolds of dimension four," *Progress in Mathematics, Birkhauser* **39** (1983) 71–125.

[179] S. Mukai, "Symplectic structure of the moduli space of sheaves on an abelian or K3 surface," *Inventiones mathematicae* **77** (1984), no. 1, 101–116.

[180] A. Beauville, "Variétés Kähleriennes dont la première classe de Chern est nulle," *Journal of Differential Geometry* **18** (1983), no. 4, 755–782 (1984).

[181] J. P. Gauntlett, G. W. Gibbons, G. Papadopoulos, and P. K. Townsend, "Hyper-Kähler manifolds and multiply intersecting branes," *Nuclear Physics* **B500** (1997) 133–162, hep-th/9702202.

[182] R. Bielawski and A. Dancer, "The geometry and topology of toric hyper-Kähler manifolds," *Communications in Analysis and Geometry* **8** (2000) 727–759.

[183] N. J. Hitchin, "Kählerian twistor spaces," *Proceedings of the London Mathematical Society* **3** (1981), no. 1, 133–150.

[184] C. Boyer and K. Galicki, *Sasakian Geometry*. Oxford University Press, 2008.

[185] R. L. Bryant and S. M. Salamon, "On the construction of some complete metrics with exceptional holonomy," *Duke Mathematical Journal* **58** (1989), no. 3, 829–850.

[186] G. W. Gibbons, D. N. Page, and C. N. Pope, "Einstein metrics on S^3, $\mathbb{R}^3$ and $\mathbb{R}^4$ bundles," *Communications in Mathematical Physics* **127** (1990) 529.

[187] L. Foscolo, M. Haskins, and J. Nordstörm, "Complete non-compact G_2-manifolds from asymptotically conical Calabi–Yau 3-folds," 2019. 1709.04904

[188] D. D. Joyce, *Compact Manifolds with Special Holonomy*. Oxford University Press on Demand, 2000.

[189] D. Joyce and S. Karigiannis, "A new construction of compact torsion-free G_2-manifolds by gluing families of Eguchi–Hanson spaces," *Journal of Differential Geometry* **117** (2021) 255–343.

[190] A. Kovalev, "Twisted connected sums and special Riemannian holonomy," *Journal für die reine und angewandte Mathematik* **565** (2003) 125–160 math/ 0012189.

[191] A. Corti, M. Haskins, J. Nordström, and T. Pacini, "G_2-manifolds and associative submanifolds via semi-Fano 3-folds," *Duke Mathematical Journal* **164** (2015), no. 10, 1971–2092, 1207.4470.

[192] C. Bär, "Real Killing spinors and holonomy," *Communications in Mathematical Physics* **154** (1993), no. 3, 509–521.

[193] J. M. Figueroa-O'Farrill, E. Hackett-Jones, and G. Moutsopoulos, "The Killing superalgebra of ten-dimensional supergravity backgrounds," *Classical and Quantum Gravity* **24** (2007) 3291–3308, hep-th/0703192.

[194] P. de Medeiros, J. Figueroa-O'Farrill, and A. Santi, "Killing superalgebras for Lorentzian four-manifolds," *Journal of High Energy Physics* **06** (2016) 106, 1605.00881.

[195] H. Lu, C. N. Pope, and J. Rahmfeld, "A construction of Killing spinors on S^n," *Journal of Mathematical Physics* **40** (1999) 4518–4526, hep-th/9805151.

[196] S. S.-T. Yau, "Kohn–Rossi cohomology and its application to the complex Plateau problem. I," *Annals of Mathematics (2)* **113** (1981) 67–110.

[197] G. Tian and S.-T. Yau, "Kähler–Einstein Metrics on Complex Surfaces With $c_1 > 0$," *Communications in Mathematical Physics* **112** (1987) 175–203.

[198] E. Witten, "Search for a Realistic Kaluza–Klein Theory," *Nuclear Physics* **B186** (1981) 412.

[199] L. Castellani, R. D'Auria, and P. Fre, "SU(3) × SU(2) × U(1) from $D = 11$ supergravity," *Nuclear Physics* **B239** (1984) 610–652.

[200] L. Castellani and L. J. Romans, "$\mathcal{N} = 3$ and $\mathcal{N} = 1$ supersymmetry in a new class of solutions for $d = 11$ supergravity," *Nuclear Physics* **B238** (1984) 683.

[201] R. D'Auria, P. Fre, and P. van Nieuwenhuizen, "$\mathcal{N} = 2$ matter coupled supergravity from compactification on a coset G/H possessing an additional Killing vector," *Physics Letters* **136B** (1984) 347–353.

[202] L. Castellani, L. J. Romans, and N. P. Warner, "A classification of compactifying solutions for $d = 11$ supergravity," *Nuclear Physics* **B241** (1984) 429–462.

[203] G. Tian, "On Kähler–Einstein metrics on certain Kähler manifolds with $c_1(m) > 0$," *Inventiones mathematicae* **89** (1987), no. 2, 225–246.

[204] A. M. Nadel, "Multiplier ideal sheaves and existence of Kähler–Einstein metrics of positive scalar curvature," *Proceedings of the National Academy of Sciences* **86** (1989), no. 19, 7299–7300.

[205] A. Futaki, H. Ono, and G. Wang, "Transverse Kahler geometry of Sasaki manifolds and toric Sasaki–Einstein manifolds," *Journal of Differential Geometry* **83** (2009) 585–636, math/0607586.

[206] J. P. Gauntlett, D. Martelli, J. Sparks, and D. Waldram, "Sasaki–Einstein metrics on $S^2 \times S^3$," *Advances in Theoretical and Mathematical Physics* **8** (2004) 711–734, hep-th/0403002.

[207] D. Martelli and J. Sparks, "Toric geometry, Sasaki–Einstein manifolds and a new infinite class of AdS/CFT duals," *Communications in Mathematical Physics* **262** (2006) 51–89, hep-th/0411238.

[208] D. Conti, "Cohomogeneity one Einstein–Sasaki 5-manifolds," *Communications in Mathematical Physics* **274** (2007), no. 3, 751–774.

[209] M. Cvetic, H. Lu, D. N. Page, and C. N. Pope, "New Einstein–Sasaki spaces in five and higher dimensions," *Physical Review Letters* **95** (2005) 071101, hep-th/0504225.

[210] J. P. Gauntlett, D. Martelli, J. F. Sparks, and D. Waldram, "A new infinite class of Sasaki–Einstein manifolds," *Advances in Theoretical and Mathematical Physics* **8** (2006) 987–1000, hep-th/0403038.

[211] D. Martelli and J. Sparks, "Notes on toric Sasaki–Einstein seven-manifolds and AdS_4/CFT_3," *Journal of High Energy Physics* **11** (2008) 016, 0808.0904.

[212] J. P. Gauntlett, D. Martelli, J. Sparks, and D. Waldram, "Supersymmetric AdS backgrounds in string and M-theory," *IRMA Lectures in Mathematics and Theoretical Physics* **8** (2005) 217–252 hep-th/0411194.

[213] W. Chen, H. Lu, C. N. Pope, and J. F. Vazquez-Poritz, "A note on Einstein–Sasaki metrics in $d \geq 7$," *Classical and Quantum Gravity* **22** (2005) 3421–3430, hep-th/0411218.

[214] A. Tomasiello and A. Zaffaroni, "Parameter spaces of massive IIA solutions," *Journal of High Energy Physics* **1104** (2011) 067, 1010.4648.

[215] D. Martelli, J. Sparks, and S.-T. Yau, "The geometric dual of a-maximisation for toric Sasaki–Einstein manifolds," *Communications in Mathematical Physics* **268** (2006) 39–65, hep-th/0503183.

[216] D. Martelli, J. Sparks, and S.-T. Yau, "Sasaki–Einstein manifolds and volume minimisation," *Communications in Mathematical Physics* **280** (2008) 611–673, hep-th/0603021.

[217] J. Sparks, "Sasaki–Einstein manifolds," *Surveys in Differential Geometry* **16** (2011) 265–324, 1004.2461.

[218] S. Yau, "Open problems in geometry," in *Proceedings of Symposia in Pure Mathematics*, vol. 54, pp. 1–28. 1993.

[219] G. Tian, "Kähler-Einstein metrics with positive scalar curvature," *Inventiones mathematicae* **130** (1997), no. 1, 1–37.

[220] X.-X. Chen, S. Donaldson, and S. Sun, "Kähler–Einstein metrics and stability," *International Mathematics Research Notices* **8** (2014) 2119–2125, `1210.7494`.

[221] T. C. Collins and G. Székelyhidi, "Sasaki–Einstein metrics and K-stability," *Geometry & Topology* **23** (2019) 1339–1413, `1512.07213`.

[222] N. Ilten and H. Süß, "K-stability for Fano manifolds with torus action of complexity one," *Duke Mathematical Journal* **166** (2017), no. 1, 177–204, `1507.04442`.

[223] P. Koerber, D. Lüst, and D. Tsimpis, "Type IIA AdS_4 compactifications on cosets, interpolations and domain walls," *Journal of High Energy Physics* **07** (2008) 017, `0804.0614`.

[224] L. Foscolo and M. Haskins, "New G_2 holonomy cones and exotic nearly Kaehler structures on the 6-sphere and the product of a pair of 3-spheres," *Annals of Mathematics* **185** (2017) 59–130, `1501.07838`.

[225] D. Gaiotto and A. Tomasiello, "Perturbing gauge/gravity duals by a Romans mass," *Journal of High Energy Physics* **01** (2010) p. 015. `0904.3959`.

[226] A. Kovalev, "Asymptotically cylindrical manifolds with holonomy spin(7). I," `1309.5027` (2013).

[227] A. P. Braun and S. Schäfer-Nameki, "Spin(7)-manifolds as generalized connected sums and 3d $\mathcal{N} = 1$ theories," *Journal of High Energy Physics* **06** (2018) 103, `1803.10755`.

[228] T. Friedrich, I. Kath, A. Moroianu, and U. Semmelmann, "On nearly parallel "G_2"-structures," *Journal of Geometry and Physics* **23** (1997), no. 3–4, 259–286.

[229] L. Randall and R. Sundrum, "A large mass hierarchy from a small extra dimension," *Physical Review Letters* **83** (1999) 3370–3373, `hep-ph/9905221`.

[230] L. Randall and R. Sundrum, "An alternative to compactification," *Physical Review Letters* **83** (1999) 4690–4693, `hep-th/9906064`.

[231] E. Ó Colgáin, M. M. Sheikh-Jabbari, J. F. Vázquez-Poritz, H. Yavartanoo, and Z. Zhang, "Warped Ricci-flat reductions," *Physical Review* **D90** (2014), no. 4, 045013, `1406.6354`.

[232] M. R. Douglas, "Effective potential and warp factor dynamics," *Journal of High Energy Physics* **03** (2010) 071, `0911.3378`.

[233] I. Chavel, *Eigenvalues in Riemannian Geometry*. Academic Press, 1984.

[234] P. Li and S. T. Yau, "Estimates of eigenvalues of a compact Riemannian manifold," in *Geometry of the Laplace operator (Proc. Sympos. Pure Math., Univ. Hawaii, Honolulu, Hawaii, 1979)*, Proceedings of Symposia in Pure Mathematics, XXXVI, pp. 205–239. American Mathematical Society, 1980.

[235] K. Hinterbichler, J. Levin, and C. Zukowski, "Kaluza–Klein towers on general manifolds," *Physical Review* **D89** (2014), no. 8, 086007, `1310.6353`.

[236] M. J. Duff, S. Ferrara, C. N. Pope, and K. S. Stelle, "Massive Kaluza–Klein modes and effective theories of superstring moduli," *Nuclear Physics* **B333** (1990) 783.

[237] M. Gabella, D. Martelli, A. Passias, and J. Sparks, "$\mathcal{N}$ = 2 supersymmetric AdS_4 solutions of M-theory," *Communications in Mathematical Physics* **325** (2014) 487–525, 1207.3082.

[238] C. Couzens, C. Lawrie, D. Martelli, S. Schafer-Nameki, and J.-M. Wong, "F-theory and AdS_3/CFT_2," *Journal of High Energy Physics* **08** (2017) 043, 1705.04679.

[239] U. Gran, J. B. Gutowski, and G. Papadopoulos, "On supersymmetric anti-de-Sitter, de-Sitter and Minkowski flux backgrounds," *Classical and Quantum Gravity* **35** (2018), no. 6, 065016, 1607.00191.

[240] J. P. Gauntlett, O. A. Mac Conamhna, T. Mateos, and D. Waldram, "AdS spacetimes from wrapped M5 branes," *Journal of High Energy Physics* **0611** (2006) 053, hep-th/0605146.

[241] K. Pilch, P. van Nieuwenhuizen, and M. F. Sohnius, "De Sitter superalgebras and supergravity," *Communications in Mathematical Physics* **98** (1985) 105.

[242] J. Lukierski and A. Nowicki, "All possible de Sitter superalgebras and the presence of ghosts," *Physical Letters*, **151B** (1985) 382–386.

[243] P. Candelas, G. T. Horowitz, A. Strominger, and E. Witten, "Vacuum configurations for superstrings," *Nuclear Physics* **B258** (1985) 46–74.

[244] S. K. Donaldson, "Anti self-dual Yang–Mills connections over complex algebraic surfaces and stable vector bundles," *Proceedings of the London Mathematical Society* **3** (1985), no. 1, 1–26.

[245] K. Uhlenbeck and S.-T. Yau, "On the existence of Hermitian-Yang–Mills connections in stable vector bundles," *Communications on Pure and Applied Mathematics* **39** (1986), no. S1, S257–S293.

[246] M. J. Duff, B. E. W. Nilsson, and C. N. Pope, "Kaluza–Klein supergravity," *Physics Reports* **130** (1986) 1–142.

[247] F. Denef, "Les Houches lectures on constructing string vacua," *Les Houches* **87** (2008) 483–610, 0803.1194.

[248] M. R. Douglas and S. Kachru, "Flux compactification," *Reviews of Modern Physics* **79** (2007) 733–796, hep-th/0610102.

[249] E. Silverstein, "Simple de Sitter solutions," *Physical Review* **D77** (2008) 106006, 0712.1196.

[250] X. Dong, B. Horn, E. Silverstein, and G. Torroba, "Micromanaging de Sitter holography," *Classical and Quantum Gravity* **27** (2010) 245020, 1005.5403.

[251] M. P. Hertzberg, S. Kachru, W. Taylor, and M. Tegmark, "Inflationary constraints on type IIA string theory," *Journal of High Energy Physics* **12** (2007) 095. 0711.2512.

[252] R. Flauger, S. Paban, D. Robbins, and T. Wrase, "Searching for slow-roll moduli inflation in massive type IIA supergravity with metric fluxes," *Physical Review* **D79** (2009) 086011, 0812.3886.

[253] O. DeWolfe, A. Giryavets, S. Kachru, and W. Taylor, "Type IIA moduli stabilization," *Journal of High Energy Physics* **07** (2005) 066, hep-th/0505160.

[254] F. F. Gautason, M. Schillo, T. Van Riet, and M. Williams, "Remarks on scale separation in flux vacua," *Journal of High Energy Physics* **03** (2016) 061, 1512.00457.

[255] R. D'Auria, S. Ferrara, M. Trigiante, and S. Vaula, "Gauging the Heisenberg algebra of special quaternionic manifolds," *Physics Letters* **B610** (2005) 147–151, hep-th/0410290.

[256] M. Graña, J. Louis, and D. Waldram, "Hitchin functionals in $\mathcal{N} = 2$ supergravity," *Journal of High Energy Physics* **01** (2006) 008, hep-th/0505264.

[257] J. Wess and J. Bagger, *Superspace and Supergravity*, Princeton University Press, 1992.

[258] D. Z. Freedman and A. Van Proeyen, *Supergravity*. Cambridge University Press, 2012.

[259] D. V. Volkov and V. P. Akulov, "Is the neutrino a Goldstone particle?" *Physics Letters* **46B** (1973) 109–110.

[260] E. A. Ivanov and A. A. Kapustnikov, "General relationship between linear and nonlinear realizations of supersymmetry," *Journal of Physics* **A11** (1978) 2375–2384.

[261] M. Rocek, "Linearizing the Volkov–Akulov model," *Physical Review Letters* **41** (1978) 451–453.

[262] R. Casalbuoni, S. De Curtis, D. Dominici, F. Feruglio, and R. Gatto, "Nonlinear realization of supersymmetry algebra from supersymmetric constraint," *Physics Letters* **B220** (1989) 569–575.

[263] Z. Komargodski and N. Seiberg, "From linear SUSY to constrained superfields," *Journal of High Energy Physics* **09** (2009) 066, 0907.2441.

[264] S. Ferrara, R. Kallosh, and A. Van Proeyen, "On the supersymmetric completion of $R + R^2$ gravity and cosmology," *Journal of High Energy Physics* **11** (2013) 134, 1309.4052.

[265] I. Antoniadis, E. Dudas, S. Ferrara, and A. Sagnotti, "The Volkov–Akulov–Starobinsky supergravity," *Physics Letters* **B733** (2014) 32–35, 1403.3269.

[266] E. A. Bergshoeff, D. Z. Freedman, R. Kallosh, and A. Van Proeyen, "Pure de Sitter Supergravity," *Physical Review* **D92** (2015), no. 8, 085040, 1507.08264.

[267] F. Hasegawa and Y. Yamada, "Component action of nilpotent multiplet coupled to matter in 4 dimensional $\mathcal{N} = 1$ supergravity," *Journal of High Energy Physics* **10** (2015) 106, 1507.08619.

[268] B. Craps, F. Roose, W. Troost, and A. Van Proeyen, "What is special Kähler geometry?," *Nuclear Physics* **B503** (1997) 565–613, hep-th/9703082.

[269] L. Andrianopoli, M. Bertolini, A. Ceresole, R. D'Auria, S. Ferrara, P. Fré, and T. Magri, "$\mathcal{N} = 2$ supergravity and $\mathcal{N} = 2$ super-Yang–Mills theory on general scalar manifolds: symplectic covariance, gaugings and the momentum map," *Journal of Geometry and Physics* **23** (1997) 111–189, hep th/9605032.

[270] D. S. Freed, "Special Kähler manifolds," *Communications in Mathematical Physics* **203** (May 1999) 3152.

[271] G. Dall'Agata, R. D'Auria, L. Sommovigo, and S. Vaulà, "$D = 4$, $\mathcal{N} = 2$ gauged supergravity in the presence of tensor multiplets," *Nuclear Physics* **B682** (2004) 243–264, hep-th/0312210.

[272] B. de Wit, H. Samtleben, and M. Trigiante, "On Lagrangians and gaugings of maximal supergravities," *Nuclear Physics* **B655** (2003) 93–126, hep-th/0212239.

[273] B. de Wit, H. Samtleben, and M. Trigiante, "Magnetic charges in local field theory," *Journal of High Energy Physics* **09** (2005) 016, hep-th/0507289.

[274] H. Samtleben, "Lectures on gauged supergravity and flux compactifications," *Classical and Quantum Gravity* **25** (2008) 214002, 0808.4076.

[275] M. Trigiante, "Gauged supergravities," *Physics Reports* **680** (2017) 1–175, 1609.09745.

[276] J. Louis, P. Smyth, and H. Triendl, "Spontaneous $\mathcal{N} = 2$ to $\mathcal{N} = 2$ supersymmetry breaking in supergravity and type II string theory," *Journal of High Energy Physics* **02** (2010) 103, 0911.5077.

[277] S. Cecotti, L. Girardello, and M. Porrati, "Two into one won't go," *Physics Letters* **145B** (1984) 61–64.

[278] D. Cassani and A. Bilal, "Effective actions and $\mathcal{N} = 1$ vacuum conditions from SU(3) × SU(3) compactifications," *Journal of High Energy Physics* **09** (2007) 076, 0707.3125.

[279] H. Stephani, D. Kramer, M. MacCallum, C. Hoenselaers, and E. Herlt, *Exact Solutions of Einsteins Field Equations*. Cambridge University Press, 2003.

[280] J. Gillard, U. Gran, and G. Papadopoulos, "The spinorial geometry of supersymmetric backgrounds," *Classical and Quantum Gravity* **22** (2005) 1033–1076, hep-th/0410155.

[281] U. Gran, J. Gutowski, and G. Papadopoulos, "Classification, geometry and applications of supersymmetric backgrounds," *Physics Reports* **794** (2019) 1–87, 1808.07879.

[282] G. Gibbons and C. Hull, "A Bogomolny bound for general relativity and solitons in $\mathcal{N} = 2$ supergravity," *Physics Letters* **B109** (1982) 190.

[283] K. Tod, "All metrics admitting supercovariantly constant spinors," *Physics Letters* **B121** (1983) 241–244.

[284] M. M. Caldarelli and D. Klemm, "Supersymmetry of anti-de Sitter black holes," *Nuclear Physics* **B545** (1999) 434–460, hep-th/9808097.

[285] S. L. Cacciatori, D. Klemm, D. S. Mansi, and E. Zorzan, "All timelike supersymmetric solutions of $\mathcal{N} = 2$, $D = 4$ gauged supergravity coupled to abelian vector multiplets," *Journal of High Energy Physics* **0805** (2008) 097, 0804.0009.

[286] P. Meessen and T. Ortín, "Supersymmetric solutions to gauged $\mathcal{N}=2$ $d=4$ sugra: the full timelike shebang," *Nuclear Physics* **B863** (2012) 65–89, 1204.0493.

[287] D. Rosa and A. Tomasiello, "Pure spinor equations to lift gauged supergravity," *Journal of High Energy Physics* **1401** (2014) 176, 1305.5255.

[288] U. Gran, J. Gutowski, and G. Papadopoulos, "Geometry of all supersymmetric four-dimensional $\mathcal{N} = 1$ supergravity backgrounds," *Journal of High Energy Physics* **06** (2008) 102, 0802.1779.

[289] J. P. Gauntlett, J. B. Gutowski, C. M. Hull, S. Pakis, and H. S. Reall, "All supersymmetric solutions of minimal supergravity in five dimensions," *Classical and Quantum Gravity* **20** (2003) 4587–4634, hep-th/0209114.

[290] J. B. Gutowski, D. Martelli, and H. S. Reall, "All supersymmetric solutions of minimal supergravity in six dimensions," *Classical and Quantum Gravity* **20** (2003) 5049–5078, hep-th/0306235.

[291] J. M. Figueroa-O'Farrill and G. Papadopoulos, "Maximally supersymmetric solutions of ten-dimensional and eleven-dimensional supergravities," *Journal of High Energy Physics* **03** (2003) 048, hep-th/0211089.

[292] U. Gran, J. Gutowski, G. Papadopoulos, and D. Roest, "$N = 31$ is not IIB," *Journal of High Energy Physics* **02** (2007) 044, hep-th/0606049.

[293] U. Gran, J. Gutowski, G. Papadopoulos, and D. Roest, "N=31, D=11," *Journal of High Energy Physics* **02** (2007) 043, hep-th/0610331.

[294] I. A. Bandos, J. A. de Azcarraga, and O. Varela, "On the absence of BPS preonic solutions in IIA and IIB supergravities," *Journal of High Energy Physics* **09** (2006) 009, hep-th/0607060.

[295] J. Figueroa-O'Farrill and N. Hustler, "The homogeneity theorem for supergravity backgrounds," *Journal of High Energy Physics* **10** (2012) 014, 1208.0553.

[296] J. Gutowski and G. Papadopoulos, "Index theory and dynamical symmetry enhancement of M-horizons," *Journal of High Energy Physics* **05** (2013) 088, 1303.0869.

[297] S. Cecotti, S. Ferrara, and L. Girardello, "Geometry of type II superstrings and the moduli of superconformal field theories," *International Journal of Modern Physics* **A4** (1989) 2475.

[298] S. Ferrara and S. Sabharwal, "Quaternionic manifolds for type II superstring vacua of Calabi–Yau spaces," *Nuclear Physics* **B332** (1990) 317–332.

[299] M. Bodner, A. C. Cadavid, and S. Ferrara, "(2, 2) vacuum configurations for type IIA superstrings: $\mathcal{N} = 2$ supergravity Lagrangians and algebraic geometry," *Classical and Quantum Gravity* **8** (1991) 789–808.

[300] J. Louis and A. Micu, "Type II theories compactified on Calabi–Yau threefolds in the presence of background fluxes," *Nuclear Physics* **B635** (2002) 395–431, hep-th/0202168.

[301] E. Witten, "New issues in manifolds of SU(3) holonomy," *Nuclear Physics* **B268** (1986) 79.

[302] A. Strominger, "Massless black holes and conifolds in string theory," *Nuclear Physics* **B451** (1995) 96–108, hep-th/9504090.

[303] J. Polchinski and A. Strominger, "New vacua for type II string theory," *Physics Letters* **B388** (1996) 736–742, hep-th/9510227.

[304] W.-Y. Chuang, S. Kachru, and A. Tomasiello, "Complex/symplectic mirrors," *Communications in Mathematical Physics* **274** (2007) 775–794, hep-th/0510042.

[305] D. Nemeschansky and A. Sen, "Conformal invariance of supersymmetric σ models on Calabi–Yau manifolds," *Physics Letters* **B178** (1986) 365–369.

[306] W. Lerche, C. Vafa, and N. P. Warner, "Chiral rings in $\mathcal{N} = 2$ superconformal theories," *Nuclear Physics* **B324** (1989) 427–474.

[307] C. Vafa and N. P. Warner, "Catastrophes and the classification of conformal theories," *Physics Letters* **B218** (1989) 51–58.

[308] E. Witten, "Phases of $\mathcal{N} = 2$ theories in two dimensions," *Nuclear Physics* **B403** (1993) 159–222, hep-th/9301042.

[309] A. Kapustin and A. Tomasiello, "The general (2,2) gauged sigma model with three-form flux," *Journal of High Energy Physics* **11** (2007) 053, hep-th/0610210.

[310] S. J. Gates, M. T. Grisaru, M. Rocek, and W. Siegel, "Superspace or one thousand and one lessons in supersymmetry," *Frontiers of Physics* **58** (1983) 1–548, hep-th/0108200.

[311] L. J. Dixon, "Some world sheet properties of superstring compactifications, on orbifolds and otherwise," in *Proceedings, Summer Workshop in High-Energy Physics and Cosmology: Superstrings, Unified Theories and Cosmology: Trieste, Italy, June 29–August 7, 1987*. World Scientific, pp. 67–26, 1987.

[312] K. Hori, R. Thomas, S. Katz, et al., *Mirror Symmetry*, vol. 1. American Mathematical Society, 2003.

[313] B. R. Greene, "String theory on Calabi–Yau manifolds," in *Theoretical Advanced Study Institute in Elementary Particle Physics (TASI 96)*, World Scientific, 1997. hep-th/9702155.

[314] B. R. Greene and M. Plesser, "Duality in Calabi–Yau moduli space," *Nuclear Physics* **B338** (1990) 15–37.

[315] P. Candelas, X. C. De La Ossa, P. S. Green, and L. Parkes, "A pair of Calabi–Yau manifolds as an exactly soluble superconformal theory," *Nuclear Physics* **B359** (1991) 21–74.

[316] D. R. Morrison, "Picard–Fuchs equations and mirror maps for hypersurfaces," hep-th/9111025. [AMS/IP Studies in Advanced Mathematics 9,185(1998)].

[317] J. Harris, "Galois groups of enumerative problems," *Duke Mathematical Journal* **46** (1979), no. 4, 685–724.

[318] A. B. Givental, "Equivariant Gromov-Witten invariants," *International Mathematics Research Notices* **1996** (1996), no. 13, 613, alg-geom/9603021.

[319] B. H. Lian, K. Liu, and S.-T. Yau, "Mirror principle I," *Asian Journal of Mathematics* **1** (1997), no. 4, 729763.

[320] H. Jockers, V. Kumar, J. M. Lapan, D. R. Morrison, and M. Romo, "Two-sphere partition functions and Gromov–Witten invariants," *Communications in Mathematical Physics* **325** (2014) 1139–1170, 1208.6244.

[321] J. D. Brown and C. Teitelboim, "Neutralization of the cosmological constant by membrane creation," *Nuclear Physics* **B297** (1988) 787–836.

[322] J. Michelson, "Compactifications of type IIB strings to four dimensions with nontrivial classical potential," *Nuclear Physics* **B495** (1997) 127–148, hep-th/9610151.

[323] R. D'Auria, S. Ferrara, and P. Fre, "Special and quaternionic isometries: general couplings in $\mathcal{N} = 2$ supergravity and the scalar potential," *Nuclear Physics* **B359** (1991) 705–740.

[324] L. Martucci, J. Rosseel, D. Van den Bleeken, and A. Van Proeyen, "Dirac actions for D-branes on backgrounds with fluxes," *Classical and Quantum Gravity* **22** (2005) 2745–2764, hep-th/0504041.

[325] G. Papadopoulos and P. K. Townsend, "Intersecting M-branes," *Physics Letters* **B380** (1996) 273–279, hep-th/9603087.

[326] A. A. Tseytlin, "Harmonic superpositions of M-branes," *Nuclear Physics* **B475** (1996) 149–163, hep-th/9604035.

[327] J. P. Gauntlett, D. A. Kastor, and J. H. Traschen, "Overlapping branes in M theory," *Nuclear Physics* **B478** (1996) 544–560, `hep-th/9604179`.

[328] N. Itzhaki, A. A. Tseytlin, and S. Yankielowicz, "Supergravity solutions for branes localized within branes," *Physics Letters* **B432** (1998) 298–304, `hep-th/9803103`.

[329] D. Youm, "Partially localized intersecting BPS branes," *Nuclear Physics* **B556** (1999) 222–246, `hep-th/9902208`.

[330] Y. Imamura, "1/4 BPS solutions in massive IIA supergravity," *Progress of Theoretical Physics* **106** (2001) 653–670, `hep-th/0105263`.

[331] K. Behrndt, E. Bergshoeff, and B. Janssen, "Intersecting D-branes in ten dimensions and six dimensions," *Physical Review* **D55** (1997) 3785–3792, `hep-th/9604168`.

[332] B. Janssen, P. Meessen, and T. Ortín, "The D8-brane tied up: string and brane solutions in massive type IIA supergravity," *Physics Letters* **B453** (1999) 229–236, `hep-th/9901078`.

[333] M. Bershadsky, C. Vafa, and V. Sadov, "D-branes and topological field theories," *Nuclear Physics* **B463** (1996) 420–434, `hep-th/9511222`.

[334] E. Witten, "Topological quantum field theory," *Communications in Mathematical Physics* **117** (1988) 353.

[335] G. Festuccia and N. Seiberg, "Rigid supersymmetric theories in curved superspace," *Journal of High Energy Physics* **1106** (2011) 114, `1105.0689`.

[336] T. T. Dumitrescu, G. Festuccia, and N. Seiberg, "Exploring curved superspace," *Journal of High Energy Physics* **1208** (2012) 141, `1205.1115`.

[337] C. Klare, A. Tomasiello, and A. Zaffaroni, "Supersymmetry on curved spaces and holography," *Journal of High Energy Physics* **1208** (2012) 061, `1205.1062`.

[338] D. Cassani, C. Klare, D. Martelli, A. Tomasiello, and A. Zaffaroni, "Supersymmetry in Lorentzian curved spaces and holography," *Communications in Mathematical Physics* **327** (2014) 577–602, `1207.2181`.

[339] H. Triendl, "Supersymmetric branes on curved spaces and fluxes," *Journal of High Energy Physics* **11** (2015) 025, `1509.02926`.

[340] R. C. McLean, "Deformations of calibrated submanifolds," *Communications in Analysis and Geometry* **6** (1998), no. 4, 705–747.

[341] M. Mariño, R. Minasian, G. W. Moore, and A. Strominger, "Nonlinear instantons from supersymmetric "*p*"-branes," *Journal of High Energy Physics* **01** (2000) 005, `hep-th/9911206`.

[342] E. Witten, "BPS bound states of D0–D6 and D0–D8 systems in a *b* field," *Journal of High Energy Physics* **2002** (April 2002) 012012.

[343] J. A. Harvey and G. W. Moore, "On the algebras of BPS states," *Communications in Mathematical Physics* **197** (1998) 489–519, `hep-th/9609017`.

[344] N. J. Hitchin, "The self-duality equations on a Riemann surface," *Proceedings of the London Mathematical Society* **3** (1987), no. 1, 59–126.

[345] M. R. Douglas, B. Fiol, and C. Romelsberger, "Stability and BPS branes," *Journal of High Energy Physics* **09** (2005) 006, `hep-th/0002037`.

[346] M. R. Douglas, "D-branes, categories and $\mathcal{N} = 1$ supersymmetry," *Journal of Mathematical Physics* **42** (2001) 2818–2843, `hep-th/0011017`.

[347] P. S. Aspinwall and M. R. Douglas, "D-brane stability and monodromy," *Journal of High Energy Physics* **05** (2002) 031, hep-th/0110071.

[348] P. S. Aspinwall, "D-branes on Calabi–Yau manifolds," in *Progress in String Theory. Proceedings, Summer School, TASI 2003, Boulder, USA, June 2–27, 2003*, pp. 1–152, World Scientific, 2004. hep-th/0403166.

[349] P. Aspinwall, T. Bridgeland, A. Craw, et al., *Dirichlet Branes and Mirror Symmetry*, vol. 4. American Mathematical Society, 2009.

[350] E. Witten and D. I. Olive, "Supersymmetry algebras that include topological charges," *Physics Letters* **78B** (1978) 97–101.

[351] Y.-K. E. Cheung and Z. Yin, "Anomalies, branes, and currents," *Nuclear Physics* **B517** (1998) 69–91, hep-th/9710206.

[352] M. B. Green, J. A. Harvey, and G. W. Moore, "I-brane inflow and anomalous couplings on D-branes," *Classical and Quantum Gravity* **14** (1997) 47–52, hep-th/9605033.

[353] A. Recknagel and V. Schomerus, "D-branes in Gepner models," *Nuclear Physics* **B531** (1998) 185–225, hep-th/9712186.

[354] E. R. Sharpe, "D-branes, derived categories, and Grothendieck groups," *Nuclear Physics* **B561** (1999) 433–450, hep-th/9902116.

[355] A. D. King, "Moduli of representations of finite dimensional algebras," *Quarterly Journal of Mathematics* **45** (1994), no. 4, 515–530.

[356] P. S. Aspinwall and A. E. Lawrence, "Derived categories and zero-brane stability," *Journal of High Energy Physics* **08** (2001) 004, hep-th/0104147.

[357] F. Denef, "Supergravity flows and D-brane stability," *Journal of High Energy Physics* **08** (2000) 050, hep-th/0005049.

[358] G. Lawlor, "The angle criterion," *Inventiones mathematicae* **95** (1989), no. 2, 437–446.

[359] D. Joyce, "On counting special Lagrangian homology three spheres," *Contemporary Mathematics* **314** (2002) 125–151, hep-th/9907013.

[360] K. Fukaya, "Morse homotopy, A_∞ category and Floer homologies," in *Proceeding of GARC Workshop on Geometry and Topology*, Seoul National University 1993.

[361] K. Fukaya, Y.-G. Oh, H. Ohta, and K. Ono, *Lagrangian Intersection Floer Theory: Anomaly and Obstruction, Part II*, vol. 2. American Mathematical Society, 2010.

[362] M. Berkooz, M. R. Douglas, and R. G. Leigh, "Branes intersecting at angles," *Nuclear Physics* **B480** (1996) 265–278, hep-th/9606139.

[363] S. Kachru and J. McGreevy, "Supersymmetric three cycles and supersymmetry breaking," *Physical Review* **D61** (2000) 026001, hep-th/9908135.

[364] A. Strominger, S.-T. Yau, and E. Zaslow, "Mirror symmetry is T-duality," *Nuclear Physics* **B479** (1996) 243–259, hep-th/9606040.

[365] M. Gross, "Topological mirror symmetry," *Inventiones mathematicae* **144** (1999), no. 1, 75–137. math/9909015.

[366] D. R. Morrison, "On the structure of supersymmetric T^3 fibrations," in *Tropical Geometry and Mirror Symmetry*, Contemporary Mathematics, vol. 527, American Mathematical Society, 2010, pp. 91–112.

[367] A. Strominger, "Superstrings with torsion," *Nuclear Physics* **B274** (1986) 253.

[368] C. M. Hull, "Superstring compactifications with torsion and space-time supersymmetry," in *First Torino Meeting on Superunification and Extra Dimensions Turin, Italy, September 22–28, 1985*, pp. 347–375. 1986.

[369] J. P. Gauntlett, D. Martelli, and D. Waldram, "Superstrings with intrinsic torsion," *Physical Review* **D69** (2004) 086002, `hep-th/0302158`.

[370] A. Sen, "Duality and orbifolds," *Nuclear Physics* **B474** (1996) 361–378, `hep-th/9604070`.

[371] A. Sen, "Stable nonBPS bound states of BPS D-branes," *Journal of High Energy Physics* **08** (1998) 010, `hep-th/9805019`.

[372] E. Goldstein and S. Prokushkin, "Geometric model for complex non-Kaehler manifolds with SU(3) structure," *Communications in Mathematical Physics* **251** (2004) 65–78, `hep-th/0212307`.

[373] J.-X. Fu and S.-T. Yau, "The theory of superstring with flux on non-Kaehler manifolds and the complex Mongempère equation," *Journal of Differential Geometry* **78** (2009). 369–428 `hep-th/0604063`.

[374] K. Becker, M. Becker, J.-X. Fu, L.-S. Tseng, and S.-T. Yau, "Anomaly cancellation and smooth non-Kaehler solutions in heterotic string theory," *Nuclear Physics* **B751** (2006) 108–128, `hep-th/0604137`.

[375] M. Garcia-Fernandez, "T-dual solutions of the Hull–Strominger system on non-Kähler threefolds," *Journal für die reine und angewandte Mathematik (Crelles Journal)* **766** (2020) 137–150.

[376] A. Fino, G. Grantcharov, and L. Vezzoni, "Solutions to the Hull–Strominger system with torus symmetry," 2019, 1901.10322.

[377] M. Rocek, K. Schoutens, and A. Sevrin, "Off-shell WZW models in extended superspace," *Physics Letters* **B265** (1991) 303–306.

[378] J. Gates, S. J., C. M. Hull, and M. Roček, "Twisted multiplets and new supersymmetric nonlinear sigma models," *Nuclear Physics* **B248** (1984) 157.

[379] M. Rocek and E. P. Verlinde, "Duality, quotients, and currents," *Nuclear Physics* **B373** (1992) 630–646, `hep-th/9110053`.

[380] T. Buscher, U. Lindström, and M. Rocek, "New supersymmetric σ models with Wess–Zumino terms," *Physics Letters* **B202** (1988) 94–98.

[381] C. Vafa, "Evidence for F-theory," *Nuclear Physics* **B469** (1996) 403–418, `hep-th/9602022`.

[382] D. R. Morrison and C. Vafa, "Compactifications of F theory on Calabi–Yau threefolds. 1," *Nuclear Physics* **B473** (1996) 74–92, `hep-th/9602114`.

[383] D. R. Morrison and C. Vafa, "Compactifications of F theory on Calabi–Yau threefolds. 2," *Nuclear Physics* **B476** (1996) 437–469, `hep-th/9603161`.

[384] T. Weigand, "F-theory," *PoS* **TASI2017** (2018) 016, `1806.01854`.

[385] B. R. Greene, A. D. Shapere, C. Vafa, and S.-T. Yau, "Stringy cosmic strings and noncompact Calabi Yau manifolds," *Nuclear Physics* **B337** (1990) 1–36.

[386] A. Sen, "F-theory and orientifolds," *Nuclear Physics* **B475** (1996) 562–578, `hep-th/9605150`.

[387] A. Sen, "Orientifold limit of F theory vacua," *Physical Review* **D55** (1997) R7345–R7349, `hep-th/9702165`.

[388] A. Collinucci, F. Denef, and M. Esole, "D-brane deconstructions in IIB orientifolds," *Journal of High Energy Physics* **02** (2009) 005, `0805.1573`.

[389] J. Polchinski and E. Silverstein, "Dual purpose landscaping tools: small extra dimensions in AdS/CFT," in *Strings, Gauge Fields, and the Geometry Behind: The Legacy of Maximilian Kreuzer*, A. Rebhan, L. Katzarkov, J. Knapp, R. Rashkov, and E. Scheidegger, eds., pp. 365–390. 2009. 0908.0756.

[390] O. Aharony, A. Fayyazuddin, and J. M. Maldacena, "The large N limit of $\mathcal{N} = 2, \mathcal{N} = 1$ field theories from three-branes in F theory," *Journal of High Energy Physics* **07** (1998) 013, hep-th/9806159.

[391] I. García-Etxebarría and D. Regalado, "$\mathcal{N} = 3$ four dimensional field theories," *Journal of High Energy Physics* **03** (2016) 083, 1512.06434.

[392] B. Haghighat, S. Murthy, C. Vafa, and S. Vandoren, "F-theory, spinning black holes and multi-string branches," *Journal of High Energy Physics* **01** (2016) 009, 1509.00455.

[393] K. Kodaira, "On compact analytic surfaces: II," *Annals of Mathematics* (1963) 563–626.

[394] A. Néron, "Modèles minimaux des variétés abéliennes sur les corps locaux et globaux," *Publications Mathématiques de l'Institut des Hautes Études Scientifiques* **21** (1964), no. 1, 5–125.

[395] W. Barth, C. Peters, and A. Van de Ven, *Compact Complex Surfaces*, vol. 2. Springer, 2004.

[396] J. de Boer, R. Dijkgraaf, K. Hori, et al., "Triples, fluxes, and strings," *Advances in Theoretical and Mathematical Physics* **4** (2002) 995–1186, hep-th/0103170.

[397] L. Bhardwaj, D. R. Morrison, Y. Tachikawa, and A. Tomasiello, "The frozen phase of F-theory," *Journal of High Energy Physics* **08** (2018) 138, 1805.09070.

[398] M. R. Gaberdiel and B. Zwiebach, "Exceptional groups from open strings," *Nuclear Physics* **B518** (1998) 151–172, hep-th/9709013.

[399] M. Bershadsky, K. A. Intriligator, S. Kachru, D. R. Morrison, V. Sadov, and C. Vafa, "Geometric singularities and enhanced gauge symmetries," *Nuclear Physics* **B481** (1996) 215–252, hep-th/9605200.

[400] S. H. Katz and C. Vafa, "Matter from geometry," *Nuclear Physics* **B497** (1997) 146–154, hep-th/9606086.

[401] A. Grassi and D. R. Morrison, "Anomalies and the Euler characteristic of elliptic Calabi–Yau threefolds," *Communications in Number Theory and Physics* **6** (2012) 51–127, 1109.0042.

[402] M. Bertolini, P. R. Merkx, and D. R. Morrison, "On the global symmetries of 6D superconformal field theories," *Journal of High Energy Physics* **07** (2016) 005, 1510.08056.

[403] P. S. Aspinwall and D. R. Morrison, "Point-like instantons on K3 orbifolds," *Nuclear Physics* **B503** (1997) 533–564, hep-th/9705104.

[404] M. Del Zotto, J. J. Heckman, A. Tomasiello, and C. Vafa, "6D conformal matter," *Journal of High Energy Physics* **1502** (2015) 054, 1407.6359.

[405] J. J. Heckman, D. R. Morrison, and C. Vafa, "On the classification of 6D SCFTs and generalized ADE orbifolds," *Journal of High Energy Physics* **05** (2014) 028, 1312.5746.

[406] J. J. Heckman, D. R. Morrison, T. Rudelius, and C. Vafa, "Atomic classification of 6D SCFTs," *Fortschritte der Physik* **63** (2015) 468–530, 1502.05405.

[407] J. J. Heckman and T. Rudelius, "Top down approach to 6D SCFTs," *Journal of Physics* **A52** (2019), no. 9, 093001, 1805.06467.

[408] P. K. Townsend, "String – membrane duality in seven dimensions," *Physics Letters* **B354** (1995) 247–255, hep-th/9504095.

[409] L. J. Romans, "The $F(4)$ gauged supergravity in six dimensions," *Nuclear Physics* **B269** (1986) 691.

[410] R. D'Auria, S. Ferrara, and S. Vaula, "Matter coupled $F(4)$ supergravity and the AdS_6/CFT_5 correspondence," *Journal of High Energy Physics* **10** (2000) 013, hep-th/0006107.

[411] K. Becker and M. Becker, "M-theory on eight-manifolds," *Nuclear Physics* **B477** (1996) 155–167, hep-th/9605053.

[412] K. Dasgupta, G. Rajesh, and S. Sethi, "M theory, orientifolds and G-flux," *Journal of High Energy Physics* **08** (1999) 023, hep-th/9908088.

[413] M. Graña and J. Polchinski, "Supersymmetric three-form flux perturbations on AdS_5," *Physical Review* **D63** (2001) 026001, hep-th/0009211.

[414] S. B. Giddings, S. Kachru, and J. Polchinski, "Hierarchies from fluxes in string compactifications," *Physical Review* **D66** (2002) 106006, hep-th/0105097.

[415] I. R. Klebanov and M. J. Strassler, "Supergravity and a confining gauge theory: duality cascades and χSB resolution of naked singularities," *Journal of High Energy Physics* **08** (2000) 052, hep-th/0007191.

[416] D. Martelli and J. Sparks, "G-structures, fluxes and calibrations in M-theory," *Physical Review* **D68** (2003) 085014, hep-th/0306225.

[417] D. Tsimpis, "M-theory on eight-manifolds revisited: $\mathcal{N} = 1$ supersymmetry and generalized Spin(7) structures," *Journal of High Energy Physics* **04** (2006) 027, hep-th/0511047.

[418] T. W. Grimm and J. Louis, "The effective action of $\mathcal{N} = 1$ Calabi–Yau orientifolds," *Nuclear Physics* **B699** (2004) 387–426, hep-th/0403067.

[419] M. Graña, "Flux compactifications in string theory: a comprehensive review," *Physics Reports* **423** (2006) 91–158, hep-th/0509003.

[420] A. R. Frey, *Warped Strings: Selfdual Flux and Contemporary Compactifications*. Ph.D. thesis, University of California, Santa Barbara, 2003. hep-th/0308156.

[421] S. B. Giddings and A. Maharana, "Dynamics of warped compactifications and the shape of the warped landscape," *Physical Review* **D73** (2006) 126003, hep-th/0507158.

[422] G. Shiu, G. Torroba, B. Underwood, and M. R. Douglas, "Dynamics of warped flux compactifications," *Journal of High Energy Physics* **06** (2008) 024, 0803.3068.

[423] S. Gukov, C. Vafa, and E. Witten, "CFT's from Calabi–Yau four-folds," *Nuclear Physics* **B584** (2000) 69–108, hep-th/9906070.

[424] E. Cremmer, S. Ferrara, C. Kounnas, and D. V. Nanopoulos, "Naturally vanishing cosmological constant in $\mathcal{N} = 1$ supergravity," *Physics Letters* **133B** (1983) 61.

[425] J. R. Ellis, A. B. Lahanas, D. V. Nanopoulos, and K. Tamvakis, "No-scale supersymmetric standard model," *Physics Letters* **134B** (1984) 429.

[426] A. R. Frey and J. Polchinski, "$\mathcal{N} = 3$ warped compactifications," *Physical Review* **D65** (2002) 126009, hep-th/0201029.

[427] E. Bergshoeff, R. Kallosh, T. Ortín, D. Roest, and A. Van Proeyen, "New formulations of $D = 10$ supersymmetry and D8–O8 domain walls," *Classical and Quantum Gravity* **18** (2001) 3359–3382, hep-th/0103233.

[428] D. Lüst and D. Tsimpis, "Supersymmetric AdS_4 compactifications of IIA supergravity," *Journal of High Energy Physics* **02** (2005) 027, hep-th/0412250.

[429] J. P. Gauntlett, D. Martelli, J. Sparks, and D. Waldram, "Supersymmetric AdS_5 solutions of type IIB supergravity," *Classical and Quantum Gravity* **23** (2006) 4693–4718, hep-th/0510125.

[430] P. Koerber and D. Tsimpis, "Supersymmetric sources, integrability and generalized structure compactifications," *Journal of High Energy Physics* **08** (2007) 082. 0706.1244.

[431] A. Legramandi, L. Martucci, and A. Tomasiello, "Timelike structures of ten-dimensional supersymmetry," *Journal of High Energy Physics* **04** (2019) 109, 1810.08625.

[432] D. Lüst, F. Marchesano, L. Martucci, and D. Tsimpis, "Generalized non-supersymmetric flux vacua," *Journal of High Energy Physics* **11** (2008) 021, 0807.4540.

[433] A. Coimbra, R. Minasian, H. Triendl, and D. Waldram, "Generalised geometry for string corrections," *Journal of High Energy Physics* **11** (2014) 160, 1407.7542.

[434] A. Tomasiello, "Generalized structures of ten-dimensional supersymmetric solutions," *Journal of High Energy Physics* **1203** (2012) 073, 1109.2603.

[435] S. Giusto, L. Martucci, M. Petrini, and R. Russo, "6D microstate geometries from 10D structures," *Nuclear Physics* **B876** (2013) 509–555, 1306.1745.

[436] F. F. Gautason, D. Junghans, and M. Zagermann, "Cosmological constant, near brane behavior and singularities," *Journal of High Energy Physics* **1309** (2013) 123, 1301.5647.

[437] N. Cribiori and D. Junghans, "No classical (anti-)de Sitter solutions with O8-planes," *Physics Letters* **B793** (2019) 54–58, 1902.08209.

[438] C. Córdova, G. B. De Luca, and A. Tomasiello, "Classical de Sitter solutions of ten-dimensional supergravity," *Physical Review Letters* **122** (2019), no. 9, 091601, 1812.04147.

[439] M. Graña, R. Minasian, M. Petrini, and A. Tomasiello, "Generalized structures of $\mathcal{N} = 1$ vacua," *Journal of High Energy Physics* **11** (2005) 020, hep-th/0505212.

[440] R. Minasian, M. Petrini, and A. Zaffaroni, "Gravity duals to deformed SYM theories and generalized complex geometry," *Journal of High Energy Physics* **12** (2006) 055, hep-th/0606257.

[441] N. Halmagyi and A. Tomasiello, "Generalized Kaehler potentials from supergravity," *Communications in Mathematical Physics* **291** (2009) 1–30, 0708.1032.

[442] L. Martucci and P. Smyth, "Supersymmetric D-branes and calibrations on general $\mathcal{N} = 1$ backgrounds," *Journal of High Energy Physics* **11** (2005) 048, hep-th/0507099.

[443] P. Koerber and L. Martucci, "D-branes on AdS flux compactifications," *Journal of High Energy Physics* **0801** (2008) 047, 0710.5530.

[444] J. Gutowski, G. Papadopoulos, and P. K. Townsend, "Supersymmetry and generalized calibrations," *Physical Review* **D60** (1999) 106006, hep-th/9905156.

[445] D. Lüst, P. Patalong, and D. Tsimpis, "Generalized geometry, calibrations and supersymmetry in diverse dimensions," *Journal of High Energy Physics* **1101** (2011) 063, 1010.5789.

[446] B. S. Acharya, M. Aganagic, K. Hori, and C. Vafa, "Orientifolds, mirror symmetry and superpotentials," (2002), hep-th/0202208.

[447] P. Koerber and L. Martucci, "From ten to four and back again: how to generalize the geometry," *Journal of High Energy Physics* **08** (2007) 059, 0707.1038.

[448] I. Benmachiche and T. W. Grimm, "Generalized $\mathcal{N} = 1$ orientifold compactifications and the Hitchin functionals," *Nuclear Physics* **B748** (2006) 200–252, hep-th/0602241.

[449] A. Ashmore, C. Strickland-Constable, D. Tennyson, and D. Waldram, "Generalising G_2 geometry: involutivity, moment maps and moduli," 1910.04795.

[450] M. Graña, J. Louis, and D. Waldram, "SU(3) × SU(3) compactification and mirror duals of magnetic fluxes," *Journal of High Energy Physics* **0704** (2007) 101, hep-th/0612237.

[451] D. Cassani and A.-K. Kashani-Poor, "Exploiting $\mathcal{N} = 2$ in consistent coset reductions of type IIA," *Nuclear Physics* **B817** (2009) 25–57, 0901.4251.

[452] D. Cassani, P. Koerber, and O. Varela, "All homogeneous $\mathcal{N} = 2$ M-theory truncations with supersymmetric AdS_4 vacua," *Journal of High Energy Physics* **11** (2012) 173, 1208.1262.

[453] A.-K. Kashani-Poor, "Nearly Kaehler reduction," *Journal of High Energy Physics* **11** (2007) 026, 0709.4482.

[454] J. P. Gauntlett and O. Varela, "Consistent Kaluza–Klein reductions for general supersymmetric AdS solutions," *Physical Review* **D76** (2007) 126007, 0707.2315.

[455] J. P. Gauntlett, S. Kim, O. Varela, and D. Waldram, "Consistent supersymmetric Kaluza–Klein truncations with massive modes," *Journal of High Energy Physics* **0904** (2009) 102, 0901.0676.

[456] D. Cassani, G. Dall'Agata, and A. F. Faedo, "Type IIB supergravity on squashed Sasaki–Einstein manifolds," *Journal of High Energy Physics* **05** (2010) 094, 1003.4283.

[457] O. Aharony, M. Berkooz, J. Louis, and A. Micu, "Non-Abelian structures in compactifications of M-theory on seven-manifolds with SU(3) structure," *Journal of High Energy Physics* **09** (2008) 108, 0806.1051.

[458] S. Gurrieri, J. Louis, A. Micu, and D. Waldram, "Mirror symmetry in generalized Calabi–Yau compactifications," *Nuclear Physics* **B654** (2003) 61–113, hep-th/0211102.

[459] A. Tomasiello, "Topological mirror symmetry with fluxes," *Journal of High Energy Physics* **06** (2005) 067, `hep-th/0502148`.

[460] R. Eager, J. Schmude, and Y. Tachikawa, "Superconformal indices, Sasaki–Einstein manifolds, and cyclic homologies," *Advances in Theoretical and Mathematical Physics* **18** (2014), no. 1, 129–175, `1207.0573`.

[461] S. Katmadas and A. Tomasiello, "AdS_4 black holes from M-theory," *Journal of High Energy Physics* **12** (2015) 111, `1509.00474`.

[462] A. Butti, M. Graña, R. Minasian, M. Petrini, and A. Zaffaroni, "The Baryonic branch of Klebanov–Strassler solution: a supersymmetric family of SU(3) structure backgrounds," *Journal of High Energy Physics* **03** (2005) 069, `hep-th/0412187`.

[463] R. Minasian, M. Petrini, and A. Zaffaroni, "New families of interpolating type IIB backgrounds," *Journal of High Energy Physics* **04** (2010) 080, `0907.5147`.

[464] T. Maxfield, J. McOrist, D. Robbins, and S. Sethi, "New examples of flux vacua," *Journal of High Energy Physics* **1312** (2013) 032, `1309.2577`.

[465] L. Martucci, J. F. Morales, and D. Ricci Pacifici, "Branes, U-folds and hyperelliptic fibrations," *Journal of High Energy Physics* **01** (2013) 145, `1207.6120`.

[466] A. P. Braun, F. Fucito, and J. F. Morales, "U-folds as K3 fibrations," *Journal of High Energy Physics* **10** (2013) 154, `1308.0553`.

[467] P. Candelas, A. Constantin, C. Damian, M. Larfors, and J. F. Morales, "Type IIB flux vacua from G-theory I," *Journal of High Energy Physics* **02** (2015) 187, `1411.4785`.

[468] P. Candelas, A. Constantin, C. Damian, M. Larfors, and J. F. Morales, "Type IIB flux vacua from G-theory II," *Journal of High Energy Physics* **02** (2015) 188, `1411.4786`.

[469] N. T. Macpherson and A. Tomasiello, "Minimal flux Minkowski classification," *Journal of High Energy Physics* **09** (2017) 126, `1612.06885`.

[470] F. Apruzzi, J. C. Geipel, A. Legramandi, N. T. Macpherson, and M. Zagermann, "$\text{Minkowski}_4 \times S^2$ solutions of IIB supergravity," *Fortschritte der Physik* **66** (2018), no. 3, 1800006, `1801.00800`.

[471] A. Legramandi and N. T. Macpherson, "$\text{Mink}_4 \times S^2$ solutions of 10 and 11 dimensional supergravity," *Journal of High Energy Physics* **07** (2019) 134, `1811.11224`.

[472] D. Andriot, "New supersymmetric flux vacua with intermediate SU(2) structure," *Journal of High Energy Physics* **08** (2008) 096, `0804.1769`.

[473] D. Andriot, "New supersymmetric vacua on solvmanifolds," *Journal of High Energy Physics* **02** (2016) 112, `1507.00014`.

[474] A. Tomasiello, "Geometrical methods for string compactifications," 2009, `http://virgilio.mib.infn.it/~atom/laces.pdf`.

[475] S. Andriolo, G. Shiu, H. Triendl, T. Van Riet, G. Venken, and G. Zoccarato, "Compact G_2 holonomy spaces from SU(3)-structures," *Journal of High Energy Physics* **03** (2019) 059, `1811.00063`.

[476] I. Brunner and K. Hori, "Orientifolds and mirror symmetry," *Journal of High Energy Physics* **11** (2004) 005, `hep-th/0303135`.

[477] L. Nirenberg, "A complex Frobenius theorem," in *Seminar on Analytic Functions*, Institute for Advanced Study, Princeton, pp. 172–189. 1957.

[478] A. Passias, G. Solard, and A. Tomasiello, "$\mathcal{N} = 2$ supersymmetric AdS_4 solutions of type IIB supergravity," *Journal of High Energy Physics* **04** (2018) 005, 1709.09669.

[479] J. Blåbäck, B. Janssen, T. Van Riet, and B. Vercnocke, "BPS domain walls from backreacted orientifolds," *Journal of High Energy Physics* **05** (2014) 040, 1312.6125.

[480] A. Legramandi and A. Tomasiello, "Breaking supersymmetry with pure spinors," *Journal of High Energy Physics* **11** (2020) 098, 1912.00001.

[481] N. Hitchin, "Generalized Calabi–Yau manifolds," *Quarterly Journal of Mathematics Oxford Series* **54** (2003) 281–308, math.dg/0209099.

[482] M. Gualtieri, "Generalized complex geometry," math/0401221. Ph.D. thesis (advisor: Nigel Hitchin).

[483] P. Koerber, "Lectures on generalized complex geometry for physicists," *Fortschritte der Physik* **59** (2011) 169–242, 1006.1536.

[484] D. Tsimpis, "Generalized geometry lectures on type II backgrounds," (2016), 1606.08674.

[485] E. M. Babalic, I. A. Coman, and C. I. Lazaroiu, "Geometric algebra techniques in flux compactifications," *Advanced High Energy Physics* **2016** (2016) 7292534, 1212.6766.

[486] V. Cortés, C. Lazaroiu, and C. S. Shahbazi, "Spinors of real type as polyforms and the generalized Killing equation," (2019), 1911.08658.

[487] M. Graña, R. Minasian, M. Petrini, and D. Waldram, "T-duality, generalized geometry and non-geometric backgrounds," *Journal of High Energy Physics* **04** (2009) 075, 0807.4527.

[488] N. J. Hitchin, "The geometry of three-forms in six and seven dimensions," *Journal of Differential Geometry* **55** (2000), no. 3, 547–576, math.dg/0010054.

[489] A. Tomasiello, "Reformulating supersymmetry with a generalized Dolbeault operator," *Journal of High Energy Physics* **02** (2008) 010, 0704.2613 [hep-th].

[490] Y. Kosmann-Schwarzbach, "Derived brackets," *Letters in Mathematical Physics* **69** (2004), no. 1, 61–87.

[491] U. Lindström, R. Minasian, A. Tomasiello, and M. Zabzine, "Generalized complex manifolds and supersymmetry," *Communications in Mathematical Physics* **257** (2005) 235–256, hep-th/0405085.

[492] S. Guttenberg, "Brackets, sigma models and integrability of generalized complex structures," *Journal of High Energy Physics* **06** (2007) 004, hep-th/0609015.

[493] N. Hitchin, "Instantons, Poisson structures and generalized Kaehler geometry," *Communications in Mathematical Physics* **265** (2006) 131–164, math/0503432.

[494] F. Saracco and A. Tomasiello, "Localized O6-plane solutions with Romans mass," *Journal of High Energy Physics* **1207** (2012) 077, 1201.5378.

[495] C. Jeschek and F. Witt, "Generalised G_2 structures and type IIB superstrings," *Journal of High Energy Physics* **03** (2005) 053, hep-th/0412280.

[496] M. Haack, D. Lüst, L. Martucci, and A. Tomasiello, "Domain walls from ten dimensions," *Journal of High Energy Physics* **0910** (2009) 089, 0905.1582.

[497] G. Dibitetto, G. Lo Monaco, A. Passias, N. Petri, and A. Tomasiello, "AdS_3 solutions with exceptional supersymmetry," *Fortschritte der Physik* **66** (2018), no. 10, 1800060, 1807.06602.

[498] A. Passias and D. Prins, "On AdS_3 solutions of Type IIB," *Journal of High Energy Physics* **05** (2020) 048, 1910.06326.

[499] D. Prins and D. Tsimpis, "IIB supergravity on manifolds with SU(4) structure and generalized geometry," *Journal of High Energy Physics* **1307** (2013) 180, 1306.2543.

[500] D. Rosa, "Generalized geometry of two-dimensional vacua," *Journal of High Energy Physics* **07** (2014) 111, 1310.6357.

[501] D. Prins and D. Tsimpis, "IIA supergravity and M-theory on manifolds with SU(4) structure," *Physical Review* **D89** (2014) 064030, 1312.1692.

[502] A. Passias, D. Prins, and A. Tomasiello, "A massive class of $\mathcal{N} = 2$ AdS_4 IIA solutions," *Journal of High Energy Physics* **10** (2018) 071, 1805.03661.

[503] J. P. Gauntlett and S. Pakis, "The geometry of $D = 11$ Killing spinors," *Journal of High Energy Physics* **04** (2003) 039, hep-th/0212008.

[504] J. P. Gauntlett, J. B. Gutowski, and S. Pakis, "The geometry of $D = 11$ null Killing spinors," *Journal of High Energy Physics* **12** (2003) 049, hep-th/0311112.

[505] P. Kaste, R. Minasian, and A. Tomasiello, "Supersymmetric M-theory compactifications with fluxes on seven-manifolds and G-structures," *Journal of High Energy Physics* **07** (2003) 004, hep-th/0303127.

[506] G. Dall'Agata and N. Prezas, "$\mathcal{N} = 1$ geometries for M-theory and type IIA strings with fluxes," *Physical Review* **D69** (2004) 066004, hep-th/0311146.

[507] A. Lukas and P. M. Saffin, "M-theory compactification, fluxes and AdS_4," *Physical Review* **D71** (2005) 046005, hep-th/0403235.

[508] E. Cremmer and B. Julia, "The SO(8) supergravity," *Nuclear Physics* **B159** (1979) 141.

[509] B. de Wit and H. Nicolai, "$d = 11$ supergravity with local SU(8) invariance," *Nuclear Physics* **B274** (1986) 363–400.

[510] C. M. Hull, "Generalised geometry for M-theory," *Journal of High Energy Physics* **07** (2007) 079, hep-th/0701203.

[511] P. Pires Pacheco and D. Waldram, "M-theory, exceptional generalised geometry and superpotentials," *Journal of High Energy Physics* **09** (2008) 123, 0804.1362.

[512] D. S. Berman and M. J. Perry, "Generalized geometry and M-theory," *Journal of High Energy Physics* **06** (2011) 074, 1008.1763.

[513] M. Graña, J. Louis, A. Sim, and D. Waldram, "$E_{7(7)}$ formulation of $\mathcal{N} = 2$ backgrounds," *Journal of High Energy Physics* **07** (2009) 104, 0904.2333.

[514] M. Graña and F. Orsi, "$\mathcal{N} = 1$ vacua in exceptional generalized geometry," *Journal of High Energy Physics* **08** (2011) 109, 1105.4855.

[515] M. Graña and F. Orsi, "$\mathcal{N} = 2$ vacua in generalized geometry," *Journal of High Energy Physics* **11** (2012) 052, 1207.3004.

[516] A. Coimbra, C. Strickland-Constable, and D. Waldram, "Supersymmetric backgrounds and generalised special holonomy," *Classical and Quantum Gravity* **33** (2016), no. 12, 125026, 1411.5721.

[517] D. Baraglia, "Leibniz algebroids, twistings and exceptional generalized geometry," *Journal of Geometry and Physics* **62** (2012) 903–934, 1101.0856.

[518] D. Cassani, O. de Felice, M. Petrini, C. Strickland-Constable, and D. Waldram, "Exceptional generalised geometry for massive IIA and consistent reductions," *Journal of High Energy Physics* **08** (2016) 074, 1605.00563.

[519] A. Coimbra, C. Strickland-Constable, and D. Waldram, "Supergravity as generalised geometry I: Type II theories," *Journal of High Energy Physics* **11** (2011) 091, 1107.1733.

[520] A. Coimbra, C. Strickland-Constable, and D. Waldram, "$E_{d(d)} \times \mathbb{R}^+$ generalised geometry, connections and M theory," *Journal of High Energy Physics* **02** (2014) 054, 1112.3989.

[521] A. Coimbra, C. Strickland-Constable, and D. Waldram, "Supergravity as generalised geometry II: $E_{d(d)} \times \mathbb{R}^+$ and M-theory," *Journal of High Energy Physics* **03** (2014) 019, 1212.1586.

[522] A. Coimbra and C. Strickland-Constable, "Generalised structures for $\mathcal{N} = 1$ AdS backgrounds," *Journal of High Energy Physics* **11** (2016) 092, 1504.02465.

[523] A. Ashmore and D. Waldram, "Exceptional Calabi–Yau spaces: the geometry of $\mathcal{N} = 2$ backgrounds with flux," *Fortschritte der Physik* **65** (2017), no. 1, 1600109, 1510.00022.

[524] A. Coimbra and C. Strickland-Constable, "Supersymmetric backgrounds, the Killing superalgebra, and generalised special holonomy," *Journal of High Energy Physics* **11** (2016) 063, 1606.09304.

[525] A. Coimbra and C. Strickland-Constable, "Supersymmetric AdS backgrounds and weak generalised holonomy," (2017), 1710.04156.

[526] P. G. O. Freund and M. A. Rubin, "Dynamics of dimensional reduction," *Physics Letters* **B97** (1980) 233–235.

[527] D. R. Morrison and M. R. Plesser, "Nonspherical horizons. 1.," *Advances in Theoretical and Mathematical Physics* **3** (1999) 1–81, hep-th/9810201.

[528] J. P. Gauntlett, "Branes, calibrations and supergravity," *Clay Mathematics Proceedings* **3** (2004) 79–126, hep-th/0305074.

[529] R. L. Bryant, S. S. Chern, R. B. Gardner, H. L. Goldschmidt, and P. A. Griffiths, *Exterior Differential Systems*, vol. 18 of *Mathematical Sciences Research Institute Publications*. Springer-Verlag, 1991.

[530] S. Beck, U. Gran, J. Gutowski, and G. Papadopoulos, "All Killing superalgebras for warped AdS backgrounds," 1710.03713.

[531] K. Sfetsos and D. C. Thompson, "On non-abelian T-dual geometries with Ramond fluxes," *Nuclear Physics* **B846** (2011) 21–42, 1012.1320.

[532] Y. Lozano, E. Ó Colgáin, K. Sfetsos, and D. C. Thompson, "Non-abelian T-duality, Ramond fields and coset geometries," *Journal of High Energy Physics* **1106** (2011) 106, 1104.5196.

[533] G. Itsios, C. Núñez, K. Sfetsos, and D. C. Thompson, "On non-abelian T-duality and new $\mathcal{N} = 1$ backgrounds," *Physics Letters* **B721** (2013) 342–346, 1212.4840.

[534] Y. Lozano and C. Núñez, "Field theory aspects of non-abelian T-duality and $\mathcal{N} = 2$ linear quivers," *Journal of High Energy Physics* **05** (2016) 107, 1603.04440.

[535] Y. Lozano, E. Ó Colgáin, D. Rodríguez-Gómez, and K. Sfetsos, "Supersymmetric AdS_6 via T-duality," *Physical Review Letters* **110** (2013) 231601, 1212.1043.

[536] J. Scherk and J. H. Schwarz, "How to get masses from extra dimensions," *Nuclear Physics* **B153** (1979) 61–88.

[537] K. Lee, C. Strickland-Constable, and D. Waldram, "Spheres, generalised parallelisability and consistent truncations," *Fortschritte der Physik* **65** (2017), no. 10–11, 1700048, 1401.3360.

[538] G. Aldazabal, W. Baron, D. Marques, and C. Nunez, "The effective action of double field theory," *Journal of High Energy Physics* **11** (2011) 052, 1109.0290.

[539] D. Geissbuhler, "Double field theory and $\mathcal{N} = 4$ gauged supergravity," *Journal of High Energy Physics* **11** (2011) 116, 1109.4280.

[540] C. Hull and B. Zwiebach, "Double field theory," *Journal of High Energy Physics* **09** (2009) 099, 0904.4664.

[541] O. Hohm and H. Samtleben, "Consistent Kaluza–Klein truncations via exceptional field theory," *Journal of High Energy Physics* **01** (2015) 131, 1410.8145.

[542] F. Ciceri, A. Guarino, and G. Inverso, "The exceptional story of massive IIA supergravity," *Journal of High Energy Physics* **08** (2016) 154, 1604.08602.

[543] E. Malek, "Half-maximal supersymmetry from exceptional field theory," *Fortschritte der Physik* **65** (2017), no. 10–11, 1700061, 1707.00714.

[544] E. Malek, H. Samtleben, and V. Vall Camell, "Supersymmetric AdS_7 and AdS_6 vacua and their minimal consistent truncations from exceptional field theory," *Physics Letters B* **786** (2018) 1711–1779 1808.05597.

[545] F. Englert, "Spontaneous compactification of eleven-dimensional supergravity," *Physics Letters* **119B** (1982) 339.

[546] C. N. Pope and N. P. Warner, "An SU(4) invariant compactification of $d = 11$ supergravity on a stretched seven sphere," *Physics Letters* **150B** (1985) 352–356.

[547] C. N. Pope and N. P. Warner, "Two new classes of compactifications of $d = 11$ supergravity," *Classical and Quantum Gravity* **2** (1985) L1.

[548] L. J. Romans, "New Compactifications of chiral $N = 2$ $d = 10$ supergravity," *Physics Letters* **153B** (1985) 392–396.

[549] D. Lüst and D. Tsimpis, "Classes of AdS_4 type IIA/IIB compactifications with SU(3)×SU(3) structure," *Journal of High Energy Physics* **04** (2009) 111, 0901.4474.

[550] D. Gaiotto and A. Tomasiello, "The gauge dual of Romans mass," *Journal of High Energy Physics* **01** (2010) 015, 0901.0969.

[551] C. Córdova, G. De Luca, and A. Tomasiello, "AdS_8 solutions in type II supergravity," *Journal of High Energy Physics* **07** (2019) 127, 1811.06987.

[552] M. J. Duff, B. E. W. Nilsson, and C. N. Pope, "The criterion for vacuum stability in Kaluza–Klein supergravity," *Physics Letters* **139B** (1984) 154–158.

[553] D. Forcella and A. Zaffaroni, "Non-supersymmetric CS-matter theories with known AdS duals," *Advanced High Energy Physics* **2011** (2011) 393645, 1103.0648.

[554] P. Breitenlohner and D. Z. Freedman, "Stability in gauged extended supergravity," *Annals of Physics* **144** (1982) 249.

[555] R. Rahman and M. Taronna, "From higher spins to strings: a primer," (2015), 1512.07932.

[556] D. N. Page and C. N. Pope, "Which compactifications of $D = 11$ supergravity are stable?" *Physics Letters* **144B** (1984) 346–350.

[557] D. N. Page and C. N. Pope, "Stability analysis of compactifications of $D = 11$ supergravity with SU(3) X SU(2) X U(1) symmetry," *Physics Letters* **145B** (1984) 337–341.

[558] A. Guarino, E. Malek, and H. Samtleben, "Stable non-supersymmetric anti-de Sitter vacua of massive IIA supergravity," *Physical Review Letters* **126** (2021), no. 6, 061601, 2011.06600.

[559] S. R. Coleman, "The fate of the false vacuum. 1. Semiclassical theory," *Physical Review* **D15** (1977) 2929–2936.

[560] S. R. Coleman and F. De Luccia, "Gravitational effects on and of vacuum decay," *Physical Review* **D21** (1980) 3305.

[561] J. M. Maldacena, J. Michelson, and A. Strominger, "Anti-de Sitter fragmentation," *Journal of High Energy Physics* **02** (1999) 011, hep-th/9812073.

[562] F. Apruzzi, G. Bruno De Luca, A. Gnecchi, G. Lo Monaco, and A. Tomasiello, "On AdS_7 stability," *Journal of High Energy Physics* **07** (2020) 033, 1912.13491.

[563] I. Bena, K. Pilch, and N. P. Warner, "Brane-jet instabilities," *Journal of High Energy Physics* **10** (2020) 091, 2003.02851.

[564] G. W. Gibbons, C. M. Hull, and N. P. Warner, "The stability of gauged supergravity," *Nuclear Physics* **B218** (1983) 173.

[565] M. Cvetic, S. Griffies, and S.-J. Rey, "Nonperturbative stability of supergravity and superstring vacua," *Nuclear Physics* **B389** (1993) 3–24, hep-th/9206004.

[566] E. Witten, "Instability of the Kaluza–Klein vacuum," *Nuclear Physics* **B195** (1982) 481–492.

[567] R. Schon and S.-T. Yau, "On the proof of the positive mass conjecture in general relativity," *Communications in Mathematical Physics* **65** (1979) 45–76.

[568] E. Witten, "A simple proof of the positive energy theorem," *Communications in Mathematical Physics* **80** (1981) 381.

[569] T. Hertog, G. T. Horowitz, and K. Maeda, "Negative energy density in Calabi–Yau compactifications," *Journal of High Energy Physics* **05** (2003) 060, hep-th/0304199.

[570] X.-Z. Dai, "A positive mass theorem for spaces with asymptotic SUSY compactification," *Communications in Mathematical Physics* **244** (2004) 335–345, math/0308249.

[571] I. García-Etxebarría, M. Montero, K. Sousa, and I. Valenzuela, "Nothing is certain in string compactifications," *Journal of High Energy Physics* **12** (2020) 032, 2005.06494.

[572] A. R. Brown and A. Dahlen, "On 'nothing' as an infinitely negatively curved spacetime," *Physical Review* **D85** (2012) 104026, 1111.0301.

[573] G. T. Horowitz, J. Orgera, and J. Polchinski, "Nonperturbative instability of $AdS_5 \times S^5/\mathbb{Z}_k$," *Physical Review* **D77** (2008) 024004, 0709.4262.

[574] H. Ooguri and L. Spodyneiko, "New Kaluza–Klein instantons and the decay of AdS vacua," *Physical Review* **D96** (2017), no. 2, 026016, 1703.03105.

[575] G. 't Hooft, "Dimensional reduction in quantum gravity," *Conference Proceedings* **C930308** (1993) 284–296, gr-qc/9310026.

[576] L. Susskind, "The world as a hologram," *Journal of Mathematical Physics* **36** (1995) 6377–6396, hep-th/9409089.

[577] J. M. Maldacena, "The large N limit of superconformal field theories and supergravity," *Advances in Theoretical and Mathematical Physics* **2** (1998) 231–252, hep-th/9711200.

[578] S. S. Gubser, I. R. Klebanov, and A. M. Polyakov, "Gauge theory correlators from noncritical string theory," *Physics Letters* **B428** (1998) 105–114, hep-th/9802109.

[579] E. Witten, "Anti-de Sitter space and holography," *Advances in Theoretical and Mathematical Physics* **2** (1998) 253–291, hep-th/9802150.

[580] F. Apruzzi, M. Fazzi, D. Rosa, and A. Tomasiello, "All AdS_7 solutions of type II supergravity," *Journal of High Energy Physics* **1404** (2014) 064, 1309.2949.

[581] J. Louis and S. Lüst, "Supersymmetric AdS_7 backgrounds in half-maximal supergravity and marginal operators of $(1,0)$ SCFTs," *Journal of High Energy Physics* **10** (2015) 120, 1506.08040.

[582] C. Córdova, T. T. Dumitrescu, and K. Intriligator, "Deformations of superconformal theories," *Journal of High Energy Physics* **11** (2016) 135, 1602.01217.

[583] J. Blåbäck, U. H. Danielsson, D. Junghans, T. Van Riet, T. Wrase, and M. Zagermann, "Smeared versus localised sources in flux compactifications," *Journal of High Energy Physics* **12** (2010) 043, 1009.1877.

[584] J. Blåbäck, U. H. Danielsson, D. Junghans, T. Van Riet, T. Wrase, and M. Zagermann, "The problematic backreaction of SUSY-breaking branes," *Journal of High Energy Physics* **1108** (2011) 105, 1105.4879.

[585] N. Bobev, G. Dibitetto, F. F. Gautason, and B. Truijen, "Holography, brane intersections and six-dimensional SCFTs," *Journal of High Energy Physics* **02** (2017) 116, 1612.06324.

[586] D. Gaiotto and E. Witten, "Supersymmetric boundary conditions in $\mathcal{N} = 4$ super Yang–Mills theory," *Journal of Statistical Physics* **135** (2009) 789–855, 0804.2902.

[587] A. Hanany and A. Zaffaroni, "Branes and six-dimensional supersymmetric theories," *Nuclear Physics* **B529** (1998) 180–206, hep-th/9712145.

[588] D. Gaiotto and A. Tomasiello, "Holography for $(1, 0)$ theories in six dimensions," *Journal of High Energy Physics* **1412** (2014) 003, 1404.0711.

[589] O. Bergman, M. Fazzi, D. Rodríguez-Gómez, and A. Tomasiello, "Charges and holography in 6d $(1, 0)$ theories," *Journal of High Energy Physics* **05** (2020) 138, 2002.04036.

[590] I. Bah, A. Passias, and A. Tomasiello, "AdS_5 compactifications with punctures in massive IIA supergravity," *Journal of High Energy Physics* **11** (2017) 050, 1704.07389.

[591] F. Apruzzi and M. Fazzi, "AdS_7/CFT_6 with orientifolds," *Journal of High Energy Physics* **01** (2018) 124, 1712.03235.

[592] H. Nastase, D. Vaman, and P. van Nieuwenhuizen, "Consistent nonlinear KK reduction of 11-d supergravity on $AdS_7 \times S^4$ and selfduality in odd dimensions," *Physics Letters* **B469** (1999) 96–102, hep-th/9905075.

[593] M. Pernici, K. Pilch, and P. van Nieuwenhuizen, "Gauged maximally extended supergravity in seven dimensions," *Physics Letters* **B143** (1984) 103.

[594] P. Townsend and P. van Nieuwenhuizen, "Gauged seven-dimensional supergravity," *Physics Letters* **B125** (1983) 41–46.

[595] A. Passias, A. Rota, and A. Tomasiello, "Universal consistent truncation for 6d/7d gauge/gravity duals," *Journal of High Energy Physics* **10** (2015) 187, 1506.05462.

[596] G. B. De Luca, A. Gnecchi, G. Lo Monaco, and A. Tomasiello, "Holographic duals of 6d RG flows," *Journal of High Energy Physics* **03** (2019) 035, 1810.10013.

[597] V. L. Campos, G. Ferretti, H. Larsson, D. Martelli, and B. E. W. Nilsson, "A study of holographic renormalization group flows in $D = 6$ and $D = 3$," *Journal of High Energy Physics* **06** (2000) 023, hep-th/0003151.

[598] F. Apruzzi, G. Dibitetto, and L. Tizzano, "A new 6D fixed point from holography," *Journal of High Energy Physics* **11** (2016) 126, 1603.06576.

[599] F. Apruzzi, M. Fazzi, A. Passias, D. Rosa, and A. Tomasiello, "AdS_6 solutions of type II supergravity," *Journal of High Energy Physics* **1411** (2014) 099, 1406.0852.

[600] A. Brandhuber and Y. Oz, "The D4–D8 brane system and five-dimensional fixed points," *Physics Letters* **B460** (1999) 307–312, hep-th/9905148.

[601] A. Passias, "A note on supersymmetric AdS_6 solutions of massive type IIA supergravity," *Journal of High Energy Physics* **1301** (2013) 113, 1209.3267.

[602] S. Ferrara, A. Kehagias, H. Partouche, and A. Zaffaroni, "Membranes and five-branes with lower supersymmetry and their AdS supergravity duals," *Physics Letters* **B431** (1998) 42–48, hep-th/9803109.

[603] N. Seiberg, "Five-dimensional SUSY field theories, nontrivial fixed points and string dynamics," *Physics Letters* **B388** (1996) 753–760, hep-th/9608111.

[604] D. R. Morrison and N. Seiberg, "Extremal transitions and five-dimensional supersymmetric field theories," *Nuclear Physics* **B483** (1997) 229–247, hep-th/9609070.

[605] O. Bergman and D. Rodríguez-Gómez, "5D quivers and their AdS_6 duals," *Journal of High Energy Physics* **1207** (2012) 171, 1206.3503.

[606] M. Cvetic, H. Lu, and C. N. Pope, "Gauged six-dimensional supergravity from massive type IIA," *Physical Review Letters* **83** (1999) 5226–5229, hep-th/9906221.

[607] L. Romans, "The $F(4)$ gauged supergravity in six dimensions," *Nuclear Physics* **B269** (1986) 691.

[608] H. Kim, N. Kim, and M. Suh, "Supersymmetric AdS_6 solutions of type IIB Supergravity," *European Physical Journal* **C75** (2015), no. 10, 484, 1506.05480.

[609] E. D'Hoker, M. Gutperle, A. Karch, and C. F. Uhlemann, "Warped $AdS_6 \times S^2$ in type IIB supergravity I: local solutions," *Journal of High Energy Physics* **08** (2016) 046, 1606.01254.

[610] E. D'Hoker, M. Gutperle, and C. F. Uhlemann, "Warped $AdS_6 \times S^2$ in type IIB supergravity II: global solutions and five-brane webs," *Journal of High Energy Physics* **05** (2017) 131, 1703.08186.

[611] Y. Lozano, N. T. Macpherson, and J. Montero, "A $\mathcal{N} = 2$ supersymmetric AdS_4 solution in M-theory with purely magnetic flux," *Journal of High Energy Physics* **10** (2015) 004, 1507.02660.

[612] Y. Lozano, E. Ó Colgáin, and D. Rodríguez-Gómez, "Hints of 5D fixed point theories from non-abelian T-duality," *Journal of High Energy Physics* **1405** (2013) 009, 1311.4842.

[613] J. Gutowski and G. Papadopoulos, "On supersymmetric AdS_6 solutions in 10 and 11 dimensions," *Journal of High Energy Physics* **12** (2017) 009, 1702.06048.

[614] E. D'Hoker, M. Gutperle, and C. F. Uhlemann, "Warped $AdS_6 \times S^2$ in type IIB supergravity III: global solutions with seven-branes," *Journal of High Energy Physics* **11** (2017) 200, 1706.00433.

[615] C. F. Uhlemann, "AdS_6/CFT_5 with O7-planes," *Journal of High Energy Physics* **04** (2020) 113, 1912.09716.

[616] O. Bergman, D. Rodríguez-Gómez, and C. F. Uhlemann, "Testing AdS_6/CFT_5 in type IIB with stringy operators," *Journal of High Energy Physics* **08** (2018) 127, 1806.07898.

[617] A. Chaney and C. F. Uhlemann, "On minimal type IIB AdS_6 solutions with commuting 7-branes," *Journal of High Energy Physics* **12** (2018) 110, 1810.10592.

[618] K. Chen, M. Gutperle, and C. F. Uhlemann, "Spin-2 operators in holographic 4d $\mathcal{N} = 2$ SCFTs," *Journal of High Energy Physics* **06** (2019) 139, 1903.07109.

[619] J. Hong, J. T. Liu, and D. R. Mayerson, "Gauged six-dimensional supergravity from warped IIB reductions," *Journal of High Energy Physics* **09** (2018) 140, 1808.04301.

[620] J. P. Gauntlett, D. Martelli, J. Sparks, and D. Waldram, "Supersymmetric AdS_5 solutions of M theory," *Classical and Quantum Gravity* **21** (2004) 4335–4366, hep-th/0402153.

[621] H. Lin, O. Lunin, and J. M. Maldacena, "Bubbling AdS space and 1/2 BPS geometries," *Journal of High Energy Physics* **0410** (2004) 025, hep-th/0409174.

[622] I. Bah, C. Beem, N. Bobev, and B. Wecht, "Four-dimensional SCFTs from M5-branes," *Journal of High Energy Physics* **1206** (2012) 005, 1203.0303.

[623] D. Gaiotto and J. Maldacena, "The gravity duals of $\mathcal{N} = 2$ superconformal field theories," *Journal of High Energy Physics* **1210** (2012) 189, 0904.4466.

[624] N. Bobev, P. Bomans, and F. F. Gautason, "Wrapped branes and punctured horizons," *Journal of High Energy Physics* **06** (2020) 011, 1912.04779.

[625] I. Bah, "AdS_5 solutions from M5-branes on Riemann surface and D6-branes sources," *Journal of High Energy Physics* **09** (2015) 163, 1501.06072.

[626] F. Apruzzi, M. Fazzi, A. Passias, and A. Tomasiello, "Supersymmetric AdS_5 solutions of massive IIA supergravity," *Journal of High Energy Physics* **06** (2015) 195, 1502.06620.

[627] O. Aharony, L. Berdichevsky, and M. Berkooz, "4D $\mathcal{N} = 2$ superconformal linear quivers with type IIA duals," *Journal of High Energy Physics* **1208** (2012) 131, 1206.5916.

[628] R. Reid-Edwards and B. j. Stefanski, "On Type IIA geometries dual to $\mathcal{N} = 2$ SCFTs," *Nuclear Physics* **B849** (2011) 549–572, 1011.0216.

[629] D. Gaiotto, "$\mathcal{N} = 2$ dualities," *Journal of High Energy Physics* **1208** (2012) 034, 0904.2715.

[630] I. Bah and F. Bonetti, "Anomaly inflow, accidental symmetry, and spontaneous symmetry breaking," *Journal of High Energy Physics* **01** (2020) 117, 1910.07549.

[631] M. R. Douglas and G. W. Moore, "D-branes, quivers, and ALE instantons," (1996), hep-th/9603167.

[632] I. R. Klebanov and E. Witten, "Superconformal field theory on three-branes at a Calabi–Yau singularity," *Nuclear Physics* **B536** (1998) 199–218, hep-th/9807080.

[633] A. Hanany and K. D. Kennaway, "Dimer models and toric diagrams," (2005), hep-th/0503149.

[634] S. Franco, A. Hanany, K. D. Kennaway, D. Vegh, and B. Wecht, "Brane dimers and quiver gauge theories," *Journal of High Energy Physics* **01** (2006) 096, hep-th/0504110.

[635] P. S. Aspinwall and D. R. Morrison, "Quivers from matrix factorizations," *Communications in Mathematical Physics* **313** (2012), no. 3, 607–633, 1005.1042.

[636] M. Fazzi and A. Tomasiello, "Holography, matrix factorizations and K-stability," *Journal of High Energy Physics* **05** (2020) 119, 1906.08272.

[637] R. G. Leigh and M. J. Strassler, "Exactly marginal operators and duality in four-dimensional $\mathcal{N} = 1$ supersymmetric gauge theory," *Nuclear Physics* **B447** (1995) 95–136, hep-th/9503121.

[638] S. Benvenuti and A. Hanany, "Conformal manifolds for the conifold and other toric field theories," *Journal of High Energy Physics* **08** (2005) 024, hep-th/0502043.

[639] O. Lunin and J. M. Maldacena, "Deforming field theories with $U(1) \times U(1)$ global symmetry and their gravity duals," *Journal of High Energy Physics* **0505** (2005) 033, hep-th/0502086.

[640] L. Martucci, "D-branes on general $\mathcal{N} = 1$ backgrounds: superpotentials and D-terms," *Journal of High Energy Physics* **06** (2006) 033, hep-th/0602129.

[641] O. Aharony, B. Kol, and S. Yankielowicz, "On exactly marginal deformations of $\mathcal{N} = 1$ SYM and type IIB supergravity on $AdS_5 \times S^5$," *Journal of High Energy Physics* **06** (2002) 039, hep-th/0205090.

[642] M. Gabella, J. P. Gauntlett, E. Palti, J. Sparks, and D. Waldram, "AdS_5 solutions of Type IIB supergravity and generalized complex geometry," *Communications in Mathematical Physics* **299** (2010) 365–408, 0906.4109.

[643] C. Couzens, "Supersymmetric AdS_5 solutions of type IIB supergravity without D3-branes," *Journal of High Energy Physics* **01** (2017) 041, 1609.05039.

[644] N. T. Macpherson, C. Núñez, L. A. Pando Zayas, V. G. J. Rodgers, and C. A. Whiting, "Type IIB supergravity solutions with AdS_5 from abelian and non-Abelian T dualities," *Journal of High Energy Physics* **02** (2015) 040, 1410.2650.

[645] M. Günaydin, L. J. Romans, and N. P. Warner, "Compact and noncompact gauged supergravity theories in five-dimensions," *Nuclear Physics* **B272** (1986) 598–646.

[646] A. Khavaev, K. Pilch, and N. P. Warner, "New vacua of gauged $\mathcal{N} = 8$ supergravity in five dimensions," *Physics Letters* **B487** (2000) 14–21, hep-th/9812035.

[647] M. Cvetic, H. Lu, C. N. Pope, A. Sadrzadeh, and T. A. Tran, "Consistent SO(6) reduction of type IIB supergravity on S^5," *Nuclear Physics* **B586** (2000) 275–286, hep-th/0003103.

[648] K. Pilch and N. P. Warner, "$\mathcal{N} = 2$ supersymmetric RG flows and the IIB dilaton," *Nuclear Physics* **B594** (2001) 209–228, hep-th/0004063.

[649] A. Baguet, O. Hohm, and H. Samtleben, "Consistent type IIB reductions to maximal 5D supergravity," *Physical Review* **D92** (2015), no. 6, 065004, 1506.01385.

[650] N. Bobev, T. Fischbacher, F. F. Gautason, and K. Pilch, "A cornucopia of AdS_5 vacua," *Journal of High Energy Physics* **07** (2020) 240, 2003.03979.

[651] C. Krishnan, V. Mohan, and S. Ray, "Machine learning $\mathcal{N} = 8$, $D = 5$ gauged supergravity," *Fortschritte der Physik* **68** (2020), no. 5, 2000027, 2002.12927.

[652] K. Pilch and N. P. Warner, "A new supersymmetric compactification of chiral IIB supergravity," *Physics Letters* **B487** (2000) 22–29, hep-th/0002192.

[653] B. S. Acharya, J. M. Figueroa-O'Farrill, C. M. Hull, and B. J. Spence, "Branes at conical singularities and holography," *Advances in Theoretical and Mathematical Physics* **2** (1999) 1249–1286, hep-th/9808014.

[654] B. S. Acharya, F. Denef, C. Hofman, and N. Lambert, "Freund–Rubin revisited," (2003), hep-th/0308046.

[655] B. E. W. Nilsson and C. N. Pope, "Hopf fibration of eleven-dimensional supergravity," *Classical and Quantum Gravity* **1** (1984) 499.

[656] S. Watamura, "Spontaneous compactification and $\mathbb{CP}^N$: SU(3)×SU(2)×U(1), $\sin^2(\theta_W)$, g_3/g_2 and SU(3) triplet chiral fermions in four dimensions," *Physics Letters* **B136** (1984) 245.

[657] D. P. Sorokin, V. I. Tkach, and D. V. Volkov, "On the relationship between compactified vacua of $d = 11$ and $d = 10$ supergravities," *Physics Letters* **B161** (1985) 301–306.

[658] D. L. Jafferis and A. Tomasiello, "A simple class of $\mathcal{N} = 3$ gauge/gravity duals," *Journal of High Energy Physics* **10** (2008) 101, `0808.0864`.

[659] A. Hanany and A. Zaffaroni, "Tilings, Chern–Simons theories and M2 Branes," *Journal of High Energy Physics* **10** (2008) 111, `0808.1244`.

[660] M. Aganagic, "A stringy origin of M2 brane Chern–Simons theories," *Nuclear Physics* **B835** (2010) 1–28, `0905.3415`.

[661] Y. Imamura and K. Kimura, "On the moduli space of elliptic Maxwell–Chern–Simons theories," *Progress of Theoretical and Experimental Physics* **120** (2008) 509–523, `0806.3727`.

[662] D. Gaiotto and D. L. Jafferis, "Notes on adding D6 branes wrapping $\mathbb{RP}^3$ in $AdS_4 \times \mathbb{CP}^3$," *Journal of High Energy Physics* **11** (2012) 015, `0903.2175`.

[663] B. de Wit and H. Nicolai, "The consistency of the S^7 truncation in $D = 11$ Supergravity," *Nuclear Physics* **B281** (1987) 211.

[664] B. de Wit and H. Nicolai, "$\mathcal{N} = 8$ supergravity," *Nuclear Physics* **B208** (1982) 323.

[665] B. de Wit and H. Nicolai, "Deformations of gauged SO(8) supergravity and supergravity in eleven dimensions," *Journal of High Energy Physics* **05** (2013) 077, `1302.6219`.

[666] N. P. Warner, "Some new extrema of the scalar potential of gauged $\mathcal{N} = 8$ supergravity," *Physics Letters* **128B** (1983) 169–173.

[667] R. Corrado, K. Pilch, and N. P. Warner, "An $\mathcal{N} = 2$ supersymmetric membrane flow," *Nuclear Physics* **B629** (2002) 74–96, `hep-th/0107220`.

[668] T. Fischbacher, K. Pilch, and N. P. Warner, "New supersymmetric and stable, non-supersymmetric phases in supergravity and holographic field theory," (2010), `1010.4910`.

[669] I. M. Comsa, M. Firsching, and T. Fischbacher, "SO(8) supergravity and the magic of machine learning," *Journal of High Energy Physics* **08** (2019) 057, `1906.00207`.

[670] N. Bobev, T. Fischbacher, and K. Pilch, "Properties of the new $\mathcal{N} = 1$ AdS_4 vacuum of maximal supergravity," *Journal of High Energy Physics* **01** (2020) 099, `1909.10969`.

[671] K. Behrndt, M. Cvetic, and T. Liu, "Classification of supersymmetric flux vacua in M-theory," *Nuclear Physics* **B749** (2006) 25 68, `hep-th/0512032`.

[672] N. Halmagyi, K. Pilch, and N. P. Warner, "On supersymmetric flux solutions of M-theory," (2012), `1207.4325`.

[673] M. Pernici and E. Sezgin, "Spontaneous compactification of seven-dimensional supergravity theories," *Classical and Quantum Gravity* **2** (1985) 673.

[674] T. Dimofte, D. Gaiotto, and S. Gukov, "Gauge theories labelled by three-manifolds," *Communications in Mathematical Physics* **325** (2014) 367–419, `1108.4389`.

[675] J. Bovy, D. Lust, and D. Tsimpis, "$\mathcal{N} = 1, 2$ supersymmetric vacua of IIA supergravity and SU(2) structures," *Journal of High Energy Physics* **08** (2005) 056, `hep-th/0506160`.

[676] C. Caviezel, P. Koerber, S. Körs, D. Lüst, D. Tsimpis, and M. Zagermann, "The effective theory of type IIA AdS_4 compactifications on nilmanifolds and cosets," *Classical and Quantum Gravity* **26** (2009) 025014 `0806.3458`.

[677] K. Behrndt and M. Cvetic, "General $\mathcal{N} = 1$ supersymmetric fluxes in massive type IIA string theory," *Nuclear Physics* **B708** (2005) 45–71, `hep-th/0407263`.

[678] J.-P. Derendinger, C. Kounnas, P. M. Petropoulos, and F. Zwirner, "Superpotentials in IIA compactifications with general fluxes," *Nuclear Physics* **B715** (2005) 211–233, `hep-th/0411276`.

[679] G. Villadoro and F. Zwirner, "$\mathcal{N} = 1$ effective potential from dual type-IIA D6/O6 orientifolds with general fluxes," *Journal of High Energy Physics* **06** (2005) 047, `hep-th/0503169`.

[680] P. G. Camara, A. Font, and L. E. Ibáñez, "Fluxes, moduli fixing and MSSM-like vacua in a simple IIA orientifold," *Journal of High Energy Physics* **09** (2005) 013, `hep-th/0506066`.

[681] G. Aldazabal and A. Font, "A second look at $\mathcal{N} = 1$ supersymmetric AdS_4 vacua of type IIA supergravity," *Journal of High Energy Physics* **02** (2008) 086, `0712.1021`.

[682] A. Guarino, J. Tarrio, and O. Varela, "Romans-mass-driven flows on the D2-brane," *Journal of High Energy Physics* **08** (2016) 168, `1605.09254`.

[683] T. House and E. Palti, "Effective action of (massive) IIA on manifolds with SU(3) structure," *Physical Review* **D72** (2005) 026004, `hep-th/0505177`.

[684] A. Tomasiello, "New string vacua from twistor spaces," *Physical Review* **D78** (2008) 046007, `0712.1396`.

[685] M. Larfors, D. Lust, and D. Tsimpis, "Flux compactification on smooth, compact three-dimensional toric varieties," *Journal of High Energy Physics* **07** (2010) 073, `1005.2194`.

[686] B. S. Acharya, F. Benini, and R. Valandro, "Fixing moduli in exact type IIA flux vacua," *Journal of High Energy Physics* **02** (2007) 018, `hep-th/0607223`.

[687] F. Marchesano, E. Palti, J. Quirant, and A. Tomasiello, "On supersymmetric AdS_4 orientifold vacua," *Journal of High Energy Physics* **08** (2020) 087, `2003.13578`.

[688] A. Rota and A. Tomasiello, "AdS_4 compactifications of AdS_7 solutions in type II supergravity," *Journal of High Energy Physics* **07** (2015) 076, `1502.06622`.

[689] D. Junghans, "O-plane backreaction and scale separation in type IIA flux vacua," *Fortschritte der Physik* **68** (2020), no. 6, 2000040, `2003.06274`.

[690] J. McOrist and S. Sethi, "M-theory and type IIA flux compactifications," *Journal of High Energy Physics* **12** (2012) 122, `1208.0261`.

[691] R. Blumenhagen, M. Cvetic, S. Kachru, and T. Weigand, "D-brane instantons in type II orientifolds," *Annual Review of Nuclear and Particle Science* **59** (2009) 269–296, 0902.3251.

[692] A. Guarino and O. Varela, "Dyonic ISO(7) supergravity and the duality hierarchy," *Journal of High Energy Physics* **02** (2016) 079, 1508.04432.

[693] A. Guarino, J. Tarrio, and O. Varela, "Flowing to $\mathcal{N} = 3$ Chern–Simons-matter theory," *Journal of High Energy Physics* **03** (2020) 100, 1910.06866.

[694] N. Bobev, T. Fischbacher, F. F. Gautason, and K. Pilch, "New AdS_4 vacua in dyonic ISO(7) gauged supergravity," *Journal of High Energy Physics* **02** (2021) 215, 2011.08542.

[695] O. Varela, "AdS_4 solutions of massive IIA from dyonic ISO(7) supergravity," *Journal of High Energy Physics* **03** (2016) 071, 1509.07117.

[696] M. Petrini and A. Zaffaroni, "$\mathcal{N} = 2$ solutions of massive type IIA and their Chern–Simons duals," *Journal of High Energy Physics* **09** (2009) 107, 0904.4915.

[697] D. Lüst and D. Tsimpis, "New supersymmetric AdS_4 type II vacua," *Journal of High Energy Physics* **0909** (2009) 098, 0906.2561.

[698] I. Bah, A. Passias, and P. Weck, "Holographic duals of five-dimensional SCFTs on a Riemann surface," *Journal of High Energy Physics* **01** (2019) 058, 1807.06031.

[699] A. Gallerati, H. Samtleben, and M. Trigiante, "The $\mathcal{N} \geq 2$ supersymmetric AdS vacua in maximal supergravity," *Journal of High Energy Physics* **12** (2014) 174, 1410.0711.

[700] Y. Pang and J. Rong, "$\mathcal{N} = 3$ solution in dyonic ISO(7) gauged maximal supergravity and its uplift to massive type IIA supergravity," *Physical Review* **D92** (2015), no. 8, 085037, 1508.05376.

[701] G. B. De Luca, G. Lo Monaco, N. T. Macpherson, A. Tomasiello, and O. Varela, "The geometry of $\mathcal{N} = 3$ AdS_4 in massive IIA," *Journal of High Energy Physics* **08** (2018) 133, 1805.04823.

[702] D. Gaiotto and A. Tomasiello, "Perturbing gauge/gravity duals by a Romans mass," *Journal of Physics* **A42** (2009) 465205, 0904.3959.

[703] N. T. Macpherson, J. Montero, and D. Prins, "$\text{Mink}_3 \times S^3$ solutions of type II supergravity," *Nuclear Physics* **B933** (2018) 185–233, 1712.00851.

[704] M. Fujita, W. Li, S. Ryu, and T. Takayanagi, "Fractional quantum hall effect via holography: Chern–Simons, edge states, and hierarchy," *Journal of High Energy Physics* **06** (2009) 066, 0901.0924.

[705] Y. Pang and J. Rong, "Evidence for the holographic dual of $\mathcal{N} = 3$ solution in massive type IIA," *Physical Review* **D93** (2016), no. 6, 065038, 1511.08223.

[706] M. Fluder and J. Sparks, "D2-brane Chern–Simons theories: F-maximization $= a$-maximization," *Journal of High Energy Physics* **01** (2016) 048, 1507.05817.

[707] K. Behrndt, M. Cvetic, and P. Gao, "General type IIB fluxes with SU(3) structures," *Nuclear Physics* **B721** (2005) 287–308, hep-th/0502154.

[708] M. Petrini, G. Solard, and T. Van Riet, "AdS vacua with scale separation from IIB supergravity," *Journal of High Energy Physics* **1311** (2013) 010, 1308.1265.

[709] G. Solard, "A method to find $\mathcal{N} = 1$ AdS_4 vacua in type IIB," *Journal of High Energy Physics* **01** (2017) 042, 1610.04237.

[710] I. Arav, K. C. M. Cheung, J. P. Gauntlett, M. M. Roberts, and C. Rosen, "Spatially modulated and supersymmetric mass deformations of $\mathcal{N} = 4$ SYM," *Journal of High Energy Physics* **11** (2020) 156, 2007.15095.

[711] C. Núñez, I. Y. Park, M. Schvellinger, and T. A. Tran, "Supergravity duals of gauge theories from f(4) gauged supergravity in six dimensions," *Journal of High Energy Physics* **4** (2001) 025.

[712] E. D'Hoker, J. Estes, and M. Gutperle, "Exact half-BPS Type IIB interface solutions. I. Local solution and supersymmetric Janus," *Journal of High Energy Physics* **0706** (2007) 021, 0705.0022.

[713] B. Assel, C. Bachas, J. Estes, and J. Gomis, "Holographic duals of $D = 3$ $\mathcal{N} = 4$ superconformal field theories," *Journal of High Energy Physics* **1108** (2011) 087, 1106.4253.

[714] B. Assel, C. Bachas, J. Estes, and J. Gomis, "IIB duals of $D = 3$ $\mathcal{N} = 4$ circular quivers," *Journal of High Energy Physics* **12** (2012) 044, 1210.2590.

[715] B. Assel and A. Tomasiello, "Holographic duals of 3D S-fold CFTs," *Journal of High Energy Physics* **06** (2018) 019, 1804.06419.

[716] G. Inverso, H. Samtleben, and M. Trigiante, "Type II supergravity origin of dyonic gaugings," *Physical Review* **D95** (2017), no. 6, 066020, 1612.05123.

[717] A. Guarino and C. Sterckx, "S-folds and (non-)supersymmetric Janus solutions," *Journal of High Energy Physics* **12** (2019) 113, 1907.04177.

[718] A. Guarino, C. Sterckx, and M. Trigiante, "$\mathcal{N} = 2$ supersymmetric S-folds," *Journal of High Energy Physics* **04** (2020) 050, 2002.03692.

[719] N. Bobev, F. F. Gautason, K. Pilch, M. Suh, and J. van Muiden, "Holographic interfaces in $\mathcal{N} = 4$ SYM: Janus and J-folds," *Journal of High Energy Physics* **05** (2020) 134, 2003.09154.

[720] A. Hanany and E. Witten, "Type IIB superstrings, BPS monopoles, and three-dimensional gauge dynamics," *Nuclear Physics* **B492** (1997) 152–190, hep-th/9611230.

[721] D. Gaiotto and E. Witten, "S-duality of boundary conditions In $\mathcal{N} = 4$ Super Yang–Mills theory," *Advances in Theoretical and Mathematical Physics* **13** (2009) 721, 0807.3720.

[722] I. Garozzo, G. Lo Monaco, and N. Mekareeya, "The moduli spaces of S-fold CFTs," *Journal of High Energy Physics* **01** (2019) 046, 1810.12323.

[723] I. Garozzo, G. Lo Monaco, and N. Mekareeya, "Variations on S-fold CFTs," *Journal of High Energy Physics* **03** (2019) 171, 1901.10493.

[724] D. Kutasov, T. Maxfield, I. Melnikov, and S. Sethi, "Constraining de Sitter space in string theory," *Physical Review Letters* **115** (2015), no. 7, 071305, 1504.00056.

[725] E. Witten, "Nonperturbative superpotentials in string theory," *Nuclear Physics* **B474** (1996) 343–360, hep-th/9604030.

[726] F. Denef and M. R. Douglas, "Distributions of flux vacua," *Journal of High Energy Physics* **05** (2004) 072, hep-th/0404116.

[727] D. Robbins and S. Sethi, "A barren landscape?" *Physical Review* **D71** (2005) 046008, hep-th/0405011.

[728] F. Denef, M. R. Douglas, and B. Florea, "Building a better racetrack," *Journal of High Energy Physics* **06** (2004) 034, hep-th/0404257.

[729] N. Goheer, M. Kleban, and L. Susskind, "The trouble with de Sitter space," *Journal of High Energy Physics* **07** (2003) 056, hep-th/0212209.

[730] S. Kachru, J. Pearson, and H. L. Verlinde, "Brane/flux annihilation and the string dual of a nonsupersymmetric field theory," *Journal of High Energy Physics* **06** (2002) 021, hep-th/0112197.

[731] C. Bachas, M. R. Douglas, and C. Schweigert, "Flux stabilization of D-branes," *Journal of High Energy Physics* **05** (2000) 048, hep-th/0003037.

[732] R. C. Myers, "Dielectric-branes," *Journal of High Energy Physics* **12** (1999) 022, hep-th/9910053.

[733] V. Balasubramanian and P. Berglund, "Stringy corrections to Kähler potentials, SUSY breaking, and the cosmological constant problem," *Journal of High Energy Physics* **11** (2004) 085, hep-th/0408054.

[734] V. Balasubramanian, P. Berglund, J. P. Conlon, and F. Quevedo, "Systematics of moduli stabilisation in Calabi–Yau flux compactifications," *Journal of High Energy Physics* **03** (2005) 007, hep-th/0502058.

[735] A. Saltman and E. Silverstein, "The scaling of the no scale potential and de Sitter model building," *Journal of High Energy Physics* **11** (2004) 066, hep-th/0402135.

[736] C. P. Burgess, R. Kallosh, and F. Quevedo, "De Sitter string vacua from supersymmetric D-terms," *Journal of High Energy Physics* **10** (2003) 056, hep-th/0309187.

[737] K. Choi, A. Falkowski, H. P. Nilles, and M. Olechowski, "Soft supersymmetry breaking in KKLT flux compactification," *Nuclear Physics* **B718** (2005) 113–133, hep-th/0503216.

[738] G. Villadoro and F. Zwirner, "De-Sitter vacua via consistent D-terms," *Physical Review Letters* **95** (2005) 231602, hep-th/0508167.

[739] S. L. Parameswaran and A. Westphal, "De Sitter string vacua from perturbative Kahler corrections and consistent D-terms," *Journal of High Energy Physics* **10** (2006) 079, hep-th/0602253.

[740] A. Westphal, "De Sitter string vacua from Kähler uplifting," *Journal of High Energy Physics* **03** (2007) 102, hep-th/0611332.

[741] M. Davidse, F. Saueressig, U. Theis, and S. Vandoren, "Membrane instantons and de Sitter vacua," *Journal of High Energy Physics* **09** (2005) 065, hep-th/0506097.

[742] B. S. Acharya, K. Bobkov, G. L. Kane, P. Kumar, and J. Shao, "Explaining the electroweak scale and stabilizing moduli in M theory," *Physical Review* **D76** (2007) 126010, hep-th/0701034.

[743] M. Becker, G. Curio, and A. Krause, "De Sitter vacua from heterotic M-theory," *Nuclear Physics* **B693** (2004) 223–260, hep-th/0403027.

[744] T. Banks, "Landskepticism or why effective potentials don't count string models," 2004. hep-th/0412129.

[745] S. Sethi, "Supersymmetry Breaking by Fluxes," *Journal of High Energy Physics* **10** (2018) 022, 1709.03554.

[746] S. Kachru and S. P. Trivedi, "A comment on effective field theories of flux vacua," *Fortschritte der Physik* **67** (2019), no. 1–2, 1800086, 1808.08971.

[747] P. McGuirk, G. Shiu, and Y. Sumitomo, "Non-supersymmetric infrared perturbations to the warped deformed conifold," *Nuclear Physics* **B842** (2011) 383–413, 0910.4581.

[748] I. Bena, M. Graña, and N. Halmagyi, "On the existence of meta-stable vacua in Klebanov–Strassler," *Journal of High Energy Physics* **09** (2010) 087, 0912.3519.

[749] I. Bena, G. Giecold, M. Graña, N. Halmagyi, and S. Massai, "The backreaction of anti-D3 branes on the Klebanov–Strassler geometry," *Journal of High Energy Physics* **06** (2013) 060, 1106.6165.

[750] J. Blåbäck, U. H. Danielsson, and T. Van Riet, "Resolving anti-brane singularities through time-dependence," *Journal of High Energy Physics* **02** (2013) 061, 1202.1132.

[751] B. Michel, E. Mintun, J. Polchinski, A. Puhm, and P. Saad, "Remarks on brane and antibrane dynamics," *Journal of High Energy Physics* **09** (2015) 021, 1412.5702.

[752] J. Polchinski, "Brane/antibrane dynamics and KKLT stability," (2015), 1509.05710.

[753] D. Cohen-Maldonado, J. Diaz, T. van Riet, and B. Vercnocke, "Observations on fluxes near anti-branes," *Journal of High Energy Physics* **01** (2016) 126, 1507.01022.

[754] R. Emparan, T. Harmark, V. Niarchos, and N. A. Obers, "World-volume effective theory for higher-dimensional black holes," *Physical Review Letters* **102** (2009) 191301, 0902.0427.

[755] J. Armas, N. Nguyen, V. Niarchos, N. A. Obers, and T. Van Riet, "Meta-stable non-extremal anti-branes," *Physical Review Letters* **122** (2019), no. 18, 181601, 1812.01067.

[756] I. Bena, E. Dudas, M. Graña, and S. Lüst, "Uplifting runaways," *Fortschritte der Physik* **67** (2019), no. 1–2, 1800100, 1809.06861.

[757] P. G. Camara, L. E. Ibáñez, and A. M. Uranga, "Flux-induced SUSY-breaking soft terms on D7–D3 brane systems," *Nuclear Physics* **B708** (2005) 268–316, hep-th/0408036.

[758] A. Dymarsky and L. Martucci, "D-brane non-perturbative effects and geometric deformations," *Journal of High Energy Physics* **04** (2011) 061, 1012.4018.

[759] B. Heidenreich, L. McAllister, and G. Torroba, "Dynamic SU(2) structure from seven-branes," *Journal of High Energy Physics* **05** (2011) 110, 1011.3510.

[760] J. Moritz, A. Retolaza, and A. Westphal, "Toward de Sitter space from ten dimensions," *Physical Review* **D97** (2018), no. 4, 046010, 1707.08678.

[761] F. F. Gautason, V. Van Hemelryck, and T. Van Riet, "The tension between 10D supergravity and dS uplifts," *Fortschritte der Physik* **67** (2019), no. 1–2, 1800091, 1810.08518.

[762] S. Kachru, M. Kim, L. McAllister, and M. Zimet, "De Sitter vacua from ten dimensions," (2019), 1908.04788.

[763] A. Saltman and E. Silverstein, "A new handle on de Sitter compactifications," *Journal of High Energy Physics* **01** (2006) 139, hep-th/0411271.

[764] G. Gur-Ari, "Brane inflation and moduli stabilization on twisted tori," *Journal of High Energy Physics* **01** (2014) 179, 1310.6787.

[765] S. S. Haque, G. Shiu, B. Underwood, and T. Van Riet, "Minimal simple de Sitter solutions," *Physical Review* **D79** (2009) 086005, 0810.5328.

[766] C. Caviezel, P. Koerber, S. Körs, D. Lüst, T. Wrase, and M. Zagermann, "On the cosmology of type IIA compactifications on SU(3)-structure manifolds," *Journal of High Energy Physics* **04** (2009) 010, 0812.3551.

[767] G. W. Gibbons and C. M. Hull, "De Sitter space from warped supergravity solutions," (2001), hep-th/0111072.

[768] U. H. Danielsson, S. S. Haque, G. Shiu, and T. Van Riet, "Towards classical de Sitter solutions in string theory," *Journal of High Energy Physics* **09** (2009) 114, 0907.2041.

[769] U. H. Danielsson, P. Koerber, and T. Van Riet, "Universal de Sitter solutions at tree-level," *Journal of High Energy Physics* **05** (2010) 090, 1003.3590.

[770] U. H. Danielsson, S. Haque, P. Koerber, G. Shiu, T. Van Riet, and T. Wrase, "De Sitter hunting in a classical landscape," *Fortschritte der Physik* **59** (2011) 897–933, 1103.4858.

[771] D. Andriot, P. Marconnet, and T. Wrase, "New de Sitter solutions of 10D type IIB supergravity," *Journal of High Energy Physics* **08** (2020) 076, 2005.12930.

[772] C. Córdova, G. B. De Luca, and A. Tomasiello, "New de Sitter solutions in ten dimensions and orientifold singularities," *Journal of High Energy Physics* **08** (2020) 093, 1911.04498.

[773] E. Silverstein, "(A)dS backgrounds from asymmetric orientifolds," *Clay Mathematics Proceedings* **1** (2002) 179, hep-th/0106209.

[774] S. Harribey and D. Tsimpis, "One-loop bosonic string and De Sitter space," *Nuclear Physics B* **948** (2019) 114768, 1810.02236.

[775] B. de Carlos, A. Guarino, and J. M. Moreno, "Flux moduli stabilisation, supergravity algebras and no-go theorems," *Journal of High Energy Physics* **01** (2010) 012, 0907.5580.

[776] J. Blåbäck, U. Danielsson, and G. Dibitetto, "Fully stable dS vacua from generalised fluxes," *Journal of High Energy Physics* **08** (2013) 054, 1301.7073.

[777] B. Souères and D. Tsimpis, "De Sitter space from dilatino condensates in (massive) IIA," *Physical Review* **D97** (2018), no. 4, 046005, 1712.07169.

[778] R. Terrisse and D. Tsimpis, "Consistent truncation with dilatino condensation on nearly Kähler and Calabi–Yau manifolds," *Journal of High Energy Physics* **02** (2019) 088, 1810.06344.

[779] T. D. Brennan, F. Carta, and C. Vafa, "The string landscape, the swampland, and the missing corner," *PoS* **TASI2017** (2017) 015, 1711.00864.

[780] E. Palti, "The swampland: introduction and review," *Fortschritte der Physik* **67** (2019), no. 6, 1900037, 1903.06239.

[781] L. Susskind, "Trouble for remnants," (1995), hep-th/9501106.

[782] T. Banks and N. Seiberg, "Symmetries and strings in field theory and gravity," *Physical Review* **D83** (2011) 084019, 1011.5120.

[783] T. Banks, M. Johnson, and A. Shomer, "A note on gauge theories coupled to gravity," *Journal of High Energy Physics* **0609** (2006) 049, hep-th/0606277.

[784] D. Harlow and H. Ooguri, "Symmetries in quantum field theory and quantum gravity," *Communications in Mathematical Physics* **383** (2021) no. 3, 1669–1804, 1810.05338.

[785] N. Arkani-Hamed, L. Motl, A. Nicolis, and C. Vafa, "The string landscape, black holes and gravity as the weakest force," *Journal of High Energy Physics* **06** (2007) 060, hep-th/0601001.

[786] T. W. Grimm, E. Palti, and I. Valenzuela, "Infinite distances in field space and massless towers of states," *Journal of High Energy Physics* **08** (2018) 143, 1802.08264.

[787] C. Cheung and G. N. Remmen, "Naturalness and the weak gravity conjecture," *Physical Review Letters* **113** (2014) 051601, 1402.2287.

[788] E. Palti, "The weak gravity conjecture and scalar fields," *Journal of High Energy Physics* **08** (2017) 034, 1705.04328.

[789] B. Heidenreich, M. Reece, and T. Rudelius, "Sharpening the weak gravity conjecture with dimensional reduction," *Journal of High Energy Physics* **02** (2016) 140, 1509.06374.

[790] H. Ooguri and C. Vafa, "Non-supersymmetric AdS and the swampland," *Advances in Theoretical and Mathematical Physics* **21** (2017) 1787–1801, 1610.01533.

[791] B. Freivogel and M. Kleban, "Vacua morghulis," (2016), 1610.04564.

[792] H. Ooguri and C. Vafa, "On the geometry of the string landscape and the swampland," *Nuclear Physics* **B766** (2007) 21–33, hep-th/0605264.

[793] D. Harlow, "Wormholes, emergent gauge fields, and the weak gravity conjecture," *Journal of High Energy Physics* **01** (2016) 122, 1510.07911.

[794] B. Heidenreich, M. Reece, and T. Rudelius, "The weak gravity conjecture and emergence from an ultraviolet cutoff," *European Physical Journal* **C78** (2018), no. 4, 337, 1712.01868.

[795] D. Lüst, E. Palti, and C. Vafa, "AdS and the swampland," *Physics Letters* **B797** (2019) 134867, 1906.05225.

[796] U. H. Danielsson and T. Van Riet, "What if string theory has no de Sitter vacua?" *International Journal of Modern Physics* **D27** (2018), no. 12, 1830007, 1804.01120.

[797] G. Obied, H. Ooguri, L. Spodyneiko, and C. Vafa, "De Sitter space and the swampland," (2018), 1806.08362.

[798] G. Dvali and C. Gomez, "Quantum exclusion of positive cosmological constant?" *Annalen der Physik* **528** (2016) 68–73.

[799] F. Denef, A. Hebecker, and T. Wrase, "De Sitter swampland conjecture and the Higgs potential," *Physical Review* **D98** (2018), no. 8, 086004, 1807.06581.

[800] H. Ooguri, E. Palti, G. Shiu, and C. Vafa, "Distance and de Sitter conjectures on the swampland," *Physics Letters* **B788** (2019) 180–184, 1810.05506.

[801] S. K. Garg and C. Krishnan, "Bounds on slow roll and the de Sitter swampland," *Journal of High Energy Physics* **11** (2019) 075, 1807.05193.

[802] C. Wetterich, "Cosmology and the fate of dilatation symmetry," *Nuclear Physics* **B302** (1988) 668–696, 1711.03844.

[803] B. Ratra and P. J. E. Peebles, "Cosmological consequences of a rolling homogeneous scalar field," *Physical Review* **D37** (1988) 3406.

[804] M. Cicoli, S. De Alwis, A. Maharana, F. Muia, and F. Quevedo, "De Sitter vs quintessence in string theory," *Fortschritte der Physik* **67** (2019), no. 1–2, 1800079, 1808.08967.

[805] S. Banerjee, U. Danielsson, G. Dibitetto, S. Giri, and M. Schillo, "Emergent de Sitter cosmology from decaying AdS," *Physical Review Letters* **121** (2018), no. 26, 261301, 1807.01570.

[806] J. J. Heckman, C. Lawrie, L. Lin, and G. Zoccarato, "F-theory and dark energy," *Fortschritte der Physik* **67** (2019), no. 10, 1900057, 1811.01959.

Index

Printed in the United States
by Baker & Taylor Publisher Services